visual

essentials of

anatomy &
physiology

Frederic H. Martini, Ph.D.
University of Hawaii at Manoa

William C. Ober, M.D.
Washington and Lee University

Edwin F. Bartholomew, M.S.

Judi L. Nath, Ph.D.
Lourdes University

Claire W. Garrison, R.N.
Medical & Scientific Illustration
Contributing Illustrator

Kathleen Welch, M.D.
Fellow, American Academy of Family Practice
Clinical Consultant

PEARSON

Boston Columbus Indianapolis New York San Francisco Upper Saddle River
Amsterdam Cape Town Dubai London Madrid Milan Munich Paris Montréal Toronto
Delhi Mexico City São Paulo Sydney Hong Kong Seoul Singapore Taipei Tokyo

Executive Editor: *Leslie Berriman*
Senior Project Editor: *Robin Pille*
Director of Development: *Barbara Yien*
Development Editor: *Alice Fugate*
Associate Editor: *Nicole McFadden*
Senior Managing Editor: *Deborah Cogan*
Assistant Managing Editor: *Nancy Tabor*
Production Management and
  Composition: *S4Carlisle Publishing Services, Inc.*
Production Manager: *Tiffany Timmerman*
Copyeditor: *Michael Rossa*

Director of Media Development: *Lauren Fogel*
Media Producer: *Joe Mochnick*
Design Manager: *Mark Ong*
Interior Designer: *Gibson Design Associates*
Cover Designer: *Riezebos Holzbaur Design Group*
Photo Lead: *Donna Kalal*
Photo Researcher: *Bill Smith Group*
Indexer: *Karen Hollister*
Senior Manufacturing Buyer: *Stacey Weinberger*
Senior Market Development Manager: *Brooke Suchomel*
Marketing Manager: *Derek Perrigo*

Cover Photo Credit: John Lund/Sam Diephuis/Blend Images/Photolibrary

Credits and acknowledgments borrowed from other sources and reproduced, with permission, in this textbook appear after the Glossary.

Notice: Our knowledge in clinical sciences is constantly changing. The authors and the publisher of this volume have taken care that the information contained herein is accurate and compatible with the standards generally accepted at the time of the publication. Nevertheless, it is difficult to ensure that all information given is entirely accurate for all circumstances. The authors and the publisher disclaim any liability, loss, or damage incurred as a consequence, directly or indirectly, of the use and application of any of the contents of this volume.

Copyright ©2013 by Frederic H. Martini, Inc., Judi L. Nath, LLC, and Edwin F. Bartholomew, Inc. Published by Pearson Education, Inc. All rights reserved. Manufactured in the United States of America. This publication is protected by Copyright, and permission should be obtained from the publisher prior to any prohibited reproduction, storage in a retrieval system, or transmission in any form or by any means, electronic, mechanical, photocopying, recording, or likewise. To obtain permission(s) to use material from this work, please submit a written request to Pearson Education, Inc., Permissions Department, 1900 E. Lake Ave., Glenview, IL 60025. For information regarding permissions, call (847) 486-2635.

Many of the designations used by manufacturers and sellers to distinguish their products are claimed as trademarks. Where those designations appear in this book, and the publisher was aware of a trademark claim, the designations have been printed in initial caps or all caps.

MasteringA&P®, A&P Flix™, Practice Anatomy Lab™ (PAL™), and Interactive Physiology® are trademarks, in the U.S. and/or other countries, of Pearson Education, Inc. or its affiliates.

Library of Congress Cataloging-in-Publication Data
  Visual essentials of anatomy & physiology / Frederic H. Martini . . . [et al.];
with Claire W. Garrison, Kathleen Welch.
    p. ; cm.
  Visual essentials of anatomy and physiology
  Includes bibliographical references and index.
  ISBN-13: 978-0-321-78077-5
  ISBN-10: 0-321-78077-9
  I. Martini, Frederic. II. Title: Visual essentials of anatomy and physiology.
  [DNLM: 1. Anatomy—Problems and Exercises. 2. Physiological
Phenomena—Problems and Exercises. QS 18.2]
  612.0076–dc23
                                        2011040690

ISBN 10: 0-321-78077-9 (Student edition)
ISBN 13: 978-0-321-78077-5 (Student edition)
ISBN 10: 0-321-79670-5 (Instructor's Review copy)
ISBN 13: 978-0-321-79670-7 (Instructor's Review copy)

1 2 3 4 5 6 7 8 9 10—CRK—15 14 13 12 11

www.pearsonhighered.com

To my son, PK, for convincing me it was time to look
at teaching and learning in new ways, and to the A&P students
and instructors who helped shape the resulting text.

— *Ric Martini*

To my sons, Todd and Carl, whose warmth and humor have
enriched my life in countless ways.

— *Bill Ober*

To my daughters, Ivy and Kate, who have made me a proud
father twice and grandfather thrice.

— *Ed Bartholomew*

To my husband, Mike, for everything, and then some.

— *Judi Nath*

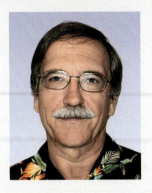

Frederic (Ric) H. Martini, Ph.D.
Author

Dr. Martini received his Ph.D. from Cornell University in comparative and functional anatomy for work on the pathophysiology of stress. In addition to professional publications that include journal articles and contributed chapters, technical reports, and magazine articles, he is the lead author of nine undergraduate texts on anatomy and physiology or anatomy. Dr. Martini is currently affiliated with the University of Hawaii at Manoa and has a long-standing bond with the Shoals Marine Laboratory, a joint venture between Cornell University and the University of New Hampshire. He has been active in the Human Anatomy and Physiology Society (HAPS) for 18 years and was a member of the committee that established the course curriculum guidelines for A&P. He is now a President Emeritus of HAPS after serving as President-Elect, President, and Past-President over 2005–2007. Dr. Martini is also a member of the American Physiological Society, the American Association of Anatomists, the Society for Integrative and Comparative Biology, the Australia/New Zealand Association of Clinical Anatomists, the Hawaii Academy of Science, the American Association for the Advancement of Science, and the International Society of Vertebrate Morphologists.

William C. Ober, M.D.
Author

Dr. Ober received his undergraduate degree from Washington and Lee University and his M.D. from the University of Virginia. He also studied in the Department of Art as Applied to Medicine at John Hopkins University. After graduation, Dr. Ober completed a residency in Family Practice and later was on the faculty at the University of Virginia in the Department of Family Medicine and in the Department of Sports Medicine. He also served as Chief of Medicine of Martha Jefferson Hospital in Charlottesville, VA. He is currently a Visiting Professor of Biology at Washington and Lee University, where he has taught several courses and led student trips to the Galapagos Islands. He also was part of the Core Faculty at Shoals Marine Laboratory for 22 years, where he taught Biological Illustration every summer. Dr. Ober has collaborated with Dr. Martini on all of his textbooks.

Edwin F. Bartholomew, M.S.
Author

Edwin F. Bartholomew received his undergraduate degree from Bowling Green State University in Ohio and his M.S. from the University of Hawaii. Mr. Bartholomew has taught human anatomy and physiology at both the secondary and undergraduate levels and a wide variety of other science courses (from botany to zoology) at Maui Community College and at historic Lahainaluna High School, the oldest high school west of the Rockies. Working with Dr. Martini, he has also coauthored *Essentials of Anatomy & Physiology*, *Structure and Function of the Human Body*, and *The Human Body in Health and Disease* (all published by Pearson). Mr. Bartholomew is a member of the Human Anatomy and Physiology Society (HAPS), the National Association of Biology Teachers, the National Science Teachers Association, the Hawaii Science Teachers Association, and the American Association for the Advancement of Science.

Judi L. Nath, Ph.D.
Author

Dr. Nath is a biology professor at Lourdes University, where she teaches anatomy & physiology, pathophysiology, medical terminology, and pharmacology. She received her B.S. and M.Ed. degrees from Bowling Green State University with majors in biology and German. She earned her Ph.D. from the University of Toledo with a research focus in immunology. Because students often view the health science vocabulary as a foreign language, Dr. Nath uses techniques applied to learning a foreign language in her teaching and her writings. She is the sole author of *Using Medical Terminology*, the first health care textbook to use a "foreign language total immersion" approach. Dr. Nath is deeply committed to teaching and student success and is part of a team implementing a First-Year Experience at Lourdes University. She has received the Faculty Excellence Award in recognition of her effective teaching, scholarship, and community service numerous times. She is active in the Human Anatomy and Physiology Society (HAPS), where she has served several terms on the board of directors. On a personal note, piano playing and bicycling are welcome diversions from authoring, and she greatly enjoys family life with her husband, Mike, and their three dogs.

Claire W. Garrison, R.N.
Contributing Illustrator

Claire W. Garrison, R.N., B.A., practiced pediatric and obstetric nursing before turning to medical illustration as a full-time career. She returned to school at Mary Baldwin College, where she received her degree with distinction in studio art. Following a five-year apprenticeship, she has worked as Dr. Ober's partner in Medical & Scientific Illustration since 1986. She was on the Core Faculty at Shoals Marine Laboratory and co-taught the Biological Illustration course with Dr. Ober. The textbooks illustrated by Medical & Scientific Illustration have won numerous design and illustration awards.

Kathleen Welch, M.D.
Clinical Consultant

Dr. Welch received her M.D. from the University of Washington in Seattle and did her residency in Family Practice at the University of North Carolina in Chapel Hill. For two years, she served as Director of Maternal and Child Health at the LBJ Tropical Medical Center in American Samoa and subsequently was a member of the Department of Family Practice at the Kaiser Permanente Clinic in Lahaina, Hawaii. She has been in private practice since 1987 and is licensed to practice in Hawaii, Washington, and New Zealand. Dr. Welch is a Fellow of the American Academy of Family Practice and a member of the Hawaii Medical Association and the Human Anatomy and Physiology Society (HAPS). With Dr. Martini, she has coauthored both a textbook on anatomy and physiology and the *A&P Applications Manual*. She and Dr. Martini were married in 1979, and they have one son, PK, to whom this book is dedicated.

Anatomy & Physiology for the 21st Century

There are already so many anatomy & physiology textbooks for the A&P essentials course, why create another one? The simple answer is that none of them seem to work for the entire range of students taking the course. *Visual Essentials of Anatomy & Physiology* is the product of a complete reanalysis of the concept of "textbook." While you will find the scope and content familiar, the organization and approach are unique.

A&P texts are still following a format that was established 500 years ago.

The anatomy text on the left, printed in 1491, conveys information in lengthy blocks of narrative text occasionally supplemented by illustrations. The same format is used in college textbooks today (although the narrative is no longer in Latin). It is certainly a "tried and true" method of instruction. But is it the best method for reaching the minds of today's students?

17.4

Filtration, reabsorption, and secretion occur in specific regions of the nephron and collecting system

1 This diagram summarizes the functions of the various segments of the nephron and collecting system in the formation of urine. Most regions perform a combination of reabsorption and secretion, but the balance between the two processes varies from one region to another. Interaction between the collecting system and the nephron loops—especially the long loops of the juxtamedullary nephrons—regulates the final volume and solute concentration of urine.

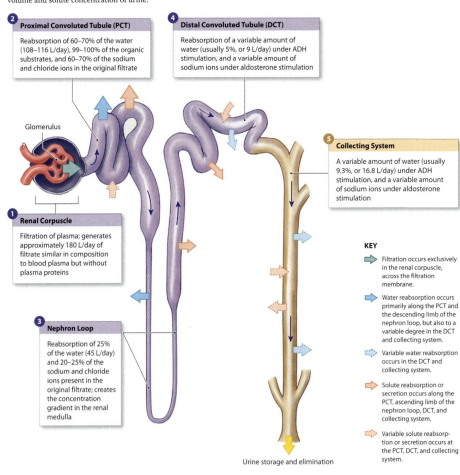

2 Proximal Convoluted Tubule (PCT)

Reabsorption of 60–70% of the water (108–116 L/day), 99–100% of the organic substrates, and 60–70% of the sodium and chloride ions in the original filtrate

4 Distal Convoluted Tubule (DCT)

Reabsorption of a variable amount of water (usually 5%, or 9 L/day) under ADH stimulation, and a variable amount of sodium ions under aldosterone stimulation

Glomerulus

1 Renal Corpuscle

Filtration of plasma; generates approximately 180 L/day of filtrate similar in composition to blood plasma but without plasma proteins

5 Collecting System

A variable amount of water (usually 9.3%, or 16.8 L/day) under ADH stimulation, and a variable amount of sodium ions under aldosterone stimulation

3 Nephron Loop

Reabsorption of 25% of the water (45 L/day) and 20–25% of the sodium and chloride ions present in the original filtrate; creates the concentration gradient in the renal medulla

Urine storage and elimination

KEY

Filtration occurs exclusively in the renal corpuscle, across the filtration membrane.

Water reabsorption occurs primarily along the PCT and the descending limb of the nephron loop, but also to a variable degree in the DCT and collecting system.

Variable water reabsorption occurs in the DCT and collecting system.

Solute reabsorption or secretion occurs along the PCT, ascending limb of the nephron loop, DCT, and collecting system.

Variable solute reabsorption or secretion occurs at the PCT, DCT, and collecting system.

Students of the 21st century have been shaped by an integrated media environment.

Everywhere you look today, information is being presented through a fusion of narrative and images. Owner's manuals for cars and electronics, emergency instruction cards on airplanes, social networking websites—all integrate text and visuals in effective ways. We have applied this contemporary presentation style to this new A&P text. By integrating text and art, this book can be clearer and easier to follow.

2 The renal corpuscle, the start of the nephron, is responsible for filtering blood. This is the vital first step in forming urine. At the renal corpuscle, the capillary knot of the glomerulus projects into the capsular space like the heart projects into the pericardial cavity. Like the pericardium, the glomerular capsule has an outer parietal layer and an inner visceral layer.

The glomerular capsule forms the outer wall of the renal corpuscle and covers the glomerular capillaries.

The **capsular space** separates the inner and outer layers of the glomerular capsule.

Initial segment of renal tubule

The **efferent arteriole** delivers blood to peritubular capillaries. It has a smaller diameter than the afferent arteriole; this elevates the blood pressure within the glomerulus to support filtration.

DCT

The **juxtaglomerular complex** consists of specialized cells that secrete renin when glomerular blood pressure falls.

The **afferent arteriole** delivers blood from a cortical radiate artery.

Outer layer
Inner layer

3 The inner layer of the glomerular capsule consists of large cells with complex processes, or "feet," that wrap around the specialized dense layer of the glomerular capillaries. These unusual cells
podos, foot + -*cyte*, cell)
at the glomerulus must
the basement membran
or **filtration slits**, betwe

A podocyte

This has been an exciting, innovative, and challenging project, and the many talented and dedicated people involved are listed in the Acknowledgments section. If you have questions, comments, or suggestions for improvement, please contact us at the e-mail address below right.

Sincerely,

Frederic H. Martini, Ph.D.   William C. Ober, M.D.

Edwin F. Bartholomew, M.S.   Judi L. Nath, Ph.D.

17.4 Review

fy the three distinct processes of
formation in the kidney.

e does filtration exclusively occur
kidney?

hormone is responsible for
ing sodium ion reabsorption in
T and collecting system?

2: Overview of Renal Physiology • 601

Visual Essentials of Anatomy & Physiology **abandons the 15th-century format and adopts a modern, integrated approach.**

The complete integration of text and art within an efficient organization makes it easier for students to access information.

Organizational overview:

• The concepts are presented in 275 modules—225 in two-page spreads and 50 as single pages, each with a focused topic statement.

• Each module ends with a series of review questions to provide reinforcement and emphasize key concepts and relationships.

• Groups of related modules are clustered in sections, and each section opens with a one-page introduction and closes with a one-page section review.

• The end-of-chapter material is organized into a highly visual review exercise capped by end-of-chapter review questions that integrate the material from the entire chapter.

These structural and organizational innovations make *Visual Essentials of Anatomy & Physiology* a completely new kind of textbook—one that students find both exciting and easier to understand. The pages that follow will walk you through the specific features of this text in more detail.

Send questions or comments to:
martini@pearson.com

What makes this book different?
The Modular Organization

The time-saving modular organization presents topics in two-page spreads. These two-page spreads give students an efficient organization for managing their time. Students can study each module during the limited time they have in their busy schedules—ten minutes for one module now, ten minutes for another module later—checking off each module as they complete it.

First, the top left page begins with a full-sentence topic heading that teaches the major point of the module. (These topic headings are correlated by number to the learning outcomes on the chapter-opening page. The learning outcomes are derived from the learning outcomes recommended by the Human Anatomy & Physiology Society.)

Next, the red-boxed numbers guide students through the presentation of the topic.

5.3

Bone has a calcified matrix associated with osteocytes, osteoblasts, and osteoclasts

Both compact bone and spongy bone contain the same three cell types.

1 Mature bone cells called **osteocytes** (OS-tē-ō-sīts; (*osteo-*, bone + *cyte*, cell) maintain the protein and mineral content of the surrounding matrix through the turnover of matrix components. Osteocytes secrete chemicals that dissolve the adjacent matrix, and the released minerals enter the circulation. The osteocytes then rebuild the matrix, stimulating the deposition of mineral crystals. Osteocytes also play a role in the repair of damaged bone.

3 **Osteoclasts** (OS-tē-ō-clasts; *clast*, to break) are cells that remove and recycle bone matrix. These are giant cells with 50 or more nuclei. They are derived from the same stem cells that produce monocytes and macrophages. Acids and proteolytic (protein-digesting) enzymes secreted by osteoclasts dissolve the matrix and release the stored minerals into interstitial fluid. This process, called **osteolysis** (os-tē-OL-i-sis; *lysis*, a loosening) or resorption, is important in the regulation of calcium and phosphate ion concentrations in body fluids.

The layers of matrix are called **lamellae** (lah-MEL-lē; singular, *lamella*, a thin plate).

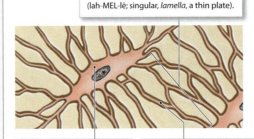

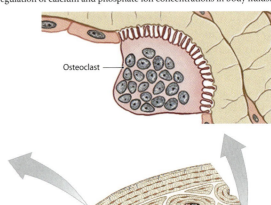

Osteoclast

Osteocytes make up most of the cells in bone. Each osteocyte occupies a **lacuna**, a pocket sandwiched between layers of matrix. Osteocytes cannot divide, and a lacuna never contains more than one osteocyte.

Processes of the osteocytes extend into passageways called **canaliculi** that penetrate the matrix. The canaliculi interconnect the lacunae and reach vascular passageways, providing a route for nutrient diffusion.

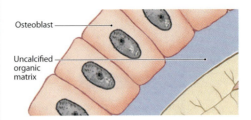

Osteoblast

Uncalcified organic matrix

2 **Osteoblasts** (OS-tē-ō-blasts; *blast*, precursor) produce new bone matrix in a process called **ossification**. Osteoblasts make and release the proteins and other organic components of the matrix. Osteoblasts also help increase the concentration of calcium phosphate to the point where calcium salts are deposited in the organic matrix forming bone. Osteocytes develop from osteoblasts that have become completely surrounded by bone matrix.

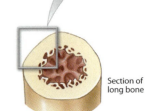

Section of long bone

150 · *Chapter 5: The Skeletal System*

4 The basic functional unit of mature compact bone is the **osteon** (OS-tē-on), or Haversian system. Many of the important features of osteons were introduced in Module 3.22, which you may wish to revisit before proceeding.

Osteon

Compact bone   LM × 375

The **central canal** generally runs parallel to the surface of the bone. Small branches of the arteries and veins within the central canal nourish the osteocytes within the osteons, and provide a route for the removal of wastes or released calcium ions.

Lamellae

The osteocytes occupy **lacunae** that lie between the lamellae. In preparing this micrograph, a small piece of bone was ground down until it was thin enough to transmit light. In this process, the lacunae and canaliculi are filled with bone dust, and thus appear black.

Then, instead of long columns of narrative text that refer to visuals, brief text is built right into the visuals. Students read while looking at the corresponding visual, which means:

- No long paragraphs
- No flipping of pages
- Everything in one place

In the intestines, calcium and phosphate ions are absorbed from the diet. The rate of absorption is hormonally regulated.

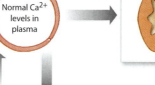

Normal Ca²⁺ levels in plasma

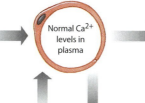

In the kidneys, the levels of calcium and phosphate ions lost in the urine are hormonally regulated.

Bone

5 **Calcium** ions play a role in a variety of physiological processes, and even small variations from the normal concentration affect cellular operations. The homeostatic regulation of calcium ion levels is a juggling act that balances activities of the intestines, bones, and kidneys.

Within the skeleton, osteoblasts are continuously depositing new bone matrix. At the same time, osteoclasts are eroding existing matrix and releasing calcium and phosphate ions that enter the circulation. The balance between osteoblast and osteoclast activity is hormonally regulated.

Module 5.3 Review

a. Define osteocyte, osteoblast, and osteoclast.

b. What is the basic functional unit of mature compact bone?

c. If osteoclast activity exceeds osteoblast activity in a bone, what would be the effect on the bone?

Finally, each two-page module ends with a set of Module Review questions that help students check their understanding before moving on.

If the calcium concentration of body fluids increases or decreases by more than 30–35 percent, neuron and muscle cell function is disrupted, with potentially lethal results. Calcium ion concentration is so closely regulated, however, that daily fluctuations of more than 10 percent are highly unusual.

What makes this book different?
The Visual Approach

The unique visual approach allows the illustrations to be the central teaching and learning element, with the text built directly around them—creating true text-art integration. This approach matches how students naturally want to use their A&P textbook. Our extensive research with A&P students—via student reviews, student focus groups, and student class tests—reveals that A&P students go first to the visuals and then to the corresponding text.

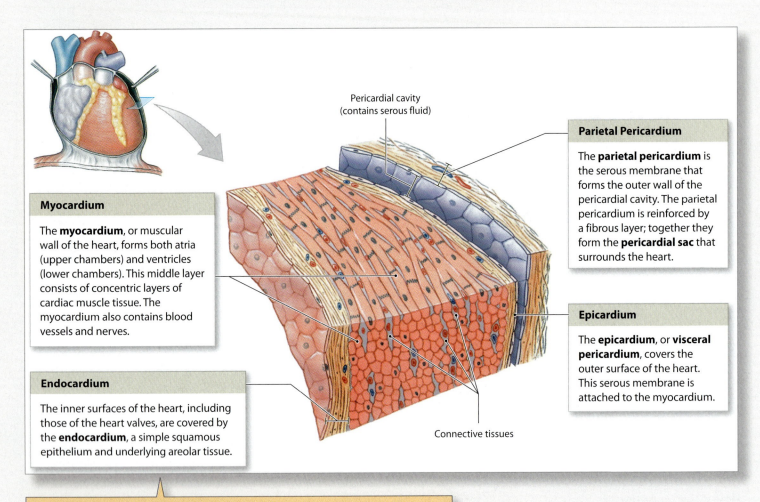

Pericardial cavity (contains serous fluid)

Parietal Pericardium

The **parietal pericardium** is the serous membrane that forms the outer wall of the pericardial cavity. The parietal pericardium is reinforced by a fibrous layer; together they form the **pericardial sac** that surrounds the heart.

Myocardium

The **myocardium**, or muscular wall of the heart, forms both atria (upper chambers) and ventricles (lower chambers). This middle layer consists of concentric layers of cardiac muscle tissue. The myocardium also contains blood vessels and nerves.

Epicardium

The **epicardium**, or **visceral pericardium**, covers the outer surface of the heart. This serous membrane is attached to the myocardium.

Endocardium

The inner surfaces of the heart, including those of the heart valves, are covered by the **endocardium**, a simple squamous epithelium and underlying areolar tissue.

Connective tissues

Descriptions and key terminology are embedded in the art.

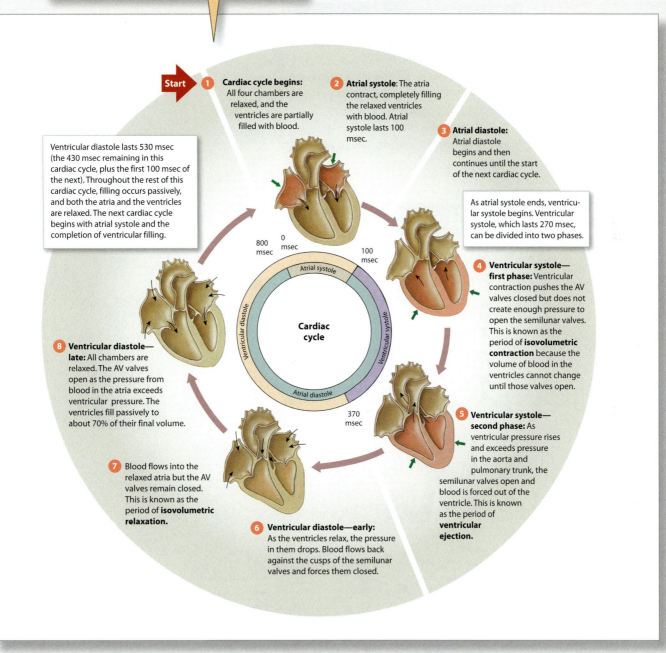

Start arrows, step numbers, and manageable "chunks" of information guide students through complex processes.

Start

1 **Cardiac cycle begins:** All four chambers are relaxed, and the ventricles are partially filled with blood.

2 **Atrial systole:** The atria contract, completely filling the relaxed ventricles with blood. Atrial systole lasts 100 msec.

3 **Atrial diastole:** Atrial diastole begins and then continues until the start of the next cardiac cycle.

As atrial systole ends, ventricular systole begins. Ventricular systole, which lasts 270 msec, can be divided into two phases.

4 **Ventricular systole— first phase:** Ventricular contraction pushes the AV valves closed but does not create enough pressure to open the semilunar valves. This is known as the period of **isovolumetric contraction** because the volume of blood in the ventricles cannot change until those valves open.

5 **Ventricular systole— second phase:** As ventricular pressure rises and exceeds pressure in the aorta and pulmonary trunk, the semilunar valves open and blood is forced out of the ventricle. This is known as the period of **ventricular ejection.**

6 **Ventricular diastole—early:** As the ventricles relax, the pressure in them drops. Blood flows back against the cusps of the semilunar valves and forces them closed.

7 Blood flows into the relaxed atria but the AV valves remain closed. This is known as the period of **isovolumetric relaxation.**

8 **Ventricular diastole— late:** All chambers are relaxed. The AV valves open as the pressure from blood in the atria exceeds ventricular pressure. The ventricles fill passively to about 70% of their final volume.

Ventricular diastole lasts 530 msec (the 430 msec remaining in this cardiac cycle, plus the first 100 msec of the next). Throughout the rest of this cardiac cycle, filling occurs passively, and both the atria and the ventricles are relaxed. The next cardiac cycle begins with atrial systole and the completion of ventricular filling.

800 msec 0 msec 100 msec 370 msec

Atrial systole

Ventricular diastole

Ventricular systole

Atrial diastole

Cardiac cycle

What makes this book different?
The Frequent Practice

Predictable places to stop and check understanding help students pace their learning throughout the chapter.

Module Reviews
appear at the end of every module for frequent and consistent self-assessment.

Module 14.3 Review

a. Describe the roles of the three types of cells associated with alveoli.

b. Describe how the structure of the respiratory membrane enhances gas diffusion.

c. What would happen to the alveoli if surfactant were not produced?

Section Reviews
appear after groups of related modules and include "workbook-style" review activities, such as labeling and concept mapping.

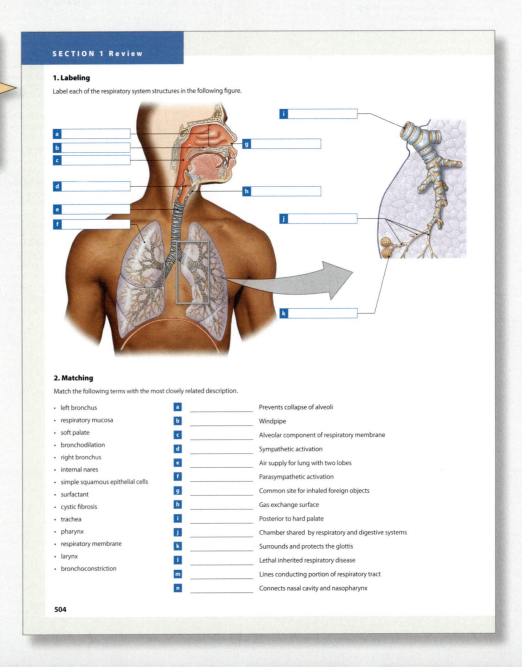

SECTION 1 Review

1. Labeling

Label each of the respiratory system structures in the following figure.

a
b
c
d
e
f
g
h
i
j
k

2. Matching

Match the following terms with the most closely related description.

- left bronchus
- respiratory mucosa
- soft palate
- bronchodilation
- right bronchus
- internal nares
- simple squamous epithelial cells
- surfactant
- cystic fibrosis
- trachea
- pharynx
- respiratory membrane
- larynx
- bronchoconstriction

a	Prevents collapse of alveoli
b	Windpipe
c	Alveolar component of respiratory membrane
d	Sympathetic activation
e	Air supply for lung with two lobes
f	Parasympathetic activation
g	Common site for inhaled foreign objects
h	Gas exchange surface
i	Posterior to hard palate
j	Chamber shared by respiratory and digestive systems
k	Surrounds and protects the glottis
l	Lethal inherited respiratory disease
m	Lines conducting portion of respiratory tract
n	Connects nasal cavity and nasopharynx

504

Visual Outline with Key Terms

Summarize the content of each module using the terms in the order provided.

SECTION 1

Functional Anatomy of the Respiratory System

- respiratory system
- upper respiratory system
- lower respiratory system
- respiratory tract
- conducting portion
- bronchioles
- respiratory portion
- alveoli
- upper respiratory tract
- lower respiratory tract

SECTION 2

Respiratory Physiology

- respiration
- external respiration
- pulmonary ventilation
- alveolar ventilation
- gas diffusion
- internal respiration
- hypoxia
- anoxia

14.1

The respiratory mucosa cleans, warms, and moistens inhaled air

- respiratory mucosa
- lamina propria
- cystic fibrosis (CF)
- pharynx
- nasopharynx
- oropharynx
- laryngopharynx
- trachea
- internal nares
- nasal vestibule
- external nares
- hard palate
- soft palate
- glottis
- larynx

14.4

Pressure changes within the pleural cavities drive pulmonary ventilation

- pneumothorax
- atelectasis
- intrapulmonary pressure
- alveolar pressure
- tidal volume (V_T)

14.2

The trachea and primary bronchi carry air to and from the lungs

- trachea
- primary bronchi
- tracheal cartilages
- trachealis muscle
- bronchioles
- bronchodilation
- bronchoconstriction
- asthma
- right lung
- horizontal and oblique fissures
- superior, middle, and inferior lobes
- hilum
- cardiac notch
- left lung
- oblique fissure
- superior and inferior lobes
- pleura

14.5

Respiratory muscles adjust the tidal volume to meet respiratory demands

- inspiratory muscles
- expiratory muscles
- primary respiratory muscles
- accessory respiratory muscles
- diaphragm
- external intercostal muscles
- respiratory cycle
- volumes
- capacities
- inspiratory reserve volume (IRV)
- tidal volume (V_T)
- dead space of the lungs
- expiratory reserve volume (ERV)
- vital capacity
- residual volume
- total lung capacity

14.3

Gas exchange occurs at the alveoli

- alveoli
- surfactant
- respiratory membrane

14.6

Gas diffusion depends on the relative concentrations and solubilities of gases

○ external respiration
○ respiratory membrane
○ internal respiration
○ hemoglobin
○ carbonic acid
○ chloride shift

● = *Term boldfaced in this module*

517

Think this is the career for you?

Key Stats:

- **Education and Training:** Associate degree is required, but a bachelor's or master's degree may be important for advancement.

- **Licensure:** All states, except Alaska and Hawaii, require respiratory therapists to be licensed.

- **Earnings:** Earnings vary but the mean annual wage is $54,280.

- **Expected Job Prospects:** Employment is expected to grow by an above-average 21% from 2008 to 2018.

Bureau of Labor Statistics, U.S. Department of Labor, *Occupational Outlook Handbook, 2010-11 Edition*, Respiratory Therapists, on the Internet at http://www.bls.gov/oco/ocos321. htm (visited September 14, 2011).

Access more review material online in the Study Area at www.masteringaandp.com.

- Chapter guides
- Chapter quizzes
- Practice tests
- Art labeling activities
- Flashcards
- A glossary with pronunciations
- Practice Anatomy Lab™ (PAL™) 3.0 virtual anatomy practice tool
- Interactive Physiology® (IP) animated tutorials
- MP3 Tutor Sessions

PAL practice anatomy lab®

For this chapter, follow these navigation paths in PAL:

- Human Cadaver>Respiratory System
- Anatomical Models>Respiratory System
- Histology>Respiratory System

iP **For this chapter, go to these topics in the Respiratory System in IP:**

- Anatomy Review: Respiratory Structures
- Pulmonary Ventilation
- Gas Exchange
- Gas Transport

Chapter 14 Review Questions

1. You spend a winter night at a friend's house. Your friend's home is quite old, and the hot-air furnace lacks a humidifier. When you wake up in the morning, you have a fair amount of nasal congestion and decide you might be coming down with a cold. After you take a steamy shower and drink some juice for breakfast, the nasal congestion disappears. Explain.

2. John breaks a rib that punctures the thoracic cavity on his left side. Which structures are potentially damaged, and what do you predict will happen to the lung as a result?

3. In emphysema, alveoli are replaced by large air spaces and elastic fibrous connective tissue. How do these changes affect the lungs?

For answers to all module, section, and chapter review questions, see the blue Answers tab at the back of the book.

519

Practice Anatomy Lab™ (PAL™) 3.0

PAL 3.0 is an indispensable virtual anatomy study and practice tool that gives students 24/7 access to the most widely used lab specimens, including the human cadaver, anatomical models, histology, cat, and fetal pig.

New! Interactive Cadaver Module

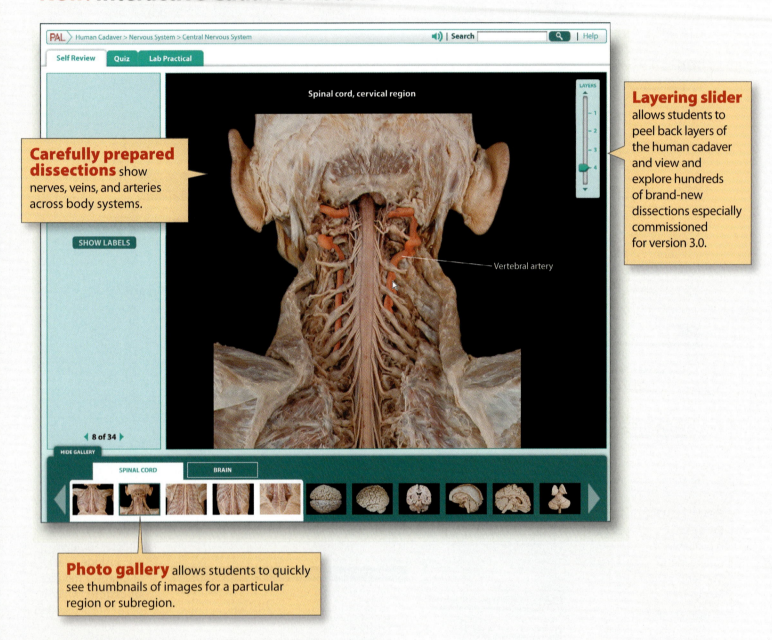

Carefully prepared dissections show nerves, veins, and arteries across body systems.

Layering slider allows students to peel back layers of the human cadaver and view and explore hundreds of brand-new dissections especially commissioned for version 3.0.

Photo gallery allows students to quickly see thumbnails of images for a particular region or subregion.

PAL 3.0 is available in the Study Area of MasteringA&P® (www.masteringaandp.com).
The PAL 3.0 DVD can be packaged with the book for no additional charge.

New! Interactive Histology Module

Magnification buttons allow students to view the same tissue slide at varying magnifications, thereby helping them identify structures and their characteristics.

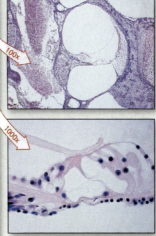

3-D Anatomy Animations

3-D Anatomy Animations of origins, insertions, actions, and innervations of over 60 muscles are now viewable in two modules: Human Cadaver and Anatomical Models. Under the Animations tab, over 50 anatomy animations of group muscle actions and joints are also viewable. A new closed-captioning option provides textual presentation of narration to help students retain information and supports ADA compliance.

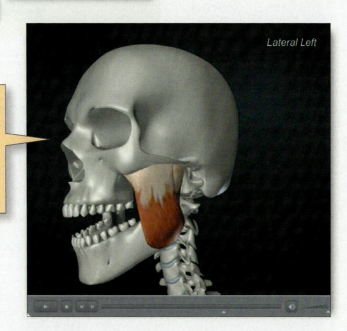

PAL 3.0 also includes:

- **NEW!** Question randomization feature
- **NEW!** Hundreds of new images and views
- **NEW!** Turn-off highlight feature
- **NEW!** IRDVD with Test Bank for PAL 3.0
- Built-in audio pronunciations
- Rotatable bones
- Simulated fill-in-the-blank lab practical exams

*See for yourself! Check out the new **PAL 3.0** at* www.masteringaandp.com.

An Online Learning and Assessment System

Get your students ready for the A&P course.

MasteringA&P allows you to assign tutorials and assessments on *Get Ready* topics:

- Study Skills
- Basic Math Review
- Terminology
- Body Basics
- Chemistry
- Cell Biology

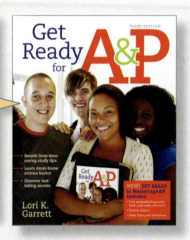

Get your students to come to class prepared.

Assignable Reading Quizzes motivate your students to read the textbook before coming to class.

Assign art from the textbook.

Assign and assess art-labeling activities based on figures from the textbook.

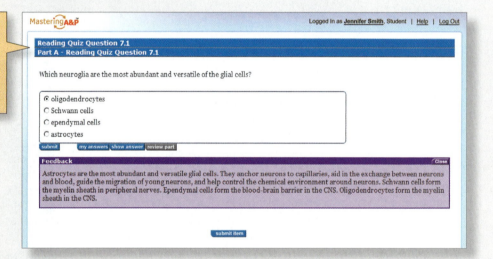

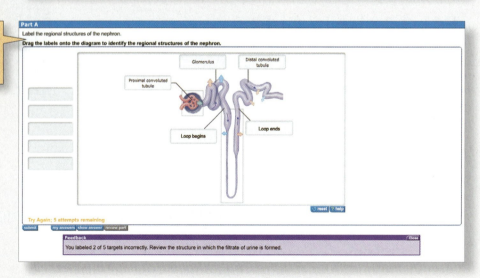

Give your students extra coaching.

Assign tutorials from your favorite media—such as Essentials of Interactive Physiology® (IP)—to help students understand and visualize tough topics. MasteringA&P provides coaching through helpful wrong-answer feedback and hints.

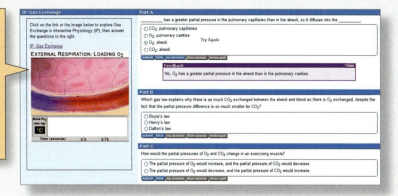

Give your students 24/7 lab practice.

Practice Anatomy Lab™ (PAL™) 3.0 is a tool that helps students study for their lab practicals outside of the lab. To learn more about PAL 3.0, see pages xiv–xv.

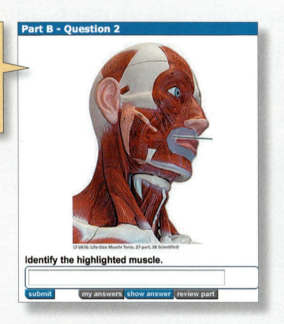

Identify struggling students before it's too late.

MasteringA&P has a color-coded gradebook that helps you identify vulnerable students at a glance. Assignments in MasteringA&P are automatically graded, and grades can be easily exported to course management systems or spreadsheets.

Go to **www.masteringaandp.com** *to learn more.*

Tools to Make the Grade

 **Study Area**

MasteringA&P includes a Study Area that will help students get ready for tests with its simple three-step approach. Students can:

1. **Take a pre-test** and obtain a personalized study plan.
2. **Learn and practice** with animations, labeling activities, and interactive tutorials.
3. **Self-test** with quizzes and a chapter post-test.

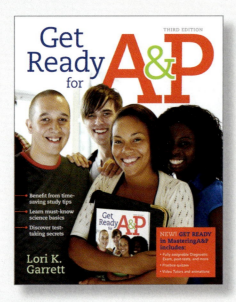

Get Ready for A&P

Students can access the **Get Ready for A&P** eText, activities, and diagnostic tests for these important topics:

- Study Skills
- Basic Math Review
- Terminology
- Body Basics
- Chemistry
- Cell Biology

Practice Anatomy Lab™ (PAL™) 3.0

Practice Anatomy Lab (PAL) 3.0 is a virtual anatomy study and practice tool that gives students 24/7 access to the most widely used lab specimens, including the human cadaver, anatomical models, histology, cat, and fetal pig. PAL 3.0 retains all of the key advantages of version 2.0, including ease-of-use, built-in audio pronunciations, rotatable bones, and simulated fill-in-the-blank lab practical exams.

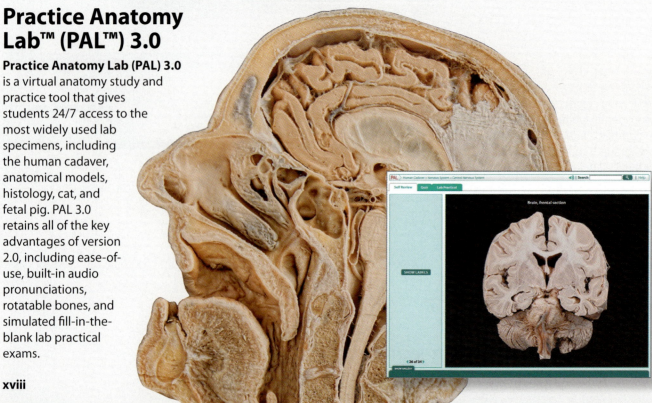

MP3 Tutor Sessions

Students can download the MP3 Tutor Sessions for specific chapters of the textbook and study wherever, whenever. They can listen to mini-lectures about the toughest topics and take audio quizzes to check their understanding.

Topics:

- Homeostasis
- Inorganic Compounds
- Membrane Transport
- Epithelial Tissue
- Layers and Associated Structures of the Integument
- How Bones React to Stress
- Types of Joints and Their Movements
- Sliding Filament Theory of Contraction
- Generation of an Action Potential
- The Differences between the Sympathetic and Parasympathetic Divisions
- The Visual Pathway
- Hypothalamic Regulation
- Hemoglobin: Function and Impact
- Cardiovascular Pressure
- Differences between Innate and Adaptive Immunity
- Hormonal Control of the Menstrual Cycle

Essentials of Interactive Physiology® (IP)

Interactive Physiology helps students understand the hardest part of A&P: physiology. Fun, interactive tutorials, games, and quizzes give students additional explanations to help them grasp difficult concepts.

Modules:

- Muscular System
- Nervous System
- Cardiovascular System
- Respiratory System
- Urinary System
- Fluids & Electrolytes
- Endocrine System
- Digestive System
- Immune System

Support for Students

eText

MasteringA&P (www.masteringaandp.com) includes an eText.
Students can access their textbook wherever and whenever they are online. eText pages look exactly like the printed text yet offer additional functionality. Students can:

- create notes
- highlight text in different colors
- create bookmarks
- zoom in and out
- view in single-page or two-page view
- click hyperlinked words and phrases to view definitions
- search quickly and easily for specific content

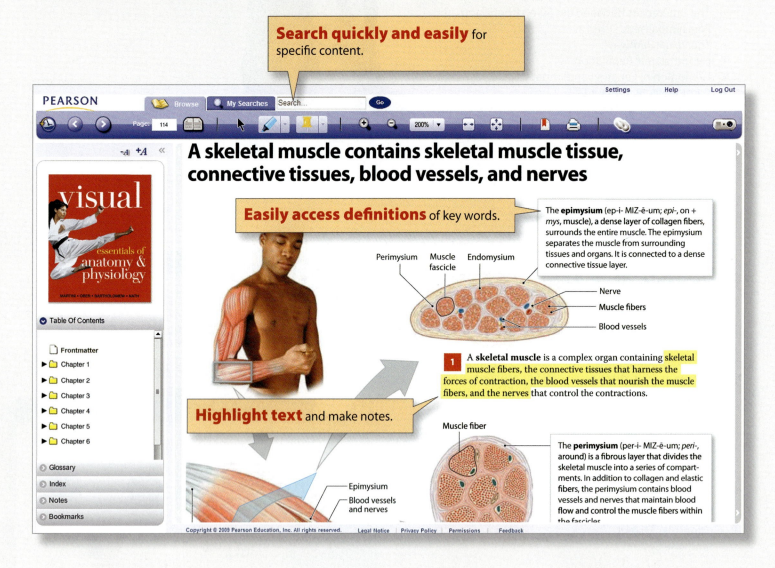

Search quickly and easily for specific content.

Easily access definitions of key words.

Highlight text and make notes.

A skeletal muscle contains skeletal muscle tissue, connective tissues, blood vessels, and nerves

Perimysium   Muscle fascicle   Endomysium

Nerve
Muscle fibers
Blood vessels

The **epimysium** (ep-i- MIZ-ē-um; *epi-*, on + *mys*, muscle), a dense layer of collagen fibers, surrounds the entire muscle. The epimysium separates the muscle from surrounding tissues and organs. It is connected to a dense connective tissue layer.

1 A **skeletal muscle** is a complex organ containing skeletal muscle fibers, the connective tissues that harness the forces of contraction, the blood vessels that nourish the muscle fibers, and the nerves that control the contractions.

Muscle fiber

The **perimysium** (per-i- MIZ-ē-um; *peri-*, around) is a fibrous layer that divides the skeletal muscle into a series of compartments. In addition to collagen and elastic fibers, the perimysium contains blood vessels and nerves that maintain blood flow and control the muscle fibers within the fascicles.

Epimysium
Blood vessels and nerves

Table Of Contents

visual
essentials of
anatomy & physiology
MARTINI • OBER • BARTHOLOMEW • NATH

Copyright © 2009 Pearson Education, Inc. All rights reserved.   Legal Notice | Privacy Policy | Permissions | Feedback

PEARSON   Browse   My Searches   Search...   Go   Settings   Help   Log Out

Page: 114   200%

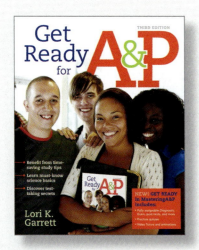

Get Ready for A&P, 3e
by Lori K. Garrett

This book and online component were created to help students be better prepared for their A&P course. Features include pre-tests, guided explanations followed by interactive quizzes and exercises, and end-of-chapter cumulative tests. Also available in the Study Area of www.masteringaandp.com.

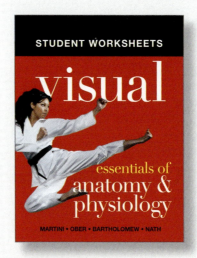

Student Worksheets for Visual Essentials of Anatomy & Physiology

This booklet contains all of the Section Review pages from the book for students who would prefer to mark their answers on separate pages rather than in the book itself. In addition, the Visual Outline with Key Terms from the end of each chapter is reprinted with space for students to summarize the content of each module using the key terms in the order provided.

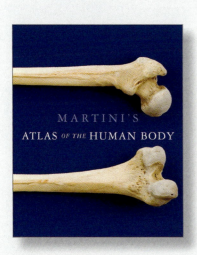

Martini's Atlas of the Human Body
by Frederic H. Martini

The Atlas offers an abundant collection of anatomy photographs, radiology scans, and embryology summaries, helping students visualize structures and become familiar with the types of images seen in a clinical setting.

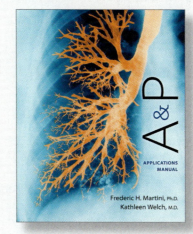

A&P Applications Manual
by Frederic H. Martini and Kathleen Welch

This manual contains extensive discussions on clinical topics and disorders to help students apply the concepts of anatomy and physiology to daily life and their future health professions.

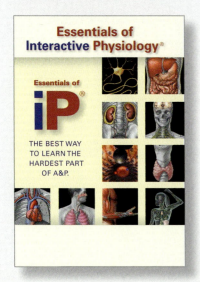

Essentials of Interactive Physiology® CD-ROM

IP helps students understand the hardest part of A&P: physiology. Fun, interactive tutorials, games, and quizzes give students additional explanations to help them grasp difficult physiological concepts.

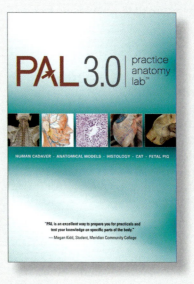

Practice Anatomy Lab™ (PAL™) 3.0 DVD

PAL 3.0 is an indispensable virtual anatomy study and practice tool that gives students 24/7 access to the most widely used lab specimens, including the human cadaver, anatomical models, histology, cat, and fetal pig.

See pages xviii–xix for the **MasteringA&P Study Area.**

Support for Instructors

eText with Whiteboard Mode

The *Visual Essentials of Anatomy & Physiology* eText within MasteringA&P comes with Whiteboard Mode, allowing instructors to use the eText for dynamic classroom presentations. Instructors can show one-page or two-page views from the book, zoom in or out to focus on select topics, and use the Whiteboard Mode to point to structures, circle parts of a process, trace pathways, and customize their presentations.

Instructors can also add notes to guide students, upload documents, and share their custom-enhanced eText with the whole class.

Instructors can find the eText with Whiteboard Mode on MasteringA&P.

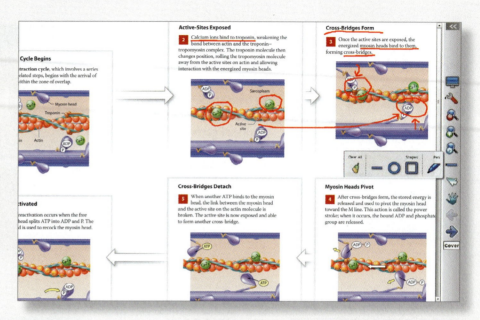

Instructor Resource DVD (IRDVD)

with Lecture Outlines by Betsy Brantley
and Clicker Questions and Quiz Shows by Margaret (Betsy) Ott
978-0-321-79276-1 / 0-321-79276-9

The IRDVD offers a wealth of instructor media resources, including presentation art, lecture outlines, test items, and answer keys—all in one convenient location. The IRDVD includes:

- Textbook images in JPEG format (in two versions—one with labels and one without)
- Customizable textbook images embedded in PowerPoint® slides (in three versions—one with editable labels, one without labels, and one as step-edit art)
- Customizable PowerPoint lecture slides, combining lecture notes, images, and tables
- Clicker Questions
- Quiz Show Questions
- **Martini's Atlas of the Human Body** images in JPEG format
- **Martini's A&P Applications Manual** images in JPEG format
- **Essentials of Interactive Physiology**® Exercise Sheets and Answer Key
- Test Bank in TestGen® and Microsoft® Word formats
- Instructor's Manual in Microsoft Word and PDF formats
- Lecture Outlines in Microsoft Word format
- Transparency Acetate masters for all figures and tables
- Separate DVD for Practice Anatomy Lab™ (PAL™) 3.0 Instructor Presentation Images and Test Bank

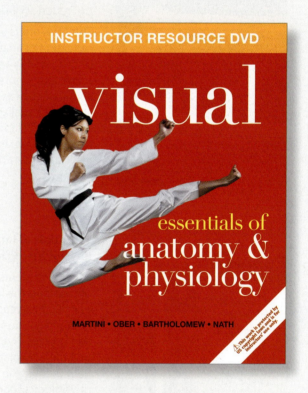

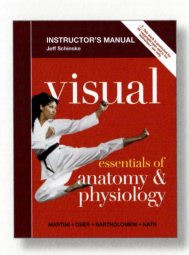

Instructor's Manual

by Jeff Schinske
978-0-321-79292-1 /
0-321-79292-0

This useful resource includes a wealth of materials to help instructors organize their lectures, such as lecture ideas, visual analogies, suggested classroom demonstrations, terminology aids, applications, and common student misconceptions/problems. It also includes sections on encouraging student talk, making learning active, incorporating diversity and the human side of A&P, and additional chapter review questions with suggested answers.

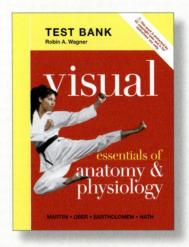

Printed Test Bank

by Robin A. Wagner
978-0-321-79275-4 /
0-321-79275-0

This test bank of close to 2,000 questions helps instructors design a variety of tests and quizzes. The test bank includes text-based and art-based questions. Each test item in the Test Bank has been tagged with a corresponding Learning Outcome from this textbook as well as a Bloom's taxonomy ranking, allowing instructors to test students on a range of learning levels. This supplement is also available in TestGen and Microsoft Word formats on the Instructor Resource DVD and in the Instructor Resources area of MasteringA&P.

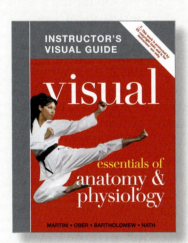

Instructor's Visual Guide

978-0-321-79280-8 /
0-321-79280-7

This guide is a printed and bound collection of thumbnails of the images and media on the IRDVD. (See previous page.) With this take-anywhere supplement, instructors can plan lectures when away from their computers.

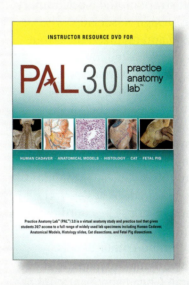

Instructor Resource DVD for Practice Anatomy Lab™ (PAL™)

978-0-321-74963-5/
0-321-74963-4

This DVD includes everything instructors need to present PAL 3.0 in lecture and lab. It includes all of the images in PowerPoint® and JPEG formats, links to animations, and access to a test bank with more than 4,000 lab practical questions.

Transparency Acetates

978-0-321-79279-2 / 0-321-79279-3

All figures and tables from the text are included in the printed Transparency Acetates. Complex figures are broken out for readable projected display. A full set of Transparency Acetate masters of all figures and tables is also available on the IRDVD. (See previous page.)

Blackboard

Preloaded book-specific content and test item files accompanying the text are available in Blackboard.

See pages xvi–xvii for MasteringA&P.

Acknowledgments

We feel very fortunate that we've been given the opportunity to use the *Visual* format to create a text for students taking the A&P essentials course. Our visual approach involves creating the text and art as a unit, and this means that the traditional sequential publishing process—manuscript generation to art development to editing to design to paging—does not apply. Over the years this project was in development, Paul Corey, President of Pearson Science, Frank Ruggirello, VP/Editorial Director of Applied Sciences, and Leslie Berriman, Executive Editor of Applied Sciences, showed unwavering support for this effort, and we thank them for their encouragement and collaboration.

The resulting text is a team effort. Ric Martini worked closely with Ed Bartholomew, Judi Nath, and Kathleen Welch as the pedagogical framework and clinical coverage evolved. Once a two-page module was drafted, Bill Ober worked with Claire Garrison and Anita Impagliazzo to produce the illustrations and a preliminary layout. Jim Gibson and Gibson Design Associates, who shaped the design, then adjusted and finalized the layout for optimal flow. This process was repeated for each module and every iteration as chapters were created, reviewed, and revised. Without the creative talents of this team, this project would not have been possible.

We would also like to thank our development editor, Alice Fugate, who assisted us by organizing and summarizing the review comments and steering us toward the necessary improvements to the modules. Throughout this process, Robin Pille, our project editor at Pearson Science, kept the wheels turning in-house by coordinating reviews, organizing manuscript development, supervising photo research, and handling a million other details. Barbara Yien oversaw the development and helped to keep the book on track. Deborah Cogan and Nancy Tabor invented new processes to shepherd this book through production and into print. Michael Rossa's careful attention to detail in his copyedit resulted in many important changes for improved consistency and accuracy. Thanks also to Janet Brodsky of Ivy Tech Community College—Lafayette for her accuracy reviews of the chapters and answer keys. We are deeply grateful to Tiffany Timmerman for her leadership of the skilled team at S4Carlisle in moving the book smoothly through composition. Thanks to Karen Hollister for her dedicated and careful work indexing the book.

We are grateful to Joe Mochnick, media producer at Pearson Science, for his careful management of the media resources for instructors and students, especially the new MasteringA&P program. Thanks also to Wendy Romaniecki, Brooke Keeney, Krysia Lazarewicz, Sade McDougal, and Pete Ratkevich for their work on MasteringA&P and the Study Area.

We are also grateful to Nicole McFadden, Associate Editor at Pearson Science, for her dedicated work on the print and media supplements and to Shannon Kong and Nancy Tabor for shepherding them smoothly through production.

Thanks also to Claire Alexander, Jeanie Chung, Hilary Thompson, Michelle Cadden, and Erik Fortier for their contributions to the editorial, media, and market development of this new book.

We are grateful to the numerous health care professionals who agreed to be interviewed for the Career Paths that appear in each chapter—thank you for taking the time to inspire the next generation of students to excel in their chosen careers.

Thanks are due to Robin A. Wagner of Lansing Community College for the Test Bank, to Betsy Brantley of Valencia College for the Lecture Presentations, to Jeff Schinske of De Anza College for the Instructor's Manual, to Margaret (Betsy) Ott of Tyler Junior College for the Clicker Questions and Quiz Shows, and to Angela Edwards of Trident Technical College for her many accuracy reviews. Thanks to Susan Heaphy of The Ohio State University at Lima, Dee Ann Sato of Cypress College, Janet Brodsky of Ivy Tech Community College—Lafayette, and Lu Clark of Lansing Community College for the assessments in MasteringA&P.

Finally, we would like to thank Brooke Suchomel, Senior Market Development Manager, Derek Perrigo, Marketing Manager, and all of the enthusiastic Pearson Science sales representatives who worked tirelessly to present this book to instructors and students over the course of its development to garner invaluable feedback and relay it to us so that we could integrate it into every page.

We are indebted to the following instructors for their comments and suggestions:

Instructor Reviewers

Teresa Alvarez
  St. Louis Community College—Forest Park

Meghan Adrikanich
  Lorain County Community College

Michelle Baragona
  Northeast Mississippi Community College

Carlena M. Benjamin
  Northeast Mississippi Community College

Mahmoud Bishr
  Metropolitan Community College—Penn Valley

Betsy Brantley
  Valencia College

Tom Carson
  Bossier Parish Community College

M. Aftab Chaudhry
  St. Louis Community College—Forest Park

Carlota B. Cinco
  Homestead Schools Inc.

Lu Anne Clark
  Lansing Community College

Estelle Coffino
  The College of Westchester

Jonathon P. Cohen
  Kankakee Community College

Janie Corbitt
  Central Georgia Technical College

Fleurdeliza Cuyco
  Preferred College of Nursing—Los Angeles

Louise DeRagon
  Mildred Elley

Andrea L. Dievendorf
  Mildred Elley

Marirose T. Ethington
  Genesee Community College

John E. Fishback
  Ozarks Technical Community College

Maria Florez
  Lone Star College—CyFair

Eric Forman
  Sauk Valley Community College

Andrew Goliszek
  North Carolina A&T State University

Roberto B. Gonzales
  Northwest Vista College

Susan Heaphy
  The Ohio State University at Lima

Peggy S. Hill
  The University of Tulsa

Dawn Hilliard
  Northeast Mississippi Community College

William Huber
  St. Louis Community College—Forest Park

Sandra Hutchinson
  Sinclair Community College

Murray Jensen
  University of Minnesota

Susanne Kalup
  Westmoreland County Community College

Vicki Kane
  YTI Career Institute—Altoona Campus

Laura Juarez de Ku
  Austin Community College

Steven Lewis
  Metropolitan Community College—Penn Valley

Lynette S. McCullough
  Southern Crescent Technical College

Melissa Meador
  Arkansas State University—Beebe

Monica M. Miklo
  Stark State College

Nathanial Mills
  Texas Woman's University—Denton

Javanika Mody
  Anne Arundel Community College

Liza Mohanty
  Olive-Harvey College

Terri Nicolau
  Coastal Bend College—Alice

Adriana Nunemaker
  Laredo Community College

Angela Porta
  Kean University

Julie M. Porterfield
  Tulsa Community College

Karla Rues
  Ozarks Technical Community College

Shaumarie Scoggins
  Texas Woman's University—Denton

Alan C. Sherrer
  Itawamba Community College

Deborah Shields
  Charter College

Betty Sims
  Coastal Bend College—Beeville

Stephen Smith
  Olive-Harvey College

Brian Stout
  Northwest Vista College

Liz Torrano
  American River College

Leticia Vosotros
  Ozarks Technical Community College

Mindy Walker
  Rockhurst University

Ronika Williams
  Coastal Bend College—Beeville

Christina Wills
  Rockhurst University

Stacy Zell
  University of Rio Grande

Class Testers

Teresa Alvarez
  St. Louis Community College—Forest Park

Michelle Baragona
  Northeast Mississippi Community College

Patty Bostwick-Taylor
  Florence-Darlington Technical College

Diep Burbridge
  Long Beach City College

Alex Cheroske
  Mesa Community College—Red Mountain

Carlota B. Cinco
  Homestead Schools Inc.

Chad Cryer
  Austin Community College

Louise DeRagon
  Mildred Elley

Andrea L. Dievendorf
  Mildred Elley

Marirose T. Ethington
  Genesee Community College

Stefanie Freese
  Northeast Mississippi Community College

Gary Fultz
  McLennan Community College

Kristie Garner
  University of Arkansas—Ft. Smith

Anne Geller
  San Diego Mesa College

Wade Hagan
  Mt. San Jacinto College—Menifee

Kristine Hicks
  University of Central Arkansas

Dawn Hilliard
  Northeast Mississippi Community College

William Huber
  St. Louis Community College—Forest Park

Sandra Hutchinson
  Santa Monica College

Murray Jensen
  University of Minnesota

Class Testers (continued)

Jill Johnson
Kirkwood Community College

Laura Juarez de Ku
Austin Community College

Tom Justice
McLennan Community College

Susanne Kalup
Westmoreland County Community College

Vicki Kane
YTI Career Institute—Altoona Campus

Charles Matsuda
Kapiolani Community College

Melissa Meador
Arkansas State University—Beebe

Sunil Nityanand
Lansing Community College

Izak Paul
Mount Royal University

Julie M. Porterfield
Tulsa Community College

David Sanchez
Austin Community College

Neal Schmidt
Pittsburg State University

Shaumarie Scoggins
Texas Woman's University—Denton

Alan C. Sherrer
Itawamba Community College

Deborah Shields
Charter College

Mary Sides
McLennan Community College

Stephen Smith
Olive-Harvey College

George Spiegel
College of Southern Maryland

Shannon Thomas
McLennan Community College

Liza Tobias-Cuyco
Preferred College of Nursing

Liz Torrano
American River College

Chase Tydell
Cerritos College

Jennifer Wester
Itawamba Community College

Kevin Young
Arizona Western College

Stacy Zell
University of Rio Grande

Contents

The Integumentary System 122

The Skeletal System 144

The Muscular System 204

The Central Nervous System 254

The Peripheral and Autonomic Nervous Systems 290

The Senses 318

The Endocrine System   358

Blood and Blood Vessels   384

The Heart and Cardiovascular Function 432

The Lymphatic System and Immunity 466

The Respiratory System   496

The Digestive System   520

Metabolism and Energetics 558

The Urinary System and Fluid, Electrolyte, and Acid-Base Balance 590

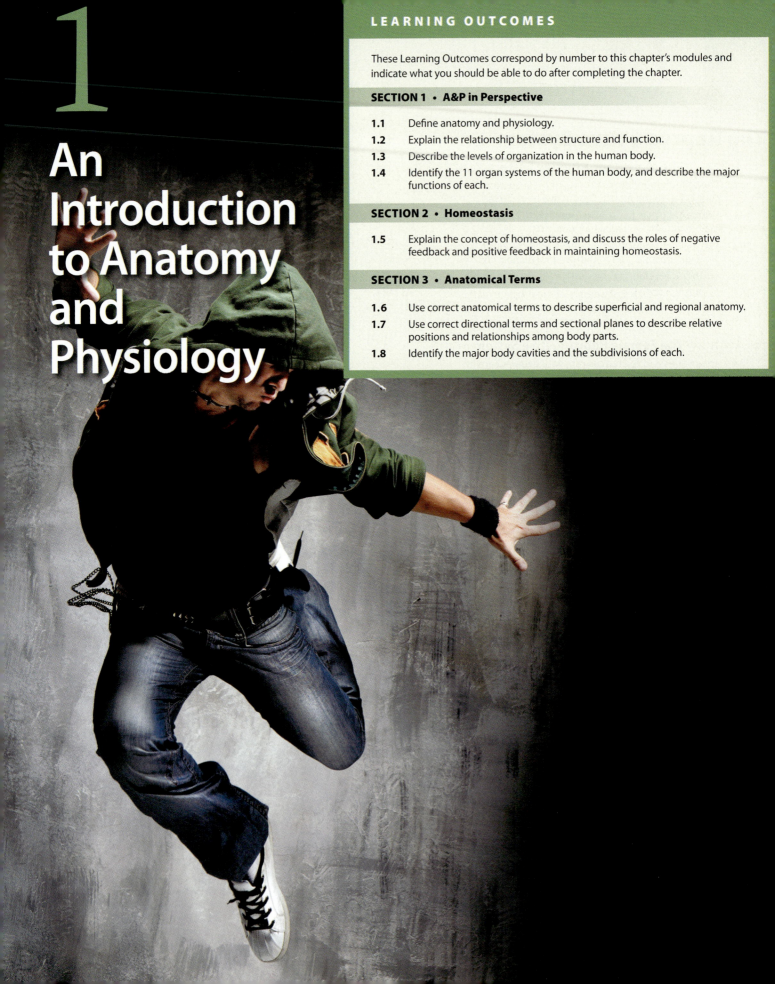

1

An Introduction to Anatomy and Physiology

A&P in Perspective

Human anatomy and physiology considers how your body performs the functions that keep you alive and alert. You will learn many interesting and important facts about the human body as we proceed. However, the approach you learn and the attitude you develop will be at least as important as the things you memorize. We can sum up the basic approach in A&P as "What is that structure, and how does it work?" The complexity of the answer depends on the level of detail you need. In science, if we know that something works but we don't know how, it's usually called a "Black Box." The more you learn, the smaller (and more numerous) those Black Boxes become; the more you learn, the more you realize how much you don't know!

We will be looking at how the body responds to normal and abnormal conditions and maintains **homeostasis**, a relatively constant internal environment. As we proceed, you will see how your body copes with injury, disease, or anything that threatens homeostasis.

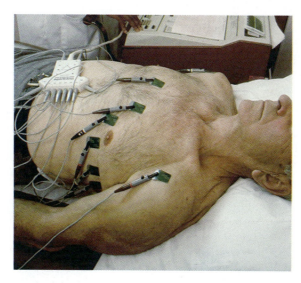

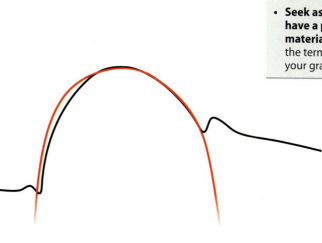

Tips on How to Succeed in Your A&P Course

- **Approach the information in different ways**. For example, you might visualize the information, talk it over with or "teach" a fellow student, or spend additional time in lab asking questions of your lab instructor.

- **Set up a study schedule** and stick to it.

- **Devote a block of time each day** to your A&P course.

- **Practice memorization.** Memorization is an important skill, and an integral part of the course. You are going to have to memorize all sorts of things—among them muscle names, directional terms, and the names of bones and brain parts. Realize that the more you practice, the better you will be at remembering terms and definitions. We will try to give you handles and tricks along the way, to help you keep the information in mind.

- **Avoid shortcuts.** Actually there are no shortcuts. (Sorry.) You won't get the grade you want if you don't put in the time and do the work. This requires preparation throughout the term.

- **Attend all lectures, labs, and study sessions**. Ask questions and participate in discussions.

- **Read your lecture and lab assignments before** coming to class.

- **Do not procrastinate**! Do not do all your studying the night before the exam! Actually STUDY the material several times throughout the week. Marathon study sessions are often counterproductive. There is no easy button; you must push yourself.

- **Seek assistance immediately if you have a problem understanding the material.** Do not wait until the end of the term when it is too late to salvage your grade.

Anatomy is the study of form . . .

Anatomy, which means "a cutting open," is the study of internal and external structures of the body and the physical relationships among body parts. Here is an overview of the anatomy of the heart, with the walls opened so that you can see the complexity of its internal structure.

1   **Gross anatomy**, or **macroscopic anatomy**, involves examining relatively large structures and features usually visible with the unaided eye. This illustration of a dissected heart is an example of gross anatomy.

2   **Microscopic anatomy** deals with structures that cannot be seen without magnification, and thus the boundaries of microscopic anatomy are established by the limits of the equipment used. With a dissecting microscope, you can see tissue structure. With a light microscope, you can see basic details of cell structure. With an electron microscope, you can see individual molecules that are only a few nanometers (billionths of a meter) across.

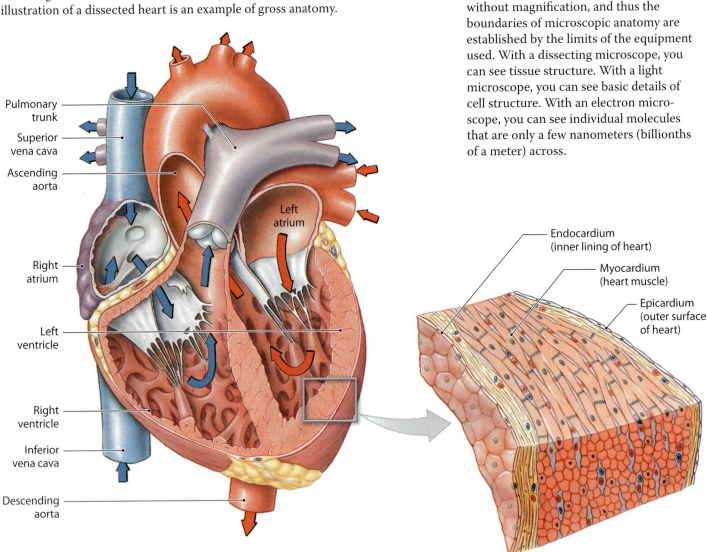

Pulmonary trunk

Superior vena cava

Ascending aorta

Left atrium

Right atrium

Left ventricle

Right ventricle

Inferior vena cava

Descending aorta

Endocardium (inner lining of heart)

Myocardium (heart muscle)

Epicardium (outer surface of heart)

All specific functions are performed by specific structures. The link between structure and function is always present, but not always understood. For example, although scientists clearly described the anatomy of the heart as early as the 15th century, almost 200 years passed before we were able to demonstrate its function—the heart is a pump.

. . . physiology is the study of function

Physiology is the study of how living organisms function. Human physiology considers the functions of the human body. These functions are complex and much more difficult to examine than most anatomical structures. A physiologist looking at the heart focuses on its functional properties, such as the timing and sequence of the heartbeat, and its effects on blood pressure in the major arteries.

3 Electrical events within the heart muscle coordinate each heartbeat. We can detect those electrical events by monitoring electrodes placed on the body surface. A record of these electrical events is called an electrocardiogram, or ECG.

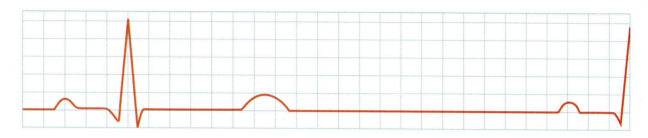

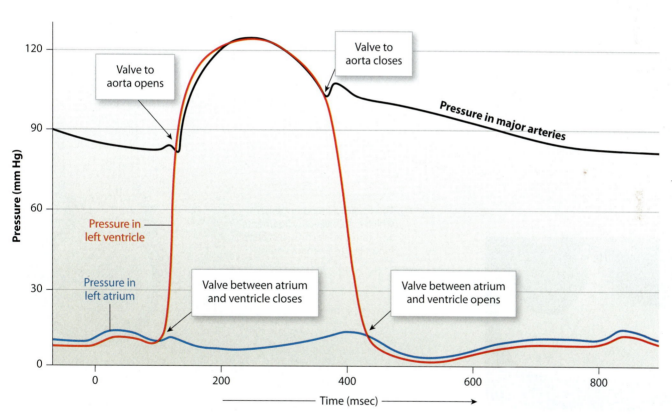

4 As the heart beats, pressure rises and falls within the major arteries and the chambers of the heart. Blood pressure in the major arteries must be maintained within normal limits to avoid damaging blood vessels (from high pressures) or collapsing the vessels (if the pressure falls too low).

Module 1.1 Review

a. Define anatomy and physiology.

b. What are the differences between gross anatomy and microscopic anatomy?

c. Explain the link between anatomy and physiology.

Form and function are interrelated

Anatomy and physiology are closely interrelated both theoretically and practically. Anatomical details are significant only because each has an effect on function, and physiological mechanisms can be fully understood only in terms of the underlying structural relationships.

1 We can easily understand this relationship at the gross anatomical level. You are well aware that your elbow joint functions like a hinge. It lets your forearm move toward or away from your shoulder, but it does not allow twisting at the joint. What imposes these functional limits? The anatomical structure of the joint imposes these limits.

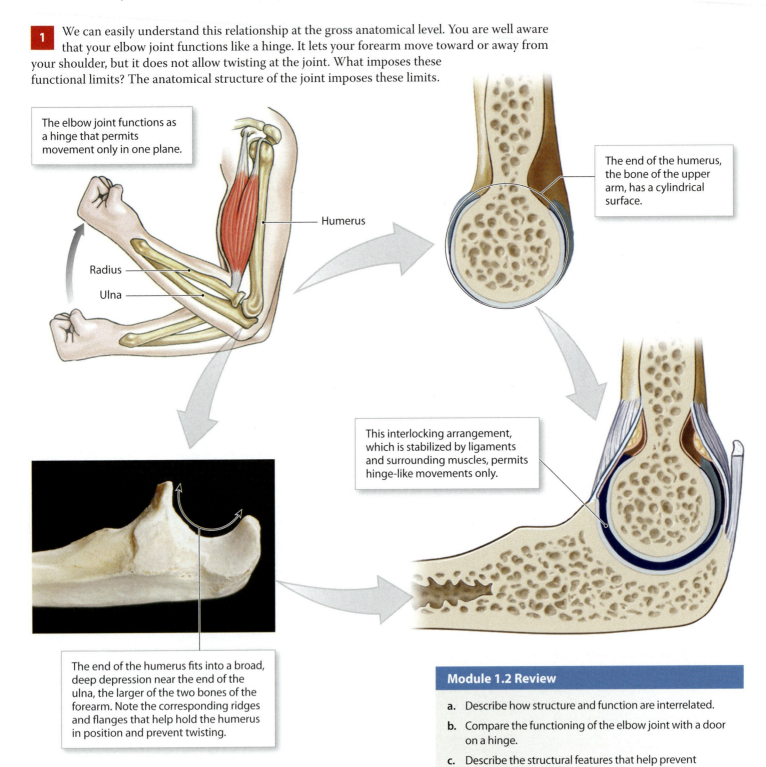

The elbow joint functions as a hinge that permits movement only in one plane.

Humerus

Radius

Ulna

The end of the humerus, the bone of the upper arm, has a cylindrical surface.

This interlocking arrangement, which is stabilized by ligaments and surrounding muscles, permits hinge-like movements only.

The end of the humerus fits into a broad, deep depression near the end of the ulna, the larger of the two bones of the forearm. Note the corresponding ridges and flanges that help hold the humerus in position and prevent twisting.

Module 1.2 Review

a. Describe how structure and function are interrelated.

b. Compare the functioning of the elbow joint with a door on a hinge.

c. Describe the structural features that help prevent twisting at the elbow joint.

The human body has multiple interdependent levels of organization

The human body is complex, but the apparent complexity represents multiple **levels of organization**. Each level is more complex than the underlying one, but all can be broken down into similar chemical and cellular components.

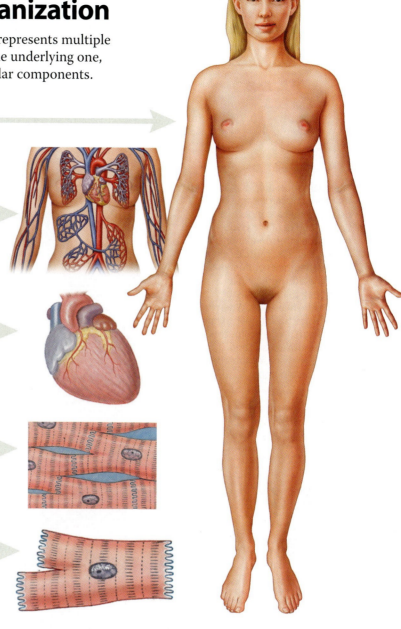

Organism Level. An **organism**—in this case, a human— is the highest level of organization. All organ systems of an organism's body work together to maintain life and health.

Organ System Level. (Chapters 4–19) Organs interact in organ systems. Each time it contracts, the heart pumps and pushes blood into a network of blood vessels. Together, the heart, blood, and blood vessels form the cardiovascular system, one of 11 **organ systems** in the body.

Organ Level. An **organ** consists of two or more tissues working in combination to perform several functions. Layers of heart muscle tissue, in combination with connective tissue (another type of tissue), form the bulk of the wall of the heart, a hollow, three-dimensional organ.

Tissue Level. (Chapter 4) A **tissue** is a group of cells and cell products working together to perform one or more specific functions. Heart muscle cells, or cardiac muscle cells (*cardium*, heart), form cardiac muscle tissue.

Cellular Level. (Chapter 3) **Cells** are the smallest living units in the body. Their functions depend on *organelles*, intracellular structures composed of complex molecules. Each organelle has a specific function; for example, one type provides the energy that powers the contractions of muscle cells in the heart.

The Chemical (or Molecular) Level. (Chapter 2) **Atoms**, the smallest stable units of matter, can combine to form molecules with complex shapes. The functional properties of a particular molecule are determined by its unique three-dimensional shape and atomic components.

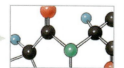
Atoms in combination

Complex protein molecules

Protein filaments

Module 1.3 Review

a. Define organ.

b. Name the lowest level of biological organization that includes the smallest living units in the body.

c. List the levels of organization between cells and organisms.

Organs and organ systems perform vital functions

1 An **organ** is a functional unit composed of more than one tissue type. The particular combination and organization of tissues within an organ both determines and limits the organ's functions. An **organ system** consists of organs that interact to perform a specific range of functions, often in a coordinated fashion.

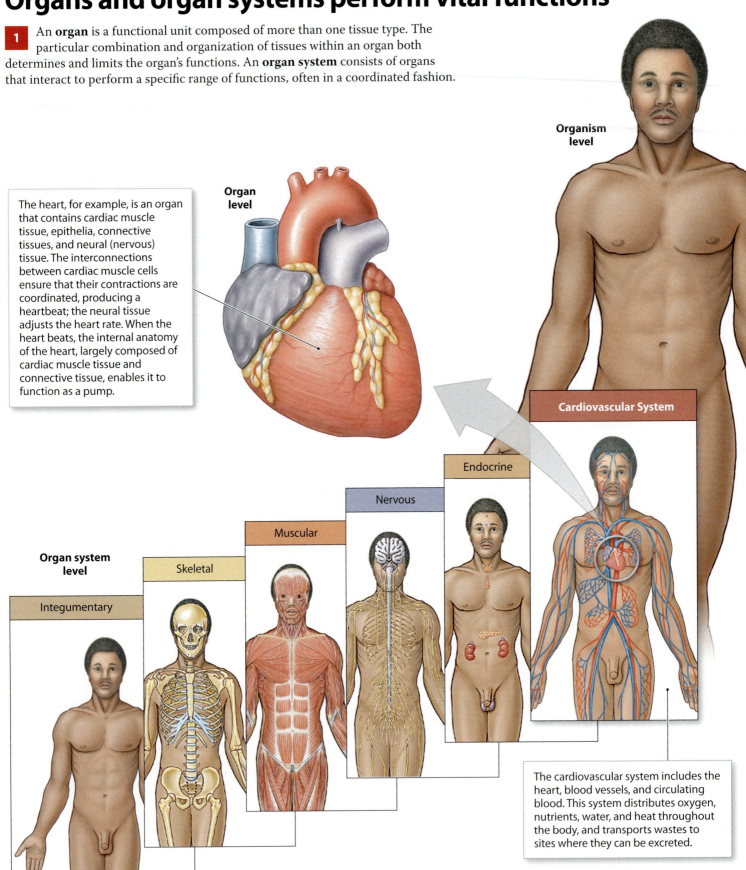

Organism level

Organ level

The heart, for example, is an organ that contains cardiac muscle tissue, epithelia, connective tissues, and neural (nervous) tissue. The interconnections between cardiac muscle cells ensure that their contractions are coordinated, producing a heartbeat; the neural tissue adjusts the heart rate. When the heart beats, the internal anatomy of the heart, largely composed of cardiac muscle tissue and connective tissue, enables it to function as a pump.

Cardiovascular System

Endocrine

Nervous

Muscular

Organ system level

Skeletal

Integumentary

The cardiovascular system includes the heart, blood vessels, and circulating blood. This system distributes oxygen, nutrients, water, and heat throughout the body, and transports wastes to sites where they can be excreted.

2 The table at right lists the 11 organ systems in the human body. Although this categorization is a convenient way to organize information, the concept of separate "organ systems" is artificial and somewhat misleading. Nothing in your body functions in isolation—not cells, not tissues, not organs, and certainly not organ systems. Organs and organ systems are interdependent, and something that affects one organ will affect the functioning of the body as a whole. For example, the heart cannot pump blood effectively after massive blood loss. If the heart cannot pump and blood cannot flow, oxygen and nutrients cannot be distributed. Very soon, cardiac muscle tissue begins to break down as individual muscle cells die from oxygen and nutrient starvation. These changes will not be restricted to the cardiovascular system; all cells, tissues, and organs in the body will be damaged, with potentially fatal results.

Organ Systems		Major Functions
	Integumentary system	Protects body from environmental hazards; controls body temperature
	Skeletal system	Supports, protects soft tissues; stores minerals; forms blood cells
	Muscular system	Moves and supports body; produces heat
	Nervous system	Direct immediate responses to stimuli, usually by coordinating the activities of other organ systems
	Endocrine system	Direct long-term changes in the activities of other organ systems
	Cardiovascular system	Transports cells and dissolved materials internally, including nutrients, wastes, and gases
	Lymphatic system	Defends against infection and disease
	Respiratory system	Delivers air to sites where gas exchange can occur between the air and circulating blood
	Digestive system	Processes food and absorps organic nutrients, minerals, vitamins, and water
	Urinary system	Eliminates excess water, salts, and waste products; control of pH
	Reproductive system	Produces sex cells and hormones

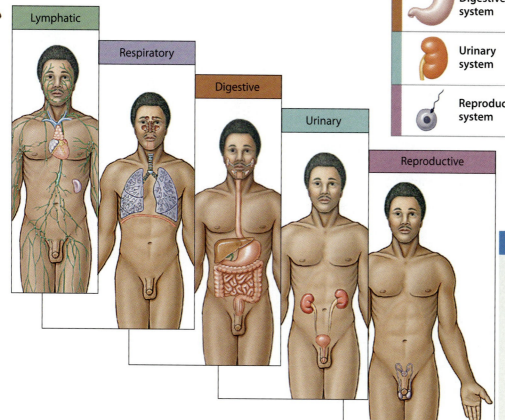

Lymphatic

Respiratory

Digestive

Urinary

Reproductive

Module 1.4 Review

a. List the 11 major organ systems of the body.

b. Explain the relationship between the skeletal system and the digestive system.

c. Using the table as a reference, describe how a compound fracture (bone break that protrudes through the skin) could affect at least six of your organ systems.

1. Short answer

Learning anatomy and physiology takes effort—there are no shortcuts. Describe eight study tips you might use to master this class.

a _____

b _____

c _____

d _____

e _____

f _____

g _____

h _____

2. Matching

Write each of the following terms under the proper heading.

	Anatomy	**Physiology**
• Right atrium		
• Myocardium	_____	_____
• Valve to aorta opens	_____	_____
• Left ventricle		
• Valve between left atrium and left ventricle closes	_____	_____
• Pressure in left atrium	_____	_____
• Electrocardiogram		
• Endocardium	_____	_____
• Superior vena cava	_____	_____
• Ulna		
• Moving forearm toward shoulder		

3. Matching

Order these six levels of organization of the human body from smallest (a) to largest (f).

_____ tissue   _____ cell   _____ organ   _____ molecule   _____ organism   _____ organ system

4. Short answer

Summarize the major functions of each of the following organ systems.

a skeletal system _____

b digestive system _____

c integumentary system _____

d urinary system _____

e nervous system _____

Homeostasis

Homeostasis (*homeo*, unchanging + *stasis*, standing) is the presence of a stable environment inside the body. Maintaining homeostasis is absolutely vital to an organism's survival; failure to maintain homeostasis soon leads to illness or even death. The principle of homeostasis is the central theme of this text and the foundation of all modern physiology. **Homeostatic regulation** is the adjustment of physiological systems to preserve homeostasis in environments that are often inconsistent, unpredictable, and potentially dangerous. An understanding of homeostatic regulation is crucial to making accurate predictions about the body's responses to both normal and abnormal conditions.

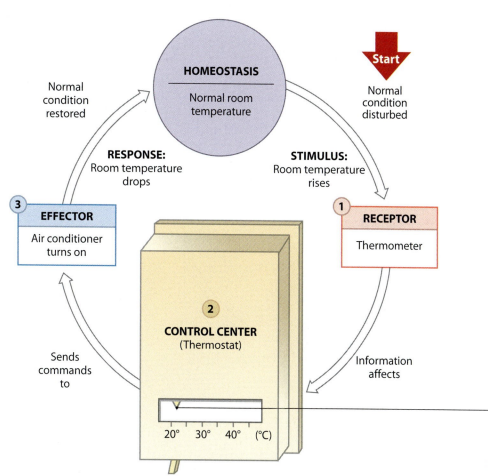

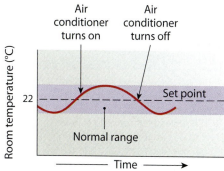

2 Homeostatic control is not precise—it maintains a normal range rather than an absolute value. The same is true for controlling room temperature—a house may have the thermostat on one wall of one room, and the air vents at multiple locations. Over time, the temperature in the house will oscillate around the set point.

The setting on a thermostat establishes the **set point**, or desired value, which in this case is the temperature you select. (In our example, the set point is 22°C, or about 72°F.) The function of the thermostat is to keep room temperature within acceptable limits, usually within a degree or so of the set point.

1 The maintenance of a relatively constant temperature in your living space is a familiar example of homeostasis. Like all homeostatic mechanisms, it consists of (1) a **receptor** or sensor—in this case, a thermometer—that is sensitive to a particular environmental change, or stimulus; (2) a **control center** or integration center—in this case, a thermostat—which receives and processes the information supplied by the receptor, and which sends out commands; and (3) an **effector**—in this case, an air conditioner—which responds to these commands by opposing the stimulus. The net effect is that any variation outside normal limits triggers a response that restores normal conditions.

Negative feedback provides stability . . .

Feedback occurs when receptor stimulation triggers a response that changes the environment at the receptor. In the case of temperature control by a thermostat, temperature variation outside the desired range triggers an automatic response that corrects the situation. This method of homeostatic regulation is called **negative feedback**, because an effector activated by the control center opposes, or negates, the original stimulus. Negative feedback thus tends to minimize change, keeping variation in key body systems within limits compatible with our long-term survival.

HOMEOSTASIS

At normal body temperature (set point: 37°C or 98.6°F), the temperature control center is relatively inactive; superficial blood flow and sweat gland activity are at normal levels.

Start

Homeostasis restored

Homeostasis disturbed

3 EFFECTORS

Increased activity in the control center targets two effectors: (1) smooth muscle in the walls of blood vessels supplying the skin and (2) sweat glands. The smooth muscle relaxes and the blood vessels dilate, increasing blood flow through vessels near the body surface; the sweat glands accelerate their secretion. The skin then acts like a radiator by losing heat to the environment, and the evaporation of sweat speeds the cooling process.

1 RECEPTORS

If body temperature rises above 37.2°C (99°F), two sets of temperature receptors, one in the skin and the other within the brain, send signals to the homeostatic control center.

Homeostasis and body temperature

2 CONTROL CENTER

The temperature control center receives information from the two sets of temperature receptors and sends commands to the effectors.

1 Negative feedback is the primary mechanism of homeostatic regulation, and it provides long-term control over the body's internal conditions and systems. Homeostatic mechanisms using negative feedback normally ignore minor variations, and they maintain a normal range rather than a fixed value. The regulatory process itself is dynamic, because the set point may vary with different activity levels and environments. For example, when you are asleep, your thermoregulatory set point is lower, whereas when you work outside on a hot day (or when you have a fever), it is higher. Thus, body temperature can vary from moment to moment or from day to day for any individual, due to either (1) small oscillations around the set point or (2) changes in the set point. Comparable variations occur in all other aspects of physiology.

...whereas positive feedback accelerates a process to completion

In **positive feedback**, an initial stimulus produces a response that exaggerates or enhances a change in the original conditions, rather than opposing it. You seldom encounter positive feedback in your daily life, simply because it tends to produce extreme responses. For example, suppose that the thermostat in your house was accidentally connected to a heater rather than to an air conditioner. Now, when room temperature exceeds the set point, the thermostat turns on the heater, causing a further rise in room temperature. Room temperature will continue to increase until someone switches off the thermostat, turns off the heater, or intervenes in some other way. This kind of escalating cycle is often called a **positive feedback loop**.

A break in a blood vessel wall causes bleeding

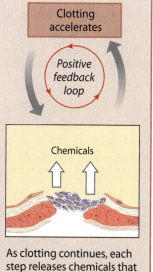

Clotting accelerates

Positive feedback loop

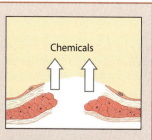

Chemicals

Damage to cells in the blood vessel wall releases chemicals that begin the process of blood clotting.

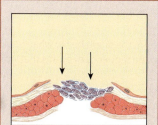

The chemicals start chain reactions in which cells, cell fragments, and soluble proteins in the blood begin to form a clot.

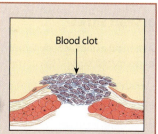

Chemicals

As clotting continues, each step releases chemicals that further accelerate the process.

Blood clot

This escalating process is a positive feedback loop that ends with the formation of a blood clot, which patches the vessel wall and stops the bleeding.

2 In the body, positive feedback loops are typically found when a potentially dangerous or stressful process must be completed quickly before homeostasis can be restored. For example, the immediate danger from a severe cut is loss of blood, which can lower blood pressure and reduce the efficiency of the heart.

Module 1.5 Review

a. Identify the components of homeostatic regulation.

b. Explain the function of negative feedback systems.

c. Why is positive feedback helpful in blood clotting but unsuitable for regulating body temperature?

1. Vocabulary

Write the term for each of the following descriptions in the space provided.

a   _____   Mechanism that increases a deviation from normal limits after an initial stimulus

b   _____   Adjustment of physiological systems to preserve homeostasis

c   _____   The maintenance of a relatively constant internal environment

d   _____   Regulation of the concentration of hormones circulating in the blood

e   _____   Corrective mechanism that opposes or cancels a variation from normal limits

2. Matching

Indicate whether each of the following processes matches with the process of negative feedback or positive feedback.

a   A rise in the level of calcium dissolved in the blood stimulates the release of a hormone that causes bone cells to absorb calcium from the blood and deposit it in bone.   _____

b   Labor contractions become increasingly forceful during childbirth.   _____

c   An increase in blood pressure triggers a nervous system response that results in lowering the blood pressure.   _____

d   Blood vessel cells damaged by a break in the vessel release chemicals that accelerate the blood clotting process.   _____

3. Short answer

Assuming a normal body temperature range of 36.7°–37.2°C (98°–99° F), identify from the graph below what would happen if body temperature increased or decreased beyond the normal limits. Use the list of descriptive terms to explain the sequence of events that would happen at (a) and (b) on the graph.

- body surface cools
- shivering occurs
- sweating increases
- temperature declines
- body heat is conserved
- blood flow to skin increases
- blood flow to skin decreases
- temperature rises

a _____

_____

_____

_____

b _____

_____

_____

_____

4. Section integration

It is a warm day but you feel a little chilled. On checking your temperature, you find that your body temperature is 1.5 degrees below normal. Suggest some possible reasons for this situation.

_____

_____

Anatomical Terms

Early anatomists created maps of the human body, and we still rely on maps for orientation. The landmarks are prominent anatomical structures; distances are measured in centimeters or inches; and we use specialized directional terms. In effect, anatomy uses a special language that must be learned almost at the start. Many terms are based on Latin or Greek words used by ancient anatomists. However, Latin and Greek terms are not the only ones that have been imported into the anatomical vocabulary over the centuries, and the vocabulary continues to expand. Many anatomical structures and clinical conditions were initially named after either the discoverer or, in the case of diseases, the most famous victim. Most of these commemorative names, or eponyms, have been replaced by more precise terms, but a few persist. The table below summarizes key word roots and derivatives that will help you understand new scientific terms as you encounter them throughout the book.

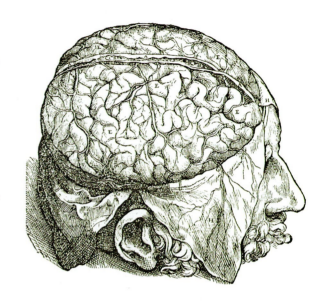

Important Word Roots, Prefixes, Suffixes, and Combining Forms in Anatomy

aer-, *aeros,* air: aerobic metabolism	**epi-,** *epi,* on: epimysium	**ost-, oste-, osteo-,** *osteon,* bone: osteal, ostealgia, osteocyte
arter-, *arteria,* artery: arterial	**hemo-,** *haima,* blood: hemopoiesis	**oto-,** *otikos,* ear: otolith
arthro-, *arthros,* joint: arthroscopy	**hemi-,** *hemi,* one-half: hemisphere	**peri-,** *peri,* around: perineurium
bio-, *bios,* life: biology	**histo-,** *histos,* tissue: histology	**phago-,** *phago,* to eat: phagocyte
-blast, *blastos,* germ: osteoblast	**hyper-,** *hyper,* above: hyperpolarization	**phot-, photo-,** *phos,* light: photalgia, photoreceptor
bronch-, *bronchus,* windpipe, airway: bronchial	**hypo-,** *hypo,* under: hypothyroid	**physio-,** *physis,* nature: physiology
cardi-, cardio-, -cardia, *kardia,* heart: cardiac, cardiopulmonary	**inter-,** *inter,* between: interventricular	**pre-,** *prae,* before: precapillary sphincter
cerebr-, *cerebrum,* brain: cerebral hemispheres	**iso-,** *isos,* equal: isotonic	**pulmo-,** *pulmo,* lung: pulmonary
cervic-, *cervicis,* neck: cervical vertebrae	**leuk-, leuko-,** *leukos,* white: leukemia, leukocyte	**retro-,** *retro,* backward: retroperitoneal
chondro-, *chondros,* cartilage: chondrocyte	**lyso-, -lysis, -lyze,** *lysis,* a loosening: hydrolysis	**sarco-,** *sarkos,* flesh: sarcomere
cranio-, *cranium,* skull: craniosacral	**myo-,** *mys,* muscle: myofilament	**scler-, sclero-,** *skleros,* hard: sclera, sclerosis
cyt-, cyto-, *kytos,* a hollow cell: cytology, cytokine	**nephr-,** *nephros,* kidney: nephron	**sub-,** *sub,* below: subcutaneous
derm-, *derma,* skin: dermatome	**neur-, neuri-, neuro-,** *neuron,* nerve: neural, neurilemma, neuromuscular	**super-,** *super,* above or beyond: superficial
end-, endo-, *endon,* within: endergonic, endometrium	**-ology,** *logos,* the study of: physiology	**vas -,** *vas,* vessel: vascular

Superficial anatomy and regional anatomy indicate locations on or in the body

1 This illustration shows the body in the **anatomical position**. In this position, the hands are at the sides with the palms facing forward, and the feet are together. Unless otherwise noted, all descriptions in this text refer to the body in the anatomical position. A person lying down in the anatomical position is said to be **supine** (soo-PĪN) when face up, and **prone** when face down. Anatomical terms are shown in boldface type and common names are in plain type.

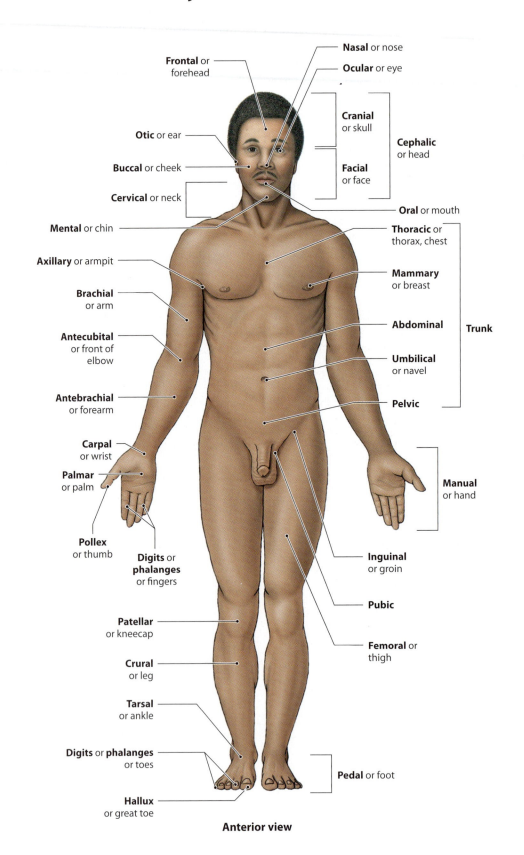

Frontal or forehead
Nasal or nose
Ocular or eye
Cranial or skull
Otic or ear
Facial or face
Cephalic or head
Buccal or cheek
Cervical or neck
Oral or mouth
Mental or chin
Thoracic or thorax, chest
Axillary or armpit
Mammary or breast
Brachial or arm
Abdominal
Trunk
Antecubital or front of elbow
Umbilical or navel
Antebrachial or forearm
Pelvic
Carpal or wrist
Palmar or palm
Manual or hand
Pollex or thumb
Digits or **phalanges** or fingers
Inguinal or groin
Pubic
Patellar or kneecap
Femoral or thigh
Crural or leg
Tarsal or ankle
Digits or **phalanges** or toes
Pedal or foot
Hallux or great toe

Anterior view

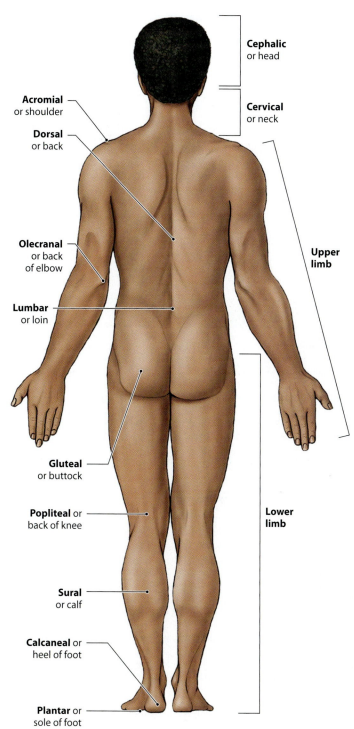

Cephalic or head

Cervical or neck

Acromial or shoulder

Dorsal or back

Olecranal or back of elbow

Lumbar or loin

Gluteal or buttock

Popliteal or back of knee

Sural or calf

Calcaneal or heel of foot

Plantar or sole of foot

Upper limb

Lower limb

Posterior view

2 Clinicians refer to four **abdominopelvic quadrants** formed by a pair of imaginary perpendicular lines that intersect at the umbilicus (navel). This simple method provides useful references for describing aches, pains, and injuries. The location can help physicians determine the possible cause.

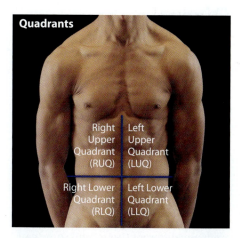

Quadrants

Right Upper Quadrant (RUQ)

Left Upper Quadrant (LUQ)

Right Lower Quadrant (RLQ)

Left Lower Quadrant (LLQ)

3 Anatomists prefer more precise terms to describe the location and orientation of internal organs. Anatomists recognize nine **abdominopelvic regions**.

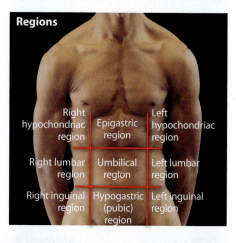

Regions

Right hypochondriac region

Epigastric region

Left hypochondriac region

Right lumbar region

Umbilical region

Left lumbar region

Right inguinal region

Hypogastric (pubic) region

Left inguinal region

4 The image at the lower right shows the relationships among quadrants, regions, and internal organs.

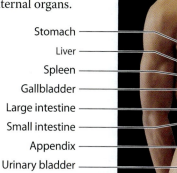

Internal organs

Stomach
Liver
Spleen
Gallbladder
Large intestine
Small intestine
Appendix
Urinary bladder

Module 1.6 Review

a. Describe a person in the anatomical position.

b. Contrast the descriptions used by clinicians and anatomists when referring to the positions of injuries or internal organs of the abdomen and pelvis.

c. A massage therapist often begins a massage by asking patrons to lie face down with their arms at their sides. What anatomical term describes that position?

Directional and sectional terms describe specific points of reference

1 The figure and table on this page introduce the principal directional terms and examples of their use. There are many different terms, and some can be used interchangeably. As you learn these directional terms, it is important to remember that all anatomical directions utilize the anatomical position as the standard point of reference.

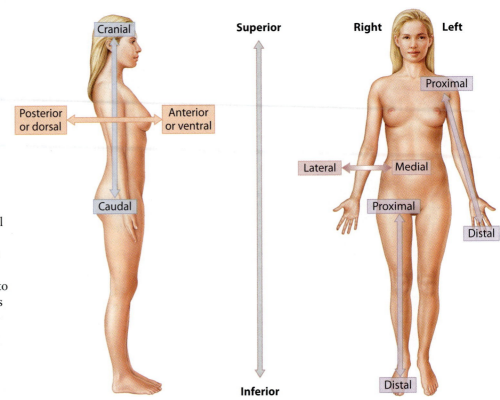

Directional Terms

Term	Region or Reference	Example
Anterior or **ventral**	The front surface of the human body	The navel is on the *anterior* surface of the trunk.
Posterior or **dorsal**	The back surface of the human body	The shoulder blade is located *posterior* to the rib cage.
Cranial or **cephalic**	The head	The *cranial*, or *cephalic*, border of the pelvis is on the side toward the head rather than toward the thigh.
Superior	Above; at a higher level (in the human body, toward the head)	In humans, the cranial border of the pelvis is *superior* to the thigh.
Caudal	The tail (coccyx in humans)	The hips are *caudal* to the waist.
Inferior	Below; at a lower level	The knees are *inferior* to the hips.
Medial	Toward the body's longitudinal axis; toward the midsagittal plane	The *medial* surfaces of the thighs may be in contact; moving medially from the arm across the chest surface brings you to the sternum.
Lateral	Away from the body's longitudinal axis; away from the midsagittal plane	The thigh bone forms a joint with the *lateral* surface of the pelvis; moving laterally from the nose brings you to the cheeks.
Proximal	Toward an attached base	The thigh is *proximal* to the foot; moving proximally from the wrist brings you to the elbow.
Distal	Away from an attached base	The fingers are *distal* to the wrist; moving distally from the elbow brings you to the wrist.
Superficial	At, near, or relatively close to the body surface	The skin is *superficial* to underlying structures.
Deep	Farther from the body surface	The bone of the thigh is *deep* to the surrounding skeletal muscles.

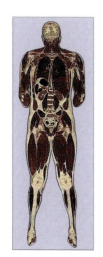 

Frontal plane

Sagittal plane

Transverse plane

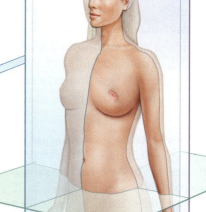

Terms That Indicate Sectional Planes

Plane	Orientation of Plane	Directional Reference	Description
Transverse or horizontal	Perpendicular to long axis	Transversely or horizontally	A **transverse**, or **horizontal**, **section** separates superior and inferior portions of the body. A cut in this plane is called a **cross section**.
Sagittal	Parallel to long axis	Sagittally	A **sagittal section** separates right and left portions. You examine a sagittal section, but you section sagittally.
Midsagittal	Parallel to long axis	Sagittally	In a **midsagittal section** or **median section**, the plane passes through the midline, dividing the body into right and left halves.
Parasagittal	Parallel to long axis	Sagittally	A **parasagittal section**, which is a cut parallel to the midsagittal plane, separates the body into right and left portions of unequal size.
Frontal or coronal	Parallel to long axis	Frontally or coronally	A **frontal**, or **coronal**, **section** separates anterior and posterior portions of the body; *coronal* usually refers to sections passing through the skull.

2 A presentation in sectional view is sometimes the only way to illustrate the relationships between the parts of a three-dimensional object. The development of medical imaging techniques has made it even more important to understand sectional planes and terms.

Module 1.7 Review

a. What is the purpose of directional and sectional terms?

b. In the anatomical position, describe an anterior view and a posterior view.

c. What type of section would separate the two eyes?

Body cavities protect internal organs and allow them to change shape

The interior of the body is often subdivided into regions established by the body wall. For example, everything deep to the chest wall is considered to be within the **thoracic cavity**, and all of the structures deep to the abdominal and pelvic walls are said to lie within the **abdominopelvic cavity**. Many vital organs within these regions are suspended within fluid-filled chambers that are called **body cavities**. Body cavities have two essential functions: (1) They protect delicate organs from shocks and impacts; and (2) they permit significant changes in the size and shape of internal organs.

1 The internal organs that are partially or completely enclosed by body cavities are called **viscera** (VIS-e-ruh) or visceral organs. Viscera do not float within the body cavities—they remain connected to the rest of the body. To understand the physical relationships, we will examine the smallest subdivision of the ventral body cavity, the **pericardial cavity** that surrounds the heart.

2 During embryological development, a single **ventral body cavity**, or **coelom** (SĒ-lōm; *koila*, cavity), forms, and it contains organs of the respiratory, cardiovascular, digestive, urinary, and reproductive systems. The ventral body cavity is later subdivided into separate body cavities whose boundaries are indicated in red. Three of these subdivisions lie within the thoracic cavity and one lies in the abdominopelvic cavity.

The relationship between the heart and the pericardial cavity resembles that of a fist pushing into a balloon. The wrist corresponds to the base (attached portion) of the heart, and the balloon corresponds to the lining of the pericardial cavity.

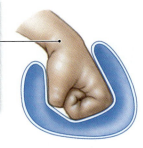

Base of heart

During each beat, the heart changes size and shape. The pericardial cavity permits these changes, and the pericardial lining, which is moist and slippery, prevents friction between the heart and adjacent structures.

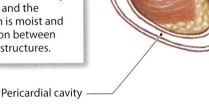

Pericardial cavity

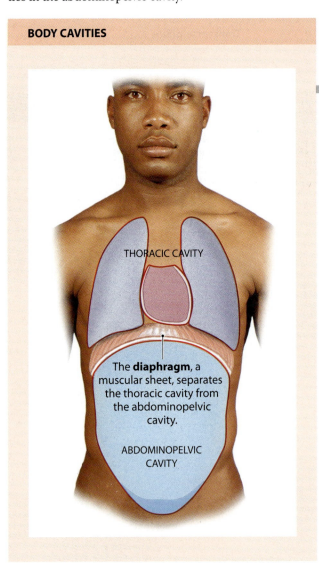

BODY CAVITIES

THORACIC CAVITY

The **diaphragm**, a muscular sheet, separates the thoracic cavity from the abdominopelvic cavity.

ABDOMINOPELVIC CAVITY

3 The thoracic cavity contains the lungs, heart, and other structures. Its boundaries are established by the chest wall and diaphragm.

THORACIC CAVITY

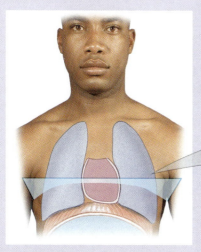

A horizontal section through the thoracic cavity shows the relationship between the subdivisions of the ventral body cavity in this region.

Each lung is surrounded by a **pleural cavity**.

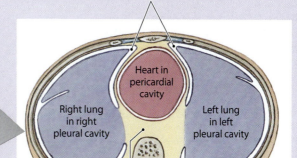

Heart in pericardial cavity

Right lung in right pleural cavity

Left lung in left pleural cavity

Note the orientation of the section. Unless otherwise noted, all cross sections are shown as inferior views. That is, just as if you are standing at the feet of a supine person and looking toward the head.

The pericardial cavity is embedded within the **mediastinum**, a mass of connective tissue that separates the two pleural cavities and stabilizes the positions of embedded organs and blood vessels.

ABDOMINOPELVIC CAVITY

During development, the portion of the original ventral body cavity extending into the abdominopelvic cavity remains intact as the **peritoneal** (per-i-tō-NĒ-al) **cavity**. A few organs, such as the kidneys and pancreas, lie between the peritoneal lining and the muscular wall of the abdominal cavity. Those organs are said to be **retroperitoneal** (re-trō-per-i-tō-NĒ-al; *retro*, behind).

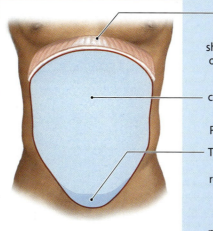

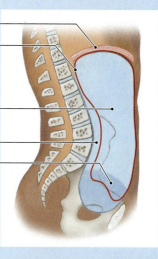

Diaphragm

Peritoneum (red) showing the boundaries of the peritoneal cavity

The abdominal cavity contains many digestive glands and organs

Retroperitoneal area

The pelvic cavity contains the urinary bladder, reproductive organs, and the last portion of the digestive tract; many of these structures lie posterior to, or inferior to, the peritoneal cavity.

4 The boundaries of the abdominopelvic cavity are established by the diaphragm, the muscles of the abdominal wall, the trunk muscles and inferior portions of the vertebral column, and the bones and muscles of the pelvis. It may be subdivided into the **abdominal cavity** and the **pelvic cavity**.

Module 1.8 Review

a. Describe two essential functions of body cavities.

b. Identify the subdivisions of the ventral body cavity.

c. If a surgeon makes an incision just inferior to the diaphragm, which body cavity will be opened?

1. Labeling

Label the directional terms in the figures at right.

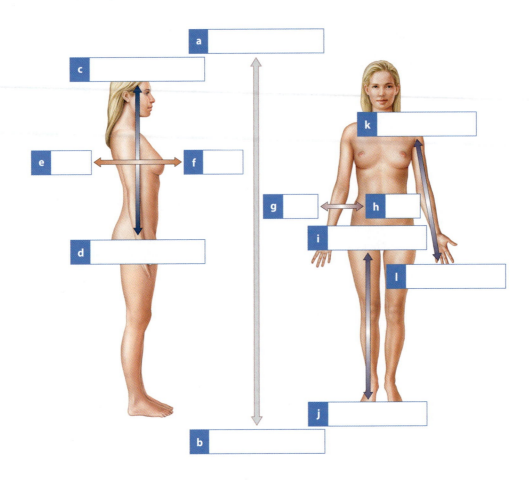

2. Concept map

Use each of the following terms once to fill in the blank boxes to correctly complete the body cavities concept map.

- digestive glands and organs
- abdominopelvic cavity
- thoracic cavity
- heart
- mediastinum
- diaphragm
- pelvic cavity
- blood vessels
- reproductive organs
- left lung
- peritoneal cavity

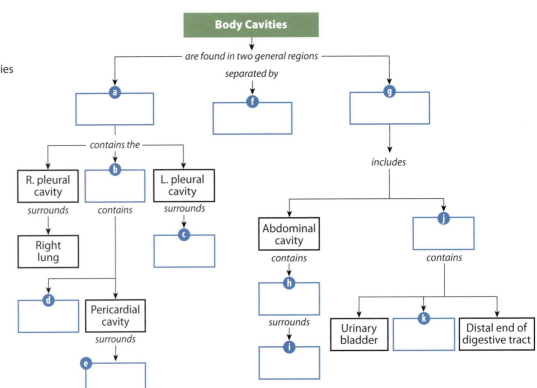

Visual Outline with Key Terms

Summarize the content of each module using the terms in the order provided.

SECTION 1

A&P in Perspective

- homeostasis

SECTION 1.4

Organs and organ systems perform vital functions

- organ
- organ system
- integumentary system
- skeletal system
- muscular system
- nervous system
- endocrine system
- cardiovascular system
- lymphatic system
- respiratory system
- digestive system
- urinary system
- reproductive system

1.1

Anatomy is the study of form; physiology is the study of function

- anatomy
- gross anatomy
- macroscopic anatomy
- microscopic anatomy
- physiology

SECTION 2

Homeostasis

- homeostasis
- homeostatic regulation
- receptor
- control center
- effector
- set point

1.2

Form and function are interrelated

- elbow joint
- humerus
- radius
- ulna

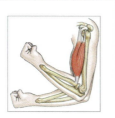

1.5

Negative feedback provides stability, whereas positive feedback accelerates a process to completion

- feedback
- negative feedback
- positive feedback
- positive feedback loop

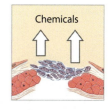

1.3

The human body has multiple interdependent levels of organization

- levels of organization
- organism
- organ systems
- organ
- tissue
- cells
- atoms

SECTION 3

Anatomical Terms

- eponyms
- word roots

● = *Term boldfaced in this module*

1.6

Superficial anatomy and regional anatomy indicate locations on or in the body

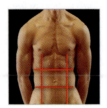

- anatomical position
- supine
- prone
- abdominopelvic quadrants
- abdominopelvic regions

1.8

Body cavities protect internal organs and allow them to change shape

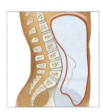

- thoracic cavity
- abdominopelvic cavity
- body cavities
- viscera
- pericardial cavity
- ventral body cavity
- coelom
- diaphragm
- pleural cavity
- mediastinum
- peritoneal cavity
- retroperitoneal
- abdominal cavity
- pelvic cavity

1.7

Directional and sectional terms describe specific points of reference

- anterior
- ventral
- posterior
- dorsal
- cranial
- cephalic
- superior
- caudal
- inferior
- medial
- lateral
- proximal
- distal
- superficial
- deep
- transverse section
- horizontal section
- cross section
- sagittal section
- midsagittal section
- median section
- parasagittal section
- frontal section
- coronal section

● = *Term boldfaced in this module*

CAREER PATHS

Diagnostic Medical Sonographer

"There is an art to ultrasound... we have to know as much anatomy as a surgeon in order to see cross sections."

— **Diana Miller**
Diagnostic Medical Sonographer,
West Coast Radiology,
Santa Ana, California

Diana Miller, a diagnostic medical sonographer, began her career as an x-ray technician in 1975. Since then, medical imaging has come a long way, and Diana has embraced the changes.

Diana became an x-ray technician because at the time she was looking for a career she could start quickly without going to college. When the opportunity arose to learn about the then-new practice of ultrasound, Diana jumped at it. "There is an art to ultrasound," she says. "You don't just go in and press a button and there's the x-ray. You have so much more responsibility; you have so much more respect. We have to know as much anatomy as a surgeon in order to see cross sections."

As a sonographer, Diana works at West Coast Radiology, a private-practice radiology

group in Santa Ana, California. She performs ultrasounds for about 14 patients a day and oversees the nine ultrasound machines in the practice. Diana does obstetric ultrasounds, but she also images every other organ in

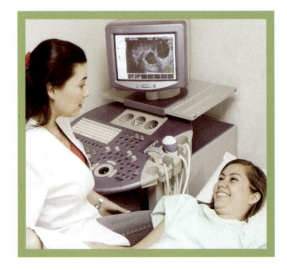

the body (except the heart, which is usually done by a cardiologist). "A lot of students who shadow me say they want to go into ultrasound because they want to look at babies," she says. "I, personally, would not want to be focused on just [obstetrics], or any one thing. I like abdominal work—the liver, pancreas, gallbladder. There are a lot of variations of normal, and there's a lot of pathology." A sonographer, for instance, is responsible for creating an image during a biopsy to show the doctor exactly where the tumor or organ to be biopsied is located.

Sonographers must interact with patients constantly, and Diana stresses the need for good people skills. "You have sick people who are anxious," she says. "You have people who aren't sick who think they are." Diana also recommends that sonographers have strong organizational skills, as well as steady nerves and hands. And of course, without a thorough knowledge of anatomy, a sonographer would be incapable of following an order to scan a patient's liver. In fact, Diana says that, on occasion, she has done additional scans based on where patients tell her they have pain. "In ultrasound, if you're very good, you learn to think outside the box, and you really help everybody," she says. "You can go beyond what they tell you to look at. I like that responsibility."

For more information, visit the website for the American Registered Diagnostic Medical Sonographers at http://www.ardms.org.

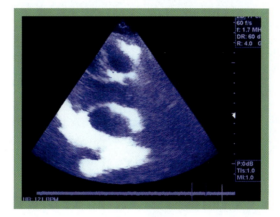

Think this is the career for you?
Key Stats:

- **Education and Training.** Formal training offered in 2-year associate degree programs and 4-year bachelor's degree programs (2-year programs are more common).

- **Licensure.** No states require licensure, but employers prefer to hire registered sonographers. To become registered, candidates must pass a national examination, and they must typically complete a required number of continuing-education hours at prescribed intervals to maintain registration.

- **Earnings.** Earnings vary but the median annual wage is $64,380.

- **Expected Job Prospects.** Employment is expected to grow faster than the national average—by 18% through 2018.

Bureau of Labor Statistics, U.S. Department of Labor, *Occupational Outlook Handbook, 2010-11 Edition,* Diagnostic Medical Sonographers, on the Internet at http://www.bls .gov/oco/ocos273.htm (visited September 14, 2011).

Access more review material online in the Study Area at www.masteringaandp.com.

- Chapter guides
- Chapter quizzes
- Practice tests
- Art labeling activities
- Flashcards
- A glossary with pronunciations
- Practice Anatomy Lab™ (PAL™) 3.0 virtual anatomy practice tool
- Interactive Physiology® (IP) animated tutorials
- MP3 Tutor Sessions

 For this chapter, go to this topic in the MP3 Tutor Sessions:

- Homeostasis

For more information about all the extra practice available to you in the Mastering A&P Study Area, turn to page xviii at the front of the book.

Chapter 1 Review Questions

1. Define homeostatic regulation in a way that states its physiological importance.

2. How does negative feedback differ from positive feedback?

3. As a surgeon, you perform an invasive procedure that necessitates cutting through the peritoneum. Are you more likely to be operating on the heart or on the stomach?

4. A hormone called calcitonin, produced by the thyroid gland, is released in response to increased levels of calcium ions in the blood. If this hormone acts through negative feedback, what effect will its release have on blood calcium levels?

5. An anatomist wishes to make detailed comparisons of medial surfaces of the left and right sides of the brain. This work requires sections that show the entire medial surface. Which kind of sections should be ordered from the lab for this investigation?

For answers to all module, section, and chapter review questions, see the blue Answers tab at the back of the book.

2

Chemical Level of Organization

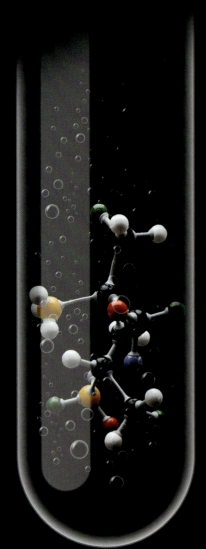

Atoms and Molecules

Our study of the human body begins at the chemical level of organization. Chemistry is the science that studies the structure of matter, which is defined as anything that takes up space and has mass. **Mass**, the amount of material in matter, is a physical property that determines the weight of an object on Earth. For our purposes, the mass of an object is the same as its weight. However, the two are not always equivalent: In orbit you would be weightless, but your mass would remain unchanged.

1 An **atom** is the smallest stable unit of matter. Atoms are composed of **subatomic particles**, three of which are especially important for understanding the basic chemical properties of matter.

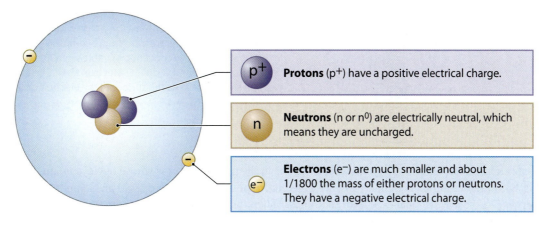

Protons (p+) have a positive electrical charge.

Neutrons (n or n0) are electrically neutral, which means they are uncharged.

Electrons (e−) are much smaller and about 1/1800 the mass of either protons or neutrons. They have a negative electrical charge.

2 An atom can be subdivided into the nucleus and the electron cloud.

The **nucleus** of an atom lies at its center. The nucleus contains one or more protons and it may also contain neutrons. The mass of the atom is primarily determined by the number of protons and neutrons in the nucleus.

The electrons in the atom whirl around the nucleus, creating an **electron cloud**.

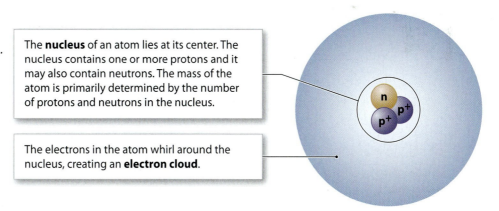

3 A **molecule** forms when atoms interact and produce larger, more complex structures.

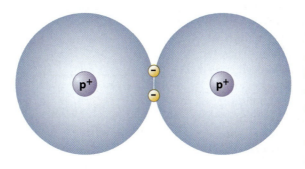

Everything around us is composed of atoms in varying combinations. The unique characteristics of each object, living or nonliving, result from the types of atoms involved and the ways those atoms combine and interact. The mass of any object represents the sum of the masses of its component atoms.

Typical atoms contain protons, neutrons, and electrons

Atoms normally contain equal numbers of protons and electrons and therefore are electrically neutral. The number of protons in an atom is known as its **atomic number**; the total number of protons and neutrons is its **mass number**. An **element** is a pure substance consisting only of atoms with the same atomic number.

1 Hydrogen (H) is the simplest atom, with an atomic number of 1. Thus, an atom of hydrogen contains one proton and one electron. Hydrogen's proton is located in the center of the atom and forms the nucleus.

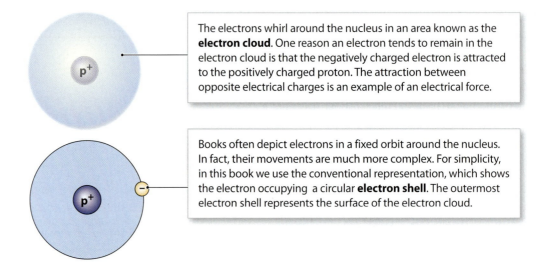

The electrons whirl around the nucleus in an area known as the **electron cloud**. One reason an electron tends to remain in the electron cloud is that the negatively charged electron is attracted to the positively charged proton. The attraction between opposite electrical charges is an example of an electrical force.

Books often depict electrons in a fixed orbit around the nucleus. In fact, their movements are much more complex. For simplicity, in this book we use the conventional representation, which shows the electron occupying a circular **electron shell**. The outermost electron shell represents the surface of the electron cloud.

2 The atoms of a single element can differ in the number of neutrons in the nucleus. Atoms whose nuclei contain the same number of protons, but different numbers of neutrons, are called **isotopes**. Different isotopes of an element have essentially identical chemical properties, and so are identical except for their mass. Therefore, we use the mass number to designate isotopes. Mass numbers are useful because they tell us the number of subatomic particles in the nuclei of different atoms.

Electron-shell model

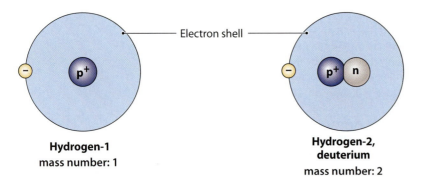

Hydrogen-1
mass number: 1

Hydrogen-2, deuterium
mass number: 2

Hydrogen-3, tritium
mass number: 3

3 The actual mass of an atom is known as its **atomic weight.** The unit used to express atomic weight is the **atomic mass unit (amu)**, also known as a dalton. One amu is very close to the weight of a single proton or neutron. Thus, the atomic weight of the most common isotope of hydrogen is very close to 1. However, the atomic weight of an element is an average mass that reflects the proportions of different isotopes. For example, the atomic number of hydrogen is 1, but the atomic weight of hydrogen is 1.0079, primarily because some hydrogen atoms (0.015 percent) have a mass number of 2, and even fewer have a mass number of 3.

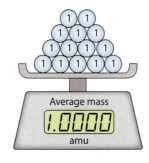

Average mass

1.0000

amu

Atomic weight of hydrogen-1 = 1

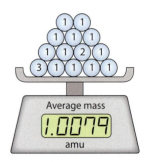

Average mass

1.0079

amu

Atomic weight of a mixture of hydrogen isotopes = 1.0079

Principal Elements of the Human Body	
Element (% of total body weight)	**Significance**
Oxygen, O (65)	A component of water and other compounds; gaseous form is essential for energy production
Carbon, C (18.6)	A component of all molecules that make up living things (see Section 4)
Hydrogen, H (9.7)	A component of water and most other compounds in the body
Nitrogen, N (3.2)	Found in proteins, nucleic acids, and other organic compounds
Calcium, Ca (1.8)	Found in bones and teeth; important for membrane function, nerve impulses, muscle contraction, and blood clotting
Phosphorus, P (1.0)	Found in bones and teeth, nucleic acids, and high-energy compounds
Potassium, K (0.4)	Important for proper membrane function, nerve impulses, and muscle contraction
Sodium, Na (0.2)	Important for blood volume, membrane function, nerve impulses, and muscle contraction
Chlorine, Cl (0.2)	Important for blood volume, membrane function, and water absorption
Magnesium, Mg (0.06)	A cofactor for many enzymes (special proteins; see module 2.4)
Sulfur, S (0.04)	Found in many proteins
Iron, Fe (0.007)	Essential for oxygen transport and energy capture
Iodine, I (0.0002)	A component of hormones of the thyroid gland
Trace elements: silicon (Si), fluorine (F), copper (Cu), manganese (Mn), zinc (Zn), selenium (Se), cobalt (Co), molybdenum (Mo), cadmium (Cd), chromium (Cr), tin (Sn), aluminum (Al), boron (B), and vanadium (V)	Some function as cofactors; the functions of many trace elements are poorly understood

4 This table shows the 13 most common elements in the human body and identifies their significance. The human body also contains atoms of another 14 elements—called **trace elements**—that are present in very small amounts. Only 92 elements exist in nature, although about two dozen additional elements have been created through nuclear reactions in research laboratories. Every element has a **chemical symbol**, an abbreviation recognized by scientists everywhere. Most of the symbols are easily connected with the English names of the elements (O for oxygen, N for nitrogen, C for carbon, and so on), but a few are abbreviations of their names in other languages. For example, the symbol for sodium, Na, comes from the Latin word *natrium*.

Module 2.1 Review

a. Define an element.

b. Describe trace elements.

c. How is it possible for two samples of hydrogen to contain the same number of atoms yet have different weights?

Electrons occupy various energy levels

Atoms are electrically neutral; every positively charged proton is balanced by a negatively charged electron. Thus, each increase in atomic number is accompanied by a comparable increase in the number of electrons traveling around the nucleus. Within the electron cloud, electrons occupy an orderly series of energy levels. Although the electrons in an energy level may travel in complex patterns around the nucleus, for our purposes the patterns can be diagrammed as a series of concentric electron shells. The first electron shell (the one closest to the nucleus) corresponds to the lowest energy level.

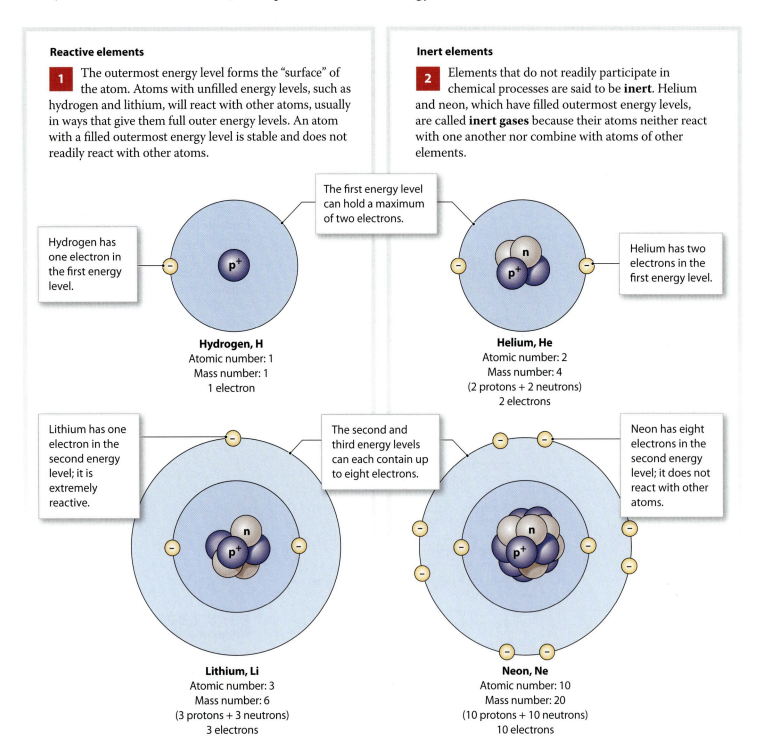

Reactive elements

1 The outermost energy level forms the "surface" of the atom. Atoms with unfilled energy levels, such as hydrogen and lithium, will react with other atoms, usually in ways that give them full outer energy levels. An atom with a filled outermost energy level is stable and does not readily react with other atoms.

Inert elements

2 Elements that do not readily participate in chemical processes are said to be **inert**. Helium and neon, which have filled outermost energy levels, are called **inert gases** because their atoms neither react with one another nor combine with atoms of other elements.

The first energy level can hold a maximum of two electrons.

Hydrogen has one electron in the first energy level.

Helium has two electrons in the first energy level.

Hydrogen, H
Atomic number: 1
Mass number: 1
1 electron

Helium, He
Atomic number: 2
Mass number: 4
(2 protons + 2 neutrons)
2 electrons

Lithium has one electron in the second energy level; it is extremely reactive.

The second and third energy levels can each contain up to eight electrons.

Neon has eight electrons in the second energy level; it does not react with other atoms.

Lithium, Li
Atomic number: 3
Mass number: 6
(3 protons + 3 neutrons)
3 electrons

Neon, Ne
Atomic number: 10
Mass number: 20
(10 protons + 10 neutrons)
10 electrons

3 Elements with unfilled outermost energy levels, such as hydrogen, lithium, or sodium, are called **reactive**, because they readily interact or combine with other atoms. In doing so, these atoms achieve stability by gaining, losing, or sharing electrons to fill their outermost energy level. When this involves the loss of electrons from the outer energy level, the result is an atom that is no longer electrically neutral—it now has more protons than electrons. The atom has a net positive charge, and it is called a positive ion or **cation**. A single missing electron gives the ion a charge of +1; some ions carry charges of +2, +3, or +4, depending on how many electrons are lost to achieve stability.

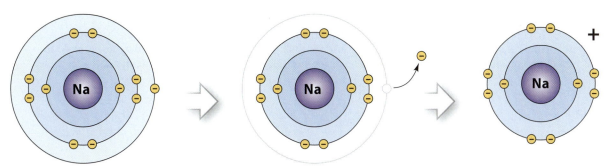

Sodium atom, Na (reactive)　　　　　　　　　　　　　　　　**Sodium ion, Na⁺** (stable)

4 At other times atoms achieve stability by filling their outer energy level with electrons obtained from other atoms. This also creates an atom that is no longer electrically neutral—it has more electrons than protons. The atom now has a net negative charge, and it is called a negative ion or **anion**. A single extra electron gives the ion a charge of −1. Some ions carry charges of −2, −3, or −4, depending on how many electrons are needed to achieve stability.

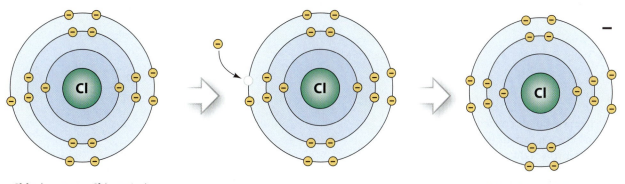

Chlorine atom, Cl (reactive)　　　　　　　　　　　　　　　**Chloride ion, Cl⁻** (stable)

The interactions that stabilize the outer energy levels of atoms often result in the formation of **chemical bonds**. These bonds hold the participating atoms together once the reaction has ended.

Module 2.2 Review

a. Indicate the maximum number of electrons that can occupy each of the first three electron shells (energy levels) of an atom.

b. Explain why the atoms of inert elements do not react with one another or combine with atoms of other elements.

c. Explain how cations and anions form.

The most common chemical bonds are ionic bonds and covalent bonds

When chemical bonding occurs, the result is the creation of new chemical entities called compounds and molecules. A **compound** is a chemical substance made up of atoms of two or more different elements, regardless of the type of bond joining them.

Step 1: Formation of sodium and chloride ions. The sodium atom loses an electron to the chlorine atom. This produces two stable ions with filled outer energy levels.

Step 2: Formation of an ionic bond. Because these ions form close together, and have opposite charges, they are attracted to one another. This creates NaCl, an ionic compound.

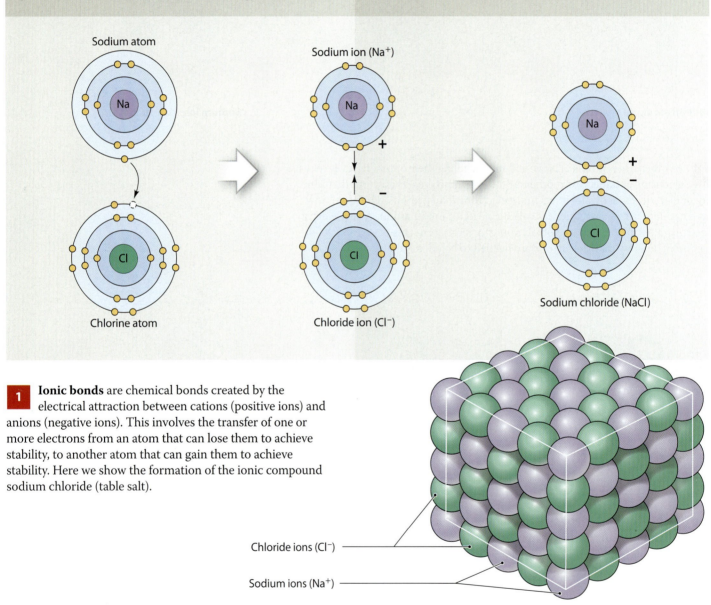

Sodium atom

Sodium ion (Na$^+$)

Chlorine atom

Chloride ion (Cl$^-$)

Sodium chloride (NaCl)

Chloride ions (Cl$^-$)

Sodium ions (Na$^+$)

1 **Ionic bonds** are chemical bonds created by the electrical attraction between cations (positive ions) and anions (negative ions). This involves the transfer of one or more electrons from an atom that can lose them to achieve stability, to another atom that can gain them to achieve stability. Here we show the formation of the ionic compound sodium chloride (table salt).

2 A crystal of sodium chloride contains a large number of sodium and chloride ions packed closely together. The formation of a chemical bond changes the properties of both reactants. In this example, a metal and a poisonous gas combine to form a crystalline solid that is edible.

3 Some atoms can complete their outer electron shells not by gaining or losing electrons, but by sharing electrons with other atoms. Such sharing creates **covalent** (kō-VĀ-lent) **bonds** between the atoms involved. A **molecule** is a chemical structure consisting of atoms of one or more elements held together by covalent bonds. In a typical covalent bond, the participating atoms share the electrons equally, and there is no electrical charge on the molecule. Such molecules are called **nonpolar molecules**.

Molecule		Description
Hydrogen (H$_2$)	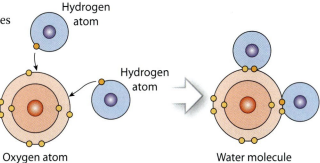	Hydrogen atoms are not found as individual atoms in nature. They exist as molecules, each containing a pair of hydrogen atoms. The two atoms share their electrons to fill their outer energy levels, and the electron pair orbits both nuclei. One electron comes from each atom, so this is called a **single covalent bond**.
Oxygen (O$_2$)		An oxygen atom has 6 electrons in its outer energy level. It may form a **double covalent bond** by sharing two pairs of electrons with another oxygen atom. This creates an oxygen molecule with a stable outer energy level.
Carbon dioxide (CO$_2$)		A carbon atom has 4 electrons in its outer energy level, so it needs to gain 4 electrons from other atoms to achieve stability. In a molecule of carbon dioxide, a carbon atom shares two pairs of electrons with each of two oxygen atoms and forms two double covalent bonds.

4 Some molecules are formed by covalent bonds that involve an unequal sharing of electrons.

Hydrogen atom

Hydrogen atom

Oxygen atom

Water molecule

5 Each hydrogen atom carries a slightly positive charge (δ^+), and the oxygen atom carries a slightly negative charge (δ^-). This forms a **polar molecule**. Covalent bonds that produce polar molecules are called **polar covalent bonds**.

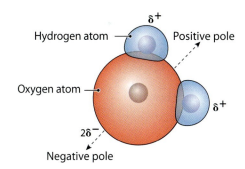

δ^+
Hydrogen atom — Positive pole
Oxygen atom —
δ^+
$2\delta^-$
Negative pole

6 The slightly positive charges on the hydrogen atoms of one polar molecule can be attracted to the slightly negative charges on another polar molecule, and this can change the shapes of the molecules or pull nearby molecules together. This weak attractive force is called a **hydrogen bond**.

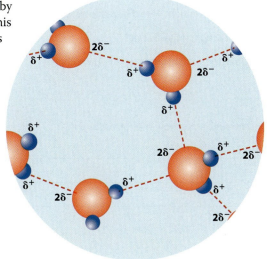

KEY

Hydrogen

Oxygen

Hydrogen bond

Module 2.3 Review

a. Name and distinguish between the two most common types of chemical bonds.

b. Describe the kind of bonds that hold the atoms in a water molecule together.

c. Relate why we can apply the term *molecule* to the smallest particle of water but not to that of table salt.

1. Short answer

Fill in the missing information in the following table.

Element	Number of protons	Number of electrons	Number of neutrons	Mass number
Helium	a	2	2	b
Hydrogen	1	c	d	1
Carbon	6	e	6	f
Nitrogen	g	7	h	14
Calcium	i	j	20	40

2. Short answer

Indicate which of the following molecules are compounds and which are elements.

H_2 (hydrogen) H_2O (water) O_2 (oxygen)   CO (carbon monoxide)

a _____   b _____   c _____   d _____

3. Matching

Match the following terms with the most closely related description.

- atomic number
- electrons
- protons
- neutrons
- isotopes
- ions
- ionic bond
- polar covalent bond
- mass number
- element
- compound
- hydrogen bond

a _____ Atoms that have gained or lost electrons

b _____ Located in the nucleus, have no charge

c _____ Atoms of two or more different elements bonded together

d _____ The number of protons in an atom

e _____ Attractive force between water molecules

f _____ Type of chemical bond within a water molecule

g _____ The number of subatomic particles in the nucleus

h _____ Substance composed only of atoms with same atomic number

i _____ Subatomic particles in the nucleus, have charge

j _____ Atoms of the same element with different masses

k _____ Type of chemical bond in table salt

l _____ Subatomic particles outside the nucleus, have charge

4. Section integration

Describe how the following pairs of terms concerning atomic interactions are similar and how they are different.

a   inert element/reactive element   _____

b   polar molecules/nonpolar molecules   _____

c   covalent bond/ionic bond   _____

Chemical Reactions

Our cells remain alive and functional by controlling chemical reactions. In a chemical reaction, new chemical bonds form between atoms, or existing bonds between atoms break. These changes occur as atoms in the reacting substances, or **reactants**, are rearranged to form different substances, or **products.** All the reactions under way in the cells and tissues of the body at any given moment are called **metabolism** (me-TAB-ō-lizm).

1 In effect, each of your cells is a chemical factory. Growth, maintenance and repair, cell division, secretion, and contraction all involve complex chemical reactions. Cells use chemical reactions to provide the energy needed to maintain homeostasis and perform essential functions.

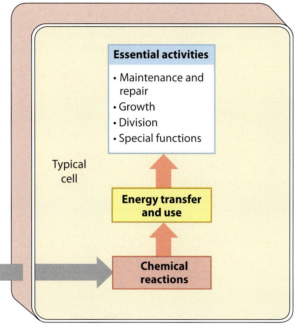

Typical cell

Essential activities
• Maintenance and repair
• Growth
• Division
• Special functions

Energy transfer and use

Substances absorbed

Chemical reactions

2 **Work** is the movement of an object or a change in the physical structure of matter. In your body, work includes movements like walking or running, and also the formation of molecules and the conversion of liquid water to water vapor (evaporation).

Energy is the capacity to perform work; movement or physical change cannot occur unless energy is provided. **Kinetic energy** is the energy of motion—energy that can be transferred to another object and perform work. Examples of kinetic energy include throwing a ball, lifting a weight, and on the cellular level, the movement or transport of structures by molecular motors. **Potential energy** is stored energy-energy that has the potential to do work. It may come from an object's position (you holding a weight overhead or standing on a ladder) or from its physical or chemical structure (a stretched spring or a charged battery).

Cells perform work as they produce complex molecules and move materials into, out of, and within the cell. The cells of a skeletal muscle at rest, for example, contain potential energy in the form of the positions of protein filaments and the covalent bonds between molecules within the cells. When the muscle contracts, it performs work, and potential energy is converted into kinetic energy. Such a conversion is never 100 percent efficient. Each time an energy exchange or transfer occurs, some of the energy is released as heat. That is why your body temperature rises when you exercise.

Chemical notation describes chemical reactions that may be helped by enzymes

1 **Chemical notation** enables us to describe complex events briefly and precisely. This table summarizes many important rules of chemical notation. It is easy to use chemical notation to calculate the weights of the reactants involved in a particular reaction.

Rules of Chemical Notation

Category	Visual Representation	Chemical Notation
Atoms The symbol of an element indicates one atom of that element. A number preceding the symbol of an element indicates the number of atoms of that element.	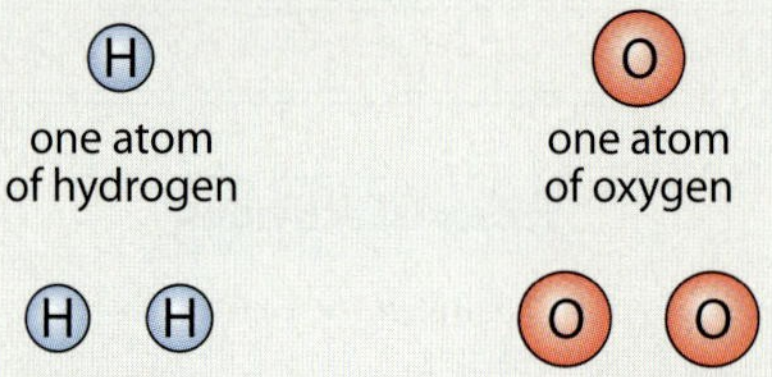 H — one atom of hydrogen O — one atom of oxygen HH — two atoms of hydrogen OO — two atoms of oxygen	H one atom of hydrogen O one atom of oxygen 2H two atoms of hydrogen 2O two atoms of oxygen
Molecules A chemical **formula**, or molecular formula, provides information about the elements and the number of their atoms in a molecule. A subscript following the symbol of an element indicates the number of atoms of that element.	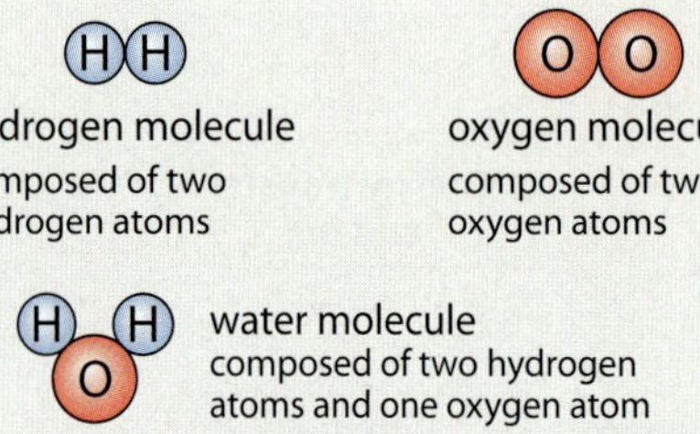 HH hydrogen molecule composed of two hydrogen atoms OO oxygen molecule composed of two oxygen atoms H H O water molecule composed of two hydrogen atoms and one oxygen atom	H_2 hydrogen molecule O_2 oxygen molecule H_2O water molecule
Reactions When we describe a chemical reaction, the participants at the start of the reaction are called reactants, and the reaction generates one or more products. An arrow indicates the direction of the reaction, from reactants (usually on the left) to products (usually on the right). In the following reaction, two atoms of hydrogen combine with one atom of oxygen to produce a single molecule of water.	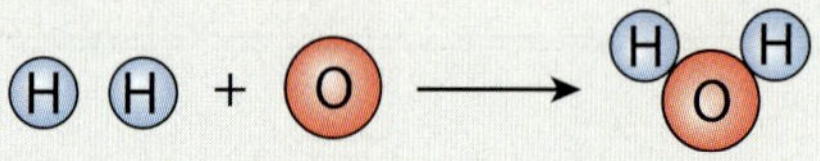 H H + O ⟶ H O H Chemical reactions neither create nor destroy atoms; they merely rearrange atoms into new combinations. Therefore, the numbers of atoms of each element must always be the same on both sides of the equation for a chemical reaction. When this is the case, we have a **balanced equation**.	$2H + O \longrightarrow H_2O$ Balanced equation $2H + 2O \longrightarrow H_2O$ Unbalanced equation
Ions A superscript plus or minus sign following the symbol of an element indicates an ion. A single plus sign indicates a cation with a charge of +1. (The original atom has lost one electron.) A single minus sign indicates an anion with a charge of −1. (The original atom has gained one electron.) If more than one electron has been lost or gained, the charge on the ion is indicated by a superscript number preceding the plus or minus sign.	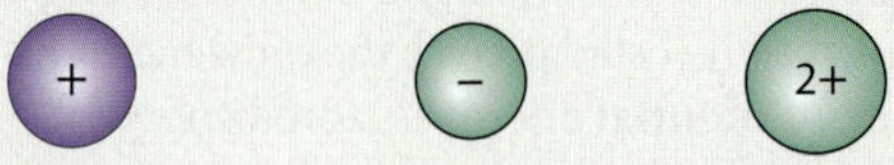 + sodium ion the sodium atom has lost one electron − chloride ion the chlorine atom has gained one electron 2+ calcium ion the calcium atom has lost two electrons **A sodium atom becomes a sodium ion** Electron lost Na → Na⁺ Sodium atom (Na) Sodium ion (Na⁺)	Na^+ sodium ion Cl^- chloride ion Ca^{2+} calcium ion

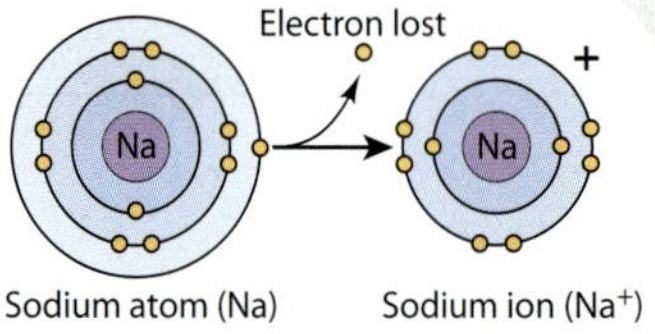

2 Most chemical reactions do not occur spontaneously, or occur so slowly that they would be of little value to cells. Before a reaction can proceed, there must be enough energy to activate the reactants. The amount of energy required to start a reaction is called the **activation energy**. Many laboratory and industrial reactions are activated by extremes in temperature, pressure, and chemical factors that are deadly to cells. Cells avoid such extremes by using special proteins called **enzymes** for most of the complex reactions in your body.

3 Enzymes promote chemical reactions by lowering the activation energy requirements. In doing so, they make it possible for chemical reactions, such as the breakdown of sugars, to proceed under conditions compatible with life. Enzymes belong to a class of substances called **catalysts** (KAT-uh-lists; *katalysis*, dissolution), compounds that accelerate chemical reactions without themselves being permanently changed or consumed. Enzymatic reactions, which are generally reversible, proceed until an equilibrium is reached.

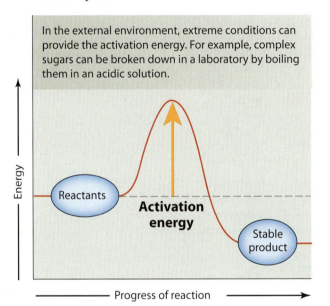

In the external environment, extreme conditions can provide the activation energy. For example, complex sugars can be broken down in a laboratory by boiling them in an acidic solution.

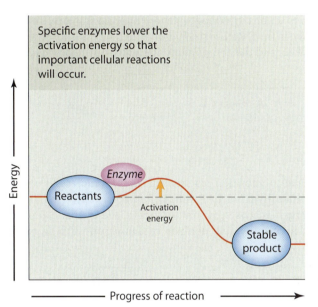

Specific enzymes lower the activation energy so that important cellular reactions will occur.

4 The complex reactions that support life proceed in a series of interlocking steps, each controlled by a specific enzyme. Such a reaction sequence is called a **metabolic pathway**. We can diagram a metabolic pathway as:

It takes activation energy to start a chemical reaction, but once it has begun, the reaction as a whole may absorb or release energy as it proceeds to completion. Reactions that release energy are **exergonic** (*exo-*, outside). If more energy is required to begin the reaction than is released as it proceeds, the reaction is **endergonic** (*endo-*, inside). Exergonic reactions are fairly common in the body; they are responsible for generating the heat that maintains your body temperature. Endergonic reactions include those required to build the molecules making up your cells.

Module 2.4 Review

a. Using the rules for chemical notation, write the molecular formula for glucose, a compound composed of 6 carbon atoms, 12 hydrogen atoms, and 6 oxygen atoms.

b. What is an enzyme?

c. Why are enzymes needed in our cells?

There are three basic types of chemical reactions

Decomposition Reactions

Decomposition is a reaction that breaks a molecule into smaller fragments. You could represent a simple decomposition reaction as

AB $\longrightarrow$ A + B

Decomposition reactions occur inside and outside cells. For example, a typical meal contains molecules of fats, sugars, and proteins that are too large and too complex for your body to absorb and use. Decomposition reactions in your digestive tract break these molecules down into smaller fragments so that absorption is possible.

Decomposition reactions involving water are important in breaking down complex molecules in the body. In **hydrolysis** (hī-DROL-i-sis; *hydro-*, water + *lysis*, a loosening), one of the bonds in a complex molecule is broken, and the components of a water molecule (H and OH) are added to the resulting fragments:

A-B + H_2O $\longrightarrow$ A-H + OH-B

Collectively, the decomposition reactions of complex molecules within the body's cells and tissues are referred to as **catabolism** (ka-TAB-ō-lizm; *katabole*, a throwing down). When a covalent bond—a form of potential energy—is broken, it releases kinetic energy that can perform work. By harnessing the energy released in this way, cells perform vital functions such as growth, movement, and reproduction.

CD $\longrightarrow$ C + D + ENERGY

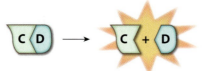

Synthesis Reactions

Synthesis (SIN-the-sis) is the opposite of decomposition. A synthesis reaction assembles smaller molecules into larger molecules. We can diagram a simple synthetic reaction as:

A + B $\longrightarrow$ AB

Synthesis reactions may involve combining atoms or molecules to form larger products. The formation of water from hydrogen and oxygen molecules is a synthesis reaction. Synthesis always involves the formation of new chemical bonds, whether the reactants are atoms or molecules.

Dehydration synthesis, or condensation, is the formation of a complex molecule by removing a water molecule:

$$\text{A-H} + \text{OH-B} \longrightarrow \text{A-B} + \text{H}_2\text{O}$$

Dehydration synthesis is therefore the opposite of hydrolysis. We will see examples of both types of reactions in later sections.

Collectively, the synthesis of new molecules within the body's cells and tissues is known as **anabolism** (a-NAB-ō-lizm; *anabole*, a raising up). Because it takes energy to create a chemical bond, anabolism is usually considered an "uphill" process. Cells must balance their energy budgets, with catabolism providing the energy to support anabolism and other vital functions.

Chemical reactions are reversible (at least theoretically), so if

$$\text{A} + \text{B} \longrightarrow \text{AB}, \text{ then } \text{AB} \longrightarrow \text{A} + \text{B}$$

Many important biological reactions are freely reversible. We can represent reversible reactions as an equation:

$$\text{A} + \text{B} \rightleftharpoons \text{AB}$$

This equation indicates that, in a sense, two reactions are occurring simultaneously: one a synthesis and the other a decomposition. At **equilibrium**, the rates at which the two reactions proceed are in balance: As fast as one molecule of AB forms, another degrades into A + B.

Exchange Reactions

In an **exchange reaction**, parts of the reacting molecules are shuffled around to produce new products:

$$\text{AB} + \text{CD} \longrightarrow \text{AD} + \text{CB}$$

Although the reactants and products contain the same components (A, B, C, and D), those components are present in different combinations. In an exchange reaction, the reactant molecules AB and CD must break apart (a decomposition) before they can interact with each other to form AD and CB (a synthesis).

Module 2.5 Review

a. Identify and describe three types of chemical reactions important in human physiology.

b. Distinguish the roles of water in hydrolysis and dehydration synthesis reactions.

c. In cells, glucose, a six-carbon molecule, is converted into two three-carbon molecules by a reaction that releases energy. What is the source of the energy?

1. Short answer

Using chemical notation, write the formula of each of the following.

a One molecule of hydrogen _____

b Two atoms of hydrogen _____

c Six molecules of water _____

d One molecule of sucrose (in this order: 12 atoms of carbon, 22 atoms of hydrogen, and 11 atoms of oxygen) _____

2. Short answer

Write the chemical equation for the following chemical reaction: one molecule of glucose combined with six molecules of oxygen produce six molecules of carbon dioxide and six molecules of water.

3. Short answer

Indicate which of the following reactions is a hydrolysis reaction, and which is a dehydration synthesis reaction.

a _____

b _____

4. Matching

Match the following terms with the most closely related description.

- exergonic
- activation energy
- products
- exchange reaction
- hydrolysis
- endergonic
- reactants
- enzyme

a _____ Catalyst

b _____ Starting substances in a chemical reaction

c _____ Chemical reaction involving water

d _____ Reactions that absorb energy

e _____ Shuffles parts of reactants

f _____ Ending substances in a chemical reaction

g _____ Reactions that release energy

h _____ Requirement for starting a chemical reaction

5. Section integration

In a metabolic pathway that consists of four steps, how would decreasing the amount of enzyme that catalyzes the second step affect the amount of product produced at the end of the pathway?

The Importance of Water in the Body

Water, H_2O, is the most important component of your body, making up about two-thirds of total body weight. A change in the body's water content can have fatal consequences because virtually all physiological systems will be affected. Although water is familiar to everyone, it has some highly unusual properties.

Important Properties of Water

Lubrication

Water is an effective lubricant because there is little friction between water molecules. Thus, even a thin layer of water between two opposing surfaces will greatly reduce friction between them. Water reduces friction within joints and in body cavities.

Reactivity

In our bodies, chemical reactions occur in water, and water molecules also participate in some reactions, including hydrolysis and dehydration synthesis.

High heat capacity

Heat capacity is the ability to absorb and retain heat. Water has an unusually high heat capacity, because water molecules in the liquid state are attracted to one another through hydrogen bonding.

- The temperature of water must be high before individual molecules have enough energy to break free to become water vapor, a gas.

- Water carries a great deal of heat away with it when it finally does change from a liquid to a gas. This feature accounts for the cooling effect of perspiration on your skin.

- A large mass of water changes temperature very slowly. This property is called **thermal inertia**.

Solubility

A remarkable number of compounds are soluble (able to dissolve) in water. The individual particles become dispersed within the water, and the result is a **solution**—a uniform mixture of two or more substances. The medium in which other atoms, ions, or molecules are dispersed is called the **solvent**; the dispersed substances are the **solutes**. In **aqueous solutions**, water is the solvent.

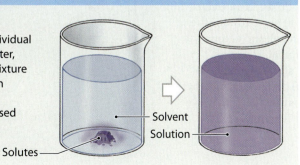

Solutes
Solvent
Solution

Physiological systems depend on water

Many compounds in the environment are held together partially or completely by ionic bonds. In water, these compounds undergo **ionization** (ī-on-ī-ZĀ-shun), or **dissociation** (di-sō-sē-Ā-shun). In this process, ionic bonds are broken as the individual ions interact with the positive or negative poles of polar water molecules.

1 A water molecule is said to be polar because it has positive and negative poles. This polarity is due to the asymmetrical positions of the hydrogen atoms that are attached by polar covalent bonds.

2 In solution, an ionic compound dissociates as water molecules break it apart. The anions are surrounded by the positive poles of water molecules, and the cations are surrounded by the negative poles of water molecules. The sheath of water molecules around an ion in solution is called a **hydration sphere.**

3 Hydration spheres also form around large molecules containing polar covalent bonds. If the molecule binds water strongly, as does glucose, it will be carried into solution—in other words, it will dissolve. Molecules that interact readily with water molecules in this way are called **hydrophilic** (hī-drō-FIL-ik; *hydro-*, water + *philos*, loving).

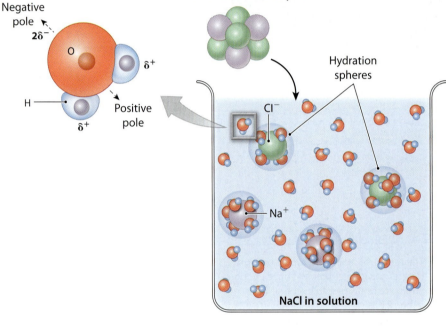

Negative pole
$2\delta^-$
O
δ^+
H
Positive pole
δ^+

Sodium chloride crystal

Hydration spheres

Cl⁻

Na⁺

NaCl in solution

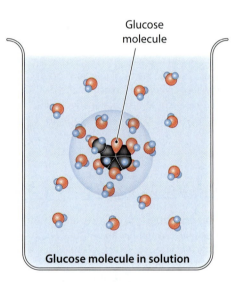

Glucose molecule

Glucose molecule in solution

4 An aqueous solution containing anions and cations will conduct an electrical current. Soluble compounds whose ions will conduct an electrical current in solution are called **electrolytes** (e-LEK-trō-līts). Sodium chloride is an important electrolyte in body fluids. In an electrical field, cations in solution will move toward the negative side, or negative terminal, and anions will move toward the positive terminal. Electrical forces across cell membranes affect the functioning of all cells. Small electrical currents carried by ions are essential to muscle contraction and nerve function, two topics that we will cover in later chapters.

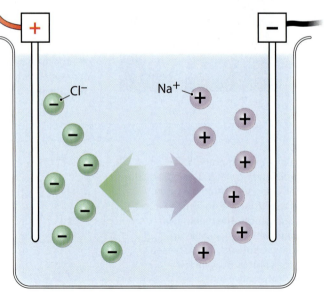

Cl⁻

Na⁺

5 This table lists the most important electrolytes. The dissociation of electrolytes in blood and other body fluids releases a variety of ions. Changes in the concentrations of electrolytes in body fluids will disturb almost every vital function. For example, declining potassium levels will paralyze muscles, and rising potassium levels will cause weak and irregular heartbeats. The concentrations of ions in body fluids are carefully regulated, primarily by the kidneys (ion excretion), digestive tract (ion absorption), and skeletal system (ion storage or release).

Important Electrolytes That Dissociate in Body Fluids

Electrolyte		Ions Released
NaCl (sodium chloride)	$\longrightarrow$	$Na^+ + Cl^-$
KCl (potassium chloride)	$\longrightarrow$	$K^+ + Cl^-$
CaPO₄ (calcium phosphate)	$\longrightarrow$	$Ca^{2+} + PO_4^{2-}$
NaHCO₃ (sodium bicarbonate)	$\longrightarrow$	$Na^+ + HCO_3^-$
MgCl₂ (magnesium chloride)	$\longrightarrow$	$Mg^{2+} + 2Cl^-$
Na₂HPO₄ (sodium hydrogen phosphate)	$\longrightarrow$	$2Na^+ + HPO_4^{2-}$
Na₂SO₄ (sodium sulfate)	$\longrightarrow$	$2Na^+ + SO_4^{2-}$

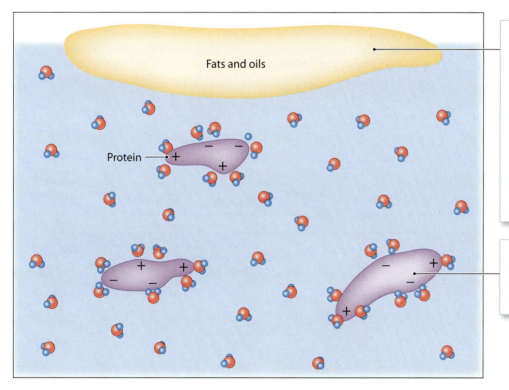

Fats and oils

Protein

Many molecules either lack polar covalent bonds or have very few. Such molecules do not have positive and negative poles and are said to be nonpolar. When nonpolar molecules are exposed to water, hydration spheres do not form and the molecules do not dissolve. Molecules that do not readily interact with water are called **hydrophobic** (hī-drō-FŌ-bik; *hydro-*, water + *phobos*, fear). Among the most familiar hydrophobic molecules are fats and oils of all kinds.

Body fluids typically contain large and complex molecules, such as proteins, that are held in solution by their association with water molecules.

6 A solution containing dispersed proteins or other large molecules is called a **colloid**. The particles or molecules in a colloid will remain in solution indefinitely. Liquid Jell-O is a familiar, viscous (thick) colloid. A **suspension** contains even larger particles that will, if undisturbed, settle out of solution due to the force of gravity. Whole blood is a temporary suspension, because the blood cells are suspended in a fluid called plasma. If clotting is prevented, the cells in a blood sample will gradually settle to the bottom of the container.

Module 2.6 Review

a. Define electrolytes.

b. Distinguish between hydrophilic and hydrophobic molecules.

c. Explain how the ionic compound sodium chloride dissolves in water.

Regulation of body fluid pH is vital for homeostasis

A hydrogen atom involved in a chemical bond or participating in a chemical reaction can easily lose its electron to become a **hydrogen ion**, H^+. Hydrogen ions are extremely reactive in solution. In excessive numbers, they will break chemical bonds, change the shapes of complex molecules, and generally disrupt cell and tissue functions. As a result, the concentration of hydrogen ions in body fluids must be regulated precisely.

1 A few hydrogen ions are normally present even in a sample of pure water, because some of the water molecules dissociate spontaneously, releasing a hydrogen ion, H^+, and a **hydroxide** (hī-DROK-sīd) **ion, OH^-**.

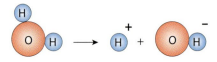

2 The hydrogen ion concentration in body fluids is important to physiological processes. It is usually reported as the **pH** of a solution. For common liquids, the pH scale ranges from 0 to 14. The increase or decrease of one unit corresponds to a tenfold change in H^+ concentration. For example, a solution with a pH of 3 is 1000 times more acidic than one at a pH of 6.

Blood

The pH of blood normally ranges from 7.35 to 7.45. Abnormal fluctuations in pH can damage cells and tissues by breaking chemical bonds, changing the shapes of proteins, and altering cellular functions. **Acidosis** is an abnormal physiological state caused by low blood pH (below 7.35); a pH below 7 can produce coma. **Alkalosis** results from an abnormally high pH (above 7.45); a blood pH above 7.8 generally causes uncontrollable and sustained skeletal muscle contractions.

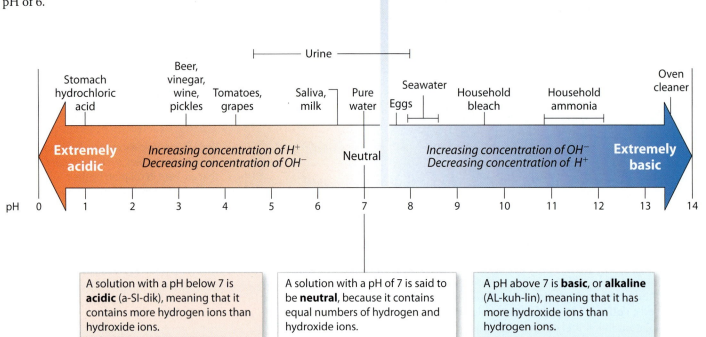

A solution with a pH below 7 is **acidic** (a-SI-dik), meaning that it contains more hydrogen ions than hydroxide ions.

A solution with a pH of 7 is said to be **neutral**, because it contains equal numbers of hydrogen and hydroxide ions.

A pH above 7 is **basic**, or **alkaline** (AL-kuh-lin), meaning that it has more hydroxide ions than hydrogen ions.

Acids and Bases

3 An **acid** is any solute that dissociates in solution and releases hydrogen ions, thereby lowering the pH. Because a hydrogen atom that loses its electron consists solely of a proton, hydrogen ions are often referred to simply as protons, and acids as proton donors. A strong acid dissociates completely in solution, and the reaction occurs essentially in one direction only. **Hydrochloric acid** (HCl) is a representative strong acid; in water, it ionizes as follows:

$$HCl \longrightarrow H^+ + Cl^-$$

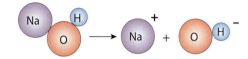

4 A **base** is a solute that removes hydrogen ions from a solution and thereby acts as a proton acceptor. In solution, many bases release a hydroxide ion. Hydroxide ions have a strong affinity for hydrogen ions and react quickly with them to form water molecules. A strong base dissociates completely in solution. **Sodium hydroxide** (NaOH) is a strong base; in solution, it releases sodium ions and hydroxide ions:

$$NaOH \longrightarrow Na^+ + OH^-$$

5 Weak acids and weak bases don't dissociate completely. At equilibrium, a significant number of molecules remains intact in the solution. For a given number of molecules in solution, weak acids and weak bases therefore have less of an impact on pH than do strong acids and strong bases. **Carbonic acid** (H_2CO_3) is a weak acid found in body fluids. In solution, carbonic acid reversibly dissociates into a hydrogen ion and a **bicarbonate ion** (HCO_3^-).

$$H_2CO_3 \rightleftharpoons H^+ + HCO_3^-$$

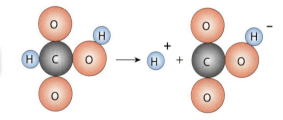

Salts

6 A **salt** is an ionic compound consisting of any cation except a hydrogen ion and any anion except a hydroxide ion. Because they are held together by ionic bonds, many salts dissociate completely in water, releasing cations and anions. For example, sodium chloride (table salt) dissociates immediately in water, releasing Na^+ and Cl^-, the most abundant ions in body fluids. The ionization of sodium chloride does not affect the local concentrations of hydrogen ions or hydroxide ions, so NaCl, like many salts, is a "neutral" solute. Other salts may indirectly affect the concentrations of H^+ and OH^-, making a solution slightly acidic or slightly basic.

$$NaCl \longrightarrow Na^+ + Cl^-$$

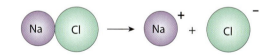

Buffers

Buffers are compounds that stabilize the pH of a solution by removing or replacing hydrogen ions. **Buffer systems** typically involve a weak acid and its related salt, which functions as a weak base. For example, the body's carbonic acid–bicarbonate buffer system consists of carbonic acid (H_2CO_3) and sodium bicarbonate, ($NaHCO_3$), otherwise known as baking soda. Buffers and buffer systems in body fluids help maintain pH within normal limits.

Module 2.7 Review

a. Define pH.

b. Explain the differences among an acid, a base, and a salt.

c. What is the significance of pH in physiological systems?

1. Short answer

List four properties of water important to the functioning of the human body.

a _____

b _____

c _____

d _____

2. Matching

Match the following terms with the most closely related description.

- solvent
- water
- buffers
- hydrophilic
- alkalosis
- hydrophobic
- acid
- solute
- alkaline
- salt

a _____ Abnormally high blood pH

b _____ A dissolved substance

c _____ A solution with a pH greater than 7

d _____ Molecules that readily interact with water

e _____ Fluid medium of a solution

f _____ Ionic compound not containing hydrogen ions or hydroxide ions

g _____ Compounds that stabilize pH in body fluids

h _____ Solution with a pH of 6.5

i _____ Molecules that do not interact with water

j _____ Makes up two-thirds of human body weight

3. Short answer

Identify the regions a–c on the pH scale below.

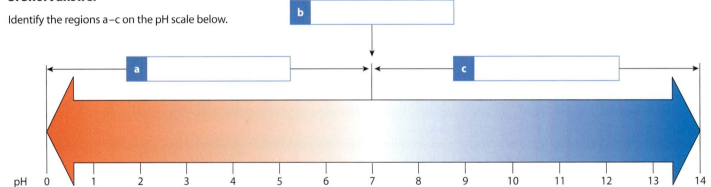

d How much more or less acidic is a solution of pH 2 compared to one with a pH of 6? _____

e Describe three negative effects of abnormal pH fluctuations in the human body.

_____

4. Section integration

The addition of table salt to pure water does not change the pH of the water. Why?

_____

_____

Metabolites and Nutrients

Metabolites (me-TAB-ō-līts; *metabole*, change) are all the molecules that can be synthesized or broken down by chemical reactions (metabolized) inside our bodies. **Nutrients** are esssential metabolites that are normally obtained from the diet. Nutrients and metabolites can be broadly categorized as either *organic* or *inorganic*.

Organic compounds always contain the elements carbon and hydrogen, and generally oxygen as well. The carbon atoms typically form a long chain held together by covalent bonds. In some cases these carbon atoms may form covalent bonds with nitrogen, phosphorus, sulfur, iron, or other elements as well as hydrogen and oxygen. **Inorganic compounds** usually do not have carbon and hydrogen as primary components, and the molecules are often held together by ionic bonds.

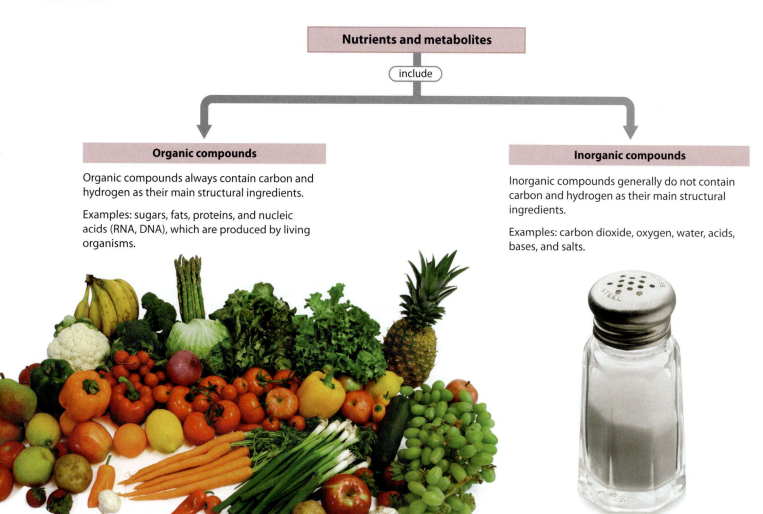

Nutrients and metabolites

include

Organic compounds

Organic compounds always contain carbon and hydrogen as their main structural ingredients.

Examples: sugars, fats, proteins, and nucleic acids (RNA, DNA), which are produced by living organisms.

Inorganic compounds

Inorganic compounds generally do not contain carbon and hydrogen as their main structural ingredients.

Examples: carbon dioxide, oxygen, water, acids, bases, and salts.

In this section we introduce the major classes of organic compounds. We also consider how enzymes aid essential reactions within living cells, and how cells capture and transfer energy with special high-energy compounds.

Carbohydrates contain carbon, hydrogen, and oxygen, usually in a 1:2:1 ratio

A **carbohydrate** is an organic molecule that contains carbon, hydrogen, and oxygen in a ratio near 1:2:1. Familiar carbohydrates include the sugars and starches that make up roughly half of the typical U.S. diet. Carbohydrates typically account for less than 1.5 percent of total body weight. Although they do have other functions, carbohydrates are most important as energy sources that are catabolized rather than stored.

Monosaccharides

1 A simple sugar, or **monosaccharide** (mon-ō-SAK-uh-rīd; *mono-*, single + *sakcharon*, sugar), is the most basic unit of a carbohydrate. A monosaccharide may contain from three to seven carbon atoms. They may then be called a triose (three-carbon), tetrose (four-carbon), pentose (five-carbon), hexose (six-carbon), or heptose (seven-carbon). The hexose **glucose** (GLOO-kōs) is the most important metabolic "fuel" in the body.

2 The three-dimensional structure of an organic molecule is an important characteristic, because it usually determines the molecule's fate or function. Some molecules have the same molecular formula—in other words, the same types and numbers of atoms—but different structures. The body usually treats these as distinct molecules.

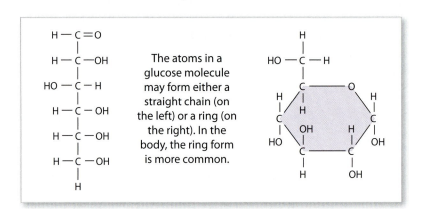

The atoms in a glucose molecule may form either a straight chain (on the left) or a ring (on the right). In the body, the ring form is more common.

Glucose

Fructose

Fructose is a hexose found in many fruits. Although its chemical formula, $C_6H_{12}O_6$, is the same as that of glucose, the arrangement of its atoms differs from that of glucose.

Carbohydrates in the Body			
Structural Class	**Examples**	**Primary Function**	**Remarks**
Monosaccharides (simple sugars)	Glucose, fructose	Energy source	Manufactured in the body and obtained from food; distributed in body fluids
Disaccharides	Sucrose, lactose, maltose	Energy source	Sucrose is table sugar, lactose is in milk, and maltose is malt sugar; all must be broken down to monosaccharides before the body can absorb them
Polysaccharides	Glycogen	Storage of glucose	Glycogen is produced in skeletal muscle cells and liver cells

Disaccharides

Two monosaccharides joined together form a **disaccharide** (dī-SAK-uh-rīd; *di-*, two). Disaccharides such as **sucrose** (table sugar) have a sweet taste and, like monosaccharides, are quite soluble in water. The formation of sucrose involves dehydration synthesis.

3 During dehydration synthesis, two molecules are joined by the removal of a water molecule.

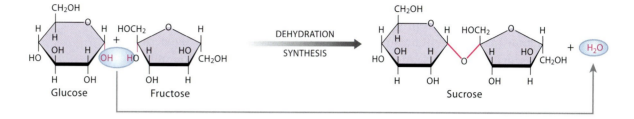

4 Hydrolysis reverses the steps of dehydration synthesis; a complex molecule is broken down by the addition of a water molecule.

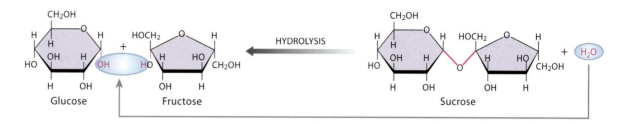

Polysaccharides

5 More-complex carbohydrates result when repeated dehydration synthesis reactions add additional monosaccharides or disaccharides. These large molecules are called **polysaccharides** (pol-ē-SAK-uh-rīdz; *poly-*, many). **Starches** are large polysaccharides formed by plants from glucose molecules. Your digestive tract can break these molecules into monosaccharides. Starches such as those in potatoes and grains are a major dietary source of energy.

Glucose molecules

The polysaccharide **glycogen** (GLĪ-kō-jen), or animal starch, has many side branches consisting of chains of glucose molecules. Muscle cells make and store glycogen. When these cells have a high demand for glucose, glycogen molecules are broken down; when the demand is low, they absorb glucose from the bloodstream and rebuild glycogen reserves.

Module 2.8 Review

a. List the three structural classes of carbohydrates, and give an example of each.

b. Cite the C:H:O ratio in carbohydrates and describe their major functions in the body.

c. Predict the reactants and the type of chemical reaction involved when muscle cells make and store glycogen.

Lipids often contain a carbon-to-hydrogen ratio of 1:2

Like carbohydrates, **lipids** (*lipos*, fat) contain carbon, hydrogen, and oxygen, and the carbon-to-hydrogen ratio is typically near 1:2. However, lipids contain much less oxygen than do carbohydrates with the same number of carbon atoms. The hydrogen-to-oxygen ratio is therefore very large; a representative lipid, such as lauric acid, has a formula of $C_{12}H_{24}O_2$. Lipids may also contain small quantities of phosphorus, nitrogen, or sulfur. Familiar lipids include fats, oils, and waxes. Most lipids are insoluble (do not dissolve) in water, but special transport mechanisms carry them in our circulating blood.

Fatty Acids

1 **Fatty acids** are long carbon chains with hydrogen atoms attached. One end of the carbon chain, called the head, always has a **carboxyl group**: —COOH.

2 In a **saturated fatty acid**, each carbon atom in the tail has four single covalent bonds.

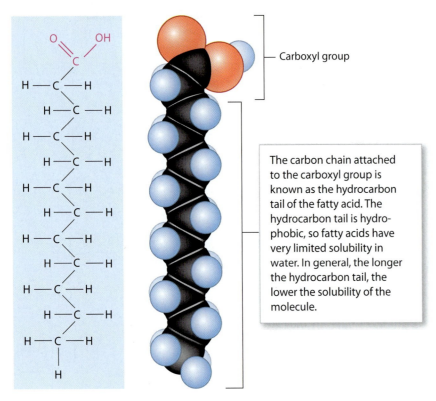

— Carboxyl group

The carbon chain attached to the carboxyl group is known as the hydrocarbon tail of the fatty acid. The hydrocarbon tail is hydrophobic, so fatty acids have very limited solubility in water. In general, the longer the hydrocarbon tail, the lower the solubility of the molecule.

Lauric acid ($C_{12}H_{24}O_2$)

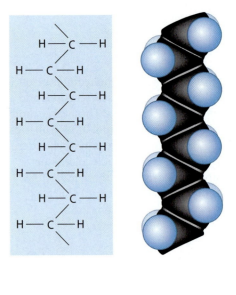

3 In an **unsaturated fatty acid**, one or more of the single covalent bonds between carbon atoms has been replaced by a double covalent bond. As a result, each carbon atom involved will bind only one hydrogen atom rather than two. This changes both the shape of the hydrocarbon tail and the way the body metabolizes that fatty acid. A monounsaturated fatty acid has one double bond in the hydrocarbon tail. A polyunsaturated fatty acid contains multiple double bonds.

Double covalent bond →

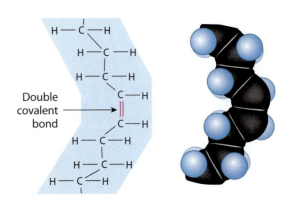

Representative Lipids in the Body

Lipid Type	Examples	Primary Functions	Remarks
Fatty acids	Lauric acid	Energy sources	Absorbed from food or synthesized in cells; transported in the blood
Fats	Monoglycerides, diglycerides, triglycerides	Energy sources, energy storage, insulation, and physical protection	Stored as fat deposits; must be broken down to fatty acids and glycerol before they can be used as an energy source
Steroids	Cholesterol	Structural components of cell membranes, hormones, digestive secretions in bile	All steroids have the same carbon ring framework
Phospholipids, glycolipids	Lecithin (a phospholipid)	Structural components of cell membranes	Derived from fatty acids and nonlipid components

Fats

4 Individual fatty acids cannot be strung together in a chain by dehydration synthesis. But they can be attached to another compound, **glycerol** (GLIS-er-ol), through a similar reaction. The most common **fats** in the body are triglycerides.

Lipids form essential structural components of all cells. In addition, lipid deposits are important as energy reserves. On average, lipids provide roughly twice as much energy as carbohydrates do, gram for gram, when broken down in the body. Lipids normally account for 12–18 percent of total body weight in adult men, and 18–24 percent in adult women. You are unable to synthesize all the lipids that your body needs, so you must obtain several fatty acids from your diet.

Dehydration synthesis can produce a **monoglyceride** (mon-ō-GLI-ser-īd), consisting of glycerol + one fatty acid. Subsequent reactions can yield a **diglyceride** (glycerol + two fatty acids) and then a **triglyceride** (glycerol + three fatty acids). Hydrolysis breaks the glycerides into fatty acids and glycerol.

Module 2.9 Review

a. Describe lipids.

b. Compare the structures of saturated and unsaturated fatty acids.

c. In the hydrolysis of a triglyceride, what are the reactants and the products?

Steroids, phospholipids, and glycolipids have diverse functions

Lipids are important as chemical messengers and as components of cellular structures.

1 **Structural lipids**, such as cholesterol, phospholipids, and glycolipids, help form and maintain the outer membrane surrounding a cell and intracellular membranes within a cell. At the cellular level, membranes are layers composed primarily of hydrophobic lipids. Functionally, a membrane is an effective barrier that can separate two aqueous solutions of differing composition. A plasma membrane is the outer boundary of a cell. It separates the intracellular contents of a cell from its extracellular surroundings.

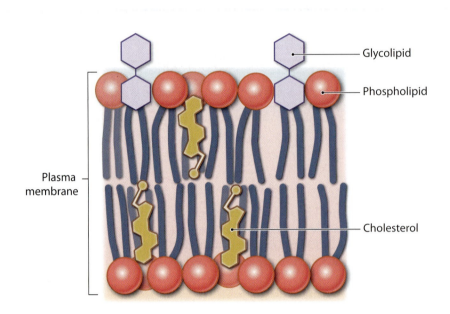

Glycolipid

Phospholipid

Plasma membrane

Cholesterol

Steroids

2 **Steroids** are large lipid molecules that share a distinctive carbon-ring framework. They differ in the functional groups that are attached to the basic ring structure.

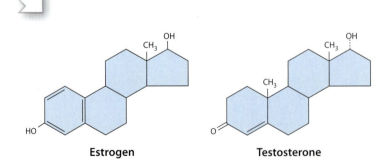

Cholesterol

All animal plasma membranes contain **cholesterol** (kōh-LES-ter-ol; *chole-*, bile + *stereos*, solid). Cells need cholesterol to maintain their plasma membranes, as well as for cell growth and division.

Cortisol

Cortisol, a hormone made in the adrenal cortex, is important in regulating of tissue metabolism.

Estrogen

Testosterone

Some steroid hormones are involved in regulating sexual function. Examples include the sex hormones estrogen and testosterone.

Phospholipids and Glycolipids

3 **Phospholipids** (FOS-fō-lip-idz) and **glycolipids** (GLĬ-kō-lip-idz) are structurally related, and our cells can synthesize both types of lipids, primarily from fatty acids.

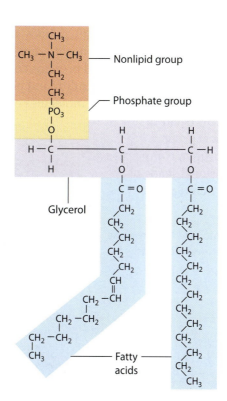

In a phospholipid, a phosphate group links a diglyceride to a nonlipid group.

Nonlipid group
Phosphate group
Glycerol
Fatty acids

Carbohydrate
Glycerol
Fatty acids

In a glycolipid, a carbohydrate is attached to a diglyceride.

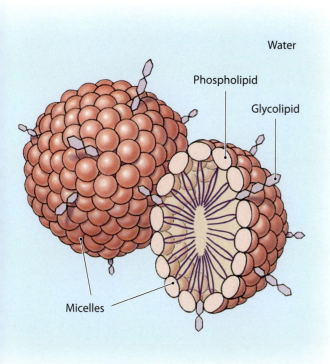

Water
Phospholipid
Glycolipid
Micelles

4 The long hydrocarbon tails of phospholipids and glycolipids are hydrophobic, but the opposite ends, the nonlipid heads, are hydrophilic. In water, large numbers of these molecules tend to form droplets, or **micelles** (mī-SELZ), with the hydrophilic portions on the outside. Most meals contain a mixture of lipids and other organic molecules, and micelles form as our digestive tract breaks down the food.

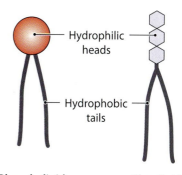

Hydrophilic heads
Hydrophobic tails

Phospholipid   **Glycolipid**

Module 2.10 Review

a. Why is cholesterol necessary in the body?

b. Describe the basic functions of steroids, phospholipids, and glycolipids.

c. Describe the orientations of phospholipids and glycolipids when they form a micelle.

Proteins are formed from amino acids

Proteins are the most abundant organic components of the human body, and in many ways the most important. The human body contains many different proteins, and they account for about 20 percent of total body weight. All proteins contain carbon, hydrogen, oxygen, and nitrogen; smaller quantities of sulfur and phosphorus may also be present. Proteins consist of long chains of organic molecules called **amino acids**. Twenty different amino acids occur in significant quantities in the body. A typical protein contains 1000 amino acids; the largest protein complexes have 100,000 or more. The three-dimensional shape of a protein plays an essential role in determining its functional properties.

Amino group
Central carbon
Carboxyl group

1 Every amino acid has the same basic structural elements.

The R group is a side chain of variable structure that is attached to the central carbon of the amino acid.

2 Dehydration synthesis can link two amino acids together.

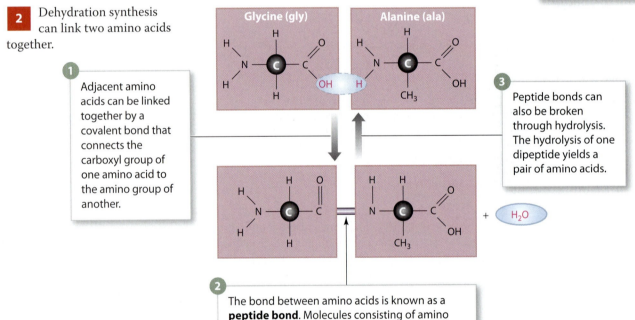

1 Adjacent amino acids can be linked together by a covalent bond that connects the carboxyl group of one amino acid to the amino group of another.

Glycine (gly)

Alanine (ala)

3 Peptide bonds can also be broken through hydrolysis. The hydrolysis of one dipeptide yields a pair of amino acids.

+ H_2O

2 The bond between amino acids is known as a **peptide bond**. Molecules consisting of amino acids held together by peptide bonds are called **peptides**. This molecule is called a **dipeptide** because it contains two amino acids.

3 The chain can be lengthened by adding more amino acids. Attaching a third amino acid produces a tripeptide. Tripeptides and larger peptide chains are called **polypeptides**. Polypeptides containing more than 100 amino acids are usually called proteins. Proteins can have four levels of structural complexity. The most basic structural level of complexity is called the primary structure.

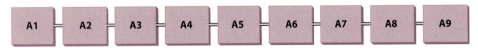

| A1 | A2 | A3 | A4 | A5 | A6 | A7 | A8 | A9 |

Primary Structure of a Protein

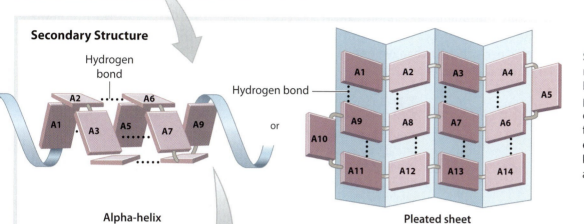

Primary structure is the sequence of amino acids along the length of a single polypeptide.

Secondary Structure

Hydrogen bond

Hydrogen bond

or

Alpha-helix

Pleated sheet

Secondary structure results from bonds between atoms at different parts of the polypeptide chain. Hydrogen bonding, for example, may create either a simple spiral, known as an alpha-helix, or a flat pleated sheet.

Tertiary Structure

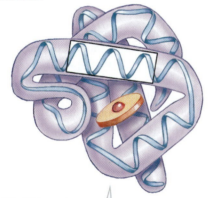

Tertiary structure results from the complex coiling and folding that gives a protein its final three-dimensional shape. Tertiary structure results primarily from interactions between the polypeptide chain and the surrounding water molecules, and to a lesser extent from interactions between the R groups of amino acids in different parts of the protein molecule.

As body temperature rises beyond its normal range, protein shape changes and enzyme function deteriorates. Eventually the protein undergoes **denaturation**, a change in tertiary or quaternary structure. Very high body temperatures (above 43°C, or 110°F) are fatal because the denaturation of structural proteins and enzymes soon causes irreparable damage to organs and organ systems.

Quaternary Structure

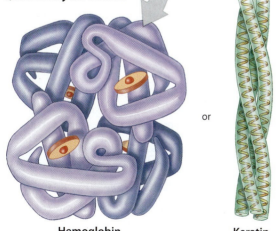

or

Hemoglobin
(globular protein)

Keratin
(fibrous protein)

Quaternary structure results from the interaction between individual polypeptide chains to form a protein complex. The protein **hemoglobin** contains four polypeptide subunits. Hemoglobin is found within red blood cells, where it binds and transports oxygen. It is an example of a globular protein. In **keratin** and **collagen**, common structural proteins in the body, three linear subunits intertwine, forming a fibrous protein.

Module 2.11 Review

a. Describe proteins.

b. What kind of bond forms during the dehydration synthesis of two amino acids?

c. How does boiling a protein affect its structural and functional properties?

Enzymes are proteins with important regulatory functions

Almost everything that happens inside the human body does so because a specific enzyme makes it possible. The reactants in enzymatic reactions are called **substrates**. As in other types of chemical reactions, the interactions among substrates yield specific products. Before an enzyme can function as a catalyst—to accelerate a chemical reaction without itself being permanently changed or consumed—the substrates must bind to a special region of the enzyme.

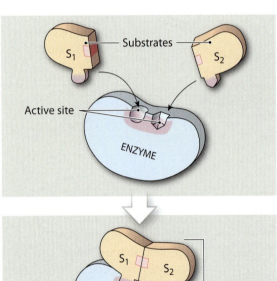

1 Substrate binding depends on the shapes of the participating molecules. Substrate binding occurs at the **active site**, typically a groove or pocket into which one or more substrates nestle, like a key fitting into a lock. Weak electrical attractive forces, such as hydrogen bonding, reinforce the physical fit. The tertiary or quaternary structure of the enzyme molecule determines the shape of the active site. Each enzyme catalyzes only one type of reaction, a characteristic called **specificity**. An enzyme's specificity is determined by the ability of its active sites to bind only to substrates with particular shapes and charges.

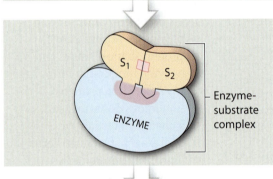

2 Substrate binding produces an **enzyme-substrate complex**. Each cell contains an assortment of enzymes, and any particular enzyme may be active under one set of conditions and inactive under another. Virtually anything that changes the tertiary or quaternary shape of an enzyme can turn it "on" or "off" by changing the properties of the active site and preventing an enzyme-substrate complex from forming. Because the change is immediate, enzyme activation or inactivation is an important method of short-term control over reaction rates and pathways.

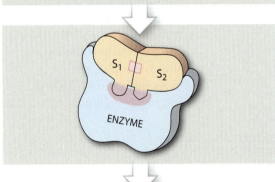

3 Substrate binding typically results in a temporary, reversible change in the shape of the enzyme. This change may further the reaction by placing physical stresses on the substrate molecules. The enzyme then promotes the formation of a product. In some cases, the change in enzyme shape that accompanies substrate binding is sufficient to catalyze the reaction. In other cases, an external source must provide the activation energy required.

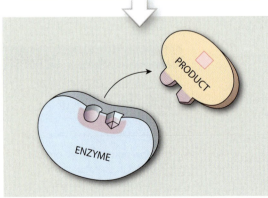

4 The completed product then detaches from the active site, and the enzyme is free to repeat the process. When every enzyme molecule is cycling through its reaction sequence at top speed, further increases in substrate concentration will not affect the rate of reaction.

Module 2.12 Review
a. Define active site.
b. What are the reactants in an enzymatic reaction called?
c. Relate an enzyme's structure to its reaction specificity.

High-energy compounds may store and transfer a portion of energy released during enzymatic reactions

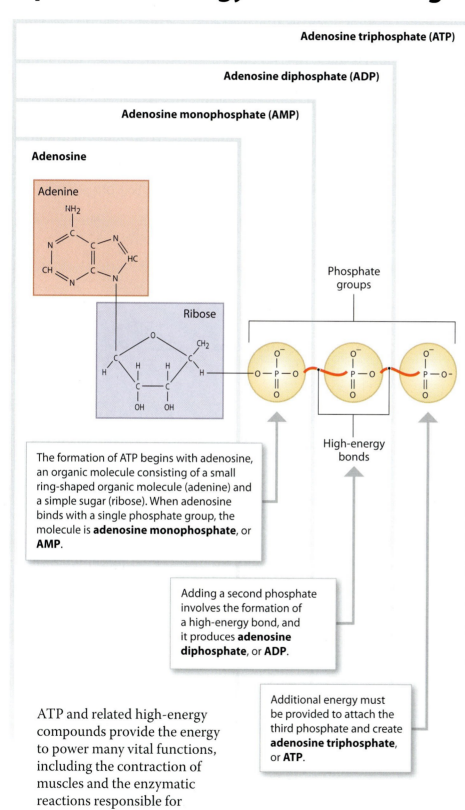

Adenosine triphosphate (ATP)

Adenosine diphosphate (ADP)

Adenosine monophosphate (AMP)

Adenosine

Adenine

NH_2

Ribose

Phosphate groups

High-energy bonds

The formation of ATP begins with adenosine, an organic molecule consisting of a small ring-shaped organic molecule (adenine) and a simple sugar (ribose). When adenosine binds with a single phosphate group, the molecule is **adenosine monophosphate**, or **AMP**.

Adding a second phosphate involves the formation of a high-energy bond, and it produces **adenosine diphosphate**, or **ADP**.

Additional energy must be provided to attach the third phosphate and create **adenosine triphosphate**, or **ATP**.

ATP and related high-energy compounds provide the energy to power many vital functions, including the contraction of muscles and the enzymatic reactions responsible for the synthesis of proteins, carbohydrates, and lipids.

1 Enzymes may catalyze synthesis, decomposition, or exchange reactions. When product formation requires an energy donor, that donor is typically a **high-energy compound**. High-energy compounds contain **high-energy bonds**, covalent bonds whose breakdown releases energy under controlled conditions. The most common high-energy compound is **adenosine triphosphate**, or **ATP**.

2 The formation of ATP from ADP is a reversible reaction. The energy stored in ATP is released when it breaks down to ADP. Cells can synthesize ATP in one location and then break it down in another, harnessing the energy released to power essential activities.

ADP + ENERGY + P ⇌ ATP

Module 2.13 Review

a. Where do cells obtain the energy needed for their vital functions?

b. Describe ATP.

c. Compare AMP with ADP.

DNA and RNA are nucleic acids

Nucleic (noo-KLĀ-ik) **acids** are large organic molecules composed of carbon, hydrogen, oxygen, nitrogen, and phosphorus. The two classes of nucleic acid molecules are **deoxyribonucleic** (dē-oks-ē-rī-bō-noo-KLĀ-ik) **acid**, or **DNA**, and **ribonucleic** (rī-bō-noo-KLĀ-ik) **acid**, or **RNA**. The primary role of nucleic acids is to store and transfer information—specifically, information essential to the synthesis of proteins by cells. A nucleic acid consists of one or two long chains that are formed by dehydration synthesis. The individual subunits of a nucleic acid are called **nucleotides**.

1 A typical nucleotide consists of a phosphate group, a sugar, and an organic molecule known as a **nitrogenous base**. Adenosine monophosphate (AMP) is an example of a nucleotide that you have already encountered.

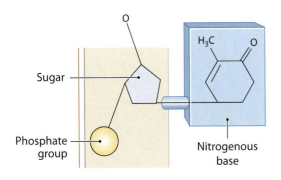

Nitrogenous bases

Adenine and **guanine** are found in both DNA and RNA.

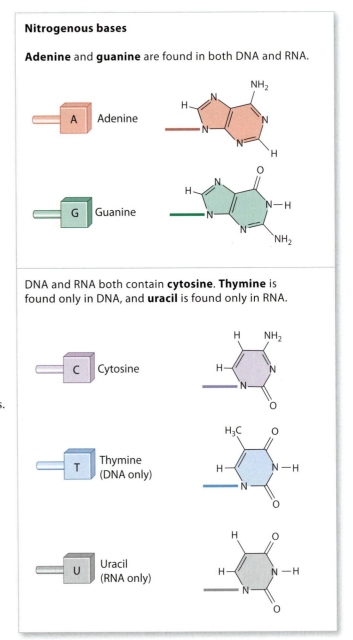

DNA and RNA both contain **cytosine**. **Thymine** is found only in DNA, and **uracil** is found only in RNA.

2 The phosphate and sugar of adjacent nucleotides can be strung together by dehydration synthesis, creating the long chains that comprise functional nucleic acids. The "backbone" of this molecule is a linear sugar-to-phosphate-to-sugar sequence, with the nitrogenous bases projecting to one side. In both DNA and RNA, it is the sequence of nitrogenous bases that carries the information for producing proteins.

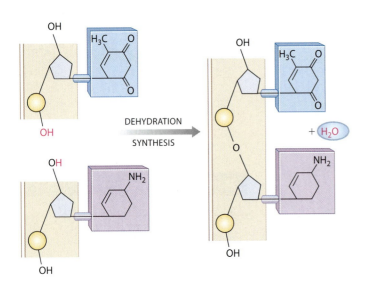

A Comparison of DNA with RNA

Characteristic	DNA	RNA
Sugar	Deoxyribose	Ribose
Nitrogenous bases	Adenine (A), guanine (G), cytosine (C), thymine (T)	Adenine, guanine, cytosine, uracil (U)
Number of nucleotides in typical molecule	Always more than 45 million	Varies from fewer than 100 to about 50,000
Shape of molecule	Paired strands coiled in a double helix	Varies with hydrogen bonding along the length of the strand of each of the three main types (mRNA, tRNA, rRNA)
Function	Stores genetic information that controls protein synthesis	Performs protein synthesis as directed by DNA

3 A DNA molecule consists of a pair of nucleotide chains. Hydrogen bonding between opposing nitrogenous bases holds the two strands together. The shapes of the nitrogenous bases allow adenine to bond only to thymine, and cytosine to bond only to guanine. As a result, the combinations adenine–thymine (A-T) and cytosine–guanine (C-G) are known as **complementary base pairs**, and the two nucleotide chains of the DNA molecule are known as **complementary strands**.

4 A molecule of RNA consists of a single chain of nucleotides. Its shape, and thus its function, depends on the order of the nucleotides and the interactions among them.

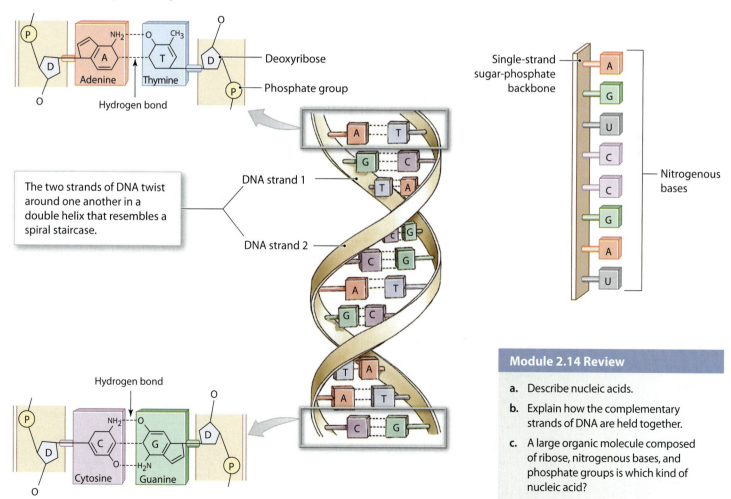

Deoxyribose
Phosphate group
Adenine
Thymine
Hydrogen bond

The two strands of DNA twist around one another in a double helix that resembles a spiral staircase.

DNA strand 1
DNA strand 2

Hydrogen bond
Cytosine
Guanine

Single-strand sugar-phosphate backbone
Nitrogenous bases

Module 2.14 Review

a. Describe nucleic acids.

b. Explain how the complementary strands of DNA are held together.

c. A large organic molecule composed of ribose, nitrogenous bases, and phosphate groups is which kind of nucleic acid?

1. Matching

Match the following terms with the most closely related description.

- monosaccharide
- ATP
- polyunsaturated
- glycerol
- cholesterol
- sucrose
- glycogen
- active site
- nucleotide
- RNA
- peptide

a _____ Polysaccharide with an energy-storage role in animal tissues

b _____ A disaccharide molecule

c _____ A fatty acid with more than one C-to-C double covalent bond

d _____ The region of an enzyme that binds the substrate

e _____ Three-carbon molecule that combines with fatty acids

f _____ A steroid essential to plasma membranes

g _____ A high-energy compound consisting of adenosine and three phosphate groups

h _____ A nucleic acid that contains the sugar ribose

i _____ The covalent bond between the carboxyl and amino groups of adjacent amino acids

j _____ Organic molecule consisting of a sugar, a phosphate group, and a nitrogenous base

k _____ A simple sugar

2. Vocabulary

In the space provided, write the boldfaced terms introduced in this section that contain the indicated word part.

Word Part	Meaning	Terms
a poly-	many	_____
b tri-	three	_____
c di-	two	_____
d glyco-	sugar	_____

3. Concept map

Use each of the following terms once to fill in the blank boxes to correctly complete the organic compounds concept map.

- lipids
- carbohydrates
- nucleic acids
- disaccharides
- RNA
- fatty acids
- phosphate groups
- glycerol
- polysaccharides
- proteins
- monosaccharides
- ATP
- amino acids
- DNA
- nucleotides

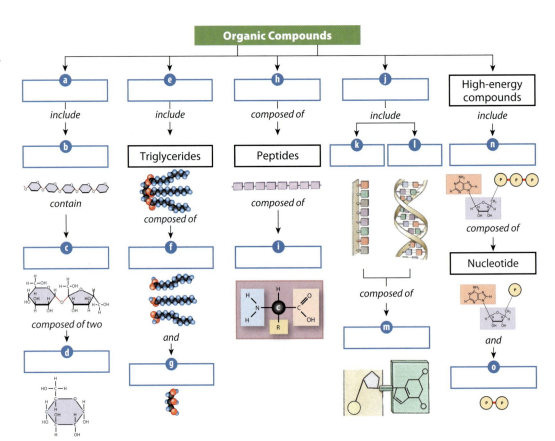

Visual Outline with Key Terms

Summarize the content of each module using the terms in the order provided.

SECTION 1

Atoms and Molecules

- mass
- atoms
- subatomic particles
- protons
- neutrons
- electrons
- nucleus
- electron cloud
- molecule

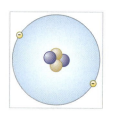

2.1

Typical atoms contain protons, neutrons, and electrons

- atomic number
- mass number
- element
- electron cloud
- electron shell
- isotopes
- atomic weight
- atomic mass unit (amu)
  - dalton
- trace elements
- chemical symbol

2.2

Electrons occupy various energy levels

- inert
- inert gases
- reactive
- cation
- anion
- chemical bonds

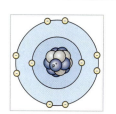

2.3

The most common chemical bonds are ionic bonds and covalent bonds

- compound
- ionic bonds
- covalent bonds
- molecule
- nonpolar molecules
- single covalent bond
- double covalent bond
- polar molecule
- polar covalent bonds
- hydrogen bond

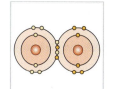

SECTION 2

Chemical Reactions

- reactants
- products
- metabolism
- work
- energy
- kinetic energy
- potential energy

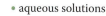

Energy transfer and use

Chemical reactions

2.4

Chemical notation describes chemical reactions that may be helped by enzymes

- chemical notation
- formula
- balanced equation
- activation energy
- enzymes
- catalysts
- metabolic pathway
- exergonic
- endergonic

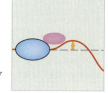

2.5

There are three basic types of chemical reactions

- decomposition
- hydrolysis
- catabolism
- synthesis
- dehydration synthesis
- anabolism
- equilibrium
- exchange reaction

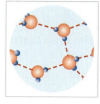

SECTION 3

The Importance of Water in the Body

- lubrication
- reactivity
- heat capacity
- thermal inertia
- solubility
- solution
- solvent
- solutes
- aqueous solutions

2.6

Physiological systems depend on water

- ionization
- dissociation
- hydration sphere
- hydrophilic
- electrolytes
- colloid
- suspension
- hydrophobic

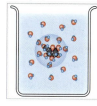

2.7

Regulation of body fluid pH is vital for homeostasis

- hydrogen ion (H⁺)
- hydroxide ion (OH⁻)
- pH
- acidosis
- alkalosis
- acidic
- neutral
- basic
- alkaline
- acid
- hydrochloric acid
- base
- sodium hydroxide
- carbonic acid
- bicarbonate ion
- salt
- buffers
- buffer systems

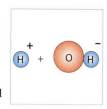

• = *Term boldfaced in this module*

SECTION 4

Metabolites and Nutrients

- metabolites
- nutrients
- organic compounds
- inorganic compounds
- high-energy compounds

2.11

Proteins are formed from amino acids

- proteins
- amino acids
- peptide bond
- peptides
- dipeptide
- polypeptides
- primary structure
- secondary structure
- tertiary structure
- quaternary structure
- hemoglobin
- keratin
- collagen
- denaturation

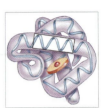

2.8

Carbohydrates contain carbon, hydrogen, and oxygen, usually in a 1:2:1 ratio

- carbohydrate
- monosaccharide
- glucose
- fructose
- disaccharide
- sucrose
- polysaccharides
- starches
- glycogen

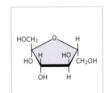

2.12

Enzymes are proteins with important regulatory functions

- substrates
- active site
- specificity
- enzyme-substrate complex

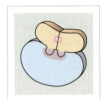

2.9

Lipids often contain a carbon-to-hydrogen ratio of 1:2

- lipids
- fatty acids
- carboxyl group
- saturated fatty acid
- unsaturated fatty acid
- glycerol
- fats
- monoglyceride
- diglyceride
- triglyceride

2.13

High-energy compounds may store and transfer a portion of energy released during enzymatic reactions

- high-energy compound
- high-energy bonds
- adenosine triphosphate (ATP)
- adenosine monophosphate (AMP)
- adenosine diphosphate (ADP)

2.10

Steroids, phospholipids, and glycolipids have diverse functions

- structural lipids
- steroids
- cholesterol
- phospholipids
- glycolipids
- micelles

2.14

DNA and RNA are nucleic acids

- nucleic acids
- deoxyribonucleic acid (DNA)
- ribonucleic acid (RNA)
- nucleotides
- nitrogenous base
- adenine
- guanine
- cytosine
- thymine
- uracil
- complementary base pairs
- complementary strands

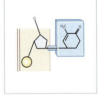

• = *Term boldfaced in this module*

CAREER PATHS

Pharmacy Technician

For Pamela Loshbaugh, no two days at work are exactly the same. As a pharmacy technician at Los Robles Hospital in Thousand Oaks, CA, Pamela works under the direct supervision of a pharmacist. She fills prescriptions, answers phones, completes paperwork, and helps the pharmacist in any other way needed.

For Pamela, a typical day begins at 6 am with preparing and distributing medication, including intravenous bags, to patients who have stayed overnight. In the afternoon, the pharmacy techs will do an update, or "catch," of patients who still need medications, as well as new patients and patients who have been discharged.

According to Pamela, communication skills are key because the pharmacy tech is often the intermediary between the doctors and nurses and the pharmacist, and medications can literally mean the difference between life and death. While chemistry—both interpersonal and scientific—is an important part of pharmacy work, A&P is equally important. "You have to know the different categories of medication," she says. "You need to look at the patient's profile and send up a red flag to the pharmacist if there are, say, two different blood-thinning medications. Or, for example, if a patient takes too many analgesics, it could cause several problems with their digestive system." Situations with incompatible medications are not common, but Pamela has seen it happen, and her knowledge of anatomy has helped prevent catastrophe.

For more information, visit the website of the National Pharmacy Technician Association at http://www.pharmacytechnician.org.

Think this is the career for you?
Key Stats:

- **Education and Training.** Employers prefer at least a high school diploma. The majority of technicians receive on-the-job training, but formal programs are available.

- **Licensure.** Most states do not require pharmacy technicians to be certified or licensed, but certification programs are available.

- **Earnings.** Earnings vary but the median hourly salary is $13.65.

- **Expected Job Prospects.** Employment is expected to grow above the national average—31% through 2018.

Bureau of Labor Statistics, U.S. Department of Labor, *Occupational Outlook Handbook, 2010-11 Edition*, Pharmacy Technicians and Aides, on the Internet at http://www.bls.gov/oco/ocos325.htm (visited September 14, 2011).

Access more review material online in the Study Area at www.masteringaandp.com.

- Chapter guides
- Chapter quizzes
- Practice tests
- Art labeling activities
- Flashcards
- A glossary with pronunciations

- Practice Anatomy Lab™ (PAL™) 3.0 virtual anatomy practice tool
- Interactive Physiology® (IP) animated tutorials
- MP3 Tutor Sessions

 For this chapter, go to this topic in the MP3 Tutor Sessions:

- Inorganic Compounds

Chapter 2 Review Questions

1. Based on the electrical properties of subatomic particles, predict how the sizes of sodium and chlorine atoms would change once they become ions.

2. An important buffer system in the human body involves carbon dioxide (CO_2) and bicarbonate ions (HCO_3^-) as shown:

$$CO_2 + H_2O \rightleftharpoons H_2CO_3 \rightleftharpoons H^+ + HCO_3^-$$

   If a person becomes excited and exhales large amounts of CO_2, how will his or her body's pH be affected?

3. A biologist analyzes a sample that contains an organic molecule and finds the following components: carbon, hydrogen, oxygen, nitrogen, and phosphorus. On the basis of this information, is the molecule a carbohydrate, a lipid, a protein, or a nucleic acid?

For answers to all module, section, and chapter review questions, see the blue Answers tab at the back of the book.

3

Cells and Tissues

An Introduction to Cells

A typical cell—the smallest living unit in the human body—is only about 0.1 mm in diameter, similar in thickness to a human hair. So, no one could examine the structure of a cell until microscopes were invented in the 17th century. Research later produced the cell theory, which we can summarize as follows:

Cells are the building blocks of all plants and animals.

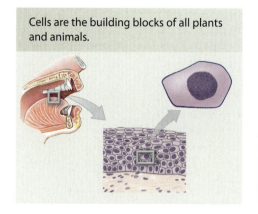

All new cells come from the division of preexisting cells.

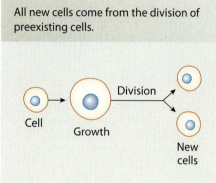

Cells are the smallest units that perform all vital physiological functions.

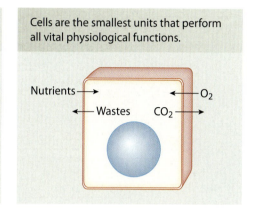

Each cell maintains homeostasis at the cellular level, but it requires the combined and coordinated actions of many cells to achieve homeostasis at higher levels of organization. Although cells vary widely in size, shape, and function, all of them are the descendants of a single cell: the fertilized ovum.

1 After a sperm fertilizes an ovum, the fertilized ovum contains the information to become any cell in the body.

2 The first few cell divisions produce *totipotent cells*, cells that are able to develop into any cell in the body. As further divisions occur, an increase in chemical signalling among the new cells begins to limit their developmental potential.

3 Differences in the chemical signals affect the DNA of the cells, turning specific genes on or off. The daughter cells begin to develop specialized structures and functions. This process of gradual specialization is called **differentiation**.

4 Differentiation produces the specialized cells that form the four basic types of body tissues. These four types of tissues are epithelial, connective, muscle, and neural.

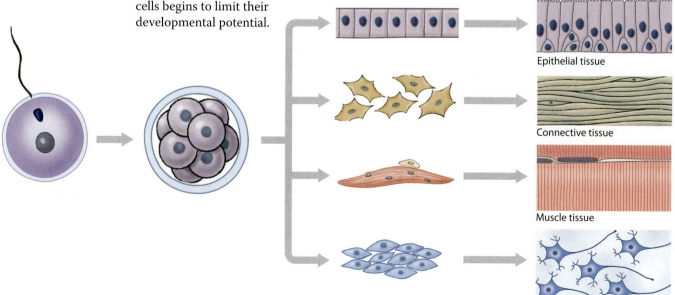

Epithelial tissue

Connective tissue

Muscle tissue

Neural tissue

Cells are the smallest living units of life

Our body cells are surrounded by a watery medium known as the **extracellular fluid**. The extracellular fluid in spaces between tissue cells is called **interstitial** (in-ter-STISH-ul) **fluid** (*interstitium*, something standing between). This module introduces the major components of our cells. Not all of these components are found in every cell of the body.

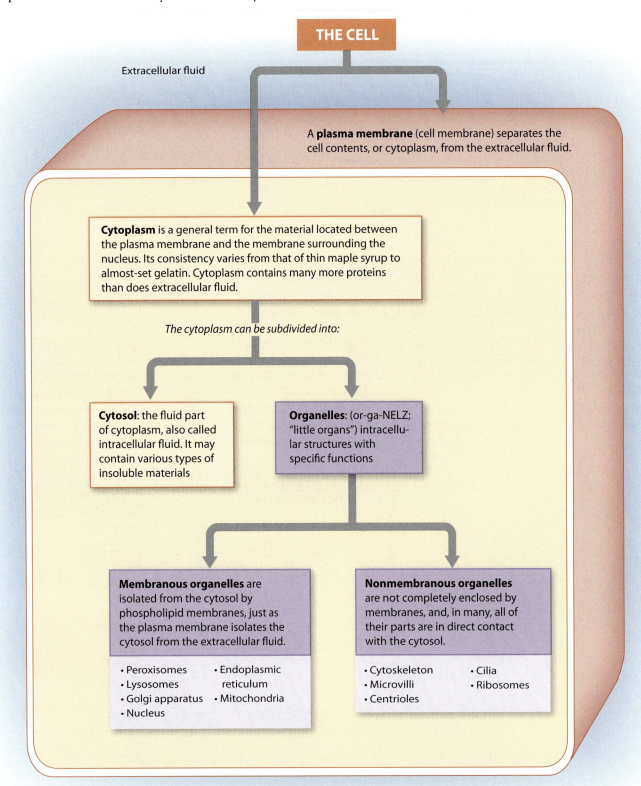

THE CELL

Extracellular fluid

A **plasma membrane** (cell membrane) separates the cell contents, or cytoplasm, from the extracellular fluid.

Cytoplasm is a general term for the material located between the plasma membrane and the membrane surrounding the nucleus. Its consistency varies from that of thin maple syrup to almost-set gelatin. Cytoplasm contains many more proteins than does extracellular fluid.

The cytoplasm can be subdivided into:

Cytosol: the fluid part of cytoplasm, also called intracellular fluid. It may contain various types of insoluble materials

Organelles: (or-ga-NELZ; "little organs") intracellular structures with specific functions

Membranous organelles are isolated from the cytosol by phospholipid membranes, just as the plasma membrane isolates the cytosol from the extracellular fluid.

• Peroxisomes
• Lysosomes
• Golgi apparatus
• Nucleus
• Endoplasmic reticulum
• Mitochondria

Nonmembranous organelles are not completely enclosed by membranes, and, in many, all of their parts are in direct contact with the cytosol.

• Cytoskeleton
• Microvilli
• Centrioles
• Cilia
• Ribosomes

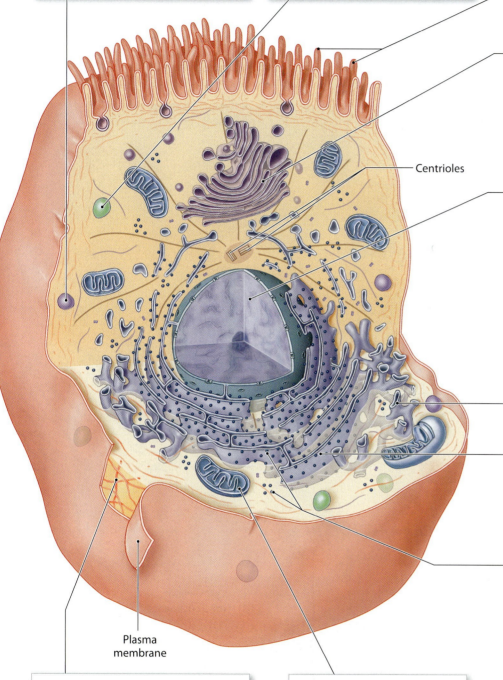

Peroxisome
STRUCTURE: Vesicles (membranous sacs) containing degradative enzymes.
FUNCTION: Break down fatty acids and organic compounds; neutralize toxic compounds generated in the process

Lysosome
STRUCTURE: Vesicles containing digestive enzymes.
FUNCTION: Break down large organic compounds, damaged organelles, and pathogens (disease-causing agents)

Microvilli
STRUCTURE: Membrane extensions containing microfilaments.
FUNCTION: Increase surface area to help absorption of extracellular materials

Golgi apparatus
STRUCTURE: Stacks of flattened membranes (cisternae) containing chambers.
FUNCTION: Modify and package proteins

Centrioles

Nucleus
STRUCTURE: A nucleoplasm (fluid in the nucleus) containing enzymes, proteins, DNA, and nucleotides; surrounded by a double membrane, the **nuclear envelope**.
FUNCTION: Control metabolism; store and process genetic information; control protein synthesis

Endoplasmic reticulum (ER)
STRUCTURE: Network of membranous sheets and channels extending throughout the cytoplasm.
FUNCTION: Synthesize secretory products; store and transport substances inside the cell

• **Smooth ER**, which has no attached ribosomes, synthesizes lipids and carbohydrates.

• **Rough ER**, which has ribosomes bound to the membranes, modifies and packages newly synthesized proteins.

Ribosomes
STRUCTURE: RNA and proteins; fixed ribosomes bound to rough ER, free ribosomes scattered in cytoplasm.
FUNCTION: Synthesize proteins

Plasma membrane

Cytoskeleton
STRUCTURE: Proteins organized in microfilaments (fine filaments) or microtubules (slender tubes); organizing center that contains a pair of centrioles.
FUNCTION: Strengthen and support cell; move cellular structures and materials

Mitochondrion
STRUCTURE: Double membrane, with inner membrane folds enclosing important metabolic enzymes.
FUNCTION: Produce 95 percent of the cell's ATP

Module 3.1 Review

a. Distinguish between cytoplasm, cytosol, and cytoskeleton.

b. Identify the membranous organelles and their functions.

c. Identify the nonmembranous organelles and describe their functions.

The plasma membrane isolates the cell from its environment and performs various functions; the cytoskeleton has structural and functional roles

1 The **plasma membrane** is a physical barrier that separates the inside of the cell from the surrounding extracellular fluid. It is a selectively permeable barrier that controls passage of materials into and out of the cell.

Superficial membrane carbohydrates form a layer known as the **glycocalyx** (glī -kō-KĀ-liks; calyx, cup). Carbohydrates account for roughly 3 percent of the weight of a plasma membrane; they are components of complex molecules such as glycoproteins (protein with some carbohydrates attached) and glycolipids (lipids with carbohydrates attached). The glycocalyx is important in cell recognition, binding to extracellular structures, and lubricating the cell surface.

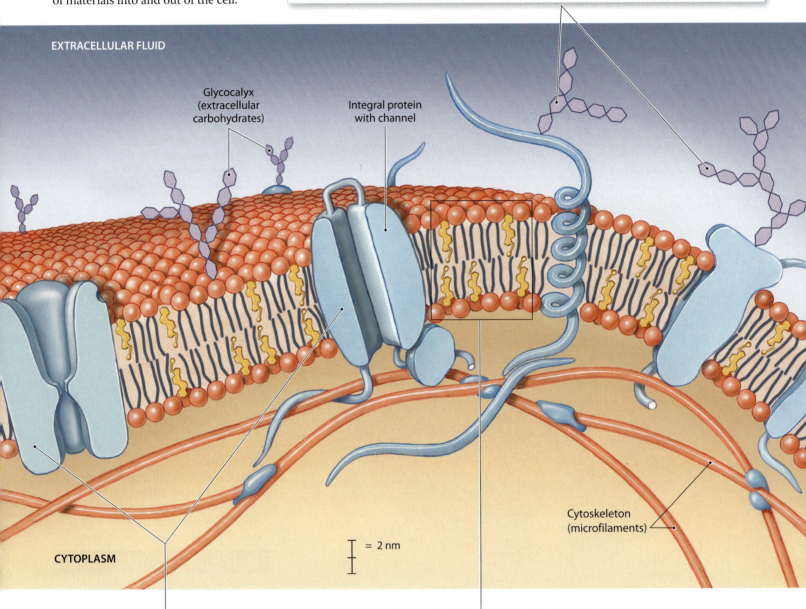

EXTRACELLULAR FLUID

Glycocalyx (extracellular carbohydrates)

Integral protein with channel

Cytoskeleton (microfilaments)

= 2 nm

CYTOPLASM

Integral proteins are part of the membrane structure and cannot be removed without damaging or destroying the membrane. Most integral proteins span the width of the membrane one or more times. Some contain pores or channels through which water and solutes may pass.

The plasma membrane is called a **phospholipid bilayer**, because the phospholipid molecules in it form two layers. In each half of the bilayer, the phospholipids lie with their hydrophilic (water-loving) heads at the membrane surface and their hydrophobic (water-fearing) tails on the inside. The hydrophobic layer in the center of the membrane isolates the cytoplasm from the extracellular fluid.

2 The cytoskeleton functions as the cell's skeleton. It provides an internal protein framework that gives the cytoplasm strength and flexibility. This table summarizes the components of the cytoskeleton and their structural role in two nonmembranous organelles, centrioles and cilia.

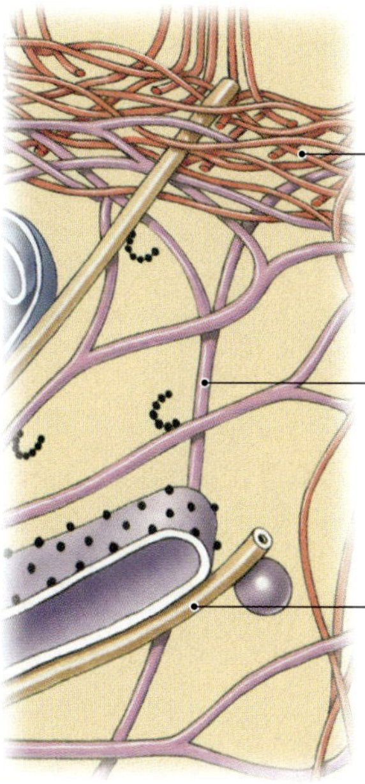

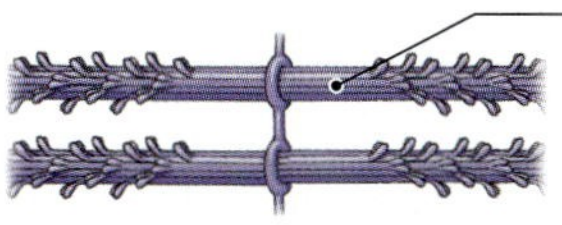

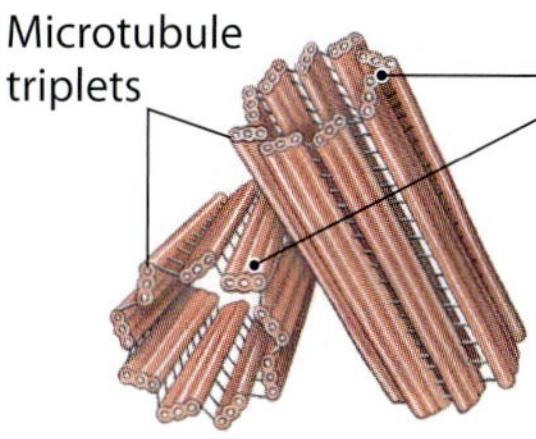

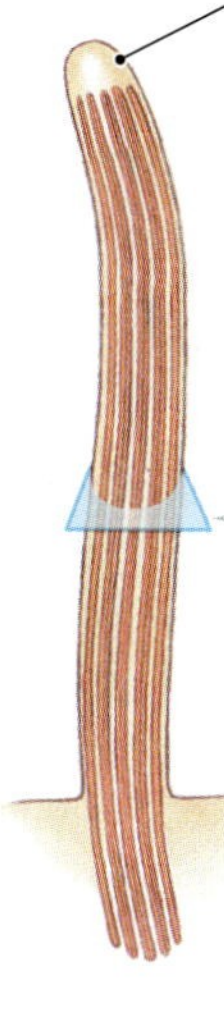

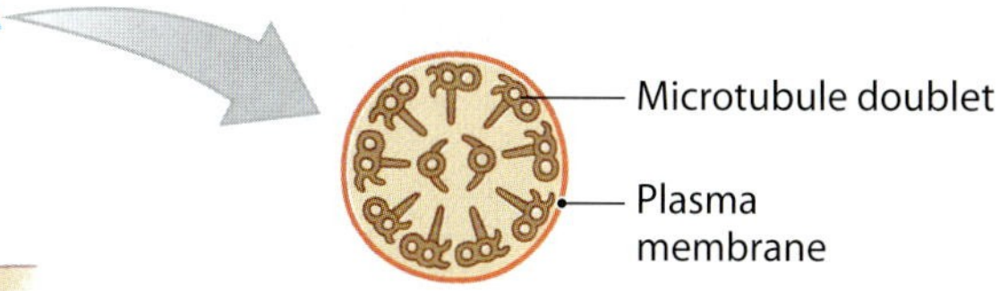

The Cytoskeleton

Structure	Remarks	Location	Functions
Microfilaments	Found in most cells; best organized in skeletal and cardiac muscle cells	In bundles beneath the cell membrane and throughout the cytoplasm	Provide strength, alter cell shape, and bind the cytoskeleton to the plasma membrane, tie the cells together, help muscles contract
Intermediate filaments	Found in most cells; at least five types known	In cytoplasm	Provide strength, move materials through cytoplasm
Microtubules	Found in most cells	In cytoplasm radiating away from centriole pair	Provide strength, move organelles; microtubules are the major components of centrioles and cilia
Thick filaments	Found in skeletal and cardiac muscle cells	In cytoplasm	Interact with actin microfilaments to contract muscles
Centrioles (organelles)	Nine groups of microtubule triplets form a short cylinder	In pairs near nucleus	Organize microtubules to move chromosomes during cell division
Cilia	Nine groups of long microtubule doublets form a cylinder around a central pair	At cell surface	Cilia beat rhythmically to move fluids or secretions across the cell surface

Microtubule triplets

Microtubule doublet

Plasma membrane

Module 3.2 Review

a. List the general functions of the plasma membrane.

b. Which structural parts of the plasma membrane are mostly responsible for its ability to isolate a cell from its external environment?

c. What is the function of cilia?

Protein synthesis in the endoplasmic reticulum and at free ribosomes depends primarily on the energy provided by mitochondrial activity

Cells are continuously expending ATP to synthesize new organic materials. They obtain that ATP by breaking down absorbed or recycled nutrients.

1 The **endoplasmic reticulum** (en-dō-PLAZ-mik re-TIK-ū-lum), or **ER**, is a network of intracellular membranes connected to the nuclear envelope, which surrounds the nucleus. The ER synthesizes and stores proteins, lipids, and carbohydates needed by the cell.

The ER forms hollow tubes, flattened sheets, and chambers called **cisternae** (sis-TUR-nē; singular, *cisterna*, a reservoir for water).

Nuclear envelope

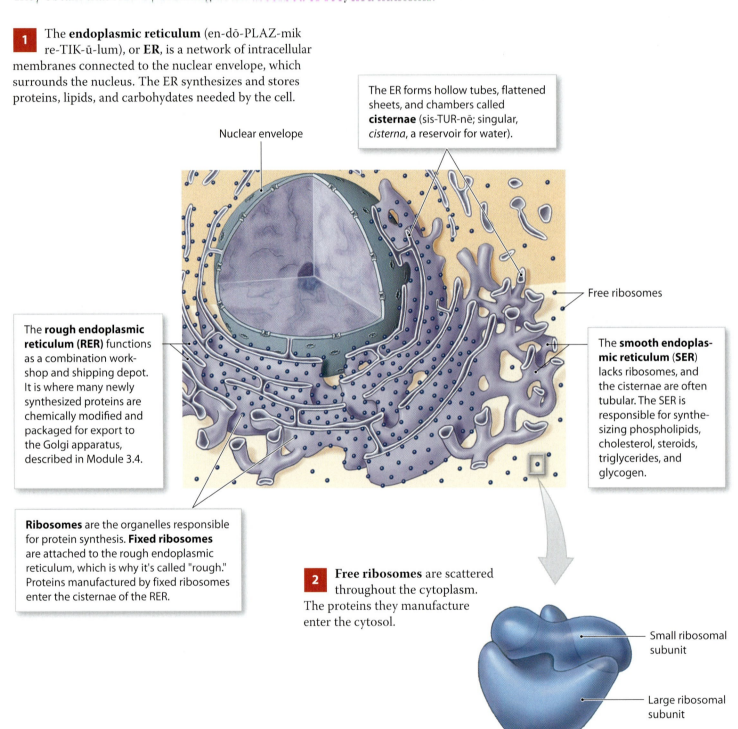

Free ribosomes

The **rough endoplasmic reticulum (RER)** functions as a combination workshop and shipping depot. It is where many newly synthesized proteins are chemically modified and packaged for export to the Golgi apparatus, described in Module 3.4.

The **smooth endoplasmic reticulum (SER)** lacks ribosomes, and the cisternae are often tubular. The SER is responsible for synthesizing phospholipids, cholesterol, steroids, triglycerides, and glycogen.

Ribosomes are the organelles responsible for protein synthesis. **Fixed ribosomes** are attached to the rough endoplasmic reticulum, which is why it's called "rough." Proteins manufactured by fixed ribosomes enter the cisternae of the RER.

2 **Free ribosomes** are scattered throughout the cytoplasm. The proteins they manufacture enter the cytosol.

Small ribosomal subunit

Large ribosomal subunit

3 All living cells require energy to carry out the functions of life. The organelles responsible for energy production are the **mitochondria** (mī-tō-KON-drē-uh; singular, mitochondrion; *mitos*, thread + *chondrion*, granule). Mitochondria vary widely in shape, from long and slender to short and fat. All share the basic structural features shown here. The number of mitochondria in a given cell varies with the cell's energy demands. Red blood cells lack mitochondria altogether, whereas mitochondria may account for 30 percent of the volume of a heart muscle cell.

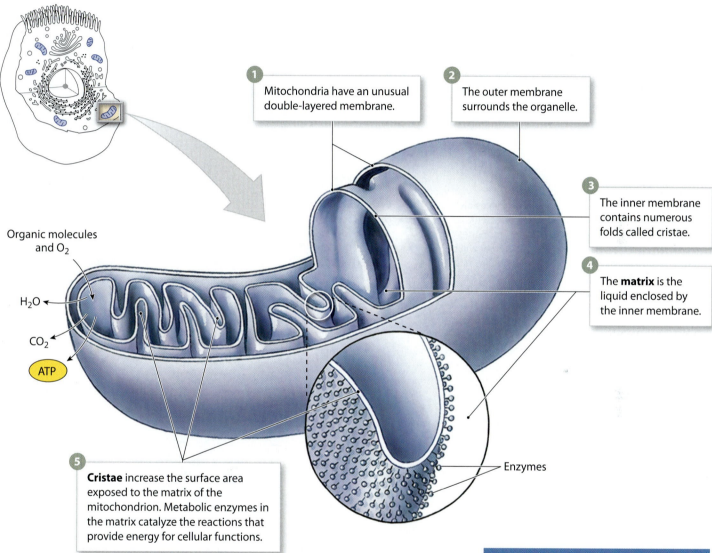

1 Mitochondria have an unusual double-layered membrane.

2 The outer membrane surrounds the organelle.

3 The inner membrane contains numerous folds called cristae.

4 The **matrix** is the liquid enclosed by the inner membrane.

Organic molecules and O_2

H_2O

CO_2

ATP

5 **Cristae** increase the surface area exposed to the matrix of the mitochondrion. Metabolic enzymes in the matrix catalyze the reactions that provide energy for cellular functions.

Enzymes

ATP production is known as **aerobic metabolism** (*aer*, air + *bios*, life), or cellular respiration. Aerobic metabolism in mitochondria produces about 95 percent of the ATP needed to keep a cell alive. (Enzymatic reactions in the cytosol produce the rest.) Note that although the chemical reactions that release energy occur in the mitochondria, most of the cellular activities that require energy occur in the surrounding cytoplasm. Think of mitochondria as little cellular batteries or fuel cells that provide the energy needed to power cellular functions.

Module 3.3 Review

a. Describe the immediate cellular destinations of newly synthesized proteins from free ribosomes and fixed ribosomes.

b. Describe the structure of smooth endoplasmic reticulum, rough endoplasmic reticulum, and a mitochondrion.

c. What does the presence of many mitochondria imply about a cell's energy requirements?

The Golgi apparatus is a packaging center

The **Golgi apparatus** (1) renews or modifies the plasma membrane; (2) modifies and packages secretions, such as hormones or enzymes, for release through exocytosis; and (3) packages special enzymes within vesicles for use in the cytosol.

1 The Golgi apparatus typically consists of five or six flattened membranous discs called **cisternae**. A single cell may contain more than one Golgi apparatus, typically situated near the nucleus.

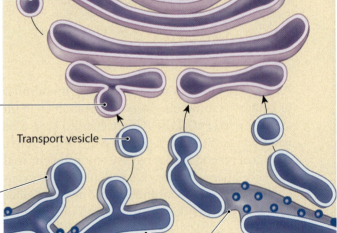

Extracellular Fluid

Cytosol

Membrane renewal

Membrane renewal vesicles add to the surface area of the plasma membrane. At the same time, other areas of the plasma membrane are being removed and recycled. The Golgi apparatus can thus change the properties of the plasma membrane, which can alter the sensitivity and functions of the cell.

Secretion

Secretory vesicles contain secretions such as hormones or enzymes that will be discharged from the cell. These vesicles fuse with the plasma membrane and empty their contents into the extracellular environment.

Enzymes for cytosol

Lysosomes (LĪ-sō-sōmz; *lyso-*, a loosening + *soma*, body) are special vesicles that provide an isolated environment for potentially dangerous chemical reactions. Lysosomes contain digestive enzymes whose varied functions are described on the facing page.

Membrane renewal

Secretion

Enzymes for cytosol

Secretory vesicle

Lysosome

Cisternae of the Golgi apparatus

4 Ultimately, the product arrives at the face oriented toward the free surface of the cell.

3 Small vesicles move material from one cisterna to the next.

2 The vesicles fuse with the Golgi membrane, emptying their contents into the cisternae. Inside the Golgi apparatus, enzymes modify the arriving proteins and glycoproteins. For example, the enzymes may change the carbohydrate structure of a glycoprotein, or they may attach a phosphate group, sugar, or fatty acid to a protein.

Transport vesicle

Start ➤ **1** **Transport vesicles** deliver some proteins and glycoproteins that were synthesized in the RER to the Golgi apparatus.

RER

72 • *Chapter 3: Cells and Tissues*

2 Cells often need to break down and recycle large organic molecules, and even complex structures like organelles. The breakdown process uses powerful enzymes, and it often generates toxic chemicals that could damage or kill the cell. Lysosomes isolate those chemical reactions from the rest of the cytoplasm. Below we show the three basic functions of the lysosomes: (1) digestion of lysosomal contents, (2) fusion with a vesicle, and (3) autolysis.

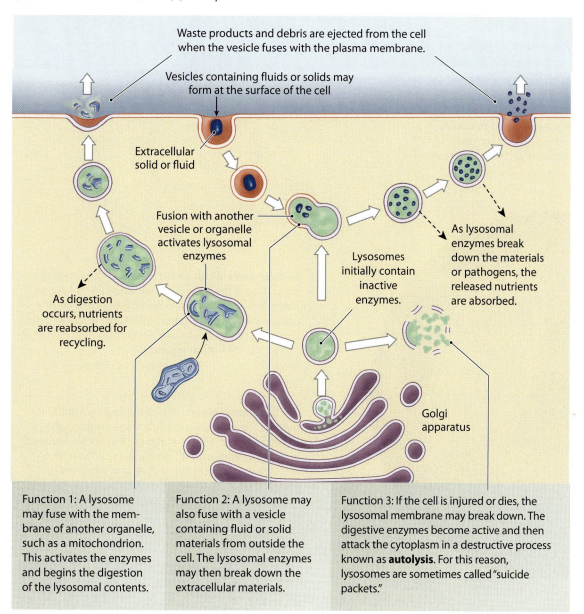

Waste products and debris are ejected from the cell when the vesicle fuses with the plasma membrane.

Vesicles containing fluids or solids may form at the surface of the cell

Extracellular solid or fluid

Fusion with another vesicle or organelle activates lysosomal enzymes

Lysosomes initially contain inactive enzymes.

As lysosomal enzymes break down the materials or pathogens, the released nutrients are absorbed.

As digestion occurs, nutrients are reabsorbed for recycling.

Golgi apparatus

Function 1: A lysosome may fuse with the membrane of another organelle, such as a mitochondrion. This activates the enzymes and begins the digestion of the lysosomal contents.

Function 2: A lysosome may also fuse with a vesicle containing fluid or solid materials from outside the cell. The lysosomal enzymes may then break down the extracellular materials.

Function 3: If the cell is injured or dies, the lysosomal membrane may break down. The digestive enzymes become active and then attack the cytoplasm in a destructive process known as **autolysis**. For this reason, lysosomes are sometimes called "suicide packets."

All membranous organelles except mitochondria are either interconnected or communicate through the movement of vesicles. This continuous movement and exchange is called membrane flow. In an actively secreting cell, an area equal to the entire membrane surface may be replaced *each hour*! Membrane flow is an example of the dynamic nature of cells. It gives cells a way to change their plasma membranes as they grow or respond to the environment.

Module 3.4 Review

a. List the three major functions of the Golgi apparatus.

b. The Golgi apparatus produces lysosomes. What do these lysosomes contain?

c. Describe three functions of lysosomes.

The nucleus contains DNA, RNA, organizing proteins, and enzymes

The nucleus is often the most prominent and visible organelle. In standard slide preparations the nucleus stains darkly and is easily identified.

1 The **nucleus** is the control center for cellular processes. The surrounding nuclear envelope, with its nuclear pores, controls chemical communication between the cytoplasm and the nucleus.

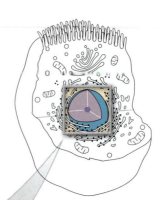

The **nuclear envelope**, which surrounds the nucleus and separates it from the cytoplasm, is a double-layered membrane.

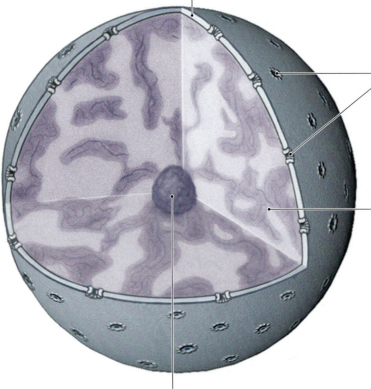

Nuclear pores, which account for about 10 percent of the surface of the nucleus, are passageways that permit chemical communication between the nucleus and the cytosol. Proteins at the pores regulate the movement of ions and small molecules. Neither proteins nor DNA can freely cross the nuclear envelope.

The term **nucleoplasm** refers to the fluid contents of the nucleus. The nucleoplasm contains the nuclear matrix, a network of fine filaments that provides structural support and may be involved with genetic activity. The nucleoplasm also contains ions, enzymes, small amounts of RNA, and DNA.

The **nucleolus** (noo-KLĒ-ō-lus) is a nuclear organelle that synthesizes ribosomal RNA. It also assembles the ribosomes, which reach the cytoplasm through the nuclear pores. Nucleoli are most prominent in cells that manufacture large amounts of proteins, such as liver, nerve, and muscle cells.

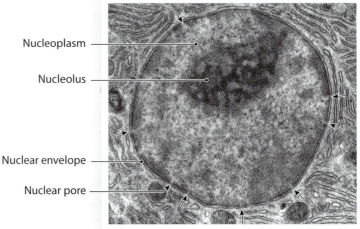

Nucleoplasm

Nucleolus

Nuclear envelope

Nuclear pore

Nucleus TEM × 34,800

2 The DNA in the nucleus stores the instructions for protein synthesis. In the nucleus, the DNA strands are coiled, rather than straight. DNA strands coil around proteins called histones. This packages a great deal of DNA in a relatively small space.

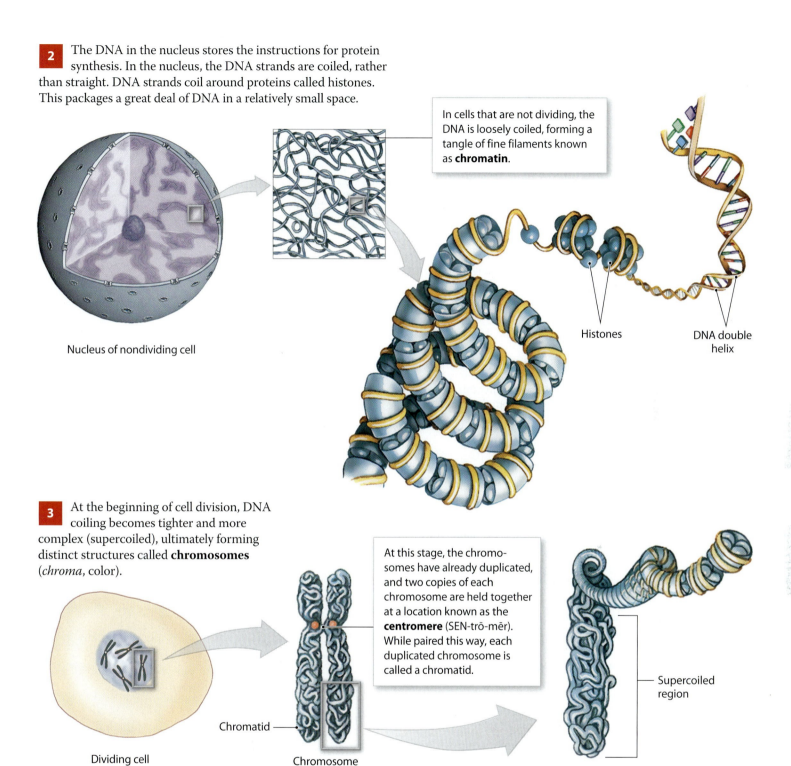

In cells that are not dividing, the DNA is loosely coiled, forming a tangle of fine filaments known as **chromatin**.

Histones

DNA double helix

Nucleus of nondividing cell

3 At the beginning of cell division, DNA coiling becomes tighter and more complex (supercoiled), ultimately forming distinct structures called **chromosomes** (*chroma*, color).

At this stage, the chromosomes have already duplicated, and two copies of each chromosome are held together at a location known as the **centromere** (SEN-trō-mēr). While paired this way, each duplicated chromosome is called a chromatid.

Supercoiled region

Chromatid

Dividing cell

Chromosome

In humans, the nuclei of somatic cells (general body cells, as opposed to sperm and oocytes, which are called sex cells) contain 23 pairs of chromosomes. One member of each pair comes from the mother, and one from the father. The DNA in these chromosomes carries the instructions for synthesizing proteins and RNA. In addition, some DNA segments have a regulatory function, and others have as yet undetermined functions.

Module 3.5 Review

a. Which molecule in the nucleus contains instructions for making proteins?

b. Describe the contents and the structure of the nucleus.

c. How many pairs of chromosomes does a typical somatic cell have?

Protein synthesis involves DNA, enzymes, and three types of RNA

In this module we introduce the key components and events of protein synthesis

1 DNA consists of long, parallel chains of nucleotides. Hydrogen bonding between complementary nitrogenous base pairs holds the two adjacent DNA chains together. The nitrogenous bases involved are adenine (A), thymine (T), cytosine (C), and guanine (G). Information is stored in the sequence of base pairs. The chemical "language" used is called the **genetic code**.

A **gene** is the functional unit of heredity. Genes contain all the DNA needed to produce specific proteins. Genes carry the information to build polypeptide chains.

A **triplet** is a sequence of three nitrogenous bases along a DNA strand that codes for a specific amino acid. For example, CTC codes for the amino acid leucine.

KEY
Nitrogenous bases

A Adenine
T Thymine
C Cytosine
G Guanine

2 In preparation for protein synthesis, the complementary DNA strands separate in the area of the active gene. An enzyme then assembles nucleotides into a single strand of **messenger RNA** (**mRNA**). Complementary base-pairing (A-U, G-C) with the DNA strand ensures that the sequence of nucleotides in the mRNA properly matches the sequence in DNA. Recall that adenine (A) can pair with either uracil (U) or thymine (T) (see Module 2.14).

The two DNA strands separate in the region containing the activated gene, exposing the triplets to the nucleoplasm.

Paired DNA strands

Enzyme

Each **codon** is a series of three RNA nucleotides; those nucleotides are complementary to those of the DNA triplets.

The mRNA strand containing the complementary codons passes through a nuclear pore and enters the cytoplasm.

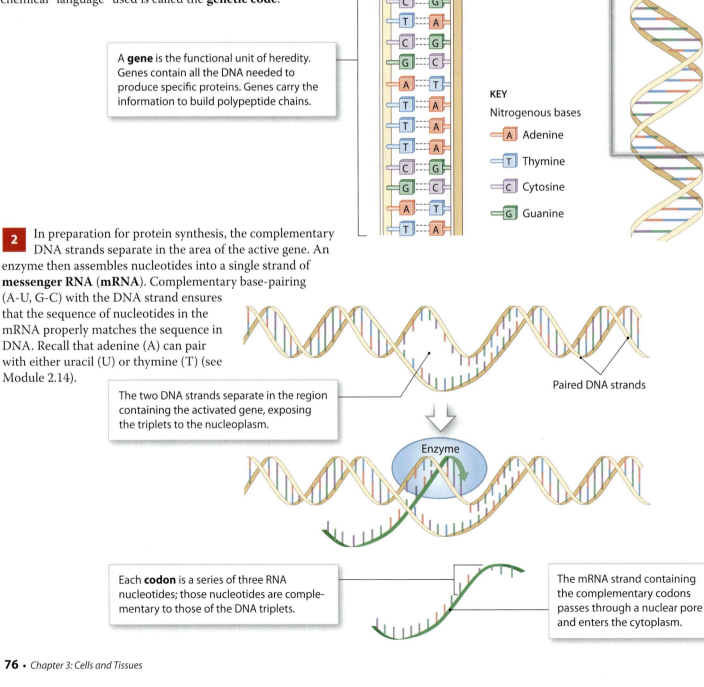

3 The complete process of protein synthesis involves two processes, transcription and translation.

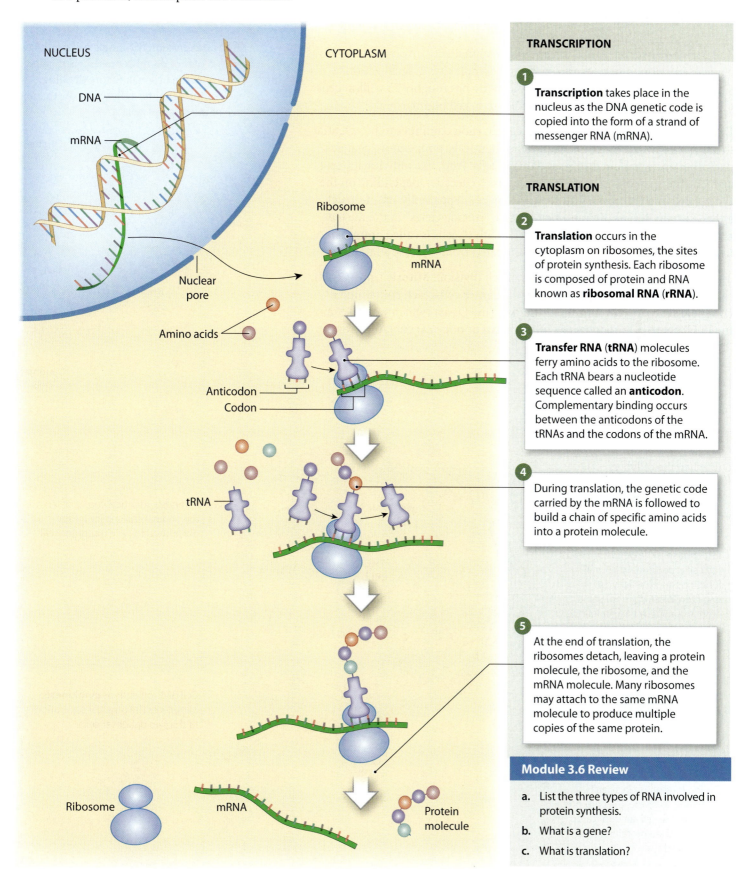

NUCLEUS

CYTOPLASM

DNA

mRNA

Nuclear pore

Amino acids

Anticodon

Codon

tRNA

Ribosome

mRNA

Ribosome

mRNA

Protein molecule

TRANSCRIPTION

1

Transcription takes place in the nucleus as the DNA genetic code is copied into the form of a strand of messenger RNA (mRNA).

TRANSLATION

2

Translation occurs in the cytoplasm on ribosomes, the sites of protein synthesis. Each ribosome is composed of protein and RNA known as **ribosomal RNA (rRNA)**.

3

Transfer RNA (**tRNA**) molecules ferry amino acids to the ribosome. Each tRNA bears a nucleotide sequence called an **anticodon**. Complementary binding occurs between the anticodons of the tRNAs and the codons of the mRNA.

4

During translation, the genetic code carried by the mRNA is followed to build a chain of specific amino acids into a protein molecule.

5

At the end of translation, the ribosomes detach, leaving a protein molecule, the ribosome, and the mRNA molecule. Many ribosomes may attach to the same mRNA molecule to produce multiple copies of the same protein.

Module 3.6 Review

a. List the three types of RNA involved in protein synthesis.

b. What is a gene?

c. What is translation?

Each cell has a life cycle that typically involves periods of growth and cell division

The period between fertilization and physical maturity involves tremendous changes in cell number, organization, and complexity. At fertilization, a single cell is all there is; at maturity, your body has roughly 75 trillion cells. **Cell division** is responsible for this increase in cell number.

Even when development is complete, cell division continues to be essential to survival. Physical wear and tear, toxic chemicals, temperature changes, and other environmental stresses damage cells. And, like individuals, cells age. The life span of a cell varies from hours to decades, depending on the type of cell and the stresses involved. Many cells apparently self-destruct after a certain period of time when specific "suicide genes" in the nucleus are activated. The genetically controlled death of cells is called **apoptosis** (ap-op-TŌ-sis or ap-ō-TŌ-sis; *apo-*, separated from + *ptosis*, a falling).

There are actually two different forms of cell division. **Mitosis** (mī-TŌ-sis), the focus of this section, produces two daughter cells, each containing a complete set of 46 chromosomes. **Meiosis** (mī-Ō-sis) (which we will examine in Chapter 18) produces sex cells (sperm or oocytes) containing only 23 chromosomes.

1 The division of a single cell produces a pair of **daughter cells**, each half the size of the original. Before dividing, each of the daughter cells will grow to the size of the original cell.

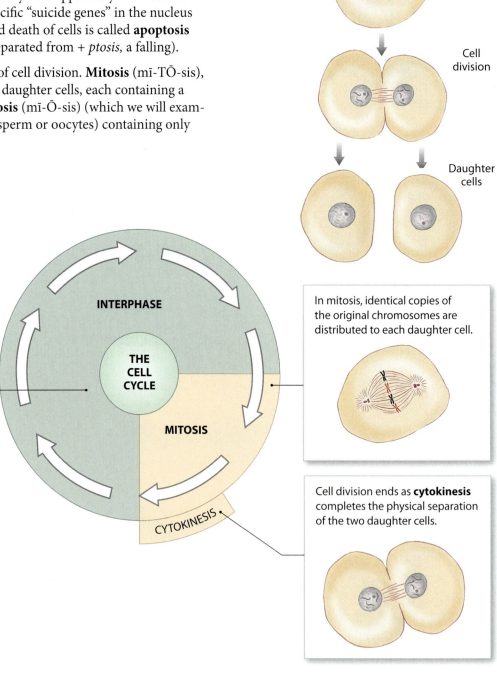

Original cell

Cell division

Daughter cells

2 This is a diagram of the life cycle of a typical cell. The cycle ends only when the cell dies.

INTERPHASE

THE CELL CYCLE

MITOSIS

CYTOKINESIS

Interphase is the period in which the cell is performing normal functions and not actively engaged in cell division. Some cells stay in interphase indefinitely; others are always either dividing or preparing to divide.

During interphase in a cell preparing to divide, the chromosomes of the cell are duplicated and associated proteins are synthesized.

In mitosis, identical copies of the original chromosomes are distributed to each daughter cell.

Cell division ends as **cytokinesis** completes the physical separation of the two daughter cells.

3 The first step in preparation for cell division is the duplication, or replication, of the DNA in the nucleus.

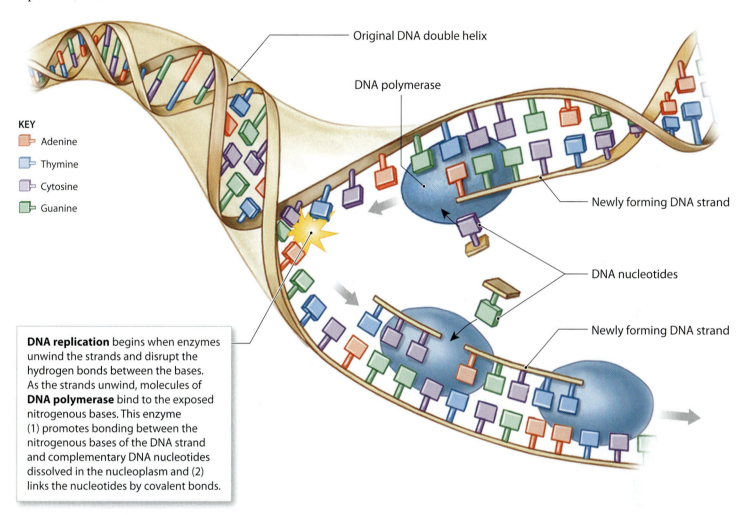

Original DNA double helix

DNA polymerase

KEY
- Adenine
- Thymine
- Cytosine
- Guanine

Newly forming DNA strand

DNA nucleotides

Newly forming DNA strand

DNA replication begins when enzymes unwind the strands and disrupt the hydrogen bonds between the bases. As the strands unwind, molecules of **DNA polymerase** bind to the exposed nitrogenous bases. This enzyme (1) promotes bonding between the nitrogenous bases of the DNA strand and complementary DNA nucleotides dissolved in the nucleoplasm and (2) links the nucleotides by covalent bonds.

4 Eventually, the unzipping completely separates the original strands. The copying ends and two identical DNA molecules have formed. Once the DNA has been replicated, the centrioles duplicated, and the necessary enzymes and proteins have been synthesized, the cell leaves interphase and proceeds to mitosis.

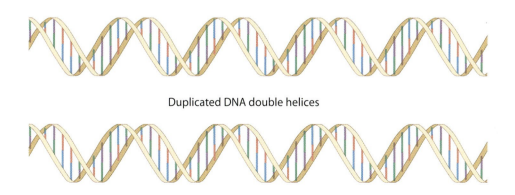

Duplicated DNA double helices

Module 3.7 Review

a. Describe what happens during interphase.

b. Which enzyme must be present for DNA replication to proceed normally?

c. What is apoptosis?

Mitosis distributes chromosomes before cytokinesis separates the daughter cells

Mitosis consists of a series of events during which the duplicated chromosomes of a cell are separated and distributed into two identical nuclei. The term mitosis specifically refers to the division and duplication of the cell's nucleus. A separate, but related, process called **cytokinesis** involves division of the cytoplasm into two distinct new cells.

MITOSIS

The centrioles have replicated, and the pairs now move to opposite sides of the nucleus.

Microtubules called **spindle fibers** interconnect the centriole pairs.

Chromatids   Centromere

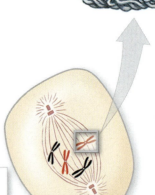

The nuclear membrane disintegrates during this period.

Centrioles   Nucleus

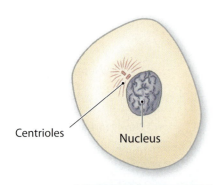

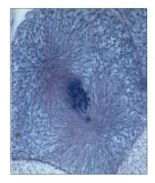

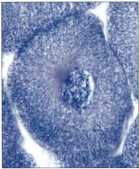

1 During interphase, the DNA strands are loosely coiled and chromosomes cannot be seen.

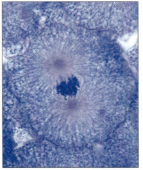

2 **Prophase** (PRŌ-fāz; *pro*, before) begins when the chromosomes coil so tightly that they can be seen under a light microscope. As a result of DNA replication, two copies of each chromosome now exist. Each copy is called a **chromatid**, and the paired chromatids are connected at a region known as the centromere.

The two chromatids are now pulled apart and drawn to opposite ends of the cell. Anaphase ends when the chromatids arrive near the centrioles at opposite ends of the cell.

As the chromatids approach the opposite sides of the dividing cell, the cytoplasm constricts along the plane of the metaphase plate, forming a **cleavage furrow**.

Metaphase plate

CYTOKINESIS

Daughter cells

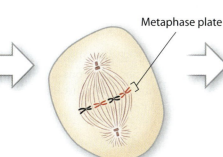

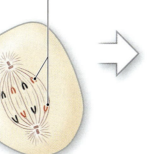
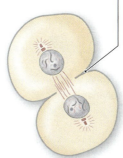

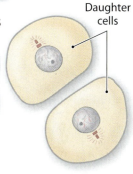

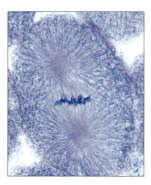
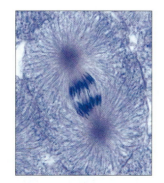
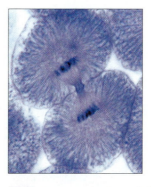
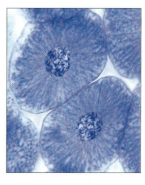

3 **Metaphase** (MET-a-fāz; *meta*, after) begins as the chromatids move to a narrow central zone called the **metaphase plate**. Metaphase ends when all the chromatids are aligned in the plane of the metaphase plate.

4 **Anaphase** (AN-a-fāz; *ana-*, apart) begins when the centromere of each chromatid pair splits and the chromatids separate. The chromatids are now pulled along the spindle fibers toward opposite sides of the dividing cell.

5 During **telophase** (TĔL-ō-fāz; *telo-*, end), the nuclear membranes re-form, the nuclei enlarge, and the chromosomes gradually uncoil to the chromatin state. The end of telophase marks the end of mitosis.

6 Cytokinesis usually begins with the formation of a cleavage furrow during anaphase and continues throughout telophase. The completion of cytokinesis marks the end of cell division.

Module 3.8 Review

a. Define mitosis, and list its four stages.

b. What is a chromatid, and how many would be present during normal mitosis in a human cell?

c. Describe the general appearance of a cell that has completed mitosis but not cytokinesis.

Tumors and cancer are characterized by abnormal cell growth and division

When the rates of cell division and growth exceed the rate of cell death, a tissue enlarges. The enlargement often results from the divisions of a single abnormal cell that no longer responds to the factors that regulate cell division. Cancer is an illness characterized by mutations—permanent changes in DNA nucleotide sequence and function—that disrupt normal control mechanisms that regulate the rates of cell division.

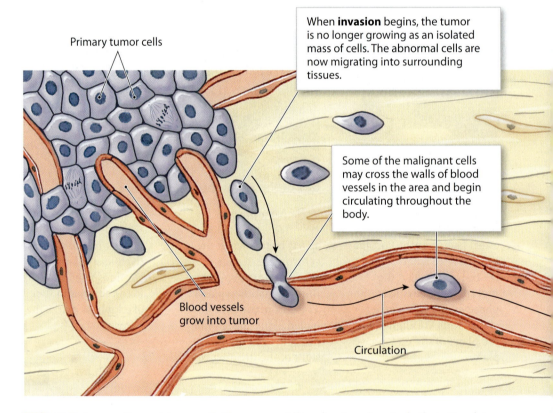

Abnormal cell

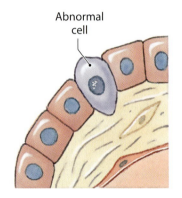

Cell divides

2 A **tumor** is a mass produced by abnormal cell growth and division. In a **benign tumor**, the cells are not cancerous and they usually remain in one place. A benign tumor seldom threatens an individual's life and can usually be surgically removed.

1 Cancer usually begins with a single abnormal cell. Every time a cell divides, there is a chance that something will go wrong with the control mechanism. As a result, cancers are most common in tissues where cells are dividing rapidly and continuously, such as the epithelium of the skin or the intestinal lining.

Primary tumor cells

When **invasion** begins, the tumor is no longer growing as an isolated mass of cells. The abnormal cells are now migrating into surrounding tissues.

Some of the malignant cells may cross the walls of blood vessels in the area and begin circulating throughout the body.

Blood vessels grow into tumor

Circulation

3 Cells in a **malignant tumor** divide very rapidly, releasing chemicals that stimulate the growth of blood vessels into the area. The availability of additional nutrients accelerates tumor growth, and malignant cells then begin migrating into surrounding tissues and nearby blood vessels. This process—**metastasis**—can produce tumors in tissues far from the site of the original tumor.

4 Malignant cells may no longer perform their original functions, or they may perform normal functions in an abnormal way. For example, endocrine cancer cells may produce normal hormones, but in excessively large amounts. This photo shows a patient with a malignant tumor of the thyroid gland. The tumor produces large amounts of thyroid hormone, causing dramatic changes in metabolic activity throughout the body.

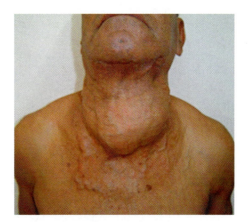

Responding to cues that are not known, cancer cells in the bloodstream ultimately escape out of blood vessels to establish tumors at other sites. These tumors are extremely active metabolically, and their presence stimulates blood vessel growth into the area. The increased blood supply provides additional nutrients to the cancer cells and further speeds up tumor growth and metastasis.

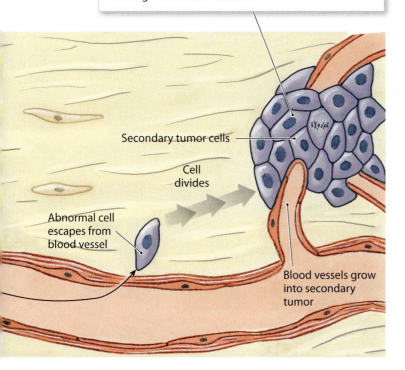

Secondary tumor cells

Cell divides

Abnormal cell escapes from blood vessel

Blood vessels grow into secondary tumor

Cancer cells do not use energy efficiently. They grow and multiply at the expense of healthy tissues, competing with normal cells for space and nutrients. Death may result when cancer cells kill or replace healthy cells, compress vital organs, or deprive normal tissues of essential nutrients. This is why many patients in the late stages of cancer appear starved.

Module 3.9 Review

a. Define cancer.

b. Distinguish between a benign tumor and a malignant tumor.

c. Why is cancer dangerous?

1. Short answer

Correctly label the indicated structures on the accompanying diagram of a cell and then describe the functions of each.

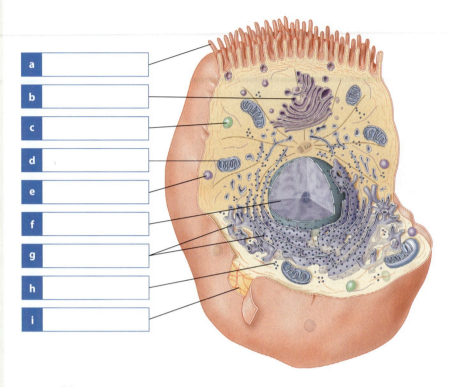

a

b

c

d

e

f

g

h

i

2. Matching

Match the following terms with the most closely related description.

- transcription
- tRNA
- chromosomes
- genetic information
- thymine
- mRNA
- gene
- uracil
- nuclear envelope
- nuclear pore
- nucleoli

a _____ DNA strands and histones

b _____ DNA nitrogen base

c _____ double membrane

d _____ RNA nitrogen base

e _____ assemble ribosomal subunits

f _____ passageway for functional mRNA

g _____ mRNA formation

h _____ functional unit of heredity

i _____ codon

j _____ anticodon

k _____ DNA nucleotide sequence

3. Section integration

The nucleus is often described as the control center of the cell. Explain the role of the nucleus in maintaining homeostasis.

_____

_____

_____

How Substances Enter and Leave the Cell

Because the plasma membrane is an effective barrier, conditions inside the cell can be considerably different from conditions outside the cell. However, the barrier cannot be absolute, because cells are not self-sufficient, and their activities must be coordinated. In this section we will consider how the plasma membrane selectively regulates the movement of materials into and out of the cell.

1 **Permeability** is the property of the plasma membrane that determines precisely which substances can enter or leave the cytoplasm.

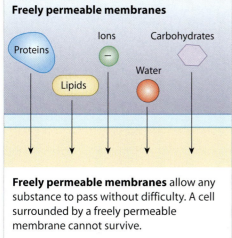

Freely permeable membranes

Proteins · Ions · Carbohydrates · Lipids · Water

Freely permeable membranes allow any substance to pass without difficulty. A cell surrounded by a freely permeable membrane cannot survive.

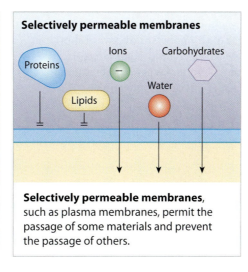

Selectively permeable membranes

Proteins · Ions · Carbohydrates · Lipids · Water

Selectively permeable membranes, such as plasma membranes, permit the passage of some materials and prevent the passage of others.

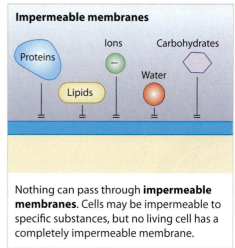

Impermeable membranes

Proteins · Ions · Carbohydrates · Lipids · Water

Nothing can pass through **impermeable membranes**. Cells may be impermeable to specific substances, but no living cell has a completely impermeable membrane.

2 Permeability is based on size, electrical charge, molecular shape, lipid solubility, or other factors. Cells differ in their permeabilities, depending on which lipids and proteins are present in the plasma membrane and how these components are arranged.

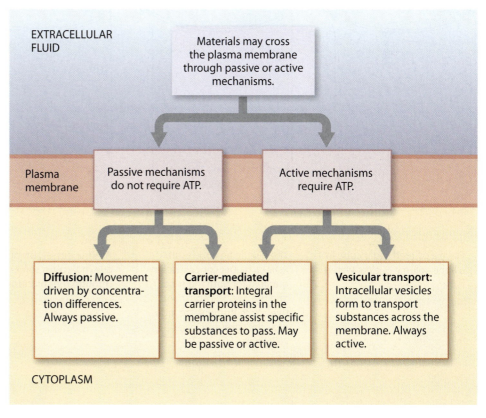

EXTRACELLULAR FLUID

Materials may cross the plasma membrane through passive or active mechanisms.

Plasma membrane

Passive mechanisms do not require ATP.

Active mechanisms require ATP.

Diffusion: Movement driven by concentration differences. Always passive.

Carrier-mediated transport: Integral carrier proteins in the membrane assist specific substances to pass. May be passive or active.

Vesicular transport: Intracellular vesicles form to transport substances across the membrane. Always active.

CYTOPLASM

Diffusion is movement driven by concentration differences

Ions and molecules in liquids and gases are constantly in motion, colliding and bouncing off one another and off obstacles in their paths. The movement is random: A molecule can bounce in any direction. One result of this continuous random motion is that, over time, the molecules in any given space tend to become evenly distributed. This distribution process is called **diffusion**. When molecules are not evenly distributed, a concentration difference, or **gradient** exists. Unless prevented from doing so, given enough time diffusion will eliminate a concentration gradient. After the gradient has been eliminated, the molecular motion continues, but net movement no longer occurs in any particular direction.

1 Diffusion in air and water is slow, and it is most important over very short distances. A simple demonstration—observing a colored sugar cube placed in a beaker of water—can give you a good mental picture of what happens at the cellular level. However, compared to a cell, a beaker of water is enormous.

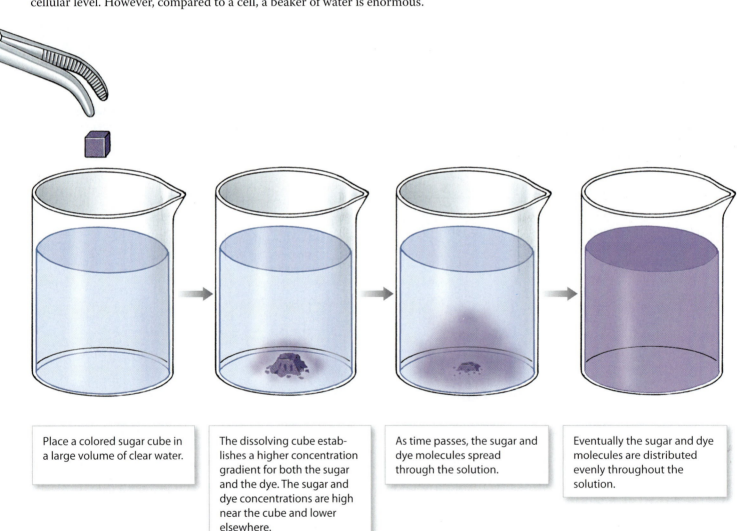

Place a colored sugar cube in a large volume of clear water.

The dissolving cube establishes a higher concentration gradient for both the sugar and the dye. The sugar and dye concentrations are high near the cube and lower elsewhere.

As time passes, the sugar and dye molecules spread through the solution.

Eventually the sugar and dye molecules are distributed evenly throughout the solution.

2 In extracellular fluids, water and dissolved solutes diffuse freely. A plasma membrane, however, acts as a barrier that selectively restricts diffusion: Some substances pass through easily, while others cannot penetrate the membrane. An ion or a molecule can diffuse across a plasma membrane only by (1) crossing the lipid portion of the membrane or (2) passing through a membrane channel.

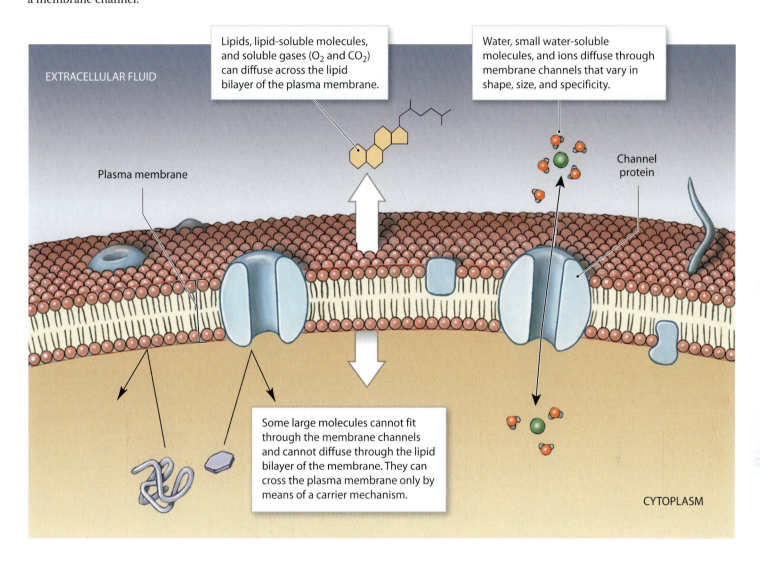

Lipids, lipid-soluble molecules, and soluble gases (O_2 and CO_2) can diffuse across the lipid bilayer of the plasma membrane.

Water, small water-soluble molecules, and ions diffuse through membrane channels that vary in shape, size, and specificity.

EXTRACELLULAR FLUID

Plasma membrane

Channel protein

Some large molecules cannot fit through the membrane channels and cannot diffuse through the lipid bilayer of the membrane. They can cross the plasma membrane only by means of a carrier mechanism.

CYTOPLASM

Intracellular and extracellular fluids are solutions that contain a variety of dissolved materials. Each solute diffuses as though it were the only material in solution. If we ignore individual solute identities and simply count ions and molecules, we find that the total concentration of dissolved ions and molecules on either side of the plasma membrane stays the same. This state of equilibrium persists because a typical plasma membrane is freely permeable to water. The diffusion of water across a membrane is so important that it is given a special name: **osmosis** (oz-MŌ-sis; *osmos*, a push).

Module 3.10 Review

a. Define diffusion.

b. Identify factors that influence diffusion.

c. How would a decrease in the oxygen concentration in the lungs affect the diffusion of oxygen into the blood?

Passive and active processes move materials in and out of cells

The plasma membrane isolates each cell from the extracellular environment, but that isolation is not complete. To survive, the cell must continuously absorb nutrients and oxygen and release carbon dioxide and waste products. This exchange involves both passive and active processes.

Passive Transport Processes (No cellular energy expended)

Process	Description	Example	
Diffusion	Diffusion is the movement of molecules from an area of high concentration to an area of low concentration. That is, the movement occurs down the concentration gradient (high to low).		When the concentration of CO_2 inside a cell is greater than outside the cell, the CO_2 diffuses out of the cell and into the extracellular fluid.
Osmosis	Osmosis is the diffusion of water molecules across a selectively permeable membrane. Movement occurs toward the higher solute concentration because that is where the concentration of water is lower. Osmosis continues until the concentration gradient is eliminated.		If the solute concentration outside a cell is greater than inside the cell, water molecules will move across the plasma membrane into the extracellular fluid.
Facilitated diffusion (carrier-mediated transport)	Facilitated diffusion is the movement of materials across a membrane by a carrier protein. Movement follows the concentration gradient.		Nutrients that are insoluble in lipids and too large to fit through membrane channels may be transported across the plasma membrane by carrier proteins. Many carrier proteins move a specific substance in one direction only, either into or out of the cell.

Process	Description	Example
Active transport (carrier-mediated transport)	**Active transport** requires carrier proteins that move specific substances across a membrane. It can continue despite an opposing concentration gradient. If the carrier moves one solute in one direction and another solute in the opposite direction, it is called an exchange pump.	EXTRACELLULAR FLUID / Sodium–potassium exchange pump / CYTOPLASM / 3 Na$^+$ / 2 K$^+$ / ATP / ADP — One of the most common active transport systems is the **sodium–potassium exchange pump.** One ATP molecule is consumed for each 3 sodium molecules that are ejected from the cell, while 2 potassium ions are reclaimed.
Vesicle formation (endocytosis)		
Pinocytosis	In **pinocytosis**, vesicles form at the plasma membrane and bring fluids and small molecules into the cell. This process is often called "cell drinking."	Pinocytic vesicle forming / CELL — Water and small molecules enter the cell across the vesicle membrane once the vesicle is inside the cytoplasm.
Receptor-mediated endocytosis	In **receptor-mediated endocytosis**, target molecules bind to receptor proteins on the membrane surface, triggering vesicle formation.	EXTRACELLULAR FLUID / Target molecules / Receptor proteins / CYTOPLASM / Vesicle containing target molecules — Each cell has specific sensitivities to extracellular materials, depending on what receptor proteins are present in the plasma membrane.
Phagocytosis	In **phagocytosis**, vesicles form at the plasma membrane to bring solid particles into the cell. This process is often called "cell eating."	Pseudopodium extends to surround object / Phagocytic vesicle / CELL — Large particles are brought into the cell by cytoplasmic extensions (called **pseudopodia**) that engulf the particle and pull it into the cell. Only specialized cells perform phagocytosis. These cells are called **phagocytes** or **macrophages**.
Exocytosis	In **exocytosis**, intracellular vesicles fuse with the plasma membrane to release fluids and/or solids from the cell.	Material ejected from cell / CELL — Cellular wastes or secretory products accumulate in vesicles and are ejected from the cell.

Module 3.11 Review

a. Which transport processes require carrier proteins?

b. Describe exocytosis.

c. Certain types of white blood cells engulf bacteria and bring them into the cell to be destroyed. What is this process called?

1. Concept map

Use each of the following terms once to fill in the blank boxes to correctly complete the membrane permeability concept map.

- exocytosis
- diffusion
- "cell eating"
- molecular size
- pinocytosis
- facilitated diffusion
- vesicular transport
- diffusion of water
- active transport
- specific substances

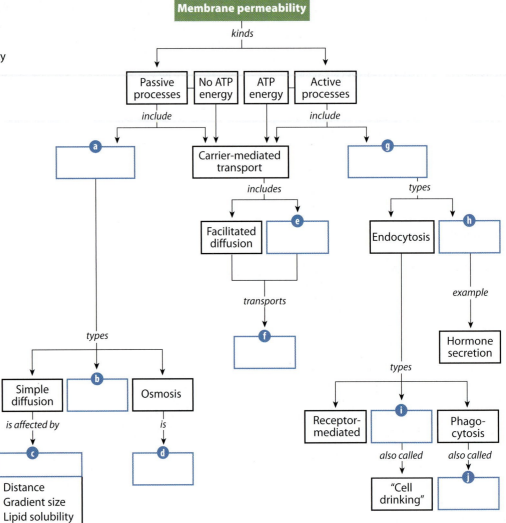

2. Short answer

Classify each of the following situations as an example of diffusion, osmosis, or neither.

- You walk into a room and smell a balsam-scented candle.
- A drop of food coloring disperses within a liquid medium.
- Water flows through a garden hose.
- A sugar cube placed in a cup of hot tea dissolves.
- Grass in the yard wilts after being exposed to excess chemical fertilizer.
- After soaking several hours in water containing sodium chloride, a stalk of celery weighs less than before it was placed in the salty water.

a _____

b _____

c _____

d _____

e _____

f _____

Tissues

Our perspective changes as we consider the tissue level of organization. For example, we require special imaging techniques and experimental procedures to examine atoms and molecules. Cellular details often escape detection unless we use an electron microscope. However, we can examine tissue structure with a light microscope, and based on your experiences in the laboratory, you may already be able to identify some tissues with the unaided eye.

The human body contains trillions of cells. For the human body to work efficiently, cells must coordinate their efforts. Cells working together form **tissues**—collections of cells and cell products that perform a relatively limited number of specialized functions. There are four basic types of tissue: (1) epithelial tissue, (2) connective tissue, (3) muscle tissue, and (4) neural tissue. In the rest of this chapter, we examine these tissue types, setting the stage for our study of organs and organ systems.

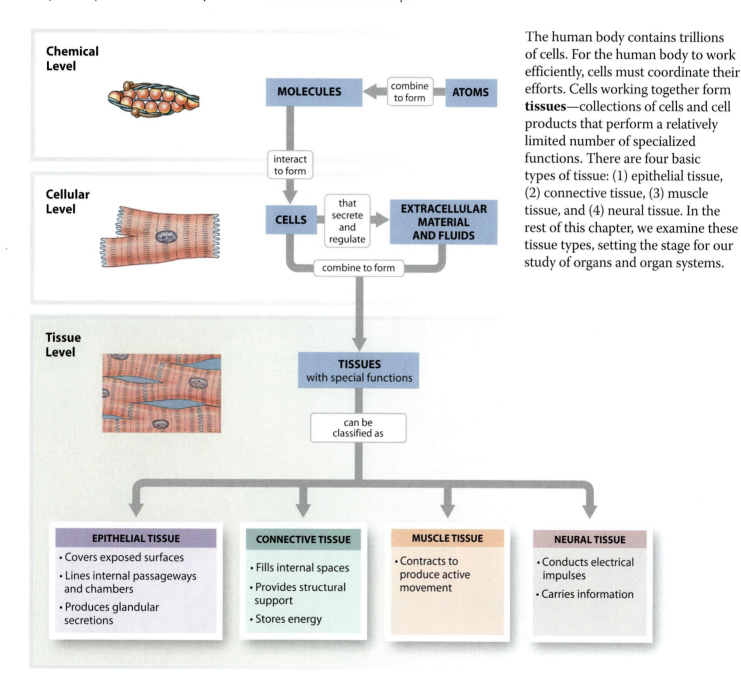

Chemical Level

MOLECULES ← combine to form ← ATOMS

interact to form

Cellular Level

CELLS → that secrete and regulate → EXTRACELLULAR MATERIAL AND FLUIDS

combine to form

Tissue Level

TISSUES with special functions

can be classified as

EPITHELIAL TISSUE	CONNECTIVE TISSUE	MUSCLE TISSUE	NEURAL TISSUE
• Covers exposed surfaces	• Fills internal spaces	• Contracts to produce active movement	• Conducts electrical impulses
• Lines internal passageways and chambers	• Provides structural support		• Carries information
• Produces glandular secretions	• Stores energy		

Tissues are specialized groups of cells and cell products

The roughly 200 different cell types in the body combine to form tissues, collections of cells and cell products that perform specific functions. **Histology** (*histos,* tissue) is the study of tissues. This module will introduce the four **primary tissue types** that, in various combinations, form the tissues of the body: epithelial tissue, connective tissue, muscle tissue, and neural tissue.

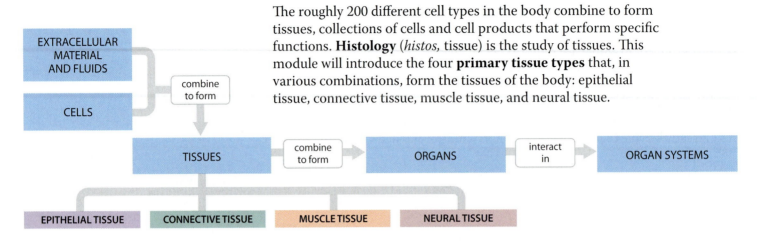

EXTRACELLULAR MATERIAL AND FLUIDS

CELLS

combine to form

TISSUES → combine to form → ORGANS → interact in → ORGAN SYSTEMS

EPITHELIAL TISSUE | CONNECTIVE TISSUE | MUSCLE TISSUE | NEURAL TISSUE

1 The most common type of **epithelial** (ep-i-THĒ-lē-ul) **tissue** is a layer of cells that forms a barrier with specific properties. Epithelia cover every exposed body surface; line the digestive, respiratory, reproductive, and urinary tracts; surround internal cavities such as the chest cavity or the fluid-filled chambers in the brain, eye, and inner ear; and line the inner surfaces of the blood vessels and heart.

2 **Connective tissue** is quite diverse in appearance. All forms of connective tissue contain specialized cells and an extracellular **matrix** that consists of protein fibers and a liquid known as the **ground substance**. The amount and consistency of the matrix depends on the type of connective tissue. In blood, the cells are suspended in a watery matrix called plasma. In bone, the solid matrix is more durable, with crystals of calcium salts organized around a fibrous framework and very little ground substance.

EPITHELIAL TISSUE

- Covers and protects exposed surfaces
- Lines internal passageways and chambers
- Produces glandular secretions

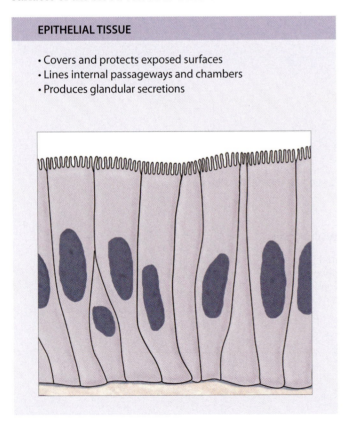

CONNECTIVE TISSUE

- Fills internal spaces
- Provides structural support
- Stores energy

Matrix

Fibers + Ground substance

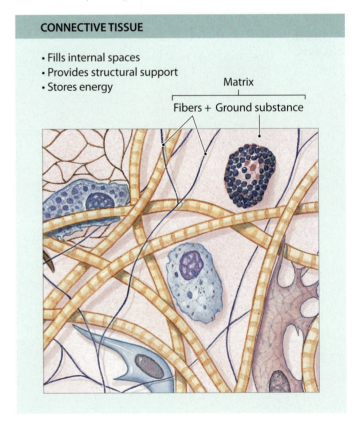

3 **Muscle tissue** is unique because individual muscle cells have the ability to contract forcefully. Major functions of muscle tissue include moving the skeleton, supporting soft tissues, maintaining blood flow, moving materials along internal passageways, and stabilizing normal body temperature. There are three different types of muscle tissue.

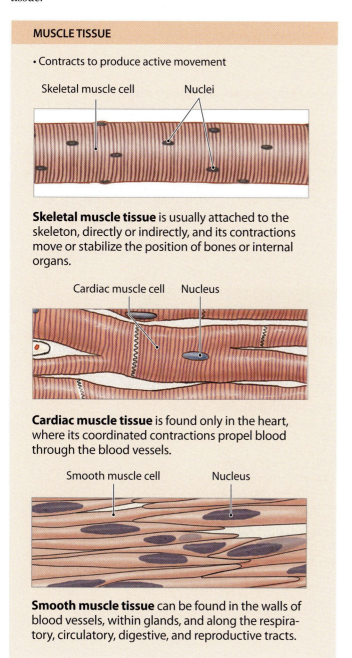

MUSCLE TISSUE

• Contracts to produce active movement

Skeletal muscle cell   Nuclei

Skeletal muscle tissue is usually attached to the skeleton, directly or indirectly, and its contractions move or stabilize the position of bones or internal organs.

Cardiac muscle cell   Nucleus

Cardiac muscle tissue is found only in the heart, where its coordinated contractions propel blood through the blood vessels.

Smooth muscle cell   Nucleus

Smooth muscle tissue can be found in the walls of blood vessels, within glands, and along the respiratory, circulatory, digestive, and reproductive tracts.

4 **Neural tissue** is specialized to carry information or instructions from one place in the body to another. Neural tissue has two basic types of cells: nerve cells, or **neurons** (NOO-rons; *neuro*, nerve), and supporting cells, or **neuroglia** (noo-ROG-lē-uh; *glia*, glue). Neurons transmit information in the form of electrical impulses. Neuroglia isolate and protect neurons while forming a supporting framework. The neural tissue in the body can be divided on anatomical grounds into the **central nervous system**, or brain and spinal cord, and the **peripheral nervous system**, which includes the nerves connecting the central nervous system with other tissues and organs.

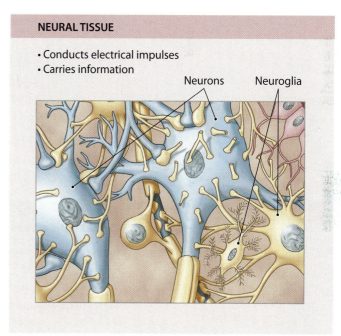

NEURAL TISSUE

• Conducts electrical impulses
• Carries information

Neurons   Neuroglia

Module 3.12 Review

a. Define histology.

b. Identify the four primary tissue types.

c. Explain the functions of each of the primary tissue types.

Epithelial tissue covers surfaces, lines structures, and forms secretory glands

1 Let's begin our discussion with epithelial tissue because it includes the surface of your skin, a familiar feature. Epithelial tissue includes epithelia and glands.

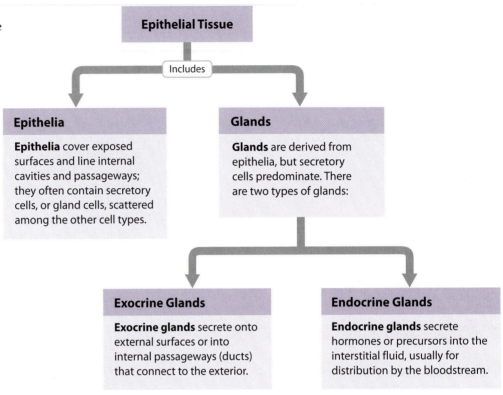

Epithelial Tissue

Includes

Epithelia

Epithelia cover exposed surfaces and line internal cavities and passageways; they often contain secretory cells, or gland cells, scattered among the other cell types.

Glands

Glands are derived from epithelia, but secretory cells predominate. There are two types of glands:

Exocrine Glands

Exocrine glands secrete onto external surfaces or into internal passageways (ducts) that connect to the exterior.

Endocrine Glands

Endocrine glands secrete hormones or precursors into the interstitial fluid, usually for distribution by the bloodstream.

2 This table describes four essential functions of epithelial tissue.

Functions of Epithelial Tissues

- **Provide physical protection**: Epithelia protect exposed and internal surfaces from abrasion, dehydration, and destruction by chemical or biological agents.

- **Control permeability**: Any substance that enters or leaves the body has to cross an epithelium. Some epithelia are relatively impermeable, whereas others are permeable to compounds as large as proteins. Most are capable of selective absorption or secretion.

- **Provide sensation**: Sensory nerves connect with most epithelia. Specialized epithelial cells can detect changes in the environment and convey information about such changes to the nervous system. For example, touch receptors respond to pressure by stimulating adjacent sensory nerves.

- **Produce specialized secretions**: Epithelial cells that produce secretions are called gland cells. Individual gland cells are often scattered among other cell types in an epithelium that may have many other functions.

3 The cells of any epithelium share a number of basic features. An epithelium has an **apical** (Ā-pi-kal) **surface**, which faces the exterior of the body or some internal space, and a **basal surface**, which is attached to adjacent tissues.

Microvilli are often found on the apical surfaces of epithelial cells that line internal passageways of the digestive, urinary, and reproductive tracts.

Cilia cover the apical surfaces in portions of the respiratory and reproductive tracts. A typical ciliated cell contains about 250 cilia that beat in a coordinated motion.

The apical surface is the region of the cell exposed to an internal or external environment. When the epithelium lines a tube, such as the intestinal tract, the apical surfaces of the epithelial cells are exposed to the space inside the tube, a passageway called the **lumen** (LOO-men).

The basal surface is where the cell attaches to underlying epithelial cells or deeper tissues.

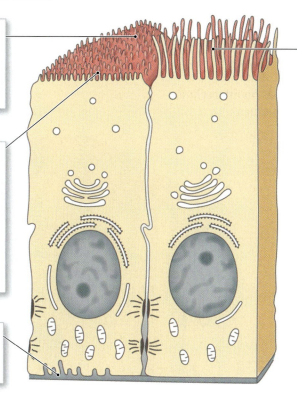

4 Epithelial cells come in three shapes: **squamous**, **cuboidal**, and **columnar**. For classification purposes, look at the superficial cells in a section. In sectional view, squamous cells appear thin and flat, cuboidal cells look like little boxes, and columnar cells look like tall, slender rectangles. If only one layer of cells is present, it is called a **simple epithelium**. In contrast, a **stratified epithelium** has several layers of cells. Stratified epithelia are generally located in areas that need protection from mechanical or chemical stresses, such as the skin surface and mouth lining.

Simple epithelia

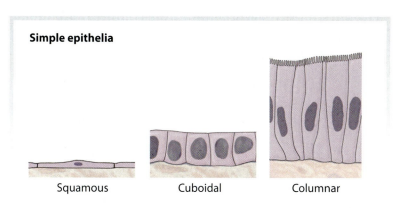

Squamous   Cuboidal   Columnar

Stratified epithelia

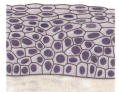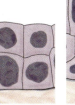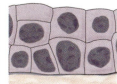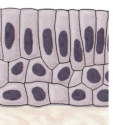

Squamous   Cuboidal   Columnar

Module 3.13 Review

a. List four essential functions of epithelial tissue.

b. Summarize the classification of an epithelium based on cell shape and number of cell layers.

c. What is the probable function of an epithelial surface whose cells have many cilia?

Epithelial cells are extensively interconnected, both structurally and functionally

To be effective as a barrier, an epithelium must form a complete cover or lining. It must also have the ability to replace lost or damaged cells through the divisions of stem cells. The physical integrity of an epithelium depends on intercellular connections and attachment to adjacent tissues.

1 The detailed structure of each form of intercellular attachment shows the linkage between structure and function at all levels.

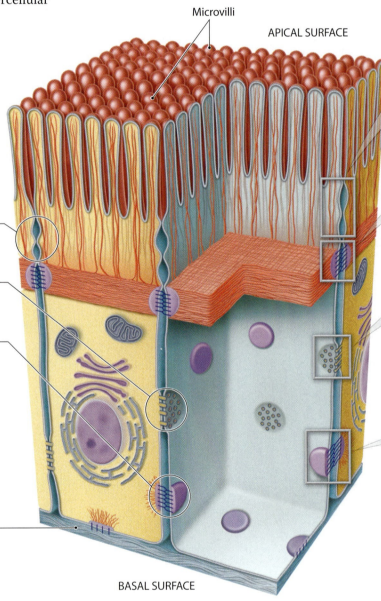

Microvilli

APICAL SURFACE

Intercellular Attachments

Tight junctions form a barrier that isolates the basolateral surfaces and deeper tissues from the contents of the lumen.

Gap junctions permit chemical communication that coordinates the activities of adjacent cells.

Desmosomes (DEZ-mō-sōms; *desmos*, ligament + *soma*, body) provide firm attachment between neighboring cells by interlocking their cytoskeletons.

Basement Membrane

The **basement membrane** is a complex structure produced by the basal surface of the epithelium and the underlying connective tissue. It contains glycoproteins and a network of fine protein filaments.

BASAL SURFACE

2 At a tight junction, the attachment is so tight that it prevents the passage of water and solutes between cells. Tight junctions are found in the digestive tract to help keep enzymes, acids, and wastes in the lumen from reaching the basolateral surfaces and digesting or damaging the underlying tissues and organs.

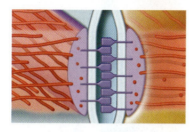

At a tight junction, interlocking membrane proteins bind the lipid portions of the two plasma membranes tightly together.

3 An adhesion belt lies inferior to the tight junctions. It is made up of a continuous band of membrane proteins that encircles cells and binds them to their neighbors. The bands are proteins that are attached to the microfilaments of the cytoskeleton.

4 At a gap junction, interlocking junctional proteins called **connexons** hold two cells together. Gap junctions between epithelial cells are common where the movement of ions helps coordinate functions such as secretion or the beating of cilia. Gap junctions occur in other tissues as well; in cardiac muscle tissue, for example, they help coordinate contractions of the heart muscle.

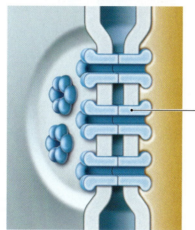

Connexons are channel proteins that form a narrow passageway and let small molecules and ions pass from cell to cell.

5 At a desmosome, the opposing plasma membranes are locked together. Desmosomes are very strong and resist stretching and twisting. Desmosomes are abundant between cells in the superficial layers of the skin. As a result, damaged skin cells are usually lost in sheets rather than as individual cells. That is why your skin peels rather than comes off as a powder after a sunburn.

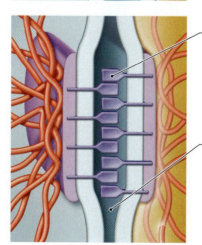

Cell adhesion molecules (CAMs) are membrane proteins that bind to each other and to extracellular materials.

The membranes of adjacent cells may also be bonded by a thin layer of **proteoglycans** made of a polysaccharide derivative.

Epithelia lack blood vessels, so they are called **avascular** (*a*, without). The cells forming the deepest layer of an epithelium must remain firmly attached to underlying connective tissues, because the blood vessels in those tissues nourish the entire epithelium.

Module 3.14 Review

a. Identify the types of epithelial intercellular connections.

b. How do epithelial tissues, which are avascular, obtain needed nutrients?

c. What are the functions of gap junctions?

The cells in a squamous epithelium are flattened and irregular in shape

The cells in a **squamous epithelium** (SKWĀ-mus; *squama,* plate or scale) are thin, flat, and irregular in shape, like pieces of a jigsaw puzzle. From the surface, the cells look like fried eggs laid side by side. In sectional view, the disc-shaped nucleus occupies the thickest portion of each cell.

1 A **simple squamous epithelium** is the body's most delicate epithelium. This type of epithelium is located in protected regions where absorption or diffusion takes place, or where a slick, slippery surface reduces friction. For example, the peritoneum forms a thin, slippery membrane that lines the abdominal cavity. Simple squamous epithelia are also found along passageways in the kidneys, in capillary walls, inside the eye, and at the gas exchange surfaces (alveoli) of the lungs.

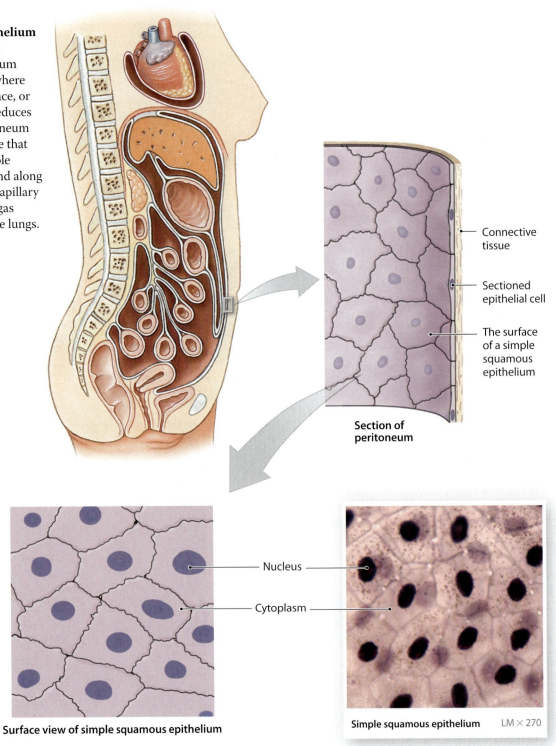

Connective tissue

Sectioned epithelial cell

The surface of a simple squamous epithelium

Section of peritoneum

Nucleus

Cytoplasm

Surface view of simple squamous epithelium

Simple squamous epithelium LM × 270

2 A **stratified squamous epithelium** is located where mechanical or chemical stresses are severe. It is composed of many layers of cells with only the superficial layers being flattened. Stratified squamous epithelia form the surface of the skin and line the mouth, throat, tongue, esophagus, rectum, anus, and vagina.

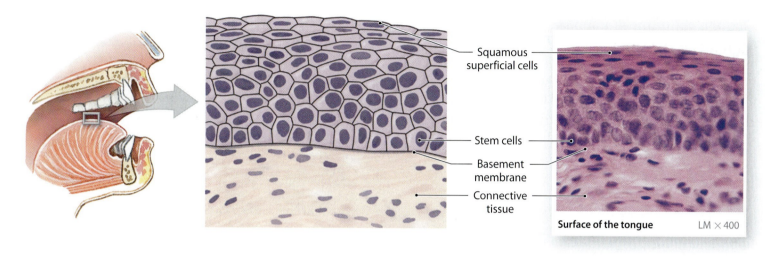

Squamous superficial cells

Stem cells

Basement membrane

Connective tissue

Surface of the tongue   LM × 400

3 On exposed body surfaces, where mechanical stress and dehydration are potential problems, apical layers of epithelial cells are packed with filaments of the protein keratin. As a result, superficial layers in a **keratinized** epithelium—such as the surface of your skin—are both tough and water resistant. A **nonkeratinized** stratified squamous epithelium resists abrasion but will dry out and deteriorate unless kept moist. More delicate non-keratinized stratified squamous epithelia are found in the oral cavity, pharynx, esophagus, anus, and vagina.

Keratinized skin cells

Keratin fibers

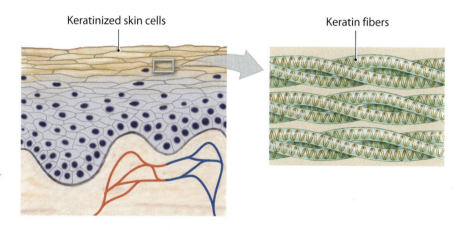

Surface of human skin

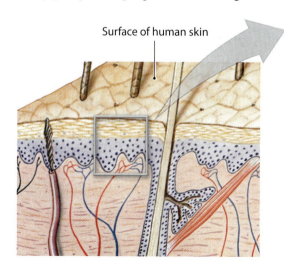

Module 3.15 Review

a. What properties are common to keratinized epithelia?

b. Why do the pharynx, esophagus, anus, and vagina have a similar epithelial organization?

c. Under a light microscope, you see simple squamous epithelium on the outer surface of a tissue. Could this be a skin surface sample? Why or why not?

Cuboidal and transitional epithelia are found along several passageways and chambers connected to the exterior

Cuboidal Epithelium

1 The cells of a **cuboidal epithelium** resemble hexagonal boxes. The spherical nuclei are near the center of each cell.

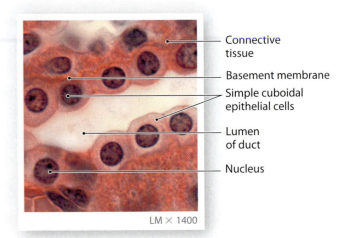

Connective tissue

Basement membrane

Simple cuboidal epithelial cells

Lumen of duct

Nucleus

LM × 1400

2 **Simple cuboidal epithelia** line exocrine glands and ducts (passageways that carry secretions). They are also found in portions of the kidneys and in the thyroid gland.

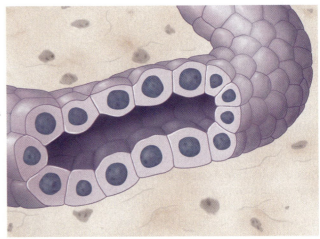

Sectioned kidney tubule

3 **Stratified cuboidal epithelia** are rare. They are most common along the ducts of sweat glands, mammary glands, and other exocrine glands.

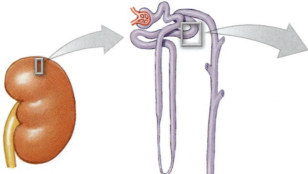

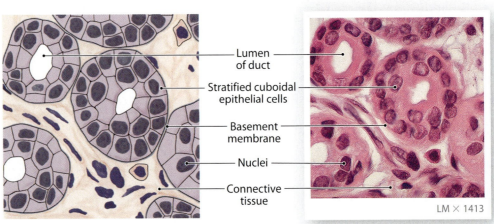

Lumen of duct

Stratified cuboidal epithelial cells

Basement membrane

Nuclei

Connective tissue

LM × 1413

Sweat gland duct

Transitional Epithelium

4 A **transitional epithelium** is an unusual stratified epithelium because it can stretch without damage. It is called *transitional* because its appearance changes as it stretches. Transitional epithelia is found only in the urinary system, where it lines the urinary bladder, the ureters, and the urine-collecting chambers within the kidneys.

Epithelium in a Relaxed Bladder

In an empty urinary bladder, the plump superficial cells are cuboidal with a dome-shaped surface.

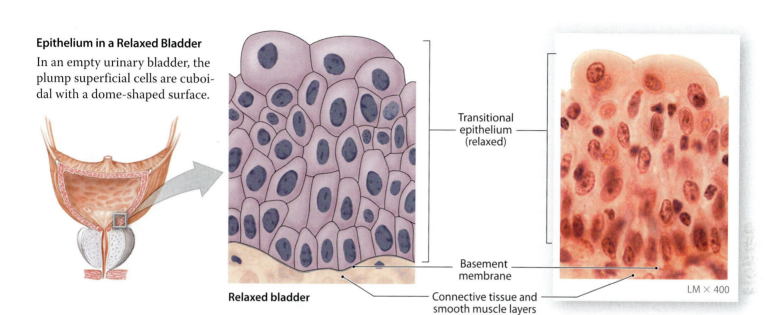

Transitional epithelium (relaxed)

Basement membrane

Relaxed bladder

Connective tissue and smooth muscle layers

LM × 400

Epithelium in a Stretched Bladder

When the urinary bladder is full, the volume of urine stretches the lining to such a degree that the epithelium appears flattened, and more like a stratified squamous epithelium.

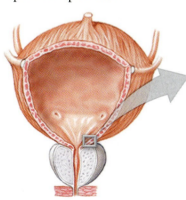

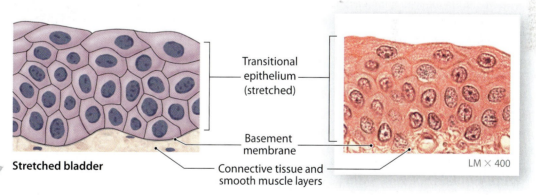

Transitional epithelium (stretched)

Basement membrane

Stretched bladder

Connective tissue and smooth muscle layers

LM × 400

Module 3.16 Review

a. Identify the epithelium that lines the urinary bladder and describe how its appearance changes as it stretches.

b. Describe the appearance of simple cuboidal epithelial cells in sectional view.

c. Stratified cuboidal epithelia are associated with which epithelial structures?

Columnar epithelia typically perform absorption or provide protection from chemical or environmental stresses

In a typical sectional view, the cells of a **columnar epithelium** appear rectangular. In reality, the densely packed cells are hexagonal, but they are taller and more slender than cells in a cuboidal epithelium. The elongated nuclei are crowded into a narrow band close to the basement membrane.

Simple Columnar Epithelium

1 **Simple columnar epithelia** line the stomach, intestine, gallbladder, uterine tubes, and ducts within the kidneys. These cells may have microvilli, which increase surface area for absorption, or cilia that move substances across the apical surface.

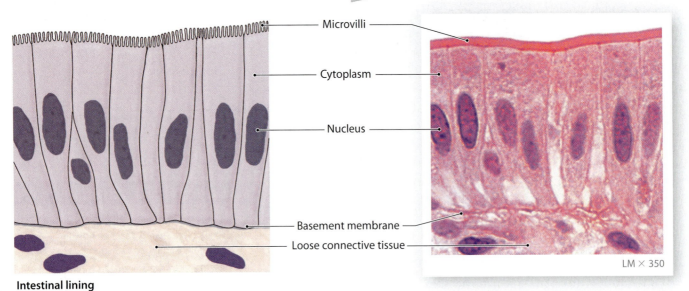

Microvilli

Cytoplasm

Nucleus

Basement membrane

Loose connective tissue

LM × 350

Intestinal lining

Pseudostratified Columnar Epithelium

Pseudostratified epithelia line the nasal cavities, the trachea, and larger airways of the lungs. They are also found along portions of the male reproductive tract.

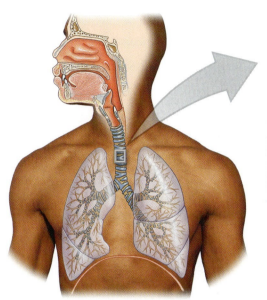

2 A **pseudostratified columnar epithelium** includes several types of cells with varying shapes and functions. The distances between the cell nuclei and the exposed surface vary, so the epithelium appears to be layered, or stratified. It is not truly stratified, though, because every epithelial cell contacts the basement membrane. Pseudostratified columnar epithelial cells usually have cilia.

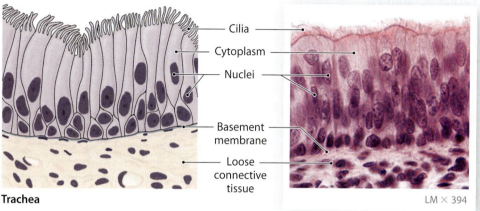

Cilia

Cytoplasm

Nuclei

Basement membrane

Loose connective tissue

Trachea

LM × 394

Stratified Columnar Epithelium

3 **Stratified columnar epithelia** are rare. These epithelia may have either two layers or multiple layers. In the latter case, only the superficial cells are columnar in shape. Stratified columnar epithelia are most often found lining large ducts such as those of the salivary glands or pancreas.

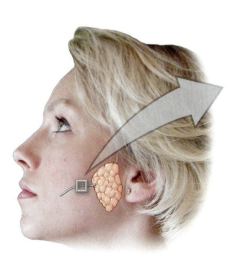

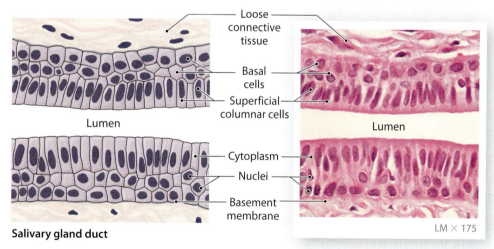

Loose connective tissue

Basal cells

Superficial columnar cells

Lumen

Lumen

Cytoplasm

Nuclei

Basement membrane

Salivary gland duct

LM × 175

Module 3.17 Review

a. Describe the appearance of simple columnar epithelial cells in a sectional view.

b. A pseudostratified columnar epithelium is not truly stratified. Explain why.

c. The columnar epithelium lining the intestine usually has _____ on its apical surface.

Glandular epithelia are specialized for secretion

Many epithelia contain gland cells that are specialized for secretion. Collections of epithelial cells (or structures derived from epithelial cells) that produce secretions are called glands. They range from scattered cells to complex glandular organs.

Types of Glands

Endocrine glands release their secretions into the interstitial fluid and plasma (fluid part of blood). **Exocrine glands** release their secretions into passageways called **ducts** that open onto an epithelial surface.

1 Endocrine glands produce *endocrine* (*endo-*, inside + *krinein*, to separate) *secretions*. Here we show a model of their basic structure. We will consider endocrine glands in a later chapter.

2 Exocrine glands produce *exocrine* (*exo-*, outside) *secretions*. Most exocrine secretions reach the surface through tubular ducts which empty onto the skin or the lining of an internal passageway that communicates with the exterior.

3 The only unicellular (one-celled) exocrine glands in the body are **mucous cells**, which release a secretion called *mucin*. These cells are scattered among other epithelial cells, such as the lining of the intestines and trachea.

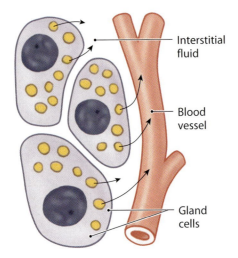

Interstitial fluid

Blood vessel

Gland cells

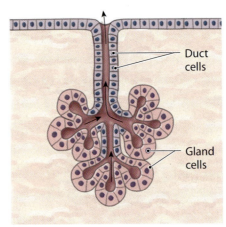

Duct cells

Gland cells

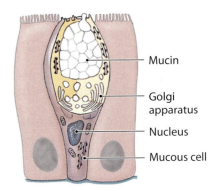

Mucin

Golgi apparatus

Nucleus

Mucous cell

Classification of Exocrine glands

4 We can classify exocrine glands by the type of secretion. *Serous glands* secrete a watery solution containing enzymes. *Mucous glands* secrete mucins that create a thick, slippery mucus. *Mixed glands* contain both serous and mucous gland cells.

Secretion Type	Description	Examples
Serous	Watery solution containing enzymes	Secretions of parotid salivary gland
Mucous	Thick, slippery mucus	Secretions of sublingual salivary gland
Mixed	Contains both serous and mucous cell secretions	Secretions of submandibular salivary gland

5 Exocrine glands can also be classified by their mechanism of secretion. Their glandular epithelial cells may release their secretions in one of three ways.

Mucin is a merocrine secretion that mixes with water to form mucus. **Mucus** is an effective lubricant, a protective barrier, and a sticky trap for foreign particles and microorganisms.

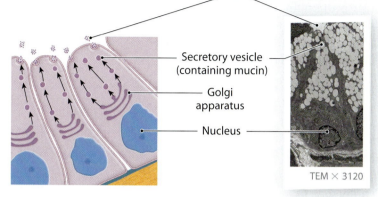

Secretory vesicle (containing mucin)

Golgi apparatus

Nucleus

TEM × 3120

In **merocrine secretion** (MER-u-krin; *meros*, part), the product is released from secretory vesicles by exocytosis. This is the most common mode of secretion. Saliva produced by salivary glands involves merocrine secretion.

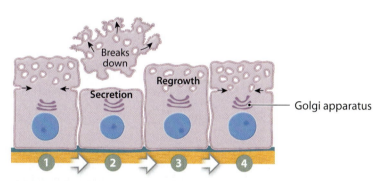

Breaks down

Regrowth

Secretion

Golgi apparatus

Apocrine secretion (AP-ō-krin; *apo-*, off) involves the loss of cytoplasm as well as the secretory product. The apical portion of the cytoplasm becomes packed with secretory vesicles and is then shed. Milk production in the mammary glands involves a combination of merocrine and apocrine secretions.

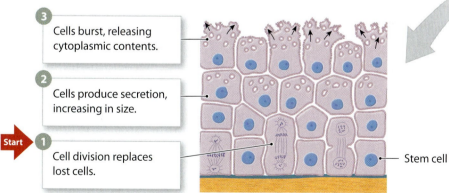

3 Cells burst, releasing cytoplasmic contents.

2 Cells produce secretion, increasing in size.

Start **1** Cell division replaces lost cells.

Stem cell

Holocrine secretion (HOL-ō-krin; *holos*, entire), by contrast, destroys the gland cell. During holocrine secretion, the entire cell becomes packed with secretory products and then bursts, releasing the secretion and killing the cell. Further secretion depends on stem cells dividing to replace destroyed gland cells.

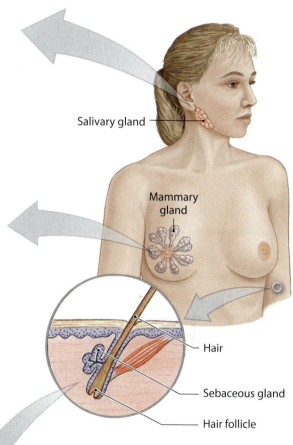

Salivary gland

Mammary gland

Hair

Sebaceous gland

Hair follicle

Module 3.18 Review

a. Name the two primary types of glands.

b. What type of secretion involves the loss of cytoplasm as well as the secretory product?

c. Describe holocrine secretion.

Loose connective tissues provide padding and support, whereas dense connective tissues provide strength

Loose and dense connective tissues contain extracellular protein fibers, a viscous ground substance, and two classes of cells. *Fixed cells* are stationary and are involved primarily with local maintenance, repair, and energy storage. *Wandering cells* are concerned primarily with defending and repairing damaged tissues.

Loose Connective Tissues

The most common forms of loose connective tissue are areolar tissue and adipose tissue.

1 **Areolar tissue** is the general packing material in the body. All of the cell types shown here can be found in areolar tissue.

Fibers

Reticular fibers are strong and form a branching network.

Collagen fibers are thick, straight or wavy, and often form bundles. They are very strong and resist stretching.

Elastic fibers are slender, unbranching, and very stretchy. They recoil to their original length after stretching or distortion.

Fixed Cells

A **melanocyte** is a fixed pigment cell that synthesizes melanin, a brownish-yellow pigment.

A **fixed macrophage** is a stationary phagocytic cell that engulfs cell debris and pathogens.

Mast cells are fixed cells that stimulate local inflammation and mobilize tissue defenses.

Fibroblasts are fixed cells that synthesize the extracellular fibers of the connective tissue.

Adipocytes (fat cells) are fixed cells that store lipids in large intracellular vesicles.

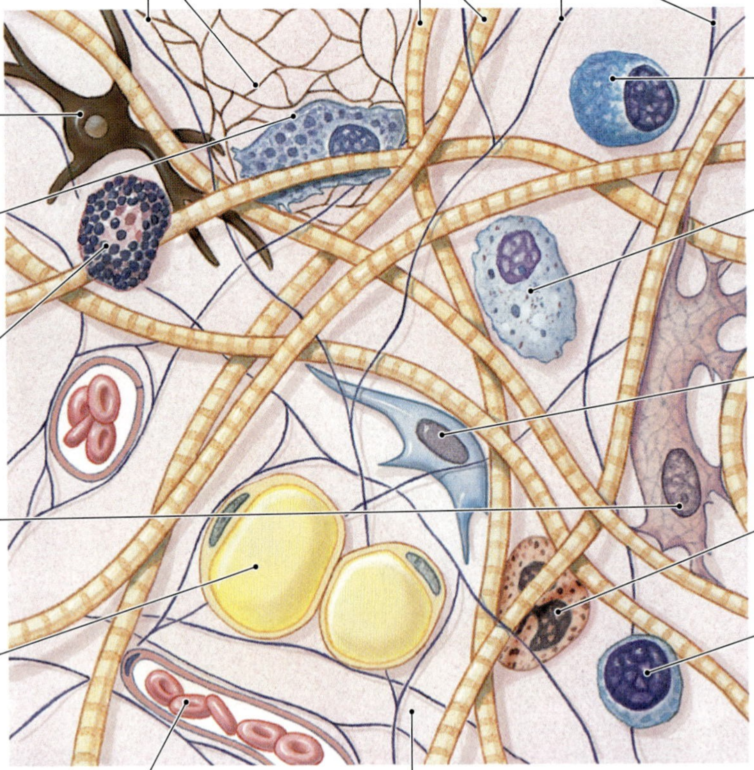

Wandering Cells

A **plasma cell** is an active, mobile immune cell that produces antibodies.

Free macrophages are wandering phagocytic cells that patrol the tissue, engulfing debris or pathogens.

Mobile **stem cells** repair damaged tissues.

Small, mobile, phagocytic blood cells enter tissues during infection or injury.

Lymphocytes are mobile cells of the immune system.

Red blood cell in vessel

Ground substance fills the spaces between cells and surrounds connective tissue fibers. Ground substance is typically clear and colorless, with a viscous (syrupy) consistency.

2 **Adipose tissue**, or fat, is found deep to the skin, especially at the flanks, buttocks, and breasts. It also forms a layer that provides padding within the orbit of the eyes, in the abdominopelvic cavity, and around the kidneys. The distinction between areolar tissue and adipose tissue is some-what arbitrary. Adipocytes account for most of the volume of adipose tissue, but only a fraction of the volume of areolar tissue.

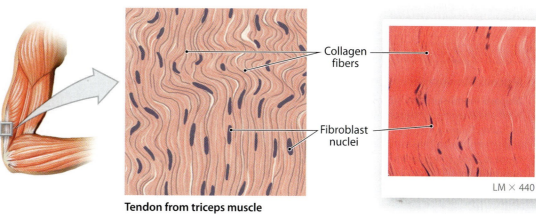

Adipocytes

LM × 300

Adipose tissue deep to the skin

Dense Connective Tissues

Most of the volume of **dense connective tissues** is occupied by extracellular fibers. The body has three types of dense connective tissues: dense regular connective tissue, dense irregular connective tissue, and elastic tissue (not shown).

3 **Dense regular connective tissue** is found in cords (tendons) or sheets connecting skeletal muscles to bone, and in cords (ligaments) that interconnect bones or stabilize the positions of internal organs. The forces applied to these cords and sheets arrive from a consistent direction: parallel to the long axis of the collagen fibers.

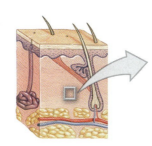

Collagen fibers

Fibroblast nuclei

LM × 440

Tendon from triceps muscle

4 In **dense irregular connective tissue**, the fibers form an interwoven meshwork in no consistent pattern. These tissues strengthen and support areas subjected to stresses from many directions. Dense irregular connective tissue forms (1) a covering, or capsule, that sheathes visceral organs; (2) a superficial layer covering bones, cartilages, and peripheral nerves; and (3) a thick supporting layer in the skin (the dermis).

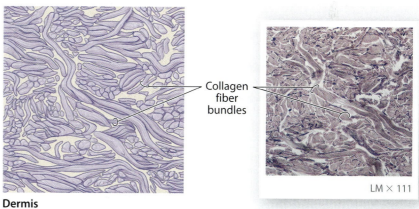

Collagen fiber bundles

LM × 111

Dermis

Module 3.19 Review

a. Identify the types of cells found in areolar tissue.

b. Lack of vitamin C in the diet interferes with the ability of fibroblasts to produce collagen. How might this affect connective tissue function?

c. Identify and describe the three types of protein fibers in connective tissue.

Membranes are physical barriers

A membrane is a physical barrier. There are many different types of anatomical membranes—you encountered plasma membranes in this chapter, and you will find other membranes in later chapters. Here we consider membranes that consist of an epithelium supported by either loose or dense connective tissue.

Synovial Membrane

A **synovial membrane** lines mobile joint cavities but does not cover the opposing joint surfaces. Although the covering of the synovial membrane is often called an epithelium, it differs from true epithelia because it lacks a basement membrane, and gaps of up to 1 mm may separate adjacent cells. Synovial fluid lubricates the joint cavity, and provides oxygen and nutrients to cartilage cells.

Cartilage covers the bone surface within the joint cavity.

The synovial membrane produces and regulates **synovial fluid**, which fills the joint cavity.

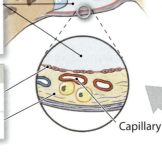

Synovial Membrane
Epithelium
Areolar tissue

Capillary

Mucous Membranes

Mucous membranes, or *mucosae* (mū-KŌ-sē), line passageways and chambers that communicate with the exterior, including those in the digestive, respiratory, reproductive, and urinary tracts. These epithelial surfaces must be kept moist to reduce friction and, in many cases, to facilitate absorption or secretion. The epithelial surfaces are lubricated either by mucus secretions or by exposure to fluids such as urine or semen.

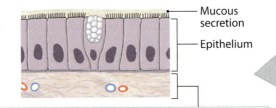

Mucous secretion

Epithelium

The **lamina propria** (PRŌ-prē-uh) is a layer of areolar tissue that supports the mucous epithelium.

Serous Membranes

Serous membranes consist of a single layer of simple squamous epithelium supported by areolar tissue. They are extremely delicate, and never directly connected to the exterior. Three serous membranes line the subdivisions of the ventral body cavity: (1) the **pleura**, which lines the pleural cavities and covers the lungs; (2) the **peritoneum**, which lines the peritoneal cavity and covers the surfaces of visceral organs; and (3) the **pericardium**, which lines the pericardial cavity and covers the heart.

A **transudate** is a liquid layer that coats the surfaces of a serous membrane and prevents friction.

Epithelium

Areolar tissue

The Cutaneous Membrane

The **cutaneous membrane**, or skin, covers the surface of the body. It consists of a stratified squamous epithelium and a layer of areolar tissue reinforced by underlying dense irregular connective tissue. In contrast to serous and mucous membranes, the cutaneous membrane is thick, relatively waterproof, and usually dry.

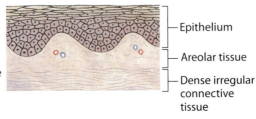

Epithelium

Areolar tissue

Dense irregular connective tissue

Module 3.20 Review

a. Identify the four types of membranes found in the body.

b. Which cavities in the body are lined by serous membranes?

c. What is the function of synovial fluid?

Cartilage provides a flexible supporting framework

Cartilage is a connective tissue with a firm, rubbery matrix that contains polysaccharide derivatives called **chondroitin sulfates** (kon-DROY-tin; *chondros,* cartilage). Together with proteins in the ground substance, they form proteoglycans. Cartilage cells, or **chondrocytes** (KON-drō-sīts), are the only cells in the cartilage matrix. They occupy small chambers called **lacunae** (la-KOO-nē; *lacus,* lake). The physical properties of cartilage depend on the proteoglycans in its matrix, and on the type and abundance of its extracellular fibers. Generally, a membrane called a **perichondrium** sets a cartilage apart from surrounding tissues. There are three types of cartilage: hyaline, elastic, and fibrocartilage.

Hyaline Cartilage

1 **Hyaline cartilage** is found between the tips of the ribs and the bones of the sternum, covering bone surfaces at mobile joints, supporting the respiratory passageways, and forming part of the nasal septum. It provides stiff but somewhat flexible support and reduces friction between bony surfaces.

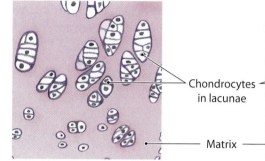

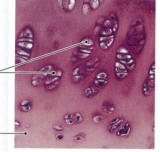

Chondrocytes in lacunae

Matrix

LM × 500

Hyaline cartilage from shoulder joint

Elastic Cartilage

2 **Elastic cartilage** supports the external ear and a number of smaller internal structures. It tolerates distortion without damage and returns to its original shape.

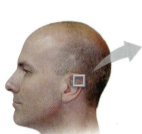

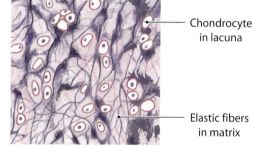

Chondrocyte in lacuna

Elastic fibers in matrix

LM × 358

Elastic cartilage from external ear

Fibrocartilage

3 **Fibrocartilage** pads are found within the knee joint, between the pubic bones of the pelvis, and in the intervertebral discs of the vertebral column. They resist compression, prevent bone-to-bone contact, and limit relative movement.

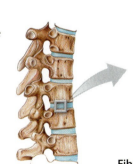

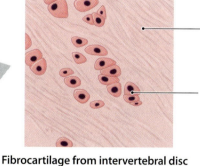

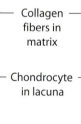

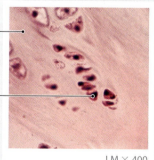

Collagen fibers in matrix

Chondrocyte in lacuna

LM × 400

Fibrocartilage from intervertebral disc

Module 3.21 Review

a. Mature cartilage cells are called _____.

b. Which connective tissue fiber is characteristic of the cartilage supporting the ear?

c. If a person has a protruding intervertebral disc, which type of cartilage has been damaged?

Bone provides a strong framework for the body

Bone, or **osseous** (OS-ē-us; *os*, bone) **tissue**, is a connective tissue with a solid, crystalline matrix. There is only a small volume of ground substance in its matrix. About two-thirds of the matrix of bone consists of a mixture of calcium salts. Collagen fibers dominate the rest of the matrix. This combination gives bone truly remarkable properties. By themselves, calcium salts are hard but rather brittle, whereas collagen fibers are strong and flexible. In bone, minerals surrounding the collagen fibers produce a strong, somewhat flexible combination that is highly resistant to shattering. In its overall properties, bone can compete with the best steel-reinforced concrete. In essence, the collagen fibers in bone act like the steel reinforcing rods, and the mineralized matrix acts like the concrete.

Compact bone also has a superficial layer of bone that was deposited during bone thickening.

Unlike cartilage, bone is highly vascular. Large vessels outside the bone are connected to smaller vessels that supply areas of compact bone and the soft tissues that fill the interior cavity.

In compact bone the matrix is organized in concentric layers around branches of blood vessels within the bone.

Compact bone

Spongy bone

1 A typical long bone is hollow, and its walls contain two different types of bone. The weight-bearing outer layer consists of well-organized **compact bone**, whereas a finer network of **spongy bone** lines the internal cavity. We will consider the structural and functional differences between the two bone types in a later chapter.

2 Cartilage and bone share a number of functions, but these two supporting connective tissues are organized very differently.

A Comparison of Cartilage and Bone

Characteristic	Cartilage	Bone
Cells	Chondrocytes in lacunae	Osteocytes in lacunae
Ground substance	Chondroitin sulfate (in proteoglycans) and water	A small volume of liquid surrounding insoluble crystals of calcium salts
Fibers	Collagen, elastic, and reticular fibers (proportions vary)	Collagen fibers predominate
Vascularity	None	Extensive
Covering	Perichondrium	Periosteum
Strength	Limited: bends easily, but hard to break	Strong: resists distortion until breaking point

3 Although the layers of bone may be arranged in varying ways, the superficial and deeper layers share common structural features.

Except in joint cavities, where they are covered by a layer of hyaline cartilage, bone surfaces are sheathed by a **periosteum** (per-ē-OS-tē-um) composed of fibrous (outer) and cellular (inner) layers.

Layers of the Periosteum

The fibrous layer helps attach a bone to surrounding tissues and to associated tendons and ligaments.

The cellular layer functions in bone growth and participates in repairs after an injury.

Lacunae in the matrix contain **osteocytes** (OS-tē-ō-sīts), or bone cells. The lacunae are typically organized around blood vessels that branch through the bony matrix.

Layers of matrix separate the lacunae. These layers are oriented along the main axis of the bone.

Canaliculi (kan-a-LIK-ū-lē; little canals) are fine passageways that form a branching network for the exchange of materials between blood vessels and osteocytes. This is important because diffusion cannot occur through the calcified matrix of bone.

A **central canal** at the center of an osteon contains the blood vessels that provide oxygen and nutrients to the osteocytes.

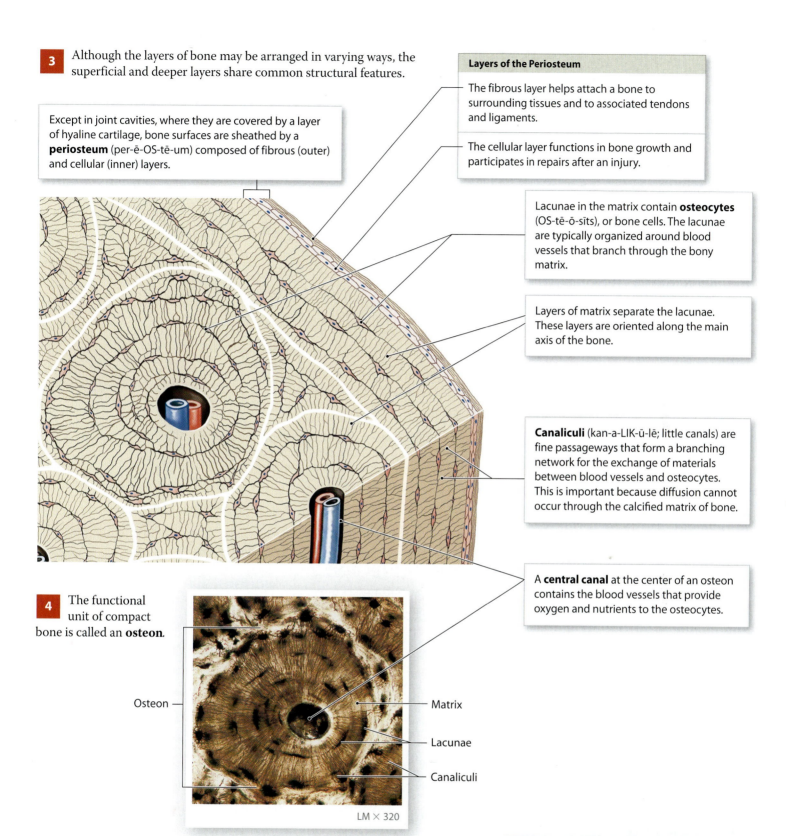

4 The functional unit of compact bone is called an **osteon**.

Osteon

Matrix

Lacunae

Canaliculi

LM × 320

Unlike cartilage, bone undergoes extensive remodeling throughout life, and complete repairs can be made even after severe damage has occurred. Bones also respond to the stresses placed on them, growing thicker and stronger with physical activity and becoming thin and brittle with inactivity.

Module 3.22 Review

a. Mature bone cells in lacunae are called _____.

b. Name the two types of bone.

c. How does exercise affect bone growth?

Muscle tissue is specialized for contraction; neural tissue is specialized for communication

Muscle Tissue

Several vital functions involve movement of one kind or another—moving materials along the digestive tract, moving blood around the cardiovascular system, or moving the body from one place to another. Movement is produced by **muscle tissue**, which is specialized for contraction. There are three types of muscle tissue: skeletal muscle, cardiac muscle, and smooth muscle.

1   **Skeletal muscle tissue** is found in skeletal muscles, organs that also contain connective tissues and neural tissue. Also known as muscle fibers, these cells are long, cylindrical, banded (**striated**), and have multiple nuclei.

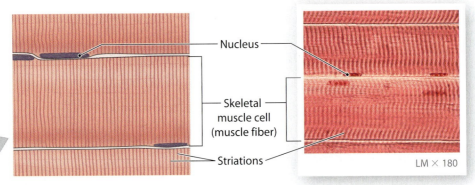

Nucleus

Skeletal muscle cell (muscle fiber)

Striations

LM × 180

Skeletal muscles move or stabilize the position of the skeleton; guard entrances and exits to the digestive, respiratory, and urinary tracts; generate heat; and protect internal organs.

2   **Cardiac muscle tissue** is found only in the heart. The cells are short, branched, and striated, usually with one or two nuclei. Cells are interconected at specialized intercellular junctions called **intercalated discs**. At an intercalcated disc, the membranes are locked together by desmosomes, proteoglycans, and gap junctions. Ion movement through gap junctions helps synchronize cardiac muscle contractions.

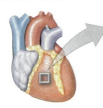

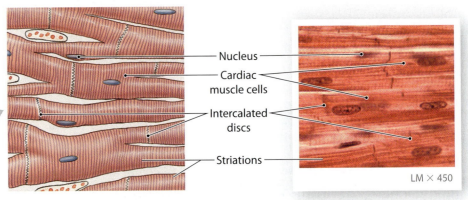

Nucleus

Cardiac muscle cells

Intercalated discs

Striations

LM × 450

Cardiac muscle moves blood through the heart and maintains blood pressure.

3   **Smooth muscle tissue** is found throughout the body. For example, smooth muscle is found in the skin, in the walls of blood vessels, and in many digestive, respiratory, urinary, and reproductive organs. The cells are short, spindle-shaped, and nonstriated, with a single, central nucleus.

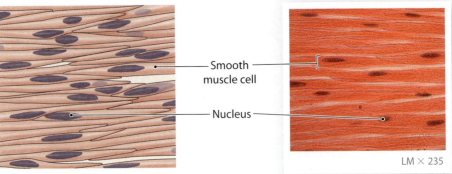

Smooth muscle cell

Nucleus

LM × 235

Smooth muscle moves food, urine, and reproductive tract secretions; controls diameter of respiratory passageways and regulates diameter of blood vessels.

Neural Tissue

Neural tissue, which is also known as nervous tissue, is specialized to conduct electrical impulses from one region of the body to another. Ninety-eight percent of the neural tissue in the body is concentrated in the brain and spinal cord, which are the control centers of the nervous system. Neural tissue contains two basic types of cells: (1) **neurons** (NOOR-onz; *neuro*, nerve) and (2) several kinds of supporting cells, collectively called **neuroglia** (noo-ROG-lē-uh), or **glial cells** (*glia*, glue).

4 Neurons transfer information from place to place and perform information processing like living computer chips. Their sizes and shapes vary widely. The longest cells in your body, found in the legs, are neurons; they can reach a meter (39 in.) in length.

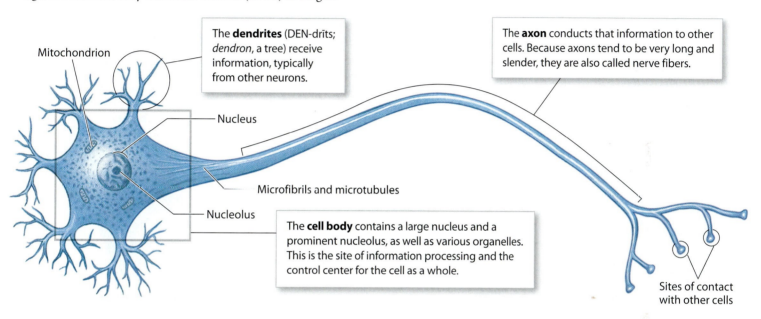

Mitochondrion

The **dendrites** (DEN-drīts; *dendron*, a tree) receive information, typically from other neurons.

The **axon** conducts that information to other cells. Because axons tend to be very long and slender, they are also called nerve fibers.

Nucleus

Microfibrils and microtubules

Nucleolus

The **cell body** contains a large nucleus and a prominent nucleolus, as well as various organelles. This is the site of information processing and the control center for the cell as a whole.

Sites of contact with other cells

5 There are several different types of neuroglia, each with specific functions. Each type has a distinctive appearance related to its primary function. In general, neuroglia protect, support, and repair neural tissue and maintain the nutrient supply to neurons.

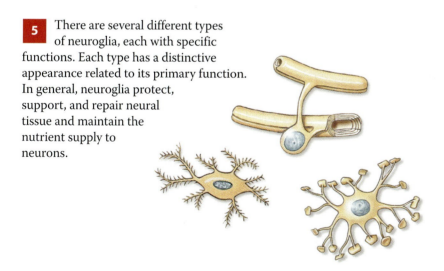

Module 3.23 Review

a. Identify the three types of muscle tissue in the body.

b. Which type of muscle tissue has small, tapering cells with single nuclei and no obvious striations?

c. Irregularly shaped cells with many fibrous projections, some several centimeters long, are probably which type of cell?

Our conscious and unconscious thought processes reflect the communication among neurons in the brain. Such communication involves the propagation of electrical impulses that are generated at the cell body and travel along the axon to reach other cells.

The response to tissue injury involves inflammation and regeneration

Tissues are not isolated; they combine to form organs with diverse functions. Therefore, any injury affects several types of tissue simultaneously. These tissues must respond in a coordinated way to preserve homeostasis. The restoration of homeostasis after an injury involves two related processes: inflammation and regeneration.

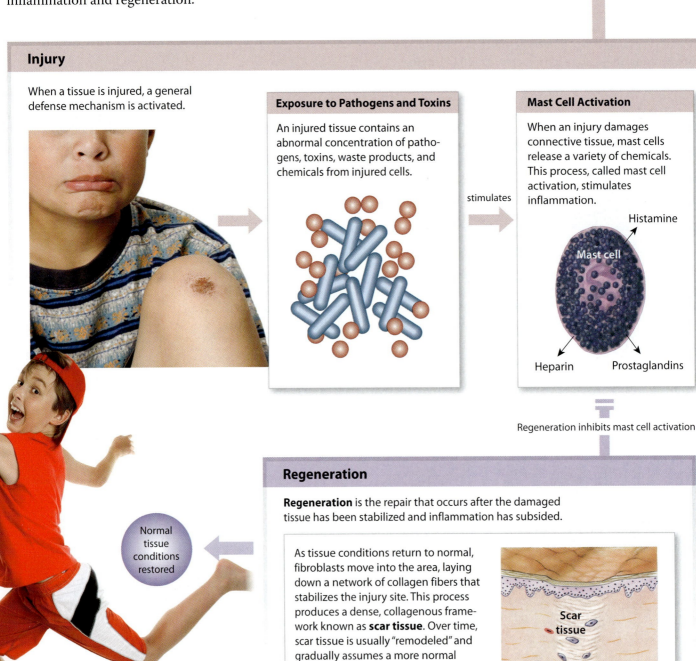

Injury

When a tissue is injured, a general defense mechanism is activated.

Exposure to Pathogens and Toxins

An injured tissue contains an abnormal concentration of pathogens, toxins, waste products, and chemicals from injured cells.

stimulates

Mast Cell Activation

When an injury damages connective tissue, mast cells release a variety of chemicals. This process, called mast cell activation, stimulates inflammation.

Histamine

Mast cell

Heparin   Prostaglandins

Regeneration inhibits mast cell activation

Regeneration

Regeneration is the repair that occurs after the damaged tissue has been stabilized and inflammation has subsided.

As tissue conditions return to normal, fibroblasts move into the area, laying down a network of collagen fibers that stabilizes the injury site. This process produces a dense, collagenous framework known as **scar tissue**. Over time, scar tissue is usually "remodeled" and gradually assumes a more normal appearance.

Scar tissue

Normal tissue conditions restored

Inflammation

Inflammation produces several familiar indications of injury, including swelling, redness, warmth, and pain. Inflammation may also result from the presence of pathogens, such as harmful bacteria, within the tissues; the presence of these pathogens constitutes an **infection**. Because all organs have connective tissues, inflammation can occur anywhere in the body.

Increased Blood Flow

In response to the released chemicals, the smooth muscle tissue that surrounds local blood vessels relaxes, and the vessels **dilate**, or enlarge in diameter. This dilation increases blood flow through the damaged tissue.

Increased Vessel Permeability

The dilation is accompanied by an increase in the permeability of the capillary walls. Plasma, including blood proteins, now diffuses into the injured tissue, so the area becomes swollen.

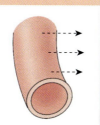

Promote

Pain

The combination of abnormal conditions within the tissue and the chemicals released by mast cells stimulates nerve endings that produce the sensation of pain.

PAIN

Increased local temperature
Increasing blood flow and permeability cause the tissue to become warm and red.

Increased oxygen and nutrients
Vessel dilation, increased blood flow, and increased vessel permeability enhance delivery of oxygen and nutrients.

O_2

Increased phagocytosis
Phagocytes in the tissue are activated, and they begin engulfing tissue debris and pathogens. Additional phagocytes migrate into the tissue from the bloodstream, drawn to the site by the abnormal local conditions.

Removal of toxins and wastes
The enhanced circulation that increases the volume of interstitial fluid also carries away toxins and waste products, distributing them to the kidneys for excretion, or to the liver for inactivation.

Toxins and wastes

Over a period of hours to days, the cleanup process generally succeeds in eliminating the inflammatory stimuli.

Each organ has a different ability to regenerate after injury—an ability that can be directly linked to the pattern of tissue organization in the injured organ. Epithelia, connective tissues (except cartilage), and smooth muscle tissue usually regenerate well. Other muscle tissues and neural tissue regenerate poorly if at all, and scar tissue often replaces damaged areas. The permanent replacement of normal tissue by scar tissue is called **fibrosis** (fī-BRŌ-sis). Fibrosis in muscle and other tissues may occur in response to injury, disease, or aging.

Module 3.24 Review

a. Identify the two processes in the response to tissue injury.

b. What are the four indications of inflammation that occur following an injury?

c. Why can inflammation occur in any organ in the body?

1. Concept map

Fill in the blank boxes with boldfaced terms and concepts introduced in this section to complete the muscle tissue concept map.

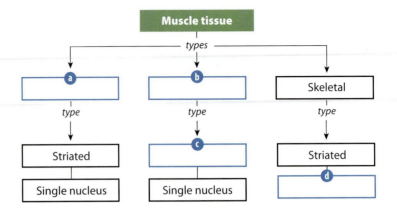

2. Vocabulary

In the space provided, write the boldfaced terms introduced in this section that contain the indicated word part.

Word Part	Meaning	Terms
peri-	around	a _____
os-	bone	b _____
chondro-	cartilage	c _____
lacus-	lake	d _____

Fill in the blanks with the appropriate term.

e _____ A type of cartilage that has a matrix with little ground substance and large amounts of collagen fibers

f _____ Cells that store lipid reserves

g _____ The membrane that lines mobile joint cavities

h _____ The membrane that covers the surface of the body

i _____ A single extension from the cell body of a neuron that carries information to other cells

j _____ A specialized intercellular junction between cardiac muscle cells

k _____ The supporting cells found in neural tissue

l _____ Muscle tissue that contains large, multinucleate, striated cells

m _____ Muscle tissue that regulates the diameter of blood vessels and respiratory passageways

n _____ The repair process that occurs after inflammation has subsided

o _____ The first process in a tissue's response to injury

3. Section integration

During inflammation, both blood flow and blood vessel permeability increase in the injured area.
Describe how these responses aid the cleanup process and eliminate the inflammatory stimuli in the injured area.

_____

_____

_____

Visual Outline with Key Terms

Summarize the content of each module using the terms in the order provided.

SECTION 1

An Introduction to Cells

- cell theory
- differentiation

3.1

Cells are the smallest living units of life

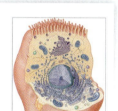

- extracellular fluid
- interstitial fluid
- plasma membrane
- cytoplasm
- cytosol
- organelles
- membranous organelles
- nonmembranous organelles
- peroxisome
- lysosome
- microvilli
- Golgi apparatus
- nucleus
- nuclear envelope
- endoplasmic reticulum (ER)
- smooth ER
- rough ER
- ribosomes
- cytoskeleton
- mitochondrion

3.2

The plasma membrane isolates the cell from its environment and performs various functions; the cytoskeleton has structural and functional roles

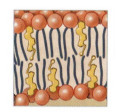

- plasma membrane
- glycocalyx
- integral proteins
- phospholipid bilayer
- microfilaments
- intermediate filaments
- microtubules
- thick filaments
- centrioles
- cilia

3.3

Protein synthesis in the endoplasmic reticulum and at free ribosomes depends primarily on the energy provided by mitochondrial activity

- endoplasmic reticulum (ER)
- cisternae
- rough endoplasmic reticulum (RER)
- smooth endoplasmic reticulum (SER)
- ribosomes
- fixed ribosomes
- free ribosomes
- mitochondria
- matrix
- cristae
- aerobic metabolism

3.4

The Golgi apparatus is a packaging center

- Golgi apparatus
- cisternae
- membrane renewal vesicles
- secretory vesicles
- lysosomes
- transport vesicles
- autolysis
- membrane flow

3.5

The nucleus contains DNA, RNA, organizing proteins, and enzymes

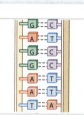

- nucleus
- nuclear envelope
- nuclear pores
- nucleoplasm
- nucleolus
- chromatin
- chromosomes
- centromere

3.6

Protein synthesis involves DNA, enzymes, and three types of RNA

- genetic code
- gene
- triplet
- messenger RNA (mRNA)
- codons
- transcription
- translation
- ribosomal RNA (rRNA)
- transfer RNA (tRNA)
- anticodon

3.7

Each cell has a life cycle that typically involves periods of growth and cell division

- cell division
- apoptosis
- mitosis
- meiosis
- daughter cells
- interphase
- cytokinesis
- DNA replication
- DNA polymerase

3.8

Mitosis distributes chromosomes before cytokinesis separates the daughter cells

- mitosis
- cytokinesis
- spindle fibers
- prophase
- chromatid
- cleavage furrow
- metaphase
- metaphase plate
- anaphase
- telophase

• = *Term boldfaced in this module*

3.9

Tumors and cancer are characterized by abnormal cell growth and division

- ○ cancer
- ○ mutations
- • tumor
- • benign tumor
- • invasion
- • malignant tumor
- • metastasis

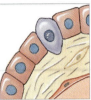

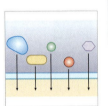

3.10

Diffusion is movement driven by concentration differences

- • diffusion
- • gradient
- • osmosis

3.11

Passive and active processes move materials in and out of cells

- ○ diffusion
- ○ osmosis
- ○ facilitated diffusion
- • active transport
- • sodium–potassium exchange pump
- • pinocytosis
- ○ cell drinking
- • receptor-mediated endocytosis
- • phagocytosis
- • pseudopodia
- • phagocytes (macrophages)
- • exocytosis

EPITHELIAL TISSUE
- Covers exposed surfaces
- Lines internal passageways and chambers
- Produces glandular secretions

3.12

Tissues are specialized groups of cells and cell products

- • histology
- • primary tissue types
- • epithelial tissue
- • connective tissue
- • matrix
- • ground substance
- • muscle tissue
- • skeletal muscle tissue
- • cardiac muscle tissue
- • smooth muscle tissue
- • neural tissue
- • neurons
- • neuroglia
- • central nervous system
- • peripheral nervous system

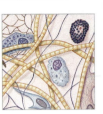

3.13

Epithelial tissue covers surfaces, lines structures, and forms secretory glands

- • epithelia
- • glands
- • exocrine glands
- • endocrine glands
- • apical surface
- • basal surface
- • lumen
- • squamous
- • cuboidal
- • columnar
- • simple epithelium
- • stratified epithelium

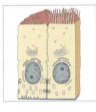

3.14

Epithelial cells are extensively interconnected, both structurally and functionally

- • tight junction
- • gap junction
- • desmosomes
- • basement membrane
- • connexons
- • cell adhesion molecules (CAMs)
- • proteoglycans
- • avascular

3.15

The cells in a squamous epithelium are flattened and irregular in shape

- • squamous epithelium
- • simple squamous epithelium
- • stratified squamous epithelium
- • keratinized
- • nonkeratinized

3.16

Cuboidal and transitional epithelia are found along several passageways and chambers connected to the exterior

- • cuboidal epithelium
- • simple cuboidal epithelium
- • stratified cuboidal epithelium
- • transitional epithelium

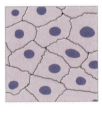

• = *Term boldfaced in this module*

3.17

Columnar epithelia typically perform absorption or provide protection from chemical or environmental stresses

- columnar epithelium
- simple columnar epithelium
- pseudostratified columnar epithelium
- stratified columnar epithelium

3.18

Glandular epithelia are specialized for secretion

- endocrine glands
- exocrine glands
- ducts
- mucous cells
- serous glands
- mucous glands
- mixed glands
- mucus
- merocrine secretion
- apocrine secretion
- holocrine secretion

3.19

Loose connective tissues provide padding and support, whereas dense connective tissues provide strength

- loose connective tissues
- areolar tissue
- reticular fibers
- collagen fibers
- elastic fibers
- melanocyte
- fixed macrophage
- mast cells
- fibroblasts
- adipocytes
- plasma cell
- free macrophages
- stem cells
- lymphocytes
- ground substance
- adipose tissue
- dense connective tissues
- dense regular connective tissue
- dense irregular connective tissue

3.20

Membranes are physical barriers

- synovial membrane
- synovial fluid
- mucous membranes
- lamina propria
- serous membranes
- pleura
- peritoneum
- pericardium
- transudate
- cutaneous membrane

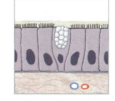

3.21

Cartilage provides a flexible supporting framework

- cartilage
- chondroitin sulfates
- chondrocytes
- lacunae
- perichondrium
- hyaline cartilage
- elastic cartilage
- fibrocartilage

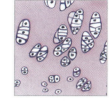

3.22

Bone provides a strong framework for the body

- osseous tissue
- compact bone
- spongy bone
- periosteum
- osteocytes
- canaliculi
- central canal
- osteon

3.23

Muscle tissue is specialized for contraction; neural tissue is specialized for communication

- muscle tissue
- skeletal muscle tissue
- striated
- cardiac muscle tissue
- intercalated discs
- smooth muscle tissue
- neural tissue
- neurons
- neuroglia (glial cells)
- dendrites
- axon
- cell body

3.24

The response to tissue injury involves inflammation and regeneration

- mast cell activation
- inflammation
- infection
- dilate
- regeneration
- scar tissue
- fibrosis

• = *Term boldfaced in this module*

CAREER PATHS

Clinical Laboratory Scientist

"I have the ability to look at patient results, put everything together and say, yeah, these lab results make sense."

— **Kathy Dagang**
Clinical Laboratory Scientist,
UC Davis Medical Center
Sacramento, CA

Kathy Dagang, a clinical laboratory scientist (CLS) at the University of California (UC) Davis Medical Center in Sacramento, jokes that hers is "a secret career." Clinical laboratory scientists have little contact with patients and are often literally hidden away in their own diagnostic labs, processing samples and checking the results. However, doctors could not diagnose patients without them. Clinical laboratory scientists process samples—of blood, urine, tissue—and release the results, helping behind-the-scenes to diagnose everything from strep throat to cancer.

Technology has advanced to the point that the processing of most samples is done by machines. In UC Davis's state-of-the-art lab, specimens come in with barcoded labels listing the tests to be run. The CLS feeds a test tube into an automation line—which Kathy describes as "kind of like a little toy train track." The machine moves the test tube through a centrifuge if necessary, then puts samples into different tubes that can be analyzed by the instruments.

A human, however, must still interpret the machine's results. If any so-called "critical" or life-threatening results emerge, the CLS must notify a doctor immediately.

Some diagnoses, like diabetic ketoacidosis, according to Kathy, are obvious from reading

a basic chemistry panel. In other cases, it is important to look at the patient's medical history. For example, Kathy may not understand an abnormal reading of pancreatic enzymes until she finds that the patient complained of upper right quadrant pain.

Kathy majored in biochemistry and always knew she wanted to go into medicine, but didn't feel compelled to be either a doctor or a nurse. Her academic adviser recommended the field, which was then called medical technology. She loved it immediately. "I have the ability to look at patient results, put everything together and say, yeah, these lab results make sense," she says.

Although anatomy and physiology is not a required course for CLS, it is recommended, and Kathy can't imagine not taking it. "To understand when something goes wrong, you

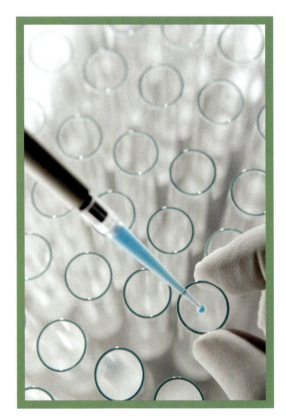

need to understand what it looks like when it goes right," she says. "You need to understand nerve transmission, liver function, kidney function."

Although most CLS don't interact with patients frequently, they do interact with doctors and each other, which makes communication and people skills important. Statistics, another recommended class, is also helpful, as laboratories work with quality controls on machines and need to understand if a margin of error is inappropriately high.

Kathy, who has been a CLS for 25 years, still helps in her lab, overseeing the flow of traffic for the thousands of specimens they process each day and ensuring that the equipment is running smoothly. She also teaches clinical chemistry to the trainees in the Davis program and trains them in the laboratory as well.

Ultimately, a CLS can go on to teach, like Kathy, or supervise a lab. Some clinical laboratory scientists work in research and development, sales, or even repair with companies that manufacture laboratory instrumentation or supplies.

For more information, visit the website of the American Society for Clinical Laboratory Science at http://www.ascls.org.

Think this is the career for you?

Key Stats:

- **Education and Training.** Bachelor's degree in a life science or medical technology required.

- **Licensure.** Some states require CLS to have a license; most employers prefer certification through one of various bodies.

- **Earnings.** Earnings vary but the median annual salary is $56,130.

- **Expected Job Prospects.** Employment is expected to grow by an above average 12% from 2008 to 2018.

Bureau of Labor Statistics, U.S. Department of Labor, *Occupational Outlook Handbook, 2010–11 Edition,* Clinical Laboratory Technicians and Technologists, on the Internet at http://www.bls.gov/oco/ocos096.htm (visited September 14, 2011).

Access more review material online in the Study Area at **www.masteringaandp.com.**

- Chapter guides
- Chapter quizzes
- Practice tests
- Art labeling activities
- Flashcards
- A glossary with pronunciations

- Practice Anatomy Lab™ (PAL™) 3.0 virtual anatomy practice tool
- Interactive Physiology® (IP) animated tutorials
- MP3 Tutor Sessions

PAL practice anatomy lab™ **For this chapter, follow these navigation paths in PAL:**

- Histology>Cytology (Cell Division)
- Histology>Epithelial Tissue
- Histology>Connective Tissue
- Histology>Muscular Tissue
- Histology>Nervous Tissue

 For this chapter, go to these topics in the MP3 Tutor Sessions:

- Membrane Transport
- Epithelial Tissue

Chapter 3 Review Questions

1. Solutions A and B are separated by a selectively permeable barrier. Over time, the level of fluid on side A increases. Which solution initially had the higher concentration of solute?

2. What is the benefit of having some of the cellular organelles enclosed by a membrane?

3. A significant structural feature in the digestive system is the presence of tight junctions near the exposed surfaces of cells lining the digestive tract. Why are these junctions so important?

4. Assuming you had the necessary materials to perform a detailed chemical analysis of body secretions, how could you determine whether a secretion was merocrine or apocrine?

5. During a lab practical, a student examines a tissue that is composed of densely packed protein fibers that are running parallel and form a cord. There are no striations, but small nuclei are visible. The student identifies the tissue as skeletal muscle. Why is the student's choice wrong, and what tissue is he probably observing?

For answers to all module, section, and chapter review questions, see the blue Answers tab at the back of the book.

4

The Integumentary System

Functional Anatomy of the Skin

The **integumentary system**, or simply the **integument** (in-TEG-U-ment), has two major components: the cutaneous membrane and accessory structures. The **cutaneous membrane**, or skin, is an organ composed of the superficial epithelium, or epidermis (epi-, above), and the underlying connective tissues of the dermis. The **accessory structures** include hair, nails, and a variety of exocrine glands. The integumentary system is the place where you and the outside world meet—and your body's first line of defense against an often hostile environment.

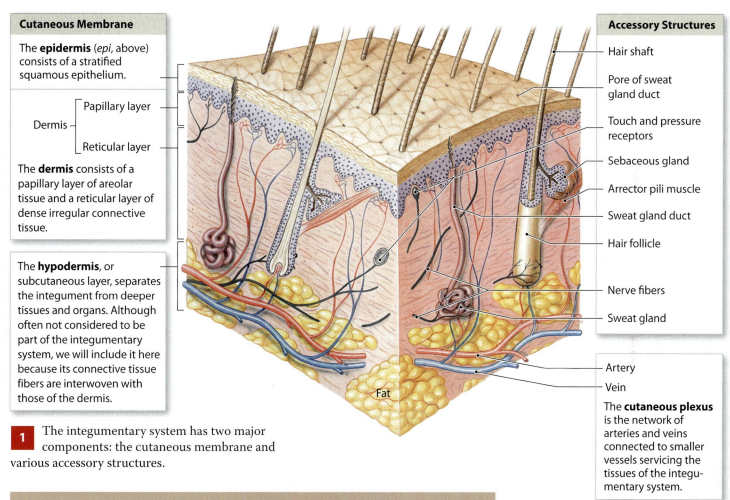

Cutaneous Membrane

The **epidermis** (*epi*, above) consists of a stratified squamous epithelium.

Dermis
 — Papillary layer
 — Reticular layer

The **dermis** consists of a papillary layer of areolar tissue and a reticular layer of dense irregular connective tissue.

The **hypodermis**, or subcutaneous layer, separates the integument from deeper tissues and organs. Although often not considered to be part of the integumentary system, we will include it here because its connective tissue fibers are interwoven with those of the dermis.

Fat

Accessory Structures

Hair shaft

Pore of sweat gland duct

Touch and pressure receptors

Sebaceous gland

Arrector pili muscle

Sweat gland duct

Hair follicle

Nerve fibers

Sweat gland

Artery

Vein

The **cutaneous plexus** is the network of arteries and veins connected to smaller vessels servicing the tissues of the integumentary system.

1 The integumentary system has two major components: the cutaneous membrane and various accessory structures.

Functions of the Integumentary System

- Protect underlying tissues and organs against impact, abrasion, fluid loss, and chemical attack
- Excrete salts, water, and organic wastes by integumentary glands
- Maintain normal body temperature through either insulation or evaporative cooling, as needed
- Produce melanin, which protects underlying tissue from ultraviolet radiation
- Produce keratin, which protects against abrasion and repels water
- Synthesize vitamin D_3, a steroid that is subsequently converted to calcitriol, a hormone important to normal calcium metabolism
- Store lipids in adipocytes in the dermis and in adipose tissue in the hypodermis
- Detect touch, pressure, pain, and temperature stimuli
- Coordinate immune response to pathogens and skin cancers

The epidermis is composed of layers with various functions

The epidermis is dominated by **keratinocytes** (ke-RAT-i-nō-sīts), the body's most abundant epithelial cells. These cells form several layers, or strata. Stem cells in the deepest layers of the epidermis continuously divide to produce keratinocytes, which are shed at the exposed surface.

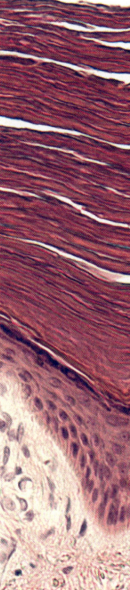

1 The deeper layers of the epidermis form epidermal ridges, which extend into the dermis. These ridges are adjacent to dermal projections called **dermal papillae** (singular, *papilla*; a nipple-shaped mound) that project into the epidermis. The ridges and papillae are significant because they greatly increase the surface area for attachment, firmly binding the epidermis to the dermis.

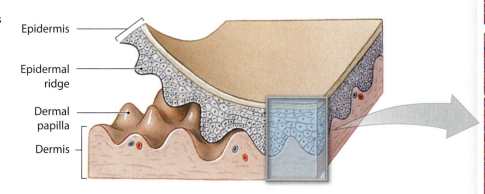

Epidermis

Epidermal ridge

Dermal papilla

Dermis

2 **Thin skin**, which covers most of the body surface, contains four strata, and is about as thick as a plastic sandwich bag (roughly 0.08 mm).

3 **Thick skin**, which occurs on the palms of the hands and the soles of the feet, contains a fifth layer. Thick skin is about as thick as a standard paper towel (roughly 0.5 mm). Note that the terms "thick" and "thin" refer to the relative thickness of the epidermis, not to the integument as a whole.

Epidermis

Dermis

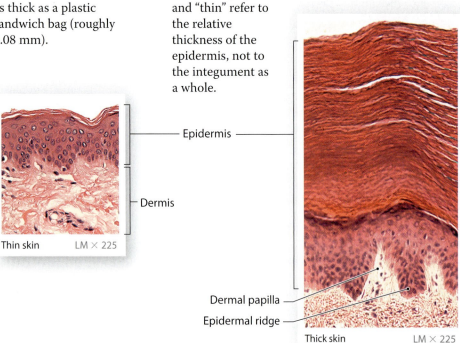

Thin skin LM × 225

Dermal papilla
Epidermal ridge

Thick skin LM × 225

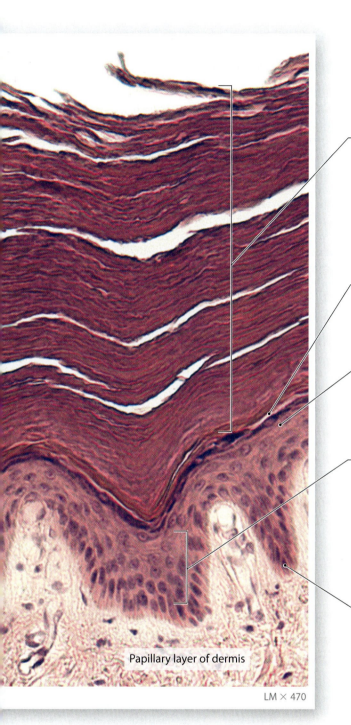

4 Like all other epithelia, the epidermis lacks local blood vessels. Epidermal cells rely on the diffusion of nutrients and oxygen from capillaries within the dermis. As a result, the cells with the highest metabolic demand are closest to the underlying dermis.

Layers of the Epidermis

Stratum corneum: At the exposed surface of both thick skin and thin skin, the stratum corneum (STRA-tum KOR-nē-um; *cornu*, horn) normally contains 15 to 30 layers of keratinized cells. Keratinization is the formation of protective, superficial layers of cells filled with the protein **keratin** (KER-a-tin; *keros*, horn). The dead cells in each layer of the stratum corneum remain tightly interconnected by desmosomes. It takes 7 to 10 days for a cell to move from the lowest layer of the epidermis (the stratum basale) to the stratum corneum. The dead cells generally remain in the exposed stratum corneum for an additional two weeks before they are shed or washed away.

Stratum lucidum: In the thick skin of the palms and soles, a stratum lucidum ("clear layer") separates the stratum corneum from deeper layers. The cells in the stratum lucidum are flattened, densely packed, largely devoid of organelles, and filled with the protein keratin. By the time they reach the stratum lucidum, they are dead and undergoing dehydration.

Stratum granulosum: This "grainy layer" consists of three to five layers of keratinocytes. By the time cells reach this layer, most have stopped dividing and have started making large amounts of keratin. As these fibers develop, the cells grow thinner and flatter, and their membranes thicken and become less permeable.

Stratum spinosum: This "spiny layer" consists of 8 to 10 layers of keratinocytes bound together by desmosomes. Its name refers to the fact that, in a standard slide, the cells look like miniature pincushions in standard histological sections. They look that way because the keratinocytes were processed with chemicals that shrank the cytoplasm but left the cytoskeletal elements and desmosomes intact. The stratum spinosum also contains **dendritic cells**, which participate in the immune response by stimulating a defense against (1) microorganisms that manage to penetrate the superficial layers of the epidermis and (2) superficial skin cancers.

Stratum basale: The stratum basale (STRA-tum buh-SAHL-āy) is the basal layer of the epidermis. Hemidesmosomes attach the cells of this layer to the basement membrane that separates the epidermis from the areolar tissue of the adjacent papillary layer of the dermis. **Basal cells** dominate the stratum basale. Basal cells are stem cells whose divisions replace the more superficial keratinocytes that are lost or shed at the epithelial surface. Skin surfaces that lack hair also contain specialized epithelial cells known as **Merkel cells** scattered among the basal cells. Merkel cells are sensitive to touch; when compressed, they release chemicals that stimulate sensory nerve endings.

Papillary layer of dermis

LM × 470

SEM × 25

5 Ridge patterns in the thick skin of the surface of fingertips produce fingerprints, which have been used to identify individuals in criminal investigations for more than a century. This scanning electron micrograph shows the tip of an index finger.

Module 4.1 Review

a. Identify the layers of the epidermis of thick skin (from deep to superficial).

b. Dandruff is caused by excessive shedding of cells from the outer layers of the skin of the scalp. Thus, dandruff is composed of cells from which epidermal layer?

c. A splinter penetrates to the third layer of the epidermis of the palm. In which layer does it lodge?

Factors influencing skin color include epidermal pigmentation and dermal circulation

Two factors influence skin color: pigments in the skin and the degree of dermal circulation.

1 The primary pigments involved in skin coloration are carotene and melanin. The micrograph at left shows a section through the thin skin of a light-skinned individual. The ratio of melanocytes to basal cells ranges between 1:4 and 1:20, depending on the region of the body. The skin covering most areas of the body has about 1000 melanocytes per square millimeter. Differences in skin pigmentation among individuals do not reflect different numbers of melanocytes, but instead different levels of pigment production. Even the melanocytes of albino individuals are distributed normally, although the cells are incapable of producing melanin. The figure below depicts a single melanocyte in the stratum basale.

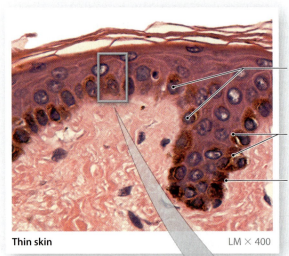

Melanocytes in stratum basale

Melanin pigment

Basement membrane

Thin skin   LM × 400

3 Melanosomes travel within the processes of melanocytes and are transferred intact to keratinocytes. In dark-skinned people, the melanosomes are larger, and the skin pigmentation is thus darker and more persistent.

Keratinocyte

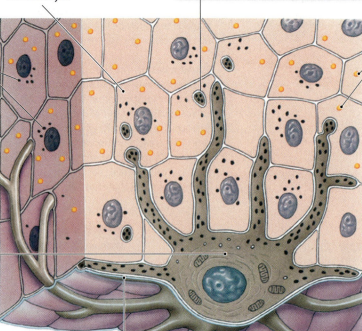

1 **Melanin** is a brown, yellow-brown, or black pigment produced by **melanocytes**.

2 The melanocytes are located in the stratum basale, squeezed between or deep to the epithelial cells. Melanocytes manufacture melanin from the amino acid tyrosine, and package it in intracellular vesicles called **melanosomes**.

4 **Carotene** (KAR-uh-tēn) is an orange-yellow pigment that normally accumulates in epidermal cells. It is most apparent in cells of the stratum corneum of light-skinned individuals, but it also accumulates in fatty tissues in the deep dermis and hypodermis. Carotene is also found in a variety of orange vegetables.

5 Skin color is genetically programmed. However, increased pigmentation, or tanning, can result in response to ultraviolet (UV) radiation.

Basement membrane

2 Dermal circulation affects skin color because blood contains red blood cells filled with the red pigment hemoglobin. When bound to oxygen, hemoglobin is bright red, giving capillaries in the dermis a reddish tint that is most apparent in lightly pigmented individuals. If those vessels dilate, the red tones become much more pronounced. For example, your skin becomes flushed and red when your body temperature rises because your superficial blood vessels dilate, allowing your skin to act like a radiator and lose heat. When blood flow decreases, oxygen levels in the tissues decline, and under these conditions hemoglobin releases oxygen and turns a much darker red. Seen from the surface, the skin then takes on a bluish color. This coloration is called **cyanosis** (sī-uh-NŌ-sis; *kyanos*, blue). In individuals of any skin color, cyanosis is most apparent in very thin skin, such as the lips or beneath the nails.

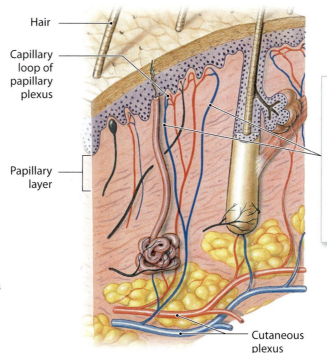

Hair

Capillary loop of papillary plexus

Papillary layer

The papillary layer of the skin contains the **papillary plexus**, a network of capillaries that provides oxygen and nutrients to the dermis and epidermis.

Cutaneous plexus

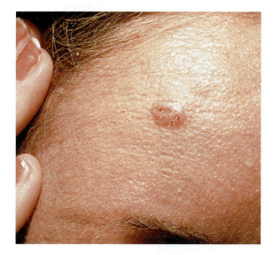

3 Skin cancers are the most common types of cancers. The most common form of skin cancer is a **basal cell carcinoma**, which often looks like a waxy bump. This is a cancer that originates in the stratum basale, due to mutations caused by overexposure to the ultraviolet (UV) radiation in sunlight and tanning beds. Basal cell carcinomas rarely metastasize (spread to other parts of the body), and most people survive these cancers. The melanin in keratinocytes provides some protection against UV radiation because the melanosomes are concentrated around the nucleus, where they act like sunshades for the enclosed DNA. **Squamous cell carcinoma** is the second most common skin cancer. It is almost totally restricted to areas of sun-exposed skin, where it forms a bump with a rough surface and reddish patches. It is more likely to metastasize than a basal cell carcinoma. Treatment for both cancers involves surgical removal of the tumor, and most people survive these cancers.

4 The most serious form of skin cancer, **malignant melanoma** (mel-a-NŌ-muh) is extremely dangerous. Cancerous melanocytes grow rapidly and metastasize through the lymphatic system. Long-term survival often depends on how early the condition is diagnosed. If detected early while still localized, the cancer can be surgically removed, and the 5-year survival rate is 99 percent. If not detected until extensive metastasis has occurred, the 5-year survival

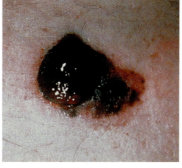

rate drops to 14 percent. To detect melanoma at an early stage, you must examine your skin, and you must know what to look for. The mnemonic ABCDE makes it easy to remember cancer's key characteristics: A is for asymmetry; melanomas tend to be irregular in shape. B is for border; the border is generally irregular. C is for color; melanoma is generally mottled. D is for diameter; any skin growth more than 5 mm in diameter is dangerous. E is for elevation; melanomas are usually raised beyond the skin surface.

Module 4.2 Review

a. Name the two pigments contained in the epidermis.

b. Why does exposure to sunlight or sunlamps darken skin?

c. Rank the three skin cancers according to their health risk.

The dermis supports the epidermis, and the hypodermis connects the dermis to the rest of the body

The **dermis** lies between the epidermis and the hypodermis. The dermis contains two types of fibers that enable it to tolerate limited stretching: collagen fibers, which are strong and resist stretching but easily bend or twist; and elastic fibers, which stretch and then recoil to their original length. The elastic fibers provide flexibility, and the collagen fibers limit that flexibility to prevent damage to the tissue. Aging, hormonal changes, and the destructive effects of ultraviolet radiation permanently reduce the amount of elastin in the dermis. The results: wrinkles and sagging skin.

1 The **papillary layer** consists of areolar tissue. This layer also contains the capillaries, lymphatic vessels, and sensory neurons that supply the surface of the skin. The papillary layer gets its name from the dermal papillae that project between the epidermal ridges. This layer also has touch receptors.

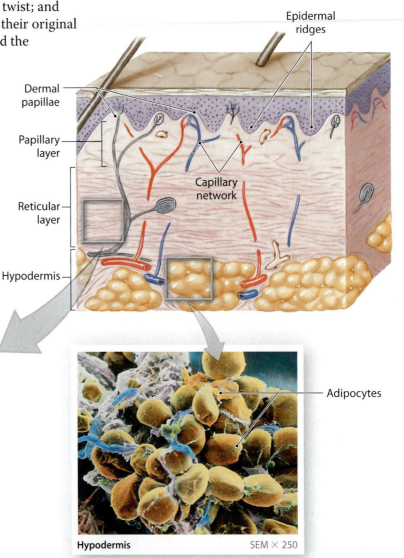

Epidermal ridges

Dermal papillae

Papillary layer

Reticular layer

Capillary network

Hypodermis

Reticular layer of dermis   SEM × 1500

Adipocytes

Hypodermis   SEM × 250

2 The **reticular layer** consists of an interwoven meshwork of dense irregular connective tissue containing both collagen and elastic fibers. Bundles of collagen fibers extend superficially beyond the reticular layer to blend into those of the papillary layer, as well as extending into the deeper hypodermis. The reticular and papillary layers of the dermis contain networks of blood vessels, lymphatic vessels, nerve fibers, and accessory organs such as hair follicles and sweat glands.

3 The **hypodermis** separates the skin from deeper structures. It stabilizes the position of the skin in relation to underlying tissues (such as skeletal muscles or other organs) while permitting independent movement. Because it is often dominated by adipose tissue, this layer also represents an important site for storing energy.

Module 4.3 Review

a. Describe the location of the dermis.

b. Where are the capillaries and sensory neurons that supply the epidermis located?

c. What accounts for the ability of the dermis to undergo repeated stretching?

The integument has endocrine functions that require ultraviolet radiation

1 Although too much sunlight can damage epithelial cells and deeper tissues, limited exposure to sunlight is beneficial because UV radiation plays a vital role in the synthesis of an important vitamin: **vitamin D₃**.

Sources of Vitamin D₃

Sunlight: When exposed to ultraviolet radiation, epidermal cells in the stratum spinosum and stratum basale convert a cholesterol-related steroid into vitamin D₃. This vitamin then diffuses across the basement membrane and enters the adjacent capillaries.

Diet: We can absorb vitamin D₃ from our diet, but few foods contain it other than fish, fish oils, and shellfish, and even then the presence and the amount of the vitamin varies. Today, many food products are "fortified with vitamin D" —most notably milk, soy milk, and orange juice.

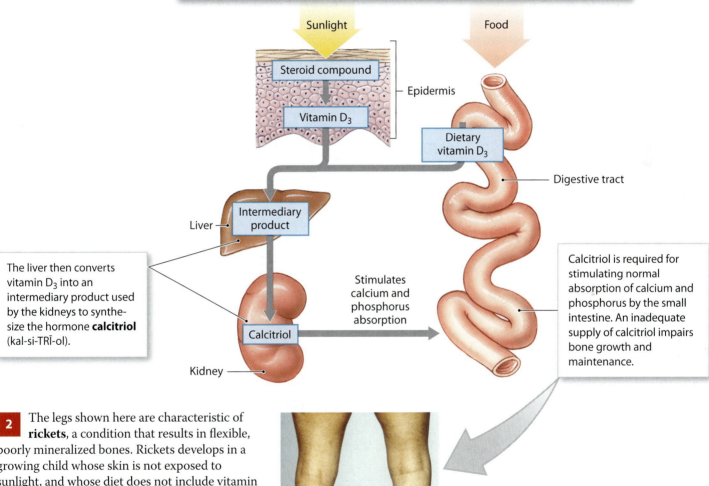

Sunlight

Steroid compound

Vitamin D₃

Epidermis

Food

Dietary vitamin D₃

Digestive tract

The liver then converts vitamin D₃ into an intermediary product used by the kidneys to synthesize the hormone **calcitriol** (kal-si-TRĪ-ol).

Liver

Intermediary product

Stimulates calcium and phosphorus absorption

Calcitriol is required for stimulating normal absorption of calcium and phosphorus by the small intestine. An inadequate supply of calcitriol impairs bone growth and maintenance.

Calcitriol

Kidney

2 The legs shown here are characteristic of **rickets**, a condition that results in flexible, poorly mineralized bones. Rickets develops in a growing child whose skin is not exposed to sunlight, and whose diet does not include vitamin D₃. The bones have the proper shape, but they lack rigidity because the bone matrix contains inadequate calcium and phosphate. Rickets has largely been eliminated in the United States because vitamin D₃ has been added to many different foods. However, vitamin D₃ is required throughout life. Adults are less likely to develop rickets due to vitamin D₃ deficiency, but bone density decreases, and this leads to a greater risk of fractures and slows the healing process. The risks are especially acute for the elderly because skin production of vitamin D₃ decreases by about 75 percent, even when exposed to sunlight.

Module 4.4 Review

a. Describe two sources of vitamin D₃.

b. Explain the relationship between sunlight exposure and vitamin D₃.

c. In some cultures, females must be covered from head to toe when they go outdoors. Explain why these women are at increased risk of developing bone problems later in life.

1. Short answer

Identify and describe the cutaneous membrane and the underlying layer of loose connective tissue in the diagram at right.

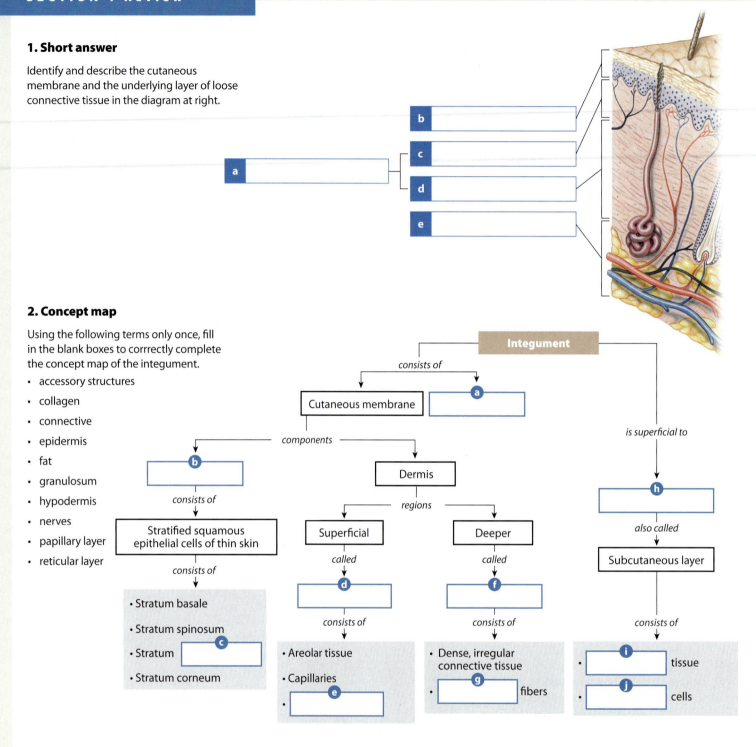

a

b

c

d

e

2. Concept map

Using the following terms only once, fill in the blank boxes to corrrectly complete the concept map of the integument.

- accessory structures
- collagen
- connective
- epidermis
- fat
- granulosum
- hypodermis
- nerves
- papillary layer
- reticular layer

Integument

consists of

Cutaneous membrane

a

components

b

consists of

Stratified squamous epithelial cells of thin skin

consists of

- Stratum basale
- Stratum spinosum
- Stratum   c
- Stratum corneum

Dermis

regions

Superficial

called

d

consists of

- Areolar tissue
- Capillaries
-    e

Deeper

called

f

consists of

- Dense, irregular connective tissue
-    g   fibers

is superficial to

h

also called

Subcutaneous layer

consists of

-    i   tissue
-    j   cells

3. Section integration

a. After taking a long bubble bath or doing dishes by hand, your fingers often appear wrinkled, or "pruny." Explain why.

_____

_____

b. Describe the sequence of events that begins with epidermal cells and UV radiation, and leads to the absorption of calcium and phosphorus by the small intestine.

_____

_____

Accessory Organs of the Skin

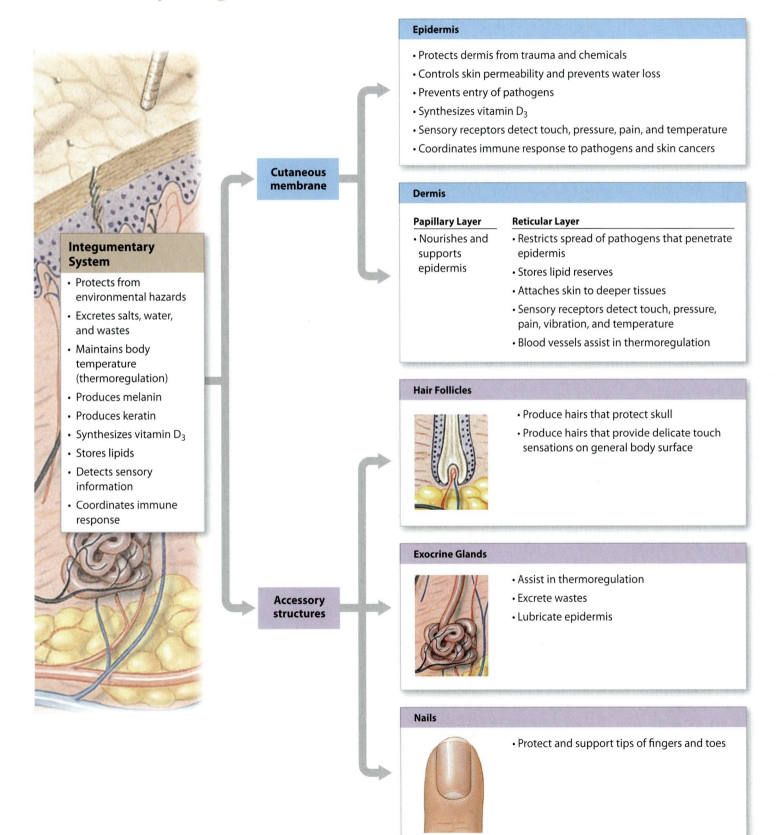

Integumentary System

- Protects from environmental hazards
- Excretes salts, water, and wastes
- Maintains body temperature (thermoregulation)
- Produces melanin
- Produces keratin
- Synthesizes vitamin D_3
- Stores lipids
- Detects sensory information
- Coordinates immune response

Cutaneous membrane

Epidermis

- Protects dermis from trauma and chemicals
- Controls skin permeability and prevents water loss
- Prevents entry of pathogens
- Synthesizes vitamin D_3
- Sensory receptors detect touch, pressure, pain, and temperature
- Coordinates immune response to pathogens and skin cancers

Dermis

Papillary Layer	Reticular Layer
• Nourishes and supports epidermis	• Restricts spread of pathogens that penetrate epidermis • Stores lipid reserves • Attaches skin to deeper tissues • Sensory receptors detect touch, pressure, pain, vibration, and temperature • Blood vessels assist in thermoregulation

Accessory structures

Hair Follicles

- Produce hairs that protect skull
- Produce hairs that provide delicate touch sensations on general body surface

Exocrine Glands

- Assist in thermoregulation
- Excrete wastes
- Lubricate epidermis

Nails

- Protect and support tips of fingers and toes

Hair is composed of dead, keratinized cells produced in a specialized hair follicle

Hairs project above the surface of the skin almost everywhere, except over the sides and soles of the feet, the palms of the hands, the sides of the fingers and toes, the lips, and portions of the external genitalia. The human body has about 2.5 million hairs, and 75 percent of them are on the general body surface, not on the head. Hairs are nonliving structures produced in organs called hair follicles.

1 A **hair follicle** is a little organ composed of epithelial and connective tissues that is responsible for the formation of a single hair.

2 Hair formation begins at the base of a hair follicle. Here a mass of epithelial cells forms a cap that surrounds a small **hair papilla**, a peg of connective tissue containing capillaries and nerves.

A **sebaceous gland** produces secretions that coat the hair and the adjacent surface of the skin.

The **hair shaft**, which we see on the surface, begins deep within the hair follicle.

The **hair root**—the portion that anchors the hair into the skin—extends from the base of the hair follicle, where hair production begins, to the point where the hair shaft loses its connection with the walls of the follicle.

A connective tissue sheath surrounds the epithelial cells of the hair follicle.

A **root hair plexus** of sensory nerves surrounds the base of each hair follicle. As a result, you can feel the movement of the shaft of even a single hair.

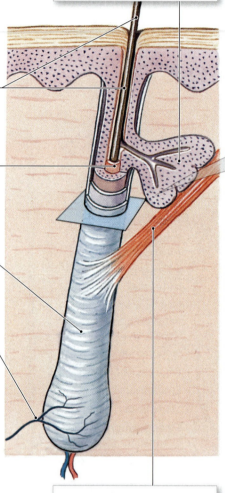

The **arrector pili** (a-REK-tor PĪ-lē; plural, arrectores pilorum) is a smooth muscle whose contraction pulls on the follicle, forcing the hair to stand erect and causing "goosebumps."

The **cuticle** of the hair consists of daughter cells produced at the edges of the hair matrix. The cuticle forms the surface of the hair.

The **cortex** of the hair is an intermediate layer of cells deep to the cuticle.

The **medulla**, or core, of the hair consists of cells at the center of the hair matrix.

As basal cells near the center of the base of the hair divide, daughter cells are gradually pushed toward the surface. This region is called the **matrix**, or growth zone of the hair.

The hair papilla is a small connective tissue peg filled with blood vessels and nerves.

Hair Structure

| The medulla of the hair contains a flexible **soft keratin**. | The cortex contains thick layers of **hard keratin**, which give the hair its stiffness. | The cuticle, although thin, contains hard keratin and is very tough. |

Hair follicle

3 A cross section through a hair follicle reveals the internal structure of both the hair and the follicle.

4 Hairs grow and are shed according to a **hair growth cycle** that includes an active phase and a resting phase. Variations in the growth rate and the duration of the hair growth cycle account for individual differences in the length of uncut hair.

2 The follicle then regresses and transitions to the resting phase.

1 The **active phase** lasts 2–5 years. During the active phase, the hair grows continuously at a rate of approximately 0.33 mm/day.

3 During the **resting phase** the hair loses its attachment to the follicle and becomes a **club hair**.

4 When follicle reactivation occurs, the club hair is shed and the hair matrix begins producing a replacement hair.

Variations in hair color reflect differences in structure and variations in the pigment produced by melanocytes at the hair papilla. Different forms of melanin give a dark brown, yellow-brown, or red color to the hair. As pigment production decreases with age, hair color lightens. White hair results from the combination of a lack of pigment and the presence of air bubbles in the medulla of the hair shaft.

Module 4.5 Review

a. Describe a typical strand of hair.

b. What happens when an arrector pili muscle contracts?

c. Pulling a hair is painful, but cutting a hair is not. Why?

Sebaceous glands and sweat glands are exocrine glands found in the skin

In this module we examine the structure and function of the accessory structures that produce exocrine secretions, with particular attention to sebaceous glands and sweat glands. These exocrine glands lubricate the epidermis, excrete wastes, and help with thermoregulation.

1 **Sebaceous** (se-BĀ-shus) **glands**, or oil glands, discharge sebum, a mixture of triglycerides, cholesterol, proteins, and electrolytes through holocrine secretion. Sebaceous glands that communicate with a single follicle share a duct and thus are classified as simple branched glands. The lipids released from gland cells enter the lumen (open passageway) of the gland. Contractions of the arrector pili muscle squeeze the sebaceous gland and force sebum into the hair follicle and onto the surface of the skin. Sebum moisturizes and conditions the hair and surrounding skin.

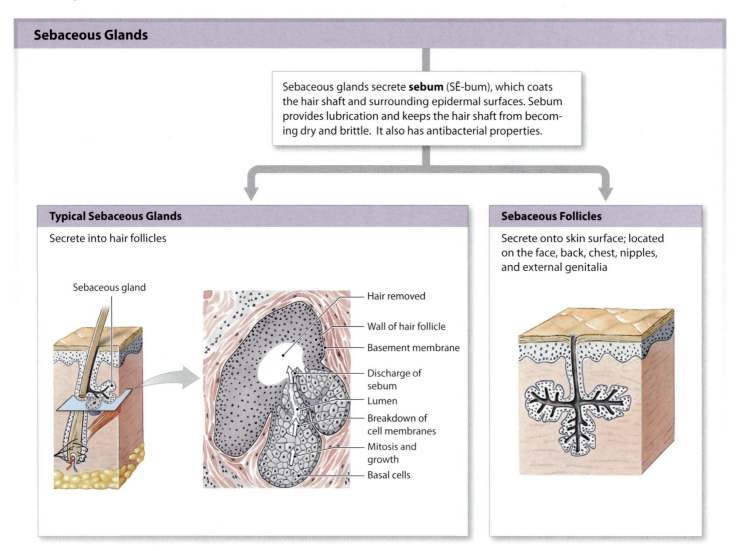

Sebaceous Glands

Sebaceous glands secrete **sebum** (SĒ-bum), which coats the hair shaft and surrounding epidermal surfaces. Sebum provides lubrication and keeps the hair shaft from becoming dry and brittle. It also has antibacterial properties.

Typical Sebaceous Glands

Secrete into hair follicles

Sebaceous gland

Hair removed

Wall of hair follicle

Basement membrane

Discharge of sebum

Lumen

Breakdown of cell membranes

Mitosis and growth

Basal cells

Sebaceous Follicles

Secrete onto skin surface; located on the face, back, chest, nipples, and external genitalia

Sweat Glands

Sweat glands produce a watery solution by merocrine secretion. It flushes the epidermal surface and performs other special functions.

Apocrine Sweat Glands

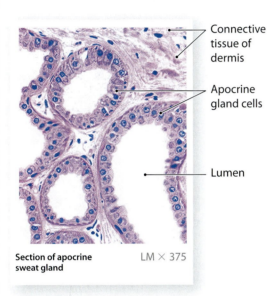

Connective tissue of dermis

Apocrine gland cells

Lumen

Section of apocrine sweat gland   LM × 375

- Limited distribution (axillae, groin, nipples)
- Produce a viscous secretion
- Include ceruminous glands of external ear and mammary glands that produce milk
- Strongly influenced by hormones
- Possible function in olfactory communication involving sense of smell

Merocrine Sweat Glands

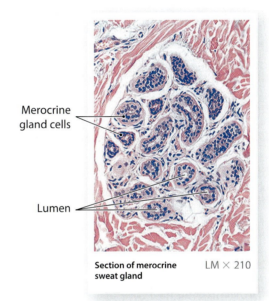

Merocrine gland cells

Lumen

Section of merocrine sweat gland   LM × 210

- Found in most areas of the skin
- Produce watery secretions containing electrolytes
- Merocrine secretion mechanism
- Controlled primarily by nervous system
- Important in thermoregulation and excretion
- Some antibacterial effects

Sweat pore

2 **Apocrine sweat glands** are found in the armpits, around the nipples, and in the pubic region, where they secrete into hair follicles. These glands produce a sticky, cloudy, and potentially odorous secretion. Despite their name, these glands rely on merocrine secretion. The adult integument also contains 2–5 million **merocrine sweat glands** that discharge their secretions directly onto the surface of the skin. The palms and soles have the highest numbers, with the palm possessing an estimated 500 merocrine sweat glands per square centimeter (3000 per square inch).

Module 4.6 Review

a. Identify two types of exocrine glands found in the skin.

b. What are the functions of sebaceous secretions?

c. Deodorants are used to mask the effects of secretions from which type of skin gland?

Nails are thick sheets of keratinized epidermal cells that protect the tips of fingers and toes

Nails protect the exposed dorsal surfaces of the tips of the fingers and toes. They also help keep your digits from being distorted by mechanical stress—for example, when you run or grasp objects.

1 These superficial and sectional views illustrate common landmarks of nail structure. The **nail body** consists of dead, tightly compressed cells packed with keratin. The nail body is recessed deep to the level of the surrounding epithelium.

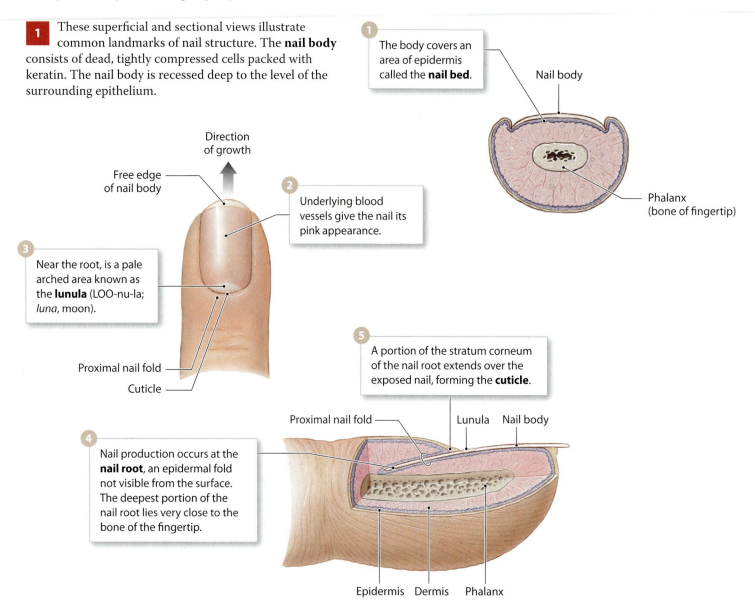

1 The body covers an area of epidermis called the **nail bed**.

Nail body

Phalanx (bone of fingertip)

Direction of growth

Free edge of nail body

2 Underlying blood vessels give the nail its pink appearance.

3 Near the root, is a pale arched area known as the **lunula** (LOO-nu-la; *luna*, moon).

Proximal nail fold

Cuticle

5 A portion of the stratum corneum of the nail root extends over the exposed nail, forming the **cuticle**.

Proximal nail fold

Lunula   Nail body

4 Nail production occurs at the **nail root**, an epidermal fold not visible from the surface. The deepest portion of the nail root lies very close to the bone of the fingertip.

Epidermis   Dermis   Phalanx

2 The cells producing the nails can be affected by health conditions that alter body metabolism, so changes in the shape, structure, or appearance of nails can provide useful diagnostic information. For example, psoriasis, a condition marked by rapid stem cell division in the stratum basale, may distort and pit nails. Some blood disorders may make the nails concave.

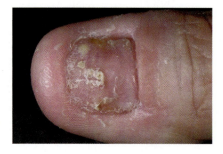

Module 4.7 Review

a. Define nail bed.

b. Describe a typical fingernail.

c. Where does nail production occur?

Age-related changes alter the integument

Fewer Melanocytes

Melanocyte activity declines, and in light-skinned individuals the skin becomes very pale. With less melanin in the skin, people become more sensitive to sun exposure and more likely to experience sunburn.

Drier Epidermis

Sebaceous glands secrete less sebum, and the skin becomes dry and often scaly.

Thinning Epidermis

The epidermis thins as basal cell activity declines, and the connections between the epidermis and dermis weaken, making older people more prone to injury, skin tears, and skin infections. The metabolic activity of the skin decreases as well. The epidermis normally produces vitamin D3, and decreased production leads to muscle weakness and brittle bones.

Diminished Immune Response

The number of dendritic cells decreases to about half the levels seen at age 21. This reduction in immune cells may decrease the sensitivity of the immune response and further encourage skin damage and infection.

Thinning Dermis

The dermis becomes thinner and has fewer elastic fibers, making the integument weaker and less resilient. The results—sagging and wrinkling—are most pronounced in body regions with the most exposure to the sun.

Decreased Perspiration

Merocrine sweat glands become less active, and with impaired perspiration, older people cannot lose heat as fast as younger people. Thus, the elderly are at greater risk of overheating in warm environments.

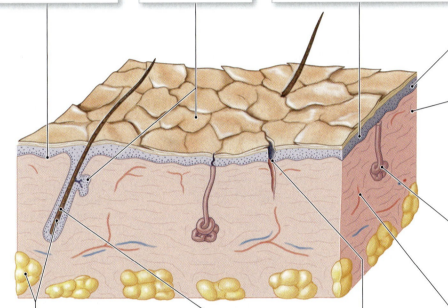

Altered Hair and Fat Distribution

With declining levels of sex hormones, differences in secondary sexual characteristics with respect to hair and body-fat distribution begin to fade. As a consequence, people age 90–100 of both sexes tend to look alike.

Fewer Active Follicles

Hair follicles stop functioning or produce thinner, finer hairs. With decreased melanocyte activity, these hairs are gray or white.

Slower Skin Repair

Skin repairs proceed more slowly. In a young adult a blister might take 3 to 4 weeks to heal. However, in a 65-year-old the same repair process could take 6 to 8 weeks.

Reduced Blood Supply

A reduction in dermal blood supply cools the skin, which can stimulate thermoreceptors and make a person feel cold even in a warm room. Reduced circulation and sweat gland function in the elderly lessen their ability to lose body heat, which can allow body temperatures to soar dangerously high with overexertion.

Aging affects all structures in the integument. This illustration summarizes the major structural and functional changes. The gradual changes in the superficial appearance of the skin provide visual cues that we use unconsciously to estimate a person's age. This accounts for the popularity of cosmetic facial surgery.

Module 4.8 Review

a. Identify some common effects of the aging process on skin.

b. Why does hair turn white or gray with age?

c. Why do we tolerate summer heat less well and become more susceptible to heat-related illness when we become older?

The integument can often repair itself, even after extensive damage

The integumentary system displays a significant degree of functional independence—it often responds directly and automatically to local influences without involving the nervous or endocrine systems. We can see a dramatic display of local regulation if the skin is injured. These four panels illustrate the steps by which the skin regenerates after injury.

1 An initial injury to the skin causes bleeding and activates mast cells.

2 After several hours, a scab has formed and cells of the stratum basale are migrating along the edges of the wound. Macrophages are removing debris, and more of these phagocytes are arriving with the enhanced circulation into the area. Clotting around the edges of the affected area partially isolates the region from adjacent undamaged tissues.

Initial Injury

Immediately after the injury, mast cells in the region trigger an inflammatory response.

Bleeding occurs at the site of injury.

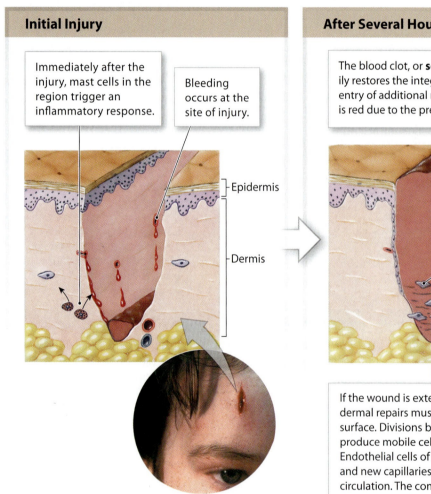

Epidermis

Dermis

After Several Hours

The blood clot, or **scab**, that forms at the surface temporarily restores the integrity of the epidermis and restricts the entry of additional microorganisms into the area. The scab is red due to the presence of trapped red blood cells.

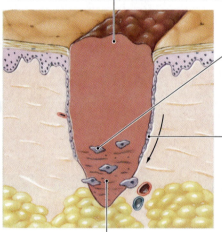

Macrophages patrol the damaged area of the dermis, phagocytizing debris and pathogens.

Cells of the stratum basale divide rapidly and migrate along the edges of the wound, to replace the missing epidermal cells.

If the wound is extensive or involves a region covered by thin skin, dermal repairs must begin before epithelial cells can cover the surface. Divisions by fibroblasts and connective tissue stem cells produce mobile cells that invade the deeper areas of injury. Endothelial cells of damaged blood vessels also begin to divide, and new capillaries grow in behind the fibroblasts, enhancing circulation. The combination of blood clot, fibroblasts, and an extensive capillary network is called **granulation tissue**.

3 One week after the injury, the scab has been undermined by epidermal cells migrating over a meshwork produced by fibroblast activity. Phagocytic activity around the site has almost ended, and the blood clot is disintegrating.

4 After several weeks, the scab has been shed, and the epidermis is complete. A shallow depression marks the injury site, but fibroblasts in the dermis continue to create scar tissue that will gradually elevate the overlying epidermis.

After One Week

Over time, deeper portions of the clot dissolve, and the number of capillaries declines. Fibroblast activity leads to the appearance of collagen fibers and typical ground substance. The repairs do not restore the integument to its original condition, however, because the dermis will contain an abnormally large number of collagen fibers and relatively few blood vessels.

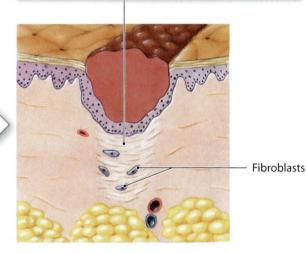

Fibroblasts

After Several Weeks

Severely damaged hair follicles, sebaceous or sweat glands, muscle cells, and nerves are seldom repaired, and they too are replaced by fibrous tissue. The formation of this rather inflexible, fibrous, noncellular **scar tissue** completes the repair process.

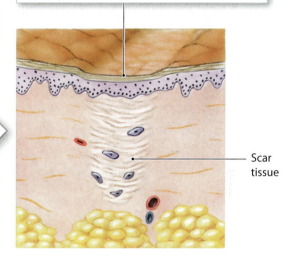

Scar tissue

5 In some adults, most often those with dark skin, scar tissue may continue forming beyond the requirements of tissue repair. The result is a **keloid** (KĒ-loyd), a raised, thickened mass of scar tissue that begins at the site of injury and grows into the surrounding dermis. Keloids are covered by a shiny, smooth epidermal surface and most commonly develop on the upper back, shoulders, anterior chest, or earlobes. They are harmless; some aboriginal cultures intentionally produce keloids as a form of body decoration.

Module 4.9 Review

a. Identify the first step in the repair of an injury to the skin.

b. Describe granulation tissue.

c. Why can skin regenerate effectively even after considerable damage?

1. Matching

Match the following terms with the most closely related description.

- malignant melanoma
- wrinkled skin
- nail root
- sebum
- apocrine sweat glands
- cuticle
- merocrine sweat glands
- scab
- reticular layer of dermis
- arrector pili

a _____ Produce watery, electrolyte-containing secretions

b _____ Epithelial fold not visible from the surface

c _____ Found in the armpit

d _____ Blood clot at skin surface

e _____ Site of hair production

f _____ Decrease in elastic fibers

g _____ Oily lipid secretion

h _____ Abnormal melanocytes metastasize through the lymphatic system

i _____ Causes hair to stand erect

j _____ Surface of a hair

2. Labeling

Label the structures of a typical nail in the accompanying figures.

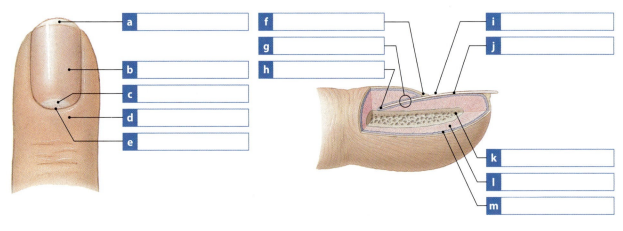

3. Labeling

Label the structures of a hair follicle in the accompanying figure.

4. Section integration

Two patients are brought to the emergency room. One has cut his finger with a knife, the other has a puncture wound from stepping on a nail. Which wound has a greater chance of becoming infected? Why?

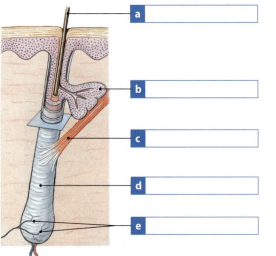

_____

_____

_____

Visual Outline with Key Terms

Summarize the content of each module using the terms in the order provided.

SECTION 1

Functional Anatomy of the Skin

- integumentary system
- integument
- cutaneous membrane
- epidermis
- dermis
- hypodermis
- accessory structures
- cutaneous plexus

SECTION 2

Accessory Organs of the Skin

- hair follicles
- exocrine glands
- nails

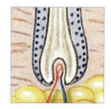

4.1

The epidermis is composed of layers with various functions

- keratinocytes
- dermal papillae
- thin skin
- thick skin
- stratum corneum
- keratin
- stratum lucidum
- stratum granulosum
- stratum spinosum
- dendritic cells
- stratum basale
- basal cells
- Merkel cells

4.5

Hair is composed of dead, keratinized cells produced in a specialized hair follicle

- hairs
- hair follicle
- hair shaft
- hair root
- root hair plexus
- sebaceous gland
- arrector pili
- hair papilla
- cuticle
- cortex
- medulla
- matrix
- soft keratin
- hard keratin
- hair growth cycle
- active phase
- resting phase
- club hair

4.2

Factors influencing skin color include epidermal pigmentation and dermal circulation

- melanin
- melanocyte
- melanosomes
- carotene
- cyanosis
- papillary plexus
- basal cell carcinoma
- squamous cell carcinoma
- malignant melanoma

4.6

Sebaceous glands and sweat glands are exocrine glands found in the skin

- sebaceous glands
- sebum
- sebaceous follicles
- sweat glands
- apocrine sweat glands
- merocrine sweat glands

4.3

The dermis supports the epidermis, and the hypodermis connects the dermis to the rest of the body

- dermis
- papillary layer
- reticular layer
- hypodermis

4.7

Nails are thick sheets of keratinized epidermal cells that protect the tips of fingers and toes

- nails
- nail body
- nail bed
- lunula
- nail root
- cuticle

4.4

The integument has endocrine functions that require ultraviolet radiation.

- vitamin D_3
- sunlight
- diet
- calcitriol
- rickets

● = Term boldfaced in this module

4.8

Age-related changes alter the integument

- fewer melanocytes
- drier epidermis
- thinning epidermis
- diminished immune response
- thinning dermis
- decreased perspiration
- altered hair and fat distribution
- fewer active hair follicles
- slower skin repair
- reduced blood supply

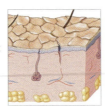

4.9

The integument can often repair itself, even after extensive damage

- scab
- granulation tissue
- scar tissue
- keloid

• = *Term boldfaced in this module*

CAREER PATHS

EMT / Paramedic

"Are we going to treat somebody differently if we know their liver is involved as opposed to the small intestine? Absolutely."

— **Sheila Moran**
Firefighter-paramedic,
Willow Springs, Illinois

Sheila Moran thinks she was destined to be a paramedic. She witnessed a stabbing at the restaurant where she worked, and in the moment of crisis, was able to administer first aid until paramedics arrived. Weeks later, she and friends were driving when they came upon a serious car accident. Sheila performed CPR on the victim, saving her life. "About a week later, I was enrolled in an Emergency Medical Technician (EMT) program," says Sheila, now a firefighter-paramedic in Willow Springs, Illinois, and an EMT instructor at Moraine Valley Community College in Palos Hills, Illinois. "It just came naturally."

The specifics vary by state, but in general, EMT-Basics receive training in caring for basic cardiac, respiratory, and trauma emergencies, and their primary job is to care for the patient on the way to the hospital. EMT-Intermediates have the same responsibilities, but can also use defibrillators in the event of cardiac arrest, administer IV fluids, and use special equipment and techniques to clear a blocked airway. Paramedics are also able to give certain drugs orally or intravenously, read electrocardiograms (ECGs or EKGs), intubate patients, and use other specialized equipment. Many EMTs and paramedics are also firefighters.

In the course of a shift, Sheila responds to calls as straightforward as a person with abdominal pains or as complicated as a multiple-car accident. And sometimes, even the abdominal pain cases get complicated. Sheila recalls one incident in which she and her partner were responding to a teenaged girl with abdominal pain at a reputed drug house and were nearly attacked by a man with a baseball bat who didn't like them questioning their patient. Fortunately, Sheila's partner was able to subdue him before he injured anyone. "Scene safety," she says, "we stress that all the time. Constantly be aware of your surroundings."

But more than the ability to expect the unexpected, Sheila says what's important is,

"radiating confidence to those around you. You need to be able to take control of the situation without being aggressive." A detailed knowledge of anatomy is also essential. "A lot of what we see is trauma–car accidents and falls–and we have to know what is what," Sheila says. "Are we going to treat somebody differently if we know their liver is involved as opposed to the small intestine? Absolutely. The liver bleeds unbelievably; with the small intestine you have a little more time." Finally, it is important to be prepared to work any shift, seven days a week. Depending on the municipality, or whether a paramedic works directly for a hospital or private ambulance company, the shifts vary. Typically, Sheila works shifts of 24 hours on duty, 48 hours off.

Sheila enjoys the camaraderie with her fellow paramedics, whom she describes as "some of the greatest people you could ever hope to know." But when talking about the satisfaction of her job, she still goes back to her first feelings of helping to save lives. "There are certain situations where you do the right thing, and you have really made a difference," she says.

For more information, visit the website for the National Association of Emergency Medical Technicians at http://www.naemt.org.

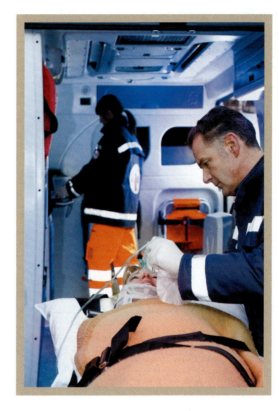

Think this is the career for you?
Key Stats:

- **Education and Training.** High school diploma required, plus additional training through an approved program at a community college or hospital.

- **Licensure.** All states require EMTs and paramedics to be licensed; licenses are currently issued by individual states but eventually will move toward a nationally standardized program.

- **Earnings.** Earnings vary but the median hourly wage is $14.60.

- **Expected Job Prospects.** Employment is expected to grow by an average of 9% from 2008 to 2018.

Bureau of Labor Statistics, U.S. Department of Labor, *Occupational Outlook Handbook, 2010–11 Edition*, Emergency Medical Technicians and Paramedics, on the Internet at http://www.bls.gov/oco/ocos101.htm (visited September 14, 2011).

Access more review material online in the Study Area at www.masteringaandp.com.

- Chapter guides
- Chapter quizzes
- Practice tests
- Art labeling activities
- Flashcards
- A glossary with pronunciations
- Practice Anatomy Lab™ (PAL™) 3.0 virtual anatomy practice tool
- Interactive Physiology® (IP) animated tutorials
- MP3 Tutor Sessions

PAL | practice anatomy lab™ — **For this chapter, follow these navigation paths in PAL:**

- Anatomical Models>Integumentary System
- Histology>Integumentary System

 For this chapter, go to this topic in the MP3 Tutor Sessions:

- Layers and Associated Structures of the Integument

Chapter 4 Review Questions

1. What two major layers make up the dermis, and what components are in each layer?

2. List the four stages in the regeneration of the skin after an injury.

3. In clinical practice, drugs can be delivered by diffusion across the skin. This delivery method is called transdermal administration. Why are fat-soluble drugs more suitable for transdermal administration than drugs that are water-soluble?

4. Exposure to optimum amounts of sunlight is necessary for proper bone maintenance and growth in children. If a child lives in an area where exposure to sunlight is rare because of pollution or overcast skies, what can be done to minimize impaired bone maintenance and growth?

5. Many people change the natural appearance of their hair by coloring it with hair dyes. Which layers of the hair do you suppose are affected by the chemical added during the hair-coloring process?

For answers to all module, section, and chapter review questions, see the blue Answers tab at the back of the book.

5

The Skeletal System

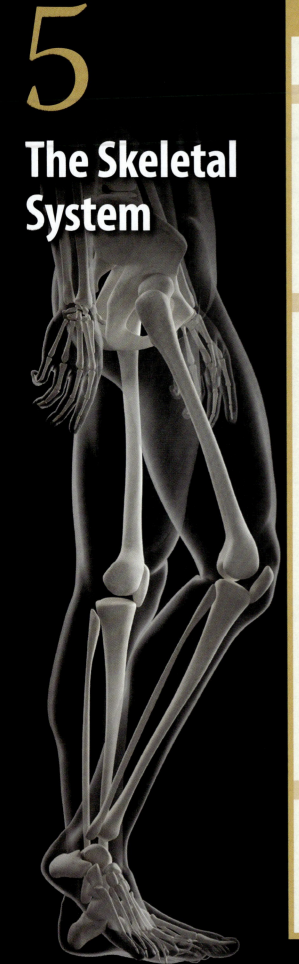

An Introduction to the Bones of the Skeletal System

The skeletal system includes the varied bones of the skeleton and the cartilages, ligaments, and other connective tissues that stabilize or interconnect them. This chapter expands upon the introduction to bone tissue (presented in Chapter 3) by considering the gross anatomy of bones and examining the mechanisms of bone growth, remodeling, and repair.

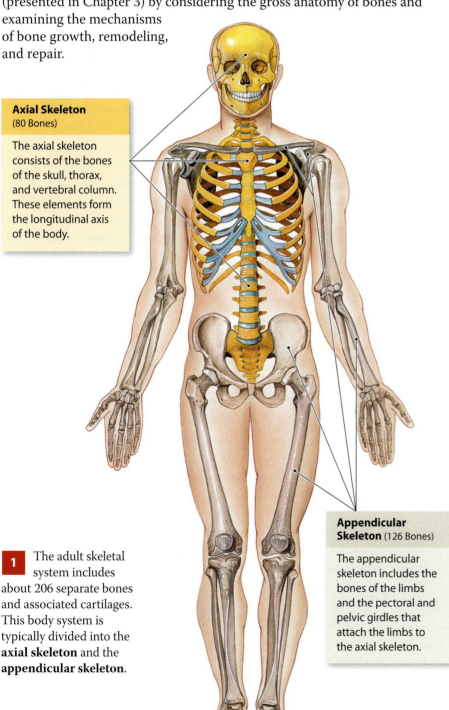

Axial Skeleton
(80 Bones)

The axial skeleton consists of the bones of the skull, thorax, and vertebral column. These elements form the longitudinal axis of the body.

Appendicular Skeleton (126 Bones)

The appendicular skeleton includes the bones of the limbs and the pectoral and pelvic girdles that attach the limbs to the axial skeleton.

1 The adult skeletal system includes about 206 separate bones and associated cartilages. This body system is typically divided into the **axial skeleton** and the **appendicular skeleton**.

Functions of the Skeletal System

- **Support:** The skeletal system provides structural support for the entire body. Individual bones or groups of bones provide a framework for the attachment of soft tissues and organs.

- **Storage of Minerals:** The calcium salts of bone represent a valuable mineral reserve that maintains normal concentrations of calcium and phosphate ions in body fluids. Calcium is the most abundant mineral in the human body. A typical body contains 1–2 kg (2.2–4.4 lb) of calcium, with more than 98 percent of it contained in the bones of the skeleton.

- **Blood Cell Production:** Red blood cells, white blood cells, and platelets are produced in the red bone marrow, which fills the internal cavities of many bones.

- **Protection:** Delicate tissues and organs are often surrounded by skeletal structures. The ribs protect the heart and lungs, the skull encloses the brain, the vertebrae shield the spinal cord, and the pelvis cradles delicate digestive and reproductive organs.

- **Movement:** Many bones of the skeleton function as levers that can change the magnitude and direction of the forces generated by skeletal muscles. The movements produced range from the delicate motions of a fingertip to powerful changes in the position of the entire body.

2 The skeleton has many vital functions that are summarized in the table above. All these functions ultimately depend on the unique and dynamic properties of bone tissue. The bone specimens that you study in lab or that you are familiar with from skeletons of dead animals are only the dry remains of this living tissue. They have the same relationship to the bone in a living organism as kiln-dried lumber does to a living tree.

Bones are classified according to shape and structure, and have a variety of surface features

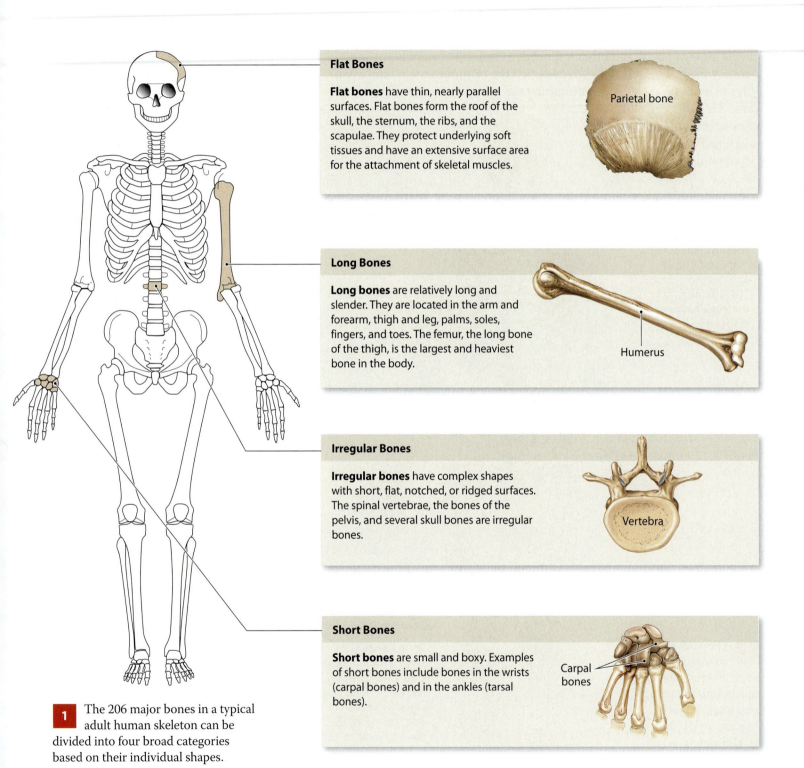

Flat Bones

Flat bones have thin, nearly parallel surfaces. Flat bones form the roof of the skull, the sternum, the ribs, and the scapulae. They protect underlying soft tissues and have an extensive surface area for the attachment of skeletal muscles.

Parietal bone

Long Bones

Long bones are relatively long and slender. They are located in the arm and forearm, thigh and leg, palms, soles, fingers, and toes. The femur, the long bone of the thigh, is the largest and heaviest bone in the body.

Humerus

Irregular Bones

Irregular bones have complex shapes with short, flat, notched, or ridged surfaces. The spinal vertebrae, the bones of the pelvis, and several skull bones are irregular bones.

Vertebra

Short Bones

Short bones are small and boxy. Examples of short bones include bones in the wrists (carpal bones) and in the ankles (tarsal bones).

Carpal bones

1 The 206 major bones in a typical adult human skeleton can be divided into four broad categories based on their individual shapes.

2 Illustrated here are the major types of **surface features**, also known as bone markings. Each bone in the body has characteristic external and internal features that are related to its particular functions. Elevations or projections form where tendons and ligaments attach, and where adjacent bones articulate (come together). Depressions, grooves, and tunnels in bone are sites where blood vessels or nerves lie alongside or penetrate the bone.

1 Surface Features of the Skull

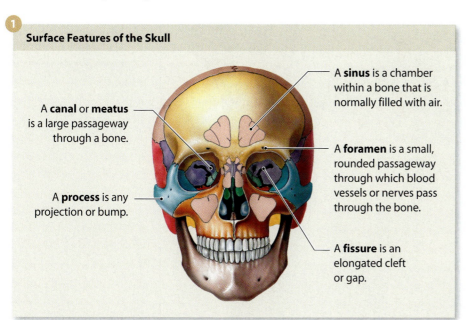

A **canal** or **meatus** is a large passageway through a bone.

A **process** is any projection or bump.

A **sinus** is a chamber within a bone that is normally filled with air.

A **foramen** is a small, rounded passageway through which blood vessels or nerves pass through the bone.

A **fissure** is an elongated cleft or gap.

2 Surface Features of the Humerus

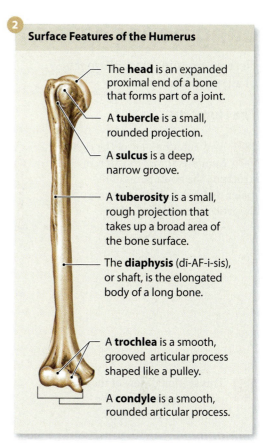

The **head** is an expanded proximal end of a bone that forms part of a joint.

A **tubercle** is a small, rounded projection.

A **sulcus** is a deep, narrow groove.

A **tuberosity** is a small, rough projection that takes up a broad area of the bone surface.

The **diaphysis** (dī-AF-i-sis), or shaft, is the elongated body of a long bone.

A **trochlea** is a smooth, grooved articular process shaped like a pulley.

A **condyle** is a smooth, rounded articular process.

3 Surface Features of the Femur

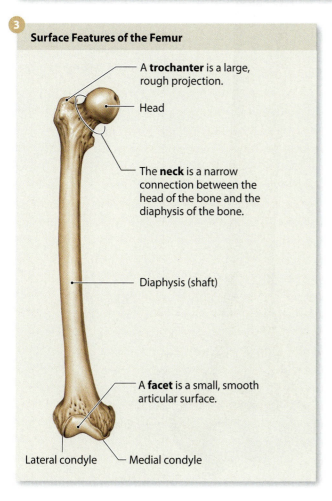

A **trochanter** is a large, rough projection.

Head

The **neck** is a narrow connection between the head of the bone and the diaphysis of the bone.

Diaphysis (shaft)

A **facet** is a small, smooth articular surface.

Lateral condyle

Medial condyle

4 Surface Features of the Pelvis

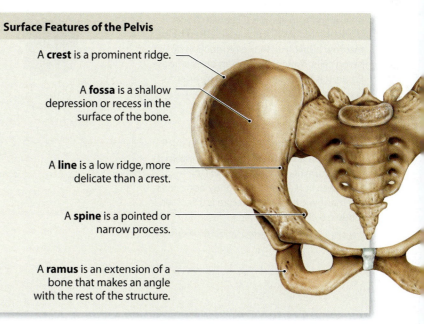

A **crest** is a prominent ridge.

A **fossa** is a shallow depression or recess in the surface of the bone.

A **line** is a low ridge, more delicate than a crest.

A **spine** is a pointed or narrow process.

A **ramus** is an extension of a bone that makes an angle with the rest of the structure.

Module 5.1 Review

a. Define *surface feature*.

b. Identify the four broad categories for classifying a bone according to shape.

c. Compare a tubercle with a tuberosity.

Long bones have a rich blood supply

Here we examine several aspects of a typical long bone.

The Humerus

1 This anterior view of the right humerus shows its external structure, the boundaries of the major regions, and the locations of the articular cartilages.

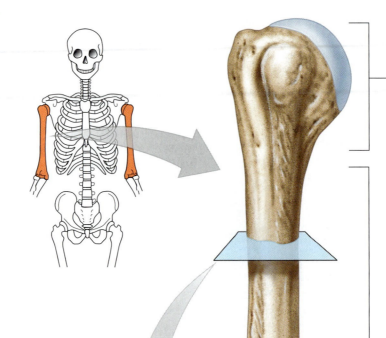

The **epiphysis** (e-PIF-i-sis) is an expanded area found at each end of the bone.

2 This sectional view of the right humerus shows the location of the marrow cavity.

The **marrow cavity** is a space within the hollow shaft. In life, it is filled with bone marrow, a highly vascular tissue. **Red bone marrow** is involved in the production of blood cells. **Yellow bone marrow** is adipose tissue. It is important as an energy reserve.

The wall of the diaphysis consists of a thick layer of dense **compact bone**. The histological structure of compact bone will be examined in Module 5.3. Spongy bone lines the marrow cavity. Its structure is discussed in Module 5.4.

The **diaphysis** (shaft) is long and tubular.

An **articular cartilage** covers portions of the epiphysis that articulate with other bones. The cartilage is avascular, and it relies primarily on diffusion from the synovial fluid to obtain oxygen and nutrients and eliminate wastes.

Epiphysis

3 In order for bones to grow and be maintained, they require an extensive blood supply.

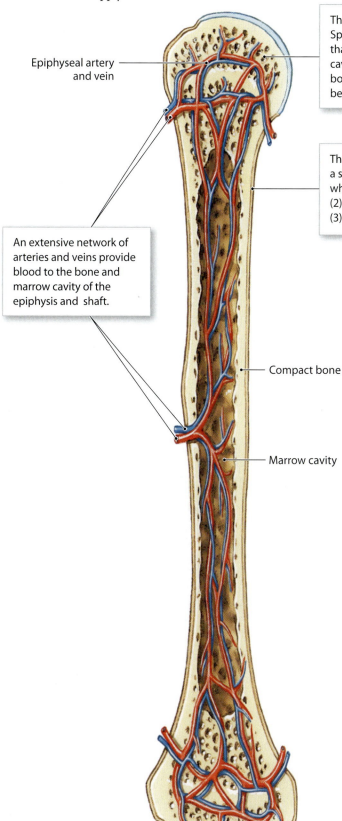

Epiphyseal artery and vein

The epiphysis consists largely of **spongy bone**. Spongy bone is an open network of struts and plates that resembles latticework. Except within joint cavities, spongy bone has a thin covering of compact bone. The histological structure of spongy bone will be examined in Module 5.3.

The superficial layer of compact bone is wrapped by a sheath of connective tissue, called the **periosteum**, which (1) isolates the bone from surrounding tissues, (2) provides a route for blood and nerves, and (3) plays a role in bone growth and repair.

An extensive network of arteries and veins provide blood to the bone and marrow cavity of the epiphysis and shaft.

Compact bone

Marrow cavity

4 A bone without a calcified matrix looks normal, but is very flexible. Roughly one-third of the weight of bone is contributed by collagen fibers, and cells account for only 2 percent of the weight of a typical bone. Calcium phosphate accounts for almost two-thirds of the weight of bone. The calcium phosphate is found in crystals, which are very hard, inflexible and brittle. Collagen fibers are strong and flexible, but they bend if they are compressed. The protein–crystal combination in bone is strong, somewhat flexible, and highly resistant to shattering. In its overall properties, bone is on a par with the best steel-reinforced concrete. In fact, bone is far superior to concrete, because it can undergo remodeling (cycles of bone formation and resorption) as needed and can repair itself after injury.

Module 5.2 Review

a. List the major parts of a long bone.

b. Describe the function of the marrow cavity.

c. How would the compressive strength of a bone be affected if the ratio of collagen to calcium phosphate increased?

Bone has a calcified matrix associated with osteocytes, osteoblasts, and osteoclasts

Both compact bone and spongy bone contain the same three cell types.

1 Mature bone cells called **osteocytes** (OS-tē-ō-sīts; (*osteo-*, bone + *cyte*, cell) maintain the protein and mineral content of the surrounding matrix through the turnover of matrix components. Osteocytes secrete chemicals that dissolve the adjacent matrix, and the released minerals enter the circulation. The osteocytes then rebuild the matrix, stimulating the deposition of mineral crystals. Osteocytes also play a role in the repair of damaged bone.

3 **Osteoclasts** (OS-tē-ō-clasts; *clast*, to break) are cells that remove and recycle bone matrix. These are giant cells with 50 or more nuclei. They are derived from the same stem cells that produce monocytes and macrophages. Acids and proteolytic (protein-digesting) enzymes secreted by osteoclasts dissolve the matrix and release the stored minerals into interstitial fluid. This process, called **osteolysis** (os-tē-OL-i-sis; *lysis*, a loosening) or resorption, is important in the regulation of calcium and phosphate ion concentrations in body fluids.

The layers of matrix are called **lamellae** (lah-MEL-lē; singular, *lamella*, a thin plate).

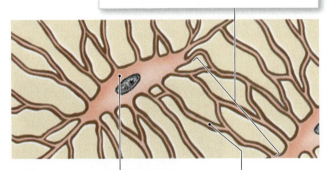

Osteocytes make up most of the cells in bone. Each osteocyte occupies a **lacuna**, a pocket sandwiched between layers of matrix. Osteocytes cannot divide, and a lacuna never contains more than one osteocyte.

Processes of the osteocytes extend into passageways called **canaliculi** that penetrate the matrix. The canaliculi interconnect the lacunae and reach vascular passageways, providing a route for nutrient diffusion.

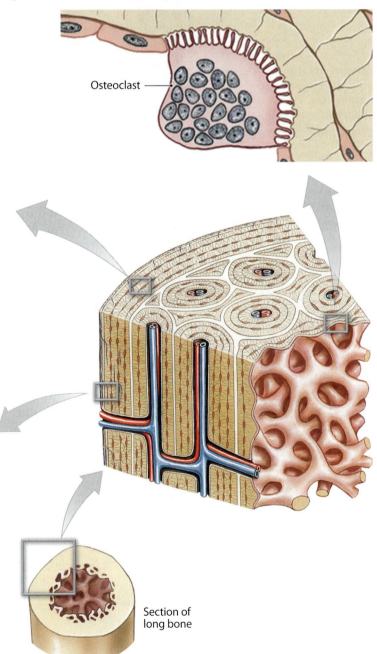

Osteoclast

Section of long bone

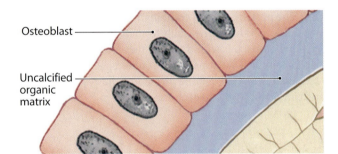

Osteoblast

Uncalcified organic matrix

2 **Osteoblasts** (OS-tē-ō-blasts; *blast*, precursor) produce new bone matrix in a process called **ossification**. Osteoblasts make and release the proteins and other organic components of the matrix. Osteoblasts also help increase the concentration of calcium phosphate to the point where calcium salts are deposited in the organic matrix forming bone. Osteocytes develop from osteoblasts that have become completely surrounded by bone matrix.

4 The basic functional unit of mature compact bone is the **osteon** (OS-tē-on), or Haversian system. Many of the important features of osteons were introduced in Module 3.22, which you may wish to revisit before proceeding.

Osteon

Compact bone   LM × 375

The **central canal** generally runs parallel to the surface of the bone. Small branches of the arteries and veins within the central canal nourish the osteocytes within the osteons, and provide a route for the removal of wastes or released calcium ions.

Lamellae

The osteocytes occupy **lacunae** that lie between the lamellae. In preparing this micrograph, a small piece of bone was ground down until it was thin enough to transmit light. In this process, the lacunae and canaliculi are filled with bone dust, and thus appear black.

In the intestines, calcium and phosphate ions are absorbed from the diet. The rate of absorption is hormonally regulated.

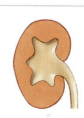

Normal Ca²⁺ levels in plasma

In the kidneys, the levels of calcium and phosphate ions lost in the urine are hormonally regulated.

5 **Calcium** ions play a role in a variety of physiological processes, and even small variations from the normal concentration affect cellular operations. The homeostatic regulation of calcium ion levels is a juggling act that balances activities of the intestines, bones, and kidneys.

Bone

Within the skeleton, osteoblasts are continuously depositing new bone matrix. At the same time, osteo-clasts are eroding existing matrix and releasing calcium and phosphate ions that enter the circulation. The balance between osteoblast and osteoclast activity is hormonally regulated.

If the calcium concentration of body fluids increases or decreases by more than 30–35 percent, neuron and muscle cell function is disrupted, with potentially lethal results. Calcium ion concentration is so closely regulated, however, that daily fluctuations of more than 10 percent are highly unusual.

Module 5.3 Review

a. Define osteocyte, osteoblast, and osteoclast.

b. What is the basic functional unit of mature compact bone?

c. If osteoclast activity exceeds osteoblast activity in a bone, what would be the effect on the bone?

Compact bone consists of parallel osteons, and spongy bone consists of a network of trabeculae

1 The lamellae of each osteon form a series of nested cylinders around the central canal. In transverse section, these lamellae resemble a target, with the **central canal** as the bull's-eye. You might think of a single osteon as a drinking straw with very thick walls: When you attempt to push the ends of the straw together or to pull them apart, the straw is quite strong. But if you hold the ends and push from the side, the straw will bend sharply with relative ease. The osteons in the diaphysis of a long bone are parallel to the long axis of the shaft. Thus, the shaft does not bend, even when extreme forces are applied to either end. The femur can withstand 10–15 times the body's weight without breaking. Yet a much smaller force applied to the side of the shaft can break any long bone, even the femur.

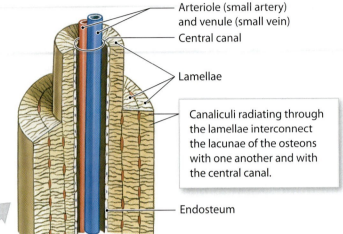

Arteriole (small artery) and venule (small vein)

Central canal

Lamellae

Canaliculi radiating through the lamellae interconnect the lacunae of the osteons with one another and with the central canal.

Endosteum

2 A section cut from the shaft of a long bone shows the organization of osteons and blood vessels, and allows a comparison between compact and spongy bone.

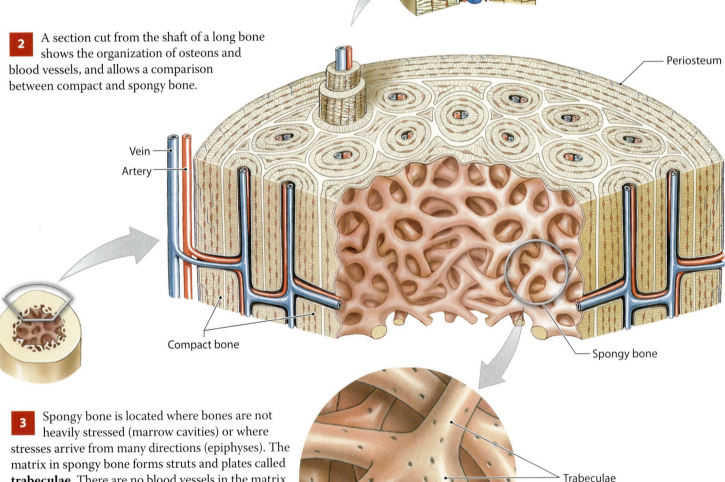

Periosteum

Vein

Artery

Compact bone

Spongy bone

3 Spongy bone is located where bones are not heavily stressed (marrow cavities) or where stresses arrive from many directions (epiphyses). The matrix in spongy bone forms struts and plates called **trabeculae**. There are no blood vessels in the matrix of spongy bone. Nutrients reach the osteocytes by diffusion along canaliculi. Because spongy bone is much lighter than compact bone, it reduces the weight of the skeleton, making it easier for muscles to move the bones.

Trabeculae

Lamellae

4 The diameter of a bone enlarges through **appositional growth** at the outer surface. In appositional growth, cells in the inner layer of the periosteum differentiate into osteoblasts and add bone matrix to the surface. This adds successive layers of lamellae to the outer surface of the bone. Osteoblasts trapped between these lamellae differentiate into osteocytes. Over time, the deeper lamellae are recycled and replaced with the osteons typical of compact bone.

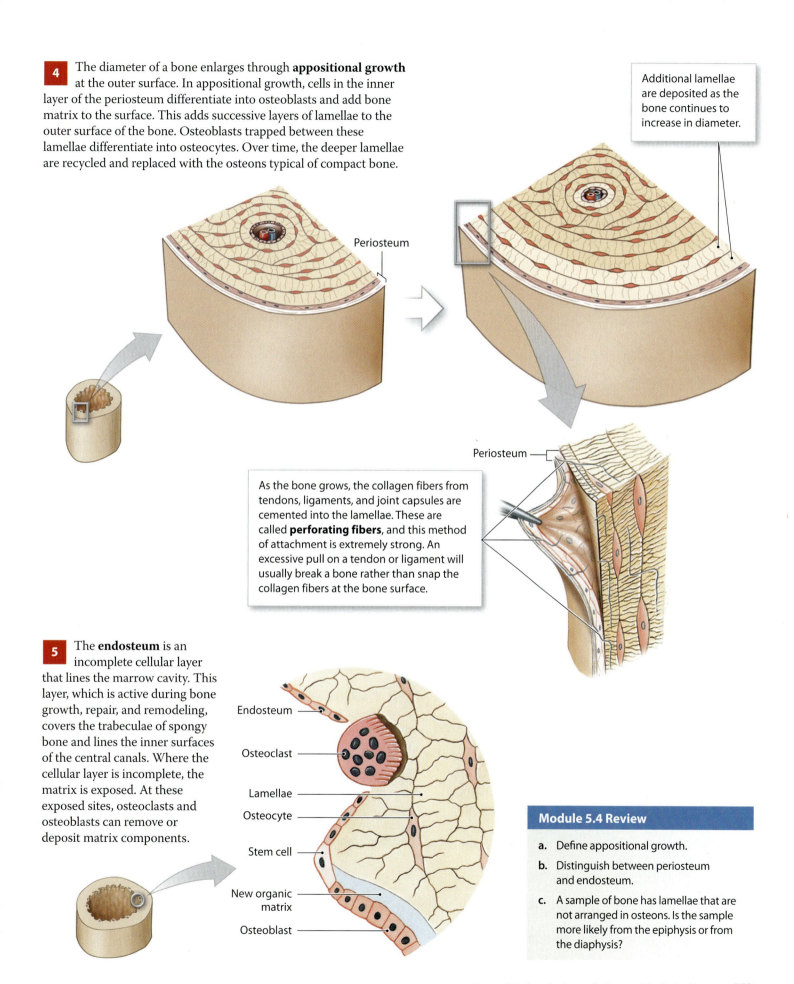

Additional lamellae are deposited as the bone continues to increase in diameter.

Periosteum

Periosteum

As the bone grows, the collagen fibers from tendons, ligaments, and joint capsules are cemented into the lamellae. These are called **perforating fibers**, and this method of attachment is extremely strong. An excessive pull on a tendon or ligament will usually break a bone rather than snap the collagen fibers at the bone surface.

5 The **endosteum** is an incomplete cellular layer that lines the marrow cavity. This layer, which is active during bone growth, repair, and remodeling, covers the trabeculae of spongy bone and lines the inner surfaces of the central canals. Where the cellular layer is incomplete, the matrix is exposed. At these exposed sites, osteoclasts and osteoblasts can remove or deposit matrix components.

Endosteum

Osteoclast

Lamellae

Osteocyte

Stem cell

New organic matrix

Osteoblast

Module 5.4 Review

a. Define appositional growth.

b. Distinguish between periosteum and endosteum.

c. A sample of bone has lamellae that are not arranged in osteons. Is the sample more likely from the epiphysis or from the diaphysis?

The most common method of bone formation involves the replacement of cartilage with bone

When bone formation begins in the embryo, all existing skeletal elements are cartilaginous. These cartilages are gradually replaced with bone through the process of **endochondral** (en-dō-KON-drul, *endo-*, inside + *chondros*, cartilage) **ossification**. This sequence illustrates the key steps in endochondral ossification. This process begins with a small cartilage that is basically a miniature model of the corresponding bone of the adult skeleton. As it forms, the bone grows in length and diameter. The increase in diameter involves appositional bone deposition.

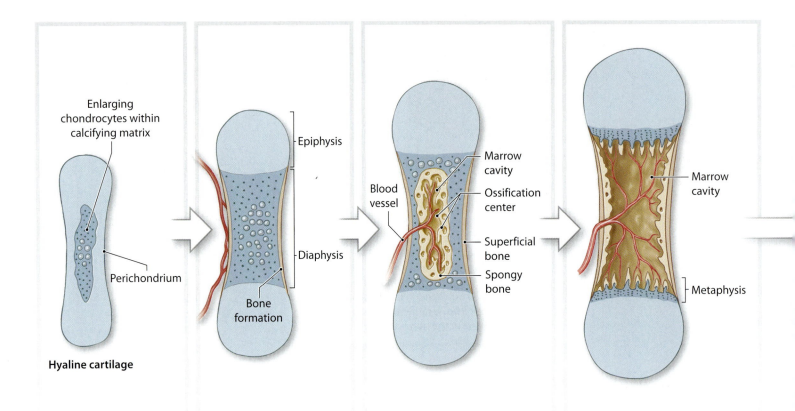

1 As the cartilage enlarges, chondrocytes near the center of the shaft increase greatly in size. The matrix is reduced to a series of small struts that soon begin to calcify. The enlarged chondrocytes then die and disintegrate, leaving cavities within the cartilage.

2 Blood vessels grow around the edges of the cartilage, and the cells of the perichondrium convert to osteoblasts. The shaft of the cartilage then becomes ensheathed in a superficial layer of bone.

3 Blood vessels penetrate the cartilage and invade the central region. Fibroblasts migrating with the blood vessels differentiate into osteoblasts and begin producing spongy bone at an **ossification center**. Bone formation then spreads along the shaft toward both ends.

4 Remodeling occurs as growth continues, creating a marrow cavity. The osseous tissue of the shaft becomes thicker, and the cartilage near each epiphysis is replaced by shafts of bone. Further growth involves increases in length and diameter.

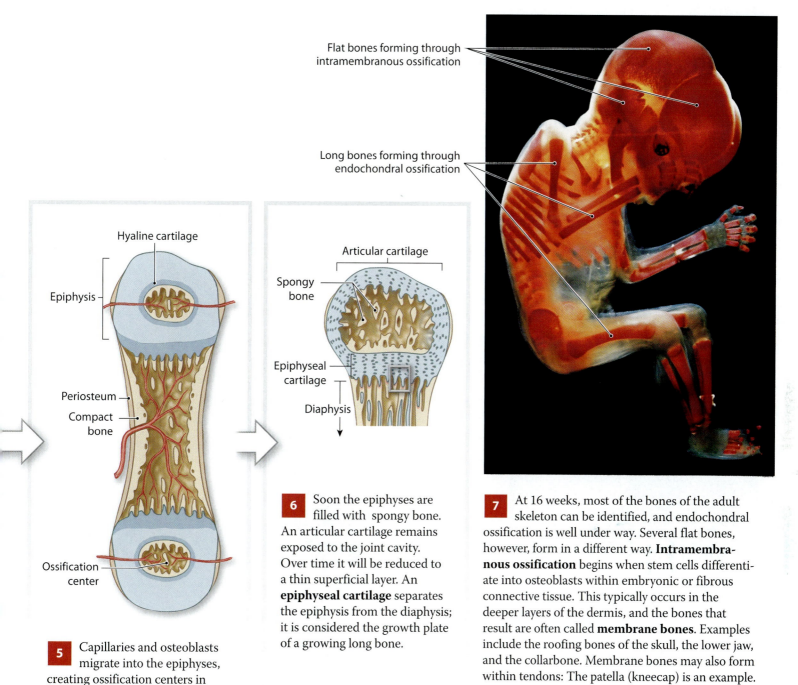

Flat bones forming through intramembranous ossification

Long bones forming through endochondral ossification

Hyaline cartilage

Epiphysis

Periosteum

Compact bone

Ossification center

Articular cartilage

Spongy bone

Epiphyseal cartilage

Diaphysis

5 Capillaries and osteoblasts migrate into the epiphyses, creating ossification centers in those areas.

6 Soon the epiphyses are filled with spongy bone. An articular cartilage remains exposed to the joint cavity. Over time it will be reduced to a thin superficial layer. An **epiphyseal cartilage** separates the epiphysis from the diaphysis; it is considered the growth plate of a growing long bone.

7 At 16 weeks, most of the bones of the adult skeleton can be identified, and endochondral ossification is well under way. Several flat bones, however, form in a different way. **Intramembranous ossification** begins when stem cells differentiate into osteoblasts within embryonic or fibrous connective tissue. This typically occurs in the deeper layers of the dermis, and the bones that result are often called **membrane bones**. Examples include the roofing bones of the skull, the lower jaw, and the collarbone. Membrane bones may also form within tendons: The patella (kneecap) is an example.

At puberty, the combination of rising levels of sex hormones, growth hormone, and thyroid hormones stimulates bone growth dramatically. Osteoblasts now begin producing bone faster than chondrocytes are producing new epiphyseal cartilage. As a result, the osteoblasts "catch up," and the epiphyseal cartilage gets narrower and narrower until it ultimately disappears. The completion of epiphyseal growth is called epiphyseal closure. In adults, the former location of this cartilage is often detectable in x-rays as a distinct **epiphyseal line**, which remains after epiphyseal growth has ended.

Module 5.5 Review

a. Describe endochondral ossification.

b. During intramembranous ossification, which type of tissue is replaced by bone?

c. How could x-rays of the femur be used to determine whether a person has reached full height?

Abnormalities of bone growth and development produce recognizable physical signs

A variety of endocrine or metabolic problems can result in atypical skeletal growth. Here we consider several conditions that affect the skeleton as a whole.

1 In **pituitary growth failure**, inadequate production of growth hormone leads to reduced epiphyseal cartilage activity and abnormally short bones. This condition is becoming increasingly rare in the United States, because children can be treated with synthetic human growth hormone.

2 **Achondroplasia** (a-kon-drō-PLĀ-zē-uh) results from abnormal epiphyseal activity. In this case the epiphyseal cartilages of the long bones grow unusually slowly and are replaced by bone early in life. As a result, the individual develops short, stocky limbs. Although other skeletal abnormalities occur, the trunk is normal in size, and sexual and mental development remain unaffected.

3 Several inherited metabolic conditions that affect many systems influence the growth and development of the skeletal system. These conditions produce characteristic variations in body proportions. For example, many individuals with **Marfan's syndrome** are very tall and have long, slender limbs, due to excessive cartilage formation at the epiphyseal cartilages. Although this is an obvious physical distinction, the characteristic body proportions are not in themselves dangerous. However, the underlying mutation, which affects the structure of connective tissue throughout the body, commonly causes life-threatening cardiovascular problems.

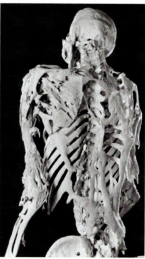
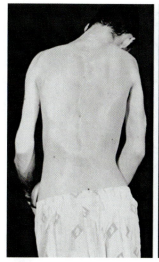

5 Under abnormal conditions, stem cells in any connective tissue can develop into osteoblasts that begin producing bone. The individual on the left has **fibrodysplasia ossificans progressiva (FOP)**, a rare single gene mutation disorder that involves the deposition of bone around skeletal muscles. The skeleton on the right shows the extent of abnormal ossification that can occur. Bones that develop in unusual places are called **heterotopic** (*hetero*, place), or **ectopic** (*ektos*, outside), **bones**. There is no effective treatment for this painful and debilitating condition, and patients seldom survive into their 40s.

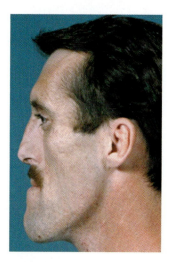

6 If growth hormone levels rise abnormally after epiphyseal cartilages close, the skeleton does not grow longer, but bones get thicker, especially those in the face, jaw, and hands. Cartilage growth and alterations in soft-tissue structure lead to changes in physical features, such as the contours of the face. These physical changes occur in the disorder called **acromegaly**.

4 **Gigantism** results from an overproduction of growth hormone before puberty. Individuals can reach heights of over 2.7 m (8 ft 11 in.) and weights of over 200 kg (440 lb). Puberty is often delayed, and the facial features in adults resemble those of acromegaly.

Module 5.6 Review

a. Describe Marfan's syndrome.

b. Compare gigantism with acromegaly.

c. Why is pituitary growth failure less common today in the United States?

A fracture is a crack or a break in a bone

Despite its mineral strength, bone can crack or even break if subjected to extreme loads, sudden impacts, or stresses from unusual directions. The damage is called a **fracture**. Most fractures heal even after severe damage, provided that the blood supply and the cellular components of the endosteum and periosteum survive.

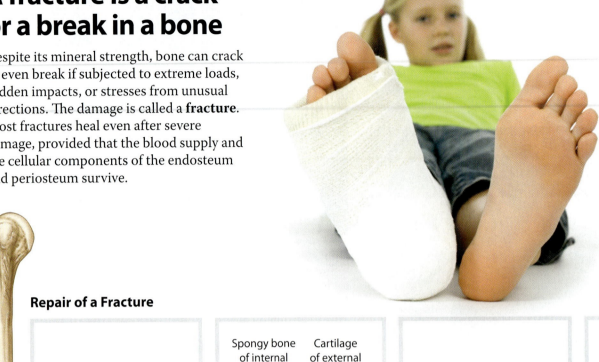

Repair of a Fracture

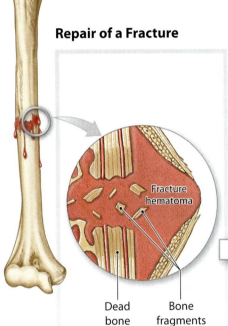

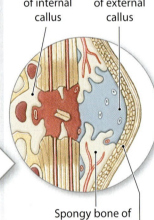

Spongy bone of internal callus

Cartilage of external callus

Fracture hematoma

Dead bone

Bone fragments

Spongy bone of external callus

Periosteum

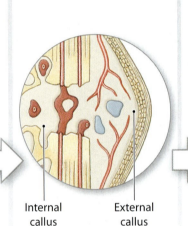

Internal callus

External callus

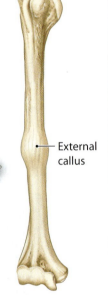

External callus

1 Immediately after the fracture, extensive bleeding occurs. Over a period of several hours, a large blood clot called a **fracture hematoma** develops.

2 An **internal callus** forms as a network of spongy bone unites the inner edges, and an **external callus** of cartilage and bone stabilizes the outer edges.

3 The cartilage of the external callus has been replaced by bone, and struts of spongy bone now unite the broken ends. Bone fragments and areas of dead bone closest to the break have been removed and replaced.

4 A swelling initially marks the location of the fracture. Over time, this region will be remodeled, and little evidence of the fracture will remain.

Types of Fractures

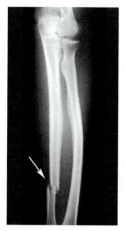

Transverse fractures, such as this fracture of the ulna, break a bone shaft across its long axis.

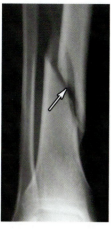

Spiral fractures, such as this fracture of the tibia, are produced by twisting stresses that spread along the length of the bone.

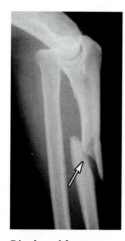

Displaced fractures produce new and abnormal bone alignments. **Nondisplaced fractures** retain the normal alignment of the bones or fragments.

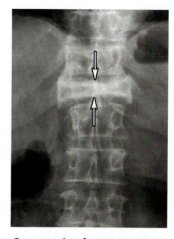

Compression fractures occur in vertebrae subjected to extreme stresses, such as those produced by the forces that arise when you land on your seat in a fall.

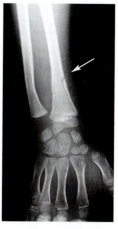

In a **greenstick fracture**, such as this fracture of the radius, only one side of the shaft is broken, and the other is bent. This type of fracture generally occurs in children, whose long bones have yet to ossify fully.

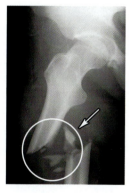

Comminuted fractures, such as this fracture of the femur, shatter the affected area into a multitude of bony fragments.

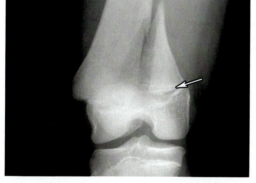

Epiphyseal fractures, such as this fracture of the femur, tend to occur where the bone matrix is undergoing calcification and chondrocytes are dying. A clean transverse fracture along this line generally heals well. Unless carefully treated, fractures between the epiphysis and the epiphyseal cartilage can permanently stop growth at this site.

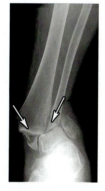

A **Pott's fracture** occurs at the ankle and affects both bones of the leg (the tibia and fibula).

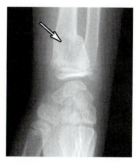

A **Colles fracture**, a break in the distal portion of the radius, is typically the result of reaching out to cushion a fall.

5 Fractures are named using various criteria, including their external appearance, their location, and the nature of the crack or break in the bone. Important types of fractures are illustrated here by representative x-rays. The broadest general categories are closed fractures and open fractures. **Closed (simple) fractures** are completely internal. They can be seen only on x-rays, because they do not involve a break in the skin. **Open (compound) fractures** project through the skin. These fractures, which are obvious on inspection, are more dangerous than closed fractures, due to the possibility of infection or uncontrolled bleeding. Many fractures fall into more than one category, because the terms overlap.

Module 5.7 Review

a. Define open fracture and closed fracture.

b. List the steps involved in fracture repair, beginning just after the fracture occurs.

c. When during fracture repair does an external callus form?

1. Vocabulary

In the space provided, write the term for each of the following definitions.

a _____ Bones with complex shapes

b _____ The expanded ends of a long bone

c _____ A shallow depression in the surface of a bone

d _____ The central space within a bone

e _____ The strut- and plate-shaped matrix of spongy bone

f _____ Cells that remove and recycle bone matrix

g _____ Type of fracture that results in many bony fragments

h _____ Tissue that separates the epiphysis and diaphysis in a growing bone

i _____ The basic functional unit of compact bone

j _____ Type of bone growth that increases bone diameter

k _____ Process by which cartilage is replaced by bone

2. Concept map

Use each of the following terms once to fill in the blank boxes to correctly complete the bone formation concept map.

- lacunae
- osteocytes
- collagen
- intramembranous ossification
- compact bone
- periosteum
- hyaline cartilage

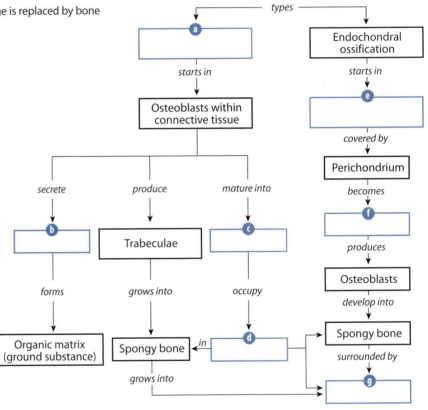

3. Section integration

While playing on her swing set, 10-year-old Rebecca falls and breaks her right leg. At the emergency room, the doctor tells her parents that the proximal end of the tibia where the epiphysis meets the diaphysis is fractured. The fracture is properly set and eventually heals. During a routine physical when she is 18, Rebecca learns that her right leg is 2 cm shorter than her left. What might account for this difference? _____

_____

_____

_____

The Skeleton

The **axial skeleton** forms the longitudinal axis of the body. This division of the skeletal system includes the skull and associated bones, the thoracic cage, the vertebral column, and various cartilages. The **appendicular skeleton** includes the bones of the limbs and the supporting elements, or girdles, that connect them to the trunk.

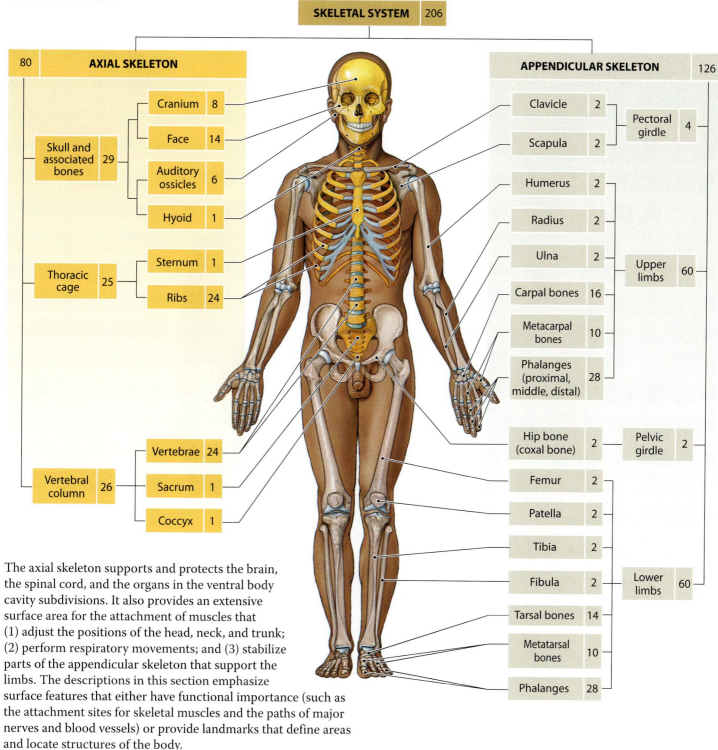

| SKELETAL SYSTEM | 206 |

AXIAL SKELETON — 80

Skull and associated bones	29	Cranium	8
		Face	14
		Auditory ossicles	6
		Hyoid	1

| Thoracic cage | 25 | Sternum | 1 |
| | | Ribs | 24 |

Vertebral column	26	Vertebrae	24
		Sacrum	1
		Coccyx	1

APPENDICULAR SKELETON — 126

| Clavicle | 2 | Pectoral girdle | 4 |
| Scapula | 2 | | |

Humerus	2	Upper limbs	60
Radius	2		
Ulna	2		
Carpal bones	16		
Metacarpal bones	10		
Phalanges (proximal, middle, distal)	28		

| Hip bone (coxal bone) | 2 | Pelvic girdle | 2 |

Femur	2	Lower limbs	60
Patella	2		
Tibia	2		
Fibula	2		
Tarsal bones	14		
Metatarsal bones	10		
Phalanges	28		

The axial skeleton supports and protects the brain, the spinal cord, and the organs in the ventral body cavity subdivisions. It also provides an extensive surface area for the attachment of muscles that (1) adjust the positions of the head, neck, and trunk; (2) perform respiratory movements; and (3) stabilize parts of the appendicular skeleton that support the limbs. The descriptions in this section emphasize surface features that either have functional importance (such as the attachment sites for skeletal muscles and the paths of major nerves and blood vessels) or provide landmarks that define areas and locate structures of the body.

Facial bones dominate the anterior skull, and cranial bones dominate the posterior surface

1 We begin by examining the skull in anterior view. If you consider the cranium as the home of the brain, the facial bones form the front porch. The skull contains 22 bones: 8 from the cranium and 14 from the face.

Facial Bones

Facial bones protect and support the entrances to the digestive and respiratory tracts. They also provide areas for the attachment of muscles that control facial expressions and assist in eating.

The **nasal bones** support the superior portion of the bridge of the nose. They are connected to cartilages that support the distal portions of the nose.

The **lacrimal bones** form part of the medial wall of the orbit (eye socket).

The **palatine bones** form the posterior portion of the hard palate and contribute to the floor of each orbit.

The **zygomatic bones** contribute to the rim and lateral wall of the orbit and form part of the cheekbone.

The **maxillae** support the upper teeth and form the inferior orbital rim, the lateral margins of the external nares, the upper jaw, and most of the hard palate.

The **inferior nasal conchae** create turbulence in air passing through the nasal cavity, and increase the epithelial surface area to promote warming and humidification of inhaled air.

The **vomer** forms the inferior portion of the bony nasal septum.

The **mandible** forms the lower jaw.

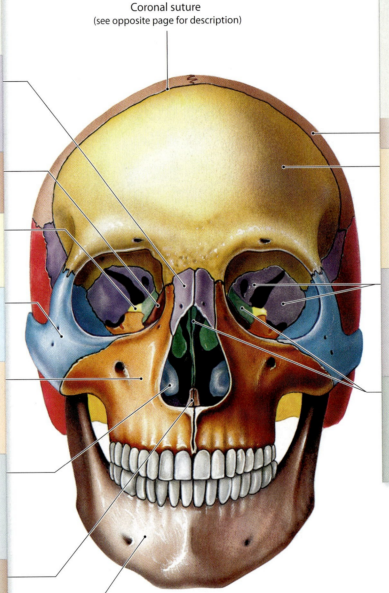

Coronal suture
(see opposite page for description)

Cranial Bones

The **cranial bones** form the **cranium**, which protects the brain. Blood vessels, nerves, and membranes that stabilize the position of the brain are attached to the inner surface of the cranium. Its outer surface provides an extensive area for the attachment of muscles that move the eyes, jaws, and head.

Parietal bone

The **frontal bone** forms the anterior portion of the cranium and the roof of the orbits. Mucous secretions of the frontal sinuses within this bone help flush the surfaces of the nasal cavities.

The **sphenoid** forms part of the floor of the cranium, unites the cranial and facial bones, and acts as a cross-brace that strengthens the sides of the skull.

The **ethmoid** forms the anteromedial floor of the cranium, the roof of the nasal cavity, and part of the nasal septum and medial orbital wall.

2 Whereas the anterior view is dominated by facial bones, the posterior view is dominated by bones of the cranium. Several prominent landmarks on the occipital and temporal bones are identified here. The occipital, parietal, and frontal bones form the skullcap.

Cranial Bones

The **parietal bone** on each side forms part of the superior and lateral surfaces of the cranium.

The **occipital bone** contributes to the posterior, lateral, and inferior surfaces of the cranium.

The **temporal bone** on either side (1) forms part of the lateral wall of the cranium and articulates with facial bones, (2) forms an articulation with the mandible, (3) surrounds and protects the sense organs of the inner ear, and (4) is an attachment site for muscles that close the jaws and move the head.

The **external occipital crest** extends inferiorly and marks the attachment of a ligament that helps stabilize the vertebrae of the neck.

Sutures

Except where the mandible contacts the cranium, the connections between the skull bones of adults are immovable joints called **sutures**. At a suture, bones are tied firmly together with dense fibrous connective tissue.

The **coronal suture** (see opposite page) attaches the frontal bone to the parietal bones of either side.

The **sagittal suture** extends from the lambdoid suture to the coronal suture, between the parietal bones.

A **squamous** (SKWĀ-mus) **suture** on each side of the skull forms the boundary between the temporal bone and the parietal bone of that side.

The **lambdoid** (LAM-doyd) **suture** (*lambda*, + *eidos*, shape) arches across the posterior surface of the skull. This suture separates the occipital bone from the two parietal bones.

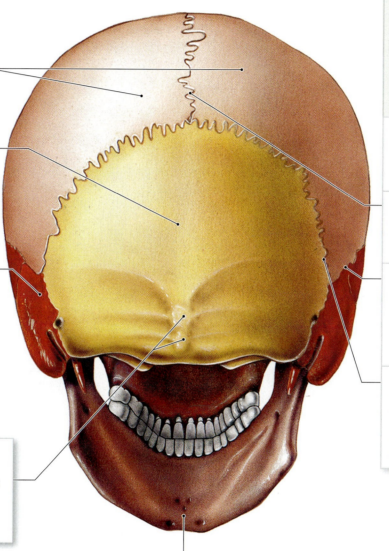

Mandible

Module 5.8 Review

a. Identify the facial bones.

b. Quincy suffers a hit to the skull that fractures the right superior lateral surface of his cranium. Which bone is fractured?

c. Identify the following bones as either a facial bone or a cranial bone: vomer, ethmoid, sphenoid, temporal, and inferior nasal conchae.

Surface features of the skull are functional landmarks

1 This lateral view of the skull shows how the large bones interconnect and reveals the prominent features of the individual cranial and facial bones.

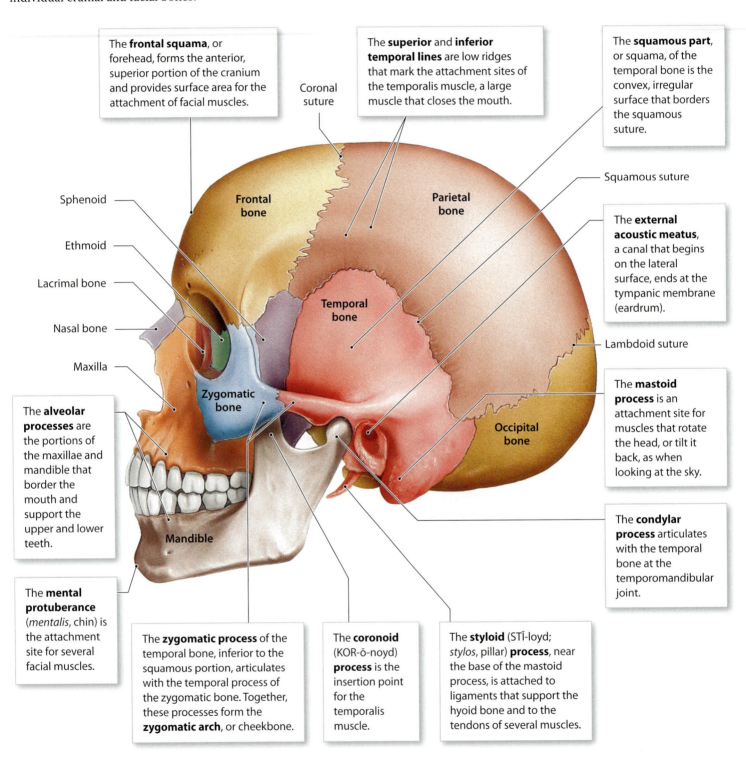

The **frontal squama**, or forehead, forms the anterior, superior portion of the cranium and provides surface area for the attachment of facial muscles.

The **superior** and **inferior temporal lines** are low ridges that mark the attachment sites of the temporalis muscle, a large muscle that closes the mouth.

The **squamous part**, or squama, of the temporal bone is the convex, irregular surface that borders the squamous suture.

Coronal suture

Sphenoid

Ethmoid

Lacrimal bone

Nasal bone

Maxilla

Frontal bone

Parietal bone

Squamous suture

The **external acoustic meatus**, a canal that begins on the lateral surface, ends at the tympanic membrane (eardrum).

Temporal bone

Lambdoid suture

Zygomatic bone

The **mastoid process** is an attachment site for muscles that rotate the head, or tilt it back, as when looking at the sky.

Occipital bone

The **alveolar processes** are the portions of the maxillae and mandible that border the mouth and support the upper and lower teeth.

The **condylar process** articulates with the temporal bone at the temporomandibular joint.

Mandible

The **mental protuberance** (*mentalis*, chin) is the attachment site for several facial muscles.

The **zygomatic process** of the temporal bone, inferior to the squamous portion, articulates with the temporal process of the zygomatic bone. Together, these processes form the **zygomatic arch**, or cheekbone.

The **coronoid** (KOR-ō-noyd) **process** is the insertion point for the temporalis muscle.

The **styloid** (STĪ-loyd; *stylos*, pillar) **process**, near the base of the mastoid process, is attached to ligaments that support the hyoid bone and to the tendons of several muscles.

2 In this inferior view you can see many of the important passageways for blood vessels and nerves, as well as the foramen used to connect the spinal cord to the brain.

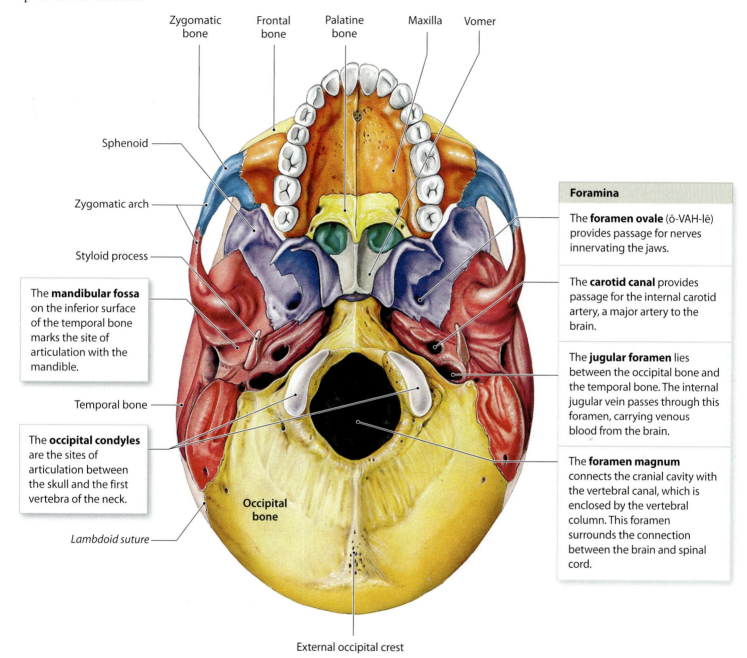

Zygomatic bone

Frontal bone

Palatine bone

Maxilla

Vomer

Sphenoid

Zygomatic arch

Styloid process

The **mandibular fossa** on the inferior surface of the temporal bone marks the site of articulation with the mandible.

Temporal bone

The **occipital condyles** are the sites of articulation between the skull and the first vertebra of the neck.

Lambdoid suture

Occipital bone

External occipital crest

Foramina

The **foramen ovale** (ō-VAH-lē) provides passage for nerves innervating the jaws.

The **carotid canal** provides passage for the internal carotid artery, a major artery to the brain.

The **jugular foramen** lies between the occipital bone and the temporal bone. The internal jugular vein passes through this foramen, carrying venous blood from the brain.

The **foramen magnum** connects the cranial cavity with the vertebral canal, which is enclosed by the vertebral column. This foramen surrounds the connection between the brain and spinal cord.

Module 5.9 Review

a. Identify the bone containing the carotid canal and name the structure that runs through this passageway.

b. Name the bone and its foramen which forms the passageway for the spinal cord.

c. The alveolar processes perform what functions in which bones?

Additional landmarks are visible in sectional views of the skull

1 This horizontal section should be compared to the inferior view in the previous module.

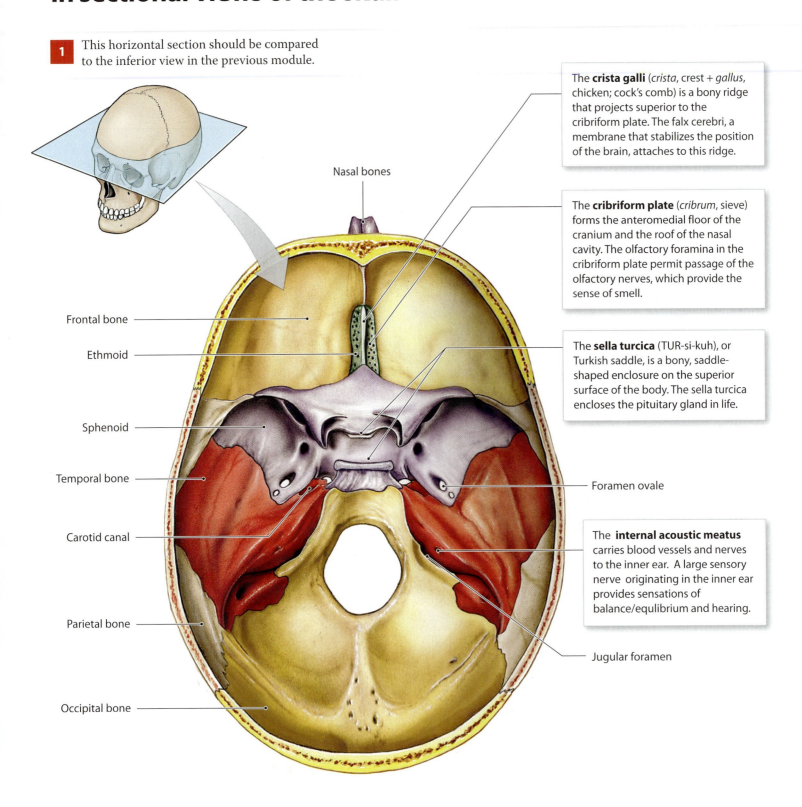

Nasal bones

Frontal bone

Ethmoid

Sphenoid

Temporal bone

Carotid canal

Parietal bone

Occipital bone

The **crista galli** (*crista*, crest + *gallus*, chicken; cock's comb) is a bony ridge that projects superior to the cribriform plate. The falx cerebri, a membrane that stabilizes the position of the brain, attaches to this ridge.

The **cribriform plate** (*cribrum*, sieve) forms the anteromedial floor of the cranium and the roof of the nasal cavity. The olfactory foramina in the cribriform plate permit passage of the olfactory nerves, which provide the sense of smell.

The **sella turcica** (TUR-si-kuh), or Turkish saddle, is a bony, saddle-shaped enclosure on the superior surface of the body. The sella turcica encloses the pituitary gland in life.

Foramen ovale

The **internal acoustic meatus** carries blood vessels and nerves to the inner ear. A large sensory nerve originating in the inner ear provides sensations of balance/equilibrium and hearing.

Jugular foramen

2 The **nasal complex** encloses the nasal cavities and the paranasal sinuses, air-filled chambers connected to the nasal cavities. The sphenoid, ethmoid, frontal bone, palatine bone, and maxillae contain the **paranasal sinuses**. (The tiny palatine sinuses, not shown, generally open into the sphenoidal sinuses.) The paranasal sinuses lighten the skull bones and provide an extensive area of mucous epithelium.

The **frontal sinuses** are extremely variable in size and time of appearance. They generally appear after age 6, but some people never develop them.

The **ethmoidal air cells** form a network whose mucus flushes the surfaces of the nasal cavities.

Sphenoid sinus

The **maxillary sinuses** produce mucus that flushes the inferior surfaces of the nasal cavities. The maxillae are the largest facial bones, and the maxillary sinuses are the largest sinuses.

Maxilla

Openings to nasal cavities

Mandible

3 This is a sagittal section through the nasal and oral cavities with the nasal septum removed. The frontal bone, sphenoid, and ethmoid form the superior wall of the nasal cavities. The lateral walls are formed by the maxillae and the lacrimal bones, the ethmoid (the superior and middle nasal conchae), and the inferior nasal conchae. Much of the anterior margin of the nasal cavity is formed by the soft tissues of the nose, but the bridge of the nose is supported by the maxillae and nasal bones.

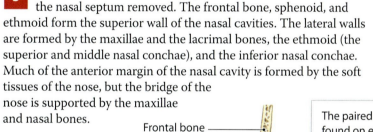

Frontal bone

Frontal sinuses

The paired **sphenoidal sinuses** are found on either side of the body of the sphenoid, inferior to the sella turcica. They are variable in size.

Crista galli

Lacrimal bone

Nasal bone

Sella turcica

Sphenoid

Ethmoid

The mass of the ethmoid projects into the nasal cavity and a central plate contributes to the nasal septum. The **superior nasal conchae** (KONG-kē; singular, *concha*, a snail shell) and the **middle nasal conchae** are delicate projections of the lateral masses of the ethmoid.

Superior nasal concha

Middle nasal concha

The **inferior nasal conchae** are separate bones that project into the nasal cavity on either side of the nasal septum.

Inferior nasal concha

Maxilla   Hard palate   Palatine bone

Module 5.10 Review

a. List at least five features of the ethmoid bone.

b. Which bones contain paranasal sinuses?

c. What roles do the paranasal sinuses play in the skull?

The associated bones of the skull perform specialized functions

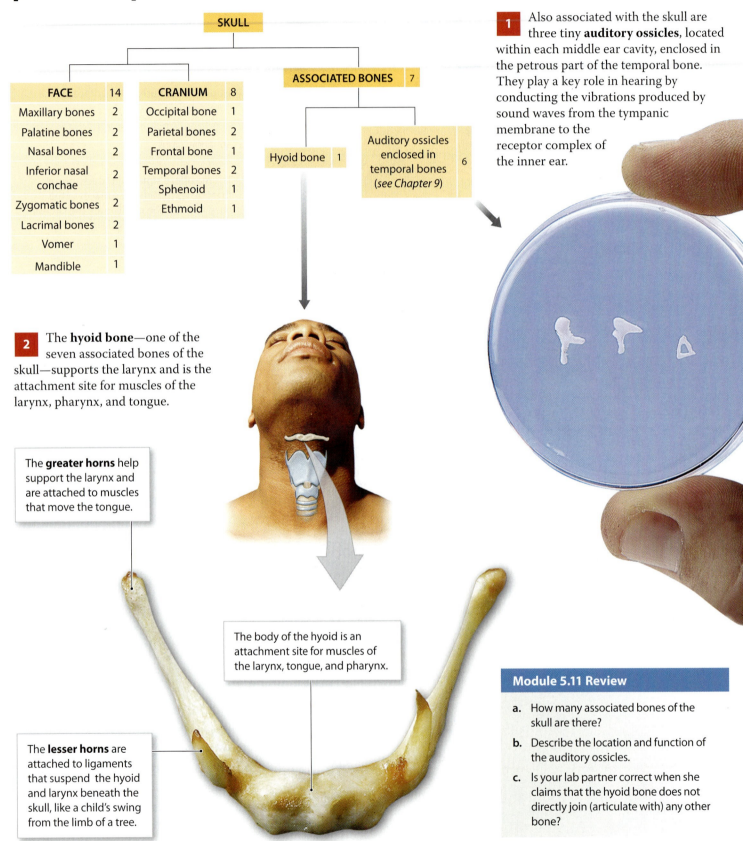

1 Also associated with the skull are three tiny **auditory ossicles**, located within each middle ear cavity, enclosed in the petrous part of the temporal bone. They play a key role in hearing by conducting the vibrations produced by sound waves from the tympanic membrane to the receptor complex of the inner ear.

2 The **hyoid bone**—one of the seven associated bones of the skull—supports the larynx and is the attachment site for muscles of the larynx, pharynx, and tongue.

The **greater horns** help support the larynx and are attached to muscles that move the tongue.

The **lesser horns** are attached to ligaments that suspend the hyoid and larynx beneath the skull, like a child's swing from the limb of a tree.

The body of the hyoid is an attachment site for muscles of the larynx, tongue, and pharynx.

Module 5.11 Review

a. How many associated bones of the skull are there?

b. Describe the location and function of the auditory ossicles.

c. Is your lab partner correct when she claims that the hyoid bone does not directly join (articulate with) any other bone?

FACE	14
Maxillary bones	2
Palatine bones	2
Nasal bones	2
Inferior nasal conchae	2
Zygomatic bones	2
Lacrimal bones	2
Vomer	1
Mandible	1

CRANIUM	8
Occipital bone	1
Parietal bones	2
Frontal bone	1
Temporal bones	2
Sphenoid	1
Ethmoid	1

ASSOCIATED BONES 7 — Hyoid bone 1 — Auditory ossicles enclosed in temporal bones (*see Chapter 9*) 6

Fontanelles permit cranial growth in infants and small children

The skull organizes around the developing brain. As the time of birth approaches, the brain enlarges rapidly. Although the bones of the skull are also growing, they don't keep pace. At birth, the cranial bones are connected by areas of flexible fibrous connective tissue that enable the skull's shape to be distorted without damage to ease the infant's passage through the birth canal. The largest fibrous areas between the cranial bones are known as **fontanelles** (fon-tuh-NELZ; sometimes spelled fontanels).

1 This lateral view shows how small an infant's facial bones are compared to the cranium.

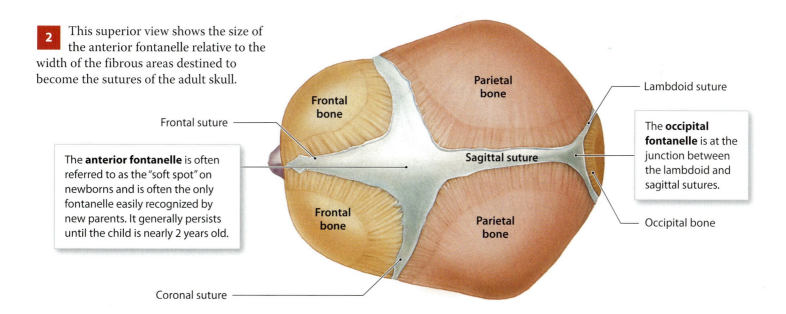

Coronal suture

The **sphenoidal fontanelle** on each side lies at the junction between the squamous suture and the coronal suture.

Frontal bone

Parietal bone

Squamous suture

Nasal bone

Maxilla

Sphenoid

Temporal bone

The **mastoid fontanelle** on each side lies at the junction between the squamous suture and the lambdoid suture.

Mandible

Lambdoid suture

Occipital bone

2 This superior view shows the size of the anterior fontanelle relative to the width of the fibrous areas destined to become the sutures of the adult skull.

Parietal bone

Frontal bone

Lambdoid suture

Frontal suture

Sagittal suture

The **occipital fontanelle** is at the junction between the lambdoid and sagittal sutures.

The **anterior fontanelle** is often referred to as the "soft spot" on newborns and is often the only fontanelle easily recognized by new parents. It generally persists until the child is nearly 2 years old.

Frontal bone

Parietal bone

Occipital bone

Coronal suture

The occipital, sphenoidal, and mastoid fontanelles disappear within a month or two after birth. Even after the fontanelles disappear, the bones of the skull remain separated by fibrous connections. The most significant growth in the skull occurs before age 5, because at that time the brain stops growing and the cranial sutures ossify.

Module 5.12 Review

a. Define fontanelle.

b. Identify the major fontanelles.

c. What purposes do fontanelles serve?

The vertebral column has four spinal curves, and vertebrae have both anatomical similarities and regional differences

1 The adult vertebral column (or spine) consists of 26 bones: the 24 vertebrae, the sacrum, and the coccyx (KOK-siks), or tailbone. The vertebrae provide a column of support that bears the weight of the head, neck, and trunk and ultimately transfers the body's weight to the appendicular skeleton of the lower limbs. The vertebrae also protect the spinal cord and help maintain an upright body position, as in sitting or standing. The total length of the vertebral column of an adult averages 71 cm (28 in.).

Spinal Curves

Primary curves develop before birth, and secondary curves after birth.

The **cervical curve**, a secondary curve, develops as the infant learns to balance the weight of the head on the vertebrae of the neck.

The **thoracic curve**, a primary curve, accommodates the thoracic organs.

The **lumbar curve**, a secondary curve, balances the weight of the trunk over the lower limbs. It develops with the ability to stand.

The **sacral curve,** a primary curve, accommodates the abdominopelvic organs.

Vertebral Regions

Regions are defined by anatomical characteristics of individual vertebrae.

Cervical (7 vertebrae)

Thoracic (12 vertebrae)

Lumbar (5 vertebrae)

Sacral

Coccygeal

Regional Differences in Vertebral Structure and Function

	Cervical Vertebrae (7)	**Thoracic Vertebrae (12)**	**Lumbar Vertebrae (5)**
Location	Neck	Chest	Inferior portion of back
Vertebral Body	Small, oval, curved faces	Medium, heart-shaped, flat faces; facets for rib articulations	Massive, oval, flat faces
Vertebral Foramen	Large	Smaller	Smallest
Spinous Process	Long; split tip; points inferiorly	Long, slender; not split; points inferiorly	Blunt, broad; points posteriorly
Transverse Processes	Have transverse foramina	All but two have facets for rib articulations; no transverse foramina	Short; no articular facets or transverse foramina
Functions	Support skull, stabilize relative positions of brain and spinal cord, and allow controlled head movements	Support weight of head, neck, upper limbs, and chest; articulate with ribs to allow changes in volume of thoracic cage	Support weight of head, neck, upper limbs, and trunk

2 Each vertebra has three basic parts: (1) articular processes, (2) a vertebral arch, and (3) a vertebral body.

3 The vertebral arch can be divided into four different parts: the spinous process, laminae, transverse processes, and pedicles.

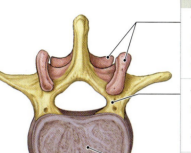

Superior view

Parts of a Vertebra

The **articular processes** extend superiorly and inferiorly to articulate with adjacent vertebrae.

The **vertebral arch** forms the posterior and lateral margins of the vertebral foramen.

The **vertebral body** is the part of a vertebra that transfers weight along the axis of the vertebral column.

The **vertebral foramen** is framed by the vertebral body and the vertebral arch.

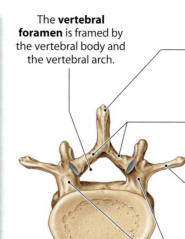

Inferior view

The Vertebral Arch

A **spinous process** projects posteriorly from the point where the vertebral laminae fuse to complete the vertebral arch.

The **laminae** (LAM-i-nē; singular, *lamina*, a thin plate) form the "roof" of the vertebral foramen.

Transverse processes project laterally on both sides from the point where the laminae join the pedicles. These processes are sites of muscle attachment, and they may also articulate with the ribs.

The **pedicles** (PED-i-kulz) form the sides of the vertebral arch.

4 This lateral view of three vertebrae shows how they combine to form the vertebral canal.

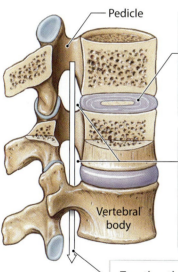

Pedicle

Vertebral body

The bodies of adjacent vertebrae are interconnected by ligaments but are separated by pads of fibrocartilage, the **intervertebral discs**.

The spaces between successive pedicles are called **intervertebral foramina**. Nerves and blood vessels running to or from the spinal cord pass through these foramina.

Together, the vertebral foramina of successive vertebrae form the **vertebral canal**, which encloses the spinal cord.

5 Each articular process has a smooth, concave surface called an **articular facet**. This posterior view shows that the vertebral arches form the roof of the vertebral canal.

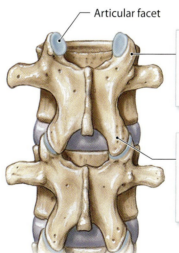

Articular facet

The **superior articular processes** articulate with the inferior articular processes of the vertebra immediately above it.

The **inferior articular processes** articulate with the superior articular processes of a more inferior vertebra (or the sacrum, in the case of the last lumbar vertebra).

Note that when referring to a specific vertebra, we use the capital letters C, T, L, S, and C_o to indicate the cervical, thoracic, lumbar, sacral, and coccygeal regions, respectively. In addition, we use a subscript number to indicate the relative position of the vertebra within that region, with 1 indicating the vertebra closest to the skull. For example, C_3 is the third cervical vertebra; C_1 is in contact with the skull. Similarly, L_4 is the fourth lumbar vertebra; L_1 is in contact with T_{12}.

Module 5.13 Review

a. Name the major components of a typical vertebra.

b. What is the importance of the secondary curves of the spine?

c. To which part of the vertebra do the intervertebral discs attach?

There are 7 cervical vertebrae and 12 thoracic vertebrae

Cervical Vertebrae

The **cervical vertebrae**, the smallest in the vertebral column, extend from the occipital bone of the skull to the thorax.

1 The vertebral foramen of a cervical vertebra is very large. At this level, the spinal cord still contains most of the axons that connect the brain to the rest of the body. However, cervical vertebrae support only the weight of the head, so the vertebral bodies can be relatively small and light. As you continue toward the sacrum, the loading increases and the vertebral bodies gradually enlarge.

The **transverse foramen,** formed by a connection of the costal and transverse processes, protects the vertebral arteries and vertebral veins that service the brain.

In a typical cervical vertebra (C_2–C_6), the tip of each spinous process has a prominent notch; such a process is a **bifid** (BĪ-fid) spinous process.

The transverse process is relatively short and stumpy.

Vertebral foramen

Vertebral body

Cervical vertebra (superior view)

2 The first two cervical vertebrae are specialized to support and stabilize the cranium while permitting head movement. The **atlas**, C_1, has no vertebral body and no spinous process, but it has a large, round vertebral foramen bounded by anterior and posterior arches. The **axis**, C_2, resembles more inferior cervical vertebrae except for the presence of the prominent dens on the superior surface of the body.

During development, the body of the atlas fuses to the body of the axis, where it forms the prominent **dens** (DENZ; *dens*, tooth).

Anterior arch of atlas

The atlas holds up the skull, articulating with the occipital condyles of the skull. This vertebra is named after Atlas, who, according to Greek myth, holds the world on his shoulders.

The articulation between the skull's occipital condyles and the atlas is a joint that permits nodding (such as when you indicate "yes").

A transverse ligament binds the dens to the inner surface of the anterior arch of the atlas. This articulation permits rotation (as when you shake your head to indicate "no").

Axis

Posterior arch of atlas

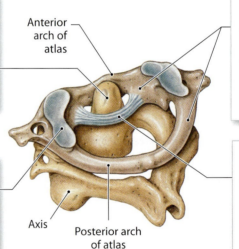

3 The last cervical vertebra has a very robust spinous process that ends in a solid tubercle that can easily be felt through the skin. A stout elastic ligament begins here and extends to an insertion along the occipital bone of the skull. When your head is upright, this ligament acts like the string on a bow to maintain the cervical curvature without muscular effort.

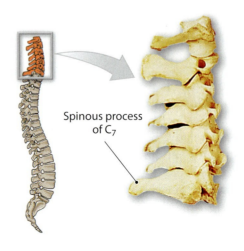

Spinous process of C_7

Thoracic Vertebrae

4 There are 12 **thoracic vertebrae**. Progressing from T_1 toward T_{12}, each vertebral body is slightly larger and more massive than the one superior to it, because the weight being transmitted along the vertebral column is steadily increasing.

5 A typical thoracic vertebra has a distinctive heart-shaped body that is much larger and more massive than that of a cervical vertebra, and the vertebral foramen is considerably smaller.

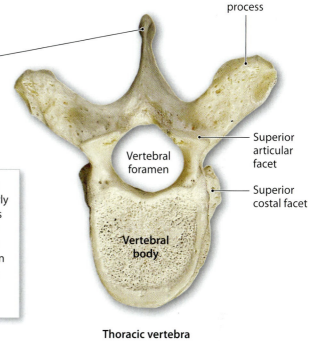

Transverse process

Superior articular facet

Vertebral foramen

Superior costal facet

Vertebral body

Thoracic vertebra (superior view)

6 Each thoracic vertebra articulates with ribs at **costal facets** on the vertebral body. The transverse processes of vertebrae T_1–T_{10} also contain costal facets. Ribs 11 and 12 do not contact the transverse processes, so they are known as "floating ribs."

The long, slender spinous process projects posteriorly and inferiorly. The spinous processes of T_{10}, T_{11}, and T_{12} increasingly resemble those of the lumbar region as the transition between the thoracic and lumbar curves approaches.

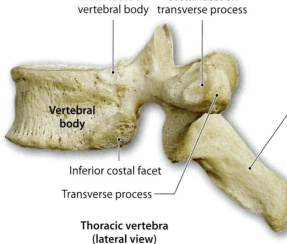

Costal facet on vertebral body

Costal facet on transverse process

Vertebral body

Inferior costal facet

Transverse process

Thoracic vertebra (lateral view)

Module 5.14 Review

a. Joe suffered a hairline fracture at the base of the dens. Which bone is fractured and where is the fractured bone located?

b. Examining a human vertebra, you notice that, in addition to the large foramen for the spinal cord, two smaller foramina are on either side of the bone in the region of the transverse processes. From which region of the vertebral column is this vertebra?

c. When you run your finger down the middle of a person's spine, what part of each vertebra are you feeling just beneath the skin?

There are five lumbar vertebrae, and the sacrum and coccyx consist of fused vertebrae

Lumbar Vertebrae

1 The five **lumbar vertebrae** are the largest vertebrae, and they transmit the most weight. Compression fractures of these vertebrae may occur at any age, but they are most often seen after aging and osteoporosis (bone loss) have reduced the strength of bones throughout the body.

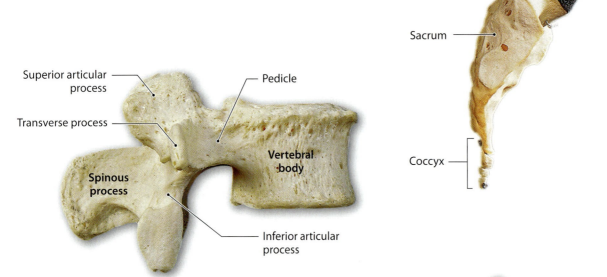

Superior articular process — Pedicle

Transverse process

Spinous process

Vertebral body

Inferior articular process

Lateral view

L₁
L₂
L₃
L₄
L₅

Sacrum

Coccyx

2 The body of a typical lumbar vertebra is thicker than that of a thoracic vertebra, and the superior and inferior surfaces are oval rather than heart shaped. Other noteworthy features are that (1) lumbar vertebrae do not have costal facets; (2) the slender transverse processes, which lack transverse costal facets, project dorsolaterally; (3) the vertebral foramen is triangular; (4) the stumpy spinous processes project posteriorly; (5) the superior articular processes face medially ("up and in"); and (6) the inferior articular processes face laterally ("down and out").

Spinous process — Superior articular process

Lamina

Transverse process

Vertebral foramen

Pedicle

Vertebral body

Superior view

Sacrum and Coccyx

The **sacrum** consists of the fused components of five sacral vertebrae. These vertebrae begin fusing shortly after puberty and, in general, are completely fused at age 25–30. The sacrum protects the reproductive, digestive, and urinary organs and, with its paired articulations, attaches the axial skeleton to the hip bones that are part of the *pelvic girdle* of the appendicular skeleton.

3 The anterior surface of the sacrum is concave, and the posterior surface is convex. The degree of curvature is more pronounced in males than in females.

At the base of the sacrum, a broad sacral **ala**, or wing, extends on either side. The anterior and superior surfaces of each ala provide an extensive area for muscle attachment.

Four pairs of **sacral foramina** extend between the posterior and anterior surfaces. The intervertebral foramina of the fused sacral vertebrae open into these passageways.

The **base** of the sacrum is the broad superior surface.

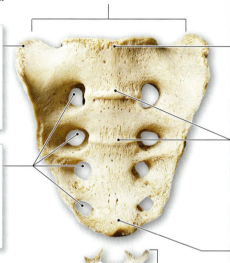

The **sacral promontory** is an important landmark in females during pelvic examinations and during labor and delivery.

Prominent transverse lines mark the former boundaries of individual vertebrae that fuse during the formation of the sacrum.

The **apex** is the narrow, inferior portion of the sacrum.

Coccyx

4 A posterior view of the sacrum shows the broad surface that provides an extensive area for the attachment of muscles.

The **sacral canal** is a passageway that extends the length of the sacrum. Nerves and membranes that line the vertebral canal in the spinal cord continue into the sacral canal.

The **superior articular process** of the sacrum articulates with the last lumbar vertebra.

The **median sacral crest** is a ridge formed by the fused spinous processes of the sacral vertebrae.

The **sacral hiatus** (hī-Ā-tus) is the opening at the inferior end of the sacral canal.

The small **coccyx** consists of three to five coccygeal vertebrae that have generally begun fusing by age 26.

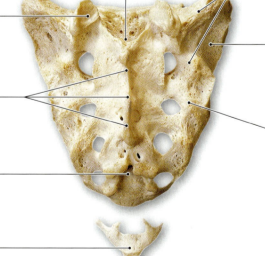

The **sacral tuberosity** is a roughened area posterior to the auricular surface. It marks the attachment site of ligaments that stabilize the sacro-iliac joint.

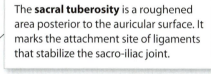

The **auricular surface** is a flat area that marks the site of articulation with the hip bones (the sacro-iliac joint).

The **lateral sacral crest** is a ridge that represents the fused transverse processes of the sacral vertebrae.

Module 5.15 Review

a. How many vertebrae are present in the lumbar region? In the sacrum?

b. What structure forms the posterior wall of the pelvic girdle?

c. Why are the bodies of the lumbar vertebrae so large?

The thoracic cage protects organs in the chest and provides sites for muscle attachment

The skeleton of the chest, or **thoracic cage**, provides bony support to the walls of the thoracic cavity. It consists of the thoracic vertebrae, the ribs, and the sternum (breastbone). The ribs and the sternum form the rib cage, whose movements are important in breathing. The thoracic cage protects the heart, lungs, thymus, and other structures in the thoracic cavity. The rib cage serves as an attachment point for muscles involved in (1) breathing, (2) maintenance of the position of the vertebral column, and (3) movements of the pectoral girdle and upper limbs.

1 An anterior view of the thoracic cage shows the sternum, the costal cartilages, and the major classes of ribs.

The **jugular notch**, located between the clavicular articulations, is a shallow indentation on the superior surface of the manubrium.

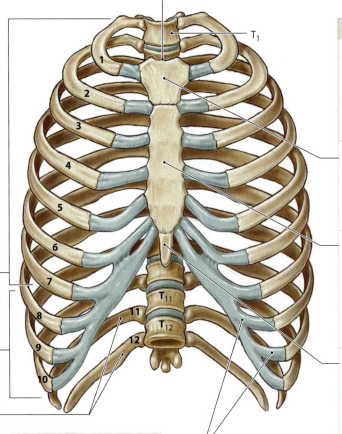

Ribs

The **ribs** reinforce the posterior and lateral walls of the thoracic cavity. Ribs 1–7 gradually increase in length and radius of curvature. Although the ribs are quite mobile and are among the most flexible of bones, they can be broken by a sharp hit or crushing impact. Fortunately, they are so stabilized by connective tissues and surrounding muscles that open fractures are rare, and splinting is unnecessary.

True ribs (ribs 1–7) are connected to the sternum by individual costal cartilages.

The **false ribs** (ribs 8–12) lack a direct connection to the sternum. (Ribs 8–10 are attached by shared costal cartilages.)

The last two pairs of ribs (11 and 12) are called **floating ribs**, because they have no connection with the sternum, or **vertebral ribs**, because they are attached only to the vertebrae and muscles of the body wall.

Costal cartilages connect ribs to the sternum, either individually or in groups.

Sternum

The adult **sternum**, or breast-bone, is a flat bone that forms in the anterior midline of the thoracic wall. It has three distinct regions that usually fuse together during adulthood.

Manubrium: The broad, trapezoid-shaped manubrium (ma-NOO-brē-um) articulates with the clavicles (collarbones) and the cartilages of the first pair of ribs.

Body: The body attaches to the inferior surface of the manubrium and extends inferiorly along the midline. Individual costal cartilages from rib pairs 2–7 are attached to this portion of the sternum.

Xiphoid process: The xiphoid (ZI-foyd) process, the smallest part of the sternum, is attached to the inferior surface of the body of the sternum.

Module 5.16 Review

a. How are true ribs distinguished from false ribs?

b. What are the three parts of the sternum?

c. In addition to the ribs and sternum, what other bones make up the thoracic cage?

Abnormalities in the axial skeleton directly affect posture and balance

The vertebral column must move, balance, and support the trunk and head. Conditions or events that damage the bones, muscles, and/or nerves can result in distorted shapes and impaired function.

1 In **kyphosis** (kī-FŌ-sis; *kyphos*, humpbacked, bent), the normal thoracic curvature becomes exaggerated posteriorly, producing a "round-back" appearance. This condition can be caused by (1) osteoporosis with compression fractures affecting the anterior portions of vertebral bodies, (2) chronic contractions in muscles that insert on the vertebrae, or (3) abnormal vertebral growth.

2 **Scoliosis** (skō-lē-Ō-sis; *scoliosis*, crookedness) is an abnormal lateral curvature of the spine. Scoliosis is the most common distortion of the vertebral curvature. This condition may result from developmental problems, damage to vertebral bodies, or muscular paralysis affecting one side of the back (as in some cases of polio). In four out of five cases, the structural or functional cause of the abnormal spinal curvature is impossible to determine. This *idiopathic* (of no known cause) scoliosis generally appears in girls during adolescence, when growth is most rapid. Small curves may later stabilize once growth is complete. For larger curves, bracing may prevent progression. Severe cases can be treated through surgical staightening with implanted metal rods or cables.

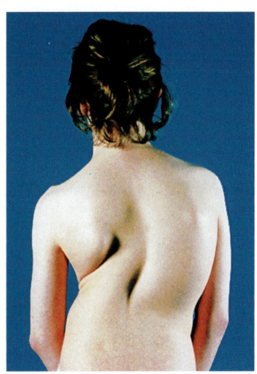

3 In **lordosis** (lor-DŌ-sis; *lordosis*, a bending backward), or "swayback," both the abdomen and buttocks protrude abnormally. The cause is an anterior exaggeration of the lumbar curvature. This may occur during pregnancy or result from abdominal obesity or weakness in the muscles of the abdominal wall.

Module 5.17 Review

a. List three causes of kyphosis.

b. How might pregnancy contribute to the development of lordosis?

c. Which condition is primarily associated with lateral distortion of the spine?

The pectoral girdles—the clavicles and scapulae—connect the upper limbs to the axial skeleton

Each arm forms a joint with the trunk at the **pectoral girdle**, or shoulder girdle. The pectoral girdle consists of two S-shaped **clavicles** (KLAV-i-kulz; collarbones) and two broad, flat **scapulae** (SKAP-ū-lē; singular, *scapula*, SKAP-ū-luh; shoulder blades).

1 The clavicles originate at the superior, lateral border of the manubrium of the sternum.

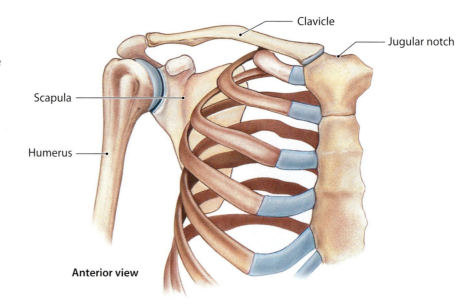

Clavicle

Jugular notch

Scapula

Humerus

Anterior view

2 From the sternal articulation at the pyramid-shaped **sternal end**, each clavicle curves laterally and posteriorly for about half its length. It then forms a smooth posterior curve to articulate with a process of the scapula, the **acromion** (a-KRŌ-mē-on).

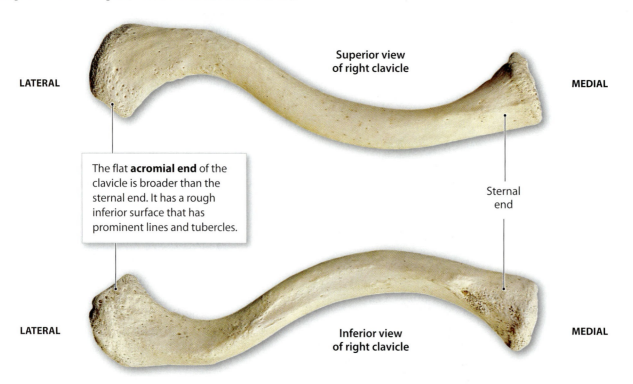

Superior view of right clavicle

LATERAL

MEDIAL

The flat **acromial end** of the clavicle is broader than the sternal end. It has a rough inferior surface that has prominent lines and tubercles.

Sternal end

LATERAL

Inferior view of right clavicle

MEDIAL

3 The anterior surface of the **body** of each scapula forms a broad, smooth triangle. The three sides of the triangle are the **superior border**, the **medial (vertebral) border**, and the **lateral (axillary) border**. Muscles that position the scapula attach along these edges. The corners of the triangle are called the **superior angle**, the **inferior angle**, and the **lateral angle**. The depression in the anterior surface is called the **subscapular fossa**.

4 The posterior surface of the scapula is convex and has prominent ridges and processes for muscle attachment.

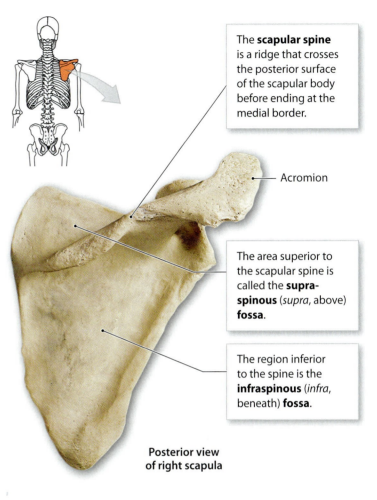

The **scapular spine** is a ridge that crosses the posterior surface of the scapular body before ending at the medial border.

Acromion

The area superior to the scapular spine is called the **supraspinous** (*supra*, above) **fossa**.

The region inferior to the spine is the **infraspinous** (*infra*, beneath) **fossa**.

Posterior view of right scapula

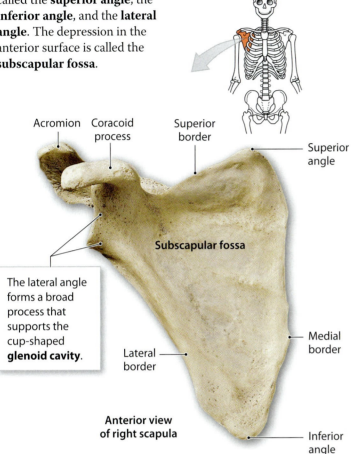

Acromion   Coracoid process

Superior border

Superior angle

Subscapular fossa

The lateral angle forms a broad process that supports the cup-shaped **glenoid cavity**.

Lateral border

Medial border

Inferior angle

Anterior view of right scapula

5 A lateral view of the scapula shows the surface of the glenoid cavity and the processes that provide attachment sites for muscles and the capsule of the shoulder joint.

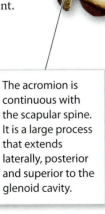

The **coracoid** (KOR-uh-koyd) **process** projects anterior and superior to the glenoid cavity.

At the glenoid cavity, the scapula articulates with the humerus, the proximal bone of the upper limb.

Lateral view of right scapula

The acromion is continuous with the scapular spine. It is a large process that extends laterally, posterior and superior to the glenoid cavity.

Module 5.18 Review

a. Name the bones of the pectoral girdle.

b. How would a broken clavicle affect the mobility and stability of the scapula?

c. Which bone articulates with the scapula at the glenoid cavity?

The humerus of the arm articulates with the radius and ulna of the forearm

The skeleton of the upper limbs consists of the bones of the arms, forearms, wrists, and hands. Note that in anatomical descriptions, the term "arm" refers only to the proximal portion of the upper limb (from shoulder to elbow), not to the entire limb. We will examine the bones of the right upper limb.

1 The arm, or brachium, contains one bone, the **humerus**, which extends from the scapula to the elbow. Near the distal articulation with the bones of the forearm, the shaft expands to either side at the **medial** and **lateral epicondyles**. Epicondyles are projections near the articular extremity above the condyle and provide additional surface area for muscle attachment.

Anterior view

Posterior view

Greater tubercle

The **intertubercular groove** lies between the greater and lesser tubercles. Both tubercles are important sites for muscle attachment; a large tendon runs along the groove.

The round **head** at the proximal end of the humerus articulates with the glenoid cavity of the scapula.

The **lesser tubercle** is a smaller projection that lies on the anterior, medial surface of the epiphysis.

The **anatomical neck** marks the extent of the joint capsule.

The **surgical neck** corresponds to the metaphysis of the growing bone. The name reflects the fact that fractures typically occur at this site.

The prominent **greater tubercle** is a rounded projection on the lateral surface of the epiphysis, near the margin of the humeral head. The greater tubercle establishes the lateral contour of the shoulder.

Shaft

The **deltoid tuberosity** is a large, rough elevation on the lateral surface of the shaft, approximately halfway along its length. It is named after the deltoid muscle, which attaches to it.

The **radial groove** crosses the inferior end of the deltoid tuberosity. This depression marks the path of the radial nerve, a large nerve that provides both sensory information from the posterior surface of the limb and motor control over the large muscles that straighten the elbow.

Coronoid fossa

Lateral epicondyle

Medial epicondyle

Trochlea

Olecranon fossa

The rounded **capitulum** forms the lateral surface of the **condyle**. At the condyle, the humerus articulates with the radius and the ulna of the forearm. The condyle is divided into two articular regions.

The **trochlea** (*trochlea*, a pulley) is the spool-shaped medial portion of the condyle. The trochlea extends from the **olecranon** (ō-LEK-ruh-non) **fossa** on the posterior surface to the **coronoid** (*corona*, crown) **fossa** on the anterior surface. These depressions accept projections from the ulna as the elbow approaches the limits of its range of motion.

2 The **ulna** and **radius** are parallel bones that support the forearm. In the anatomical position, the ulna lies medial to the radius.

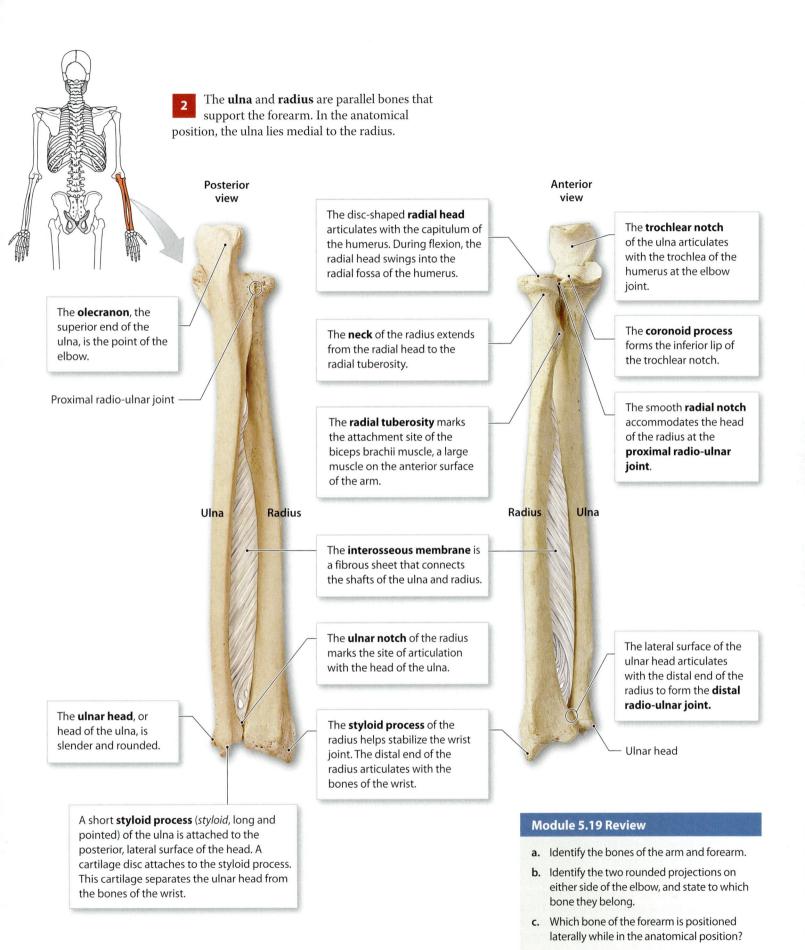

Posterior view

Anterior view

The disc-shaped **radial head** articulates with the capitulum of the humerus. During flexion, the radial head swings into the radial fossa of the humerus.

The **trochlear notch** of the ulna articulates with the trochlea of the humerus at the elbow joint.

The **olecranon**, the superior end of the ulna, is the point of the elbow.

Proximal radio-ulnar joint

The **neck** of the radius extends from the radial head to the radial tuberosity.

The **coronoid process** forms the inferior lip of the trochlear notch.

The **radial tuberosity** marks the attachment site of the biceps brachii muscle, a large muscle on the anterior surface of the arm.

The smooth **radial notch** accommodates the head of the radius at the **proximal radio-ulnar joint**.

Ulna   Radius

Radius   Ulna

The **interosseous membrane** is a fibrous sheet that connects the shafts of the ulna and radius.

The **ulnar notch** of the radius marks the site of articulation with the head of the ulna.

The lateral surface of the ulnar head articulates with the distal end of the radius to form the **distal radio-ulnar joint.**

The **ulnar head**, or head of the ulna, is slender and rounded.

The **styloid process** of the radius helps stabilize the wrist joint. The distal end of the radius articulates with the bones of the wrist.

Ulnar head

A short **styloid process** (*styloid*, long and pointed) of the ulna is attached to the posterior, lateral surface of the head. A cartilage disc attaches to the styloid process. This cartilage separates the ulnar head from the bones of the wrist.

Module 5.19 Review

a. Identify the bones of the arm and forearm.

b. Identify the two rounded projections on either side of the elbow, and state to which bone they belong.

c. Which bone of the forearm is positioned laterally while in the anatomical position?

The wrist is composed of carpal bones, and the hand consists of metacarpal bones and phalanges

1 The **carpus**, or wrist, contains eight carpal bones arranged in two rows: a row of four proximal carpal bones, and a row containing four distal carpal bones.

2 Five **metacarpal** (met-uh-KAR-pul; *metacarpus*, hand) **bones** articulate with the distal carpal bones and support the hand. The metacarpal bones are identified by Roman numerals I–V, beginning with the lateral metacarpal bone, which articulates with the trapezium. Distally, the metacarpal bones articulate with the proximal finger bones. The enlarged dorsal surfaces of the metacarpals at these joints are often called "knuckles".

3 Distally, the metacarpal bones articulate with the proximal finger bones. Each hand has 14 finger bones, or **phalanges** (fa-LAN-jēz; singular, *phalanx*). Metacarpal I articulates with the proximal bone of the thumb, or **pollex** (POL-eks). The pollex has two phalanges (proximal and distal). Each of the other fingers has three phalanges (proximal, middle, and distal).

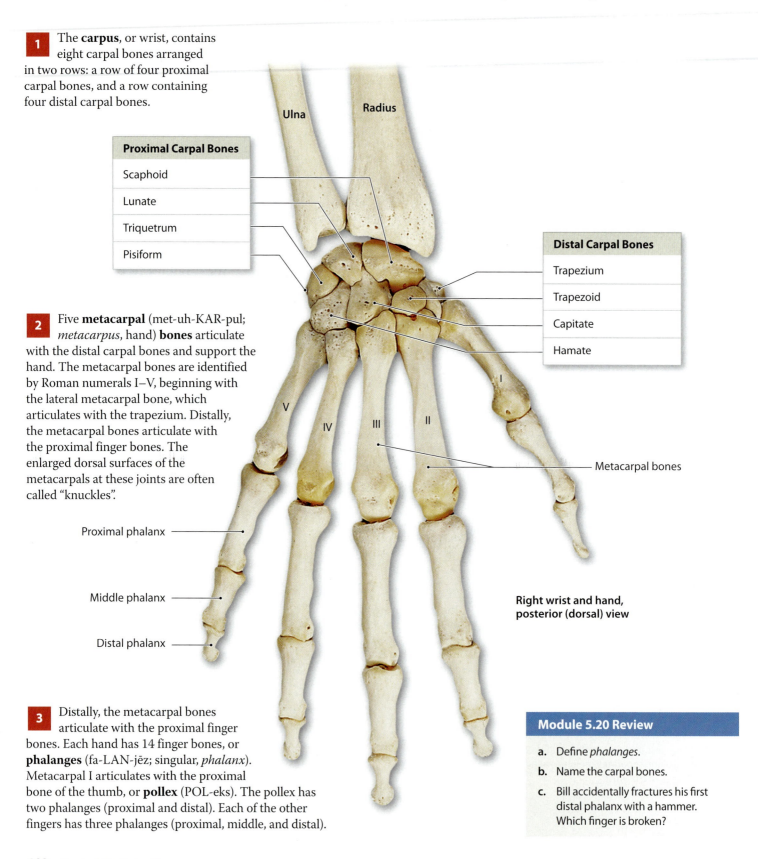

Ulna

Radius

Proximal Carpal Bones
Scaphoid
Lunate
Triquetrum
Pisiform

Distal Carpal Bones
Trapezium
Trapezoid
Capitate
Hamate

Metacarpal bones

Proximal phalanx

Middle phalanx

Distal phalanx

Right wrist and hand, posterior (dorsal) view

Module 5.20 Review

a. Define *phalanges*.

b. Name the carpal bones.

c. Bill accidentally fractures his first distal phalanx with a hammer. Which finger is broken?

5.21

The hip bone forms by the fusion of the ilium, ischium, and pubis

1 The **pelvic girdle** consists of the paired **hip bones**, which are also called the coxal bones. Each hip bone forms by the fusion of three bones: an **ilium** (IL-ē-um; plural, *ilia*), an **ischium** (IS-kē-um; plural, *ischia*), and a **pubis** (PŪ-bis).

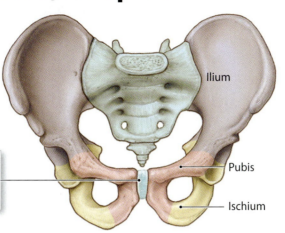

Ilium

At the **pubic symphysis,** the right and left pubic bones are attached to a median fibrocartilage pad.

Pubis

Ischium

2 In lateral view, the ilium dominates the hip bone. Landmarks along the margin of the ilium include the **iliac spines**, which mark the attachment sites of important muscles and ligaments; the **gluteal lines**, which mark the attachment of large hip muscles; and the **greater sciatic** (sī-AT-ik) **notch**, through which a major nerve (the sciatic nerve) reaches the lower limb.

3 A medial view shows the roughened articular surfaces that stabilize the pelvis and many additional features of the ischium and pubis.

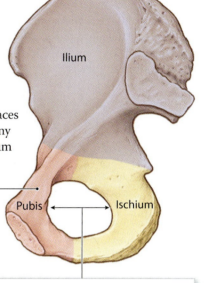

Ilium

Superior pubic ramus

Pubis

Ischium

ANTERIOR

The **obturator** (OB-too-rā-tor) **foramen** is a space that is closed by a sheet of collagen fibers. It provides a firm base for the attachment of muscles of the hip. This foramen is bounded by the **ischial ramus**, the **inferior pubic ramus**, and the **superior pubic ramus**.

The **iliac crest** is an important ridge for muscle attachment.

Anterior superior iliac spine

Gluteal lines

ANTERIOR

Ilium

Posterior superior iliac spine

Posterior inferior iliac spine

Greater sciatic notch

The prominent **ischial spine** projects inferior to the greater sciatic notch.

The **acetabulum** (as-e-TAB-ū-lum; *acetabulum*, vinegar cup), a concave socket, articulates with the head of the femur. The ilium, ischium, and pubis meet inside the acetabulum, in an arrangement resembling a pie sliced into three pieces.

Ischium

Pubis

Ischial ramus

The **ischial tuberosity,** a roughened projection, bears the body's weight when you are seated.

Module 5.21 Review

a. Describe the acetabulum.

b. Which three bones fuse to make up the hip bone?

c. When you are seated, which part of the hip bone bears your body's weight?

The pelvis consists of the two hip bones plus the sacrum and the coccyx

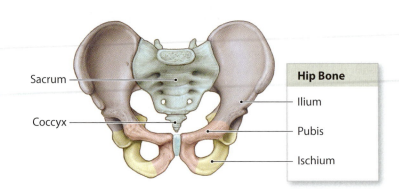

Hip Bone
Ilium
Pubis
Ischium

Sacrum
Coccyx

1 The **pelvis** consists of the two hip bones (coxal bones), the sacrum, and the coccyx. An extensive network of ligaments connects the lateral borders of the sacrum with the iliac crest, the ischial tuberosity, and the ischial spine. Other ligaments tie the ilia to the posterior lumbar vertebrae. These interconnections increase the stability of the pelvis.

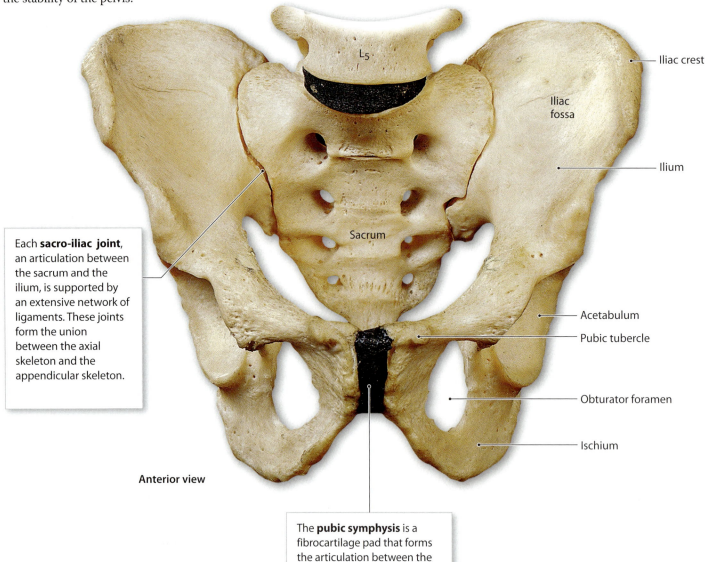

Each **sacro-iliac joint**, an articulation between the sacrum and the ilium, is supported by an extensive network of ligaments. These joints form the union between the axial skeleton and the appendicular skeleton.

The **pubic symphysis** is a fibrocartilage pad that forms the articulation between the two pubic bones.

L₅
Iliac crest
Iliac fossa
Ilium
Sacrum
Acetabulum
Pubic tubercle
Obturator foramen
Ischium

Anterior view

2 The pelvis may be divided into the **false** (greater) **pelvis** and the **true** (lesser) **pelvis.** The false pelvis encloses organs within the inferior portion of the abdominal cavity. The true pelvis (shaded in purple) encloses the pelvic cavity, a subdivision of the abdominopelvic cavity.

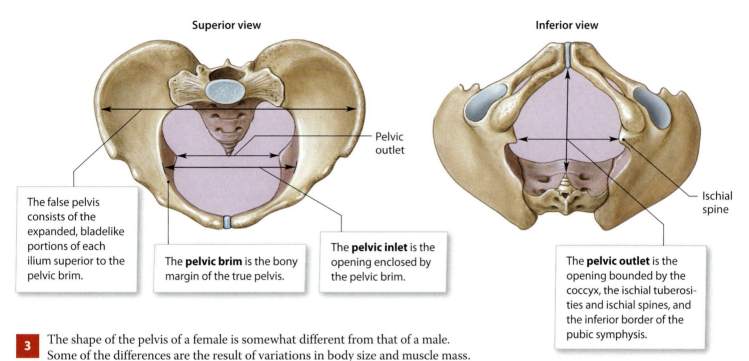

Superior view

Inferior view

Pelvic outlet

The false pelvis consists of the expanded, bladelike portions of each ilium superior to the pelvic brim.

The **pelvic brim** is the bony margin of the true pelvis.

The **pelvic inlet** is the opening enclosed by the pelvic brim.

Ischial spine

The **pelvic outlet** is the opening bounded by the coccyx, the ischial tuberosities and ischial spines, and the inferior border of the pubic symphysis.

3 The shape of the pelvis of a female is somewhat different from that of a male. Some of the differences are the result of variations in body size and muscle mass. For example, in females, the pelvis is generally smoother and lighter and has less prominent markings. Females have other skeletal adaptations for childbearing, including:

- An enlarged pelvic outlet.
- A broader pubic angle (the inferior angle between the pubic bones), greater than 100°.
- Less curvature on the sacrum and coccyx, which, in males, arcs into the pelvic outlet.
- A wider, more circular pelvic inlet.
- A relatively broad pelvis that does not extend as far superiorly (a "low pelvis").
- Ilia that project farther laterally but do not extend as far superior to the sacrum.

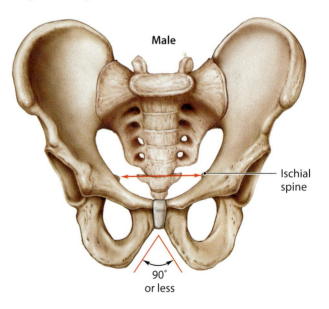

Male

Ischial spine

90° or less

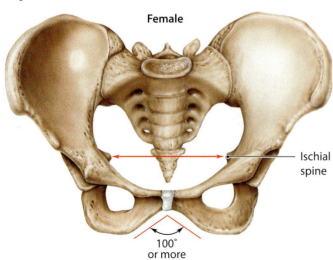

Female

Ischial spine

100° or more

The femur articulates with the patella and tibia

The skeleton of each lower limb consists of a **femur** (thigh), a **patella** (kneecap), a **tibia** and a **fibula** (leg), and the **tarsal bones**, **metatarsal bones**, and **phalanges** of the foot. Once again, anatomical terminology differs from common usage. In anatomical terms, "leg" refers only to the distal portion of the limb, not to the entire lower limb. Thus, we will use thigh and leg, rather than upper leg and lower leg. The functional anatomy of the lower limbs differs from that of the upper limbs, mostly because the lower limbs transfer the body weight to the ground.

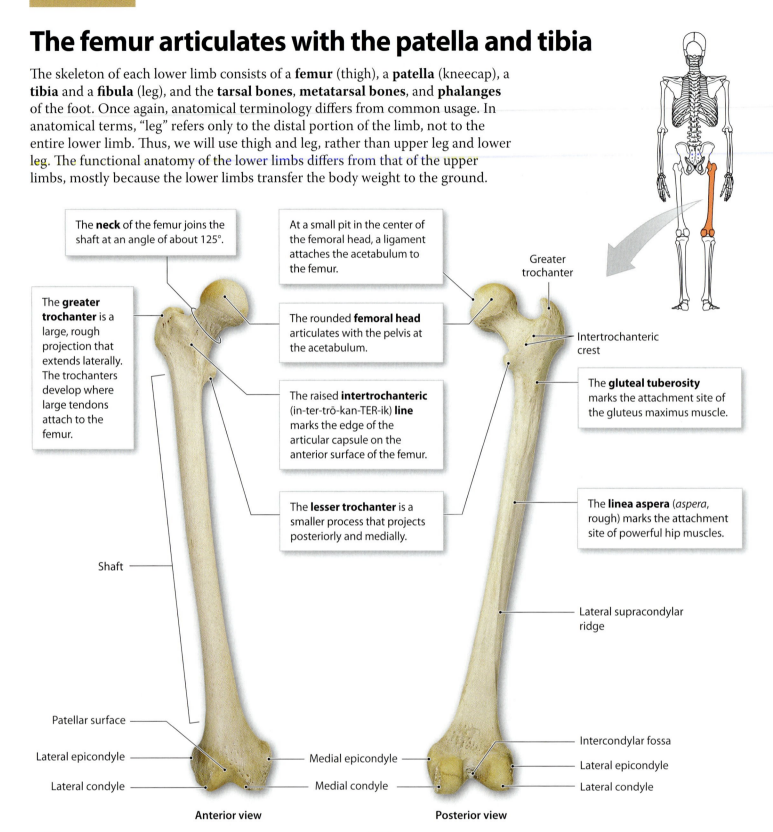

The **neck** of the femur joins the shaft at an angle of about 125°.

At a small pit in the center of the femoral head, a ligament attaches the acetabulum to the femur.

Greater trochanter

The **greater trochanter** is a large, rough projection that extends laterally. The trochanters develop where large tendons attach to the femur.

The rounded **femoral head** articulates with the pelvis at the acetabulum.

Intertrochanteric crest

The **gluteal tuberosity** marks the attachment site of the gluteus maximus muscle.

The raised **intertrochanteric** (in-ter-trō-kan-TER-ik) **line** marks the edge of the articular capsule on the anterior surface of the femur.

The **lesser trochanter** is a smaller process that projects posteriorly and medially.

The **linea aspera** (*aspera*, rough) marks the attachment site of powerful hip muscles.

Shaft

Lateral supracondylar ridge

Patellar surface

Lateral epicondyle — — Medial epicondyle

Intercondylar fossa

Lateral epicondyle

Lateral condyle — — Medial condyle

Lateral condyle

Anterior view　　　　**Posterior view**

1 The femur is the longest and heaviest bone in the body. It articulates with the hip bone at the hip joint and with the tibia of the leg at the knee joint. Major landmarks are shown here on the anterior and posterior surfaces of the right femur.

2 At the distal end of the femur, the **medial** and **lateral condyles** participate in the knee joint. On the anterior and inferior surfaces, the two condyles are separated by the **patellar surface**, a smooth articular surface over which the patella glides. On the posterior surface, the medial and lateral condyles are separated by a deep **intercondylar fossa** that does not extend onto the anterior surface.

3 The **tibia** (TIB-ē-uh), or shinbone, is the large medial bone of the leg. At the proximal end of the tibia, the **medial** and **lateral tibial condyles** articulate with the medial and lateral condyles of the femur. The slender **fibula** (FIB-ū-luh) parallels the lateral border of the tibia but does not participate in the knee joint and bears no weight. However, the fibula is important as a site for the attachment of muscles that move the foot and toes. In addition, the distal tip of the fibula extends lateral to the ankle, providing important stability to that joint.

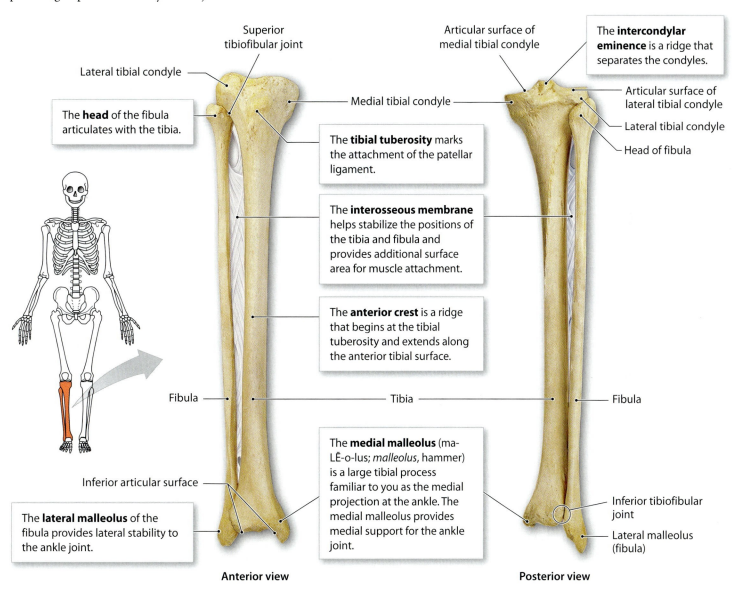

Superior tibiofibular joint

Lateral tibial condyle

The **head** of the fibula articulates with the tibia.

Medial tibial condyle

The **tibial tuberosity** marks the attachment of the patellar ligament.

The **interosseous membrane** helps stabilize the positions of the tibia and fibula and provides additional surface area for muscle attachment.

The **anterior crest** is a ridge that begins at the tibial tuberosity and extends along the anterior tibial surface.

Fibula — Tibia — Fibula

Inferior articular surface

The **lateral malleolus** of the fibula provides lateral stability to the ankle joint.

The **medial malleolus** (ma-LĒ-o-lus; *malleolus*, hammer) is a large tibial process familiar to you as the medial projection at the ankle. The medial malleolus provides medial support for the ankle joint.

Anterior view

Articular surface of medial tibial condyle

The **intercondylar eminence** is a ridge that separates the condyles.

Articular surface of lateral tibial condyle

Lateral tibial condyle

Head of fibula

Inferior tibiofibular joint

Lateral malleolus (fibula)

Posterior view

Module 5.23 Review

a. Identify the bones of the lower limb.

b. Which structure articulates with the acetabulum?

c. The fibula neither participates in the knee joint nor bears weight. Yet, when it is fractured, walking becomes difficult. Why?

The ankle and foot contain tarsal bones, metatarsal bones, and phalanges

1 The bones of the **ankle** transfer the body weight from the leg to the ground by distributing it through the bones of the foot. The combination of ankle and foot must be strong enough yet flexible enough to deal with the changes in loading that occur during walking, running, and jumping.

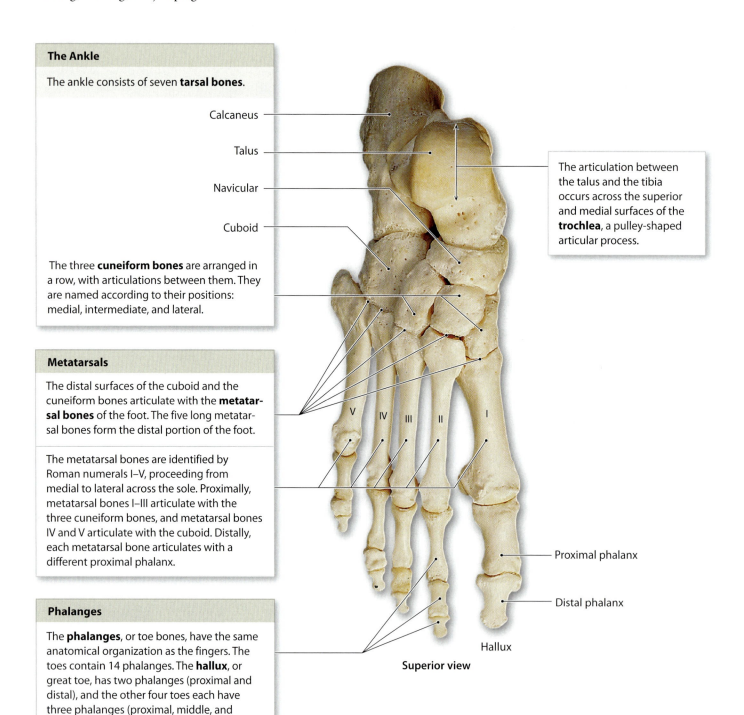

The Ankle

The ankle consists of seven **tarsal bones**.

Calcaneus

Talus

Navicular

Cuboid

The three **cuneiform bones** are arranged in a row, with articulations between them. They are named according to their positions: medial, intermediate, and lateral.

The articulation between the talus and the tibia occurs across the superior and medial surfaces of the **trochlea**, a pulley-shaped articular process.

Metatarsals

The distal surfaces of the cuboid and the cuneiform bones articulate with the **metatarsal bones** of the foot. The five long metatarsal bones form the distal portion of the foot.

The metatarsal bones are identified by Roman numerals I–V, proceeding from medial to lateral across the sole. Proximally, metatarsal bones I–III articulate with the three cuneiform bones, and metatarsal bones IV and V articulate with the cuboid. Distally, each metatarsal bone articulates with a different proximal phalanx.

V   IV   III   II   I

Proximal phalanx

Distal phalanx

Hallux

Superior view

Phalanges

The **phalanges**, or toe bones, have the same anatomical organization as the fingers. The toes contain 14 phalanges. The **hallux**, or great toe, has two phalanges (proximal and distal), and the other four toes each have three phalanges (proximal, middle, and distal).

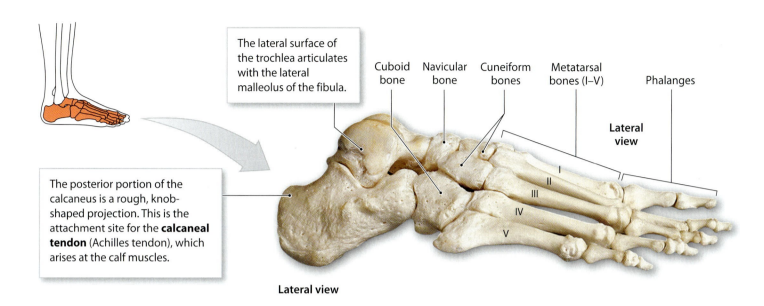

The lateral surface of the trochlea articulates with the lateral malleolus of the fibula.

Cuboid bone

Navicular bone

Cuneiform bones

Metatarsal bones (I–V)

Phalanges

Lateral view

I
II
III
IV
V

The posterior portion of the calcaneus is a rough, knob-shaped projection. This is the attachment site for the **calcaneal tendon** (Achilles tendon), which arises at the calf muscles.

Lateral view

Phalanges

Metatarsal bones

Medial cuneiform bone

Navicular bone

Talus

Medial view

I

Calcaneus

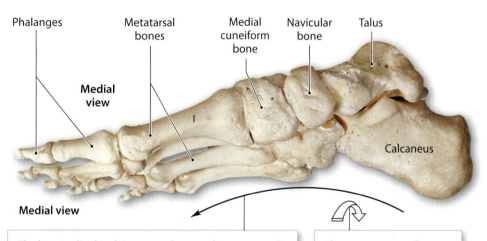

Medial view

2 Weight transfer occurs along the **longitudinal arch** of the foot. The amount of weight transferred forward depends on the position of the foot and the placement of your body weight. When you "dig in your heels" all your body weight rests on the calcaneus, but when you are on tiptoes all your weight is transferred to the metatarsal bones and phalanges. When you stand normally, your body weight is distributed evenly between the calcaneus and the distal ends of the metatarsal bones.

The longitudinal arch is present because ligaments and tendons connect the calcaneus to the distal portions of the metatarsal bones. However, the lateral (calcaneal) portion of the longitudinal arch has much less curvature than the medial (talar) portion, in part because the talar portion has considerably more elasticity. As a result, the medial plantar surface of the foot remains elevated, so that the muscles, nerves, and blood vessels that supply the inferior surface are not squeezed between the metatarsal bones and the ground.

The **transverse arch** exists because the degree of longitudinal curvature changes from the medial border to the lateral border of the foot.

The arches of the foot are usually present at birth. Sometimes, however, they don't develop properly. In **congenital talipes equinovarus** (clubfoot), abnormal muscle development distorts growing bones and joints. One or both feet may be involved, and the condition can be mild, moderate, or severe. In most cases, the tibia, ankle, and foot are affected; the longitudinal arch is exaggerated, and the feet are turned medially and inverted. If both feet are involved, the soles face one another. This condition, which affects 1 in 1000 births, is about twice as common in boys as girls. Prompt treatment with casts or other supports in infancy helps alleviate the problem, and fewer than half the cases require surgery.

Module 5.24 Review

a. Identify the tarsal bones.

b. Which foot bone transmits the weight of the body from the tibia toward the toes?

c. While jumping off the back steps at his house, 10-year old Joey lands on his right heel and breaks his foot. Which foot bone is most likely broken?

1. Concept map

Use each of the following terms once to fill in the blank boxes to correctly complete the skeleton concept map.

- floating ribs
- temporal
- mandible
- axial
- hyoid
- sacral
- vertebral column
- lacrimal
- xiphoid process
- occipital
- sternum
- skull
- thoracic
- longitudinal
- lumbar

Skeleton

a _____ division

forms the

b _____ axis of body

subdivisions

c _____

consists of

Face	Cranium	Associated bones
bones	*bones*	

Face *bones*:
- Zygomatic (2)
- Maxilla (2)
- **d** _____ (1)
- Nasal (2)
- Palatine (2)
- **e** _____ (2)
- Vomer (1)
- Inferior nasal conchae (2)

Cranium *bones*:
- **f** _____ (1)
- Parietal (2)
- Frontal (1)
- **g** _____ (2)
- Sphenoid (1)
- Ethmoid (1)

Associated bones:
- Auditory ossicles (6)
- **h** _____ (1)

i _____

consists of

Vertebrae

types
- Cervical (7)
- **j** _____ (12)
- **k** _____ (5)
- **l** _____ (1)
- Coccyx (1)

Ribs

types
- True ribs (7 pr)
- False ribs (5 pr)
- **m** _____ (2 pr)

n _____

parts

Manubrium

connected to

Body

connected to

o _____

2. Labeling

Label the bones of the appendicular skeleton in the diagram at right.

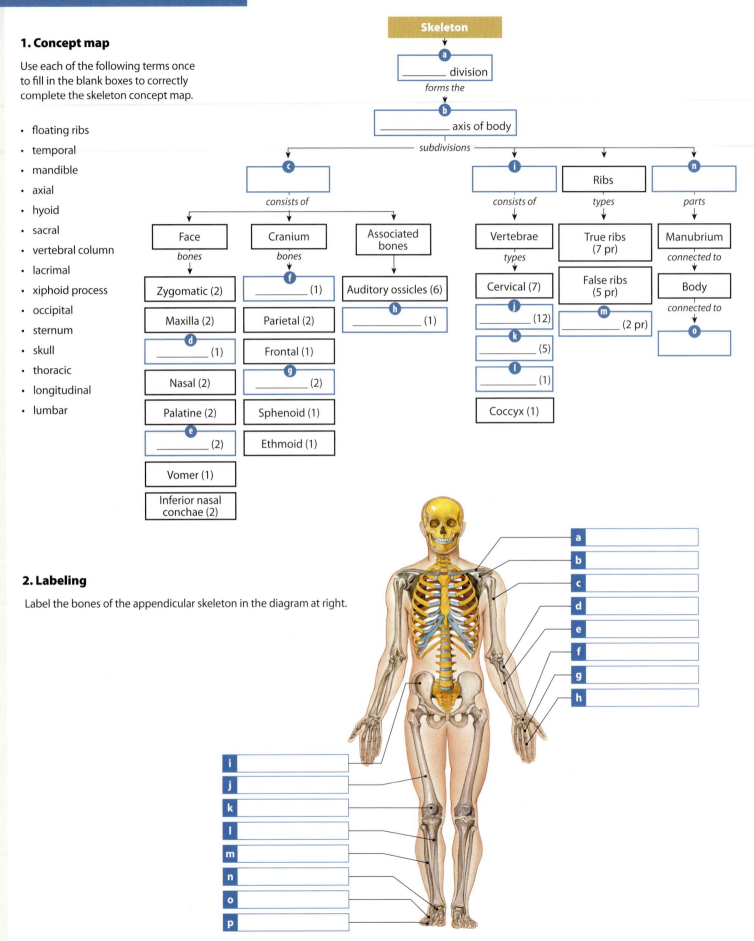

a _____
b _____
c _____
d _____
e _____
f _____
g _____
h _____

i _____
j _____
k _____
l _____
m _____
n _____
o _____
p _____

Joint Structure and Movement

Because the bones of the skeleton are relatively inflexible, movements can occur only at **articulations**, or joints, where two bones interconnect. The anatomical structure of a joint determines the type and amount of movement that may occur. Each joint reflects a compromise between the need for strength and the need for mobility. As a result, articulations differ in the amount of movement permitted, and this property is known as **range of motion (ROM)**. Articulations are often categorized by their range of motion, with subgroups based on anatomical structure.

Functional and Structural Classifications of Articulations

Functional Category (degree of movement)	Structural Category (connection method) and Type		Description	Example
Synarthrosis (no movement) At a synarthrosis, the bony edges are quite close together and may even interlock. These extremely strong joints are located where movement between the bones must be prevented.	**Fibrous** Suture		A **suture** (*sutura*, a sewing together) is a fibrous (dense connective tissue) connection plus interlocked surfaces.	Between the bones of the skull
	Gomphosis		A **gomphosis** (gom-FŌ-sis; *gomphosis*, a bolting together) is a fibrous connection plus insertion in a bony socket.	Between the teeth and jaws
	Cartilaginous Synchondrosis		A **synchondrosis** (sin-kon-DRŌ-sis; *syn*, together + *chondros*, cartilage) is an interposition of a cartilage plate.	Between the first rib and sternum
Amphiarthrosis (little movement) An amphiarthrosis permits more movement than a synarthrosis, but is much stronger than a freely movable joint. The articulating bones are connected by collagen fibers or cartilage.	**Fibrous** Syndesmosis		At a **syndesmosis** (sin-dez-MŌ-sis; *desmos*, a band or ligament), bones are connected by a ligament.	Between the tibia and fibula
	Cartilaginous Symphysis		At a **symphysis**, the articulating bones are connected by a wedge or pad of fibrocartilage.	Between the right and left halves of the pelvis
Diarthrosis (free movement)	**Synovial**		**Diarthroses**, or **synovial** (si-NŌ-ve-ul) **joints**, are complex joints bounded by joint capsules and containing synovial fluid.	Numerous; subdivided by range of motion

Synarthrotic and amphiarthrotic joints are simple in structure, with direct connections between the articulating bones. Diarthrotic joints are quite complex in structure, and they have the greatest range of motion. In this section, we will consider the structure and function of synovial joints.

Synovial joints (diarthroses) are freely movable and contain synovial fluid

In this module we discuss the general structure of a synovial joint. We focus our attention on the key components of synovial joints: the articular cartilages and accessory structures.

1 Under normal conditions, the opposing bony surfaces within a synovial joint cannot contact one another, because these surfaces are covered by special articular cartilages, which are slick and smooth. This feature alone can reduce friction during movement at the joint. However, even when pressure is applied across a joint, the smooth articular cartilages do not touch one another, because they are separated by a thin film of synovial fluid within the joint cavity.

Components of Synovial Joints

Articular cartilages resemble hyaline cartilages elsewhere in the body. However, articular cartilages have no perichondrium, and the matrix contains more water than that of other cartilages.

The **joint capsule**, or articular capsule, is dense and fibrous, and it may be reinforced with various accessory structures such as tendons or ligaments.

Synovial fluid within the joint cavity lubricates, cushions shocks, prevents abrasion, and supports the chondrocytes of the articular cartilages.

Marrow cavity

The periosteum of each bone is continuous with the capsule of the joint. This adds strength and helps to stabilize the joint.

Synovial membrane

Spongy bone of epiphysis

Compact bone

2 This table summarizes the major functions of synovial fluid. Synovial fluid is produced by the synovial membrane that lines the joint cavity. During normal movement, synovial fluid circulates from the areolar tissue into the joint cavity and percolates through the articular cartilages, providing oxygen and nutrients to the chondrocytes and carrying away their metabolic wastes.

Functions of Synovial Fluid

- **Lubrication**. When part of an articular cartilage is compressed during movement, some of the synovial fluid is squeezed out of the cartilage and into the space between the opposing surfaces. This thin layer of fluid markedly reduces friction between moving surfaces, just as a thin film of water reduces friction between a car's tires and a highway.

- **Nutrient Distribution**. The synovial fluid in a joint must circulate continuously to provide nutrients and waste disposal for the chondrocytes of the articular cartilages. It circulates whenever the joint moves, and the compression and reexpansion of the articular cartilages pump synovial fluid into and out of the cartilage matrix.

- **Shock Absorption**. Synovial fluid cushions shocks in joints that are compressed. For example, when you jog, your knees are severely compressed and the synovial fluid distributes that force evenly across the articular surfaces and outward to the joint capsule.

3 In complex synovial joints, such as the knee, a variety of accessory structures provide support and additional stability. Many of these accessory structures can be seen in this diagrammatic view of a sagittal section of the knee joint.

The tendon of the quadriceps muscles attaches to the base of the patella. Although not part of the articulation itself, tendons passing across or around a joint can limit the joint's range of motion and provide mechanical support for it.

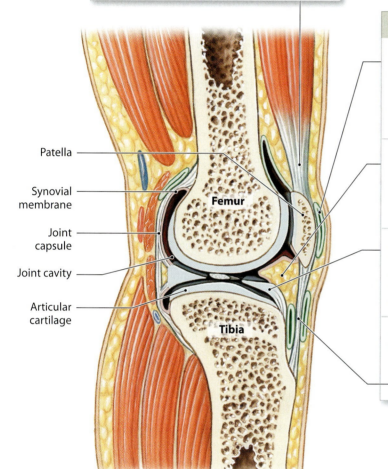

Patella

Synovial membrane

Joint capsule

Joint cavity

Articular cartilage

Femur

Tibia

Accessory Structures Supporting the Knee

A **bursa** (BUR-sa; a pouch; plural, *bursae*) is a small, fluid-filled pocket that forms in connective tissue. It contains synovial fluid and is lined by a synovial membrane. Bursae often form where a tendon or ligament rubs against other tissues. Located around most synovial joints, including the knee joint, bursae reduce friction and act as shock absorbers.

Fat pads are localized masses of adipose tissue covered by a layer of synovial membrane. They are commonly superficial to the joint capsule. Fat pads protect the articular cartilages and act as packing material for the joint. When the bones move, the fat pads fill in the spaces created as the joint cavity changes shape.

A **meniscus** (me-NIS-kus; a crescent; plural, *menisci*) is a pad of fibrocartilage between opposing bones within a synovial joint. Menisci, or articular discs, may subdivide a synovial cavity, channel the flow of synovial fluid, or allow for variations in the shapes of the articular surfaces.

Accessory ligaments support, strengthen, and reinforce synovial joints.

Patellar ligament

A joint cannot be both highly mobile and very strong. The greater the range of motion at a joint, the weaker it becomes. A synarthrosis, the strongest type of joint, permits no movement, whereas a diarthrosis, such as the shoulder, is far weaker but permits a broad range of motion. Any diarthrosis will be damaged by movement beyond its normal range of motion. When reinforcing structures cannot protect a joint from extreme stresses, a **dislocation** results. In a dislocation, the articulating surfaces are forced out of position. The displacement can damage the articular cartilages, tear ligaments, or distort the joint capsule. Although the inside of a joint has no pain receptors, nerves that monitor the capsule, ligaments, and tendons are quite sensitive, so dislocations are very painful.

Module 5.25 Review

a. Define dislocation.

b. Describe the components of a synovial joint, and identify the function of each.

c. Why would improper circulation of synovial fluid lead to the degeneration of articular cartilages in the affected joint?

Anatomical organization determines the functional properties of synovial joints

An accurate description of the functions of a joint includes terms that indicate the types of motion permitted. This visual summary gives representative examples of the various anatomical classes of synovial joints based on the shapes of the articulating surfaces. It relates articular structure to simplified joint models and lists examples of each group of joints.

Types of Synovial Joints	Models of Joint Motion	Examples
Gliding joint — Permits a sliding motion in any direction on a relatively flat surface	Clavicle, Manubrium	• Acromioclavicular and sternoclavicular joints • Intercarpal and intertarsal joints • Vertebrocostal joints • Sacro-iliac joints
Hinge joint — Permits movement in one plane only, like the hinge on a door	Humerus, Ulna	• Elbow joints • Knee joints • Ankle joints • Interphalangeal joints
Pivot joint — Permits rotation around a fixed axis	Atlas, Axis	• Atlas/axis • Proximal radio-ulnar joints
Condylar joint — Permits movement in two planes, but prevents rotation	Scaphoid bone, Ulna, Radius	• Radiocarpal joints • Metacarpophalangeal joints 2–5 • Metatarsophalangeal joints
Saddle joint — Permits more extensive motion in two planes, while preventing rotation	III II, Metacarpal bone of thumb, Trapezium	• First carpometacarpal joints
Ball-and-socket joint — Permits movements in multiple directions plus rotation	Scapula, Humerus	• Shoulder joints • Hip joints

Module 5.26 Review

a. Identify the types of synovial joints.

b. What type of synovial joint permits the widest range of motion?

c. Indicate the type of synovial joint for each of the following: shoulder, elbow, and thumb.

Adjacent vertebrae are separated by intervertebral discs that are compressed by the weight of the trunk

From axis to sacrum, the bodies of adjacent vertebrae are separated and cushioned by pads of fibrocartilage called **intervertebral discs**. Each disc consists of a fibrous pad with a gelatinous core that acts like a shock absorber.

1 Extreme stresses may distort an intervertebral disc, forcing it partway into the vertebral canal. This condition, seen here in lateral view, is called a **bulging disc**. The tough outer layer of cartilage is actually bulging, and this is considered a normal part of aging.

2 If the disc projects posteriorly into the vertebral canal, spinal nerves are distorted, producing pain and potentially causing problems with sensation and muscle control. This condition, called a **herniated disc**, shows the gelatinous core breaking through the outer fibrocartilage layer.

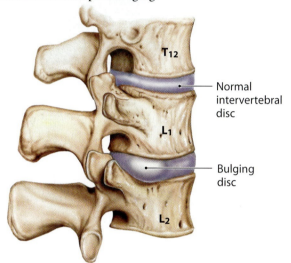

Normal intervertebral disc

Bulging disc

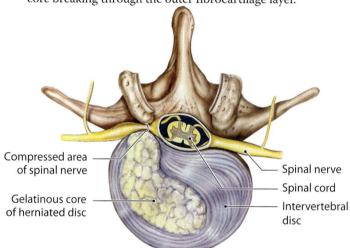

Compressed area of spinal nerve

Gelatinous core of herniated disc

Spinal nerve

Spinal cord

Intervertebral disc

3 The bones of the skeleton become thinner and weaker as a normal part of the aging process. This reduction in bone mass begins between the ages of 30 and 40 as osteoblast activity begins to decline, while osteoclast activity continues at previous levels. Thereafter, women lose roughly 8 percent of their skeletal mass every decade; men lose roughly 3 percent per decade. When the reduction in bone mass begins to compromise normal function, the condition is known as **osteoporosis** (os-tē-ō-po-RŌ-sis; *porosus*, porous). The combination of osteoporosis and the reduction in the cushioning properties of the intervertebral discs makes vertebral fractures a common problem among the elderly.

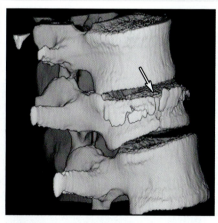

Clinical scan of a compression fracture in a lumbar vertebra

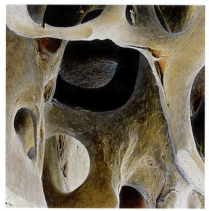

Normal spongy bone   SEM × 25

Spongy bone with osteoporosis   SEM × 21

Module 5.27 Review

a. What is the function of an intervertebral disc?

b. Compare a bulging disc and a herniated disc.

c. What is osteoporosis?

Arthritis can disrupt normal joint structure and function

Joints are subjected to heavy wear and tear throughout our lifetimes, and problems with joint function are relatively common, especially as we age. **Rheumatism** (ROO-muh-tiz-um) is a general term that indicates pain and stiffness affecting bones, muscles, or both. Several major forms of rheumatism exist. **Arthritis** (ar-THRĪ-tis) encompasses all the rheumatic diseases that affect synovial joints. Arthritis always involves damage to the articular cartilages, but the specific cause can vary. **Osteoarthritis** (os-tē-ō-ar-THRĪ-tis) generally affects individuals age 60 or older. Osteoarthritis can result from the cumulative effects of wear and tear at joint surfaces or from genetic factors affecting collagen formation. In the U.S. population, 25 percent of women and 15 percent of men over age 60 show signs of this disease.

Normal Joint

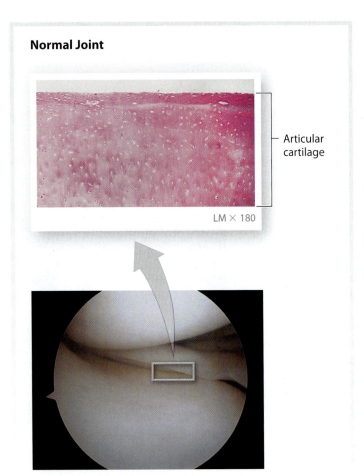

Articular cartilage

LM × 180

Arthroscopic view of normal cartilage

1 This is a normal articular cartilage. The cartilage is thick and the surface smooth and slick.

Arthritic Joint

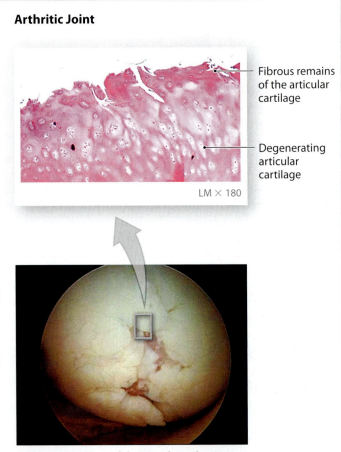

Fibrous remains of the articular cartilage

Degenerating articular cartilage

LM × 180

Arthroscopic view of damaged cartilage

2 This is an articular cartilage damaged by osteoarthritis. The exposed surfaces change from a slick, smooth-gliding surface to a surface composed of a rough feltwork of bristly collagen fibers. Such a change drastically increases friction at the joint, which then promotes further degeneration of the articular cartilage.

3 There are several options available that enable clinicians to examine the structure of problematic joints. This is an arthroscopic view of the interior of the left knee, showing injuries to the ligaments. An **arthroscope** uses fiber optics within a narrow tube to explore a joint without major surgery. Optical fibers are thin threads of glass or plastic that conduct light. The fibers can be bent around corners, so they can be introduced into a knee or other joint and moved around, enabling the physician to see what is going on inside the joint. If necessary, a second small incision can be made to insert flexible instruments that permit surgery inside the joint, within view of the arthroscope. This procedure, called **arthroscopic surgery**, has greatly improved the treatment of knee and other joint injuries.

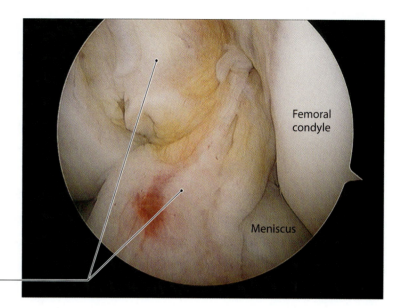

Femoral
condyle

Meniscus

Ligaments within
the joint cavity

4 Artificial joints such as these may be the method of last resort when regular exercise, physical therapy, and anti-inflammatory drugs (such as aspirin) fail to slow the progress of arthritis. Artificial joints can restore mobility and relieve pain. Unfortunately, artificial joints typically have a service life of only about 10 years, and replacing an artificial joint in a patient 80–90 years old is a very stressful procedure.

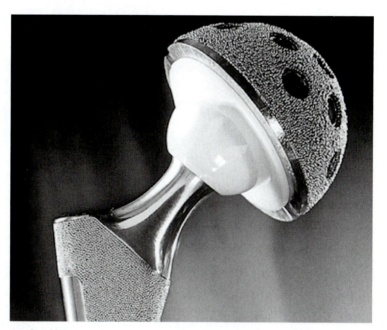

Artificial hip

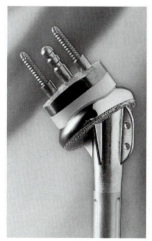

Artificial shoulder

Artificial knee

Module 5.28 Review

a. Compare rheumatism to osteoarthritis.

b. Explain the use of an arthroscope.

c. What can a person do to slow the progression of arthritis?

1. Matching

Match the following terms with the most closely related description.

- amphiarthrosis
- arthritis
- synovial fluid
- synarthrosis
- fluid-filled pouch
- diarthrosis
- fibrocartilage discs
- hip joint
- osteoporosis
- articular cartilages

a	_____	Freely movable joint
b	_____	Lacks a perichondrium
c	_____	Ball and socket
d	_____	Menisci
e	_____	Immovable joint
f	_____	Reduced bone mass
g	_____	Bursa
h	_____	Slightly movable joint
i	_____	Articular cartilage damage
j	_____	Provides lubrication and nutrients

2. Labeling

Label the structures in the synovial joint figure at right.

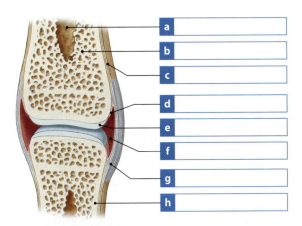

a	
b	
c	
d	
e	
f	
g	
h	

3. Short Answer

Identify the type of synovial joints represented by each model of joint motion.

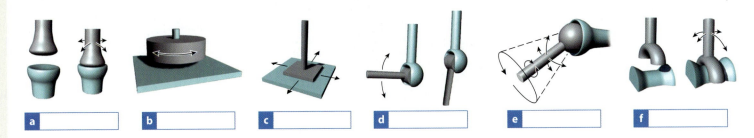

a	
b	
c	
d	
e	
f	

4. Short Answer

Why are vertebral fractures a common problem in the elderly?

_____

_____

Visual Outline with Key Terms

Summarize the content of each module using the terms in the order provided.

SECTION 1

An Introduction to the Bones of the Skeletal System

- axial skeleton
- appendicular skeleton

5.1

Bones are classified according to shape and structure, and have a variety of surface features

- flat bones
- long bones
- irregular bones
- short bones
- surface features
- canal (meatus)
- process
- sinus
- foramen
- fissure
- head
- tubercle
- sulcus
- tuberosity
- diaphysis
- trochlea
- condyle
- trochanter
- neck
- facet
- crest
- fossa
- line
- spine
- ramus

5.2

Long bones have a rich blood supply

- epiphysis
- diaphysis
- marrow cavity
- red bone marrow
- yellow bone marrow
- compact bone
- articular cartilage
- spongy bone
- periosteum

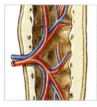

5.3

Bone has a calcified matrix associated with osteocytes, osteoblasts, and osteoclasts

- osteocytes
- lamellae
- lacuna
- canaliculi
- osteoblasts
- ossification
- osteoclasts
- osteolysis
- osteon
- central canal
- lacunae
- calcium

5.4

Compact bone consists of parallel osteons, and spongy bone consists of a network of trabeculae

- central canal
- trabeculae
- appositional growth
- perforating fibers
- endosteum

5.5

The most common method of bone formation involves the replacement of cartilage with bone

- endochondral ossification
- ossification center
- epiphyseal cartilage
- intramembranous ossification
- membrane bones
- epiphyseal line

5.6

Abnormalities of bone growth and development produce recognizable physical signs

- pituitary growth failure
- achondroplasia
- Marfan's syndrome
- gigantism
- fibrodysplasia ossificans progressiva (FOP)
- heterotopic (ectopic) bones
- acromegaly

5.7

A fracture is a crack or a break in a bone

- fracture
- fracture hematoma
- internal callus
- external callus
- closed (simple) fractures
- open (compound) fractures
- transverse fractures
- spiral fractures
- displaced fractures
- nondisplaced fractures
- compression fractures
- greenstick fracture
- comminuted fractures
- epiphyseal fractures
- Pott's fracture
- Colles fracture

● = *Term boldfaced in this module*

199

SECTION 2

The Skeleton

- axial skeleton
  - skull and associated bones
  - thoracic cage
  - vertebral column
- appendicular skeleton
  - pectoral girdle
  - upper limbs
  - pelvic girdle
  - lower limbs

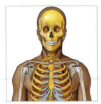

5.8

Facial bones dominate the anterior skull, and cranial bones dominate the posterior surface

- facial bones
- nasal bones
- lacrimal bones
- palatine bones
- zygomatic bones
- maxillae
- inferior nasal conchae
- vomer
- mandible
- cranial bones
- cranium
- frontal bone
- sphenoid
- ethmoid
- parietal bones
- occipital bone
- temporal bones
- external occipital crest
- sutures

- coronal suture
- sagittal suture
- squamous suture
- lambdoid suture

5.9

Surface features of the skull are functional landmarks

- frontal squama
- superior temporal lines
- inferior temporal lines
- squamous part
- external acoustic meatus
- mastoid process
- condylar process
- styloid process
- coronoid process
- zygomatic process
- zygomatic arch
- mental protuberance
- alveolar processes
- mandibular fossa
- occipital condyles
- foramen ovale
- carotid canal
- jugular foramen
- foramen magnum

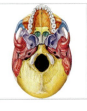

5.10

Additional landmarks are visible in sectional views of the skull

- crista galli
- cribiform plate
- sella turcica
- internal acoustic meatus
- nasal complex
- paranasal sinuses
- frontal sinuses
- ethmoidal air cells
- maxillary sinuses
- sphenoidal sinuses
- superior nasal conchae
- middle nasal conchae
- inferior nasal conchae

5.11

The associated bones of the skull perform specialized functions

- auditory ossicles
- hyoid bone
- greater horns
- lesser horns

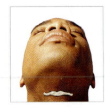

5.12

Fontanelles permit cranial growth in infants and small children

- fontanelles
- sphenoidal fontanelles
- mastoid fontanelles
- anterior fontanelle
- occipital fontanelle

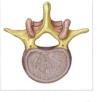

5.13

The vertebral column has four spinal curves, and vertebrae have both anatomical similarities and regional differences

- cervical curve
- thoracic curve
- lumbar curve
- sacral curve
- articular processes
- vertebral arch
- vertebral body
- vertebral foramen
- spinous process
- laminae
- transverse processes
- pedicles
- intervertebral discs
- intervertebral foramina
- vertebral canal
- articular facet
- superior articular processes
- inferior articular processes

5.14

There are 7 cervical vertebrae and 12 thoracic vertebrae

- cervical vertebrae
- transverse foramen
- bifid
- atlas
- axis
- dens
- thoracic vertebrae
- costal facets

5.15

There are five lumbar vertebrae, and the sacrum and coccyx consist of fused vertebrae

- lumbar vertebrae
- sacrum
- ala
- sacral foramina
- base (of sacrum)
- sacral promontory
- apex (of sacrum)
- sacral canal
- superior articular processes
- median sacral crest
- sacral hiatus
- coccyx
- sacral tuberosity
- auricular surface
- lateral sacral crest

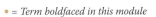
= *Term boldfaced in this module*

5.16

The thoracic cage protects organs in the chest and provides sites for muscle attachment

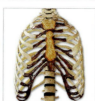

- thoracic cage
- ribs
- true ribs
- costal cartilages
- false ribs
- floating ribs
- vertebral ribs
- jugular notch
- sternum
- manubrium
- body (of sternum)
- xiphoid process

5.17

Abnormalities in the axial skeleton directly affect posture and balance

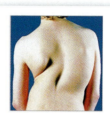

- kyphosis
- scoliosis
- lordosis

5.18

The pectoral girdles—the clavicles and scapulae—connect the upper limbs to the axial skeleton

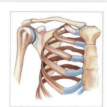

- pectoral girdle
- clavicles
- scapulae
- sternal end
- acromion
- acromial end (of clavicle)
- body (of scapula)
- superior border
- medial border
- lateral border
- superior angle
- inferior angle
- lateral angle
- subscapular fossa
- glenoid cavity
- scapular spine
- supraspinous fossa
- infraspinous fossa
- coracoid process

5.19

The humerus of the arm articulates with the radius and ulna of the forearm

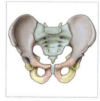

- humerus
- medial epicondyle
- lateral epicondyle
- intertubercular groove
- capitulum
- condyle
- head (of humerus)
- lesser tubercle
- anatomical neck
- surgical neck
- deltoid tuberosity
- greater tubercle
- radial groove
- trochlea
- olecranon fossa
- coronoid fossa
- ulna
- radius
- olecranon
- ulnar head
- styloid process (ulna)
- radial head
- neck (of radius)
- radial tuberosity
- interosseous membrane
- ulnar notch
- styloid process (radius)
- trochlear notch
- coronoid process
- radial notch
- proximal radio-ulnar joint
- distal radio-ulnar joint

5.20

The wrist is composed of carpal bones, and the hand consists of metacarpal bones and phalanges

- carpus
- scaphoid
- lunate
- triquetrum
- pisiform
- trapezium
- trapezoid
- capitate
- hamate
- metacarpal bones
- phalanges
- pollex

5.21

The hip bone forms by the fusion of the ilium, ischium, and pubis

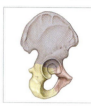

- pelvic girdle
- hip bones
- ilium
- ischium
- pubis
- pubic symphysis
- iliac spines
- gluteal lines
- greater sciatic notch
- iliac crest
- acetabulum
- ischial tuberosity
- ischial spine
- obturator foramen
- ischial ramus
- inferior pubic ramus
- superior pubic ramus

5.22

The pelvis consists of the two hip bones plus the sacrum and the coccyx

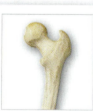

- pelvis
- sacro-iliac joint
- pubic symphysis
- false pelvis
- true pelvis
- pelvic brim
- pelvic inlet
- pelvic outlet

5.23

The femur articulates with the patella and tibia

- femur
- patella
- tibia
- fibula
- tarsal bones
- metatarsal bones
- phalanges
- neck (femur)
- greater trochanter
- femoral head
- intertrochanteric line
- lesser trochanter
- gluteal tuberosity
- linea aspera
- medial condyle
- lateral condyle
- patellar surface
- intercondylar fossa
- tibia
- medial tibial condyle
- lateral tibial condyle
- fibula
- head (fibula)
- lateral malleolus
- tibial tuberosity
- interosseous membrane
- anterior crest (tibia)
- medial malleolus
- intercondylar eminence

• = *Term boldfaced in this module*

5.24

The ankle and foot contain tarsal bones, metatarsal bones, and phalanges

- ankle
- tarsal bones
  - calcaneus
  - talus
  - navicular
  - cuboid
- cuneiform bones
- trochlea
- metatarsal bones
- phalanges
- hallux
- calcaneal tendon
- longitudinal arch
- transverse arch
- congenital talipes equinovarus

SECTION 3

Joint Structure and Movement

- articulations
- range of motion (ROM)
- synarthrosis
- suture
- gomphosis
- synchondrosis
- amphiarthrosis
- syndesmosis
- symphysis
- diarthrosis
- diarthroses (synovial joints)

5.25

Synovial joints (diarthroses) are freely movable and contain synovial fluid

- articular cartilages
- joint capsule
- synovial fluid
- bursa
- fat pads
- meniscus
- accessory ligaments
- dislocation

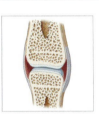

- = Term boldfaced in this module

5.26

Anatomical organization determines the functional properties of synovial joints

- gliding joint
- hinge joint
- pivot joint
- condylar joint
- saddle joint
- ball-and-socket joint

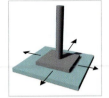

5.27

Adjacent vertebrae are separated by intervertebral discs that are compressed by the weight of the trunk

- intervertebral discs
- bulging disc
- herniated disc
- osteoporosis

5.28

Arthritis can disrupt normal joint structure and function

- rheumatism
- arthritis
- osteoarthritis
- arthroscope
- arthroscopic surgery

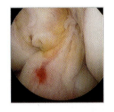

CAREER PATHS

Physical Therapist

"You not only need knowledge of anatomy, but you also need to understand the physiology of healing."

— Derrick Isa
Telesis Physical Therapy
Thousand Oaks, CA

For Derrick Isa, physical therapy is about "seeing the small victories you make with each patient"–the woman with a spinal cord injury, for instance, whom Derrick helped walk so that she could visit her son's grave, or the grandfather who had shoulder surgery so he could throw the baseball with his grandkids. "You can improve their quality of life," he says.

Physical therapists work with patients who have medical conditions, illnesses, or injuries that limit their ability to move and perform basic activities. While some therapists specialize in one group—such as athletes, stroke victims, or patients recovering from surgery—Derrick's private practice sees a variety of patients, from babies born with torticollis, where their heads are turned to one side, to high school athletes recovering from injuries, to patients rehabilitating after joint replacement surgery. Derrick's youngest patient is 4 months old; his oldest is 97.

Given that the physical therapist helps the body to move correctly, Derrick says anatomy and physiology are essential. "You not only need knowledge of anatomy, but you also need to understand the physiology of healing," he says.

Understanding physics and the impact of stress and torque on the body is also helpful.

Derrick also stresses that physical therapists must be sensitive to the patients, for whom the challenges of rehabilitation can seem insurmountable. Derrick says he needs to be skilled in day-to-day small talk, as well as sometimes just listening. To build their sensitivity, most physical therapists take courses in child development as well as general and abnormal psychology. For Derrick, the most effective path to empathy is to experience what his patients will experience. Most physical therapy programs will put students in wheelchairs for a day. Derrick had to go down a flight of 12 stairs in a wheelchair, and notes that the first time he had to do a wheelie to get up onto a curb, "I landed on my backside."

For more information, visit the website for American Physical Therapy Association at http://www.apta.org/.

Think this is the career for you?
Key Stats:

- **Education and Training.** Physical therapists must have a graduate degree—either a master's or doctorate. Undergraduate prerequisites for admission are similar to those for medical school. Some graduate programs also require volunteer experience at a local hospital or clinic.

- **Licensure.** All 50 states require physical therapists to be licensed.

- **Earnings.** Earnings vary but the median annual salary is $76,310.

- **Expected Job Prospects.** Employment is expected to grow faster than the national average—by 30% through 2018.

Bureau of Labor Statistics, U.S. Department of Labor, *Occupational Outlook Handbook, 2010-11 Edition,* Physical Therapists, on the Internet at http://www.bls.gov/oco/ocos080.htm (visited September 14, 2011).

Access more review material online in the Study Area at www.masteringaandp.com.

- Chapter guides
- Chapter quizzes
- Practice tests
- Art labeling activities
- Flashcards
- A glossary with pronunciations
- Practice Anatomy Lab™ (PAL™) 3.0 virtual anatomy practice tool
- Interactive Physiology® (IP) animated tutorials
- MP3 Tutor Sessions

For this chapter, follow these navigation paths in PAL:

- Human Cadaver>Axial Skeleton
- Human Cadaver>Appendicular Skeleton
- Human Cadaver>Joints
- Anatomical Models>Axial Skeleton
- Anatomical Models>Appendicular Skeleton
- Anatomical Models>Joints

 For this chapter, go to these topics in the MP3 Tutor Sessions:

- How Bones React to Stress
- Types of Joints and their Movements

Chapter 5 Review Questions

1. If spongy bone has no osteons, how do nutrients reach the osteocytes?

2. Why are stresses or impacts to the side of the shaft in a long bone more dangerous than stress applied to the long axis of the shaft?

3. Why would a physician concerned about the relative growth of a young child request an x-ray of the hand?

4. While working at an excavation site, an anthropologist finds several small skull bones. She examines the frontal, parietal, and occipital bones and concludes that the skulls are those of children not yet 1 year old. How can she tell their ages from an examination of these bones?

5. Why would a person suffering from osteoporosis be more likely to suffer a broken hip than a broken shoulder?

For answers to all module, section, and chapter review questions, see the blue Answers tab at the back of the book.

6

The Muscular System

Functional Anatomy of Skeletal Muscle Tissue

Muscle tissue, one of the four primary tissue types, consists chiefly of muscle cells that are highly specialized for contraction. Without the three types of muscle tissue—skeletal muscle, cardiac muscle, and smooth muscle—nothing in the body would move, and the body itself could not move. In this chapter we consider the structure and function of skeletal muscle tissue. Each cell in skeletal muscle tissue is a single muscle fiber. Skeletal muscles are organs composed primarily of skeletal muscle tissue plus connective tissues, nerves, and blood vessels. Skeletal muscles are directly or indirectly attached to bones and have several functions, as detailed in the list of functions at right.

Muscle Tissue

Skeletal Muscle Tissue

Skeletal muscle tissue contractions move the body by pulling on bones of the skeleton, making it possible for us to walk, dance, bite an apple, or play the ukulele.

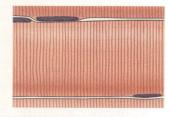

Cardiac Muscle Tissue

Cardiac muscle tissue contractions in the heart propel blood through the blood vessels.

(See Chapter 12.)

Smooth Muscle Tissue

Smooth muscle tissue contractions move fluids and solids along the digestive tract and regulate the diameters of small arteries, among other functions.

(See Chapter 15.)

In Module 6.1 we examine the functional anatomy of a typical skeletal muscle. Throughout this section we will emphasize the microscopic structural features that make contractions possible.

Functions of Skeletal Muscle Tissue

- **Produce Skeletal Movement**. Skeletal muscle contractions pull on tendons and move the bones of the skeleton. The effects range from simple motions such as extending the arm or breathing, to the highly coordinated movements of swimming, skiing, or typing.

- **Maintain Posture and Body Position**. Tension in skeletal muscles maintains body posture—for example, holding your head still when you read a book, or balancing your body weight above your feet when you walk. Without constant muscular activity, you could neither sit upright nor stand.

- **Support Soft Tissues**. The abdominal wall and the floor of the pelvic cavity consist of layers of skeletal muscle. These muscles support the weight of visceral organs and shield internal tissues from injury.

- **Guard Entrances and Exits**. Skeletal muscles encircle the openings of the digestive and urinary tracts. These muscles (sphincters) provide voluntary control over swallowing, defecation, and urination.

- **Maintain Body Temperature**. Muscle contractions require energy; whenever energy is used in the body, some of it is converted to heat. The heat released by working muscles keeps body temperature in the range required for normal functioning.

- **Provide Nutrient Reserves**. When the diet contains inadequate proteins or calories, the contractile proteins in skeletal muscles are broken down into amino acids, which are released into the circulation. The liver can use some of these amino acids to synthesize glucose. Others can be broken down to provide energy.

A skeletal muscle contains skeletal muscle tissue, connective tissues, blood vessels, and nerves

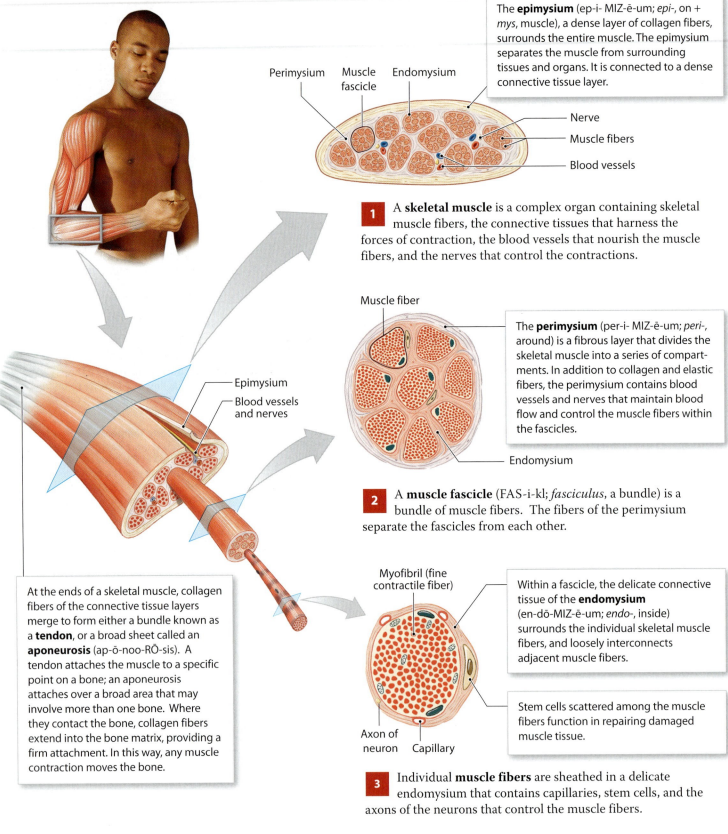

The **epimysium** (ep-i- MIZ-ē-um; *epi-*, on + *mys*, muscle), a dense layer of collagen fibers, surrounds the entire muscle. The epimysium separates the muscle from surrounding tissues and organs. It is connected to a dense connective tissue layer.

Perimysium   Muscle fascicle   Endomysium

Nerve
Muscle fibers
Blood vessels

1 A **skeletal muscle** is a complex organ containing skeletal muscle fibers, the connective tissues that harness the forces of contraction, the blood vessels that nourish the muscle fibers, and the nerves that control the contractions.

Muscle fiber

The **perimysium** (per-i- MIZ-ē-um; *peri-*, around) is a fibrous layer that divides the skeletal muscle into a series of compartments. In addition to collagen and elastic fibers, the perimysium contains blood vessels and nerves that maintain blood flow and control the muscle fibers within the fascicles.

Endomysium

Epimysium
Blood vessels and nerves

2 A **muscle fascicle** (FAS-i-kl; *fasciculus*, a bundle) is a bundle of muscle fibers. The fibers of the perimysium separate the fascicles from each other.

At the ends of a skeletal muscle, collagen fibers of the connective tissue layers merge to form either a bundle known as a **tendon**, or a broad sheet called an **aponeurosis** (ap-ō-noo-RŌ-sis). A tendon attaches the muscle to a specific point on a bone; an aponeurosis attaches over a broad area that may involve more than one bone. Where they contact the bone, collagen fibers extend into the bone matrix, providing a firm attachment. In this way, any muscle contraction moves the bone.

Myofibril (fine contractile fiber)

Within a fascicle, the delicate connective tissue of the **endomysium** (en-dō-MIZ-ē-um; *endo-*, inside) surrounds the individual skeletal muscle fibers, and loosely interconnects adjacent muscle fibers.

Stem cells scattered among the muscle fibers function in repairing damaged muscle tissue.

Axon of neuron   Capillary

3 Individual **muscle fibers** are sheathed in a delicate endomysium that contains capillaries, stem cells, and the axons of the neurons that control the muscle fibers.

4 Mature skeletal muscle fibers are enormous. A muscle fiber from a thigh muscle could have a diameter of 100 μm and a length of up to 30 cm, or 12 inches. Each skeletal muscle fiber contains hundreds of nuclei. The genes in these nuclei control the production of enzymes and structural proteins required for normal muscle contraction. Because skeletal muscle fibers are so unusual in size and appearance, special terms are used to describe them. The plasma membrane is called the **sarcolemma** (sar-kō-LEM-uh; *sarkos*, flesh + *lemma*, husk), and the cytoplasm is called **sarcoplasm** (SAR-kō-plazm).

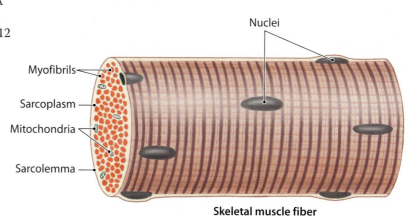

Skeletal muscle fiber

5 A **myofibril** (*myo-*, muscle) is a cylindrical structure 1–2 μm in diameter and as long as the entire muscle fiber. The sarcoplasm of a single skeletal muscle fiber may contain hundreds to thousands of myofibrils. Because each myofibril has a banded appearance and the fiber is packed with myofibrils lying side-by-side, the entire muscle fiber appears to have bands, or striations.

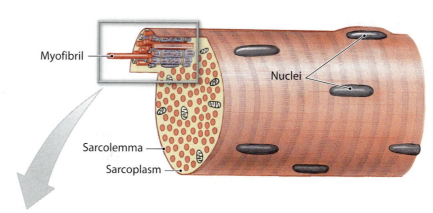

6 Myofibrils consist of bundles of protein filaments called **myofilaments**. The most abundant myofilaments are **thin filaments** composed primarily of actin, and **thick filaments** composed primarily of myosin.

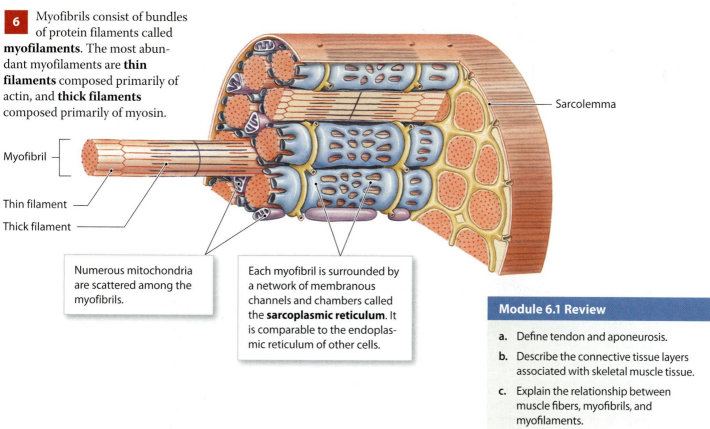

Numerous mitochondria are scattered among the myofibrils.

Each myofibril is surrounded by a network of membranous channels and chambers called the **sarcoplasmic reticulum**. It is comparable to the endoplasmic reticulum of other cells.

Module 6.1 Review

a. Define tendon and aponeurosis.

b. Describe the connective tissue layers associated with skeletal muscle tissue.

c. Explain the relationship between muscle fibers, myofibrils, and myofilaments.

Skeletal muscle fibers have contractile myofibrils containing hundreds to thousands of sarcomeres

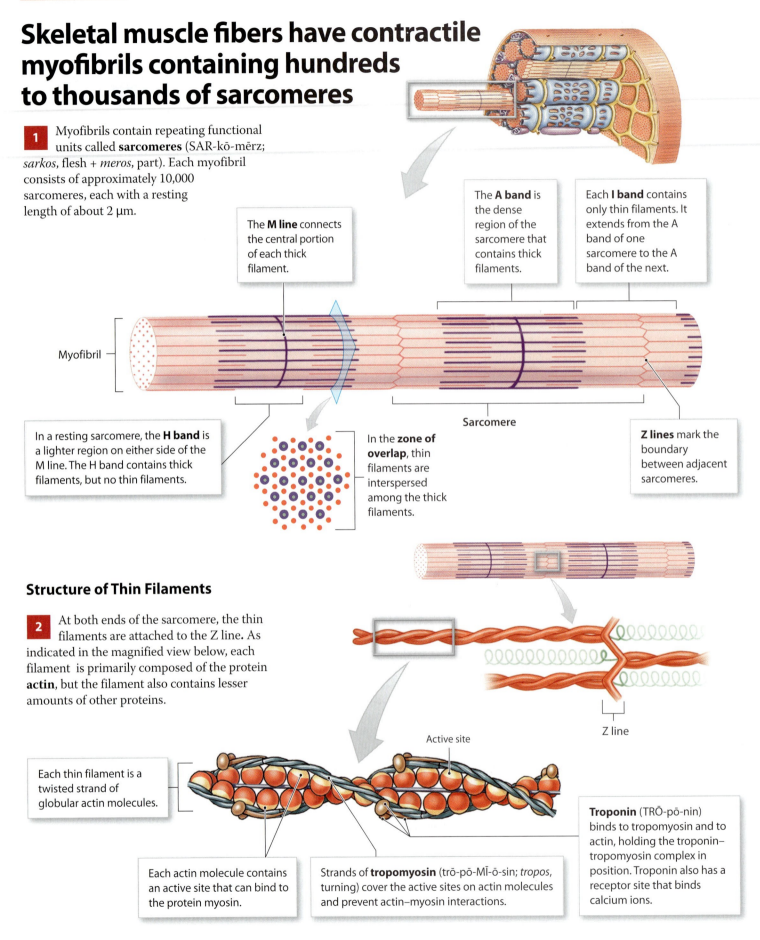

1 Myofibrils contain repeating functional units called **sarcomeres** (SAR-kō-mērz; *sarkos*, flesh + *meros*, part). Each myofibril consists of approximately 10,000 sarcomeres, each with a resting length of about 2 μm.

The **M line** connects the central portion of each thick filament.

The **A band** is the dense region of the sarcomere that contains thick filaments.

Each **I band** contains only thin filaments. It extends from the A band of one sarcomere to the A band of the next.

Myofibril

In a resting sarcomere, the **H band** is a lighter region on either side of the M line. The H band contains thick filaments, but no thin filaments.

In the **zone of overlap**, thin filaments are interspersed among the thick filaments.

Sarcomere

Z lines mark the boundary between adjacent sarcomeres.

Structure of Thin Filaments

2 At both ends of the sarcomere, the thin filaments are attached to the Z line. As indicated in the magnified view below, each filament is primarily composed of the protein **actin**, but the filament also contains lesser amounts of other proteins.

Z line

Active site

Each thin filament is a twisted strand of globular actin molecules.

Each actin molecule contains an active site that can bind to the protein myosin.

Strands of **tropomyosin** (trō-pō-MĪ-ō-sin; *tropos*, turning) cover the active sites on actin molecules and prevent actin–myosin interactions.

Troponin (TRŌ-pō-nin) binds to tropomyosin and to actin, holding the troponin–tropomyosin complex in position. Troponin also has a receptor site that binds calcium ions.

Arriving action potential

Sarcolemma of motor end plate

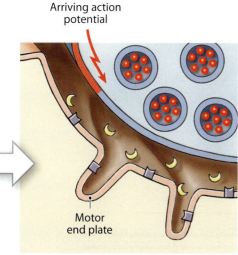

Motor end plate

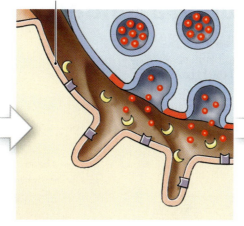

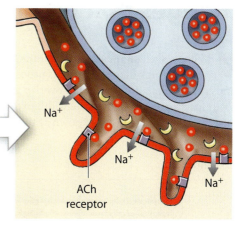

Na⁺

Na⁺

Na⁺

ACh receptor

3 The stimulus for ACh release is the arrival of an electrical impulse, or **action potential**, at the axon terminal. An action potential is a sudden change in the membrane potential that travels along the length of the axon.

4 When the action potential reaches the neuron's axon terminal, permeability changes in the membrane trigger the exocytosis of ACh into the synaptic cleft.

5 ACh molecules diffuse across the synaptic cleft and bind to ACh receptors on the surface of the motor end plate. ACh binding alters the membrane's permeability to sodium ions. Because the extracellular fluid contains a high concentration of sodium ions, and sodium ion concentration inside the cell is very low, sodium ions rush into the sarcoplasm.

Action potential

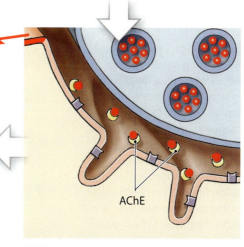

AChE

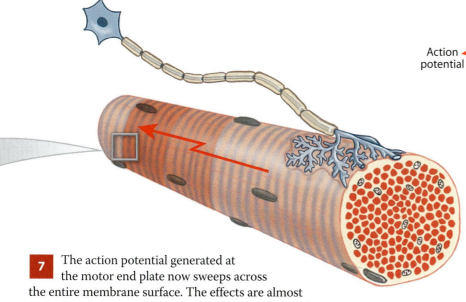

7 The action potential generated at the motor end plate now sweeps across the entire membrane surface. The effects are almost immediate because an action potential is an electrical event that flashes like a spark across the sarcolemma. However, the effects are brief because the ACh has been broken down. No further stimulus acts upon the motor end plate until another action potential arrives at the axon terminal.

6 The sudden inrush of the positively charged sodium ions generates an action potential in the sarcolemma. AChE quickly breaks down the ACh in the synaptic cleft, thus inactivating the ACh receptors.

Module 6.3 Review

a. Describe the neuromuscular junction.

b. How would a drug that blocked acetylcholine release affect muscle contraction?

c. Predict what would happen if there were no AChE in the synaptic cleft.

The sliding filament theory explains the physical changes that occur during a contraction

1 When a skeletal muscle fiber contracts, thin filaments slide past the thick filaments. In this process: (1) the H bands and I bands get smaller, (2) the zones of overlap get larger, (3) the Z lines move closer together, and (4) the width of the A band remains constant. This explanation is known as the **sliding filament theory**.

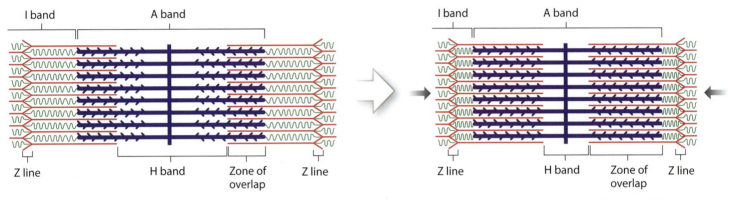

Sarcomere at rest

Sarcomere contraction and filament sliding

2 During a contraction, sliding occurs in every sarcomere along a myofibril, so the myofibril gets shorter. Because myofibrils are attached to the sarcolemma at each Z line and at either end of the muscle fiber, when myofibrils shorten, so does the muscle fiber. When both ends are free to move, the ends of a contracting muscle fiber move toward the center of the muscle fiber.

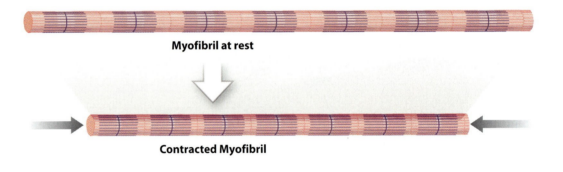

Myofibril at rest

Contracted Myofibril

When one end of a myofibril is fixed in position and the other end is free to move, the free end is pulled toward the fixed end.

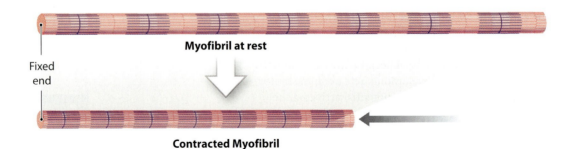

Fixed end

Myofibril at rest

Contracted Myofibril

3 Each end of a myofibril connects to one end of the muscle fiber, and all of the muscle fibers in a skeletal muscle are usually attached to tendons that are connected to bone. As a result, when the sarcomeres in the muscle fibers of the skeletal muscle contract, they pull on the attached bones. This can move an associated joint.

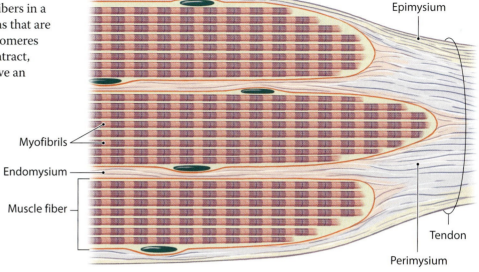

Epimysium

Myofibrils

Endomysium

Muscle fiber

Tendon

Perimysium

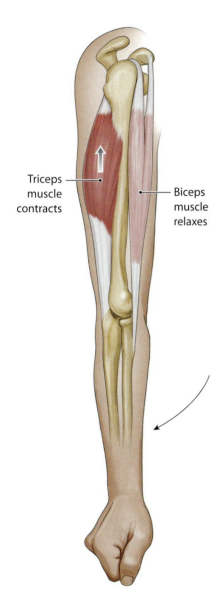

Triceps muscle contracts

Biceps muscle relaxes

4 Muscle contraction is an active process that uses ATP as a power source. A skeletal muscle contains hundreds to thousands of skeletal muscle fibers, and the strength of the contraction depends on how many of those muscle fibers are active at any given moment.

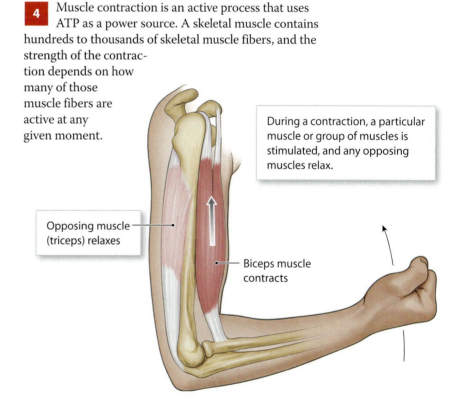

During a contraction, a particular muscle or group of muscles is stimulated, and any opposing muscles relax.

Opposing muscle (triceps) relaxes

Biceps muscle contracts

5 There is no active mechanism for lengthening a muscle fiber. After a skeletal muscle contracts, it returns to its original length primarily through a combination of gravity, the contraction of opposing muscles, and elasticity in the tissues stretched by the contraction.

Module 6.4 Review

a. Define the sliding filament theory.

b. Summarize the changes a sarcomere undergoes during a contraction.

c. If there is no active mechanism for lengthening a muscle, how does a muscle regain its initial length after a contraction?

A muscle fiber contraction uses ATP in a cycle that repeats for the duration of the contraction

Resting Sarcomere

In a resting sarcomere, each myosin head is already "energized"—charged with the energy that will be used to power a contraction. Each myosin head points away from the M line. In this position, the myosin head is "cocked" like the spring in a mousetrap. Cocking the myosin head requires energy, which is obtained by breaking down ATP. In doing so, the myosin head functions as an enzyme that breaks down ATP. At the start of the contraction cycle, the breakdown products—ADP and phosphate (P)—remain bound to the myosin head.

Contraction Cycle Begins

1 The **contraction cycle**, which involves a series of interrelated steps, begins with the arrival of calcium ions within the zone of overlap.

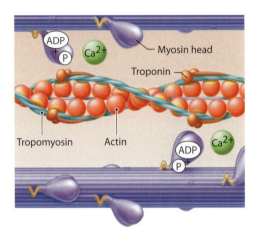

Contracted Sarcomere

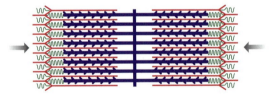

The entire cycle repeats several times each second, as long as Ca²⁺ levels are high and there is enough ATP. Once the stimulus is removed, the SR can pump Ca²⁺ from the sarcoplasm and store it for release in the future. Troponin molecules then shift position, swinging the tropomyosin strands over the active sites and preventing further cross-bridge formation.

Myosin Reactivated

6 Myosin reactivation occurs when the free myosin head splits ATP into ADP and P. The energy released is used to recock the myosin head.

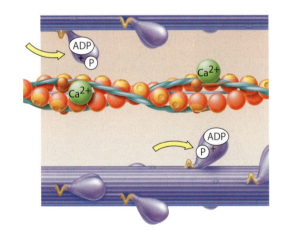

Active-Sites Exposed

2 Calcium ions bind to troponin, weakening the bond between actin and the troponin–tropomyosin complex. The troponin molecule then changes position, rolling the tropomyosin molecule away from the active sites on actin and allowing interaction with the energized myosin heads.

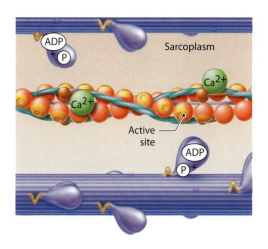

Cross-Bridges Form

3 Once the active sites are exposed, the energized myosin heads bind to them, forming cross-bridges.

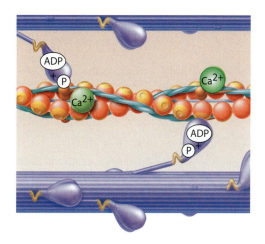

Cross-Bridges Detach

5 When another ATP binds to the myosin head, the link between the myosin head and the active site on the actin molecule is broken. The active site is now exposed and able to form another cross-bridge.

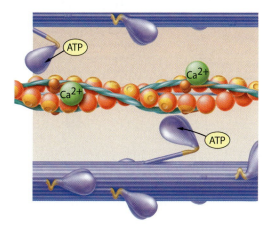

Myosin Heads Pivot

4 After cross-bridges form, the stored energy is released and used to pivot the myosin head toward the M line. This action is called the power stroke; when it occurs, the bound ADP and phosphate group are released.

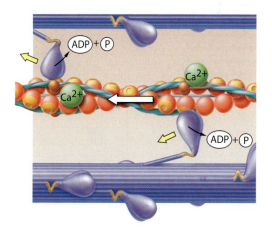

Module 6.5 Review

a. What molecule supplies energy for a muscle contraction?

b. List the five interrelated steps that occur once the contraction cycle has begun.

c. What triggers myosin reactivation?

1. Labeling

Label the structures in the following figure of a skeletal muscle fiber.

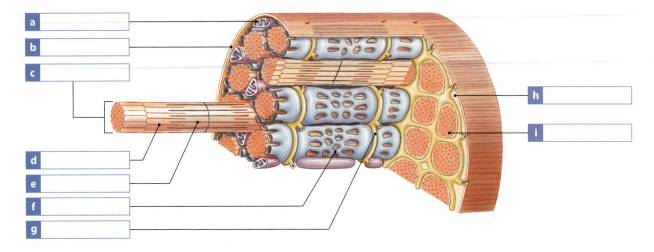

a

b

c

d

e

f

g

h

i

2. Labeling

Label the structures in the following diagram of adjacent sarcomeres along a myofibril.

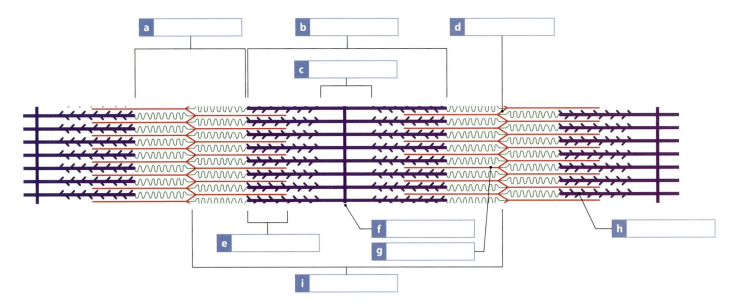

a

b

c

d

e

f

g

h

i

3. Vocabulary

In the space provided, write the boldfaced terms introduced in this section that contain the indicated word part.

Word Part	Meaning	Terms
myo-	muscle	a _____
sarko-	flesh	b _____

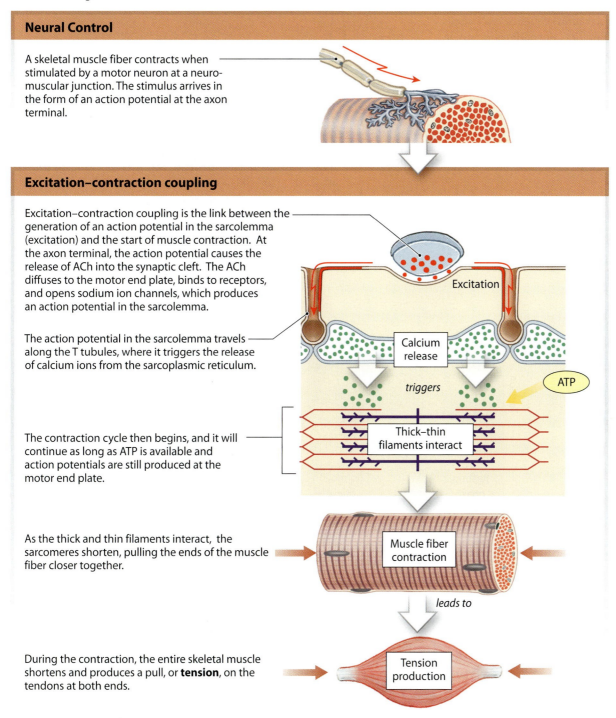

Functional Properties of Skeletal Muscle Tissue

Before we examine how skeletal muscle tissue works, let's recall the sequence of events we have considered thus far.

Neural Control

A skeletal muscle fiber contracts when stimulated by a motor neuron at a neuro-muscular junction. The stimulus arrives in the form of an action potential at the axon terminal.

Excitation–contraction coupling

Excitation–contraction coupling is the link between the generation of an action potential in the sarcolemma (excitation) and the start of muscle contraction. At the axon terminal, the action potential causes the release of ACh into the synaptic cleft. The ACh diffuses to the motor end plate, binds to receptors, and opens sodium ion channels, which produces an action potential in the sarcolemma.

The action potential in the sarcolemma travels along the T tubules, where it triggers the release of calcium ions from the sarcoplasmic reticulum.

The contraction cycle then begins, and it will continue as long as ATP is available and action potentials are still produced at the motor end plate.

As the thick and thin filaments interact, the sarcomeres shorten, pulling the ends of the muscle fiber closer together.

During the contraction, the entire skeletal muscle shortens and produces a pull, or **tension**, on the tendons at both ends.

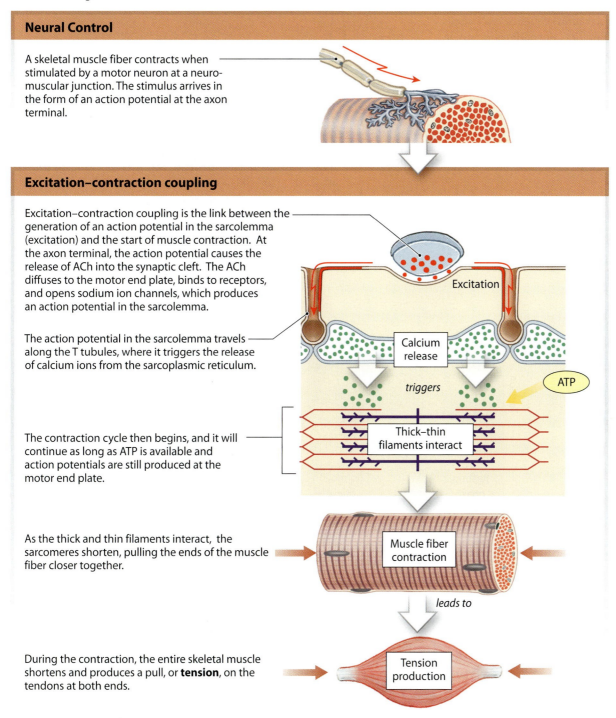

In this section we focus on how tension is produced, how muscle contractions are classified and fueled, and how different muscle fibers respond from one moment to the next.

Tension production rises to maximum levels as the rate of muscle stimulation increases

1 As a muscle fiber shortens, it exerts a pull, or tension, on the connective tissue fibers attached to it. This graph of tension development in muscle fibers shows two examples of a **twitch**, a single stimulus-contraction-relaxation sequence in a muscle fiber. Twitches vary in duration, depending on muscle type and location, internal and external environmental conditions, and other factors. Although electrical stimulation can produce muscle twitches in a laboratory, they are too brief to be part of any normal activity. However, they reveal some interesting differences in the functional properties of various skeletal muscles.

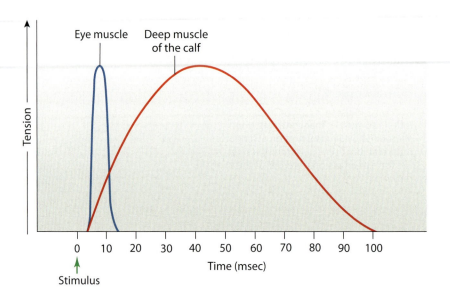

2 This graph shows the phases of a 40-msec twitch in a muscle fiber from the gastrocnemius muscle, a prominent muscle of the calf.

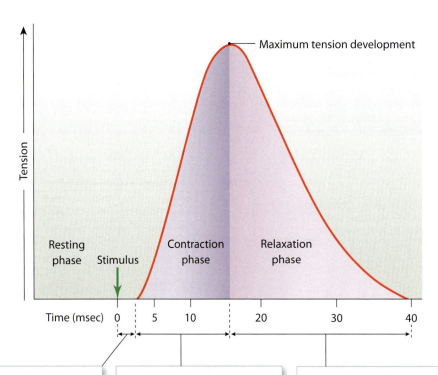

The **latent period** begins at stimulation and typically lasts about 2 msec. During this period, an action potential sweeps across the sarcolemma, and the sarcoplasmic reticulum releases calcium ions. The muscle fiber does not produce tension during the latent period, because the contraction cycle has yet to begin.

In the **contraction phase**, lasting about 13 msec, tension rises to a peak. As the tension rises, calcium ions are binding to troponin, active sites on thin filaments are being exposed, and cross-bridge interactions are occurring.

The **relaxation phase** lasts about 25 msec. During this period, calcium levels are falling, tropomyosin is covering active sites, and the number of cross-bridges is declining as they detach. As a result, tension returns to resting levels.

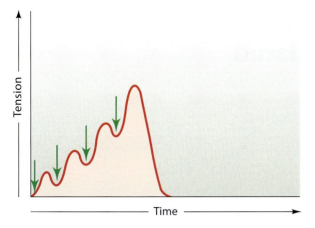

3 If a second stimulus arrives before the relaxation phase has ended, a second, more powerful contraction occurs. The addition of one twitch to another in this way constitutes **wave summation**. The duration of a single twitch determines the maximum time available to produce wave summation.

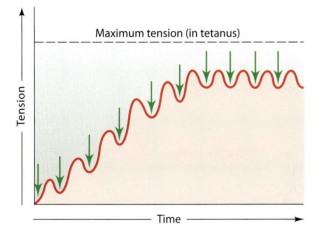

Maximum tension (in tetanus)

4 A muscle producing almost peak tension during rapid cycles of contraction and relaxation is said to be in **incomplete tetanus** (*tetanos*, convulsive tension). (Note that the dashed line in this graph and the one below represents maximum, or peak, tension developed in tetanus.)

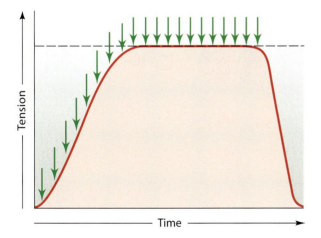

5 **Complete tetanus** occurs when a higher stimulation frequency eliminates the relaxation phase. During complete tetanus, action potentials arrive so rapidly that the sarcoplasmic reticulum cannot reclaim calcium ions. The high Ca^{2+} concentration in the cytoplasm prolongs the contraction, making it continuous. Although muscles can be forced into tetanus in the laboratory, they seldom if ever develop peak tension in the course of normal activities. This is because normal movements require precise control over, and continuous variation in, the amount of tension produced.

KEY

↓ = Stimulus

Module 6.6 Review

a. Define a twitch.

b. Describe the events occurring during the relaxation phase of a twitch.

c. Contrast complete and incomplete tetanus.

A skeletal muscle controls muscle tension by the number of motor units stimulated

1 A typical skeletal muscle contains thousands of muscle fibers. Although some motor neurons control just a few muscle fibers, most control hundreds of them. The amount of tension produced is controlled at the subconscious level through variations in the number of muscle fibers stimulated.

Spinal cord

Cell bodies of motor neurons

Axons of motor neurons

Motor nerve (collection of motor neuron axons)

A **motor unit** is a single motor neuron and all the muscle fibers it innervates. The size of a motor unit indicates how fine the control of movement can be. In the muscles of the eye, where precise control is extremely important, a motor neuron may control only 4–6 muscle fibers. We have much less precise control over our leg muscles, where a single motor neuron may control 1000–2000 muscle fibers.

The muscle fibers of each motor unit are intermingled with those of other motor units. Because of this intermingling, the direction of pull on the tendon does not change when the number of activated motor units changes. If you decide to perform a specific movement, the contraction begins when the smallest motor units in the stimulated muscle are activated. As the movement continues, larger motor units containing faster and more powerful muscle fibers are activated, and tension production rises steeply. The smooth but steady increase in muscular tension produced by increasing the number of active motor units is called **recruitment**.

KEY

- Motor unit 1
- Motor unit 2
- Motor unit 3

A variable number of motor units is always active, even when the entire muscle is not contracting. Their contractions do not produce enough tension to cause movement, but they do tense and firm the muscle. This resting tension in a skeletal muscle is called **muscle tone**, and it is regulated at the subconscious level. The activity level of each motor neuron changes constantly, and individual muscle fibers can relax while a constant tension is maintained in the attached tendon. Activated muscle fibers use energy, so the greater the muscle tone, the higher the "resting" rate of metabolism. Elevated muscle tone increases resting energy consumption by a small amount, but the effects are cumulative, and they continue 24 hours per day.

Module 6.7 Review

a. Define motor unit.

b. What is recruitment?

c. Describe the relationship between the number of fibers in a motor unit and the precision of body movements.

Muscle contractions may be isotonic or isometric

Isotonic contraction

1 In an **isotonic contraction** (*iso-*, equal + *tonos*, tension), tension rises and the skeletal muscle's length changes. Lifting an object off a desk, walking, and running involve isotonic contractions. Consider a skeletal muscle that is 1 cm² in cross-sectional area and can produce roughly 4 kg (8.8 lb) of tension in complete tetanus. If we hang a load of 2 kg (4.4 lb) from that muscle and stimulate it, the muscle will shorten.

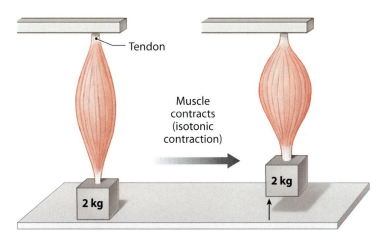

2 Before the muscle can shorten, the cross-bridges must produce enough tension to overcome the load—in this case, the 2-kg weight. During this initial period, tension in the muscle fibers rises until the tension in the tendon exceeds the load. As the muscle shortens, the tension in the skeletal muscle remains constant.

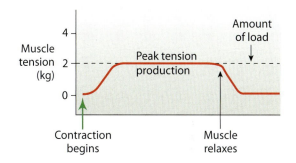

Isometric contraction

3 In contrast to an isotonic contraction, in an **isometric contraction** (*metric*, measure), the muscle as a whole does not change length, and the tension produced never exceeds the load. If the load is greater than or equals the peak tension, the load won't move when the muscle contracts. The contracting muscle bulges, but not as much as it does during an isotonic contraction. In an isometric contraction, although the muscle as a whole does not shorten, the individual muscle fibers shorten as connective tissues stretch. The muscle fibers cannot shorten further, because the tension does not exceed the load.

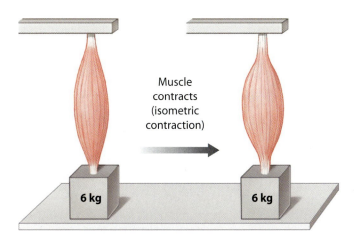

Many of the reflexive muscle contractions that keep your body upright when you stand or sit involve isometric contractions of muscles that oppose the force of gravity.

4 Because the muscle cannot produce enough tension to overcome the 6-kg load, the weight does not move.

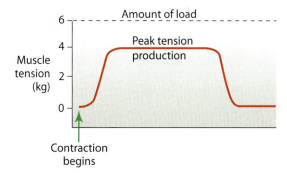

Module 6.8 Review

a. Compare an isotonic contraction and an isometric contraction.

b. Can a skeletal muscle contract without shortening? Why or why not?

c. In the two graphs above, which contraction produced the greater tension?

Muscle contraction requires a large amount of ATP that may be produced aerobically or anaerobically

1 In a resting skeletal muscle, the demand for ATP is low. More than enough oxygen is available for the mitochondria to meet that demand, and they produce a surplus of ATP. The extra ATP is used to build up reserves of **creatine phosphate** (**CP**), another high-energy compound, and glycogen. Resting muscle fibers absorb fatty acids and glucose delivered by the bloodstream. The fatty acids are broken down in the mitochondria, generating ATP that is used to convert creatine to creatine phosphate and glucose to glycogen.

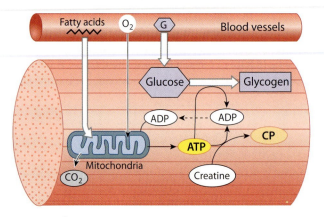

Muscle at rest

2 At moderate levels of activity, the demand for ATP increases. This demand is met by the mitochondria, and the rate of oxygen consumption increases as a result. The skeletal muscle now relies primarily on the aerobic (with oxygen) metabolism of **pyruvate** in mitochondria. Two 3-carbon pyruvate molecules are produced each time enzymes break down a single 6-carbon glucose molecule. Although the production of pyruvate generates a small amount of ATP, the contribution is dwarfed by that of the mitochondria. As long as mitochondrial activity meets the demand for ATP, the muscle will not become fatigued until glycogen, lipid, and amino acid reserves are exhausted. This type of fatigue affects the muscles of endurance athletes, such as marathon runners, after hours of exertion.

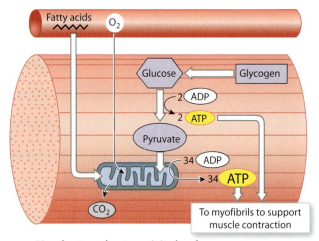

Muscle at moderate activity levels

3 At peak levels of activity, ATP demands are enormous and mitochondrial ATP production rises to a maximum rate determined by the availability of oxygen. At peak exertion, mitochondrial activity can provide only about one-third of the ATP needed. The remainder is produced anaerobically (without oxygen) by breaking down glucose to pyruvate. As pyruvate levels climb, pyruvate is converted to **lactic acid**, a related 3-carbon molecule that dissociates into a 3-carbon **lactate** molecule and a hydrogen ion (H^+). The production of lactic acid during peak activity lowers the intracellular and extracellular pH. After only a few seconds of peak activity, changes in pH will alter the characteristics of key enzymes so that the muscle fiber can no longer contract. Sprinters usually experience this type of muscle fatigue.

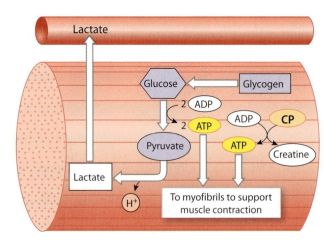

Muscle at peak activity levels

Peak Activity

Much of the large amounts of lactate produced during peak exertion diffuses out of the muscle fibers and into the bloodstream. The liver absorbs this lactate and begins converting it back into pyruvate.

Recovery

This process continues after exertion has ended, because lactate levels within muscle fibers remain relatively high, and lactate continues to diffuse into the bloodstream. After the liver converts the reabsorbed lactate to pyruvate, roughly 30 percent of the new pyruvate molecules are broken down in the mitochondria. This provides the ATP needed to convert the remaining 70 percent of pyruvate molecules back into glucose. The glucose molecules are then released into the circulation, where they are absorbed by skeletal muscle fibers and used to rebuild their glycogen reserves.

4   Much of the lactate released by muscle fibers during strenuous activity diffuses from the muscle tissue and into the bloodstream. The liver absorbs and recycles it, and releases glucose into the circulation. This shuffling of lactate to the liver and of glucose back to muscle cells is called the **Cori cycle**. Throughout the recovery period, the body's oxygen demand remains elevated above normal resting levels. The more ATP required, the more oxygen will be needed. The amount of oxygen required to restore normal, pre-exertion conditions is called the **oxygen debt**.

Module 6.9 Review

a. Identify the two compounds in which energy is stored in resting muscle fibers.

b. Under what conditions do muscle fibers produce lactate?

c. Define oxygen debt.

Many factors can result in muscle hypertrophy, atrophy, or paralysis

1 As a result of repeated, exhaustive stimulation, muscle fibers develop more mitochondria, a higher concentration of glycolytic enzymes, and larger glycogen reserves. Such muscle fibers have more myofibrils than do fibers that are less used, and each myofibril contains more thick and thin filaments. The net effect is **hypertrophy**, or an enlargement of the stimulated muscle. The number of muscle fibers does not change significantly, but the muscle as a whole enlarges because each muscle fiber increases in diameter. The muscle also becomes stronger, since tension production is proportional to the cross-sectional area of the muscle. The muscles of a bodybuilder are excellent examples of muscular hypertrophy. Although hypertrophy can be promoted by using steroid hormones, exhaustive training is still required, and the side effects of steroid use can be very dangerous.

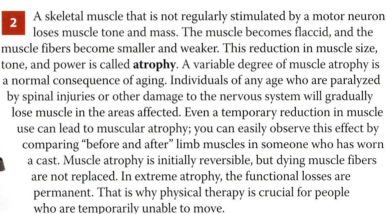

2 A skeletal muscle that is not regularly stimulated by a motor neuron loses muscle tone and mass. The muscle becomes flaccid, and the muscle fibers become smaller and weaker. This reduction in muscle size, tone, and power is called **atrophy**. A variable degree of muscle atrophy is a normal consequence of aging. Individuals of any age who are paralyzed by spinal injuries or other damage to the nervous system will gradually lose muscle in the areas affected. Even a temporary reduction in muscle use can lead to muscular atrophy; you can easily observe this effect by comparing "before and after" limb muscles in someone who has worn a cast. Muscle atrophy is initially reversible, but dying muscle fibers are not replaced. In extreme atrophy, the functional losses are permanent. That is why physical therapy is crucial for people who are temporarily unable to move.

3 This figure considers a few of the many clinical conditions that can affect skeletal muscles. Because skeletal muscles depend on motor neurons for stimulation, disorders that affect the nervous system can indirectly affect the muscular system.

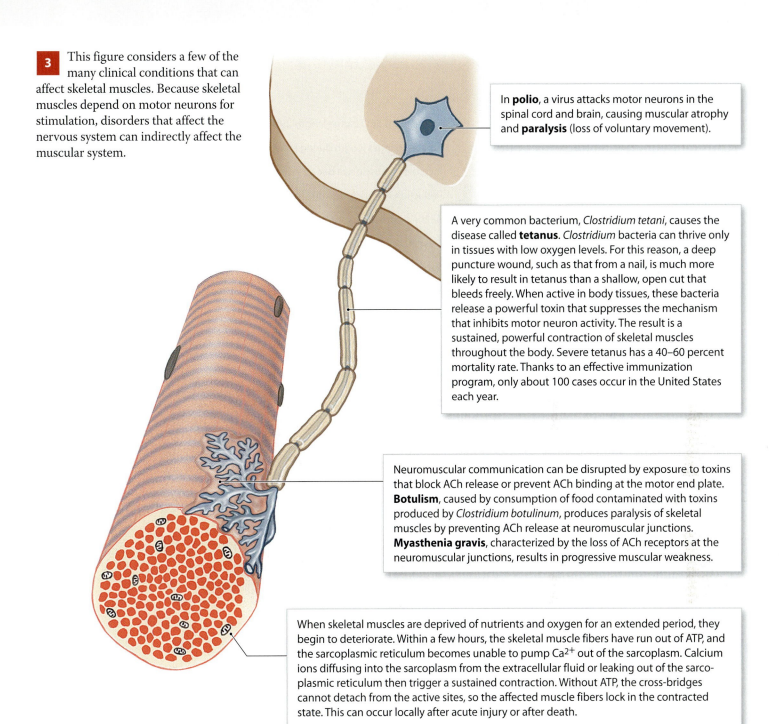

In **polio**, a virus attacks motor neurons in the spinal cord and brain, causing muscular atrophy and **paralysis** (loss of voluntary movement).

A very common bacterium, *Clostridium tetani*, causes the disease called **tetanus**. *Clostridium* bacteria can thrive only in tissues with low oxygen levels. For this reason, a deep puncture wound, such as that from a nail, is much more likely to result in tetanus than a shallow, open cut that bleeds freely. When active in body tissues, these bacteria release a powerful toxin that suppresses the mechanism that inhibits motor neuron activity. The result is a sustained, powerful contraction of skeletal muscles throughout the body. Severe tetanus has a 40–60 percent mortality rate. Thanks to an effective immunization program, only about 100 cases occur in the United States each year.

Neuromuscular communication can be disrupted by exposure to toxins that block ACh release or prevent ACh binding at the motor end plate. **Botulism**, caused by consumption of food contaminated with toxins produced by *Clostridium botulinum*, produces paralysis of skeletal muscles by preventing ACh release at neuromuscular junctions. **Myasthenia gravis**, characterized by the loss of ACh receptors at the neuromuscular junctions, results in progressive muscular weakness.

When skeletal muscles are deprived of nutrients and oxygen for an extended period, they begin to deteriorate. Within a few hours, the skeletal muscle fibers have run out of ATP, and the sarcoplasmic reticulum becomes unable to pump Ca^{2+} out of the sarcoplasm. Calcium ions diffusing into the sarcoplasm from the extracellular fluid or leaking out of the sarcoplasmic reticulum then trigger a sustained contraction. Without ATP, the cross-bridges cannot detach from the active sites, so the affected muscle fibers lock in the contracted state. This can occur locally after acute injury or after death.

Shortly after death, a generalized skeletal muscle contraction, called **rigor mortis**, occurs throughout the body, beginning with the smaller muscles of the face, neck, and arms. As the SR deteriorates, calcium ions are released and a sustained contraction begins. As ATP reserves are exhausted, the muscles become locked in the contracted state. Because all skeletal muscles are involved during rigor mortis, the individual becomes "stiff as a board." It typically begins 2–7 hours after death and disappears after 1–6 days or when decomposition begins; but the timing depends on environmental factors such as temperature. Forensic pathologists can estimate time of death based on environmental conditions and the degree of rigor mortis.

Module 6.10 Review

a. Define muscle hypertrophy and muscle atrophy.

b. Six weeks after Fred broke his leg the cast is removed, and as he steps down from the exam table, his leg gives way and he falls. Propose a logical explanation.

c. Explain how the flexibility or rigidity of a dead body can provide a clue to a murder victim's time of death.

1. Matching

Match the following terms with their descriptions.

- isometric contraction
- atrophy
- oxygen
- rigor mortis
- motor unit
- lactic acid
- isotonic contraction
- hypertrophy

a _____ Muscle does not change length during contraction

b _____ Muscle changes length during contraction

c _____ Increase in muscle size due to increased use

d _____ Reduction in muscle size due to disuse

e _____ All the muscle fibers controlled by a single motor neuron

f _____ Overall contraction of body muscles after death

g _____ Required for mitochondrial ATP production

h _____ Produced during peak muscular activity

2. Labeling

Use the following terms to correctly label the structures in the diagram representing a blood vessel and a skeletal muscle at rest.

- creatine
- glycogen
- CP
- glucose
- O_2
- fatty acids

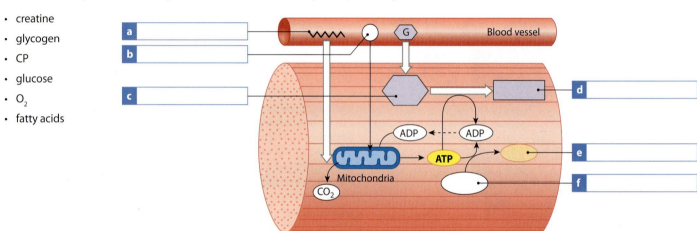

3. Section Integration

a Explain what happens to the lactate produced during strenuous muscle activity.

_____

_____

_____

b Some clinical conditions of the muscular system are due to disorders of the nervous system.

Describe how disorders of the nervous system produce polio, tetanus, botulism, and myasthenia gravis.

_____

_____

_____

_____

c Explain the physiological events that lead to rigor mortis.

_____

_____

_____

Functional Organization of the Muscular System

The skeletal muscles of the **muscular system** account for almost half of the weight of your body.

2 The muscular system can be divided into axial and appendicular divisions. **Axial muscles** support and position the axial skeleton. **Appendicular muscles** support, move, and brace the limbs.

1 The muscular system contributes more to body weight than does any other organ system.

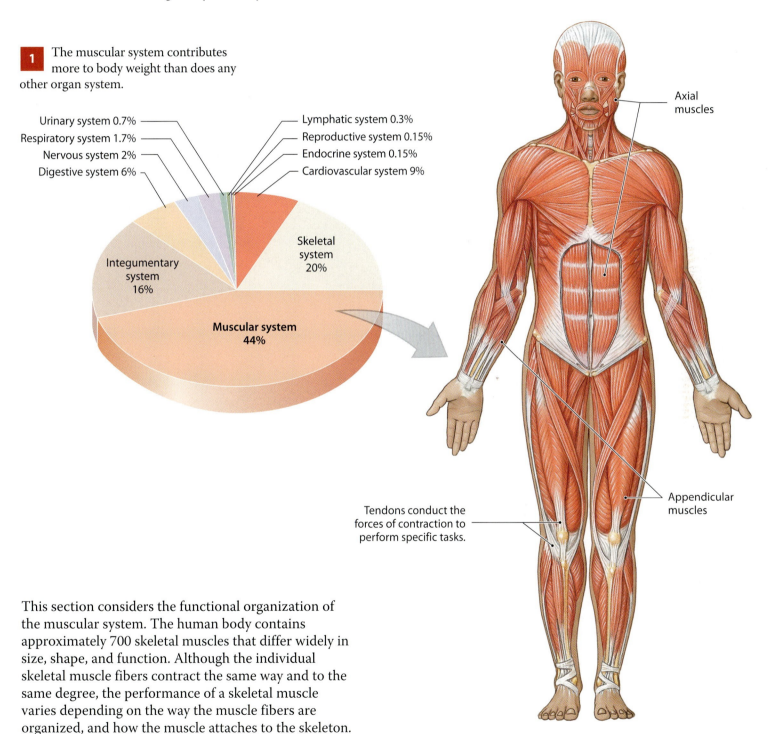

Urinary system 0.7%
Respiratory system 1.7%
Nervous system 2%
Digestive system 6%

Lymphatic system 0.3%
Reproductive system 0.15%
Endocrine system 0.15%
Cardiovascular system 9%

Skeletal system 20%

Integumentary system 16%

Muscular system 44%

Axial muscles

Appendicular muscles

Tendons conduct the forces of contraction to perform specific tasks.

This section considers the functional organization of the muscular system. The human body contains approximately 700 skeletal muscles that differ widely in size, shape, and function. Although the individual skeletal muscle fibers contract the same way and to the same degree, the performance of a skeletal muscle varies depending on the way the muscle fibers are organized, and how the muscle attaches to the skeleton.

The names of muscles provide clues to their appearance and/or function

1 In most cases one end of a muscle is fixed in position, and the other end moves during a contraction. The place where the fixed end attaches is the **origin** of the muscle. Most muscles originate at a bone, but some originate at a connective tissue sheath or band. The site where the movable end attaches to another structure is the **insertion** of the muscle. The origin is typically proximal to the insertion when the body is in the anatomical position. However, knowing which end is the origin and which is the insertion is ultimately less important than knowing where the two ends attach and what the muscle does when it contracts. When a muscle contracts, it produces a specific movement or **action**. In general, we describe actions in terms of movement at specific joints.

Action

Origins of biceps brachii muscle

Insertion of biceps brachii muscle

2 With complex movements, muscles work in groups rather than individually. Their cooperation makes a particular movement more efficient. Based on their functions, muscles may be described as agonists, antagonists, or synergists.

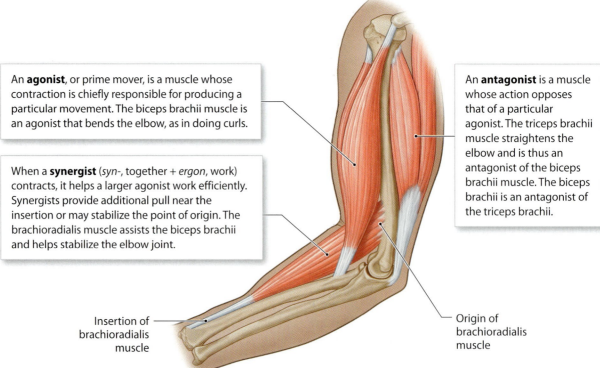

An **agonist**, or prime mover, is a muscle whose contraction is chiefly responsible for producing a particular movement. The biceps brachii muscle is an agonist that bends the elbow, as in doing curls.

When a **synergist** (*syn-*, together + *ergon*, work) contracts, it helps a larger agonist work efficiently. Synergists provide additional pull near the insertion or may stabilize the point of origin. The brachioradialis muscle assists the biceps brachii and helps stabilize the elbow joint.

An **antagonist** is a muscle whose action opposes that of a particular agonist. The triceps brachii muscle straightens the elbow and is thus an antagonist of the biceps brachii muscle. The biceps brachii is an antagonist of the triceps brachii.

Insertion of brachioradialis muscle

Origin of brachioradialis muscle

Muscle Terminology

Terms Indicating Specific Regions of the Body*	Terms Indicating Position, Direction, or Fascicle Organization	Terms Indicating Structural Characteristics of the Muscle	Terms Indicating Actions
Abdominal (abdomen)	Anterior (front)	**Nature of Origin**	**General**
Ancon (elbow)	External (on the outside)	Biceps (two heads)	Abductor (movement away)
Auricular (ear)	Extrinsic (originating outside)	Triceps (three heads)	Adductor (movement toward)
Brachial (arm)	Inferior (below)	Quadriceps (four heads)	Depressor (lowering movement)
Capitis (head)	Internal (away from the surface)		Extensor (straightening movement)
Carpi (wrist)	Intrinsic (inside)	**Shape**	Flexor (bending movement)
Cervicis (neck)	Lateral (on the side)	Deltoid (triangle)	Levator (raising movement)
Coccygeal (coccyx)	Medial (middle)	Orbicularis (circle)	Pronator (turning into prone position)
Costal (rib)	Oblique (slanting)	Pectinate (comblike)	Supinator (turning into supine position)
Cutaneous (skin)	Posterior (back)	Piriformis (pear-shaped)	Tensor (tensing movement)
Femoris (thigh)	Profundus (deep)	Platy- (flat)	
Glossal (tongue)	Rectus (straight)	Pyramidal (pyramid)	
Hallux (great toe)	Superficial (toward the surface)	Rhomboid (parallelogram)	**Specific**
Ilium (groin)	Superior (toward the head)	Serratus (serrated)	Buccinator (trumpeter)
Inguinal (groin)	Transverse (crosswise)	Splenius (bandage)	Risorius (laugher)
Lumbar (lumbar region)		Teres (round and long)	Sartorius (like a tailor)
Nasalis (nose)		Trapezius (trapezoid)	
Nuchal (back of neck)			
Ocular (eye)		**Other Striking Features**	
Oris (mouth)		Alba (white)	
Palpebra (eyelid)		Brevis (short)	
Pollex (thumb)		Gracilis (slender)	
Popliteal (posterior to knee)		Latae (wide)	
Psoas (loin)		Latissimus (widest)	
Radial (forearm)		Longissimus (longest)	
Scapular (scapula)		Longus (long)	
Temporal (temple)		Magnus (large)	
Thoracic (thorax)		Major (larger)	
Tibial (tibia; shin)		Maximus (largest)	
Ulnar (ulna)		Minimus (smallest)	
		Minor (smaller)	
		Vastus (great)	

* For other regional terms, refer to Module 1.6.

3 This table includes a useful summary of the most important terms used in naming skeletal muscles. Becoming familiar with these terms will help you identify and remember specific muscles. The complete names of nearly all skeletal muscles include the term "muscle" (the only exceptions are the platysma and the diaphragm). Although the full name, such as the biceps brachii muscle, will usually appear in the text, for simplicity we will use only the descriptive name (biceps brachii) in figures and tables.

Module 6.11 Review

a. Define the term *synergist* as it relates to muscle actions.

b. What is the relationship between the biceps brachii and the triceps brachii?

c. What does the name *flexor carpi radialis longus* tell you about this muscle?

Broad descriptive terms are used to describe movements with reference to the anatomical position

It is easiest to understand the terms used to describe body movements when you see the actions under way. You should become very familiar with the descriptive terms presented in Modules 6.12 and 6.13, because we will use them when we consider the actions of skeletal muscles.

Flexion and Extension

1 Flexion and extension are usually applied to the movements of the long bones of the limbs, but they are also used to describe movements of the axial skeleton. For example, when you bring your head toward your chest, you flex the intervertebral joints of the neck.

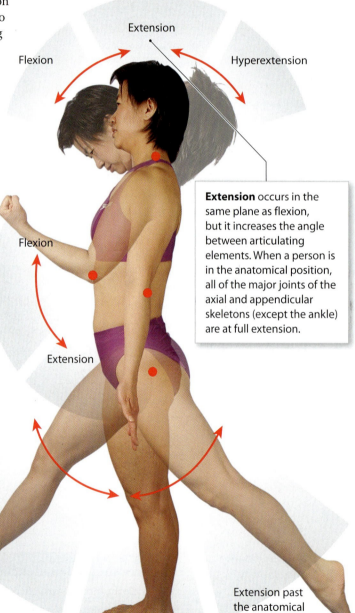

Extension

Flexion

Hyperextension

Flexion

Flexion (FLEK-shun) is movement in the anterior–posterior plane that reduces the angle between the articulating elements. Here you see flexion at the neck, the elbow, and the hip.

Extension

Extension occurs in the same plane as flexion, but it increases the angle between articulating elements. When a person is in the anatomical position, all of the major joints of the axial and appendicular skeletons (except the ankle) are at full extension.

Extension past the anatomical position is called **hyperextension**.

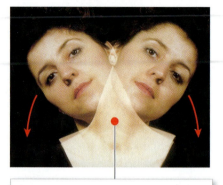

Lateral flexion occurs when your vertebral column bends to the side. This movement is most pronounced in the cervical and thoracic regions.

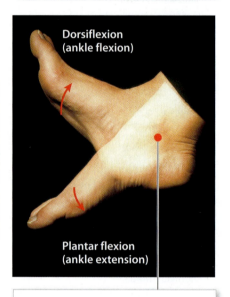

Dorsiflexion (ankle flexion)

Plantar flexion (ankle extension)

Dorsiflexion is flexion at the ankle joint and elevation of the sole, as when you dig in your heel. **Plantar flexion** (*planta*, sole), the opposite movement, extends the ankle joint and elevates the heel, as when you stand on tiptoe.

Abduction and Adduction

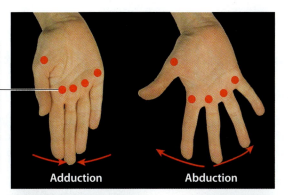

Spreading the fingers or toes apart abducts them, because they move away from a central digit. Bringing them together is adduction. (Fingers move toward or away from the middle finger; toes move toward or away from the second toe.)

Adduction   **Abduction**

Abduction   **Adduction**

2 Abduction and adduction always refer to the movements of the appendicular skeleton, not to those of the axial skeleton.

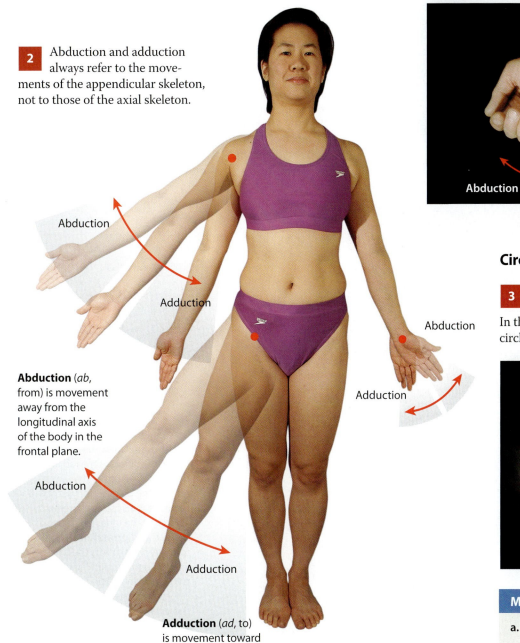

Abduction

Adduction

Abduction

Adduction

Abduction (*ab*, from) is movement away from the longitudinal axis of the body in the frontal plane.

Abduction

Adduction

Adduction (*ad*, to) is movement toward the longitudinal axis of the body in the frontal plane.

Circumduction

3 Moving your arm as if to draw a big circle on the wall is **circumduction**. In this movement your hand moves in a circle, but your arm does not rotate.

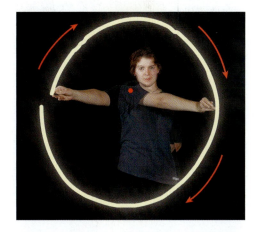

Module 6.12 Review

a. When doing jumping jacks, which lower limb movements are necessary?

b. Which movements are associated with hinge joints?

c. Compare dorsiflexion to plantar flexion.

Terms of more limited application describe rotational movements and special movements

Rotation

1 Rotational movements of the trunk are described as left or right rotation. Rotation of the limbs can be described as medial or lateral rotation; special terms are used to describe the rotational movements of the forearm.

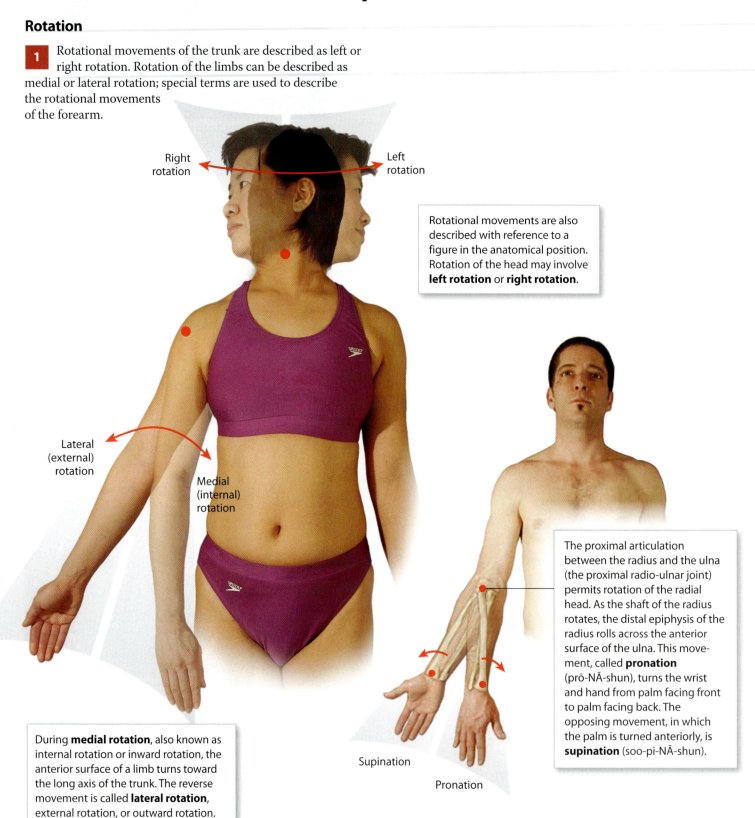

Right rotation | Left rotation

Rotational movements are also described with reference to a figure in the anatomical position. Rotation of the head may involve **left rotation** or **right rotation**.

Lateral (external) rotation

Medial (internal) rotation

During **medial rotation**, also known as internal rotation or inward rotation, the anterior surface of a limb turns toward the long axis of the trunk. The reverse movement is called **lateral rotation**, external rotation, or outward rotation.

Supination

Pronation

The proximal articulation between the radius and the ulna (the proximal radio-ulnar joint) permits rotation of the radial head. As the shaft of the radius rotates, the distal epiphysis of the radius rolls across the anterior surface of the ulna. This movement, called **pronation** (prō-NĀ-shun), turns the wrist and hand from palm facing front to palm facing back. The opposing movement, in which the palm is turned anteriorly, is **supination** (soo-pi-NĀ-shun).

Special Movements

2 There are several special terms that apply to specific articulations or unusual types of movement.

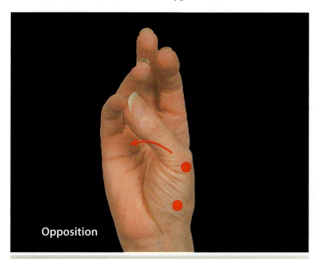

Opposition

Opposition is movement of the thumb toward the palm or the pads of other fingers. Opposition enables you to hold objects between your thumb and palm. It involves movement at the first carpometacarpal and metacarpophalangeal joints. Flexion at the fifth metacarpophalangeal joint can assist this movement.

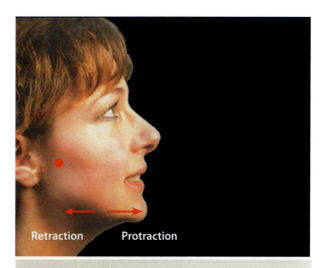

Retraction   Protraction

Protraction entails moving a part of the body anteriorly in the horizontal plane. **Retraction** is the reverse movement. You protract your jaw when you grasp your upper lip with your lower teeth, and you protract your clavicles when you cross your arms.

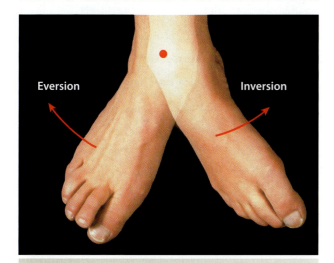

Eversion   Inversion

Inversion (*in*, into + *vertere*, to turn) is a twisting motion of the foot that turns the sole inward, elevating the medial edge of the sole. The opposite movement is called **eversion** (ē-VER-zhun; *e*, out).

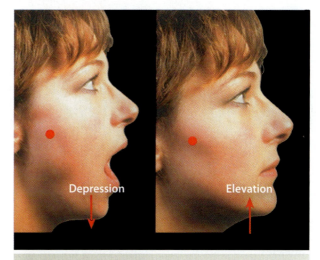

Depression   Elevation

Depression and **elevation** occur when a structure moves in an inferior or a superior direction, respectively.

Module 6.13 Review

a. Snapping your fingers involves what movement with the thumb and third metacarpophalangeal joint?

b. What movements are made possible by the rotation of the radius head?

c. What hand movements occur when wriggling into tight-fitting gloves?

Skeletal muscles are grouped in the axial division or appendicular division based on origins and functions

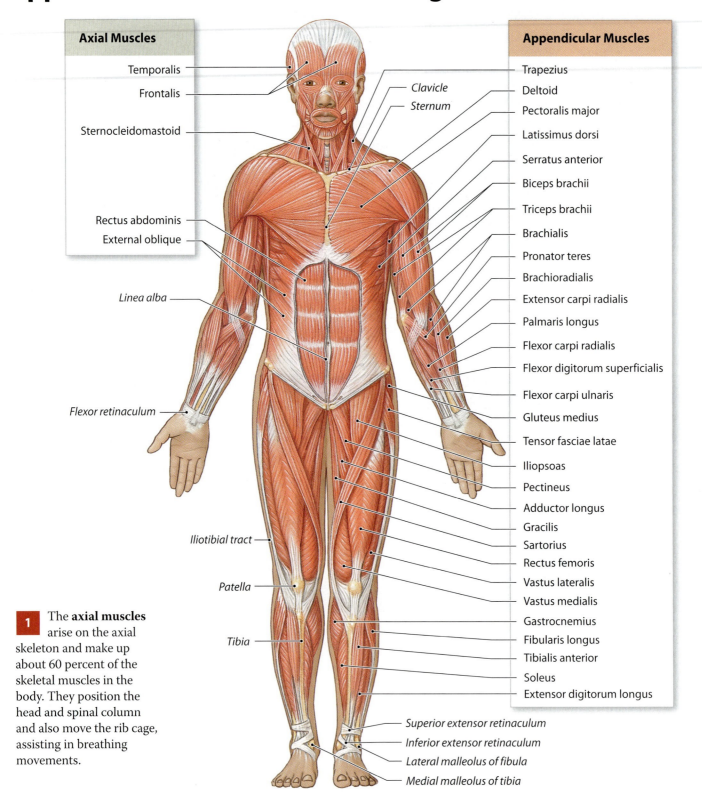

Axial Muscles

Temporalis
Frontalis
Sternocleidomastoid
Rectus abdominis
External oblique
Linea alba
Flexor retinaculum
Iliotibial tract
Patella
Tibia

Clavicle
Sternum

Appendicular Muscles

Trapezius
Deltoid
Pectoralis major
Latissimus dorsi
Serratus anterior
Biceps brachii
Triceps brachii
Brachialis
Pronator teres
Brachioradialis
Extensor carpi radialis
Palmaris longus
Flexor carpi radialis
Flexor digitorum superficialis
Flexor carpi ulnaris
Gluteus medius
Tensor fasciae latae
Iliopsoas
Pectineus
Adductor longus
Gracilis
Sartorius
Rectus femoris
Vastus lateralis
Vastus medialis
Gastrocnemius
Fibularis longus
Tibialis anterior
Soleus
Extensor digitorum longus

Superior extensor retinaculum
Inferior extensor retinaculum
Lateral malleolus of fibula
Medial malleolus of tibia

1 The **axial muscles** arise on the axial skeleton and make up about 60 percent of the skeletal muscles in the body. They position the head and spinal column and also move the rib cage, assisting in breathing movements.

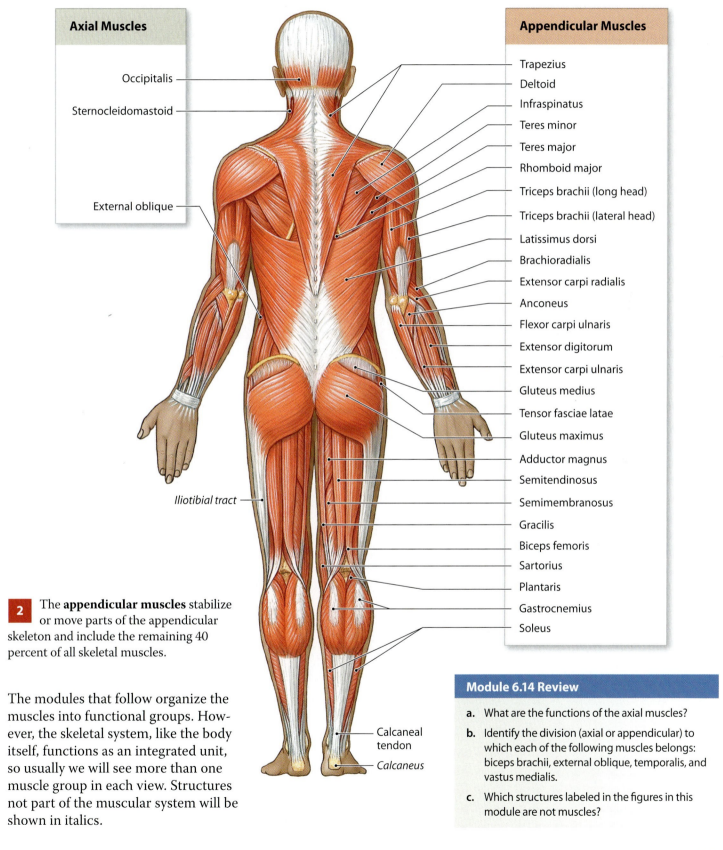

Axial Muscles

- Occipitalis
- Sternocleidomastoid
- External oblique

Appendicular Muscles

- Trapezius
- Deltoid
- Infraspinatus
- Teres minor
- Teres major
- Rhomboid major
- Triceps brachii (long head)
- Triceps brachii (lateral head)
- Latissimus dorsi
- Brachioradialis
- Extensor carpi radialis
- Anconeus
- Flexor carpi ulnaris
- Extensor digitorum
- Extensor carpi ulnaris
- Gluteus medius
- Tensor fasciae latae
- Gluteus maximus
- Adductor magnus
- Semitendinosus
- Semimembranosus
- Gracilis
- Biceps femoris
- Sartorius
- Plantaris
- Gastrocnemius
- Soleus

Iliotibial tract

Calcaneal tendon

Calcaneus

2 The **appendicular muscles** stabilize or move parts of the appendicular skeleton and include the remaining 40 percent of all skeletal muscles.

The modules that follow organize the muscles into functional groups. However, the skeletal system, like the body itself, functions as an integrated unit, so usually we will see more than one muscle group in each view. Structures not part of the muscular system will be shown in italics.

Module 6.14 Review

a. What are the functions of the axial muscles?

b. Identify the division (axial or appendicular) to which each of the following muscles belongs: biceps brachii, external oblique, temporalis, and vastus medialis.

c. Which structures labeled in the figures in this module are not muscles?

The muscles of the head and neck are important in eating and useful for communication

The muscles of facial expression originate on the surface of the skull. At their insertions, the fibers of the epimysium are woven into those of the connective tissue and the dermis of the skin. Thus, when they contract, the skin moves.

1 This anterior view shows superficial muscles on the right side of the face and deeper muscles on the left side of the face.

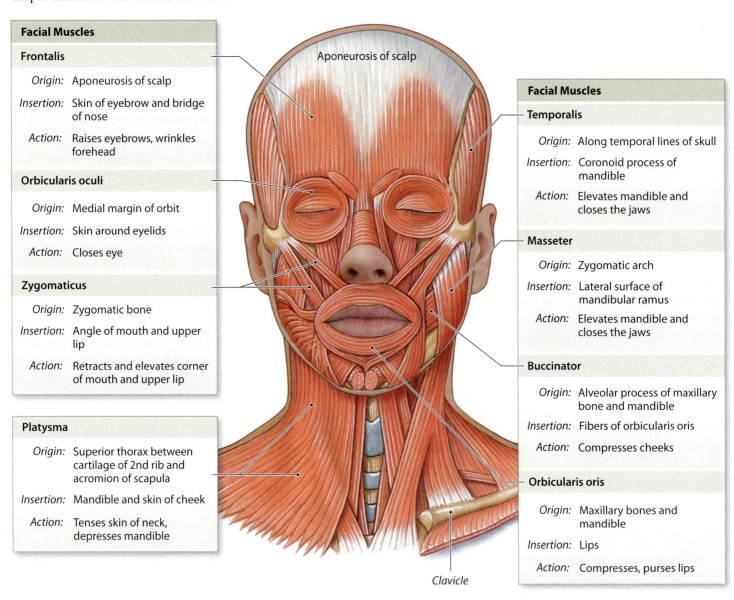

Facial Muscles

Frontalis

Origin: Aponeurosis of scalp

Insertion: Skin of eyebrow and bridge of nose

Action: Raises eyebrows, wrinkles forehead

Orbicularis oculi

Origin: Medial margin of orbit

Insertion: Skin around eyelids

Action: Closes eye

Zygomaticus

Origin: Zygomatic bone

Insertion: Angle of mouth and upper lip

Action: Retracts and elevates corner of mouth and upper lip

Platysma

Origin: Superior thorax between cartilage of 2nd rib and acromion of scapula

Insertion: Mandible and skin of cheek

Action: Tenses skin of neck, depresses mandible

Aponeurosis of scalp

Facial Muscles

Temporalis

Origin: Along temporal lines of skull

Insertion: Coronoid process of mandible

Action: Elevates mandible and closes the jaws

Masseter

Origin: Zygomatic arch

Insertion: Lateral surface of mandibular ramus

Action: Elevates mandible and closes the jaws

Buccinator

Origin: Alveolar process of maxillary bone and mandible

Insertion: Fibers of orbicularis oris

Action: Compresses cheeks

Orbicularis oris

Origin: Maxillary bones and mandible

Insertion: Lips

Action: Compresses, purses lips

Clavicle

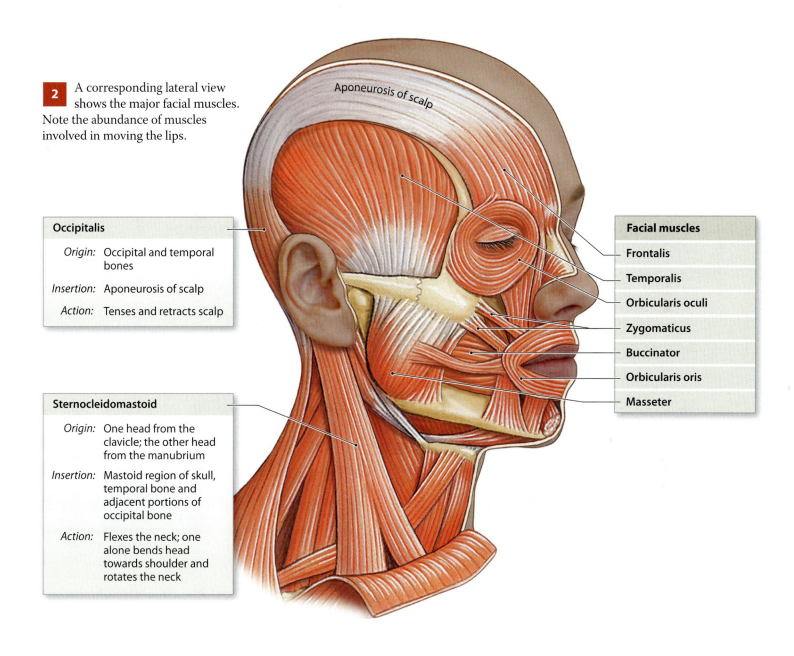

2 A corresponding lateral view shows the major facial muscles. Note the abundance of muscles involved in moving the lips.

Aponeurosis of scalp

Occipitalis

Origin: Occipital and temporal bones

Insertion: Aponeurosis of scalp

Action: Tenses and retracts scalp

Sternocleidomastoid

Origin: One head from the clavicle; the other head from the manubrium

Insertion: Mastoid region of skull, temporal bone and adjacent portions of occipital bone

Action: Flexes the neck; one alone bends head towards shoulder and rotates the neck

Facial muscles

Frontalis

Temporalis

Orbicularis oculi

Zygomaticus

Buccinator

Orbicularis oris

Masseter

Module 6.15 Review

a. Where do the muscles of facial expression originate?

b. You bite into an apple. Name the muscles involved.

c. Which muscles do you contract to produce a smile?

The muscles of the vertebral column support and align the axial skeleton

1 The **muscles of the vertebral column** are arranged in several layers. They include muscles originating or inserting on the ribs and the processes of the vertebrae. Although this mass of muscles extends from the sacrum to the skull, each muscle group is composed of numerous separate muscles of various lengths.

Spinal Extensors, Superficial Layer

Splenius capitis

Erector Spinae

The erector spinae muscles are subdivided based on proximity to the vertebral column.

Spinalis

Origin: Spinous and transverse processes of thoracic and superior lumbar vertebrae

Insertion: Spinous processes of superior thoracic vertebrae

Action: Extend vertebral column

Longissimus

Origin: Transverse processes of inferior cervical, thoracic, and lumbar vertebrae

Insertion: Mastoid process of temporal bone; transverse processes of middle and superior cervical and thoracic vertebrae; inferior surfaces of ribs

Action: Together, the two sides extend the vertebral column; alone, each rotates and laterally flexes it to that side

Iliocostalis

Origin: Superior borders of ribs, iliac crest, sacral crests, and spinous processes of lumbar vertebrae

Insertion: Transverse processes of middle and inferior cervical vertebrae; inferior surfaces of inferior seven ribs

Action: Extend or laterally flex vertebral column, elevate or depress ribs

Quadratus lumborum

Origin: Iliac crest

Insertion: Last rib and transverse processes of lumbar vertebrae

Action: Together, they depress ribs; alone each side laterally flexes vertebral column

Thoracodorsal fascia

Posterior view

Module 6.16 Review

a. Name the erector spinae muscles from medial to lateral in relation to the vertebral column.

b. Which vertebral column muscles originate on the iliac crest?

c. Name the spinal extensor muscles that have insertions on the skull.

The oblique and rectus muscles form the muscular walls of the trunk

1 *Oblique* muscles lie at an angle to the body axis, whereas *rectus* muscles are parallel to the body axis.

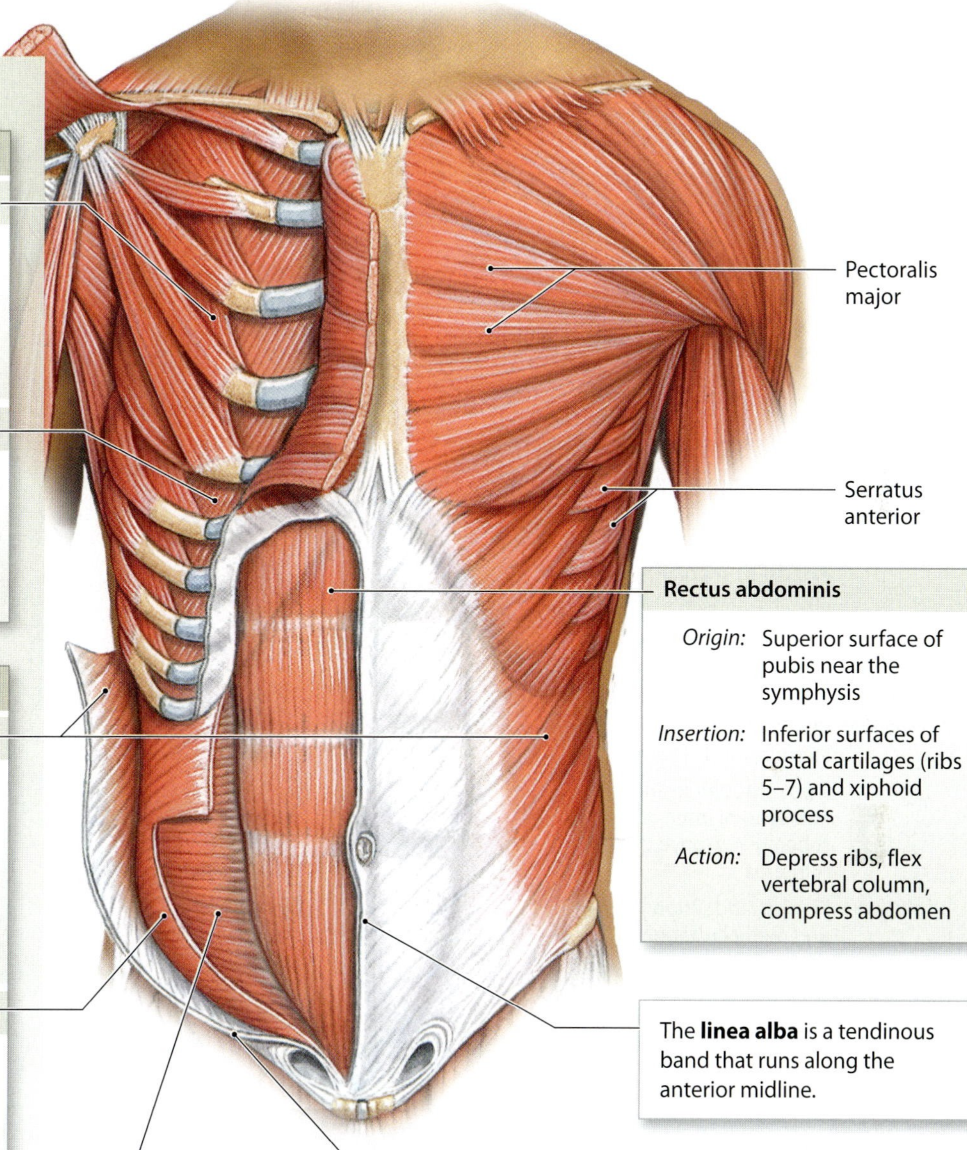

Oblique Group

Thoracic Region

External intercostals

Origin: Inferior border of each rib

Insertion: Superior border of more inferior rib

Action: Elevate ribs

Internal intercostals

Origin: Superior border of each rib

Insertion: Inferior border of the adjacent superior rib

Action: Depress ribs

Abdominal Region

External oblique

Origin: External and inferior borders of ribs 5–12

Insertion: Linea alba and iliac crest

Action: Compress abdomen; depress ribs, flex or bend vertebral column

Internal oblique

Origin: Lumbodorsal fascia and iliac crest

Insertion: Inferior ribs, xiphoid process, and linea alba

Action: Compress abdomen; depress ribs, flex or bend vertebral column

Transversus abdominis

Origin: Cartilages of ribs 6–12, iliac crest, and lumbodorsal fascia

Insertion: Linea alba and pubis

Action: Compress abdomen

Pectoralis major

Serratus anterior

Rectus abdominis

Origin: Superior surface of pubis near the symphysis

Insertion: Inferior surfaces of costal cartilages (ribs 5–7) and xiphoid process

Action: Depress ribs, flex vertebral column, compress abdomen

The **linea alba** is a tendinous band that runs along the anterior midline.

Cut edge of rectus sheath

Anterior view

Deep muscles are shown on the right side of the body, and superficial muscles of the oblique and rectus groups are shown on the left side.

Module 6.17 Review

a. Name the abdominal muscles from superficial to deep.

b. Damage to the external intercostal muscles would interfere with what important process?

c. If someone hit you in the rectus abdominis muscle, how would your body position change?

Large axial and appendicular muscles originate on the trunk

1 Modules 6.15–6.17 have considered the axial muscles, many of which are included here. In general, muscles originating on the trunk control gross movements of the limbs. These muscles are often large and powerful. Distally, the limb muscles get smaller and more numerous, and their movements become more precise.

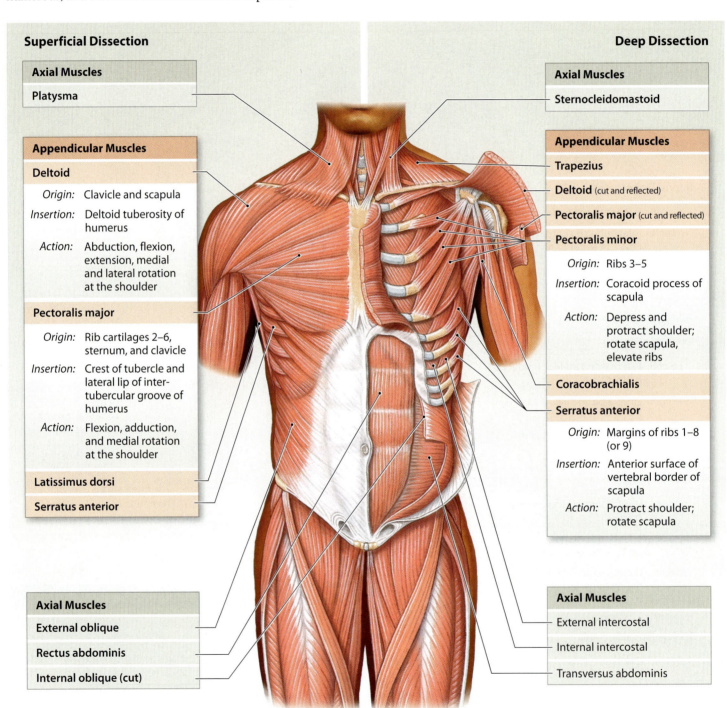

Superficial Dissection

Axial Muscles

Platysma

Appendicular Muscles

Deltoid

Origin:	Clavicle and scapula
Insertion:	Deltoid tuberosity of humerus
Action:	Abduction, flexion, extension, medial and lateral rotation at the shoulder

Pectoralis major

Origin:	Rib cartilages 2–6, sternum, and clavicle
Insertion:	Crest of tubercle and lateral lip of inter-tubercular groove of humerus
Action:	Flexion, adduction, and medial rotation at the shoulder

Latissimus dorsi

Serratus anterior

Axial Muscles

External oblique

Rectus abdominis

Internal oblique (cut)

Deep Dissection

Axial Muscles

Sternocleidomastoid

Appendicular Muscles

Trapezius

Deltoid (cut and reflected)

Pectoralis major (cut and reflected)

Pectoralis minor

Origin:	Ribs 3–5
Insertion:	Coracoid process of scapula
Action:	Depress and protract shoulder; rotate scapula, elevate ribs

Coracobrachialis

Serratus anterior

Origin:	Margins of ribs 1–8 (or 9)
Insertion:	Anterior surface of vertebral border of scapula
Action:	Protract shoulder; rotate scapula

Axial Muscles

External intercostal

Internal intercostal

Transversus abdominis

2 The posterior surface of the trunk is dominated by appendicular muscles that originate on the large bones of the limb girdles and the proximal bones of the limbs.

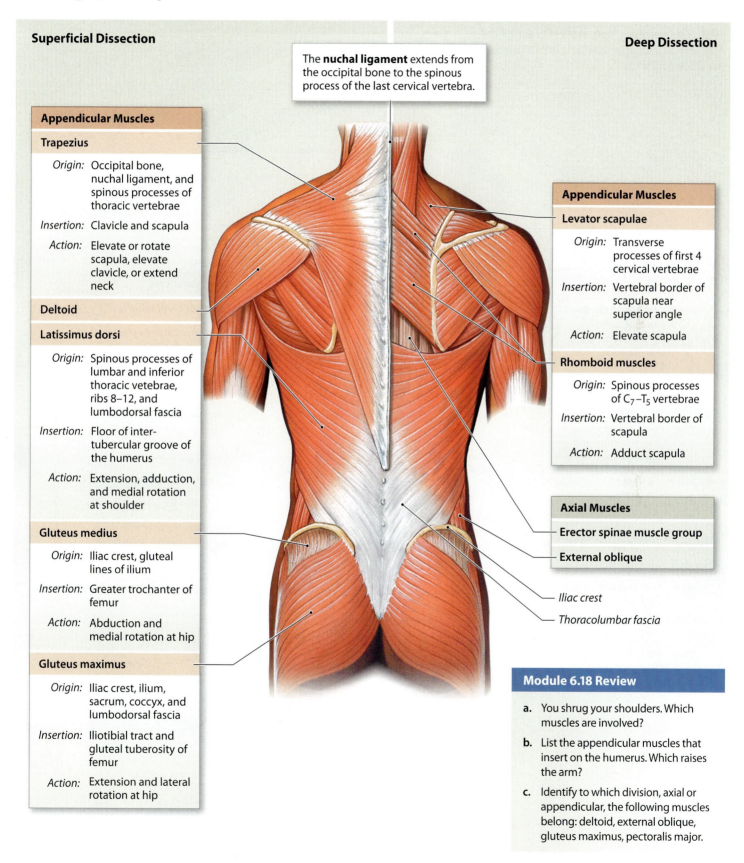

Superficial Dissection

Deep Dissection

The **nuchal ligament** extends from the occipital bone to the spinous process of the last cervical vertebra.

Appendicular Muscles

Trapezius

Origin: Occipital bone, nuchal ligament, and spinous processes of thoracic vertebrae

Insertion: Clavicle and scapula

Action: Elevate or rotate scapula, elevate clavicle, or extend neck

Deltoid

Latissimus dorsi

Origin: Spinous processes of lumbar and inferior thoracic vetebrae, ribs 8–12, and lumbodorsal fascia

Insertion: Floor of inter-tubercular groove of the humerus

Action: Extension, adduction, and medial rotation at shoulder

Gluteus medius

Origin: Iliac crest, gluteal lines of ilium

Insertion: Greater trochanter of femur

Action: Abduction and medial rotation at hip

Gluteus maximus

Origin: Iliac crest, ilium, sacrum, coccyx, and lumbodorsal fascia

Insertion: Iliotibial tract and gluteal tuberosity of femur

Action: Extension and lateral rotation at hip

Appendicular Muscles

Levator scapulae

Origin: Transverse processes of first 4 cervical vertebrae

Insertion: Vertebral border of scapula near superior angle

Action: Elevate scapula

Rhomboid muscles

Origin: Spinous processes of C_7–T_5 vertebrae

Insertion: Vertebral border of scapula

Action: Adduct scapula

Axial Muscles

Erector spinae muscle group

External oblique

Iliac crest

Thoracolumbar fascia

Module 6.18 Review

a. You shrug your shoulders. Which muscles are involved?

b. List the appendicular muscles that insert on the humerus. Which raises the arm?

c. Identify to which division, axial or appendicular, the following muscles belong: deltoid, external oblique, gluteus maximus, pectoralis major.

Proximal limb muscles are larger, stronger, fewer, and less precise in their actions than distal limb muscles

1 This posterior view shows the superficial layer of muscles involved in extension at the elbow and wrist.

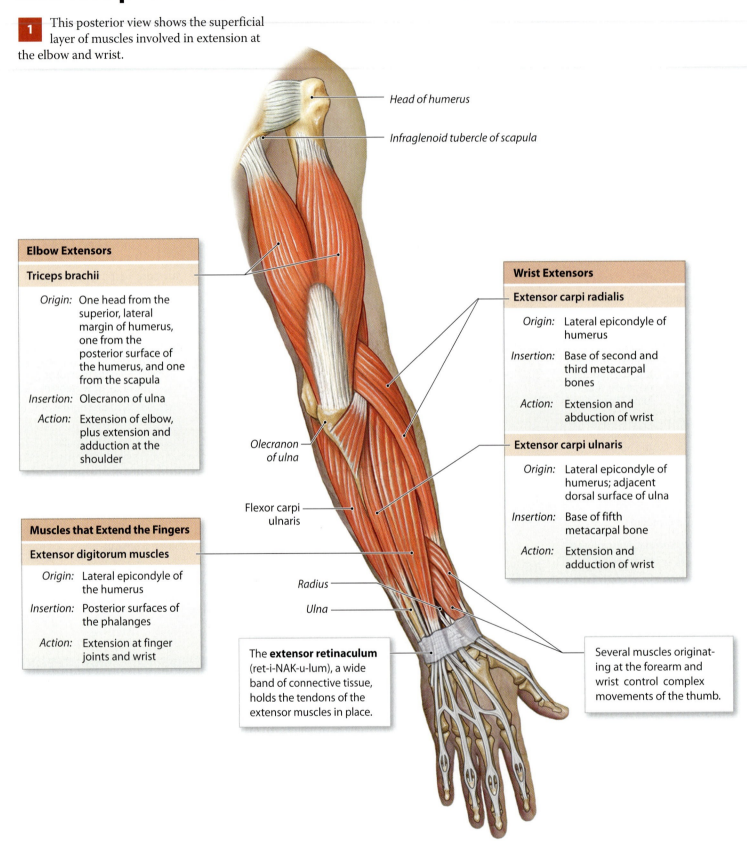

Head of humerus

Infraglenoid tubercle of scapula

Elbow Extensors

Triceps brachii

Origin: One head from the superior, lateral margin of humerus, one from the posterior surface of the humerus, and one from the scapula

Insertion: Olecranon of ulna

Action: Extension of elbow, plus extension and adduction at the shoulder

Olecranon of ulna

Flexor carpi ulnaris

Muscles that Extend the Fingers

Extensor digitorum muscles

Origin: Lateral epicondyle of the humerus

Insertion: Posterior surfaces of the phalanges

Action: Extension at finger joints and wrist

Wrist Extensors

Extensor carpi radialis

Origin: Lateral epicondyle of humerus

Insertion: Base of second and third metacarpal bones

Action: Extension and abduction of wrist

Extensor carpi ulnaris

Origin: Lateral epicondyle of humerus; adjacent dorsal surface of ulna

Insertion: Base of fifth metacarpal bone

Action: Extension and adduction of wrist

Radius

Ulna

The **extensor retinaculum** (ret-i-NAK-u-lum), a wide band of connective tissue, holds the tendons of the extensor muscles in place.

Several muscles originating at the forearm and wrist control complex movements of the thumb.

2 This anterior view shows the superficial muscles involved in flexion at the elbow and wrist.

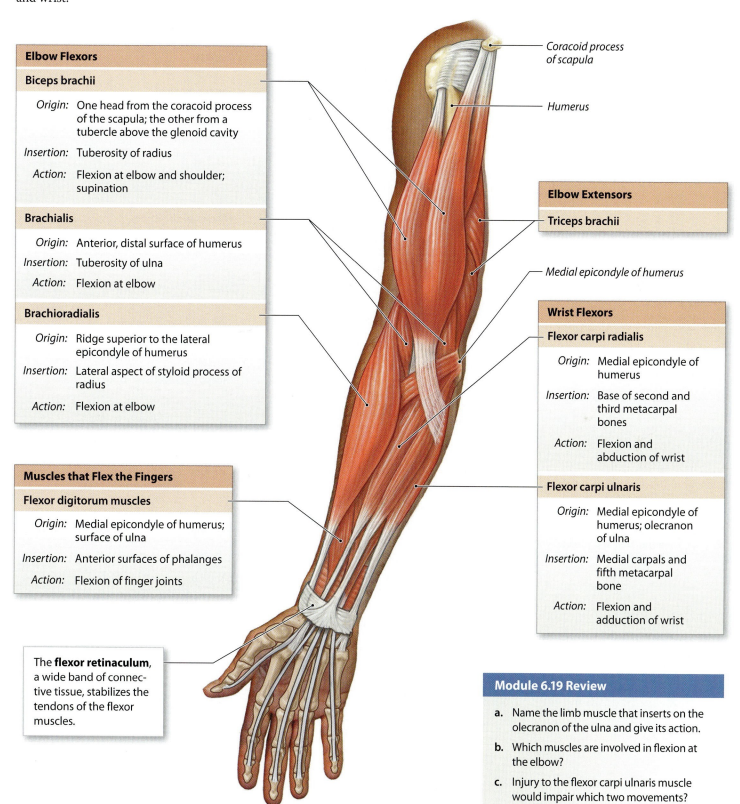

Elbow Flexors

Biceps brachii

Origin:	One head from the coracoid process of the scapula; the other from a tubercle above the glenoid cavity
Insertion:	Tuberosity of radius
Action:	Flexion at elbow and shoulder; supination

Brachialis

Origin:	Anterior, distal surface of humerus
Insertion:	Tuberosity of ulna
Action:	Flexion at elbow

Brachioradialis

Origin:	Ridge superior to the lateral epicondyle of humerus
Insertion:	Lateral aspect of styloid process of radius
Action:	Flexion at elbow

Muscles that Flex the Fingers

Flexor digitorum muscles

Origin:	Medial epicondyle of humerus; surface of ulna
Insertion:	Anterior surfaces of phalanges
Action:	Flexion of finger joints

The **flexor retinaculum**, a wide band of connective tissue, stabilizes the tendons of the flexor muscles.

Coracoid process of scapula

Humerus

Elbow Extensors

Triceps brachii

Medial epicondyle of humerus

Wrist Flexors

Flexor carpi radialis

Origin:	Medial epicondyle of humerus
Insertion:	Base of second and third metacarpal bones
Action:	Flexion and abduction of wrist

Flexor carpi ulnaris

Origin:	Medial epicondyle of humerus; olecranon of ulna
Insertion:	Medial carpals and fifth metacarpal bone
Action:	Flexion and adduction of wrist

Module 6.19 Review

a. Name the limb muscle that inserts on the olecranon of the ulna and give its action.

b. Which muscles are involved in flexion at the elbow?

c. Injury to the flexor carpi ulnaris muscle would impair which two movements?

The muscles that move the leg originate on the pelvis and femur

1 **Flexors of the knee** originate on the pelvic girdle and extend along the posterior and medial surfaces of the thigh. The semitendinosus, biceps femoris, and semimembranosus at the back of the thigh are commonly called the **hamstring muscles**. A strain in any of these is called a pulled hamstring.

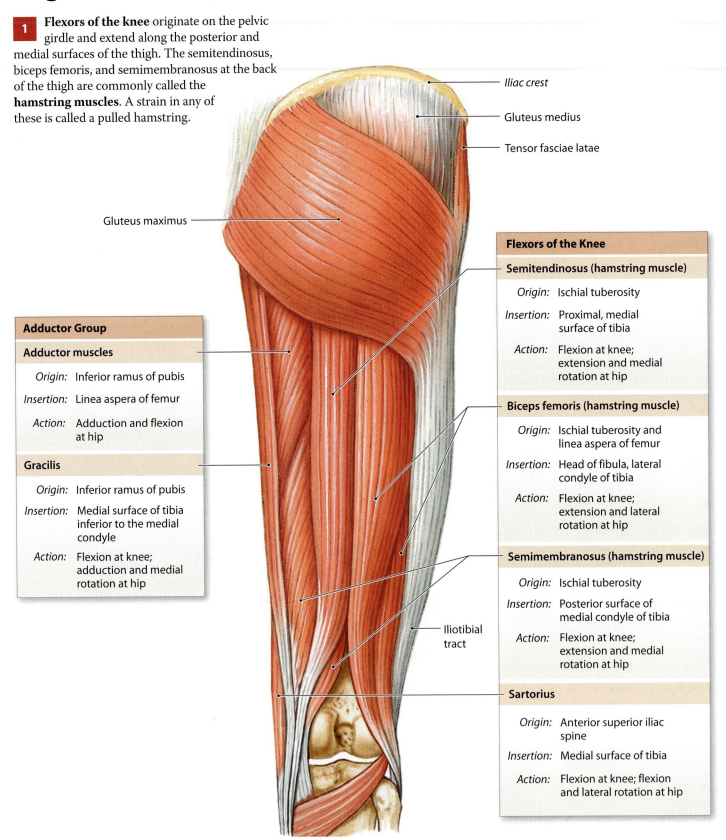

Iliac crest

Gluteus medius

Tensor fasciae latae

Gluteus maximus

Iliotibial tract

Adductor Group

Adductor muscles

Origin: Inferior ramus of pubis

Insertion: Linea aspera of femur

Action: Adduction and flexion at hip

Gracilis

Origin: Inferior ramus of pubis

Insertion: Medial surface of tibia inferior to the medial condyle

Action: Flexion at knee; adduction and medial rotation at hip

Flexors of the Knee

Semitendinosus (hamstring muscle)

Origin: Ischial tuberosity

Insertion: Proximal, medial surface of tibia

Action: Flexion at knee; extension and medial rotation at hip

Biceps femoris (hamstring muscle)

Origin: Ischial tuberosity and linea aspera of femur

Insertion: Head of fibula, lateral condyle of tibia

Action: Flexion at knee; extension and lateral rotation at hip

Semimembranosus (hamstring muscle)

Origin: Ischial tuberosity

Insertion: Posterior surface of medial condyle of tibia

Action: Flexion at knee; extension and medial rotation at hip

Sartorius

Origin: Anterior superior iliac spine

Insertion: Medial surface of tibia

Action: Flexion at knee; flexion and lateral rotation at hip

2 Most of the **extensors of the knee** originate on the femoral surface and extend along the anterior and lateral surfaces of the thigh. Collectively the knee extensors are called the **quadriceps muscles**.

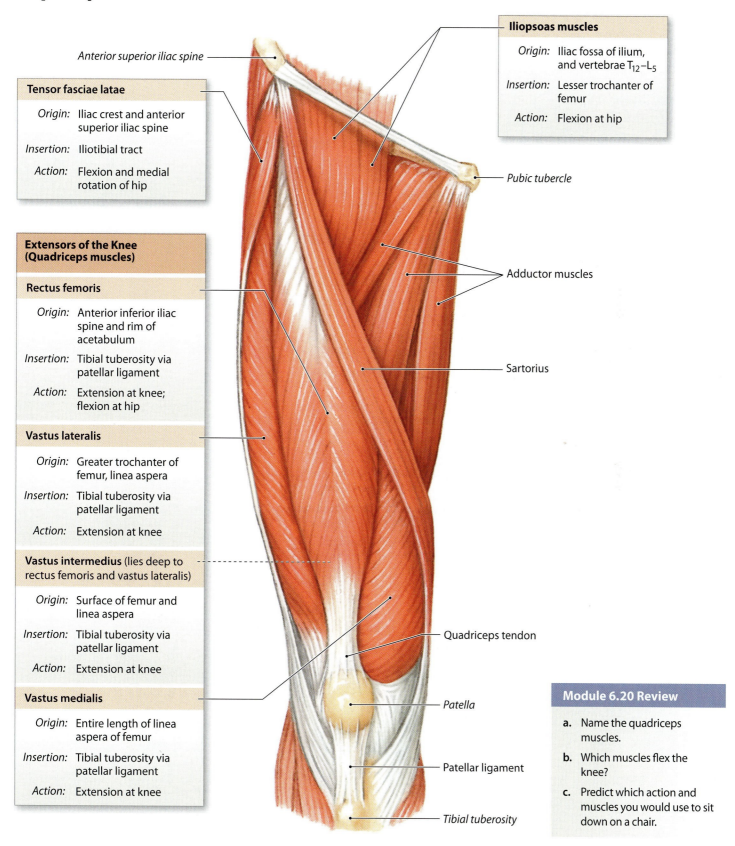

Anterior superior iliac spine

Tensor fasciae latae

Origin:	Iliac crest and anterior superior iliac spine
Insertion:	Iliotibial tract
Action:	Flexion and medial rotation of hip

Iliopsoas muscles

Origin:	Iliac fossa of ilium, and vertebrae T_{12}–L_5
Insertion:	Lesser trochanter of femur
Action:	Flexion at hip

Pubic tubercle

Adductor muscles

Sartorius

Extensors of the Knee (Quadriceps muscles)

Rectus femoris

Origin:	Anterior inferior iliac spine and rim of acetabulum
Insertion:	Tibial tuberosity via patellar ligament
Action:	Extension at knee; flexion at hip

Vastus lateralis

Origin:	Greater trochanter of femur, linea aspera
Insertion:	Tibial tuberosity via patellar ligament
Action:	Extension at knee

Vastus intermedius (lies deep to rectus femoris and vastus lateralis)

Origin:	Surface of femur and linea aspera
Insertion:	Tibial tuberosity via patellar ligament
Action:	Extension at knee

Vastus medialis

Origin:	Entire length of linea aspera of femur
Insertion:	Tibial tuberosity via patellar ligament
Action:	Extension at knee

Quadriceps tendon

Patella

Patellar ligament

Tibial tuberosity

Module 6.20 Review

a. Name the quadriceps muscles.

b. Which muscles flex the knee?

c. Predict which action and muscles you would use to sit down on a chair.

The primary muscles that move the foot and toes originate on the tibia and fibula

1 These two posterior views show the superficial and deep muscle layers of the right leg.

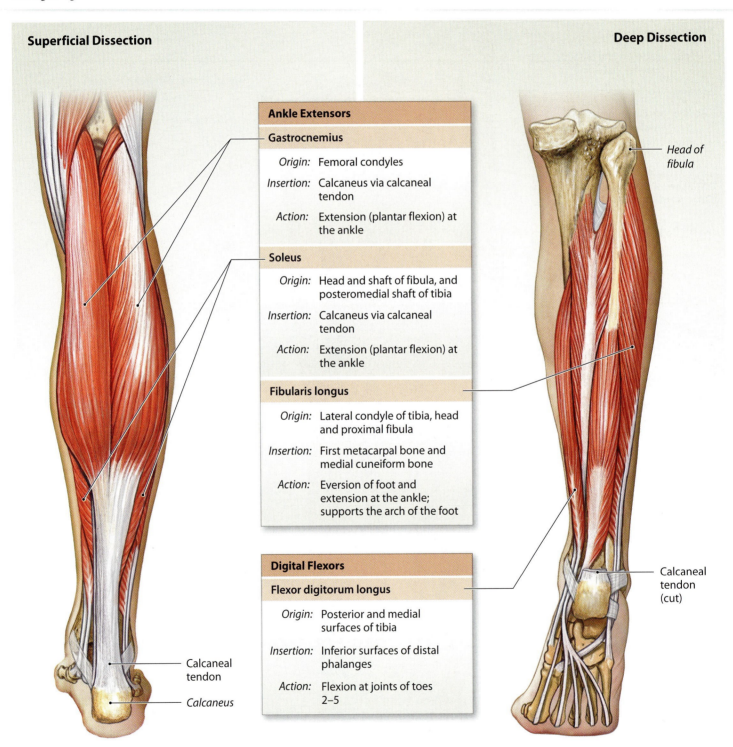

Superficial Dissection

Deep Dissection

Ankle Extensors

Gastrocnemius

Origin:	Femoral condyles
Insertion:	Calcaneus via calcaneal tendon
Action:	Extension (plantar flexion) at the ankle

Soleus

Origin:	Head and shaft of fibula, and posteromedial shaft of tibia
Insertion:	Calcaneus via calcaneal tendon
Action:	Extension (plantar flexion) at the ankle

Fibularis longus

Origin:	Lateral condyle of tibia, head and proximal fibula
Insertion:	First metacarpal bone and medial cuneiform bone
Action:	Eversion of foot and extension at the ankle; supports the arch of the foot

Digital Flexors

Flexor digitorum longus

Origin:	Posterior and medial surfaces of tibia
Insertion:	Inferior surfaces of distal phalanges
Action:	Flexion at joints of toes 2–5

Head of fibula

Calcaneal tendon (cut)

Calcaneal tendon

Calcaneus

2 These lateral and medial views show the arrangement of the major superficial muscles.

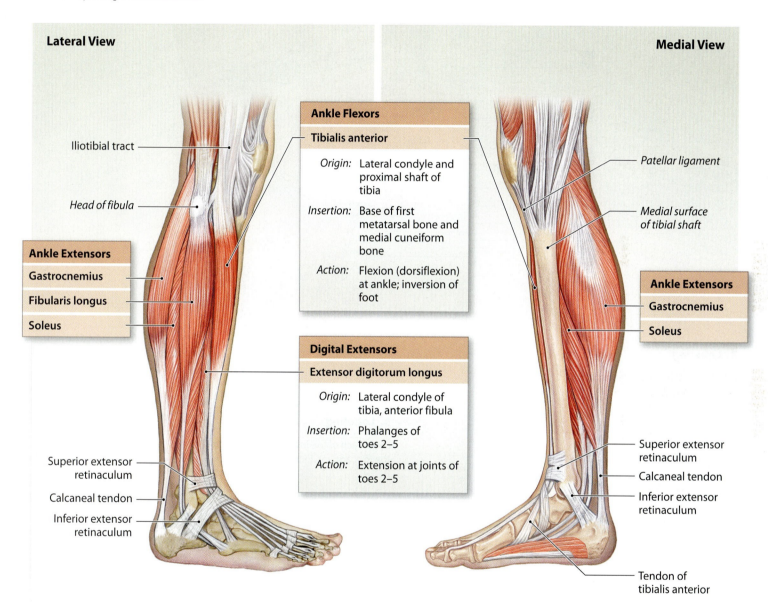

Lateral View

Iliotibial tract

Head of fibula

Ankle Extensors

Gastrocnemius

Fibularis longus

Soleus

Superior extensor retinaculum

Calcaneal tendon

Inferior extensor retinaculum

Ankle Flexors

Tibialis anterior

Origin:	Lateral condyle and proximal shaft of tibia
Insertion:	Base of first metatarsal bone and medial cuneiform bone
Action:	Flexion (dorsiflexion) at ankle; inversion of foot

Digital Extensors

Extensor digitorum longus

Origin:	Lateral condyle of tibia, anterior fibula
Insertion:	Phalanges of toes 2–5
Action:	Extension at joints of toes 2–5

Medial View

Patellar ligament

Medial surface of tibial shaft

Ankle Extensors

Gastrocnemius

Soleus

Superior extensor retinaculum

Calcaneal tendon

Inferior extensor retinaculum

Tendon of tibialis anterior

The largest muscles associated with ankle movement are the **gastrocnemius** and **soleus muscles**. These muscles produce extension (plantar flexion at the ankle), a movement essential to walking and running. The muscles that move the toes are much smaller, and they originate on the surface of the tibia, the fibula, or both. Large tendon sheaths surround tendons where they cross the ankle joint. The **superior** and **inferior retinacula**, tough supporting bands of collagen fibers, stabilize the positions of these sheaths.

Module 6.21 Review

a. Name the muscles that produce plantar flexion.

b. You let up on the gas pedal while driving. Identify the action and muscles involved.

c. How would a torn calcaneal tendon affect movement of the foot?

1. Labeling

Label each of the indicated muscles of the face and neck in the following diagram.

a

b

c

d

e

f

g

h

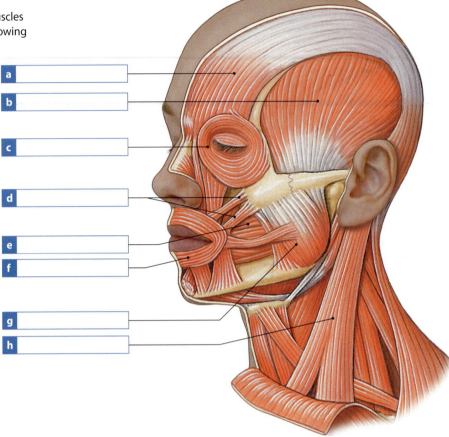

2. Labeling

Label each of the indicated muscles that move the thigh and leg in the diagram below.

a

b

c

d

e

f

g

h

i

g

3. Labeling

Label each of the indicated muscles that move the foot and toe in the diagram below.

a

b

c

d

e

f

h

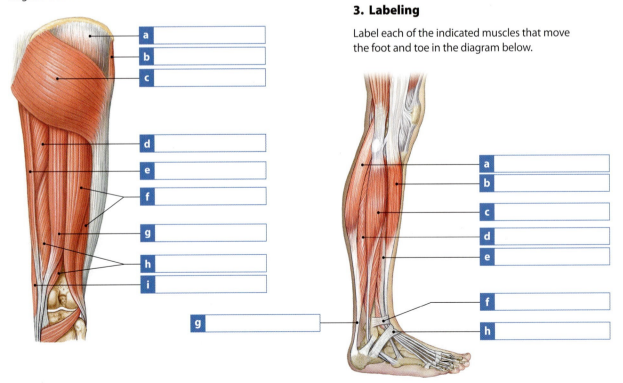

Visual Outline with Key Terms

Summarize the content of each module using the terms in the order provided.

SECTION 1

Functional Anatomy of Skeletal Muscle Tissue

- skeletal muscles
- skeletal muscle tissue
- cardiac muscle tissue
- smooth muscle tissue

6.1

A skeletal muscle contains skeletal muscle tissue, connective tissues, blood vessels, and nerves

- epimysium
- skeletal muscle
- perimysium
- muscle fascicle
- endomysium
- muscle fibers
- tendon
- aponeurosis
- sarcolemma
- sarcoplasm
- myofibril
- myofilaments
- thin filaments
- thick filaments
- sarcoplasmic reticulum

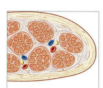

6.2

Skeletal muscle fibers have contractile myofibrils containing hundreds to thousands of sarcomeres

- ○ skeletal muscle fiber
- sarcomeres
- H band
- M line
- zone of overlap
- A band
- I band
- Z lines
- actin
- tropomyosin
- troponin
- ○ thick filaments
- myosin molecules
- tail
- head
- membrane potential
- transverse tubules (T tubules)
- sarcoplasmic reticulum (SR)

6.3

A skeletal muscle fiber contracts when stimulated by a motor neuron

- neuromuscular junction (NMJ)
- axon terminal
- motor end plate
- synaptic cleft
- acetylcholine (ACh)
- acetylcholinesterase (AChE)
- action potential

6.4

The sliding filament theory explains the physical changes that occur during a contraction

- sliding filament theory
- ○ sarcomere
- ○ myofibril at rest
- ○ contracted myofibril

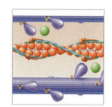

6.5

A muscle fiber contraction uses ATP in a cycle that repeats for the duration of the contraction

- contraction cycle
- ○ active-site exposure
- ○ cross-bridge formation
- ○ myosin head pivoting
- ○ cross-bridge detachment
- ○ myosin reactivation

SECTION 2

Functional Properties of Skeletal Muscle Tissue

- ○ neural control
- ○ excitation–contraction coupling
- tension

6.6

Tension production rises to maximum levels as the rate of muscle stimulation increases

- twitch
- latent period
- contraction phase
- relaxation phase
- wave summation
- incomplete tetanus
- complete tetanus

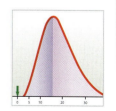

• = *Term boldfaced in this module*

6.7

A skeletal muscle controls muscle tension by the number of motor units stimulated

- motor unit
- recruitment
- muscle tone

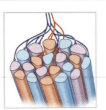

SECTION 3

Functional Organization of the Muscular System

- muscular system
- axial muscles
- appendicular muscles

Muscular system

6.8

Muscle contractions may be isotonic or isometric

- isotonic contraction
- isometric contraction

6.11

The names of muscles provide clues to their appearance and/or function

- origin
- insertion
- action
- agonist
- synergist
- antagonist

6.9

Muscle contraction requires a large amount of ATP that may be produced aerobically or anaerobically

- creatine phosphate (CP)
- pyruvate
- lactic acid
- lactate
- Cori cycle
- oxygen debt

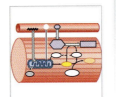

6.12

Broad descriptive terms are used to describe movements with reference to the anatomical position

- flexion
- extension
- hyperextension
- lateral flexion
- dorsiflexion
- plantar flexion
- abduction
- adduction
- circumduction

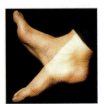

6.10

Many factors can result in muscle hypertrophy, atrophy, or paralysis

- hypertrophy
- atrophy
- polio
- paralysis
- tetanus
- botulism
- myasthenia gravis
- rigor mortis

6.13

Terms of more limited application describe rotational movements and special movements

- left rotation
- right rotation
- medial rotation
- lateral rotation
- pronation
- supination
- opposition
- protraction
- retraction
- inversion
- eversion
- depression
- elevation

● = *Term boldfaced in this module*

6.14

Skeletal muscles are grouped in the axial division or appendicular division based on origins and functions

- axial muscles
- appendicular muscles

6.18

Large axial and appendicular muscles originate on the trunk

- appendicular muscles
- deltoid
- pectoralis major
- latissimus dorsi
- serratus anterior
- trapezius
- pectoralis minor
- coracobrachialis
- nuchal ligament
- gluteus medius
- gluteus maximus
- levator scapulae
- rhomboid muscles

6.15

The muscles of the head and neck are important in eating and useful for communication

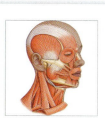

- facial muscles
- frontalis
- orbicularis oculi
- zygomaticus
- temporalis
- platysma
- masseter
- buccinator
- orbicularis oris
- occipitalis
- sternocleidomastoid

6.19

Proximal limb muscles are larger, stronger, fewer, and less precise in their actions than distal limb muscles

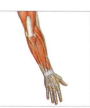

- extensors
- triceps brachii
- extensor digitorum muscles
- extensor carpi radialis
- extensor carpi ulnaris
- extensor retinaculum
- flexors
- biceps brachii
- brachialis
- brachioradialis
- flexor digitorum muscles
- flexor retinaculum
- flexor carpi radialis
- flexor carpi ulnaris

6.16

The muscles of the vertebral column support and align the axial skeleton

- muscles of the vertebral column
- spinal extensors
- splenius capitis
- erector spinae muscles
- spinalis
- longissimus
- iliocostalis
- quadratus lumborum

6.20

The muscles that move the leg originate on the pelvis and femur

- flexors of the knee
- hamstring muscles
- adductor group
- adductor muscles
- gracilis
- semitendinosus
- biceps femoris
- semimembranosus
- sartorius
- extensors of the knee
- quadriceps muscles
- tensor fasciae latae
- rectus femoris
- vastus lateralis
- vastus intermedius
- vastus medialis
- iliopsoas muscles

6.17

The oblique and rectus muscles form the muscular walls of the trunk

- oblique group
- thoracic region
- external intercostals
- internal intercostals
- abdominal region
- external oblique
- internal oblique
- transversus abdominis
- rectus abdominis
- linea alba

6.21

The primary muscles that move the foot and toes originate on the tibia and fibula

- ankle extensors
- gastrocnemius
- soleus
- fibularis longus
- digital flexors
- flexor digitorum longus
- ankle flexors
- tibialis anterior
- digital extensors
- extensor digitorum longus
- gastrocnemius muscles
- soleus muscles
- superior retinacula
- inferior retinacula

• = *Term boldfaced in this module*

"If you're going to claim you're working with someone's body, you need to understand muscles—not just where they are but what they do."

— **Pauline Hui**
Personal Trainer
San Francisco, CA

CAREER PATHS

Personal Trainer

Many of the professions profiled in this book focus on people who have been hurt or ill. Pauline Hui, a personal trainer who works primarily at the Bay Club in San Francisco, does help people who are recovering from injury or surgery, but she also works with healthy people who are trying to stay that way. "I love helping people, whether it's for the treatment part or the personal training part," she says.

"I like seeing them make good changes, whether they're beginners or advanced. It's a feel-good job."

Personal trainers do not need to have any special certification to seek employment, yet certification is offered through the American College of Sports Medicine. However, a thorough understanding of how the body works is crucial, as doctors and physical therapists may refer clients as part of rehabilitation. Pauline is certified in Muscle Activation Technique, which focuses on strengthening muscles to reduce the risk of injury. It also is intended to help people recover more quickly from injury or surgery and develop greater stability in their movements.

Pauline's oldest client is 83 and has had two strokes. Her sessions begin with treatment, then some strength exercises. In some sessions, the client can only move her arm 90 degrees at the beginning, but by the end of the session she has almost the full range of motion.

Pauline emphasizes how important it is for personal trainers to understand anatomy and physiology. "If you're going to claim you're working with someone's body, you need to understand muscles—not just where they are but what they do."

In addition to anatomy, Pauline recommends that personal trainers be well-versed in physics, to understand how the forces of different exercise machines affect the body.

Because personal trainers work one-on-one with their clients, they also must have good people skills. They must be comfortable not only explaining the value of doing certain exercises, but also standing and making small talk with clients. They must be able to read clients and understand how much each client wants to be pushed—and when not to push even when the client asks.

"I had a person with sciatica who wanted to do a certain exercise, and I said, 'I'll just be upfront with you: You're not ready for this,'" Pauline says. "A lot of people have that mentality that pain is good, but that's not always the case. A good trainer knows that."

An average training session for Pauline lasts about 50 minutes, with the specific activities tailored to the client. "When somebody picks me, right off the bat, I ask what his or her goals are," she says. Pauline has worked with some clients for her entire 14-year career and has watched their goals change over time: from simply losing weight to running a marathon or adding muscle.

The stereotypical image of personal trainers is that of young, hard-bodied men and women, but Pauline emphasizes that trainers can be older, and don't have to look like they stepped out of a fitness magazine. "You have to be physically fit," she says. "You should be able to run a mile, if you have no joint issues. Other than that, you don't have to look like a bodybuilder."

In addition to health clubs and fitness centers, personal trainers also work for civic organizations, parks, and school districts, as well as being self-employed.

For more information, visit the website of the American College of Sports Medicine at http://www.acsm.org.

Think this is the career for you?

Key Stats:

- **Education and Training.** A high school diploma and additional training is usually sufficient for entry level, but a bachelor's degree in exercise science, physical education, kinesiology, or a related area is usually required for advancement.

- **Licensure.** States do not require licensing or certification, but most employers do.

- **Earnings.** Earnings vary but the median annual salary is $31.090. Many personal trainers work part-time in addition to other jobs.

- **Expected Job Prospects.** Employment is expected to grow by an above average 29% from 2008 to 2018.

Bureau of Labor Statistics, U.S. Department of Labor, *Occupational Outlook Handbook, 2010-11 Edition*, Fitness Workers, on the Internet at http://www.bls.gov/oco/ocos296.htm (visited September 14, 2011).

Access more review material online in the Study Area at www.masteringaandp.com.

- Chapter guides
- Chapter quizzes
- Practice tests
- Art labeling activities
- Flashcards
- A glossary with pronunciations

- Practice Anatomy Lab™ (PAL™) 3.0 virtual anatomy practice tool
- Interactive Physiology® (IP) animated tutorials
- MP3 Tutor Sessions

 practice anatomy lab™   **For this chapter, follow these navigation paths in PAL:**

- Human Cadaver>Muscular System
- Anatomical Models>Muscular System
- Histology>Muscular System

iP   **For this chapter, go to these topics in the Muscular System in IP:**

- Anatomy Review: Skeletal Muscle Tissue
- The Neuromuscular Junction
- Sliding Filament Theory
- Muscle Metabolism
- Contraction of Whole Muscle

   For this chapter, go to this topic in the MP3 Tutor Sessions:

- Sliding Filament Theory of Contraction

Chapter 6 Review Questions

1. Describe the basic sequence of events that occurs at a neuromuscular junction.

2. Which two mechanisms are used to generate ATP from glucose in muscle cells?

3. Many potent insecticides contain toxins called organophosphates, which interfere with the action of the enzyme acetylcholinesterase. Terry is using an insecticide containing organophosphates and is very careless. He does not use gloves or a mask, so he absorbs some of the chemical through his skin and inhales a large amount as well. What signs would you expect to observe in Terry as a result of organophosphate poisoning?

4. The time of a murder victim's death is commonly estimated by the flexibility/stiffness of the body. Explain why this is possible.

5. Jared is interested in building up his thigh muscles, specifically the quadriceps group. What exercises would you recommend to help him accomplish his goal?

For answers to all module, section, and chapter review questions, see the blue Answers tab at the back of the book.

7

The Central Nervous System

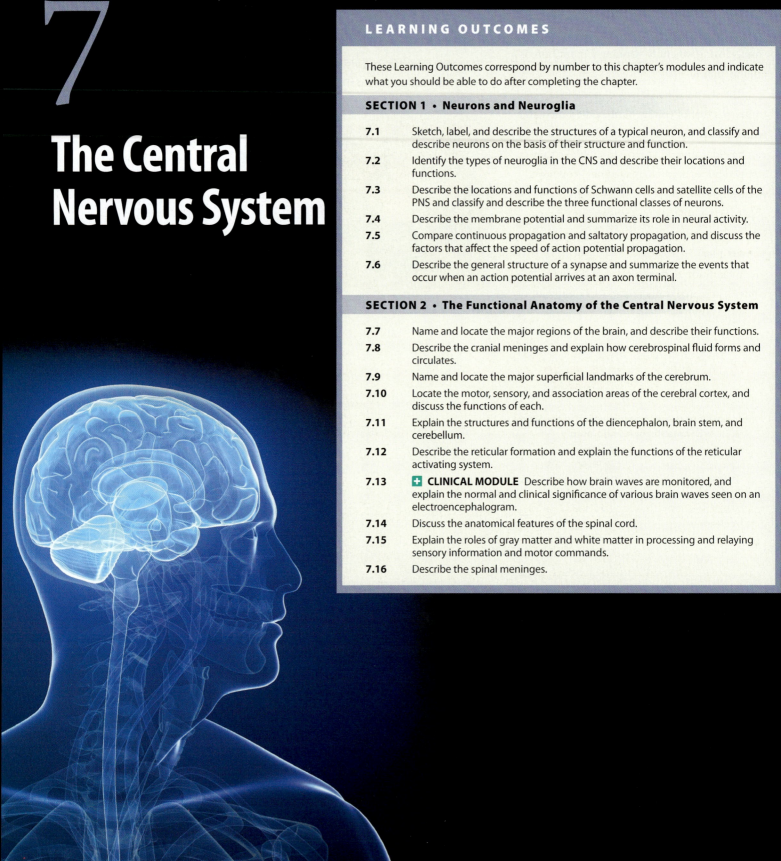

Neurons and Neuroglia

In this chapter we consider the structure of neural tissue and introduce basic principles of neurophysiology. Our discussion of neural function requires a basic understanding of the functional organization of the nervous system. This flowchart indicates major components and functions of the nervous system.

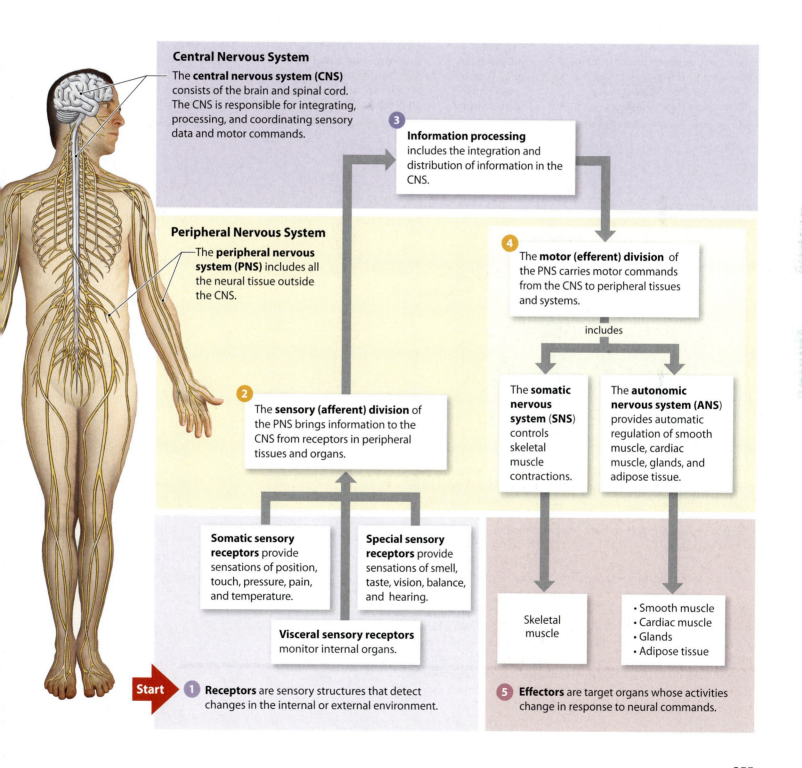

Central Nervous System

The **central nervous system (CNS)** consists of the brain and spinal cord. The CNS is responsible for integrating, processing, and coordinating sensory data and motor commands.

3 **Information processing** includes the integration and distribution of information in the CNS.

Peripheral Nervous System

The **peripheral nervous system (PNS)** includes all the neural tissue outside the CNS.

4 The **motor (efferent) division** of the PNS carries motor commands from the CNS to peripheral tissues and systems.

includes

2 The **sensory (afferent) division** of the PNS brings information to the CNS from receptors in peripheral tissues and organs.

The **somatic nervous system (SNS)** controls skeletal muscle contractions.

The **autonomic nervous system (ANS)** provides automatic regulation of smooth muscle, cardiac muscle, glands, and adipose tissue.

Somatic sensory receptors provide sensations of position, touch, pressure, pain, and temperature.

Special sensory receptors provide sensations of smell, taste, vision, balance, and hearing.

Visceral sensory receptors monitor internal organs.

Skeletal muscle

• Smooth muscle
• Cardiac muscle
• Glands
• Adipose tissue

Start **1** **Receptors** are sensory structures that detect changes in the internal or external environment.

5 **Effectors** are target organs whose activities change in response to neural commands.

255

Neurons are nerve cells specialized for intercellular communication

In this module we examine the structural features of neurons. Neurons have three general regions: dendrites, which receive stimuli from the environment or from other neurons; a cell body, which contains the nucleus and other organelles; and an axon, which carries information toward other cells.

Dendrites

Typical dendrites are highly branched, with each branch bearing fine 0.5- to 1-μm-long processes. CNS neurons receive most of their information at the dendrites.

Axon

The **axon hillock** is the origin of the axon from the nerve cell body.

Axons may produce branches known as **collaterals**. These branches enable a single neuron to communicate with several other cells.

The main axon trunk ends in a series of fine extensions. At the end of each extension is an **axon terminal.**

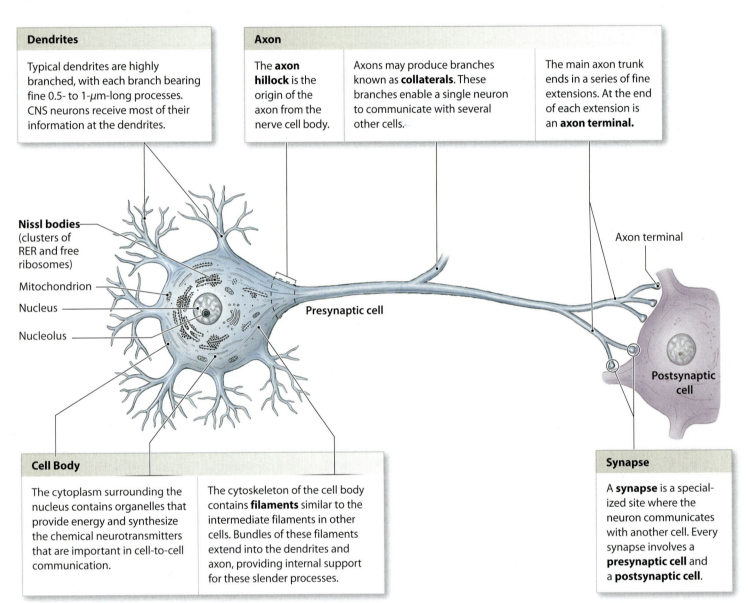

Nissl bodies (clusters of RER and free ribosomes)

Mitochondrion

Nucleus

Nucleolus

Presynaptic cell

Axon terminal

Postsynaptic cell

Cell Body

The cytoplasm surrounding the nucleus contains organelles that provide energy and synthesize the chemical neurotransmitters that are important in cell-to-cell communication.

The cytoskeleton of the cell body contains **filaments** similar to the intermediate filaments in other cells. Bundles of these filaments extend into the dendrites and axon, providing internal support for these slender processes.

Synapse

A **synapse** is a specialized site where the neuron communicates with another cell. Every synapse involves a **presynaptic cell** and a **postsynaptic cell**.

1 Here is a diagrammatic view of a representative neuron. The cell body contains most of the organelles of the neuron. The axon is a long cytoplasmic process that extends away from the cell body. Many materials, including enzymes and lysosomes, travel the length of the axon in both directions along **neurotubules** (neuron microtubules). If debris or unusual chemicals appear in the axon terminal, this transport mechanism soon delivers them to the cell body.

Stuctural Variations among Neurons

There are three major anatomical classes of neurons.

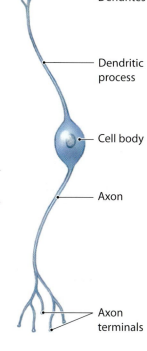

Dendrites

Dendritic process

Cell body

Axon

Axon terminals

2 **Bipolar neurons** have two processes—a dendritic process and an axon—separated by the cell body. Bipolar neurons are uncommon, but occur in special sense organs, where they relay information about sight, smell, or hearing from receptor cells to other neurons. Bipolar neurons are small; the largest measure less than 30 μm from end to end.

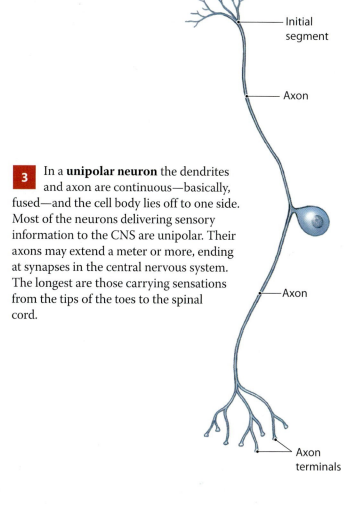

Dendrites

Initial segment

Axon

Axon

Axon terminals

3 In a **unipolar neuron** the dendrites and axon are continuous—basically, fused—and the cell body lies off to one side. Most of the neurons delivering sensory information to the CNS are unipolar. Their axons may extend a meter or more, ending at synapses in the central nervous system. The longest are those carrying sensations from the tips of the toes to the spinal cord.

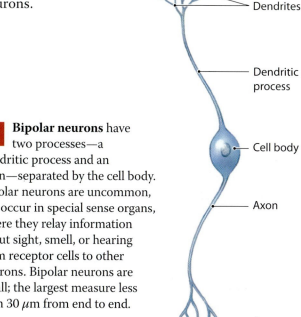

Dendrites

Axon

Axon terminals

4 **Multipolar neurons** have two or more dendrites and a single axon. These are the most common neurons in the CNS. All motor neurons that control skeletal muscles, for example, are multipolar neurons. The axons of multipolar neurons can be as long as those of unipolar neurons; the longest carry motor commands from the spinal cord to small muscles that move the toes.

Most CNS neurons lack centrioles and cannot divide. As a result, neurons lost to injury or disease are seldom replaced. Although neural stem cells persist in the adult nervous system, these cells are typically inactive (except in the epithelium responsible for our sense of smell, in the retina of the eye, and in the hippocampus, a portion of the brain involved with storing memories).

Module 7.1 Review

a. Name the structural components of a typical neuron.

b. Classify neurons according to their structure.

c. Why is a CNS neuron not usually replaced after it is injured?

Oligodendrocytes, astrocytes, ependymal cells, and microglia are neuroglia of the CNS

Neuroglia (or **glial cells**)—cells that support and protect neurons in the PNS and CNS—are abundant and diverse, accounting for about half the volume of the nervous system. Neural tissue in the CNS is organized differently than in the PNS, primarily because the CNS has a greater variety of glial cell types. This illustration summarizes information about neuroglia in the CNS.

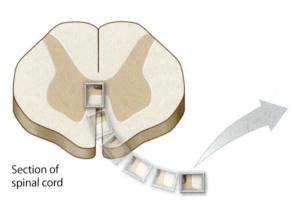

Section of
spinal cord

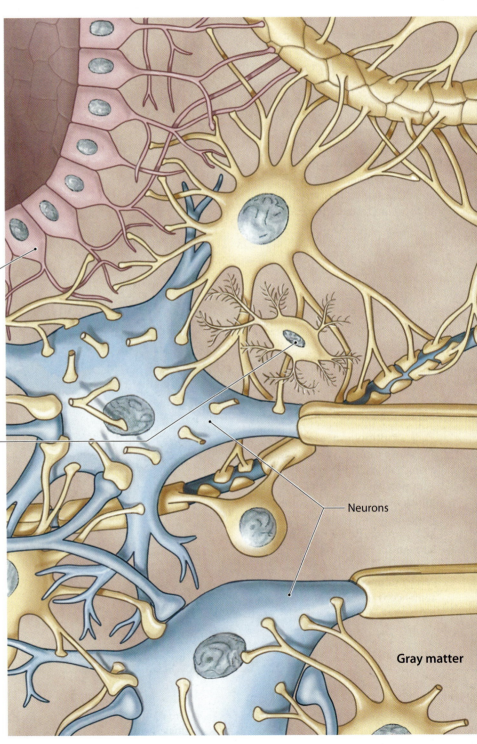

Ependymal cells form an epithelium known as the **ependyma** (ep-EN-di-muh), which lines a fluid-filled passageway within the spinal cord and brain. This passageway is filled with **cerebrospinal fluid (CSF)**, which also surrounds the brain and spinal cord. Ependymal cells help produce, monitor, and circulate the CSF; some cells in the ventricles of the brain may be ciliated.

Microglia (mī-KROG-lē-uh) are embryologically related to monocytes and macrophages. Microglia migrate into the CNS as the nervous system forms and they persist as mobile cells, constantly moving through the neural tissue, removing cellular debris, waste products, and pathogens by phagocytosis.

Neurons

Gray matter

Astrocytes maintain the **blood–brain barrier** that isolates the CNS from the chemicals and hormones circulating in the blood. They also provide structural support within neural tissue; regulate ion, nutrient, and dissolved gas concentrations in the interstitial fluid surrounding the neurons; absorb and recycle neurotransmitters that are not broken down or reabsorbed at synapses; and form scar tissue after CNS injury.

Oligodendrocytes (ol-i-gō-DEN-drō-sīts; *oligo-*, few) provide a structural framework within the CNS by stabilizing the positions of axons. They also produce **myelin** (MĪ-e-lin), a membranous lipid-rich wrapping that coats axons and speeds up the transmission of nerve impulses. When myelinating an axon, the tip of an oligodendrocyte process expands to form an enormous membranous pad containing very little cytoplasm. This flattened "pancake" somehow winds around the axon, forming concentric layers of plasma membrane. These layers constitute a **myelin sheath**.

Many oligodendrocytes form the myelin sheath along the length of an axon. Such an axon is said to be **myelinated**. Each oligodendrocyte myelinates segments of several axons. The relatively large areas of the axon that are thus wrapped in myelin are called **internodes** (*inter*, between).

The small gaps of a few micrometers that separate adjacent internodes are called **nodes**. In dissection, myelinated axons appear glossy white, primarily because of the lipids within the myelin. As a result, regions dominated by myelinated axons make up the **white matter** of the CNS.

Not all axons in the CNS are myelinated. **Unmyelinated axons** may not be completely covered by the processes of neuroglia. Such axons are common where relatively short axons and collaterals form synapses with densely packed neuron cell bodies. Areas containing neuron cell bodies, dendrites, and unmyelinated axons have a dusky gray color, and they make up the **gray matter** of the CNS.

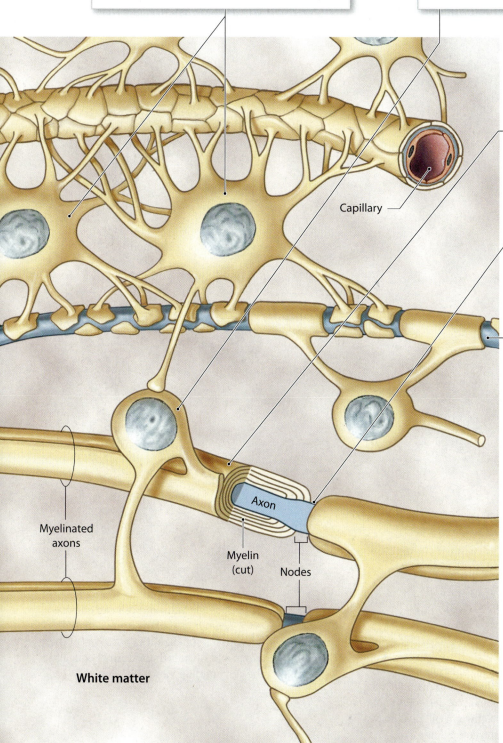

Capillary

Myelinated axons

Axon

Myelin (cut)

Nodes

White matter

Module 7.2 Review

a. Identify the neuroglia of the central nervous system.

b. Which glial cell protects the CNS from chemicals and hormones circulating in the blood?

c. Which type of neuroglia would occur in increased numbers in the brain tissue of a person with a CNS infection?

Schwann cells and satellite cells protect the axons and cell bodies of sensory and motor neurons in the PNS

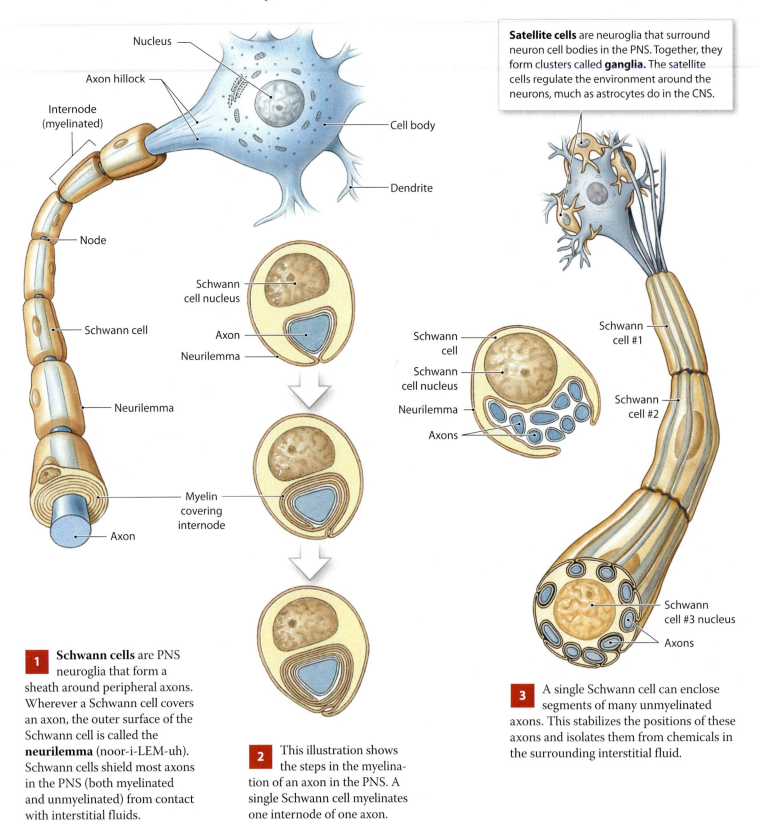

Nucleus

Axon hillock

Internode (myelinated)

Cell body

Dendrite

Satellite cells are neuroglia that surround neuron cell bodies in the PNS. Together, they form clusters called **ganglia.** The satellite cells regulate the environment around the neurons, much as astrocytes do in the CNS.

Node

Schwann cell nucleus

Axon

Neurilemma

Schwann cell

Schwann cell nucleus

Neurilemma

Axons

Schwann cell

Neurilemma

Schwann cell #1

Schwann cell #2

Myelin covering internode

Axon

Schwann cell #3 nucleus

Axons

1 **Schwann cells** are PNS neuroglia that form a sheath around peripheral axons. Wherever a Schwann cell covers an axon, the outer surface of the Schwann cell is called the **neurilemma** (noor-i-LEM-uh). Schwann cells shield most axons in the PNS (both myelinated and unmyelinated) from contact with interstitial fluids.

2 This illustration shows the steps in the myelination of an axon in the PNS. A single Schwann cell myelinates one internode of one axon.

3 A single Schwann cell can enclose segments of many unmyelinated axons. This stabilizes the positions of these axons and isolates them from chemicals in the surrounding interstitial fluid.

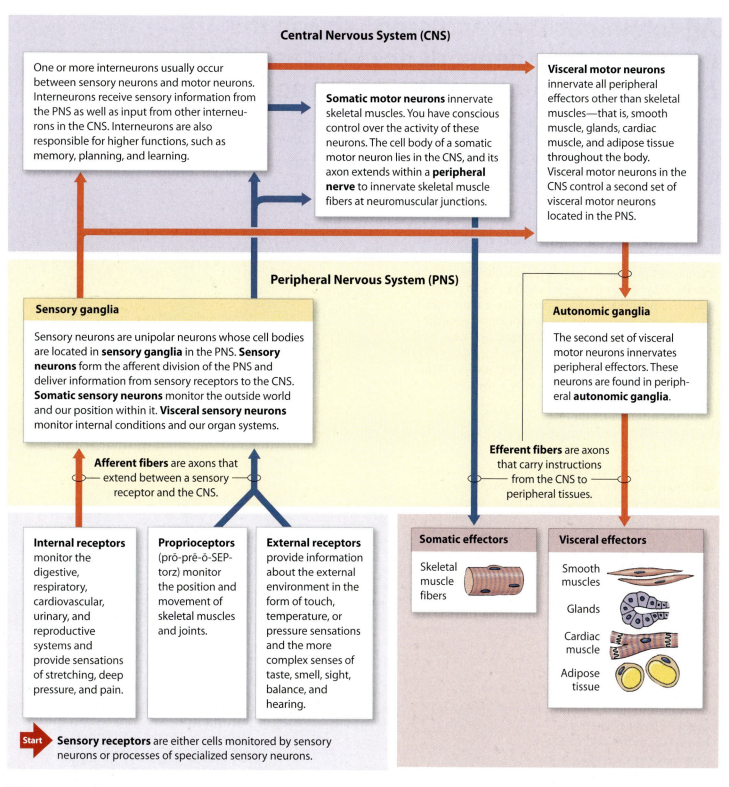

Central Nervous System (CNS)

One or more interneurons usually occur between sensory neurons and motor neurons. Interneurons receive sensory information from the PNS as well as input from other interneurons in the CNS. Interneurons are also responsible for higher functions, such as memory, planning, and learning.

Somatic motor neurons innervate skeletal muscles. You have conscious control over the activity of these neurons. The cell body of a somatic motor neuron lies in the CNS, and its axon extends within a **peripheral nerve** to innervate skeletal muscle fibers at neuromuscular junctions.

Visceral motor neurons innervate all peripheral effectors other than skeletal muscles—that is, smooth muscle, glands, cardiac muscle, and adipose tissue throughout the body. Visceral motor neurons in the CNS control a second set of visceral motor neurons located in the PNS.

Peripheral Nervous System (PNS)

Sensory ganglia

Sensory neurons are unipolar neurons whose cell bodies are located in **sensory ganglia** in the PNS. **Sensory neurons** form the afferent division of the PNS and deliver information from sensory receptors to the CNS. **Somatic sensory neurons** monitor the outside world and our position within it. **Visceral sensory neurons** monitor internal conditions and our organ systems.

Autonomic ganglia

The second set of visceral motor neurons innervates peripheral effectors. These neurons are found in peripheral **autonomic ganglia**.

Afferent fibers are axons that extend between a sensory receptor and the CNS.

Efferent fibers are axons that carry instructions from the CNS to peripheral tissues.

Internal receptors monitor the digestive, respiratory, cardiovascular, urinary, and reproductive systems and provide sensations of stretching, deep pressure, and pain.

Proprioceptors (prō-prē-ō-SEP-torz) monitor the position and movement of skeletal muscles and joints.

External receptors provide information about the external environment in the form of touch, temperature, or pressure sensations and the more complex senses of taste, smell, sight, balance, and hearing.

Somatic effectors

Skeletal muscle fibers

Visceral effectors

Smooth muscles

Glands

Cardiac muscle

Adipose tissue

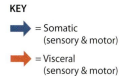

Start **Sensory receptors** are either cells monitored by sensory neurons or processes of specialized sensory neurons.

---

4 This simplified flowchart indicates the basic relationships among the three functional classes of neurons: **sensory neurons**, **interneurons**, and **motor neurons**. The human body has about 10 million sensory neurons, 20 billion interneurons, and half a million motor neurons.

KEY

➡ = Somatic (sensory & motor)

➡ = Visceral (sensory & motor)

Module 7.3 Review

a. Compare Schwann cells and satellite cells.

b. Describe the neurilemma.

c. Classify neurons into three categories according to their function.

All communication and processing in the nervous system depends on changes in the membrane potential of individual neurons

All of the diverse functions of the nervous system depend on cell-to-cell communication. Each neuron communicates with another neuron or other cell types. Changes in the permeability of the neuron's plasma membrane determine the nature of that communication.

1 Plasma membranes are selectively permeable, and the cytosol and extracellular fluid differ in composition. Under normal circumstances, the inside of the plasma membrane has a slight negative charge with respect to the outside. Why? Because there is a slight excess of positively charged ions outside the plasma membrane, and a slight excess of negatively charged ions and proteins inside the plasma membrane. This unequal charge distribution is called the **membrane potential**, and it is a characteristic of all living cells. It results from differences in the permeability of the membrane to various ions, as well as by active transport mechanisms.

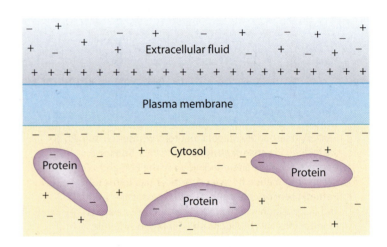

2 The extracellular fluid (ECF) contains high concentrations of sodium ions (Na^+) and chloride ions (Cl^-), whereas the cytosol contains high concentrations of potassium ions (K^+) and negatively charged proteins (Pr^-). The ions cannot freely cross the lipid portions of the plasma membrane; they can enter or leave the cell only through membrane channels or by active transport mechanisms. The membrane potential exists primarily because plasma membranes contain large numbers of sodium and potassium ion channels that are always open. The size, shape, and structure of a channel determines which ions can pass through it. The membrane also contains sodium and potassium ion channels that are closed in the resting cell. One set of channels opens when exposed to specific neurotransmitters, whereas another set of channels opens or closes in response to large changes in the membrane potential.

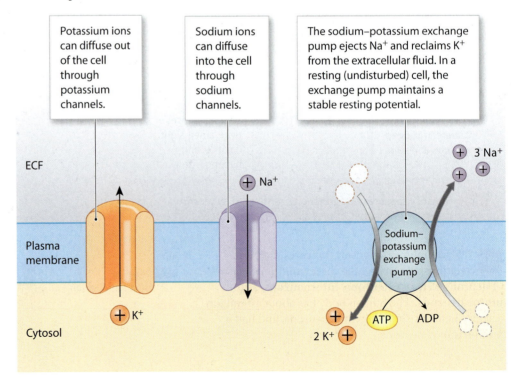

Potassium ions can diffuse out of the cell through potassium channels.

Sodium ions can diffuse into the cell through sodium channels.

The sodium–potassium exchange pump ejects Na^+ and reclaims K^+ from the extracellular fluid. In a resting (undisturbed) cell, the exchange pump maintains a stable resting potential.

3 This figure summarizes the role of the membrane potential in neural activity. Changes in the membrane potential have many important functions. For example, they can trigger muscle contraction and gland secretion as well as transfer information in the nervous system.

1 The membrane potential of a resting cell is called the **resting potential**. All neural activities begin with a change in the resting potential of a neuron.

2 A typical stimulus produces a temporary, localized change in the resting potential. The effect, which decreases with distance from the stimulus, is called a **graded potential**.

3 If the graded potential is sufficiently large, it triggers an **action potential** in the membrane of the axon. An action potential is an electrical event that involves one location on the membrane. Once an action potential develops in one location, it propagates (spreads) along the surface of an axon toward the axon terminals.

4 Typically, **synaptic activity** involves the release of chemical compounds by the presynaptic cell. These compounds, called **neurotransmitters**, bind to receptors on the plasma membrane of the postsynaptic cell, changing its permeability.

5 Each neuron cell body is carpeted with thousands to tens of thousands of axon terminals. Any number of them may be active at any given moment; some may release chemicals that stimulate the neuron, and others may release chemicals that inhibit the cell. It is the overall net effect that determines whether or not an action potential develops in the adjacent axon. This is the simplest form of **information processing** in the nervous system.

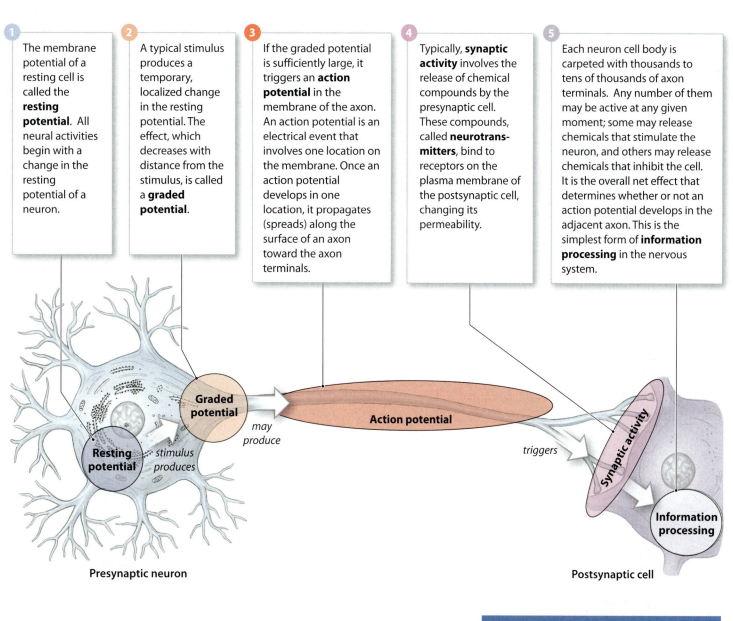

Resting potential   *stimulus produces*   **Graded potential**   *may produce*   **Action potential**   *triggers*   **Synaptic activity**   **Information processing**

Presynaptic neuron   **Postsynaptic cell**

Module 7.4 Review

a. Define membrane potential.

b. What happens at the sodium–potassium exchange pump?

c. List three body functions that result from changes in the membrane potential of a cell.

An action potential can affect other portions of the membrane through continuous or saltatory propagation

1 This graph shows the changes in the membrane potential that occur over time when different chemical stimuli are applied to the cell body of a neuron. The resting membrane potential is negative, because there are more negative charges inside the plasma membrane than outside. This unequal distribution of charges is called **polarization**.

A neurotransmitter chemical such as ACh opens sodium ion channels that are closed in the resting plasma membrane. As Na$^+$ ions cross the plasma membrane, the excess of negative charges inside the membrane is reduced and the membrane becomes less polarized. This change is called a **depolarization**.

When the chemical stimulus is removed, the sodium–potassium exchange pump returns the membrane to its normal resting potential. This process is called **repolarization**.

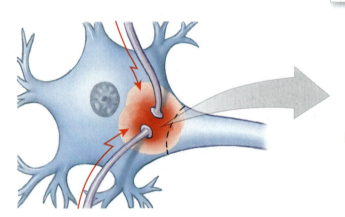

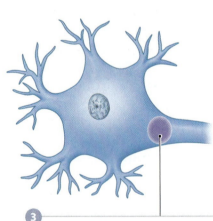

2 An action potential develops when a graded potential depolarizes the membrane to a point called the threshold potential. At that point, some membrane channels begin to respond as the membrane becomes less polarized. Three major events then occur.

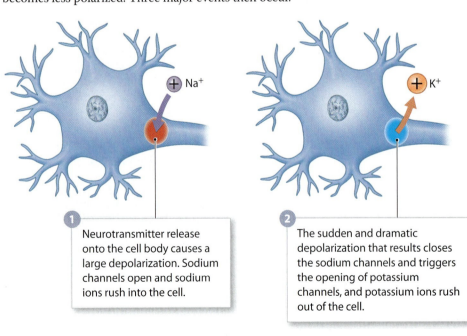

1 Neurotransmitter release onto the cell body causes a large depolarization. Sodium channels open and sodium ions rush into the cell.

2 The sudden and dramatic depolarization that results closes the sodium channels and triggers the opening of potassium channels, and potassium ions rush out of the cell.

3 The rapid loss of potassium ions begins to repolarize the plasma membrane. Over time the sodium-potassium exchange pump will eject the sodium ions and recapture the potassium ions.

3 When an action potential develops at one location, the change in the membrane potential at that site triggers an action potential at the adjacent portion of the plasma membrane, even though that area was not exposed to the neurotransmitter.

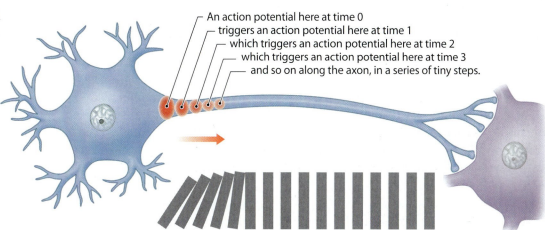

An action potential here at time 0
triggers an action potential here at time 1
which triggers an action potential here at time 2
which triggers an action potential here at time 3
and so on along the axon, in a series of tiny steps.

This process, called **continuous propagation**, is like the falling of closely spaced dominos. The action potential propagates slowly along the axon as each domino topples the next in line.

4 In a myelinated axon, the axonal membrane is exposed only at the nodes. As a result, action potentials can only develop at the nodes. This means that the action potential seems to jump from node to node along the axon.

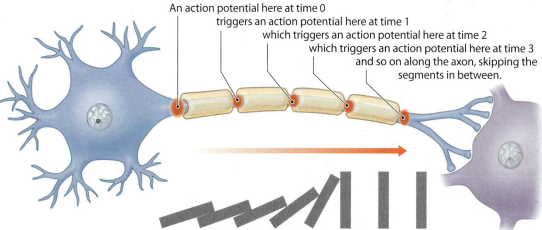

An action potential here at time 0
triggers an action potential here at time 1
which triggers an action potential here at time 2
which triggers an action potential here at time 3
and so on along the axon, skipping the segments in between.

This process, called **saltatory propagation** (*saltare*, leaping), is much faster than continuous propagation. Because the dominos are far apart, the distance covered per falling domino is much greater.

Axon diameter also affects propagation speed; action potentials travel fastest along large-diameter myelinated axons. So, why isn't every axon in our nervous system large and myelinated? The most likely reason is that it would be physically impossible. If all sensory information and motor commands were carried by large myelinated fibers, your peripheral nerves would be the size of garden hoses, and your spinal cord would be the diameter of a garbage can. Instead, information transfer in the nervous system is prioritized: Urgent news—sensory information about things that threaten survival and motor commands that prevent injury—travels over myelinated fibers (the equivalent of instant messaging). Unmyelinated fibers relay less urgent sensory information and motor commands.

Module 7.5 Review

a. Describe depolarization and repolarization.

b. Compare continuous and saltatory propagation.

c. What is the relationship between myelin and the propagation speed of action potentials?

At a synapse, information travels from presynaptic cell to postsynaptic cell

In the nervous system, messages are transmitted from one location to another along axons in the form of action potentials, also known as "nerve impulses." To be effective, messages must be not only propagated along an axon but also transferred in some way to another neuron or an effector cell. That transfer occurs at a **synapse**. At a synapse involving two neurons, information is relayed from a presynaptic neuron to a postsynaptic neuron. Except in a few special cases involving sensory receptors, a presynaptic cell is always a neuron.

1 Communication between presynaptic and postsynaptic cells most commonly involves the release of chemicals called **neurotransmitters** into the **synaptic cleft**, a narrow space separating the two cells. Axon terminals can reabsorb and reassemble fragments of neurotransmitters broken down in the synaptic cleft.

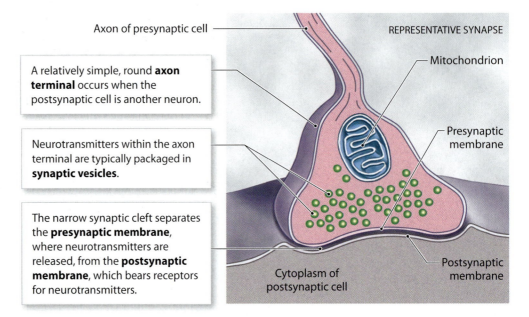

Axon of presynaptic cell

REPRESENTATIVE SYNAPSE

Mitochondrion

A relatively simple, round **axon terminal** occurs when the postsynaptic cell is another neuron.

Neurotransmitters within the axon terminal are typically packaged in **synaptic vesicles**.

The narrow synaptic cleft separates the **presynaptic membrane**, where neurotransmitters are released, from the **postsynaptic membrane**, which bears receptors for neurotransmitters.

Presynaptic membrane

Postsynaptic membrane

Cytoplasm of postsynaptic cell

2 One of the reasons the nervous system is so complex and versatile is that it uses more than 100 different neurotransmitters, each of which works in a different way. This table introduces a few of the important neurotransmitters.

Selected Neurotransmitters

Neurotransmitter	Chemical Structure	Comments
Acetylcholine (ACh)	$CH_3-N^+(CH_3)(CH_3)-CH_2-CH_2-O-C(=O)-CH_3$	Widespread in CNS and PNS; released at neuromuscular junctions (NMJs); best known and most studied neurotransmitter
Norepinephrine (NE)	$NH_2-CH_2-CH(OH)-$ (benzene ring with OH, OH)	Involved in attention and consciousness, control of body temperature, and regulation of pituitary gland secretion
Epinephrine (E)	$CH_2-NH-CH_2-CH(OH)-$ (benzene ring with OH, OH)	Generally excitatory effect along autonomic pathways
Serotonin	$NH_2-CH_2-CH_2-$ (indole ring with OH)	Important in emotional states, moods, and body temperature; several illicit hallucinogenic drugs, such as Ecstasy, target serotonin receptors
Glutamine	$HO-C(=O)-CH(NH_2)-CH_2-CH_2-C(=O)-OH$	Direct inhibitory effects: opens Cl^- channels; indirect effects: opens K^+ channels and blocks entry of Ca^{2+}

Step 1

An action potential arrives at the presynaptic cell's axon terminal. It depolarizes the axon terminal, which stimulates the release of neurotransmitter.

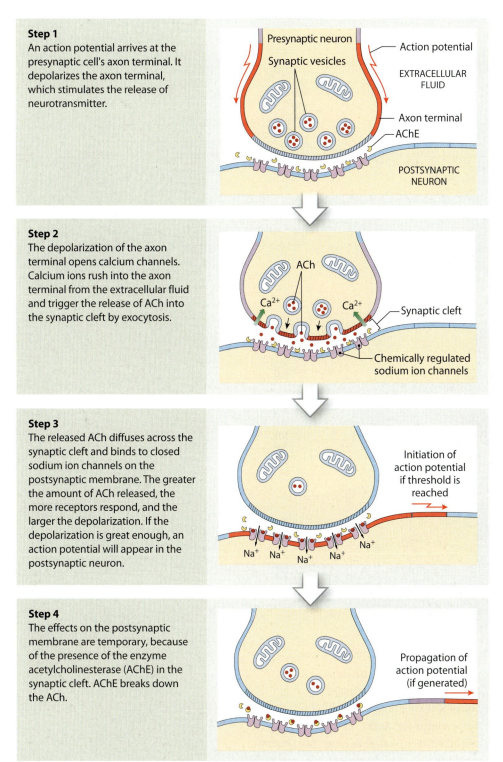

Presynaptic neuron

Synaptic vesicles

Action potential

EXTRACELLULAR FLUID

Axon terminal

AChE

POSTSYNAPTIC NEURON

Step 2

The depolarization of the axon terminal opens calcium channels. Calcium ions rush into the axon terminal from the extracellular fluid and trigger the release of ACh into the synaptic cleft by exocytosis.

ACh

Ca^{2+} Ca^{2+}

Synaptic cleft

Chemically regulated sodium ion channels

Step 3

The released ACh diffuses across the synaptic cleft and binds to closed sodium ion channels on the postsynaptic membrane. The greater the amount of ACh released, the more receptors respond, and the larger the depolarization. If the depolarization is great enough, an action potential will appear in the postsynaptic neuron.

Initiation of action potential if threshold is reached

Na^+ Na^+ Na^+ Na^+ Na^+

Step 4

The effects on the postsynaptic membrane are temporary, because of the presence of the enzyme acetylcholinesterase (AChE) in the synaptic cleft. AChE breaks down the ACh.

Propagation of action potential (if generated)

3 This figure illustrates the events that occur at an ACh-releasing synapse when an action potential arrives at an axon terminal. The same steps are involved in communication across a neuromuscular junction (a process described in Module 6.3).

A synaptic delay of 0.2–0.5 msec occurs between the arrival of the action potential at the axon terminal and the effect on the postsynaptic membrane. Although a delay of 0.5 msec is not very long, in that time an action potential may travel more than 7 cm (about 3 in.) along a myelinated axon. When information is being passed along a chain of neurons, the cumulative synaptic delay really adds up. This is why reflexes are important for survival—they can provide rapid and automatic responses to stimuli because very few synapses are involved.

Module 7.6 Review

a. Describe the structure of a synapse.

b. Describe the events that occur at a synapse when an action potential arrives at the axon terminal.

c. Why is the depolarization on the postsynaptic cell temporary?

1. Labeling

Label each of the structures in the following diagram of a neuron.

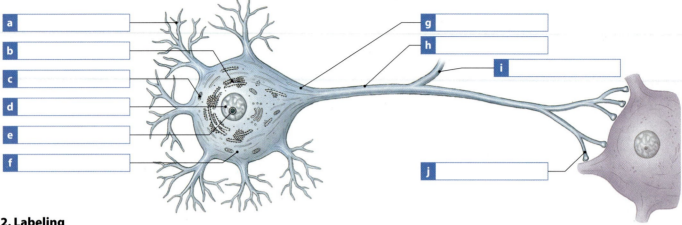

a _____

b _____

c _____

d _____

e _____

f _____

g _____

h _____

i _____

j _____

2. Labeling

Label each of the structures in the following diagram of a synapse.

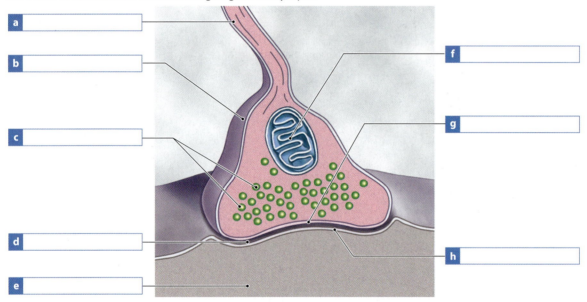

a _____

b _____

c _____

d _____

e _____

f _____

g _____

h _____

3. Vocabulary

Write the boldfaced term introduced in this section in the blank next to the term's definition.

a _____ A propagated change in the membrane potential

b _____ Smallest of the three anatomical classes of neurons

c _____ The membrane potential of a nonstimulated cell

d _____ Neuroglial cells that myelinate CNS axons

e _____ The structure where information is relaye between two neurons

f _____ A shift in the membrane potential to a less polarized state

g _____ Most numerous of the three functional classes of neurons

h _____ A shift in the membrane potential to a more polarized state when the chemical stimulus is removed

4. Section integration

Guillain-Barré *(ghee-yan bah-ray)* syndrome is a degeneration of myelin sheaths that ultimately may result in paralysis. Propose a mechanism by which myelin sheath degeneration can cause muscular paralysis.

_____

_____

The Functional Anatomy of the Central Nervous System

This section introduces the functional anatomy of the brain and spinal cord. In adult humans, the brain contains almost 97 percent of the body's neural tissue. A "typical" brain weighs 1.4 kg (3 lb) and has a volume of 1200 mL (71 in.³). Brain size varies considerably among individuals. On average, the brains of males are about 10 percent larger than those of females, due to differences in average body size. No correlation exists between brain size and intelligence. Individuals with the smallest brains (750 mL) and the largest brains (2100 mL) are functionally normal.

1 This lateral view of the brain and spinal cord of an embryo after four weeks of development shows the **neural tube**, the hollow cylinder that is the beginning of the central nervous system. In the cephalic portion, three areas enlarge rapidly through expansion of the internal cavity.

The midbrain is an expansion next to the forebrain.

The hindbrain is continuous with the spinal cord.

Spinal cord

The forebrain is at the tip of the neural tube.

2 By week 5 of development, the brain has changed position and the forebrain and hindbrain have subdivided, forming the regions found in the adult brain.

Forebrain	Hindbrain	
The **diencephalon** (dī-en-SEF-a-lon; *dia*, through + *encephalos*, brain) becomes the major relay and processing center for information headed to and from the cerebrum.	The region closest to the midbrain will form the **cerebellum** and the **pons** of the adult brain.	The region closest to the spinal cord will become the **medulla oblongata**.
The **cerebrum** will grow to become the largest part of the adult brain.		

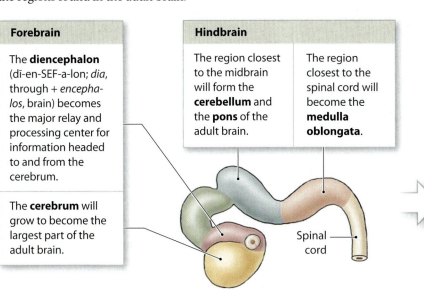

Spinal cord

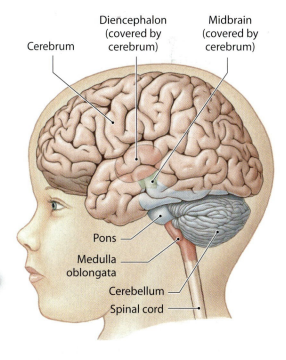

Cerebrum

Diencephalon (covered by cerebrum)

Midbrain (covered by cerebrum)

Pons

Medulla oblongata

Cerebellum

Spinal cord

3 As development continues, the cerebrum enlarges to the point where it covers other portions of the brain.

Each region of the brain has distinct structural and functional characteristics

1 This diagrammatic view of the brain introduces the major regions of the brain and their general functions. The regions are color coded, and we will use these colors in illustrations throughout the chapter.

Cerebrum

The **cerebrum** (se-RĒ-brum) of the adult brain is divided into a pair of large **cerebral hemispheres.** The surfaces of the cerebral hemispheres are highly folded and covered by a superficial layer of gray matter called the **cerebral cortex** (*cortex*, rind or bark). Functions include conscious thought, memory storage and processing, sensory processing, and the regulation of skeletal muscle contractions.

Fissures are deep grooves that subdivide the cerebral hemisphere.

Gyri (JĪ-rī; singular, *gyrus*) are folds in the cerebral cortex that increase its surface area.

Sulci (SUL-sī; singular, *sulcus*) are shallow depressions in the cerebral cortex that separate adjacent gyri.

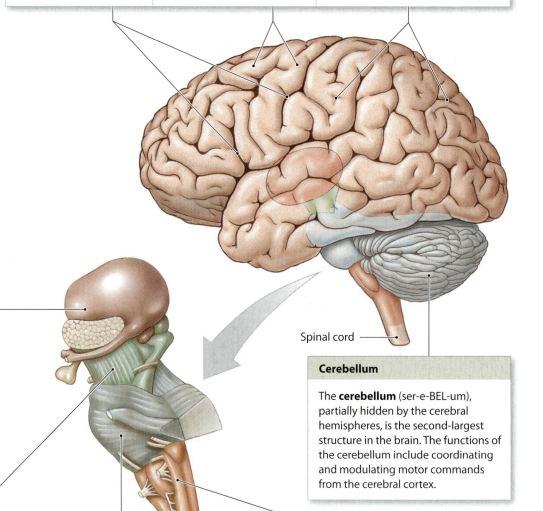

Diencephalon

The **diencephalon** is the structural and functional link between the cerebral hemispheres and the rest of the CNS.

The **thalamus** (THAL-a-mus) contains relay and processing centers for sensory information.

The **hypothalamus** (*hypo-*, below), or floor of the diencephalon, contains centers involved with emotions, autonomic function, and hormone production.

Spinal cord

Cerebellum

The **cerebellum** (ser-e-BEL-um), partially hidden by the cerebral hemispheres, is the second-largest structure in the brain. The functions of the cerebellum include coordinating and modulating motor commands from the cerebral cortex.

Brain stem

The **brain stem** includes the midbrain, pons, and medulla oblongata.

The **midbrain** processes visual and auditory (hearing) information and controls reflexes triggered by these stimuli. It also contains centers that help maintain consciousness.

The **pons** (*pons*, bridge) connects the cerebellum to the brain stem. In addition to tracts and relay centers, the pons also functions in somatic and visceral motor control.

The **medulla oblongata** relays sensory information to other portions of the brain stem, and to the thalamus. The medulla oblongata also contains major centers that regulate autonomic function, such as heart rate and blood pressure.

2 During development, the passageway within the neural tube in the cerebral hemispheres, diencephalon, pons and cerebellum, and medulla oblongata expands to form chambers called **ventricles** (VEN-tri-kls). The ventricles are filled with cerebrospinal fluid and lined by ependymal cells.

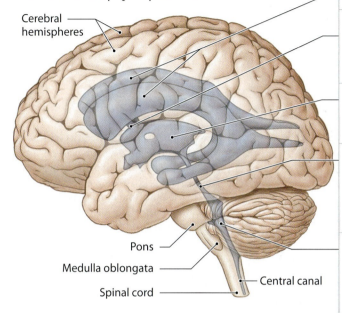

Cerebral hemispheres

Pons

Medulla oblongata

Spinal cord

Central canal

Ventricular system, lateral view

Ventricles of the Brain

Each cerebral hemisphere contains a large **lateral ventricle**.

Each lateral ventricle communicates with the third ventricle through an **interventricular foramen**.

The **third ventricle** is located in the diencephalon.

The **cerebral aqueduct** is a slender canal within the midbrain that connects the third ventricle to the fourth ventricle.

The **fourth ventricle** begins in the pons and extends into the superior portion of the medulla oblongata. It then narrows and becomes the central canal of the spinal cord.

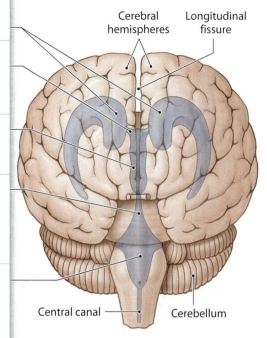

Cerebral hemispheres

Longitudinal fissure

Central canal

Cerebellum

Ventricular system, anterior view

3 This coronal section of the brain shows the interconnections between the ventricles.

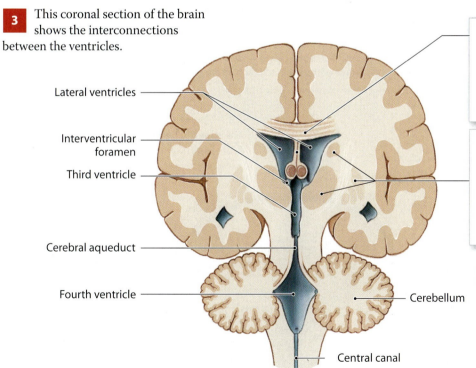

Lateral ventricles

Interventricular foramen

Third ventricle

Cerebral aqueduct

Fourth ventricle

Cerebellum

Central canal

The **corpus callosum** is a thick tract of white matter that interconnects the two cerebral hemispheres. It contains more than 200 million axons carrying some 4 billion impulses per second.

Nuclei are groups of nerve cell bodies within the CNS. The **basal nuclei** lie within each cerebral hemisphere. They provide subconscious control of muscle tone and help direct complex learned movements, such as walking or running.

Module 7.7 Review

a. Name the major regions of the brain and the distinct structures of each.

b. Describe the role of the medulla oblongata.

c. Which ventricles would lose communication by a blocked cerebral aqueduct?

The cranial meninges and the cerebrospinal fluid protect and support the brain

The delicate tissues of the brain are protected from mechanical forces by the bones of the cranium, the cranial meninges, and the cerebrospinal fluid. In addition, the blood–brain barrier biochemically isolates the brain's neural tissue from the general circulation.

1 The three layers that make up the **cranial meninges**—the cranial **dura mater**, **arachnoid mater**, and **pia mater**—are continuous with those of the spinal meninges. Knowing the Latin meanings for these terms helps in understanding their names. *Mater*, which means mother, is a covering. The outer *dura* means hard, the middle *arachnoid* means like a cobweb, and the inner *pia* means tender.

2 Arachnoid mater

The arachnoid mater consists of an outer epithelial layer supported by a fibrous meshwork that connects it to the pia mater. The arachnoid mater does not follow the brain's underlying folds. The fluid-filled subarachnoid space separates the outer layer of the arachnoid mater from the pia mater.

Subdural space

Cranium (skull)

Arachnoid mater

Subarachnoid space

1 Dura mater

The dura mater consists of outer and inner fibrous layers. The outer layer is fused to the periosteum of the cranial bones. As a result, there is no epidural space. The outer and inner layers of the cranial dura mater are typically separated by a slender gap that contains tissue fluids and blood vessels, including several large dural sinuses, which collect blood from the veins of the brain.

Dura mater (outer layer)
Dural sinus
Dura mater (inner layer)

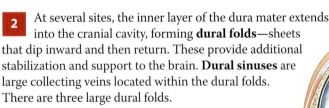

Cerebral cortex

3 Pia mater

The pia mater sticks to the surface of the brain. It extends into every fold and accompanies the branches of cerebral blood vessels as they penetrate the surface of the brain to reach internal structures.

2 At several sites, the inner layer of the dura mater extends into the cranial cavity, forming **dural folds**—sheets that dip inward and then return. These provide additional stabilization and support to the brain. **Dural sinuses** are large collecting veins located within the dural folds. There are three large dural folds.

The **falx cerebri** (FALKS SER-e-brī; *falx*, sickle shaped) is a fold of dura mater that projects between the cerebral hemispheres. Its inferior portions attach anteriorly to the crista galli and posteriorly to the internal occipital crest of the occipital bone. The superior and inferior sagittal sinuses lie within this dural fold.

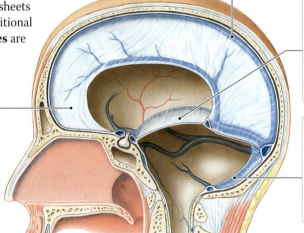

The **superior sagittal sinus** is the largest dural sinus.

The **tentorium cerebelli** (ten-TŌ-rē-um ser-e-BEL-ī; *tentorium*, a tent) separates the cerebral hemispheres from the cerebellum.

The **falx cerebelli** separates the two cerebellar hemispheres along the midsagittal line inferior to the tentorium cerebelli.

3 **Cerebrospinal fluid (CSF)** completely surrounds and bathes the exposed surfaces of the CNS. Each of the ventricles contains an area of **choroid plexus**, which consists of a combination of specialized ependymal cells and capillaries involved in producing and maintaining cerebrospinal fluid. The CSF circulates from the choroid plexus through the ventricles and fills the central canal of the spinal cord. As it circulates, materials diffuse between the CSF and the interstitial fluid of the CNS across the ependymal cells.

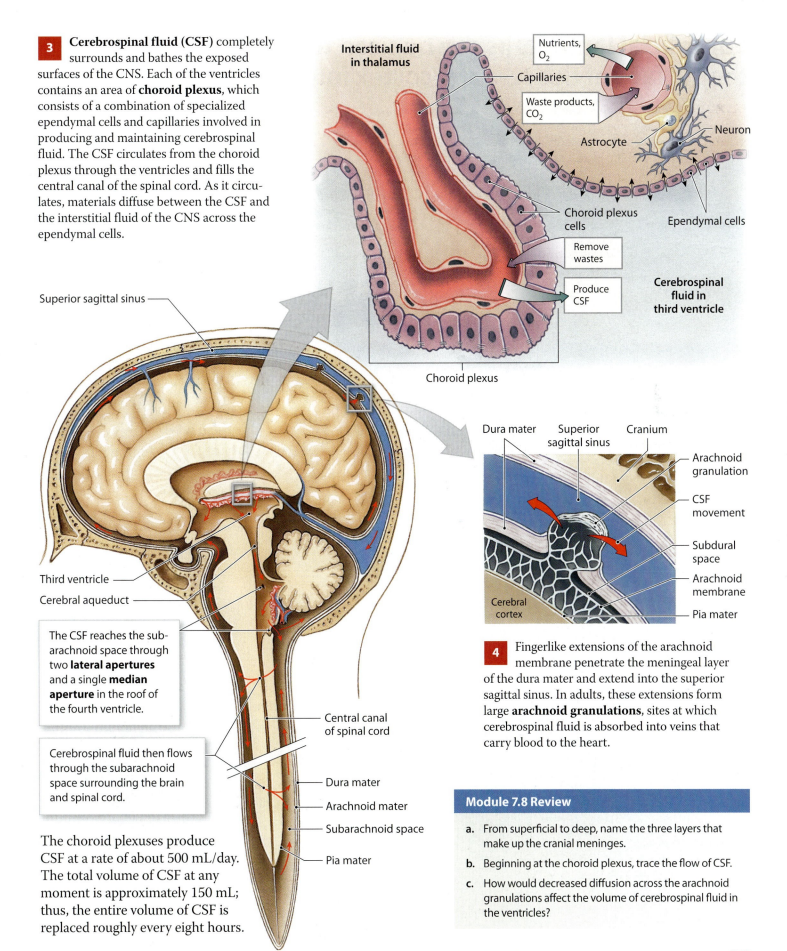

Interstitial fluid in thalamus

Nutrients, O_2

Capillaries

Waste products, CO_2

Neuron

Astrocyte

Choroid plexus cells

Ependymal cells

Remove wastes

Produce CSF

Cerebrospinal fluid in third ventricle

Choroid plexus

Superior sagittal sinus

Third ventricle

Cerebral aqueduct

The CSF reaches the sub-arachnoid space through two **lateral apertures** and a single **median aperture** in the roof of the fourth ventricle.

Cerebrospinal fluid then flows through the subarachnoid space surrounding the brain and spinal cord.

Central canal of spinal cord

Dura mater

Arachnoid mater

Subarachnoid space

Pia mater

The choroid plexuses produce CSF at a rate of about 500 mL/day. The total volume of CSF at any moment is approximately 150 mL; thus, the entire volume of CSF is replaced roughly every eight hours.

Dura mater

Superior sagittal sinus

Cranium

Arachnoid granulation

CSF movement

Subdural space

Arachnoid membrane

Pia mater

Cerebral cortex

4 Fingerlike extensions of the arachnoid membrane penetrate the meningeal layer of the dura mater and extend into the superior sagittal sinus. In adults, these extensions form large **arachnoid granulations**, sites at which cerebrospinal fluid is absorbed into veins that carry blood to the heart.

Module 7.8 Review

a. From superficial to deep, name the three layers that make up the cranial meninges.

b. Beginning at the choroid plexus, trace the flow of CSF.

c. How would decreased diffusion across the arachnoid granulations affect the volume of cerebrospinal fluid in the ventricles?

Superficial landmarks can be used to divide the surface of the cerebral cortex into lobes

1 Each cerebral hemisphere can be divided into regions called **lobes**. Your brain has a unique pattern of sulci and gyri, as individual as a fingerprint, but the boundaries between lobes are reliable landmarks. Lobes on the external surfaces are named after the overlying bones of the skull.

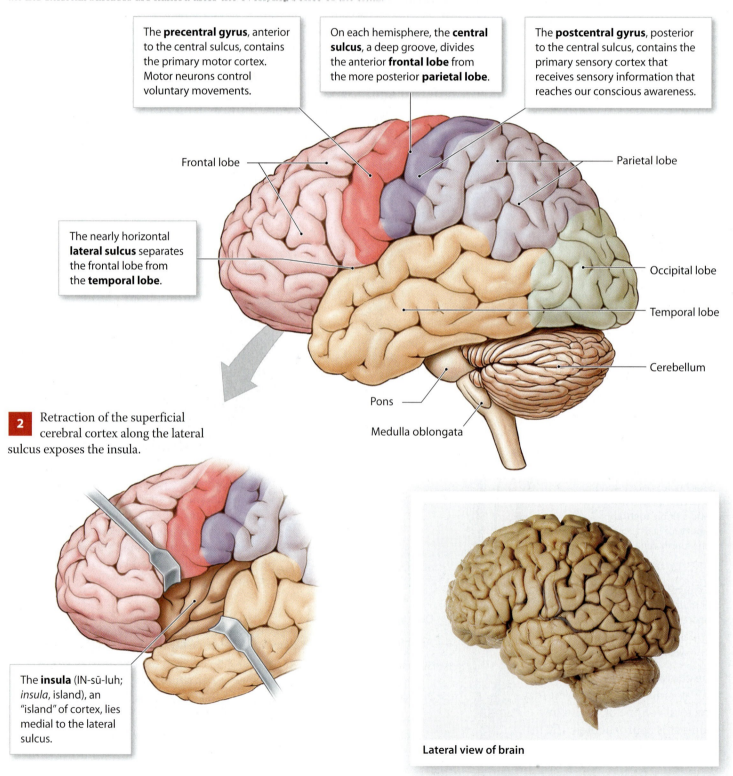

The **precentral gyrus**, anterior to the central sulcus, contains the primary motor cortex. Motor neurons control voluntary movements.

On each hemisphere, the **central sulcus**, a deep groove, divides the anterior **frontal lobe** from the more posterior **parietal lobe**.

The **postcentral gyrus**, posterior to the central sulcus, contains the primary sensory cortex that receives sensory information that reaches our conscious awareness.

Frontal lobe

Parietal lobe

The nearly horizontal **lateral sulcus** separates the frontal lobe from the **temporal lobe**.

Occipital lobe

Temporal lobe

Cerebellum

Pons

Medulla oblongata

2 Retraction of the superficial cerebral cortex along the lateral sulcus exposes the insula.

The **insula** (IN-sū-luh; *insula*, island), an "island" of cortex, lies medial to the lateral sulcus.

Lateral view of brain

3 This midsagittal view indicates the inner boundaries of the lobes and highlights the way the cerebral hemispheres cover the rest of the brain. For clarity, we have labeled structures and regions outside the cerebrum in italics.

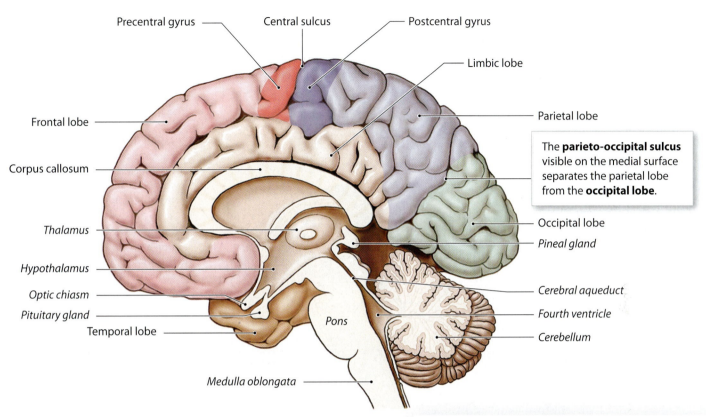

The **parieto-occipital sulcus** visible on the medial surface separates the parietal lobe from the **occipital lobe**.

As you proceed, remember the following facts about the cerebral hemispheres:

- Each cerebral hemisphere receives sensory information from, and sends motor commands to, the opposite side of the body. Thus the motor areas of your left cerebral hemisphere control muscles on your right side, and motor areas of your right cerebral hemisphere control muscles on your left side. This crossing over, which occurs in the brain stem and spinal cord, has no known functional significance.

- Even though the two hemispheres may look identical and have many similar functions, important differences exist.

- The correspondence between a specific function and a specific region of the cerebral cortex is imprecise. The boundaries are indistinct and have considerable overlap, and some cortical functions, such as consciousness, cannot easily be assigned to any single region. However, we know that healthy individuals use all portions of the brain.

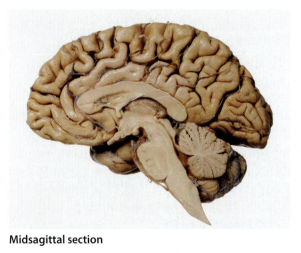

Midsagittal section

Module 7.9 Review

a. Identify the lobes of the cerebrum and indicate the basis for their names.

b. Describe the insula.

c. What effect would damage to the left postcentral gyrus produce?

The lobes of the cerebral cortex contain regions with specific functions

1 The **primary motor cortex** issues voluntary commands to skeletal muscles, and the **primary sensory cortex** receives general somatic sensory information. The special senses of sight, hearing, smell, and taste reach other portions of the cerebral cortex. Each sensory and motor region of the cortex is connected to a nearby **association area**. Association areas are regions of the cortex that interpret incoming data or coordinate a motor response.

Motor Cortex

Neurons of the primary motor cortex are called **pyramidal cells**, because their cell bodies resemble little pyramids.

The **somatic motor association area** is responsible for the coordination of learned movements.

The **gustatory cortex** of the insula receives information from taste receptors.

The **olfactory cortex** receives sensory information from olfactory (smell) receptors.

Auditory Cortex

Primary auditory cortex is responsible for monitoring auditory (sound) information.

The **auditory association area** monitors sensory activity in the auditory cortex and recognizes sounds, such as spoken words.

Sensory Cortex

Neurons in the primary sensory cortex receive somatic sensory information from receptors for touch, pressure, pain, vibration, taste, or temperature.

The **somatic sensory association area** monitors activity in the primary sensory cortex. It allows you to recognize a light touch, such as a mosquito landing on your arm.

Visual Cortex

The **primary visual cortex** receives information from the lateral thalamic nuclei.

The **visual association area** monitors the patterns of activity in the visual cortex and interprets the results. When you see the symbols c, a, and r, your visual association area recognizes that they form the word "car."

Central sulcus

PARIETAL LOBE

FRONTAL LOBE

OCCIPITAL LOBE

Lateral sulcus

TEMPORAL LOBE

2 **Integrative centers** concerned with performing complex processes, such as speech, writing, mathematics, and understanding spatial relationships, are restricted to either the left or the right hemisphere.

The **speech center,** also called the Broca area or the motor speech area, lies in the same hemisphere as the general interpretive area. The speech center regulates the patterns of breathing and vocalization needed for normal speech.

The **prefrontal cortex** coordinates information relayed from the association areas of the cortex. In the process, it performs such abstract intellectual functions as predicting the consequences of events or actions.

The **frontal eye field** controls learned eye movements, such as when you scan these lines of text.

The **general interpretive area** receives information from all the sensory association areas. This analytical center is present in only one hemisphere (typically the left). This region plays an essential role in your personality by integrating sensory information and coordinating access to complex visual and auditory memories.

3 Each of the two cerebral hemispheres is responsible for specific functions that are not ordinarily performed by the opposite hemisphere. This regional specialization is called **hemispheric lateralization**.

Left Cerebral Hemisphere

In most people, the left hemisphere contains the general interpretive and speech centers and is responsible for language-based skills. Reading, writing, and speaking, for example, depend on processing done in the left cerebral hemisphere. In addition, the premotor cortex that controls hand movements is larger on the left side for right-handed individuals than for left-handed individuals. The left hemisphere is also important in performing analytical tasks, such as mathematics and logic.

Right Cerebral Hemisphere

The right cerebral hemisphere analyzes sensory information and relates the body to the sensory environment. Interpretive centers in this hemisphere enable you to identify familiar objects by touch, smell, sight, taste, or feel. For example, the right hemisphere plays a dominant role in recognizing faces and in understanding three-dimensional relationships. It is also important in analyzing the emotional context of a conversation—for instance, distinguishing between the threat "Get lost!" and the question "Get lost?"

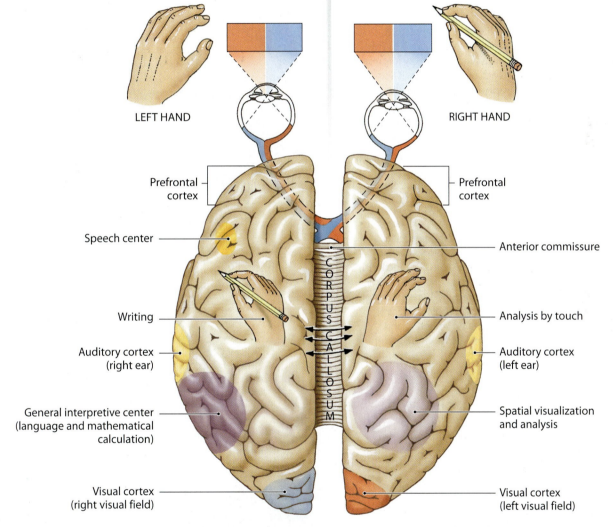

Left-handed people represent about 9 percent of the human population. Although in most cases the primary motor cortex of the right hemisphere controls motor function for the dominant left hand, the centers involved with speech and analytical function are in the left hemisphere. Interestingly, an unusually high percentage of musicians and artists are left-handed. Such a link may exist because the primary motor cortex and association areas on the right cerebral hemisphere are near the association areas involved with spatial visualization and emotions.

Module 7.10 Review

a. Where is the primary motor cortex located?

b. Which senses are affected by damage to the temporal lobes?

c. A stroke patient is unable to speak. Which part of the brain has been affected?

The diencephalon, brain stem, and cerebellum contain relay stations that process information outside our awareness

1 The **diencephalon**, which surrounds the third ventricle, consists of the left and right thalamus and the hypothalamus. An area of the thalamus forms the roof of the diencephalon. Its anterior portion contains an extensive area of choroid plexus. The posterior portion contains the **pineal gland**, an endocrine structure that secretes the hormone melatonin. Among other functions, melatonin is important in regulating day–night cycles.

Diencephalon

Thalamus

The third ventricle separates the left thalamus from the right thalamus. Each contains a rounded mass of thalamic nuclei. The thalamus is the final relay point for all ascending sensory information, other than olfactory, that will reach our awareness. It acts as a filter, passing on to the primary sensory cortex only a small portion of the arriving sensory information. The rest is relayed to subconscious centers in the brain. The thalamus also plays a role in coordinating voluntary and involuntary motor commands.

Hypothalamus

The hypothalamus contains important control and integrative centers associated with: (1) the subconscious centers involved with rage, pleasure, pain, and sexual arousal, as part of a larger network known as the *limbic system*; (2) adjusting the activities of autonomic centers in the pons and medulla oblongata (such as heart rate, blood pressure, respiration, and digestive functions); (3) coordinating activities of the nervous and endocrine systems; (4) secreting a variety of hormones, including antidiuretic hormone (ADH) and oxytocin (OXT); (5) producing the behavioral "drives" involved in hunger and thirst; (6) coordinating voluntary and autonomic functions; (7) regulating normal body temperature; and (8) coordinating the daily cycles of activity.

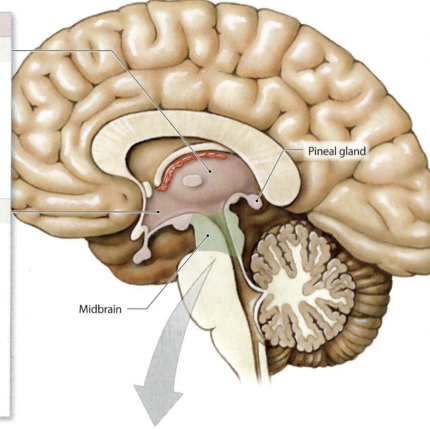

Pineal gland

Midbrain

2 The **midbrain** contains two pairs of sensory nuclei, the **colliculi** (ko-LIK-u-lī; singular: *colliculus*, a small hill), involved in processing visual and auditory sensations. The midbrain also contains (1) motor nuclei for two of the cranial nerves (N III, IV) involved in controlling eye movements, (2) the headquarters of the reticular formation [see next module], (3) nuclei involved in maintaining muscle tone and posture, and (4) the **substantia nigra**, a nucleus that regulates the motor output of the basal nuclei.

The **cerebral peduncles** (*peduncles*, little feet) contain descending bundles of nerve fibers. Some of the descending fibers go to the cerebellum by way of the pons, and others carry voluntary motor commands issued by the primary motor cortex of each cerebral hemisphere.

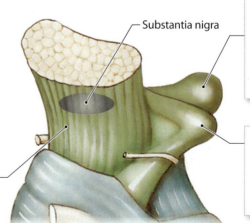

Substantia nigra

The *superior colliculi* control the reflex movements of the eyes, head, and neck in response to visual stimuli, such as a blinding flash of light.

The *inferior colliculi* control reflex movements of the head, neck, and trunk in response to auditory stimuli, such as a loud noise.

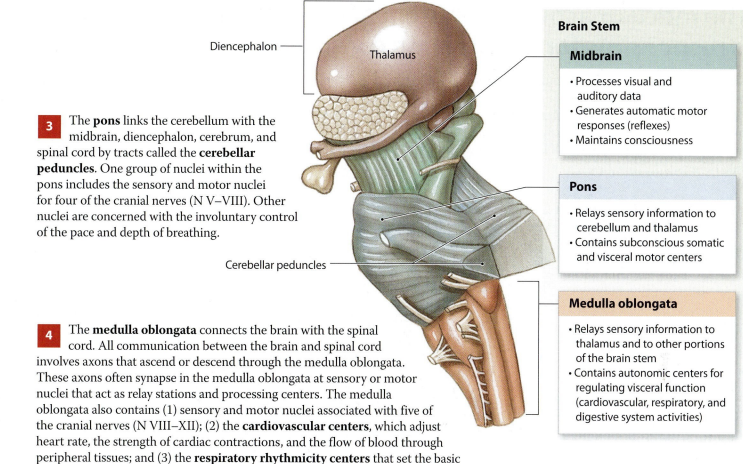

Brain Stem

Midbrain

- Processes visual and auditory data
- Generates automatic motor responses (reflexes)
- Maintains consciousness

Diencephalon

Thalamus

3 The **pons** links the cerebellum with the midbrain, diencephalon, cerebrum, and spinal cord by tracts called the **cerebellar peduncles**. One group of nuclei within the pons includes the sensory and motor nuclei for four of the cranial nerves (N V–VIII). Other nuclei are concerned with the involuntary control of the pace and depth of breathing.

Pons

- Relays sensory information to cerebellum and thalamus
- Contains subconscious somatic and visceral motor centers

Cerebellar peduncles

Medulla oblongata

- Relays sensory information to thalamus and to other portions of the brain stem
- Contains autonomic centers for regulating visceral function (cardiovascular, respiratory, and digestive system activities)

4 The **medulla oblongata** connects the brain with the spinal cord. All communication between the brain and spinal cord involves axons that ascend or descend through the medulla oblongata. These axons often synapse in the medulla oblongata at sensory or motor nuclei that act as relay stations and processing centers. The medulla oblongata also contains (1) sensory and motor nuclei associated with five of the cranial nerves (N VIII–XII); (2) the **cardiovascular centers**, which adjust heart rate, the strength of cardiac contractions, and the flow of blood through peripheral tissues; and (3) the **respiratory rhythmicity centers** that set the basic pace for breathing movements.

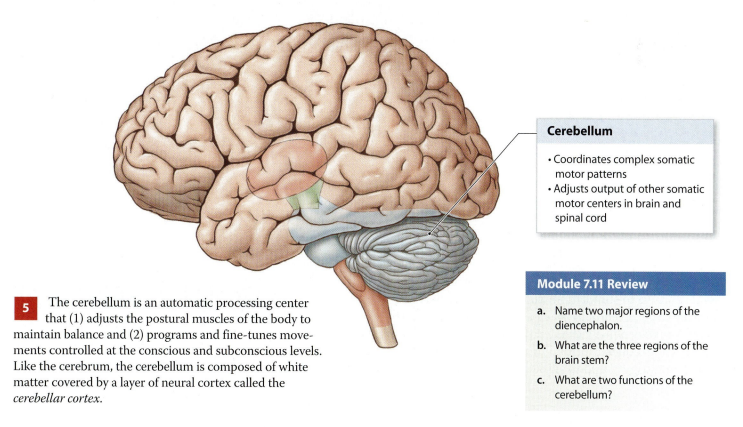

Cerebellum

- Coordinates complex somatic motor patterns
- Adjusts output of other somatic motor centers in brain and spinal cord

5 The cerebellum is an automatic processing center that (1) adjusts the postural muscles of the body to maintain balance and (2) programs and fine-tunes movements controlled at the conscious and subconscious levels. Like the cerebrum, the cerebellum is composed of white matter covered by a layer of neural cortex called the *cerebellar cortex*.

Module 7.11 Review

a. Name two major regions of the diencephalon.

b. What are the three regions of the brain stem?

c. What are two functions of the cerebellum?

The reticular activating system of the midbrain is responsible for maintaining consciousness

The brain stem contains the **reticular formation**, an interconnected neural network that regulates many involuntary functions. The reticular formation of the midbrain contains the **reticular activating system (RAS)**. The output of this system directly affects the activity of the cerebral cortex. When the RAS is inactive, so are we; when the RAS is stimulated, so is our state of attention or wakefulness.

1 This diagrammatic midsagittal section of the brain shows how the reticular formation extends through much of the brain stem. Inputs from a variety of sensory pathways are carried through the reticular formation to the reticular activating system, which in turn stimulates large areas of the cerebral cortex.

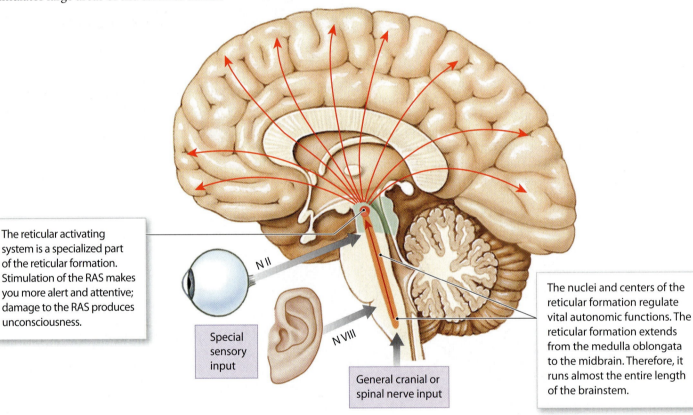

The reticular activating system is a specialized part of the reticular formation. Stimulation of the RAS makes you more alert and attentive; damage to the RAS produces unconsciousness.

N II

Special sensory input

N VIII

General cranial or spinal nerve input

The nuclei and centers of the reticular formation regulate vital autonomic functions. The reticular formation extends from the medulla oblongata to the midbrain. Therefore, it runs almost the entire length of the brainstem.

After many hours of activity, the reticular formation becomes less responsive to stimulation. You become less alert and more lethargic. Eventually you lose consciousness and fall asleep. Your sleep may be ended by any stimulus sufficient to activate your reticular formation and RAS. Over time the size of the stimulus required to awaken you gradually decreases.

Module 7.12 Review

a. Describe the reticular formation.

b. Describe the reticular activating system (RAS).

c. You are sleeping and your RAS is suddenly activated. What will happen?

Brain activity can be monitored using external electrodes; the record is called an electroencephalogram, or EEG

1 Neural function depends on electrical impulses, and the brain contains billions of neurons and the axons of the central white matter. Brain activity at any given moment generates an electrical field that can be measured by placing electrodes on the scalp. The electrical activity changes constantly, as nuclei and cortical areas are stimulated or quieted down. An **electroencephalogram** (**EEG**) is a printed report of the brain's electrical activity. The electrical patterns observed are called **brain waves**. The different patterns of brain waves are distinguished by their frequency (number of waves per unit of time) and amplitude (wave height).

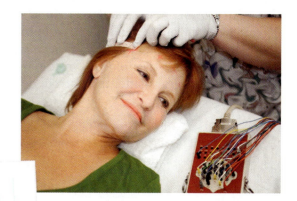

Alpha waves occur in the brains of healthy, awake adults who are resting with their eyes closed. Alpha waves disappear during sleep, but they also vanish when the individual begins to concentrate on some specific task.

Beta waves are higher frequency waves typical of individuals who are either concentrating on a task, under stress, or in a state of psychological tension.

Theta waves may appear for a short time during sleep in normal adults but are mostly seen in children and in intensely frustrated adults. The presence of theta waves under other circumstances may indicate the presence of a brain disorder, such as a tumor.

Delta waves are very-large-amplitude, low-frequency waves. They are normally seen during deep sleep in individuals of all ages. Delta waves are also seen in the brains of infants (whose brains are still developing) and in awake adults when a tumor, vascular blockage, or inflammation has damaged portions of the brain.

Seconds   1   2   3   4

Electrical activity in the two hemispheres is generally synchronized by a "pacemaker" mechanism that appears to involve the thalamus. A lack of synchrony between the hemispheres can therefore indicate localized damage or other cerebral abnormalities. A tumor or injury affecting one hemisphere, for example, typically changes the pattern in that hemisphere, and the patterns of the two hemispheres are no longer aligned. A **seizure** is a temporary cerebral disorder accompanied by abnormal movements, unusual sensations, inappropriate behavior, or some combination of these signs and symptoms. Clinical conditions characterized by seizures are known as seizure disorders, or **epilepsies**. Seizures of all kinds show remarkable changes on an electroencephalogram. The change begins in one portion of the cerebral cortex but may then spread across the entire cortical surface, like a wave on the surface of a pond.

Module 7.13 Review

a. Define electroencephalogram (EEG) and describe the four wave types associated with it.

b. You are reading this textbook. If you had an EEG right now, which brain wave(s) would you expect to see?

c. Differentiate between a seizure and epilepsy.

The spinal cord contains gray matter and white matter

The adult spinal cord measures approximately 45 cm (18 in.) in length and has a maximum width of roughly 14 mm (0.55 in.). In sectional view, it has an outer layer of white matter and an inner layer of gray matter that surrounds a small central canal.

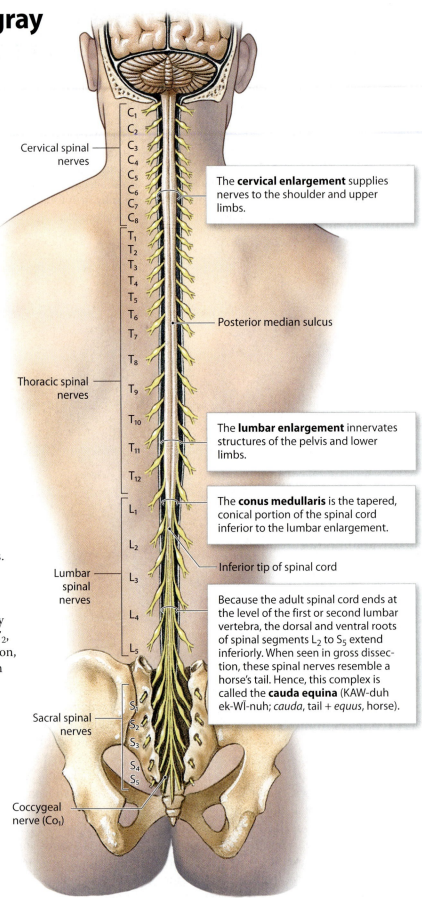

Cervical spinal nerves

C₁
C₂
C₃
C₄
C₅
C₆
C₇
C₈

The **cervical enlargement** supplies nerves to the shoulder and upper limbs.

T₁
T₂
T₃
T₄
T₅
T₆
T₇
T₈

Thoracic spinal nerves

T₉
T₁₀
T₁₁
T₁₂

Posterior median sulcus

The **lumbar enlargement** innervates structures of the pelvis and lower limbs.

L₁
L₂

The **conus medullaris** is the tapered, conical portion of the spinal cord inferior to the lumbar enlargement.

Lumbar spinal nerves

L₃
L₄
L₅

Inferior tip of spinal cord

Because the adult spinal cord ends at the level of the first or second lumbar vertebra, the dorsal and ventral roots of spinal segments L₂ to S₅ extend inferiorly. When seen in gross dissection, these spinal nerves resemble a horse's tail. Hence, this complex is called the **cauda equina** (KAW-duh ek-WĪ-nuh; *cauda*, tail + *equus*, horse).

Sacral spinal nerves

S₁
S₂
S₃
S₄
S₅

Coccygeal nerve (Co₁)

1 The entire spinal cord can be divided into 31 segments, each giving rise to a pair of spinal nerves. Every pair is identified by its association with adjacent vertebrae. Each spinal nerve inferior to the first thoracic vertebra takes its name from the vertebra immediately superior to it. Thus, spinal nerve T₁ emerges immediately inferior to vertebra T₁, spinal nerve T₂ follows vertebra T₂, and so forth. The arrangement differs in the cervical region, because the first pair of spinal nerves, C₁, passes between the skull and the first cervical vertebra. For this reason, each cervical nerve takes its name from the vertebra immediately inferior to it. In other words, cervical nerve C₂ precedes vertebra C₂, and the same system is used for the rest of the cervical series. The transition from one numbering system to another occurs between the last cervical vertebra and the first thoracic vertebra. The spinal nerve found at this location has been designated C₈. Therefore, although there are only seven cervical vertebrae, there are eight cervical nerves.

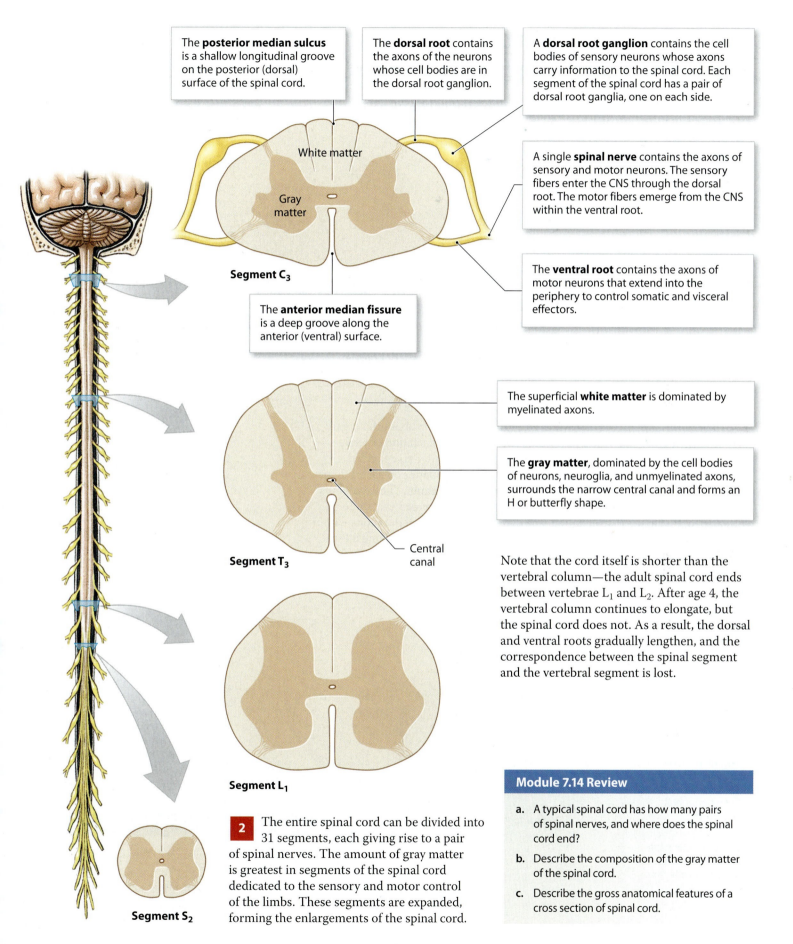

The **posterior median sulcus** is a shallow longitudinal groove on the posterior (dorsal) surface of the spinal cord.

The **dorsal root** contains the axons of the neurons whose cell bodies are in the dorsal root ganglion.

A **dorsal root ganglion** contains the cell bodies of sensory neurons whose axons carry information to the spinal cord. Each segment of the spinal cord has a pair of dorsal root ganglia, one on each side.

White matter

Gray matter

A single **spinal nerve** contains the axons of sensory and motor neurons. The sensory fibers enter the CNS through the dorsal root. The motor fibers emerge from the CNS within the ventral root.

Segment C₃

The **anterior median fissure** is a deep groove along the anterior (ventral) surface.

The **ventral root** contains the axons of motor neurons that extend into the periphery to control somatic and visceral effectors.

The superficial **white matter** is dominated by myelinated axons.

The **gray matter**, dominated by the cell bodies of neurons, neuroglia, and unmyelinated axons, surrounds the narrow central canal and forms an H or butterfly shape.

Central canal

Segment T₃

Note that the cord itself is shorter than the vertebral column—the adult spinal cord ends between vertebrae L_1 and L_2. After age 4, the vertebral column continues to elongate, but the spinal cord does not. As a result, the dorsal and ventral roots gradually lengthen, and the correspondence between the spinal segment and the vertebral segment is lost.

Segment L₁

2 The entire spinal cord can be divided into 31 segments, each giving rise to a pair of spinal nerves. The amount of gray matter is greatest in segments of the spinal cord dedicated to the sensory and motor control of the limbs. These segments are expanded, forming the enlargements of the spinal cord.

Segment S₂

Module 7.14 Review

a. A typical spinal cord has how many pairs of spinal nerves, and where does the spinal cord end?

b. Describe the composition of the gray matter of the spinal cord.

c. Describe the gross anatomical features of a cross section of spinal cord.

Gray matter is the region of integration, and white matter carries information

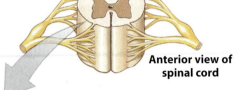

Anterior view of spinal cord

1 This cross section shows most of the anatomical landmarks of the spinal cord. Compare this view to the diagrammatic view below.

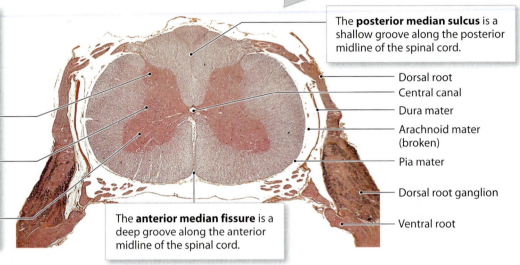

Structural Organization of Gray Matter

The projections of gray matter toward the outer surface of the spinal cord are called **horns**.

The **posterior gray horn** contains somatic and visceral sensory nuclei.

The **lateral gray horn**, located only in thoracic and lumbar segments, contains visceral motor nuclei.

The **anterior gray horn** contains somatic motor nuclei.

The **posterior median sulcus** is a shallow groove along the posterior midline of the spinal cord.

Dorsal root
Central canal
Dura mater
Arachnoid mater (broken)
Pia mater
Dorsal root ganglion
Ventral root

The **anterior median fissure** is a deep groove along the anterior midline of the spinal cord.

2 This diagrammatic view provides additional information about the organization of gray matter, and introduces the regional organization of white matter in the spinal cord. Like the gray horns, the white matter is organized according to the region of the body innervated. The white matter on each side of the spinal cord can be divided into three regions called **columns**. These columns contain tracts. A **tract** is a bundle of axons in the CNS that is relatively uniform with respect to diameter, myelination, and conduction speed. All the axons within a tract relay the same type of information (sensory or motor) in the same direction. **Ascending tracts** carry sensory information toward the brain, and **descending tracts** convey motor commands to the spinal cord.

Structural and Functional Organization of White Matter

The **posterior white column** lies between the posterior gray horns and the posterior median sulcus.

The **lateral white column** includes the white matter on either side of the spinal cord, between the anterior and posterior columns.

The **anterior white column** lies between the anterior gray horns and the anterior median fissure.

Functional Organization of Gray Matter

The cell bodies of neurons in the gray matter of the spinal cord are organized into functional groups called **nuclei**.

Sensory nuclei receive and relay sensory information from peripheral receptors.

Motor nuclei issue motor commands to peripheral effectors.

Somatic
Visceral
Visceral
Somatic

Dorsal root

Dorsal root ganglion

Ventral root

Module 7.15 Review

a. Name the three horns of the spinal cord gray matter.

b. Differentiate between sensory and motor nuclei.

c. What are the three columns in the white matter?

The spinal cord is surrounded by the meninges, which consist of the dura mater, arachnoid mater, and pia mater

The delicate neural tissues must be protected from shocks, including damaging contact with the surrounding bony walls of the vertebral canal. The **spinal meninges** (me-NIN-jēz; singular, *meninx,* membrane), three layers of membranes surrounding the spinal cord, provide the necessary physical stability and shock absorption. Blood vessels branching within these layers deliver oxygen and nutrients to the spinal cord. The spinal meninges consist of the dura mater, the arachnoid mater, and the pia mater. At the foramen magnum of the skull, the spinal meninges are continuous with the cranial meninges, which surround the brain.

Between the dura mater and the walls of the vertebral canal lies the **epidural space**, a region that contains blood vessels and a protective padding of adipose tissue.

1 This posterior view of the dissected spinal cord shows the basic relationships among the spinal meninges.

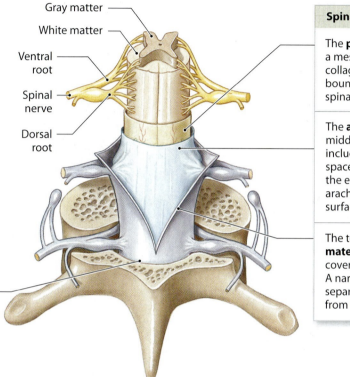

Gray matter

White matter

Ventral root

Spinal nerve

Dorsal root

Spinal meninges

The **pia mater** consists of a meshwork of elastic and collagen fibers that is firmly bound to the surface of the spinal cord.

The **arachnoid mater** is the middle meningeal layer. It includes the subarachnoid space that extends between the epithelial layer of the arachnoid and the outer surface of the pia mater.

The tough, fibrous **dura mater** is the outermost covering of the spinal cord. A narrow subdural space separates the dura mater from the arachnoid mater.

2 Small samples of cerebrospinal fluid (CSF) can be collected in a procedure called a **spinal tap**, or **lumbar puncture**. To avoid damaging the spinal cord—which extends only to the level of L_1 or L_2 vertebra—a needle is inserted between L_2 and the sacrum. In this region, the spinal meninges enclose only the relatively sturdy cauda equina and a significant volume of CSF. A small amount (3–9 mL) of fluid is collected by allowing the fluid to drip out under its own pressure. Spinal taps are performed to administer spinal anesthesia, when CNS infection is suspected, or when severe back pain, headaches, disc problems, and some types of strokes are diagnosed.

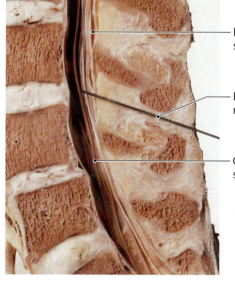

Epidural space

Lumbar puncture needle

Cauda equina in subarachnoid space

Module 7.16 Review

a. What are the three layers of the spinal meninges?

b. Where is the epidural space located and what does it contain?

c. Why is a spinal tap done below the level of the L_2 vertebra?

1. Labeling

Label the structures in the accompanying figure of a
lateral view of the human brain.

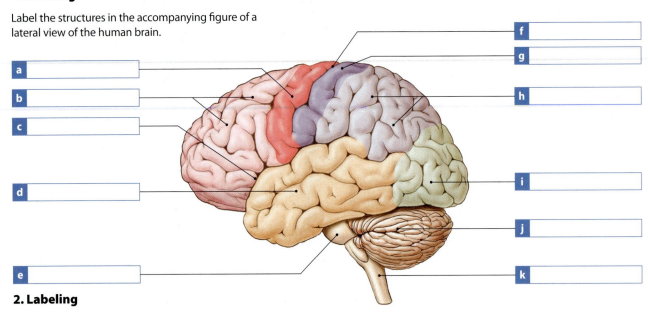

a _____
b _____
c _____
d _____
e _____
f _____
g _____
h _____
i _____
j _____
k _____

2. Labeling

Label each of the structures in the following cross-sectional diagram of the
spinal cord.

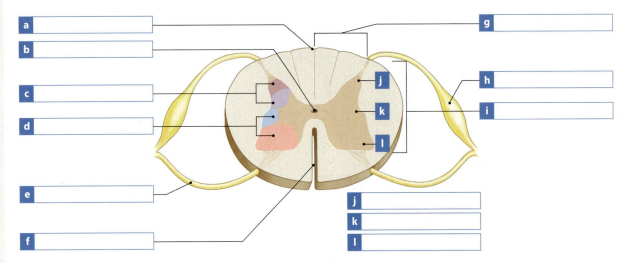

a _____
b _____
c _____
d _____
e _____
f _____
g _____
h _____
i _____
j _____
k _____
l _____

3. Vocabulary

Write the boldfaced term introduced in this section that matches the following definitions.

a _____ The regions of white matter in the spinal cord

b _____ Cerebral region containing the primary motor cortex

c _____ Region of the brain stem containing centers that adjust the heart and breathing rate

d _____ Specialized membranes that provide stability and support for the spinal cord and brain

e _____ The complex made up of the long dorsal and ventral spinal nerve roots inferior to the spinal cord

f _____ Final relay point for all ascending sensory information (other than olfactory) reaching our awareness

g _____ The outermost covering of the spinal cord

h _____ Cerebral cortex integrative center which predicts consequences of events or actions

i _____ Fluid that surrounds and bathes the CNS

Visual Outline with Key Terms

Summarize the content of each module using the terms in the order provided.

SECTION 1

Neurons and Neuroglia

- receptors
- visceral sensory receptors
- somatic sensory receptors
- special sensory receptors
- sensory (afferent) division
- peripheral nervous system (PNS)
- information processing
- central nervous system (CNS)
- motor (efferent) division
- somatic nervous system (SNS)
- autonomic nervous system (ANS)
- effectors

7.1

Neurons are nerve cells specialized for intercellular communication

- ○ dendrites
- ○ cell body
- ○ axon
- neurotubules
- axon hillock
- collaterals
- axon terminal
- nissl bodies
- filaments
- synapse
- presynaptic cell
- postsynaptic cell
- bipolar neuron
- unipolar neuron
- multipolar neuron

7.2

Oligodendrocytes, astrocytes, ependymal cells, and microglia are neuroglia of the CNS

- neuroglia (glial cells)
- ependymal cells
- ependyma
- cerebrospinal fluid (CSF)
- microglia
- astrocytes
- blood–brain barrier
- oligodendrocytes
- myelin
- myelin sheath
- myelinated
- internodes
- nodes
- white matter
- unmyelinated axons
- gray matter

7.3

Schwann cells and satellite cells protect the axons and cell bodies of sensory and motor neurons in the PNS

- Schwann cells
- neurilemma
- satellite cells
- ganglia
- sensory neurons
- interneurons
- motor neurons
- sensory receptors
- internal receptors
- proprioceptors
- external receptors
- afferent fibers
- sensory ganglia
- sensory neurons
- somatic sensory neurons
- visceral sensory neurons
- somatic motor neurons
- peripheral nerve
- visceral motor neurons
- efferent fibers
- autonomic ganglia
- ○ somatic effectors
- ○ visceral effectors

7.4

All communication and processing in the nervous system depends on changes in the membrane potential of individual neurons

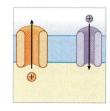

- membrane potential
- resting potential
- graded potential
- action potential
- synaptic activity
- neurotransmitters
- information processing

7.5

An action potential can affect other portions of the membrane through continuous or saltatory propagation

- polarization
- depolarization
- repolarization
- continuous propagation
- saltatory propagation

7.6

At a synapse, information travels from presynaptic cell to postsynaptic cell

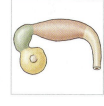

- synapse
- neurotransmitters
- synaptic cleft
- axon terminal
- synaptic vesicles
- presynaptic membrane
- postsynaptic membrane

SECTION 2

The Functional Anatomy of the Central Nervous System

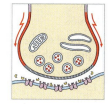

- neural tube
- diencephalon
- cerebrum
- cerebellum
- pons
- medulla oblongata

7.7

Each region of the brain has distinct structural and functional characteristics

- cerebrum
- cerebral hemispheres
- cerebral cortex
- fissures
- gyri
- sulci
- diencephalon
- thalamus
- hypothalamus
- brain stem
- midbrain
- pons
- medulla oblongata
- ventricles
- lateral ventricle
- interventricular foramen
- third ventricle
- cerebral aqueduct
- fourth ventricle
- corpus callosum
- nuclei
- basal nuclei

• = Term boldfaced in this module

7.8

The cranial meninges and the cerebrospinal fluid protect and support the brain

- cranial meninges
- dura mater
- arachnoid mater
- pia mater
- dural folds
- dural sinuses
- superior sagittal sinus
- tentorium cerebelli
- falx cerebelli
- falx cerebri
- cerebrospinal fluid (CSF)
- choroid plexus
- lateral apertures
- median aperture
- arachnoid granulations

7.9

Superficial landmarks can be used to divide the surface of the cerebral cortex into lobes

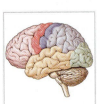

- lobes
- precentral gyrus
- central sulcus
- frontal lobe
- parietal lobe
- postcentral gyrus
- lateral sulcus
- temporal lobe
- insula
- parieto-occipital sulcus
- occipital lobe

7.10

The lobes of the cerebral cortex contain regions with specific functions

- primary motor cortex
- primary sensory cortex
- association area
- ○ motor cortex
- pyramidal cells
- somatic motor association area
- gustatory cortex
- olfactory cortex
- ○ auditory cortex
- primary auditory cortex
- auditory association area
- ○ sensory cortex
- somatic sensory association area
- ○ visual cortex
- primary visual cortex
- visual association area
- integrative centers
- speech center
- prefrontal cortex
- frontal eye field
- general interpretive area
- hemispheric lateralization

7.11

The diencephalon, brain stem, and cerebellum contain relay stations that process information outside our awareness

- diencephalon
- pineal gland
- thalamus
- hypothalamus
- midbrain
- colliculi
- substantia nigra
- cerebral peduncles
- pons
- cerebellar peduncles
- medulla oblongata
- cardiovascular centers
- respiratory rhythmicity centers

7.12

The reticular activating system of the midbrain is responsible for maintaining consciousness

- reticular formation
- reticular activating system (RAS)

7.13

Brain activity can be monitored using external electrodes; the record is called an electroencephalogram, or EEG

- electroencephalogram (EEG)
- brain waves
- alpha waves
- beta waves
- theta waves
- delta waves
- seizure
- epilepsies

7.14

The spinal cord contains gray matter and white matter

- cervical enlargement
- lumbar enlargement
- conus medullaris
- cauda equina
- posterior median sulcus
- dorsal root
- dorsal root ganglion
- spinal nerve
- ventral root
- anterior median fissure
- white matter
- gray matter

7.15

Gray matter is the region of integration, and white matter carries information

- horns
- posterior gray horn
- lateral gray horn
- anterior gray horn
- posterior median sulcus
- anterior median fissure
- columns
- tract
- ascending tracts
- descending tracts
- ○ gray matter
- nuclei
- sensory nuclei
- motor nuclei
- ○ white matter
- posterior white column
- lateral white column
- anterior white column

7.16

The spinal cord is surrounded by meninges, which consist of the dura mater, arachnoid mater, and pia mater

- spinal meninges
- pia mater
- arachnoid mater
- dura mater
- epidural space
- spinal tap (lumbar puncture)

● = Term boldfaced in this module

CAREER PATHS

Physician Assistant

Physician Assistants, or PAs, work in almost every branch of medicine. "We're not unlike stem cells," says Stephen Lummus, a PA at an urgent care clinic in Longview, Texas. "We PAs start out the same, but then we can evolve into all kinds of different functions."

As a PA, Stephen sees patients, formulates treatment plans, orders tests, and writes prescriptions. Sometimes, he will refer patients to a specialist or the emergency room. In other words, he does everything a doctor would do, except he works under a doctor's supervision.

Before becoming a PA, Stephen worked as a firefighter, then as an emergency medical

technician (EMT). He became a PA because he wanted to do more for the people he treated. "It has changed my life," he says. "I love helping people. I knew I wanted a career where I could look myself in the mirror at the end of the day and feel good about what I accomplished."

Because their training spans the breadth of medicine, understanding A&P is as important to a PA as basic mechanics is to someone who fixes cars. "Medicine relies on it," Stephen says. "Without anatomy and physiology, you couldn't take an in-depth medical history, and formulating a diagnosis would be completely impossible." Like any member of a medical team, PAs also need good communication skills, deductive reasoning ability, and the willingness to work long hours.

For additional information, visit the website for the American Academy of Physician Assistants at http://www.aapa.org.

Think this is the career for you?
Key Stats:

- **Education and Training.** A degree from an accredited PA program is required. (Admission requirements vary, but most applicants have a BA and health-related work experience.)

- **Licensure.** PA candidates must pass an exam to become licensed, and complete continuing medical education every two years to remain certified. Every six years, PAs must pass a recertification examination.

- **Earnings.** Earnings vary but the median annual wage is $86,410.

- **Expected Job Prospects.** Employment is expected to grow faster than the national average—39% from 2008 to 2018.

Bureau of Labor Statistics, U.S. Department of Labor, *Occupational Outlook Handbook, 2010–11 Edition*, Physician Assistants, on the Internet at http://www.bls.gov/oco/ocos081.htm (visited September 14, 2011)

Access more review material online in the Study Area at www.masteringaandp.com.

- Chapter guides
- Chapter quizzes
- Practice tests
- Art labeling activities
- Flashcards
- A glossary with pronunciations

- Practice Anatomy Lab™ (PAL™) 3.0 virtual anatomy practice tool
- Interactive Physiology® (IP) animated tutorials
- MP3 Tutor Sessions

 practice anatomy lab™ **For this chapter, follow these navigation paths in PAL:**

- Human Cadaver>Nervous System>Central Nervous System
- Anatomical Models>Nervous System>Central Nervous System
- Histology>Nervous Tissue

 For this chapter, go to these topics in the Nervous System in IP:

- Anatomy Review
- The Membrane Potential
- Synaptic Transmission

 For this chapter, go to this topic in the MP3 Tutor Sessions:

- Generation of an Action Potential

Chapter 7 Review Questions

1. List all of the CNS sites where cerebrospinal fluid (CSF) is located. What are the functions of CSF?

2. Predict the effects on the body of a spinal cord transection at C_7. How would these effects differ from those of a spinal cord transection at T_{10}?

3. What are the main functional differences between the right and left hemispheres of the cerebrum?

4. If neurons in the central nervous system lack centrioles and are unable to divide, how can a person develop brain cancer?

5. Which part of the brain is associated with respiratory and cardiac activity?

For answers to all module, section, and chapter review questions, see the blue Answers tab at the back of the book.

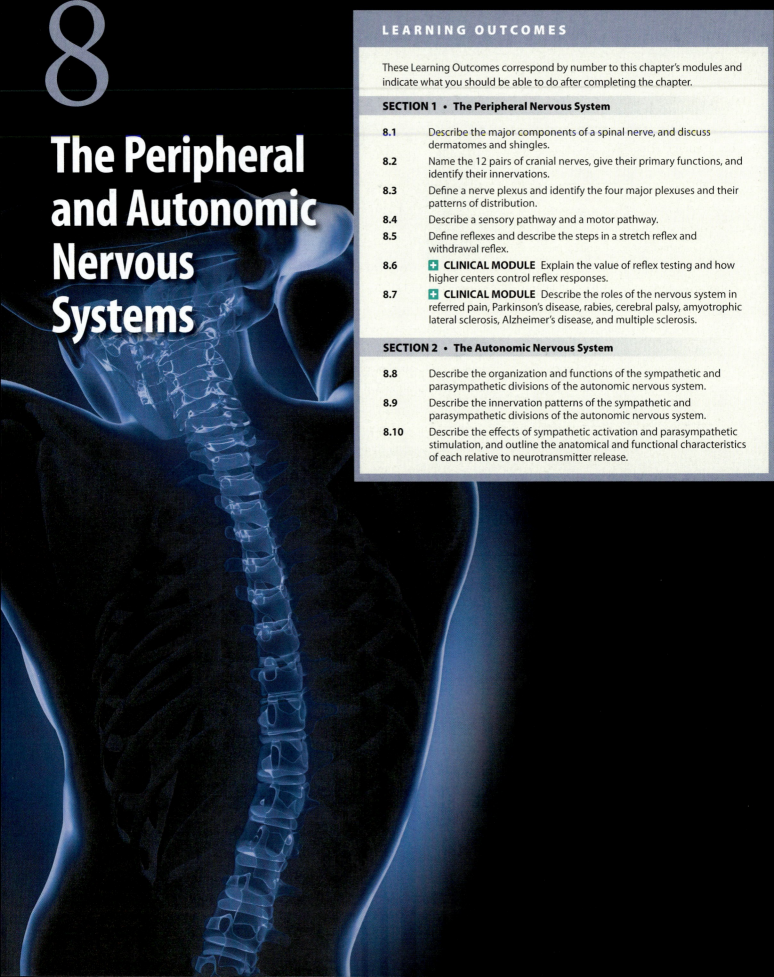

8

The Peripheral and Autonomic Nervous Systems

The Peripheral Nervous System

The central nervous system processes sensory information and conveys motor commands over a network of peripheral nerves called the **peripheral nervous system (PNS)**. Based on their origin, the **peripheral nerves** can be divided into the **cranial nerves** of the brain, and the **spinal nerves** of the spinal cord.

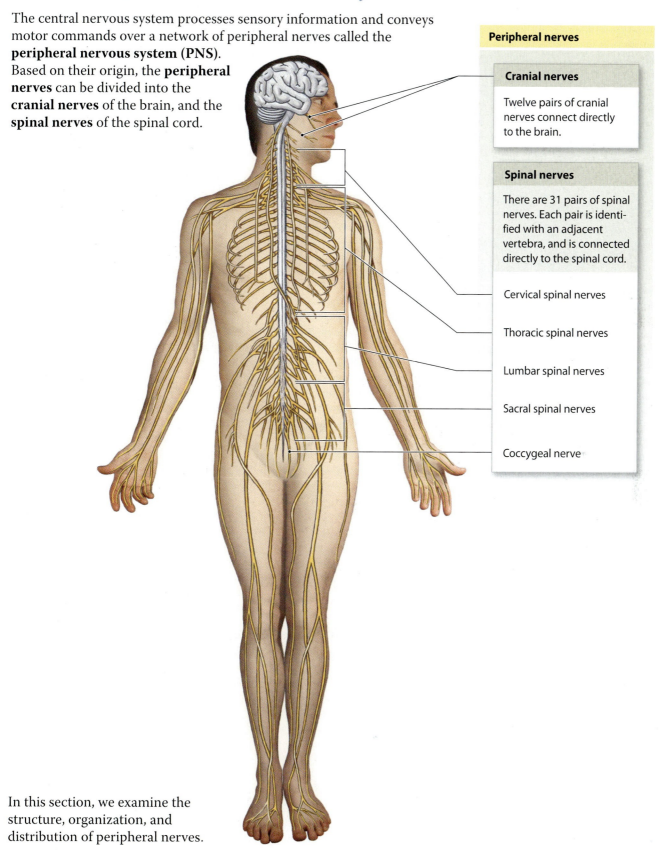

Peripheral nerves

Cranial nerves

Twelve pairs of cranial nerves connect directly to the brain.

Spinal nerves

There are 31 pairs of spinal nerves. Each pair is identified with an adjacent vertebra, and is connected directly to the spinal cord.

Cervical spinal nerves

Thoracic spinal nerves

Lumbar spinal nerves

Sacral spinal nerves

Coccygeal nerve

In this section, we examine the structure, organization, and distribution of peripheral nerves.

Spinal nerves have a consistent anatomical structure and pattern of distribution

1 Every segment of the spinal cord is connected to a pair of spinal nerves. Surrounding each spinal nerve is a series of connective tissue layers that is continuous with those of their associated peripheral nerves. These layers, best seen in sectional view, are comparable to those associated with skeletal muscles.

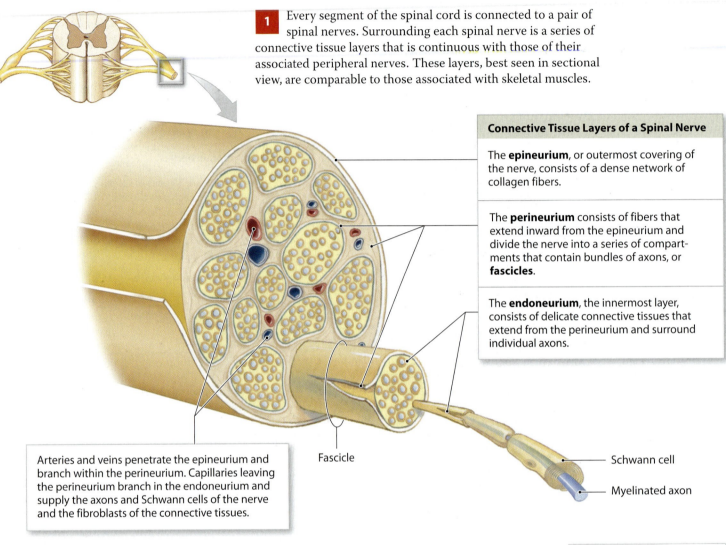

Connective Tissue Layers of a Spinal Nerve

The **epineurium**, or outermost covering of the nerve, consists of a dense network of collagen fibers.

The **perineurium** consists of fibers that extend inward from the epineurium and divide the nerve into a series of compartments that contain bundles of axons, or **fascicles**.

The **endoneurium**, the innermost layer, consists of delicate connective tissues that extend from the perineurium and surround individual axons.

Arteries and veins penetrate the epineurium and branch within the perineurium. Capillaries leaving the perineurium branch in the endoneurium and supply the axons and Schwann cells of the nerve and the fibroblasts of the connective tissues.

Fascicle

Schwann cell

Myelinated axon

2 Each spinal nerve branches to form **rami** (RĀ-mī; singular *ramus*, a branch). Some of these rami carry visceral motor fibers of the autonomic nervous system (ANS). Spinal nerves in the thoracic and upper lumbar segments of the spinal cord carry the motor output of the **sympathetic division** that is responsible for the "fight or flight" response.

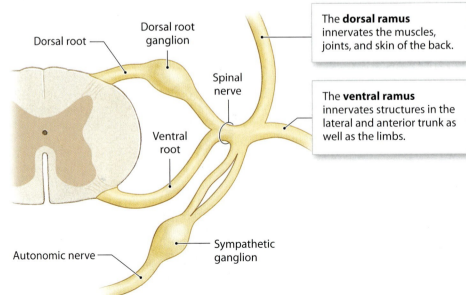

Dorsal root

Dorsal root ganglion

Spinal nerve

Ventral root

Autonomic nerve

Sympathetic ganglion

The **dorsal ramus** innervates the muscles, joints, and skin of the back.

The **ventral ramus** innervates structures in the lateral and anterior trunk as well as the limbs.

3 The specific bilateral region of the skin surface monitored by a single pair of spinal nerves is known as a **dermatome**. Each pair of spinal nerves serves its own dermatome, but the boundaries of adjacent dermatomes overlap to some degree. Spinal nerve C_1 typically lacks a sensory branch to the skin; when present, it innervates the scalp with C_2 and C_3. The face is innervated by a pair of cranial nerves. The most inferior spinal nerve is the coccygeal nerve (C_0). Its dermatome is not visible on this diagram.

4 Dermatomes are clinically important because damage or infection of a spinal nerve or dorsal root ganglion produces a characteristic loss of sensation in the corresponding region of the skin. Additionally, characteristic signs may appear on the skin supplied by that specific nerve. The skin eruptions shown here are characteristic of **shingles**, a viral infection of dorsal root ganglia. Shingles (derived from the Latin *cingulum*, girdle) is caused by the varicella-zoster virus (VZV), the same herpes virus that causes chickenpox. This herpes virus attacks neurons within the dorsal roots of spinal nerves and sensory ganglia of cranial nerves. This disorder produces a painful rash and blisters whose distribution corresponds to that of the affected sensory nerve and its associated dermatome. Any person who has had chickenpox is at risk of developing shingles, because the virus can remain dormant within the anterior gray horns of the spinal cord. It is not known what triggers the reactivation of the virus. In 2006, the U.S. Food and Drug Administration approved a VZV vaccine (Zostavax) for use in people ages 60 and above who have had chickenpox.

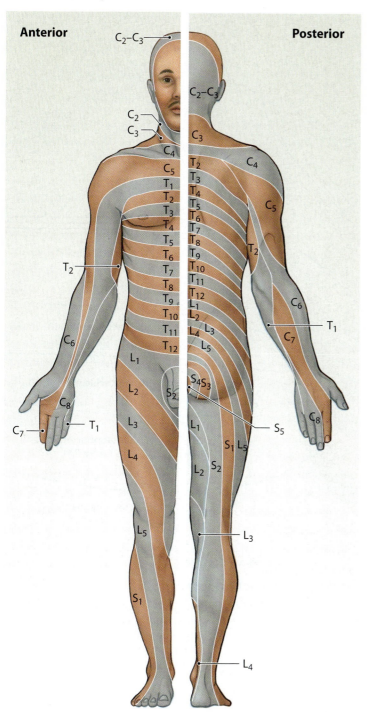

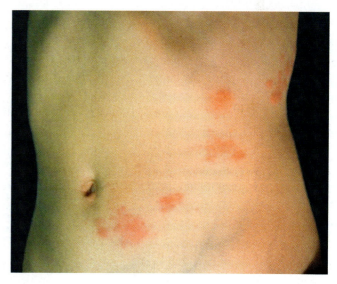

Module 8.1 Review

a. Identify the three layers of connective tissue of a spinal nerve and identify the major peripheral branches of a spinal nerve.

b. Describe a dermatome.

c. Explain the etiology (cause) of shingles.

The twelve pairs of cranial nerves are classified as sensory, special sensory, motor, or mixed nerves

The 12 pairs of cranial nerves are part of the PNS, but they connect directly to the brain and are visible on its ventral surface. Like spinal nerves, cranial nerves are wrapped in layers of connective tissue. Each nerve has a name related to its appearance or its function. The Roman numeral given to each nerve corresponds to its position on the brain, beginning with the cerebrum.

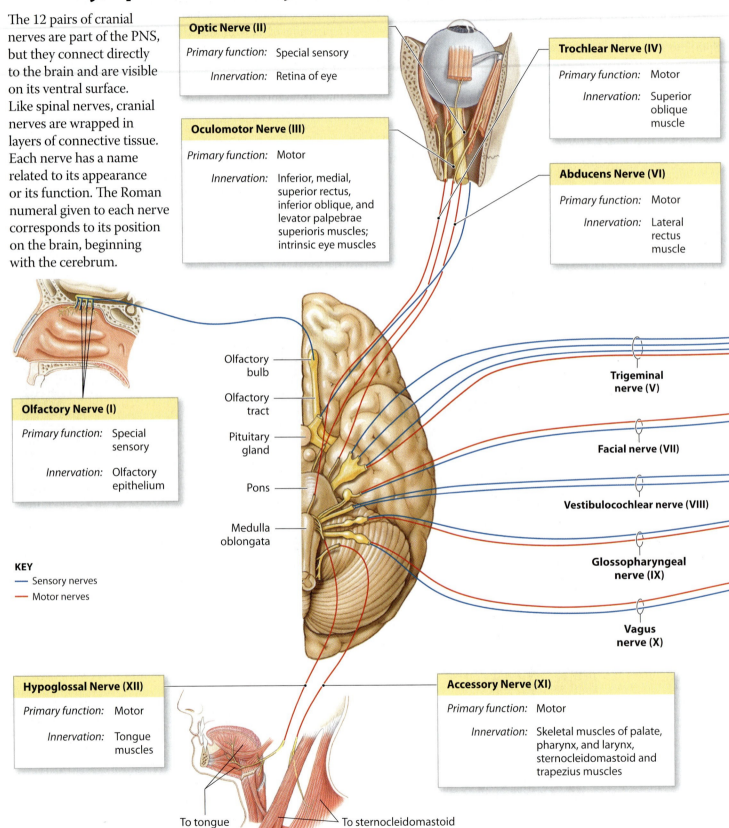

Optic Nerve (II)

Primary function: Special sensory

Innervation: Retina of eye

Oculomotor Nerve (III)

Primary function: Motor

Innervation: Inferior, medial, superior rectus, inferior oblique, and levator palpebrae superioris muscles; intrinsic eye muscles

Trochlear Nerve (IV)

Primary function: Motor

Innervation: Superior oblique muscle

Abducens Nerve (VI)

Primary function: Motor

Innervation: Lateral rectus muscle

Olfactory Nerve (I)

Primary function: Special sensory

Innervation: Olfactory epithelium

Olfactory bulb
Olfactory tract
Pituitary gland
Pons
Medulla oblongata

KEY
— Sensory nerves
— Motor nerves

Trigeminal nerve (V)

Facial nerve (VII)

Vestibulocochlear nerve (VIII)

Glossopharyngeal nerve (IX)

Vagus nerve (X)

Hypoglossal Nerve (XII)

Primary function: Motor

Innervation: Tongue muscles

Accessory Nerve (XI)

Primary function: Motor

Innervation: Skeletal muscles of palate, pharynx, and larynx, sternocleidomastoid and trapezius muscles

To tongue muscles

To sternocleidomastoid and trapezius muscles

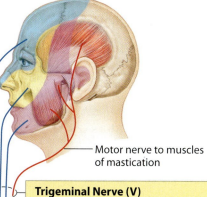

Motor nerve to muscles
of mastication

Trigeminal Nerve (V)

Primary function: Mixed

Innervation: **Sensory:** regions of the
face, lips, part of palate,
and part of tongue
Motor: muscles
associated with the jaws
and mastication

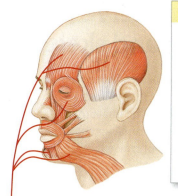

Facial Nerve (VII)

Primary function: Mixed

Innervation: **Sensory:** taste
receptors on anterior
two-thirds of tongue
Motor: muscles of
facial expression,
lacrimal gland,
submandibular and
sublingual salivary
glands

Motor nerve
to facial muscles

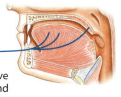

Sensory nerve
to tongue and
soft palate

Vestibulocochlear Nerve (VIII)

Primary function: Special sensory

Innervation: Cochlea (receptors
for hearing)
Vestibule (receptors
for motion and
balance)

Cochlear branch

Vestibular branch

Glossopharyngeal Nerve (IX)

Primary function: Mixed

Innervation: **Sensory:** posterior
third of tongue;
pharynx and part
of palate; receptors
for blood pressure,
pH, oxygen, and
carbon dioxide
concentrations
Motor: pharyngeal
muscles and parotid
salivary gland

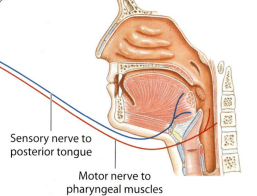

Sensory nerve to
posterior tongue

Motor nerve to
pharyngeal muscles

Vagus Nerve (X)

Primary function: Mixed

Innervation: **Sensory:** pharynx; auricle and
external acoustic meatus; diaphragm;
visceral organs in thoracic and
abdominopelvic cavities
Motor: palatal and pharyngeal
muscles and visceral organs in
thoracic and abdominopelvic cavities

Module 8.2 Review

a. Identify the cranial nerves by name and
number.

b. Which cranial nerves have motor
functions only?

c. Which cranial nerves are mixed nerves?

8.3

Spinal nerves form nerve plexuses that innervate the skin and skeletal muscles

During development, small skeletal muscles innervated by different ventral rami typically fuse to form larger muscles with compound origins. The anatomical distinctions between the component muscles may disappear, but separate ventral rami continue to provide sensory innervation and motor control to each part of the compound muscle. As they converge, the ventral rami of adjacent spinal nerves blend their fibers, producing a series of compound nerve trunks. Such a complex interwoven network of nerves is called a **nerve plexus** (PLEK-sus; *plexus*, braid).

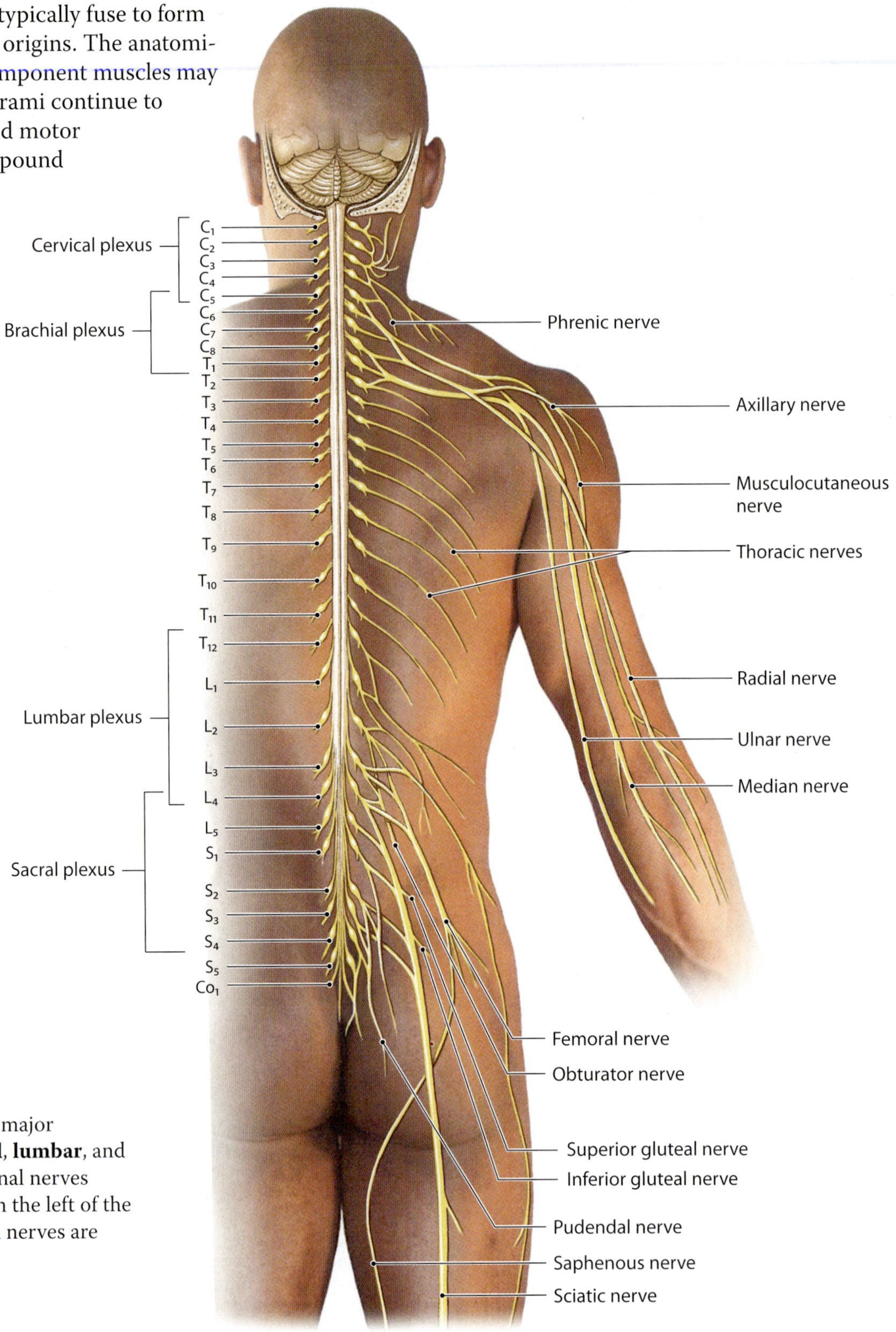

Cervical plexus
Brachial plexus
Lumbar plexus
Sacral plexus
C_1
C_2
C_3
C_4
C_5
C_6
C_7
C_8
T_1
T_2
T_3
T_4
T_5
T_6
T_7
T_8
T_9
T_{10}
T_{11}
T_{12}
L_1
L_2
L_3
L_4
L_5
S_1
S_2
S_3
S_4
S_5
Co_1
Phrenic nerve
Axillary nerve
Musculocutaneous nerve
Thoracic nerves
Radial nerve
Ulnar nerve
Median nerve
Femoral nerve
Obturator nerve
Superior gluteal nerve
Inferior gluteal nerve
Pudendal nerve
Saphenous nerve
Sciatic nerve

1 The ventral rami form four major plexuses: **cervical**, **brachial**, **lumbar**, and **sacral**. In this illustration, the spinal nerves forming each plexus are shown on the left of the figure, while the major peripheral nerves are shown on the right.

2 The cervical plexus consists of the ventral rami of spinal nerves C_1–C_5. The branches of the cervical plexus innervate the muscles of the neck and extend into the thoracic cavity. The **phrenic nerve**, the major nerve of the cervical plexus, provides the entire nerve supply to the diaphragm, a key respiratory muscle. Other branches of this nerve plexus are distributed to the skin of the neck and the superior part of the chest.

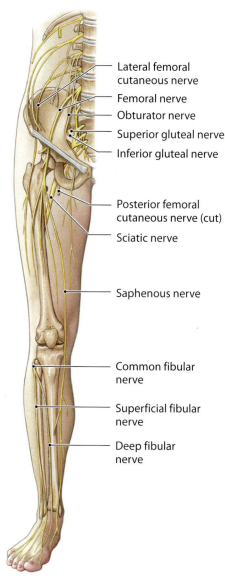

Nerve Roots of Cervical Plexus

C_1
C_2
C_3
C_4
C_5

Phrenic nerve

3 The brachial plexus innervates the pectoral girdle and upper limbs, with contributions from the ventral rami of spinal nerves C_5–T_1.

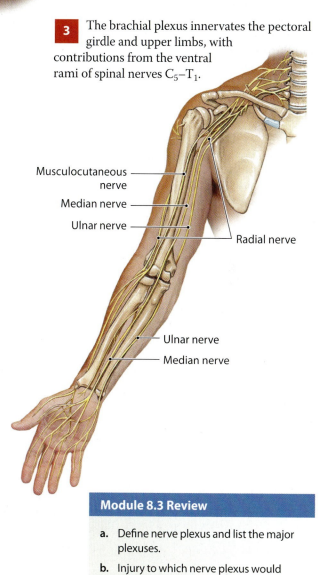

Musculocutaneous nerve
Median nerve
Ulnar nerve
Radial nerve

Ulnar nerve
Median nerve

4 The lumbar plexus and the sacral plexus arise from the lumbar and sacral segments of the spinal cord, respectively. The nerves arising at these plexuses innervate the pelvic girdle and lower limbs.

Lateral femoral cutaneous nerve
Femoral nerve
Obturator nerve
Superior gluteal nerve
Inferior gluteal nerve

Posterior femoral cutaneous nerve (cut)
Sciatic nerve

Saphenous nerve

Common fibular nerve

Superficial fibular nerve

Deep fibular nerve

Module 8.3 Review

a. Define nerve plexus and list the major plexuses.

b. Injury to which nerve plexus would interfere with the ability to breathe?

c. List the major nerves of the brachial plexus.

Sensory and motor pathways carry information between the PNS and processing centers in the CNS

Sensory Pathways

The posterior column pathway carries the sensations of highly localized (fine) touch, pressure, vibration, and proprioception (the sense of body movements and position, especially the limbs). The pathway that delivers this information begins at a peripheral receptor and ends at the primary sensory cortex of the cerebral hemispheres.

1 In a typical sensory pathway, sensory neurons deliver sensations to the CNS. The axons of these **first-order neurons** synapse on interneurons known as **second-order neurons,** which may be located in the spinal cord or brain stem. If the sensation is to reach our awareness, the axons of these interneurons synapse on a **third-order neuron** in the thalamus. Somewhere along its length, the axon of the second-order neuron crosses over to the opposite side of the CNS. Third-order neurons synapse in the primary sensory cortex. The **posterior column pathway** is shown here.

A **sensory homunculus** ("little man") is a map created by determining the area of the primary sensory cortex that is devoted to a particular part of the body. The proportions of the sensory homunculus reflect the *number of sensory receptors* in each region. The greater the number of receptors, the greater the number of cortical neurons required to process the information.

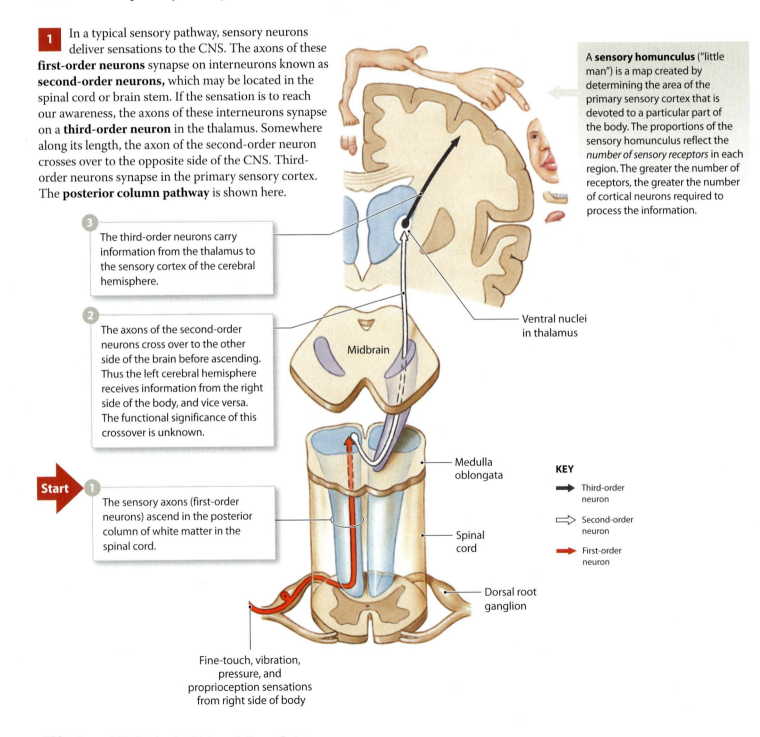

3 The third-order neurons carry information from the thalamus to the sensory cortex of the cerebral hemisphere.

2 The axons of the second-order neurons cross over to the other side of the brain before ascending. Thus the left cerebral hemisphere receives information from the right side of the body, and vice versa. The functional significance of this crossover is unknown.

Start **1** The sensory axons (first-order neurons) ascend in the posterior column of white matter in the spinal cord.

Ventral nuclei in thalamus

Midbrain

Medulla oblongata

Spinal cord

Dorsal root ganglion

Fine-touch, vibration, pressure, and proprioception sensations from right side of body

KEY

→ Third-order neuron

⇨ Second-order neuron

→ First-order neuron

Motor Pathways

Motor pathways always involve at least two motor neurons: an **upper motor neuron**, whose cell body lies in a CNS processing center, and a **lower motor neuron**, whose cell body lies in a nucleus of the brain stem or in the spinal cord.

2 The **corticospinal pathway** provides voluntary control over skeletal muscles. This pathway is sometimes called the pyramidal system because it begins at neurons with pyramid-shaped cell bodies in the primary motor cortex. The axons of these upper motor neurons descend into the brain stem and spinal cord to synapse on lower motor neurons that control skeletal muscles.

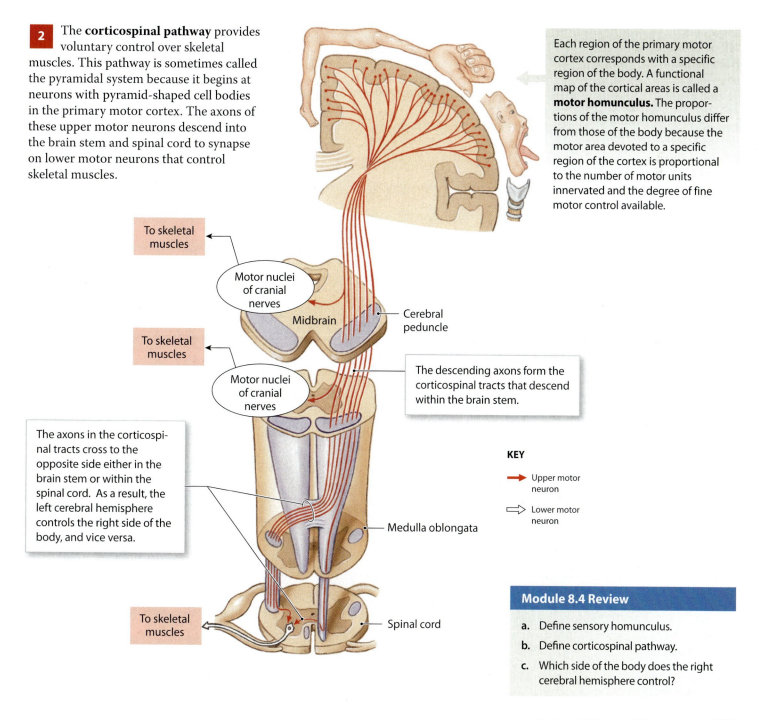

Each region of the primary motor cortex corresponds with a specific region of the body. A functional map of the cortical areas is called a **motor homunculus.** The proportions of the motor homunculus differ from those of the body because the motor area devoted to a specific region of the cortex is proportional to the number of motor units innervated and the degree of fine motor control available.

To skeletal muscles

Motor nuclei of cranial nerves

Midbrain

Cerebral peduncle

To skeletal muscles

Motor nuclei of cranial nerves

The descending axons form the corticospinal tracts that descend within the brain stem.

The axons in the corticospinal tracts cross to the opposite side either in the brain stem or within the spinal cord. As a result, the left cerebral hemisphere controls the right side of the body, and vice versa.

KEY

→ Upper motor neuron

⇨ Lower motor neuron

Medulla oblongata

To skeletal muscles

Spinal cord

Module 8.4 Review

a. Define sensory homunculus.

b. Define corticospinal pathway.

c. Which side of the body does the right cerebral hemisphere control?

Reflexes are rapid, automatic responses to stimuli

Reflexes are rapid, automatic responses to specific stimuli. Reflexes preserve homeostasis by making rapid adjustments in the function of organs or organ systems. The response shows little variability: Each time a particular reflex is activated, it usually produces the same motor response.

1 The wiring of a single reflex is called a **reflex arc**. A reflex arc begins at a receptor and ends at a peripheral effector, such as a muscle fiber. The simplest and fastest reflex arcs are those of **stretch reflexes**, such as the **patellar reflex** shown here. These reflexes involve just two neurons— a sensory neuron and a motor neuron.

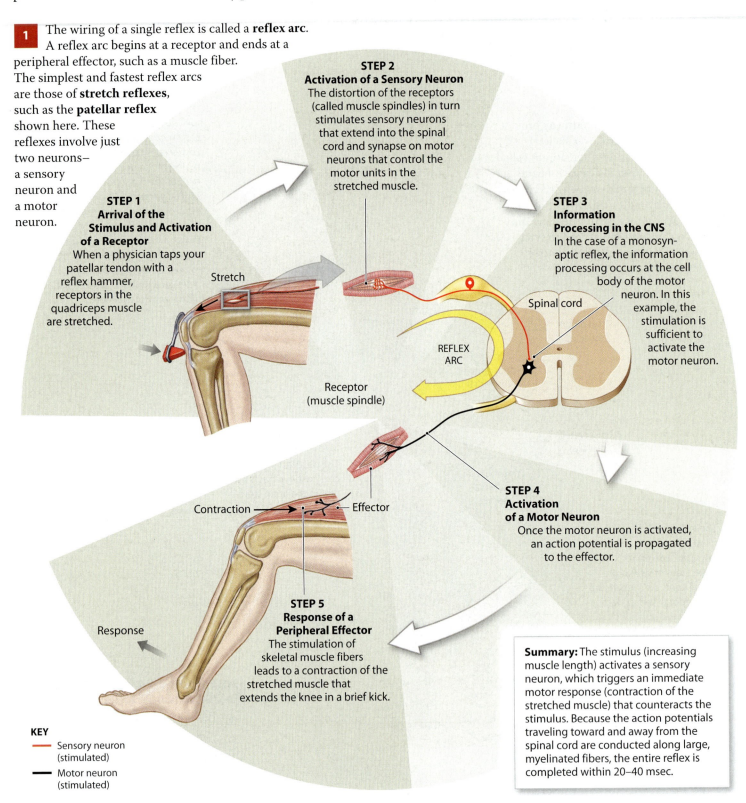

STEP 1
Arrival of the Stimulus and Activation of a Receptor
When a physician taps your patellar tendon with a reflex hammer, receptors in the quadriceps muscle are stretched.

Stretch

STEP 2
Activation of a Sensory Neuron
The distortion of the receptors (called muscle spindles) in turn stimulates sensory neurons that extend into the spinal cord and synapse on motor neurons that control the motor units in the stretched muscle.

STEP 3
Information Processing in the CNS
In the case of a monosynaptic reflex, the information processing occurs at the cell body of the motor neuron. In this example, the stimulation is sufficient to activate the motor neuron.

Spinal cord

REFLEX ARC

Receptor (muscle spindle)

STEP 4
Activation of a Motor Neuron
Once the motor neuron is activated, an action potential is propagated to the effector.

Contraction — Effector

Response

STEP 5
Response of a Peripheral Effector
The stimulation of skeletal muscle fibers leads to a contraction of the stretched muscle that extends the knee in a brief kick.

Summary: The stimulus (increasing muscle length) activates a sensory neuron, which triggers an immediate motor response (contraction of the stretched muscle) that counteracts the stimulus. Because the action potentials traveling toward and away from the spinal cord are conducted along large, myelinated fibers, the entire reflex is completed within 20–40 msec.

KEY
— Sensory neuron (stimulated)
— Motor neuron (stimulated)

2 More complex reflexes involve multiple neurons, which can relay information about the sensation and the reflex response to other neurons and centers in the spinal cord and brain. A **withdrawal reflex** moves affected parts of the body away from a stimulus.

STEP 1
The Arrival of a Stimulus and Activation of a Receptor
A receptor is either a specialized cell or the dendrites of a sensory neuron. Receptors are sensitive to physical or chemical changes in the body or to changes in the external environment. In this example, leaning on a tack stimulates pain receptors in the hand. These receptors respond to stimuli that cause tissue damage.

STEP 2
The Activation of a Sensory Neuron
The stimulation of dendrites leads to the formation and propagation of action potentials along the axons of the sensory neurons. This information reaches the spinal cord by way of a dorsal root ganglion.

STEP 3
Information Processing
Information processing begins when the sensory neuron releases neurotransmitter molecules that may stimulate a second neuron. This neuron is called an **interneuron** because it lies between the sensory neuron and the motor neuron.

The neurotransmitter produces a graded depolarization that, if suffiicient, triggers action potentials in the interneuron.

Dorsal root ganglion

To higher centers

REFLEX ARC

Receptor

Stimulus

Effector

STEP 4
The Activation of a Motor Neuron
Activation of the interneuron leads to stimulation of motor neurons whose axons carry action potentials to peripheral effectors.

STEP 5
The Response of a Peripheral Effector
The release of neurotransmitters at axon terminals then leads to a response—in this case, a skeletal muscle whose contraction pulls your hand away from the tack. A reflex response generally removes or opposes the original stimulus; in this case, the contracting muscle pulls your hand away from a painful stimulus. This reflex arc is therefore an example of negative feedback.

KEY
— Sensory neuron (stimulated)
═ Excitatory interneuron
— Motor neuron (stimulated)

Module 8.5 Review

a. What are the common characteristics of reflexes?

b. Define reflex and list the components of a reflex arc.

c. In the patellar reflex, identify the response observed and the effectors involved.

Reflexes can be used to determine the location and severity of damage to the CNS

Activities in the brain can have a profound effect on the performance of a reflex by facilitating (assisting) or inhibiting the motor neurons or interneurons involved. The facilitation of motor neurons involved in reflexes is called **reinforcement**. For example, a voluntary effort to pull apart clasped hands can produce reinforcement of stretch reflexes so that a light patellar tap produces a big kick rather than a twitch.

1 The **biceps reflex**, **triceps reflex**, and **ankle-jerk reflex** are stretch reflexes often tested during a physical exam. Each reflex is controlled by specific segments of the spinal cord. As a result, testing these reflexes provides information about the status of the corresponding spinal segments. The table on the right presents information on several reflexes commonly used in physical exams.

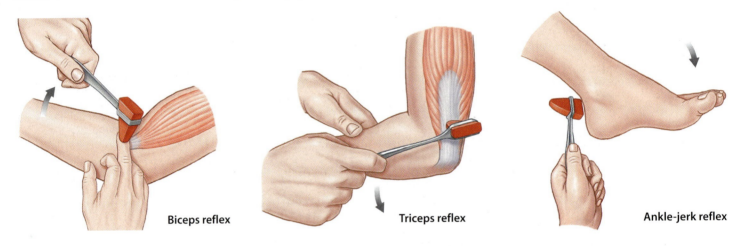

Biceps reflex   Triceps reflex   Ankle-jerk reflex

2 Descending fibers may also have an inhibitory effect on spinal reflexes. Stroking an infant's foot on the lateral side of the sole produces a fanning of the toes known as the **Babinski sign**, or positive Babinski reflex. This response disappears as descending motor pathways develop, because those pathways inhibit this reflex response.

3 In normal adults, stroking the lateral side of the sole produces a curling of the toes, called a **plantar reflex** (or negative Babinski reflex) after about a 1-second delay. If either the higher centers or the descending tracts are damaged, the Babinski sign will reappear in an adult. As a result, this reflex is often tested if CNS injury is suspected.

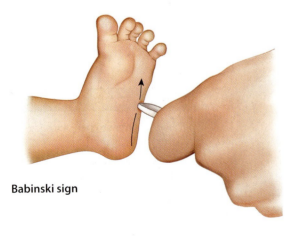

Babinski sign

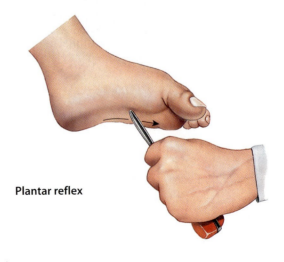

Plantar reflex

4 The **abdominal reflex** depends on descending facilitation, rather than descending inhibition. In this reflex, seen in normal adults, a light stroking of the skin produces a reflexive twitch in the abdominal muscles that moves the navel toward the stimulus. The absence of an abdominal reflex may indicate damage to the descending tracts.

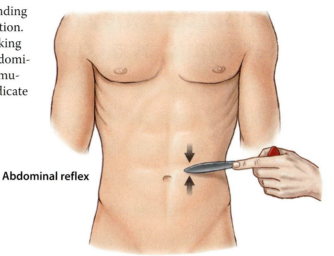

Abdominal reflex

5 This table presents the common reflexes tested during a clinical evaluation, along with the nerves and spinal cord regions those reflexes examine.

Reflexes Used in Diagnostic Testing

Reflex	Stimulus	Afferent Nerve(s)	Spinal Segment	Efferent Nerve(s)	Normal Response
Abdominal reflex	Light stroking of skin of abdomen	Thoracic nerves	T_7–T_{12} at level of arrival	Thoracic nerves	Contractions of abdominal muscles that pull navel toward the stimulus
Plantar reflex	Longitudinal stroking of lateral side of sole of foot	Tibial nerve	S_1, S_2	Tibial nerve	Flexion at toe joints
Biceps reflex	Tap to tendon of biceps brachii muscle near its insertion	Musculocutaneous nerve	C_5, C_6	Musculocutaneous nerve	Flexion at elbow
Triceps reflex	Tap to tendon of triceps brachii muscle near its insertion	Radial nerve	C_6, C_7	Radial nerve	Extension at elbow
Brachioradialis reflex	Tap to forearm near styloid process of the radius	Radial nerve	C_5, C_6	Radial nerve	Flexion at elbow, supination, and flexion at finger joints
Patellar reflex	Tap to patellar tendon	Femoral nerve	L_2–L_4	Femoral nerve	Extension at knee
Ankle-jerk reflex	Tap to calcaneal tendon	Tibial nerve	S_1, S_2	Tibial nerve	Extension (plantar flexion) at ankle

Module 8.6 Review

a. Define reinforcement as it pertains to spinal reflexes.

b. What purpose does reflex testing serve?

c. After injuring her back, 22-year-old Tina exhibits a positive Babinski reflex. What does this imply about her injury?

Nervous system disorders may result from problems with neurons, pathways, or a combination of the two

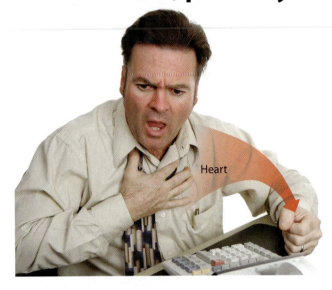

Heart

Referred Pain

1 **Referred pain** is the sensation of pain in a part of the body other than its actual source. A familiar example is the pain of a heart attack, which is frequently felt in the left arm. Strong visceral pain sensations arriving at a segment of the spinal cord can stimulate interneurons that are part of the spinothalamic pathway. Activity in these interneurons leads to the stimulation of the primary sensory cortex, so the individual feels pain in a specific part of the body surface.

Parkinson's Disease

2 **Parkinson's disease** results when neurons of the substantia nigra are damaged or they secrete less dopamine. The basal nuclei become more active, which raises skeletal muscle tone and produces rigidity and stiffness. Individuals with Parkinson's disease have difficulty starting voluntary movements, because opposing muscle groups do not relax; they must be overpowered. Once a movement is under way, every aspect must be voluntarily controlled through intense effort and concentration.

Normal substantia nigra   Diminished substantia nigra in Parkinson patient

Rabies

3 **Rabies** is a dramatic example of a clinical condition directly related to communication between the PNS and CNS involving peripheral axons. A bite from a rabid animal injects the rabies virus into peripheral tissues, where virus particles quickly enter axon terminals, then travel along the axon and into the CNS—with potentially fatal results. Many toxins (including heavy metals), some pathogenic bacteria, and other viruses also bypass CNS defenses in this way.

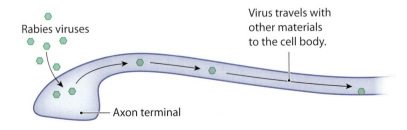

Rabies viruses

Virus travels with other materials to the cell body.

Axon terminal

Cerebral Palsy

4 The term **cerebral palsy (CP)** refers to a number of disorders that affect voluntary motor performance; they appear during infancy or childhood and persist throughout the life of affected individuals. The cause may be trauma associated with premature or unusually stressful birth, maternal exposure to drugs (including alcohol), or a genetic defect that causes the improper development of motor pathways.

ALS

5 **Amyotrophic lateral sclerosis (ALS)** is a progressive, degenerative disorder that affects motor neurons in the spinal cord, brain stem, and cerebral hemispheres. The degeneration affects both upper and lower motor neurons. A defect in axonal transport is thought to underlie the disease. Because a motor neuron and its dependent muscle fibers are so intimately related, the destruction of CNS neurons causes atrophy of the associated skeletal muscles. It is commonly known as *Lou Gehrig disease*, named after the famous New York Yankees player who died of the disorder. Noted physicist Stephen Hawking is also afflicted with this condition.

Alzheimer's Disease

6 **Alzheimer's disease (AD)** is a progressive disorder characterized by the loss of higher-order cerebral functions. It is the most common cause of **senile dementia**, or senility. Symptoms may appear at 50–60 years of age or later, although the disease occasionally affects younger individuals. An estimated 2 million people in the United States—including roughly 15 percent of those over age 65, and nearly half of those over age 85—have some form of the condition, and it causes approximately 100,000 deaths each year. Microscopic examination of the brains of AD patients reveals intracellular and extracellular abnormalities in brain regions such as the hippocampus, specifically associated with memory processing.

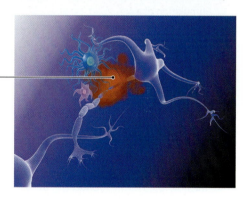

Abnormal dendrites, axons, and extracellular proteins form complexes known as Alzheimer's plaques.

Multiple Sclerosis

7 **Multiple sclerosis** (skler-Ō-sis; *sclerosis*; hardness), or **MS**, is a disease characterized by recurrent incidents of demyelination that affects axons in the optic nerve, brain, and spinal cord. Common signs and symptoms include partial loss of vision and problems with speech, balance, and general motor coordination, including bowel and urinary bladder control. The time between incidents and the degree of recovery vary from case to case. In about one-third of all cases, the disorder is progressive, and functional impairment increases following each new incident. The first attack typically occurs in individuals 30–40 years old; the incidence among women is 1.5 times that among men.

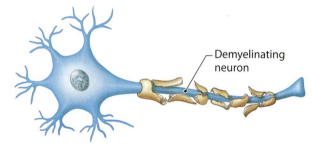

Demyelinating neuron

Module 8.7 Review

a. Define referred pain.

b. Describe how rabies is contracted.

c. Describe amyotrophic lateral sclerosis (ALS).

1. Short answer

Identify the cranial nerves in the accompanying figure, and indicate the function of each: M = motor, S = sensory, or B = both motor and sensory.

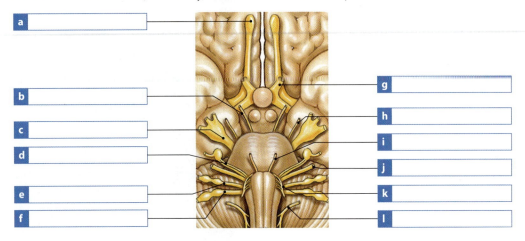

a _____

b _____

c _____

d _____

e _____

f _____

g _____

h _____

i _____

j _____

k _____

l _____

2. Labeling

Label the indicated structures in the accompanying diagram of a reflex arc.

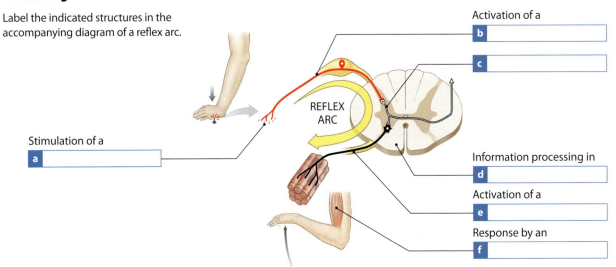

REFLEX ARC

Activation of a

b _____

c _____

Stimulation of a

a _____

Information processing in

d _____

Activation of a

e _____

Response by an

f _____

3. Vocabulary

Write the boldfaced term introduced in this section that matches each of the following definitions.

a _____ Condition with skin eruptions along a dermatome

b _____ Outermost connective tissue layer covering a spinal nerve

c _____ Reflex often tested if CNS injury is suspected

d _____ Nerve that supplies the diaphragm

e _____ Complex interwoven network of nerves

f _____ Motor pathway leading from the cerebral cortex to the spinal cord

g _____ Sensory pathway leading from the spinal cord to the cerebral cortex

h _____ Innermost connective tissue layer of a spinal nerve; covers Schwaan cells

i _____ Condition appearing in infancy or childhood that affects voluntary motor performance

j _____ The enhancement of reflexes due to facilitation of motor neurons

The Autonomic Nervous System

So far, we have focused on the organization of the **somatic nervous system**, or **SNS**. The SNS provides conscious and subconscious control over the skeletal muscles of the body. In this section we consider the **autonomic nervous system**, or **ANS**, which controls visceral function largely outside of our awareness.

1 In the SNS, motor neurons of the central nervous system exert direct control over skeletal muscles. The lower motor neurons may be controlled by reflexes based in the spinal cord or brain, or by upper motor neurons whose cell bodies lie within nuclei of the brain or at the primary motor cortex.

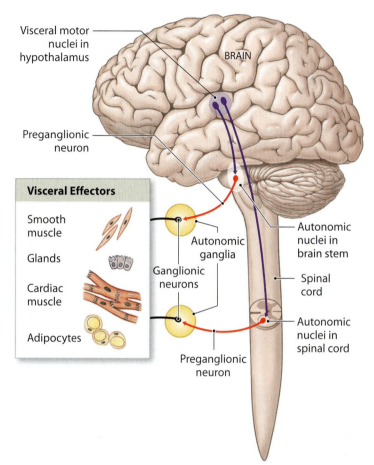

Visceral motor nuclei in hypothalamus

BRAIN

Preganglionic neuron

Visceral Effectors

Smooth muscle

Glands

Cardiac muscle

Adipocytes

Autonomic ganglia

Ganglionic neurons

Autonomic nuclei in brain stem

Spinal cord

Autonomic nuclei in spinal cord

Preganglionic neuron

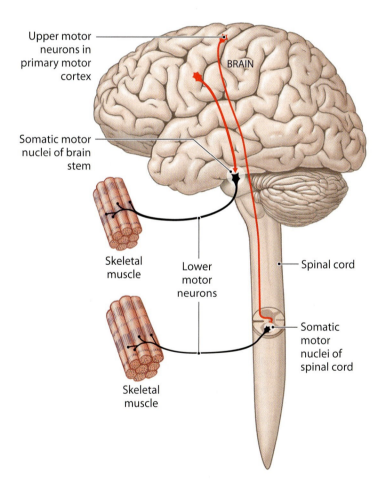

Upper motor neurons in primary motor cortex

BRAIN

Somatic motor nuclei of brain stem

Skeletal muscle

Lower motor neurons

Spinal cord

Somatic motor nuclei of spinal cord

Skeletal muscle

2 In the autonomic nervous system (ANS), integrative centers for autonomic activity are located in the hypothalamus. The neurons in these autonomic centers are comparable to the upper motor neurons in the SNS. However, instead of a single lower motor neuron, there are two motor neurons in series. Visceral motor neurons whose cell bodies lie in the brain stem and spinal cord are known as **preganglionic neurons**. These neurons are part of visceral reflex arcs, and most of their activities represent direct reflex responses, rather than responses to commands from the hypothalamus. The axons of preganglionic neurons leave the CNS and synapse on **ganglionic neurons**—visceral motor neurons in peripheral ganglia. These ganglia, which contain hundreds to thousands of ganglionic neurons, are called **autonomic ganglia**. Ganglionic neurons innervate visceral effectors such as cardiac muscle, smooth muscle, glands, and adipose tissue.

The sympathetic division has chain ganglia, collateral ganglia, and the adrenal medullae ...

Organization of the Sympathetic Division

1 In the sympathetic division of the ANS, preganglionic neurons from the thoracic and superior lumbar segments of the spinal cord synapse on ganglionic neurons that may be located (1) within the sympathetic chain of ganglia near the spinal cord, (2) in collateral ganglia that lie within the abdominopelvic cavity, or (3) within modified ganglion cells in the adrenal medullae. Because the ganglionic neurons are relatively close to the vertebral column, the axons (fibers) of the preganglionic neurons (**preganglionic fibers**) are short, and the axons of the ganglionic neurons (**postganglionic fibers**) are long. The axon terminals of preganglionic fibers release **acetylcholine (ACh)** whereas those of postganglionic fibers release the neurotransmitter **norepinephrine (NE)**.

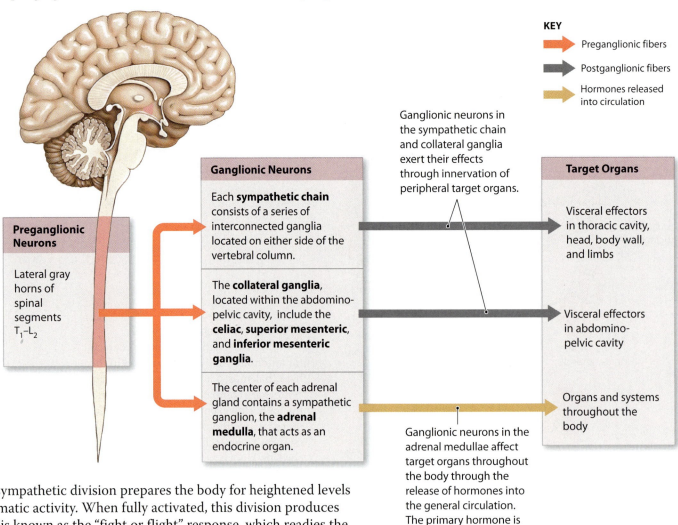

KEY

➡ Preganglionic fibers

➡ Postganglionic fibers

➡ Hormones released into circulation

Preganglionic Neurons

Lateral gray horns of spinal segments T_1–L_2

Ganglionic Neurons

Each **sympathetic chain** consists of a series of interconnected ganglia located on either side of the vertebral column.

The **collateral ganglia**, located within the abdomino-pelvic cavity, include the **celiac**, **superior mesenteric**, and **inferior mesenteric ganglia**.

The center of each adrenal gland contains a sympathetic ganglion, the **adrenal medulla**, that acts as an endocrine organ.

Ganglionic neurons in the sympathetic chain and collateral ganglia exert their effects through innervation of peripheral target organs.

Ganglionic neurons in the adrenal medullae affect target organs throughout the body through the release of hormones into the general circulation. The primary hormone is epinephrine, but norepinephrine is also released.

Target Organs

Visceral effectors in thoracic cavity, head, body wall, and limbs

Visceral effectors in abdomino-pelvic cavity

Organs and systems throughout the body

The sympathetic division prepares the body for heightened levels of somatic activity. When fully activated, this division produces what is known as the "fight or flight" response, which readies the body for a crisis that may require sudden, intense physical activity. Increased levels of sympathetic activity cause the following changes: (1) heightened mental alertness, (2) increased metabolic rate, (3) reduced digestive and urinary functions, (4) activation of energy reserves, (5) increased respiratory rate and dilation of respiratory passageways, (6) elevated heart rate and blood pressure, and (7) activation of sweat glands.

... whereas the parasympathetic division has terminal or intramural ganglia

Organization of the Parasympathetic Division

2 In the parasympathetic division of the ANS, a typical preganglionic neuron synapses on six to eight ganglionic neurons. These neurons may be situated in **terminal ganglia**, located near the target organ, or in **intramural ganglia** (*murus*, wall), which are embedded in the tissues of the target organ. Terminal ganglia are usually paired; examples include the parasympathetic ganglia associated with the cranial nerves. Intramural ganglia typically consist of clusters of ganglion cells. Both preganglionic and postganglionic axon terminals of the parasympathetic division release acetylcholine.

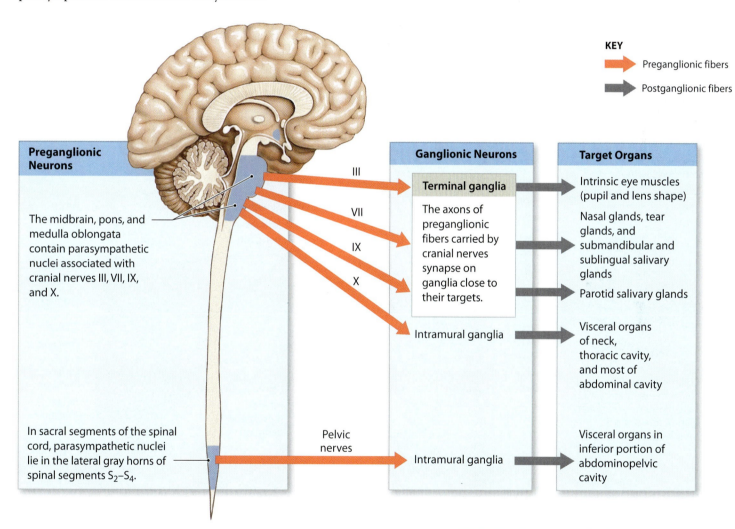

KEY

→ Preganglionic fibers

→ Postganglionic fibers

Preganglionic Neurons

The midbrain, pons, and medulla oblongata contain parasympathetic nuclei associated with cranial nerves III, VII, IX, and X.

In sacral segments of the spinal cord, parasympathetic nuclei lie in the lateral gray horns of spinal segments S_2–S_4.

III
VII
IX
X

Pelvic nerves

Ganglionic Neurons

Terminal ganglia

The axons of preganglionic fibers carried by cranial nerves synapse on ganglia close to their targets.

Intramural ganglia

Intramural ganglia

Target Organs

Intrinsic eye muscles (pupil and lens shape)

Nasal glands, tear glands, and submandibular and sublingual salivary glands

Parotid salivary glands

Visceral organs of neck, thoracic cavity, and most of abdominal cavity

Visceral organs in inferior portion of abdominopelvic cavity

The parasympathetic division regulates visceral function and energy conservation. It is known as the "rest and digest" system. Responses to increased levels of parasympathetic activity include the following: (1) decreased metabolic rate, (2) decreased heart rate and blood pressure, (3) increased secretion by salivary and digestive glands, (4) increased motility and blood flow in the digestive tract, and (5) stimulation of urination and defecation.

Module 8.8 Review

a. List general responses to increased sympathetic activity and to parasympathetic activity.

b. Describe an intramural ganglion.

c. Starting in the spinal cord, trace the path of a nerve impulse through the sympathetic ANS to its target organ in the abdominopelvic cavity.

The two ANS divisions innervate many of the same structures, but the innervation patterns are different

Innervation in the Sympathetic Division

1 The left side of this image shows the distribution to the skin (and to skeletal muscles and other tissues of the body wall), whereas the right side shows the innervation of visceral organs.

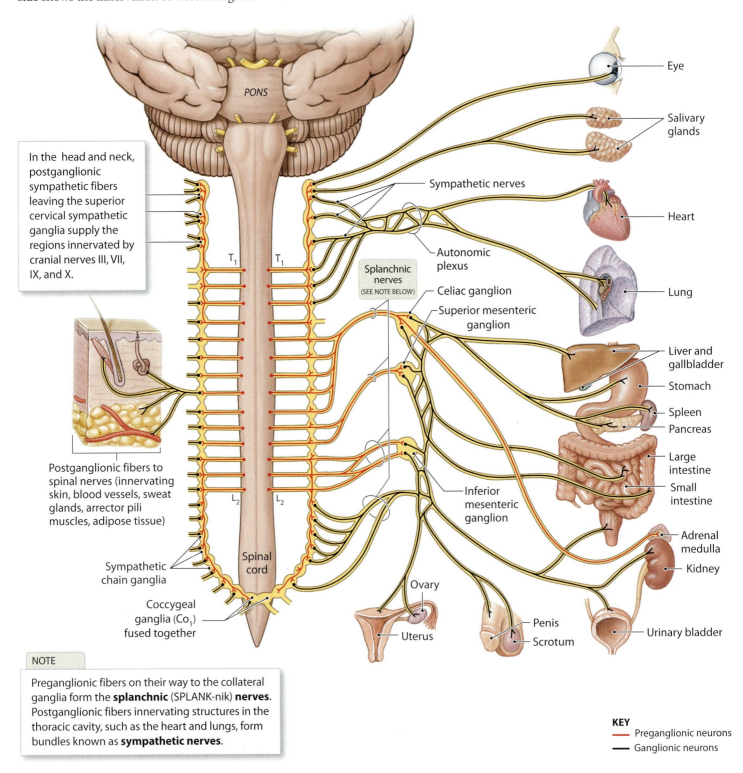

PONS

In the head and neck, postganglionic sympathetic fibers leaving the superior cervical sympathetic ganglia supply the regions innervated by cranial nerves III, VII, IX, and X.

T_1

Splanchnic nerves (SEE NOTE BELOW)

Sympathetic nerves

Autonomic plexus

Celiac ganglion

Superior mesenteric ganglion

Inferior mesenteric ganglion

Postganglionic fibers to spinal nerves (innervating skin, blood vessels, sweat glands, arrector pili muscles, adipose tissue)

L_2

Sympathetic chain ganglia

Spinal cord

Coccygeal ganglia (Co_1) fused together

Ovary

Uterus

Penis

Scrotum

Eye

Salivary glands

Heart

Lung

Liver and gallbladder

Stomach

Spleen

Pancreas

Large intestine

Small intestine

Adrenal medulla

Kidney

Urinary bladder

NOTE

Preganglionic fibers on their way to the collateral ganglia form the **splanchnic** (SPLANK-nik) **nerves**. Postganglionic fibers innervating structures in the thoracic cavity, such as the heart and lungs, form bundles known as **sympathetic nerves**.

KEY
— Preganglionic neurons
— Ganglionic neurons

Innervation in the Parasympathetic Division

2 This image shows the parasympathetic innervation on one side of the body; the innervation on the opposite side (not shown) follows the same pattern.

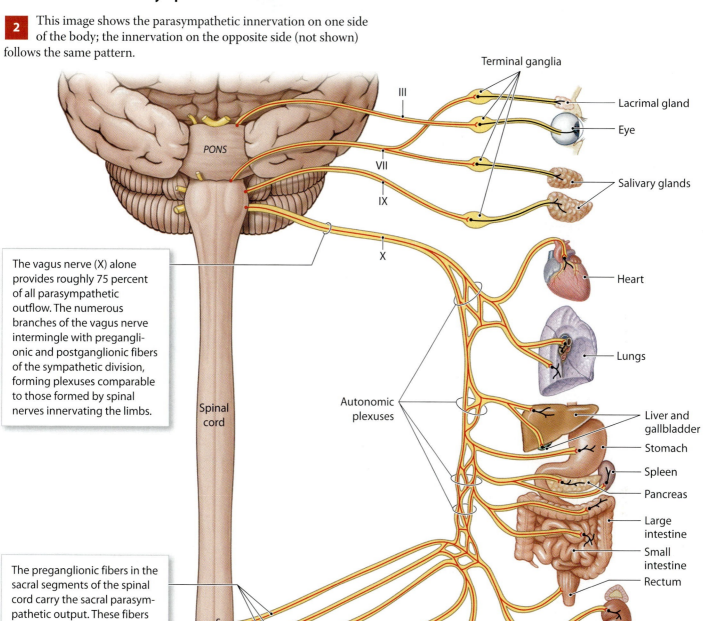

Terminal ganglia

III

VII

IX

X

PONS

Spinal cord

Lacrimal gland

Eye

Salivary glands

Heart

Lungs

Autonomic plexuses

Liver and gallbladder

Stomach

Spleen

Pancreas

Large intestine

Small intestine

Rectum

Kidney

Penis

Scrotum

Urinary bladder

Uterus

Ovary

S_2
S_3
S_4

The vagus nerve (X) alone provides roughly 75 percent of all parasympathetic outflow. The numerous branches of the vagus nerve intermingle with preganglionic and postganglionic fibers of the sympathetic division, forming plexuses comparable to those formed by spinal nerves innervating the limbs.

The preganglionic fibers in the sacral segments of the spinal cord carry the sacral parasympathetic output. These fibers form distinct **pelvic nerves**, which innervate intramural ganglia in the walls of the kidneys, urinary bladder, terminal portions of the large intestine, and sex organs.

Module 8.9 Review

a. Define splanchnic nerves.

b. Which nerve carries the majority of the parasympathetic outflow?

c. Describe sympathetic nerves.

The functional differences between the two ANS divisions reflect their divergent anatomical and physiological characteristics

Functional Characteristics of the Sympathetic Division

1 Unlike the parasympathetic division, the entire sympathetic division can be quickly activated. This table summarizes the effects of sympathetic activation.

2 This illustration and the table below it summarize the anatomical characteristics of the sympathetic division of the ANS.

Effects of Sympathetic Activation

The sympathetic division can change the activities of tissues and organs by releasing NE at peripheral synapses, and by distributing E and NE throughout the body in the bloodstream. The visceral motor fibers that target specific effectors, such as smooth muscle fibers in blood vessels of the skin, can be activated in reflexes that do not involve other visceral effectors.

In a crisis, however, the entire division responds. This event, called **sympathetic activation**, is controlled by sympathetic centers in the hypothalamus. The effects are not limited to peripheral tissues; sympathetic activation also alters CNS activity. During sympathetic activation, the following changes occur in an individual:

- Increased alertness via stimulation of the reticular activating system, causing the individual to feel "on edge."

- A feeling of energy, often associated with a disregard for danger and a temporary insensitivity to painful stimuli.

- Increased activity in the cardiovascular and respiratory centers of the pons and medulla oblongata, leading to elevations in blood pressure, heart rate, breathing rate, and depth of respiration.

- A general elevation in muscle tone, so the person looks tense and may begin to shiver.

- The mobilization of energy reserves, through the accelerated breakdown of glycogen in muscle and liver cells and the release of lipids by adipose tissues. These changes, plus the peripheral changes already noted, complete the preparations necessary for the individual to cope with a stressful situation.

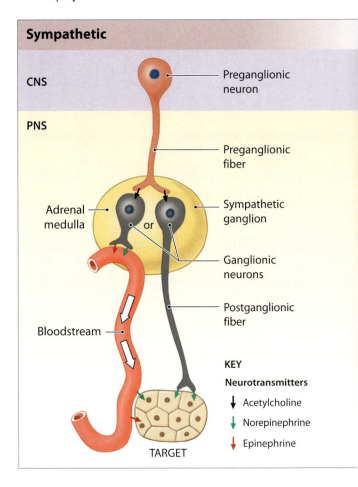

Sympathetic

CNS — Preganglionic neuron

PNS — Preganglionic fiber

Adrenal medulla — or — Sympathetic ganglion

Ganglionic neurons

Postganglionic fiber

Bloodstream

TARGET

KEY

Neurotransmitters

↓ Acetylcholine

↓ Norepinephrine

↓ Epinephrine

Characteristic	Sympathetic Division
Location of CNS visceral motor neurons	Lateral gray horns of spinal segments T_1–L_2
Location of PNS ganglia	Near vertebral column (sympathetic chain ganglia)
Preganglionic fibers Neurotransmitter	Short ACh
Postganglionic fibers Neurotransmitter	Long Usually norepinephrine (NE)
General functions	Stimulates metabolism; increases alertness; prepares for emergency ("fight or flight")

Functional Characteristics of the Parasympathetic Division

3 This illustration and the table below it summarize the anatomical characteristics of the parasympathetic division of the ANS.

4 The parasympathetic division does not release neurotransmitters directly into the bloodstream, so it does not undergo a division-wide activation. This table summarizes the effects of parasympathetic stimulation.

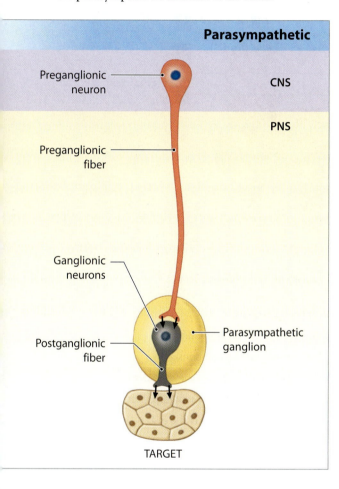

Parasympathetic

Preganglionic neuron — CNS

PNS

Preganglionic fiber

Ganglionic neurons

Postganglionic fiber — Parasympathetic ganglion

TARGET

Characteristic	Parasympathetic Division
Location of CNS visceral motor neurons	Brain stem and spinal segments S_2–S_4
Location of PNS ganglia	Typically intramural
Preganglionic fibers Neurotransmitter	Relatively long ACh
Postganglionic fibers Neurotransmitter	Relatively short ACh
General functions	Promotes relaxation, nutrient uptake, energy storage ("rest and digest")

Effects of Parasympathetic Stimulation

Under normal conditions, the entire parasympathetic division—unlike the sympathetic division—is neither controlled nor activated as a whole. Although it is active continuously, the activities are reflex responses to conditions within specific structures or regions. Examples of the major effects produced by the parasympathetic division include the following:

- Constriction of the pupils (to restrict the amount of light that enters the eyes) and focusing of the lenses of the eyes on nearby objects.

- Secretion by digestive glands, including salivary glands, gastric glands, duodenal glands, intestinal glands, the pancreas (exocrine and endocrine), and the liver.

- Secretion of hormones that promote the absorption and utilization of nutrients by peripheral cells.

- Changes in blood flow and glandular activity associated with sexual arousal.

- Increased smooth muscle activity along the digestive tract.

- Stimulation and coordination of defecation.

- Contraction of the urinary bladder during urination.

- Constriction of the respiratory passageways.

- Reduction in heart rate and in the force of contraction.

These functions center on relaxation, food processing, and energy absorption. Stimulation of the parasympathetic division leads to a general increase in the nutrient content of the blood. Cells throughout the body respond to this increase by absorbing nutrients and using them to support anabolic activities such as growth, cell division, and the creation of energy reserves in the form of lipids or glycogen.

Module 8.10 Review

a. What neurotransmitter is released by all parasympathetic neurons?

b. Why is the parasympathetic division called the anabolic system?

c. What physiological changes are typical in tense (anxious) individuals?

SECTION 2 Review

1. Labeling

Fill in the missing labels in this diagram of the sympathetic division of the ANS. Also indicate the distribution of the sympathetic innervation using red for the preganglionic fibers and black for the postganglionic fibers.

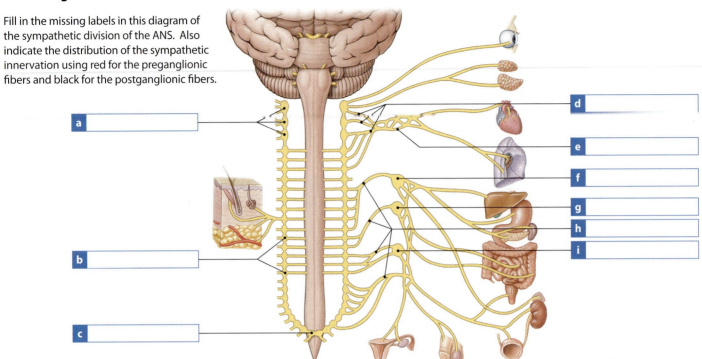

a _____

b _____

c _____

d _____

e _____

f _____

g _____

h _____

i _____

2. Matching

Write S (for sympathetic) or P (for parasympathetic) to indicate the ANS division responsible for each of the following effects.

a _____ decreased metabolic rate

b _____ increased salivary and digestive secretions

c _____ increased metabolic rate

d _____ stimulation of urination and defecation

e _____ activation of sweat glands

f _____ heightened mental alertness

g _____ decreased heart rate and blood pressure

h _____ activation of energy reserves

i _____ increased heart rate and blood pressure

j _____ reduced digestive and urinary functions

k _____ increased motility and blood flow in the digestive tract

l _____ increased respiratory rate and dilation of respiratory passages

m _____ constriction of the pupils and focus of the eyes on nearby objects

3. Section integration

Compare and contrast the following characteristics of the sympathetic and parasympathetic divisions of the ANS: a. Location of CNS visceral motor neurons; b. Location of PNS ganglia; c. Relative length of preganglionic fibers and the neurotransmitter they release; d. Relative length of postganglionic fibers and the neurotransmitter they release; and, e. General functions.

_____

_____

_____

_____

_____

_____

Visual Outline with Key Terms Summarize the content of each module using the terms in the order provided.

SECTION 1

The Peripheral Nervous System

- peripheral nervous system (PNS)
- peripheral nerves
- cranial nerves
- spinal nerves

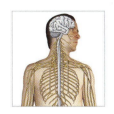

8.1

Spinal nerves have a consistent anatomical structure and pattern of distribution

- epineurium
- perineurium
- fascicles
- endoneurium
- rami
- sympathetic division
- dorsal ramus
- ventral ramus
- dermatome
- shingles

8.2

The 12 pairs of cranial nerves are classified as sensory, special sensory, motor, or mixed nerves

- olfactory nerve (I)
- optic nerve (II)
- oculomotor nerve (III)
- trochlear nerve (IV)
- trigeminal nerve (V)
- abducens nerve (VI)
- facial nerve (VII)
- vestibulocochlear nerve (VIII)
- glossopharyngeal nerve (IX)
- vagus nerve (X)
- accessory nerve (XI)
- hypoglossal nerve (XII)

8.3

Spinal nerves form nerve plexuses that innervate the skin and skeletal muscles

- nerve plexus
- cervical plexus
- brachial plexus
- lumbar plexus
- sacral plexus
- phrenic nerve

8.4

Sensory and motor pathways carry information between the PNS and processing centers in the CNS

- first-order neurons
- second-order neurons
- third-order neurons
- posterior column pathway
- sensory homunculus
- upper motor neuron
- lower motor neuron
- corticospinal pathway
- motor homunculus

8.5

Reflexes are rapid, automatic responses to stimuli

- reflexes
- reflex arc
- stretch reflexes
- patellar reflex
- withdrawal reflex
- interneuron

8.6

Reflexes can be used to determine the location and severity of damage to the CNS

- reinforcement
- biceps reflex
- triceps reflex
- ankle-jerk reflex
- Babinski sign
- plantar reflex
- abdominal reflex

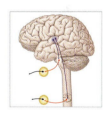

8.7

Nervous system disorders may result from problems with neurons, pathways, or a combination of the two

- referred pain
- Parkinson disease
- rabies
- cerebral palsy (CP)
- amyotrophic lateral sclerosis (ALS)
- Alzheimer disease (AD)
- senile dementia
- multiple sclerosis (MS)

SECTION 2

The Autonomic Nervous System

- somatic nervous system (SNS)
- autonomic nervous system (ANS)
- preganglionic neurons
- ganglionic neurons
- autonomic ganglia

8.8

The sympathetic division has chain ganglia, collateral ganglia, and the adrenal medullae, whereas the parasympathetic division has terminal or intramural ganglia

- preganglionic fibers
- postganglionic fibers
- acetylcholine (ACh)
- norepinephrine (NE)
- sympathetic chain
- collateral ganglia
- celiac ganglia
- superior mesenteric ganglia
- inferior mesenteric ganglia
- adrenal medulla
- terminal ganglia
- intramural ganglia

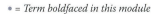

● = *Term boldfaced in this module*

● = *Term boldfaced in this module*

"I really like working with the patients. It is gratifying to know that you are helping someone else."

— **Barbara Davis**
Occupational Therapist
Castle Medical Center
Kailua, Hawaii

CAREER PATHS

Occupational Therapist

You take for granted most activities you do every day to take care of yourself: showering, brushing your teeth, getting dressed, making breakfast. But imagine if you couldn't do them anymore. That is where people like Barbara Davis, an occupational therapist at Castle Medical Center in Kailua, Hawaii, come in.

Barbara works primarily with the hospital's acute care inpatients. "I really like working with the patients," she says. "It is gratifying to know that you are helping someone else."

Occupational therapy (OT) is similar to physical therapy, but differs in that the treatment focuses on getting patients to perform specific everyday activities, such as preparing a meal or handling objects. Occupational therapy focuses on adapting the environment to fit the patient.

Occupational therapists will often specialize in one type of patient—special-needs children, patients with chronic lung disease, senior citizens, and so on. Barbara works with general adults, about 60 percent of whom are over 65.

A typical day for Barbara begins when she gets her list of patients. She reviews the prior OT evaluation for patients with whom she is not familiar. For patients who are new to occupational therapy, she reviews their medical history and any other pertinent information.

The therapy sessions which can run from 30 to 60 minutes, focus on the activities of daily life (ADL) such as dressing, bathing, and sitting on and getting up from toilets and chairs. Barbara guides her patients through these activities, but also supervises some muscle-strengthening exercises. Most important, she advises the patients and their families on how to adapt their environments to their specific needs. For example, occupational therapists will talk to hip-replacement patients about chairs and grab-bars for the shower, as well as teach patients how to use these devices.

Barbara, however, would not be able to instruct patients on how to adapt to their changing needs without the necessary education.

"One class I found extremely interesting and useful for OT was a neuroanatomy course, as we have quite a few patients with a diagnosis of stroke, TIA, traumatic brain injury, etc.," she says. "An active knowledge of anatomy and physiology is also essential when working with our orthopedic patients as well as for general patient and family education. Many of our patients have other extensive medical conditions—diabetes, HIV, hypertension, hypotension, glaucoma—which we have to understand to know how these conditions may affect the patient's overall function and rehabilitation in general."

Because occupational therapists work closely with doctors, nurses, patients, and patients' families, Barbara emphasizes the importance of interpersonal skills. "We have to work well as a team," she says. "Anything less is really unacceptable when you work in a fast-paced environment and want to provide the best-quality services for the patients."

Occupational therapists work in hospitals, rehabilitation centers, and in private practice. Occasionally, they work in schools. Some occupational therapists work part-time, and some work at multiple facilities. As the Baby Boomer generation ages and needs more help, the demand for occupational therapists is expected to rise.

For more information, visit the website of the American Occupational Therapy Association at http://www.aota.org.

Think this is the career for you?

Key Stats:

- **Education and Training.** A master's degree is required for a career in occupational therapy.

- **Licensure.** All 50 states require occupational therapists to be licensed; board certification is optional.

- **Earnings.** Earnings vary but the median annual salary is $72,320.

- **Expected Job Prospects.** Employment is expected to grow faster than the national average—by 26% through 2018.

Bureau of Labor Statistics, U.S. Department of Labor, *Occupational Outlook Handbook, 2010–11 Edition,* Occupational Therapists, on the Internet at http://www.bls.gov/oco/ocos078.htm (visited September 14, 2011).

Access more review material online in the Study Area at www.masteringaandp.com.

- Chapter guides
- Chapter quizzes
- Practice tests
- Art labeling activities
- Flashcards
- A glossary with pronunciations
- Practice Anatomy Lab™ (PAL™) 3.0 virtual anatomy practice tool
- Interactive Physiology® (IP) animated tutorials
- MP3 Tutor Sessions

 practice anatomy lab™ **For this chapter, follow these navigation paths in PAL:**

- Human Cadaver>Nervous System>Peripheral Nervous System
- Human Cadaver>Nervous System>Autonomic Nervous System
- Anatomical Models>Nervous System>Peripheral Nervous System
- Anatomical Models>Nervous System>Autonomic Nervous System

 For this chapter, go to this topic in the MP3 Tutor Sessions:

- The Differences between the Sympathetic and Parasympathetic Divisions

Chapter 8 Review Questions

1. Few people are able to remember the names, numbers, and functions of the cranial nerves without some effort. Many people use mnemonic phrases, such as Oh, Once One Takes The Anatomy Final, Very Good Vacations Are Heavenly, in which the first letter of each word represents the names of the cranial nerves. Using this mnemonic, list the 12 pairs of cranial nerves in numerical order.

2. Describe the relationship among first-, second-, and third-order neurons in a sensory pathway.

3. What is a motor homunculus? How does it differ from a sensory homunculus?

4. The response time in the patellar reflex is much faster than response time in a withdrawal reflex. Explain.

5. How does the emergence of sympathetic fibers along the spinal cord differ from the emergence of parasympathetic fibers?

For answers to all module, section, and chapter review questions, see the blue Answers tab at the back of the book.

9

The Senses

The General Senses

1 The term **general senses** is used to describe our sensitivity to temperature, pain, touch, pressure, vibration and proprioception. Sensory information is picked up by **sensory receptors**, specialized cells or cell processes that monitor conditions inside or outside the body. The simplest receptors are the dendrites (cell processes) of sensory neurons. As shown at the right, the branching tips of these dendrites, called free nerve endings, are not protected by accessory structures. Free nerve endings extend through a tissue the way grass roots extend into the soil. They can be stimulated by many different stimuli and therefore exhibit little receptor specificity. For example, free nerve endings that respond to tissue damage by providing pain sensations may be stimulated by chemical stimulation, pressure, temperature changes, or trauma.

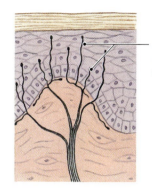

Free nerve endings

2 General sensory receptors are simple in structure and widely distributed in the body. As indicated in this concept map, they can be classified according to the nature of the primary stimulus.

A Functional Classification of General Sensory Receptors

Nociceptors

Nociceptors are pain receptors. They are free nerve endings with large receptive fields and broad sensitivity. Two types of axons—Type A and Type C fibers—carry pain sensations.

Thermoreceptors

Thermoreceptors, or temperature receptors, are free nerve endings located in the dermis, skeletal muscles, liver, and hypothalamus. Cold receptors are three or four times more numerous than warm receptors. No structural differences between warm and cold thermoreceptors have been identified.

Mechanoreceptors

Mechanoreceptors are sensitive to stimuli that distort their plasma membranes. These membranes contain mechanically gated ion channels whose gates open or close in response to stretching, compression, twisting, or other distortions of the membrane.

Chemoreceptors

Chemoreceptors respond to water-soluble and lipid-soluble substances that are dissolved in body fluids (interstitial fluid, plasma, and CSF).

Proprioceptors monitor the positions of joints and muscles. They are the most structurally and functionally complex of the general sensory receptors.

Baroreceptors (bar-ō-rē-SEP-torz; *baro-*, pressure) detect pressure changes in the walls of blood vessels and in portions of the digestive, respiratory, and urinary tracts.

Tactile receptors provide the sensations of touch, pressure, and vibration. Touch sensations provide information about shape or texture, whereas pressure sensations indicate the degree and frequency of mechanical distortion.

General sensory receptors in the skin vary widely in form and function

There are millions of general sensory receptors in the body. Not surprisingly, the greatest diversity is found in the skin, which is in constant contact with the external environment and its associated threats to homeostasis. This overview introduces the basic structural and functional characteristics of the important receptor types in the skin.

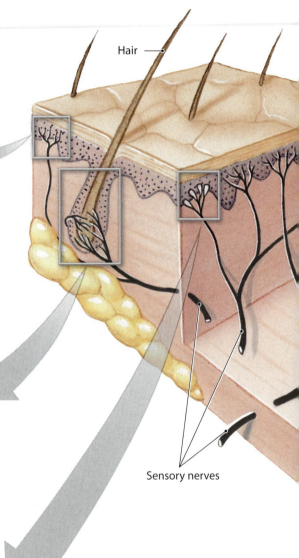

Hair

Sensory nerves

Free Nerve Endings

Free nerve endings are the branching tips of sensory neurons. They are not protected by any accessory structures, and they are nonspecific: They can respond to tactile, pain, and temperature stimuli. Free nerve endings are the most common receptors in the skin.

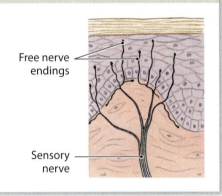

Free nerve endings

Sensory nerve

Root Hair Plexus

Wherever hairs are located, the nerve endings of the **root hair plexus** monitor distortions and movements across the body surface. When a hair is displaced, the movement of the follicle distorts the sensory dendrites and produces action potentials.

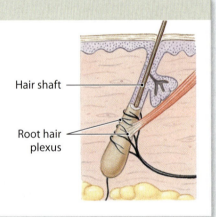

Hair shaft

Root hair plexus

Tactile Discs

Tactile discs are fine touch and pressure receptors. The dendritic processes of a single myelinated afferent fiber make close contact with unusually large epithelial cells in the stratum basale of the epidermis of the skin.

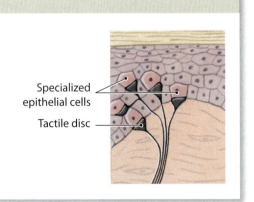

Specialized epithelial cells

Tactile disc

Tactile Corpuscles

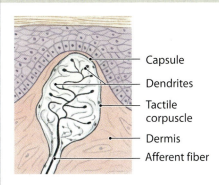

- Capsule
- Dendrites
- Tactile corpuscle
- Dermis
- Afferent fiber

Corpuscles are sensory end structures. **Tactile corpuscles**, or Meissner (MĪS-ner) corpuscles, provide sensations of fine touch and pressure and low-frequency vibration. These receptors become less sensitive to stimulation within a second after contact. These receptors are most abundant in the eyelids, lips, fingertips, nipples, and external genitalia. The dendrites are highly coiled and interwoven, and they are surrounded by modified Schwann cells. A fibrous capsule surrounds the entire complex and anchors it within the dermis.

Lamellated Corpuscles

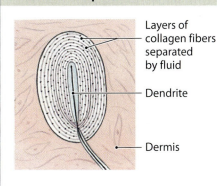

- Layers of collagen fibers separated by fluid
- Dendrite
- Dermis

Lamellated (LAM-e-lāt-ed; *lamella*, thin plate) **corpuscles**, or pacinian (pa-SIN-ē-an) corpuscles, are sensitive to deep pressure. They are most sensitive to pulsing or high-frequency vibrating stimuli. A single dendrite lies within a series of concentric layers of collagen fibers and specialized fibroblasts. The concentric layers, separated by interstitial fluid, shield the dendrite from virtually every source of stimulation other than direct pressure. Somatic sensory information is provided by lamellated corpuscles located throughout the dermis, notably in the fingers, mammary glands, and external genitalia. Visceral sensory information is provided by lamellated corpuscles in mesenteries, in the pancreas, and in the walls of the urethra and urinary bladder.

Ruffini Corpuscles

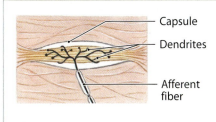

- Capsule
- Dendrites
- Afferent fiber

Ruffini (roo-FĒ-nē) **corpuscles** are sensitive to pressure and distortion of the reticular (deep) dermis. These receptors show little, if any, **adaptation** (reduction in sensitivity to a constant stimulus). A capsule surrounds a core of collagen fibers that are continuous with those of the surrounding dermis. Within the capsule, a network of dendrites is intertwined with the collagen fibers. Any tension or distortion of the dermis tugs or twists the capsular fibers, stretching or compressing the attached dendrites and altering the activity in the myelinated afferent fiber.

Sensitivity to tactile sensations can be altered by infection, disease, and damage to sensory neurons or pathways. The locations of tactile responses may have diagnostic significance. For example, sensory loss along the boundary of a dermatome can help identify the affected spinal nerve or nerves.

Module 9.1 Review

a. Identify the six types of tactile receptors located in the skin, and describe their sensitivities.

b. Which types of tactile receptors are located only in the dermis?

c. Which is likely to be more sensitive to continuous deep pressure: a lamellated corpuscle or a Ruffini corpuscle?

Baroreceptors and chemoreceptors start important autonomic reflexes involving visceral sensory pathways

Baroreceptors and chemoreceptors play key roles in the reflexive control of organ function by the autonomic nervous system.

1 **Baroreceptors** are stretch receptors that monitor changes in pressure. The receptor consists of free nerve endings that branch within the elastic tissues in the walls of hollow organs, blood vessels, and tubes in the respiratory, digestive, and urinary tract. When the pressure changes, the elastic walls of these structures stretch or recoil. These changes in shape distort the receptor's dendritic branches and alter the rate of action-potential generation. Baroreceptors monitor blood pressure in the walls of major vessels, including the carotid artery (at the carotid sinus, an expanded portion of the internal carotid artery) and the aorta (at the aortic sinus). The information provided by these baroreceptors plays a major role in regulating cardiac function and adjusting blood flow to vital tissues. Baroreceptors in the lungs monitor the degree of lung expansion. This information is relayed to the respiratory rhythmicity centers in the medulla oblongata, which set the pace of respiration. Baroreceptors in the urinary and digestive tracts trigger a variety of visceral reflexes, such as urination.

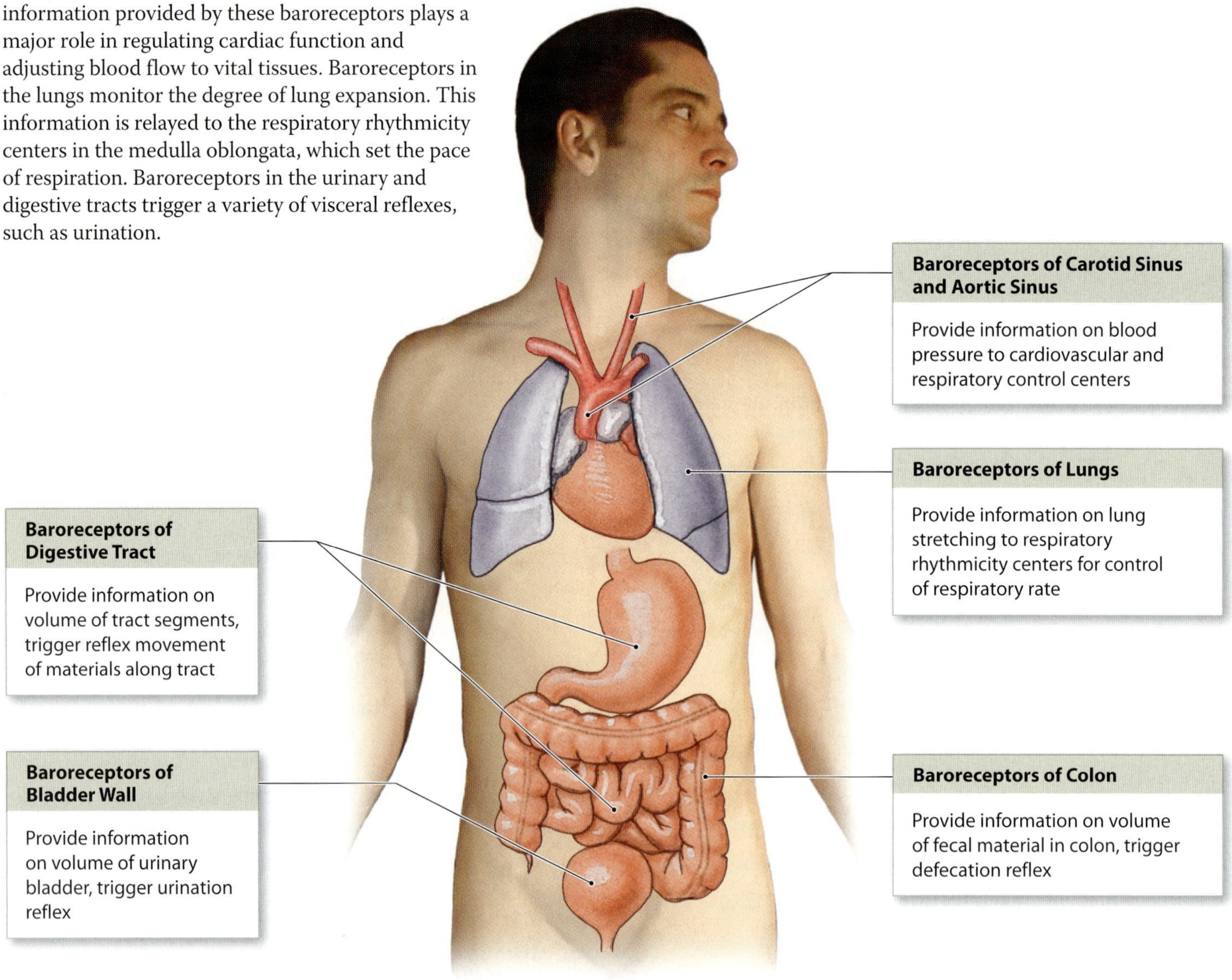

Baroreceptors of Carotid Sinus and Aortic Sinus

Provide information on blood pressure to cardiovascular and respiratory control centers

Baroreceptors of Lungs

Provide information on lung stretching to respiratory rhythmicity centers for control of respiratory rate

Baroreceptors of Digestive Tract

Provide information on volume of tract segments, trigger reflex movement of materials along tract

Baroreceptors of Bladder Wall

Provide information on volume of urinary bladder, trigger urination reflex

Baroreceptors of Colon

Provide information on volume of fecal material in colon, trigger defecation reflex

2 **Chemoreceptors** are specialized neurons that can detect small changes in the concentrations of specific chemicals or compounds. Chemoreceptive neurons are found (1) within the medulla oblongata and elsewhere in the brain, (2) in the **carotid bodies**, near the origin of the internal carotid arteries on each side of the neck, and (3) in the **aortic bodies** between the major branches of the aortic arch. The neurons in the respiratory centers in the medulla oblongata monitor the pH and carbon dioxide levels in cerebrospinal fluid (CSF). The carotid and aortic bodies monitor the pH and levels of carbon dioxide and oxygen in arterial blood. These chemoreceptors play an important role in the reflexive control of respiration and cardiovascular function.

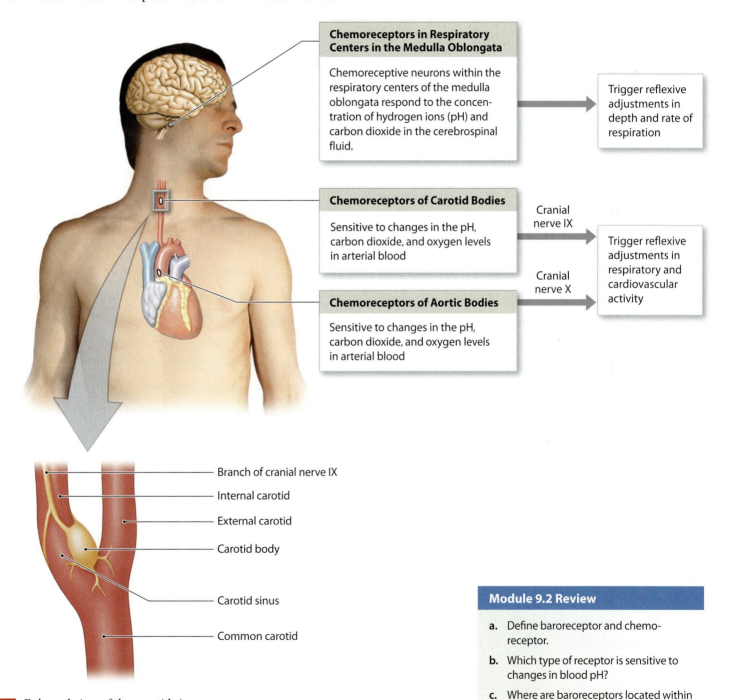

Chemoreceptors in Respiratory Centers in the Medulla Oblongata

Chemoreceptive neurons within the respiratory centers of the medulla oblongata respond to the concentration of hydrogen ions (pH) and carbon dioxide in the cerebrospinal fluid.

Trigger reflexive adjustments in depth and rate of respiration

Chemoreceptors of Carotid Bodies

Sensitive to changes in the pH, carbon dioxide, and oxygen levels in arterial blood

Cranial nerve IX

Trigger reflexive adjustments in respiratory and cardiovascular activity

Cranial nerve X

Chemoreceptors of Aortic Bodies

Sensitive to changes in the pH, carbon dioxide, and oxygen levels in arterial blood

Branch of cranial nerve IX

Internal carotid

External carotid

Carotid body

Carotid sinus

Common carotid

3 Enlarged view of the carotid sinus and the location of the carotid body.

Module 9.2 Review

a. Define baroreceptor and chemoreceptor.

b. Which type of receptor is sensitive to changes in blood pH?

c. Where are baroreceptors located within the body?

1. Labeling

Label each type of tactile receptor found in the skin.

 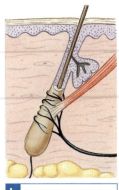 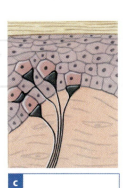 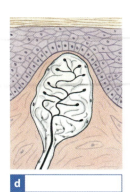 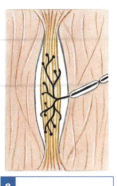 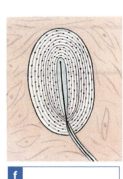

a _____
b _____
c _____
d _____
e _____
f _____

2. Matching

Match the following terms with the most closely related description (some terms are used more than once).

- thermoreceptors
- nociceptors
- baroreceptors
- tactile receptors
- chemoreceptors
- proprioceptors
- free nerve endings

a	_____	Most common receptors in the skin
b	_____	Sensitive to substances dissolved in body fluids
c	_____	Provide sensations of touch, pressure, and vibration
d	_____	Mechanoreceptors in the walls of major blood vessels
e	_____	Type of receptors located in the hypothalamus
f	_____	Type of receptors located in the medulla oblongata
g	_____	Sensitive to temperature
h	_____	Type of receptors located in the carotid and aortic bodies
i	_____	Type of receptors located in the carotid and aortic sinuses
j	_____	Mechanoreceptors that monitor muscle and joint positions
k	_____	Provide sensations of pain

3. Short answer

a What are the general senses?

_____

_____

b List the general sensory receptors that may be stimulated by a mosquito landing on and biting your arm.

_____

_____

The Special Senses

The **special senses** are **olfaction** (smell), **vision** (sight), **gustation** (taste), **equilibrium** (balance), and **hearing**. These sensations are provided by receptors that are structurally more complex than those of the general senses. Special sensory receptors are located in **sense organs** such as the eye or ear, where surrounding tissues protect the receptors. The information these receptors provide is distributed to specific areas of the cerebral cortex (the auditory cortex, the visual cortex, and so forth) and to centers throughout the brain stem.

1 In olfaction, the sensory receptors are modifed neurons; in gustation, the receptors are specialized receptor cells that communicate with sensory neurons. In each case, the sensory receptors are located within epithelia exposed to the external environment, and the information is routed directly to the CNS for processing.

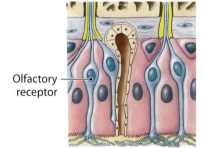

Olfactory receptor

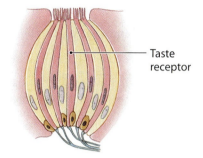

Taste receptor

2 In contrast, the receptors responsible for the sensations of equilibrium and hearing are isolated and protected from the external environment. These receptors are located within the internal ear, a complex sense organ, and the sensory information may be integrated and organized before it is forwarded to the CNS.

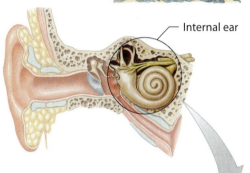

Internal ear

3 The receptors of the internal ear are called **hair cells** because their free surfaces are covered with specialized processes similar to the cilia and microvilli of other cells. They are basically mechanoreceptors sensitive to contact or movement, and they are always surrounded by supporting cells and monitored by the dendrites of sensory neurons. When an external force pushes against the processes of a hair cell, the distortion of the plasma membrane alters the rate at which the hair cell releases chemical transmitters. In this way, hair cells provide information about the direction and strength of mechanical stimuli.

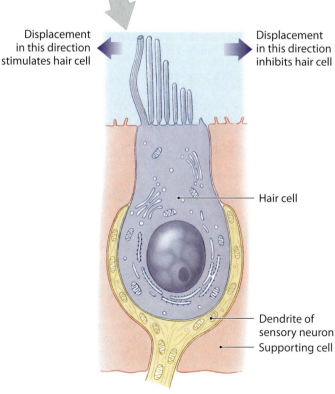

Displacement in this direction stimulates hair cell

Displacement in this direction inhibits hair cell

Hair cell

Dendrite of sensory neuron

Supporting cell

325

Olfaction involves specialized chemoreceptive neurons whereas taste receptors are specialized epithelial cells

Olfaction

1 The sense of smell, more precisely called **olfaction**, is provided by paired **olfactory organs**. These organs are located in the nasal cavity on either side of the nasal septum. They cover the inferior surface of the cribriform plate, the superior portion of the perpendicular plate, and the superior nasal conchae of the ethmoid bone.

Olfactory Pathway to the Cerebrum

The sensory neurons within the olfactory organ are stimulated by chemicals in the air. Between 10 and 20 million olfactory receptors are packed into an area of roughly 5 cm².	Axons leaving the olfactory epithelium collect into 20 or more bundles that penetrate the cribriform plate of the ethmoid bone.	The first synapse occurs in the **olfactory bulb**, which is located just superior to the cribriform plate.	Axons leaving the olfactory bulb travel along the **olfactory tract** to reach the olfactory cortex, the hypothalamus, and portions of the limbic system.	The distribution of olfactory information to the limbic system and hypothalamus explains the profound emotional and behavioral responses, as well as the memories, that can be triggered by certain smells.

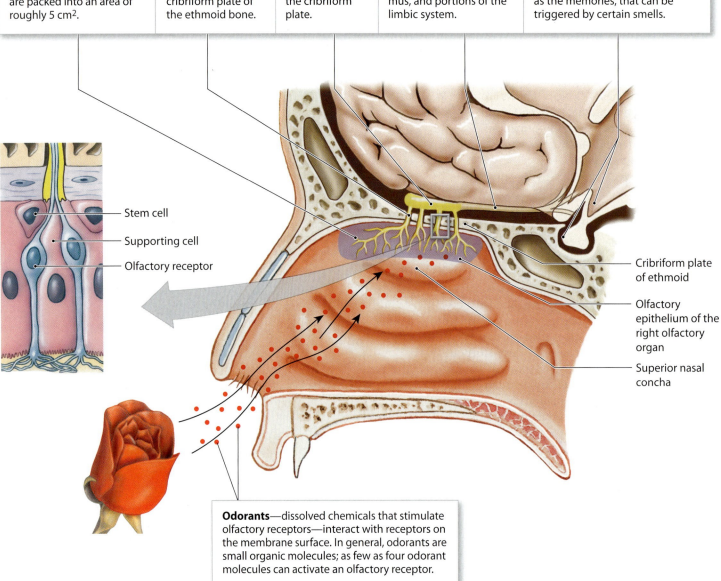

Stem cell

Supporting cell

Olfactory receptor

Cribriform plate of ethmoid

Olfactory epithelium of the right olfactory organ

Superior nasal concha

Odorants—dissolved chemicals that stimulate olfactory receptors—interact with receptors on the membrane surface. In general, odorants are small organic molecules; as few as four odorant molecules can activate an olfactory receptor.

Gustation

Taste, or **gustation**, provides information about the foods and liquids we consume. **Taste receptors**, or **gustatory** (GUS-ta-tor-ē) **receptors**, are distributed over the superior surface of the tongue and adjacent portions of the pharynx and larynx. The most important taste receptors are on the tongue.

2 The superior surface of the tongue bears epithelial projections called **lingual papillae** (pa-PIL-ē; *papilla*, a nipple-shaped mound). The human tongue bears three types of lingual papillae: (1) circumvallate (sir-kum-VAL-āt; *circum-*, around + *vallum*, wall) papillae, (2) fungiform (*fungus*, mushroom) papillae, and (3) filiform (*filum*, thread) papillae. Circumvallate and fungiform papillae contain taste receptors and specialized epithelial cells in sensory structures called **taste buds**. An adult has about 5000 taste buds.

Circumvallate Papillae

Circumvallate papillae are relatively large, shaped like the tip of a pencil eraser, and surrounded by deep epithelial folds. Each of these papillae contains as many as 100 taste buds.

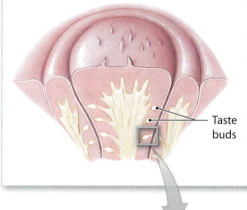

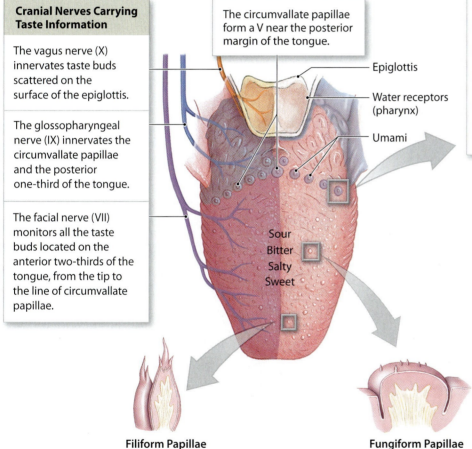

Cranial Nerves Carrying Taste Information
The vagus nerve (X) innervates taste buds scattered on the surface of the epiglottis.
The glossopharyngeal nerve (IX) innervates the circumvallate papillae and the posterior one-third of the tongue.
The facial nerve (VII) monitors all the taste buds located on the anterior two-thirds of the tongue, from the tip to the line of circumvallate papillae.

The circumvallate papillae form a V near the posterior margin of the tongue.

Epiglottis

Water receptors (pharynx)

Umami

Sour
Bitter
Salty
Sweet

Filiform Papillae

Fungiform Papillae

Taste buds

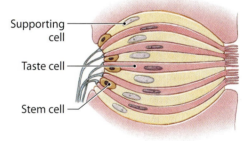

Supporting cell

Taste cell

Stem cell

Diagrammatic view of a taste bud

3 There is some evidence that sensitivity to the four primary taste sensations—sweet, salty, sour, and bitter—varies along the long axis of the tongue. However, there are no differences in the structure of the taste buds, and taste buds in all portions of the tongue provide all four primary taste sensations. There are also two other taste sensations. **Umami** (oo-MAH-mē) is a pleasant taste that is characteristic of beef broth, chicken broth, and Parmesan cheese. This taste is detected by receptors sensitive to the presence of amino acids, small peptides, and nucleotides. These receptors are present in taste buds of the circumvallate papillae. **Water receptors** are especially concentrated in the pharynx. The hypothalamus processes the sensory output of these receptors and affects water balance and the regulation of blood volume.

Module 9.3 Review

a. Describe olfaction and its receptors.

b. Describe gustation and its receptors.

c. Trace the olfactory pathway, beginning at the olfactory epithelium.

The ear is divided into the external ear, the middle ear, and the internal ear

1 The ear is divided into three anatomical regions. The external and middle ear play a role in hearing, while the internal ear plays a role in both hearing and equilibrium.

External Ear	Middle Ear	Internal Ear
The **external ear**—the visible portion of the ear—collects and directs sound waves toward the middle ear.	The **middle ear**, or tympanic cavity, is an air-filled chamber separated from the external acoustic meatus by the tympanic membrane. It is connected to the pharynx by the auditory tube.	The **internal ear** contains the sensory organs for hearing and equilibrium. It receives amplified sound waves from the middle ear.

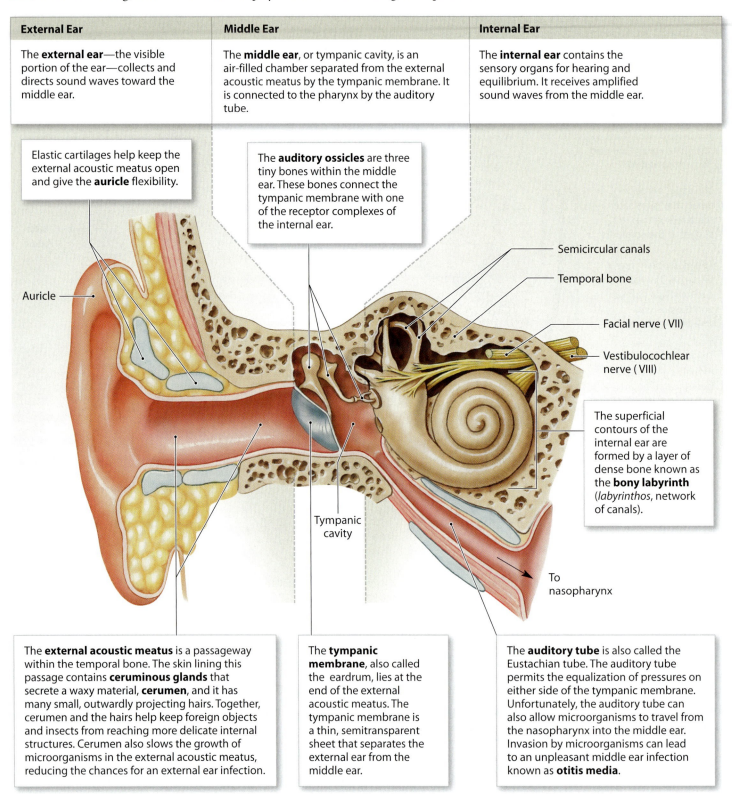

Elastic cartilages help keep the external acoustic meatus open and give the **auricle** flexibility.

The **auditory ossicles** are three tiny bones within the middle ear. These bones connect the tympanic membrane with one of the receptor complexes of the internal ear.

Auricle

Semicircular canals

Temporal bone

Facial nerve (VII)

Vestibulocochlear nerve (VIII)

The superficial contours of the internal ear are formed by a layer of dense bone known as the **bony labyrinth** (*labyrinthos*, network of canals).

Tympanic cavity

To nasopharynx

The **external acoustic meatus** is a passageway within the temporal bone. The skin lining this passage contains **ceruminous glands** that secrete a waxy material, **cerumen**, and it has many small, outwardly projecting hairs. Together, cerumen and the hairs help keep foreign objects and insects from reaching more delicate internal structures. Cerumen also slows the growth of microorganisms in the external acoustic meatus, reducing the chances for an external ear infection.

The **tympanic membrane**, also called the eardrum, lies at the end of the external acoustic meatus. The tympanic membrane is a thin, semitransparent sheet that separates the external ear from the middle ear.

The **auditory tube** is also called the Eustachian tube. The auditory tube permits the equalization of pressures on either side of the tympanic membrane. Unfortunately, the auditory tube can also allow microorganisms to travel from the nasopharynx into the middle ear. Invasion by microorganisms can lead to an unpleasant middle ear infection known as **otitis media**.

2 The middle ear contains the three auditory ossicles and communicates with both the superior portion of the pharynx (the nasopharynx), through the auditory tube, and with the mastoid air cells, through a number of small connections. The articulations between the auditory ossicles are the smallest synovial joints in the body. Each ossicle has a tiny joint capsule and supporting extracapsular ligaments.

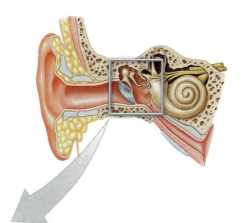

Auditory Ossicles

The **malleus** (*malleus*, hammer) attaches at three points to the interior surface of the tympanic membrane.

The **incus** (*incus*, anvil), the middle ossicle, attaches the malleus to the stapes.

The edges of the base of the **stapes** (*stapes*, stirrup) are bound to the edges of the oval window, an opening in the bone that surrounds the internal ear.

Temporal bone (petrous part)

Connections to mastoid air cells

Stabilizing ligament

Branch of facial nerve VII (cut)

External acoustic meatus

Tympanic cavity (middle ear)

Oval window

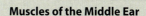

Muscles of the Middle Ear

Small skeletal muscles insert on the malleus and the stapes. Contractions of these muscles reduce the amount of vibration transferred from the tympanic membrane to the oval window.

Auditory tube

Round window

Arriving sound waves vibrate the tympanic membrane, thereby converting the waves into mechanical movements. The auditory ossicles conduct those vibrations to the internal ear, because they are connected in such a way that an in–out movement of the tympanic membrane produces a rocking motion of the stapes. The ossicles thus function as a lever system that collects the force applied to the tympanic membrane and focuses it on the oval window. Because the tympanic membrane is 22 times larger and heavier than the oval window, considerable amplification occurs, so we can hear very faint sounds. But that degree of amplification can be a problem when we are exposed to very loud noises. Contractions of the muscles inserting on the malleus and stapes protect the tympanic membrane and ossicles from violent movements under very noisy conditions.

Module 9.4 Review

a. Name the three tiny bones located in the middle ear.

b. What is the function of the auditory tube?

c. Why are external ear infections relatively uncommon?

The bony labyrinth protects the membranous labyrinth

1 The internal ear contains the receptors for the special senses of equilibrium and hearing. It can be subdivided into a bony labyrinth and a membranous labyrinth.

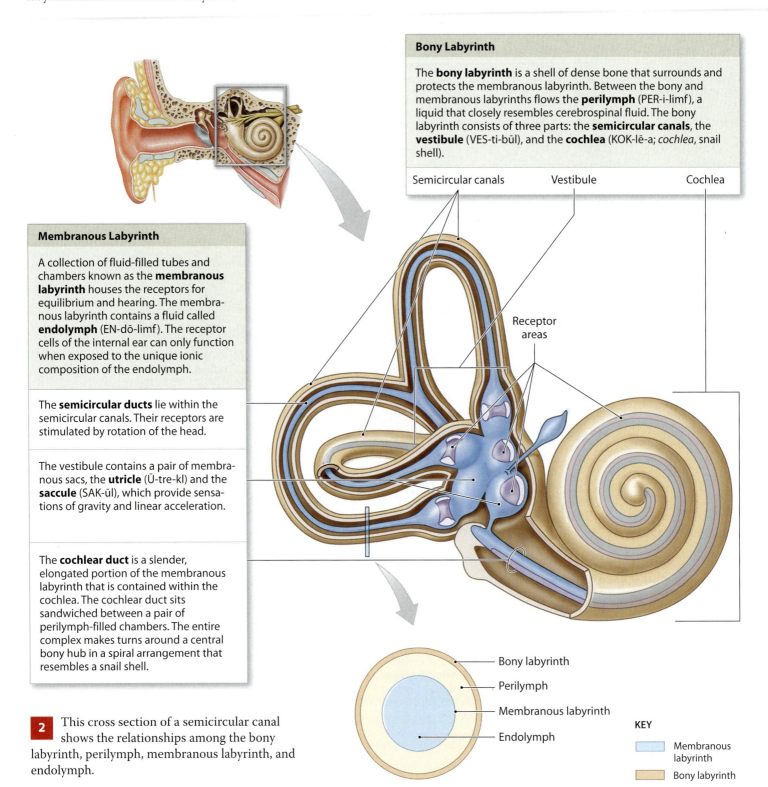

Bony Labyrinth

The **bony labyrinth** is a shell of dense bone that surrounds and protects the membranous labyrinth. Between the bony and membranous labyrinths flows the **perilymph** (PER-i-limf), a liquid that closely resembles cerebrospinal fluid. The bony labyrinth consists of three parts: the **semicircular canals**, the **vestibule** (VES-ti-būl), and the **cochlea** (KOK-lē-a; *cochlea*, snail shell).

Semicircular canals   Vestibule   Cochlea

Membranous Labyrinth

A collection of fluid-filled tubes and chambers known as the **membranous labyrinth** houses the receptors for equilibrium and hearing. The membranous labyrinth contains a fluid called **endolymph** (EN-dō-limf). The receptor cells of the internal ear can only function when exposed to the unique ionic composition of the endolymph.

The **semicircular ducts** lie within the semicircular canals. Their receptors are stimulated by rotation of the head.

The vestibule contains a pair of membranous sacs, the **utricle** (Ū-tre-kl) and the **saccule** (SAK-ūl), which provide sensations of gravity and linear acceleration.

The **cochlear duct** is a slender, elongated portion of the membranous labyrinth that is contained within the cochlea. The cochlear duct sits sandwiched between a pair of perilymph-filled chambers. The entire complex makes turns around a central bony hub in a spiral arrangement that resembles a snail shell.

Receptor areas

Bony labyrinth
Perilymph
Membranous labyrinth
Endolymph

KEY

Membranous labyrinth

Bony labyrinth

2 This cross section of a semicircular canal shows the relationships among the bony labyrinth, perilymph, membranous labyrinth, and endolymph.

3 Here is a simple concept map that summarizes information about the regions and functions of the membranous labyrinth.

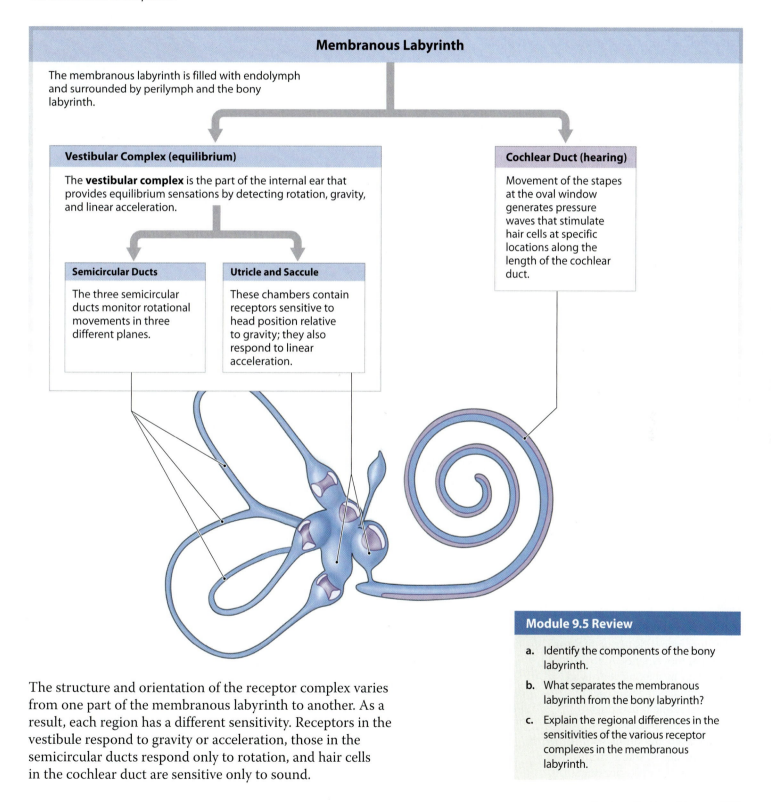

Membranous Labyrinth

The membranous labyrinth is filled with endolymph and surrounded by perilymph and the bony labyrinth.

Vestibular Complex (equilibrium)

The **vestibular complex** is the part of the internal ear that provides equilibrium sensations by detecting rotation, gravity, and linear acceleration.

Semicircular Ducts

The three semicircular ducts monitor rotational movements in three different planes.

Utricle and Saccule

These chambers contain receptors sensitive to head position relative to gravity; they also respond to linear acceleration.

Cochlear Duct (hearing)

Movement of the stapes at the oval window generates pressure waves that stimulate hair cells at specific locations along the length of the cochlear duct.

The structure and orientation of the receptor complex varies from one part of the membranous labyrinth to another. As a result, each region has a different sensitivity. Receptors in the vestibule respond to gravity or acceleration, those in the semicircular ducts respond only to rotation, and hair cells in the cochlear duct are sensitive only to sound.

Module 9.5 Review

a. Identify the components of the bony labyrinth.

b. What separates the membranous labyrinth from the bony labyrinth?

c. Explain the regional differences in the sensitivities of the various receptor complexes in the membranous labyrinth.

Hair cells in the semicircular ducts respond to rotation, while those in the utricle and saccule respond to gravity and linear acceleration

The **anterior**, **posterior**, and **lateral semicircular ducts** are continuous with the utricle. Each semicircular duct contains an **ampulla**, an expanded region that contains the receptors.

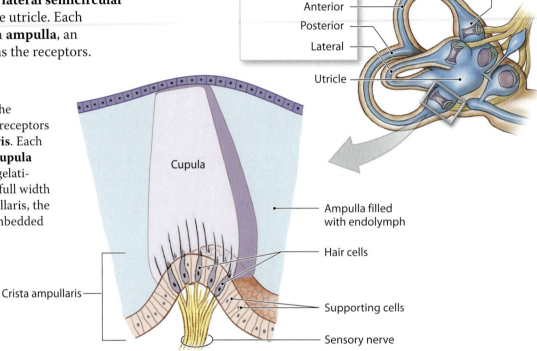

Semicircular Ducts

Anterior
Posterior
Lateral

Utricle

Ampulla

1 The region in the wall of the ampulla that contains the receptors is known as the **crista ampullaris**. Each crista ampullaris is bound to a **cupula** (KŪ-pū-luh), a flexible, elastic, gelatinous structure that extends the full width of the ampulla. At a crista ampullaris, the processes of the hair cells are embedded in the cupula.

Cupula

Ampulla filled with endolymph

Hair cells

Crista ampullaris

Supporting cells

Sensory nerve

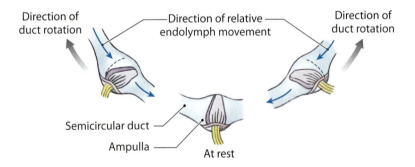

Direction of duct rotation

Direction of relative endolymph movement

Direction of duct rotation

Semicircular duct

Ampulla

At rest

2 The cupula has a density very close to that of the surrounding endolymph, so it essentially floats above the receptor surface. When your head rotates in the plane of a semicircular duct, the movement of endolymph along the length of the duct pushes the cupula to the side, distorting the receptor processes. Movement of fluid in one direction stimulates the hair cells, and movement in the opposite direction inhibits them. When the endolymph stops moving, the elastic cupula rebounds to its normal position.

3 Even the most complex movement can be analyzed in terms of motion in three rotational planes. Each semicircular duct responds to one of these rotational movements. A horizontal rotation, as in shaking your head "no," stimulates the hair cells of the lateral semicircular duct. Nodding "yes" excites the anterior duct, and tilting your head from side to side activates receptors in the posterior duct.

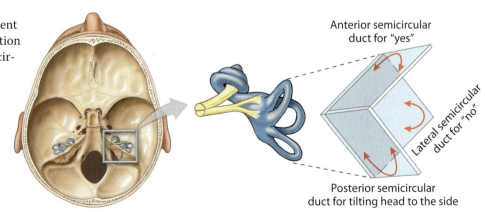

Anterior semicircular duct for "yes"

Lateral semicircular duct for "no"

Posterior semicircular duct for tilting head to the side

4 The utricle and saccule each contain a sensory structure called a **macula** that provides equilibrium sensations, whether the body is moving or stationary. The macula in the utricle is sensitive to changes in horizontal movement. The macula in the saccule is sensitive to vertical movement.

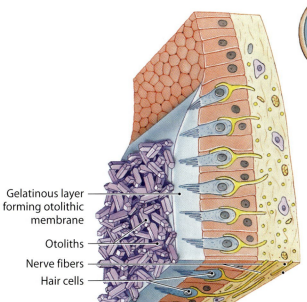

Gelatinous layer forming otolithic membrane

Otoliths

Nerve fibers

Hair cells

5 The hair cell processes are embedded in a gelatinous otolithic membrane whose surface contains densely packed calcium carbonate crystals. These calcium carbonate crystals are called **otoliths** ("ear stones").

Utricle

Endolymphatic sac

Endolymphatic duct

Saccule

6 Changes in the position of the head cause distortion of the hair cell processes in the maculae and send signals to the brain.

When your head is in the normal, upright position, the otoliths sit atop the otolithic membrane of the macula. Their weight presses on the macular surface, pushing the hair cell processes down rather than to one side or another.

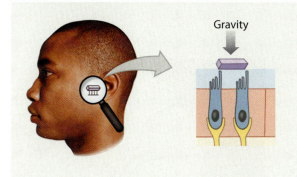

Gravity

When your head is tilted, the pull of gravity on the otoliths shifts them to the side, thereby distorting the hair cell processes and stimulating the macular receptors. This mechanism accounts for your perception of linear acceleration, as when your car speeds up suddenly. The otoliths lag behind, and the effect on the hair cells is comparable to tilting your head back.

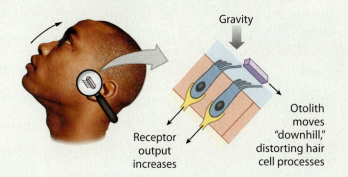

Gravity

Receptor output increases

Otolith moves "downhill," distorting hair cell processes

Module 9.6 Review

a. Define otoliths.

b. Cite the function of receptors in the saccule and utricle.

c. Damage to the cupula of the lateral semicircular duct would interfere with what perception?

The cochlear duct contains the hair cells of the spiral organ

1 The **cochlear duct** lies between a pair of perilymphatic chambers: the **scala vestibuli** and the **scala tympani**. The bony labyrinth encases the outer surfaces of these ducts everywhere except at the **oval window** (the base of the scala vestibuli) and the **round window** (the base of the scala tympani). Because the scala vestibuli and scala tympani are connected at the tip of the cochlear spiral, they really form one long and continuous perilymphatic chamber. This chamber begins at the oval window, extends through the scala vestibuli, and proceeds around the top of the cochlea and back along the scala tympani; it ends at the round window.

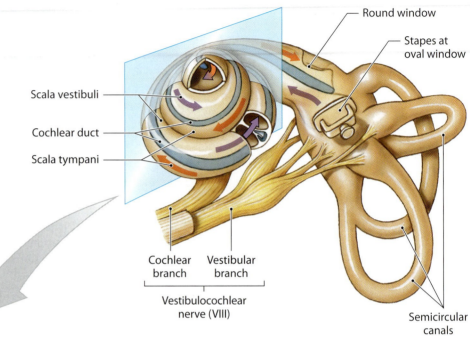

Round window

Stapes at oval window

Scala vestibuli

Cochlear duct

Scala tympani

Cochlear branch   Vestibular branch

Vestibulocochlear nerve (VIII)

Semicircular canals

KEY

From oval window to tip of spiral

From tip of spiral to round window

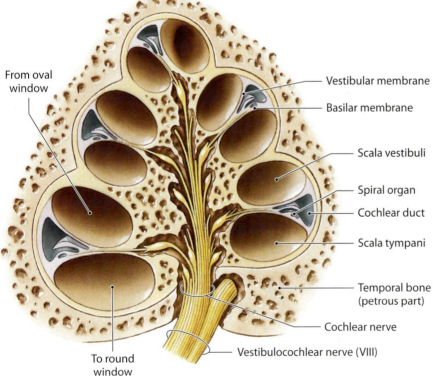

From oval window

Vestibular membrane

Basilar membrane

Scala vestibuli

Spiral organ

Cochlear duct

Scala tympani

Temporal bone (petrous part)

Cochlear nerve

Vestibulocochlear nerve (VIII)

To round window

2 The cochlear duct is a long, coiled tube suspended between the scala vestibuli and the scala tympani. The hair cells of the cochlear duct are located in a structure called the **spiral organ**, or organ of Corti. The spiral organ is located on the basilar membrane. Like the chambers of the cochlea, the spiral organ runs through the entire length of the cochlea.

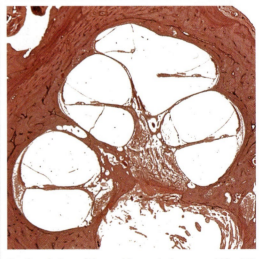

Sectional view of the cochlear spiral   LM × 200

3 This sectional view shows a single turn of the cochlea. The **vestibular membrane** separates the cochlear duct from the scala vestibuli, and the **basilar membrane** separates the cochlear duct from the scala tympani. The scala vestibuli and scala tympani are filled with perilymph. The cochlear duct, which contains the spiral organ, is filled with endolymph.

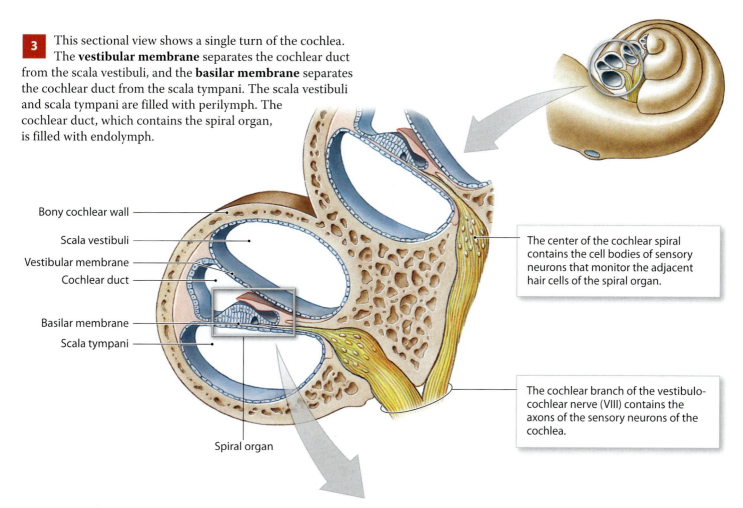

Bony cochlear wall

Scala vestibuli

Vestibular membrane

Cochlear duct

Basilar membrane

Scala tympani

Spiral organ

The center of the cochlear spiral contains the cell bodies of sensory neurons that monitor the adjacent hair cells of the spiral organ.

The cochlear branch of the vestibulo-cochlear nerve (VIII) contains the axons of the sensory neurons of the cochlea.

4 The hair cells of the spiral organ are arranged in a series of longitudinal rows. Their processes are in contact with the overlying **tectorial** (tek-TOR-ē-al; *tectum*, roof) **membrane**. This membrane is firmly attached to the inner wall of the cochlear duct. When a portion of the basilar membrane bounces up and down in response to pressure fluctuations within the perilymph, the processes of the hair cells are pressed against the tectorial membrane and distorted. If the amount of movement increases, more hair cells—and more rows of hair cells—are stimulated. These pressure changes within the perilymph are triggered by sound waves arriving at the tympanic membrane.

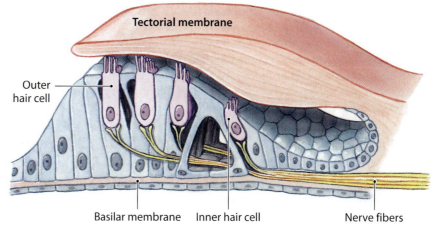

Tectorial membrane

Outer hair cell

Basilar membrane   Inner hair cell   Nerve fibers

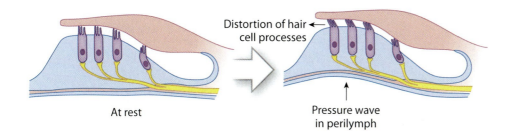

Distortion of hair cell processes

At rest

Pressure wave in perilymph

Module 9.7 Review

a. Where is the spiral organ located?

b. Name the fluids found within the scala vestibuli, scala tympani, and cochlear duct.

c. Identify the structures visible in the light micrograph of the cochlear spiral in sectional view.

The sensations of pitch and volume depend on movements of the basilar membrane

1 Hearing is the perception of sound, which consists of waves of pressure conducted through a medium such as air or water. In air, each pressure wave consists of a region where the air molecules are compressed together and one where they are farther apart.

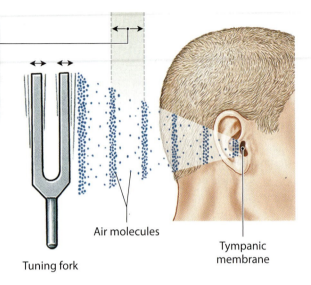

The **wavelength** of sound is the distance between two adjacent wave crests (peaks) or, equivalently, the distance between two adjacent wave troughs.

Air molecules

Tympanic membrane

Tuning fork

2 Sound waves can be graphed as S-shaped curves that repeat in a regular pattern. At sea level, sound waves travel through the air at about 1235 km/h (768 mph). The **frequency** is the number of waves that pass a fixed reference point—such as the tympanic membrane—in a given time. Physicists use the term "cycles" rather than waves. Hence, the frequency of a sound is measured in terms of the number of cycles per second (cps), a unit called **hertz (Hz)**. What we perceive as the pitch of a sound is our sensory response to its frequency. A high-frequency sound (high pitch, short wavelength) might have a frequency of 15,000 Hz or more; a very low-frequency sound (low pitch, long wavelength) could have a frequency of 100 Hz or less.

Because all sound waves travel at the same speed, as the frequency increases, the wavelength must become shorter.

The **amplitude** is determined by the amount of energy carried by the wave. The louder the sound, the higher the amplitude.

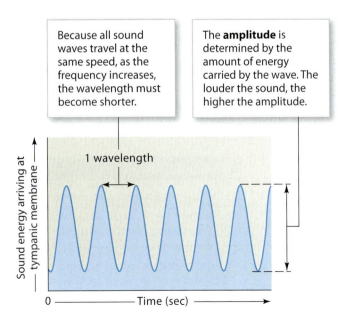

Sound energy arriving at tympanic membrane

1 wavelength

0 ——— Time (sec) ———→

3 This table lists the **intensity**—or energy in sound waves—of familiar sounds. Intensity determines how loud it seems; the greater the energy content, the larger the amplitude, and the louder the sound. Sound energy is reported in **decibels** (DES-i-belz) (**dB**).

The Intensity of Representative Sounds

Typical Decibel Level	Example	Dangerous Time Exposure
0	Lowest audible sound	
30	Quiet library; soft whisper	
40	Quiet office; living room; bedroom away from traffic	
50	Light traffic at a distance; refrigerator; gentle breeze	
60	Air conditioner from 20 feet; conversation; sewing machine in operation	
70	Busy traffic; noisy restaurant	Some damage if continuous
80	Subway; heavy city traffic; alarm clock at 2 feet; factory noise	More than 8 hours
90	Truck traffic; noisy home appliances; shop tools; gas lawn mower	Less than 8 hours
100	Chain saw; boiler shop; pneumatic drill	2 hours
120	"Heavy metal" rock concert; sandblasting; thunderclap nearby	Immediate danger
140	Gunshot; jet plane	Immediate danger
160	Rocket launching pad	Hearing loss inevitable

4 The energy of sound waves is a physical pressure. When sound waves strike a flexible object, the object responds to that pressure. Given the right combination of frequencies and amplitudes, the object will begin to vibrate at the same frequency as the sound, a phenomenon called **resonance**. The higher the amplitude, the greater the amount of vibration. For you to be able to hear a sound, your tympanic membrane must vibrate in resonance with the sound waves. Pressure waves at the tympanic membrane generate movement of the stapes at the oval window. The flexibility of the basilar membrane on which the spiral organ rests varies along its length. As a result, pressure waves of different frequencies affect different parts of the membrane. The location of the vibration is interpreted as **pitch**. The number of stimulated hair cells is interpreted as **volume**.

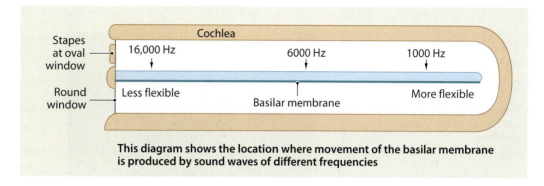

This diagram shows the location where movement of the basilar membrane is produced by sound waves of different frequencies

5 This illustration summarizes the events involved in hearing.

Events Involved in Hearing

1	2	3	4	5	6
Sound waves arrive at the tympanic membrane.	Movement of the tympanic membrane causes displacement of the auditory ossicles.	Movement of the stapes at the oval window establishes pressure waves in the perilymph of the scala vestibuli.	The pressure waves distort the basilar membrane on their way to the round window of the scala tympani.	Vibration of the basilar membrane causes vibration of hair cells against the tectorial membrane.	Information about the region and the intensity of stimulation is relayed to the CNS over the cochlear branch of cranial nerve VIII.

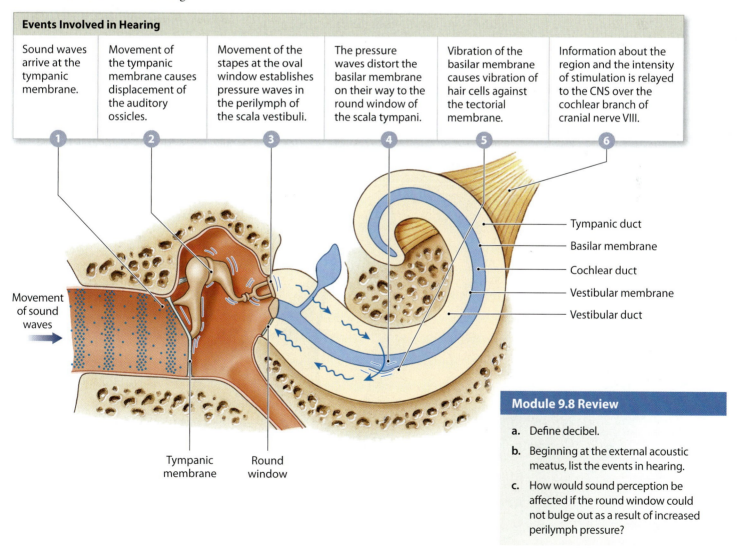

Module 9.8 Review

a. Define decibel.

b. Beginning at the external acoustic meatus, list the events in hearing.

c. How would sound perception be affected if the round window could not bulge out as a result of increased perilymph pressure?

The accessory structures of the eye provide protection while allowing light to reach the interior of the eye

1 The **accessory structures** of the eye include the eyelids, eyelashes, the superficial epithelium of the eye, and the structures associated with the production, secretion, and removal of tears.

The **cornea** is a transparent area on the anterior surface of the eye.

Laterally, the two eyelids are connected at the **lateral canthus**.

Light enters the eye by passing through the cornea and then through the **pupil**, an opening at the center of the colored **iris**.

Eyelids and Eyelashes

The **eyelashes**, along the margins of the eyelids, are very robust hairs that help prevent foreign matter from reaching the eye surface.

The eyelid, or **palpebra** (pal-PĒ-bra), is a continuation of the skin. The continual blinking of the palpebrae keeps the surface of the eye lubricated, and removes dust and debris. The eyelids can also close firmly to protect the delicate surface of the eye.

Medially, the two eyelids are connected at the **medial canthus** (KAN-thus).

The **palpebral fissure** is the gap that separates the free margins of the upper and lower eyelids.

Glands in the **lacrimal caruncle** (KAR-ung-kul) produce the thick secretions that cause the gritty deposits that sometimes appear after a good night's sleep.

2 The epithelium covering the inner surfaces of the eyelids and the outer surface of the eye is called the **conjunctiva** (kon-junk-TĪ-vuh). It is a mucous membrane covered by a specialized stratified squamous epithelium.

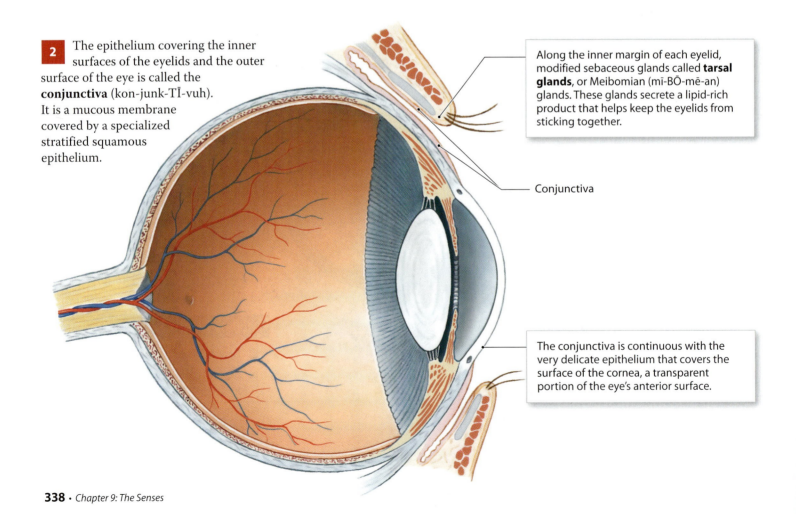

Along the inner margin of each eyelid, modified sebaceous glands called **tarsal glands**, or Meibomian (mī-BŌ-mē-an) glands. These glands secrete a lipid-rich product that helps keep the eyelids from sticking together.

Conjunctiva

The conjunctiva is continuous with the very delicate epithelium that covers the surface of the cornea, a transparent portion of the eye's anterior surface.

3 A constant flow of tears keeps conjunctival surfaces moist and clean. Tears reduce friction, remove debris, prevent bacterial infection, and provide nutrients and oxygen to the conjunctival epithelium. The **lacrimal apparatus** produces, distributes, and removes tears. The lacrimal apparatus of each eye consists of (1) a lacrimal gland (or tear gland) with associated ducts, (2) paired lacrimal canaliculi, (3) a lacrimal sac, and (4) a naso-lacrimal duct.

Components of the Lacrimal Apparatus

The almond-shaped **lacrimal gland** produces about 1 mL of watery, slightly alkaline tears per day. The tears provide lubrication and supply the nutrient and oxygen demands of the corneal cells. Lacrimal gland secretions contain the antibacterial enzyme **lysozyme** and antibodies that attack pathogens before they enter the body.

Tear ducts deliver tears from the lacrimal gland to the space behind the upper eyelid.

The **lacrimal puncta** (singular, *punctum*) are two small pores that drain the lacrimal lake (the area where tears collect).

The **lacrimal canaliculi** are small canals that connect the lacrimal puncta to the lacrimal sac.

The **lacrimal sac** is a small chamber that nestles within the lacrimal sulcus of the orbit.

The **nasolacrimal duct** originates at the inferior tip of the lacrimal sac. It passes through the naso-lacrimal canal to deliver tears to the nasal cavity.

The nasolacrimal duct empties into the inferior meatus, a narrow passageway inferior and lateral to the inferior nasal concha.

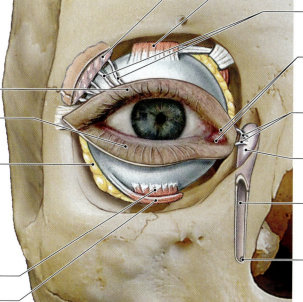

Superior rectus muscle

Upper eyelid

Lower eyelid

A layer of resilient orbital fat posterior to the eyeball provides padding while permitting eye movements.

Inferior rectus muscle

Inferior oblique muscle

4 **Conjunctivitis**, or pinkeye, results from damage to, and irritation of, the conjunctival surface. The most obvious sign, redness, is due to the dilation of blood vessels deep to the conjunctival epithelium. This condition may be caused by pathogenic infection or by physical, allergic, or chemical irritation of the conjunctival surface.

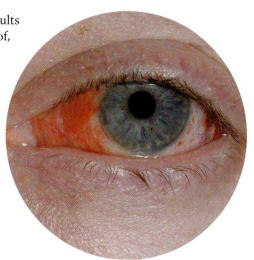

Module 9.9 Review

a. List the accessory structures associated with the eye.

b. Explain conjunctivitis.

c. Which layer of the eye would be the first affected by inadequate tear production?

The eye has a layered wall; it is hollow, with fluid-filled anterior and posterior cavities

1 The wall of the eye has three layers.

Fibrous Layer

The **fibrous layer**, the outermost layer of the eye, consists of the cornea and the **sclera** (SKLER-uh). These two components are continuous, and the border between the two is called the **corneal limbus**. The fibrous layer (1) provides mechanical support and some degree of physical protection, (2) serves as an attachment site for the extrinsic eye muscles, and (3) contains the cornea. The cornea allows the passage of light, and its curvature aids in the focusing process.

Vascular Layer

The **vascular layer** contains numerous blood vessels, lymphatic vessels, and the intrinsic (smooth) muscles of the eye. The functions of this middle layer include (1) providing a route for blood vessels and lymphatics that supply tissues of the eye; (2) regulating the amount of light that enters the eye; (3) secreting and reabsorbing the aqueous humor that circulates within the chambers of the eye; and (4) controlling the shape of the lens, an essential part of the focusing process.

The **iris** is visible through the transparent corneal surface. It contains blood vessels, pigment cells, and layers of smooth muscle fibers. When these muscles contract, they change the diameter of the pupil and the amount of light entering the eye.

The **ciliary body** is a thickened region that bulges into the interior of the eye. It consists of a ring of smooth muscle and epithelial cells that secrete a fluid called **aqueous humor**. Ligaments extending from the ciliary body hold the lens in position posterior to the iris and pupil.

The **choroid** is a vascular layer that is covered by the sclera. The choroid contains an extensive capillary network that delivers oxygen and nutrients to the neural tissue within the neural layer.

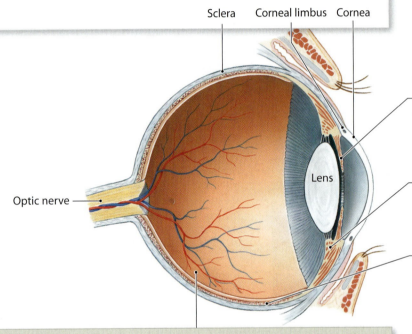

Sclera   Corneal limbus   Cornea

Optic nerve

Lens

Neural Layer

The **neural layer**, or **retina**, is the innermost layer of the eye. The retina consists of a thin outer layer (the pigmented layer) that absorbs light, and a thick inner layer (the neural layer) that contains the **photoreceptors**—the cells that are sensitive to light.

2 This superior sectional view reveals that the ciliary body and the lens divide the interior of the eye into a small **anterior cavity** and a large **posterior cavity**.

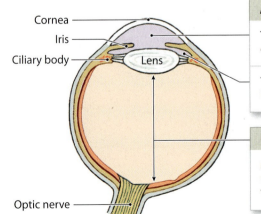

Cornea
Iris
Ciliary body
Lens
Optic nerve

Anterior Cavity

The **anterior chamber** extends from the cornea to the iris.

The **posterior chamber** extends between the iris and the ciliary body and lens.

Posterior Cavity

Most of the posterior cavity's volume is taken up by a gelatinous substance known as the **vitreous body**, or vitreous humor.

Extrinsic Eye Muscles

The extrinsic eye muscles position the eye.

3 Five of the six extrinsic eye muscles are visible in this lateral view of the right eye.

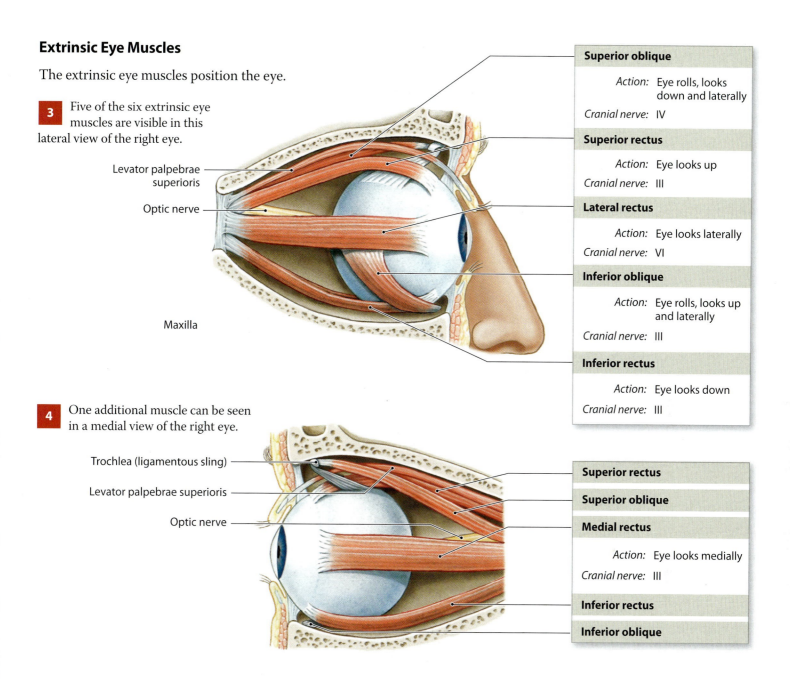

Levator palpebrae superioris

Optic nerve

Maxilla

Superior oblique	
Action:	Eye rolls, looks down and laterally
Cranial nerve:	IV
Superior rectus	
Action:	Eye looks up
Cranial nerve:	III
Lateral rectus	
Action:	Eye looks laterally
Cranial nerve:	VI
Inferior oblique	
Action:	Eye rolls, looks up and laterally
Cranial nerve:	III
Inferior rectus	
Action:	Eye looks down
Cranial nerve:	III

4 One additional muscle can be seen in a medial view of the right eye.

Trochlea (ligamentous sling)

Levator palpebrae superioris

Optic nerve

Superior rectus	
Superior oblique	
Medial rectus	
Action:	Eye looks medially
Cranial nerve:	III
Inferior rectus	
Inferior oblique	

5 This anterior view of the right eye shows the direction of eye movements produced by the contraction of each extrinsic eye muscle operating independently.

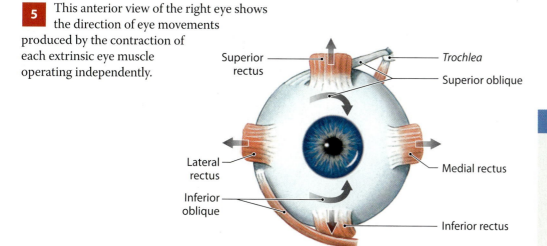

Superior rectus

Trochlea

Superior oblique

Lateral rectus

Medial rectus

Inferior oblique

Inferior rectus

Module 9.10 Review

a. Name the three layers of the eye.

b. What gives the eyes their characteristic color?

c. Name the extrinsic eye muscles and describe the way in which each moves the eye.

The eye is highly organized and has a consistent visual axis that directs light to the fovea of the retina

1 The sectional view below presents key aspects of eye anatomy that are associated with positioning the eye and allowing light to reach the photoreceptors of the retina.

The cornea permits the entry of light into the eye, so its transparency and clarity are vital to eye function. The cornea consists primarily of a dense matrix containing multiple layers of collagen fibers, organized so as not to interfere with the passage of light. The cornea has no blood vessels; the superficial epithelial cells must obtain oxygen and nutrients from the tears that flow across their free surfaces.

The lens lies posterior to the cornea, held in place by the **suspensory ligaments** that originate on the ciliary body. The lens consists of concentric layers of cells surrounded by a dense fibrous capsule. The cells are slender, elongated, and filled with transparent proteins responsible for both the clarity and the focusing power of the lens. Around the edges of the lens, its capsular fibers intermingle with those of the suspensory ligaments. The primary function of the lens is to focus the visual image on the photoreceptors, and the lens accomplishes this by changing shape.

Tension in the suspensory ligaments resists the tendency of the lens to assume a spherical shape.

The ciliary body supports the lens and controls its shape.

The retina contains the photoreceptors, pigment cells, supporting cells, and neurons.

The blood vessels of the choroid directly or indirectly provide nutrients to all structures within the eye.

The sclera, or "white of the eye," consists of dense fibrous connective tissue containing both collagen and elastic fibers. This layer is thickest over the posterior surface of the eye, near the exit of the optic nerve, and thinnest over the anterior surface. The sclera stabilizes the shape of the eye during eye movements. The six extrinsic eye muscles insert on the sclera, blending their collagen fibers with those of the sclera.

The **optic nerve** (N II) carries visual information to the brain.

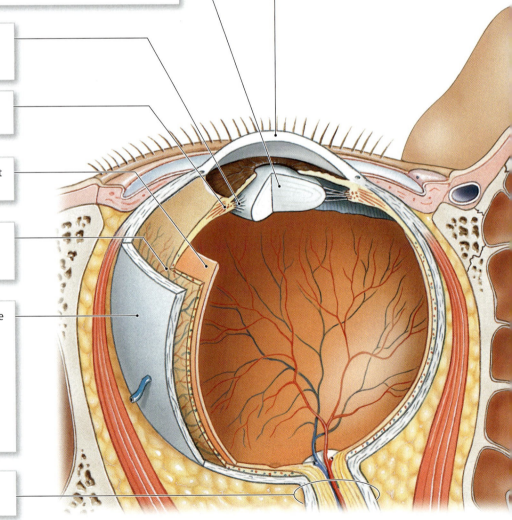

2 The amount of light entering the eye and passing through the lens is controlled by the two layers of the pupillary muscles of the iris. When these smooth muscles contract, they change the diameter of the pupil. Both muscle layers are controlled by the autonomic nervous system. Parasympathetic activation in response to bright light causes the pupils to constrict, whereas sympathetic activation in response to dim light causes the pupils to dilate.

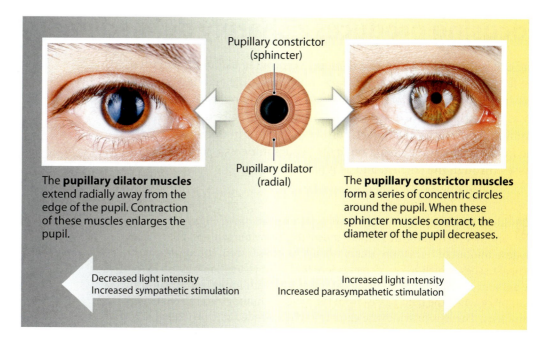

Pupillary constrictor (sphincter)

Pupillary dilator (radial)

The **pupillary dilator muscles** extend radially away from the edge of the pupil. Contraction of these muscles enlarges the pupil.

The **pupillary constrictor muscles** form a series of concentric circles around the pupil. When these sphincter muscles contract, the diameter of the pupil decreases.

Decreased light intensity
Increased sympathetic stimulation

Increased light intensity
Increased parasympathetic stimulation

3 Light passing through the center of the cornea and the center of the lens strikes a specific location that contains the highest density of photoreceptors anywhere in the eye.

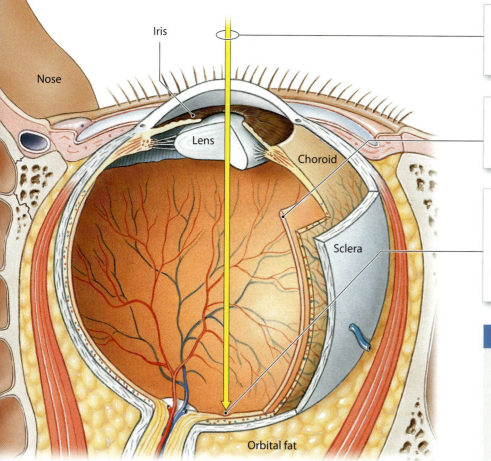

Iris

Nose

Lens

Choroid

Sclera

Orbital fat

The **visual axis** of the eye is an imaginary line drawn from the center of an object you are looking at directly, through the center of the cornea and the center of the lens to the retina.

The photoreceptors are located in the inner, neural portion of the retina. The type and density of receptors varies from one portion of the retina to another.

The very highest concentration of photoreceptors occurs at the center of an area called the **macula** (MAK-ū-luh; spot). This central area is called the **fovea** (FŌ-vē-uh; shallow depression). The fovea is the site of sharpest vision: When you look directly at an object, its image falls on this portion of the retina.

Module 9.11 Review

a. Which eye structure does not contain blood vessels?

b. List the structures and fluids that light passes through from the cornea to the retina.

c. What happens to the pupils when light intensity decreases?

Focusing produces a sharply defined image at the retina

The eye is often compared to a camera. To provide useful information, the lens of the eye, like a camera lens, must focus the arriving image. For an image to be "in focus," the rays of light arriving from an object must strike the sensitive surface of the retina in precise order, forming a miniature image of the object. If the rays are not perfectly focused, the image is blurry. Focusing typically occurs in two steps: as the light passes through the cornea, and as it passes through the lens.

1 Light is **refracted**, or bent, when it passes from one medium to another medium with a different density. In the human eye, the greatest amount of refraction occurs when light passes from the air into the corneal tissues, which have a density close to that of water. You cannot vary the amount of refraction that occurs at the cornea. Additional refraction takes place when the light passes from the aqueous humor into the relatively dense lens. The lens provides the extra refraction needed to focus the light rays from an object toward a **focal point**—a specific point of intersection on the retina. The distance between the center of the lens and its focal point is the **focal distance** of the lens. Whether in the eye or in a camera, the focal distance is determined by the distance from the object to the lens, and the shape of the lens.

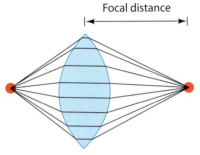

Focal distance

Light from distant source (object)

Focal distance

Close source

Lens

Focal point

The closer the light source, the longer the focal distance

Focal distance

The rounder the lens, the shorter the focal distance

2 A camera focuses an image by moving the lens toward or away from the digital image sensor. This method of focusing cannot work in our eyes, because the distance from the lens to the macula cannot change. We focus images on the retina by changing the shape of the lens to keep the focal distance constant, a process called **accommodation**.

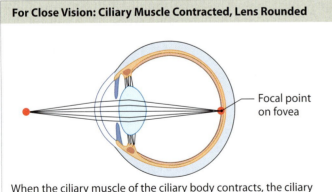

For Close Vision: Ciliary Muscle Contracted, Lens Rounded

Focal point on fovea

When the ciliary muscle of the ciliary body contracts, the ciliary body moves toward the lens, thereby reducing the tension in the suspensory ligaments. The elastic capsule of the lens then pulls it into a more spherical shape that increases the refractive power of the lens, enabling it to bring light from nearby objects into focus on the retina.

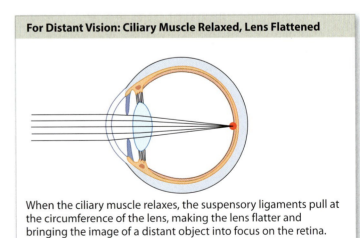

For Distant Vision: Ciliary Muscle Relaxed, Lens Flattened

When the ciliary muscle relaxes, the suspensory ligaments pull at the circumference of the lens, making the lens flatter and bringing the image of a distant object into focus on the retina.

3 An object in view isn't a single point; the image consists of a large number of individual points, like the pixels on a computer screen. In the eye, light from each point is focused on the retina, creating a miniature image of the original. However, the image is inverted and reversed. The brain compensates for this image reversal, and we are not aware of any difference between the orientation of the image on the retina and that of the object. The compensation is learned by experience—a person wearing glasses that reverse and invert the visual image can adapt to the change relatively quickly.

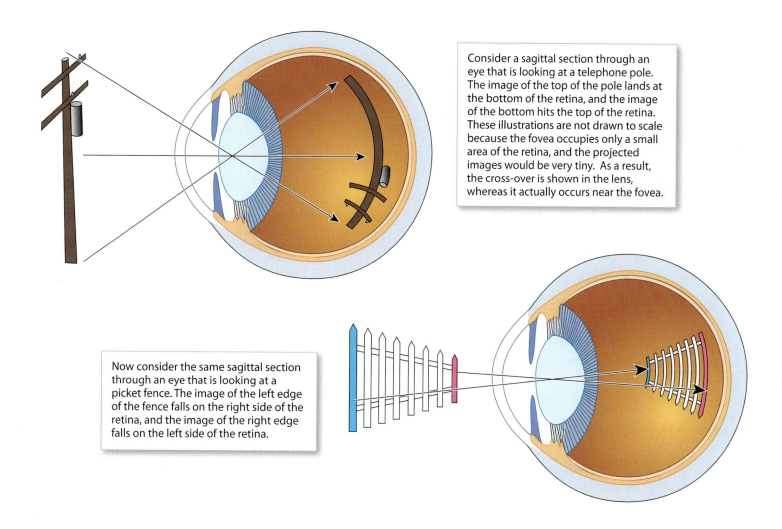

Consider a sagittal section through an eye that is looking at a telephone pole. The image of the top of the pole lands at the bottom of the retina, and the image of the bottom hits the top of the retina. These illustrations are not drawn to scale because the fovea occupies only a small area of the retina, and the projected images would be very tiny. As a result, the cross-over is shown in the lens, whereas it actually occurs near the fovea.

Now consider the same sagittal section through an eye that is looking at a picket fence. The image of the left edge of the fence falls on the right side of the retina, and the image of the right edge falls on the left side of the retina.

The greatest amount of refraction is required to view objects that are very close to the lens. The inner limit of clear vision is determined by the degree of elasticity in the lens. Children can usually focus on something 7–9 cm from the eye, but over time the lens tends to become stiffer and less responsive. A young adult can usually focus on objects 15–20 cm away. As aging proceeds, this distance gradually increases; the near point at age 60 is typically about 83 cm.

Module 9.12 Review

a. Define focal point.

b. When the ciliary muscles are relaxed, are you viewing something close up or something in the distance?

c. Why does the near point of vision typically increase with age?

The neural tissue of the retina contains multiple layers of specialized photoreceptors, neurons, and supporting cells

1 This diagrammatic sectional view through the eye shows the retina near the origin of the optic nerve.

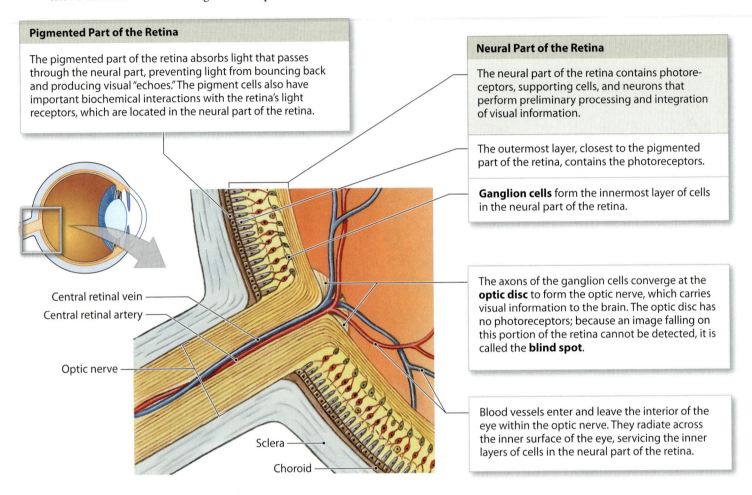

Pigmented Part of the Retina

The pigmented part of the retina absorbs light that passes through the neural part, preventing light from bouncing back and producing visual "echoes." The pigment cells also have important biochemical interactions with the retina's light receptors, which are located in the neural part of the retina.

Neural Part of the Retina

The neural part of the retina contains photoreceptors, supporting cells, and neurons that perform preliminary processing and integration of visual information.

The outermost layer, closest to the pigmented part of the retina, contains the photoreceptors.

Ganglion cells form the innermost layer of cells in the neural part of the retina.

The axons of the ganglion cells converge at the **optic disc** to form the optic nerve, which carries visual information to the brain. The optic disc has no photoreceptors; because an image falling on this portion of the retina cannot be detected, it is called the **blind spot**.

Blood vessels enter and leave the interior of the eye within the optic nerve. They radiate across the inner surface of the eye, servicing the inner layers of cells in the neural part of the retina.

Central retinal vein

Central retinal artery

Optic nerve

Sclera

Choroid

2 This is a photograph of the retina, taken through the cornea, pupil, and lens of the right eye.

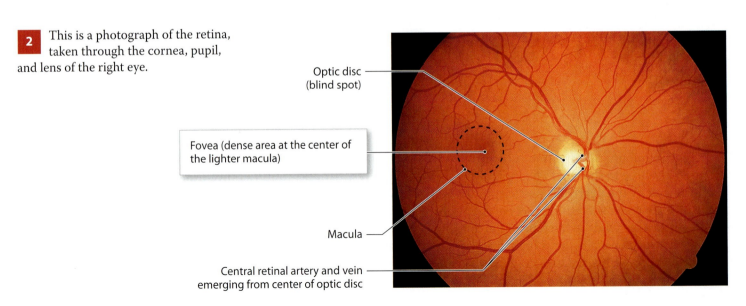

Optic disc (blind spot)

Fovea (dense area at the center of the lighter macula)

Macula

Central retinal artery and vein emerging from center of optic disc

3 This sectional view shows that the retina contains multiple layers of specialized cells, including two types of photoreceptors: rods and cones.

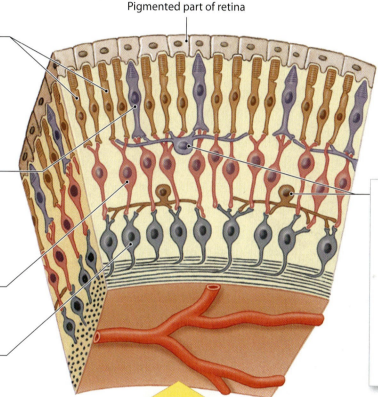

Photoreceptors of the Retina

Rods do not discriminate among colors of light. Highly sensitive, they enable us to see in dimly lit rooms, at twilight, and in pale moonlight. The retina of each eye contains approximately 125 million rods. The density of rods is highest at the periphery of the retina, where there are very few cones.

Cones provide us with color vision. They give us sharper, clearer images than rods do, but cones require more intense light. The retina of each eye contains approximately 6 million cones. The density of cones reaches its maximum at the fovea of the macula, where there are no rods.

Rods and cones synapse with neurons called **bipolar cells**.

Bipolar cells synapse on ganglion cells.

Pigmented part of retina

Other cells in the retina can facilitate or inhibit communication between photoreceptors and ganglion cells, thereby altering the sensitivity of the retina. The effect is comparable to adjusting the contrast on a television set. These cells play an important role in the eye's adjustment to dim or brightly lit environments.

LIGHT

When you look directly at an object, the image falls on the fovea, the center of color vision and image sharpness. However, in very dim light, cones cannot function. That is why you can't see a dim star if you stare directly at it, but you can see it if you shift your gaze to one side or the other so that the image falls on the more sensitive rods.

Module 9.13 Review

a. Define rods and cones and briefly state their functions.

b. If you enter a dimly lit room, will you be able to see clearly? Why or why not?.

c. If you had been born without cones in your eyes, explain why you would or would not be able to see.

Photoreception, which occurs in the outer segment of rods and cones, involves the activation of visual pigments

The rods and cones of the retina are called photoreceptors because they detect photons, the basic units of visible light. Light energy is a form of radiant energy that travels in waves with a characteristic wavelength (distance between wave peaks). Our eyes are sensitive to wavelengths of 400–700 nm, the spectrum of visible light. A nanometer (nm) is one billionth of a meter.

1 This illustration shows the major structural features of rods and cones, and the adjacent pigment epithelium and bipolar cells. The outer segments of both rods and cones have membranous plates, or discs, that contain special organic compounds called **visual pigments**.

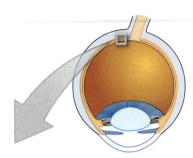

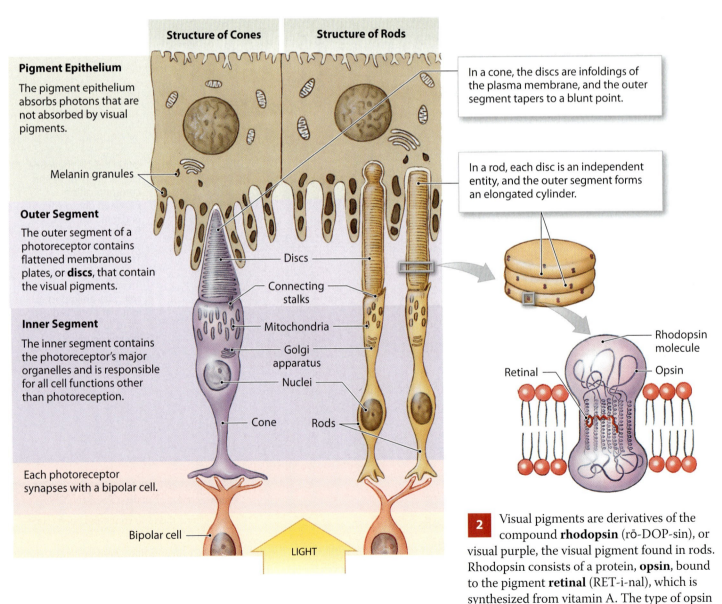

Structure of Cones

Structure of Rods

Pigment Epithelium

The pigment epithelium absorbs photons that are not absorbed by visual pigments.

Melanin granules

Outer Segment

The outer segment of a photoreceptor contains flattened membranous plates, or **discs**, that contain the visual pigments.

Inner Segment

The inner segment contains the photoreceptor's major organelles and is responsible for all cell functions other than photoreception.

Each photoreceptor synapses with a bipolar cell.

Discs

Connecting stalks

Mitochondria

Golgi apparatus

Nuclei

Cone

Rods

Bipolar cell

LIGHT

In a cone, the discs are infoldings of the plasma membrane, and the outer segment tapers to a blunt point.

In a rod, each disc is an independent entity, and the outer segment forms an elongated cylinder.

Rhodopsin molecule

Retinal

Opsin

2 Visual pigments are derivatives of the compound **rhodopsin** (rō-DOP-sin), or visual purple, the visual pigment found in rods. Rhodopsin consists of a protein, **opsin**, bound to the pigment **retinal** (RET-i-nal), which is synthesized from vitamin A. The type of opsin present determines the wavelength of light that can be absorbed by retinal.

3 When light strikes a visual pigment, the retinal molecule changes shape, which changes the permeability of the outer segment. This permeability change is the key to transduction in the eye, as it converts light energy into a nerve impulse.

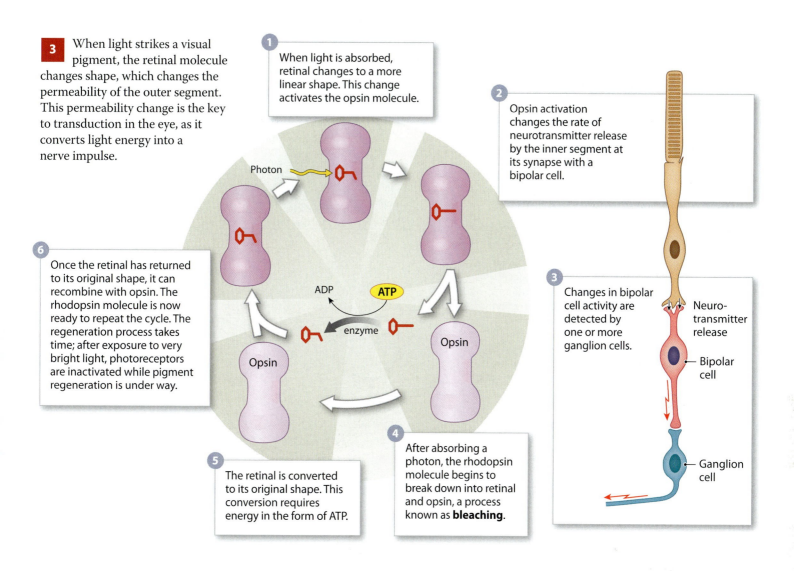

1 When light is absorbed, retinal changes to a more linear shape. This change activates the opsin molecule.

2 Opsin activation changes the rate of neurotransmitter release by the inner segment at its synapse with a bipolar cell.

Photon

6 Once the retinal has returned to its original shape, it can recombine with opsin. The rhodopsin molecule is now ready to repeat the cycle. The regeneration process takes time; after exposure to very bright light, photoreceptors are inactivated while pigment regeneration is under way.

ADP

ATP

enzyme

Opsin

Opsin

3 Changes in bipolar cell activity are detected by one or more ganglion cells.

Neuro-transmitter release

Bipolar cell

Ganglion cell

5 The retinal is converted to its original shape. This conversion requires energy in the form of ATP.

4 After absorbing a photon, the rhodopsin molecule begins to break down into retinal and opsin, a process known as **bleaching**.

4 There are three types of cones: **blue cones**, **green cones**, and **red cones**. Each type has a different form of opsin that is sensitive to a different range of wavelengths. Their stimulation in various combinations is the basis for color vision. In an individual with normal vision, the cone population consists of 16 percent blue cones, 10 percent green cones, and 74 percent red cones. Although their wavelength sensitivities overlap, each type is most sensitive to a specific portion of the visual spectrum. If all three cone populations are stimulated, we perceive the color as white. We also perceive white if rods (but not cones) are stimulated, which is why everything appears black-and-white when we enter dimly lit surroundings or walk by starlight.

NOTE

If a person lacks one or more cone pigments, **color blindness** results. The common forms are sex-linked and the genetic basis will be considered in a later chapter. Total color blindness (no cone pigments) is very rare, but 2% of males lack either red or green cone pigments and are partially color blind.

Module 9.14 Review

a. Identify the three types of cones.

b. Explain why your vision is momentarily impaired after viewing a camera's flash.

c. How could a diet deficient in vitamin A affect vision?

The visual pathways distribute visual information from each eye to both cerebral hemispheres

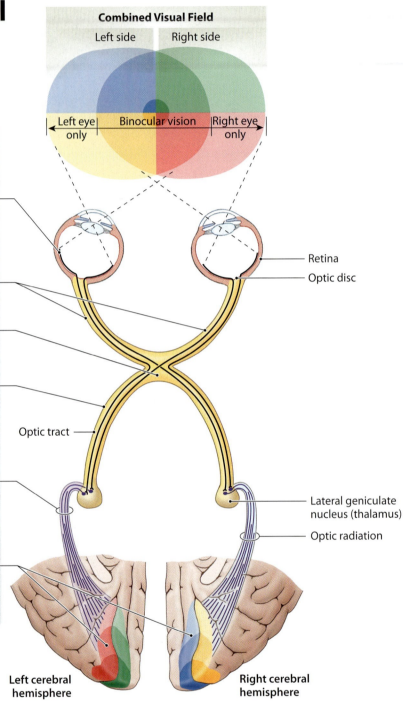

Combined Visual Field

Left side   Right side

Left eye only   Binocular vision   Right eye only

Retina

Optic disc

Optic tract

Lateral geniculate nucleus (thalamus)

Optic radiation

Left cerebral hemisphere

Right cerebral hemisphere

The Visual Pathways

The visual pathways begin at the photoreceptors in the retina. Each photoreceptor monitors a specific receptive field, and when stimulated, passes the information through a bipolar cell and to a ganglion cell.

Axons from the approximately 1 million ganglion cells converge on the optic disc, penetrate the wall of the eye, and proceed toward the diencephalon as the optic nerve (II).

The two optic nerves, one from each eye, reach the diencephalon at the optic chiasm.

From that point, approximately half the fibers proceed toward the lateral geniculate nucleus of the same side of the brain, whereas the other half cross over to reach the lateral geniculate nucleus of the opposite side.

From each lateral geniculate nucleus, visual information travels to the occipital cortex of the cerebral hemisphere on that side. The bundle of projection fibers linking each lateral geniculate nucleus with the visual cortex is known as the **optic radiation**.

The perception of a visual image reflects the integration of information that arrives at the visual cortex of the occipital lobes. Each eye receives a slightly different visual image, because (1) the foveae are 5–7.5 cm (2–3.0 in.) apart, and (2) the nose and eye socket block the view of the opposite side.

1   The visual images from the left and right eyes overlap, and the visual cortex of each cerebral hemisphere receives information from both eyes. The information is sorted, however, so that the left visual cortex gets information on the right half of the visual field, and the right visual cortex receives information on the left half of the visual field. The brain achieves **depth perception**, an interpretation of the three-dimensional relationships among objects in view, by comparing the relative positions of objects within the images received by the two eyes. The map in the visual cortex is upside down and backward, duplicating the orientation of the visual image at the retina.

Module 9.15 Review

a. Define optic radiation.

b. Where are visual images perceived?

c. Trace the visual pathway, beginning at the photoreceptors in the retina.

Accommodation problems result from abnormalities in the cornea or lens, or in the shape of the eye

Emmetropia

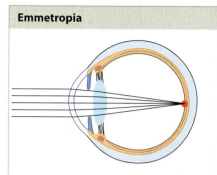

In the normal healthy eye, when the ciliary muscle is relaxed and the lens is flattened, the image of a distant object will be focused on the retina's surface. This condition is called **emmetropia** (*emmetro-*, proper + *-opia*, vision), or normal vision.

Myopia

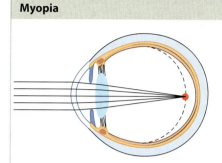

If the eyeball is too deep or the resting curvature of the lens is too great, the image of a distant object is projected in front of the retina. Such individuals are said to be nearsighted because vision at close range is clear but distant objects are out of focus. Their condition is more formally termed **myopia** (*myo-*, to shut + *-opia*, vision).

Diverging lens

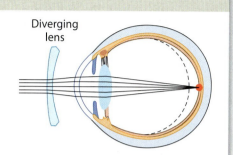

Myopia can be treated by placing a diverging lens in front of the eye. Diverging lenses have at least one concave surface and spread the light rays apart as if the object were closer to the viewer.

1 Small variations in the performance of the lens or the structure of the eye can be corrected with external lenses (glasses or contact lenses).

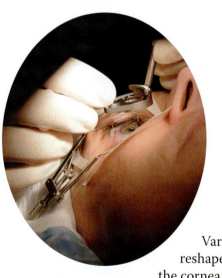

Hyperopia

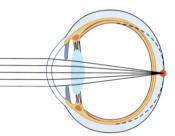

If the eyeball is too shallow or the lens is too flat, **hyperopia** results. The ciliary muscle must contract to focus even a distant object on the retina, and at close range the lens cannot provide enough refraction to focus an image on the retina. Individuals with this problem are said to be farsighted, because they can see distant objects most clearly.

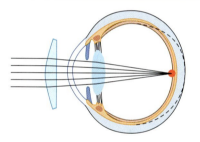

Hyperopia can be corrected by placing a converging lens in front of the eye. Converging lenses have at least one convex surface and provide the additional refraction needed to bring nearby objects into focus.

Variable success at correcting myopia and hyperopia has been achieved by surgery that reshapes the cornea. In **photorefractive keratectomy (PRK)** a computer-guided laser shapes the cornea to exact specifications. Tissue is removed only to a depth of 10–20 μm—no more than about 10 percent of the cornea's thickness. The entire procedure can be done in less than a minute. A variation on PRK is called **laser-assisted in-situ keratomileusis (LASIK)**. In this procedure the interior layers of the cornea are reshaped and covered by a flap of the normal corneal epithelium. Roughly 70 percent of LASIK patients achieve normal vision; it is now the most common form of refractive surgery. Each year, an estimated 100,000 people undergo PRK therapy in the United States. Corneal scarring is rare, and approximately 10 million Americans have had corneal refractive surgery. However, many still need reading glasses, and both immediate and long-term visual problems can occur.

Module 9.16 Review

a. Define emmetropia.

b. Discuss two surgical procedures for correcting myopia and hyperopia.

c. Which type of lens would correct hyperopia?

Aging is associated with many disorders of the special senses; trauma, infection, and abnormal stimuli may cause problems at any age

Olfaction

1 Disorders of the sense of smell may result from a head injury or from normal, age-related changes. If an injury to the head damages the olfactory nerves (N I), then the sense of smell may be impaired. Unlike other populations of neurons, olfactory receptor cells are regularly replaced by the division of stem cells. Despite this process, the total number of receptors declines with age, and the remaining receptors become less sensitive. As a result, the elderly have difficulty detecting odors in low concentrations. This explains why your grandmother may overdo her perfume, and your grandfather's aftershave may seem so strong—they must use more of the odorous solution to be able to smell it themselves.

Gustation

2 Disorders of the sense of taste can be caused by problems with olfactory receptors, damage to taste buds, damage to cranial nerves, and age-related changes. The sense of smell also makes a large contribution to our sense of taste, so conditions that affect the olfactory receptors—such as the common cold—can also dull your sense of taste. A reduced sense of taste may also result from damage to taste buds by inflammation or infections of the mouth. Alternatively, the cranial nerves carrying taste sensations (VII, IX, X) may be damaged through trauma or compression by a tumor.

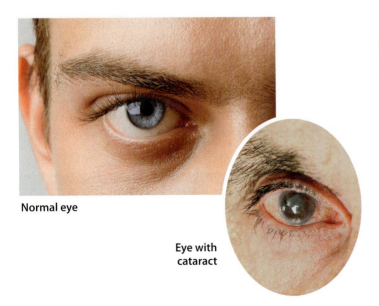

Normal eye

Eye with cataract

Vision

3 Several common eye problems were introduced earlier in the chapter. A condition in which the lens loses its transparency—called a **cataract**—can result from injuries, radiation, or reaction to drugs, but **senile cataracts**, a natural consequence of aging, are the most common form. As the condition advances, the individual needs brighter and brighter light for reading, and visual acuity may eventually decrease to the point of blindness. A damaged or nonfunctional lens can be replaced by an artificial substitute, and vision fine-tuned with glasses or contact lenses.

Equilibrium

4 **Vertigo** is an illusion of movement (within oneself or within one's surroundings). Vertigo is caused by conditions that alter the function of the internal ear receptor complex, the vestibular branch of the vestibulocochlear nerve, or sensory nuclei and pathways in the central nervous system. Any event that sets endolymph into motion can stimulate the equilibrium receptors and produce vertigo. For example, flushing the external acoustic meatus with cold water may chill the endolymph in the outermost portions of the labyrinth and establish a temperature-related circulation of fluid that produces mild and temporary vertigo. Excessive consumption of alcohol or exposure to certain drugs can also produce vertigo by changing endolymph composition or disturbing hair cells. Perhaps the most common cause of vertigo is **motion sickness**. Its unpleasant signs and symptoms include headache, sweating, flushing of the face, nausea, and vomiting. The drugs commonly administered to prevent motion sickness appear to depress activity in the brain stem.

Hearing

5 An estimated 6 million Americans have at least a partial hearing deficit, or deafness. **Conductive deafness** results from interference with the normal transfer of vibrations from the tympanic membrane to the oval window. Causes include excess wax or trapped water in the external acoustic meatus, scarring or perforation of the tympanic membrane, or immobilization of one or more auditory ossicles by fluid, a tumor, or abnormal bone growth that restricts movement of the ossicles (*otosclerosis*). In **nerve deafness**, the problem lies within the cochlea or somewhere along the auditory pathway. Vibrations reach the oval window, but the receptors either cannot respond or their response cannot reach the central nervous system. Very loud noises can cause nerve deafness by damaging the sensory cilia on the receptor cells. Bacterial or viral infections may also kill receptor cells and damage sensory nerves. Young children have the greatest hearing range: They can detect sounds ranging from a 20-Hz buzz to a 20,000-Hz whine. With age, damage due to loud noises or other injuries accumulates: The tympanic membrane gets less flexible, the articulations between the auditory ossicles stiffen, and the round window may begin to ossify. As a result, older individuals typically show some degree of hearing loss.

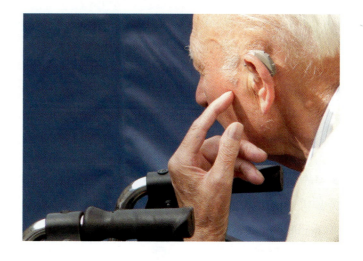

Module 9.17 Review

a. Which cranial nerves provide taste sensations from the tongue?

b. Identify two common classes of hearing-related disorders.

c. What causes vertigo?

1. Labeling

Label the structures in the following
diagram of a sagittal section of the left eye.

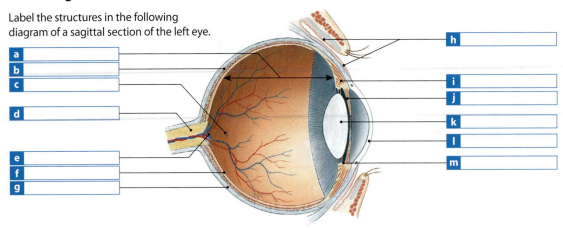

a
b
c
d
e
f
g
h
i
j
k
l
m

2. Labeling

Label the structures in the following
diagram of the right ear.

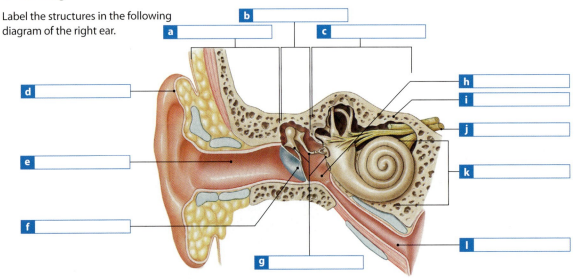

a
b
c
d
e
f
g
h
i
j
k
l

3. Matching

Match the following terms with the most closely related description.

- optic disc
- lingual papillae
- olfactory bulb
- stapes
- spiral organ
- taste bud
- endolymph
- cerebral cortex
- cones

a _____ Epithelial projections of the tongue

b _____ Cluster of gustatory receptors

c _____ Photoreceptors that provide the perception of color

d _____ Receives all special senses stimuli

e _____ Region of retina called the "blind spot"

f _____ Attached to oval window

g _____ Site of first synapse by olfactory receptors

h _____ Contains hair cells in the cochlear duct

i _____ Fluid within certain chambers and canals of the inner ear

4. Section integration

A bright flash of light from nearby exploding fireworks blinds Rachel's eyes. The result is a "ghost" image
that temporarily remains on her retinas. What might account for the images and their subsequent disappearance?

_____

_____

Visual Outline with Key Terms

Summarize the content of each module using the terms in the order provided.

SECTION 1

The General Senses

- sensory receptors
- nociceptors
- thermoreceptors
- mechanoreceptors
- proprioceptors
- baroreceptors
- tactile receptors
- chemoreceptors

9.1

General sensory receptors in the skin vary widely in form and function

- free nerve endings
- root hair plexus
- tactile discs
- tactile corpuscles
- lamellated corpuscles
- Ruffini corpuscles
- adaptation

9.2

Baroreceptors and chemoreceptors start important autonomic reflexes involving visceral sensory pathways

- baroreceptors
- chemoreceptors
- carotid bodies
- aortic bodies

SECTION 2

The Special Senses

- special senses
- olfaction
- vision
- gustation
- equilibrium
- hearing
- ○ internal ear
- hair cells

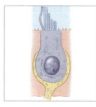

9.3

Olfaction involves specialized chemoreceptive neurons whereas taste receptors are specialized epithelial cells

- olfactory organs
- olfactory bulb
- olfactory tract
- odorants
- gustation
- taste receptors
- gustatory receptors
- lingual papillae
- taste buds
- circumvallate papillae
- umami
- water receptors

9.4

The ear is divided into the external ear, the middle ear, and the internal ear

- external ear
- auricle
- external acoustic meatus
- ceruminous glands
- cerumen
- middle ear
- auditory ossicles
- tympanic membrane
- internal ear
- bony labyrinth
- auditory tube
- otitis media
- malleus
- incus
- stapes

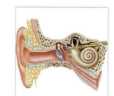

9.5

The bony labyrinth protects the membranous labyrinth

- internal ear
- bony labyrinth
- perilymph
- semicircular canals
- vestibule
- cochlea
- membranous labyrinth
- endolymph
- semicircular ducts
- utricle
- saccule
- cochlear duct
- vestibular complex

9.6

Hair cells in the semicircular ducts respond to rotation, while those in the utricle and saccule respond to gravity and linear acceleration

- anterior semicircular duct
- posterior semicircular duct
- lateral semicircular duct
- ampulla
- crista ampullaris
- cupula
- otolith
- maculae

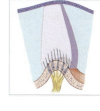

9.7

The cochlear duct contains the hair cells of the spiral organ

- cochlear duct
- scala vestibuli
- scala tympani
- oval window
- round window
- vestibular membrane
- basilar membrane
- spiral organ
- tectorial membrane

● = *Term boldfaced in this module*

9.8

The sensations of pitch and volume depend on movements of the basilar membrane

- wavelength
- frequency
- hertz (Hz)
- amplitude
- intensity
- decibels (dB)
- resonance
- pitch
- volume

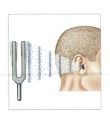

9.9

The accessory structures of the eye provide protection while allowing light to reach the interior of the eye

- accessory structures
- cornea
- lateral canthus
- pupil
- iris
- lacrimal caruncle
- eyelashes
- palpebra
- medial canthus
- palpebral fissure
- conjunctiva
- tarsal glands
- lacrimal apparatus
- lacrimal gland
- lysozyme
- tear ducts
- lacrimal puncta
- lacrimal canaliculi
- lacrimal sac
- nasolacrimal duct
- conjunctivitis

9.10

The eye has a layered wall; it is hollow, with fluid-filled anterior and posterior cavities

- fibrous layer
  - cornea
- sclera
- corneal limbus
- vascular layer
- iris
- ciliary body
- choroid
- neural layer (retina)
- photoreceptors
- anterior cavity
- anterior chamber
- posterior chamber
- posterior cavity
- vitreous body

9.11

The eye is highly organized and has a consistent visual axis that directs light to the fovea of the retina

- suspensory ligaments
- optic nerve
- pupillary dilator muscles
- pupillary constrictor muscles
- visual axis
- macula
- fovea

9.12

Focusing produces a sharply defined image at the retina

- refracted
- focal point
- focal distance
- accommodation

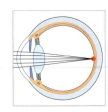

9.13

The neural tissue of the retina contains multiple layers of specialized photoreceptors, neurons, and supporting cells

- ganglion cells
- optic disc
- blind spot
- rods
- cones
- bipolar cells

9.14

Photoreception, which occurs in the outer segment of rods and cones, involves the activation of visual pigments

- visual pigments
- pigment epithelium
- photoreceptor
- outer segment
- discs
- inner segment
- rhodopsin
- opsin
- retinal
- bleaching
- blue cones
- green cones
- red cones
- color blindness

9.15

The visual pathways distribute visual information from each eye to both cerebral hemispheres

- depth perception
- optic radiation
  - lateral geniculate nucleus

9.16

Accommodation problems result from abnormalities in the cornea or lens, or in the shape of the eye

- emmetropia
- myopia
- hyperopia
- photorefractive keratectomy (PRK)
- laser-assisted in-situ keratomileusis (LASIK)

9.17

Aging is associated with many disorders of the special senses; trauma, infection, and abnormal stimuli may cause problems at any age

- cataract
- senile cataracts
- vertigo
- motion sickness
- conductive deafness
- nerve deafness

• = *Term boldfaced in this module*

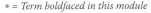

CAREER PATHS

Massage Therapist

Stephanie Miller, a certified massage therapist, says that a working knowledge of anatomy and physiology is essential 0to her career: "Learning relevant sciences is becoming more of an integrative part of studying massage therapy. It gives this field of work more weight as a medical practice."

Stephanie estimates that about half of her clients have chronic pains or ailments and need therapeutic massage, and the other half have no specific health needs and are more interested in a luxury or "spa" massage. She sees four to five clients a day and takes 30 minutes between each client. Sometimes, she has to spend time icing the irritated tendons in her arms.

"Because you use your body as your tool, you really have to start committing yourself to a healthier lifestyle," she says. "A massage therapist has to learn body mechanics: how to use your body so you can continue using it for the next 15 to 20 years."

With each new client, Stephanie does an intake interview and asks, in addition to health history, what kind of work he or she does. "If you know somebody's doing a lot of desk work, you know which muscles are used to create those actions," she says. As a result, a knowledge of the body and how it works is important for Stephanie to address her clients' needs.

For more information visit the website of the American Massage Therapy Association at http://www.amtamassage.org.

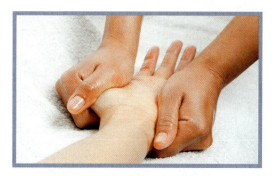

Think this is the career for you?

Key Stats:

- **Education and Training.** Usually a vocational degree, but requirements vary by state.

- **Licensure.** Most states require massage therapists to be licensed.

- **Earnings.** Earnings vary but the median hourly wage is $16.78; because of the physical work required, most massage therapists work fewer than 40 hours per week.

- **Expected Job Prospects.** Employment is expected to grow by an above average 18.9% from 2008 to 2018.

Bureau of Labor Statistics, U.S. Department of Labor, *Occupational Outlook Handbook, 2010–11 Edition*, Massage Therapists, on the Internet at http://www.bls.gov/oco/ocos295.htm (visited September 14, 2011).

Mastering A&P®

Access more review material online in the Study Area at www.masteringaandp.com.

- Chapter guides
- Chapter quizzes
- Practice tests
- Art labeling activities
- Flashcards
- A glossary with pronunciations
- Practice Anatomy Lab™ (PAL™) 3.0 virtual anatomy practice tool
- Interactive Physiology® (IP) animated tutorials
- MP3 Tutor Sessions

 practice anatomy lab™ **For this chapter, follow these navigation paths in PAL:**

- Human Cadaver>Nervous System>Special Senses
- Anatomical Models>Nervous System>Special Senses
- Histology>Special Senses

 For this chapter, go to this topic in the MP3 Tutor Sessions:

- The Visual Pathway

Chapter 9 Review Questions

1. Distinguish between the general senses and the special senses of the human body.

2. Why are olfactory sensations long lasting and an important part of our memories and emotions?

3. You are at a park watching some deer 35 feet away from you. Your friend taps you on the shoulder to ask a question. As you turn to look at your friend, who is standing 2 feet away, what changes will occur regarding the pupils, muscles, and lenses of your eyes?

4. After attending a Fourth of July fireworks extravaganza, Mona finds it difficult to hear normal conversation, and her ears keep "ringing." What is causing her hearing problems?

5. After riding the express elevator from the twentieth floor to the ground floor, for a few seconds you feel as if you are still descending, even though you have obviously come to a stop. Why?

For answers to all module, section, and chapter review questions, see the blue Answers tab at the back of the book.

10

The Endocrine System

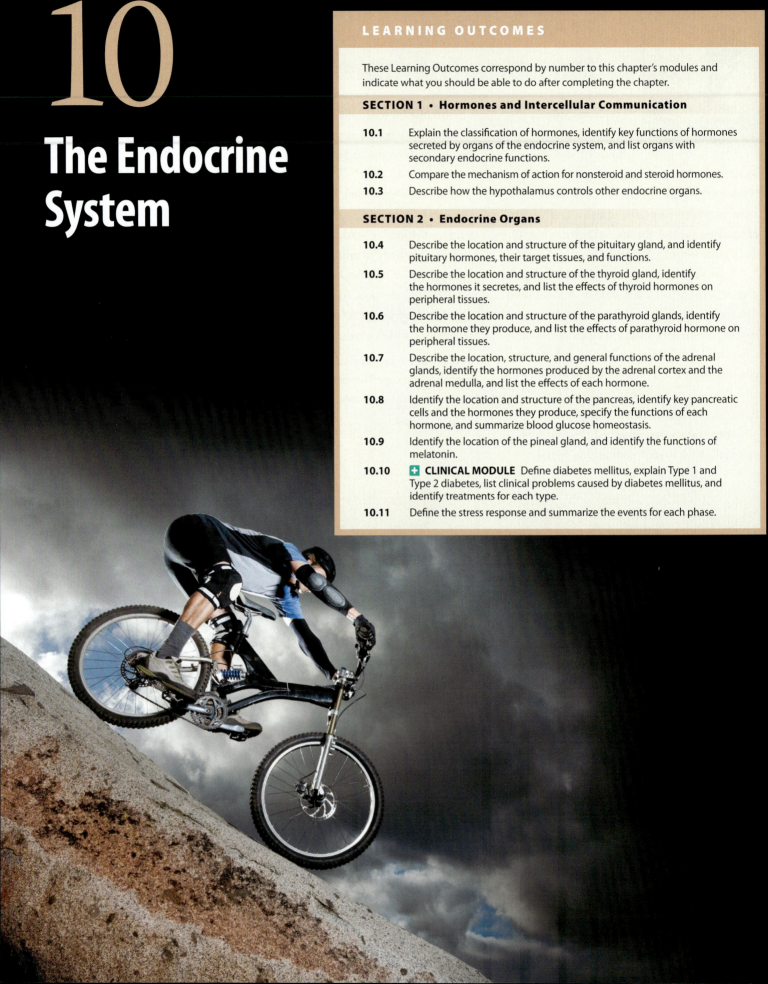

Hormones and Intercellular Communication

To preserve homeostasis, cellular activities must be coordinated throughout the body. Most of this communication involves the release of chemical messengers into extracellular fluid (interstitial fluid and blood). Chemical communication commonly occurs between neighboring cells. The table below summarizes how cellular communication over greater distances is coordinated by the nervous and endocrine systems and their chemical messengers.

Intercellular Communication by the Endocrine and Nervous Systems			
Cell Type	**Transmission**	**Chemical Messengers**	**Effects**
Endocrine glandular cells	Through the bloodstream	**Hormones** (chemical messengers that are released by one tissue and transported in the bloodstream to reach target cells in other tissues of the body)	Target cells are primarily in other tissues and organs and must have appropriate receptors. Provides long-term communication, such as growth and development.
Neurons	Across synaptic clefts	Neurotransmitters	Limited to very specific area; target cells must have appropriate receptors. Provides short, quick communication, such as reflexes.

The differences between the nervous and endocrine systems seem relatively clear. In fact, the distinctions noted above are the basis for treating them as two separate systems. On the other hand, when we consider them in detail, we see similarities in the ways they are organized.

• Both systems rely on the release of chemicals that bind to specific receptors on their target cells.

• The two systems share many chemical messengers. For example, norepinephrine and epinephrine are called hormones when released into the bloodstream, but they are called neurotransmitters when released across synapses.

• Regulation of their activities typically involves negative feedback. The neural activity or hormone released either blocks or opposes the original stimulus.

• The two systems share a common goal: to preserve homeostasis by coordinating and regulating the activities of other cells, tissues, organs, and systems.

In this section we consider the structure and functions of the endocrine system, as well as the integration of neural and endocrine activities.

Hormones may be amino acid derivatives, peptides, or lipid derivatives

1 Hormones can be divided into three groups based on their chemical structure: (1) **amino acid derivatives**, (2) **peptide hormones**, and (3) **lipid derivatives**.

Amino Acid Derivatives

Amino acid derivatives are small molecules that are structurally related to amino acids, the building blocks of proteins.

Derivatives of Tyrosine

Thyroid Hormones

Thyroxine (T_4)

Catecholamines

Epinephrine

Sources of tyrosine include meat, dairy, and fish.

Derivative of Tryptophan

Melatonin

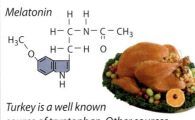

Turkey is a well known source of tryptophan. Other sources include chocolate, oats, bananas, dried dates, milk, cottage cheese, and peanuts.

Peptide Hormones

Peptide hormones are chains of amino acids. Most peptide hormones are synthesized as **prohormones**—inactive molecules that are converted to active hormones before or after they are secreted.

Glycoproteins

These proteins are more than 200 amino acids long and have carbohydrate side chains. Glycoproteins include *thyroid-stimulating hormone* (TSH), *luteinizing hormone* (LH), and *follicle-stimulating hormone* (FSH) from the anterior lobe of the pituitary gland, as well as several hormones produced in other organs.

Short Polypeptides/Small Proteins

This group of peptide hormones is large and diverse. It includes hormones that range from **short chain polypeptides**, such as *antidiuretic hormone* (ADH) and *oxytocin* (OXT) (each 9 amino acids long), to **small proteins**, such as *growth hormone* (GH; 191 amino acids) and *prolactin* (PRL; 198 amino acids). This group includes all the hormones secreted by the hypothalamus, heart, thymus, digestive tract, pancreas, and posterior lobe of the pituitary gland, as well as several hormones produced in other organs.

Lipid Derivatives

There are two classes of lipid derivatives: **eicosanoids**, derived from arachidonic acid, a 20-carbon fatty acid; and **steroid hormones**, derived from cholesterol.

Eicosanoids

Eicosanoids (ī-kō-sa-noydz) are factors that coordinate cellular activities and affect enzymatic processes (such as blood clotting) in extracellular fluids. One group of eicosanoids—**prostaglandins**—is involved primarily in coordinating local cellular activities.

Prostaglandin E

Aspirin suppresses the production of prostaglandins.

Steroid Hormones

Steroid hormones are released by the reproductive organs (androgens by the testes in males, estrogen and progesterone by the ovaries in females), by the cortex of the adrenal glands (corticosteroids), and by the kidneys (calcitriol). Because circulating steroid hormones are bound to specific transport proteins in the plasma, they remain in circulation longer than do secreted peptide hormones.

Estrogen

2 The **endocrine system** includes only those organs (indicated in purple in this figure) whose primary function is the production of hormones. Many other organs contain tissues that secrete hormones, but they are not considered part of the endocrine system because their endocrine functions are secondary. Examples include the heart, kidneys, intestines, thymus, and reproductive organs. We will consider the endocrine functions of these organs in later chapters.

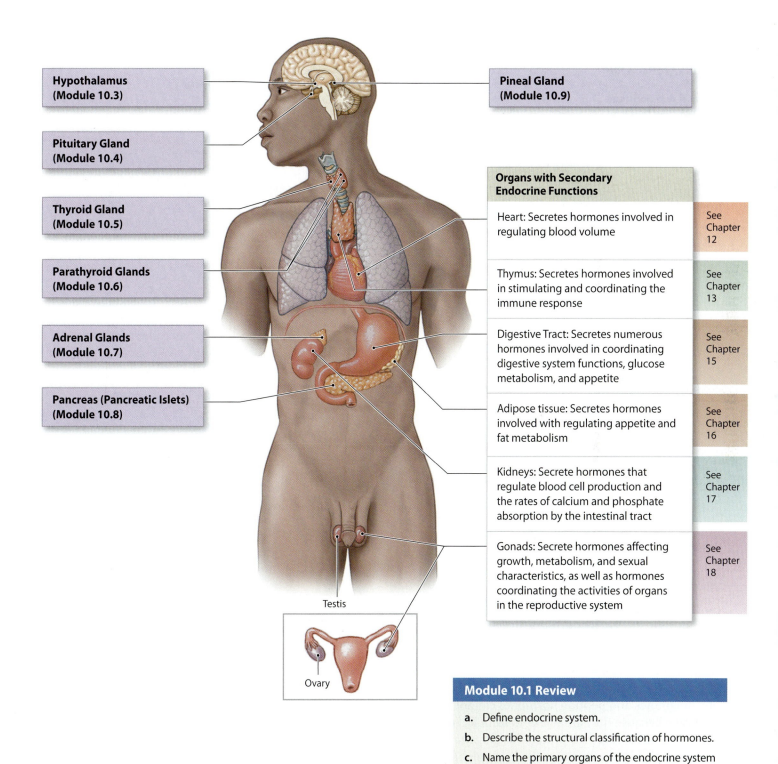

Hypothalamus
(Module 10.3)

Pituitary Gland
(Module 10.4)

Thyroid Gland
(Module 10.5)

Parathyroid Glands
(Module 10.6)

Adrenal Glands
(Module 10.7)

Pancreas (Pancreatic Islets)
(Module 10.8)

Pineal Gland
(Module 10.9)

Organs with Secondary Endocrine Functions

Heart: Secretes hormones involved in regulating blood volume — See Chapter 12

Thymus: Secretes hormones involved in stimulating and coordinating the immune response — See Chapter 13

Digestive Tract: Secretes numerous hormones involved in coordinating digestive system functions, glucose metabolism, and appetite — See Chapter 15

Adipose tissue: Secretes hormones involved with regulating appetite and fat metabolism — See Chapter 16

Kidneys: Secrete hormones that regulate blood cell production and the rates of calcium and phosphate absorption by the intestinal tract — See Chapter 17

Gonads: Secrete hormones affecting growth, metabolism, and sexual characteristics, as well as hormones coordinating the activities of organs in the reproductive system — See Chapter 18

Testis

Ovary

Module 10.1 Review

a. Define endocrine system.

b. Describe the structural classification of hormones.

c. Name the primary organs of the endocrine system and those organs and tissues with secondary endocrine functions.

There are two major mechanisms by which hormones act on target cells

1 Nonsteroid hormones bind to receptors on the plasma membrane and activate G proteins. They exert their effects on target cells through a **second messenger**, an intracellular intermediate that alters the activity of enzymes present in the cell. Second messengers are often high-energy compounds such as **cyclic-AMP** (**cAMP**).

2 Steroid hormones pass directly through the target cell's plasma membrane because they are lipid-soluble. They can bind to receptors in both the cytoplasm and nucleus and affect gene activity and protein synthesis.

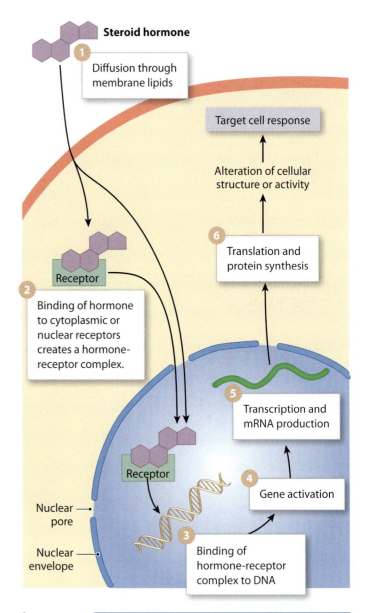

Nonsteroid hormone
such as epinephrine or norepinephrine (first messenger)

Plasma membrane

Receptor

G protein

ATP

cAMP

1 Binding of hormone to plasma membrane receptor

2 G protein activated

3 Second messenger system activated

CYTOPLASM

4 Cellular enzyme activity changed

Target cell response

NUCLEUS

Nuclear envelope

Nuclear pore

DNA

Steroid hormone

1 Diffusion through membrane lipids

Target cell response

Receptor

2 Binding of hormone to cytoplasmic or nuclear receptors creates a hormone-receptor complex.

Alteration of cellular structure or activity

6 Translation and protein synthesis

5 Transcription and mRNA production

Receptor

4 Gene activation

3 Binding of hormone-receptor complex to DNA

Nuclear pore

Nuclear envelope

Receptors for catecholamines (E and NE), peptide hormones, and eicosanoids are in the plasma membranes of their target cells. Catecholamines and peptide hormones cannot penetrate a plasma membrane because they are not lipid soluble. Instead, these hormones bind to receptor proteins at the *outer* surface of the plasma membrane. Eicosanoids *are* lipid soluble. They diffuse across the plasma membrane to reach receptor proteins on the *inner* surface of the membrane.

Module 10.2 Review

a. Describe two mechanisms of hormone action.

b. Which type of hormone binds to a plasma membrane receptor and why?

c. Which type of hormone diffuses across the plasma membrane and binds to receptors in the cytoplasm?

The hypothalamus exerts direct or indirect control over the activities of many different endocrine organs

1 The **hypothalamus** provides the highest level of endocrine control; it integrates the activities of the nervous and endocrine systems. The hypothalamus accomplishes this integration through three mechanisms.

1 Hypothalamic neurons synthesize two hormones —antidiuretic hormone (ADH) and oxytocin (OXT)—and transport them along axons that extend to the posterior lobe of the pituitary gland.

2 The hypothalamus secretes **regulatory hormones**, called releasing hormones and inhibitory hormones. A **releasing hormone (RH)** stimulates secretion of hormones at the anterior lobe. An **inhibiting hormone (IH)** prevents secretion of hormones from the anterior lobe.

3 The hypothalamus contains autonomic centers that exert direct neural control over the endocrine cells of the adrenal medullae. When the sympathetic division is activated, the adrenal medullae are stimulated directly and immediately.

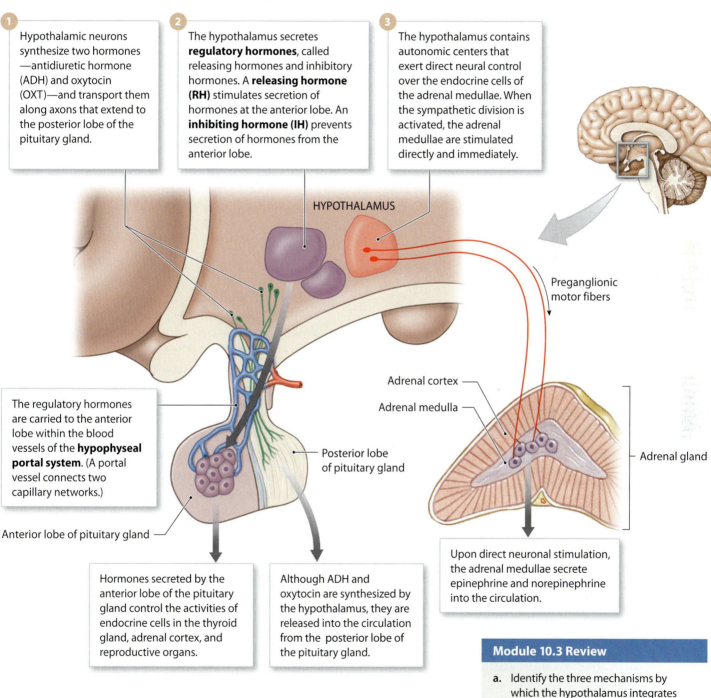

HYPOTHALAMUS

Preganglionic motor fibers

The regulatory hormones are carried to the anterior lobe within the blood vessels of the **hypophyseal portal system**. (A portal vessel connects two capillary networks.)

Adrenal cortex

Adrenal medulla

Posterior lobe of pituitary gland

Adrenal gland

Anterior lobe of pituitary gland

Hormones secreted by the anterior lobe of the pituitary gland control the activities of endocrine cells in the thyroid gland, adrenal cortex, and reproductive organs.

Although ADH and oxytocin are synthesized by the hypothalamus, they are released into the circulation from the posterior lobe of the pituitary gland.

Upon direct neuronal stimulation, the adrenal medullae secrete epinephrine and norepinephrine into the circulation.

Module 10.3 Review

a. Identify the three mechanisms by which the hypothalamus integrates neural and endocrine function.

b. Define regulatory hormone.

c. Contrast releasing hormones with inhibiting hormones.

1. Concept map

Place each of the following terms in the appropriate box to correctly complete the hormones concept map.

- steroid hormones
- tryptophan derivatives
- glycoproteins
- short polypeptides
- catecholamines
- peptide hormones
- thyroid hormones
- transport proteins
- lipid derivatives
- small proteins
- eicosanoids

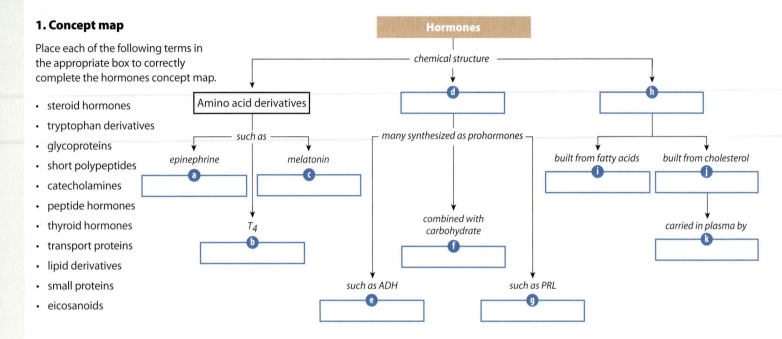

2. Multiple Choice

Choose the bulleted item that best completes each statement.

a The endocrine system regulates processes such as growth and development that require long-term coordination by using:

- physical changes to control homeostatic regulation
- DNA synthesis to control cellular metabolic activities
- chemical messengers to relay information and instructions between cells
- nuclear control of cells to produce additional mitochondria

b The release of hormones by endocrine cells alters the:

- rate at which neurotransmitters are released
- very specific responses to environmental stimuli
- simultaneous metabolic activities of many tissues and organs bearing the same receptors
- anatomical boundary between the nervous and endocrine systems

c A functional similarity between the endocrine and nervous systems is:

- both systems secrete hormones into the bloodstream
- the cells of both systems are functionally the same
- both produce very rapid responses to environmental stimuli
- both systems rely on chemical messengers that bind to target cells with specific receptors

d The two classes of hypothalamic regulatory hormones are:

- humoral and neural
- releasing and inhibiting
- cyclic-AMP and G proteins
- steroid and peptide

e Epinephrine, norepinephrine, and peptide hormones affect target organ cells by:

- binding to target receptors in the nucleus
- second messengers released when receptor binding occurs at the plasma membrane surface
- enzymatic reactions that occur in the ribosomes
- binding to receptors in the cytoplasm

f Steroid hormones affect target organ cells by:

- binding to target receptors in the cytoplasm or nucleus
- binding to target receptors in the plasma membrane
- binding to target receptors in peripheral tissues
- releasing second messengers at plasma membrane receptors

g Steroid hormones remain in circulation longer than peptide hormones because they are:

- produced as prohormones
- side chains attached to the basic carbon ring structure
- bound to specific transport proteins in the blood plasma
- attached to the hemoglobin molecule for cellular use

h Hormones alter the operations of target cells by changing the:

- plasma membrane permeability properties
- types, amounts, or activities of important enzymes and structural proteins
- arrangement of the molecules making up the plasma membrane
- rate at which hormones affect the target organ cells

Endocrine Organs

This section focuses on six organs whose primary function is to release hormones.

Pituitary Gland

The **pituitary gland** secretes multiple hormones that regulate the endocrine activities of the adrenal cortex, thyroid gland, and reproductive organs, and a hormone that stimulates melanin production.

Pineal Gland

The **pineal gland** secretes melatonin, which affects reproductive function and helps establish circadian (day/night) rhythms.

Thyroid Gland

The **thyroid gland** secretes hormones that affect metabolic rate and calcium levels in body fluids.

Parathyroid Glands

The **parathyroid glands** secrete a hormone important for regulating calcium ion concentrations in body fluids.

Adrenal Glands

The **adrenal glands** secrete hormones involved with mineral balance, metabolic control, and resistance to stress. The adrenal medullae release E and NE during sympathetic activation.

Pancreas (Pancreatic Islets)

The **pancreatic islets** secrete hormones regulating glucose uptake and utilization by body tissues.

Hormones of the pituitary gland, parathyroid gland, pineal gland, pancreatic islets, and some adrenal hormones exert their effects through second messenger systems, primarily impacting enzyme activities. Thyroid hormones and other adrenal hormones work primarily by altering genetic activities. As a result, their effects take longer to appear.

The pituitary gland acts as a master gland

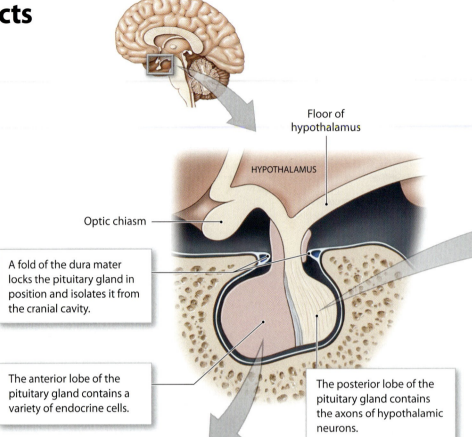

1 The pituitary gland is a small, oval gland that lies nestled within the sella turcica, a depression in the sphenoid bone. The pituitary gland releases nine important peptide hormones—the anterior lobe secretes seven hormones and the posterior lobe releases two. Posterior lobe hormones are made by the hypothalamus. All nine hormones bind to membrane receptors, and all nine use the same second messenger. The hormones of the anterior lobe are also called **tropic hormones** (*trope*, a turning), because they "turn on" endocrine glands or support the functions of other organs.

Floor of hypothalamus

HYPOTHALAMUS

Optic chiasm

A fold of the dura mater locks the pituitary gland in position and isolates it from the cranial cavity.

The anterior lobe of the pituitary gland contains a variety of endocrine cells.

The posterior lobe of the pituitary gland contains the axons of hypothalamic neurons.

Hormones of the Anterior Lobe of the Pituitary Gland

TSH	ACTH	Gonadotropins (FSH and LH)

TSH

Thyroid-stimulating hormone (TSH), also called thyrotropin, targets the thyroid gland, where it triggers the release of thyroid hormones.

Thyroid gland

ACTH

Adrenocorticotropic hormone (ACTH), also known as corticotropin, stimulates the release of steroid hormones by the adrenal cortex, the outer portion of the adrenal gland. ACTH specifically targets cells that produce hormones affecting glucose metabolism.

Adrenal gland

Gonadotropins (FSH and LH)

The hormones called **gonadotropins** (gō-nad-ō-TRŌ-pinz) regulate the activities of the gonads. (These organs—the testes in males and ovaries in females—produce reproductive cells as well as hormones.)

Follicle-stimulating hormone (FSH) promotes ovarian follicle development in females and, in combination with luteinizing hormone, stimulates the secretion of **estrogens** (ES-trō-jenz) by ovarian cells. In males, FSH promotes the physical maturation of developing sperm. FSH production is inhibited by **inhibin**, a peptide hormone released by cells in the testes and ovaries.

Luteinizing (LOO-tē-in-ī-zing) **hormone (LH)** induces ovulation, the release of reproductive cells in females. It also promotes the secretion, by the ovaries, of estrogens and progestins (such as progesterone), which prepare the body for possible pregnancy. In males, this gonadotropin stimulates the production of sex hormones by the interstitial cells of the testes. These sex hormones are called **androgens** (AN-drō-jenz; *andros*, man); the most important androgen is testosterone.

Ovary

Testis

Hormones of the Posterior Lobe of the Pituitary Gland

ADH

Antidiuretic hormone (ADH), also known as **vasopressin**, is released in response to a variety of stimuli, most notably a rise in the solute concentration in the blood or a fall in blood volume or blood pressure. Rising solute levels stimulate osmoreceptors, specialized hypothalamic neurons that respond to changes in the osmotic concentration of body fluids. The osmoreceptors then stimulate the neurosecretory neurons that release ADH.

- The primary function of ADH is to decrease the amount of water lost at the kidneys. With losses minimized, the body retains any water absorbed from the digestive tract, reducing the concentrations of electrolytes in extracellular fluid.
- In high concentrations, ADH also causes vasoconstriction, a constriction of peripheral blood vessels that helps raise blood pressure.
- Alcohol inhibits ADH release, which explains the increased fluid excretion that follows the consumption of alcoholic beverages.

Kidney

OXT

In women, **oxytocin** (*okytokos,* swift birth), or **OXT**, stimulates smooth muscle contraction in the wall of the uterus, promoting labor and delivery. After delivery, oxytocin stimulates the contraction of myoepithelial cells around the secretory alveoli and the ducts of the mammary glands, promoting the ejection of milk. Although the functions of oxytocin in sexual activity remain unclear, it is known that circulating concentrations of oxytocin rise during sexual arousal and peak at orgasm in both sexes.

Uterus

GH

Growth hormone (GH) stimulates cell growth and reproduction by accelerating the rate of protein synthesis. Skeletal muscle cells and chondrocytes are particularly sensitive to GH.

- In epithelia and connective tissues, GH stimulates stem cell divisions and the differentiation of daughter cells.
- In adipose tissue, GH stimulates adipocytes (fat cells) to break down stored triglycerides and release fatty acids into the blood. Many tissues then stop breaking down glucose and use fatty acids instead to generate ATP. This is termed a **glucose-sparing effect**.
- In the liver, GH stimulates the breakdown of glycogen reserves, leading to the release of glucose into the bloodstream.

Muscular and skeletal systems

PRL

Prolactin (*pro*-, before + *lac*, milk) **(PRL)** works with other hormones to stimulate mammary gland development. In pregnancy and during the nursing period that follows delivery, PRL also stimulates milk production by the mammary glands.

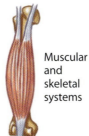

MSH

The pars intermedia, a narrow portion of the anterior lobe closest to the posterior lobe, may secrete **melanocyte-stimulating hormone (MSH)**. MSH stimulates the melanocytes of the skin to increase their production of melanin. In adults, this portion of the anterior lobe is virtually nonfunctional, and the circulating blood usually does not contain MSH.

Module 10.4 Review

a. Name the two lobes of the pituitary gland and the cellular sources of their secreted hormones.

b. Identify the nine pituitary hormones and their target tissues.

c. In a dehydrated person, how would the amount of ADH released by the posterior pituitary change?

Thyroid gland hormones regulate metabolism or calcium ion levels

1 The **thyroid gland** curves across the anterior surface of the trachea just inferior to the thyroid ("shield-shaped") cartilage, which forms most of the anterior surface of the larynx. The two lobes of the thyroid gland are united by a slender connection, the isthmus (IS-mus). You can easily feel the gland with your fingers. The size of the gland is quite variable, depending on heredity and environmental and nutritional factors, but its average weight is about 34 g (1.2 oz). An extensive blood supply gives the thyroid gland a deep red color.

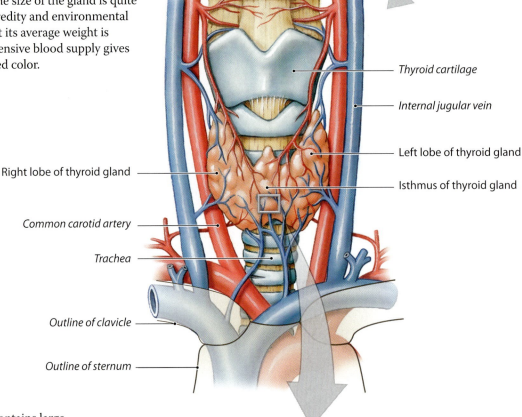

Thyroid cartilage

Internal jugular vein

Left lobe of thyroid gland

Right lobe of thyroid gland

Isthmus of thyroid gland

Common carotid artery

Trachea

Outline of clavicle

Outline of sternum

Simple cuboidal epithelium of follicle

Thyroid follicle

Thyroglobulin in colloid

Section of thyroid gland LM × 260

2 The thyroid gland contains large numbers of **thyroid follicles**, hollow spheres lined by a simple cuboidal epithelium. The follicle cells surround a follicle cavity that holds **colloid**, a viscous fluid containing large quantities of dissolved proteins. A network of capillaries surrounds each follicle, delivering nutrients and regulatory hormones to the glandular cells and accepting their secretory products and metabolic wastes. The follicle cells synthesize a globular protein called **thyroglobulin** (thī-rō-GLOB-ū-lin). Thyroglobulin molecules contain the amino acid tyrosine. Iodide ions are added to the tyrosine molecules to form thyroid hormones. The thyroglobulin-hormone complex is then stored in the thyroid follicle.

A second population of endocrine cells lies sandwiched between the basement membranes of the follicle cells. These large, pale cells—called **C (clear) cells** or parafollicular cells—produce the hormone **calcitonin (CT)**, which helps regulate calcium ion concentrations in body fluids.

3 Thyroid follicle cells break down thyroglobulin into two thyroid hormones, **T₃** (triiodothyronine) and **T₄** (thyroxine), which are released into the bloodstream. (The subscripts 3 and 4 refer to how many iodide ions each hormone contains.) Roughly 75 percent of thyroid hormones entering the bloodstream become attached to transport proteins called **thyroid-binding globulins** (**TBGs**). The transport proteins release thyroid hormones only gradually, and the bound thyroid hormones represent a substantial reserve: The bloodstream normally contains more than a week's supply of thyroid hormones.

4 Thyroid hormones are primarily transported across the plasma membrane by carrier-mediated processes. Once in the cells, these hormones bind to receptors on mitochondria and receptors within the nucleus. Thyroid hormones bound to mitochondria increase the mitochondrial rates of ATP production. Hormone-receptor complexes in the nucleus activate specific genes or change the rate of transcription, which affect the metabolic activities of the cell by increasing or decreasing the concentrations of specific enzymes.

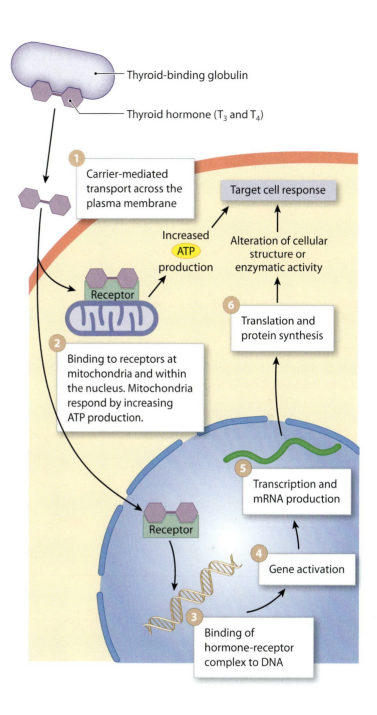

Thyroid-binding globulin

Thyroid hormone (T₃ and T₄)

1 Carrier-mediated transport across the plasma membrane

Target cell response

Increased ATP production

Alteration of cellular structure or enzymatic activity

Receptor

2 Binding to receptors at mitochondria and within the nucleus. Mitochondria respond by increasing ATP production.

6 Translation and protein synthesis

5 Transcription and mRNA production

Receptor

4 Gene activation

3 Binding of hormone-receptor complex to DNA

5 This table summarizes the important effects that thyroid hormones have on peripheral tissues.

Effects of T₃ and T₄ on Peripheral Tissues

- Raise metabolic rate (increased oxygen and energy consumption); in children, may raise body temperature
- Increase heart rate and force of contraction; generally raise blood pressure
- Increase sensitivity to sympathetic stimulation
- Maintain normal sensitivity of respiratory centers to changes in oxygen and carbon dioxide concentrations
- Stimulate red blood cell formation, enhancing oxygen delivery
- Stimulate activity in other endocrine tissues
- Accelerate turnover of minerals in bone

Module 10.5 Review

a. Name the three hormones secreted by the thyroid gland.

b. List five effects that thyroid hormones have on peripheral tissues.

c. After a thyroidectomy (surgical removal of the thyroid gland), symptoms of decreased thyroid hormone concentrations take about a week to appear. Why?

Parathyroid hormone, produced by the parathyroid glands, is the primary regulator of calcium ion levels in body fluids

Two pairs of **parathyroid glands** are embedded in the posterior surfaces of the thyroid gland.

1 The parathyroid glands are shaped like small peas. Altogether, the four parathyroid glands weigh a mere 1.6 g (0.06 oz). The capsule of the thyroid gland covers the parathyroid glands and holds them in position.

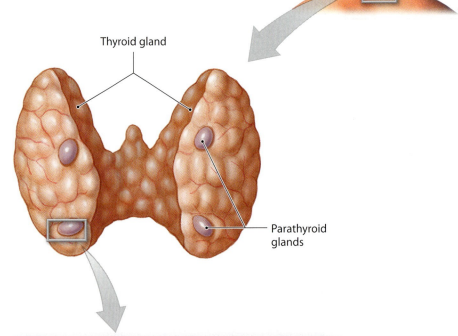

Thyroid gland

Parathyroid glands

2 The parathyroid glands have at least two cell populations: (1) **Parathyroid cells** produce parathyroid hormone. (2) **Oxyphils**, the other cell type, have no known function. Like the C cells of the thyroid gland, the parathyroid cells monitor the circulating concentration of calcium ions. When the Ca^{2+} concentration of the blood falls below normal, the parathyroid cells secrete **parathyroid hormone (PTH)**. PTH secretion raises Ca^{2+} concentration in body fluids.

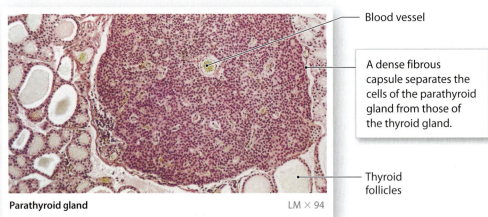

Blood vessel

A dense fibrous capsule separates the cells of the parathyroid gland from those of the thyroid gland.

Thyroid follicles

Parathyroid gland   LM × 94

3 Parathyroid hormone and calcitonin (secreted by the thyroid) have opposing effects on levels of calcium ions in body fluids. However, in healthy adults PTH, aided by calcitriol secreted by the kidneys, is the primary regulator of circulating calcium ion concentrations. Removal of the thyroid gland seldom affects calcium ion homeostasis because dietary intake and metabolic demand are so closely balanced that elevated blood calcium levels are very rare. However, calcitonin can be administered clinically to treat several metabolic disorders that raise calcium levels and cause excessive bone formation.

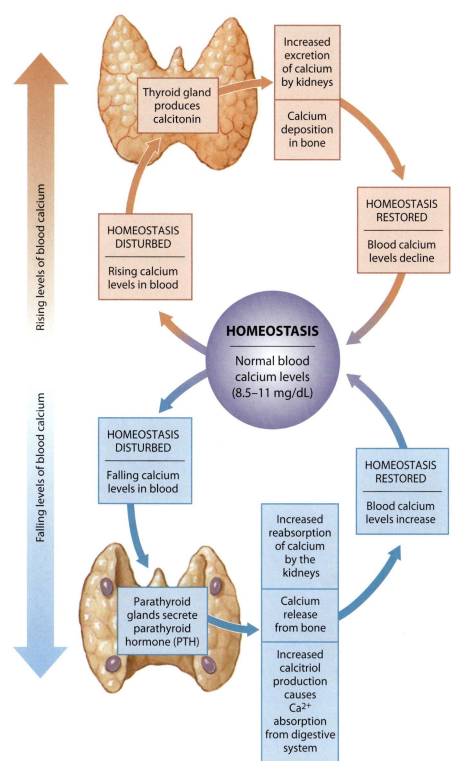

Rising levels of blood calcium

Thyroid gland produces calcitonin

Increased excretion of calcium by kidneys

Calcium deposition in bone

HOMEOSTASIS RESTORED

Blood calcium levels decline

HOMEOSTASIS DISTURBED

Rising calcium levels in blood

HOMEOSTASIS

Normal blood calcium levels (8.5–11 mg/dL)

Falling levels of blood calcium

HOMEOSTASIS DISTURBED

Falling calcium levels in blood

HOMEOSTASIS RESTORED

Blood calcium levels increase

Parathyroid glands secrete parathyroid hormone (PTH)

Increased reabsorption of calcium by the kidneys

Calcium release from bone

Increased calcitriol production causes Ca^{2+} absorption from digestive system

Effects of Parathyroid Hormone on Peripheral Tissues

- *Mobilizes calcium from bone by affecting osteoblast and osteoclast activity*. PTH inhibits osteoblasts, reducing the rate of calcium deposition in bone. Osteoclast activity predominates, and as bone matrix erodes, plasma Ca^{2+} levels rise.

- *Enhances reabsorption of Ca^{2+} by the kidneys*

- *Stimulates formation and secretion of calcitriol at the kidneys*. In general, the effects of calcitriol complement or enhance those of PTH, but one major effect of calcitriol is to enhance Ca^{2+} and PO_4^{3-} absorption by the digestive tract.

Module 10.6 Review

a. Describe the locations of the parathyroid glands.

b. Explain how parathyroid hormone raises blood calcium levels.

c. Increased blood calcium levels would result in increased secretion of which hormone?

The adrenal glands produce hormones involved in metabolic regulation

1 A yellow, pyramid-shaped **adrenal gland** sits on the superior border of each kidney. The adrenal glands are retroperitoneal, as are the kidneys, and only their anterior surfaces are covered by a layer of parietal peritoneum. Like other endocrine glands, the adrenal glands are richly supplied with blood vessels.

2 The adrenal cortex has a yellow color due to stored lipids, especially cholesterol and various fatty acids. The adrenal cortex produces more than two dozen steroid hormones, collectively called **corticosteroids**. Like other steroid hormones, corticosteroids exert their effects by determining which genes in the nuclei of their target cells are transcribed, and at what rate. The resulting changes in the nature and concentration of enzymes in the cytoplasm affect cellular metabolism. Corticosteroids are vital: If the adrenal glands are destroyed or removed, the individual will die unless corticosteroids are administered.

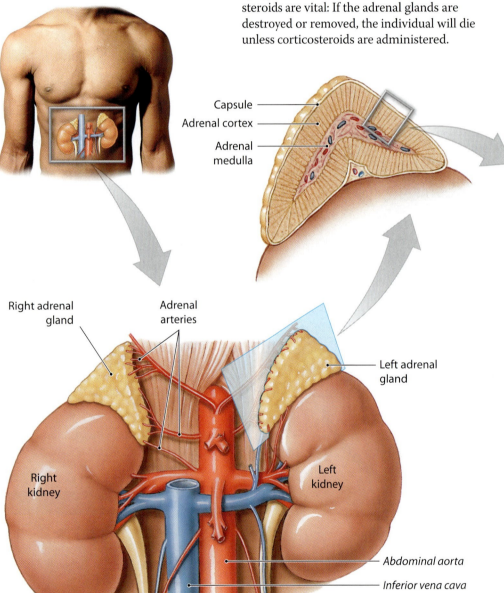

Capsule
Adrenal cortex
Adrenal medulla

Right adrenal gland
Adrenal arteries
Left adrenal gland
Right kidney
Left kidney
Abdominal aorta
Inferior vena cava

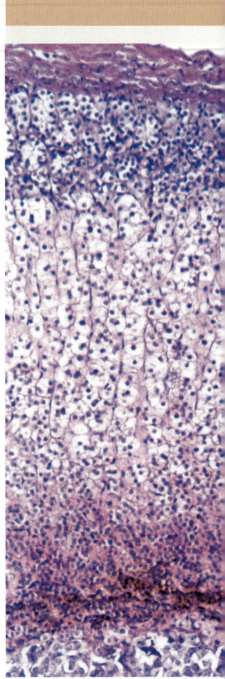

Adrenal gland LM × 250

3 Deep to the adrenal capsule are three distinct regions, or zones, in the adrenal cortex. Each zone synthesizes specific steroid hormones. Deep to the cortex lies the adrenal medulla, which synthesizes epinephrine and norepinephrine.

The Adrenal Hormones

Region/Zone	Hormones	Primary Targets	Hormonal Effects	Regulatory Control
Capsule				
ADRENAL CORTEX				
Outer zone of the adrenal cortex	**Mineralocorticoids (MCs)**, primarily **aldosterone**	Kidneys	Aldosterone increases renal reabsorption of Na$^+$ and water, especially in the presence of ADH. It also accelerates urinary loss of K$^+$.	Mineralocorticoid secretion is stimulated by the activation of the renin-angiotensin system (Module 12.11) and inhibited by hormones opposing that system.
Large, central zone of the adrenal cortex	**Glucocorticoid**s **(GCs)** are steroid hormones that affect glucose metabolism. The primary hormones are **cortisol** (KOR-ti-sol), also called hydrocortisone, and smaller amounts of the related steroid **corticosterone** (kor-ti-KOS-te-rōn). The liver converts some of the circulating cortisol to **cortisone**, another metabolically active glucocorticoid.	Most cells	Glucocorticoids increase rates of glucose and glycogen formation by the liver. They also stimulate the release of amino acids from skeletal muscles, and lipids from adipose tissues, and they promote lipid catabolism within peripheral cells. These actions supplement the glucose-sparing effect of growth hormone (Module 10.4). Cortisol also reduces inflammation (an **anti-inflammatory effect**).	Glucocorticoid secretion is stimulated by adrenocorticotropic hormone (ACTH) from the anterior lobe of the pituitary gland.
Narrow zone bordering each adrenal medulla	Small quantities of androgens (male sex hormones) that may be converted to estrogens in the bloodstream	Skin, bones, and other tissues, but minimal effects in normal adults	Adrenal androgens stimulate the development of pubic hair in boys and girls before puberty.	Androgen secretion is stimulated by ACTH.
ADRENAL MEDULLA	Epinephrine (E) and norepinephrine (NE)	Most cells	Epinephrine and norepinephrine increase cardiac activity, blood pressure, glycogen breakdown, and blood glucose levels.	Epinephrine and norepinephrine secretion is stimulated by sympathetic preganglionic fibers during sympathetic activation.

Module 10.7 Review

a. Identify the two regions of an adrenal gland, and cite the hormones secreted by each.

b. Identify the target tissue for aldosterone.

c. How would elevated cortisol levels affect blood glucose levels?

The pancreatic islets secrete insulin and glucagon and regulate glucose use by most cells

1 The **pancreas** lies within the abdominopelvic cavity in the loop formed between the inferior border of the stomach and the proximal portion of the small intestine. It is a slender, pale organ with a lumpy consistency. In adults, the pancreas is 20–25 cm (8–10 in.) long and weighs about 80 g (2.8 oz). The pancreas has both exocrine and endocrine functions. The **exocrine pancreas**, roughly 99 percent of the organ's volume, consists of clusters of gland cells and their attached ducts. The gland and duct cells secrete large quantities of an alkaline, enzyme-rich fluid that reaches the lumen of the digestive tract through one or two pancreatic ducts.

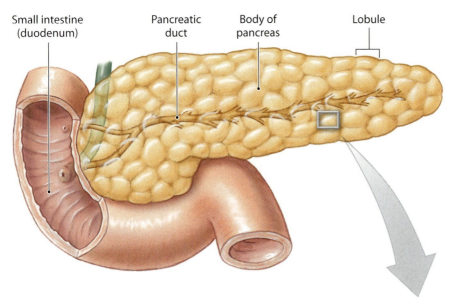

Small intestine (duodenum) · Pancreatic duct · Body of pancreas · Lobule

2 The **endocrine pancreas** consists of small groups of cells scattered among the exocrine cells. The endocrine clusters are known as **pancreatic islets**, or islets of Langerhans (LAN-ger-hanz). Pancreatic islets account for only about 1 percent of all cells in the pancreas. Nevertheless, a typical pancreas contains roughly 2 million pancreatic islets, and their secretions are vital to our survival.

Alpha cells produce the hormone **glucagon** (GLOO-ka-gon). Glucagon raises blood glucose levels by increasing the rates of glycogen breakdown and glucose release by the liver.

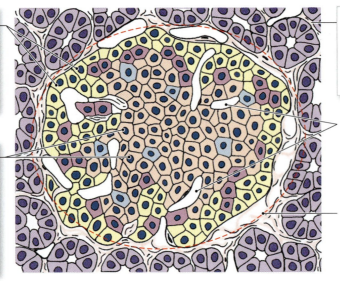

Pancreatic exocrine cells form small clusters that secrete into a lumen continuous with the pancreatic duct. The glandular clusters are called **pancreatic acini** (singular, *acinus*).

Capillaries

Beta cells produce the hormone **insulin** (IN-suh-lin). Insulin lowers blood glucose levels by increasing the rate of glucose uptake and utilization by cells, and by increasing glycogen synthesis in skeletal muscles and the liver.

Pancreatic islet

3 Insulin and glucagon are the primary hormones that regulate blood glucose levels. When blood glucose levels rise, beta cells secrete insulin, which then stimulates the transport of glucose across plasma membranes into target cells. When blood glucose levels decline, alpha cells secrete glucagon, which stimulates glycogen breakdown and glucose release by the liver.

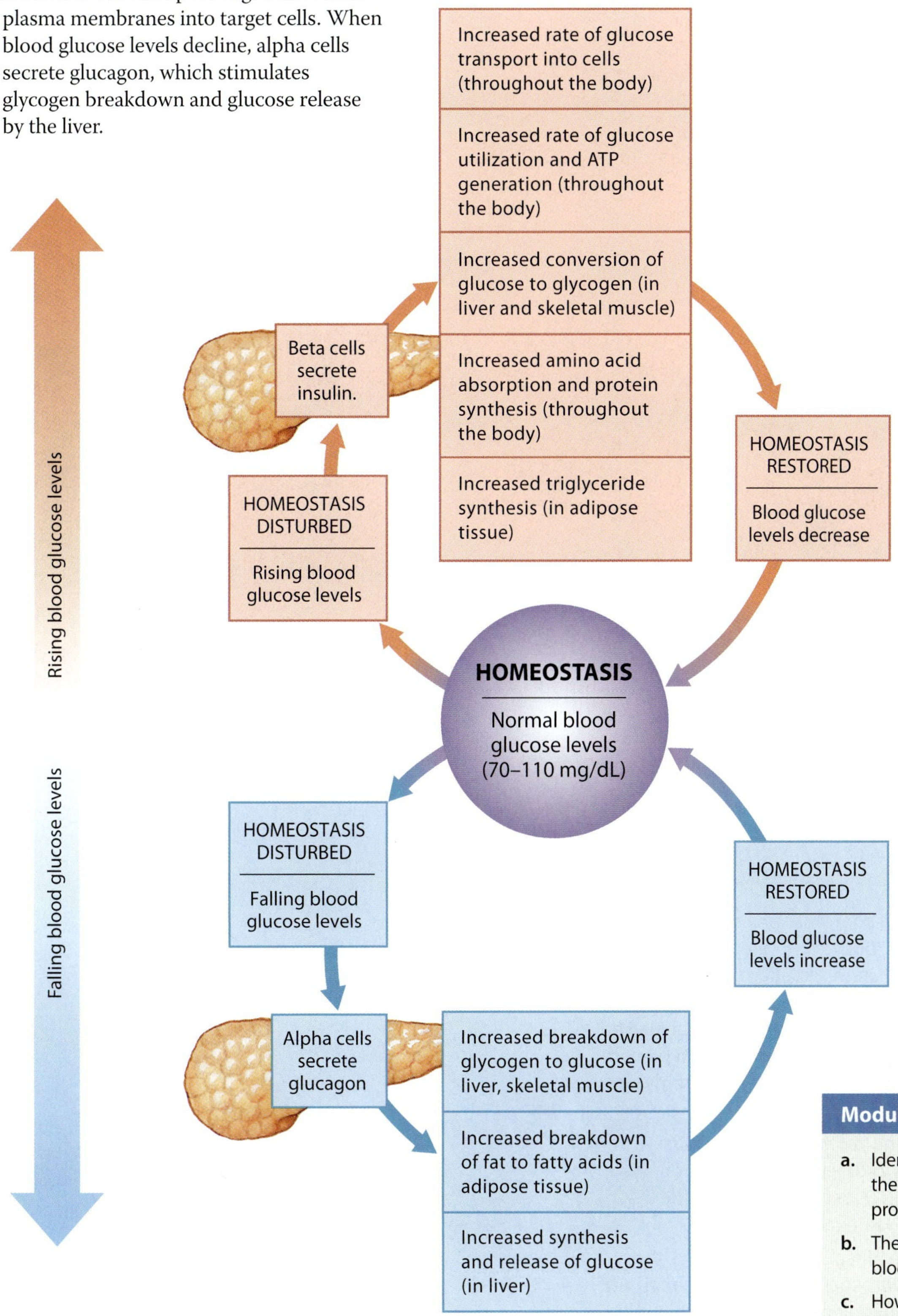

Rising blood glucose levels

Increased rate of glucose transport into cells (throughout the body)

Increased rate of glucose utilization and ATP generation (throughout the body)

Increased conversion of glucose to glycogen (in liver and skeletal muscle)

Beta cells secrete insulin.

Increased amino acid absorption and protein synthesis (throughout the body)

HOMEOSTASIS RESTORED
___
Blood glucose levels decrease

HOMEOSTASIS DISTURBED
___
Rising blood glucose levels

Increased triglyceride synthesis (in adipose tissue)

HOMEOSTASIS
___
Normal blood glucose levels (70–110 mg/dL)

Falling blood glucose levels

HOMEOSTASIS DISTURBED
___
Falling blood glucose levels

HOMEOSTASIS RESTORED
___
Blood glucose levels increase

Alpha cells secrete glucagon

Increased breakdown of glycogen to glucose (in liver, skeletal muscle)

Increased breakdown of fat to fatty acids (in adipose tissue)

Increased synthesis and release of glucose (in liver)

Module 10.8 Review

a. Identify two important types of cells in the pancreatic islets and the hormones produced by each.

b. The secretion of which hormone lowers blood glucose concentrations?

c. How do rising glucagon levels affect the amount of glycogen stored in the liver?

The pineal gland of the epithalamus secretes melatonin

1 The **pineal gland**, part of the epithalamus, lies in the posterior portion of the roof of the third ventricle. The pineal gland contains neurons, neuroglia, and special secretory cells that synthesize the hormone **melatonin**. Collaterals from the visual pathways enter the pineal gland and affect the rate of melatonin production, which is lowest during daylight hours and highest at night.

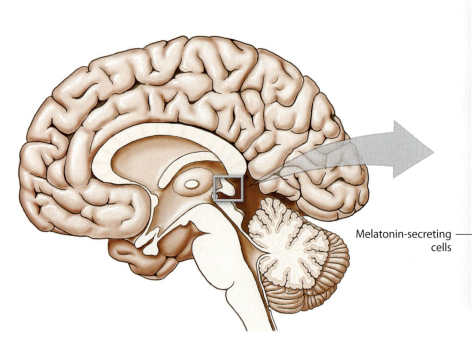

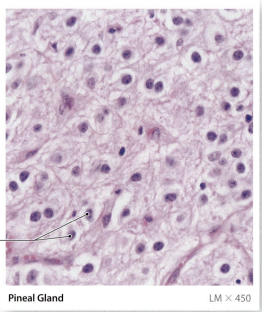

Melatonin-secreting cells

Pineal Gland   LM × 450

Functions of Melatonin in Humans

• *Inhibit reproductive functions.* In some mammals, melatonin slows the maturation of sperm, oocytes, and reproductive organs by reducing the rate of GnRH secretion. The significance of this effect in humans remains unclear, but evidence suggests that melatonin may play a role in timing human sexual maturation. Blood levels of melatonin decline at puberty, and pineal tumors that eliminate melatonin production cause premature puberty in young children.

• *Protect against tissue damage by free radicals.* Melatonin is a very effective antioxidant that may protect CNS neurons from **free radicals**, such as nitric oxide (NO) or hydrogen peroxide (H_2O_2), that may be generated in active neural tissue.

• *Set circadian rhythms.* Because pineal activity is cyclical, the pineal gland is also involved with maintaining basic **circadian rhythms**—daily changes in physiological processes that follow a regular day–night pattern.

Module 10.9 Review

a. Describe the location of the pineal gland.

b. How would longer hours of daylight, as during the summer, affect the production of melatonin?

c. List three functions of melatonin.

Diabetes mellitus is an endocrine disorder characterized by excessively high blood glucose levels

Diabetes Mellitus

Diabetes mellitus (mel-Ī-tus; *mellitum*, honey) is characterized by glucose concentrations that are high enough to overwhelm the reabsorption capabilities of the kidneys. (The presence of abnormally high blood glucose levels is called **hyperglycemia** [hī-per-glī-SĒ-mē-ah].) In diabetes mellitus, glucose appears in the urine (**glycosuria**; glī-kō-SOO-rē-a), and urine volume generally becomes excessive (**polyuria**). Diabetes mellitus can be caused by genetic abnormalities or mutations that result in inadequate insulin production, the synthesis of abnormal insulin molecules, or the production of defective insulin-receptor proteins.

Type 1 (insulin dependent) Diabetes

In **Type 1 (insulin dependent) diabetes**, the pancreatic beta cells do not produce enough insulin. Individuals with type 1 diabetes must take insulin to live—typically multiple injections daily, or continuous infusion through an insulin pump or other device. Type 1 diabetes accounts for only about 5–10 percent of diabetes cases and often develops in childhood.

Type 2 (non-insulin dependent) Diabetes

Type 2 (non-insulin dependent) diabetes is the most common form of diabetes mellitus. Most individuals with Type 2 diabetes produce normal amounts of insulin, at least initially, but their tissues do not respond properly—a condition known as insulin resistance. Type 2 diabetes is associated with obesity, and weight loss through diet and exercise can be an effective treatment, especially when coupled with drugs that alter rates of glucose synthesis and release by the liver.

1   Untreated diabetes mellitus disrupts metabolic activities throughout the body. Clinical problems arise because the tissues are experiencing an energy crisis—in essence, most of the tissues are responding as they would during chronic starvation, breaking down lipids and even proteins because they are unable to absorb glucose from their surroundings. The byproducts of this metabolic shift, molecules called ketone bodies, can cause body fluids to become dangerously acidic. This condition is called **diabetic ketoacidosis**. In addition, abnormal changes in blood vessel structure are particularly dangerous. An estimated 23.6 million people in the United States have some form of diabetes.

Clinical Problems Caused by Diabetes Mellitus

Proliferation of capillaries and hemorrhaging at the retina may cause partial or complete blindness. This condition is called **diabetic retinopathy.**

Degenerative blockages in cardiac circulation can lead to early heart attacks. For a given age group, heart attacks are three to five times more likely in individuals with diabetes than in people who do not have the condition.

Diabetic nephropathy—degenerative changes in the kidneys—can lead to kidney failure.

Abnormal blood flow to neural tissues is probably responsible for a variety of problems with peripheral nerves, including abnormal autonomic function. As a group, these disorders are termed **diabetic neuropathy**.

Blood flow to the distal portions of the limbs is reduced, damaging peripheral tissues. Reduced blood flow to the feet, for example, can lead to tissue death, ulceration, infection, and the loss of toes or a major portion of one or both feet.

Module 10.10 Review

a. Define diabetes mellitus.

b. Identify and describe the two types of diabetes mellitus.

c. Describe three clinical problems caused by diabetes mellitus.

The stress response is a predictable response to any significant threat to homeostasis

Any condition—physical or emotional—that threatens homeostasis is a form of **stress**. Many stresses are opposed by specific homeostatic adjustments. For example, a decline in body temperature leads to shivering or changes in blood flow, which can restore body temperature to normal. In addition, the body has a general response to stress that can occur while other, more specific responses are under way. Exposure to a wide variety of stress-causing factors will produce the same general pattern of hormonal and physiological adjustments. These responses are part of the **stress response**, also known as the **general adaptation syndrome (GAS)**. The stress response has three stages: the alarm, resistance, and exhaustion phases.

Alarm Phase ("Fight or Flight")

During the **alarm phase**, an immediate response to the stress occurs. The sympathetic division of the autonomic nervous system directs this response. In the alarm phase, (1) energy reserves are mobilized, mainly in the form of glucose, and (2) the body prepares to deal with the stress-causing factor through "fight or flight" responses. Epinephrine is the dominant hormone of the alarm phase, and its secretion accompanies a generalized sympathetic activation.

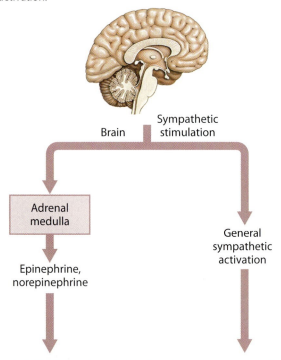

Brain | Sympathetic stimulation

Adrenal medulla

Epinephrine, norepinephrine

General sympathetic activation

Immediate Short-Term Responses to Crises

- Mobilize glycogen and lipid reserves to form glucose
- Increase mental alertness
- Increase energy use by all cells
- Change circulation patterns
- Reduce digestive activity and urine production
- Increase sweat gland secretion
- Increase heart rate and respiratory rate

Resistance Phase

If a stress lasts longer than a few hours, the individual enters the **resistance phase** of the stress response. Glucocorticoids are the dominant hormones of the resistance phase. Epinephrine, GH, and thyroid hormones are also involved. Energy demands in the resistance phase remain higher than normal, due to the combined effects of these hormones. Neural tissue has a high demand for energy, and neurons must have a reliable supply of glucose. Glycogen reserves are adequate to maintain normal glucose concentrations during the alarm phase but are nearly exhausted after several hours. The hormones of the resistance phase mobilize the body's metabolic reserves while shifting tissue metabolism away from glucose, sparing whatever glucose becomes available for use by neural tissues.

Brain

Sympathetic stimulation

ACTH

Pancreas

Renin-angiotensin system → Adrenal cortex

Mineralocorticoids (with ADH)   Glucocorticoids   Glucagon   Growth hormone

Long-Term Metabolic Adjustments

- Mobilize remaining energy reserves: Adipose tissue releases lipids; skeletal muscle releases amino acids
- Conserve glucose: Peripheral tissues (except neural) break down lipids to obtain energy
- Raise blood glucose concentrations: Liver synthesizes glucose from other carbohydrates, amino acids, and lipids
- Conserve salts and water; lose K^+ and H^+

Exhaustion Phase

The body's lipid reserves are sufficient to maintain the resistance phase for weeks or even months. But the resistance phase cannot be sustained indefinitely. When the resistance phase ends, homeostatic regulation breaks down and the **exhaustion phase** begins. Unless corrective actions are taken almost immediately, the failure of one or more organ systems will prove fatal. Mineral imbalances contribute to the existing problems with major systems. The production of aldosterone throughout the resistance phase conserves Na^+ at the expense of K^+. As the body's K^+ content declines, a variety of cells—notably neurons and muscle fibers—begin to malfunction. Although a single cause (such as heart failure) may be listed as the cause of death, the underlying problem is the body's inability to sustain the endocrine and metabolic adjustments of the resistance phase.

Factors That Can Trigger the Exhaustion Phase

- Exhaustion of lipid reserves and the breakdown of structural proteins as the body's primary energy source, damaging vital organs
- Infections that develop due to suppression of inflammation and the immune response, a secondary effect of the glucocorticoids that are essential to the metabolic activities of the resistance phase
- Cardiovascular damage from the ADH and aldosterone-related elevations in blood pressure and blood volume
- Inability of the adrenal cortex to continue producing glucocorticoids, which results in a failure to maintain acceptable blood glucose concentrations
- Failure to maintain adequate fluid and electrolyte balance

Module 10.11 Review

a. List the three phases of the stress response.

b. Describe the resistance phase.

c. During which phase of the stress response is there a collapse of vital systems?

1. Short answer

Identify each endocrine gland based on the major effects produced by its secreted hormone(s).

a _____ establishes day–night cycles

b _____ secretes insulin and regulates glucose uptake and utilization

c _____ regulates mineral balance, metabolic control, and resistance to stress

d _____ regulates secretions of adrenal cortex, thyroid gland, and reproductive organs

e _____ regulates metabolic rate and calcium levels in body fluids

f _____ two pairs of glands that play an important role in the response to decreasing calcium levels in body fluids

2. Matching

Match the following list of hormones with their effects in the body.

- parathyroid hormone (PTH)
- cortisol
- glucagon
- calcitonin
- aldosterone
- insulin
- thyroid hormones (T$_3$, T$_4$)
- growth hormone (GH)
- prolactin (PRL)
- oxytocin (OXT)
- antidiuretic hormone (ADH)
- calcitriol

a _____ stimulates mammary gland milk production

b _____ stimulates cell growth and cell division

c _____ stimulates greater oxygen and energy consumption

d _____ decreases water loss at the kidneys and causes vasoconstriction

e _____ increases blood glucose levels

f _____ stimulates uterine muscle contractions that promote labor and delivery

g _____ increases calcium and phosphate absorption by digestive tract

h _____ reduces inflammation

i _____ lowers blood glucose levels

j _____ stimulates increase in blood calcium levels

k _____ decreases blood calcium levels

l _____ stimulates the reabsorption of sodium and water by the kidneys

3. Matching

Match the following terms with the most closely related description.

- iodide ions
- glucocorticoids
- pituitary gland
- homeostasis threat
- sympathetic activation
- hyperglycemia
- FSH
- diabetes mellitus
- ACTH
- osmoreceptors

a _____ detect(s) changes in solute concentration in the blood

b _____ secrete(s) tropic hormones

c _____ gonadotropin

d _____ stimulates adrenal cortex to release steroid hormones

e _____ abnormally high blood glucose level

f _____ stress

g _____ hyperglycemia, glycosuria, and polyuria

h _____ alarm phase of GAS

i _____ resistance phase of GAS

j _____ essential for making thyroid hormones

4. Section Integration

Hormones may produce different, but complementary, effects in specific tissues and organs. Describe the differing effects that calcitriol and parathyroid hormone (PTH) have on tissues involved in calcium metabolism.

_____

_____

_____

Visual Outline with Key Terms
Summarize the content of each module using the terms in the order provided.

SECTION 1

Hormones and Intercellular Communication

- ○ endocrine system
- • hormones
- ○ nervous system
- ○ neurotransmitters

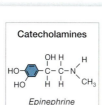

10.1

Hormones may be amino acid derivatives, peptides, or lipid derivatives

- • amino acid derivatives
- • peptide hormones
- • lipid derivatives
- • catecholamines
- • prohormones
- • short chain peptides
- • small proteins
- • eicosanoids

- • prostaglandins
- • steroid hormones
- • endocrine system
- ○ hypothalamus
- ○ pituitary gland
- ○ thyroid gland
- ○ adrenal glands
- ○ pancreas (pancreatic islets)

- ○ pineal gland
- ○ parathyroid glands

Catecholamines

Epinephrine

10.2

There are two major mechanisms by which hormones act on target cells

- • second messenger
- • cyclic-AMP (cAMP)
- ○ nonsteroid hormone

- ○ steroid hormones
- ○ thyroid hormones

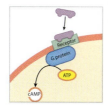

10.3

The hypothalamus exerts direct or indirect control over the activities of many different endocrine organs

- • hypothalamus
- • regulatory hormones
- • releasing hormones (RH)

- • inhibiting hormone (IH)
- • hypophyseal portal system

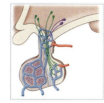

SECTION 2

Endocrine Organs

- • pituitary gland
- • thyroid gland
- • adrenal glands
- • pineal gland
- • parathyroid glands
- • pancreatic islets

10.4

The pituitary gland acts as a master gland

- • tropic hormones
- • antidiuretic hormone (ADH)
- • vasopressin
- ○ osmoreceptors
- • oxytocin (OXT)
- • thyroid-stimulating hormone (TSH)
- • adrenocorticotropic hormone (ACTH)
- • gonadotropins
- • follicle-stimulating hormone (FSH)

- • estrogens
- • inhibin
- • luteinizing hormone (LH)
- • androgens
- • growth hormone (GH)
- • glucose-sparing effect
- • prolactin (PRL)
- • melanocyte-stimulating hormone (MSH)

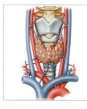

10.5

The thyroid gland hormones regulate metabolism or calcium ion levels

- • thyroid gland
- • thyroid follicles
- • colloid
- • thyroglobulin
- • C (clear) cells
- ○ parafollicular cells

- • calcitonin (CT)
- • T_3 (triiodothyronine) and T_4 (thyroxine)
- • thyroid-binding globulins (TBGs)

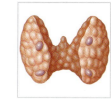

10.6

Parathyroid hormone, produced by the parathyroid glands, is the primary regulator of calcium ion levels in body fluids

- • parathyroid glands
- • parathyroid cells
- ○ oxyphils
- • parathyroid hormone (PTH)
- ○ calcitriol

10.7

The adrenal glands produce hormones involved in metabolic regulation

- • adrenal gland
- ○ retroperitoneal
- ○ adrenal cortex
- • corticosteroids
- • mineralocorticoids (MCs)
- • aldosterone
- • glucocorticoids (GCs)
- • cortisol
- • corticosterone

- • cortisone
- ○ glucose-sparing effect
- • anti-inflammatory effect
- ○ androgens
- ○ adrenal medulla
- ○ epinephrine (E)
- ○ norepinephrine (NE)

• = *Term boldfaced in this module*

10.8

The pancreatic islets secrete insulin and glucagon and regulate glucose use by most cells

- pancreas
- exocrine pancreas
- endocrine pancreas
- pancreatic islets
- alpha cells
- glucagon
- beta cells
- insulin
- pancreatic acini

10.10

Diabetes mellitus is an endocrine disorder characterized by excessively high blood glucose levels

- diabetes mellitus
- hyperglycemia
- glycosuria
- polyuria
- type 1 (insulin dependent) diabetes
- type 2 (non-insulin dependent) diabetes
- diabetic ketoacidosis
- diabetic retinopathy
- diabetic nephropathy
- diabetic neuropathy

10.9

The pineal gland of the epithalamus secretes melatonin

- pineal gland
- melatonin
- free radicals
- circadian rhythms

10.11

The stress response is a predictable response to any significant threat to homeostasis

- stress
- stress response
- general adaptation syndrome (GAS)
- alarm phase
- resistance phase
- exhaustion phase

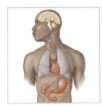

• = *Term boldfaced in this module*

CAREER PATHS

Sports Trainer

"Anatomy and physiology is essential. It teaches you the parts of the engine you're trying to improve, and it's the foundation on which all sports disciplines are based."

— **Andy Walshe**
Sports Trainer
Park City, Utah

Andy Walshe is the high-performance manager for the athletic program of the U.S. Ski and Snowboard Association. He works closely with over 150 athletes who are training to become Olympic champions. Andy supervises a staff of physiologists, nutritionists, dieticians, biomechanicists, and sports psychologists.

For trainers at this level, there is no such thing as a typical day. "This profession is full of variety. One day you may be running training camps, and the next you may be on the snow in South America shooting high-speed video for movement analysis."

In addition to supervising the training team, Andy develops elite sports performance models—frameworks that allow a trainer to analyze athletes in relation to their sport to determine their strengths and weaknesses. His expertise is in biomechanical physiological analysis. This requires a strong background in anatomy and physiology. "You need to understand and analyze movement patterns, and you can't do that without a good understanding of muscles and bones. And you

need to understand energy systems as well, so the physiology element is very important."

In fact, says Andy, for all of his staff, "anatomy and physiology is essential. It teaches you the parts of the engine you're trying to improve, and it's the foundation on which all sports disciplines are based."

Andy is one of a small group of elite-level trainers working with the best athletes in the world. Elite trainers bring to the job a combination of advanced academic work and practical experience. A master's degree is required for this type of job.

Andy worked first as a volunteer assistant physiologist, traveling with teams in an unpaid position, and worked his way up to his current job, which he's had for seven years. "To become a sports trainer, it's a good idea to have as much practical training as possible in a variety of sports, working with athletes and coaches. Volunteer your time. Do what it takes to get your foot in the door."

Sports trainers can work at the junior level, college level, or elite level (with Olympic or professional athletes). There are jobs in the United States and chances to work in other countries. At the elite level, most entry-level jobs are unpaid internships—opportunities to learn while showing the trainers what you can do. Andy suggests, "Find out what organizations are associated with sports that interest you. The U.S. Olympic Committee has internship programs, and there are many organizations to investigate for each sport."

Andy says the rewards of the job are great. "The athletes and the coaches are the best part of this job. It's great being a part of the team that helps athletes chase their dreams. Along the way, you make great friends."

And there's a larger goal as well. "Our number one goal is to get more kids into the sport. Of course, we're here to get our athletes medals, but the impact of those medals is that we get more kids involved in the sport."

For more information, contact the National Athletic Trainers' Association at http://www.nata.org.

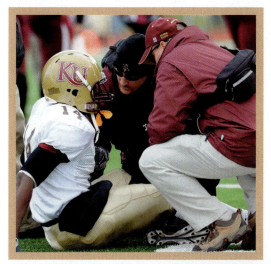

Think this is the career for you?
Key Stats:

- **Education and Training.** A bachelor's degree is required; most sports trainers also earn a master's or doctoral degree.

- **Licensure.** Certification from the Board of Certification, Inc. (BOC) is required in 47 states. Earning certification in states where it is not required is helpful when seeking job placement.

- **Earnings.** Earnings vary but the median annual wage is $41,600.

- **Expected Job Prospects.** Employment is expected to grow by an above average 37 % from 2008 to 2018.

Bureau of Labor Statistics, U.S. Department of Labor, *Occupational Outlook Handbook, 2010–11 Edition*, Athletic Trainers, on the Internet at http://www.bls.gov/oco/ocos294.htm (visited September 14, 2011).

Access more review material online in the Study Area at www.masteringaandp.com.

- Chapter guides
- Chapter quizzes
- Practice tests
- Art labeling activities
- Flashcards
- A glossary with pronunciations

- Practice Anatomy Lab™ (PAL™) 3.0 virtual anatomy practice tool
- Interactive Physiology® (IP) animated tutorials
- MP3 Tutor Sessions

PAL practice anatomy lab™   **For this chapter, follow these navigation paths in PAL:**

- Human Cadaver>Endocrine System
- Anatomical Models>Endocrine System
- Histology>Endocrine System

 For this chapter, go to these topics in the Endocrine System in IP:

- Orientation
- Endocrine System Review
- Biochemistry, Secretion, and Transport of Hormones
- The Actions of Hormones on Target Cells
- The Hypothalamic-Pituitary Axis
- Response to Stress

 For this chapter, go to this topic in the MP3 Tutor Sessions:

- Hypothalamic Regulation

Chapter 10 Review Questions

1. A fellow student in your anatomy and physiology class claims that the hypothalamus is exclusively a part of the CNS. What argument(s) would you make to document that the hypothalamus integrates the nervous system and endocrine system?

2. How does control of the adrenal medulla differ from control of the adrenal cortex?

3. What are two benefits of having a portal system connect the hypothalamus with the anterior lobe of the pituitary gland?

4. Julie is pregnant but is not receiving any prenatal care. She has a poor diet consisting mostly of fast food. She drinks no milk, preferring colas instead. How would this situation affect Julie's level of parathyroid hormone?

5. a. As a diagnostic clinician, which endocrine malfunctions would you associate with the following signs and symptoms in patients A and B?

   Patient A: (↑) urine output, (↑) blood sugar level, dehydration, (↑) thirst, (↑) body fluid acidity

   Patient B: Subcutaneous swelling, hair loss, dry skin, low body temperature, muscle weakness, and slowed reflexes

   b. Which glands are associated with each disorder?

For answers to all module, section, and chapter review questions, see the blue Answers tab at the back of the book.

11

Blood and Blood Vessels

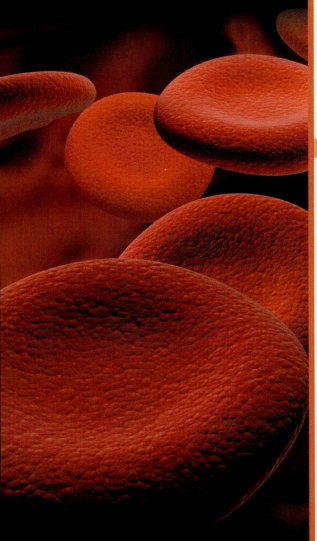

Blood

1 The **cardiovascular system** includes a fluid (blood), a series of conducting tubes (the blood vessels) that distribute the fluid throughout the body, and a pump (the heart) that keeps the fluid in motion.

The Components of the Cardiovascular System

BLOOD distributes oxygen, carbon dioxide, and blood cells; delivers nutrients and hormones; transports wastes; and assists in temperature regulation and defense against disease.

BLOOD VESSELS distribute blood around the body

Capillaries	permit diffusion between blood and interstitial fluids
Arteries	carry blood away from the heart to the capillaries
Veins	return blood from capillaries to the heart

THE HEART propels blood and maintains blood pressure

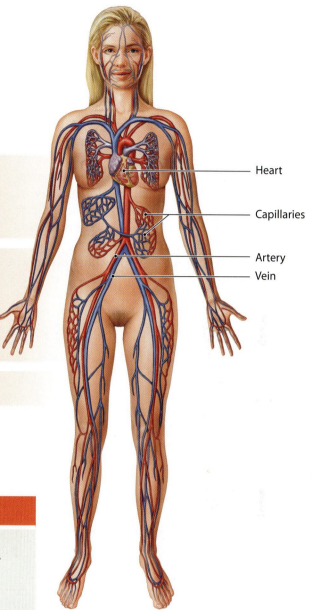

Heart

Capillaries

Artery

Vein

Functions of the Blood

- **Transports Dissolved Gases, Nutrients, Hormones, and Metabolic Wastes.** Blood carries oxygen from the lungs to peripheral tissues, and carbon dioxide from those tissues to the lungs. Blood distributes nutrients absorbed at the digestive tract or released from storage in adipose tissue or in the liver. It carries hormones from endocrine glands toward their target cells, and it absorbs and carries the wastes produced by cells to the kidneys for excretion.

- **Regulates the pH and Ion Composition of Interstitial Fluids.** Blood absorbs and neutralizes acids generated by active tissues, such as lactic acid produced by skeletal muscles. Diffusion between interstitial fluids and blood eliminates local deficiencies or excesses of ions, such as calcium or potassium.

- **Restricts Fluid Losses at Injury Sites.** Blood contains enzymes and other substances that respond to breaks in vessel walls by initiating the clotting process. A blood clot acts as a temporary patch that prevents further blood loss.

- **Defends against Toxins and Pathogens.** Blood transports white blood cells, specialized cells that migrate into peripheral tissues to fight infections or remove debris. Blood also transports antibodies, proteins that specifically attack invading microbes.

- **Stabilizes Body Temperature.** Blood absorbs the heat generated by active skeletal muscles and redistributes it to other tissues. If body temperature is already high, that heat will be lost across the skin surface. If body temperature is too low, the warm blood is directed to the brain and other temperature-sensitive organs.

2 This section examines the structure and function of blood, a fluid with multiple functions and remarkable properties. In adults, circulating blood provides each of the body's roughly 75 trillion cells with nutrients, oxygen, and chemical messengers, and a way of transporting wastes to their sites of removal. The blood also transports specialized cells that defend peripheral tissues from infection and disease. These services are so essential that a body region deprived of blood dies in a matter of minutes.

Blood is a fluid connective tissue containing plasma and formed elements

Blood is a fluid connective tissue with a unique composition. It consists of **plasma** (PLAZ-muh), a liquid matrix, and **formed elements** (cells and cell fragments). The cardiovascular system of an adult male contains 5–6 liters (5.3–6.4 quarts) of blood; that of an adult female contains 4–5 liters (4.2–5.3 quarts). The difference in blood volume between the sexes primarily reflects differences in average body size. After blood is removed for analysis or storage, the term **whole blood** is used to indicate that the blood composition has not been altered. The components of whole blood can, however, be separated, or **fractionated**, if only one component is of interest.

1 Plasma forms about 55 percent of the volume of whole blood. In many respects, the composition of plasma resembles that of interstitial fluid. This similarity exists because water, ions, and small solutes are continuously exchanged between plasma and interstitial fluids across the walls of capillaries.

PLASMA COMPOSITION	
Plasma proteins	7%
Other solutes	1%
Water	92%

Transports organic and inorganic molecules, formed elements, and heat

Whole blood

Plasma

55%
(Range: 46–63%)

+

Formed elements

45%
(Range: 37–54%)

The **hematocrit** (he-MAT-ō-krit) is the percentage of whole blood volume contributed by formed elements, 99.9 percent of which are red blood cells. In adult males, the normal hematocrit, or **packed cell volume (PCV)**, averages 47% (range: 40–54%); the average for adult females is 42% (range: 37–47%). The difference in hematocrit between the sexes primarily reflects the fact that androgens (male hormones) stimulate red blood cell production, whereas estrogens (female hormones) do not.

FORMED ELEMENTS	
Platelets	< .1%
White blood cells	< .1%
Red blood cells	99.9%

2 Formed elements are blood cells and cell fragments suspended in plasma, and they account for about 45 percent of the volume of whole blood.

Properties of Whole Blood

- Blood temperature is roughly 38°C (100.4°F), slightly above normal body temperature.

- Blood is five times as viscous as water—that is, five times as thick. Blood's high viscosity results from interactions among dissolved proteins, formed elements, and water molecules in the plasma.

- Blood is slightly alkaline, with an average pH of 7.40 (range: 7.35–7.45)

Plasma Proteins

Plasma proteins are in solution rather than forming insoluble fibers like those in other connective tissues, such as loose connective tissue or cartilage. On average, each 100 mL of plasma contains 7.6 g of protein, almost five times the concentration in interstitial fluid. The large size and globular shapes of most blood proteins usually prevent them from leaving the bloodstream. The liver synthesizes and releases more than 90 percent of all plasma proteins.

Albumins (al-BŪ-minz) make up about 60 percent of the plasma proteins. As the most abundant plasma proteins, these solutes are major contributors to the osmotic pressure of plasma.

Globulins (GLOB-ū-linz) account for approximately 35 percent of the proteins in plasma. Important plasma globulins include antibodies and transport globulins. Antibodies, also called **immunoglobulins** (i-mū-nō-GLOB-ū-linz), attack foreign proteins and pathogens. **Transport globulins** bind small ions, hormones, lipids, and other compounds.

Fibrinogen (fī-BRIN-ō-jen) functions in clotting and normally accounts for roughly 4 percent of plasma proteins. Under certain conditions, fibrinogen molecules interact to form large, insoluble strands of **fibrin** (FĪ-brin) that form the basic framework for a blood clot.

The plasma also contains active and inactive enzymes and hormones with varying concentrations.

Other Solutes

Other solutes are generally present in plasma in concentrations similar to those in the interstitial fluids. However, differences in the concentrations of nutrients and waste products can exist between arterial blood and venous blood.

Electrolytes: Normal extracellular ion composition is essential for vital cellular activities. The major plasma electrolytes are Na^+, K^+, Ca^{2+}, Mg^{2+}, Cl^-, HCO_3^-, HPO_4^-, and SO_4^{2-}.

Organic nutrients: Organic nutrients are used for ATP production, growth, and cell maintenance. This category includes lipids (fatty acids, cholesterol, and glycerides), carbohydrates (primarily glucose), and amino acids.

Organic wastes: Waste products are carried to sites of breakdown or excretion. Examples of organic wastes include urea, uric acid, creatinine, bilirubin, and ammonium ions.

Platelets

Platelets (PLĀT-lets) are small, membrane-bound cell fragments that contain enzymes and other substances important to the clotting process.

White Blood Cells

White blood cells (**WBCs**), or leukocytes (LOO-kō-sīts; *leukos*, white + *-cyte*, cell), participate in the body's defense mechanisms. There are five classes of leukocytes, each with slightly different functions that will be explored later in the chapter.

 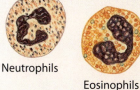

Neutrophils

Eosinophils

Basophils

Lymphocytes

Monocytes

Red Blood Cells

Red blood cells (**RBCs**), or erythrocytes (e-RITH-rō-sīts; *erythros*, red + *-cyte*, cell), are the most abundant blood cells. These specialized cells are essential for oxygen transport in the blood.

Module 11.1 Review

a. Define hematocrit.

b. Identify the two components making up whole blood, and list the composition of each.

c. During an infection, which components of blood would be elevated?

Red blood cells, the most common formed elements, contain hemoglobin

1 The human body contains an enormous number of red blood cells (RBCs). A standard blood test—the **red blood cell count**—reports the number of RBCs per microliter (μL) of whole blood. In adult males, 1 μL, or 1 cubic millimeter (mm^3), of whole blood contains 4.5–6.3 million RBCs; in adult females, 1 μL contains 4.2–5.5 million. A single drop of whole blood has about 260 million RBCs, and the blood of an average adult has 25 trillion RBCs. So RBCs make up about one-third of all cells in the human body.

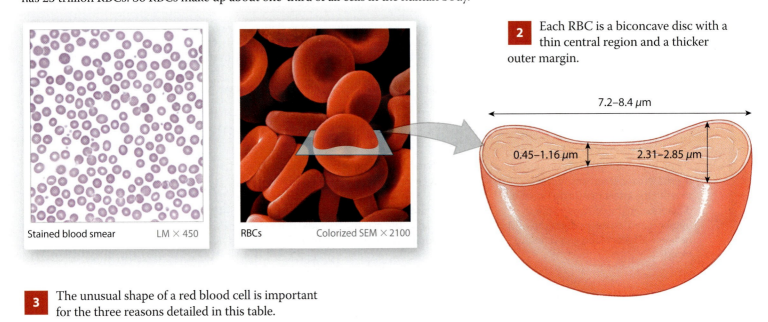

Stained blood smear   LM × 450

RBCs   Colorized SEM × 2100

2 Each RBC is a biconcave disc with a thin central region and a thicker outer margin.

7.2–8.4 μm

0.45–1.16 μm   2.31–2.85 μm

3 The unusual shape of a red blood cell is important for the three reasons detailed in this table.

Functional Aspects of Red Blood Cells

- **Large surface area-to-volume ratio.** Each RBC carries oxygen bound to hemoglobin, an intracellular protein, and that oxygen must be absorbed or released quickly as the RBC passes through the capillaries. The greater the surface area per unit volume, the faster the exchange of oxygen between the RBC's interior and the surrounding plasma.

- **RBCs can form stacks.** Like dinner plates, RBCs can form stacks that ease the flow through narrow blood vessels. An entire stack can pass along a blood vessel only slightly larger than the diameter of a single RBC, whereas individual cells would bump the walls, bang together, and form logjams that could restrict or prevent blood flow.

- **Flexibility.** Red blood cells are very flexible and can bend and flex when entering small capillaries and branches. By changing shape, individual RBCs can squeeze through capillaries as narrow as 4 μm.

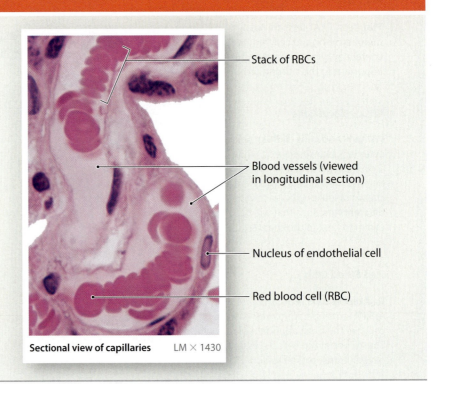

Stack of RBCs

Blood vessels (viewed in longitudinal section)

Nucleus of endothelial cell

Red blood cell (RBC)

Sectional view of capillaries   LM × 1430

4 A red blood cell is very different from the "typical cell" we discussed in Chapter 3. As our RBCs develop, they lose most of their organelles, including nuclei; they retain only the cytoskeleton. Because mature RBCs lack nuclei and ribosomes, they cannot divide or synthesize structural proteins or enzymes. As a result, RBCs cannot repair themselves, and their life span is normally less than 120 days. In effect, a developing RBC loses any organelle not directly associated with the transport of respiratory gases. That function is performed by molecules of **hemoglobin (Hb)**, which account for more than 95 percent of an RBC's intracellular proteins. The hemoglobin content of whole blood is reported in grams of Hb per deciliter (100 mL) of whole blood (g/dL). Normal ranges are 14–18 g/dL in males and 12–16 g/dL in females.

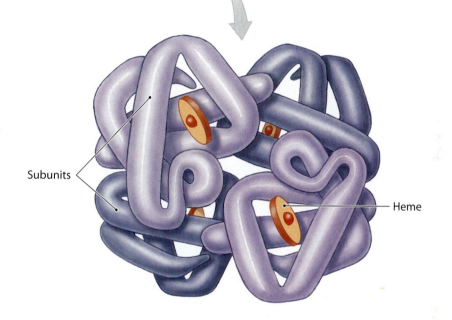

5 Hemoglobin has a complex structure with four globin subunits. Each of these subunits contains a single molecule of **heme**, a non-protein pigment complex.

Subunits

Heme

6 Each heme unit holds an iron ion in such a way that the iron can interact with an oxygen molecule, forming **oxyhemoglobin**, HbO_2. Blood containing RBCs filled with oxyhemoglobin is bright red. A hemoglobin molecule whose iron is not bound to oxygen is called **deoxyhemoglobin**. Blood containing RBCs filled with deoxyhemoglobin is dark red—almost burgundy. Roughly 98.5 percent of the oxygen carried by the blood travels through the bloodstream bound to Hb molecules inside RBCs; the rest is dissolved in the plasma.

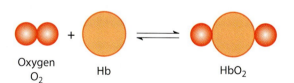

Oxygen O_2 +   Hb   ⇌ HbO_2

RBCs are continuously produced and recycled. About 1 percent of the circulating RBCs are replaced each day, and in the process approximately 3 million new RBCs enter the bloodstream each second! Such a rapid rate of replacement is necessary because a typical RBC has a relatively short life span. After it travels about 700 miles in 120 days, either its plasma membrane ruptures or it is engulfed by macrophages in the liver, spleen, or red bone marrow. The continuous elimination of RBCs usually goes unnoticed, as long as new ones enter the bloodstream at a comparable rate.

Module 11.2 Review

a. Why is it important for RBCs to have a large surface area-to-volume ratio?

b. Describe hemoglobin.

c. Compare oxyhemoglobin with deoxyhemoglobin.

Blood type is determined by the presence or absence of specific surface antigens on RBCs

Antigens are substances that can trigger a protective defense mechanism called an **immune response**. Most antigens are proteins, although some other types of organic molecules are antigens as well. The plasma membranes of your cells contain **surface antigens**, substances your immune system recognizes as "normal" or "self." In other words, your immune system ignores these substances rather than attacking them as "foreign."

1 Your **blood type** is a classification determined by specific surface antigens in RBC plasma membranes. These surface antigens are genetically determined. Your immune system ignores the surface antigens on your own RBCs. However, your plasma may contain **antibodies** that will attack the antigens on "foreign" RBCs. Although red blood cells have at least 50 kinds of surface antigens, three surface antigens are of particular importance: A, B, and Rh (or D). There are four blood types based on the presence or absence of the A and B surface antigens.

Type A	Type B	Type AB	Type O
Type A blood has RBCs with surface antigen A only.	**Type B** blood has RBCs with surface antigen B only.	**Type AB** blood has RBCs with both A and B surface antigens.	**Type O** blood has RBCs lacking both A and B surface antigens. In an emergency, Type O blood can be administered with little fear of a cross reaction. Type O individuals are often called **universal donors**.
Surface antigen A	Surface antigen B		
If you have Type A blood, your plasma contains anti-B antibodies, which will attack Type B surface antigens.	If you have Type B blood, your plasma contains anti-A antibodies, which will attack Type A surface antigens.	If you have Type AB blood, your plasma has neither anti-A nor anti-B antibodies. A person with Type AB blood can receive blood of any type; such individuals are often called **universal recipients**.	If you have Type O blood, your plasma contains both anti-A and anti-B antibodies.

2 If surface antigens on RBCs of one blood type are exposed to the corresponding antibodies from another blood type, the RBCs will **agglutinate**, or clump together. This process is called **agglutination**. The cells may also undergo **hemolysis** (RBC rupture). Such a **cross-reaction** is very dangerous because clumps and fragments of RBCs can plug small blood vessels in the kidneys, lungs, heart, or brain, damaging or destroying affected tissues. Accidental cross-reactions may occur if a person being treated for severe blood loss is accidentally given a transfusion of the wrong blood type.

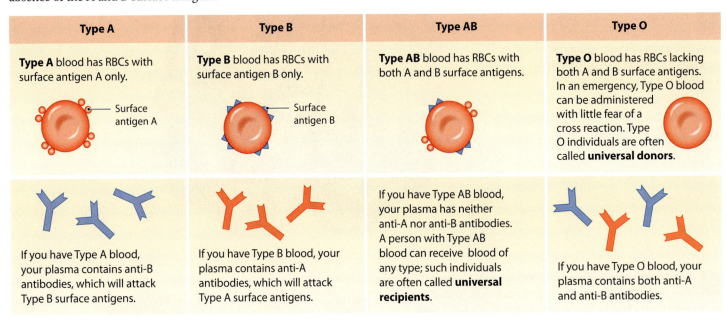

Surface antigens + Opposing antibodies → Agglutination (clumping) → Hemolysis

3 As this table indicates, the various blood types are not evenly distributed throughout the population. The term **Rh positive** (**Rh**$^+$) indicates the presence of the **Rh surface antigen**. The absence of this antigen is indicated as **Rh negative** (**Rh**$^-$). When the complete blood type is recorded, "Rh" is usually omitted and blood type is reported as O negative (O$^-$), A positive (A$^+$), and so on.

Differences in Blood Type Distribution

Population	Percentage with Each Blood Type				
	O	**A**	**B**	**AB**	**Rh**$^+$
U.S. (Average)	46	40	10	4	85
African American	49	27	20	4	95
Caucasian	45	40	11	4	85
Chinese American	42	27	25	6	100
Filipino American	44	22	29	6	100
Hawaiian	46	46	5	3	100
Japanese American	31	39	21	10	100
Korean American	32	28	30	10	100
Native North American	79	16	4	<1	100
Native South American	100	0	0	0	100
Australian Aborigine	44	56	0	0	100

4 Shown here are the results of blood typing tests on blood samples from four individuals. Drops of blood are mixed with solutions containing antibodies to the surface antigens A, B, and Rh. Clumping (agglutination) occurs when the sample contains the corresponding surface antigen or antigens. The blood type for each individual is shown at right. Because cross-reactions, or **transfusion reactions**, are so dangerous, care must be taken to ensure that the blood types of a blood donor and the recipient are **compatible**—that is, the donor's blood cells and the recipient's plasma will not cross-react.

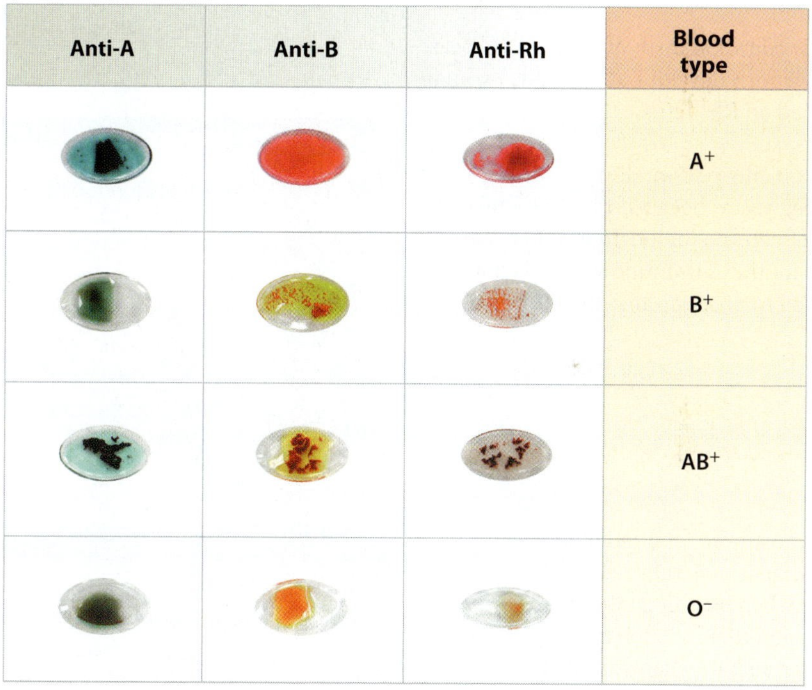

Anti-A	Anti-B	Anti-Rh	Blood type
			A$^+$
			B$^+$
			AB$^+$
			O$^-$

The presence of anti-A and/or anti-B antibodies is genetically determined and remains constant throughout life, regardless of whether the individual has ever been exposed to foreign RBCs. In contrast, the plasma of an Rh-negative individual does not necessarily contain anti-Rh antibodies. These antibodies are present only if the individual has been sensitized by previous exposure to Rh-positive RBCs.

Module 11.3 Review

a. What is the function of surface antigens on RBCs?

b. Which blood type(s) can be safely transfused into a person with Type O blood?

c. Why can't a person with Type A blood safely receive blood from a person with Type B blood?

Hemolytic disease of the newborn is an RBC-related disorder caused by a cross-reaction between fetal and maternal blood types

Genes controlling the presence or absence of any surface antigen in the plasma membrane of a red blood cell are provided by both parents. A child may inherit a combination of genes different from either parent and so may have a blood type different from that of either parent. During pregnancy, when fetal and maternal vascular systems are closely intertwined, the mother's antibodies may cross the placenta, attacking and destroying fetal RBCs. The resulting condition, called **hemolytic disease of the newborn (HDN)**, has many forms, some so mild as to remain undetected. Those involving the Rh surface antigen are quite dangerous, because unlike anti-A and anti-B antibodies, anti-Rh antibodies are able to cross the placenta and enter the fetal bloodstream.

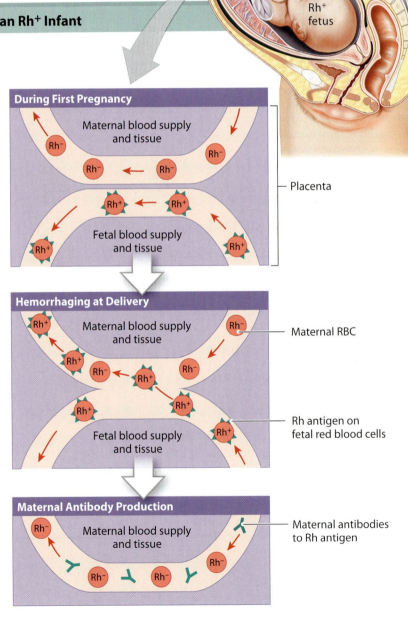

Rh⁻ mother

Rh⁺ fetus

Placenta

Maternal RBC

Rh antigen on fetal red blood cells

Maternal antibodies to Rh antigen

First Pregnancy of an Rh⁻ Mother with an Rh⁺ Infant

The most common form of hemolytic disease of the newborn develops after an Rh⁻ woman has carried an Rh⁺ fetus.

1 Problems seldom develop during a first pregnancy, because very few fetal cells enter the maternal bloodstream then, and thus the mother's immune system is not stimulated to produce anti-Rh antibodies.

During First Pregnancy
Maternal blood supply and tissue
Fetal blood supply and tissue

2 Exposure to fetal red blood cell antigens generally occurs during delivery, when bleeding takes place at the placenta and uterus. Such mixing of fetal and maternal blood can stimulate the mother's immune system to produce anti-Rh antibodies. This leads to **sensitization**, a reaction in which specific antibodies develop in response to an antigen.

Hemorrhaging at Delivery
Maternal blood supply and tissue
Fetal blood supply and tissue

3 Roughly 20 percent of Rh⁻ mothers who carried Rh⁺ children become sensitized within 6 months of delivery. Because the anti-Rh antibodies are not produced in significant amounts until after delivery, a woman's first infant is not affected.

Maternal Antibody Production
Maternal blood supply and tissue

Rh⁻
mother

Second Pregnancy of an Rh⁻ Mother with an Rh⁺ Infant

4 If a future pregnancy involves an Rh⁺ fetus, maternal anti-Rh antibodies produced after the first delivery cross the placenta and enter the fetal bloodstream. These antibodies destroy fetal RBCs, producing a dangerous anemia (RBC deficiency). The fetal demand for blood cells increases, and they begin leaving the red bone marrow and entering the bloodstream before completing their development. Because these immature RBCs are erythroblasts, HDN is also known as **erythroblastosis fetalis** (e-rith-rō-blas-TŌ-sis fē-TAL-is). Without treatment, the fetus will probably die before delivery or the infant will die shortly thereafter. Because the maternal antibodies remain active in the newborn for one to two months after delivery, the infant's entire blood volume may require replacement to remove the maternal anti-Rh antibodies, as well as the damaged RBCs. Fortunately, the mother's anti-Rh antibody production can be prevented if anti-Rh antibodies (available under the name RhoGAM) are administered to the mother in weeks 26–28 of pregnancy and during and after delivery. These antibodies destroy any fetal RBCs that cross the placenta before they can stimulate a maternal immune response. Because maternal sensitization does not occur, no anti-Rh antibodies are produced. In the United States, this relatively simple procedure has almost entirely eliminated HDN mortality caused by Rh incompatibilities.

Rh⁺
fetus

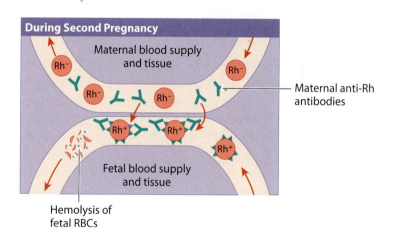

During Second Pregnancy

Maternal blood supply and tissue

Maternal anti-Rh antibodies

Fetal blood supply and tissue

Hemolysis of fetal RBCs

Module 11.4 Review

a. Define hemolytic disease of the newborn (HDN).

b. Why is RhoGAM administered to Rh⁻ mothers?

c. Does an Rh⁺ mother carrying an Rh⁻ fetus require a RhoGAM injection? Explain your answer.

White blood cells defend the body against pathogens, toxins, cellular debris, and abnormal or damaged cells

White blood cells (WBCs), or leukocytes, share the properties described below.

1 **Granular leukocytes** have abundant cytoplasmic granules that absorb histological stains, such as Wright stain or Giemsa stain.

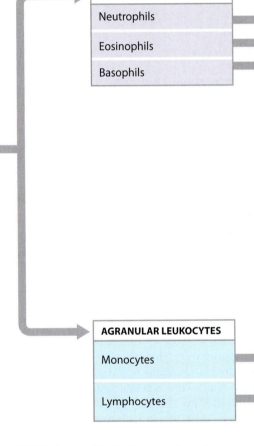

GRANULAR LEUKOCYTES
Neutrophils
Eosinophils
Basophils

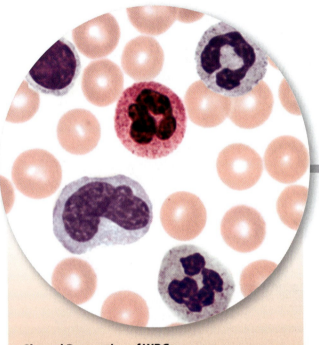

Shared Properties of WBCs

- WBCs circulate for only a short portion of their life span, using the bloodstream primarily to travel between organs and to rapidly reach areas of infection or injury. White blood cells spend most of their time migrating through loose and dense connective tissues throughout the body.

- All WBCs can migrate out of the bloodstream. When circulating white blood cells in the bloodstream become activated, they contact and adhere to the vessel walls and squeeze between adjacent endothelial cells to enter the surrounding tissue. This process is called **diapedesis** (*dia*, through; *pedesis*, a leaping).

- All WBCs are attracted to specific chemical stimuli. This characteristic, called **positive chemotaxis** (kē-mō-TAK-sis), guides WBCs to invading pathogens, damaged tissues, and other active WBCs.

- Neutrophils, eosinophils, and monocytes are capable of phagocytosis. These phagocytes can engulf pathogens, cell debris, or other materials. Macrophages are monocytes that have moved out of the bloodstream and have become actively phagocytic.

AGRANULAR LEUKOCYTES
Monocytes
Lymphocytes

2 **Agranular leukocytes** have few, if any, cytoplasmic granules that absorb histological stain.

White Blood Cells

Cell Type	Average Amount per μL	Appearance in a Stained Blood Smear	Functions
Neutrophils	4150 (range 1800–7300) Differential count: 50–70%	Round cell; nucleus lobed and may resemble a string of beads; cytoplasm contains large, pale inclusions	Phagocytic: engulf pathogens or debris in injured or infected tissues; release cytotoxic enzymes and chemicals
Eosinophils	165 (range 0–700) Differential count: 2–4%	Round cell; nucleus generally has two lobes; cytoplasm contains large granules that generally stain bright red	Phagocytic: engulf antibody-labeled materials; release cytotoxic enzymes; reduce inflammation; increase in number in allergic reactions and parasitic infections
Basophils	44 (range: 0–150) Differential count: <1%	Round cell; nucleus generally cannot be seen through dense, blue-stained granules in cytoplasm	Enter damaged tissues and release histamine and other chemicals that promote inflammation
Monocytes	456 (range: 200–950) Differential count: 2–8%	Very large cell; nucleus kidney bean- to horseshoe-shaped; abundant cytoplasm	Phagocytic: enter tissues and become macrophages; engulf pathogens or debris
Lymphocytes	2185 (range: 1500–4000) Differential count: 20–40%	Generally round cell, slightly larger than RBC; round nucleus; very little cytoplasm	Cells of lymphatic system; provide defense against specific pathogens or toxins

On average, there are 7000 WBCs per μL (range: 5000-10,000). A variety of conditions, including pathogenic infections, inflammation, and allergic reactions, cause characteristic changes in the populations of circulating WBCs. By examining a stained blood smear, we can obtain a **differential count** of the WBC population. The values reported indicate the number of each type of cell in a sample of 100 WBCs.

Module 11.5 Review

a. Identify the five types of white blood cells.

b. Which types of white blood cells would you find in the greatest numbers in an infected cut?

c. How do basophils respond during inflammation?

Formed elements are produced by stem cells derived from hemocytoblasts

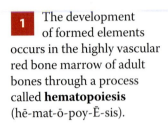

1 The development of formed elements occurs in the highly vascular red bone marrow of adult bones through a process called **hematopoiesis** (hē-mat-ō-poy-Ē-sis).

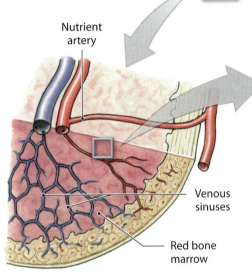

Nutrient artery

Venous sinuses

Red bone marrow

Hemocytoblasts

Hemocytoblasts (*hemo-*, blood + *cyte*, cell + *blastos*, precursor), or multipotent stem cells, are found in the red bone marrow of adults. Their divisions give rise to two types of stem cells that produce all formed elements.

Lymphoid Stem Cells

Lymphoid stem cells, which are responsible for the production of lymphocytes, originate in the red bone marrow. Some remain there, while others migrate from the bone marrow to other **lymphoid tissues**, including the thymus, spleen, and lymph nodes. As a result, lymphocytes are produced in these organs as well as in the red bone marrow.

Myeloid Stem Cells

Myeloid stem cells are stem cells in red bone marrow that divide to give rise to all types of formed elements other than lymphocytes.

2 The term "formed elements" is appropriate because platelets are cell fragments, rather than specialized cells. This table summarizes important information about platelets.

Structure and Function of Platelets

Appearance in a Stained Blood Smear	Average Number per μL	Function	Remarks
Platelets (PLĀT-lets) are flattened discs that appear round when viewed from above, and spindle-shaped in section or in a blood smear. 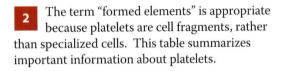	350,000 (range: 150,000–500,000)	Platelets clump together and stick to damaged vessel walls, and they release chemicals that stimulate blood clotting.	Platelets are continuously replaced. Each platelet circulates for 9–12 days before being removed by phagocytes, mainly in the spleen.

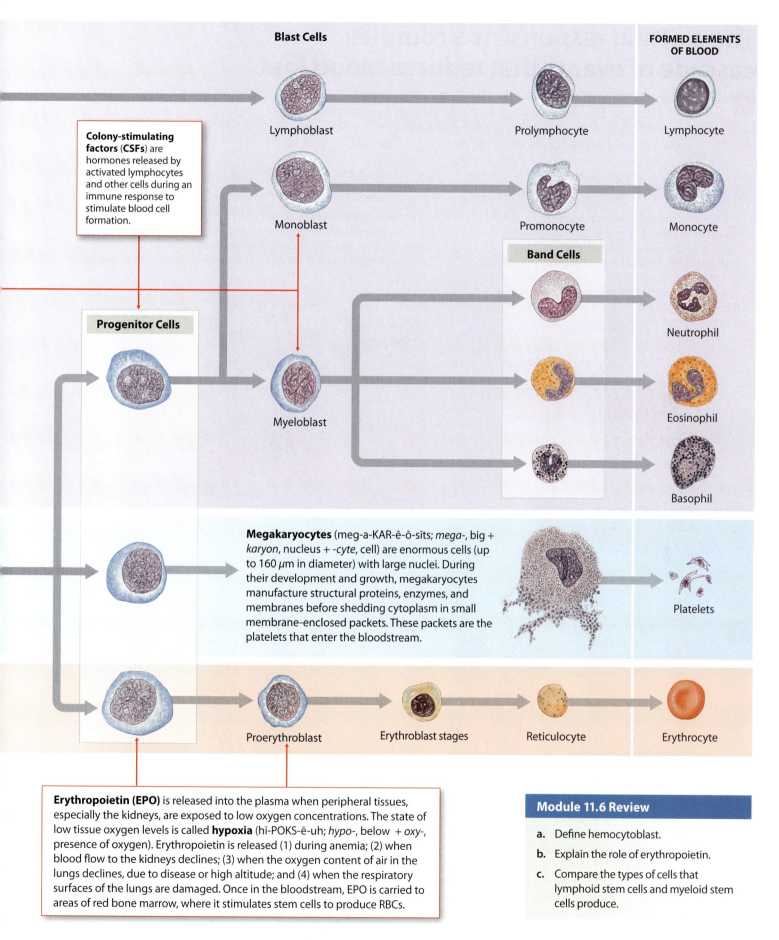

Blast Cells

FORMED ELEMENTS OF BLOOD

Lymphoblast

Prolymphocyte

Lymphocyte

Colony-stimulating factors (CSFs) are hormones released by activated lymphocytes and other cells during an immune response to stimulate blood cell formation.

Monoblast

Promonocyte

Monocyte

Band Cells

Progenitor Cells

Myeloblast

Neutrophil

Eosinophil

Basophil

Megakaryocytes (meg-a-KAR-ē-ō-sīts; *mega-*, big + *karyon*, nucleus + *-cyte*, cell) are enormous cells (up to 160 μm in diameter) with large nuclei. During their development and growth, megakaryocytes manufacture structural proteins, enzymes, and membranes before shedding cytoplasm in small membrane-enclosed packets. These packets are the platelets that enter the bloodstream.

Platelets

Proerythroblast

Erythroblast stages

Reticulocyte

Erythrocyte

Erythropoietin (EPO) is released into the plasma when peripheral tissues, especially the kidneys, are exposed to low oxygen concentrations. The state of low tissue oxygen levels is called **hypoxia** (hī-POKS-ē-uh; *hypo-*, below + *oxy-*, presence of oxygen). Erythropoietin is released (1) during anemia; (2) when blood flow to the kidneys declines; (3) when the oxygen content of air in the lungs declines, due to disease or high altitude; and (4) when the respiratory surfaces of the lungs are damaged. Once in the bloodstream, EPO is carried to areas of red bone marrow, where it stimulates stem cells to produce RBCs.

Module 11.6 Review

a. Define hemocytoblast.

b. Explain the role of erythropoietin.

c. Compare the types of cells that lymphoid stem cells and myeloid stem cells produce.

The clotting response is a complex cascade of events that reduces blood loss

1 The process of **hemostasis** (*haima*, blood + *stasis*, halt) is responsible for stopping the loss of blood through the walls of damaged vessels. At the same time, it establishes a framework for tissue repairs. Although usually divided into three phases, hemostasis is a complex **cascade**, a series of interactions in which one phase triggers the next.

Vascular Phase

The **vascular phase** of hemostasis lasts for roughly 30 minutes after the injury occurs. It is dominated by endothelial cell response and vascular spasm. Endothelial cells contract and expose the underlying basement membrane to the bloodstream. Vascular spasm is local contraction of smooth muscle within the vessel walls.

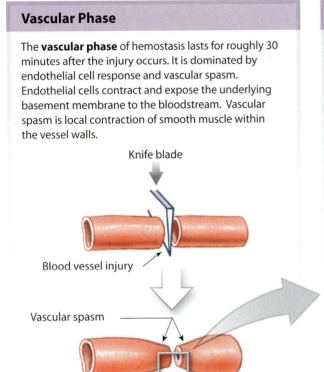

Knife blade

Blood vessel injury

Vascular spasm

Platelet Phase

The **platelet phase** of hemostasis begins with the attachment of platelets to sticky endothelial surfaces, to the basement membrane, to exposed collagen fibers, and to each other. Platelets release clotting factors that accelerate the number of aggregating platelets and stimulate further local vessel contractions.

Release of clotting factors

Plasma in vessel lumen

Platelet aggregation

Platelet adhesion to damaged vessel

Endothelium

Basement membrane

Vessel wall

Platelet plug may form

Cut edge of vessel wall

Contracted smooth muscle cells

2 If the clotting response is inadequately controlled, blood clots will begin to form in the bloodstream rather than at the site of an injury. These blood clots may not stick to the wall of the vessel, but continue to drift around until either plasmin digests them or they become lodged in a small blood vessel. A drifting blood clot is called an **embolus** (EM-bō-lus; *embolos*, plug). An embolus that becomes stuck in a blood vessel blocks circulation to the area downstream, killing the affected tissues. The sudden blockage is called an **embolism**, and the area of tissue damage caused by the circulatory interruption is called an **infarction**, often shortened to infarct. Infarctions in the brain are known as *strokes*; infarctions in the heart are called *myocardial infarctions*, or heart attacks.

Embolism blocks blood vessel

Embolus

3 A **thrombus** (*thrombos*, clot) is a blood clot attached to a vessel wall. It begins to form when platelets stick to the wall of an intact blood vessel. Often, the platelets are attracted to areas called *plaques*, where endothelial and smooth muscle cells within the vessel wall contain large quantities of lipids and constrict the vessel's diameter.

Thrombus

Plaque

Coagulation Phase

The **coagulation** (cō-ag-ū-LĀ-shun) **phase** of hemostasis does not start until 30 seconds or more after the vessel has been damaged. **Coagulation**, or blood clotting, involves a complex sequence of steps leading to the conversion of circulating fibrinogen into the insoluble protein **fibrin**. As the fibrin network grows, blood cells and additional platelets are trapped in the fibrous tangle, forming a blood clot that seals off the damaged portion of the vessel. **Procoagulants** (clotting factors) in the plasma play a key role in this phase. Important clotting factors include Ca^{2+} and 11 different proteins (identified by Roman numerals). Many clotting factors are proenzymes, which, when converted to active enzymes, direct essential reactions in the clotting response. The activation of one proenzyme commonly creates a chain reaction.

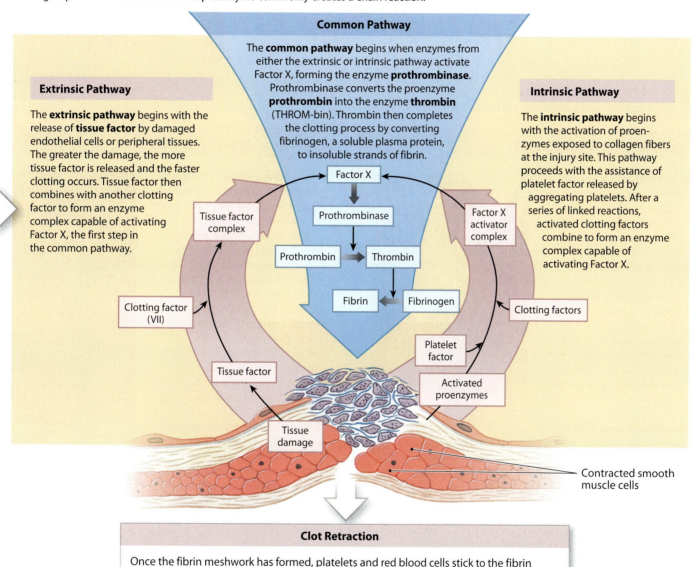

Common Pathway

The **common pathway** begins when enzymes from either the extrinsic or intrinsic pathway activate Factor X, forming the enzyme **prothrombinase**. Prothrombinase converts the proenzyme **prothrombin** into the enzyme **thrombin** (THROM-bin). Thrombin then completes the clotting process by converting fibrinogen, a soluble plasma protein, to insoluble strands of fibrin.

Extrinsic Pathway

The **extrinsic pathway** begins with the release of **tissue factor** by damaged endothelial cells or peripheral tissues. The greater the damage, the more tissue factor is released and the faster clotting occurs. Tissue factor then combines with another clotting factor to form an enzyme complex capable of activating Factor X, the first step in the common pathway.

Intrinsic Pathway

The **intrinsic pathway** begins with the activation of proenzymes exposed to collagen fibers at the injury site. This pathway proceeds with the assistance of platelet factor released by aggregating platelets. After a series of linked reactions, activated clotting factors combine to form an enzyme complex capable of activating Factor X.

Factor X

Tissue factor complex

Prothrombinase

Factor X activator complex

Prothrombin → Thrombin

Clotting factor (VII)

Fibrin ← Fibrinogen

Clotting factors

Platelet factor

Tissue factor

Activated proenzymes

Tissue damage

Contracted smooth muscle cells

Clot Retraction

Once the fibrin meshwork has formed, platelets and red blood cells stick to the fibrin strands. The platelets then contract, and the entire clot begins to undergo **clot retraction**, a process that continues over a period of 30–60 minutes and pulls the cut edges together.

As repairs proceed, the clot gradually dissolves. This process, called **fibrinolysis** (fī-bri-NOL-i-sis), begins with the activation of the proenzyme **plasminogen** by thrombin, produced by the common pathway, and **tissue plasminogen activator (t-PA)**, released by damaged tissues. The activation of plasminogen produces the enzyme **plasmin** (PLAZ-min), which erodes the foundation of the clot.

Module 11.7 Review

a. Define hemostasis and name its three phases.

b. Briefly describe the vascular, platelet, and coagulation phases of hemostasis.

c. Compare an embolus with a thrombus.

Blood disorders can be classified by their origins and the changes in blood characteristics

1 The procedure called **venipuncture** (VĒN-i-punk-chur; *vena*, vein + *punctura*, a piercing) is crucial in diagnosing blood disorders. Fresh whole blood is generally collected from a superficial vein, such as the median cubital vein on the anterior surface of the elbow. Venipuncture is commonly used because (1) superficial veins are easy to locate, (2) the walls of veins are thinner than those of comparably sized arteries, and (3) blood pressure in the venous system is relatively low, so the puncture wound seals quickly. The most common clinical procedures examine venous blood.

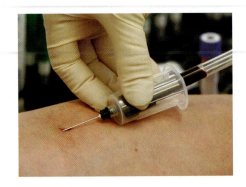

2 Blood disorders can have a variety of causes. The major categories of blood disorders are introduced here.

Nutritional Blood Disorders

In **iron deficiency anemia**, normal hemoglobin synthesis cannot occur because iron reserves or the dietary intake of iron is inadequate. Because developing RBCs cannot synthesize functional hemoglobin, they are unusually small and do not transport as much oxygen as a normal RBCs. Women are especially dependent on a normal dietary supply of iron because their iron reserves are about one-half that of a typical man.

A deficiency in **vitamin B$_{12}$** prevents normal stem cell divisions in the red bone marrow, which can result in **pernicious** (per-NISH-us) **anemia**. Fewer red blood cells are produced, and those that are produced are abnormally large and may develop bizarre shapes.

Calcium ions and **vitamin K** affect almost every aspect of the clotting process. Clotting pathways require Ca^{2+}, so any disorder that lowers plasma Ca^{2+} concentrations can impair blood clotting. Additionally, the liver requires adequate amounts of vitamin K for synthesizing four of the clotting factors, including prothrombin.

Congenital Blood Disorders

Sickle cell anemia results from a mutation affecting the amino acid sequence of the globin subunits of the Hb molecule. Affected RBCs take on a sickled shape when they release bound oxygen. This makes the RBCs fragile and easily damaged. Moreover, a sickled RBC can become stuck in a narrow capillary. A circulatory blockage results, and nearby tissues become starved for oxygen. To develop sickle cell anemia, an individual must have two copies of the sickling gene—one from each parent. If only one sickling gene is present, the individual has the **sickling trait**. In such cases, most of the hemoglobin is of the normal form, and the RBCs function normally. However, having the sickling trait gives an individual some resistance to malaria, a deadly mosquito-borne parasitic disease. Infection of RBCs by the parasites induces sickling, and sickled cells containing the parasites are engulfed and destroyed by macrophages.

Normal and sickled RBCs

Hemophilia (hē-mō-FĒ-lē-a) is an inherited bleeding disorder. About 1 person in 10,000 is a hemophiliac, and of those, 80–90 percent are males. In most cases, hemophilia is caused by the reduced production of a single clotting factor. The severity of hemophilia varies, depending on how little clotting factor is produced. In severe cases, extensive bleeding occurs with relatively minor contact, and bleeding occurs at joints and around muscles.

The **thalassemias** (thal-ah-SĒ-mē-uhs) are a diverse group of inherited blood disorders caused by an inability to produce adequate amounts of normal protein subunits of hemoglobin. The severity of different types of thalassemias depends on which and how many protein subunits are abnormal.

Infections of the Blood

Blood is normally free of microorganisms, but they can enter the blood through a wound or infection. **Bacteremia** is a condition in which bacteria circulate in blood, but do not multiply there. **Viremia** is a similar condition associated with viruses. **Sepsis** (SEP-sis) is a widespread pathogenic infection of body tissues. Sepsis of the blood, or **septicemia** (formerly known as "blood poisoning"), results when pathogens multiply in the blood and spread throughout the body.

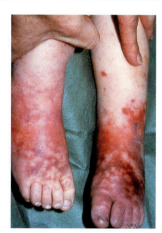

Infant with septicemia from meningococcal bacteria

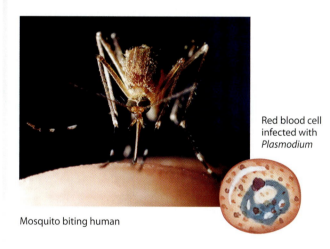

Mosquito biting human

Red blood cell infected with *Plasmodium*

Malaria is a parasitic disease caused by several species of the protozoan *Plasmodium*. It is one of the most severe diseases in tropical countries, killing 1.5–3 million people per year, of which up to half are children under the age of 5. Malaria is transmitted by a mosquito. The parasite initially infects liver cells, then enlarges and fragments into smaller forms that infect red blood cells. Periodically, at intervals of 2–3 days, all of the infected RBCs rupture simultaneously and release more parasites that infect additional RBCs. The release and reinfection of RBCs corresponds to the cycles of fever and chills that characterize malaria. Dead RBCs can block blood vessels leading to vital organs, such as the kidney and brain, resulting in tissue death.

Cancers of the Blood

The **leukemias** are cancers of blood-forming tissues. The cancerous cells of leukemia do not form a compact tumor, but instead are spread throughout the body from their origin in red bone marrow. Both of the two types of leukemia— myeloid and lymphoid—are characterized by elevated levels of circulating WBCs. **Myeloid leukemia** is characterized by the presence of abnormal granulocytes (neutrophils, eosinophils, and basophils) or other cells of the red bone marrow. **Lymphoid leukemia** involves lymphocytes and their stem cells. The first symptoms appear when immature and abnormal white blood cells appear in the circulation. Untreated leukemia is fatal.

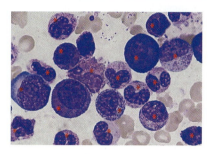

Abnormal WBCs (∗) seen in a blood smear of a patient with myeloid leukemia

Degenerative Blood Disorders

In **disseminated intravascular coagulation (DIC)**, bacterial toxins activate several steps in the coagulation process that converts fibrinogen to fibrin within the circulating blood. Although much of the fibrin is removed by phagocytes or is dissolved by plasmin, small clots may block small vessels and damage nearby tissues. If the liver cannot produce enough circulating fibrinogen to keep pace with the rate at which fibrinogen is being removed, clotting declines and uncontrolled bleeding may occur.

Module 11.8 Review

a. Define venipuncture.

b. Identify the two types of leukemia.

c. Compare iron deficiency anemia with pernicious anemia.

1. Concept map

Use the terms in the following list to complete the whole blood concept map.

- globulins
- electrolytes, glucose, urea
- protein
- albumins
- leukocytes
- fibrinogen
- solutes
- basophils
- erythrocytes
- formed elements
- plasma
- eosinophils
- lymphocytes
- water
- platelets
- monocytes
- neutrophils

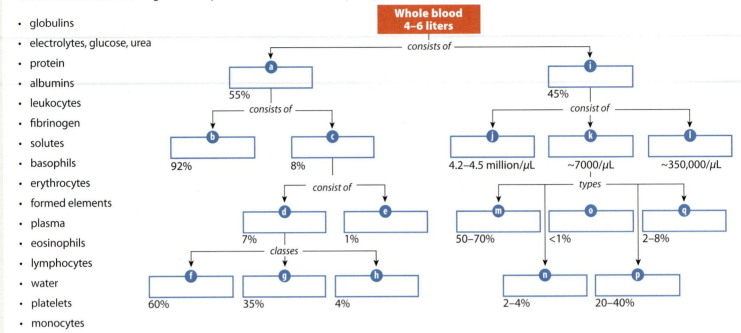

2. Matching

Match the following terms with the most closely related description.

- liquid matrix
- transport protein
- RBCs rupture
- venipuncture
- drifting blood clot
- mature RBCs
- pigment complex
- platelets
- erythropoietin
- cross-reaction
- monocytes
- lymphocytes

a	_____	embolus
b	_____	lack nuclei
c	_____	plasma
d	_____	macrophages
e	_____	globulin
f	_____	agglutination
g	_____	specific immunity
h	_____	malaria
i	_____	median cubital vein
j	_____	heme
k	_____	hormone
l	_____	blood clotting

3. Section integration

Why are mature red blood cells in humans incapable of protein synthesis and mitosis?

_____

_____

The Functional Anatomy of Blood Vessels

Blood flows through a network of blood vessels that extend between the heart and peripheral tissues. Those blood vessels can be organized into a **pulmonary circuit**, which carries blood to and from the gas exchange surfaces of the lungs, and a **systemic circuit**, which transports blood to and from the rest of the body. Each circuit begins and ends at the heart, and blood travels through these circuits in sequence. Thus, blood returning to the heart from the systemic circuit must complete the pulmonary circuit before reentering the systemic circuit. Blood is carried away from the heart by **arteries**, or efferent vessels, and returns to the heart by way of **veins**, or afferent vessels. Microscopic thin-walled vessels called **capillaries** interconnect the smallest arteries and the smallest veins. Capillaries are called exchange vessels, because their thin walls permit the exchange of nutrients, dissolved gases, and waste products between the blood and surrounding tissues.

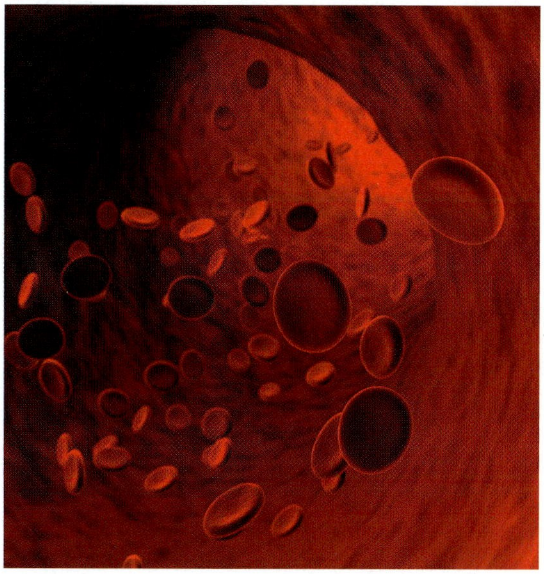

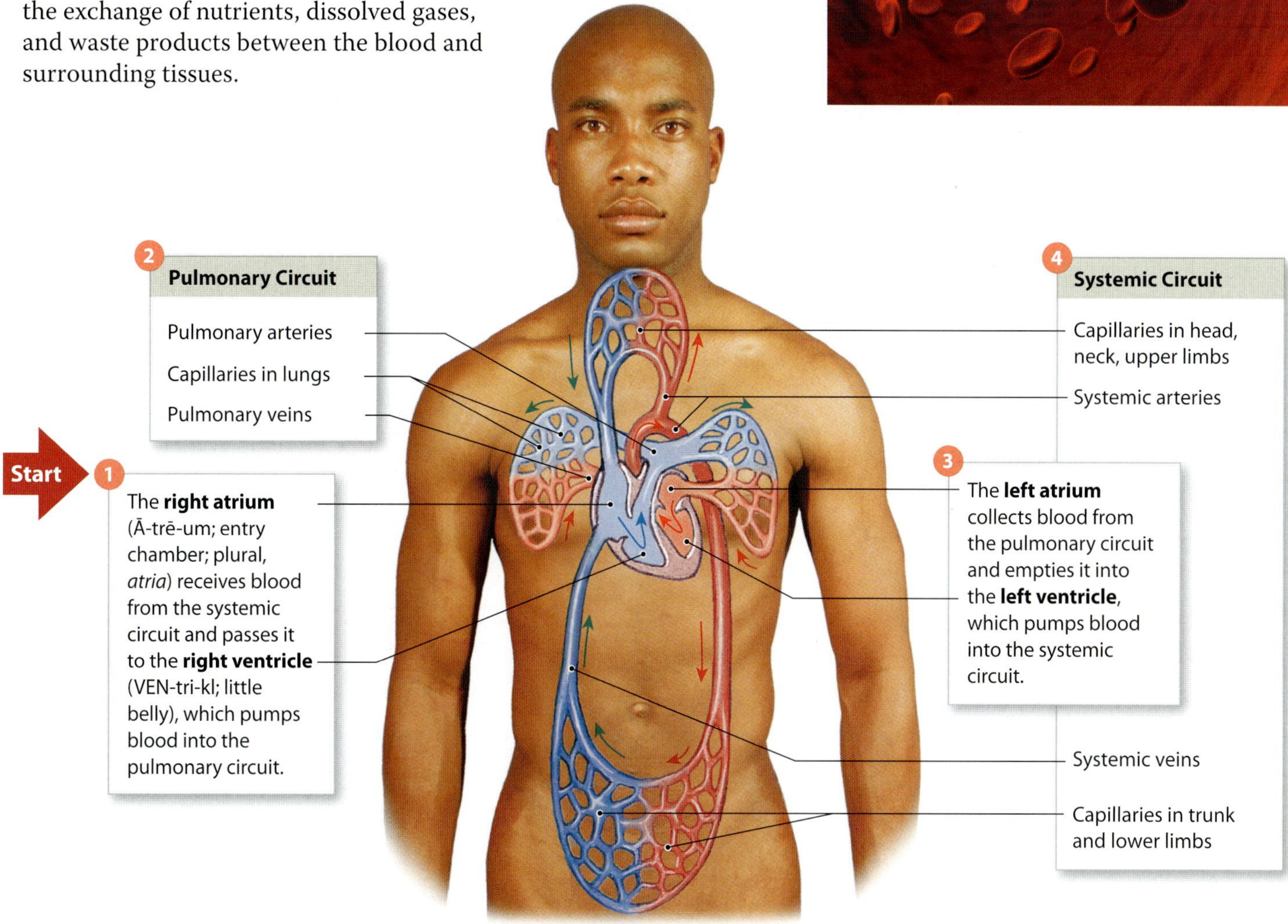

2
Pulmonary Circuit

Pulmonary arteries

Capillaries in lungs

Pulmonary veins

4
Systemic Circuit

Capillaries in head, neck, upper limbs

Systemic arteries

Start

1 The **right atrium** (Ā-trē-um; entry chamber; plural, *atria*) receives blood from the systemic circuit and passes it to the **right ventricle** (VEN-tri-kl; little belly), which pumps blood into the pulmonary circuit.

3 The **left atrium** collects blood from the pulmonary circuit and empties it into the **left ventricle**, which pumps blood into the systemic circuit.

Systemic veins

Capillaries in trunk and lower limbs

In this section we will take a close look at the structure and distribution of the major blood vessels in the body.

Arteries and veins differ in the structure and thickness of their walls

1 The walls of arteries and veins contain three distinct layers: the tunica intima, tunica media, and tunica externa.

The **tunica intima** (IN-ti-muh), or tunica interna, is the innermost layer of a blood vessel. This layer includes the endothelial lining and an underlying layer of connective tissue containing elastic fibers. In arteries, the outer margin of the tunica intima contains a thick layer of elastic fibers called the **internal elastic membrane**.

The **tunica media**, the middle layer, contains concentric sheets of smooth muscle in a framework of loose connective tissue. When these smooth muscles contract, the vessel decreases in diameter; this is called **vasoconstriction**. When the smooth muscles relax, the diameter increases; this is called **vasodilation**. An external elastic membrane is found only in arteries.

The **tunica externa** (eks-TER-nuh) or tunica adventitia, the outermost layer of a blood vessel, is a connective tissue sheath. In arteries, this layer contains collagen fibers with scattered bands of elastic fibers. In veins, it is generally thicker than the tunica media and contains networks of elastic fibers and bundles of smooth muscle cells. The connective tissue fibers of the tunica externa typically blend into those of adjacent tissues, stabilizing and anchoring the blood vessel.

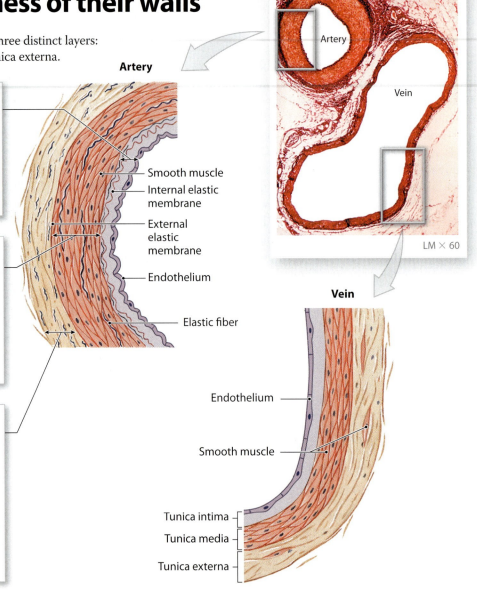

Artery

Smooth muscle
Internal elastic membrane
External elastic membrane
Endothelium
Elastic fiber

Artery

Vein

LM × 60

Vein

Endothelium
Smooth muscle

Tunica intima
Tunica media
Tunica externa

Features of Typical Arteries and Veins

Feature	Typical Artery	Typical Vein
General appearance in sectional view	Usually round, with relatively thick wall	Usually flattened or collapsed, with relatively thin wall
Tunica Intima		
Endothelium	Usually rippled, due to vessel constriction	Often smooth
Internal elastic membrane	Present	Absent
Tunica Media	Thick, dominated by smooth muscle cells and elastic fibers	Thin, dominated by smooth muscle cells and collagen fibers
External elastic membrane	Present	Absent
Tunica Externa	Collagen and elastic fibers	Collagen, elastic fibers, and smooth muscle cells

Large Vein

Large veins include the superior and inferior venae cavae and their tributaries. All three vessel wall layers are present in all large veins.

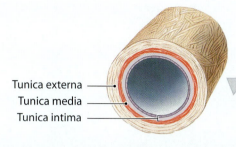

Tunica externa
Tunica media
Tunica intima

Elastic Artery

Elastic arteries are large vessels that transport blood away from the heart. The pulmonary trunk and aorta, as well as their major arterial branches, are elastic arteries. These vessels are capable of stretching and recoiling as the heart beats and arterial pressures change.

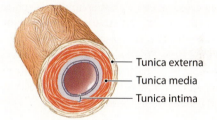

Internal elastic membrane
Tunica intima
Tunica media
Tunica externa

Medium-sized Vein

Medium-sized veins have a thin tunica media that contains smooth muscle cells and collagen fibers. The thickest layer is the tunica externa, which contains elastic and collagen fibers.

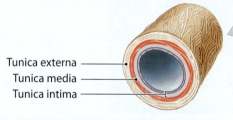

Tunica externa
Tunica media
Tunica intima

Muscular Artery

Muscular arteries, or medium-sized arteries, distribute blood to the body's skeletal muscles and internal organs.

Tunica externa
Tunica media
Tunica intima

Venule

Venules are small veins that collect blood from capillary beds. Venules are the smallest venous vessels.

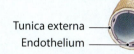

Tunica externa
Endothelium

Arteriole

Arterioles have a tunica media that consists of only one or two layers of smooth muscle cells.

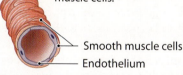

Smooth muscle cells
Endothelium

Capillaries

Capillaries are the only blood vessels whose walls permit exchange between the blood and the surrounding interstitial fluids. Because capillary walls are thin, diffusion distances are short, so exchange can occur quickly. The pores allow rapid exchange of water and solutes between plasma and the interstitial fluid.

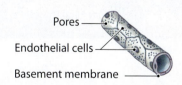

Pores
Endothelial cells
Basement membrane

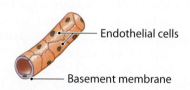

Endothelial cells
Basement membrane

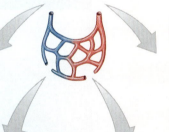

---

2 There are five general classes of blood vessels: arteries, arterioles, capillaries, venules, and veins. **Arteries** carry blood away from the heart. As arteries enter peripheral tissues they branch repeatedly, and the branches decrease in diameter. The smallest arterial branches are called arterioles (ar-TĒR-ē-ōls). From the arterioles, blood moves into capillaries, where diffusion occurs between blood and interstitial fluid. From the capillaries, blood enters small veins, or venules (VEN-ūls), which unite to form larger **veins** that return blood to the heart.

Module 11.9 Review

a. List the five general classes of blood vessels.

b. Describe a capillary.

c. A cross section of tissue shows several small, thin-walled vessels with very little smooth muscle tissue in the tunica media. Which type of vessels are these?

Arteriosclerosis can restrict blood flow and damage vital organs

Arteriosclerosis (ar-tē-rē-ō-skle-RŌ-sis; *arterio-*, artery + *sklerosis*, hardness) is a thickening and toughening of arterial walls. This condition may not sound life-threatening, but complications related to arteriosclerosis account for roughly half of all deaths in the United States. The effects of arteriosclerosis are varied. For example, arteriosclerosis of arteries within the heart is responsible for coronary artery disease (CAD) and potential heart attacks, and arteriosclerosis of arteries supplying the brain can lead to strokes.

1 **Atherosclerosis** (ath-er-ō-skler-Ō-sis; *athero-*, fatty degeneration) is the formation of lipid deposits in the arterial tunica media associated with damage to the endothelial lining. Atherosclerosis, the most common form of arteriosclerosis, tends to develop in people whose blood contains elevated levels of plasma lipids—specifically, cholesterol. The result is an atherosclerotic **plaque**, a fatty mass of tissue that projects into the lumen of the vessel and restricts blood flow. Elderly individuals—especially elderly men—are most likely to develop atherosclerotic plaques. Other important risk factors for atherosclerosis include high blood pressure and cigarette smoking.

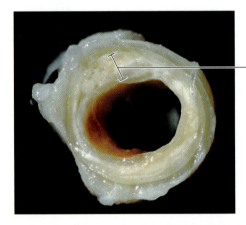

Plaque deposit in vessel wall

2 Plaques can be treated by removing the damaged segment of the vessel and replacing it (often with a superficial vein removed from the leg), but such surgery can be difficult and dangerous. In **balloon angioplasty** (AN-jē-ō-plas-tē; *angeion*, vessel), the tip of a catheter containing an inflatable balloon is inserted into an artery. Once in position, the balloon is inflated, pressing the plaque against the vessel walls. This photo shows the catheter within an artery of a cadaver; the artery has been opened to show the orientation and relative sizes of the catheter and artery. Balloon angioplasty is most effective in treating small, soft plaques. Several factors make this a highly attractive treatment: (1) the mortality rate during surgery is only about 1 percent; (2) the success rate is over 90 percent; and (3) the procedure can be performed on an outpatient basis.

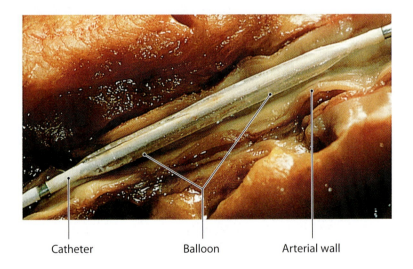

Catheter   Balloon   Arterial wall

Module 11.10 Review

a. Compare arteriosclerosis with atherosclerosis.

b. Identify risk factors for the development of atherosclerosis.

c. Describe balloon angioplasty.

Capillaries function as part of an interconnected network called a capillary bed

1 A **capillary bed** contains several relatively direct connections between arterioles and venules. The wall in the initial part of such a passageway possesses smooth muscle capable of changing its diameter. This segment is called a **metarteriole** (met-ar-TĒR-ē-ōl). The rest of the passageway resembles a typical capillary in structure. Although blood normally flows at a constant rate from the arteriole to the venule across the capillary bed, the flow within each capillary is quite variable. Precapillary sphincters adjust the flow of blood to each capillary by alternately contracting and relaxing about a dozen times per minute. Such rhythmic changes in vessel diameter are called **vasomotion**. Vasomotion causes the blood flow within the associated capillary to occur in pulses rather than as a steady and constant stream.

A capillary bed may receive blood from more than one artery. These **collateral arteries** enter the region and fuse before giving rise to arterioles. The fusion of two collateral arteries that supply a capillary bed is an example of an **arterial anastomosis**. An arterial anastomosis is in effect an insurance policy: If one artery is compressed or blocked, capillary circulation will continue.

A single arteriole generally gives rise to dozens of capillaries that empty into several venules.

The entrance to each capillary is guarded by a band of smooth muscle called a **precapillary sphincter**. Contraction or relaxation of the smooth muscle cells changes the diameter of the capillary entrance, thereby controlling the flow of blood into the capillary.

An **arteriovenous** (ar-tēr-ē-ō-VĒ-nus) **anastomosis** is a direct connection between an arteriole and a venule. When this anastomosis is dilated, blood will bypass the capillary bed and flow directly into the venous circulation. The pattern of blood flow through these anastomoses is regulated primarily by sympathetic innervation under the control of the cardiovascular centers of the medulla oblongata.

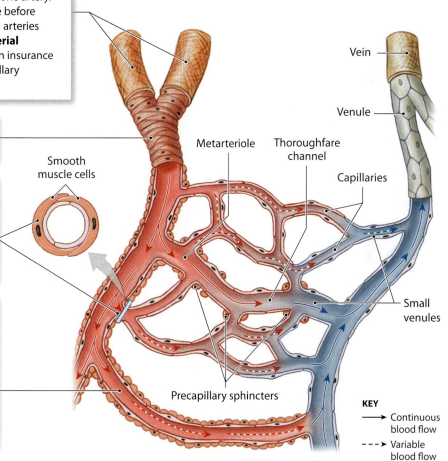

Smooth muscle cells

Metarteriole

Thoroughfare channel

Capillaries

Vein

Venule

Small venules

Precapillary sphincters

KEY

→ Continuous blood flow

---→ Variable blood flow

Module 11.11 Review

a. What is the role of precapillary sphincters?

b. Define vasomotion.

c. Describe blood flow through an arteriovenous anastomosis.

The venous system has low pressures and contains almost two-thirds of the body's blood volume

The arterial system is a high-pressure system: Almost all the force developed by the heart is required to push blood along the network of arteries and through miles of capillaries. Blood pressure in a peripheral venule is only about 10 percent of that in the ascending aorta, and pressures continue to fall along the venous system.

1 The blood pressure in venules and medium-sized veins is so low that it cannot overcome the force of gravity. In the limbs, veins of this size contain **valves**, folds of the tunica intima that project from the vessel wall and point in the direction of blood flow. Venous valves permit blood flow in one direction only to prevent the backflow of blood toward the capillaries. If the walls of the veins near the valves weaken or become stretched and distorted, the valves may not work properly. Blood then pools in the veins, which become grossly distended. The effects range from mild discomfort and a cosmetic problem, as in superficial **varicose veins** in the thighs and legs, to painful distortion of adjacent tissues, as in **hemorrhoids**.

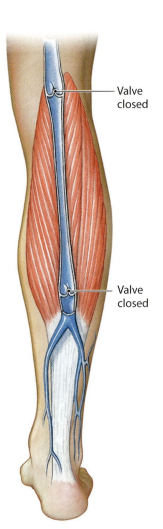

Valve closed

Valve closed

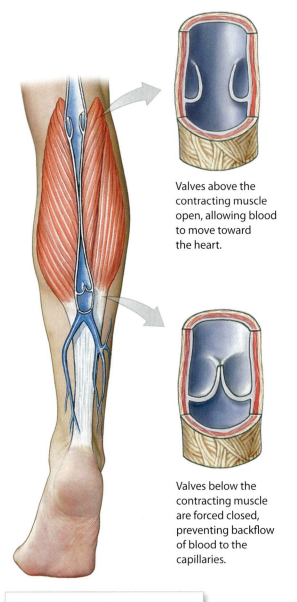

Valves above the contracting muscle open, allowing blood to move toward the heart.

Valves below the contracting muscle are forced closed, preventing backflow of blood to the capillaries.

When you are standing, venous blood from your feet must overcome the pull of gravity to ascend to the heart. Valves compartmentalize the blood within the veins, thereby dividing the blood between the compartments.

Any contraction of the surrounding skeletal muscles squeezes the blood toward the heart. Although you are probably not aware of it, when you stand, rapid cycles of contraction and relaxation are occurring within your leg muscles, helping to push blood toward the trunk.

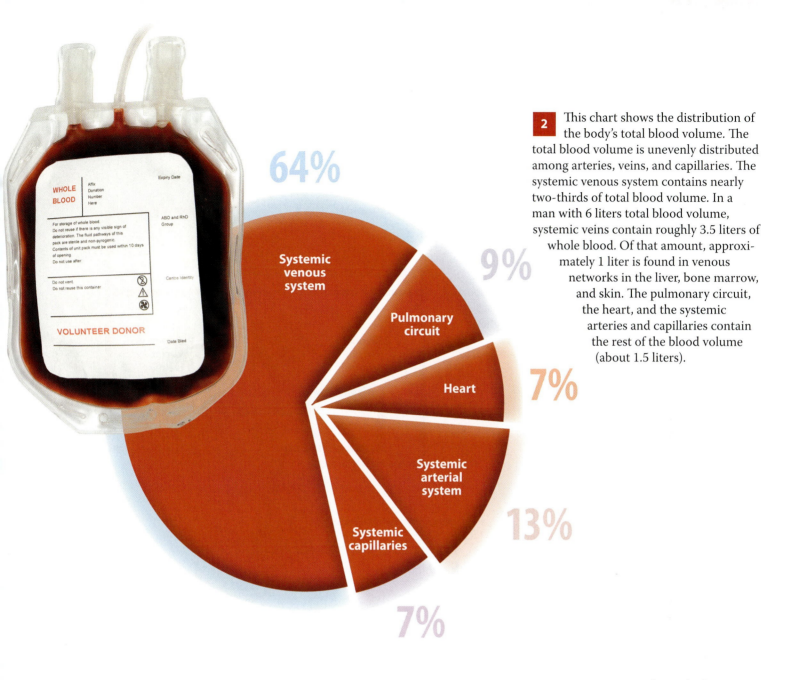

64%

Systemic venous system

9%

Pulmonary circuit

Heart

7%

Systemic arterial system

13%

Systemic capillaries

7%

WHOLE BLOOD

Affix
Donation
Number
Here

Expiry Date

ABO and RhD Group

For storage of whole blood.
Do not reuse if there is any visible sign of deterioration. The fluid pathways of this pack are sterile and non-pyrogenic.
Contents of unit pack must be used within 10 days of opening.
Do not use after:

Do not vent.
Do not reuse this container.

Centre Identity

VOLUNTEER DONOR

Date Bled

2 This chart shows the distribution of the body's total blood volume. The total blood volume is unevenly distributed among arteries, veins, and capillaries. The systemic venous system contains nearly two-thirds of total blood volume. In a man with 6 liters total blood volume, systemic veins contain roughly 3.5 liters of whole blood. Of that amount, approximately 1 liter is found in venous networks in the liver, bone marrow, and skin. The pulmonary circuit, the heart, and the systemic arteries and capillaries contain the rest of the blood volume (about 1.5 liters).

3 If serious hemorrhaging occurs, the body maintains blood volume within the arterial system at near-normal levels by reducing the volume of blood in the venous system. In the mechanism involved, the vasomotor center in the medulla oblongata stimulates sympathetic nerves innervating smooth muscle cells in the walls of medium-sized veins. The resulting contraction produces **venoconstriction**, reducing the diameter of the veins and the amount of blood contained in the venous system. In addition, blood enters the general circulation from venous networks in the liver, bone marrow, and skin. Reducing the amount of blood in the venous system can maintain the volume within the arterial system at near-normal levels despite a significant blood loss.

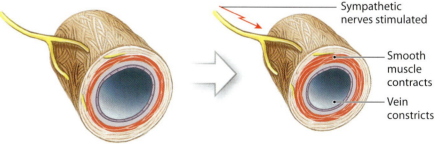

Sympathetic nerves stimulated

Smooth muscle contracts

Vein constricts

Module 11.12 Review

a. Define varicose veins.

b. Why are valves located in veins, but not in arteries?

c. How is blood flow maintained in veins to counter the pull of gravity?

The pulmonary circuit, which is relatively short, carries deoxygenated blood from the right ventricle to the lungs and returns oxygenated blood to the left atrium

1 This flow chart provides an overview of the organization of the cardiovascular system. The pulmonary circuit is composed of arteries and veins that transport blood between the heart and lungs. This circuit begins at the right ventricle and ends at the left atrium. From the left ventricle, the arteries of the systemic circuit transport oxygenated blood and nutrients to all organs and tissues, ultimately returning deoxygenated blood to the right atrium.

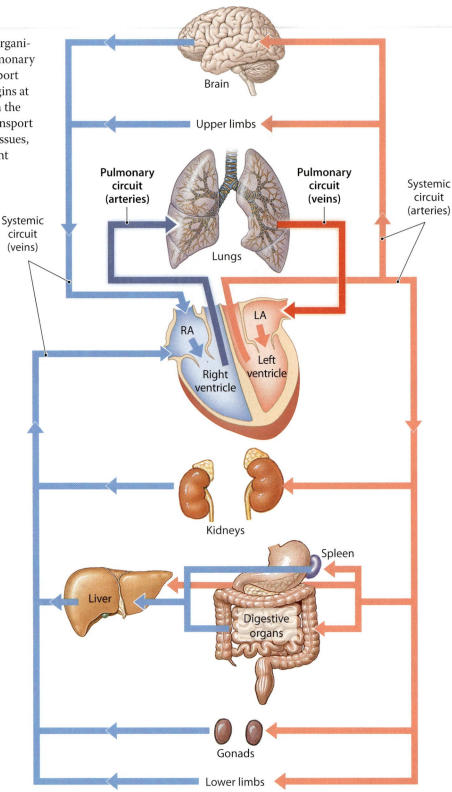

2 The table below summarizes the overall blood vessel organization in the body.

Major Patterns of Blood Vessel Organization

1. The peripheral distributions of arteries and veins on the body's left and right sides are generally identical, except near the heart, where the largest vessels connect to the atria or ventricles.

2. A single vessel may have several names as it crosses specific anatomical boundaries, making accurate anatomical descriptions possible when the vessel extends far into the periphery. For example, the external iliac artery becomes the femoral artery as it leaves the trunk and enters the lower limb.

3. Tissues and organs are usually serviced by several arteries and veins. Often, anastomoses between adjacent arteries or veins reduce the impact of a temporary or even permanent **occlusion** (blockage) of a single blood vessel.

3 Arteries of the pulmonary circuit differ from those of the systemic circuit in that they carry deoxygenated blood. (This is why most color-coded diagrams show the pulmonary arteries in blue, the same color as systemic veins.) By convention, several large arteries are called **trunks**; the **pulmonary trunk** is one important example. As the pulmonary trunk curves over the superior border of the heart, it gives rise to the left and right **pulmonary arteries**. These large arteries enter the lungs before branching repeatedly, giving rise to smaller and smaller arteries. The smallest branches, the **pulmonary arterioles**, provide blood to **alveolar capillaries** that surround small air pockets called **alveoli** (al-VĒ-ō-lī; singular, *alveolus*). The walls of the alveoli are thin enough for gas to be exchanged between the capillary blood and inspired air; the blood absorbs oxygen and releases carbon dioxide. Oxygenated blood leaving the alveolar capillaries enters venules that in turn unite to form larger vessels carrying blood toward the **pulmonary veins**. These four veins, two from each lung, deliver oxygenated blood into the left atrium, completing the pulmonary circuit.

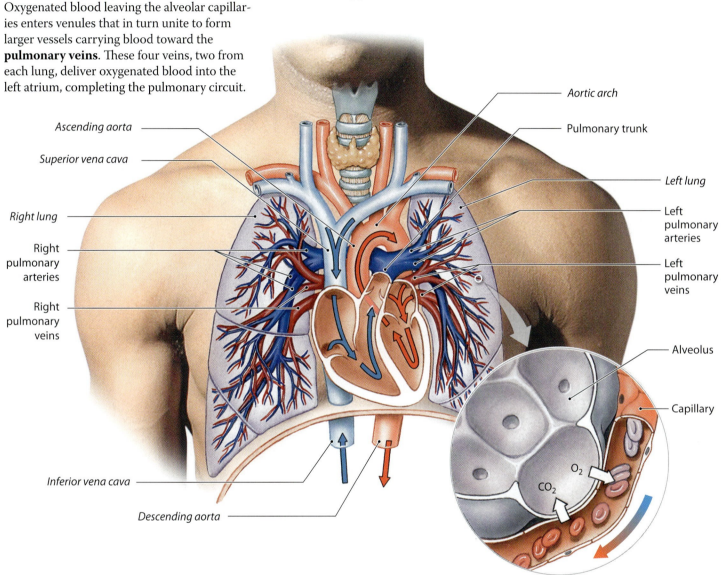

Ascending aorta

Superior vena cava

Right lung

Right pulmonary arteries

Right pulmonary veins

Inferior vena cava

Descending aorta

Aortic arch

Pulmonary trunk

Left lung

Left pulmonary arteries

Left pulmonary veins

Alveolus

Capillary

O_2

CO_2

Module 11.13 Review

a. Identify the two circulatory circuits of the cardiovascular system.

b. Briefly describe the three major patterns of blood vessel organization.

c. Trace the path of a drop of blood through the lungs, beginning at the right ventricle and ending at the left atrium.

The systemic arterial and venous systems operate in parallel, and the major vessels often have similar names

1 This figure provides an overview of the systemic **arterial system**. Note that all the vessels of the systemic arterial system originate from the aorta, the large elastic artery extending from the left ventricle of the heart. In each of the six modules that follow, the left-hand page will illustrate and discuss the systemic arterial vessels and their branches. To reduce clutter, "artery" will not be repeated in every label. Because most of the major arteries are paired, with one artery of each pair on either side of the body, the terms "right" and "left" will appear in figure labels only when the arteries on both sides are labeled.

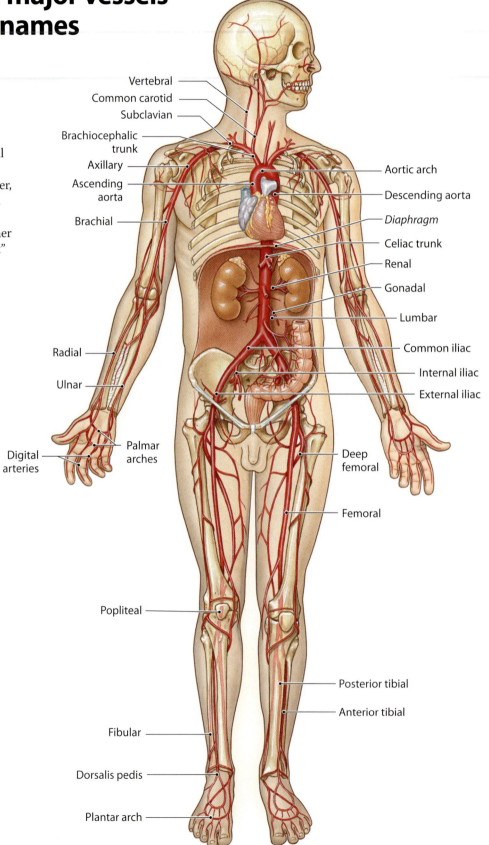

Vertebral
Common carotid
Subclavian
Brachiocephalic trunk
Axillary
Ascending aorta
Brachial

Aortic arch
Descending aorta
Diaphragm
Celiac trunk
Renal
Gonadal
Lumbar
Common iliac
Internal iliac
External iliac

Radial
Ulnar

Digital arteries
Palmar arches

Deep femoral

Femoral

Popliteal

Posterior tibial
Anterior tibial

Fibular

Dorsalis pedis

Plantar arch

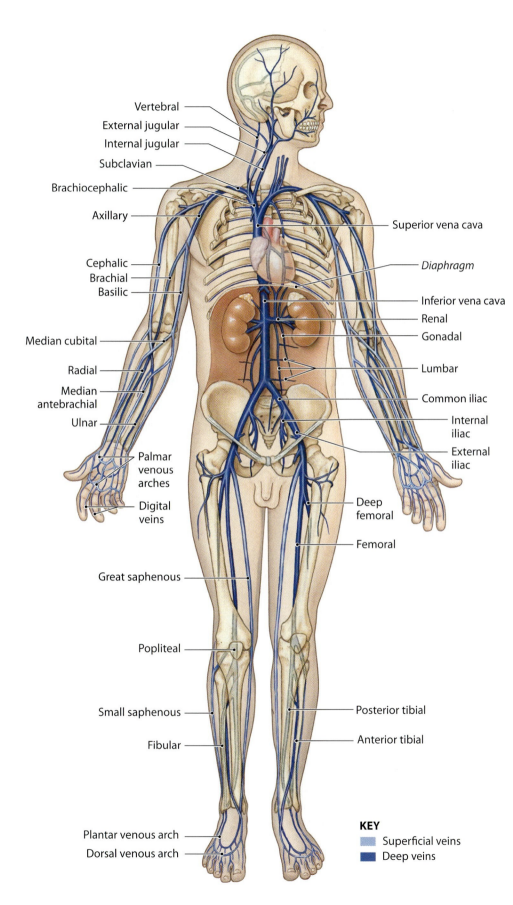

Vertebral
External jugular
Internal jugular
Subclavian
Brachiocephalic
Axillary

Cephalic
Brachial
Basilic

Median cubital

Radial

Median antebrachial

Ulnar

Palmar venous arches

Digital veins

Great saphenous

Popliteal

Small saphenous

Fibular

Plantar venous arch
Dorsal venous arch

Superior vena cava

Diaphragm

Inferior vena cava
Renal
Gonadal

Lumbar

Common iliac

Internal iliac

External iliac

Deep femoral

Femoral

Posterior tibial

Anterior tibial

KEY
Superficial veins
Deep veins

2 This figure provides an overview of the systemic **venous system**. Note that all of the vessels of the systemic venous system merge into two large veins: the **superior vena cava**, which collects systemic blood from the head, chest, and upper limbs, and the **inferior vena cava**, which collects systemic blood from all structures inferior to the diaphragm. In each of the six modules that follow, the right-hand page will illustrate and discuss the systemic venous vessels and their tributaries. The term "vein" will not be repeated in every label. The same name often applies to both the artery and the vein servicing a particular structure or region. Additionally, "right" and "left" will appear in figure labels only when the veins on both sides are shown.

One significant difference between the arterial and venous systems concerns the distribution of major veins in the neck and limbs. Arteries in these areas are located deep beneath the skin, protected by bones and surrounding soft tissues. In contrast, the neck and limbs generally have two sets of peripheral veins, one superficial and the other deep. This dual venous drainage is important for controlling body temperature. In hot weather, venous blood flows through superficial veins, where heat loss can occur; in cold weather, blood is routed to the deep veins to minimize heat loss.

Module 11.14 Review

a. Name the two large veins that collect blood from the systemic circuit.

b. Identify the largest artery in the body.

c. Besides containing valves, cite another major difference between the arterial and venous systems.

The branches of the aortic arch supply structures ...

1 The aorta is a large artery that is the main trunk of the systemic arterial system. It arises from the base of the heart's left ventricle and is subdivided into the ascending aorta, aortic arch, and descending aorta.

Start

Branches of the Aortic Arch

Three elastic arteries originate along the aortic arch and deliver blood to the head, neck, shoulders, and upper limbs:

The **brachiocephalic** (brā-kē-ō-se-FAL-ik) **trunk** is the first branch off the aortic arch. It ascends for a short distance before branching to form the **right subclavian artery** and the **right common carotid artery**.

The **left common carotid artery**, the second vessel arising from the aortic arch, supplies the left side of the head and neck.

The **left subclavian artery**, the third artery arising from the aortic arch, has the same distribution pattern as the right subclavian artery.

The Right Subclavian Artery

Two major branches arise before a subclavian artery leaves the thoracic cavity: (1) the **internal thoracic artery**, supplying the pericardium (heart) and anterior wall of the chest; and (2) the **vertebral artery**, which provides blood to the brain and spinal cord.

Arteries of the Arm

After leaving the thoracic cavity and passing across the superior border of the first rib, the subclavian becomes the **axillary artery**. This artery crosses the axilla to enter the arm, where it becomes the **brachial artery**, which supplies blood to the upper limb. The brachial artery gives rise to the deep brachial artery, which supplies deep structures on the posterior aspect of the arm, and the ulnar collateral arteries, which supply the area around the elbow.

Arteries of the Forearm

As it approaches the coronoid fossa of the humerus, the brachial artery divides into the **radial artery**, which follows the radius, and the **ulnar artery**, which follows the ulna to the wrist.

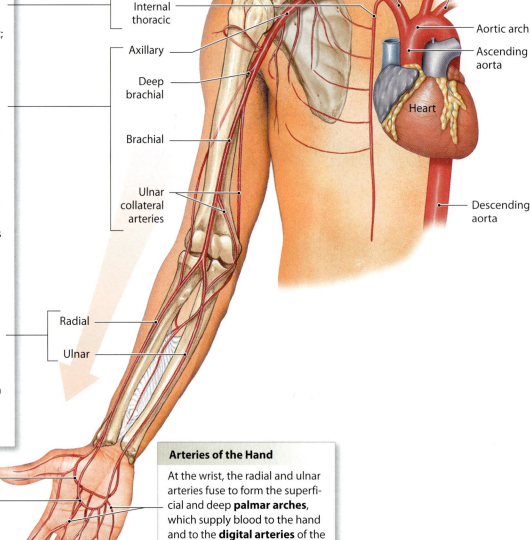

Vertebral
Internal thoracic
Axillary
Deep brachial
Brachial
Ulnar collateral arteries
Radial
Ulnar

Aortic arch
Ascending aorta
Heart
Descending aorta

Deep palmar arch
Superficial palmar arch

Arteries of the Hand

At the wrist, the radial and ulnar arteries fuse to form the superficial and deep **palmar arches**, which supply blood to the hand and to the **digital arteries** of the thumb and fingers.

... that are drained by the superior vena cava

2 The superior vena cava collects systemic blood from the head, chest, and upper limbs. Note the alternate routes of venous blood flow in the upper limb.

Veins of the Neck

| The **external jugular vein** drains superficial structures of the head and neck. | The **vertebral vein** drains the cervical spinal cord and the posterior surface of the skull. | The **internal jugular vein** drains deep structures of the head and neck. |

The Right Subclavian Vein

The axillary vein is joined by the cephalic vein on the lateral surface of the first rib, forming the **subclavian vein**, which continues into the chest.

Veins of the Arm

As the **brachial vein** merges with the **basilic vein** it becomes the **axillary vein**, which enters the axilla.

The **cephalic vein** extends along the lateral side of the arm.

Veins of the Forearm

The **median cubital vein** interconnects the cephalic and basilic veins. (The median cubital is the vein from which venous blood samples are typically collected.)

The **ulnar vein** and the **radial vein** drain the **deep palmar arch**. After crossing the elbow, these veins fuse to form the brachial vein.

The cephalic vein, the **median antebrachial vein**, and the basilic vein drain the **superficial palmar arch**.

The **brachiocephalic vein** forms as the jugular veins empty into the axillary vein. It receives blood from the vertebral vein and the internal thoracic vein draining the anterior chest wall.

The **superior vena cava (SVC)** carries blood from the two brachiocephalic veins to the right atrium of the heart.

The **internal thoracic vein** collects blood from the intercostal veins and delivers it to the brachiocephalic vein.

Brachial

Basilic

KEY

▨	Superficial veins
▨	Deep veins

Cephalic

Radial

Basilic

Ulnar

Deep palmar arch

Superficial palmar arch

Start

Veins of the Hand

The **digital veins** empty into superficial and deep veins of the hand, which are interconnected to form the palmar venous arches.

Module 11.15 Review

a. Name the two arteries formed by the division of the brachiocephalic trunk.

b. A blockage of which branch from the aortic arch would interfere with blood flow to the left arm?

c. Whenever Thor gets angry, a large vein bulges in the lateral region of his neck. Which vein is this?

The external carotid arteries supply the neck, lower jaw, and face, and the internal carotid and vertebral arteries supply the brain ...

1 The **common carotid arteries** ascend deep in the tissues of the neck and supply blood to the structures of the face, neck, and brain. To locate the carotid artery, gently press a finger along either side of the windpipe (trachea) until you feel a strong pulse. Details of the arterial supply of the brain are covered in Module 11.17.

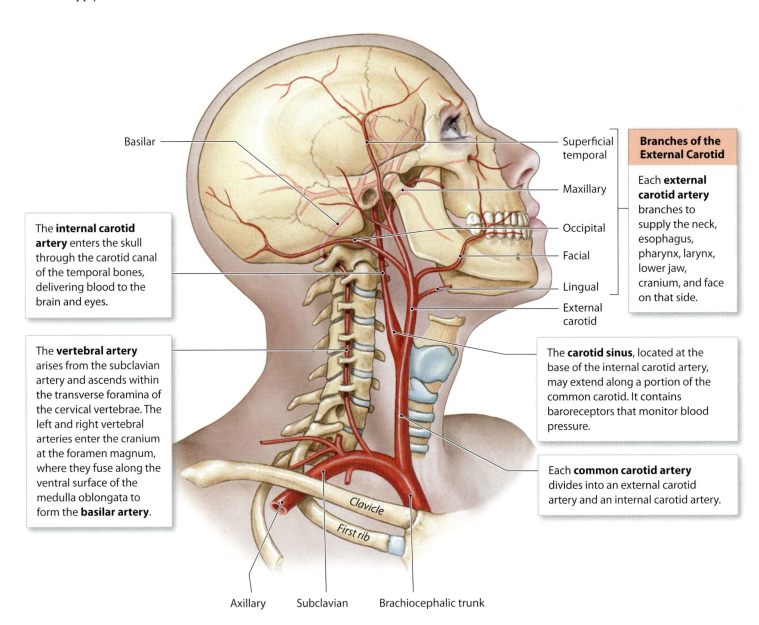

Basilar

Superficial temporal

Maxillary

Occipital

Facial

Lingual

External carotid

Branches of the External Carotid

Each **external carotid artery** branches to supply the neck, esophagus, pharynx, larynx, lower jaw, cranium, and face on that side.

The **internal carotid artery** enters the skull through the carotid canal of the temporal bones, delivering blood to the brain and eyes.

The **vertebral artery** arises from the subclavian artery and ascends within the transverse foramina of the cervical vertebrae. The left and right vertebral arteries enter the cranium at the foramen magnum, where they fuse along the ventral surface of the medulla oblongata to form the **basilar artery**.

The **carotid sinus**, located at the base of the internal carotid artery, may extend along a portion of the common carotid. It contains baroreceptors that monitor blood pressure.

Each **common carotid artery** divides into an external carotid artery and an internal carotid artery.

Clavicle

First rib

Axillary   Subclavian   Brachiocephalic trunk

... while the external jugular veins drain the regions supplied by the external carotids, and the internal jugular veins drain the brain

2 The **external jugular veins** are formed by the maxillary and temporal veins, while the **internal jugular veins** drain the blood from the various venous sinuses within the cranium (see Module 11.17 for additional details). The external and internal jugular veins combine with the vertebral and subclavian to form the **brachiocephalic vein**.

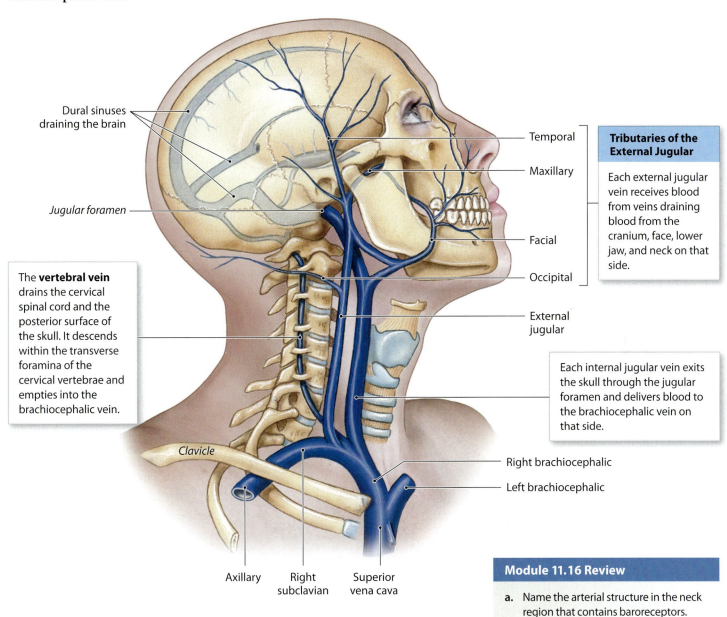

Dural sinuses draining the brain

Jugular foramen

The **vertebral vein** drains the cervical spinal cord and the posterior surface of the skull. It descends within the transverse foramina of the cervical vertebrae and empties into the brachiocephalic vein.

Clavicle

Temporal

Maxillary

Tributaries of the External Jugular

Each external jugular vein receives blood from veins draining blood from the cranium, face, lower jaw, and neck on that side.

Facial

Occipital

External jugular

Each internal jugular vein exits the skull through the jugular foramen and delivers blood to the brachiocephalic vein on that side.

Right brachiocephalic

Left brachiocephalic

Axillary   Right subclavian   Superior vena cava

Module 11.16 Review

a. Name the arterial structure in the neck region that contains baroreceptors.

b. Identify branches of the external carotid artery.

c. Identify the veins that combine to form the brachiocephalic vein.

The internal carotid arteries and the vertebral arteries supply the brain ...

1 This lateral view shows the major arteries supplying the brain. The internal carotid arteries normally supply the arteries of the anterior half of the cerebrum, and the rest of the brain receives blood from the vertebral and basilar arteries. The internal carotid artery ascends to the level of the optic nerves, where each artery divides into three branches: (1) an **ophthalmic artery**, which supplies the eyes; (2) an **anterior cerebral artery**, which supplies the frontal and parietal lobes of the brain; and (3) a **middle cerebral artery**, which supplies the midbrain and the lateral surfaces of the cerebral hemispheres.

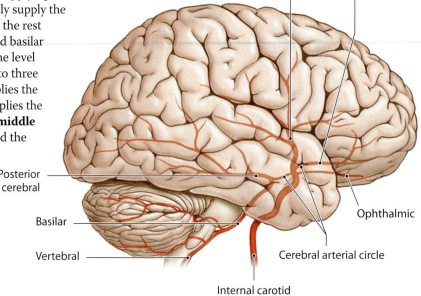

Middle cerebral

Anterior cerebral

Posterior cerebral

Basilar

Vertebral

Ophthalmic

Cerebral arterial circle

Internal carotid

2 The internal carotid arteries and the basilar artery are interconnected in a ring-shaped anastomosis called the **cerebral arterial circle**.

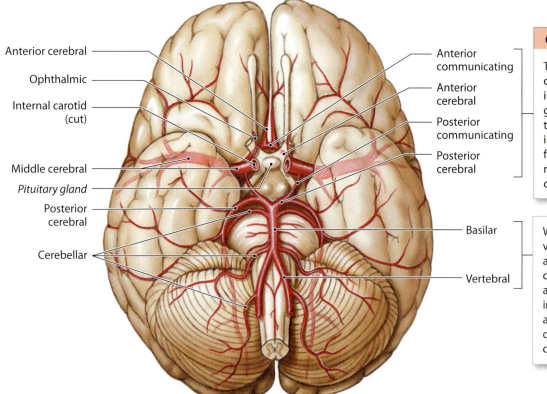

Anterior cerebral

Ophthalmic

Internal carotid (cut)

Middle cerebral

Pituitary gland

Posterior cerebral

Cerebellar

Anterior communicating

Anterior cerebral

Posterior communicating

Posterior cerebral

Basilar

Vertebral

Cerebral Arterial Circle

The cerebral arterial circle, or circle of Willis, encircles the infundibulum of the pituitary gland. This arrangement reduces the likelihood of a serious interruption of cerebral blood flow, because the brain can receive blood from either the carotid or the vertebral arteries.

Within the cranium, the vertebral arteries and the basilar artery supply blood to the spinal cord, medulla oblongata, pons, and cerebellum before dividing into the posterior cerebral arteries, which in turn branch off into the posterior communicating arteries.

... which is drained by the dural sinuses and the internal jugular veins

3 The superficial cerebral veins and small veins of the brain stem empty into a network of dural sinuses. Most of the deep cerebral veins converge within the brain to form the **great cerebral vein**, which delivers blood from the interior of the cerebral hemispheres and the choroid plexus to the **straight sinus**. Numerous small veins from the orbit and other cerebral veins drain into the **cavernous sinus**.

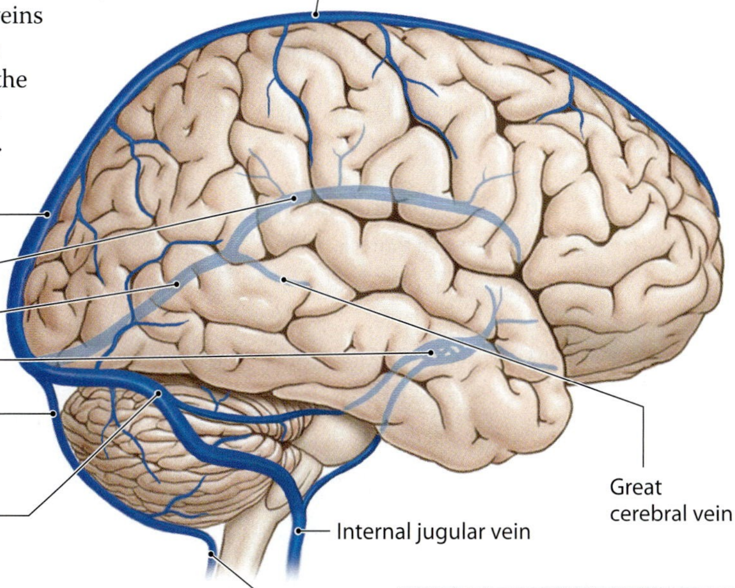

The **superior sagittal sinus**, in the falx cerebri, is the largest dural sinus.

- Superior sagittal sinus
- Inferior sagittal sinus
- Straight sinus
- Cavernous sinus
- Occipital sinus
- Right transverse sinus
- Great cerebral vein
- Internal jugular vein

4 The transverse sinuses, the straight sinus, and the superior sagittal sinus converge, penetrate the jugular foramina, and leave the skull as the internal jugular veins.

The **vertebral vein** on each side receives blood from the dural sinuses as well as superficial veins of the skull and veins draining the cervical vertebrae.

- Superior sagittal sinus (cut)
- Cavernous sinus
- Internal jugular
- Cerebral veins
- Petrosal sinus
- Sigmoid sinus
- Cerebellar veins
- Straight sinus
- Transverse sinus
- Occipital sinus

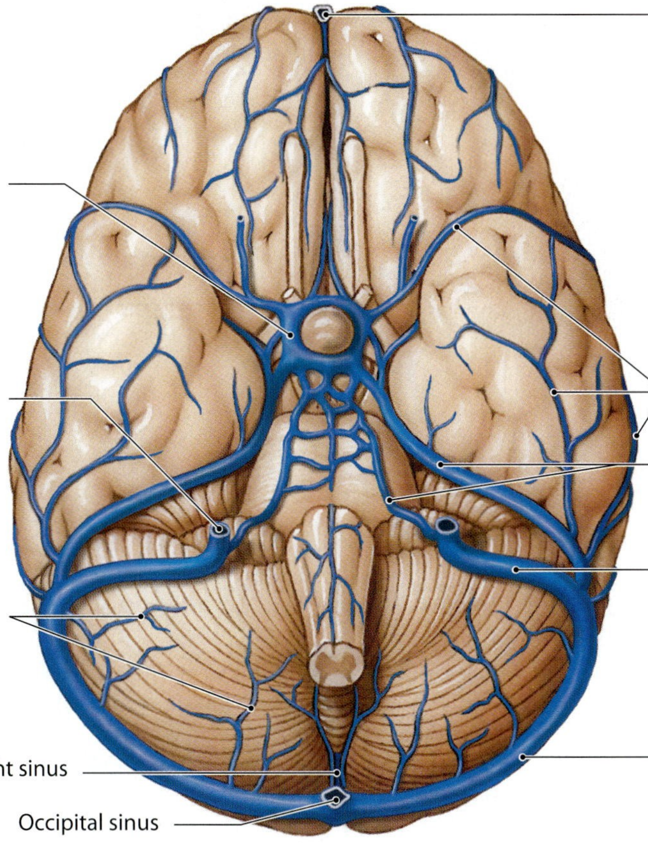

Module 11.17 Review

a. Name the veins that drain the dural sinuses of the brain.

b. Name the three branches of the internal carotid artery and the structures they supply.

c. Describe the structure and function of the cerebral arterial circle.

The regions supplied by the descending aorta ...

1 The descending aorta is continuous with the aortic arch. The diaphragm divides the descending aorta into a superior **thoracic aorta** and an inferior **abdominal aorta.** The figure details the branches of the thoracic aorta and introduces the branches of the abdominal aorta that are detailed further in the table below.

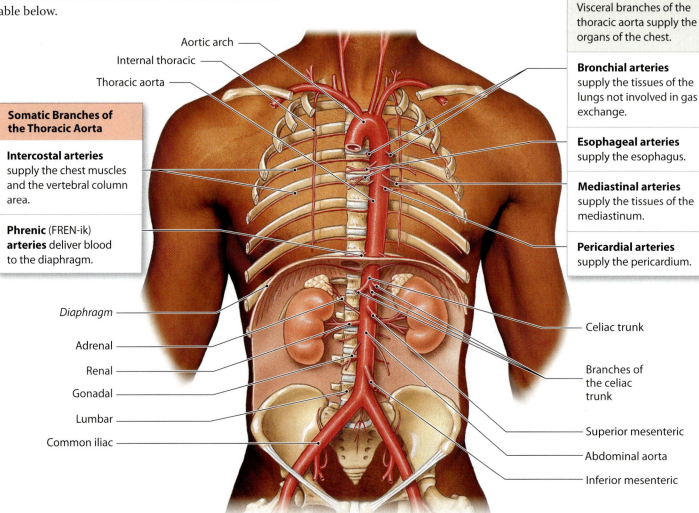

Somatic Branches of the Thoracic Aorta

Intercostal arteries supply the chest muscles and the vertebral column area.

Phrenic (FREN-ik) **arteries** deliver blood to the diaphragm.

Visceral Branches of the Thoracic Aorta

Visceral branches of the thoracic aorta supply the organs of the chest.

Bronchial arteries supply the tissues of the lungs not involved in gas exchange.

Esophageal arteries supply the esophagus.

Mediastinal arteries supply the tissues of the mediastinum.

Pericardial arteries supply the pericardium.

Labels on figure:
- Aortic arch
- Internal thoracic
- Thoracic aorta
- Diaphragm
- Adrenal
- Renal
- Gonadal
- Lumbar
- Common iliac
- Celiac trunk
- Branches of the celiac trunk
- Superior mesenteric
- Abdominal aorta
- Inferior mesenteric

Major Paired Branches of the Abdominal Aorta

- The **phrenic arteries** supply the diaphragm and the inferior portion of the esophagus.
- The **adrenal arteries** supply the adrenal glands.
- The short **renal arteries** supply the kidneys.
- The **gonadal** (gō-NAD-al) **arteries** are called testicular arteries in males and ovarian arteries in females.
- Small **lumbar arteries** supply the vertebrae, spinal cord, and abdominal wall.

Major Unpaired Branches of the Abdominal Aorta

- The **celiac** (SĒ-lē-ak) **trunk** divides into three branches that supply the inferior portion of the esophagus, spleen, pancreas, stomach, liver, gallbladder, and proximal portion of the small intestine.
- The **superior mesenteric** (mez-en-TER-ik) **artery** supplies the pancreas, duodenum, small intestine, and most of the large intestine.
- The **inferior mesenteric artery** delivers blood to the terminal portions of the colon and the rectum.

... are drained by the superior and inferior venae cavae

2 The **superior vena cava** drains blood from the head, neck, shoulders, chest, and upper limbs. The chief collecting vessels of the thorax are the **azygos** (AZ-i-gos) **vein** and the **hemiazygos vein**. These veins receive blood from **intercostal veins**, which in turn receive blood from the chest muscles; **esophageal veins**, which drain blood from the inferior portion of the esophagus; **bronchial veins** drain the passageways of the lungs; and **mediastinal veins** drain other mediastinal structures. The **inferior vena cava** collects most of the blood inferior to the diaphragm.

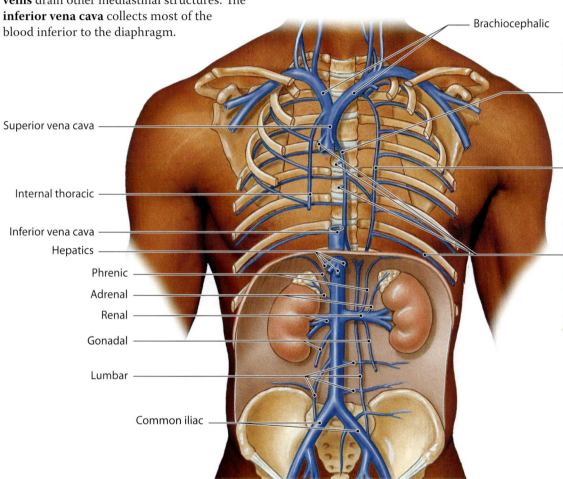

Brachiocephalic

Superior vena cava

Internal thoracic

Inferior vena cava

Hepatics

Phrenic

Adrenal

Renal

Gonadal

Lumbar

Common iliac

The Azygos and Hemiazygos Veins

The azygos vein is a major tributary of the superior vena cava.

The smaller hemiazygos vein drains into the azygos vein and may also drain into the left brachiocephalic vein.

Together these veins collect blood from the esophagus, mediastinum, conducting passageways of the lungs, and muscles of the chest wall.

Major Tributaries of the Inferior Vena Cava

- **Lumbar veins** drain the spinal cord and muscles of the body wall.
- **Gonadal** (ovarian or testicular) **veins** drain the ovaries or testes.
- **Hepatic veins** drain the sinusoids (channels) of the liver.
- **Renal veins** collect blood from the kidneys.
- **Adrenal veins** drain the adrenal glands.
- **Phrenic veins** drain the diaphragm.

Module 11.18 Review

a. Which vessel collects most of the venous blood inferior to the diaphragm?

b. Identify the major tributaries of the inferior vena cava.

c. Grace is in an automobile accident, and her celiac trunk is ruptured. Which organs will be affected most directly by this injury?

The viscera supplied by the celiac trunk and mesenteric arteries ...

1 The abdominal aorta begins immediately inferior to the diaphragm. Three unpaired branches supply the abdominal viscera: the celiac trunk, the superior mesenteric artery, and the inferior mesenteric artery.

The Celiac Trunk

The **celiac trunk** divides into the common hepatic artery, the left gastric artery, and the splenic artery.

The **common hepatic artery** branches to supply the liver, stomach, gallbladder, and the duodenum (the proximal segment of the small intestine).

The **left gastric artery** supplies the stomach. Its anastomosis with the right gastric artery ensures a continuous blood supply to the stomach.

The **splenic artery** supplies the spleen and sends branches to the stomach and pancreas.

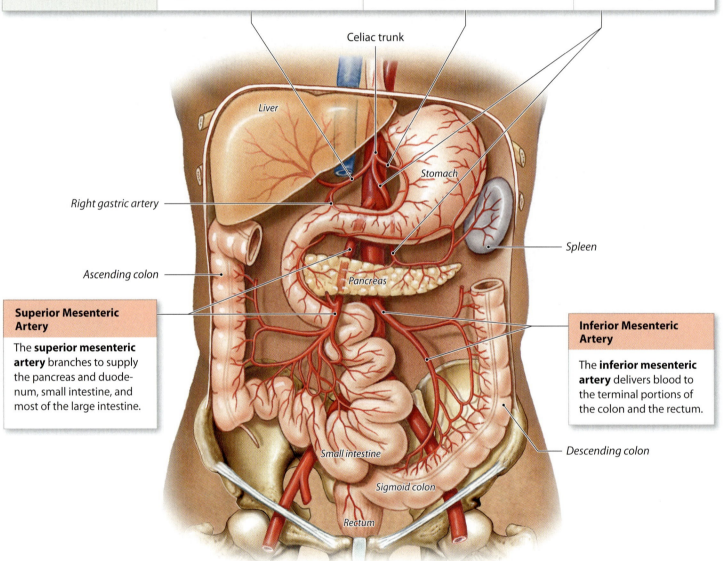

Celiac trunk

Liver

Stomach

Right gastric artery

Spleen

Ascending colon

Pancreas

Superior Mesenteric Artery

The **superior mesenteric artery** branches to supply the pancreas and duodenum, small intestine, and most of the large intestine.

Inferior Mesenteric Artery

The **inferior mesenteric artery** delivers blood to the terminal portions of the colon and the rectum.

Descending colon

Small intestine

Sigmoid colon

Rectum

... are drained by the tributaries of the hepatic portal vein

2 The **hepatic portal vein** forms through the fusion of the superior mesenteric, inferior mesenteric, and splenic veins. The largest volume of blood (and most of the nutrients) flows through the superior mesenteric vein. The hepatic portal vein receives blood from the left and right **gastric veins**, which drain the medial border of the stomach, and from the **cystic vein**, from the gallbladder.

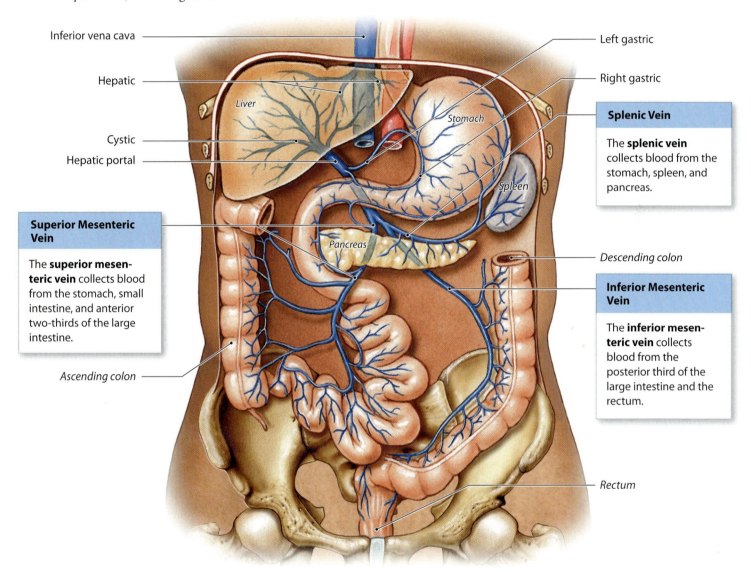

Inferior vena cava

Hepatic

Cystic

Hepatic portal

Left gastric

Right gastric

Splenic Vein

The **splenic vein** collects blood from the stomach, spleen, and pancreas.

Liver

Stomach

Spleen

Pancreas

Superior Mesenteric Vein

The **superior mesenteric vein** collects blood from the stomach, small intestine, and anterior two-thirds of the large intestine.

Ascending colon

Descending colon

Inferior Mesenteric Vein

The **inferior mesenteric vein** collects blood from the posterior third of the large intestine and the rectum.

Rectum

Module 11.19 Review

a. List the unpaired branches of the abdominal aorta that supply blood to the visceral organs.

b. Identify the three veins that merge to form the hepatic portal vein.

c. Identify the branches of the celiac trunk.

The pelvis and lower limb are supplied by branches of the common iliac arteries ...

1 Near the level of vertebra L_4, the abdominal aorta divides to form a pair of elastic arteries: the **right** and **left common iliac** (IL-ē-ak) **arteries**. At the level of the lumbosacral joint, each common iliac divides to form an **internal iliac artery** and an **external iliac artery**.

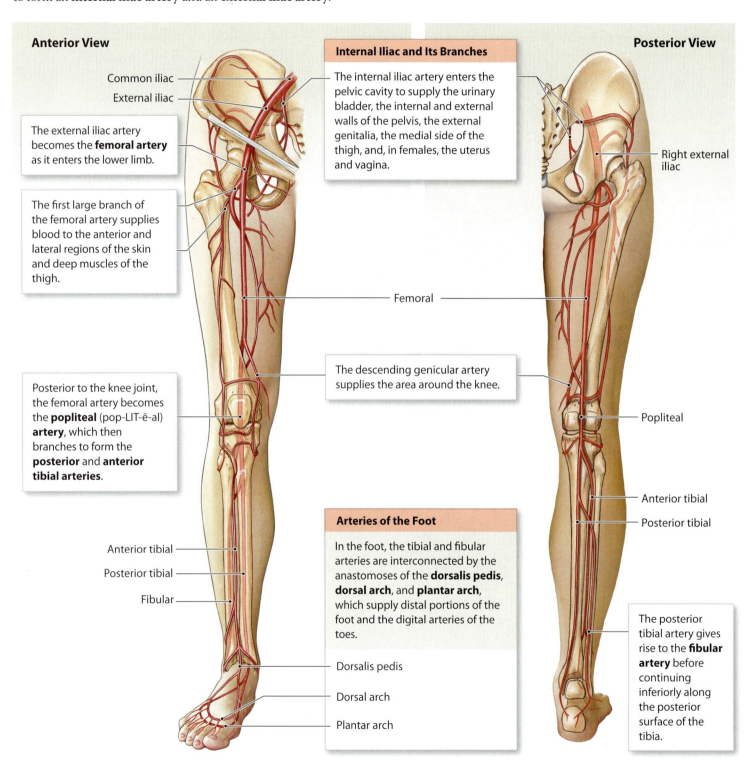

Anterior View

Common iliac

External iliac

The external iliac artery becomes the **femoral artery** as it enters the lower limb.

The first large branch of the femoral artery supplies blood to the anterior and lateral regions of the skin and deep muscles of the thigh.

Internal Iliac and Its Branches

The internal iliac artery enters the pelvic cavity to supply the urinary bladder, the internal and external walls of the pelvis, the external genitalia, the medial side of the thigh, and, in females, the uterus and vagina.

Posterior View

Right external iliac

Femoral

Posterior to the knee joint, the femoral artery becomes the **popliteal** (pop-LIT-ē-al) **artery**, which then branches to form the **posterior** and **anterior tibial arteries**.

The descending genicular artery supplies the area around the knee.

Popliteal

Anterior tibial

Posterior tibial

Arteries of the Foot

In the foot, the tibial and fibular arteries are interconnected by the anastomoses of the **dorsalis pedis**, **dorsal arch**, and **plantar arch**, which supply distal portions of the foot and the digital arteries of the toes.

Anterior tibial

Posterior tibial

Fibular

Dorsalis pedis

Dorsal arch

Plantar arch

The posterior tibial artery gives rise to the **fibular artery** before continuing inferiorly along the posterior surface of the tibia.

... and drained by tributaries of the common iliac veins

2 The external iliac veins receive blood from the lower limbs, the pelvis, and the lower abdomen. As the left and right external iliac veins cross the inner surface of the ilium, they are joined by the internal iliac veins, which drain the pelvic organs. The internal iliac veins are formed by the fusion of the gluteal, internal pudendal, obturator, and lateral sacral veins. The union of the external and internal iliac veins forms the **common iliac vein**.

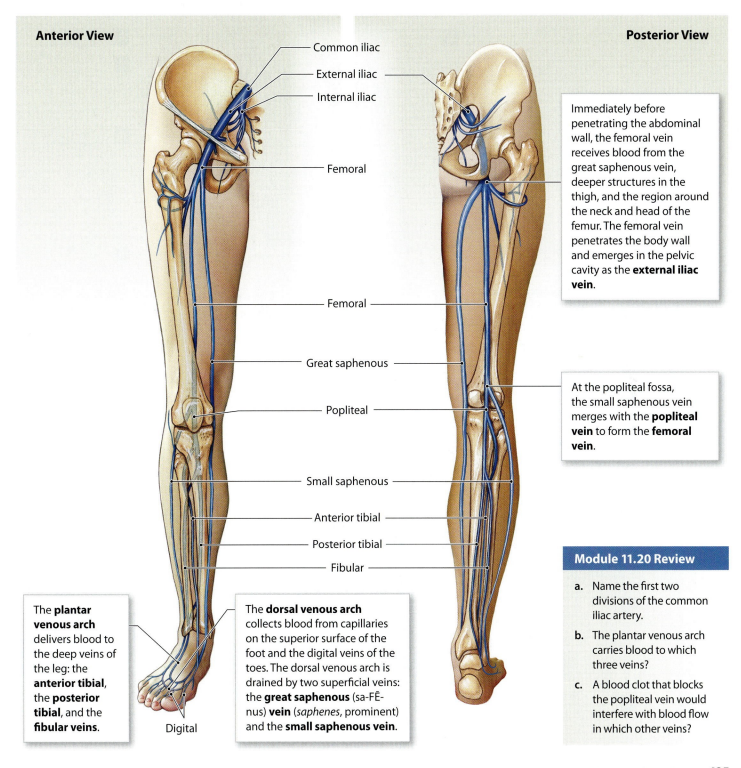

Anterior View

Posterior View

Common iliac

External iliac

Internal iliac

Femoral

Femoral

Great saphenous

Popliteal

Small saphenous

Anterior tibial

Posterior tibial

Fibular

Digital

Immediately before penetrating the abdominal wall, the femoral vein receives blood from the great saphenous vein, deeper structures in the thigh, and the region around the neck and head of the femur. The femoral vein penetrates the body wall and emerges in the pelvic cavity as the **external iliac vein**.

At the popliteal fossa, the small saphenous vein merges with the **popliteal vein** to form the **femoral vein**.

The **plantar venous arch** delivers blood to the deep veins of the leg: the **anterior tibial**, the **posterior tibial**, and the **fibular veins**.

The **dorsal venous arch** collects blood from capillaries on the superior surface of the foot and the digital veins of the toes. The dorsal venous arch is drained by two superficial veins: the **great saphenous** (sa-FĒ-nus) **vein** (*saphenes*, prominent) and the **small saphenous vein**.

Module 11.20 Review

a. Name the first two divisions of the common iliac artery.

b. The plantar venous arch carries blood to which three veins?

c. A blood clot that blocks the popliteal vein would interfere with blood flow in which other veins?

1. Labeling

Label the major arteries in the diagram at right.

a

b

c

d

e

f

g

h

i

j

k

l

m

n

2. Labeling

Label the major veins in the accompanying diagram.

a

b

c

d

e

f

g

h

i

j

k

l

m

n

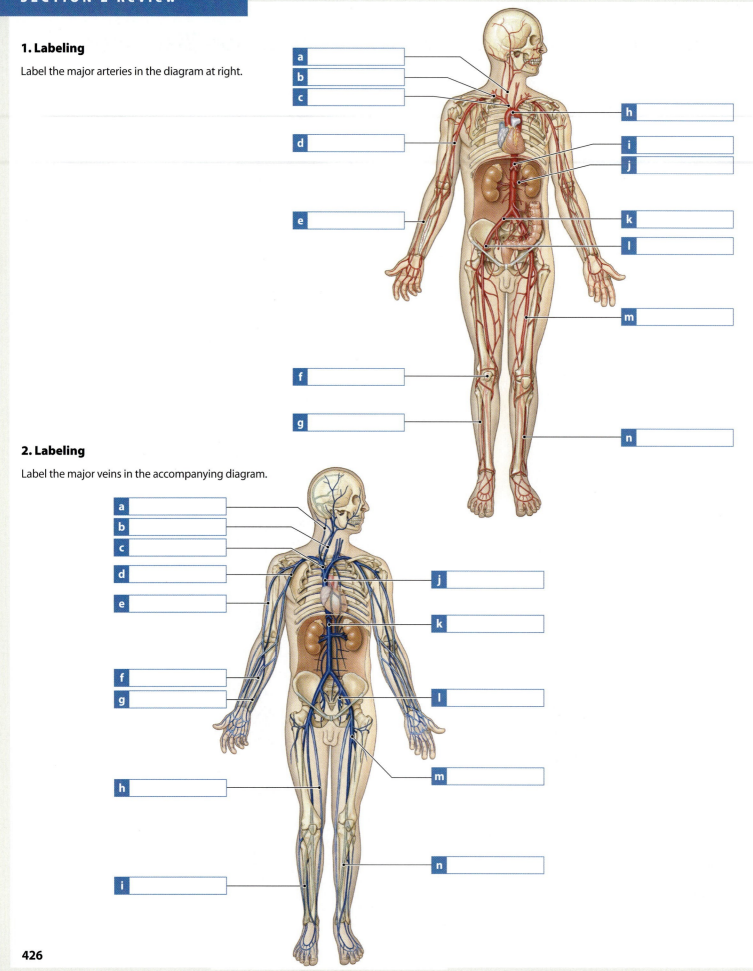

Visual Outline with Key Terms   Summarize the content of each module using the terms in the order provided.

SECTION 1

Blood

- cardiovascular system
  ○ blood
  ○ blood vessels
  ○ heart

11.1

Blood is a fluid connective tissue containing plasma and formed elements

- plasma
- formed elements
- whole blood
- fractionated
- hematocrit
- packed cell volume (PCV)
- albumin
- globulins

- immunoglobulins
- transport globulins
- fibrinogen
- fibrin
- electrolytes
- organic nutrients
- organic wastes
- platelets

- white blood cells (WBCs)
- red blood cells (RBCs)

11.2

Red blood cells, the most common formed elements, contain hemoglobin

- red blood cell count
- hemoglobin (Hb)
- heme
- oxyhemoglobin (HbO₂)
- deoxyhemoglobin

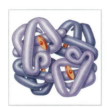

11.3

Blood type is determined by the presence or absence of specific surface antigens on RBCs

- antigens
- immune response
- surface antigens
- blood type
- antibodies
- Type A
- Type B
- Type AB

- universal recipients
- Type O
- universal donors
- agglutinate
- agglutination
- hemolysis
- cross-reaction
- Rh positive (Rh⁺)

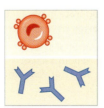

- Rh surface antigen
- Rh negative (Rh⁻)
- transfusion reactions
- compatible

11.4

Hemolytic disease of the newborn is an RBC-related disorder caused by a cross-reaction between fetal and maternal blood types

- hemolytic disease of the newborn (HDN)
- sensitization
- erythroblastosis fetalis

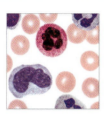

11.5

White blood cells defend the body against pathogens, toxins, cellular debris, and abnormal or damaged cells

- white blood cells (WBCs)
- diapedesis
- positive chemotaxis
- granular leukocytes
- neutrophils
- eosinophils

- basophils
- agranular leukocytes
- monocytes
- lymphocytes
- differential count

11.6

Formed elements are produced by stem cells derived from hemocytoblasts

- hematopoiesis
- hemocytoblasts
- lymphoid stem cells
- lymphoid tissues
- colony-stimulating factors (CSFs)
  ○ blast cells
- myeloid stem cells

  ○ progenitor cells
  ○ band cells
- megakaryocytes
- erythropoietin (EPO)
- hypoxia
- platelets

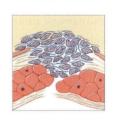

11.7

The clotting response is a complex cascade of events that reduces blood loss

- hemostasis
- cascade
- vascular phase
- platelet phase
- coagulation phase
- coagulation
  ○ fibrinogen
- fibrin
- procoagulants
- embolus

- embolism
- infarction
- thrombus
- clot retraction
- fibrinolysis
- plasminogen
- tissue plasminogen activator (t-PA)
- plasmin

• = *Term boldfaced in this module*

11.8

Blood disorders can be classified by their origins and the changes in blood characteristics

- venipuncture
- iron deficiency anemia
- vitamin B$_{12}$
- pernicious anemia
- vitamin K
- sickle cell anemia
- sickling trait
- hemophilia
- thalassemias
- bacteremia
- viremia
- sepsis
- septicemia
- malaria
- leukemias
- myeloid leukemia
- lymphoid leukemia
- disseminated intravascular coagulation (DIC)

SECTION 2

The Functional Anatomy of Blood Vessels

- pulmonary circuit
- systemic circuit
- arteries
- veins
- capillaries
- right atrium
- right ventricle
- left atrium
- left ventricle

11.9

Arteries and veins differ in the structure and thickness of their walls

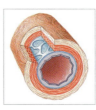

- tunica intima
- internal elastic membrane
- tunica media
- vasoconstriction
- vasodilation
- tunica externa
- large veins
- medium-sized veins
- venules
- elastic arteries
- muscular arteries
- arterioles
- capillaries
- arteries
- veins

11.10

Arteriosclerosis can restrict blood flow and damage vital organs

- arteriosclerosis
- atherosclerosis
- plaque
- balloon angioplasty

11.11

Capillaries function as part of an interconnected network called a capillary bed

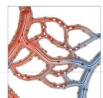

- capillary bed
- metarteriole
- vasomotion
- collateral arteries
- arterial anastomosis
- precapillary sphincters
- arteriovenous anastomosis

11.12

The venous system has low pressures and contains almost two-thirds of the body's blood volume

- valves
- varicose veins
- hemorrhoids
- venoconstriction

11.13

The pulmonary circuit, which is relatively short, carries deoxygenated blood from the right ventricle to the lungs and returns oxygenated blood to the left atrium

- occlusion
- trunks
- pulmonary trunk
- pulmonary arteries
- pulmonary arterioles
- alveolar capillaries
- alveoli
- pulmonary veins

11.14

The systemic arterial and venous systems operate in parallel, and the major vessels often have similar names

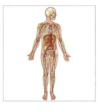

- arterial system
- venous system
- superior vena cava
- inferior vena cava
- dual venous drainage

• = *Term boldfaced in this module*

11.15

The branches of the aortic arch supply structures that are drained by the superior vena cava

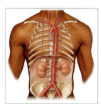

- brachiocephalic trunk
- right subclavian artery
- right common carotid artery
- left common carotid artery
- left subclavian artery
- internal thoracic artery
- vertebral artery
- axillary artery
- brachial artery
- radial artery
- ulnar artery
- palmar arches
- digital arteries
- digital veins
- superficial palmar arch
- deep palmar arch
- median antebrachial vein
- radial vein
- ulnar vein
- median cubital vein
- cephalic vein
- axillary vein
- basilic vein
- brachial vein
- subclavian vein
- external jugular vein
- internal jugular vein
- vertebral vein
- brachiocephalic vein
- superior vena cava (SVC)
- internal thoracic vein

11.16

The external carotid arteries supply the neck, lower jaw, and face, and the internal carotid and vertebral arteries supply the brain, while the external jugular veins drain the regions supplied by the external carotids, and the internal jugular veins drain the brain

- common carotid arteries
- internal carotid artery
- vertebral artery
- basilar artery
- external carotid artery
- carotid sinus
- common carotid artery
- external jugular veins
- internal jugular vein
- brachiocephalic vein
- vertebral vein

11.17

The internal carotid arteries and the vertebral arteries supply the brain, which is drained by the dural sinuses and the internal jugular veins

- ophthalmic artery
- anterior cerebral artery
- middle cerebral artery
- cerebral arterial circle
- great cerebral vein
- straight sinus
- cavernous sinus
- superior sagittal sinus
- vertebral vein

11.18

The regions supplied by the descending aorta are drained by the superior and inferior venae cavae

- thoracic aorta
- abdominal aorta
- intercostal arteries
- phrenic arteries
- bronchial arteries
- esophageal arteries
- mediastinal arteries
- pericardial arteries
- adrenal arteries
- renal arteries
- gonadal arteries
- lumbar arteries
- celiac trunk
- superior mesenteric artery
- inferior mesenteric artery
- superior vena cava
- azygos vein
- hemiazygos vein
- intercostal veins
- esophageal veins
- bronchial veins
- inferior vena cava
- mediastinal veins
- lumbar veins
- gonadal veins
- hepatic veins
- renal veins
- adrenal veins
- phrenic veins

11.19

The viscera supplied by the celiac trunk and mesenteric arteries are drained by the tributaries of the hepatic portal vein

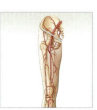

- celiac trunk
- common hepatic artery
- left gastric artery
- splenic artery
- superior mesenteric artery
- gastric veins
- cystic vein
- inferior mesenteric artery
- hepatic portal vein
- splenic vein
- superior mesenteric vein
- inferior mesenteric vein

11.20

The pelvis and lower limbs are supplied by branches of the common iliac arteries and drained by tributaries of the common iliac veins

- right common iliac artery
- left common iliac artery
- internal iliac artery
- external iliac artery
- femoral artery
- popliteal artery
- posterior tibial artery
- anterior tibial artery
- dorsalis pedis
- dorsal arch
- plantar arch
- fibular artery
- common iliac vein
- plantar venous arch
- anterior tibial vein
- posterior tibial vein
- fibular vein
- dorsal venous arch
- great saphenous vein
- small saphenous vein
- external iliac vein
- popliteal vein
- femoral vein

• = *Term boldfaced in this module*

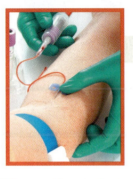

"The specimens that a phlebotomist collects will relate to every single body system. If a physician orders a CBC, you need to know what specimen tube to use, what department in the lab it goes to, and what the clinical significance is."

— **Sue Phelan**
Associate Professor, Phlebotomy
Moraine Valley Community College
Palos Hills, Illinois

CAREER PATHS

Phlebotomist

Phlebotomists are a type of clinical laboratory technician who draw blood for testing and for blood donation. Sue Phelan has worked as a phlebotomist and is currently a full-time faculty member in phlebotomy and chair of the Career Programs Department at Moraine Valley Community College in Palos Hills, Illinois.

When Sue trained as a clinical laboratory scientist (CLS) at Christ Hospital in Oak Lawn, Illinois, one component of her twelve-month internship was phlebotomy. Once the interns were trained in phlebotomy, they often assisted with the "morning draws," a roster of hospital patients who all needed their blood drawn for testing. Later on, while working as a CLS, Sue filled in on phlebotomy shifts for overtime, and eventually worked on the hospital's blood donor mobile. She became the hospital's first pheresis tech—a person who handles blood platelet donations—and went on to specialize in bloodbanking immunohematology.

A typical day for a phlebotomist working at a hospital involves drawing blood from patients all around the hospital and taking it to the lab for testing. Prepping, sterilizing and

drawing blood from each patient takes about five minutes. Because doctors prefer blood from a patient in a basal state—one who has rested, fasted, and is lying down—the early morning is often the busiest, but phlebotomists also work evening and overnight shifts. Phlebotomists who work for doctors' offices or blood banks may have more traditional 9-5 hours.

Sue estimates that phlebotomists spend 60 percent of their time on patient interaction. She notes that sticking someone with a needle requires a great deal of bedside manner, especially with children.

"With children, the interpersonal communication becomes even more complicated," Sue says. Not only are children more likely to be frightened, the actual blood draw is more difficult because their veins are smaller. "You need communication strategies for working with patients across the lifespan," she says, in addition to being able to recognize and work with patients who have special needs.

Sue says that phlebotomists also need good fine motor skills and steady hands, noting that many of her colleagues cross-stitch as a hobby. "Football is a game of inches—this is a game of millimeters," she says.

Knowledge of anatomy and physiology is also important. "The specimens that the phlebotomist collects will relate to every single body system," Sue says. "If a physician orders a CBC (complete blood count), you need to know what specimen tube to use, what department in the lab it goes to, and what the clinical significance is." Above all, a phlebotomist needs to be able to find veins and arteries and know the differences between venous and arterial blood, or what it means when a specimen is rejected because it is hemolyzed, she explains.

Sue finds lab work fascinating, but her favorite aspect of phlebotomy is the people. "When you're working in the lab with your specimens, they all look the same in the rack," she explains. But outside of the lab, "the specimen has a face." Sue especially loved working blood drives and interacting with all the people who came to give blood. "Folks who are donors are generally very nice people," she says. "They're generally healthy, so they have great veins. You get to talk to them and find out all kinds of interesting things."

While phlebotomy itself is a career, Sue notes that many of her former students have gone on to become nurses, respiratory therapists, doctors, radiologic technologists, physician assistants, and clinical laboratory scientists. Because phlebotomists travel throughout the hospital, they have a chance to see a multitude of professions and departments on a regular basis.

For more information, visit the website for the Center for Phlebotomy Education at http://www.phlebotomy.com/.

Think this is the career for you?

Key Stats:

- **Education and Training.** High school diploma and training are required; some phlebotomists go on to become clinical laboratory scientists, nurses, or other medical professionals.

- **Licensure.** Most states do not require phlebotomists to be certified or licensed, though more and more employers do. Multiple organizations certify phlebotomists.

- **Earnings.** Earnings vary but the median hourly salary is $17.44.

- **Expected Job Prospects.** Employment is expected to grow faster than the national average—16% from 2008 to 2018.

Bureau of Labor Statistics, U.S. Department of Labor, *Occupational Outlook Handbook, 2010–11 Edition*, Clinical Laboratory Technologists and Technicians, on the Internet at http://www.bls.gov/oco/ocos096.htm (visited September 14, 2011).

Mastering A&P®

Access more review material online in the Study Area at www.masteringaandp.com.

- Chapter guides
- Chapter quizzes
- Practice tests
- Art labeling activities
- Flashcards
- A glossary with pronunciations

- Practice Anatomy Lab™ (PAL™) 3.0 virtual anatomy practice tool
- Interactive Physiology® (IP) animated tutorials
- MP3 Tutor Sessions

 practice anatomy lab™ **For this chapter, follow these navigation paths in PAL:**

- Human Cadaver>Cardiovascular System>Blood Vessels
- Anatomical Models>Cardiovascular System>Veins
- Anatomical Models>Cardiovascular System>Arteries

MP3 tutor sessions **For this chapter, go to this topic in the MP3 Tutor Sessions:**

- Hemoglobin: Function and Impact

Chapter 11 Review Questions

1. Relate the anatomical differences between arteries and veins to their functions.

2. What happens to the smooth muscle lining the arterioles during running?

3. What is the role of blood in stabilizing and maintaining body temperature?

4. Describe some cellular features of a red blood cell (RBC) that limit it to a fairly short life span.

5. Why is blood flow to the brain relatively continuous and constant?

For answers to all module, section, and chapter review questions, see the blue Answers tab at the back of the book.

12

The Heart and Cardiovascular Function

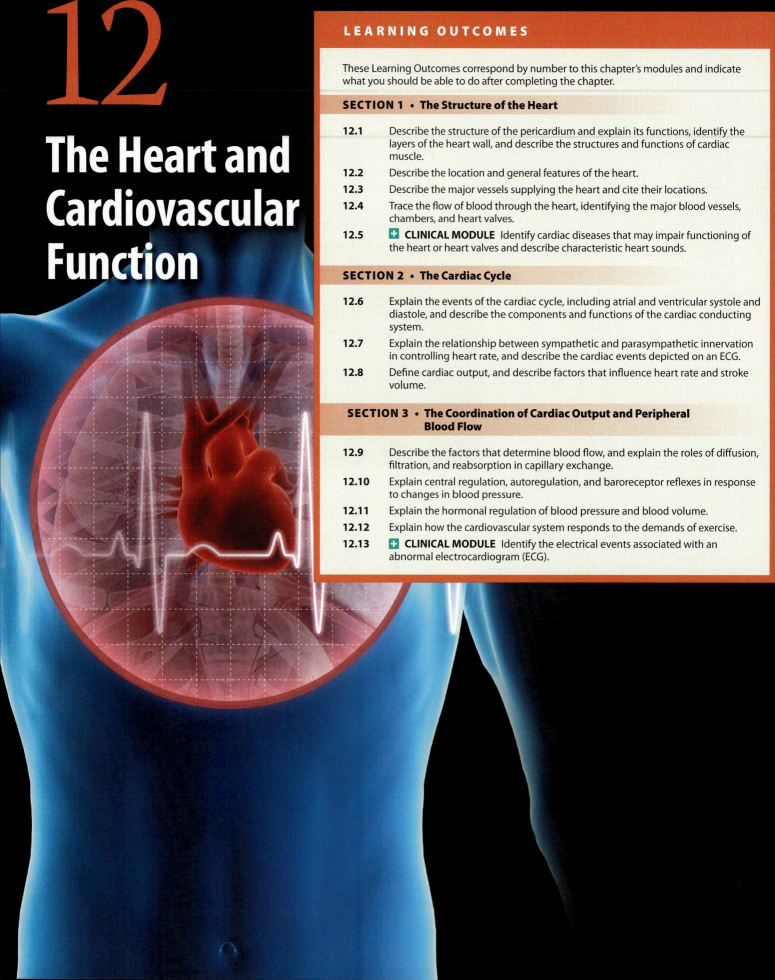

The Structure of the Heart

1 The heart is located near the anterior chest wall, directly posterior to the sternum. A midsagittal section through the trunk does not divide the heart into two equal halves, because the center lies slightly to the left of the midline. The entire heart is rotated to the left, so that the right atrium and right ventricle dominate an anterior view of the heart.

The **base** of the heart is at the superior surface, where the great veins and arteries are attached. The base sits posterior to the sternum at the level of the third costal cartilage, centered about 1.2 cm (0.5 in.) to the left side.

Ribs

The **apex** (A-peks) is the inferior, pointed tip of the heart. A typical adult heart measures approximately 12.5 cm (5 in.) from the base to the apex, which reaches the fifth intercostal space approximately 7.5 cm (3 in.) to the left of the midline.

2 This anterior view shows the borders of the heart. The base forms the **superior border**. The **right border** of the heart is formed by the right atrium; the **left border** is formed by the left ventricle and a small portion of the left atrium. The left border extends to the apex, where it meets the inferior border. The **inferior border** is formed mainly by the inferior wall of the right ventricle.

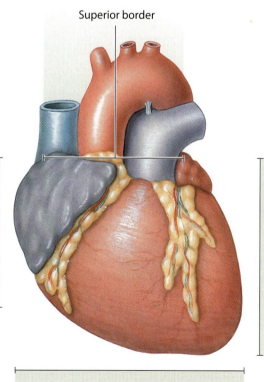

Superior border

Right border

Left border

Inferior border

This section examines the anatomy of the heart. We will then consider the regulation of cardiac (heart) function (Section 2) before exploring how cardiac activity and blood flow are regulated under changing conditions (Section 3).

433

The wall of the heart contains concentric layers of cardiac muscle tissue

1 This view shows a section taken from the heart wall and the surrounding pericardium. The heart wall contains three layers: epicardium, myocardium, and endocardium.

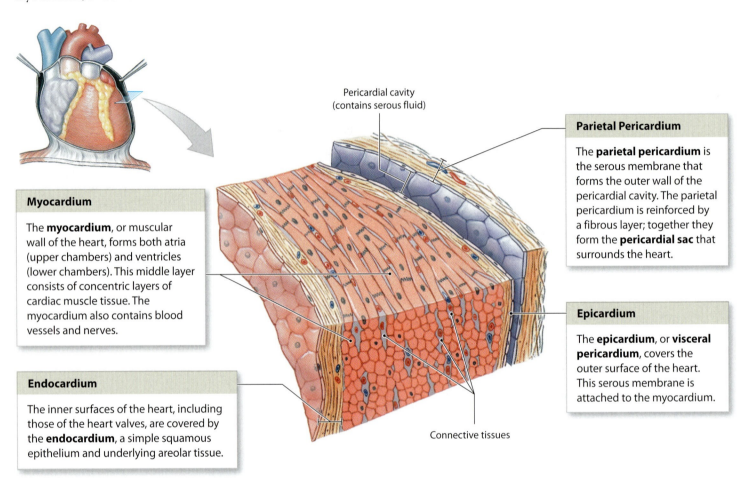

Pericardial cavity (contains serous fluid)

Parietal Pericardium

The **parietal pericardium** is the serous membrane that forms the outer wall of the pericardial cavity. The parietal pericardium is reinforced by a fibrous layer; together they form the **pericardial sac** that surrounds the heart.

Myocardium

The **myocardium**, or muscular wall of the heart, forms both atria (upper chambers) and ventricles (lower chambers). This middle layer consists of concentric layers of cardiac muscle tissue. The myocardium also contains blood vessels and nerves.

Epicardium

The **epicardium**, or **visceral pericardium**, covers the outer surface of the heart. This serous membrane is attached to the myocardium.

Endocardium

The inner surfaces of the heart, including those of the heart valves, are covered by the **endocardium**, a simple squamous epithelium and underlying areolar tissue.

Connective tissues

2 The atrial myocardium contains muscle bundles that wrap around the atria and form figure-eights that encircle the large arteries and veins at the base of the heart. Superficial ventricular muscles wrap around both ventricles; deeper muscle layers spiral around and between the ventricles toward the apex in a figure-eight pattern.

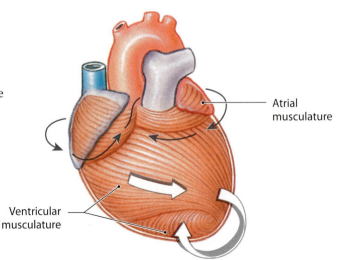

Atrial musculature

Ventricular musculature

3 This light micrograph shows the histological characteristics that distinguish cardiac muscle tissue from skeletal muscle tissue: small cell size, a single, centrally located nucleus, branching interconnections between cells, and specialized intercellular connections.

Each cardiac muscle cell is connected to several others at specialized sites known as **intercalated** (in-TER-ka-lā-ted) **discs**.

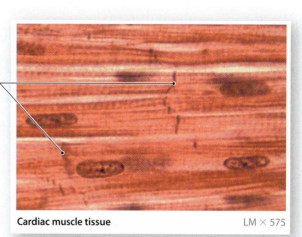

Cardiac muscle tissue   LM × 575

4 Cardiac muscle cells are found only in the heart. Like skeletal muscle fibers (cells), cardiac muscle cells contain organized myofibrils, and the presence of many aligned sarcomeres gives the cells a striated appearance. They depend almost totally on aerobic metabolism to obtain the energy they need to continue contracting. The sarcoplasm of a cardiac muscle cell contains large numbers of mitochondria and abundant reserves of myoglobin that store oxygen. Because these cells are metabolically very active and have a high demand for oxygen and nutrients, cardiac muscle tissues are richly supplied with capillaries.

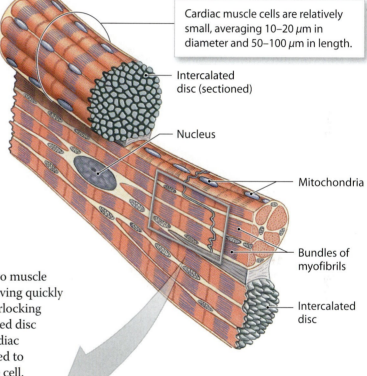

Cardiac muscle cells are relatively small, averaging 10–20 μm in diameter and 50–100 μm in length.

Intercalated disc (sectioned)

Nucleus

Mitochondria

Bundles of myofibrils

Intercalated disc

5 At an intercalated disc, the plasma membranes of two adjacent cardiac muscle cells are extensively intertwined and bound together by gap junctions and desmosomes. These connections help stabilize the positions of adjacent cells. The gap junctions allow ions and small molecules to move from one cell to another. This creates a direct electrical connection between the two muscle cells. An action potential can travel across an intercalated disc, moving quickly from one cardiac muscle cell to another. Myofibrils in the two interlocking muscle cells are firmly anchored to the membrane at the intercalated disc and can "pull together" with maximum efficiency. Because the cardiac muscle cells are mechanically, chemically, and electrically connected to one another, the entire tissue resembles a single, enormous muscle cell.

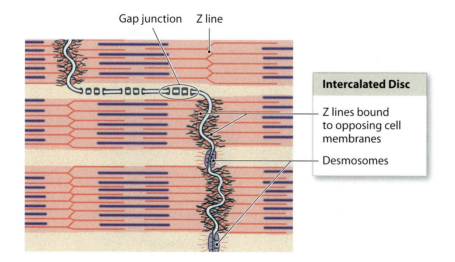

Gap junction   Z line

Intercalated Disc

Z lines bound to opposing cell membranes

Desmosomes

Module 12.1 Review

a. From superficial to deep, name the layers of the heart wall.

b. Describe the serous membrane lining the pericardial cavity.

c. Why is it important that cardiac tissue be richly supplied with mitochondria and capillaries?

The heart is suspended within the pericardial cavity; the boundaries of the internal chambers are visible on the external surface

1 This anterior view shows the position and orientation of the heart relative to the major vessels and the ribs, sternum, and lungs. The heart, surrounded by the pericardial sac, sits in the anterior portion of the **mediastinum** (mē-dē-AS-ti-num or mē-dē-a-STĪ-num), the region between the two pleural (lung) cavities. The mediastinum also contains blood vessels and the thymus, esophagus, and trachea.

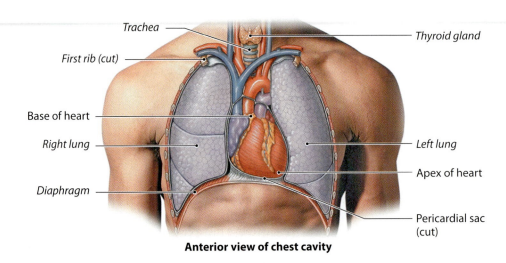

Anterior view of chest cavity

2 To visualize the relationship between the heart and the pericardial cavity, imagine pushing your fist toward the center of a large, partially inflated balloon. The balloon represents the pericardium, and your fist is the heart. Your wrist, where the balloon folds back on itself, corresponds to the base of the heart, to which the **great vessels**, the largest veins and arteries in the body, are attached. The air space inside the balloon corresponds to the pericardial cavity.

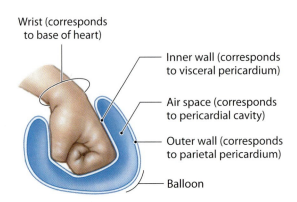

The pericardial cavity contains 15–50 mL of pericardial fluid, secreted by the visceral serous pericardium. This fluid acts as a lubricant that reduces friction between the opposing surfaces as the heart beats. Pathogens can infect the pericardium, producing the condition **pericarditis**. The inflamed pericardial surfaces rub against one another, producing a distinctive scratching sound that can be heard through a stethoscope.

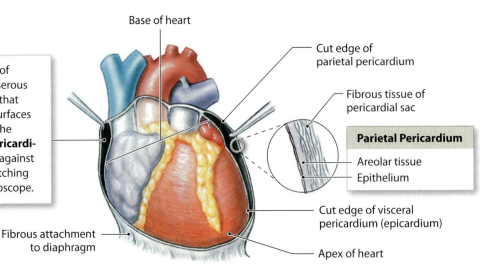

Parietal Pericardium

3 We can easily identify the four cardiac chambers in a superficial view of the **anterior surface** of the heart. The right and left atria have relatively thin muscular walls and are highly expandable. When not filled with blood, the outer portion of each atrium deflates and becomes a lumpy, wrinkled flap. Shallow grooves, or **sulci** (singular: *sulcus*), mark the boundaries between the atria and ventricles and between the left and right ventricles.

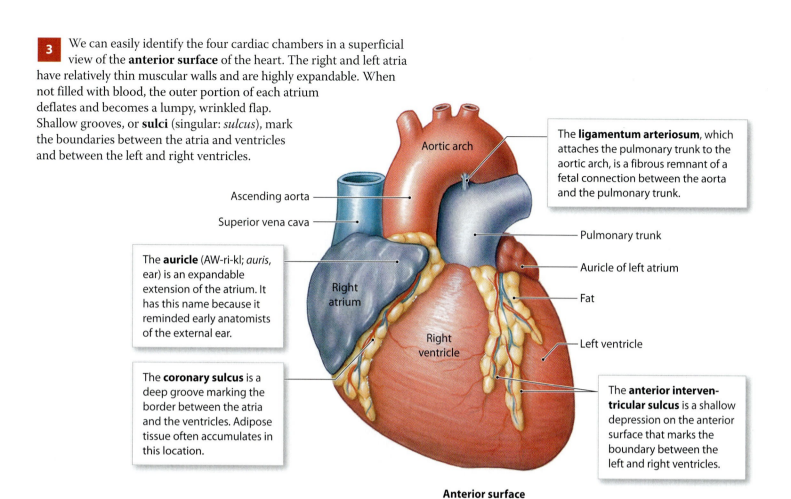

Aortic arch

The **ligamentum arteriosum**, which attaches the pulmonary trunk to the aortic arch, is a fibrous remnant of a fetal connection between the aorta and the pulmonary trunk.

Ascending aorta

Superior vena cava

Pulmonary trunk

Auricle of left atrium

The **auricle** (AW-ri-kl; *auris*, ear) is an expandable extension of the atrium. It has this name because it reminded early anatomists of the external ear.

Right atrium

Fat

Right ventricle

Left ventricle

The **coronary sulcus** is a deep groove marking the border between the atria and the ventricles. Adipose tissue often accumulates in this location.

The **anterior interventricular sulcus** is a shallow depression on the anterior surface that marks the boundary between the left and right ventricles.

Anterior surface

4 This view of the **posterior surface** of the heart shows the left atrium and its connection to the pulmonary veins. It also shows the right atrium and its connection to the coronary veins and the venae cavae.

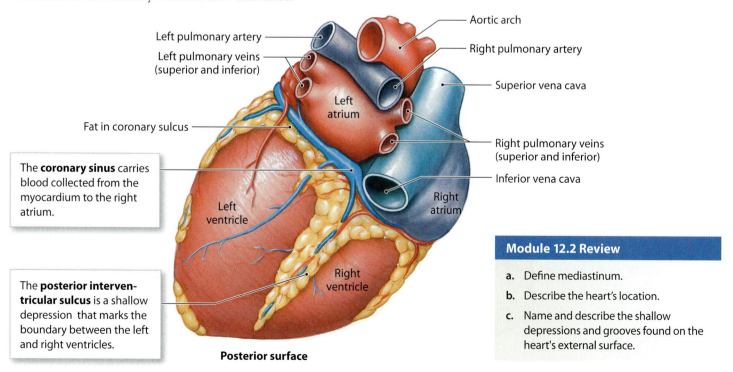

Left pulmonary artery

Left pulmonary veins (superior and inferior)

Aortic arch

Right pulmonary artery

Superior vena cava

Fat in coronary sulcus

Left atrium

The **coronary sinus** carries blood collected from the myocardium to the right atrium.

Right pulmonary veins (superior and inferior)

Inferior vena cava

Left ventricle

Right atrium

The **posterior interventricular sulcus** is a shallow depression that marks the boundary between the left and right ventricles.

Right ventricle

Posterior surface

Module 12.2 Review

a. Define mediastinum.

b. Describe the heart's location.

c. Name and describe the shallow depressions and grooves found on the heart's external surface.

The heart has an extensive blood supply

The heart works continuously, so cardiac muscle cells require reliable supplies of oxygen and nutrients. Although a great volume of blood flows through the chambers of the heart, the myocardium needs its own, separate blood supply. The coronary circulation supplies that blood to the muscle tissue of the heart. During maximum exertion, blood flow to the myocardium may increase to nine times that of resting levels.

1 The left and right **coronary arteries** originate at the base of the ascending aorta, where blood pressure is the highest in the systemic circuit. However, myocardial blood flow is not steady; it peaks while the heart muscle is relaxed, and almost ceases while the heart muscle contracts.

Left Coronary Artery

The **left coronary artery** supplies blood to the left ventricle, left atrium, and interventricular septum (wall between the ventricles).

Circumflex artery

The large **anterior interventricular artery**, or left anterior descending artery, runs along the surface within the anterior interventricular sulcus.

Right Coronary Artery

The **right coronary artery** follows the coronary sulcus around the heart. It supplies blood to the right atrium and portions of both ventricles.

Marginal arteries from the right coronary artery supply the surface of the right ventricle.

Pulmonary trunk

Aortic arch

Left atrium

Right atrium

Right ventricle

Left ventricle

Anterior view

Arterial anastomoses between the anterior and posterior interventricular arteries maintain continuous blood flow despite pressure fluctuations in the left and right coronary arteries.

2 Branches of the left and right coronary arteries continue onto the posterior surface of the heart.

The **circumflex artery** is a branch of the left coronary artery that curves to the left around the coronary sulcus, eventually meeting and fusing with small branches of the right coronary artery.

Marginal artery

Left atrium

Left ventricle

Right atrium

Right ventricle

Right coronary artery

The right coronary artery continues across the posterior surface of the heart, supplying the **posterior interventricular artery**, or posterior descending artery, which runs toward the apex within the posterior interventricular sulcus. This vessel supplies blood to the interventricular septum and adjacent portions of the ventricles.

Posterior view

3 This view identifies the major collecting vessels on the anterior surface of the heart.

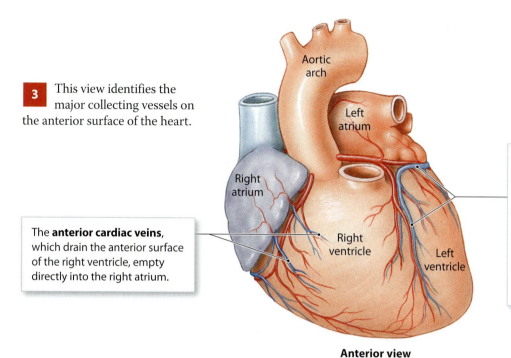

Aortic arch

Left atrium

Right atrium

Right ventricle

Left ventricle

The **anterior cardiac veins**, which drain the anterior surface of the right ventricle, empty directly into the right atrium.

The **great cardiac vein** begins on the anterior surface of the ventricles, along the interventricular sulcus. This vein drains blood from the region supplied by the anterior interventricular artery. The great cardiac vein reaches the level of the atria and then curves around the left side of the heart within the coronary sulcus to empty into the coronary sinus.

Anterior view

4 This view identifies the major collecting vessels on the posterior surface of the heart.

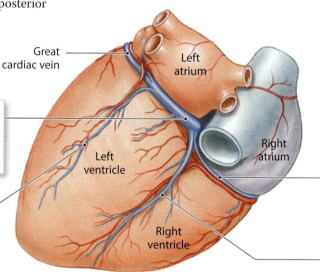

Great cardiac vein

Left atrium

The **coronary sinus** is an expanded vein that opens into the right atrium near the base of the inferior vena cava.

Left ventricle

Right atrium

The **posterior cardiac vein** drains the area supplied by the circumflex artery.

Right ventricle

The **small cardiac vein** receives blood from the posterior surfaces of the right atrium and ventricle. It empties into the coronary sinus in company with the middle cardiac vein.

The **middle cardiac vein**, draining the area supplied by the posterior interventricular artery, empties into the coronary sinus.

Posterior view

Each time the left ventricle contracts, it forces blood into the aorta. The arrival of additional blood at elevated pressures stretches the elastic walls of the aorta. When the left ventricle relaxes, pressure declines, and the walls of the aorta recoil. This recoil, called **elastic rebound**, pushes blood both forward, into the systemic circuit, and backward, into the coronary arteries. Thus, the combination of blood pressure and elastic rebound ensures a continuous flow of blood to meet the demands of active cardiac muscle tissue.

Module 12.3 Review

a. List the arteries and veins of the coronary circulation.

b. Describe what happens to blood flow in the aorta during elastic rebound.

c. Identify the main vessel that drains blood from the myocardial capillaries.

Internal valves control the direction of blood flow between the heart chambers

1 In a sectional view, you can see that the right atrium opens into the right ventricle, and the left atrium with the left ventricle. The **interatrial septum** (*septum*, wall) separates the two atria. The much thicker **interventricular septum** separates the two ventricles. Each septum is a muscular partition. **Atrioventricular (AV) valves**, folds of fibrous tissue, extend into the openings between the atria and ventricles. The AV valves permit blood flow in one direction only: from the atria to the ventricles.

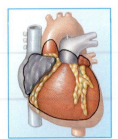

Right Atrium

The right atrium receives deoxygenated blood from the superior and inferior venae cavae. It also receives blood from the cardiac veins through the coronary sinus.

The **fossa ovalis** marks the location of the foramen ovale in the embryo. The foramen ovale closes at birth, and this opening between the atria permanently seals off within the first year.

The anterior atrial wall and the inner surface of the auricle contain prominent muscular ridges called the **pectinate muscles** (*pectin*, comb).

The opening of the coronary sinus carries blood from the cardiac veins.

Right Ventricle

Blood travels from the right atrium into the right ventricle through a broad opening bounded by the **right atrioventricular (AV) valve**, also known as the **tricuspid** (trī- KUS-pid; *tri*, three) **valve**.

The free edge of each valve consists of three flaps, or **cusps**, attached to tendinous connective tissue fibers called the **chordae tendineae** (KOR-dē TEN-di-nē-ē; tendinous cords).

The fibers originate at conical muscular projections called the **papillary** (PAP-i-ler-ē) **muscles**.

The superior portion of the right ventricle tapers toward the **pulmonary valve** (pulmonary semilunar valve). Blood leaving the right ventricle passes through the pulmonary valve to enter the pulmonary trunk.

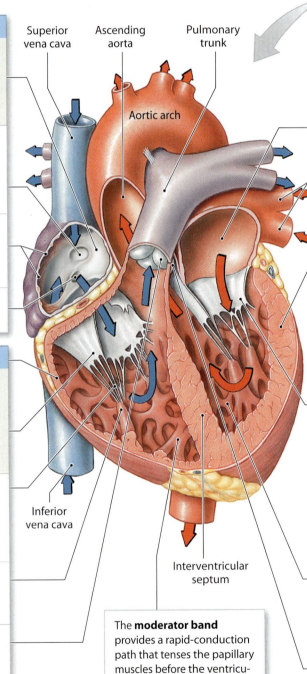

Superior vena cava

Ascending aorta

Pulmonary trunk

Aortic arch

Inferior vena cava

Interventricular septum

Left Atrium

The left atrium receives oxygenated blood from the pulmonary veins.

Left pulmonary veins

Left Ventricle

The left ventricle wall is much thicker than the right ventricle wall, enabling the left ventricle to develop enough pressure to push blood through the large systemic circuit, whereas the right ventricle only needs to develop enough pressure to pump blood through the nearby lungs.

The **left atrioventricular (AV) valve**, or **bicuspid** (bī-KUS-pid) **valve**, allows blood to flow from the left atrium into the left ventricle but prevents backflow during ventricular contraction. As the name bicuspid implies, the left AV valve contains two cusps. Clinicians often call this valve the **mitral** (MĪ-tral; *mitre*, a bishop's hat) **valve**.

The **trabeculae carneae** (tra-BEK-ū-lē KAR-nē-ē; *carneus*, fleshy) are a series of muscular ridges on the inner surfaces of the right and left ventricles.

Blood leaves the left ventricle by passing through the **aortic valve** (aortic semilunar valve) and into the ascending aorta.

The **moderator band** provides a rapid-conduction path that tenses the papillary muscles before the ventricular myocardium contracts. This prevents "slamming" of the right AV valve cusps.

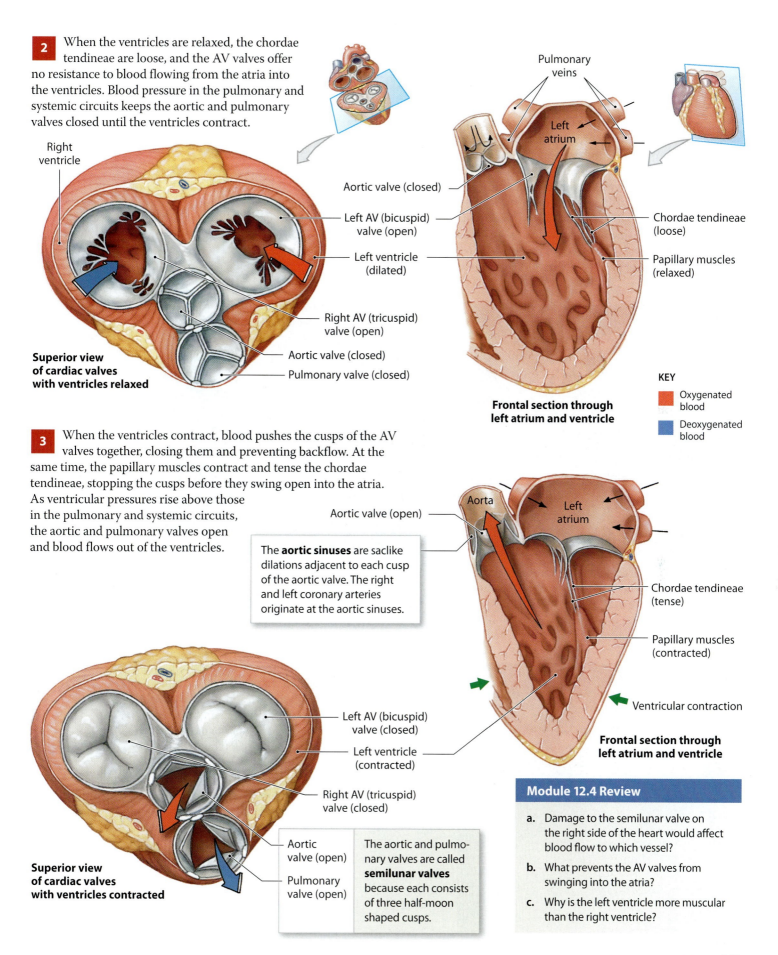

2 When the ventricles are relaxed, the chordae tendineae are loose, and the AV valves offer no resistance to blood flowing from the atria into the ventricles. Blood pressure in the pulmonary and systemic circuits keeps the aortic and pulmonary valves closed until the ventricles contract.

Right ventricle

Aortic valve (closed)

Left AV (bicuspid) valve (open)

Left ventricle (dilated)

Right AV (tricuspid) valve (open)

Aortic valve (closed)

Pulmonary valve (closed)

Superior view of cardiac valves with ventricles relaxed

Pulmonary veins

Left atrium

Chordae tendineae (loose)

Papillary muscles (relaxed)

Frontal section through left atrium and ventricle

KEY

Oxygenated blood

Deoxygenated blood

3 When the ventricles contract, blood pushes the cusps of the AV valves together, closing them and preventing backflow. At the same time, the papillary muscles contract and tense the chordae tendineae, stopping the cusps before they swing open into the atria. As ventricular pressures rise above those in the pulmonary and systemic circuits, the aortic and pulmonary valves open and blood flows out of the ventricles.

Aorta

Left atrium

Aortic valve (open)

The **aortic sinuses** are saclike dilations adjacent to each cusp of the aortic valve. The right and left coronary arteries originate at the aortic sinuses.

Chordae tendineae (tense)

Papillary muscles (contracted)

Ventricular contraction

Frontal section through left atrium and ventricle

Left AV (bicuspid) valve (closed)

Left ventricle (contracted)

Right AV (tricuspid) valve (closed)

Aortic valve (open)

Pulmonary valve (open)

The aortic and pulmonary valves are called **semilunar valves** because each consists of three half-moon shaped cusps.

Superior view of cardiac valves with ventricles contracted

Module 12.4 Review

a. Damage to the semilunar valve on the right side of the heart would affect blood flow to which vessel?

b. What prevents the AV valves from swinging into the atria?

c. Why is the left ventricle more muscular than the right ventricle?

Damage to the myocardium or heart valves may compromise heart function

1   **Coronary artery disease (CAD)** is a condition in which the coronary circulation is partially or completely blocked. Cardiac muscle cells need a constant supply of oxygen and nutrients, so any reduction in blood flow to the heart muscle produces a corresponding reduction in cardiac performance. Reduced circulatory supply, known as **coronary ischemia** (is-KĒ-mē-uh), generally results from partial or complete blockage of the coronary arteries. The usual cause is an atherosclerotic plaque in a coronary artery. The plaque, or an associated **thrombus** (blood clot), narrows the passageway and reduces blood flow. Spasms in the smooth muscles of the vessel wall can further decrease or even stop blood flow. Plaques may be visible in clinical scans or high-resolution ultrasound images.

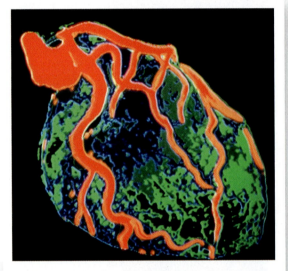

This is a color-enhanced **digital subtraction angiography (DSA)** scan of a normal heart. The major branches of the left and right coronary arteries are clearly visible.

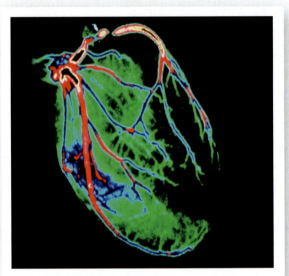

This is a color-enhanced DSA scan of the heart of an individual with advanced CAD. Blood flow to the ventricular myocardium is severely restricted.

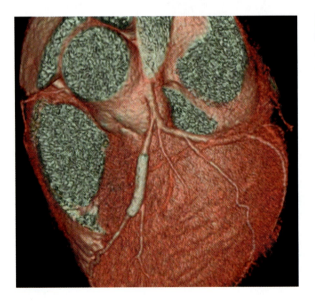

2   Balloon angioplasty removes most plaques, but they can redevelop. To prevent this, cardiac specialists routinely insert a fine wire-mesh tube called a **stent** into the vessel. The stent pushes against the vessel wall and holds it open, making angioplasty more effective and reducing long-term complications. This 3D spiral scan shows a stent that has been placed in the anterior interventricular artery. (This imaging technique reveals the lumen of the blood vessels, rather than the vessel, so the stent appears to surround the vessel.) If the circulatory blockage is extensive, mutiple stents can be inserted along the length of the vessel.

3 The pulmonary and aortic valves are composed of a set of three semilunar cusps of fibrous connective tissue. As a result, the semilunar valves do not require muscular braces. When the semilunar valves close, the three symmetrical cusps support one another like the legs of a tripod.

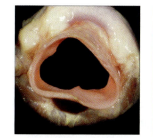 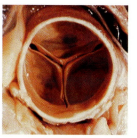

Open   Closed

Superior views

4 Defects or damage to any of the four valves of the heart can compromise cardiac function. If valve function deteriorates to the point at which the heart cannot maintain adequate blood flow, symptoms of **valvular heart disease (VHD)** appear. Congenital malformations may be responsible, but in many cases the condition develops as a consequence of **carditis**, an inflammation of the heart. In severe cases, the only option may be to replace the damaged valve with an artificial substitute, such as a bioprosthetic valve.

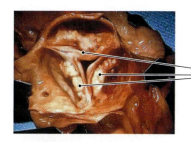

This is a superior view of a damaged aortic valve. The cusps are irregular in shape; they are also stiff and inflexible. A valve like this cannot open properly or close completely.

This bioprosthetic valve is an artificial valve that uses the cusps from a pig's heart. Pig or cow valves do not stimulate blood clotting, but they may wear out after roughly 10 years of service.

Heart sounds — S_4 S_1 "Lubb" S_2 "Dupp" S_3 S_4

5 Heart valves closing, blood rushing through the heart, and the heart muscle contracting—all of these events produce noises called heart sounds. There are four **heart sounds**, designated S_1 through S_4. If you listen to your own heart with a stethoscope, you will clearly hear the first and second heart sounds. The first heart sound, known as "lubb" (S_1), lasts a little longer than the second, called "dupp" (S_2). S_1, which marks the start of ventricular contraction, is produced as the AV valves close; S_2 occurs when the semilunar valves close. Third and fourth heart sounds are usually very faint and not heard in healthy adults. These sounds are associated with blood flowing into the ventricles (S_3) and atrial contraction (S_4), rather than with valve action.

Module 12.5 Review

a. What is coronary ischemia, and what danger does it pose?

b. Describe the purpose of a stent.

c. Summarize the characteristic heart sounds that can be heard with a stethoscope, and the events that cause them.

1. Labeling

Label each of the structures in the accompanying figure.

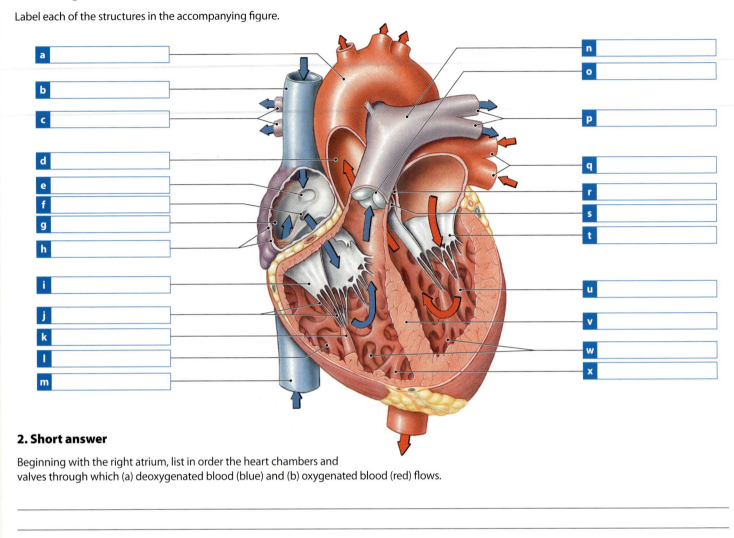

a _____
b _____
c _____
d _____
e _____
f _____
g _____
h _____
i _____
j _____
k _____
l _____
m _____

n _____
o _____
p _____
q _____
r _____
s _____
t _____
u _____
v _____
w _____
x _____

2. Short answer

Beginning with the right atrium, list in order the heart chambers and
valves through which (a) deoxygenated blood (blue) and (b) oxygenated blood (red) flows.

_____

_____

_____

3. Matching

Match the following terms with the most closely related description.

- apex
- fossa ovalis
- intercalated discs
- coronary arteries
- tricuspid valve
- aortic valve
- endocardium
- aorta
- myocardium
- elastic rebound
- coronary sinus
- pericardial cavity

a _____ Blood to systemic arteries

b _____ Contains serous fluid

c _____ Muscular wall of heart

d _____ Returns blood from heart wall back to heart

e _____ Depression in interatrial septum

f _____ Right atrioventricular valve

g _____ Provides an extra push to blood flow in the systemic circuit and coronary circulation

h _____ Cardiac muscle fiber connections

i _____ Provide blood to heart muscle tissue

j _____ Inner surface of heart

k _____ Inferior tip of heart

l _____ Semilunar valve

The Cardiac Cycle

1 Each heartbeat is followed by a brief resting phase, which allows time for all four chambers to relax and prepare for the next heartbeat. The period between the start of one heartbeat and the beginning of the next is a single **cardiac cycle**. The cardiac cycle, therefore, includes alternating periods of contraction and relaxation.

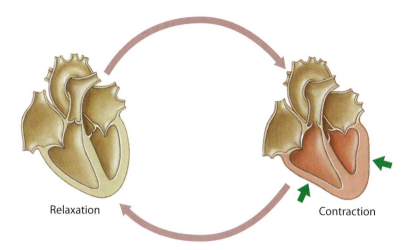

Relaxation   Contraction

2 Although we think of the heart as a pump, it is really four pumps that work in pairs, and thus a heartbeat is a complicated event. If all four chambers contracted at once, normal blood flow couldn't occur. Instead, the two atria contract first, pushing blood into the ventricles, and then the two ventricles contract, pushing blood through the pulmonary and systemic circuits and then back into the atria. The elaborate pacemaking and conducting systems within the heart normally provide the required spacing between atrial and ventricular contractions.

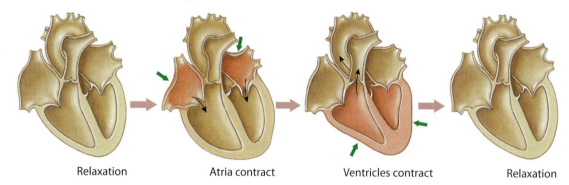

Relaxation   Atria contract   Ventricles contract   Relaxation

3 For any one chamber in the heart, the cardiac cycle can be divided into two phases: systole and diastole. During **systole** (SIS-tō-lē), or contraction, the chamber contracts and pushes blood into an adjacent chamber or into an arterial trunk. Systole is followed by **diastole** (dī-AS-tō-lē), or relaxation. During diastole, the chamber fills with blood and prepares for the next contraction. At a representative heart rate of 75 beats per minute (bpm), a sequence of systole and diastole in either the atria or the ventricles lasts 800 msec. This section examines a representative cardiac cycle—how it is started and coordinated and how the pressures generated by the contracting chambers move blood in one direction. For convenience, we will assume that the atria determine the cardiac cycle, and that it includes one cycle of atrial systole and atrial diastole.

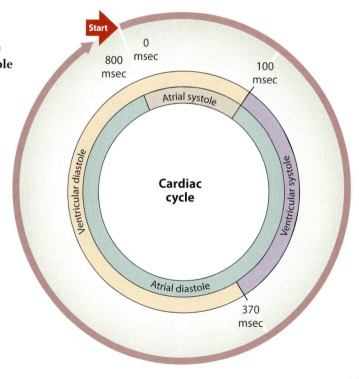

The cardiac cycle, which creates pressure gradients that maintain blood flow, is coordinated by specialized cardiac muscle cells

1 Here we diagram the phases of the cardiac cycle for a heart rate of 75 beats per minute (bpm). When the heart rate increases, all the phases of the cardiac cycle shorten. The greatest reduction occurs in the amount of time spent in diastole. When the heart rate climbs from 75 bpm to 200 bpm, the duration of systole drops by less than 40 percent, but the duration of diastole falls by almost 75 percent.

Start

1 Cardiac cycle begins: All four chambers are relaxed, and the ventricles are partially filled with blood.

2 Atrial systole: The atria contract, completely filling the relaxed ventricles with blood. Atrial systole lasts 100 msec.

3 Atrial diastole: Atrial diastole begins and then continues until the start of the next cardiac cycle.

Ventricular diastole lasts 530 msec (the 430 msec remaining in this cardiac cycle, plus the first 100 msec of the next). Throughout the rest of this cardiac cycle, filling occurs passively, and both the atria and the ventricles are relaxed. The next cardiac cycle begins with atrial systole and the completion of ventricular filling.

As atrial systole ends, ventricular systole begins. Ventricular systole, which lasts 270 msec, can be divided into two phases.

4 Ventricular systole— first phase: Ventricular contraction pushes the AV valves closed but does not create enough pressure to open the semilunar valves. This is known as the period of **isovolumetric contraction** because the volume of blood in the ventricles cannot change until those valves open.

8 Ventricular diastole— late: All chambers are relaxed. The AV valves open as the pressure from blood in the atria exceeds ventricular pressure. The ventricles fill passively to about 70% of their final volume.

800 msec · 0 msec · 100 msec

Atrial systole

Ventricular diastole

Cardiac cycle

Ventricular systole

Atrial diastole

370 msec

5 Ventricular systole— second phase: As ventricular pressure rises and exceeds pressure in the aorta and pulmonary trunk, the semilunar valves open and blood is forced out of the ventricle. This is known as the period of **ventricular ejection.**

7 Blood flows into the relaxed atria but the AV valves remain closed. This is known as the period of **isovolumetric relaxation.**

6 Ventricular diastole—early: As the ventricles relax, the pressure in them drops. Blood flows back against the cusps of the semilunar valves and forces them closed.

2 Cardiac muscle tissue contracts on its own, even without neural or hormonal stimulation. This property is called **automaticity**. The **conducting system** is a network of specialized cardiac muscle cells responsible for starting and distributing the stimulus for contraction. This illustration introduces the components of the heart's conducting system.

1 Each heartbeat begins with an action potential generated at the **sinoatrial** (sī-nō-Ā-trē-al) **node**, or simply the **SA node**. The SA node is embedded in the posterior wall of the right atrium, near the entrance to the superior vena cava. The electrical impulse generated by this cardiac pacemaker is then distributed by other cells of the conducting system.

2 In the atria, conducting cells in **internodal pathways** distribute the contractile stimulus to atrial muscle cells as the electrical impulse travels toward the ventricles.

3 The **atrioventricular (AV) node** is located at the junction between the atria and ventricles. The AV node also contains pacemaker cells, but they do not ordinarily affect the heart rate. However, if the SA node or internodal pathways are damaged, the heart will continue to beat because without commands from the SA node, the AV node will generate impulses at a rate of 40–60 beats per minute.

5 **Purkinje fibers** are large-diameter conducting cells that propagate action potentials very rapidly—as fast as small myelinated axons. Purkinje fibers are the final link in the distribution network, and they are responsible for the depolarization of the ventricular myocardial cells that triggers ventricular systole.

4 The AV node delivers the stimulus to the **AV bundle**, located within the interventricular septum. The AV bundle is normally the only electrical connection between the atria and the ventricles.

The AV bundle leads to the right and left **bundle branches**. The left bundle branch, which supplies the massive left ventricle, is much larger than the right bundle branch. Both branches extend toward the apex of the heart, turn, and fan out deep to the endocardial surface.

Moderator band

The cells of the AV node can generate impulses at a maximum rate of 230 per minute. Because each impulse results in a ventricular contraction, this value is the maximum normal heart rate. Even if the SA node generates impulses at an even faster rate, the ventricles will not contract faster than 230 bpm. Such high heart rates occur only when the heart or the conducting system has been damaged or stimulated by drugs.

Module 12.6 Review

a. Provide the alternate terms for the contraction and relaxation of heart chambers.

b. Describe the phases of the cardiac cycle.

c. Is the heart always pumping blood when pressure in the left ventricle is rising? Explain.

The ANS innervates the heart, and an ECG can display heart activity

Cardiac Innervation

1 The cardiac centers of the medulla oblongata contain the autonomic headquarters for cardiac control. The autonomic nervous system (ANS) adjusts both heart rate and the force of contraction.

The **cardioinhibitory center** controls the parasympathetic neurons that slow the heart rate.

The **cardioaccelerary center** controls sympathetic neurons that increase the heart rate.

Sympathetic

Sympathetic innervation arrives by postganglionic fibers within the cardiac nerves. These fibers innervate the conducting system, and the atrial and ventricular myocardium.

Cervical sympathetic ganglion

Sympathetic preganglionic fiber

Sympathetic postganglionic fiber

Cardiac nerve

Vagal nucleus

Medulla oblongata

Vagus nerve (N X)

Spinal cord

Parasympathetic

Parasympathetic innervation arrives by the vagus nerve and synapses with postganglionic neurons in the cardiac plexus. Postganglionic fibers innervate the SA node, AV node, and atrial musculature; innervation of the ventricular musculature is very limited.

Parasympathetic preganglionic fiber

Synapses in cardiac plexus

Parasympathetic postganglionic fibers

2 Each of us has a characteristic resting heart rate that varies with age, general health, and physical conditioning. According to the American Heart Association, the normal range of resting heart rates is 60–100 bpm.

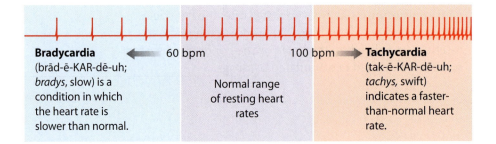

Bradycardia (brād-ē-KAR-dē-uh; *bradys,* slow) is a condition in which the heart rate is slower than normal.

60 bpm

Normal range of resting heart rates

100 bpm

Tachycardia (tak-ē-KAR-dē-uh; *tachys,* swift) indicates a faster-than-normal heart rate.

The Electrocardiogram (ECG)

3 The electrical events occurring in the heart are powerful enough to be detected by electrodes placed on the surface of the body. A recording of these events over time is an **electrocardiogram** (e-lek-trō-KAR-dē-ō-gram), also called an **ECG** or **EKG**. The appearance of the ECG varies with the placement of the monitoring electrodes, or leads. The photo shows the leads in one of the standard configurations and the graph indicates the important features of an ECG obtained using that configuration.

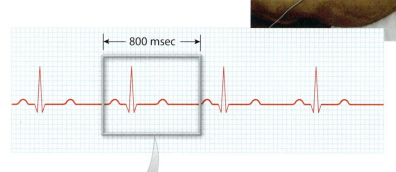

← 800 msec →

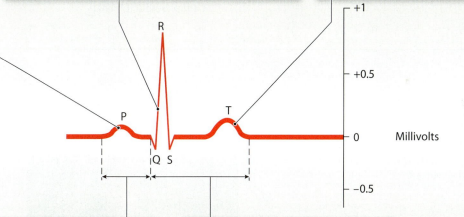

The small **P wave** accompanies the depolarization of the atria. The atria begin contracting about 25 msec after the P wave starts.

The **QRS complex** appears as the ventricles depolarize. This is a relatively strong electrical signal, because the ventricular muscle is much more massive than the atrial muscle. It is also a complex signal, largely because of the complex pathway that the spread of depolarization takes through the ventricles. The ventricles begin contracting shortly after the peak of the **R wave**.

The smaller **T wave** coincides with ventricular repolarization. A wave corresponding to atrial repolarization is not apparent, because it occurs while the ventricles are depolarizing, and the electrical events are masked by the QRS complex.

R

+1

+0.5

P   T

0   Millivolts

Q S

−0.5

The **P–R interval** extends from the start of atrial depolarization to the start of the QRS complex (ventricular depolarization) rather than to R, because in abnormal ECGs the peak at R can be difficult to determine. A P–R interval that lasts longer than 200 msec can indicate damage to the conducting pathways or AV node.

The **Q–T interval** indicates the time required for the ventricles to undergo a single cycle of depolarization and repolarization. It is usually measured from the end of the P–R interval rather than from the bottom of the Q wave.

Clinicians use an ECG to assess the performance of specific nodal, conducting, and contractile components. When a heart attack has damaged part of the heart, for example, the ECG will reveal an abnormal pattern of impulse conduction.

Module 12.7 Review

a. Describe the sites and actions of the cardioinhibitory and cardioacceler\atory centers.

b. Compare bradycardia with tachycardia.

c. List five important features of the ECG, and indicate what each represents.

Adjustments in heart rate and stroke volume regulate cardiac output

1 The goal of cardiovascular regulation is to maintain adequate blood flow to vital tissues. The best indicator of peripheral blood flow is **cardiac output (CO)**, the amount of blood pumped by the left ventricle in one minute. Cardiac output depends on two factors: the heart rate and the **stroke volume**, the amount of blood pumped out of the ventricle during a single heartbeat.

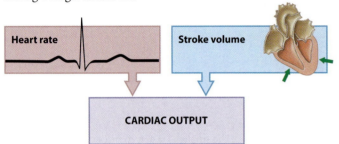

2 The calculation of cardiac output is straightforward: you multiply the heart rate (HR) by the average stroke volume (SV). For example, if the heart rate is 75 bpm and stroke volume is 80 mL/beat:

| HR: 75 beats/min | x | SV: 80 mL/beat | = | CO: 6000 mL/min |

The body precisely adjusts cardiac output such that peripheral tissues receive an adequate circulatory supply under a variety of conditions. When necessary, the heart rate can increase by 250 percent, and stroke volume in a normal heart can almost double.

3 As this figure indicates, cardiac output varies widely to meet metabolic demands, and the body must closely monitor and tightly control those variations.

Cardiac output (L/min)

0   5   10   15   20   25   30   35   40

Average resting cardiac output

Normal range of cardiac output during heavy exercise

Cardiac output in some forms of **heart failure**, a condition that exists when the heart can no longer meet the demands of peripheral tissues

Maximum for trained athletes exercising at peak levels

4 Anything that changes heart rate or stroke volume can change cardiac output. This diagram summarizes important factors that affect cardiac output. Review this carefully before proceeding to the next section.

Factors affecting heart rate (HR)

Body Temperature

When body temperature rises, heart rate goes up; that's why your heart seems to race when you have a fever. A drop in body temperature lowers heart rate; during heart surgery, body temperature may be artificially lowered.

Neural Regulation

Sympathetic stimulation increases HR; parasympathetic stimulation decreases HR.

Hormonal Regulation

Many hormones increase heart rate (notably, epinephrine and thyroid hormones).

Factors affecting stroke volume (SV)

Venous Return

Venous return is the amount of venous blood delivered to the heart by the systemic circulation each minute. Blood volume, muscular activity, and the rate of blood flow through peripheral capillaries are the major factors that determine venous return.

The **filling time** is the duration of ventricular diastole. The longer the filling time, the more blood the ventricles will contain when they start to contract. The greater the volume of blood in the ventricle, the stronger the contraction. This relationship of "more blood in → more blood out" is known as **Starling's law of the heart**. Both an increased filling time and increased ventricular blood volume lead to an increase in the stroke volume.

| Muscular contractions compress veins and help valves direct venous blood toward the right atrium. | Large reductions in blood volume due to bleeding or dehydration reduce venous return. | Changes in peripheral blood flow patterns can increase or decrease venous return. |

In general, when venous return increases, SV increases; when venous return decreases, SV decreases.

Contractility

Contractility is the amount of force produced during a ventricular contraction. Many medicines are intended to reduce contractility; examples include the "beta-blockers" such as propranolol or timolol and calcium channel blockers such as nifedipine or verapamil.

Neural Regulation	Hormonal Regulation
Sympathetic stimulation increases contractility; parasympathetic stimulation decreases it.	Many hormones—including epinephrine, norepinephrine, and thyroid hormones—increase contractility.

HEART RATE (HR)

STROKE VOLUME (SV)

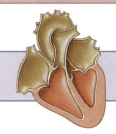

CARDIAC OUTPUT (CO) = HR × SV

Module 12.8 Review

a. Define cardiac output.

b. How would an increase in venous return affect stroke volume?

c. Why is it a potential problem if the heart beats too rapidly?

1. Short answer

Refer to the discussion of the cardiac cycle in Module 12.6 in answering questions a through g below.

a A contracting heart chamber is in _____ (diastole/systole), and a relaxed heart chamber is in _____ (diastole/systole).

b As the cardiac cycle begins, the ventricles are _____ (partially/completely) filled with blood.

c What happens to the left AV valve when the pressure in the left ventricle rises above that in the left atrium? _____

d During most of ventricular diastole, the pressure in the left ventricle is _____ (greater than/the same as/less than) the pressure in the left atrium.

e What happens to the aortic valve when the pressure within the left ventricle becomes greater than the pressure within the aorta? _____

f During isovolumetric contraction, the semilunar valves are _____ (open/closed).

g During isovolumetric relaxation, the AV and semilunar valves are _____ (open/closed).

2. Short answer

Starting with the SA node, name and describe the roles of the structures that carry an action potential through the conducting network of the heart.

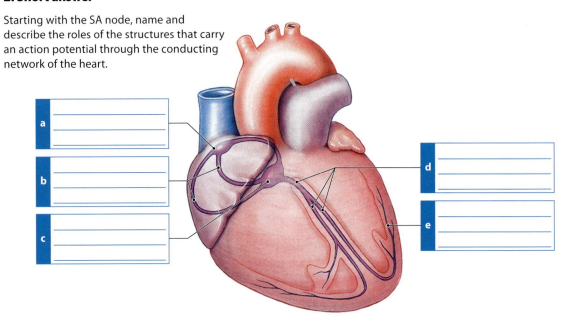

a _____

b _____

c _____

d _____

e _____

3. Matching

Match the following terms with the most closely related description.

- P wave
- cardiac output
- automaticity
- Purkinje fibers
- contractility
- filling time
- sympathetic neurons
- QRS complex
- stroke volume
- tachycardia
- bradycardia
- parasympathetic neurons

a _____ Amount of force produced during ventricular contraction

b _____ Self-stimulated cardiac muscle contractions

c _____ Atrial depolarization

d _____ Amount of blood ejected by the left ventricle during a single beat

e _____ Ventricular depolarization

f _____ Decrease the heart rate

g _____ Term for slower-than-normal heart rate

h _____ Increase the heart rate

i _____ Depolarize the ventricular myocardial cells

j _____ Term for faster-than-normal heart rate

k _____ HR × SV

l _____ Duration of ventricular diastole

The Coordination of Cardiac Output and Peripheral Blood Flow

You are now familiar with the factors involved in regulating cardiac output. Both cardiac output and the distribution of blood within the pulmonary and systemic circuits must constantly be adjusted to meet the demands of active tissues. Here we summarize the sites and mechanisms of cardiovascular regulation.

Cardiac Output

To maintain **cardiac output**, the heart must generate enough pressure to force blood through thousands of miles of peripheral capillaries, most scarcely larger in diameter than a single red blood cell.

Arterial Blood Pressure

Blood pressure is the pressure within the cardiovascular system as a whole. **Arterial pressure** is much higher than **venous pressure**. Without a high arterial pressure, blood could not be pushed through smaller and smaller arteries and then through innumerable capillary networks.

Venous Return

The **venous return** is the amount of blood arriving at the right ventricle each minute. On average, venous return is equal to cardiac output.

Regulation (Neural and Hormonal)

Neural and hormonal regulation make coordinated adjustments in heart rate, stroke volume, peripheral resistance, and venous pressure so that cardiac output is sufficient to meet the demands of peripheral tissues.

Venous Pressure

Valves and muscular compression of peripheral veins play an important role in maintaining venous pressure and venous blood flow. As blood moves toward the heart, the vessels get larger and larger in diameter, and resistance decreases continuously.

Peripheral Resistance

Resistance is a force that opposes movement. The **peripheral resistance** is the resistance of the arterial system as a whole. Resistance increases as the arterial branches get smaller and smaller.

Capillary Pressure

Because capillary pressures are very low, blood flows slowly, allowing plenty of time for diffusion between the blood and the surrounding interstitial fluid. This interplay, called **capillary exchange**, is the primary focus of the entire cardiovascular system.

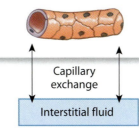

Capillary exchange

Interstitial fluid

In this section we will consider the functioning of the cardiovascular system as a whole, and learn how neural and hormonal mechanisms regulate cardiovascular function.

Arterial pressure and peripheral resistance determine blood flow and affect capillary exchange

Blood flow is the volume of blood flowing per unit of time through a vessel or group of vessels. It is directly proportional to **blood pressure** or **arterial pressure** (increased pressure results in increased flow), and inversely proportional to **peripheral resistance** (increased resistance results in decreased flow). However, the absolute blood pressure is less important than the **pressure gradient**—the difference in pressure from one end of a vessel to the other. The largest pressure gradient is found between the base of the aorta and the proximal ends of peripheral capillary beds. Cardiovascular control centers can alter this gradient, and thereby change the rate of capillary blood flow, by adjusting cardiac output and peripheral resistance.

1 Arterial pressure is not constant; it rises during ventricular systole and falls during ventricular diastole as the elastic arterial walls stretch and recoil. The peak blood pressure measured during ventricular systole is called **systolic pressure**, and the minimum blood pressure at the end of ventricular diastole is called **diastolic pressure**. In recording blood pressure, we separate systolic and diastolic pressures with a slash. So, the blood pressure illustrated here is written as 120/90 (read as "one-twenty over ninety").

The difference between the systolic and diastolic pressures is the **pulse pressure**.

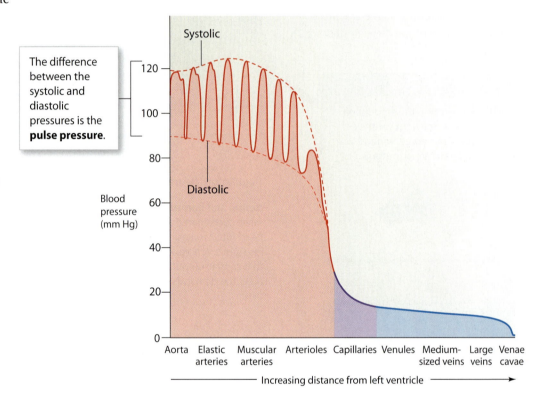

2 **Capillary exchange** involves a combination of filtration, diffusion, and osmosis. **Capillary hydrostatic pressure (CHP)** is the blood pressure within capillary beds, and it provides the driving force for filtration. Along the length of a typical capillary, CHP pushes water and soluble molecules out of the bloodstream and into the interstitial fluid. Only small solutes can cross the endothelium. Larger molecules, such as plasma proteins, must stay in the bloodstream. This size-selective process is called **filtration**.

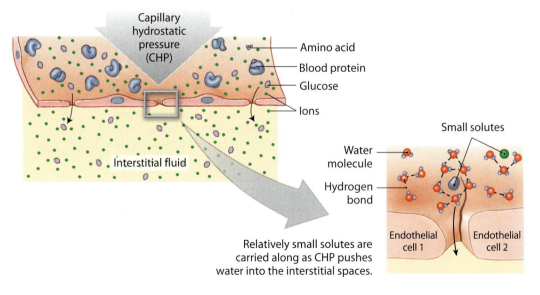

Relatively small solutes are carried along as CHP pushes water into the interstitial spaces.

3 Diffusion is the net movement of ions or molecules from an area where their concentration is higher to an area where their concentration is lower. Diffusion occurs most rapidly when the distances involved are short, the concentration gradient is large, and the ions or molecules involved are small. Diffusion occurs continuously across capillary walls, but different substances use different routes, as noted in this table.

Routes of Diffusion across Capillary Walls

- Water, ions, and small organic molecules, such as glucose, amino acids, and urea: Usually enter or leave the bloodstream by diffusion either between adjacent endothelial cells or through the pores of special capillaries.
- Many ions, including sodium, potassium, calcium, and chloride: Diffuse across endothelial cells by passing through channels in plasma membranes.
- Large water-soluble compounds: Cannot enter or leave the bloodstream except at special capillaries, such as those in the hypothalamus, the kidneys, many endocrine organs, and the intestinal tract.
- Lipids (such as fatty acids and steroids) and lipid-soluble materials (including soluble gases such as oxygen and carbon dioxide): Cross capillary walls by diffusion through the endothelial plasma membranes.
- Plasma proteins: Normally cannot cross the endothelial lining anywhere except in sinusoids, such as those of the liver, where plasma proteins enter the bloodstream.

4 In a capillary, blood pressure declines as blood flows from the arterial end to the venous end. As a result, the rates of filtration and reabsorption gradually change as blood passes along the length of a capillary. The figure below diagrams the factors involved.

Filtration Predominates

Plasma protein concentrations increase.

At a point roughly two-thirds of the way along the capillary, blood pressure has fallen to the point where there is no net movement of fluid into or out of the capillary.

Reabsorption Predominates

In the final third of the capillary, blood pressure is lowest and blood osmotic concentration is highest. Water flows back into the capillary due to osmosis. Note that more water leaves the bloodstream during filtration than gets retrieved through reabsorption. The difference, roughly 3.6 L/day, flows through body tissues and enters nearby lymphatic vessels that eventually return it to the venous system.

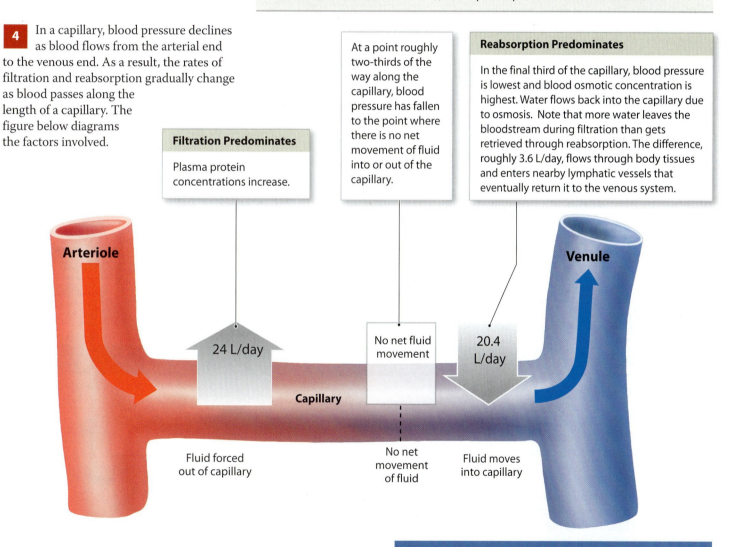

Arteriole

Venule

24 L/day

No net fluid movement

20.4 L/day

Capillary

Fluid forced out of capillary

No net movement of fluid

Fluid moves into capillary

Module 12.9 Review

a. Define blood flow, and describe its relationship to blood pressure and peripheral resistance.

b. In a healthy individual, is blood pressure higher in the aorta or in the inferior vena cava? Explain.

c. More water leaves a capillary than is reabsorbed along its length. Where does this water go?

Cardiovascular regulatory mechanisms respond to changes in blood pressure

1 Homeostatic mechanisms regulate cardiovascular activity to ensure that **tissue perfusion**—blood flow through tissues—meets the tissue demands for oxygen and nutrients. There are two regulatory pathways: (1) autoregulation and, if that is ineffective, (2) central regulation by neural and endocrine mechanisms.

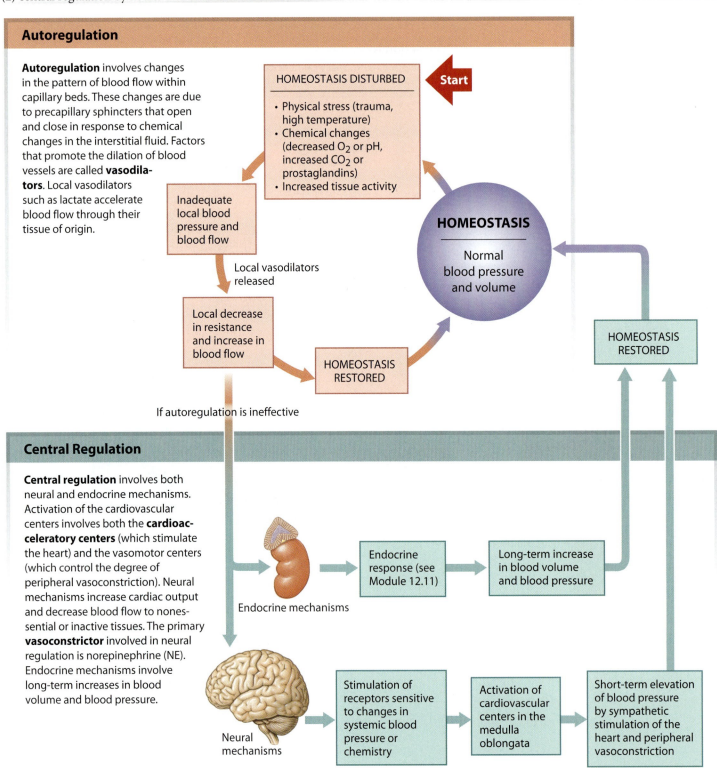

Autoregulation

Autoregulation involves changes in the pattern of blood flow within capillary beds. These changes are due to precapillary sphincters that open and close in response to chemical changes in the interstitial fluid. Factors that promote the dilation of blood vessels are called **vasodilators**. Local vasodilators such as lactate accelerate blood flow through their tissue of origin.

Start

HOMEOSTASIS DISTURBED

- Physical stress (trauma, high temperature)
- Chemical changes (decreased O_2 or pH, increased CO_2 or prostaglandins)
- Increased tissue activity

Inadequate local blood pressure and blood flow

Local vasodilators released

Local decrease in resistance and increase in blood flow

HOMEOSTASIS RESTORED

HOMEOSTASIS

Normal blood pressure and volume

HOMEOSTASIS RESTORED

If autoregulation is ineffective

Central Regulation

Central regulation involves both neural and endocrine mechanisms. Activation of the cardiovascular centers involves both the **cardioacceleratory centers** (which stimulate the heart) and the vasomotor centers (which control the degree of peripheral vasoconstriction). Neural mechanisms increase cardiac output and decrease blood flow to nonessential or inactive tissues. The primary **vasoconstrictor** involved in neural regulation is norepinephrine (NE). Endocrine mechanisms involve long-term increases in blood volume and blood pressure.

Endocrine mechanisms

Endocrine response (see Module 12.11)

Long-term increase in blood volume and blood pressure

Neural mechanisms

Stimulation of receptors sensitive to changes in systemic blood pressure or chemistry

Activation of cardiovascular centers in the medulla oblongata

Short-term elevation of blood pressure by sympathetic stimulation of the heart and peripheral vasoconstriction

2 The **baroreceptor reflexes** (*baro-*, pressure) respond to changes in blood pressure. The baroreceptors are located in: (1) the walls of the carotid sinuses (expanded chambers near the bases of the internal carotid arteries in the neck); (2) the aortic sinuses (pockets in the walls of the ascending aorta adjacent to the heart); and (3) the right atrium.

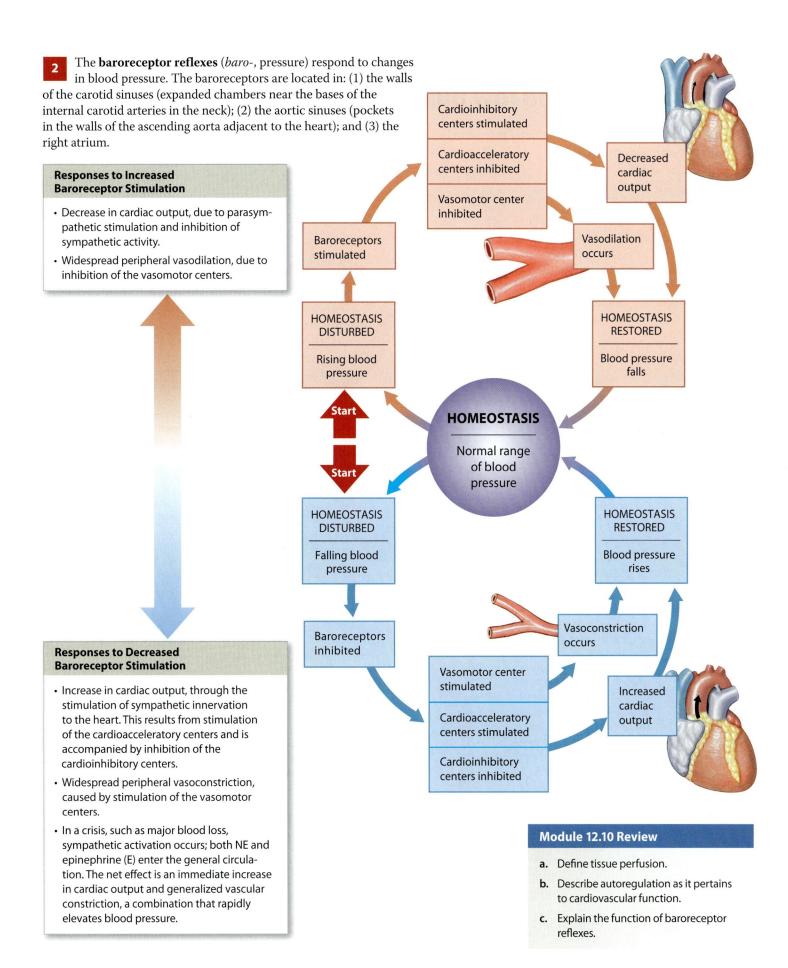

Responses to Increased Baroreceptor Stimulation

- Decrease in cardiac output, due to parasympathetic stimulation and inhibition of sympathetic activity.
- Widespread peripheral vasodilation, due to inhibition of the vasomotor centers.

Cardioinhibitory centers stimulated

Cardioacceleratory centers inhibited

Vasomotor center inhibited

Decreased cardiac output

Vasodilation occurs

Baroreceptors stimulated

HOMEOSTASIS DISTURBED
Rising blood pressure

HOMEOSTASIS RESTORED
Blood pressure falls

Start

HOMEOSTASIS
Normal range of blood pressure

Start

HOMEOSTASIS DISTURBED
Falling blood pressure

HOMEOSTASIS RESTORED
Blood pressure rises

Baroreceptors inhibited

Vasoconstriction occurs

Vasomotor center stimulated

Cardioacceleratory centers stimulated

Cardioinhibitory centers inhibited

Increased cardiac output

Responses to Decreased Baroreceptor Stimulation

- Increase in cardiac output, through the stimulation of sympathetic innervation to the heart. This results from stimulation of the cardioacceleratory centers and is accompanied by inhibition of the cardioinhibitory centers.
- Widespread peripheral vasoconstriction, caused by stimulation of the vasomotor centers.
- In a crisis, such as major blood loss, sympathetic activation occurs; both NE and epinephrine (E) enter the general circulation. The net effect is an immediate increase in cardiac output and generalized vascular constriction, a combination that rapidly elevates blood pressure.

Module 12.10 Review

a. Define tissue perfusion.

b. Describe autoregulation as it pertains to cardiovascular function.

c. Explain the function of baroreceptor reflexes.

The endocrine responses to low blood pressure and low blood volume ...

The endocrine system provides both short-term and long-term regulation of cardiovascular performance through the endocrine functions of the heart and kidneys, in addition to the actions of antidiuretic hormone (ADH) released from the pituitary gland.

1 When blood pressure and blood volume fall below normal, the adrenal medullae immediately release E and NE, stimulating cardiac output and peripheral vasoconstriction. Other hormones important in the long-term response include ADH, angiotensin II, erythropoietin (EPO), and aldosterone. All four are concerned primarily with long-term regulation of blood volume. ADH and angiotensin II also affect blood pressure.

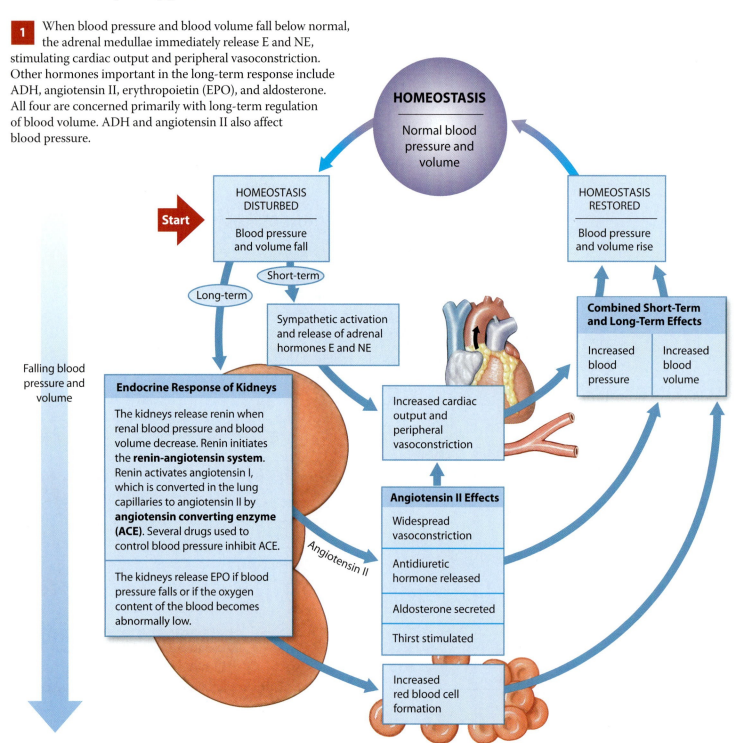

HOMEOSTASIS

Normal blood pressure and volume

HOMEOSTASIS DISTURBED

Blood pressure and volume fall

Start

Short-term

Long-term

HOMEOSTASIS RESTORED

Blood pressure and volume rise

Sympathetic activation and release of adrenal hormones E and NE

Combined Short-Term and Long-Term Effects

| Increased blood pressure | Increased blood volume |

Falling blood pressure and volume

Endocrine Response of Kidneys

The kidneys release renin when renal blood pressure and blood volume decrease. Renin initiates the **renin-angiotensin system**. Renin activates angiotensin I, which is converted in the lung capillaries to angiotensin II by **angiotensin converting enzyme (ACE)**. Several drugs used to control blood pressure inhibit ACE.

The kidneys release EPO if blood pressure falls or if the oxygen content of the blood becomes abnormally low.

Increased cardiac output and peripheral vasoconstriction

Angiotensin II

Angiotensin II Effects

Widespread vasoconstriction

Antidiuretic hormone released

Aldosterone secreted

Thirst stimulated

Increased red blood cell formation

... are very different from those to high blood pressure and high blood volume

2   Excessive blood volume triggers a response through its effects on the walls of the heart. When the heart walls are abnormally stretched during diastole, cardiac muscle cells release a peptide hormone. **Atrial natriuretic peptide** (nā-trē-ū-RET-ik; *natrium*, sodium + *ouresis*, urination), or **ANP**, is produced by cardiac muscle cells in the wall of the right atrium. Falling blood volume and blood pressure reduce the stresses on the walls of the heart, and ANP production ceases.

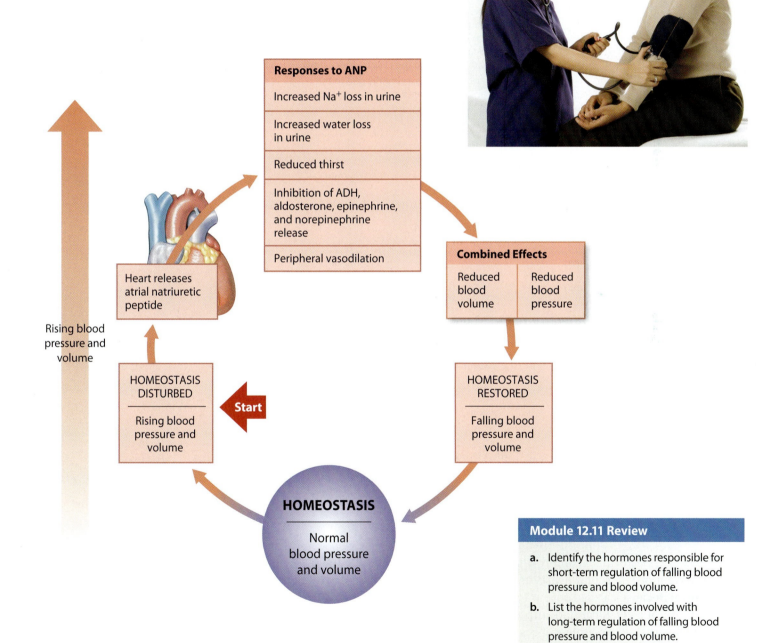

Responses to ANP

Increased Na⁺ loss in urine

Increased water loss in urine

Reduced thirst

Inhibition of ADH, aldosterone, epinephrine, and norepinephrine release

Peripheral vasodilation

Combined Effects

| Reduced blood volume | Reduced blood pressure |

Heart releases atrial natriuretic peptide

Rising blood pressure and volume

HOMEOSTASIS DISTURBED
___
Rising blood pressure and volume

Start

HOMEOSTASIS RESTORED
___
Falling blood pressure and volume

HOMEOSTASIS
___
Normal blood pressure and volume

Module 12.11 Review

a. Identify the hormones responsible for short-term regulation of falling blood pressure and blood volume.

b. List the hormones involved with long-term regulation of falling blood pressure and blood volume.

c. In what way would a kidney respond to vasoconstriction of its renal artery?

The cardiovascular centers make extensive adjustments to cardiac output and blood distribution during exercise

At Rest

At rest, cardiac output averages around 5.8 L/min per ventricle. Note the pattern of blood distribution to major organs, especially the skeletal muscles, brain, and abdominal viscera.

Light Exercise

As you begin light exercise, three interrelated changes take place:

1. Vasodilation occurs, peripheral resistance drops, and blood flow through the capillaries increases.
2. Venous return increases as skeletal muscle contractions squeeze blood along the peripheral veins. At the same time, each inhalation creates negative pressure in the thoracic cavity that pulls blood into the venae cavae from their branches. This mechanism is called the **respiratory pump**.
3. Cardiac output rises, primarily due to the increased venous return.

Heavy Exercise

At higher levels of exertion, cardiac output increases toward maximal levels. Major changes in the peripheral distribution of blood allow a massive increase in blood flow to skeletal muscles while preventing a potentially disastrous decline in systemic blood pressure. Under massive sympathetic stimulation, the vasomotor centers severely restrict blood flow to "nonessential" organs, such as the abdominal viscera. Although blood flow to most tissues declines, body temperature rises and skin perfusion increases to promote heat loss. Only the blood supply to the brain remains unchanged.

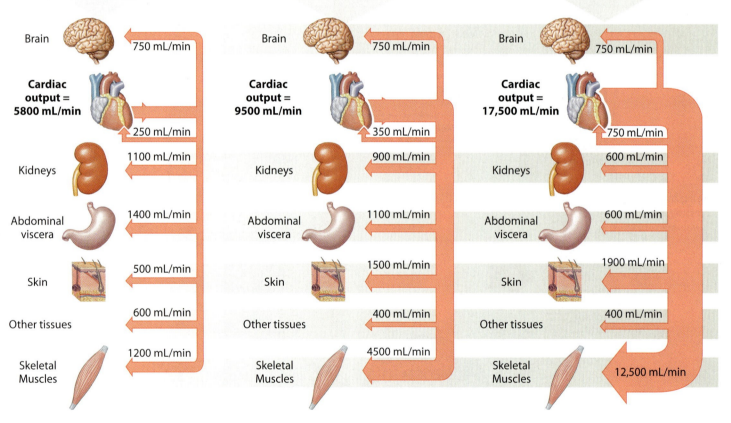

At Rest	Light Exercise	Heavy Exercise
Brain — 750 mL/min	Brain — 750 mL/min	Brain — 750 mL/min
Cardiac output = 5800 mL/min	Cardiac output = 9500 mL/min	Cardiac output = 17,500 mL/min
250 mL/min	350 mL/min	750 mL/min
Kidneys — 1100 mL/min	Kidneys — 900 mL/min	Kidneys — 600 mL/min
Abdominal viscera — 1400 mL/min	Abdominal viscera — 1100 mL/min	Abdominal viscera — 600 mL/min
Skin — 500 mL/min	Skin — 1500 mL/min	Skin — 1900 mL/min
Other tissues — 600 mL/min	Other tissues — 400 mL/min	Other tissues — 400 mL/min
Skeletal Muscles — 1200 mL/min	Skeletal Muscles — 4500 mL/min	Skeletal Muscles — 12,500 mL/min

Cardiovascular performance improves significantly with training. Trained athletes have more muscular hearts and larger stroke volumes than do nonathletes. Because cardiac output is equal to the stroke volume times the heart rate, at a given cardiac output, the larger the stroke volume, the slower the heart rate. An athlete at rest can maintain normal blood flow to peripheral tissues with a heart rate as low as 32 beats per minute, and, when necessary, the cardiac output of an athlete in peak condition can increase to levels 50 percent higher than those of nonathletes.

Module 12.12 Review

a. Describe the respiratory pump.

b. Describe the changes in cardiac output and blood flow during exercise.

c. Why does blood flow to visceral organs decrease during exercise?

An ECG can reveal cardiac arrhythmias

Despite the variety of sophisticated equipment available to assess cardiac function, in most cases the ECG provides the most important diagnostic information. ECG analysis is especially useful in detecting and diagnosing **cardiac arrhythmias** (ā-RITH-mē-az)—abnormal patterns of cardiac electrical activity. Momentary arrhythmias are not necessarily dangerous; about 5 percent of healthy individuals experience a few abnormal heartbeats each day. Clinical problems appear when arrhythmias reduce the pumping efficiency of the heart. Here we describe several important arrhythmias.

Normal ECG

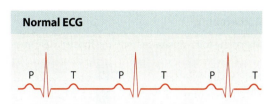

Premature Atrial Contractions (PACs)

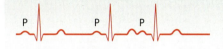

Premature atrial contractions may occur in healthy individuals. In a PAC, the normal atrial rhythm is momentarily interrupted by a "surprise" atrial contraction. Stress, caffeine, and various drugs may increase the incidence of PACs, presumably by increasing the membrane permeabilities of the SA pacemakers. The impulse spreads along the conduction pathway, and a normal ventricular contraction follows the atrial beat.

Paroxysmal Atrial Tachycardia (PAT)

In **paroxysmal** (par-ok-SIZ-mal) **atrial tachycardia**, or **PAT**, a premature atrial contraction triggers a flurry of atrial activity. The ventricles are still able to keep pace, and the heart rate jumps to about 180 beats per minute.

Atrial Fibrillation (AF)

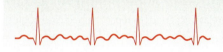

During **atrial fibrillation** (fib-ri-LĀ-shun), or **AF**, the impulses move over the atrial surface at rates of perhaps 500 beats per minute. The atrial wall quivers instead of producing an organized contraction. The ventricular rate cannot follow the atrial rate and may remain within normal limits. Even though the atria are now nonfunctional, their normal contribution to ventricular blood volume (amount of blood in the ventricle just before contraction begins) is so small that the condition may go unnoticed in older individuals.

Premature Ventricular Contractions (PVCs)

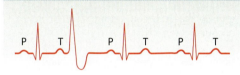

Premature ventricular contractions (PVCs) occur when a Purkinje fiber or ventricular myocardial cell depolarizes to threshold and triggers a premature contraction. Single PVCs are a common type of abnormal heartbeat and not dangerous. The cell responsible is called an **ectopic pacemaker.** An ectopic pacemaker is any pacemaker other than the sinoatiral node. Epinephrine, stimulatory drugs, caffeine, and ionic changes that depolarize cardiac muscle cell membranes can increase the frequency of PVCs.

Ventricular Tachycardia (VT)

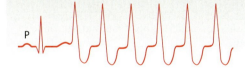

Ventricular tachycardia is defined as four or more PVCs without intervening normal beats. It is also known as **VT** or **V-tach**. Multiple PVCs and VT may indicate that serious cardiac problems exist.

Ventricular Fibrillation (VF)

Ventricular fibrillation (VF) is responsible for the condition known as **cardiac arrest**. VF is rapidly fatal, because the ventricles quiver and stop pumping blood.

Module 12.13 Review

a. Define cardiac arrhythmias.

b. Why is ventricular fibrillation fatal?

c. Which condition is characterized by premature atrial contraction that triggers a flurry of atrial activity?

1. Short answer

What three primary factors influence blood pressure and blood flow? _____

2. Matching

Match the following terms with the most closely related description.

- autoregulation
- venous return
- local vasodilators
- atrial natriuretic peptide
- vasoconstriction
- baroreceptors
- ACE
- capillary hydrostatic pressure
- medulla oblongata
- cardiac arrest
- pulse pressure
- vasodilation

a	_____	Detect changes in blood pressure
b	_____	Vasomotor center inhibited
c	_____	Aided by thoracic pressure changes due to breathing and skeletal muscle activity
d	_____	Causes immediate, local homeostatic responses
e	_____	Decreased tissue O_2 and increased CO_2
f	_____	Inhibited by several drugs to control high blood pressure
g	_____	Vasomotor center stimulated
h	_____	Difference between systolic and diastolic pressures
i	_____	Ventricular fibrillation
j	_____	Forces water out of a capillary
k	_____	Release triggered by excessive blood volume when heart walls are abnormally stretched
l	_____	Cardiovascular centers

3. Multiple choice

Choose the bulleted item that best completes each statement.

a Of the following blood vessels, the greatest drop in blood pressure occurs in the _____.

- capillaries
- veins
- venules
- arterioles

b The neural regulation of cardiac output primarily involves the _____.

- somatic nervous system
- central nervous system
- autonomic nervous system
- all of these

c Hormonal regulation by ADH, epinephrine, angiotensin II, and norepinephrine results in _____.

- decreasing peripheral vasoconstriction
- increasing peripheral vasoconstriction
- increasing peripheral vasodilation
- all of these

d The three primary interrelated changes that occur as exercise begins are _____.

- decreased vasodilation, increased venous return, increased cardiac output
- increased vasodilation, decreased venous return, increased cardiac output
- decreased vasodilation, decreased venous return, decreased cardiac output
- increased vasodilation, increased venous return, increased cardiac output

e The only part of the body where the blood supply is unchanged during exercise at maximal levels is the _____.

- heart
- brain
- kidney
- skin

f The systems responsible for modifying heart rate and regulating blood pressure are the _____ systems.

- respiratory and muscular
- respiratory and nervous
- muscular and urinary
- nervous and endocrine

4. Section integration

Explain why arterial pressure is higher than venous pressure.

_____

_____

Visual Outline with Key Terms

Summarize the content of each module using the terms in the order provided.

The Structure of the Heart

- base
- apex
- superior border
- right border
- left border
- inferior border

12.1

The wall of the heart contains concentric layers of cardiac muscle tissue

- myocardium
- endocardium
- parietal pericardium
- pericardial sac
- epicardium (visceral pericardium)
- intercalated disc

12.2

The heart is suspended within the pericardial cavity; the boundaries of the internal chambers are visible on the external surface

- mediastinum
- great vessels
- pericarditis
- anterior surface
- sulci
- auricle
- coronary sulcus
- ligamentum arteriosum
- anterior interventricular sulcus
- posterior surface
- coronary sinus
- posterior interventricular sulcus

12.3

The heart has an extensive blood supply

- coronary arteries
- right coronary artery
- marginal arteries
- left coronary artery
- anterior interventricular artery
- circumflex artery
- posterior interventricular artery
- anterior cardiac veins
- great cardiac vein
- coronary sinus
- posterior cardiac vein
- small cardiac vein
- middle cardiac vein
- elastic rebound

12.4

Internal valves control the direction of blood flow between the heart chambers

- interatrial septum
- interventricular septum
- atrioventricular (AV) valves
- fossa ovalis
- pectinate muscles
- right atrioventricular (AV) valve (tricuspid valve)
- cusps
- chordae tendineae
- papillary muscles
- pulmonary valve
- moderator band
- left atrioventricular (AV) valve (bicuspid valve)
- mitral valve
- trabeculae carneae
- aortic valve
- aortic sinuses
- semilunar valves

12.5

Damage to the myocardium or heart valves may compromise heart function

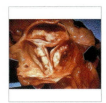

- arteriosclerosis
- atherosclerosis
- plaque
- balloon angioplasty
- coronary artery disease (CAD)
- coronary ischemia
- thrombus
- digital subtraction angiography (DSA)
- stent
- valvular heart disease (VHD)
- carditis
- heart sounds (S_1–S_4)

The Cardiac Cycle

- cardiac cycle
- systole
- diastole

12.6

The cardiac cycle, which creates pressure gradients that maintain blood flow, is coordinated by specialized cardiac muscle cells

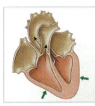

- cardiac cycle
- atrial systole
- atrial diastole
- ventricular systole—first phase
- isovolumetric contraction
- ventricular systole—second phase
- ventricular ejection
- ventricular diastole—early
- isovolumetric relaxation
- ventricular diastole—late
- automaticity
- conducting system
- sinoatrial node (SA node)
- internodal pathways
- atrioventricular node (AV node)
- AV bundle
- bundle branches
- Purkinje fibers

12.7

The ANS innervates the heart and an ECG can display heart activity

- cardioinhibitory center
- cardioacceleratory center
- bradycardia
- tachycardia
- electrocardiogram (ECG or EKG)
- P wave
- QRS complex
- R wave
- T wave
- P–R interval
- Q–T interval

12.8

Adjustments in heart rate and stroke volume regulate cardiac output

- cardiac output (CO)
- stroke volume
- heart failure
- factors affecting heart rate (HR)
- body temperature
- neural regulation
- hormonal regulation
- venous return
- filling time
- Starling's law of the heart
- contractility

• = *Term boldfaced in this module*

SECTION 3

The Coordination of Cardiac Output and Peripheral Blood Flow

- cardiac output
- blood pressure
- arterial pressure
- venous pressure
- resistance
- peripheral resistance
- capillary exchange
- venous return
- neural and hormonal regulation

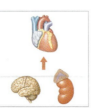

12.9

Arterial pressure and peripheral resistance determine blood flow and affect capillary exchange

- blood flow
- blood pressure
- arterial pressure
- peripheral resistance
- pressure gradient
- systolic pressure
- diastolic pressure
- pulse pressure
- capillary exchange
- capillary hydrostatic pressure (CHP)

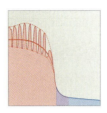

- filtration
- diffusion

12.10

Cardiovascular regulatory mechanisms respond to changes in blood pressure

- tissue perfusion
- autoregulation
- vasodilators
- central regulation
- cardioacceleratory centers
- vasoconstrictor
- baroreceptor reflexes

12.11

The endocrine responses to low blood pressure and low blood volume are very different from those to high blood pressure and high blood volume

- renin-angiotensin system
- angiotensin converting enzyme (ACE)
- atrial natriuretic peptide (ANP)

12.12

The cardiovascular centers make extensive adjustments to cardiac output and blood distribution during exercise

- cardiac output
- respiratory pump
- light exercise
- heavy exercise

12.13

An ECG can reveal cardiac arrythmias

- cardiac arrhythmias
- premature atrial contractions (PACs)
- paroxysmal atrial tachycardia (PAT)
- atrial fibrillation (AF)
- premature ventricular contractions (PVCs)
- ectopic pacemaker
- ventricular tachycardia (VT or V-tach)
- ventricular fibrillation (VF)
- cardiac arrest

● = *Term boldfaced in this module*

CAREER PATHS

Radiologic Technologist

"You don't know which area you may end up working in, so get as much A&P as you can."

— **Allen Nakagawa**
Radiologic Technologist,
San Francisco, California

If you have ever broken a bone, had a mammogram, or undergone radiation treatment, you have met a radiologic technologist. Radiologic technologists, or imaging technologists as they are often called, are the professionals who manage the machines and position the patients for scans and treatments involving radiation.

Allen Nakagawa is a senior diagnostic radiologic technologist at University of California San Francisco (UCSF) Medical Center. Allen specializes in spinal imaging, orthopedics, and pediatric orthopedics.

"A typical day for me begins at about 6:15 a.m.," says Allen. "I check on the incoming patient workload to see what will be happening that day. I also coordinate the workload of the students that I train." On most days, Allen is assigned to the operating room (OR). "In the OR, we have MRI machines that are built into the room."

To become a radiologic technologist, you must study in a two- or four-year program. A&P is a required course. "You will take intensive anatomy and additional anatomy and physiology in each modality you are considering working

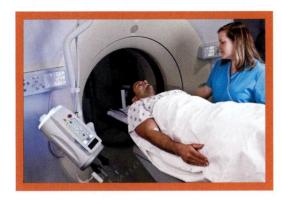

in," says Allen. "You don't know which area you may end up working in, so get as much A&P as you can."

Regardless of where a technologist works, contact with patients is the core of the job. "You have to enjoy working with people. It can be hard, because many of your patients are really

suffering. We get patients who are terminally ill, and we get kids who are really sick."

Allen enjoys working in one of the best hospitals in the world, using a variety of equipment to assist physicians and help diagnose their patients' conditions. As an imaging technologist, he is an integral part of the patient care team. "Physicians need to see what is going on inside the body. Even when the body is opened up, physicians still need information they can only get from imaging. For example, patients who have been in auto accidents may behave normally, but they may have a hematoma or a skull fracture that isn't apparent until it is revealed on an imaging scan."

For more information, contact the American Society of Radiologic Technologists at http:// www.asrt.org.

Think this is the career for you?
Key Stats:

- **Education and Training.** A certificate, an associate degree, or a bachelor's degree, earned through a formal training program in radiography.

- **Licensure.** Most states require licensure, though the requirements vary by state. The American Registry of Radiologic Technologists (ARRT) offers voluntary certification. Many employers prefer to hire certified radiologic technologists.

- **Earnings.** Earnings vary but the median annual wage is $53,240.

- **Expected Job Prospects.** Employment is expected to grow by an above average 17% from 2008 to 2018.

Bureau of Labor Statistics, U.S. Department of Labor, *Occupational Outlook Handbook, 2010–11 Edition,* Radiologic Technologists and Technicians, on the Internet at http://www .bls.gov/oco/ocos105.htm (visited September 14, 2011).

MasteringA&P®

Access more review material online in the Study Area at www.masteringaandp.com.

- Chapter guides
- Chapter quizzes
- Practice tests
- Art labeling activities
- Flashcards
- A glossary with pronunciations

- Practice Anatomy Lab™ (PAL™) 3.0 virtual anatomy practice tool
- Interactive Physiology® (IP) animated tutorials
- MP3 Tutor Sessions

PAL practice anatomy lab™ **For this chapter, follow these navigation paths in PAL:**

- Human Cadaver>Cardiovascular System>Heart
- Anatomical Models>Cardiovascular System>Heart
- Histology>Cardiovascular System

iP **For this chapter, go to these topics in the Cardiovascular System in IP:**

- Anatomy Review: The Heart
- Intrinsic Conduction System
- Cardiac Cycle
- Cardiac Output
- Blood Vessel Structure and Function
- Measuring Blood Pressure
- Factors that Affect Blood Pressure

 For this chapter, go to this topic in the MP3 Tutor Sessions:

- Cardiovascular Pressure

Chapter 12 Review Questions

1. Describe the relationships of the four chambers of the heart to the pulmonary and systemic circuits.

2. Why can't cardiac output increase indefinitely?

3. A patient's ECG tracing shows a consistent pattern of two P waves followed by a normal QRS complex and T wave. What is the cause of this abnormal wave pattern?

4. The following measurements were made on two individuals (the values recorded remained stable for 1 hour):
   Person A: heart rate 75 bpm; stroke volume 60 mL
   Person B: heart rate 90 bpm; stroke volume 95 mL
   Which person has the greater venous return? Which person has the longer ventricular filling time?

5. Karen is taking the medication verapamil, a heart drug that causes a decrease in the force of contraction. What effect should this medication have on Karen's stroke volume?

For answers to all module, section, and chapter review questions, see the blue Answers tab at the back of the book.

13

The Lymphatic System and Immunity

Anatomy of the Lymphatic System

The **lymphatic system** includes the cells, tissues, and organs that defend the body against both environmental hazards (such as various microbes) and internal threats (such as cancer cells).

1 We introduced **lymphoycytes**, the primary cells of the lymphatic system, in earlier chapters. Lymphocytes respond to invading pathogens (such as bacteria or viruses), abnormal body cells (such as virus-infected cells or cancer cells), and foreign proteins (such as the toxins released by some bacteria). Within the lymphatic system, lymphocytes are surrounded by **lymph**, a liquid that resembles interstitial fluid.

2 The lymphatic system includes a network of **lymphatic vessels**, often called **lymphatics**, which begin in peripheral tissues and end at connections to veins. The lymphatic system also includes an array of lymphoid tissues and lymphoid organs scattered throughout the body. Most of the body's lymphocytes are produced and stored within these lymphoid tissues and organs. However, lymphocytes are also produced in areas of red bone marrow, along with other defense cells, such as monocytes and macrophages.

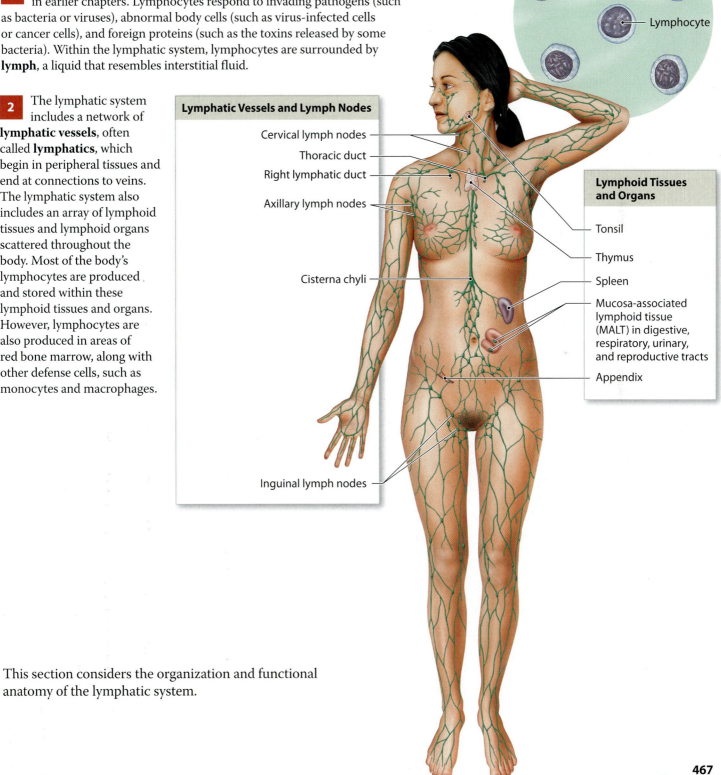

Lymphatic Vessels and Lymph Nodes

Cervical lymph nodes
Thoracic duct
Right lymphatic duct
Axillary lymph nodes
Cisterna chyli
Inguinal lymph nodes

Lymph
Lymphocyte

Lymphoid Tissues and Organs

Tonsil
Thymus
Spleen
Mucosa-associated lymphoid tissue (MALT) in digestive, respiratory, urinary, and reproductive tracts
Appendix

This section considers the organization and functional anatomy of the lymphatic system.

Interstitial fluid flows into lymphatic capillaries to become lymph within lymphatic vessels

Lymphatic vessels carry lymph from peripheral tissues to the venous system. The lymphatic network begins with **lymphatic capillaries**, which branch through peripheral tissues.

1 Lymphatic capillaries are found in almost every tissue and organ in the body. As shown here, they are closely associated with blood capillaries within these tissues. Interstitial fluid flows continuously into lymphatic capillaries, where it is called **lymph**. The arrows show the movement of fluid out of capillaries and the net flow of interstitial fluid and lymph.

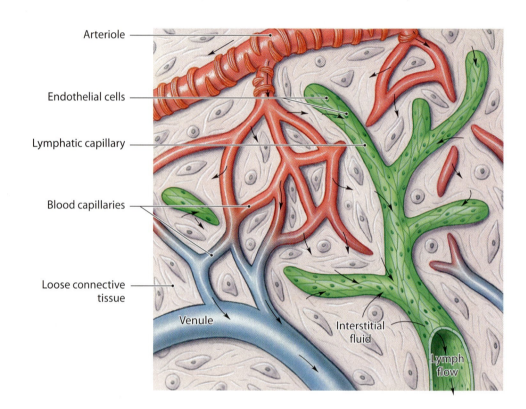

Arteriole

Endothelial cells

Lymphatic capillary

Blood capillaries

Loose connective tissue

Venule

Interstitial fluid

Lymph flow

2 Lymphatic capillaries differ from blood capillaries in that lymphatic capillaries:
1) originate as pockets rather than forming continuous tubes
2) have larger diameters
3) have thinner walls
4) typically look flattened or irregular in sectional view

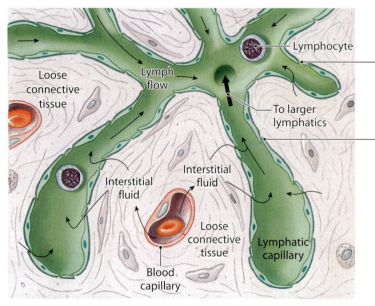

Loose connective tissue

Lymph flow

Lymphocyte

To larger lymphatics

Interstitial fluid

Interstitial fluid

Loose connective tissue

Lymphatic capillary

Blood capillary

Sectional view

Although lymphatic capillaries are lined by endothelial cells, the basement membrane is incomplete or missing entirely.

The endothelial cells of a lymphatic capillary overlap. The region of overlap acts as a one-way valve, permitting the entry of fluids and solutes (including proteins), as well as viruses, bacteria, and cellular debris, but preventing their return to the intercellular spaces.

3 From the lymphatic capillaries, lymph flows into larger lymphatic vessels that lead toward the body's trunk. These lymphatic vessels commonly occur in association with arteries and veins.

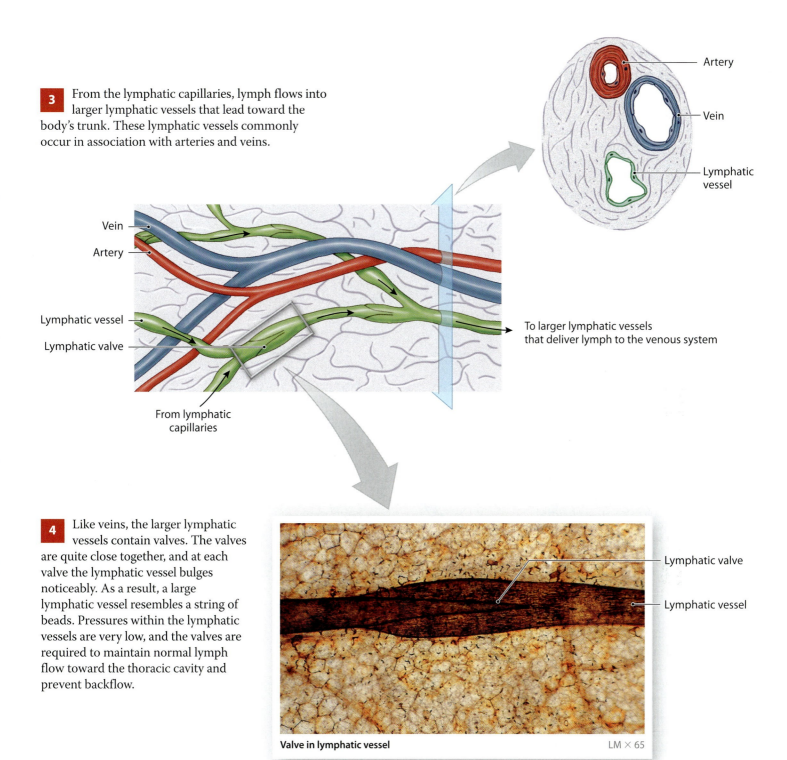

Artery

Vein

Lymphatic vessel

Vein

Artery

Lymphatic vessel

Lymphatic valve

From lymphatic capillaries

To larger lymphatic vessels that deliver lymph to the venous system

4 Like veins, the larger lymphatic vessels contain valves. The valves are quite close together, and at each valve the lymphatic vessel bulges noticeably. As a result, a large lymphatic vessel resembles a string of beads. Pressures within the lymphatic vessels are very low, and the valves are required to maintain normal lymph flow toward the thoracic cavity and prevent backflow.

Lymphatic valve

Lymphatic vessel

Valve in lymphatic vessel   LM × 65

Areas of the body without a blood supply, such as the cornea of the eye, have no lymphatic capillaries. The red bone marrow and the central nervous system also lack lymphatic vessels.

Module 13.1 Review

a. What is the function of lymphatic vessels?

b. What is lymph?

c. What is the function of overlapping endothelial cells in lymphatic capillaries?

Small lymphatic vessels merge to form lymphatic ducts that empty lymph into subclavian veins

1 **Blood** is normally confined to the vessels of the cardiovascular system, and contractions of the heart keep it in motion. As blood flows through body tissues, water and solutes move from the plasma into the surrounding interstitial fluid. Lymph forms as interstitial fluid drains into lymphatic vessels that begin in peripheral tissues and empty into the venous system. This continuous recirculation of interstitial fluid is essential to homeostasis because: 1) it helps eliminate local differences in levels of nutrients, wastes, and toxins, 2) maintains blood volume, and 3) alerts the immune system to infections that may be under way in peripheral tissues.

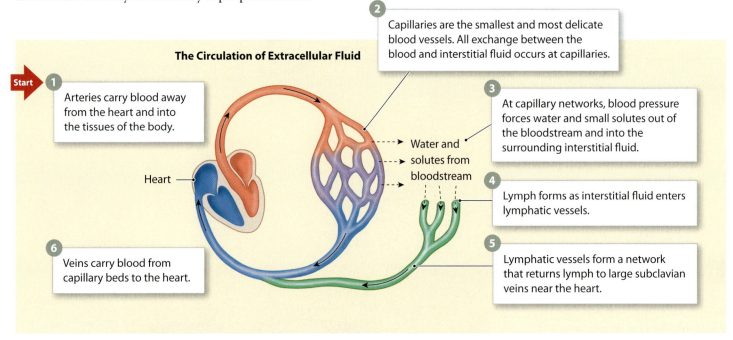

The Circulation of Extracellular Fluid

2 Capillaries are the smallest and most delicate blood vessels. All exchange between the blood and interstitial fluid occurs at capillaries.

Start **1** Arteries carry blood away from the heart and into the tissues of the body.

3 At capillary networks, blood pressure forces water and small solutes out of the bloodstream and into the surrounding interstitial fluid.

Water and solutes from bloodstream

4 Lymph forms as interstitial fluid enters lymphatic vessels.

Heart

5 Lymphatic vessels form a network that returns lymph to large subclavian veins near the heart.

6 Veins carry blood from capillary beds to the heart.

2 Superficial and deep lymphatics converge to form larger vessels called **lymphatic trunks**, which in turn empty into two large collecting vessels: the thoracic duct and the right lymphatic duct. The **thoracic duct** collects lymph from the body inferior to the diaphragm and from the left side of the body superior to the diaphragm. The smaller **right lymphatic duct** collects lymph from the right side of the body superior to the diaphragm.

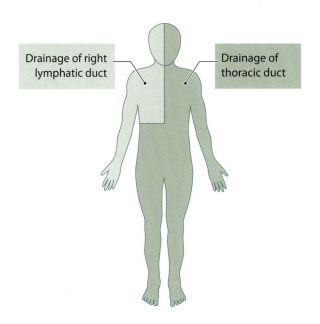

Drainage of right lymphatic duct

Drainage of thoracic duct

3 This illustration shows the relationship between the lymphatic ducts and the venous system. The thoracic duct empties into the left subclavian vein. The right lymphatic duct drains into the right subclavian vein.

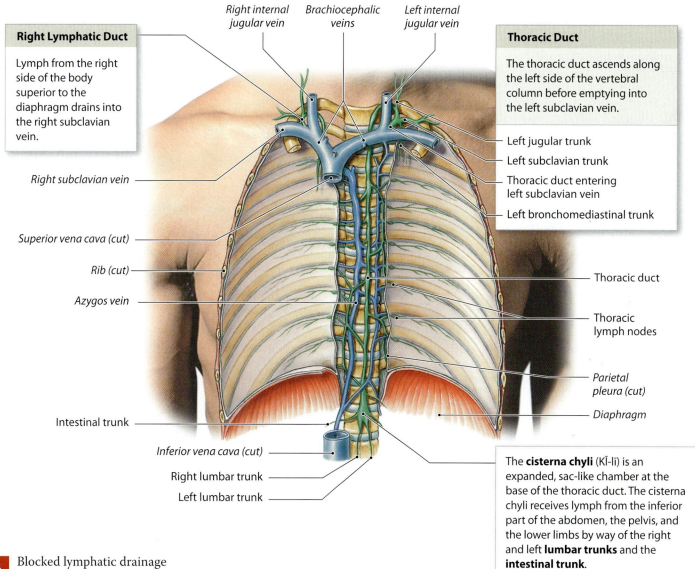

Right Lymphatic Duct

Lymph from the right side of the body superior to the diaphragm drains into the right subclavian vein.

Right internal jugular vein

Brachiocephalic veins

Left internal jugular vein

Thoracic Duct

The thoracic duct ascends along the left side of the vertebral column before emptying into the left subclavian vein.

Left jugular trunk

Left subclavian trunk

Thoracic duct entering left subclavian vein

Left bronchomediastinal trunk

Right subclavian vein

Superior vena cava (cut)

Rib (cut)

Azygos vein

Thoracic duct

Thoracic lymph nodes

Parietal pleura (cut)

Diaphragm

Intestinal trunk

Inferior vena cava (cut)

Right lumbar trunk

Left lumbar trunk

The **cisterna chyli** (KĪ-lī) is an expanded, sac-like chamber at the base of the thoracic duct. The cisterna chyli receives lymph from the inferior part of the abdomen, the pelvis, and the lower limbs by way of the right and left **lumbar trunks** and the **intestinal trunk**.

4 Blocked lymphatic drainage produces **lymphedema** (limf-e-DĒ-muh), a condition in which interstitial fluids accumulate and the affected area gradually becomes swollen and grossly distended. Lymphedema most often affects a limb, as in this photo, although it can occur in other locations. If the condition persists, the connective tissues lose their elasticity and the swelling becomes permanent. Because the interstitial fluids are essentially stagnant, toxins and pathogens can accumulate and overwhelm local defenses without fully activating the immune system.

Module 13.2 Review

a. Name the two large lymphatic ducts into which the lymphatic trunks empty.

b. Describe the drainage of the right lymphatic duct and the thoracic duct.

c. Explain lymphedema.

Lymphocytes are responsible for the immune functions of the lymphatic system

Lymphocytes account for 20–40 percent of circulating leukocytes. However, circulating lymphocytes are only a small fraction of the total lymphocyte population. The body contains some 10^{12} lymphocytes, with a combined weight of more than a kilogram (2.2 lb).

1 Three classes of lymphocytes circulate in blood. While each class has distinctive biochemical and functional characteristics, all are sensitive to specific chemicals called **antigens**. Most antigens are either pathogens, parts or products of pathogens, or other foreign substances. Most antigens are proteins, but some lipids, polysaccharides, and nucleic acids can also act as antigens. An antigen stimulates an immune response that ultimately destroys the substance or pathogen associated with that antigen.

Classes of Lymphocytes

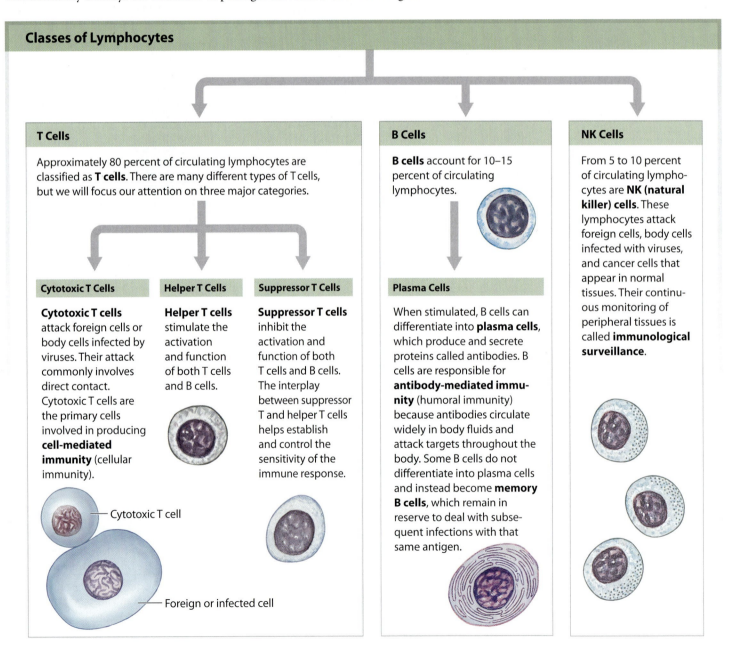

T Cells

Approximately 80 percent of circulating lymphocytes are classified as **T cells**. There are many different types of T cells, but we will focus our attention on three major categories.

Cytotoxic T Cells

Cytotoxic T cells attack foreign cells or body cells infected by viruses. Their attack commonly involves direct contact. Cytotoxic T cells are the primary cells involved in producing **cell-mediated immunity** (cellular immunity).

— Cytotoxic T cell

— Foreign or infected cell

Helper T Cells

Helper T cells stimulate the activation and function of both T cells and B cells.

Suppressor T Cells

Suppressor T cells inhibit the activation and function of both T cells and B cells. The interplay between suppressor T and helper T cells helps establish and control the sensitivity of the immune response.

B Cells

B cells account for 10–15 percent of circulating lymphocytes.

Plasma Cells

When stimulated, B cells can differentiate into **plasma cells**, which produce and secrete proteins called antibodies. B cells are responsible for **antibody-mediated immunity** (humoral immunity) because antibodies circulate widely in body fluids and attack targets throughout the body. Some B cells do not differentiate into plasma cells and instead become **memory B cells**, which remain in reserve to deal with subsequent infections with that same antigen.

NK Cells

From 5 to 10 percent of circulating lymphocytes are **NK (natural killer) cells**. These lymphocytes attack foreign cells, body cells infected with viruses, and cancer cells that appear in normal tissues. Their continuous monitoring of peripheral tissues is called **immunological surveillance**.

2 In adults, erythropoiesis (red blood cell formation) is normally confined to red bone marrow. However, lymphocyte production, or **lymphopoiesis** (lim-fō-poy-Ē-sis), involves the red bone marrow, thymus, and peripheral lymphoid tissues. Red bone marrow plays the primary role in maintaining normal lymphocyte populations.

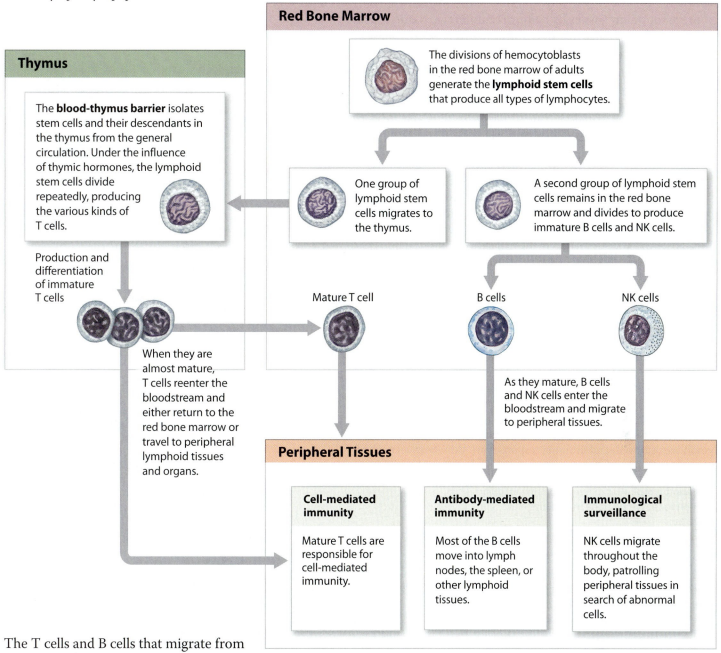

Red Bone Marrow

The divisions of hemocytoblasts in the red bone marrow of adults generate the **lymphoid stem cells** that produce all types of lymphocytes.

One group of lymphoid stem cells migrates to the thymus.

A second group of lymphoid stem cells remains in the red bone marrow and divides to produce immature B cells and NK cells.

Thymus

The **blood-thymus barrier** isolates stem cells and their descendants in the thymus from the general circulation. Under the influence of thymic hormones, the lymphoid stem cells divide repeatedly, producing the various kinds of T cells.

Production and differentiation of immature T cells

When they are almost mature, T cells reenter the bloodstream and either return to the red bone marrow or travel to peripheral lymphoid tissues and organs.

Mature T cell

B cells

NK cells

As they mature, B cells and NK cells enter the bloodstream and migrate to peripheral tissues.

Peripheral Tissues

Cell-mediated immunity

Mature T cells are responsible for cell-mediated immunity.

Antibody-mediated immunity

Most of the B cells move into lymph nodes, the spleen, or other lymphoid tissues.

Immunological surveillance

NK cells migrate throughout the body, patrolling peripheral tissues in search of abnormal cells.

The T cells and B cells that migrate from their sites of origin retain the ability to divide. Their divisions produce daughter cells of the same type. For example, a dividing B cell produces other B cells, but not T cells or NK cells. The ability of specific types of lymphocytes to increase in number is crucial to the success of the immune response.

Module 13.3 Review

a. Identify the three main classes of lymphocytes.

b. Which cells are responsible for antibody-mediated immunity?

c. Which tissues are involved in lymphopoiesis?

Lymphocytes within lymphoid tissues and lymphoid organs protect the body from pathogens and toxins

Lymphoid tissue consists of connective tissue dominated by lymphocytes. Lymphoid nodules are spherical masses of lymphoid tissue. In contrast, lymphoid organs (lymph nodes, thymus, and spleen) are separated from surrounding tissues by a fibrous connective tissue capsule.

1 The digestive, respiratory, urinary, and reproductive tracts communicate with the exterior environment, which contains dangerous pathogens and toxins. Clusters of lymphoid nodules are found deep to the epithelia lining these passageways. These clusters, which protect the passageways from pathogens and toxins, form the **mucosa-associated lymphoid tissue (MALT)**.

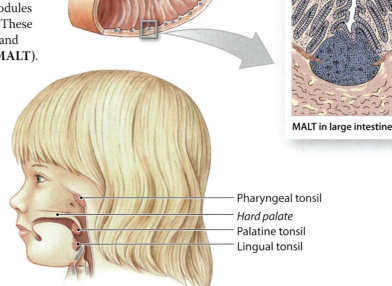

MALT in large intestine

2 The **tonsils** are large lymphoid nodules in the walls of the pharynx. A single **pharyngeal tonsil**, often called the adenoid, lies in the posterior superior wall of the nasopharynx. **Palatine tonsils** are located at the posterior, inferior margin of the oral cavity, along the boundary with the pharynx. A pair of **lingual tonsils** lie deep to the mucous epithelium covering the base of the tongue. Because of their location, the lingual tonsils are not usually visible unless they become infected and swollen, a condition known as **tonsillitis**.

Pharyngeal tonsil
Hard palate
Palatine tonsil
Lingual tonsil

3 **Lymph nodes** are small lymphoid organs shaped like a kidney bean. A lymph node functions like a kitchen water filter, purifying lymph before it reaches the venous circulation. As lymph flows through a lymph node, at least 99 percent of the antigens in the lymph are removed, and immune responses are stimulated as needed.

Lymph node
Lymphatic vessel

Lymph nodes

Afferent lymphatics (*afferens*, to bring to) carry lymph to the lymph node from peripheral tissues.

The afferent vessels deliver lymph to a meshwork of reticular fibers, macrophages, and dendritic cells. **Dendritic cells** are involved in the initiation of the immune response.

Within the lymph node, lymph flows through regions containing both B cells and plasma cells.

Efferent lymphatics (*efferens*, to bring out) leave the lymph node and carry lymph toward the venous circulation.

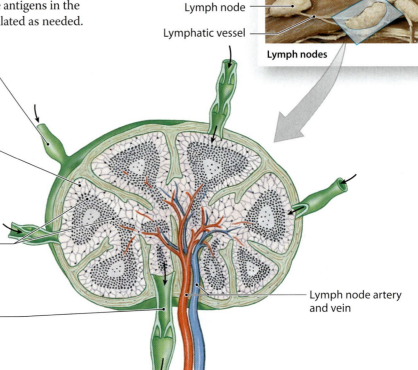

Lymph node artery and vein

4 The thymus produces several hormones that are important to the development of functional T cells, and thus to maintaining normal immunological defenses. *Thymosin* (THĪ-mō-sin) is the name originally given to an extract from the thymus that promotes the development and maturation of lymphocytes. This extract actually contains several complementary hormones collectively known as **thymosins**. The gradual decrease in the size and secretory abilities of the thymus that occurs with age is correlated with increased susceptibility to disease.

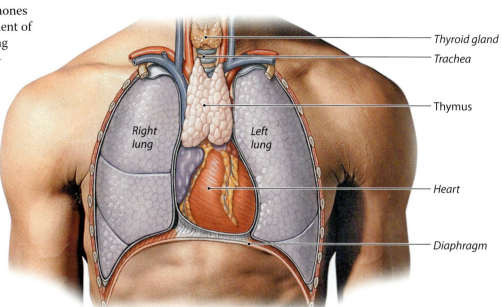

Thyroid gland

Trachea

Thymus

Right lung

Left lung

Heart

Diaphragm

5 In adults, the spleen contains the largest mass of lymphoid tissue in the body. In essence, the spleen performs the same functions for blood that lymph nodes perform for lymph. The spleen: 1) removes abnormal red blood cells by phagocytosis, 2) stores iron recycled from red blood cells, and 3) initiates immune responses by B cells and T cells.

Diaphragm

Stomach

Rib

Liver

Gastrosplenic ligament

Gastric area

Spleen

Pancreas

Inferior vena cava

Aorta

Spleen

Hilum

Kidneys

Renal area

The lateral surface of the spleen is smooth and convex, conforming to the shape of the diaphragm and body wall.

Lateral surface of the spleen

Module 13.4 Review

a. Name the lymphoid tissue that protects epithelia lining the digestive, respiratory, urinary, and reproductive tracts.

b. Define tonsils, and name the three types of tonsils.

c. Describe the functions of the spleen.

1. Labeling

Label the structures of the lymphatic system in the accompanying figure.

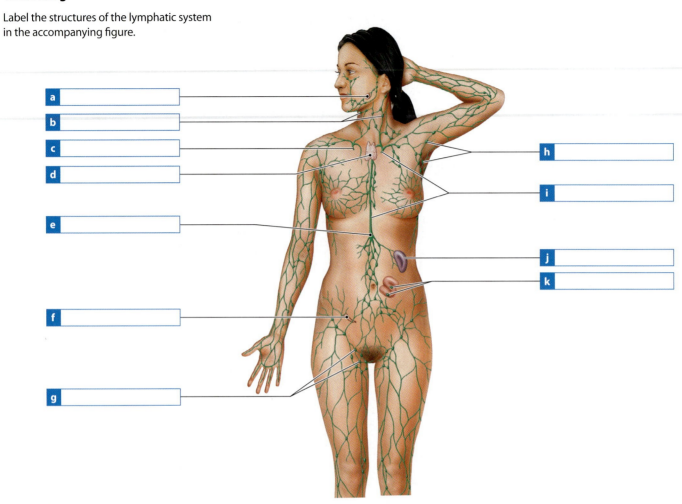

a _____

b _____

c _____

d _____

e _____

f _____

g _____

h _____

i _____

j _____

k _____

2. Matching

Match the following terms with the most closely related description.

- B cells
- spleen
- lymphatic capillaries
- cytotoxic T cells
- efferent lymphatics
- tonsils
- lymphopoiesis
- lymphedema
- lymphoid organs
- antigens
- afferent lymphatics
- right subclavian vein
- thoracic duct
- lymph nodes

a _____ Beginning of lymphatic system

b _____ Results from blocked lymphatic drainage

c _____ Receives lymph from right lymphatic duct

d _____ Receives lymph from left half of head

e _____ Thymus, spleen, and lymph nodes

f _____ Smallest lymphoid organs

g _____ Carry lymph away from lymph nodes

h _____ Largest mass of lymphoid tissue in body

i _____ Specific chemicals to which lymphocytes are sensitive

j _____ Occurs in red bone marrow, thymus, and lymphoid tissues

k _____ Cell-mediated immunity

l _____ Lymphoid nodules in walls of pharynx

m _____ Antibody-mediated immunity

n _____ Carry lymph to lymph nodes

Nonspecific Immunity

Immunity is the ability to resist infection and disease, often caused by microorganisms called pathogens. We have two forms of immunity that work independently or together. These forms are *nonspecific (innate) immunity* and *specific (adaptive) immunity*. The body has several anatomical barriers and defense mechanisms. They either prevent or slow the entry of infectious organisms, or attack them if they do enter. In this section we consider nonspecific (innate) defenses.

Immunity

Nonspecific (Innate) Defenses

Nonspecific defenses do not distinguish one type of pathogen from another. Their response is the same, regardless what the invading agent is. These defenses, which are innate (present at birth), provide a defensive capability known as **nonspecific resistance**.

Physical barriers keep hazardous organisms and materials outside the body. For example, a mosquito that lands on your head may be unable to reach the surface of the scalp if you have a full head of hair.

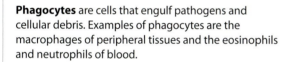

Phagocytes are cells that engulf pathogens and cellular debris. Examples of phagocytes are the macrophages of peripheral tissues and the eosinophils and neutrophils of blood.

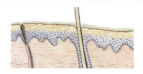

Immunological surveillance is the destruction of abnormal cells by natural killer (NK) cells in peripheral tissues.

Destruction of abnormal cells

Interferons are chemical messengers that coordinate the defenses against viral infections.

The **complement system** is a group of circulating proteins that helps antibodies destroy pathogens.

The **inflammatory response** is a localized, tissue-level response that tends to limit the spread of an injury or infection.

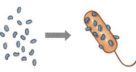
Inflammation

Fever is an elevated body temperature that accelerates tissue metabolism and the activity of defenses.

Specific (Adaptive) Defenses

Specific defenses are defense mechanisms that protect against particular pathogens. For example, a specific defense may protect against infection by one type of bacterium but be ineffective against other bacteria and viruses. Specific defenses depend on the activities of specific lymphocytes. The body's specific defenses produce a state of protection known as **specific resistance**. We will consider specific immunity in Section 3.

13.5

Physical barriers and phagocytes play a role in nonspecific defenses

To cause trouble, an antigenic compound or a pathogen must enter body tissues. This requires crossing an epithelium—either at the skin or across a mucous membrane.

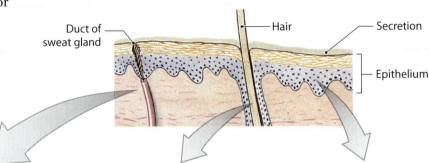

Duct of sweat gland · Hair · Secretion · Epithelium

1 The integumentary system provides the major physical barrier to the external environment.

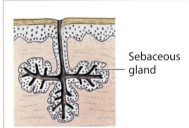

Sebaceous gland

Specialized accessory structures and secretions protect most epithelia. The epidermal surface also receives the secretions of sebaceous and sweat glands. These secretions, which flush the surface to wash away microorganisms and chemical agents, may also contain bactericidal chemicals, destructive enzymes (lysozymes), and antibodies.

The hair on most areas of your body protects against mechanical abrasion (especially on the scalp), and hair often prevents hazardous materials or insects from contacting your skin.

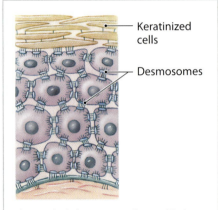

Keratinized cells · Desmosomes

The epithelial covering of your skin has multiple layers, a coating of keratinized cells, and a network of desmosomes that lock adjacent cells together, blocking pathogens from entering.

2 The epithelia lining the digestive, respiratory, urinary, and reproductive tracts also form an important barrier against antigenic compounds and pathogens. As noted in Module 13.4, MALT in these locations provides nonspecific defense.

Mucus bathes most surfaces of your digestive tract, and your stomach contains a powerful acid that can destroy many pathogens. Mucus moves across the lining of the respiratory tract, urine flushes the urinary passageways, and glandular secretions do the same for the reproductive tract. Special enzymes, antibodies, and an acidic pH make these secretions even more effective.

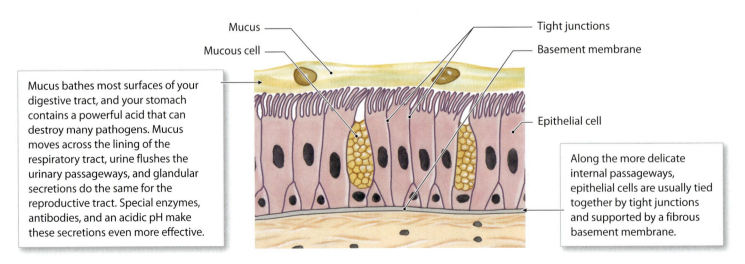

Mucus · Mucous cell · Tight junctions · Basement membrane · Epithelial cell

Along the more delicate internal passageways, epithelial cells are usually tied together by tight junctions and supported by a fibrous basement membrane.

3 Phagocytes perform janitorial and police services in peripheral tissues, removing cellular debris and responding to invasion by foreign compounds or pathogens. Phagocytes represent the "first line of cellular defense" against pathogenic invasion. Many phagocytes attack and remove microorganisms even before lymphocytes detect their presence. Although different types of phagocytes target different threats, all phagocytic cells function in the same basic way as described below.

Types of Phagocytes

12 μm

8–10 μm

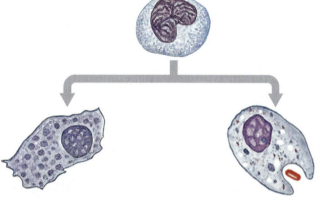

There are two major classes of **macrophages** derived from the monocytes of the circulating blood. This collection of phagocytic cells is called the **monocyte–macrophage system**, or the reticuloendothelial system.

Neutrophils are abundant, mobile, and quick to phagocytize cellular debris or invading bacteria. They circulate in the bloodstream and roam through peripheral tissues, especially at sites of injury or infection.

Eosinophils, which are less abundant than neutrophils, phagocytize foreign compounds or pathogens that have been coated with antibodies.

Fixed macrophages are permanent residents of specific tissues and organs and are scattered among connective tissues. They normally do not move within these tissues.

Free macrophages travel throughout the body, arriving at the site of an injury by migrating through adjacent tissues or by recruitment from the circulating blood.

4 All phagocytes share several capabilities, as summarized in the table below.

Functional Characteristics of Phagocytes

- Phagocytes can leave capillaries by squeezing between adjacent endothelial cells, a process known as **diapedesis**.

- Phagocytes may be attracted to or repelled by chemicals in the surrounding fluids, a phenomenon called **chemotaxis**. They are particularly sensitive to chemicals released by either body cells or pathogens.

- Phagocytosis always begins with the phagocyte attaching to its target. Receptors on the plasma membrane of the phagocyte bind to the surface of the target.

- After attaching, the phagocyte either destroys the target itself or promotes its destruction by activating specific defenses.

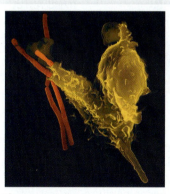
Phagocyte (yellow) engulfing bacteria (orange)   SEM × 2900

Module 13.5 Review

a. How does the integumentary system protect the body?

b. Identify the types of phagocytes in the body, and differentiate between fixed macrophages and free macrophages.

c. Define chemotaxis.

Inflammation is a localized tissue response to injury; fever is a generalized response to tissue damage and infection

Inflammation

1 **Inflammation**, or the inflammatory response, is a localized tissue response to injury. Inflammation produces local swelling, redness, heat, and pain. Many stimuli—impact, abrasion, distortion, chemical irritation, infection by pathogens, and extreme temperatures (hot or cold)—can cause inflammation. Each of these stimuli kills cells, damages connective tissue fibers, or injures the tissue in some other way. The changes alter the chemical composition of the interstitial fluid. Damaged cells release prostaglandins, proteins, and potassium ions, and the injury itself may introduce foreign proteins or pathogens. The changes in the interstitial environment trigger the complex process of inflammation.

Fever

2 **Fever** is the rise in core body temperature above 37.2°C (99°F). Circulating proteins called **pyrogens** (PĪ-rō-jenz; *pyro-*, fever or heat + *-gen*, substance) can reset the temperature thermostat in the hypothalamus and raise body temperature. Within limits, a fever can be beneficial. Higher-than-normal body temperatures may inhibit some viruses and bacteria, but the most likely beneficial effect is on body metabolism. For each 1°C rise in body temperature, metabolic rate jumps by 10 percent. The result: Quicker mobilization of tissue defenses and an accelerated repair process.

Tissue Damage

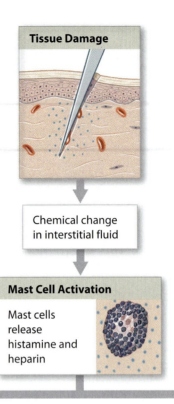

Chemical change in interstitial fluid

Mast Cell Activation

Mast cells release histamine and heparin

Redness, Swelling, Heat, and Pain

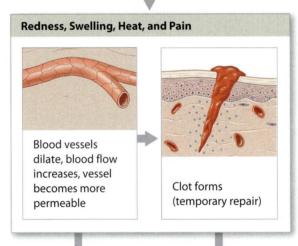

Blood vessels dilate, blood flow increases, vessel becomes more permeable

Clot forms (temporary repair)

Phagocyte Attraction

Phagocytes, especially neutrophils, are attracted to area

Release of cytokines (chemical messengers affecting immune defenses)

Neutrophils and macrophages remove debris; fibroblasts are stimulated

Specific defenses are activated

Tissue Repair

Pathogens are removed, clot erodes, scar tissue forms

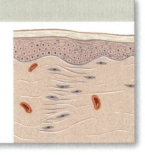

Summary of Nonspecific Immunity

3 This table summarizes various nonspecific defenses.

Physical Barriers

Prevent approach of and deny access to pathogens

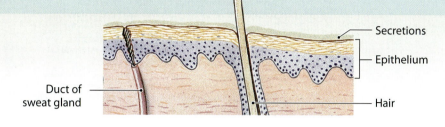

Secretions

Epithelium

Duct of sweat gland

Hair

Phagocytes

Remove debris and pathogens

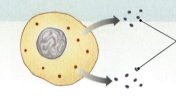

Neutrophil Eosinophil Monocyte Free macrophage Fixed macrophage

Immune Surveillance by NK cells

Continuously monitor normal tissues and destroy abnormal cells

Natural killer cell

Lysed abnormal cell

When an abnormal cell is detected, the NK cell releases **perforins**, cytotoxic chemicals that destroy the abnormal cell's membrane.

Interferons

Increase resistance of cells to viral infection; slow the spread of disease

Interferons are released by activated lymphocytes, macrophages, or virus-infected cells. Interferon triggers the production of antiviral proteins in the cytoplasm of normal cells. It also stimulates the activities of macrophages and NK cells.

Complement System

When activated, attacks and breaks down the surfaces of cells, bacteria, and viruses; attracts phagocytes; stimulates inflammation

Complement

Lysed pathogen

Inflammation (Inflammatory Response)

Multiple effects make nonspecific and specific defenses more effective

Mast cell

- Blood flow increased
- Phagocytes activated
- Damaged area isolated by clotting reaction
- Capillary permeability increased
- Complement activated
- Regional temperature increased
- Specific defenses activated

Fever

Mobilizes defenses; accelerates repairs; inhibits pathogens

Body temperature rises above 37.2°C in response to pyrogens

Module 13.6 Review

a. What is the result of mast cell activation?

b. Summarize the body's nonspecific defenses.

c. A rise in the level of interferons in the body suggests what kind of infection?

1. Labeling

Fill in the spaces with the name of the nonspecific defense described at right.

a _____ _____

b _____

c _____ _____

d _____

e _____ _____

f _____

g _____

a keep hazardous organisms and materials outside the body. For example, a mosquito that lands on your head may be unable to reach the surface of the scalp if you have a full head of hair.	
b are cells that engulf pathogens and cell debris. Examples are the macrophages of peripheral tissues and the eosinophils and neutrophils of blood.	
c is the destruction of abnormal cells by NK cells in peripheral tissues.	 Destruction of abnormal cells
d are chemical messengers that coordinate the defenses against viral infections.	
e is a group of circulating proteins that helps antibodies destroy pathogens.	
f is a localized, tissue-level response that tends to limit the spread of an injury or infection.	 Swelling, redness, heat, and pain
g is elevated body temperature that accelerates tissue metabolism and the activity of defenses.	

2. Multiple choice

Choose the bulleted item that best completes each statement.

a A physical barrier such as the skin provides a nonspecific body defense due to its makeup, which includes _____.

- multiple layers
- a coating of keratinized cells
- a network of desmosomes locking adjacent cells together
- all of these

b NK (natural killer) cells sensitive to the presence of abnormal plasma membranes are primarily involved in _____.

- defenses against specific pathogens
- phagocytic activity for defense
- complex, time-consuming defense mechanisms
- immune surveillance

c The nonspecific defense that breaks down cells, attracts phagocytes, and stimulates inflammation is _____.

- the inflammatory response
- the action of interferons
- the complement system
- immune surveillance

d The protein(s) that interfere(s) with the replication of viruses is (are) _____.

- complement proteins
- heparin
- pyrogens
- interferons

e Circulating proteins that reset the thermostat in the hypothalamus, causing body temperature to rise, are called _____.

- pyrogens
- interferons
- lysosomes
- complement proteins

f The "first line of cellular defense" against pathogenic invasion is _____.

- phagocytes
- mucus
- hair
- interferon

3. Short answer

We usually associate a fever with illness. Can a fever ever be beneficial? Explain your answer.

_____

Specific Immunity

1 The coordinated activities of T cells and B cells provide **specific defenses** (adaptive defenses), which protect against specific pathogens. These activities produce **specific immunity**, a specific resistance against potentially dangerous antigens. In general, T cells are responsible for cell-mediated immunity, which defends against abnormal cells and pathogens inside cells, and B cells provide antibody-mediated immunity, which defends against antigens and pathogens in body fluids. However, as indicated in this flowchart, there are many different forms of immunity.

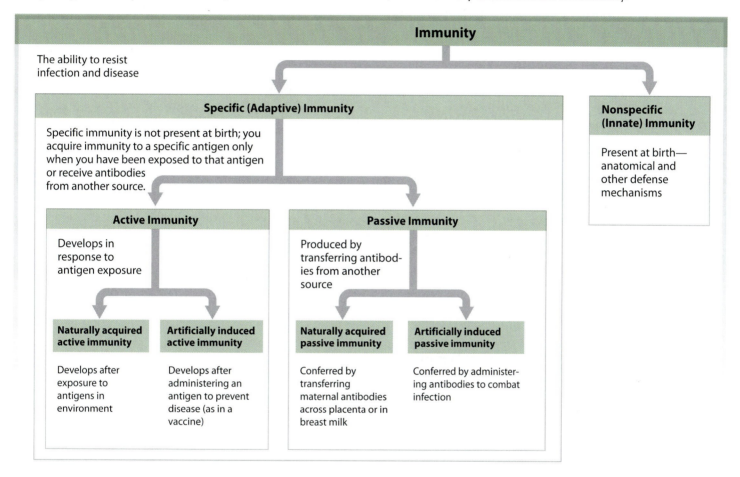

2 Regardless of the form, immunity exhibits four general properties, which we summarize below.

Specific Properties of Specific Immunity

- **Specificity** results from the activation of appropriate lymphocytes and the production of antibodies with targeted effects. Each T cell or B cell has receptors that bind to one specific antigen, but ignore all others. The response of an activated T cell or B cell is equally specific, and leaves other antigens unaffected.

- **Versatility** results from the large diversity of lymphocytes in the body. There are millions of different lymphocyte populations, each sensitive to a different antigen. When activated by an appropriate antigen, a lymphocyte divides, producing more lymphocytes with the same specificity. All the cells produced by the division of an activated lymphocyte constitute a **clone**.

- **Immunologic memory** exists because cell divisions of activated lymphocytes produce two groups of cells: one group that attacks the invader immediately, and another that remains inactive unless it is exposed to the same antigen at a later date. These inactive **memory cells** enable your immune system to "remember" antigens it has previously encountered, and to launch a faster, stronger, and longer-lasting counterattack if the same antigen reappears.

- **Tolerance** exists because the immune response ignores normal ("self") tissues but targets abnormal and foreign ("non-self") cells as well as toxins. Tolerance can also develop over time in response to chronic exposure to an antigen in the environment. Such tolerance generally lasts only as long as the exposure continues.

Specific defenses are triggered by exposure to antigenic fragments

1 This figure provides a "big picture" overview of the immune response: Exposure to an antigen activates phagocyes. These phagocytes stimulate: 1) cell-mediated events involving attacks by T cells, and 2) antibody-mediated events involving antibodies produced by cells derived from B cells. We will examine each of these processes more closely in later modules. This module focuses on the first step: how antigens trigger an immune response.

Antigen Presentation

Most antigens must either infect cells or be "processed" by phagocytes before specific defenses are activated. The trigger is the appearance of antigens or antigenic fragments in plasma membranes. This is called **antigen presentation**.

Specific Defenses

Antigen presentation triggers specific defenses, or an immune response.

Virus

Bacterium

1 Cell-Mediated Immunity

Phagocytes activated → T cells activated

Direct Physical and Chemical Attack

Activated T cells find the pathogens and attack them through phagocytosis or the release of chemical toxins.

Communication and feedback

2 Antibody-Mediated Immunity

Activated B cells give rise to cells that produce antibodies.

Attack by Circulating Antibodies

Destroy antigens

2 **Antigen-presenting cells (APCs)** are specialized cells that include monocytes, macrophages, and the dendritic cells of the skin and lymphoid organs. Here we show the process of antigen presentation by a phagocytic cell.

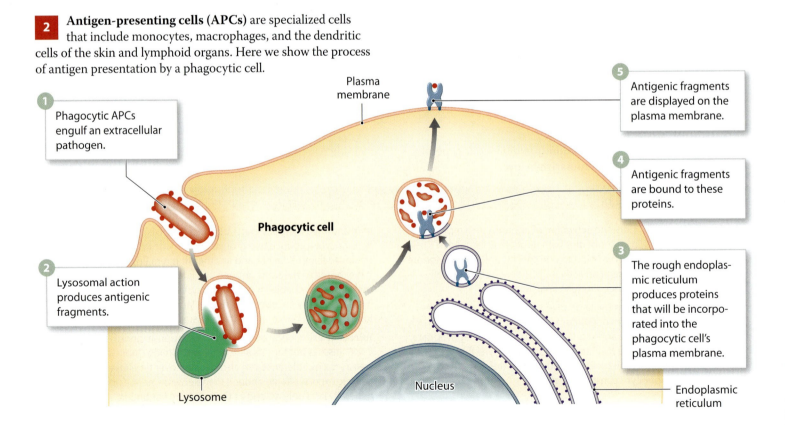

Plasma membrane

5 Antigenic fragments are displayed on the plasma membrane.

1 Phagocytic APCs engulf an extracellular pathogen.

4 Antigenic fragments are bound to these proteins.

Phagocytic cell

2 Lysosomal action produces antigenic fragments.

3 The rough endoplasmic reticulum produces proteins that will be incorporated into the phagocytic cell's plasma membrane.

Lysosome

Nucleus

Endoplasmic reticulum

3 Defenses against bacterial pathogens are usually initiated by active macrophages.

4 Defenses against viruses involve direct contact with virus-infected cells and antigen presentation by APCs.

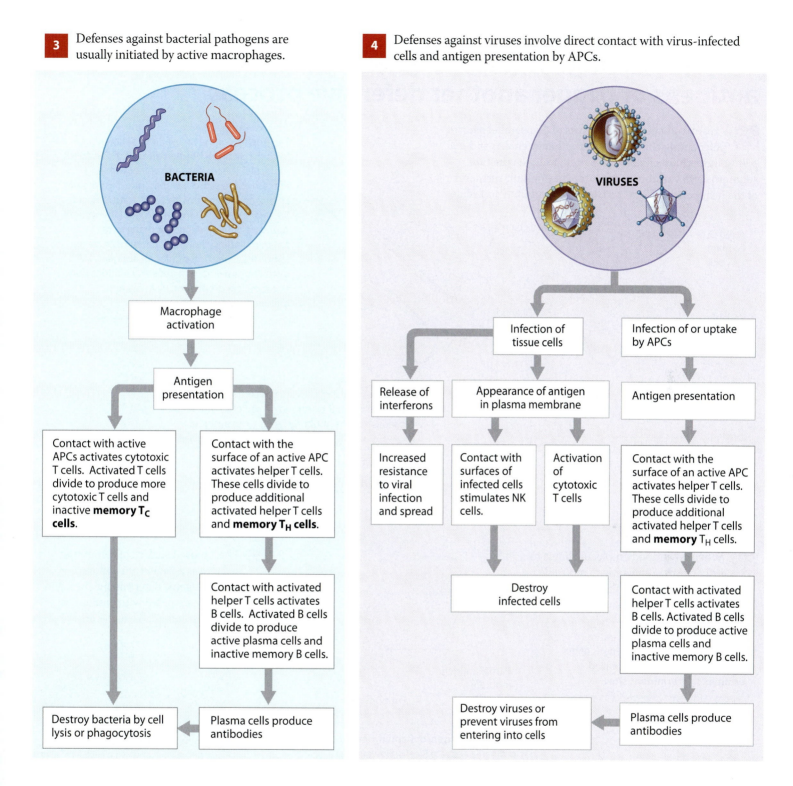

BACTERIA

Macrophage activation

Antigen presentation

Contact with active APCs activates cytotoxic T cells. Activated T cells divide to produce more cytotoxic T cells and inactive **memory T_C cells**.

Contact with the surface of an active APC activates helper T cells. These cells divide to produce additional activated helper T cells and **memory T_H cells**.

Contact with activated helper T cells activates B cells. Activated B cells divide to produce active plasma cells and inactive memory B cells.

Destroy bacteria by cell lysis or phagocytosis

Plasma cells produce antibodies

VIRUSES

Infection of tissue cells

Infection of or uptake by APCs

Release of interferons

Appearance of antigen in plasma membrane

Antigen presentation

Increased resistance to viral infection and spread

Contact with surfaces of infected cells stimulates NK cells.

Activation of cytotoxic T cells

Contact with the surface of an active APC activates helper T cells. These cells divide to produce additional activated helper T cells and **memory T_H cells**.

Destroy infected cells

Contact with activated helper T cells activates B cells. Activated B cells divide to produce active plasma cells and inactive memory B cells.

Destroy viruses or prevent viruses from entering into cells

Plasma cells produce antibodies

Memory cells play no role in overcoming the initial infection. But if the same pathogen appears at a later date, the memory cells dramatically reduce the time required to respond to the infection. When exposed to that pathogen, these cells immediately differentiate into: 1) cytotoxic T cells (memory T_C cells), 2) helper T cells (memory T_H cells), and 3) plasma cells (memory B cells), and begin fighting the infection. The response is so rapid and effective that we may never even realize that an infection has occurred.

Module 13.7 Review

a. Describe antigen presentation.

b. Which cells can be activated by direct contact with virus-infected cells?

c. Which cells produce antibodies?

Antibodies are small soluble proteins that bind to specific antigens; they may inactivate the antigens or trigger another defensive process

1 An antibody molecule consists of two parallel pairs of polypeptide chains: one pair of **heavy chains** and one pair of **light chains**. Each chain contains both **constant segments** and **variable segments**. The constant segments of the heavy chains form the base of the antibody molecule.

The free tips of the two variable segments form the **antigen binding sites** of the antibody molecule. These sites can interact with an antigen in the same way that the active site of an enzyme interacts with a substrate molecule. Small differences in the structure of the variable segments affect the precise shape of the antigen binding site. These differences make antibodies specific for different antigens. The distinctions result from genetic variations that occur during production, division, and differentiation of B cells.

Antigen binding site

Heavy chain

Variable segment

Disulfide bond

Light chain

Binding sites that can activate the complement system are covered when the antibody is secreted but become exposed when the antibody binds to an antigen.

Constant segments of light and heavy chains

Binding sites may also be present that attach the secreted antibody to the surfaces of macrophages, basophils, or mast cells.

2 When an antibody molecule binds to its corresponding antigen molecule, they form an **antigen-antibody complex**.

1 Antibodies bind not to the entire antigen, but to specific portions of its exposed surface—regions called **antigenic determinant sites**.

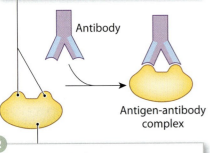

Antibody

Antigen-antibody complex

2 A **complete antigen** is an antigen with at least two antigenic determinant sites, one for each of the antigen binding sites on an antibody molecule.

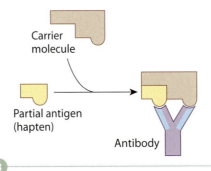

Carrier molecule

Partial antigen (hapten)

Antibody

3 A **partial antigen**, or hapten, does not ordinarily activate B cells. However, haptens may become attached to carrier molecules, forming combinations that can function as complete antigens. The antibodies produced will attack both the hapten and the carrier molecule. If the carrier molecule is normally present in the tissues, the antibodies may attack and destroy normal cells. This is the basis for several drug reactions, including allergies to penicillin.

3 The exposed surface of something as large as a bacterium contains millions of antigenic determinant sites, and it may become carpeted with antibodies, eventually destroying the bacterium.

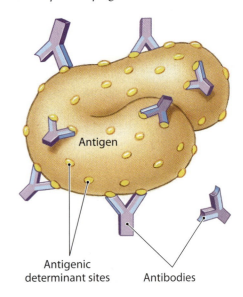

Antigen

Antigenic determinant sites

Antibodies

4 There are five different classes of antibodies, or **immunoglobulins** (**Igs**). The classes are determined by differences in the structure of the heavy-chain constant segments and so have no effect on the antibody's specificity, which is determined by its antigen binding sites.

Classes of Antibodies

   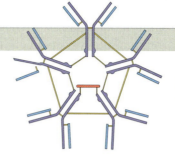 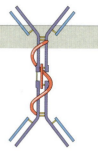

IgG antibodies are responsible for resistance against many viruses, bacteria, and bacterial toxins. They account for 80 percent of all antibodies.

IgE attaches to basophils and mast cells, releasing histamine and speeding up inflammation.

IgD is an individual molecule on the surfaces of B cells, where it can bind antigens in the extracellular fluid. This binding can play a role in sensitizing the B cell.

IgM is the first class of antibody secreted after an antigen is encountered. IgM concentration declines as IgG production accelerates. The anti-A and anti-B antibodies responsible for the agglutination of incompatible blood types are IgM antibodies.

IgA is found primarily in glandular secretions such as mucus, tears, saliva, and semen. These antibodies attack pathogens before they gain access to internal tissues.

5 The initial response to exposure to an antigen is called the **primary response**. The primary response takes time to develop, because the antigen must activate the appropriate B cells, which must then differentiate into antibody-secreting plasma cells. During the primary response, the **antibody titer**, or antibody concentration in the plasma, does not peak until one to two weeks after initial exposure. If the individual is no longer exposed to the antigen, the antibody titer declines thereafter.

6 When an antigen is encountered a second time, it triggers the **secondary response** which is more extensive and lasts longer. During the secondary response, antibody titers increase more rapidly and reach levels many times higher than they did in the primary response. This reflects the presence of large numbers of memory cells that are already primed for this antigen. These memory B cells respond immediately—much faster than the B cells stimulated during the initial exposure. The secondary response appears even if the second exposure occurs years after the first, because memory cells may survive for 20 years or more.

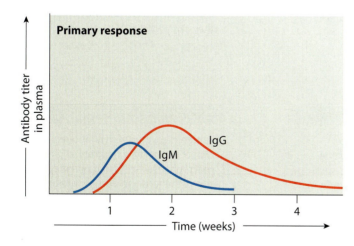

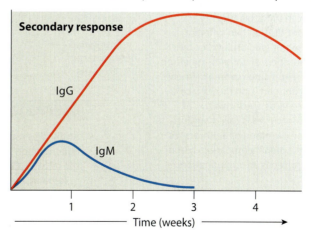

Module 13.8 Review

a. Define antigenic determinant site.

b. Describe the structure of an antibody.

c. Name the five classes of immunoglobulins and cite a function of each.

Antibodies use many different mechanisms to destroy target antigens

As a group, the antibodies of the various classes provide a versatile and effective defense against a variety of pathogens. The formation of an antigen-antibody complex can eliminate an antigen in seven different ways.

Neutralization

Both viruses and bacterial toxins must bind to the plasma membranes of body cells before they can enter or injure those cells. Binding occurs at superficial sites on the bacteria or toxins. Antibodies may bind to those sites, making the virus or toxin incapable of attaching itself to a cell. This mechanism is known as **neutralization**.

Prevention of Pathogen Adhesion

Antibodies dissolved in saliva, mucus, tears, and perspiration coat epithelia, providing an additional layer of defense. A covering of antibodies makes it difficult for bacteria or viruses to adhere to and penetrate body surfaces.

Activation of Complement

Upon binding to an antigen, portions of the antibody molecule change shape, exposing areas that bind complement proteins. The bound complement molecules then activate the complement system, which destroys the antigen.

Opsonization

Phagocytes bind more easily to surfaces that are covered with antibodies and complement proteins. As a result, a coating of antibodies and complement proteins makes phagocytosis more effective, an effect called **opsonization**. Some bacteria have slick plasma membranes or capsules, and phagocytes must be able to hang onto their prey before they can engulf it.

Agglutination

If antigens are close together, an antibody can bind to antigenic determinant sites on two different antigens. In this way, antibodies can tie large numbers of antigens together, creating an **immune complex**. When the target antigen is on the surface of a cell or a virus, the formation of immune complexes is called **agglutination**. The clumping of erythrocytes that occurs when incompatible blood types are mixed is an agglutination reaction.

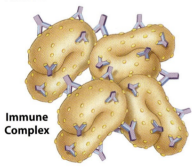

Immune Complex

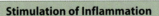

Antigenic determinant sites

Stimulation of Inflammation

Antibodies may promote inflammation by stimulating the release of chemicals, such as heparin and histamine, from basophils and mast cells.

Attraction of Phagocytes

Antigens covered with antibodies attract eosinophils, neutrophils, and macrophages—cells that phagocytize pathogens and destroy foreign or abnormal plasma membranes.

Module 13.9 Review

a. Describe the ways that antigen-antibody complexes can destroy target antigens.

b. Define opsonization.

c. Which cells are involved in inflammation?

Antibody responses can cause allergies and anaphylaxis

Allergies are inappropriate or excessive immune responses to antigens. The sudden increase in cellular activity or antibody titers can have several unpleasant side effects. For example, neutrophils or cytotoxic T cells may destroy normal cells while attacking the antigen, or the antigen-antibody complex may trigger massive inflammation. Antigens that trigger allergic reactions are called **allergens**.

1 Sensitization to an allergen during the initial exposure leads to the production of large quantities of IgE. The tendency to produce IgE antibodies in response to specific allergens may be genetically determined.

First Exposure

Allergen fragment

Allergens → Macrophage → T$_H$ cell activation

B cell sensitization and activation

Plasma cell

IgE antibodies

2 **Immediate hypersensitivity** is a rapid and especially severe response to the presence of an antigen. One form, **allergic rhinitis**, includes hay fever and other environmental allergies and may affect 15 percent of the U.S. population. Allergic rhinitis, characterized by inflammation of the nasal membranes, is one example of a **hypersensitivity reaction** that is restricted to the body surface. If the allergen enters the bloodstream, however, the response could be lethal. In **anaphylaxis** (an-a-fi-LAK-sis; *ana-*, again + *phylaxis*, protection), a circulating allergen stimulates mast cells throughout the body to release histamines. In severe cases extensive peripheral vasodilation occurs, producing a fall in blood pressure that can lead to a circulatory collapse. This response is called **anaphylactic shock**.

Subsequent Exposure

Allergen

IgE

Granules

Sensitization of mast cells and basophils

Massive stimulation of mast cells and basophils

Release of histamines and other chemicals that cause pain and inflammation

Localized Allergic Reactions	Systemic Allergic Reactions
If the allergen is at the body surface: localized inflammation, pain, and itching Example: allergic rhinitis	If the allergen is in the bloodstream: itching, swelling, and difficulty breathing (due to constricted airway) Example: anaphylaxis

Module 13.10 Review

a. Define allergy and allergen.

b. What is anaphylaxis?

c. Which chemicals do mast cells and basophils release when stimulated in an allergic reaction?

Immune disorders involving both overactivity and underactivity can be harmful

Excessive or Misdirected Immune Response

Autoimmune Disorders

In an **autoimmune disorder**, the immune system mistakenly attacks the body's own tissues. Usually the immune system recognizes but ignores self-antigens—normal antigens found in the body. When the recognition system malfunctions, however, activated B cells may make antibodies against normal body cells and tissues. These "misguided" antibodies are called **autoantibodies**. The resulting autoimmune disorder depends on the specific antigen attacked by auto-antibodies. Examples include:

- **Thyroiditis** is inflammation resulting from the release of autoantibodies against thyroglobulin.

- **Rheumatoid arthritis** occurs when autoanti-bodies form immune complexes within connective tissues around the joints. These complexes cause marked inflammation and an excessive immune response, eventually destroying the joint.

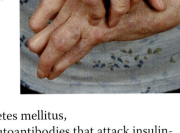

- **Type 1 diabetes**, formerly known as insulin-dependent diabetes mellitus, is generally caused by autoantibodies that attack insulin-producing cells in the pancreatic islets.

Autoimmune disorders affect an estimated 5 percent of adults in North America and Europe. Many of these disorders appear to be cases of mistaken identity. For example, proteins associated with the measles, Epstein–Barr, influenza, and other viruses contain amino acid sequences that are similar to those of myelin proteins. As a result, antibodies that target these viruses may also attack myelin sheaths. This mechanism is likely responsible for multiple sclerosis.

Graft Rejection

After an organ transplant, the major problem is **graft rejection**. In graft rejection, contact with proteins on plasma membranes in the donated tissue activates the recipient's T cells. The cytotoxic T cells that develop then attack and destroy the foreign cells. **Immunosuppression**—reducing the sensitivity of the immune system—can significantly improve transplant success. An understanding of the communication among T cells, macrophages, and B cells has led to the development of drugs with more selective effects. **Cyclosporin A (CsA)**, a compound derived from a fungus, is the most important immunosuppressive drug. This compound suppresses the immune response primarily by inhibiting helper T cell activity while leaving suppressor T cells relatively unaffected.

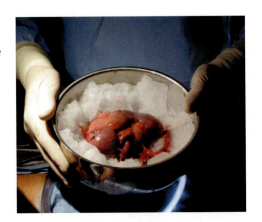

Allergies

The effects of the many forms of allergies range from mild to potentially lethal. (For examples, see Module 13.12)

Immunodeficiency Diseases

Immunodeficiency diseases result from (1) problems with the embryological development of lymphoid organs and tissues; (2) an infection with a virus that depresses immune function; or (3) treatment with, or exposure to, immunosuppressive agents, such as radiation or drugs.

Acquired immune deficiency syndrome (AIDS), the most common immunodeficiency disease, is caused by the **human immunodeficiency virus (HIV)**. The virus binds to specific membrane proteins, known as CD4 proteins, and infects helper T cells. The infected cells begin synthesizing and shedding viral proteins. Cells infected with HIV are ultimately destroyed either by the virus or immune defenses. The gradual destruction of helper T cells impairs both cell-mediated and antibody-mediated responses to antigens. Making matters worse, suppressor T cells are relatively unaffected by the virus, and over time the excess of suppressing factors "turns off" the normal immune response. Circulating antibody levels decline, cell-mediated immunity is reduced, and the body is left vulnerable to microbial invaders. With immune function suppressed, ordinarily harmless microorganisms can initiate lethal **opportunistic infections**. Because immune surveillance is also depressed, the risk of cancer increases. Infection with HIV occurs through intimate contact with the body fluids of infected individuals. The major routes of transmission involve contact with blood, semen, or vaginal secretions, although all body fluids may contain the virus. An estimated 33 million people are infected worldwide; 22 million of them are in sub-Saharan Africa, and 1.2 million in North America. AIDS causes about 17,000 deaths each year in the United States, and 2 million deaths worldwide.

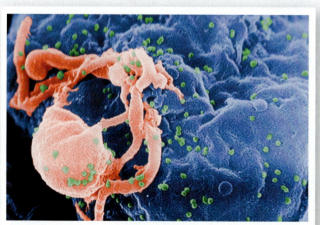

HIV (green) budding from an infected T$_H$ cell   SEM × 40,000

Age-Related Reductions in Immune Activity

With advancing age, the immune system becomes less effective at combating disease. T cells become less responsive to antigens, so fewer cytotoxic T cells respond to an infection. This effect may, at least in part, be associated with the gradual shrinkage of the thymus and a reduction in circulating levels of thymic hormones. Because the number of helper T cells also falls, B cells are less responsive, so antibody levels do not rise as quickly after antigen exposure. The net result is an increased susceptibility to viral and bacterial infections. For this reason, vaccinations for acute viral diseases such as the flu (influenza) and pneumococcal pneumonia, are strongly recommended for elderly individuals. The increased incidence of cancer in the elderly reflects declining immune surveillance, so tumor cells are not eliminated as effectively.

Module 13.11 Review

a. Define autoimmune disorders.

b. Describe immunosuppression.

c. Provide a plausible explanation for the increased incidence of cancer in the elderly.

1. Matching

Match the following terms with the most closely related description.

- opsonization
- helper T cells
- antibody
- complete antigen
- neutralization
- IgM
- agglutination
- IgG
- passive immunity
- anaphylaxis
- graft rejection
- acquired immunity
- B cells

a _____ Two parallel pairs of polypeptide chains

b _____ Major problem after an organ transplant

c _____ Active and passive

d _____ Involves the transfer of antibodies

e _____ Attacked by HIV

f _____ Enhances phagocytosis

g _____ Clumping of RBCs when incompatible blood types are mixed

h _____ Differentiate into memory and plasma cells

i _____ Antigen with at least two antigenic determinant sites

j _____ Antibodies used to determine blood type

k _____ Antibody mechanism that prevents a virus or toxin from binding to a body cell

l _____ Abundant antibody that provides resistance against many viruses, bacteria, and bacterial toxins

m _____ Circulating allergen stimulates mast cells throughout body

2. Matching

Match the following terms with the most closely related description.

- cytotoxic T cells
- viruses
- B cells
- antibodies
- helper T cells
- macrophages
- natural killer (NK) cells
- suppressor T cells
- memory T cells and B cells

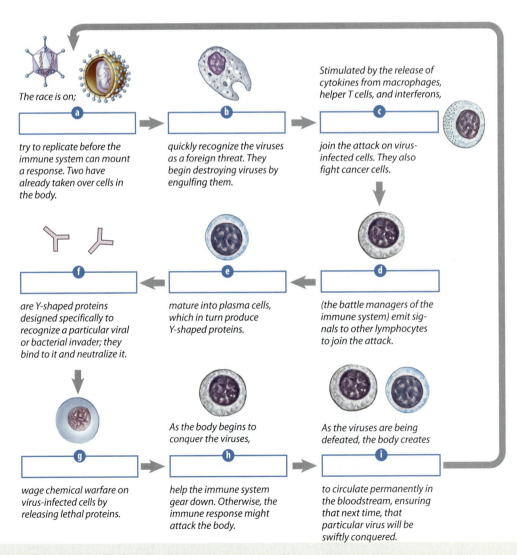

The race is on;

a _____

try to replicate before the immune system can mount a response. Two have already taken over cells in the body.

b _____

quickly recognize the viruses as a foreign threat. They begin destroying viruses by engulfing them.

Stimulated by the release of cytokines from macrophages, helper T cells, and interferons,

c _____

join the attack on virus-infected cells. They also fight cancer cells.

f _____

are Y-shaped proteins designed specifically to recognize a particular viral or bacterial invader; they bind to it and neutralize it.

e _____

mature into plasma cells, which in turn produce Y-shaped proteins.

d _____

(the battle managers of the immune system) emit signals to other lymphocytes to join the attack.

g _____

wage chemical warfare on virus-infected cells by releasing lethal proteins.

As the body begins to conquer the viruses,

h _____

help the immune system gear down. Otherwise, the immune response might attack the body.

As the viruses are being defeated, the body creates

i _____

to circulate permanently in the bloodstream, ensuring that next time, that particular virus will be swiftly conquered.

Visual Outline with Key Terms

Summarize the content of each module using the terms in the order provided.

SECTION 1

Anatomy of the Lymphatic System

- lymphatic system
- lymphocytes
- lymph
- lymphatic vessels
- lymphatics

13.1

Interstitial fluid flows into lymphatic capillaries to become lymph within lymphatic vessels

- lymphatic vessels
- lymphatic capillaries
- lymph

13.2

Small lymphatic vessels merge to form lymphatic ducts that empty lymph into subclavian veins

- blood
- lymphatic trunks
- thoracic duct
- right lymphatic duct
- cisterna chyli
- lumbar trunks
- intestinal trunk
- lymphedema

13.3

Lymphocytes are responsible for the immune functions of the lymphatic system

- antigens
- T cells
- cytotoxic T cells
- cell-mediated immunity
- helper T cells
- suppressor T cells
- B cells
- plasma cells
- antibody-mediated immunity
- memory B cells
- NK (natural killer) cells
- immunological surveillance
- lymphopoiesis
- red bone marrow
- lymphoid stem cells
- thymus
- blood–thymus barrier

13.4

Lymphocytes within lymphoid tissues and lymphoid organs protect the body from pathogens and toxins

- lymphoid tissues
- lymphoid nodule
- lymphoid organs
- mucosa-associated lymphoid tissue (MALT)
- tonsils
- pharyngeal tonsil
- palatine tonsils
- lingual tonsils
- tonsillitis
- lymph nodes
- efferent lymphatics
- dendritic cells
- afferent lymphatics
- thymosins

SECTION 2

Nonspecific immunity

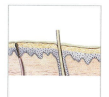

- immunity
- nonspecific defenses
- nonspecific resistance
- physical barriers
- phagocytes
- immune surveillance
- interferons
- complement system
- inflammatory response
- fever
- specific defenses
- specific resistance

13.5

Physical barriers and phagocytes play a role in nonspecific defenses

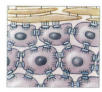

- integumentary system
- epithelia
- phagocytes
- neutrophils
- eosinophils
- macrophages
- monocyte–macrophage system
- fixed macrophages
- free macrophages
- diapedesis
- chemotaxis

13.6

Inflammation is a localized tissue response to injury; fever is a generalized response to tissue damage and infection

- inflammation
- fever
- pyrogens
- perforins

SECTION 3

Specific immunity

- specific defenses
- immunity
- acquired immunity
- passive immunity
- naturally acquired passive immunity
- artificially acquired passive immunity
- active immunity (immune response)
- naturally acquired active immunity
- artificially acquired active immunity
- innate immunity
- specificity
- versatility
- clone
- immunologic memory

Specific Defenses (Immunity)
Respond to threats on an individualized basis

- memory cells
- tolerance

13.7

Specific defenses are triggered by exposure to antigenic fragments

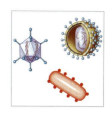

- antigen presentation
- antigen-presenting cells (APCs)
- memory T_C cells
- memory T_H cells

● = *Term boldfaced in this module*

13.8

Antibodies are small soluble proteins that bind to specific antigens; they may inactivate the antigens or trigger another defensive process

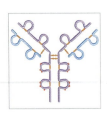

- ○ antibody molecule
- heavy chains
- light chains
- constant segments
- variable segments
- antigen binding sites
- antigen-antibody complex
- antigenic determinant sites
- complete antigen
- partial antigen
- immunoglobulins (Igs)
- IgG, IgE, IgD, IgM, IgA
- primary response
- antibody titer
- secondary response

13.9

Antibodies use many different mechanisms to destroy target antigens

- neutralization
- prevention of adhesion
- activation of complement
- opsonization
- attraction of phagocytes
- stimulation of inflammation
- agglutination
- immune complex

13.10

Antibody responses can cause allergies and anaphylaxis

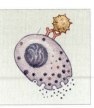

- allergies
- allergens
- immediate hypersensitivity
- allergic rhinitis
- hypersensitivity reaction
- anaphylaxis
- anaphylactic shock
- ○ localized allergic reactions
- ○ systemic allergic reactions

13.11

Immune disorders involving both overactivity and underactivity can be harmful

- autoimmune disorder
- autoantibodies
- thyroiditis
- rheumatoid arthritis
- Type 1 diabetes
- graft rejection
- immunosuppression
- cyclosporin A (CsA)
- ○ allergies
- immuno-deficiency diseases
- acquired immune deficiency syndrome (AIDS)
- human immunodeficiency virus (HIV)
- opportunistic infections

● = *Term boldfaced in this module*

CAREER PATHS

Burn Nurse Specialist

"There are patients in need here, and the work is important. The rewards of working with burns—seeing patients get better—are great."

— **Cynthia Pronze**
Burn Nurse Specialist
Ann Arbor, Michigan

Medical professionals who work in trauma units must be ready to handle whatever comes through the door. The patients are very ill or severely injured, and the days are long and intense.

Cynthia Pronze is a clinical nurse supervisor for the trauma burn intensive care unit and acute care unit of the University of Michigan Hospital. As a supervisor, Cynthia manages a staff of nurses and is in charge of patient care. She begins work early, around 6:15 a.m., checking in with the night-shift nurses before they leave so she can catch up on the status of all the patients. She goes unit to unit until she is fully up-to-date. Much of the rest of her day is filled with meetings, either with her direct managers or with doctors or other nurses on the unit. "We're a level 1 trauma burn center, which means this is a vast place. I always check in with staff again, usually before

noon, to find out how things are going. What are their needs? Do they need help?"

Working in the burn ward has special challenges. "The number of burns that occur every year in this country is huge. Still, this is not an area that everyone wants to go into." Many people are uncomfortable with treating burns, so they choose other fields.

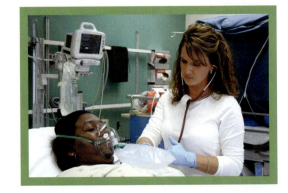

Cynthia chose to work in trauma because it began to interest her while she was a nursing student. "There are patients in need here, and the work is important. The rewards of working with burns—seeing patients get better—are great. Depending on the percentage of burn coverage, some of these patients are here for many months. We get to know them and their families. We help them deal with stress and the huge gaps in their lives."

Cynthia began her nursing career after working for 20 years as a hairdresser and raising three children. "I come from a very large family, and college was not an option for me when I was young. But at one point my husband turned to me and said, 'You've always wanted to become a nurse. What are you waiting for?' I enrolled first in the local community college and earned an A.D.N.—an associate's degree in nursing. Later I took more course work and obtained a bachelor's and then a master's degree."

Cynthia took anatomy and physiology as an undergraduate and again in her master's program. "A&P course work is very challenging. But I couldn't get enough of it. It's so relevant to nursing. You have to have an in-depth knowledge base when you care for another human being, understanding how the body works, how it responds. A&P answers questions that may be swirling around in your head. And once you're treating people, you realize, 'Oh, that's why this may or may not have happened with this patient.'"

What qualities make a successful nurse? "Working well with people, good organizational skills, following directions," she says, "and being able to give of yourself."

For more information, visit the website for the National League for Nursing at http://www .nin.org.

Think this is the career for you?
Key Stats:

- **Education and Training.** Most burn nurse specialists are registered nurses (RNs), which requires an associate's or bachelor's degree in nursing, though more and more are pursuing graduate degrees.

- **Licensure.** All states require nurses to graduate from an approved nursing program and pass a national licensing exam.

- **Earnings.** Earnings vary but the median annual salary is $64,690.

- **Expected Job Prospects.** Employment is expected to grow faster than the national average—by 22 % through 2018.

Bureau of Labor Statistics, U.S. Department of Labor, *Occupational Outlook Handbook, 2010–11 Edition,* Registered Nurses, on the Internet at http://www.bls.gov/oco/ocos083.htm (visited September 14, 2011).

Access more review material online in the Study Area at www.masteringaandp.com.

- Chapter guides
- Chapter quizzes
- Practice tests
- Art labeling activities
- Flashcards
- A glossary with pronunciations
- Practice Anatomy Lab™ (PAL™) 3.0 virtual anatomy practice tool
- Interactive Physiology® (IP) animated tutorials
- MP3 Tutor Sessions

 practice anatomy lab™   **For this chapter, follow these navigation paths in PAL:**

- Human Cadaver>Lymphatic System
- Anatomical Models>Lymphatic System
- Histology>Lymphatic System

 For this chapter, go to these topics in the Immune System in IP:

- Immune System Overview
- Anatomy Review
- Innate Host Defenses
- Common Characteristics of B and T Lymphocytes
- Humoral Immunity

 For this chapter, go to this topic in the MP3 Tutor Sessions:

- Differences between Innate and Adaptive Immunity

Chapter 13 Review Questions

1. List and explain four general properties of immunity.

2. In what ways can the formation of an antigen-antibody complex eliminate an antigen?

3. Kathy's grandfather is diagnosed with lung cancer. When his physician takes tissue samples (biopsies) of several lymph nodes from neighboring regions of the body, Kathy wonders why, since his cancer is in his lungs. What would you tell her?

4. Tony finds out that he has been exposed to the measles and is concerned that he might have contracted the disease. His physician takes a blood sample and sends it to a lab to measure antibody levels. The results show an elevated level of IgM antibodies to measles virus, but very few IgG antibodies to the virus. Did Tony contract the disease?

5. An investigator at a crime scene discovers some body fluid on the victim's clothing. The investigator carefully takes a sample and sends it to the crime lab for analysis. On the basis of the analysis of immunoglobulins, could the crime lab determine whether the sample is blood plasma or semen? Explain your answer.

For answers to all module, section, and chapter review questions, see the blue Answers tab at the back of the book.

14

The Respiratory System

Functional Anatomy of the Respiratory System

The **respiratory system** is made up of structures involved in moving air to and from the lungs, and exchanging gases between air and blood. In this section we consider this body system's functions and structural organization.

1 The respiratory system is anatomically divided into an **upper respiratory system** and **lower respiratory system**. The passageways of both divisions that carry air to and from the gas exchange surfaces of the lungs make up the **respiratory tract**. The tract is composed of a conducting portion and a respiratory portion. The **conducting portion** begins at the nasal cavity and extends to the narrow passageways called **bronchioles**. The **respiratory portion** includes the smallest bronchioles and the air-filled pockets called **alveoli**. Gas exchange between air and blood occurs at the alveoli.

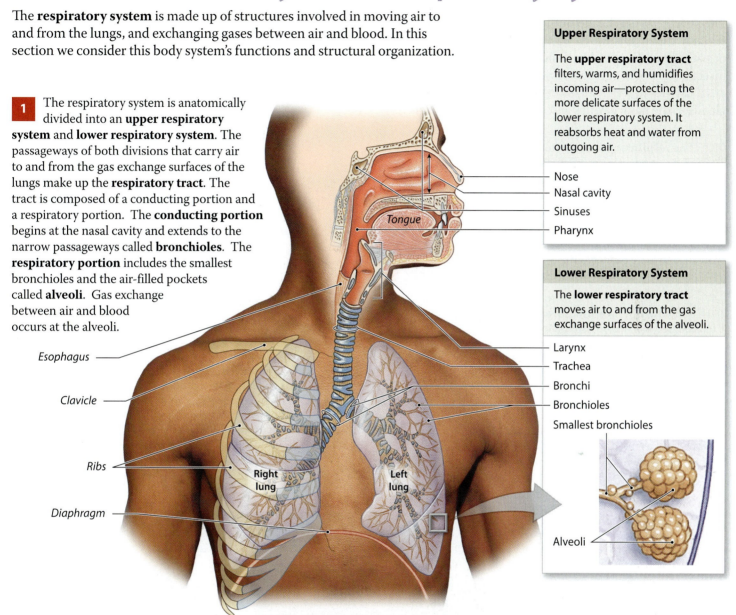

Tongue

Esophagus

Clavicle

Ribs

Right lung

Left lung

Diaphragm

Upper Respiratory System

The **upper respiratory tract** filters, warms, and humidifies incoming air—protecting the more delicate surfaces of the lower respiratory system. It reabsorbs heat and water from outgoing air.

Nose
Nasal cavity
Sinuses
Pharynx

Lower Respiratory System

The **lower respiratory tract** moves air to and from the gas exchange surfaces of the alveoli.

Larynx
Trachea
Bronchi
Bronchioles
Smallest bronchioles

Alveoli

2 This table summarizes the major functions of the respiratory system.

Major Functions of the Respiratory System

- Provide a large surface area for gas exchange between air and circulating blood.
- Move air to and from the exchange surfaces of the lungs along the respiratory tract.
- Protect respiratory surfaces from dehydration and temperature changes, and defend against invasion by pathogens.
- Produce sounds involved in speaking, singing, and other forms of communication.
- Aid the sense of smell by olfactory receptors in the superior portions of the nasal cavity.

The respiratory mucosa cleans, warms, and moistens inhaled air

The delicate gas exchange surfaces of your respiratory system can be severely damaged if the air you inhale (breathe in) contains debris or pathogens. The nasal hairs prevent the entry of large debris or insects. Networks of blood vessels in the walls radiate heat into the nasal cavities to warm incoming air, and the secretions of mucous glands in the nasal cavities and sinuses humidify the air.

1 The **respiratory mucosa** (mū-KŌ-suh) lines most of the conducting portion of the respiratory tract. The structure of the respiratory epithelium changes along the respiratory tract. A pseudostratified ciliated columnar epithelium with numerous mucous cells lines the nasal cavity, the superior portion of the pharynx, and the trachea, bronchi, and large bronchioles.

The beating of cilia sweeps mucus and any trapped debris or microorganisms toward the pharynx, where they can be coughed out or swallowed and destroyed by the acids and enzymes of the stomach. This ciliary movement continuously cleans and protects the respiratory surfaces. By the time air reaches the respiratory portion of the tract, particles larger than around 5 μm have been trapped and removed.

Mucous cell

Ciliated columnar epithelial cell

Mucus layer

The **lamina propria** (LAM-in-nuh PRŌ-prē-uh) is the underlying layer of areolar tissue that supports the respiratory epithelium. In the upper respiratory system, trachea, and bronchi, the lamina propria contains mucous glands that discharge secretions onto the epithelial surface.

2 **Cystic fibrosis (CF)** is the most common lethal inherited disease among Caucasians of Northern European descent. There are 1000 new cases of CF diagnosed each year. The most dangerous signs and symptoms result from the production of abnormally thick and sticky mucus. The dense mucus accumulates, and restricts airflow. Potentially lethal infections may develop in the respiratory passageways and lungs if bacteria colonize the stagnant mucus. Over time, chronic inflammation and infections may damage the respiratory system permanently. This CF patient is receiving a breathing treatment through her inhaler, which will thin the mucus temporarily.

3 We can see the features of the upper respiratory system and those leading into the lower respiratory system best in sagittal section.

The nasal cavity opens into the nasopharynx through a connection known as the **internal nares**.

The **nasal vestibule** is the space contained within the flexible tissues of the nose. Its epithelium contains coarse hairs that extend across the external nares. These hairs trap large airborne particles and prevent them from entering the nasal cavity.

Pharynx

The **pharynx** (FAR-inks) is a chamber shared by the digestive and respiratory systems. The curving superior and posterior walls of the pharynx are closely bound to the axial skeleton, but the lateral walls are flexible and muscular.

The **nasopharynx** (nā-zō-FAR-inks) is the superior portion of the pharynx located between the soft palate and the internal nares.

The **oropharynx** (*oris*, mouth) extends between the soft palate and the base of the tongue at the level of the hyoid bone. At the boundary between the nasopharynx and the oropharynx, the epithelium changes from pseudostratified columnar to stratified squamous epithelium.

The narrow **laryngopharynx** (la-rin-gō-FAR-inks) includes that portion of the pharynx between the hyoid bone and the entrance to the larynx and esophagus. Like the oropharynx, the laryngopharynx is lined with a stratified squamous epithelium.

The **trachea**, or windpipe, conducts air toward the lungs.

Air normally enters the nose through the paired **external nares** (NĀ-res), or nostrils, which open into the nasal cavity.

Nasal cavity

A bony **hard palate** forms the floor of the nasal cavity and separates it from the oral cavity.

Tongue

A fleshy **soft palate** extends posterior to the hard palate.

Hyoid bone

Inhaled air leaves the pharynx and enters the larynx through a narrow opening called the **glottis** (GLOT-is).

The **larynx** (LAR-inks) is a cartilaginous structure that surrounds and protects the glottis. The larynx marks the beginning of the lower respiratory system and contains the vocal cords.

The lining of the nasal cavity contains an extensive network of highly expandable veins that can release heat like a radiator. As cool, dry air passes inward over the exposed surfaces of the nasal cavity, the air warms and water in the mucus evaporates. Air moving from your nasal cavity to your lungs is thus heated almost to body temperature, and it is nearly saturated with water vapor. This protects more delicate respiratory surfaces from chilling or drying out. Breathing through your mouth eliminates much of the conditioning of inhaled air and increases heat and water loss every time you exhale (breathe out).

Module 14.1 Review

a. What is the role of cilia lining the respiratory mucosa?

b. Why can cystic fibrosis become lethal?

c. How is inhaled air warmed?

The trachea and primary bronchi carry air to and from the lungs

1 The **trachea** (TRĀ-kē-uh), or windpipe, is a tough, flexible tube with a diameter of about 2.5 cm (1 in.). The trachea branches to form the right and left **primary bronchi** (BRONG-kī; singular, *bronchus*). The branching pattern of bronchi is often called the bronchial tree.

2 This is a sectional view of the trachea. An elastic ligament and the **trachealis muscle** connect the ends of each C-shaped tracheal cartilage. When the trachealis muscle contracts, the trachea gets smaller in diameter, which increases the resistance to airflow. Such constriction of the trachea also increases the force of air expelled during a cough, and aids the expulsion of irritants. When the trachealis muscle relaxes, the trachea becomes larger in diameter, which makes it easier to move air along the respiratory passageways.

Hyoid bone

Larynx

The trachea contains 15–20 **tracheal cartilages**, which stiffen the tracheal walls and protect the airway.

Esophagus

Because the tracheal cartilages are incomplete posteriorly, the posterior tracheal wall can easily change shape when large masses of food pass along the esophagus.

The right primary bronchus is larger in diameter than the left, and descends toward the lung at a steeper angle. Thus, most foreign objects that enter the trachea find their way into the right bronchus.

Trachealis muscle

Respiratory epithelium

Trachea

Lumen of trachea

Tracheal cartilage

Mucous gland

Left primary bronchus

Smaller bronchi

Right lung

Cartilage plates

Visceral pleura

Left lung

Bronchiole

Respiratory epithelium

The walls of bronchioles contain relatively thick layers of smooth muscle. Involuntary contraction and relaxation of this smooth muscle alters bronchial diameter and resistance to airflow. Sympathetic activation leads to **bronchodilation**, the enlargement of airway diameter. Parasympathetic stimulation leads to **bronchoconstriction**, a reduction in the diameter of the airway. Extreme bronchoconstriction may occur during allergic reactions such as **asthma** (AZ-muh), in which swelling and bronchoconstriction can severely restrict—or even prevent—airflow.

Smooth muscle

3 Each lung is shaped like a blunt cone, with the tip extending to the first rib. The broad concave inferior portion of each lung rests on the superior surface of the diaphragm.

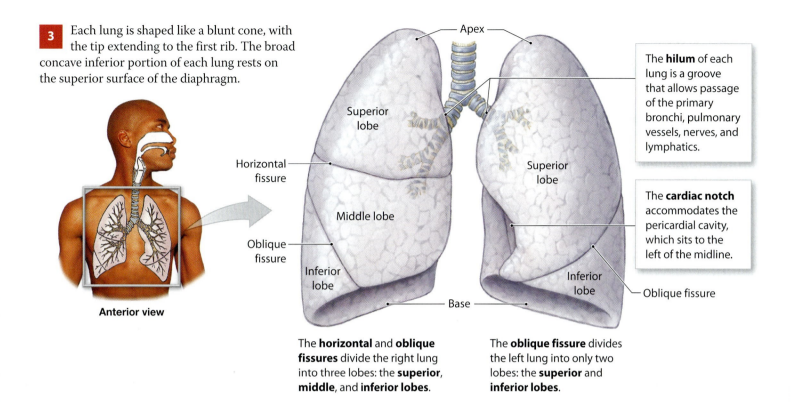

Anterior view

Apex

The **hilum** of each lung is a groove that allows passage of the primary bronchi, pulmonary vessels, nerves, and lymphatics.

Superior lobe

Horizontal fissure

Superior lobe

The **cardiac notch** accommodates the pericardial cavity, which sits to the left of the midline.

Middle lobe

Oblique fissure

Inferior lobe

Inferior lobe

Base

Oblique fissure

The **horizontal** and **oblique fissures** divide the right lung into three lobes: the **superior, middle,** and **inferior lobes**.

The **oblique fissure** divides the left lung into only two lobes: the **superior** and **inferior lobes**.

4 This superior view of the thoracic cavity shows the relationship between the lungs, heart, and mediastinum. The pleural cavities surround the lungs, and the pericardial cavity surrounds the heart. The lining of the pleural cavity is a serous membrane called the **pleura** (PLOO-rah). The parietal pleura lines the inner surface of the chest wall, the visceral pleura covers the surfaces of the lungs, and the pleural cavity lies between them.

Parietal pleura

Right pleural cavity

Visceral pleura

Mediastinum

Right Lung

Left Lung

Heart

Left pleural cavity

Pericardial cavity

Module 14.2 Review

a. Why are the tracheal cartilages C-shaped rather than complete rings?

b. If food accidentally enters the trachea, in which bronchus is it more likely to lodge? Why?

c. Which lung features the cardiac notch? What is the purpose of the cardiac notch?

Gas exchange occurs at the alveoli

1 The conducting portion of the respiratory tract enters the lungs as large-diameter primary bronchi, supported by cartilage rings. As the bronchi branch and become more narrow, the support changes first to cartilage blocks and then, in the bronchioles, to a layer of smooth muscle. In the smallest bronchioles, the walls contain bands of smooth muscle cells (see below). These delicate bronchioles lead to tiny sacs called **alveoli** (al-VĒ-ō-lī; singular, alveolus). Each lung contains about 150 million alveoli, which gives the lung an open, spongy appearance.

Primary bronchus

Secondary bronchi

Small bronchi

Cluster of alveoli

Bronchioles

2 Each alveolus is surrounded by a capillary network that is supported by a meshwork of connective tissue fibers. Alveoli most often form clusters, like bunches of grapes, with adjacent clusters sharing a connection to a single bronchiole.

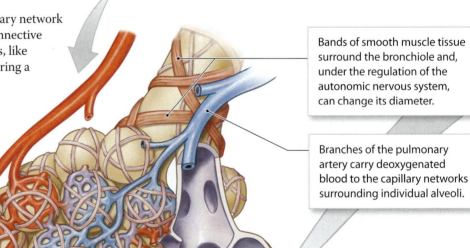

Bands of smooth muscle tissue surround the bronchiole and, under the regulation of the autonomic nervous system, can change its diameter.

Branches of the pulmonary vein collect oxygenated blood from the alveolar capillary networks and carry it to the heart.

Branches of the pulmonary artery carry deoxygenated blood to the capillary networks surrounding individual alveoli.

The alveolar capillary networks are the sites of gas exchange between the air in the alveoli and the circulating blood.

The elastic fibers that surround each alveolus allow it to expand and recoil as air moves in and out of the respiratory tract. The recoil of these fibers during exhalation both reduces the size of the alveoli and helps push air out of the lungs.

Alveoli

3 The alveolar epithelium is primarily a simple squamous epithelium. The surrounding capillaries differ functionally from other capillaries because they dilate when alveolar oxygen levels are high, and constrict when alveolar oxygen levels are low. This response to oxygen levels directs blood flow to the alveoli containing the most oxygen.

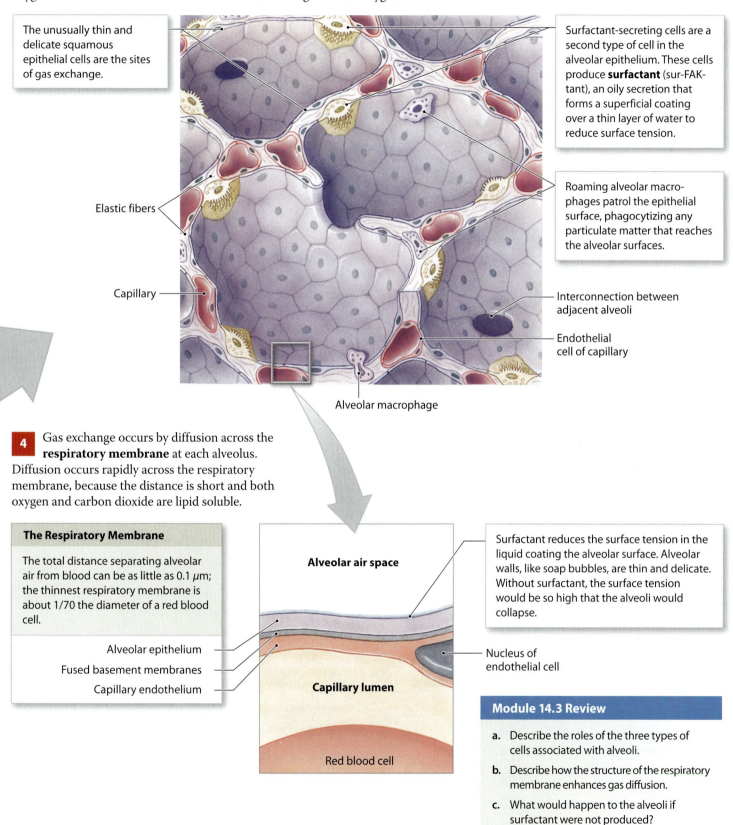

The unusually thin and delicate squamous epithelial cells are the sites of gas exchange.

Surfactant-secreting cells are a second type of cell in the alveolar epithelium. These cells produce **surfactant** (sur-FAK-tant), an oily secretion that forms a superficial coating over a thin layer of water to reduce surface tension.

Roaming alveolar macrophages patrol the epithelial surface, phagocytizing any particulate matter that reaches the alveolar surfaces.

Elastic fibers

Capillary

Interconnection between adjacent alveoli

Endothelial cell of capillary

Alveolar macrophage

4 Gas exchange occurs by diffusion across the **respiratory membrane** at each alveolus. Diffusion occurs rapidly across the respiratory membrane, because the distance is short and both oxygen and carbon dioxide are lipid soluble.

The Respiratory Membrane

The total distance separating alveolar air from blood can be as little as 0.1 μm; the thinnest respiratory membrane is about 1/70 the diameter of a red blood cell.

Alveolar epithelium

Fused basement membranes

Capillary endothelium

Alveolar air space

Capillary lumen

Red blood cell

Surfactant reduces the surface tension in the liquid coating the alveolar surface. Alveolar walls, like soap bubbles, are thin and delicate. Without surfactant, the surface tension would be so high that the alveoli would collapse.

Nucleus of endothelial cell

Module 14.3 Review

a. Describe the roles of the three types of cells associated with alveoli.

b. Describe how the structure of the respiratory membrane enhances gas diffusion.

c. What would happen to the alveoli if surfactant were not produced?

1. Labeling

Label each of the respiratory system structures in the following figure.

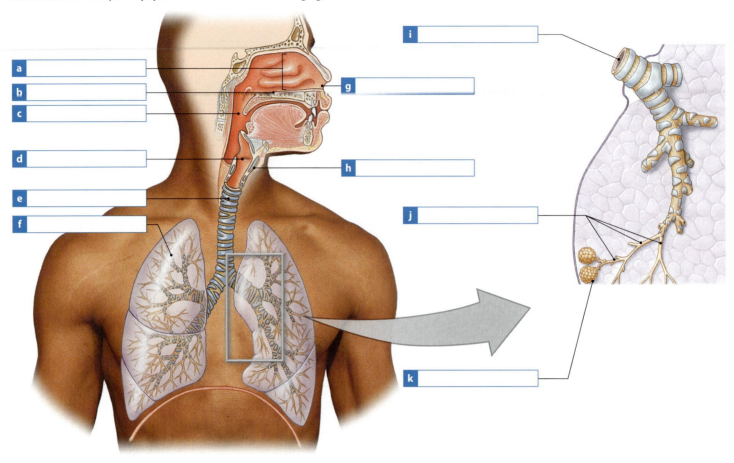

2. Matching

Match the following terms with the most closely related description.

- left bronchus
- respiratory mucosa
- soft palate
- bronchodilation
- right bronchus
- internal nares
- simple squamous epithelial cells
- surfactant
- cystic fibrosis
- trachea
- pharynx
- respiratory membrane
- larynx
- bronchoconstriction

a	_____	Prevents collapse of alveoli
b	_____	Windpipe
c	_____	Alveolar component of respiratory membrane
d	_____	Sympathetic activation
e	_____	Air supply for lung with two lobes
f	_____	Parasympathetic activation
g	_____	Common site for inhaled foreign objects
h	_____	Gas exchange surface
i	_____	Posterior to hard palate
j	_____	Chamber shared by respiratory and digestive systems
k	_____	Surrounds and protects the glottis
l	_____	Lethal inherited respiratory disease
m	_____	Lines conducting portion of respiratory tract
n	_____	Connects nasal cavity and nasopharynx

Respiratory Physiology

The general term **respiration** refers to two integrated processes: external respiration and internal respiration.

1 This illustration gives an overview of respiration. It shows the relationships between external respiration (pulmonary ventilation, gas diffusion, and gas transport), and internal respiration.

Respiration

External Respiration

External respiration is involved with the exchange of oxygen and carbon dioxide between the body's tissues and the external environment. The purpose of external respiration, and the primary function of the respiratory system, is meeting the respiratory demands of cells.

Internal Respiration

Internal respiration is the absorption of O_2 and the release of CO_2 by tissue cells. (We will consider the biochemical pathways responsible for consuming O_2 and generating CO_2 in Chapter 16.)

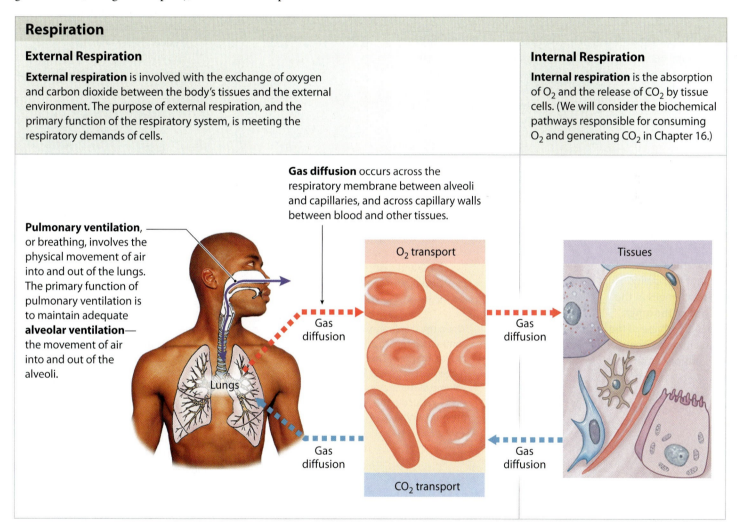

Gas diffusion occurs across the respiratory membrane between alveoli and capillaries, and across capillary walls between blood and other tissues.

Pulmonary ventilation, or breathing, involves the physical movement of air into and out of the lungs. The primary function of pulmonary ventilation is to maintain adequate **alveolar ventilation**— the movement of air into and out of the alveoli.

Lungs

O_2 transport

Gas diffusion

Gas diffusion

CO_2 transport

Tissues

Gas diffusion

Gas diffusion

Abnormalities affecting any aspects of external respiration will ultimately affect the oxygen concentrations available to tissues. If the oxygen content declines, the affected tissues will become starved for oxygen. **Hypoxia**, or low tissue oxygen levels, severely limits the metabolic activities of the affected area. If the supply of oxygen is cut off completely, the condition called **anoxia** (an-OK-sē-a; *a-*, without + *ox-*, oxygen) results. Much of the damage caused by strokes and heart attacks results from localized anoxia, which kills cells in that area.

Pressure changes within the pleural cavities drive pulmonary ventilation

In a gas such as air, the molecules bounce around as independent objects. At normal atmospheric pressures, gas molecules are much farther apart than the molecules in a liquid, so the density of air is relatively low. If a gas is enclosed in a container, the pressure exerted by the gas results from its molecules colliding with the container walls. The greater the number of collisions, the higher the gas pressure.

1 These diagrams will help you understand the relationships between pressure and volume in a gas. For a gas in a closed container and at a constant temperature, if you decrease the volume of the gas, its pressure will rise. If you increase the volume of the gas, its pressure will fall. This is an inverse relationship between volume (V) and pressure (P), so V↓, P↑.

V↓ causes P↑
If you decrease the volume of the container, collisions occur more frequently per unit time, elevating the pressure of the gas.

V↑ causes P↓
If you increase the volume of the container, fewer collisions occur per unit time, because it takes longer for a gas molecule to travel from one wall to another. As a result, the gas pressure inside the container declines.

2 What does this have to do with breathing? Movements of the diaphragm and rib cage change the volume of the thoracic cavity. This causes changes in the volume of the lungs. When the shape of the thoracic cavity changes, it expands or compresses the lungs, changing the air (gas) pressure within the respiratory tract.

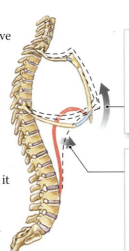

Superior movement of the rib cage increases the depth and width of the thoracic cavity, increasing its volume and reducing pressure within it.

When the diaphragm contracts, it tenses and moves inferiorly. This movement increases the volume of the thoracic cavity, reducing the pressure within it.

3 When you start to take a breath, the pressures (P) inside and outside the thoracic cavity are identical, and no air moves into or out of your lungs.

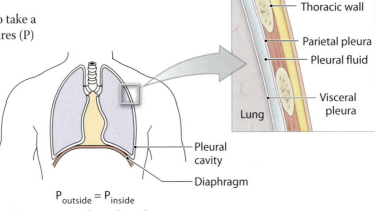

Thoracic wall

Parietal pleura
Pleural fluid

Visceral pleura

Lung

Pleural cavity

Diaphragm

$P_{outside} = P_{inside}$
Pressure outside and inside are equal, so no air movement occurs

The layer of pleural fluid contained in the pleural cavity makes the lungs stick to the inner walls of the thorax. This kind of fluid bond is responsible for making a coaster stick to the bottom of a wet glass. The elastic tissues of the lungs are always trying to recoil and reduce lung volume to about 5 percent of its normal size. The fluid bond between the parietal and visceral pleura prevents this collapse. If an injury allows air into the pleural cavity, this bond is broken and the lung collapses. Air in the pleural cavity is called a **pneumothorax** (nū-mō-THOR-aks), and the resulting lung collapse is called **atelectasis** (a-te-LEK-ta-sis).

4 Air will flow from an area of higher pressure to an area of lower pressure. When the thoracic cavity enlarges during inhalation, pressure falls inside the lungs and air flows in. When the thoracic cavity decreases in volume during exhalation, pressure rises inside the lungs, forcing air out of the respiratory tract.

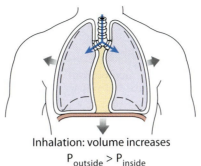

Inhalation: volume increases
$P_{outside} > P_{inside}$
Pressure inside falls, so air flows in

Exhalation: volume decreases
$P_{outside} < P_{inside}$
Pressure inside rises, so air flows out

5 What determines the direction of airflow? The difference between atmospheric pressure and **intrapulmonary** (in-tra-PUL-mo-nār-ē) **pressure**, the pressure inside the respiratory tract. **Alveolar pressure** is the intrapulmonary pressure measured in the alveoli.

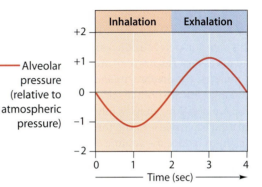

Alveolar pressure (relative to atmospheric pressure)

A pressure difference of 0 mm Hg exists when atmospheric and intrapulmonary pressures are equal. Positive intrapulmonary pressures will push air out of the lungs; negative intrapulmonary pressures will cause air to rush into the lungs.

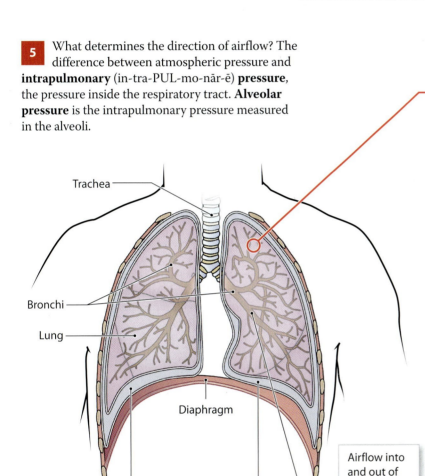

Trachea

Bronchi

Lung

Diaphragm

Right pleural cavity

Left pleural cavity

Airflow into and out of the bronchial tree of the lungs.

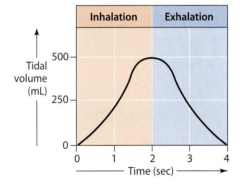

Tidal volume (mL)

The **tidal volume (V_T)** is the amount of air moved into the lungs during inhalation or out of the lungs during exhalation. At rest, the tidal volume is approximately 500 mL.

Module 14.4 Review

a. Describe the relationship between volume and pressure for a gas.

b. What physical changes affect the volume of the lungs?

c. What pressures determine the direction of airflow within the respiratory tract?

Respiratory muscles adjust the tidal volume to meet respiratory demands

Respiratory muscles may be involved with either inhalation or exhalation. Those involved with inhalation are called **inspiratory muscles**. Those muscles involved in exhalation are called **expiratory muscles**.

1 This anterior view introduces the **primary** and **accessory respiratory muscles**. The primary muscles, the diaphragm and external intercostal muscles, are both inspiratory muscles. When you are breathing quietly, inhalation is active but exhalation is passive—elastic forces and gravity are sufficient to reduce the volume of the lungs.

Accessory Inspiratory Muscles

Sternocleido-mastoid muscle

Scalene muscles

Pectoralis minor muscle

Serratus anterior muscle

Primary Inspiratory Muscle

Diaphragm

Primary Inspiratory Muscle

External intercostal muscles

Accessory Expiratory Muscles

Internal intercostal muscles

Transversus thoracis muscle

External oblique muscle

Rectus abdominis

Internal oblique muscle

2 This lateral view during inhalation shows the inspiratory muscles that elevate the ribs and depress the diaphragm to enlarge the thoracic cavity.

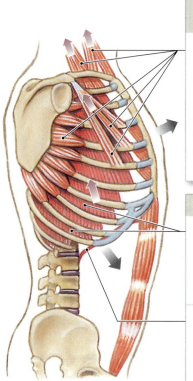

Accessory Inspiratory Muscles (active when needed)

The contraction of accessory muscles assists the external intercostal muscles in elevating the ribs. The muscles increase the speed and amount of rib movement when the primary respiratory muscles are unable to move enough air to meet the oxygen demands of tissues.

Primary Inspiratory Muscles

Contraction of the external intercostal muscles elevates the ribs. This action contributes roughly 25 percent to the volume of air in the lungs at rest.

Contraction of the diaphragm flattens the floor of the thoracic cavity, increasing its volume and drawing air into the lungs. This is responsible for roughly 75 percent of the air movement in normal breathing at rest.

3 This corresponding lateral view during forced exhalation shows the accessory expiratory muscles that depress the ribs and push the relaxed diaphragm into the thoracic cavity. The abdominal muscles that assist in forced exhalation are represented by a single muscle (the rectus abdominis).

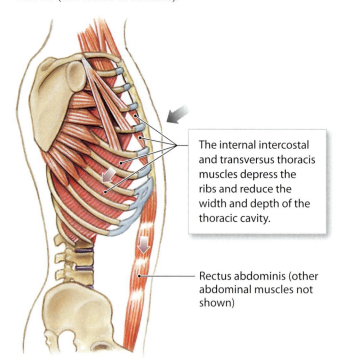

The internal intercostal and transversus thoracis muscles depress the ribs and reduce the width and depth of the thoracic cavity.

Rectus abdominis (other abdominal muscles not shown)

4 A **respiratory cycle** consists of one inhalation and one exhalation. Only a small amount of the air in the lungs is exhanged during a single quiet respiratory cycle. We can increase the tidal volume by inhaling more vigorously and exhaling more completely. We can divide the total volume of the lungs into a series of **volumes** and **capacities** (each capacity is the sum of various volumes) based on different degrees of effort, as indicated in this graph. The red line shows the volume of air within the lungs during respiration. These volumes and capacities are useful in medicine, to aid in diagnosing various respiratory disorders, or in assessing the state of illness or recovery from a lung disease.

Pulmonary Volumes and Capacities (adult male)

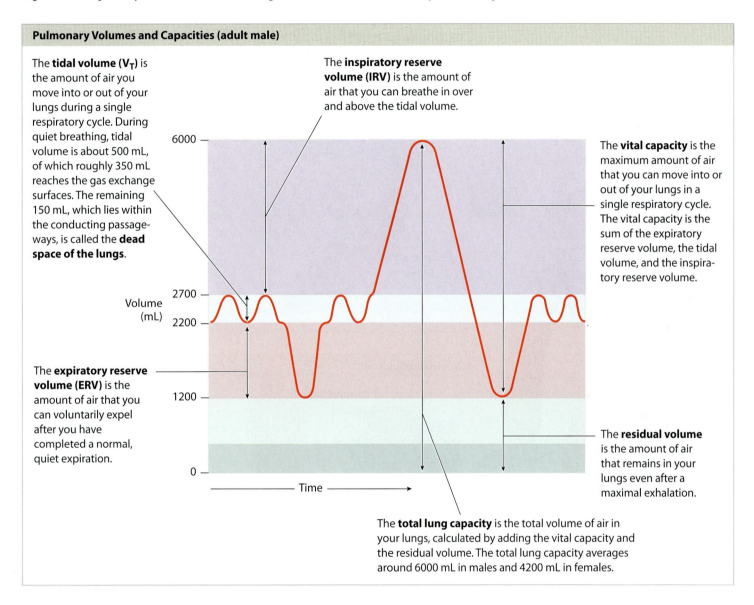

The **tidal volume (V$_T$)** is the amount of air you move into or out of your lungs during a single respiratory cycle. During quiet breathing, tidal volume is about 500 mL, of which roughly 350 mL reaches the gas exchange surfaces. The remaining 150 mL, which lies within the conducting passageways, is called the **dead space of the lungs**.

The **inspiratory reserve volume (IRV)** is the amount of air that you can breathe in over and above the tidal volume.

The **vital capacity** is the maximum amount of air that you can move into or out of your lungs in a single respiratory cycle. The vital capacity is the sum of the expiratory reserve volume, the tidal volume, and the inspiratory reserve volume.

The **expiratory reserve volume (ERV)** is the amount of air that you can voluntarily expel after you have completed a normal, quiet expiration.

The **residual volume** is the amount of air that remains in your lungs even after a maximal exhalation.

The **total lung capacity** is the total volume of air in your lungs, calculated by adding the vital capacity and the residual volume. The total lung capacity averages around 6000 mL in males and 4200 mL in females.

Volume (mL) axis values: 6000, 2700, 2200, 1200, 0

Time →

5 This table compares the pulmonary volumes in healthy males and females.

Pulmonary Volumes		Males	Females
Vital capacity	IRV	3300	1900
	V$_T$	500	500
	ERV	1000	700
Residual volume		1200	1100
Total lung capacity		6000 mL	4200 mL

Module 14.5 Review

a. Name the various measurable pulmonary volumes.

b. Identify the primary inspiratory muscles.

c. When do the accessory respiratory muscles become active?

Gas diffusion depends on the relative concentrations and solubilities of gases

1 This illustration shows the diffusion of oxygen and carbon dioxide during external respiration in the pulmonary circuit and during internal respiration in the systemic circuit.

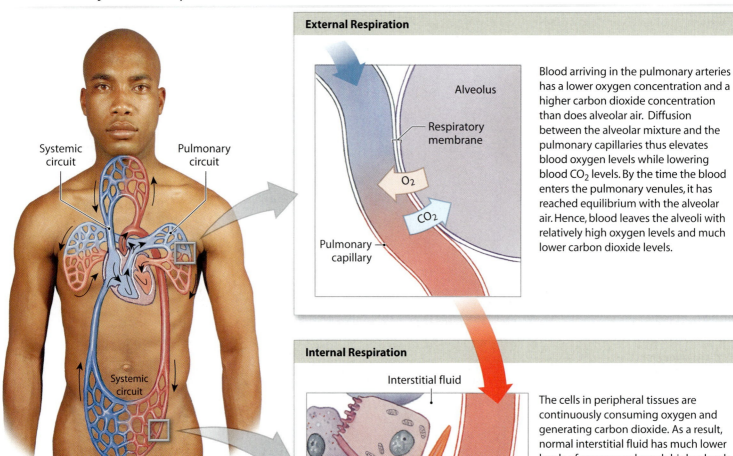

Systemic circuit

Pulmonary circuit

Systemic circuit

External Respiration

Alveolus

Respiratory membrane

O_2

CO_2

Pulmonary capillary

Blood arriving in the pulmonary arteries has a lower oxygen concentration and a higher carbon dioxide concentration than does alveolar air. Diffusion between the alveolar mixture and the pulmonary capillaries thus elevates blood oxygen levels while lowering blood CO_2 levels. By the time the blood enters the pulmonary venules, it has reached equilibrium with the alveolar air. Hence, blood leaves the alveoli with relatively high oxygen levels and much lower carbon dioxide levels.

Internal Respiration

Interstitial fluid

O_2

CO_2

Systemic capillary

The cells in peripheral tissues are continuously consuming oxygen and generating carbon dioxide. As a result, normal interstitial fluid has much lower levels of oxygen and much higher levels of carbon dioxide than the blood arriving in the capillaries. Driven by these concentration differences, oxygen diffuses out of the blood and carbon dioxide diffuses into the blood.

Every oxygen molecule entering peripheral tissues is balanced by an oxygen molecule absorbed at the alveoli, and the absorbed oxygen molecule will be replaced in the next respiratory cycle. But if tissue oxygen demand accelerates—suppose you are running to a class—that equilibrium is disturbed. Your respiratory rate and tidal volume must then increase.

2 This illustration summarizes how the bloodstream transports oxygen and carbon dioxide in the lungs (at left) and in peripheral tissues (at right).

Oxygen pickup from the lungs

In the pulmonary capillaries, most of the oxygen diffusing into the bloodstream is absorbed by red blood cells (RBCs) and reversibly bound to hemoglobin (Hb). A small percentage of oxygen dissolves in the plasma.

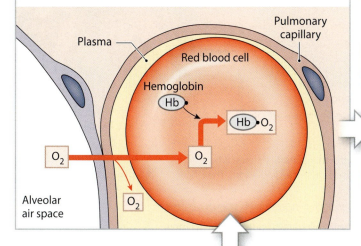

Oxygen delivery to the tissues

In systemic capillaries, hemoglobin releases the oxygen, which diffuses into the surrounding tissues.

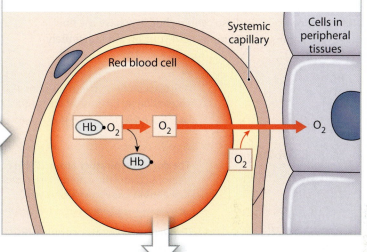

Carbon dioxide delivery to the lungs

At the lungs, the sequence reverses, and carbon dioxide diffuses out of the capillaries and into the alveolar air spaces.

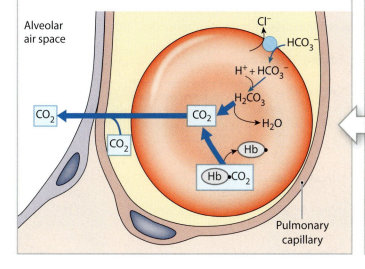

Carbon dioxide pickup from the tissues

In systemic capillaries, some of the carbon dioxide dissolves in the plasma, but most is absorbed by red blood cells. Most of that CO_2 is converted to carbonic acid (H_2CO_3) and the rest is bound to hemoglobin.

1 The carbonic acid dissociates, releasing H^+ and a bicarbonate ion (HCO_3^-).

2 The bicarbonate ion leaves the RBC in exchange for a chloride ion (Cl^-). This is known as the **chloride shift**.

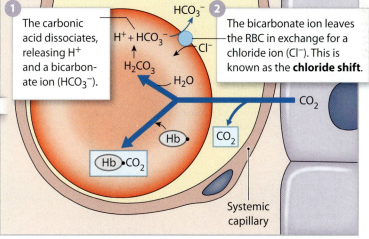

Module 14.6 Review

a. Describe the forces that move oxygen and carbon dioxide across the respiratory membrane.

b. What molecule binds oxygen in the RBC?

c. Describe the three ways that carbon dioxide is transported in the bloodstream.

Respiratory control involves interacting centers in the brain stem

Breathing continues throughout life, without requiring your conscious attention. Subconscious centers in the pons and medulla oblongata control the respiratory cycles.

1 Respiratory control involves multiple levels of regulation in the brain stem. Conscious effort can temporarily influence or override these centers, but they have the final say—you cannot, for example, hold your breath indefinitely.

Higher centers in the brain can alter the activity of the respiratory centers in the brain stem, but essentially normal respiratory cycles continue even if the brain stem above the pons has been severely damaged.

Pons

The respiratory centers in the pons adjust the pace of respiration set by those of the medulla oblongata.

Pons

Medulla oblongata

The **dorsal respiratory group (DRG)** contains an inspiratory center whose neurons control the external intercostal muscles and the diaphragm. The DRG functions in every respiratory cycle.

The **ventral respiratory group (VRG)** has both inspiratory and expiratory centers, but they function only when oxygen demands increase and accessory respiratory muscles become involved.

Medulla oblongata

2 This flowchart shows the events involved in **quiet breathing** in a resting individual.

3 **Forced breathing** increases pulmonary ventilation toward maximal levels by involving accessory respiratory muscles.

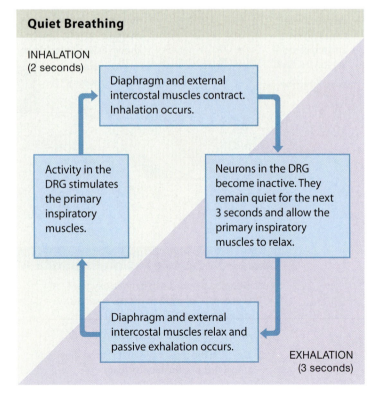

Quiet Breathing

INHALATION (2 seconds)

Diaphragm and external intercostal muscles contract. Inhalation occurs.

Activity in the DRG stimulates the primary inspiratory muscles.

Neurons in the DRG become inactive. They remain quiet for the next 3 seconds and allow the primary inspiratory muscles to relax.

Diaphragm and external intercostal muscles relax and passive exhalation occurs.

EXHALATION (3 seconds)

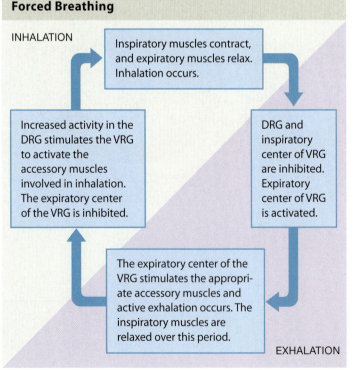

Forced Breathing

INHALATION

Inspiratory muscles contract, and expiratory muscles relax. Inhalation occurs.

Increased activity in the DRG stimulates the VRG to activate the accessory muscles involved in inhalation. The expiratory center of the VRG is inhibited.

DRG and inspiratory center of VRG are inhibited. Expiratory center of VRG is activated.

The expiratory center of the VRG stimulates the appropriate accessory muscles and active exhalation occurs. The inspiratory muscles are relaxed over this period.

EXHALATION

Under normal conditions, carbon dioxide levels are the most important factor influencing respiratory activity. A rise of just 10 percent in arterial CO_2 causes the respiratory rate to double, even if oxygen levels are completely normal. In contrast, even a relatively large drop in arterial oxygen concentrations has little effect on the respiratory centers.

4 This diagram shows how changes in carbon dioxide levels affect the respiratory rate.

An elevated arterial blood CO_2 level constitutes **hypercapnia.** The most common cause of hypercapnia is an abnormally low respiratory rate that is insufficient to meet normal demands for delivering oxygen and removing carbon dioxide. The most obvious sign of this imbalance, called **hypoventilation**, is the accumulation of carbon dioxide in the blood.

Hyperventilation, when the rate and depth of respiration exceed the demands for delivering oxygen and removing carbon dioxide, gradually leads to **hypocapnia**, an abnormally low carbon dioxide level. Snorkelers sometimes hyperventilate to extend their time underwater; this works because it is carbon dioxide that stimulates respiratory activity. If the carbon dioxide levels are driven down too far, a snorkeler may become unconscious from oxygen starvation in the brain without ever feeling the urge to breathe. This dangerous condition is called **shallow water blackout**.

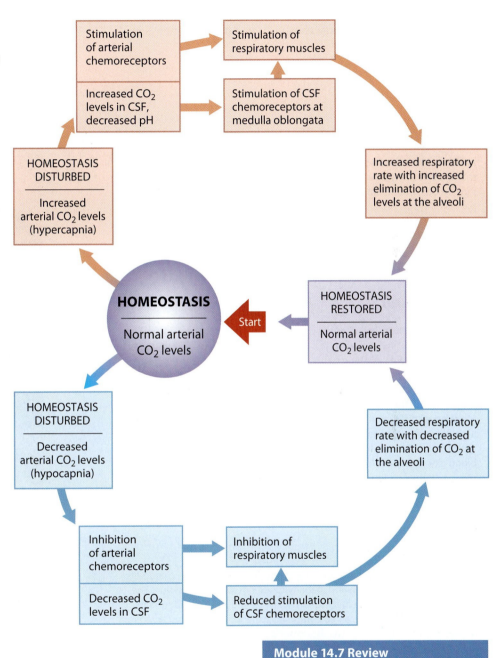

Module 14.7 Review

a. Where are the centers that adjust the pace of respiration located?

b. Describe the two respiratory groups found in the medulla oblongata.

c. What conditions are caused by hyperventilation and hypoventilation?

Respiratory disorders and aging both reduce lung function; smoking makes matters worse

Chronic Obstructive Pulmonary Disease

1 **Chronic obstructive pulmonary disease (COPD)** is a general term for three progressive disorders of the airways that restrict airflow and ventilation: asthma, chronic bronchitis, and emphysema.

Asthma

Asthma (AZ-muh) is a condition characterized by respiratory passageways that are extremely sensitive to irritation. The airways respond by constricting smooth muscles all along the airways. The mucosa of the respiratory passageways swells, and mucus production accelerates. The combination makes breathing very difficult. Asthma attacks can be triggered by allergies, toxins, exercise, cold weather, or even stress.

Chronic Bronchitis

Chronic bronchitis (brong-KĪ-tis) is a long-term inflammation and swelling of the bronchial lining, leading to overproduction of mucus. The characteristic sign is frequent coughing with copious sputum production. Cigarette smoking is the leading cause of chronic bronchitis. Other environmental irritants, such as chemical vapors, can also cause it. Over time, increased mucus production can block smaller airways, increasing resistance and reducing respiratory efficiency. Chronic bacterial infections leading to more lung damage are common. Individuals with chronic bronchitis may have symptoms of heart failure, including widespread edema (swelling). Their blood oxygenation is low, and their skin may have a bluish color. The combination of widespread edema and bluish coloration has led to the descriptive term **blue bloaters** for individuals with this condition.

Emphysema

Emphysema (em-fi-ZĒ-muh) is a chronic, progressive condition characterized by shortness of breath and inability to tolerate physical exertion. The underlying problem is the destruction of alveoli and inadequate surface area for gas exchange. The alveoli gradually expand, and adjacent alveoli merge to form larger air spaces supported by fibrous tissue without alveolar capillary networks. As elastic connective tissues are destroyed, the loss of respiratory surface area restricts oxygen absorption, so the individual becomes short of breath. The respiratory muscles work hard, and these individuals tend to be thin because they use a lot of energy just breathing. On chest x-rays, such as the one at right, the lungs appear overexpanded. People with emphysema breathe heavily and are able to maintain near-normal blood oxygen levels, so their skin usually appears pink. The combination of heavy breathing and pink coloration has led to the descriptive term **pink puffers** for these individuals.

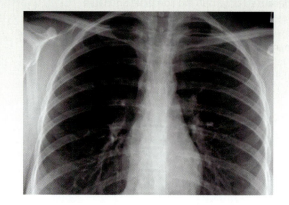

Lung Cancer

2 **Lung cancers** now account for 12.6 percent of new cancer cases in both men and women. It kills more people each year than colon, breast, and prostate cancer combined. Over 50 percent of lung cancer patients die within a year of diagnosis. Studies show that 85–90 percent of all lung cancers are the direct result of cigarette smoking. Even secondhand smoke is harmful: A nonsmoker who lives with a smoker has a 20–30 percent higher risk of developing lung cancer. Before about 1970, this disease affected primarily middle-aged men, but as the number of women smokers has increased, so has the number of women who develop lung cancer. This graph shows that the incidence of lung cancer among men has gradually declined since 1992. In contrast, there hasn't been a comparable decrease in the incidence of lung cancer among women.

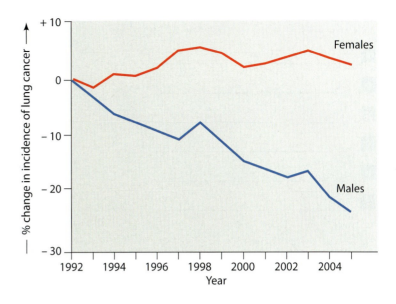

Effects of Aging

Aging affects all aspects of respiratory function. As elastic tissue deteriorates throughout the body, vital capacity decreases and it becomes harder to move air in and out of the lungs. These respiratory limitations make strenuous exercise more difficult.

3 In addition to skeletal and connective tissue changes associated with age, some degree of emphysema is normal in individuals over age 50. However, the extent varies widely depending on lifetime exposure to cigarette smoke and other respiratory irritants. This graph compares the respiratory performance of individuals who have never smoked with individuals who have smoked for various periods of time. The message is clear: Although some decrease in respiratory performance is inevitable, you can prevent serious respiratory problems by stopping smoking or never starting. And it's never too late to quit!

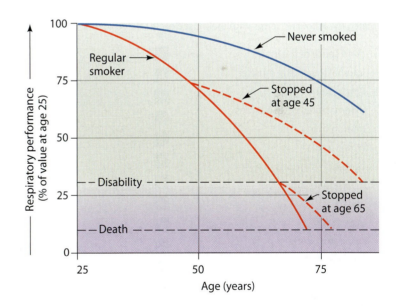

Module 14.8 Review

a. Describe chronic obstructive pulmonary disease (COPD).

b. List two important risk factors for developing lung cancer.

c. Name several age-related factors that affect the respiratory system.

1. Short answer

Identify and describe the various pulmonary volumes and capacities indicated in the following graph:

a _____

b _____

c _____

d _____

e _____

f _____

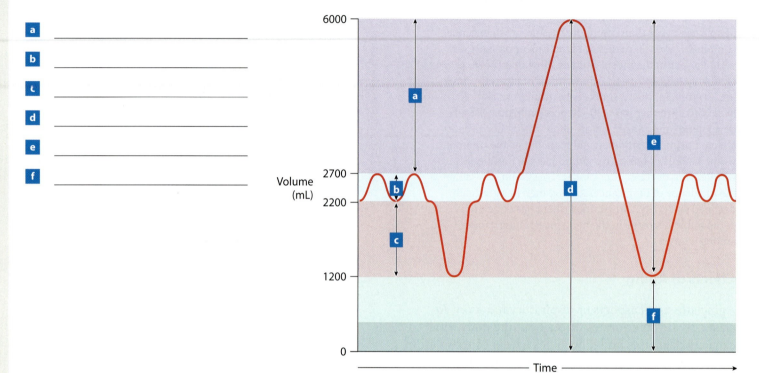

2. Matching

Match the following terms with the most closely related description.

- dead space of the lungs
- internal intercostal muscles
- emphysema
- external intercostal muscles
- hemoglobin
- hypercapnia
- centers in the pons
- bicarbonate ion
- anoxia
- atelectasis
- respiratory cycle
- chloride shift
- DRG and VRG centers
- hypocapnia

a _____ RBC-plasma ion exchange

b _____ A cause of tissue death

c _____ Set the pace of respiration

d _____ Act to depress ribs

e _____ Transports CO_2 in blood

f _____ Pink puffers

g _____ Act to elevate ribs

h _____ Binds oxygen in RBC

i _____ Result of hyperventilation

j _____ Portion of tidal volume

k _____ Adjust the pace of respiration

l _____ Result of hypoventilation

m _____ Collapsed lung

n _____ An inhalation and exhalation

3. Section integration

Compare and contrast external respiration, pulmonary ventilation, and internal respiration. _____

_____

_____

Visual Outline with Key Terms

Summarize the content of each module using the terms in the order provided.

SECTION 1

Functional Anatomy of the Respiratory System

- respiratory system
- upper respiratory system
- lower respiratory system
- respiratory tract
- conducting portion
- bronchioles
- respiratory portion
- alveoli
- upper respiratory tract
- lower respiratory tract

SECTION 2

Respiratory Physiology

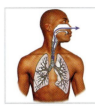

- respiration
- external respiration
- pulmonary ventilation
- alveolar ventilation
- gas diffusion
- internal respiration
- hypoxia
- anoxia

14.1

The respiratory mucosa cleans, warms, and moistens inhaled air

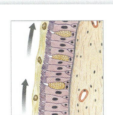

- respiratory mucosa
- lamina propria
- cystic fibrosis (CF)
- pharynx
- nasopharynx
- oropharynx
- laryngopharynx
- trachea
- internal nares
- nasal vestibule
- external nares
- hard palate
- soft palate
- glottis
- larynx

14.4

Pressure changes within the pleural cavities drive pulmonary ventilation

- pneumothorax
- atelectasis
- intrapulmonary pressure
- alveolar pressure
- tidal volume (V_T)

14.2

The trachea and primary bronchi carry air to and from the lungs

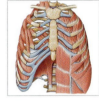

- trachea
- primary bronchi
- tracheal cartilages
- trachealis muscle
- bronchioles
- bronchodilation
- bronchoconstriction
- asthma
- right lung
- horizontal and oblique fissures
- superior, middle, and inferior lobes
- hilum
- cardiac notch
- left lung
- oblique fissure
- superior and inferior lobes
- pleura

14.5

Respiratory muscles adjust the tidal volume to meet respiratory demands

- inspiratory muscles
- expiratory muscles
- primary respiratory muscles
- accessory respiratory muscles
- diaphragm
- external intercostal muscles
- respiratory cycle
- volumes
- capacities
- inspiratory reserve volume (IRV)
- tidal volume (V_T)
- dead space of the lungs
- expiratory reserve volume (ERV)
- vital capacity
- residual volume
- total lung capacity

14.3

Gas exchange occurs at the alveoli

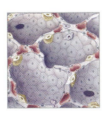

- alveoli
- surfactant
- respiratory membrane

14.6

Gas diffusion depends on the relative concentrations and solubilities of gases

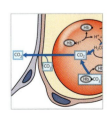

- external respiration
- respiratory membrane
- internal respiration
- hemoglobin
- carbonic acid
- chloride shift

• = *Term boldfaced in this module*

14.7

Respiratory control involves interacting centers in the brain stem

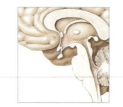

- ○ pons
- ○ medulla oblongata
- ● dorsal respiratory group (DRG)
- ● ventral respiratory group (VRG)
- ● quiet breathing

- ● forced breathing
- ● hypercapnia
- ● hypoventilation
- ● hyperventilation
- ● hypocapnia
- ● shallow water blackout

● = *Term boldfaced in this module*

14.8

Respiratory disorders and aging both reduce lung function; smoking makes matters worse

- ● chronic obstructive pulmonary disease (COPD)
- ● asthma
- ● chronic bronchitis

- ● blue bloaters
- ● emphysema
- ● pink puffers
- ● lung cancers

CAREER PATHS

"I love seeing the patients and being able to help them understand the process. I enjoy interacting with the families. I just love what I do. You can fix people. You can make them better."

— Deborah Kimball
Registered Respiratory Therapist,
Mountain Vista Medical Center,
Mesa, Arizona

Respiratory Therapist

Respiratory therapists work under the supervision of a doctor to help patients breathe, or, if they are having trouble doing so on their own, assist them in getting enough oxygen. Most respiratory therapists work in hospitals—everywhere from the neonatal unit to post-surgery. Because a patient who cannot breathe can suffer brain damage or die in a matter of seconds, quick, calm reactions are essential.

Deborah Kimball, a registered respiratory therapist at Mountain Vista Medical Center in Mesa, Arizona, rotates among the intensive care unit, emergency room, and regular wards. She assists with the insertion and removal of breathing tubes (known as intubation and extubation), responds to distress codes as part of the rapid response team, performs chest compressions, and puts oxygen masks on patients who are not breathing properly (called "bagging"). She also administers medications via a nebulizer (which vaporizes liquid medicine so that it can be inhaled), monitors patients' arterial blood gases, and changes fittings on ventilators as needed.

Deborah used to be a private investigator and paralegal, but after the birth of her twin boys, she decided she needed a secure,

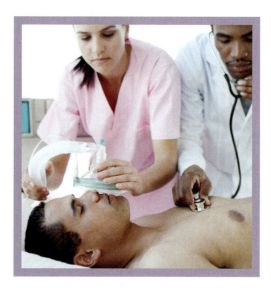

full-time job. Interested in medicine, she considered radiology, but found it too slow. "I'm an adrenaline junkie," she says. "I love seeing the patients and being able to help them understand the process. I enjoy interacting with the families. I just love what I do. You can fix people. You can make them better." The most difficult part of her job is the fact that sometimes you can't make them better.

"You have to be able to deal with death," she says. "The baby codes are the worst. There are times when everybody working a code is in tears, but you have to keep it together."

In addition to an ability to handle stress, Deborah points out that knowledge of anatomy and physiology is an important part of her work. "We're the ABC team: Airway, Breathing, and Circulation. You're pumping the heart—you need to know where it is. For example, when you're doing your assessments, and you have someone coming in with pneumonia, you need to know how the body works." Respiratory therapists should also be flexible and possess strong communication skills. They convey important information to doctors, nurses, and patients on a daily basis, and must be prepared to work nights and weekends to accommodate the needs of their patients. Those who work in home care may have daytime hours only, with the understanding that they are on call.

Although most respiratory therapists work in hospitals, there is a need for them anywhere patients may have difficulty breathing—nursing homes, doctors' offices, sleep centers (where they work with patients with sleep apnea), home care, and air transport and ambulance programs. Increasingly, they are employed by asthma education or smoking cessation programs.

For more information, visit the website for American Association for Respiratory Care at http://www.aarc.org.

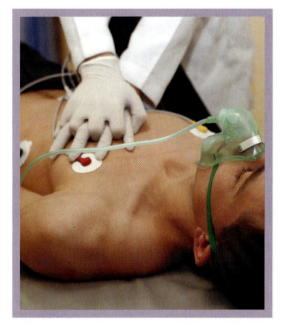

Think this is the career for you?

Key Stats:

- **Education and Training.** Associate degree is required, but a bachelor's or master's degree may be important for advancement.

- **Licensure.** All states, except Alaska and Hawaii, require respiratory therapists to be licensed.

- **Earnings.** Earnings vary but the mean annual wage is $54,280.

- **Expected Job Prospects.** Employment is expected to grow by an above-average 21% from 2008 to 2018.

Bureau of Labor Statistics, U.S. Department of Labor, *Occupational Outlook Handbook, 2010-11 Edition*, Respiratory Therapists, on the Internet at http://www.bls.gov/oco/ocos321. htm (visited September 14, 2011).

MasteringA&P®

Access more review material online in the Study Area at www.masteringaandp.com.

- Chapter guides
- Chapter quizzes
- Practice tests
- Art labeling activities
- Flashcards
- A glossary with pronunciations
- Practice Anatomy Lab™ (PAL™) 3.0 virtual anatomy practice tool
- Interactive Physiology® (IP) animated tutorials
- MP3 Tutor Sessions

PAL practice anatomy lab™ **For this chapter, follow these navigation paths in PAL:**

- Human Cadaver>Respiratory System
- Anatomical Models>Respiratory System
- Histology>Respiratory System

 For this chapter, go to these topics in the Respiratory System in IP:

- Anatomy Review: Respiratory Structures
- Pulmonary Ventilation
- Gas Exchange
- Gas Transport

Chapter 14 Review Questions

1. You spend a winter night at a friend's house. Your friend's home is quite old, and the hot-air furnace lacks a humidifier. When you wake up in the morning, you have a fair amount of nasal congestion and decide you might be coming down with a cold. After you take a steamy shower and drink some juice for breakfast, the nasal congestion disappears. Explain.

2. John breaks a rib that punctures the thoracic cavity on his left side. Which structures are potentially damaged, and what do you predict will happen to the lung as a result?

3. In emphysema, alveoli are replaced by large air spaces and elastic fibrous connective tissue. How do these changes affect the lungs?

For answers to all module, section, and chapter review questions, see the blue Answers tab at the back of the book.

15

The Digestive System

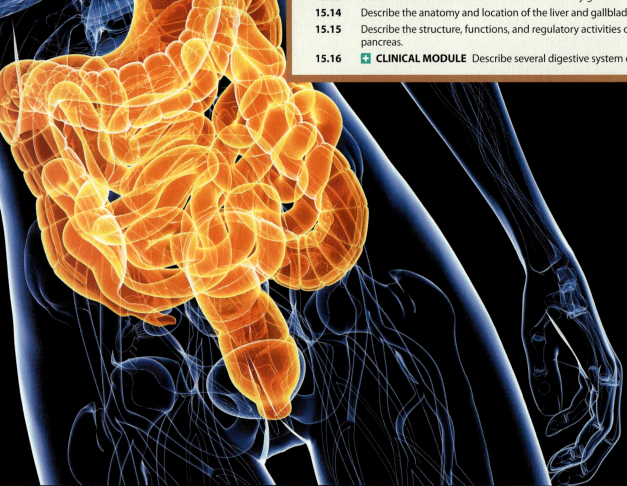

General Organization of the Digestive System

The **digestive system** consists of a muscular tube, the digestive tract— also called the gastrointestinal (GI) tract—plus various accessory organs. Food enters the mouth and passes through the digestive tract. Along the way, accessory organs produce secretions containing water, enzymes, buffers, and other components that help prepare organic and inorganic nutrients for absorption.

1 The digestive system works with other systems to support tissues that have no direct connection with the outside environment and no other means of obtaining nutrients.

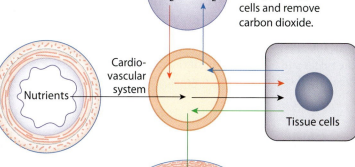

The digestive system provides the nutrients cells need for maintenance and growth.

The respiratory system works with the cardiovascular system to supply oxygen to cells and remove carbon dioxide.

O_2 and CO_2

Cardio-vascular system

Nutrients

Tissue cells

Wastes

The urinary system removes the organic wastes generated by cell activities.

2 The **digestive tract** begins at the mouth and continues through the oral cavity, pharynx, esophagus, stomach, small intestine, and large intestine, which opens to the exterior at the anus. These structures have distinctive structural and functional characteristics, but all share an underlying pattern of histological organization (tissue structure) that we will consider in this section. The accessory organs shown here are glandular organs that secrete their products into the digestive tract.

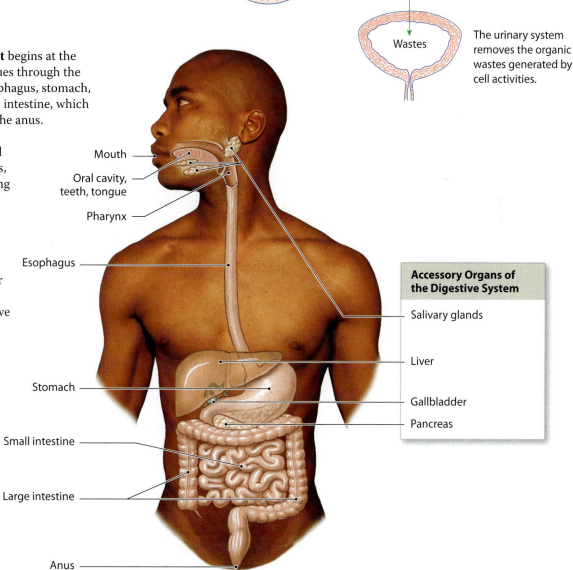

Mouth

Oral cavity, teeth, tongue

Pharynx

Esophagus

Stomach

Small intestine

Large intestine

Anus

Accessory Organs of the Digestive System

Salivary glands

Liver

Gallbladder

Pancreas

The digestive tract is a muscular tube lined by a mucous membrane

The structure of the digestive tract varies from one region to another. The composite view shown in this module most closely resembles the small intestine, the longest segment of the digestive tract.

1 The digestive tract is basically a long muscular tube lined by a mucous membrane. The lining often contains permanent ridges as well as temporary folds that disappear as the passageway fills. Together the ridges and folds dramatically increase the surface area available for absorbing nutrients.

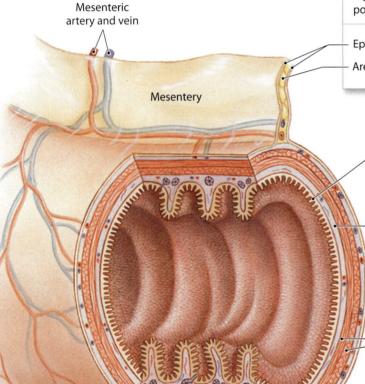

Mesenteric artery and vein

Mesentery

Mesentery

A **mesentery** is a double sheet of peritoneal membrane. The areolar tissue between the epithelial surfaces provides an access route for blood vessels, nerves, and lymphatic vessels to and from the digestive tract. Mesenteries also stabilize the attached organs and prevent the intestines from becoming entangled during digestive movements or sudden changes in body position.

Epithelial surface

Areolar tissue

Major Layers of the Digestive Tract

Mucosa

The inner lining, or **mucosa**, of the digestive tract is a mucous membrane consisting of an epithelium, which is moistened by glandular secretions, and a lamina propria of areolar tissue.

Submucosa

The **submucosa** is a layer of dense irregular connective tissue that surrounds the mucosa. The submucosa has large blood vessels and lymphatic vessels, and in some regions it also contains exocrine glands that secrete buffers and enzymes into the lumen of the digestive tract.

Muscularis Externa

The **muscularis externa** is dominated by smooth muscle cells in two layers, an inner circular layer and an outer longitudinal layer. These layers play an essential role in mechanical processing and in moving materials along the digestive tract.

Serosa

Along most portions of the digestive tract within the peritoneal cavity, the muscularis externa is covered by a layer of visceral peritoneum known as the **serosa**. There is no serosa covering the muscularis externa of the oral cavity, pharynx, esophagus, and rectum. Instead, a dense network of collagen fibers forms a sheath that firmly attaches the digestive tract to adjacent structures.

2 Each layer in the wall of the digestive tract has a different set of functions. The mucosa has several structural features that enhance its ability to absorb nutrients from the lumen of the tract. This diagram summarizes those features.

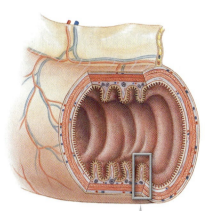

Secretory Glands

Ducts carry the secretions of gland cells located in the mucosa and submucosa—or in accessory glandular organs—to the epithelial surfaces.

Mucosal glands

Submucosal gland

Circular folds (plicae circulares) are permanent transverse folds in the intestinal lining.

Artery and vein

Lymphatic vessel

Mucosa

Submucosa

Muscularis Externa

Circular muscle layer

Longitudinal muscle layer

Serosa

Components of the Mucosa

The mucosal epithelium of the oral cavity, pharynx, and esophagus are lined by stratified squamous epithelium. The stomach, the small intestine, and almost the entire length of the large intestine are lined by simple columnar epithelium that contains mucous cells.

Villi (singular, *villus*) are small mucosal projections, like tiny fingers, that stick into the lumen of the small intestine. Villi are permanent features that further increase the surface area available for absorption.

The lamina propria consists of a layer of areolar tissue that also contains blood vessels, sensory nerve endings, lymphatic vessels, smooth muscle cells, scattered areas of lymphoid tissue, and, in some regions, mucous glands.

The **muscularis** (mus-kū-LAIR-is) **mucosae** (mū-KŌ-sē) consists of two concentric layers of smooth muscle. The inner layer encircles the lumen (the circular muscle layer), and the outer layer contains muscle cells oriented parallel to the long axis of the tract (the longitudinal muscle layer). Contractions in these layers change the shape of the lumen and move the circular folds and villi.

Superficial to the muscularis externa, the submucosa contains the **submucosal plexus**, a nerve network that contains sensory neurons, parasympathetic ganglionic neurons, and sympathetic postganglionic fibers that innervate the mucosa and submucosa.

The muscularis externa contains the **myenteric** (mī-en-TER-ik) **plexus** (*mys*, muscle + *enteron*, intestine). This network of parasympathetic ganglia, sensory neurons, interneurons, and sympathetic postganglionic fibers lies sandwiched between the circular and longitudinal muscle layers. The myenteric plexus and the submucosal plexus contain neurons involved with the local control of digestive activities. In general, parasympathetic stimulation increases digestive muscle tone and activity; sympathetic stimulation relaxes muscles and inhibits muscle activity.

Module 15.1 Review

a. Name the four layers of the digestive tract from superficial to deep.

b. What is the importance of the mesenteries?

c. Compare the submucosal plexus with the myenteric plexus.

Smooth muscle contractions mix and propel materials through the digestive tract

Smooth muscle tissue forms sheets, bundles, or sheaths around other tissues in almost every organ. Smooth muscles around blood vessels regulate blood flow through vital organs. In the digestive and urinary systems, ring-shaped smooth muscles called sphincters regulate the movement of materials along internal passageways.

1 Wherever smooth muscle tissue forms layers, the cells are aligned parallel to one another. In the digestive tract, there is usually an inner circular layer and an outer longitudinal layer. In a longitudinal section of the digestive tract, the muscle cells in the circular layer of the muscularis externa look like little round balls, whereas those in the longitudinal layer look like long spindles.

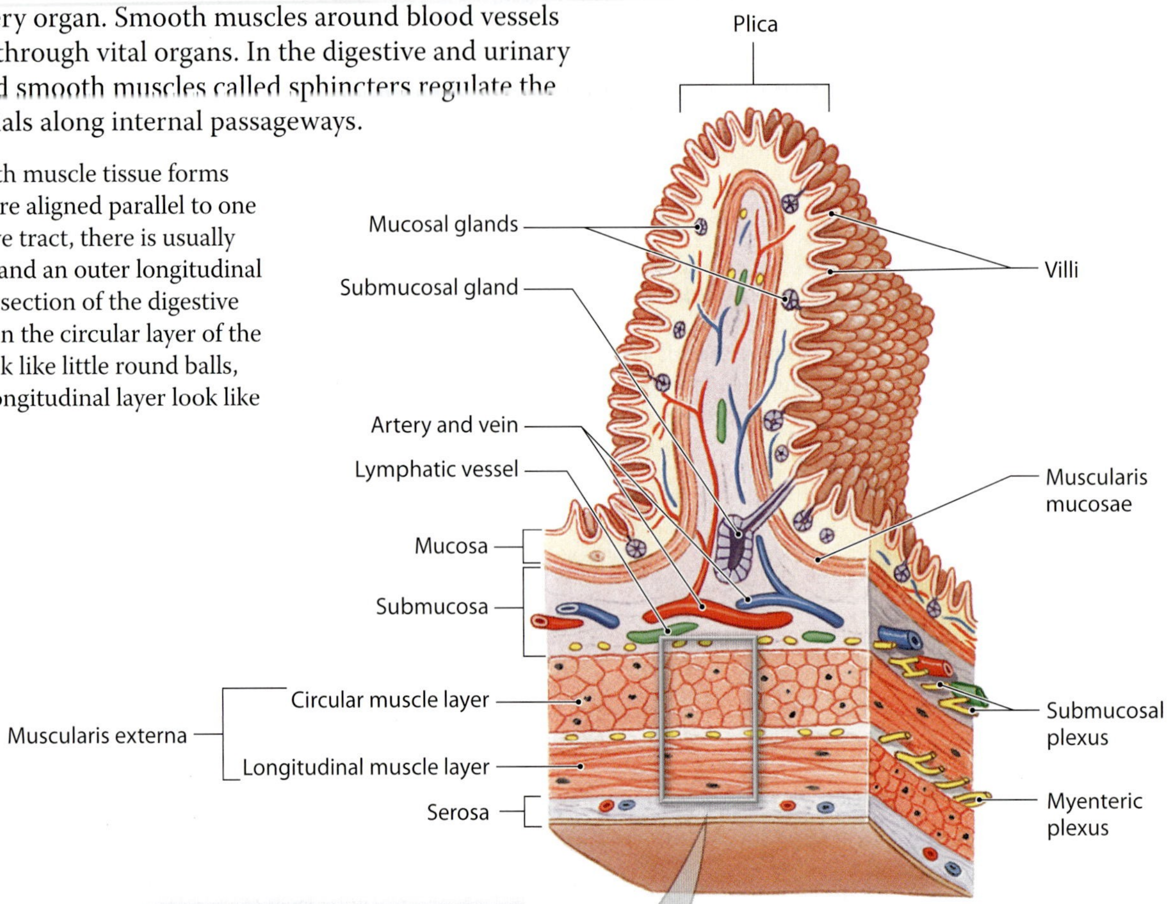

Plica

Mucosal glands

Submucosal gland

Villi

Artery and vein

Lymphatic vessel

Muscularis mucosae

Mucosa

Submucosa

Circular muscle layer

Muscularis externa

Longitudinal muscle layer

Submucosal plexus

Serosa

Myenteric plexus

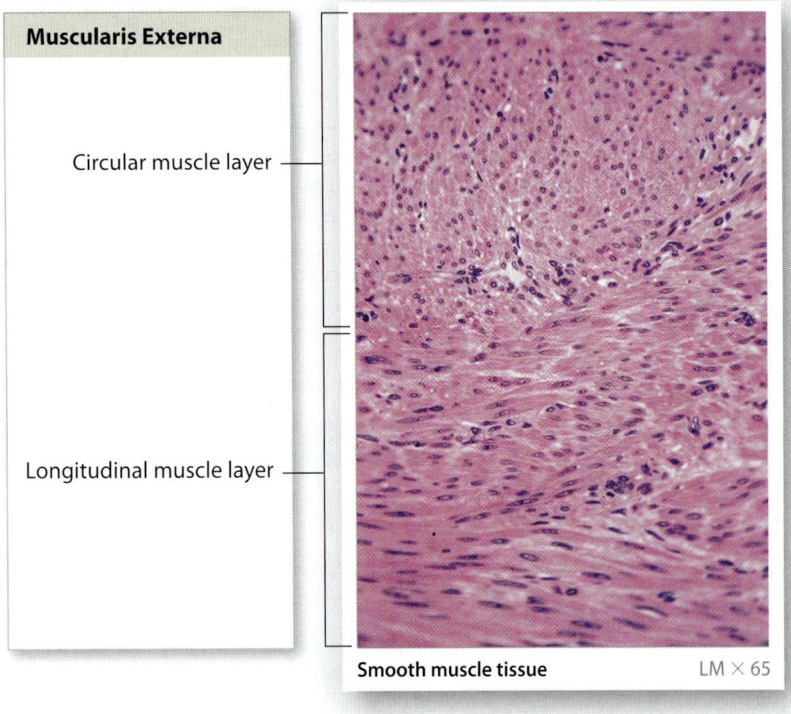

Muscularis Externa

Circular muscle layer

Longitudinal muscle layer

Smooth muscle tissue   LM × 65

Peristalsis

2 Food enters the digestive tract as a moist, compact mass known as a **bolus** (BŌ-lus). The muscularis externa propels materials from one portion of the digestive tract to another by contractions known as **peristalsis** (per-i-STAL-sis). During a peristaltic movement, a wave of contraction in the circular muscles forces the bolus forward.

Segmentation

3 Most areas of the small intestine and some portions of the large intestine undergo cycles of contraction that churn and fragment the bolus, mixing its contents with intestinal secretions. This activity is called **segmentation**. These rhythmic cycles of contraction do not follow a set pattern and thus do not push materials along the tract in any one direction.

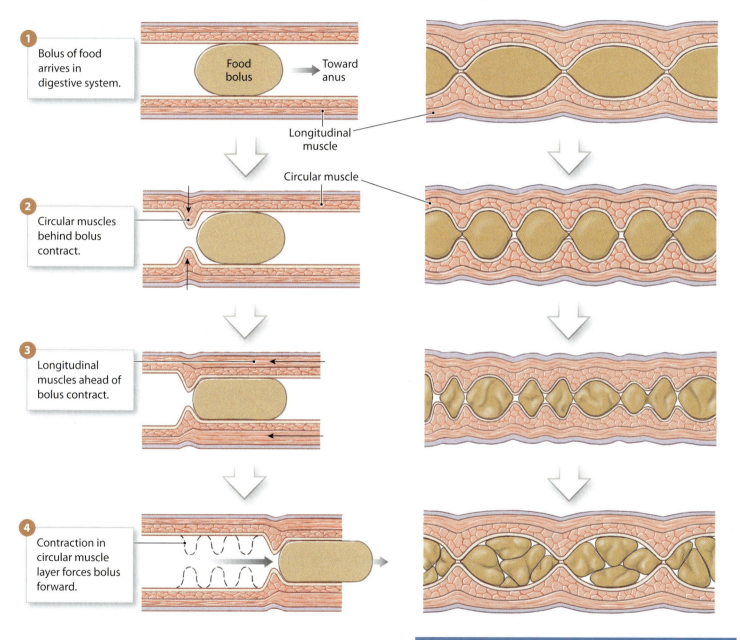

1 Bolus of food arrives in digestive system.

Food bolus → Toward anus

Longitudinal muscle

Circular muscle

2 Circular muscles behind bolus contract.

3 Longitudinal muscles ahead of bolus contract.

4 Contraction in circular muscle layer forces bolus forward.

Module 15.2 Review

a. Describe the orientation of smooth muscle fibers in the muscularis externa of the digestive tract.

b. Define bolus.

c. Compare peristalsis and segmentation. Which is more efficient in propelling intestinal contents along the digestive tract?

1. Labeling

Label each of the structures of the digestive tract in the following figure.

a _____

b _____

c _____

d _____

e _____

f _____

g _____

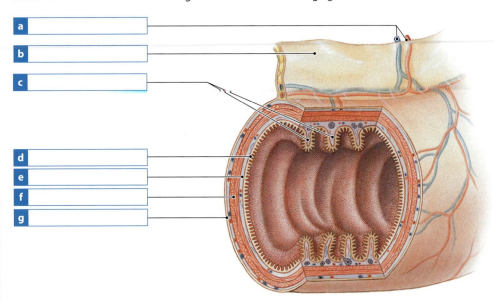

2. Matching

Match the following terms with the most closely related description.

- lamina propria
- peristalsis
- submucosal plexus
- esophagus
- muscularis mucosae
- segmentation
- circular folds
- sphincter
- myenteric plexus
- smooth
- submucosa
- liver
- muscularis externa
- bolus

a _____ Digestive tube between the pharynx and stomach

b _____ Moves circular folds and villi

c _____ Areolar tissue layer containing blood vessels, nerve endings, and lymphatics

d _____ Permanent transverse folds in the digestive tract lining

e _____ Waves of muscular contractions that propel materials along digestive tract

f _____ Layer of digestive tract containing large blood vessels and lymphatics

g _____ Nerve network within the muscularis externa

h _____ Muscular layer of digestive tract

i _____ Digestive system accessory organ

j _____ Rhythmic muscular contractions that mix materials in digestive tract

k _____ Mass of food entering the digestive tract

l _____ Type of muscle tissue in digestive tract wall

m _____ Nerve network superficial to the muscularis externa

n _____ Ring of muscle tissue

3. Section integration

How would a decrease in smooth muscle tone affect the digestive processes and possibly promote constipation (infrequent bowel movements)?

_____

_____

_____

The Digestive Tract

The digestive tract is a muscular tube approximately 10 m (33 ft) long. It can be divided into regions that differ in histological structure and function. In this section we consider each of the major subdivisions of the digestive tract, which are summarized below. In Section 3, we will discuss the accessory organs of the digestive system.

Major Subdivisions of the Digestive Tract
Oral Cavity, Teeth, Tongue
Mechanical processing (crushing and shredding of food), moistening, mixing food with saliva
Pharynx
Muscular propulsion of food into the esophagus
Esophagus
Transport of swallowed food to the stomach
Stomach
Chemical breakdown of food by acid and enzymes; mechanical processing (mixing and churning) through muscular contractions
Small Intestine
Enzymatic digestion and absorption of water, organic substrates, vitamins, and ions
Large Intestine
Dehydration and compaction of indigestible materials in preparation for elimination

Mouth

Accessory Organs of the Digestive System

Salivary glands

Liver

Gallbladder

Pancreas

Anus

1 This table summarizes the general functions of the digestive tract. The lining of the digestive tract also protects surrounding tissues against digestive acids and enzymes, abrasion, and pathogens within the digestive tract.

General Functions of the Digestive Tract

- **Ingestion**: Food and liquids enter the digestive tract through the mouth.
- **Mechanical processing**: Occurs to most ingested solids either before they are swallowed or in the proximal portions of the digestive tract. It includes the crushing and shredding of food in the oral cavity and the mixing and churning of swallowed food in the stomach.
- **Digestion**: The chemical and enzymatic breakdown of food into small organic molecules that can be absorbed by the digestive epithelium.

- **Secretion**: Performed along most of the digestive tract. (However, the accessory digestive organs provide most of the acids, enzymes, and buffers required for digestion.)
- **Absorption**: The movement of organic molecules, electrolytes, vitamins, and water across the digestive epithelium and into the interstitial fluid of the digestive tract.
- **Compaction**: The progressive dehydration of indigestible foods and organic wastes prior to elimination from the body. The compacted material is called **feces**. The elimination of feces from the body is called **defecation** (def-e-KĀ-shun).

The oral cavity contains the tongue, salivary glands, and teeth, and receives saliva from the salivary glands

The **oral cavity** is lined by the **oral mucosa**, which has a stratified squamous epithelium. A layer of keratinized cells covers regions exposed to severe abrasion, such as the superior surface of the tongue and the opposing surface of the hard palate. The epithelial lining of the cheeks, lips, and inferior surface of the tongue is relatively thin, nonkeratinized, and delicate. Although nutrients are not absorbed in the oral cavity, digestion of carbohydrates and lipids begins here.

1 This sagittal section introduces the major components forming the boundaries of the oral cavity.

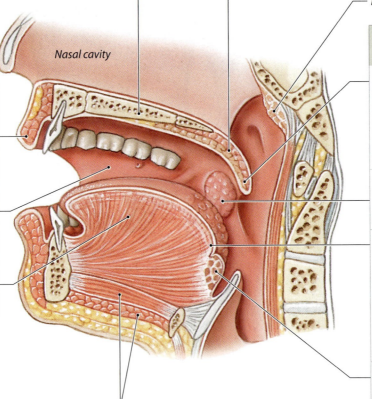

Nasal cavity

Pharyngeal tonsil

Superior Boundary of the Oral Cavity

The **hard palate** is formed by the palatine processes of the maxillary bones and the horizontal plates of the palatine bones.

The muscular **soft palate** lies posterior to the hard palate.

Anterior and Lateral Boundary of the Oral Cavity

Anteriorly, the mucosa of each cheek is continuous with the lips, or **labia** (LĀ-bē-uh; singular, *labium*).

The mucosae of the **cheeks**, or lateral walls of the oral cavity, are supported by pads of fat and the buccinator muscles.

The **body** of the tongue is the anterior, mobile portion.

Posterior Boundary of the Oral Cavity

The posterior margin of the soft palate supports the **uvula** (Ū-vū-luh), a dangling process that helps prevent food from entering the pharynx prematurely.

A palatine tonsil lies on either side of the entrance to the pharynx.

The **root** of the tongue is the fixed portion that projects into the pharynx. A V-shaped line of circumvallate papillae roughly marks the boundary between the root and the body of the tongue.

A lingual tonsil is embedded in the root of the tongue.

Inferior Boundary of the Oral Cavity

The floor of the oral cavity inferior to the tongue receives extra support from the underlying muscles.

2 An anterior view of the oral cavity shows additional details not visible in the sagittal section at left.

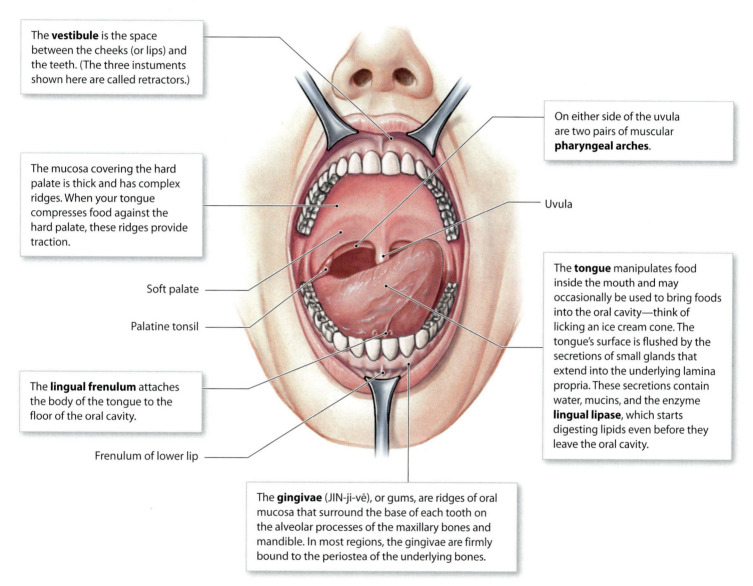

The **vestibule** is the space between the cheeks (or lips) and the teeth. (The three instuments shown here are called retractors.)

The mucosa covering the hard palate is thick and has complex ridges. When your tongue compresses food against the hard palate, these ridges provide traction.

Soft palate

Palatine tonsil

The **lingual frenulum** attaches the body of the tongue to the floor of the oral cavity.

Frenulum of lower lip

On either side of the uvula are two pairs of muscular **pharyngeal arches**.

Uvula

The **tongue** manipulates food inside the mouth and may occasionally be used to bring foods into the oral cavity—think of licking an ice cream cone. The tongue's surface is flushed by the secretions of small glands that extend into the underlying lamina propria. These secretions contain water, mucins, and the enzyme **lingual lipase**, which starts digesting lipids even before they leave the oral cavity.

The **gingivae** (JIN-ji-vē), or gums, are ridges of oral mucosa that surround the base of each tooth on the alveolar processes of the maxillary bones and mandible. In most regions, the gingivae are firmly bound to the periostea of the underlying bones.

In the oral cavity, the teeth crush and shred food. This process, called **mastication** (chewing), increases the surface area of the food exposed to salivary gland secretions. There are three pairs of salivary glands, accessory glands that begin the process of digesting carbohydrates. We will consider salivary glands in a later section.

Module 15.3 Review

a. Name the structure that forms the roof of the oral cavity.

b. Which type of epithelium lines the oral cavity, and why?

c. Name the enzyme that begins digesting lipids in the oral cavity.

Teeth in different regions of the jaws vary in size, shape, and function

1 This sectional view shows a representative adult tooth. The bulk of each tooth consists of **dentin**, a mineralized matrix similar to that of bone. However, unlike bone, dentin does not contain cells. Instead, cytoplasmic processes extend into the dentin from cells in the central **pulp cavity**, an interior chamber.

1 The **crown** of the tooth projects into the oral cavity from the surface of the gingivae.

2 The **neck** of the tooth marks the boundary between the crown and the root.

3 The **root** of each tooth sits in a bony socket called an **alveolus**.

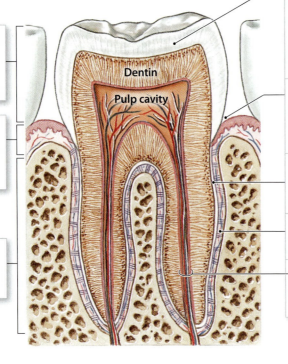

Dentin

Pulp cavity

Components of a Tooth

A layer of **enamel** covers the dentin of the crown. Enamel, which contains calcium phosphate in a crystalline form, is the hardest biologically manufactured substance. Adequate amounts of calcium, phosphate, and vitamin D during childhood are essential for the enamel coating to become complete and resistant to decay.

A shallow groove surrounds the neck of each tooth. The epithelium of the gingiva is bound to the tooth at the base of this groove, blocking bacteria from the deeper tissues around the root.

A layer of **cementum** (se-MEN-tum) covers the dentin of the root. Cementum is less resistant to erosion than is dentin.

Collagen fibers of the **periodontal ligament** extend from the dentin of the root to the bone of the alveolus.

The pulp cavity receives blood vessels and nerves through the **root canal**, a narrow tunnel within the root of the tooth.

2 There are four different types of teeth, each with a distinctive shape and root pattern that enable us to process food in different ways.

The Four Types of Teeth

Incisors (in-SĪ-zerz) are blade-shaped teeth located at the front of the mouth. Incisors are useful for clipping or cutting. Incisors have a single root.

The **cuspids** (KUS-pidz), or canine teeth, are conical, with a sharp ridgeline and a pointed tip. They are used for tearing or slashing. Cuspids have a single root.

Bicuspids (bī-KUS-pidz), or premolars, have flattened crowns with prominent ridges. They crush, mash, and grind. Bicuspids have one or two roots.

Molars have very large, flattened crowns with prominent ridges adapted for crushing and grinding. Molars typically have three or more roots.

Upper jaw

Lower jaw

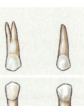

3 Two sets of teeth form during embryonic development. The first to erupt through the gums are the **deciduous teeth** (de-SID-ū-us; *deciduus*, falling off), the temporary teeth of the **primary dentition**. Deciduous teeth are also called primary teeth, milk teeth, or baby teeth. At two years of age, children have 20 deciduous teeth—five on each side of the upper and lower jaws: two incisors, one cuspid, and a pair of deciduous molars.

4 As the jaws grow larger, the primary dentition is gradually replaced by the **secondary dentition**. Three additional molars appear on each side of the upper and lower jaws as the individual ages, bringing the count of permanent teeth to 32. As replacement proceeds, the periodontal ligaments and roots of the primary teeth erode until the deciduous teeth either fall out or are pushed aside by the secondary teeth as they erupt.

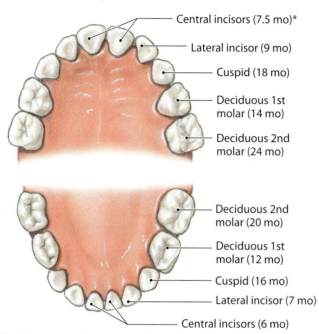

Central incisors (7.5 mo)*
Lateral incisor (9 mo)
Cuspid (18 mo)
Deciduous 1st molar (14 mo)
Deciduous 2nd molar (24 mo)

Deciduous 2nd molar (20 mo)
Deciduous 1st molar (12 mo)
Cuspid (16 mo)
Lateral incisor (7 mo)
Central incisors (6 mo)

Deciduous teeth, upper and lower jaws

*Time indicates age at eruption

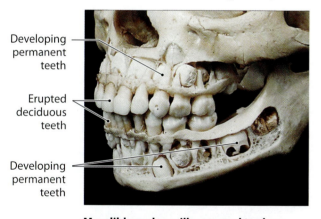

Developing permanent teeth

Erupted deciduous teeth

Developing permanent teeth

Mandible and maxilla exposed to show developing permanent teeth

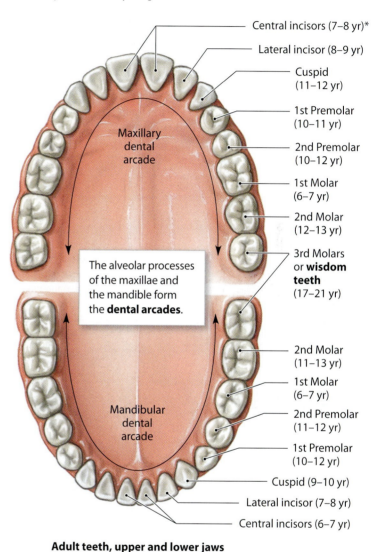

Central incisors (7–8 yr)*
Lateral incisor (8–9 yr)
Cuspid (11–12 yr)
1st Premolar (10–11 yr)
2nd Premolar (10–12 yr)
1st Molar (6–7 yr)
2nd Molar (12–13 yr)
3rd Molars or **wisdom teeth** (17–21 yr)

Maxillary dental arcade

The alveolar processes of the maxillae and the mandible form the **dental arcades**.

2nd Molar (11–13 yr)
1st Molar (6–7 yr)
2nd Premolar (11–12 yr)
1st Premolar (10–12 yr)
Cuspid (9–10 yr)
Lateral incisor (7–8 yr)
Central incisors (6–7 yr)

Mandibular dental arcade

Adult teeth, upper and lower jaws

*Time indicates age at eruption

Tooth decay generally results from the action of bacteria that normally inhabit your mouth. Bacteria adhering to the surfaces of the teeth produce a sticky matrix that traps food particles and creates deposits known as **dental plaque**. If the gingival attachment at the neck of a tooth weakens, the result can be **gingivitis**, inflammation of the gingivae. Severe gingivitis, usually caused by bacterial infection, erodes the gums and damages the root, eventually causing the tooth to fall out. Regular brushing and flossing are important to remove dental plaque and massage the gums, stimulating and strengthening the epithelial attachment of each tooth.

Module 15.4 Review

a. Name the four types of teeth and the three main parts of a typical tooth.

b. What is the name sometimes given to the third molars?

c. Distinguish between the primary dentition and the secondary dentition.

The muscular walls of the pharynx and esophagus play a key role in swallowing

1 The **pharynx** (FAR-inks), a membrane-lined cavity posterior to the nose and mouth, is a common passageway for solid food, liquids, and air. Food passes through the oropharynx and laryngopharynx on its way to the esophagus. The illustration on the right reviews the major landmarks and boundaries of the pharynx.

2 The **esophagus** is a hollow muscular tube approximately 25 cm (10 in.) long with a diameter of about 2 cm (0.80 in.) at its widest point. The primary function of the esophagus is to transport food and liquid from the pharynx to the stomach.

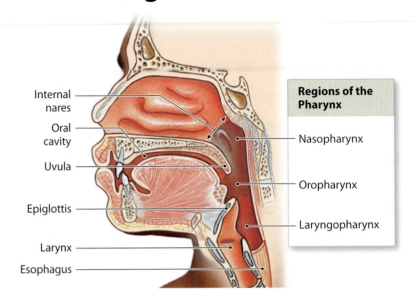

Internal nares
Oral cavity
Uvula
Epiglottis
Larynx
Esophagus

Regions of the Pharynx

Nasopharynx

Oropharynx

Laryngopharynx

The mucosa and submucosa are thrown into large folds that extend the length of the esophagus. These folds allow the esophagus to expand when a large bolus passes through. Muscle tone in the walls keeps the lumen closed, except when you swallow.

Mucosa

Submucosa

In the superior third of the esophagus, the circular and longitudinal layers of the muscularis externa contain skeletal muscle fibers. The middle third contains a mixture of skeletal and smooth muscle tissue. Along the inferior third, only smooth muscle occurs.

The epithelium lining the mucosa of the esophagus is a nonkeratinized, stratified squamous epithelium similar to that of the pharynx and oral cavity.

Lamina propria

The muscularis mucosae consists of an irregular layer of smooth muscle.

LM × 60

3 The layers of the esophageal wall are comparable to those in other portions of the digestive tract. However, the shape of the lumen and the structure of the muscular externa are unique to the esophagus. There is no serosa, but an **adventitia** of connective tissue outside the muscularis externa anchors the esophagus to the posterior body wall.

4 This light micrograph illustrates the extreme thickness of the epithelium making up the esophageal mucosa layer.

5 Swallowing, or **deglutition** (dē-gloo-TISH-un), is a complex process that can be initiated voluntarily but proceeds automatically once it begins. Although you take conscious control over swallowing when you eat or drink, swallowing is also controlled at the subconscious level. Each day you swallow approximately 2400 times. We can divide swallowing into three phases: buccal, pharyngeal, and esophageal.

Buccal Phase

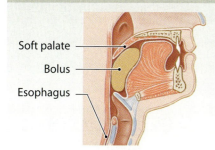

Soft palate
Bolus
Esophagus

The **buccal phase** begins when the bolus is compressed against the hard palate. The tongue retracts, forcing the bolus into the oropharynx and helping to elevate the soft palate, thereby sealing off the nasopharynx. The buccal phase is strictly voluntary. Once the bolus enters the oropharynx, reflex responses are initiated that move the bolus toward the stomach.

Pharyngeal Phase

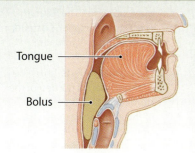

Tongue
Bolus

The **pharyngeal phase** begins when tactile receptors on the pharyngeal arches and uvula are stimulated. In response, motor commands from the swallowing center in the medulla oblongata direct the pharyngeal muscles to contract in a coordinated pattern. Contractions of the pharyngeal muscles elevate the larynx and fold the epiglottis, while the palatal muscles elevate the uvula and soft palate to block the entrance to the nasopharynx. Pharyngeal constrictors then force the bolus through the pharynx, past the folded epiglottis, and into the esophagus.

Esophageal Phase

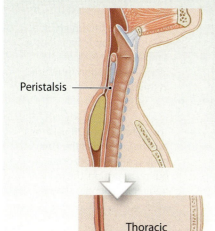

Peristalsis

Thoracic cavity

Lower esophageal sphincter

Stomach

The **esophageal phase** of swallowing begins as the contraction of pharyngeal muscles forces the bolus through the entrance to the esophagus. Once in the esophagus, peristalsis pushes the bolus toward the stomach. The approach of the bolus opens the lower esophageal sphincter, and the bolus then continues into the stomach. For a typical bolus, the entire trip takes about 9 seconds. Liquids may travel faster, flowing ahead of the peristaltic contractions with the assistance of gravity. A dry or poorly lubricated bolus travels more slowly, and a series of local reflexes, called **secondary peristaltic waves,** may be required to push it all the way to the stomach.

The esophagus begins posterior to the cricoid cartilage. From this point, where it is at its narrowest, the esophagus descends toward the thoracic cavity posterior to the trachea. It passes inferiorly along the posterior wall of the mediastinum and enters the abdominopelvic cavity through the **esophageal hiatus** (hī-Ā-tus), an opening in the diaphragm. Parasympathetic and sympathetic fibers from the esophageal plexus innervate the esophagus. Resting muscle tone in the circular muscle layer in the superior 3 cm of the esophagus normally prevents air from entering the esophagus. The band of smooth muscle involved functions as an upper esophageal sphincter. A comparable area of smooth muscle at the inferior end of the esophagus forms the **lower esophageal sphincter**. It normally remains in a state of active contraction, which prevents the backflow of materials from the stomach into the esophagus.

Module 15.5 Review

a. Describe the structure and function of the pharynx.

b. Name the structure connecting the pharynx to the stomach.

c. Describe the major event in each of the three phases of swallowing.

The stomach is a muscular, expandable, J-shaped organ with three layers in its muscularis externa

1 This illustration presents the major surfaces and regions of the stomach. Its shape is actually highly variable. When empty, the stomach resembles a muscular J-shaped tube with a narrow, constricted lumen. When full, it can contain 1–1.5 liters of material. That material, which combines ingested food with saliva and the secretions of gastric glands, is a viscous, highly acidic, soupy mixture called **chyme** (kīm).

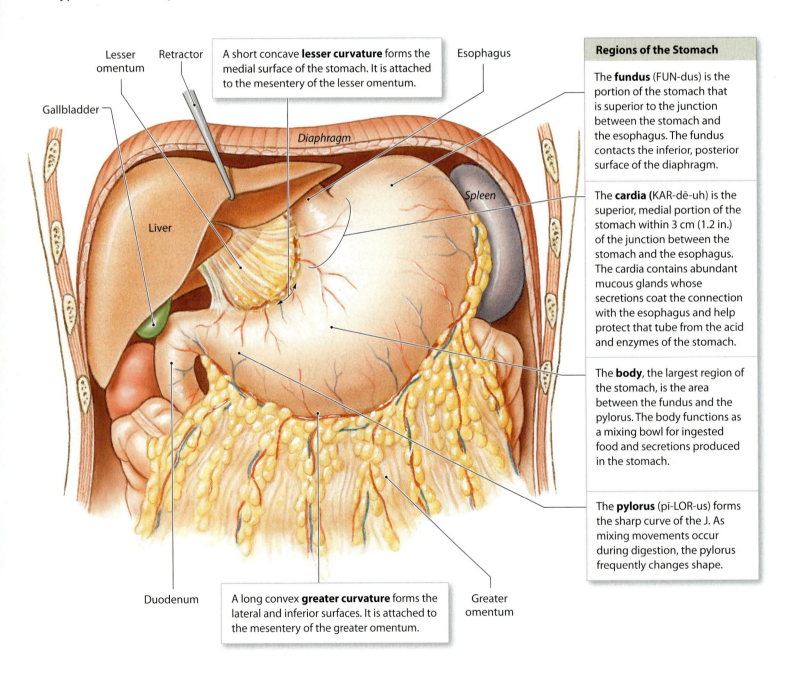

Lesser omentum

Retractor

A short concave **lesser curvature** forms the medial surface of the stomach. It is attached to the mesentery of the lesser omentum.

Esophagus

Gallbladder

Diaphragm

Spleen

Liver

Duodenum

A long convex **greater curvature** forms the lateral and inferior surfaces. It is attached to the mesentery of the greater omentum.

Greater omentum

Regions of the Stomach

The **fundus** (FUN-dus) is the portion of the stomach that is superior to the junction between the stomach and the esophagus. The fundus contacts the inferior, posterior surface of the diaphragm.

The **cardia** (KAR-dē-uh) is the superior, medial portion of the stomach within 3 cm (1.2 in.) of the junction between the stomach and the esophagus. The cardia contains abundant mucous glands whose secretions coat the connection with the esophagus and help protect that tube from the acid and enzymes of the stomach.

The **body**, the largest region of the stomach, is the area between the fundus and the pylorus. The body functions as a mixing bowl for ingested food and secretions produced in the stomach.

The **pylorus** (pī-LOR-us) forms the sharp curve of the J. As mixing movements occur during digestion, the pylorus frequently changes shape.

2 This cadaver dissection exposes the stomach and the lesser and greater omenta.

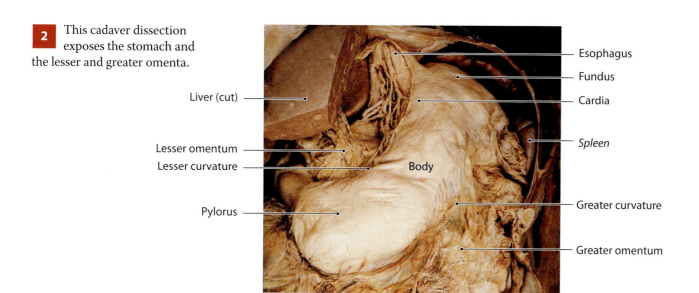

Esophagus

Fundus

Cardia

Liver (cut)

Spleen

Lesser omentum

Lesser curvature

Body

Pylorus

Greater curvature

Greater omentum

3 A partially sectioned and dissected stomach reveals additional details about the internal structure of this organ.

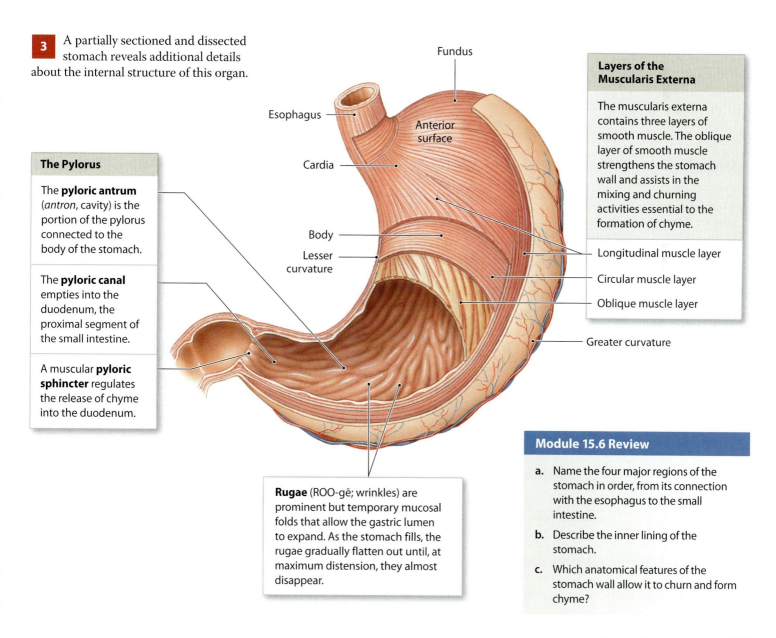

Fundus

Esophagus

Anterior surface

Cardia

Body

Lesser curvature

Greater curvature

The Pylorus

The **pyloric antrum** (*antron,* cavity) is the portion of the pylorus connected to the body of the stomach.

The **pyloric canal** empties into the duodenum, the proximal segment of the small intestine.

A muscular **pyloric sphincter** regulates the release of chyme into the duodenum.

Layers of the Muscularis Externa

The muscularis externa contains three layers of smooth muscle. The oblique layer of smooth muscle strengthens the stomach wall and assists in the mixing and churning activities essential to the formation of chyme.

Longitudinal muscle layer

Circular muscle layer

Oblique muscle layer

Rugae (ROO-gē; wrinkles) are prominent but temporary mucosal folds that allow the gastric lumen to expand. As the stomach fills, the rugae gradually flatten out until, at maximum distension, they almost disappear.

Module 15.6 Review

a. Name the four major regions of the stomach in order, from its connection with the esophagus to the small intestine.

b. Describe the inner lining of the stomach.

c. Which anatomical features of the stomach wall allow it to churn and form chyme?

The stomach breaks down the organic nutrients in food

1 The wall of the stomach is relatively thick and muscular, and its mucosa has deep folds that form gastric glands.

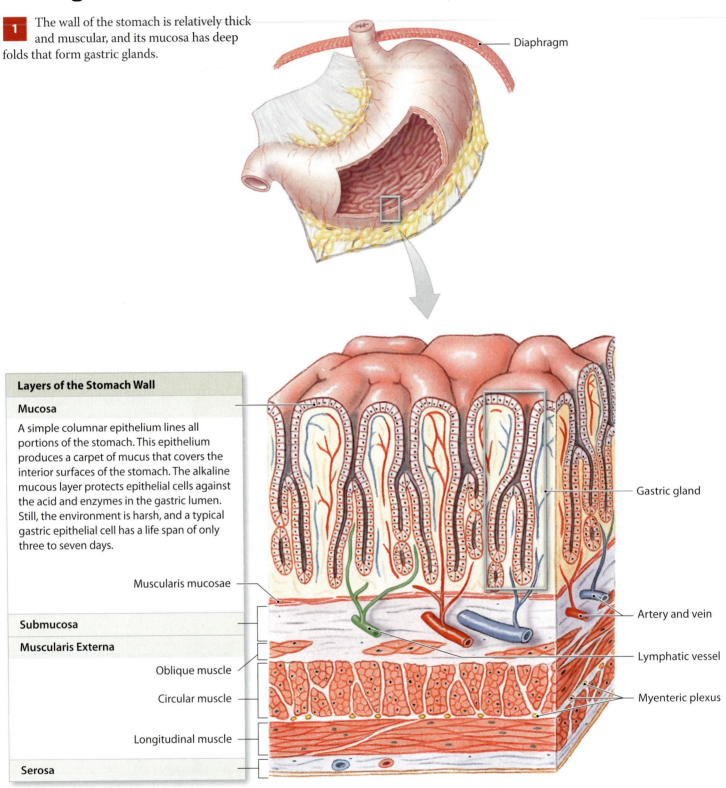

Diaphragm

Gastric gland

Artery and vein

Lymphatic vessel

Myenteric plexus

Layers of the Stomach Wall

Mucosa

A simple columnar epithelium lines all portions of the stomach. This epithelium produces a carpet of mucus that covers the interior surfaces of the stomach. The alkaline mucous layer protects epithelial cells against the acid and enzymes in the gastric lumen. Still, the environment is harsh, and a typical gastric epithelial cell has a life span of only three to seven days.

Muscularis mucosae

Submucosa

Muscularis Externa

Oblique muscle

Circular muscle

Longitudinal muscle

Serosa

2 **Gastric glands** in the fundus and body secrete most of the acid and enzymes involved in gastric digestion. The gastric glands in these areas are dominated by parietal cells and chief cells. Together, these cells secrete about 1500 mL of gastric juice each day. **Gastric juice** is a thin, colorless, acidic fluid that contains enzymes. Gastric glands in the pylorus secrete mucus and hormones involved in coordinating and controlling digestive activity.

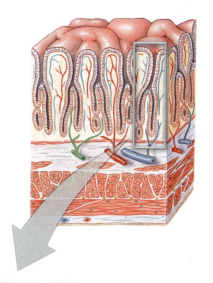

Shallow depressions called **gastric pits** open onto the interior surface of the stomach. Stem cells at the base of each gastric pit actively divide, replacing epithelial cells that are shed into the chyme.

Each gastric pit communicates with several gastric glands that extend deep into the lamina propria.

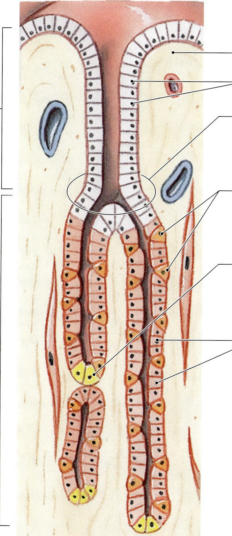

Lamina propria

Mucous epithelial cells

Neck

Cells of Gastric Glands

Parietal cells secrete **intrinsic factor**, a glycoprotein that aids the absorption of vitamin B_{12} across the intestinal lining. Parietal cells also secrete hydrochloric acid (HCl).

Enteroendocrine cells produce a variety of hormones. We consider these hormones and their functions in Module 15.10.

Chief cells secrete **pepsinogen** (pep-SIN-ō-jen), an inactive proenzyme. Hydrochloric acid in the gastric lumen converts pepsinogen to **pepsin**, an active proteolytic (protein-digesting) enzyme. In addition, the stomachs of newborn infants (but not of adults) produce **rennin** and **gastric lipase**, enzymes important for digesting milk.

Module 15.7 Review

a. What is the function of parietal cells?

b. Explain the significance of the alkaline mucous layer lining the interior surface of the stomach.

c. Describe the relationship between pepsinogen and pepsin.

The intestinal tract is specialized for absorbing nutrients

1 The intestinal lining bears a series of transverse folds called **circular folds**. Unlike the rugae in the stomach, the circular folds are permanent features that do not disappear when the small intestine fills. The small intestine contains roughly 800 circular folds, most of them within the jejunum, the middle segment of the small intestine. Their presence greatly increases the surface area available for absorbing nutrients.

Small intestine

2 If the small intestine were a simple tube with smooth walls, it would have a total absorptive area of only about 3300 cm^2 (3.6 ft^2). But instead, its mucosa contains circular folds. Each circular fold supports a forest of fingerlike projections, the **intestinal villi**. Each villus is covered by epithelial cells whose exposed surfaces are covered with microvilli. This arrangement increases the total area for absorption by a factor of more than 600, to approximately 2 million cm^2 (over 2200 ft^2).

Circular fold

Villi

3 This sectional diagram of the intestinal wall shows features common to all segments of the small intestine.

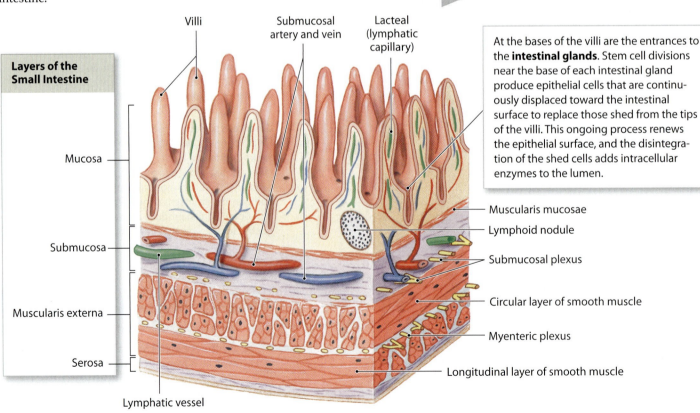

Villi

Submucosal artery and vein

Lacteal (lymphatic capillary)

At the bases of the villi are the entrances to the **intestinal glands**. Stem cell divisions near the base of each intestinal gland produce epithelial cells that are continuously displaced toward the intestinal surface to replace those shed from the tips of the villi. This ongoing process renews the epithelial surface, and the disintegration of the shed cells adds intracellular enzymes to the lumen.

Layers of the Small Intestine

Mucosa

Submucosa

Muscularis externa

Serosa

Lymphatic vessel

Muscularis mucosae

Lymphoid nodule

Submucosal plexus

Circular layer of smooth muscle

Myenteric plexus

Longitudinal layer of smooth muscle

4 Each villus has a complex internal structure. The lamina propria of each villus contains an extensive network of capillaries that originate in a vascular network within the submucosa. These capillaries carry absorbed nutrients to the hepatic portal circulation for delivery to the liver, which adjusts the nutrient concentrations in blood before the blood reaches the general systemic circulation.

5 The surface of each villus consists of a simple columnar epithelium that is carpeted with microvilli. Because the microvilli project from the epithelium like the bristles on a brush, these cells are said to have a **brush border**. Brush border enzymes are integral membrane proteins located on the surfaces of intestinal microvilli. These enzymes break down materials that come in contact with the brush border. The epithelial cells then absorb the breakdown products.

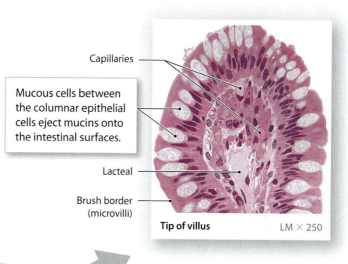

Capillaries

Mucous cells between the columnar epithelial cells eject mucins onto the intestinal surfaces.

Lacteal

Brush border (microvilli)

Tip of villus   LM × 250

Each villus contains a lymphatic capillary called a **lacteal** (LAK-tē-ul; *lacteus*, milky). Lacteals transport materials that cannot enter blood capillaries. For example, absorbed fatty acids are assembled into protein–lipid packages that are too large to diffuse into the bloodstream. After entering the lacteals, these lipid-rich packets reach the venous circulation as the thoracic duct delivers lymph to the left subclavian vein.

Columnar epithelial cell

Mucous cell

Nerve

Capillary network

Lamina propria

Arteriole

Lymphatic vessel

Venule

Contractions of the muscularis mucosae and smooth muscle cells within the intestinal villi move the villi back and forth, exposing their epithelial surfaces to the liquefied intestinal contents. This movement makes absorption more efficient by quickly eliminating local differences in nutrient concentration. Movements of the villi also squeeze the lacteals, thereby helping lymph move out of the villi.

Muscularis mucosae

Module 15.8 Review

a. Name the layers of the small intestine from superficial to deep.

b. Describe the anatomy of the intestinal mucosa.

c. Explain the function of lacteals.

The small intestine is divided into the duodenum, jejunum, and ileum

The small intestine plays the key role in digesting and absorbing nutrients. Ninety percent of nutrient absorption occurs in the small intestine; the remainder occurs in the large intestine.

1 The small intestine fills much of the peritoneal cavity. The small intestine averages 6 m (19.7 ft) in length. It has a diameter ranging from 4 cm (1.6 in.) at the stomach to about 2.5 cm (1 in.) at its junction with the large intestine.

Regions of the Small Intestine

The **duodenum** (doo-AH-de-num or doo-ō-DĒ-num), 25 cm (10 in.) in length, is the segment closest to the stomach. This portion of the small intestine receives chyme from the stomach and digestive secretions from the pancreas and liver.

An abrupt bend marks the boundary between the duodenum and the **jejunum** (je-JOO-num). The jejunum is about 2.5 meters (8.2 ft) long. Most chemical digestion and nutrient absorption occur in the jejunum.

The **ileum** (IL-ē-um), the final segment of the small intestine, is also the longest, averaging 3.5 meters (11.5 ft) in length. The ileum ends at the **ileocecal** (il-ē-o-SĒ-kal) **valve**, a sphincter that controls the flow of material from the ileum into the large intestine.

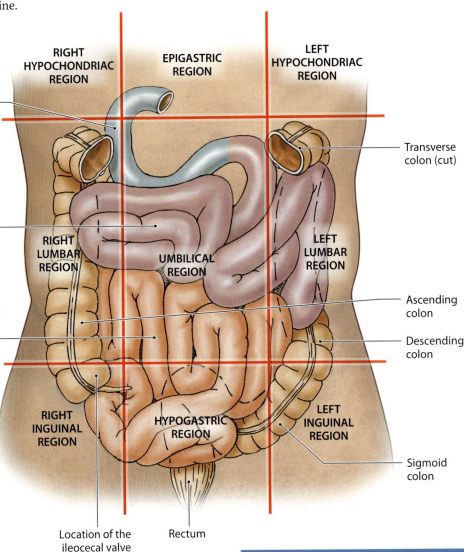

RIGHT HYPOCHONDRIAC REGION

EPIGASTRIC REGION

LEFT HYPOCHONDRIAC REGION

Transverse colon (cut)

RIGHT LUMBAR REGION

UMBILICAL REGION

LEFT LUMBAR REGION

Ascending colon

Descending colon

RIGHT INGUINAL REGION

HYPOGASTRIC REGION

LEFT INGUINAL REGION

Sigmoid colon

Location of the ileocecal valve

Rectum

Module 15.9 Review

a. Name the three regions of the small intestine from proximal to distal.

b. Identify the region of the small intestine that can be found within the epigastric region.

c. What is the primary function of the duodenum?

Five hormones are involved in regulating digestive activities

1 Five hormones help regulate digestive function: gastrin, secretin, gastric inhibitory peptide, cholescystokinin, and vasoactive intestinal peptide. Four of the five are produced by the duodenum, which receives partially digested materials from the stomach. The duodenum adjusts gastric activity and coordinates the secretions of accessory digestive organs according to the characteristics of the arriving chyme.

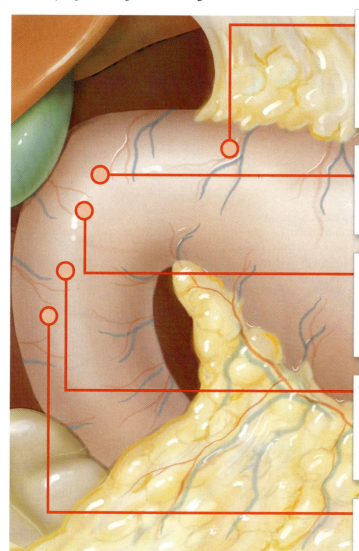

Gastrin is secreted by G cells in the pyloric antrum and enteroendocrine cells in the duodenum. The pyloric antrum secretes gastrin when stimulated by the vagus nerves or when food arrives in the stomach. Duodenal cells release gastrin when they are exposed to large quantities of incompletely digested proteins. Gastrin increases stomach motility (ability to move) and stimulates the production of gastric acid and enzymes.

Secretin is released when chyme arrives in the duodenum. Secretin's primary effect is to increase in the secretion of bile (by the liver) and buffers (by the pancreas), which in turn increase the pH of the chyme. Among its secondary effects, secretin reduces gastric motility and secretory rates.

Gastric inhibitory peptide (GIP) is secreted when fats and carbohydrates—especially glucose—enter the small intestine. The inhibition of gastric activity is accompanied by the stimulation of insulin release by the pancreatic islets. GIP has several secondary effects, including stimulating duodenal gland activity, stimulating lipid synthesis in adipose tissue, and increasing glucose use by skeletal muscles.

Cholecystokinin (CCK) is secreted when chyme arrives in the duodenum, especially when the chyme contains lipids and partially digested proteins. In the pancreas, CCK accelerates the production and secretion of all types of digestive enzymes. It also relaxes the hepatopancreatic sphincter and contracts the gallbladder, resulting in the ejection of bile and pancreatic juice into the duodenum.

Vasoactive intestinal peptide (VIP) stimulates intestinal gland secretion, dilates regional capillaries, and inhibits acid production in the stomach. By dilating capillaries in active areas of the intestinal tract, VIP provides an efficient mechanism for removing absorbed nutrients.

Module 15.10 Review

a. Name the five major hormones that regulate digestive activities and summarize their effects.

b. Does a high-fat meal raise or lower the level of cholecystokinin (CCK) in the blood?

c. Suppose the duodenum of the small intestine did not produce secretin. How would this affect the pH of the intestinal contents?

The large intestine stores and concentrates fecal material

The **large intestine**, also known as the large bowel, averages about 1.5 meters (4.9 ft) long and 7.5 cm (3 in) wide. The major functions of the large intestine are to: (1) absorb important vitamins produced by bacteria in the colon, (2) reabsorb water from intestinal contents, (3) compact indigestible intestinal contents into feces, and (4) store feces prior to defecation. The large intestine consists of three segments: the cecum, colon, and rectum.

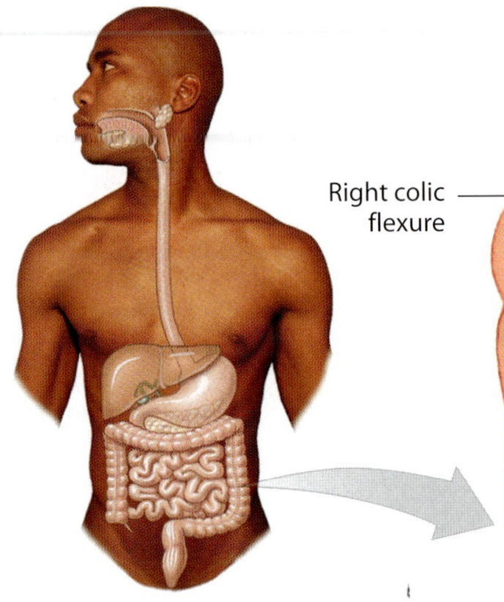

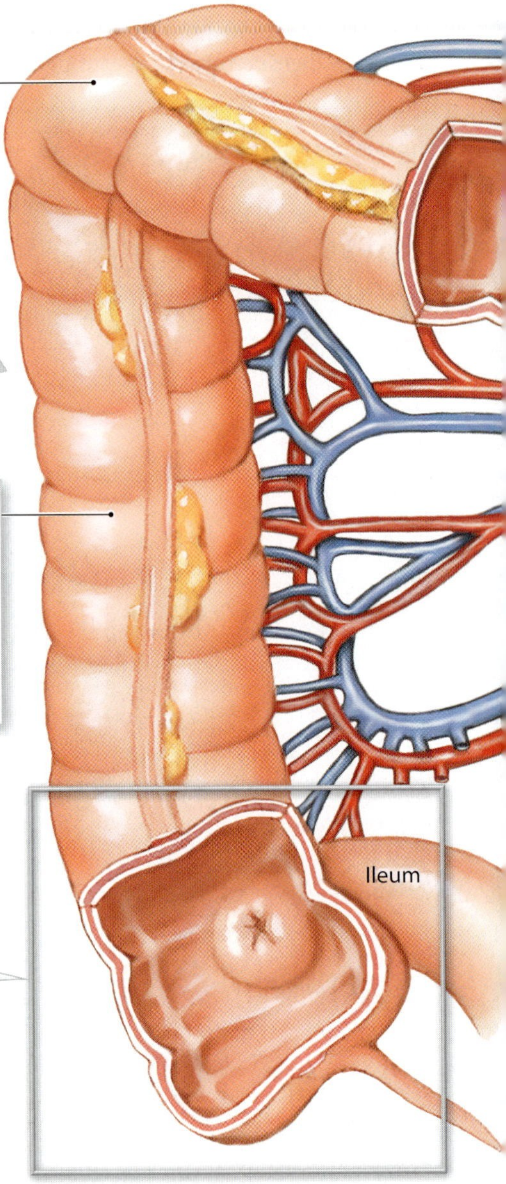

1 Material arriving from the ileum first enters an expanded pouch called the **cecum** (SĒ-kum). The cecum collects and stores materials from the ileum and begins the process of **compaction** (compressing intestinal contents to form feces).

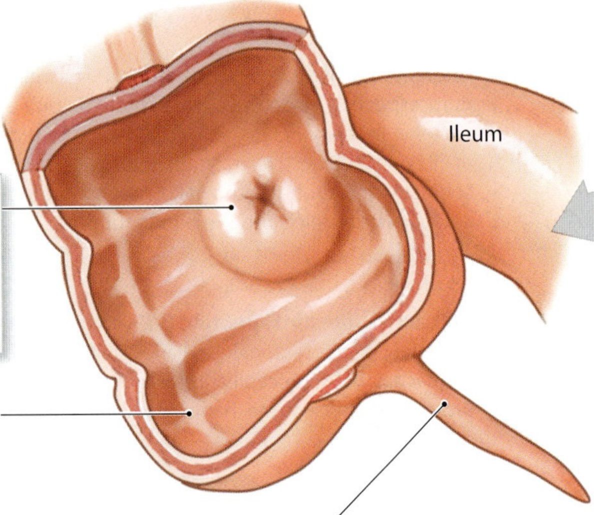

Right colic flexure

The **ascending colon** begins at the cecum and ascends along the right margin of the peritoneal cavity to the inferior surface of the liver. There, the colon bends sharply to the left at the **right colic flexure**; this marks the end of the ascending colon.

Ileum

The ileum attaches to the medial surface of the cecum and opens into the cecum at the **ileocecal valve**.

Ileum

Cecum

The slender, hollow **appendix** is attached to the cecum. The appendix is generally about 9 cm (3.6 in.) long, but its size and shape are quite variable. The mucosa and submucosa of the appendix are dominated by lymphoid nodules, and the appendix functions primarily as an organ of the lymphatic system. Inflammation of the appendix is known as **appendicitis**.

2 The **colon** has a larger diameter and a thinner wall than the small intestine. We can subdivide the colon into four regions: the ascending colon, transverse colon, descending colon, and sigmoid colon. The ascending and descending colon are firmly attached to the abdominal wall. The transverse colon and sigmoid colon are suspended by mesenteries.

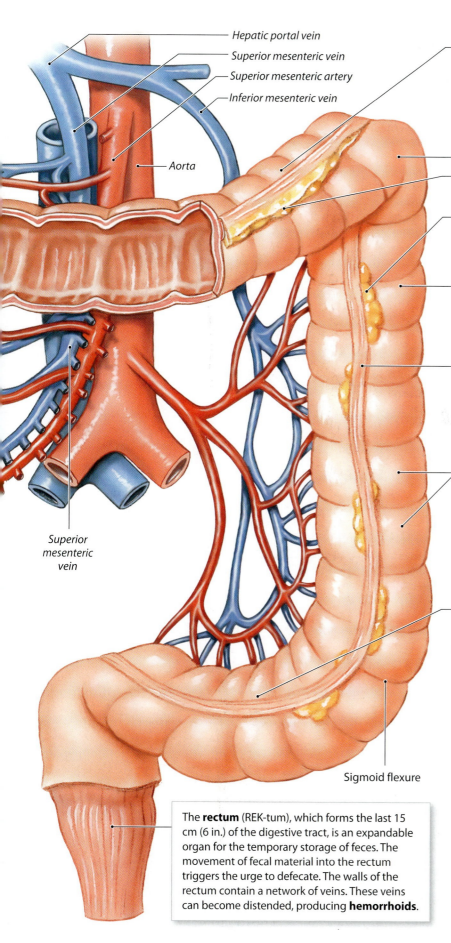

Hepatic portal vein

Superior mesenteric vein

Superior mesenteric artery

Inferior mesenteric vein

Aorta

Superior mesenteric vein

The **transverse colon** crosses the abdomen from right to left. As the transverse colon reaches the left side of the body, the colon makes a 90° turn at the **left colic flexure**.

Left colic flexure

Greater omentum (cut)

The serosa of the colon contains numerous teardrop-shaped sacs of fat called **fatty appendices**.

The **descending colon** proceeds inferiorly along the body's left side until reaching the iliac fossa. At the iliac fossa, the descending colon ends at the **sigmoid flexure**.

Three separate longitudinal bands of smooth muscle—called the **taeniae coli** (TĒ-nē-ē KŌ-lē)—run along the outer surfaces of the colon just deep to the serosa. These bands correspond to the outer layer of the muscularis externa in other portions of the digestive tract.

Muscle tone within the taeniae coli is what creates **haustra** (HAWS-truh), a series of pouches in the wall of the colon. Cutting into the intestinal lumen reveals that the creases between the haustra affect the mucosal lining as well, producing a series of internal folds. Haustra permit the colon to expand and elongate, rather like the bellows that allow an accordion to lengthen.

The sigmoid flexure is the start of the **sigmoid** (SIG-moyd; *sigmeidos*, Greek letter S) **colon**, an S-shaped segment that is about 15 cm (6 in.) long and empties into the rectum.

Sigmoid flexure

Powerful peristaltic contractions called **mass movements** occur a few times each day in response to distension of the stomach and duodenum. These contractions begin at the transverse colon and push materials along the distal portion of the large intestine.

The **rectum** (REK-tum), which forms the last 15 cm (6 in.) of the digestive tract, is an expandable organ for the temporary storage of feces. The movement of fecal material into the rectum triggers the urge to defecate. The walls of the rectum contain a network of veins. These veins can become distended, producing **hemorrhoids**.

Module 15.11 Review

a. Identify the three segments of the large intestine and the four regions of the colon.

b. Name the major functions of the large intestine.

c. Describe mass movements.

Neural reflexes and hormones work together to control digestive activities

1 This flowchart summarizes the pattern of hormone release and the effects of those hormones within the digestive system.

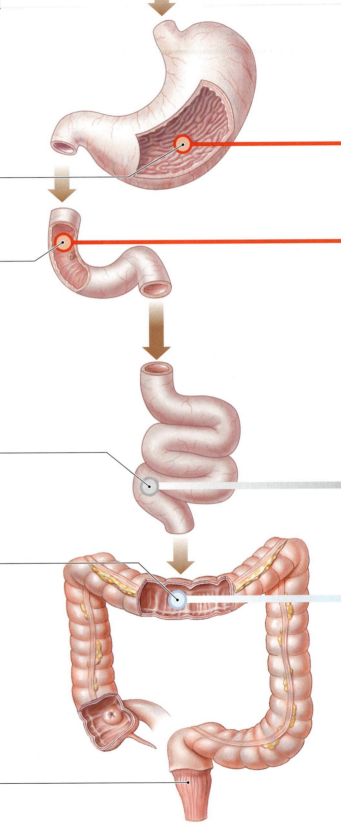

Ingested food

1 Gastric secretion begins before food arrives in the stomach. When you see, smell, taste, or think of food, the parasympathetic division of the ANS prepares the stomach for a meal. The vagus nerves innervate the mucous cells, chief cells, parietal cells, and enteroendocrine cells of the stomach. In response to stimulation, the production of gastric juice rapidly accelerates.

2 When food arrives in the stomach, the physical distortion stimulates the production of gastrin.

3 When chyme arrives in the duodenum, the physical and chemical stimuli promote neural events and the secretion of multiple hormones that coordinate the activities of the intestinal tract and the pancreas, liver, and gallbladder. The local distension also stimulates stretch receptors that trigger the **enterogastric reflex**. This reflex inhibits both gastrin production and gastric contractions, thereby controlling the rate at which chyme enters the duodenum.

4 The material arriving in the jejunum is further broken down by digestive enzymes. The intestinal epithelium absorbs the products of digestion.

5 On average, 1500 mL of material enters the colon each day. Over 1 L of water is reabsorbed through osmosis, and only 200 mL of feces is ejected. In addition to preventing dehydration by reabsorbing water, the epithelium absorbs three vitamins produced by the normal bacterial residents of the colon: vitamin K, vitamin B_5, and vitamin B_7 (biotin). The fecal material ejected is 75 percent water, 5 percent bacteria, and the rest a mixture of indigestible materials, inorganic matter, and the remains of epithelial cells.

6 The rectum is usually empty, except when a powerful peristaltic contraction forces feces out of the sigmoid colon. Distension of the rectal wall then starts the **defecation reflex**. However, feces will only be ejected if the external anal sphincter is voluntarily relaxed.

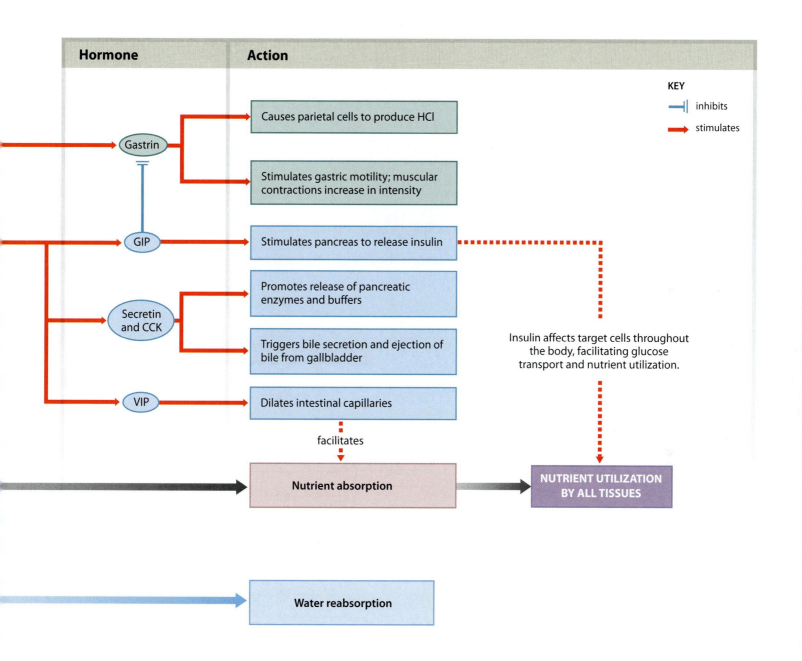

Hormone	Action

Gastrin → Causes parietal cells to produce HCl

Gastrin → Stimulates gastric motility; muscular contractions increase in intensity

GIP → Stimulates pancreas to release insulin

Secretin and CCK → Promotes release of pancreatic enzymes and buffers

Secretin and CCK → Triggers bile secretion and ejection of bile from gallbladder

VIP → Dilates intestinal capillaries

KEY
⊣ inhibits
→ stimulates

Insulin affects target cells throughout the body, facilitating glucose transport and nutrient utilization.

facilitates

Nutrient absorption → **NUTRIENT UTILIZATION BY ALL TISSUES**

Water reabsorption

Module 15.12 Review

a. Describe the two reflexes triggered by stimulation of stretch receptors.

b. What is the role of GIP?

c. Describe the components of feces.

1. Labeling

Label the structures of a typical tooth in the following figure.

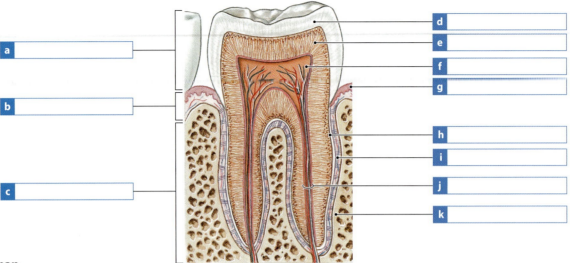

a [ ]

b [ ]

c [ ]

d [ ]

e [ ]

f [ ]

g [ ]

h [ ]

i [ ]

j [ ]

k [ ]

2. Concept map

Using the following terms, fill in the blank boxes to complete the hormones of digestive activity concept map.

- acid production
- gastrin
- VIP
- insulin
- intestinal capillaries
- GIP
- material in jejunum
- gallbladder
- bile
- inhibits
- secretin and CCK
- nutrient utilization by tissues

Hormones of Digestive Activity

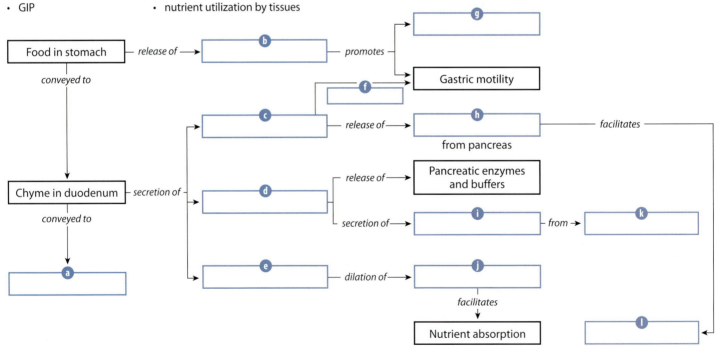

Food in stomach — *release of* → [b]

conveyed to ↓

Chyme in duodenum — *secretion of* → [d]

conveyed to ↓

[a]

[c]

[e]

[g]

promotes

Gastric motility

[f]

release of → [h] from pancreas — *facilitates*

release of → Pancreatic enzymes and buffers

secretion of → [i] — *from* → [k]

dilation of → [j]

facilitates ↓

Nutrient absorption

[l]

3. Short answer

Briefly describe the similarities and differences between parietal cells and chief cells in the stomach wall. _____

_____

_____

Accessory Digestive Organs

The major accessory digestive organs are the salivary glands, liver, gallbladder, and pancreas. The salivary glands and pancreas produce and store enzymes and buffers that are essential to normal digestive function. In addition to their roles in digestion, the salivary glands, liver, and pancreas have vital metabolic and endocrine functions.

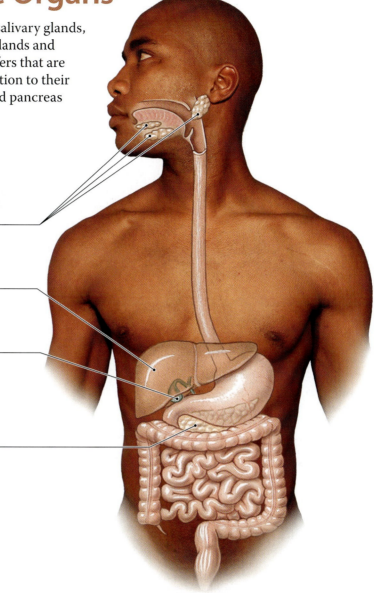

Accessory Digestive Organs

Salivary Glands

Three pairs of salivary glands produce saliva that contains mucins and enzymes.

Liver

The liver has almost 200 known functions. The table below lists the most important.

Gallbladder

The gallbladder stores and concentrates bile secreted by the liver.

Pancreas

Exocrine cells secrete buffers and digestive enzymes. Endocrine cells secrete insulin and glucagon, hormones introduced in Module 10.8.

Digestive and Metabolic Functions of the Liver

- Synthesize and secrete bile
- Store glycogen and lipid reserves
- Maintain normal concentrations of glucose, amino acids, and fatty acids in the bloodstream
- Synthesize and convert nutrient types (for instance, converting carbohydrates to lipids)
- Synthesize and release cholesterol bound to transport proteins
- Inactivative toxins
- Store iron
- Store fat-soluble vitamins

Other Major Functions

- Synthesize plasma proteins
- Synthesize clotting factors
- Destroy damaged red blood cells by phagocytosis (by Kupfer cells)
- Store blood
- Absorb and break down circulating hormones and immunoglobulins
- Absorb and inactivate lipid-soluble drugs

Secretions from three pairs of salivary glands produce saliva

1 Three pairs of salivary glands secrete into the oral cavity. Each pair has a distinctive cellular organization and produces saliva with slightly different properties. Any object in your mouth can trigger a **salivary reflex** controlled by the parasympathetic division of the ANS. This reflex accelerates secretion by all the salivary glands, producing large amounts of saliva.

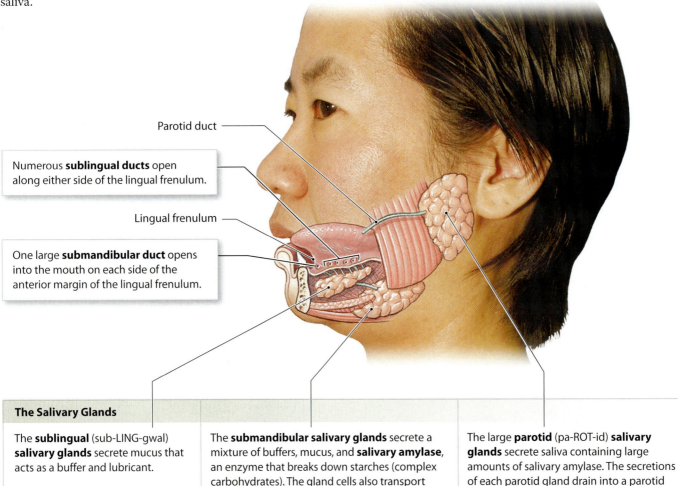

Parotid duct

Numerous **sublingual ducts** open along either side of the lingual frenulum.

Lingual frenulum

One large **submandibular duct** opens into the mouth on each side of the anterior margin of the lingual frenulum.

The Salivary Glands

The **sublingual** (sub-LING-gwal) **salivary glands** secrete mucus that acts as a buffer and lubricant.	The **submandibular salivary glands** secrete a mixture of buffers, mucus, and **salivary amylase**, an enzyme that breaks down starches (complex carbohydrates). The gland cells also transport antibodies into the saliva, to protect against pathogens in food.	The large **parotid** (pa-ROT-id) **salivary glands** secrete saliva containing large amounts of salivary amylase. The secretions of each parotid gland drain into a parotid duct that empties into the vestibule at the level of the second upper molar.

A continuous background level of saliva flushes the oral surfaces, helping to keep them clean. Buffers in the saliva keep the pH of your mouth near 7.0 and prevent the buildup of acids produced by bacteria. In addition, saliva contains antibodies that help control populations of oral bacteria.

Module 15.13 Review

a. Name the three pairs of salivary glands and list their secretions.

b. List several functions of saliva.

c. If the parotid salivary glands were damaged, the digestion of which nutrient would be affected?

The liver, the largest visceral organ, is divided into left, right, caudate, and quadrate lobes

1 The **liver**, the largest visceral organ, weighs about 1.5 kg (3.3 lb). These anterior and posterior views of the liver surface show its four lobes (right, left, caudate, and quadrate) and its major anatomical landmarks. The table on page 545 summarizes some of the liver's functions.

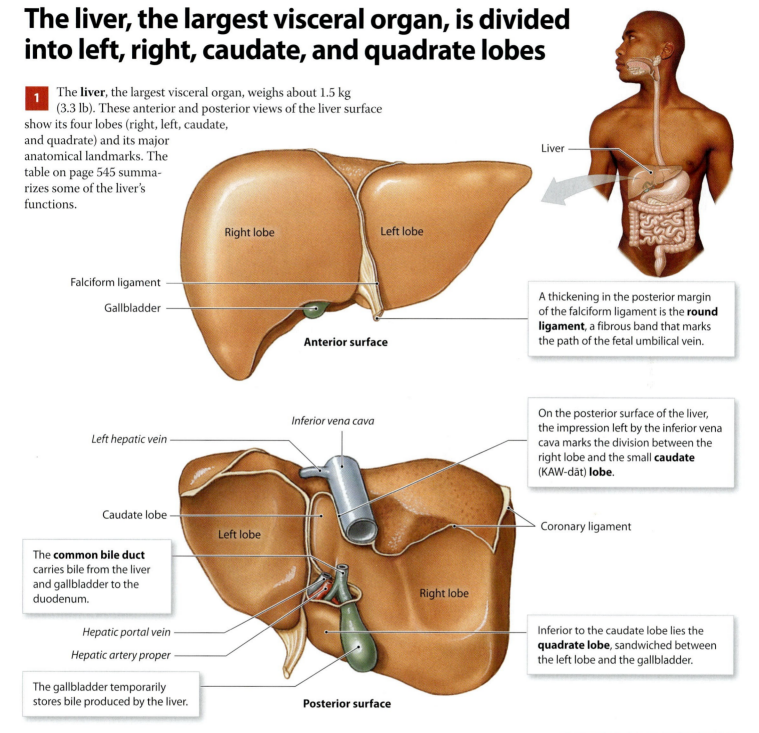

Liver

Right lobe

Left lobe

Falciform ligament

Gallbladder

Anterior surface

A thickening in the posterior margin of the falciform ligament is the **round ligament**, a fibrous band that marks the path of the fetal umbilical vein.

Inferior vena cava

Left hepatic vein

On the posterior surface of the liver, the impression left by the inferior vena cava marks the division between the right lobe and the small **caudate** (KAW-dāt) **lobe**.

Caudate lobe

Left lobe

Coronary ligament

The **common bile duct** carries bile from the liver and gallbladder to the duodenum.

Right lobe

Hepatic portal vein

Hepatic artery proper

Inferior to the caudate lobe lies the **quadrate lobe**, sandwiched between the left lobe and the gallbladder.

The gallbladder temporarily stores bile produced by the liver.

Posterior surface

Nutrient-rich blood from the digestive tract drains into the liver through the hepatic portal vein. Normally pressures in the hepatic portal system are low, averaging 10 mm Hg or less. Liver infections and conditions such as alcoholism can cause degenerative changes in the liver that interfere with normal blood flow. If blood flow through the liver becomes restricted, pressure in the hepatic portal system increses markedly, leading to **portal hypertension**. As pressures rise, small peripheral veins and capillaries in the portal system become distended. If they rupture, extensive bleeding can take place. Portal hypertension can also force fluid into the peritoneal cavity across the serosal surfaces of the liver and viscera, producing ascites.

Module 15.14 Review

a. Name the lobes of the liver.

b. Which structure marks the division between the left lobe and right lobe of the liver?

c. Explain why degenerative changes in the liver may cause extensive bleeding.

The gallbladder stores and concentrates bile ...

The **gallbladder** is a hollow, pear-shaped organ that stores and concentrates bile prior to its excretion into the small intestine. This muscular sac is located in a recess in the posterior surface of the liver's right lobe.

1 In this view, the liver has been pulled upward to show the gallbladder and its associated ducts.

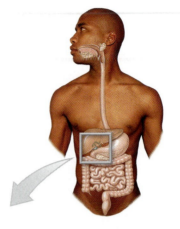

The **right** and **left hepatic ducts** collect bile from all the bile ducts of the liver lobes.

The hepatic ducts unite to form the **common hepatic duct**. Bile in the common hepatic duct either flows into the common bile duct or enters the **cystic duct**, which leads to the gallbladder.

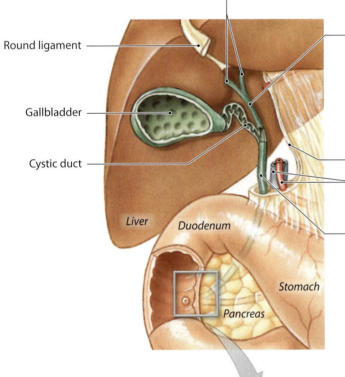

Round ligament

Gallbladder

Cystic duct

Liver

Duodenum

Stomach

Pancreas

Cut edge of lesser omentum

Hepatic portal vein and hepatic artery

The common hepatic duct and the cystic duct unite to form the **common bile duct**. The common bile duct empties into the duodenum.

2 The common bile duct passes within the lesser omentum toward the stomach, turns, and penetrates the wall of the duodenum. There it meets the pancreatic duct at the entrance to the **duodenal ampulla** (am-PUL-a), a chamber within the **duodenal papilla** that projects into the intestinal lumen.

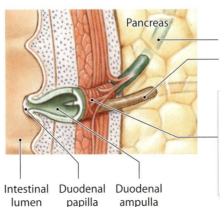

Pancreas

Common bile duct

Pancreatic duct

A muscular **hepatopancreatic sphincter** encircles the common bile duct, pancreatic duct, and duodenal ampulla. Resting tension in the sphincter prevents bile from flowing into the duodenum except at mealtimes.

Intestinal lumen   Duodenal papilla   Duodenal ampulla

... and the pancreas has vital endocrine and exocrine functions

The **pancreas** lies posterior to the stomach, extending laterally from the duodenum toward the spleen. (We considered the endocrine functions of the pancreas in Chapter 10.)

3 The pancreas lies within the loop formed by the duodenum as it leaves the pylorus.

The large **pancreatic duct** delivers the exocrine secretions to the duodenum. The epithelial cells lining the duct and its smaller tributaries secrete water and ions that mix with the secretions of the exocrine gland cells to form a watery **pancreatic juice**.

Partitions of connective tissue divide the interior of the pancreas into distinct **pancreatic lobules**.

Common bile duct

In 3–10 percent of the population, a small **accessory pancreatic duct** branches from the pancreatic duct and empties separately into the duodenum.

Body of pancreas

Tail of pancreas

Head of pancreas

Duodenal papilla

Duodenum

The pancreatic duct meets the common bile duct at the entrance to the duodenal ampulla.

4 Each day, the pancreas secretes about 1000 mL (1 qt) of pancreatic juice containing a variety of enzymes and a watery buffer solution. This table introduces the primary pancreatic enzymes produced; we discuss their functions in the next chapter.

Major Pancreatic Enzymes

- **Pancreatic alpha-amylase** is a carbohydrase (kar-bō-HĪ-drās)—an enzyme that breaks down certain starches. Pancreatic alpha-amylase is almost identical to salivary amylase.
- **Pancreatic lipase** breaks down lipids, releasing products (such as fatty acids) that can be easily absorbed.
- **Nucleases** break down RNA or DNA.
- **Proteolytic enzymes** break proteins apart. They are secreted as inactive proenzymes to protect the pancreas against the action of its own enzymes. The proenzymes become active once they are in the duodenal lumen. The active enzymes include trypsin, chymotrypsin, carboxypeptidase, and elastase. Together, they break down proteins into a mixture of tripeptides, dipeptides, and amino acids.

Module 15.15 Review

a. Trace a drop of bile from the hepatic ducts to the duodenal lumen.

b. Describe pancreatic juice.

c. Name the major pancreatic enzymes and cite their primary functions.

Disorders of the digestive system are diverse and relatively common

Oral Cavity

Periodontal disease, the most common cause for losing teeth, occurs when dental plaque forms in the area between the gums and teeth. The bacterial activity may cause **gingivitis** (shown here) and tooth decay, and eventual breakdown of the periodontal ligament and surrounding bone.

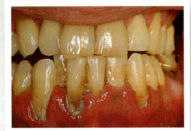

Periodontal disease

Salivary Glands

The **mumps virus** causes **mumps**, an infection of the salivary glands. The infection most often occurs in the parotid salivary gland, as seen here, but it may also infect other salivary glands and other organs, including the gonads and the meninges. Infection typically occurs at 5 to 9 years of age. In postadolescent males, the mumps virus infecting the testes may cause sterility. There is an effective mumps vaccine, usually combined with the measles and rubella vaccines to form the MMR vaccine. It is administered to infants after the age of 15 months.

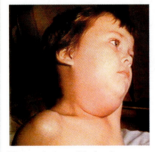

Mumps in parotid gland

Esophagus

Esophagitis (ē-sof-a-JĪ-tis) is an inflammation of the esophagus. This painful condition, which can be seen in this endoscopic view (internal image using an instrument called an endoscope) of the esophagus, usually results from stomach acids that leak through a weakened or permanently relaxed lower esophageal sphincter. This backflow, or **gastroesophageal reflux**, is responsible for the symptoms of heartburn.

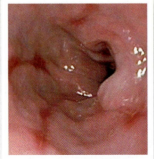

Esophagitis seen on endoscopy

Liver

Any condition that severely damages the liver is a threat to life. **Hepatitis**, an inflammation of the liver, can be caused by excessive alcohol consumption, drugs, or infection. **Cirrhosis** (sir-RŌ-sis) is a form of hepatitis characterized by the degeneration of liver cells and their replacement by fibrous connective tissue (a process called scarring). The surviving liver cells divide, but the fibrous tissue prevents normal tissue structure from being reestablished. As a result, liver function declines and a variety of other complications develop.

There are many different forms of viral hepatitis: The most common are **hepatitis A**, **B,** and **C**. The hepatitis viruses disrupt liver function by attacking and destroying liver cells. An infected individual may develop a high fever, and the liver may become inflamed and tender. In a condition called **jaundice** (JAWN-dis), the skin and eyes develop a yellow color because the bilirubin normally excreted in bile accumulates in body fluids.

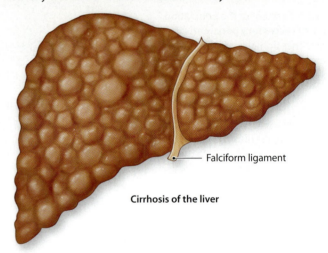

Falciform ligament

Cirrhosis of the liver

Gallbladder

If bile becomes too concentrated, crystals of insoluble minerals and salts called **gallstones** form. Small gallstones are not a problem if they can be flushed down the bile duct and excreted. In **cholecystitis** (kō-lē-sis-TĪ-tis; *chole*, bile + *kystis*, bladder + *itis*, inflammation), the gallstones become so large that they damage the wall of the gallbladder or block the cystic duct or common bile duct. In that case, the gallbladder may need to be surgically removed. This removal does not seriously impair digestion, because bile production continues at normal levels.

Gallstones

Stomach

Inflammation of the mucous membrane lining the stomach is called **gastritis** (gas-TRĪ-tis). This condition may develop after ingesting drugs, including aspirin and alcohol. It may also appear after severe emotional or physical stress, bacterial infection of the gastric wall, or the ingestion of strong chemicals. Gastritis may cause ulcers to form. A **peptic ulcer** develops when gastric enzymes and acids erode through the stomach or duodenal lining. A **gastric ulcer** is located in the stomach, and a **duodenal ulcer** is located in the duodenum. Infection by a bacterium, *Helicobacter pylori*, (HE-li-kō-bak-ter pī-LŌR-ī) is responsible for over 80 percent of peptic ulcers. Treatment for ulcers involves the administration of drugs, such as **cimetidine** (si-MET-i-dēn) (Tagamet), that inhibit acid production by gastric glands, combined with antibiotics if *Helicobacter pylori* is present.

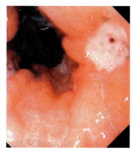

Gastric ulcer

Pancreas

Pancreatitis (pan-krē-a-TĪ-tis) is an inflammation of the pancreas. Causes of pancreatitis include gallstones that block the excretory ducts, viral infections, and toxic drugs, such as alcohol. Any of these stimuli may injure exocrine cells in a portion of the organ. Lysosomes then activate digestive enzymes within the cells, which begin to break down. In about one-eighth of the cases, death results when the process does not stop, and the released lysosomal enzymes destroy the pancreas.

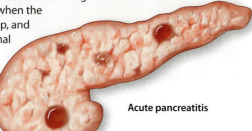

Acute pancreatitis

Small Intestine

Enteritis is inflammation of the intestine (usually applied to the small intestine). Enteritis typically causes watery bowel movements, or **diarrhea** (dī-a-RĒ-uh). One cause of diarrhea due to enteritis is the protozoan *Giardia lamblia* (shown here). **Dysentery** (dis-en-TER-ē) is inflammation of the small and large intestine that usually produces diarrhea containing blood and mucus. **Gastroenteritis** is inflammation of the stomach and the intestines due to bacterial, viral, protozoan, or parasitic worm infections. Most of these conditions are prevalent in areas that have poor sanitation and low water quality.

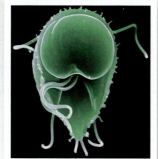

Giardia lamblia   SEM × 4000

Large Intestine

Colitis (kō-LĪ-tis) is a general term referring to inflammation of the colon, often involving diarrhea or constipation. Diarrhea results when the lining of the colon becomes unable to reabsorb normal amounts of water, or when so much fluid enters the colon that its water reabsorption capacity is exceeded. **Constipation** is infrequent bowel movement (or defecation), generally involving dry, hard feces. It results when fecal material moves through the colon so slowly that excessive water is reabsorbed.

Each year roughly 152,000 new cases of **colorectal cancer** are diagnosed in the United States, and 57,000 deaths result from this condition. Most common among persons over 50 years of age, risk factors for colorectal cancer include a diet rich in animal fats and low in fiber, and a number of inherited disorders that promote the formation of epithelial tumors along the intestines. Most colorectal cancers begin as small localized tumors, or **polyps** (POL-ips), that grow from the mucosa lining the intestinal wall. The prognosis improves dramatically if cancerous polyps are removed before metastasis has occurred.

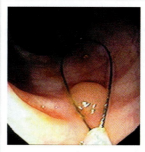

Colon polyp seen on colonoscopy

Module 15.16 Review

a. Describe periodontal disease.

b. Describe cholecystitis.

c. What bacterium is responsible for the vast majority of peptic ulcers?

1. Labeling

Label each of the anatomical structures in the following diagram.

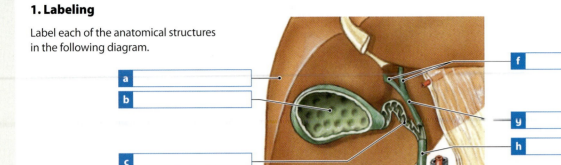

a _____

b _____

c _____

d _____

e _____

f _____

y _____

h _____

i _____

j _____

k _____

2. Matching

Match the following terms with the most closely related description.

- saliva
- pancreatic juice
- gallstones
- portal hypertension
- pancreatic lipase
- liver
- starch
- insulin
- hepatopancreatic sphincter
- salivary reflex
- gallbladder
- mumps
- common bile duct
- peptic ulcer

a	_____	Substrate of pancreatic alpha-amylase
b	_____	Regulates bile flow into duodenum
c	_____	Drains liver and gallbladder
d	_____	Contains antibodies that reduce bacterial populations
e	_____	Viral infection of salivary glands
f	_____	Condition caused by epithelial damage in the stomach or intestines by acids.
g	_____	Condition resulting from restricted blood flow through the liver
h	_____	Pancreatic enzyme that breaks down complex lipids
i	_____	Organ that synthesizes and secretes bile
j	_____	Endocrine secretion produced by the pancreas
k	_____	Exocrine secretion produced by the pancreas
l	_____	Stimulated by any object in the mouth
m	_____	Organ that stores bile
n	_____	Cause cholecystitis

3. Short answer

Describe the beneficial roles of saliva. _____

_____

4. Section integration

Predict the consequences of a tumor blocking the duodenal ampulla. _____

_____

Visual Outline with Key Terms

Summarize the content of each module using the terms in the order provided.

SECTION 1

General Organization of the Digestive System

- digestive system
- digestive tract
- gastrointestinal (GI) tract

15.1

The digestive tract is a muscular tube lined by a mucous membrane

- mesentery
- mucosa
- submucosa
- muscularis externa
- serosa
- secretory glands
- circular folds
- villi
- muscularis mucosae
- submucosal plexus
- myenteric plexus

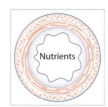

15.2

Smooth muscle contractions mix and propel materials through the digestive tract.

- bolus
- peristalsis
- segmentation

SECTION 2

The Digestive Tract

- oral cavity, teeth, tongue
- pharynx
- esophagus
- stomach
- small intestine
- large intestine
- ingestion
- mechanical processing
- digestion
- secretion
- absorption
- compaction
- feces
- defecation

15.3

The oral cavity contains the tongue, salivary glands, and teeth, and receives saliva from the salivary glands

- oral cavity
- oral mucosa
- hard palate
- soft palate
- uvula
- root (of the tongue)
- labia
- cheeks
- body (of the tongue)
- vestibule
- lingual frenulum
- pharyngeal arches
- tongue
- lingual lipase
- gingivae
- mastication

15.4

Teeth in different regions of the jaws vary in size, shape, and function

- dentin
- pulp cavity
- crown
- neck
- root
- alveolus
- enamel
- cementum
- periodontal ligament
- root canal
- incisors
- cuspids
- bicuspids
- molars
- deciduous teeth
- primary dentition
- secondary dentition
- wisdom teeth
- dental arcades
- tooth decay
- dental plaque
- gingivitis

15.5

The muscular walls of the pharynx and esophagus play a key role in swallowing

- pharynx
- esophagus
- adventitia
- deglutition
- buccal phase
- pharyngeal phase
- esophageal phase
- secondary peristaltic waves
- esophageal hiatus
- lower esophageal sphincter

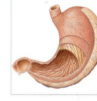

15.6

The stomach is a muscular, expandable, J-shaped organ with three layers in its muscularis externa

- chyme
- lesser curvature
- greater curvature
- fundus
- cardia
- body
- pylorus
- pyloric antrum
- pyloric canal
- pyloric sphincter
- rugae

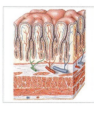

15.7

The stomach breaks down the organic nutrients in food

- mucosa
- submucosa
- muscularis externa
- serosa
- gastric glands
- gastric juice
- gastric pits
- parietal cells
- intrinsic factor
- enteroendocrine cells
- chief cells
- pepsinogen
- pepsin
- rennin
- gastric lipase

• = *Term boldfaced in this module*

15.8

The intestinal tract is specialized for absorbing nutrients

- plicae circulares
- intestinal villi
- intestinal glands
- lacteal
- brush border

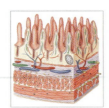

15.9

The small intestine is divided into the duodenum, jejunum, and ileum

- duodenum
- jejunum
- ileum
- ileocecal valve

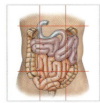

15.10

Five hormones are involved in regulating digestive activities

- gastrin
- secretin
- gastric inhibitory peptide (GIP)
- cholecystokinin (CCK)
- vasoactive intestinal peptide (VIP)

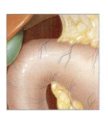

15.11

The large intestine stores and concentrates fecal material

- large intestine
- cecum
- compaction
- ileocecal valve
- appendix
- appendicitis
- ascending colon
- right colic flexure
- colon
- transverse colon
- left colic flexure
- fatty appendices
- descending colon
- sigmoid flexure
- taeniae coli
- haustra
- sigmoid colon
- rectum
- hemorrhoids
- mass movements

15.12

Neural reflexes and hormones work together to control digestive activities

- ○ gastrin
- enterogastric reflex
- defecation reflex

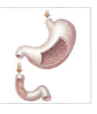

SECTION 3

Accessory Digestive Organs

- ○ salivary glands
- ○ liver
- ○ gallbladder
- ○ pancreas

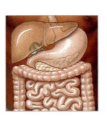

15.13

Secretions from three pairs of salivary glands produce saliva

- salivary reflex
- sublingual ducts
- submandibular duct
- sublingual salivary glands
- submandibular salivary glands
- salivary amylase
- parotid salivary glands
- parotid duct
- salivary reflex

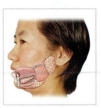

15.14

The liver, the largest visceral organ, is divided into left, right, caudate, and quadrate lobes

- liver
- round ligament
- ○ right lobe
- ○ left lobe
- caudate lobe
- quadrate lobe
- common bile duct
- ○ gallbladder
- portal hypertension

15.15

The gallbladder stores and concentrates bile, and the pancreas has vital endocrine and exocrine functions

- gallbladder
- right and left hepatic ducts
- common hepatic duct
- cystic duct
- common bile duct
- duodenal ampulla
- duodenal papilla
- hepatopancreatic sphincter
- pancreas
- pancreatic duct
- pancreatic juice
- pancreatic lobules
- accessory pancreatic duct
- pancreatic alpha-amylase
- pancreatic lipase
- nucleases
- proteolytic enzymes

15.16

Disorders of the digestive system are diverse and relatively common

- periodontal disease
- gingivitis
- mumps virus
- mumps
- esophagitis
- gastroesophageal reflux
- hepatitis
- cirrhosis
- hepatitis A, B, and C
- jaundice
- gallstones
- cholecystitis
- gastritis
- peptic ulcer
- gastric ulcer
- duodenal ulcer
- cimetidine
- pancreatitis
- enteritis
- diarrhea
- dysentery
- gastroenteritis
- colitis
- constipation
- colorectal cancer
- polyps

• = *Term boldfaced in this module*

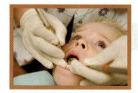

CAREER PATHS

Dental Hygienist

"The mouth is kind of a window to your entire body," says dental hygienist Mary Cattadoris. "I can look at people's teeth and I can tell if they clench or grind from stress." She once had a patient whose mouth looked abnormal. Mary recommended that the patient see a physician—it turned out she had leukemia.

Of the patients she sees each day, most of Mary's work is preventive: scaling for tartar, cleanings, fluoride treatments, sealants, x-rays, whitening, and, more recently, laser periodontal therapy, which is a specialty in which she had to become certified. She does an oral health assessment and cancer screening of each patient, then works out a treatment plan. If the patient requires more than just regular cleanings, she refers him or her to the dentist in the practice, or sometimes directly to an orthodontist.

A particular passion of hers within the job is education: teaching her patients that good oral health involves more than just brushing your teeth twice a day. "It always shocks me to recognize how little people know when it comes to disease prevention and diet," Mary says. "There are areas in the world where people have never seen a dentist but they are cavity-free because they don't eat processed sugar. Some very smart people don't have a good dental IQ, and it's exciting to change their thought process."

"Knowledge of anatomy and physiology is vital," Mary says, noting that her education included an entire semester of head and neck anatomy. "You need to know what normal looks like," she says. "You can recognize when things are not healthy, and you can compliment patients on what they're doing well."

For more information, visit the website of the American Dental Hygienists Association at http:// www.adha.org.

Think this is the career for you?
Key Stats:

- **Education and Training.** A degree from an accredited dental hygiene school is required. Most programs offer an associate's degree, although some offer a certificate, a BS, or an MS.

- **Licensure.** All states require dental hygienists to be licensed, and nearly all states require candidates to pass a written and clinical examination.

- **Earnings.** Earnings vary but the median annual wage is $68,250.

- **Expected Job Prospects.** Employment is expected to grow faster than the national average—36% from 2008 to 2018.

Bureau of Labor Statistics, U.S. Department of Labor, *Occupational Outlook Handbook, 2010–11 Edition,* Dental Hygienists, on the Internet at http://www.bls.gov/oco/ocos097 .htm (visited September 14, 2011).

MasteringA&P®

Access more review material online in the Study Area at www.masteringaandp.com.

- Chapter guides
- Chapter quizzes
- Practice tests
- Art labeling activities
- Flashcards
- A glossary with pronunciations
- Practice Anatomy Lab™ (PAL™) 3.0 virtual anatomy practice tool
- Interactive Physiology® (IP) animated tutorials
- MP3 Tutor Sessions

PAL practice anatomy lab™

For this chapter, follow these navigation paths in PAL:

- Human Cadaver>Digestive System
- Anatomical Models>Digestive System
- Histology>Digestive System

 For this chapter, go to these topics in the Digestive System in IP:

- Orientation
- Anatomy Review
- Motility
- Secretion
- Digestion and Absorption

Chapter 15 Review Questions

1. How does the stomach promote and assist in the digestive process?

2. Explain why some patients with gallstones develop pancreatitis.

3. Barb suffers from Crohn's disease, a regional inflammation of the intestine that is thought to have some genetic basis, although the actual cause remains unknown. When the disease flares up, she experiences abdominal pain, weight loss, and anemia. Which parts of the intestine are probably involved, and what are the causes of her signs and symptoms?

4. Through which layers of a molar would an oral surgeon drill to perform a root canal (removal of the nerves, blood vessels, and pulp tissue in a severely infected tooth)?

5. Recently, more people have turned to surgery to help them lose weight. One form of weight control surgery involves stapling a portion of the stomach shut, creating a smaller volume. How would such a surgery result in weight loss?

For answers to all module, section, and chapter review questions, see the blue Answers tab at the back of the book.

16

Metabolism and Energetics

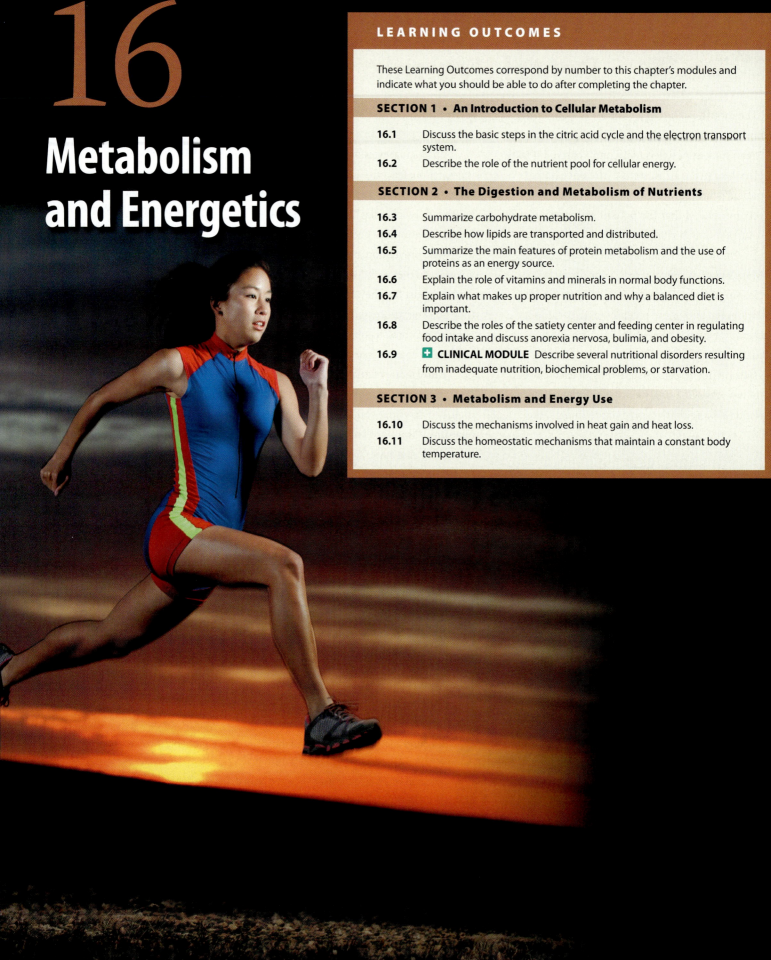

An Introduction to Cellular Metabolism

The term **metabolism** (me-TAB-ō-lizm) refers to all the chemical reactions that occur in an organism. Chemical reactions within cells, collectively known as **cellular metabolism**, provide the energy needed to maintain homeostasis and perform essential functions. **Catabolism** is the breakdown of organic substances in the body, whereas **anabolism** is the synthesis of new organic molecules (as noted in Chapter 2).

An Overview of Cellular Metabolism

In the process of **metabolic turnover**, cells continuously break down and replace all their organic components except DNA. Catabolism releases energy in the form of ATP. All of the cell's organic building blocks form a **nutrient pool**—an accessible reserve of organic substrates that can be used for metabolic turnover or energy production.

Cells continuously absorb organic molecules from the surrounding interstitial fluids, supplementing those broken down through catabolism. The components of the nutrient pool can be either used for anabolism or broken down further to produce ATP. This section considers the origins and fate of the nutrient pool, beginning with an overview of the catabolic pathways that provide ATP.

In mitochondria, catabolism releases significant amounts of energy. Roughly 40 percent of the energy released in these catabolic reactions can be captured and used to convert ADP to ATP. The other 60 percent escapes as heat that warms the interior of the cell and the surrounding tissues.

The mitochondria produce ATP for anabolism and other cell functions.

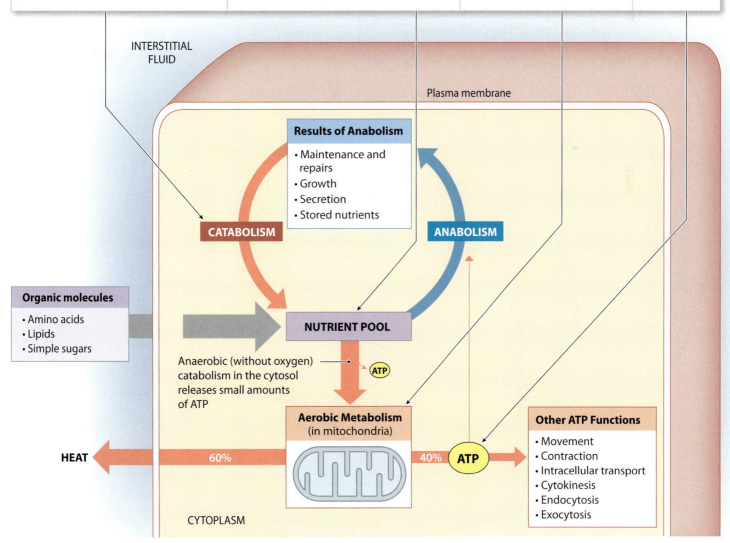

Cells obtain most of their ATP from the electron transport system, which is linked to the citric acid cycle

1 Mitochondria provide almost all of the energy that supports cellular operations. The cell "feeds" its mitochondria from its nutrient pool, and in return, the cell gets the ATP it needs. However, mitochondria are picky eaters: They will accept only specific organic molecules for processing and energy production.

Start Chemical reactions in the cytoplasm take organic nutrients from the nutrient pool and break them down into smaller fragments that the mitochondria can process. Once absorbed by mitochondria, it doesn't matter whether these fragments were obtained from a carbohydrate, a lipid, or an amino acid. They are all broken down through the catabolic pathway shown here.

1 The common substrate for mitochondria is a 2-carbon molecule of **acetate** attached to coenzyme A to form **acetyl-CoA**. **Coenzymes** are organic molecules that must bind to enzymes to make the enzymes functional. They also act as hydrogen carriers. Many coenzymes are derived from organic substances in food called vitamins.

2 Coenzyme A is released in a reaction that produces **citric acid**. This is the first step in what is known as the **citric acid cycle**, or Kreb's cycle.

3 The citric acid cycle removes hydrogen atoms from organic molecules and transfers them to coenzymes. In addition, it produces a single molecule of ATP for every turn in the cycle.

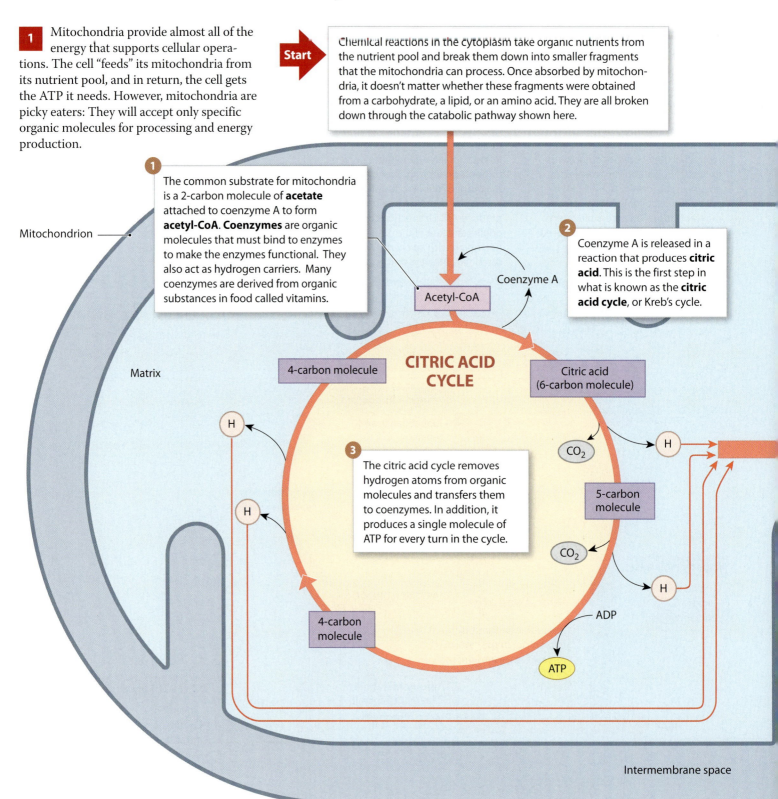

Mitochondrion

Matrix

Coenzyme A

Acetyl-CoA

CITRIC ACID CYCLE

4-carbon molecule

Citric acid (6-carbon molecule)

CO_2

H

5-carbon molecule

CO_2

H

4-carbon molecule

ADP

ATP

H

H

Intermembrane space

2 Within each mitochondrion, the hydrogen atoms are used to generate ATP through the process of **oxidative phosphorylation**. *Oxidation* refers to the transfer of electrons. *Phosphorylation* refers to the attachment of a high-energy phosphate group to ADP, producing ATP. Oxidative phosphorylation requires H atoms from the citric acid cycle, coenzymes, special electron transfer molecules, and oxygen to generate ATP. It involves the transfer of electrons in a series such that each step releases a small amount of energy. In contrast, the release of energy in one step would result in the destruction of a cell. Oxidative phosphorylation produces roughly 95 percent of the ATP used by body cells.

4 Oxidative Phosphorylation

Coenzymes deliver hydrogen atoms with high-energy electrons from the citric acid cycle to the **electron transport system (ETS)**. The ETS generates ATP, consumes oxygen, and produces water.

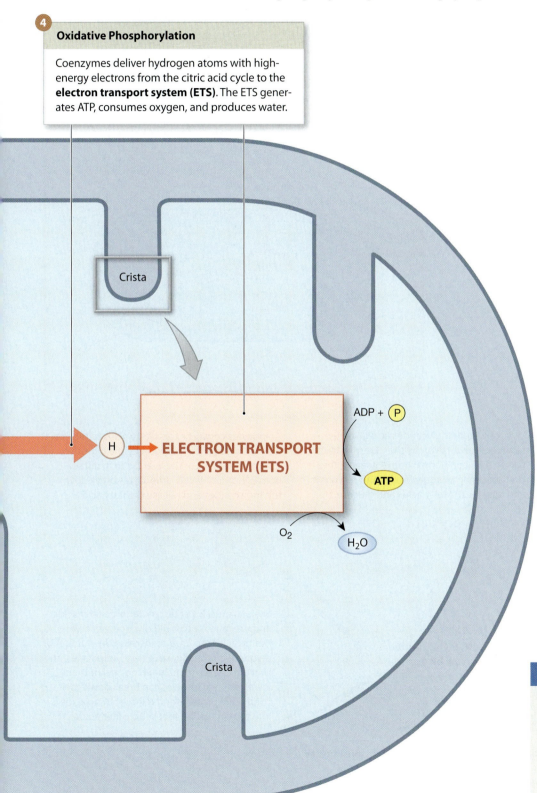

Mitochondria are very efficient but aerobic metabolism depends on oxygen. Oxygen is needed because it is the final acceptor of the electrons transferred along the ETS. After accepting electrons from the ETS, the oxygen joins with hydrogen ions (protons) to form water. However, if the supply of oxygen is restricted, then the transfer of electrons cannot continue and the ETS stops working. When the ETS stops, the citric acid cycle halts as well. If the problem persists, the cell will die because it lacks the ATP to support vital functions. Poisons like cyanide prevent the flow of electrons to oxygen. This has the same effect on the cell as oxygen starvation.

Module 16.1 Review

a. Briefly describe the citric acid cycle and explain its role in cellular metabolism.

b. Explain the role of the ETS in cellular metabolism.

c. Explain the link between the ETS and oxygen.

Cells can break down any available substrate from the nutrient pool to obtain the energy they need

The nutrient pool is the source of the substrates for both catabolism and anabolism. Cells tend to conserve the materials needed to build new compounds and break down the excess.

1 The nutrient pool is the key to a cell's survival because it contains the organic materials required for both anabolism and catabolism. Cells must continuously replace membranes, organelles, enzymes, and structural proteins as well as perform their specialized functions. All of these functions require ATP, so the cell's mitochondria must have a continuous supply of two-carbon substrate molecules. In this diagram, the thickness of the arrow indicates the relative importance of that pathway in an inactive cell.

The bloodstream distributes organic compounds for cells throughout the body to absorb.

KEY

= Catabolic pathway

= Anabolic pathway

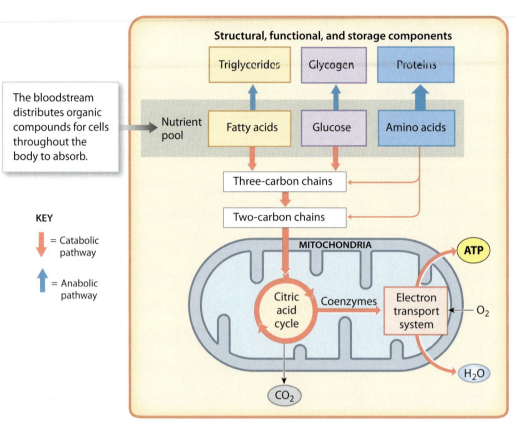

Structural, functional, and storage components

Triglycerides Glycogen Proteins

Nutrient pool

Fatty acids Glucose Amino acids

Three-carbon chains

Two-carbon chains

MITOCHONDRIA

ATP

Citric acid cycle Coenzymes Electron transport system

O_2

H_2O

CO_2

2 The nutrient pool of each cell contributes to the metabolic reserves of the body as a whole. When absorption across the digestive tract is insufficient to maintain normal nutrient levels in the blood, the body mobilizes those reserves.

Liver cells store triglycerides and glycogen reserves. If absorption by the digestive tract fails to maintain normal nutrient levels, the liver breaks down triglycerides and glycogen and releases the fatty acids and glucose.

Adipocytes convert excess fatty acids to triglycerides for storage. If absorption by the digestive tract and reserves in the liver fail to maintain normal nutrient levels, adipocytes break down the triglycerides and release the fatty acids.

Nutrients obtained through digestion and absorption

Nutrients distributed in the blood

Skeletal muscles at rest metabolize fatty acids and use glucose to build glycogen reserves. Amino acids are used to increase the number of myofibrils. If the digestive tract, adipocytes, and liver are unable to maintain normal nutrient levels, skeletal muscles can break down their contractile proteins and release amino acids into the circulation for use by other tissues.

Neural tissue requires a continuous supply of glucose. During starvation, other tissues shift to fatty acid or amino acid catabolism, conserving glucose for neural tissue.

Cells in most tissues continuously absorb and catabolize glucose.

3 Neither the diet nor the nutrient pool provides everything needed to build every protein, carbohydrate, and lipid a cell might require. With few exceptions, this is not a problem because the cell also contains the enzymes necessary to synthesize what it needs from available substrates. This diagram summarizes the catabolic and anabolic pathways involved.

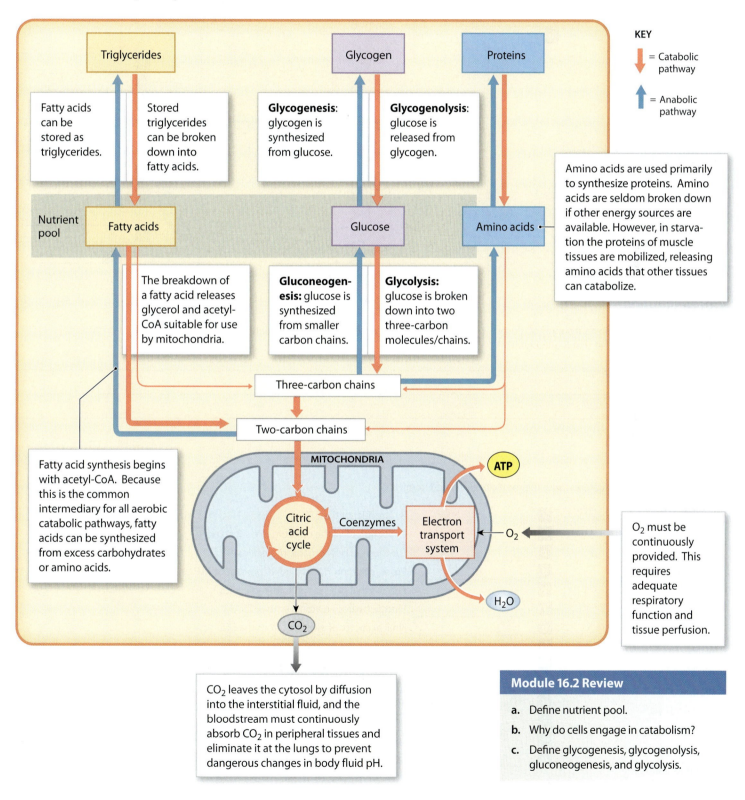

KEY

= Catabolic pathway

= Anabolic pathway

Triglycerides

Glycogen

Proteins

Fatty acids can be stored as triglycerides.

Stored triglycerides can be broken down into fatty acids.

Glycogenesis: glycogen is synthesized from glucose.

Glycogenolysis: glucose is released from glycogen.

Amino acids are used primarily to synthesize proteins. Amino acids are seldom broken down if other energy sources are available. However, in starvation the proteins of muscle tissues are mobilized, releasing amino acids that other tissues can catabolize.

Nutrient pool

Fatty acids

Glucose

Amino acids

The breakdown of a fatty acid releases glycerol and acetyl-CoA suitable for use by mitochondria.

Gluconeogenesis: glucose is synthesized from smaller carbon chains.

Glycolysis: glucose is broken down into two three-carbon molecules/chains.

Three-carbon chains

Two-carbon chains

Fatty acid synthesis begins with acetyl-CoA. Because this is the common intermediary for all aerobic catabolic pathways, fatty acids can be synthesized from excess carbohydrates or amino acids.

MITOCHONDRIA

ATP

Citric acid cycle

Coenzymes

Electron transport system

O_2

O_2 must be continuously provided. This requires adequate respiratory function and tissue perfusion.

H_2O

CO_2

CO_2 leaves the cytosol by diffusion into the interstitial fluid, and the bloodstream must continuously absorb CO_2 in peripheral tissues and eliminate it at the lungs to prevent dangerous changes in body fluid pH.

Module 16.2 Review

a. Define nutrient pool.

b. Why do cells engage in catabolism?

c. Define glycogenesis, glycogenolysis, gluconeogenesis, and glycolysis.

1. Matching

Use the following terms to fill in the blanks in the cellular metabolism figure to the right.

- glucose
- electron transport system
- O_2
- fatty acids
- proteins
- citric acid cycle
- ATP
- CO_2
- H_2O
- coenzymes
- two-carbon chains

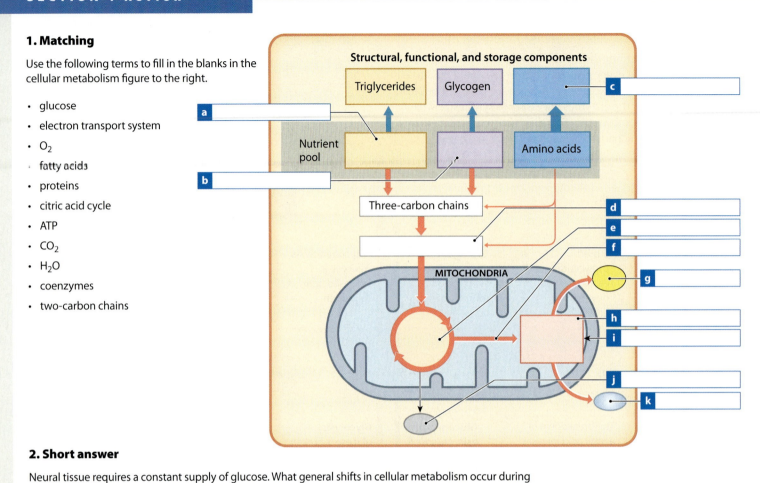

Structural, functional, and storage components

Triglycerides   Glycogen   **c**

a

Nutrient pool   Amino acids

b

Three-carbon chains   **d**

e

f

MITOCHONDRIA   **g**

h

i

j

k

2. Short answer

Neural tissue requires a constant supply of glucose. What general shifts in cellular metabolism occur during fasting or starvation to meet that requirement?

_____

_____

_____

3. Matching

Match each of the following terms with the most closely related item.

- coenzymes
- glycogenolysis
- glycogenesis
- nutrients scarce
- nutrient pool
- anabolism
- triglycerides
- water
- acetyl CoA
- oxygen
- citric acid cycle
- catabolism
- oxidative phosphorylation
- nutrients abundant

a	_____	By-product of the ETS
b	_____	Collection of all the cell's organic substances
c	_____	Synthesis of new organic molecules
d	_____	Process that produces over 90 percent of ATP used by body cells
e	_____	Process that releases glucose molecules
f	_____	Shuttle hydrogen atoms to the ETS
g	_____	Final acceptor of electrons from the ETS
h	_____	Breakdown of organic molecules
i	_____	Condition in which cells preferentially break down carbohydrates
j	_____	Fatty acid reserves
k	_____	Reaction sequence in which mitochondrial CO_2 production occurs
l	_____	Process that synthesizes glycogen from glucose molecules
m	_____	Condition in which cells preferentially break down lipids
n	_____	Common substrate for mitochondrial ATP production

The Digestion and Metabolism of Nutrients

Nutrients are essential elements in food required for normal physiology. **Metabolites** (meh-TAB-eh-līt; *metabole*, change), a much larger group, include all the molecules (nutrients included) that can be synthesized or broken down by chemical reactions inside our bodies. The food we eat has an organized physical structure, and the organic compounds it contains are large and often insoluble. During digestion the physical structure is broken down, and a combination of chemical and enzymatic attack breaks the complex organic compounds into simpler components that can be absorbed by the digestive tract and subsequently distributed by the bloodstream. Cells throughout the body rely on the availability of these organic molecules for energy production and to replenish the intracellular nutrient pool.

This section will provide an overview of the digestion, absorption, and fates of carbohydrates, lipids, and proteins. The modules will emphasize general patterns that will make it easier for you to understand the specifics presented in biochemistry or nutrition courses. We will focus attention on how nutrients are absorbed, stored or interconverted, or catabolized to yield the energy needed to support vital activities. (The major anabolic pathways leading to the synthesis of carbohydrates, lipids, and proteins are not discussed here; they were considered in Chapter 2.)

Steps in the Process of Digestion

In the oral cavity, saliva dissolves some organic nutrients, and mechanical processing by the teeth and tongue disrupts the physical structure of the material and provides access for digestive enzymes. Those enzymes begin the digestion of complex carbohydrates (polysaccharides) and lipids.

In the stomach, the material is further broken down physically and chemically by stomach acid and by enzymes that can operate at an extremely low pH. These enzymes begin the digestion of protein.

In the duodenum, buffers from the pancreas and liver moderate the pH of the arriving chyme, and various digestive enzymes are secreted by the pancreas that catalyze the catabolism of carbohydrates, lipids, proteins, and nucleic acids.

Nutrient absorption then occurs in the small intestine, primarily in the jejunum, and the nutrients enter the bloodstream.

Indigestible materials and wastes enter the large intestine, where water is reabsorbed and bacterial action generates both organic nutrients and vitamins. These organic products are absorbed before the residue is ejected at the anus.

Most of the nutrients absorbed by the digestive tract end up in a tributary of the hepatic portal vein that transports them into the liver. The liver absorbs nutrients as needed to maintain normal levels in the systemic circuit.

Within peripheral tissues, cells absorb the nutrients needed to maintain their nutrient pool and ongoing operations.

Carbohydrates are usually the preferred substrates for catabolism and ATP production under resting conditions

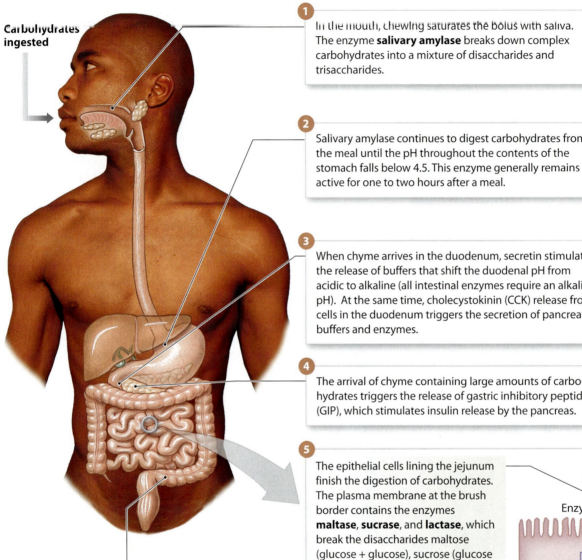

Carbohydrates ingested

1 In the mouth, chewing saturates the bolus with saliva. The enzyme **salivary amylase** breaks down complex carbohydrates into a mixture of disaccharides and trisaccharides.

2 Salivary amylase continues to digest carbohydrates from the meal until the pH throughout the contents of the stomach falls below 4.5. This enzyme generally remains active for one to two hours after a meal.

3 When chyme arrives in the duodenum, secretin stimulates the release of buffers that shift the duodenal pH from acidic to alkaline (all intestinal enzymes require an alkaline pH). At the same time, cholecystokinin (CCK) release from cells in the duodenum triggers the secretion of pancreatic buffers and enzymes.

4 The arrival of chyme containing large amounts of carbohydrates triggers the release of gastric inhibitory peptide (GIP), which stimulates insulin release by the pancreas.

5 The epithelial cells lining the jejunum finish the digestion of carbohydrates. The plasma membrane at the brush border contains the enzymes **maltase**, **sucrase**, and **lactase**, which break the disaccharides maltose (glucose + glucose), sucrose (glucose + fructose), and lactose (glucose + galactose) into simple sugars that are then absorbed. As these enzymes function, they transport the monosaccharides across the plasma membrane and release them into the cytosol.

The simple sugars that are transported into the cell at its apical surface diffuse through the cytoplasm and reach the interstitial fluid by facilitated diffusion. These monosaccharides then diffuse into the capillaries of the intestinal villi for eventual transport to the liver by the hepatic portal vein.

6 Intestinal enzymes do not alter indigestible carbohydrates such as cellulose, so they arrive in the colon virtually intact. This provides a reliable nutrient source for colonic bacteria, whose metabolic activities generate small quantities of **flatus**, or intestinal gas. Foods containing large amounts of indigestible carbohydrates (such as beans) stimulate bacterial gas production, distending the colon and causing cramps and the frequent discharge of intestinal gases.

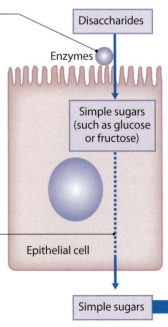

Hepatic portal vein

Disaccharides

Enzymes

Simple sugars (such as glucose or fructose)

Epithelial cell

Simple sugars

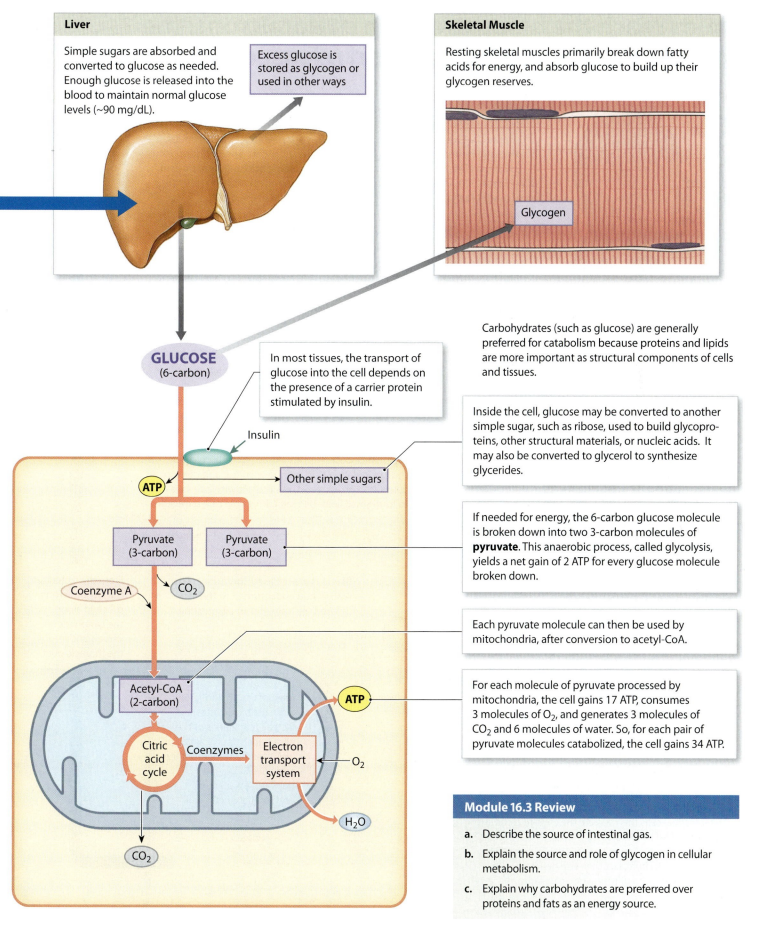

Liver

Simple sugars are absorbed and converted to glucose as needed. Enough glucose is released into the blood to maintain normal glucose levels (~90 mg/dL).

Excess glucose is stored as glycogen or used in other ways

Skeletal Muscle

Resting skeletal muscles primarily break down fatty acids for energy, and absorb glucose to build up their glycogen reserves.

Glycogen

GLUCOSE
(6-carbon)

In most tissues, the transport of glucose into the cell depends on the presence of a carrier protein stimulated by insulin.

Carbohydrates (such as glucose) are generally preferred for catabolism because proteins and lipids are more important as structural components of cells and tissues.

Insulin

ATP

Other simple sugars

Inside the cell, glucose may be converted to another simple sugar, such as ribose, used to build glycoproteins, other structural materials, or nucleic acids. It may also be converted to glycerol to synthesize glycerides.

Pyruvate
(3-carbon)

Pyruvate
(3-carbon)

Coenzyme A

CO_2

If needed for energy, the 6-carbon glucose molecule is broken down into two 3-carbon molecules of **pyruvate**. This anaerobic process, called glycolysis, yields a net gain of 2 ATP for every glucose molecule broken down.

Each pyruvate molecule can then be used by mitochondria, after conversion to acetyl-CoA.

Acetyl-CoA
(2-carbon)

ATP

Citric acid cycle

Coenzymes

Electron transport system

O_2

H_2O

CO_2

For each molecule of pyruvate processed by mitochondria, the cell gains 17 ATP, consumes 3 molecules of O_2, and generates 3 molecules of CO_2 and 6 molecules of water. So, for each pair of pyruvate molecules catabolized, the cell gains 34 ATP.

Module 16.3 Review

a. Describe the source of intestinal gas.

b. Explain the source and role of glycogen in cellular metabolism.

c. Explain why carbohydrates are preferred over proteins and fats as an energy source.

Lipids reach the bloodstream in chylomicrons; the cholesterol is then extracted and released as lipoproteins

Lipids ingested

1 Mechanical processing in the mouth breaks food into smaller chunks and disrupts connective tissue organization. As this is under way, the bolus becomes saturated with saliva containing a **lipase**. This enzyme attacks triglycerides, breaking them down into monoglycerides and fatty acids. Body cells cannot build every fatty acid they can break down. **Linolenic acid** and **linoleic acid**, which are required to synthesize and maintain plasma membranes, are called **essential fatty acids** because they must be included in the diet.

2 Most dietary lipids are not water soluble. The mixing of chyme in the stomach creates large drops containing a variety of lipids. Lipase continues to function in the acid environment, but can only attack triglycerides at the surfaces of these drops. As a result, only about 20 percent of the triglycerides have been broken down by the time the chyme leaves the stomach.

3 When chyme arrives in the duodenum, CCK is released, triggering the secretion of pancreatic enzymes, including another lipase, and also stimulating the gallbladder to contract and eject bile into the duodenum. Bile contains **bile salts** that break the large lipid drops into tiny droplets, a process called **emulsification**, which provides better access for pancreatic lipase. Pancreatic lipase then breaks apart the triglycerides to form a mixture of fatty acids, monoglycerides, and glycerol. As these molecules are released, they interact with bile salts in the lumen to form small lipid–bile salt complexes called **micelles** (mī-SELZ).

Thoracic duct

4 When a micelle contacts the intestinal epithelium, the lipids diffuse across the plasma membrane and enter the cytoplasm.

The intestinal cells synthesize new triglycerides from the monoglycerides, fatty acids, and glycerol. These triglycerides, in company with absorbed cholesterol, phospholipids, and other lipid-soluble materials, are then coated with proteins, creating complexes known as **chylomicrons** (kī-lō-MĪ-kronz; chylos, milky lymph; mikros, small). Chylomicrons are **lipoproteins**—lipid-protein complexes that contain insoluble lipids. The superficial coating of phospholipids and proteins makes the entire complex water soluble.

The intestinal cells then secrete the chylomicrons into interstitial fluid by exocytosis. The superficial protein coating of the chylomicrons keeps them suspended in the interstitial fluid, but their size generally prevents them from diffusing into capillaries.

Bile salts are released; most are reabsorbed and returned to the liver before they leave the small intestine.

Micelle

Monoglycerides, fatty acids

Triglycerides + other lipids and proteins

Exocytosis

Chylomicron

Most of the chylomicrons diffuse into the intestinal lacteals, capillaries of the lymphatic system, which lack basement membranes and have large gaps between adjacent endothelial cells.

5 From the lacteals, the chylomicrons proceed along the lymphatic vessels and into the thoracic duct.

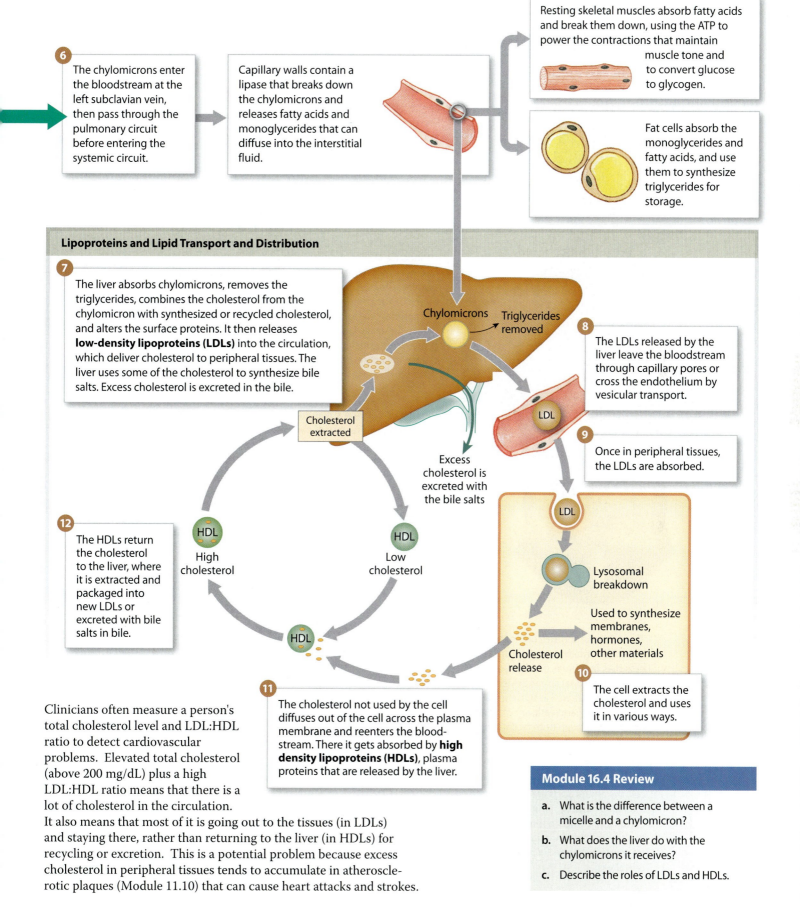

6 The chylomicrons enter the bloodstream at the left subclavian vein, then pass through the pulmonary circuit before entering the systemic circuit.

Capillary walls contain a lipase that breaks down the chylomicrons and releases fatty acids and monoglycerides that can diffuse into the interstitial fluid.

Resting skeletal muscles absorb fatty acids and break them down, using the ATP to power the contractions that maintain muscle tone and to convert glucose to glycogen.

Fat cells absorb the monoglycerides and fatty acids, and use them to synthesize triglycerides for storage.

Lipoproteins and Lipid Transport and Distribution

7 The liver absorbs chylomicrons, removes the triglycerides, combines the cholesterol from the chylomicron with synthesized or recycled cholesterol, and alters the surface proteins. It then releases **low-density lipoproteins (LDLs)** into the circulation, which deliver cholesterol to peripheral tissues. The liver uses some of the cholesterol to synthesize bile salts. Excess cholesterol is excreted in the bile.

Chylomicrons — Triglycerides removed

Cholesterol extracted

Excess cholesterol is excreted with the bile salts

8 The LDLs released by the liver leave the bloodstream through capillary pores or cross the endothelium by vesicular transport.

LDL

9 Once in peripheral tissues, the LDLs are absorbed.

LDL

Lysosomal breakdown

Used to synthesize membranes, hormones, other materials

Cholesterol release

10 The cell extracts the cholesterol and uses it in various ways.

12 The HDLs return the cholesterol to the liver, where it is extracted and packaged into new LDLs or excreted with bile salts in bile.

HDL — High cholesterol

HDL — Low cholesterol

HDL

11 The cholesterol not used by the cell diffuses out of the cell across the plasma membrane and reenters the bloodstream. There it gets absorbed by **high density lipoproteins (HDLs)**, plasma proteins that are released by the liver.

Clinicians often measure a person's total cholesterol level and LDL:HDL ratio to detect cardiovascular problems. Elevated total cholesterol (above 200 mg/dL) plus a high LDL:HDL ratio means that there is a lot of cholesterol in the circulation. It also means that most of it is going out to the tissues (in LDLs) and staying there, rather than returning to the liver (in HDLs) for recycling or excretion. This is a potential problem because excess cholesterol in peripheral tissues tends to accumulate in atherosclerotic plaques (Module 11.10) that can cause heart attacks and strokes.

Module 16.4 Review

a. What is the difference between a micelle and a chylomicron?

b. What does the liver do with the chylomicrons it receives?

c. Describe the roles of LDLs and HDLs.

An amino acid not needed for protein synthesis may be broken down or converted to a different amino acid

1 Proteins have very complex structures, so protein digestion is both complex and time-consuming.

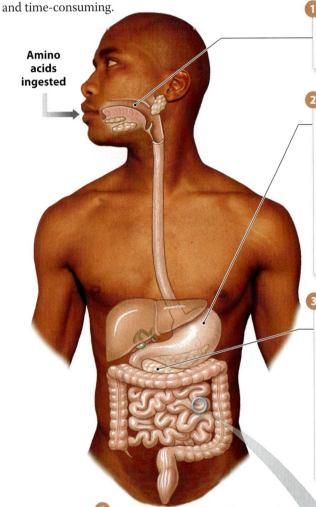

Amino acids ingested

1 The first step is to disrupt the tough three-dimensional structure of the food. This begins in the mouth as the food is thoroughly chewed and mixed with saliva.

2 Additional mechanical processing occurs in the stomach through churning and mixing. Exposure of the bolus to a strongly acidic environment kills pathogens and breaks down connective tissues and plant cell walls. Stomach acids also denature most proteins and disrupt tertiary and secondary protein structure. This exposes peptide bonds to enzymatic attack by the proteolytic enzyme **pepsin**, which is secreted by the parietal cells as the inactive proenzyme pepsinogen. Pepsin generally has enough time to break down complex proteins into smaller peptide and polypeptide chains before the chyme enters the duodenum.

Hepatic portal vein

3 When acidic chyme arrives in the duodenum, CCK stimulates production and release of pancreatic enzymes. These enzymes are secreted as inactive proenzymes that are then activated within the duodenum. **Enteropeptidase**, an enzyme released by the duodenal epithelium, begins the process, converting the proenzyme trypsinogen to the proteolytic enzyme **trypsin**. Trypsin then converts the other proenzymes to yield **chymotrypsin, carboxypeptidase**, and **elastase**. Each enzyme attacks peptide bonds linking specific amino acids and ignores others. Together, they break down proteins into a mixture of dipeptides, tripeptides, and amino acids.

4 The epithelial surfaces of the small intestine contain several **peptidases**, enzymes that break peptide bonds and release individual amino acids. These amino acids, as well as those released by the action of pancreatic enzymes, are absorbed through both facilitated diffusion and cotransport mechanisms.

After diffusing to the basal surface of the cell, the amino acids are released into interstitial fluid by facilitated diffusion and cotransport. Once in the interstitial fluid, the amino acids enter intestinal capillaries for transport to the liver by the hepatic portal vein.

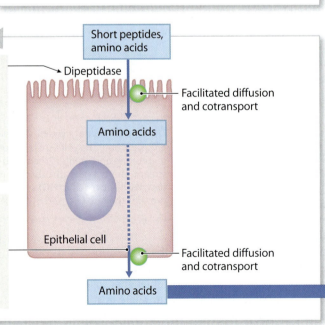

Short peptides, amino acids

Dipeptidase

Facilitated diffusion and cotransport

Amino acids

Epithelial cell

Facilitated diffusion and cotransport

Amino acids

5

The liver does not control circulating levels of amino acids as precisely as it does glucose concentrations. The liver itself uses many amino acids to synthesize plasma proteins, and it has all of the enzymes needed to synthesize, convert, or catabolize amino acids. Amino acids that can be broken down to 3-carbon molecules can be used for gluconeogenesis when other sources of glucose are unavailable.

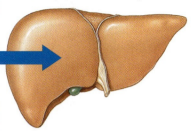

Amino Acid Synthesis

Liver cells and other body cells can readily produce half of the total 20 amino acids needed to synthesize their proteins. Because these 10 amino acids do not have to be obtained in the diet, they are known as **nonessential amino acids**. In contrast, **essential amino acids** cannot be synthesized at all by the body (or cannot be made in amounts sufficient to meet physiological needs) and so must be obtained in the diet.

Meat, eggs, and dairy foods are good sources of protein

2 The diagram below shows how your body uses the organic components of food under resting conditions.

Carbohydrates, fats, and proteins in food

↓

Digested to simple sugars, fatty acids, and amino acids that can be absorbed and distributed in the bloodstream

↓

Absorbed by cells throughout the body

↓

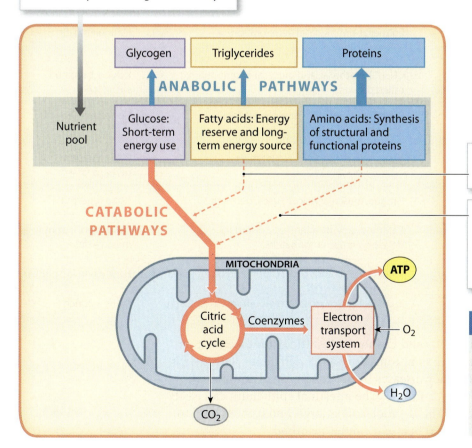

Most fatty acids are broken down to acetyl-CoA and processed by the citric acid cycle.

A cell surviving by amino acid catabolism is like someone in a winter cabin burning the walls to stay warm. It can help temporarily, but it isn't a permanent solution, and in the long run it only makes matters worse.

Module 16.5 Review

a. What is the role of enteropeptidase?

b. What is the difference between essential amino acids and nonessential amino acids?

c. What happens to nutrients in the nutrient pool under resting conditions?

Vitamins and minerals are essential to the function of many metabolic pathways

The ingestion and metabolism of nutrients from food is called **nutrition**. **Vitamins** are organic compounds required in very small quantities that play an essential role in specific metabolic pathways. **Minerals** are inorganic ions released through the dissociation of electrolytes.

Water-Soluble Vitamins

1 The **water-soluble vitamins** are the B vitamins and vitamin C. Most of them are components of coenzymes. The B vitamins are found in meat, eggs, and dairy products, while citrus fruits are good sources of vitamin C.

2 The table below summarizes information about water-soluble vitamins. These water-soluble vitamins are rapidly exchanged between the fluid compartments of the digestive tract and the circulating blood, and excessive amounts are readily excreted in urine. For this reason, conditions involving an excess of water-soluble vitamins is unlikely unless you take megadoses of vitamin supplements.

The Water-Soluble Vitamins

Vitamin	Significance	Effects of Deficiency	Effects of Excess
B_1 (thiamine)	Coenzyme in many pathways	Muscle weakness, CNS and cardiovascular problems, including heart disease; called *beriberi*	Hypotension (low blood pressure)
B_2 (riboflavin)	Part of coenzyme, involved in many pathways (glycolysis and citric acid cycle)	Epithelial and mucosal deterioration	Itching, tingling
B_3 (niacin)	Part of coenzyme, involved in many pathways (glycolysis and citric acid cycle)	CNS, GI, epithelial, and mucosal deterioration; called *pellagra*	Itching, burning; vasodilation
B_5 (pantothenic acid)	Component of coenzyme A, in many pathways	Retarded growth, anemia, convulsions, epithelial changes	None reported
B_6 (pyridoxine)	Coenzyme in amino acid and lipid metabolism	Retarded growth, CNS disturbances	CNS alterations, perhaps fatal
B_7 (biotin)	Coenzyme in many pathways	Fatigue, muscular pain, nausea, dermatitis	None reported
B_9 (folate/folic acid)	Coenzyme in amino acid and nucleic acid metabolism	Retarded growth, anemia, GI disorders, developmental abnormalities	Few noted, except in massive doses
B_{12} (cobalamin)	Coenzyme in nucleic acid metabolism	Impaired RBC production, causing pernicious anemia	Polycythemia ($\uparrow$ RBCs)
C (ascorbic acid)	Coenzyme in many pathways	Epithelial and mucosal deterioration; called *scurvy*	Kidney stones

The body stores significant amounts of vitamins B_{12} and C only. So insufficient intake of other water-soluble vitamins can lead to initial signs and symptoms of vitamin deficiency within days to weeks. Intestinal bacteria help prevent deficiency diseases by producing small amounts of four of the nine water-soluble vitamins (B_5, B_7, B_9, and B_{12}). The intestinal epithelium can easily absorb all the water-soluble vitamins except B_{12}. The B_{12} molecule is large, and it must bind to **intrinsic factor** (synthesized by the gastric mucosa) before it can be absorbed.

Fat-Soluble Vitamins

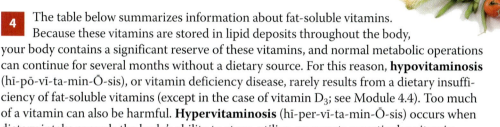

3 The **fat-soluble vitamins** are vitamins A, D_3, E, and K. These vitamins are absorbed primarily from the digestive tract along with the lipid contents of micelles. Vegetables are good sources of fat-soluble vitamins. In addition, intestinal bacteria produce some vitamin K and, when exposed to sunlight, your skin can synthesize some vitamin D_3.

4 The table below summarizes information about fat-soluble vitamins. Because these vitamins are stored in lipid deposits throughout the body, your body contains a significant reserve of these vitamins, and normal metabolic operations can continue for several months without a dietary source. For this reason, **hypovitaminosis** (hī-pō-vī-ta-min-Ō-sis), or vitamin deficiency disease, rarely results from a dietary insufficiency of fat-soluble vitamins (except in the case of vitamin D_3; see Module 4.4). Too much of a vitamin can also be harmful. **Hypervitaminosis** (hī-per-vī-ta-min-Ō-sis) occurs when dietary intake exceeds the body's ability to store, utilize, or excrete a particular vitamin.

The Fat-Soluble Vitamins

Vitamin	Significance	Effects of Deficiency	Effects of Excess
A	Maintains epithelia; required to synthesize visual pigments; supports immune system and bone growth	Retarded growth, night blindness, deterioration of epithelial membranes	Liver damage, pale skin, CNS effects (nausea, anorexia)
D_3	Required for bone growth, intestinal calcium and phosphorus absorption, and retention of these ions	Rickets (in children), osteomalacia (in adults), skeletal deterioration	Calcium deposits in many tissues, disrupting functions
E	Prevents breakdown of vitamin A and fatty acids	Anemia, other problems suspected	Nausea, stomach cramps, blurred vision, fatigue
K	Essential for liver to synthesize clotting factors	Bleeding disorders	Liver dysfunction, jaundice

Minerals

5 Minerals have three important roles: (1) ions such as sodium and chloride determine the osmotic concentrations of body fluids, (2) ions in various combinations play major roles in buffering body fluids, and (3) ions are essential in a variety of enzymatic reactions. Minerals are **cofactors** that act with enzymes to catalyze a chemical reaction. This table lists the major minerals and their significance, and some of the effects of deficiency or excess.

The Major Minerals

Mineral	Significance	Effects of Deficiency or Excess
Sodium	Major cation in body fluids; essential for normal membrane function; cofactor of enzymes	*Deficiency:* muscle cramps *Excess:* hypertension
Potassium	Major cation in cytosol; essential for normal membrane function; cofactor of enzymes	*Deficiency:* irregular heartbeat *Excess:* irregular heartbeat
Chloride	Major anion in body fluids; functions in forming HCl	*Deficiency:* irregular heartbeat
Calcium	Essential for muscle and neuron function, and normal bone structure; cofactor of enzymes	*Deficiency:* osteoporosis *Excess:* kidney stones
Phosphorus	In high-energy compounds, nucleic acids, and bone matrix	*Deficiency:* muscle weakness *Excess:* muscle spasms
Magnesium	Cofactor of enzymes	*Deficiency:* muscle spasms, nausea

Module 16.6 Review

a. Define nutrition.

b. If vitamins do not provide a source of energy, what is their role in the body?

c. What are the roles of minerals in the body?

Proper nutrition depends on eating a balanced diet

A **balanced diet** contains all the ingredients needed to maintain homeostasis, including adequate substrates to produce ATP, essential amino acids and fatty acids, and vitamins. In addition, the diet must include electrolytes and enough water to replace losses in urine, feces, and evaporation. A balanced diet prevents **malnutrition**, an unhealthy state resulting from inadequate or excessive absorption of one or more nutrients.

1 How do you know if you're eating a balanced diet? Follow the updated food recommendations on the website **www.choosemyplate.gov**. The United States Department of Agriculture created this diagram to offer personalized eating plans and food assessments based on the current Dietary Guidelines for Americans. The color-coded food group slices indicate the proportions of food we should consume from each of the five basic food groups: grains (orange), vegetables (green), fruits (red), dairy products (blue), and meat and beans (purple). In addition, use oils sparingly.

Know the Limits on Fats, Sugars, and Salt (Sodium)

- Get most of your fat from fish, nuts, and vegetable oils, which are key sources of essential fatty acids.

- Limit solid fats like butter, shortening, and lard that are high in saturated fats and cholesterol.

- Check the Nutrition Facts label to keep saturated fats, trans fats, and sodium low.

- Choose food and beverages low in added sugars. Added sugars contribute to excessive caloric intake.

Find Your Balance between Food and Physical Activity

- Stay within your daily calorie needs.

- Adults should be physically active for at least 30 minutes most days of the week.

- About 60 minutes a day of physical activity may be needed to prevent weight gain.

- To sustain weight loss, at least 60 to 90 minutes a day of physical activity may be required.

- Children and teenagers should be physically active for 60 minutes most days.

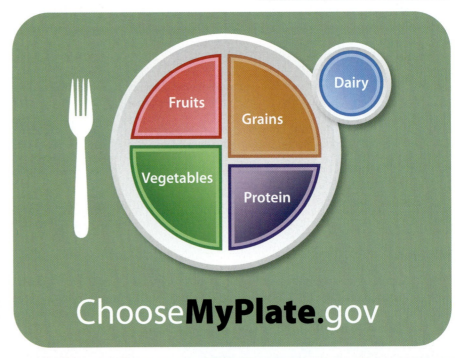

GRAINS	VEGETABLES	FRUITS	O I L S	DAIRY	MEAT & BEANS
Make half your grains whole	Vary your veggies	Focus on fruits		Get your calcium-rich foods	Go lean with proteins

2 In addition to considering the types of food you eat, think about how much energy that food contains relative to how much energy you use on a daily basis. We express the energy content of food in **calories**. One calorie is the energy needed to raise the temperature of 1 g of water by 1°C. We use the term **kilocalorie** (KIL-ō-kal-o-rē) (kcal) or **Calorie** (Cal) when talking about the metabolism of the entire body. One Calorie is the amount of energy needed to raise the temperature of 1 kilogram of water 1°C. The energy yield of the components of your diet differ; the values are 4.18 Cal/g for carbohydrates, 4.32 Cal/g for proteins, and 9.46 Cal/g for lipids. Depending on the level of activity, the average adult needs between 2000 to 3000 Cal each day to maintain a stable weight. That's a lot of mitochondrial activity—1 Cal is roughly equivalent to 83 million trillion ATP.

Food name	Serving	Cal
Breakfast bar	1 bar	368
Long-grain rice	1.5 cup	308
Bread, whole wheat	4 slices	277
Butter	0.5 tbsp	51
Beer, regular	12 fl oz	160
Cola, regular	12 fl oz	140

3 Here is additional information about the various food groups. The food groups in themselves are less important than making intelligent choices about what (and how much) you eat. Poor choices can lead to malnutrition even if all five groups are represented.

Five Basic Food Groups and Their Effects on Health

Nutrient Group	Provides	Health Effects
Grains (recommended: at least half of total eaten should be whole grains)	Carbohydrates; vitamins E, thiamine, niacin, folate; calcium; phosphorus; iron; sodium; dietary fiber	Whole grains prevent rapid rise in blood glucose levels, and consequent rapid rise in insulin levels
Vegetables (recommended: especially dark-green and orange vegetables)	Carbohydrates; vitamins A, C, E, folate; dietary fiber; potassium	Reduce risk of cardiovascular disease; protect against colon cancer (folate) and prostate cancer (lycopene in tomatoes)
Fruits (recommended: a variety of fruit each day)	Carbohydrates; vitamins A, C, E, folate; dietary fiber; potassium	Reduce risk of cardiovascular disease; protect against colon cancer (folate)
Dairy (recommended: low-fat or fat-free milk, yogurt, and cheese)	Complete proteins; fats; carbohydrates; calcium; potassium; magnesium; sodium; phosphorus; vitamins A, B_{12}, pantothenic acid, thiamine, riboflavin	Good source of calcium, which strengthens bones. Whole milk: High in calories, may cause weight gain; saturated fats associated with heart disease
Meat and Beans (recommended: lean meats, fish, poultry, eggs, dry beans, nuts, legumes)	Complete proteins; fats; calcium; potassium; phosphorus; iron; zinc; vitamins E, thiamine, B_6	Fish and poultry lower risk of heart disease and colon cancer (compared to red meat). Consumption of up to one egg per day does not appear to increase incidence of heart disease; nuts and legumes improve blood cholesterol ratios, lower risk of heart disease and diabetes

4 Some members of the dairy products and meat and beans groups—for example milk, yogurt, beef, fish, poultry, and eggs—provide all the essential amino acids in sufficient quantities. They are said to contain **complete proteins**.

5 Many plants supply adequate *amounts* of protein but they are **incomplete proteins**, which lack one or more of the essential amino acids. Vegetarians, who largely restrict themselves to grains, vegetables, and fruits groups (with or without dairy products), must become adept at varying their food choices to include combinations of ingredients that meet all their amino acid requirements. Even with a proper balance of amino acids, people who do not consume animal products face a vitamin B_{12} deficiency. This is because the only sources of vitamin B_{12} are animal products or fortified foods.

Module 16.7 Review

a. Define a balanced diet.

b. Distinguish between a complete protein and an incomplete protein.

c. Of these three —carbohydrates, lipids, or proteins—which one releases the greatest number of Calories per gram during catabolism?

The control of appetite is complex and involves both short-term and long-term mechanisms

1 The **feeding center** and the **satiety center** are hypothalamic nuclei involved with the control of appetite. Multiple factors influence these centers; here we diagram the most important. In addition, social factors, psychological pressures, and dietary habits can play a role, although the mechanisms and pathways involved are not known.

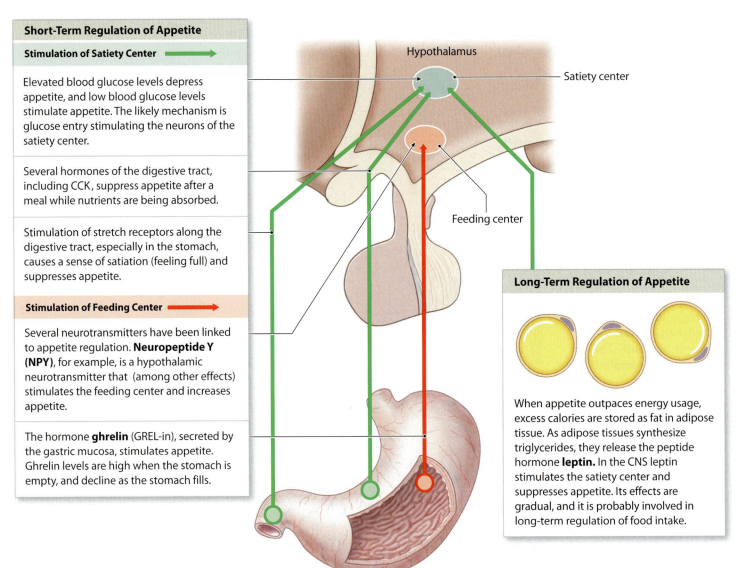

Short-Term Regulation of Appetite

Stimulation of Satiety Center ⟶

Elevated blood glucose levels depress appetite, and low blood glucose levels stimulate appetite. The likely mechanism is glucose entry stimulating the neurons of the satiety center.

Several hormones of the digestive tract, including CCK, suppress appetite after a meal while nutrients are being absorbed.

Stimulation of stretch receptors along the digestive tract, especially in the stomach, causes a sense of satiation (feeling full) and suppresses appetite.

Stimulation of Feeding Center ⟶

Several neurotransmitters have been linked to appetite regulation. **Neuropeptide Y (NPY)**, for example, is a hypothalamic neurotransmitter that (among other effects) stimulates the feeding center and increases appetite.

The hormone **ghrelin** (GREL-in), secreted by the gastric mucosa, stimulates appetite. Ghrelin levels are high when the stomach is empty, and decline as the stomach fills.

Hypothalamus

Satiety center

Feeding center

Long-Term Regulation of Appetite

When appetite outpaces energy usage, excess calories are stored as fat in adipose tissue. As adipose tissues synthesize triglycerides, they release the peptide hormone **leptin.** In the CNS leptin stimulates the satiety center and suppresses appetite. Its effects are gradual, and it is probably involved in long-term regulation of food intake.

In general, activation of the satiety center inhibits the feeding center. In addition, factors that primarily stimulate one center often have a secondary, inhibitory effect on the other.

2 **Eating disorders** are psychological problems that result in either inadequate or excessive food consumption. There are many different forms; a common thread in these conditions is an obsessive concern about food and body weight.

Anorexia Nervosa

Anorexia (an-o-REK-sē-ah) is the lack or loss of appetite. It may also accompany disorders that involve other systems. **Anorexia nervosa** is a form of self-induced starvation that appears to result from severe psychological problems. It is most common in adolescent Caucasian females whose weight is about 30 percent below normal. Although obviously underweight, patients are convinced that they are too fat and refuse to eat normal amounts of food. Death rates in severe cases range from 10 to 15 percent.

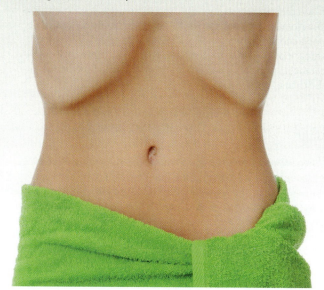

Bulimia

In **bulimia** (bu-LĒM-ē-ah) an individual goes on an "eating binge" that may include 20,000 or more calories. Afterward the person induces vomiting and often uses laxatives (to move materials more quickly through the digestive tract), and diuretics (drugs that promote fluid loss in the urine). Bulimia is more common than anorexia nervosa, and generally affects adolescent females.

3 **Obesity** is defined as being 20 percent over ideal weight, because it is at this point that serious health risks appear. The U.S. Centers for Disease Control and Prevention estimate that 32 percent of men and 35 percent of women in the United States can be considered obese.

Obesity

The most widely accepted measure of overweight and obesity is the **body mass index (BMI)**, calculated by dividing a person's weight in kilograms by his or her height in meters squared. Generally, desirable weight–height ratios correspond to a BMI between 19 and 25. Somewhat arbitrarily, overweight is defined as a BMI of 25 or over and obesity as a BMI of 30 or over. Basically, obese individuals take in more food energy than they use. There are two major categories of obesity: regulatory obesity and metabolic obesity. **Regulatory obesity**, the most common form, results from a failure to regulate food intake so that appetite, diet, and activity are in balance. In **metabolic obesity**, the condition is secondary to some underlying bodily malfunction that affects cell and tissue metabolism. Cases of metabolic obesity are relatively rare.

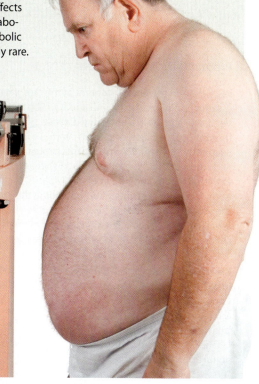

Module 16.8 Review

a. Which hormone inhibits the satiety center and stimulates appetite in the short-term?

b. Describe leptin and its effect on appetite.

c. Compare anorexia nervosa and bulimia.

Metabolic disorders may result from inadequate nutrition, biochemical problems, or starvation

Phenylketonuria (PKU)

Several inherited metabolic disorders result from an inability to produce specific enzymes involved in amino acid metabolism. Individuals with **phenylketonuria** (fen-il-kē-tō-NOO-rē-uh), or **PKU**, for example, cannot convert phenylalanine to tyrosine. This reaction is an essential step in the synthesis of norepinephrine, epinephrine, dopamine, and melanin. If PKU is not detected at birth, central nervous system development is inhibited, and severe brain damage results. The condition is common enough that a warning is printed on the packaging of products, such as diet drinks, that contain phenylalanine. Individuals with PKU must eat a low phenylalanine diet and take protein supplements.

Protein Deficiency Diseases

Regardless of the energy content of the diet, if it is deficient in essential amino acids, the individual will be malnourished to some degree. In a **protein deficiency disease**, protein synthesis decreases throughout the body. As protein synthesis in the liver fails to keep pace with the breakdown of plasma proteins, plasma osmolarity falls. Reduced osmolarity results in a fluid shift as more water moves out of the capillaries and into interstitial spaces, the peritoneal cavity, or both. The longer the individual remains in this state, the more severe the ascites (accumulating fluid in the peritoneal cavity) and edema (swelling) that result. It is estimated that more than 100 million children worldwide suffer from protein deficiency diseases.

Kwashiorkor (kwash-ē-OR-kor) occurs in children whose protein intake is inadequate, even if total caloric intake is acceptable. Complications include damage to the developing brain. The term is from the Ghana language and its literal translation means "first-second", describing the development of the disease in an older child who had been weaned from his mother's breast when a younger sibling was born.

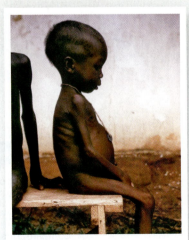

Kwashiorkor

Ketoacidosis

When glucose supplies are limited, the breakdown of fatty acids and some amino acids in liver cells elevates acetyl-CoA levels and results in the production of small organic acids called **ketone bodies**. Most of these compounds diffuse out of the liver and accumulate in the bloodstream. One of the ketone bodies, acetone, can often be smelled in the breath of someone who has skipped one or more meals. This condition is called **ketosis** (ke-TŌ-sis). Over time ketone bodies can accumulate, lowering blood pH and acidifying the blood, a condition called **ketoacidosis** (kē-tō-as-i-DŌ-sis). In severe cases, blood pH may drop below 7.05, and this may cause coma, cardiac arrhythmias, and death. An individual with poorly controlled or undiagnosed diabetes mellitus is at serious risk of developing ketoacidosis because the liver responds as if the person is starving, catabolizing proteins and lipids and dumping ketone bodies into the circulation.

Gout

When RNA is recycled as part of metabolic turnover, two of its nitrogenous bases (adenine and guanine) cannot be used as a nutrient source and are catabolized into **uric acid**. Like urea, uric acid is a relatively nontoxic waste product, but it is far less soluble than urea. Urea and uric acid are called **nitrogenous wastes**, because they are waste products that contain nitrogen atoms. Normal uric acid concentrations in plasma average 2.7–7.4 mg/dL, depending on gender and age. At concentrations over 7.4 mg/dL, body fluids become saturated with uric acid. Insoluble uric acid crystals (sodium urate) may begin to form, leading to the condition called **gout**. Initially, the joints of the limbs, especially the metatarsal–phalangeal joint of the great toe, are likely to be affected. This intensely painful condition, called **gouty arthritis**, may persist for several days and then disappear for a period of days to years. Meat, seafood, and beer are rich in purines, so limiting the consumption of these foods is recommended.

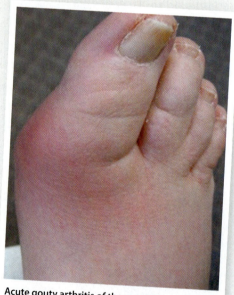

Acute gouty arthritis of the great toe

Starvation and Metabolic Reserves

Metabolic reserves are organic substrates in the body that can be catabolized to obtain the ATP needed to sustain life. The metabolic reserves of a typical 70-kg (154-lb) individual include lipids, proteins, and carbohydrates. Due to its high energy content, adipose tissue represents a disproportionate percentage of the total reserve in the form of triglycerides. Most of the available protein reserve is located in the contractile proteins of skeletal muscle. Carbohydrate reserves are relatively small and sufficient for only a few hours or, at most, overnight.

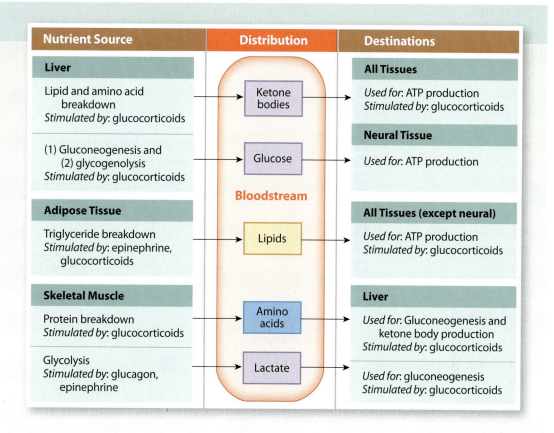

Nutrient Source	Distribution	Destinations
Liver		**All Tissues**
Lipid and amino acid breakdown *Stimulated by*: glucocorticoids	Ketone bodies	*Used for*: ATP production *Stimulated by*: glucocorticoids
		Neural Tissue
(1) Gluconeogenesis and (2) glycogenolysis *Stimulated by*: glucocorticoids	Glucose	*Used for*: ATP production
	Bloodstream	
Adipose Tissue		**All Tissues (except neural)**
Triglyceride breakdown *Stimulated by*: epinephrine, glucocorticoids	Lipids	*Used for*: ATP production *Stimulated by*: glucocorticoids
Skeletal Muscle		**Liver**
Protein breakdown *Stimulated by*: glucocorticoids	Amino acids	*Used for*: Gluconeogenesis and ketone body production *Stimulated by*: glucocorticoids
Glycolysis *Stimulated by*: glucagon, epinephrine	Lactate	*Used for*: gluconeogenesis *Stimulated by*: glucocorticoids

This graph follows changes in the metabolic stores of a 70-kg individual during prolonged starvation. At the point where proteins are being used for energy, the kidneys begin to assist the liver by catabolizing amino acids and generating additional glucose molecules.

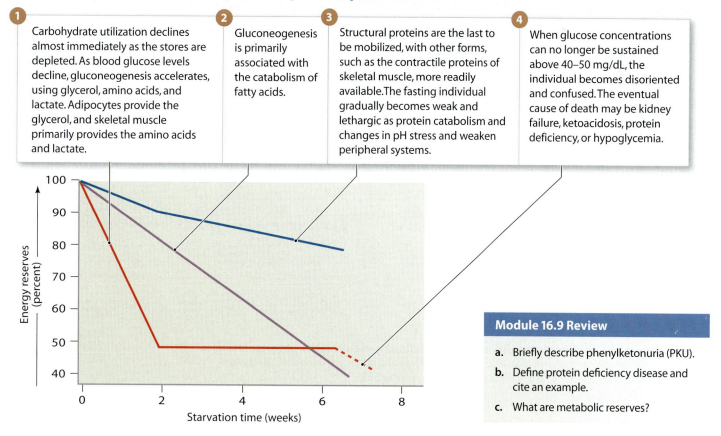

1 Carbohydrate utilization declines almost immediately as the stores are depleted. As blood glucose levels decline, gluconeogenesis accelerates, using glycerol, amino acids, and lactate. Adipocytes provide the glycerol, and skeletal muscle primarily provides the amino acids and lactate.

2 Gluconeogenesis is primarily associated with the catabolism of fatty acids.

3 Structural proteins are the last to be mobilized, with other forms, such as the contractile proteins of skeletal muscle, more readily available. The fasting individual gradually becomes weak and lethargic as protein catabolism and changes in pH stress and weaken peripheral systems.

4 When glucose concentrations can no longer be sustained above 40–50 mg/dL, the individual becomes disoriented and confused. The eventual cause of death may be kidney failure, ketoacidosis, protein deficiency, or hypoglycemia.

Module 16.9 Review

a. Briefly describe phenylketonuria (PKU).

b. Define protein deficiency disease and cite an example.

c. What are metabolic reserves?

1. Matching

Match each of the following terms with the most closely related item.

- lipogenesis
- hypovitaminosis
- lipolysis
- A, D, E, K
- malnutrition
- leptin
- ketone bodies
- calorie
- uric acid
- B complex and C
- inhibits feeding center
- lipoproteins
- pyruvate
- skeletal muscle

a _____	Adipose tissue hormone
b _____	Source of glycogen reserves
c _____	Water-soluble vitamins
d _____	Product of fatty acid and amino acid catabolism in liver
e _____	Lipid synthesis
f _____	Condition due to a vitamin deficiency
g _____	Fat-soluble vitamins
h _____	Lipid transport
i _____	Product of the anaerobic breakdown of glucose
j _____	Lipid breakdown
k _____	Gout
l _____	Unit of energy
m _____	Inadequate or excessive nutrient absorption
n _____	Role of satiety center

2. Multiple choice

Choose the bulleted item that best completes each statement.

a Intestinal absorption of nutrients occurs mainly in the _____.

- duodenum
- ileocecum
- ileum
- jejunum

b When blood glucose concentrations are elevated, the glucose molecules are _____.

- catabolized for energy or converted into glycogen for storage
- used to build proteins
- used for tissue repair
- all of these

c Most of the lipids absorbed by the digestive tract are immediately transferred to the _____.

- liver
- red blood cells
- hepatocytes for storage
- venous circulation by the thoracic duct

d Hypervitaminosis involving water-soluble vitamins is relatively uncommon because _____.

- the excess amount is stored in adipose tissue
- the excess amount is readily excreted in the urine
- the excess amount is stored in the bones
- excess amounts are readily absorbed by skeletal muscle tissue

3. Short answer

a What is the difference between an essential amino acid and a non-essential amino acid? _____
_____
_____
_____

b Describe three reasons why protein catabolism is an impractical source of quick energy. _____
_____
_____
_____

c A cofactor is an ion or organic molecule (coenzyme) that attaches to the active site of an enzyme. List some examples of both types of cofactors. _____
_____
_____
_____

d Why is the liver the focal point for metabolic regulation and control? _____
_____
_____
_____

4. Section integration

Darla suffers from anorexia nervosa. One afternoon she is rushed to the emergency room because of cardiac arrhythmias. Her breath has the smell of an aromatic hydrocarbon, and blood and urine samples contain high levels of ketone bodies. Why do you think she is having the arrhythmias?

_____
_____

Metabolism and Energy Use

Nutrition and metabolism have a direct impact on body temperature. This illustration provides an overview of the processes involved.

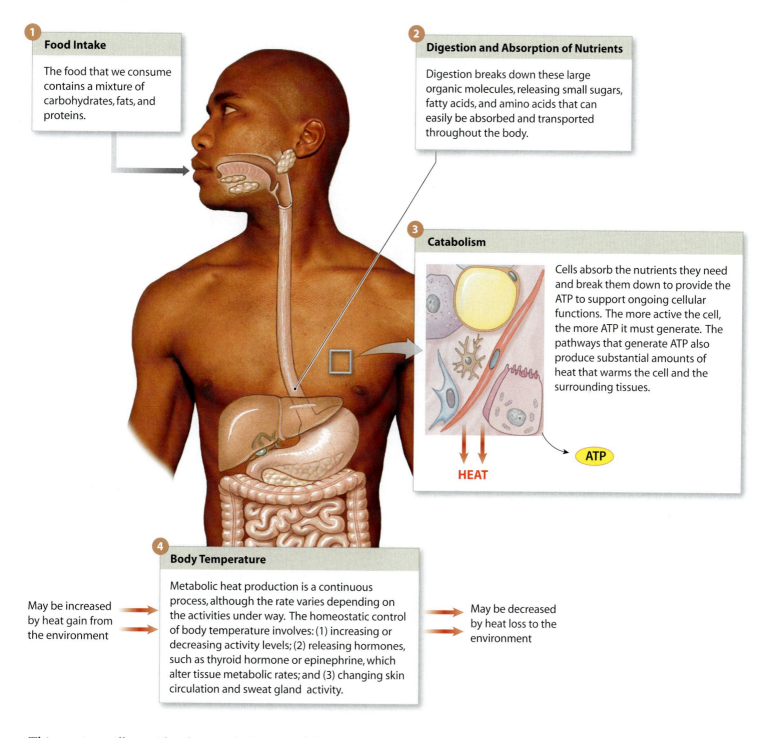

1 Food Intake

The food that we consume contains a mixture of carbohydrates, fats, and proteins.

2 Digestion and Absorption of Nutrients

Digestion breaks down these large organic molecules, releasing small sugars, fatty acids, and amino acids that can easily be absorbed and transported throughout the body.

3 Catabolism

Cells absorb the nutrients they need and break them down to provide the ATP to support ongoing cellular functions. The more active the cell, the more ATP it must generate. The pathways that generate ATP also produce substantial amounts of heat that warms the cell and the surrounding tissues.

HEAT

ATP

4 Body Temperature

Metabolic heat production is a continuous process, although the rate varies depending on the activities under way. The homeostatic control of body temperature involves: (1) increasing or decreasing activity levels; (2) releasing hormones, such as thyroid hormone or epinephrine, which alter tissue metabolic rates; and (3) changing skin circulation and sweat gland activity.

May be increased by heat gain from the environment

May be decreased by heat loss to the environment

This section will consider the metabolic rate of the body as a whole. We will also examine the basic factors involved in controlling body temperature.

The body uses energy continuously, and to maintain a constant body temperature, heat gain and heat loss must be in balance

Energetics

The amount of energy needed to support ongoing activities varies from moment to moment. The study of the flow of energy and its change(s) from one form to another is called energetics. A common benchmark in energetics studies is the **basal metabolic rate** (**BMR**), the minimum resting energy expenditure of an awake, alert person.

1 A direct method of determining the BMR involves monitoring respiratory activity. If we assume that average amounts of carbohydrates, lipids, and proteins are being catabolized, 4.825 Calories are expended per liter of oxygen consumed.

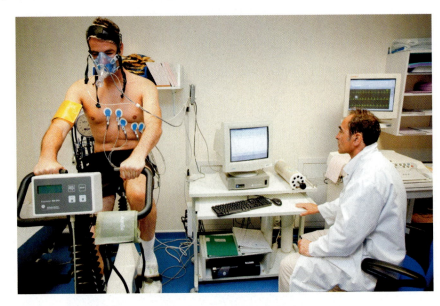

2 An average individual has a BMR of 70 Cal per hour, or about 1680 Cal per day. The actual energy consumption per day may be several times that amount, depending on a person's size, gender, weight, and level of physical activity. This graph shows the approximate number of Calories expended per hour at various levels of physical exertion.

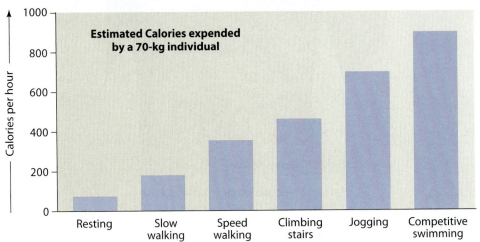

Estimated Calories expended by a 70-kg individual

Calories per hour

Resting — Slow walking — Speed walking — Climbing stairs — Jogging — Competitive swimming

To maintain energy balance, food intake must be adequate to support the activities under way. All of the reactions that generate ATP also generate heat. Only about 40 percent of the energy released through catabolism can be used to form ATP, and the rest warms the surrounding cytoplasm. When activity levels increase, ATP production accelerates, generating more heat. Enzymes will only function normally over a relatively narrow range of temperatures, and metabolic pathways are at risk unless heat is lost as quickly as it is produced. So this section ends with the topic of **thermoregulation**—the homeostatic control of body temperature.

Heat Gain and Heat Loss

About 60 percent of energy released by catabolism is lost as heat that warms surrounding tissues. If body temperature is to remain constant, heat production and heat loss must be kept in balance despite wide variations in activity levels and environmental conditions.

3 This illustration introduces the primary mechanisms of heat transfer between your body and the surrounding environment.

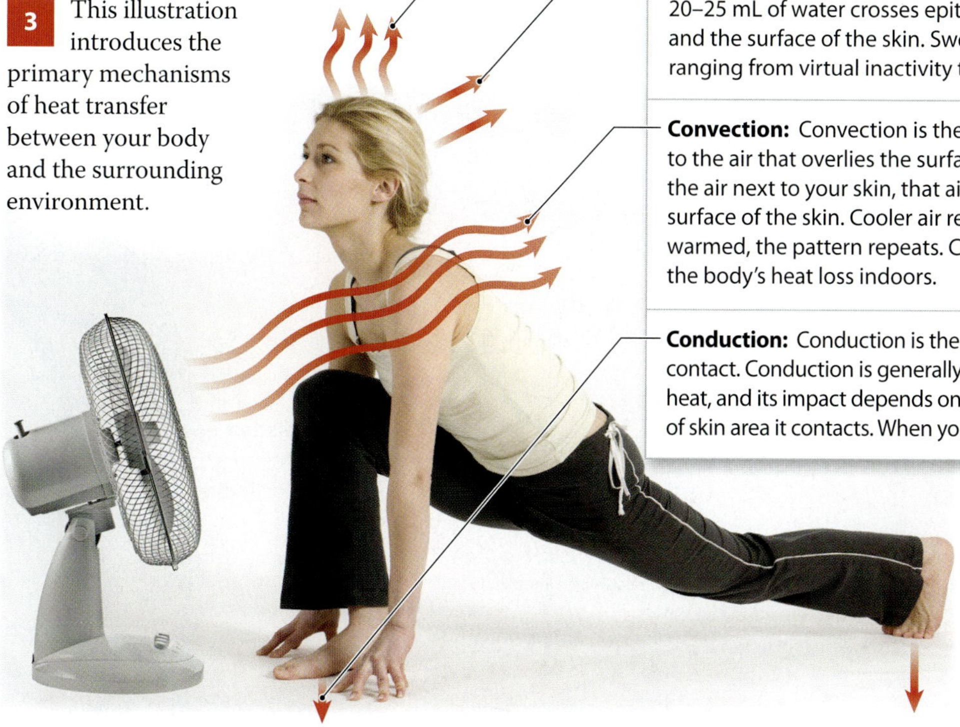

Primary Mechanisms of Heat Transfer

Radiation: Warm objects lose heat energy as infrared radiation. When you feel the heat from the sun, you are detecting that radiation. Your body loses heat the same way, but in proportionately smaller amounts. More than 50 percent of the heat you lose indoors is attributable to radiation; the exact amount varies with both body temperature and skin temperature.

Evaporation: When water evaporates, it changes from a liquid to a vapor. Evaporation absorbs energy—roughly 0.58 Cal per gram of water evaporated—and cools the surface where evaporation occurs. Each hour, 20–25 mL of water crosses epithelia and evaporates from the alveolar surfaces and the surface of the skin. Sweat glands have a tremendous scope of activity, ranging from virtual inactivity to perspiration rates of 2–4 liters per hour.

Convection: Convection is the result of conductive heat loss to the air that overlies the surface of the body. As your body conducts heat to the air next to your skin, that air warms and rises, moving away from the surface of the skin. Cooler air replaces it, and as this air in turn becomes warmed, the pattern repeats. Convection accounts for roughly 15 percent of the body's heat loss indoors.

Conduction: Conduction is the direct transfer of energy through physical contact. Conduction is generally not an effective mechanism for gaining or losing heat, and its impact depends on the temperature of the object and the amount of skin area it contacts. When you are standing, conductive losses are negligible.

4 This diagram shows the effects of failing to control body temperature.

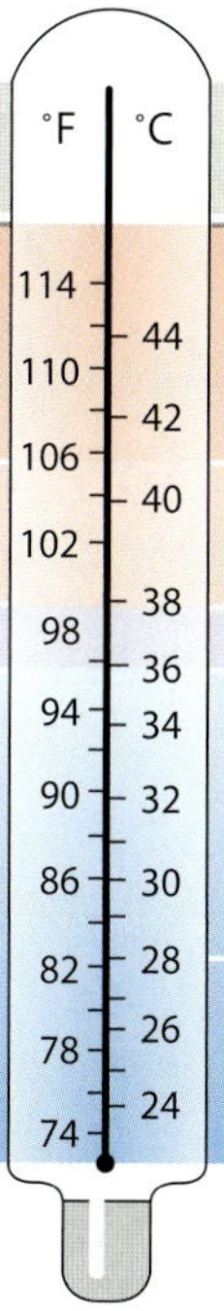

Underlying physical or environmental condition	°F	°C	Thermoregulatory capabilities	Major physiological effects
	114		Severely impaired	Death Proteins denature
CNS damage	110	44		Convulsions
Heat stroke		42	Impaired	Cell damage
	106			
Disease-related fevers		40		Disorientation
Extreme exercise	102		Effective	
Active children		38		
Normal range (oral)	98	36		Systems normal
Early mornings in cold weather	94	34	Impaired	Disorientation
	90	32		Loss of muscle control
Severe exposure	86	30	Severely impaired	Loss of consciousness
Hypothermia for open heart surgery	82	28		Cardiac arrest
	78	26	Lost	
	74	24		Death

Module 16.10 Review

a. Define basal metabolic rate.

b. Which heat transfer process accounts for about one-half of an individual's heat loss when indoors?

c. Distinguish between conduction and convection. How does heat loss differ between the two mechanisms?

Thermoregulatory centers in the hypothalamus adjust heat loss and heat gain

Heat loss and heat gain involve the activities of many systems. Those activities are coordinated by the **heat-loss center** and **heat-gain center**, respectively, in the preoptic area of the hypothalamus.

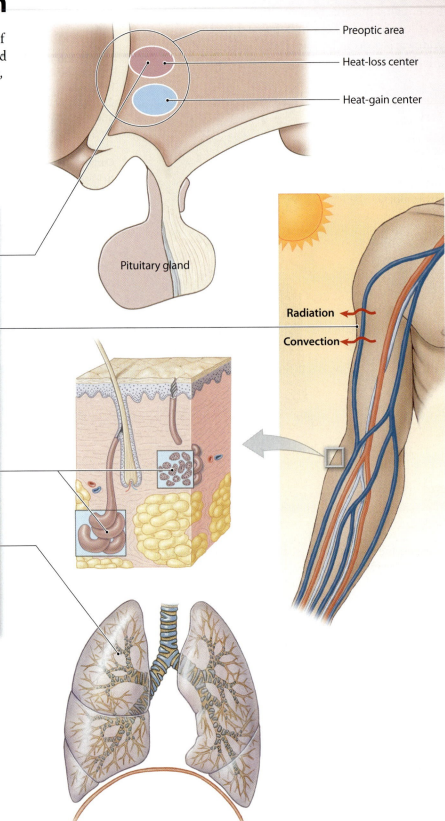

Preoptic area

Heat-loss center

Heat-gain center

Pituitary gland

Radiation

Convection

Heat-Loss Center Coordinates Responses to Rising Body Temperature

Behavioral Changes: A sense of discomfort leads to behavioral responses—getting into shade, going into water, or taking other voluntary steps that reduce body temperature.

Vasodilation and Shunting Blood to Skin Surface: Inhibiting the vasomotor center causes peripheral vasodilation, and warm blood flows to the surface of the body. The skin takes on a reddish color, skin temperatures rise, and radiational and convective losses increase.

Sweat Production: As blood flow to the skin increases, sweat glands are stimulated to increase sweat production. The perspiration flows across the body surface, and evaporative heat losses accelerate. Maximal secretion, if completely evaporated, can remove 2320 Cal of heat per hour.

Respiratory Heat Loss: The respiratory centers are stimulated, and respiration becomes deeper. Often, the individual begins breathing through an open mouth rather than through the nasal passageways, increasing evaporative heat losses through the lungs.

When Body Temperature Falls

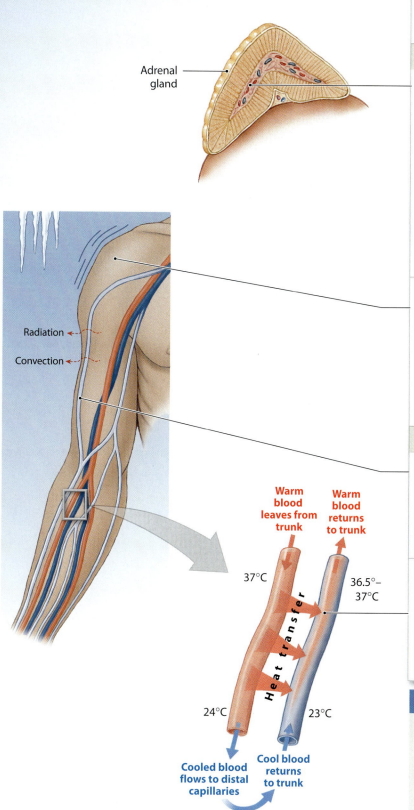

Adrenal gland

Radiation

Convection

Warm blood leaves from trunk

Warm blood returns to trunk

37°C

36.5°–37°C

Heat transfer

24°C

23°C

Cooled blood flows to distal capillaries

Cool blood returns to trunk

Heat-Gain Center Coordinates Responses to Falling Body Temperature

The heat-gain center responds to low body temperature in two ways.

Increased Generation of Body Heat

Nonshivering thermogenesis (ther-mō-JEN-e-sis): Hormones are released that raise the metabolic activity of all tissues. Sympathetic stimulation of the adrenal medullae releases epinephrine, which quickly increases the rates of glycogenolysis in liver and skeletal muscle and the metabolic rate of most tissues. In children, but not usually in adults, the heat-gain center can increase secretion of thyrotropin-releasing hormone (TRH) by the hypothalamus when body temperature falls below normal. TRH stimulates the release of thyroid-stimulating hormone (TSH) by the anterior lobe of the pituitary gland, and as the rate of thyroid hormone release increases, so does the rate of catabolism throughout the body.

Shivering thermogenesis: A gradual increase in muscle tone raises the energy consumption of skeletal muscle tissue throughout your body. Both agonists and antagonists are involved, and muscle tone gradually increases to the point at which stretch receptor stimulation will produce brief, oscillatory contractions of antagonistic muscles. In other words, you begin to shiver. Shivering can elevate body temperature quite effectively, increasing the rate of heat generation by as much as 400 percent.

Conservation of Body Heat

Vasomotor center: This center decreases blood flow to the dermis, reducing losses by radiation and convection. The skin cools, and with blood flow restricted, it may look bluish or pale. The epithelial cells are not damaged, because they can tolerate extended periods at temperatures as low as 25°C (77°F) or as high as 49°C (120°F).

Countercurrent exchange: The deep veins lie alongside the deep arteries, and heat is conducted from the warm blood flowing outward to the limbs to the cooler blood returning from the periphery. This arrangement traps the heat close to the body core and reduces heat loss. The transfer of heat, water, or solutes between fluids moving in opposite directions is called **countercurrent exchange**.

Module 16.11 Review

a. Name the heat conservation mechanism that conducts heat from deep arteries to adjacent deep veins in the limbs.

b. Describe the role of nonshivering thermogenesis in regulating body temperature.

c. Predict the effect of peripheral vasodilation on an individual's body temperature.

1. Matching

Match each of the following terms with the most closely related item.

- conduction   **a** _____ Study of the flow of energy and its changes
- basal metabolic rate   **b** _____ Most important heat-loss process when indoors
- 40 percent   **c** _____ Homeostatic control of body temperature
- countercurrent exchange   **d** _____ Heat-loss process associated with sweating
- hypothalamus   **e** _____ Release of hormones; increased metabolism
- evaporation   **f** _____ Percent of catabolic energy released as heat
- shivering thermogenesis   **g** _____ Stimulation of vasomotor center
- peripheral vasoconstriction   **h** _____ Heat transfer mechanism between deep blood vessels in the limbs
- thermoregulation   **i** _____ Resting energy expenditure
- energetics   **j** _____ Direct transfer of heat energy by physical contact
- radiation   **k** _____ Percent of catabolic energy captured as ATP
- 60 percent   **l** _____ Inhibition of vasomotor center
- peripheral vasodilation   **m** _____ Coordinates heat-loss center and heat-gain center
- nonshivering thermogenesis   **n** _____ Result of increased skeletal muscle tone

2. Multiple choice

Choose the bulleted item that best completes each statement.

a An individual's BMR is influenced by _____ .

- gender
- body weight
- age
- all of these

b The four processes involved in heat exchange with the environment are _____ .

- energetics, heat conservation, heat loss, and heat gain
- radiation, conduction, convection, and evaporation
- shivering, sweating, thermoregulation, and behavioral modifications
- energetics, evaporation, hormones, and heat conservation

c The primary mechanisms for increasing heat loss from the body include _____ .

- vasomotor and respiratory
- perspiring and shivering
- physiological responses and behavioral modifications
- acclimatization and vasomotor

d All of the following are responses to an increase in body temperature, except _____ .

- stimulation of the respiratory centers
- stimulation of sweat glands
- peripheral vasoconstriction
- peripheral vasodilation

e If daily food intake exceeds total energy demands, the excess energy is stored primarily as _____ .

- triglycerides in adipose tissue
- lipoproteins in the liver
- glycogen in the liver
- glucose in the bloodstream

f Your body uses energy _____ .

- only during the day
- only during strenuous activities
- only when exposed to cold
- continuously

3. Short answer

a How can energy consumption at rest be estimated by monitoring oxygen utilization? _____

_____

_____

b Describe the responses generated by the heat-gain center. _____

_____

_____

_____

c Describe the heat-gain mechanisms, both short-term and long-term, involved in nonshivering thermogenesis. _____

_____

_____

_____

_____

Visual Outline with Key Terms

Summarize the content of each module using the terms in the order provided.

An Introduction to Cellular Metabolism

- metabolism
- cellular metabolism
- catabolism
- anabolism
- metabolic turnover
- nutrient pool

NUTRIENT POOL

ATP

Aerobic Metabolism (in mitochondria)

16.1

Cells obtain most of their ATP from the electron transport system, which is linked to the citric acid cycle

- acetate
- acetyl-CoA
- coenzymes
- citric acid
- citric acid cycle
- oxidative phosphorylation
- electron transport system (ETS)

16.2

Cells can break down any available substrate from the nutrient pool to obtain the energy they need

- nutrient pool
- catabolic pathway
- anabolic pathway
- glycogenesis
- glycogenolysis
- gluconeogenesis
- glycolysis

SECTION 2

The Digestion and Metabolism of Nutrients

- oral cavity
- stomach
- duodenum
- jejunum
- hepatic portal vein
- liver

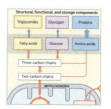

16.3

Carbohydrates are usually the preferred substrates for catabolism and ATP production under resting conditions

- salivary amylase
- maltase
- sucrase
- lactase
- flatus
- liver
- skeletal muscle
- glucose
- glycogen
- insulin
- pyruvate

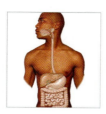

16.4

Lipids reach the bloodstream in chylomicrons; the cholesterol is then extracted and released as lipoproteins

- lipase
- linolenic acid
- linoleic acid
- essential fatty acids
- bile salts
- emulsification
- micelles
- chylomicrons
- lipoproteins
- lacteals
- thoracic duct
- cholesterol
- low-density lipoproteins (LDLs)
- high-density lipoproteins (HDLs)

16.5

An amino acid not needed for protein synthesis may be broken down or converted to a different amino acid

- pepsin
- enteropeptidase
- trypsin
- chymotrypsin
- carboxypeptidase
- elastase
- peptidases
- hepatic portal vein
- liver
- nonessential amino acids
- essential amino acids

16.6

Vitamins are essential to the function of many metabolic pathways

- nutrition
- vitamins
- minerals
- water-soluble vitamins
- intrinsic factor
- fat-soluble vitamins
- hypovitaminosis
- hypervitaminosis
- cofactors

16.7

Proper nutrition depends on eating a balanced diet

- balanced diet
- malnutrition
- food pyramid
- www.choosemyplate .gov
- calories
- kilocalorie
- Calorie
- complete proteins
- incomplete proteins

16.8

The control of appetite is complex and involves both short-term and long-term mechanisms

- feeding center
- satiety center
- neuropeptide Y (NPY)
- ghrelin
- leptin
- eating disorders
- anorexia
- anorexia nervosa
- bulimia
- obesity
- body mass index (BMI)
- regulatory obesity
- metabolic obesity

• = *Term boldfaced in this module*

16.9

Metabolic disorders may result from inadequate nutrition, biochemical problems, or starvation

- phenylketonuria (PKU)
- protein deficiency disease
- kwashiorkor
- ketone bodies
- ketosis
- ketoacidosis
- uric acid
- nitrogenous wastes
- gout
- gouty arthritis
- ○ metabolic reserves

16.10

The body uses energy continuously, and to maintain a constant body temperature, heat gain and heat loss must be in balance

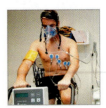

- energetics
- basal metabolic rate (BMR)
- thermoregulation
- ○ heat transfer
- radiation
- convection
- evaporation
- conduction

SECTION 3

Metabolism and Energy Use

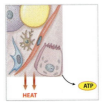

- ○ food intake
- ○ digestion and absorption of nutrients
- ○ catabolism
- ○ body temperature

16.11

Thermoregulatory centers in the hypothalamus adjust heat loss and heat gain

- heat-loss center
- heat-gain center
- ○ preoptic area
- behavioral changes
- vasodilation and shunting of blood to skin surface
- sweat production
- respiratory heat loss
- nonshivering thermogenesis
- shivering thermogenesis
- vasomotor center
- countercurrent exchange

● = *Term boldfaced in this module*

"Being able to see a tangible change in people's lives is so rewarding."

— **Mandy Murphy**
Registered Dietitian
Marin, California

CAREER PATHS

Registered Dietitian

Mandy Murphy became a dietitian in part because she loved to cook but also because she was interested in health and well-being. What she came to love about the job, however, is the difference she makes every day to her clients. "Being able to see a tangible change in people's lives is so rewarding," says Mandy, who works for the County of Marin in California. "That's what I love: seeing the lifelong changes."

In general, dietitians counsel people about the best ways to improve or maintain their health through diet. Mandy's job has three primary components during a normal 5-day work week. She spends two days at a Women, Infants, and Children (WIC) facility, two days at an obstetrics clinic with pregnant women who have gestational diabetes, and one day working with a community nutrition program. At the WIC facility, she meets with women,

infants, and children who have been diagnosed as malnourished for nutritional counseling sessions of 30 to 45 minutes, during which they will discuss everything from possible changes in diet to specific recipes and strategies for making meals. Her role at the obstetrics clinic is similar, though more narrowly focused. Her work with the community nutrition program involves a lot of community outreach, including

working with grocery stores to advertise and offer more healthy foods. She provides grocery store tours, nutrition workshops, and a weekly class for pregnant women.

Although Mandy's official job categorization is "nutritionist, bilingual," she is technically a registered dietitian. She speaks fluent Spanish, having honed her skills during a year studying in Argentina, and then an internship in Puerto Rico. She estimates that 80 percent of her work is in Spanish. While clearly advantageous, knowledge of a language other than English is not a requirement. On the other hand, the sciences, especially anatomy and physiology, are absolutely necessary. Mandy says anatomy and physiology "give us a great base" for explaining concepts to clients. A thorough understanding of digestion—the mechanical, chemical, and enzymatic processes whereby food is converted into materials the body can use—is essential to help clients understand their nutritional needs. For Mandy's work with pregnant women, knowing the basics of pregnancy and lactation is also crucial.

Many dietitians work in public health settings, like Mandy, but they can also work for school districts, hospitals, and in private practice. Although Mandy occasionally sees

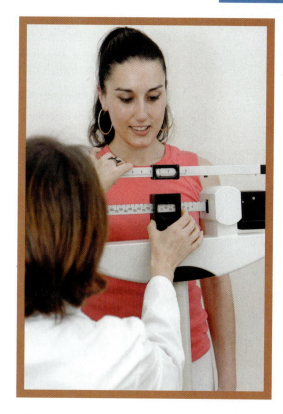

clients on an evening or weekend, most dietitians work a traditional 40-hour week, though some who work in hospitals may work weekends.

For more information, visit the website of the American Dietetic Association at http://www.eatright.org.

Think this is the career for you?
Key Stats:

- **Education and Training.** At minimum, a bachelor's degree in nutritional science is required.

- **Licensure.** Certification requirements for dietitians vary by state: Some states require licensure, some require statutory certification, and one only requires registration. The Registered Dietitian (RD) credential is achieved by passing an exam and participating in a yearlong internship approved by the American Dietetic Association.

- **Earnings.** Earnings vary but the median annual salary is $53,250.

- **Expected Job Prospects.** Employment is expected to grow by an average of 9% through 2018.

Bureau of Labor Statistics, U.S. Department of Labor, *Occupational Outlook Handbook, 2010–11 Edition,* Dietitians and Nutritionists, on the Internet at http://www.bls.gov/oco/ocos077 .htm (visited September 14, 2011).

MasteringA&P®

Access more review material online in the Study Area at www.masteringaandp.com.

- Chapter guides
- Chapter quizzes
- Practice tests
- Art labeling activities
- Flashcards
- A glossary with pronunciations

- Practice Anatomy Lab™ (PAL™) 3.0 virtual anatomy practice tool
- Interactive Physiology® (IP) animated tutorials
- MP3 Tutor Sessions

Chapter 16 Review Questions

1. Why are vitamins and minerals essential components of the diet?

2. Charlie has a blood test that shows a normal level of LDLs but an elevated level of HDLs in this blood. Given that his family has a history of cardiovascular disease, he wonders if he should modify his lifestyle. What would you tell him?

3. What effect do vasoconstriction and vasodilation of skin blood vessels have on body temperature?

4. How is the brain involved in the regulation of body temperature?

5. Why is a starving person more susceptible to infectious disease than a well-nourished person?

For answers to all module, section, and chapter review questions, see the blue Answers tab at the back of the book.

17

The Urinary System and Fluid, Electrolyte, and Acid-Base Balance

Anatomy of the Urinary System

The **urinary system** eliminates excess water, salts, and physiological wastes by producing urine. In this section we consider the functional anatomy of urinary system components: the kidneys, ureters, urinary bladder, and urethra. Later sections consider how the kidneys remove metabolic wastes from the bloodstream to produce urine, how the process is regulated, and how urine is eliminated through the passageways of the **urinary tract**.

The two **kidneys** are metabolically very active. Even at rest they receive roughly 25 percent of the cardiac output. The kidneys perform the excretory functions of the urinary system. They produce **urine**, a fluid containing water, ions, and small soluble compounds.

The **ureters** receive the urine from the kidneys and conduct it to the urinary bladder. Urine movement involves a combination of gravity and peristaltic contractions of smooth muscle in the walls of the ureters.

The **urinary bladder** receives and stores urine prior to its elimination from the body. The contraction of smooth muscle in the walls of the urinary bladder drives **urination**, the ejection of urine.

The **urethra** is a passageway that conducts urine from the urinary bladder to the outside of the body.

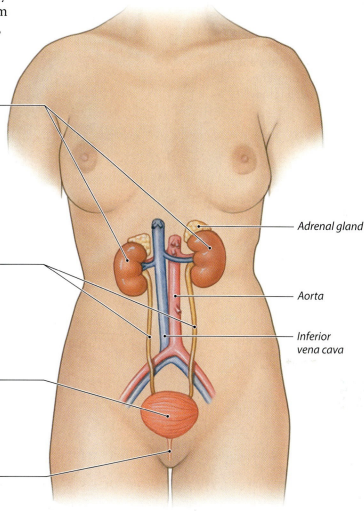

Adrenal gland

Aorta

Inferior
vena cava

Functions of the Urinary System

- Adjust blood volume and blood pressure
- Regulate plasma concentrations of sodium, potassium, chloride, and other ions
- Stabilize blood pH
- Conserve valuable nutrients
- Remove drugs, toxins, and metabolic wastes from the bloodstream

The kidneys are paired retroperitoneal organs

In this module you will learn the location of the kidneys and their placement in relation to the axial skeleton and the organs of the abdominopelvic cavity. You will also see how the kidneys are supported, protected, and stabilized in position. (In our discussions we will often encounter the term "renal." Renal means "relating to the kidneys.")

1 This anterior view shows the kidneys, ureters, urinary bladder, and associated blood vessels. Because the kidneys are in a retroperitoneal position, they are clearly visible in an anterior view only after other abdominal organs have been removed.

The **renal fascia** is a dense, fibrous outer layer that anchors the kidney to surrounding structures.

A typical adult **kidney** is reddish-brown and about 10 cm (4 in.) long, 5.5 cm (2.2 in.) wide, and 3 cm (1.2 in.) thick. Each kidney weighs about 150 g (5.25 oz).

The **hilum**, a prominent medial indentation, is the point of entry for the renal artery and renal nerves, and the point of exit for the renal vein and the ureter.

The **ureters** pass inferiorly and cross the anterior surfaces of the external iliac artery and vein before emptying into the posterior, inferior surface of the urinary bladder.

This is the cut edge of the posterior peritoneum, which has been removed. The kidneys, adrenal glands, and ureters lie between the muscles of the posterior body wall and the parietal peritoneum, in a retroperitoneal position.

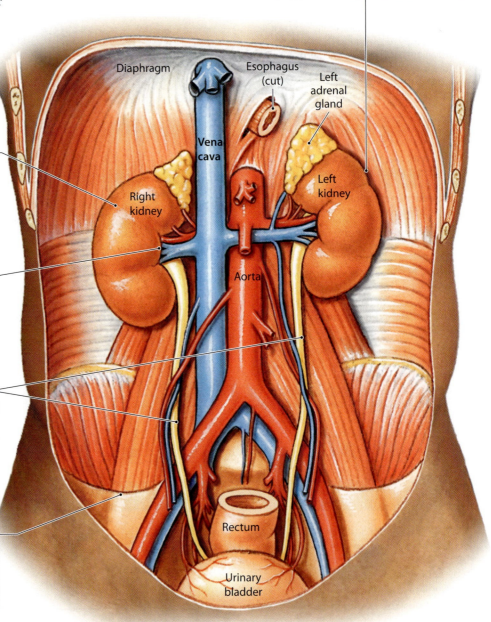

Diaphragm

Esophagus (cut)

Left adrenal gland

Vena cava

Left kidney

Right kidney

Aorta

Rectum

Urinary bladder

2 This diagram shows a sectioned kidney with its major landmarks and features. The blood vessels servicing the kidney enter through the hilum.

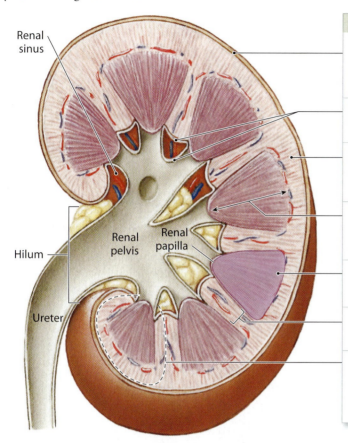

Renal sinus

Hilum

Renal pelvis

Renal papilla

Ureter

Major Structural Landmarks of the Kidney

The **fibrous capsule** covering the outer surface of the kidney also lines the **renal sinus**, an internal cavity within the kidney. The outer and inner linings are continuous at the hilum.

Within the renal sinus, the fibrous capsule stabilizes the positions of the ureter, renal blood vessels, and renal nerves.

The **renal cortex** is the superficial region of the kidney, in contact with the fibrous capsule. The cortex is light reddish-brown and granular.

The **renal medulla** extends from the renal cortex to the renal sinus. The renal medulla is the inner darker, reddish-brown region of the kidney.

A **renal pyramid** is a conical structure extending from the cortex to a tip called the **renal papilla**.

A **renal column** is a band of granular tissue that separates adjacent pyramids.

A **kidney lobe** consists of a renal pyramid, the overlying area of renal cortex, and adjacent tissues of the renal columns. Each kidney contains 6–18 kidney lobes. Urine is produced in the kidney lobes.

3 Here is a frontal section of a human kidney. The labels show the pathway of newly formed urine from within the kidney into the ureter.

A **minor calyx** collects the urine produced by a single kidney lobe.

A **major calyx** forms through the fusion of 4–5 minor calyces [KĀ-li-sēz].

Hilum

The **renal pelvis** is a large, funnel-shaped structure that collects urine from the major calyces. The renal pelvis is continuous with the ureter.

Ureter

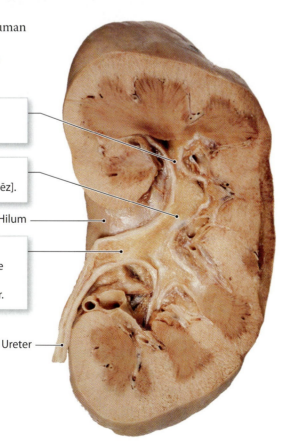

Module 17.1 Review

a. Describe the location of the kidneys.

b. Describe the structural landmarks of the kidney.

c. Which structure is a conical mass within the renal medulla that ends at the renal papilla?

A nephron can be divided into regions; each region has specific functions

1 A **nephron** is a microscopic structure that performs the essential functions of the kidney. A nephron consists of a **renal corpuscle** and a **renal tubule**. At the renal corpuscle, blood pressure forces water and dissolved solutes out of the glomerular capillaries and into a chamber—the **capsular space**—that is continuous with the lumen of the renal tubule. Filtration produces an essentially protein-free solution, known as **filtrate**, that is otherwise similar to blood plasma. After being modified by the renal tubule and collecting system, the filtrate leaves the kidneys as urine.

Nephron

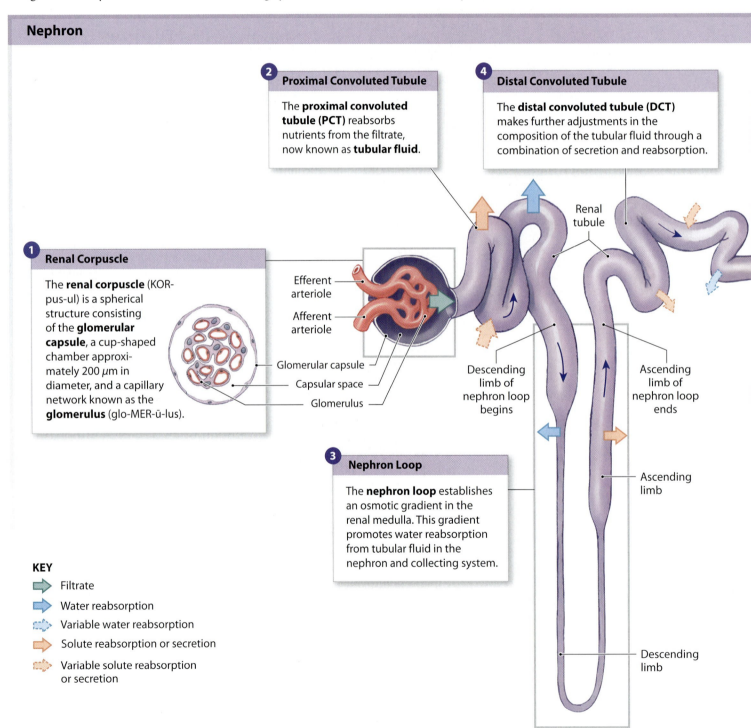

2 Proximal Convoluted Tubule

The **proximal convoluted tubule (PCT)** reabsorbs nutrients from the filtrate, now known as **tubular fluid**.

4 Distal Convoluted Tubule

The **distal convoluted tubule (DCT)** makes further adjustments in the composition of the tubular fluid through a combination of secretion and reabsorption.

Renal tubule

1 Renal Corpuscle

The **renal corpuscle** (KOR-pus-ul) is a spherical structure consisting of the **glomerular capsule**, a cup-shaped chamber approximately 200 μm in diameter, and a capillary network known as the **glomerulus** (glo-MER-ū-lus).

Efferent arteriole

Afferent arteriole

Glomerular capsule

Capsular space

Glomerulus

Descending limb of nephron loop begins

Ascending limb of nephron loop ends

Ascending limb

3 Nephron Loop

The **nephron loop** establishes an osmotic gradient in the renal medulla. This gradient promotes water reabsorption from tubular fluid in the nephron and collecting system.

Descending limb

KEY

Filtrate

Water reabsorption

Variable water reabsorption

Solute reabsorption or secretion

Variable solute reabsorption or secretion

2 Each nephron empties into the **collecting system**, a series of tubes that carry tubular fluid away from the nephron. Collecting ducts receive this fluid from many nephrons. Each collecting duct begins in the renal cortex and descends into the renal medulla, carrying fluid to a papillary duct that drains into a minor calyx.

3 Roughly 85 percent of all nephrons are **cortical nephrons**, located almost entirely within the superficial cortex of the kidney. The remaining 15 percent are **juxtamedullary** (juks-tuh-MED-ular-ē; *juxta*, near) **nephrons**, which have long nephron loops that extend deep into the renal medulla. The juxtamedullary nephrons establish conditions in the renal medulla that are essential for conserving water and producing concentrated urine.

Collecting System

The collecting system receives the urine from individual nephrons and performs final adjustments in urine volume and composition before delivering it to a minor calyx.

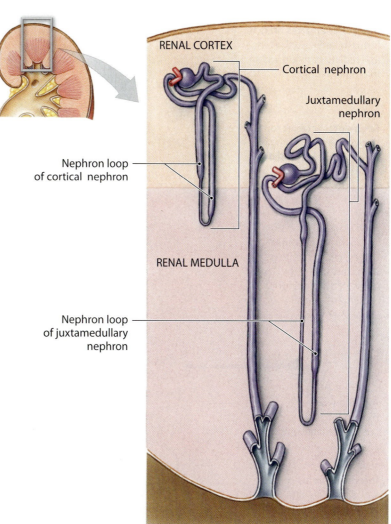

RENAL CORTEX

Cortical nephron

Juxtamedullary nephron

Nephron loop of cortical nephron

RENAL MEDULLA

Nephron loop of juxtamedullary nephron

5 **Collecting Duct**

A **collecting duct** carries tubular fluid through the osmotic gradient in the renal medulla.

6 **Papillary Duct**

A **papillary duct** collects tubular fluid from multiple collecting ducts and delivers it to a minor calyx.

As it travels along the renal tubule, tubular fluid gradually changes in composition. The characteristics of the urine that enters the minor calyx vary from moment to moment depending on the activities under way in each segment of the nephron and collecting system.

Module 17.2 Review

a. List the primary structures of the nephron and collecting system.

b. Identify the components of the renal corpuscle.

c. What is the difference between fluid formed in the glomerulus and blood plasma?

The kidneys are highly vascular, and the circulation patterns are complex

1 These diagrams show the **arterial** system supplying the kidney (top, red arrows) and the **venous** system draining the kidney (bottom, blue arrows).

Interlobar arteries branch from the segmental arteries and radiate outward within the renal columns.

Segmental arteries form through the branching of the renal artery inside the renal sinus.

Each kidney receives blood through a **renal artery**, which originates at the aorta near the origin of the superior mesenteric artery.

Arcuate arteries originate from interlobar arteries and arch along the boundary between the renal cortex and renal medulla.

Cortical radiate arteries supply the cortical portions of adjacent kidney lobes.

Afferent arterioles that branch off the cortical radiate arteries supply blood to individual nephrons.

Each afferent arteriole delivers blood to a capillary knot called a **glomerulus**. Blood is then distributed to the capillaries of the nephron as detailed in **2**.

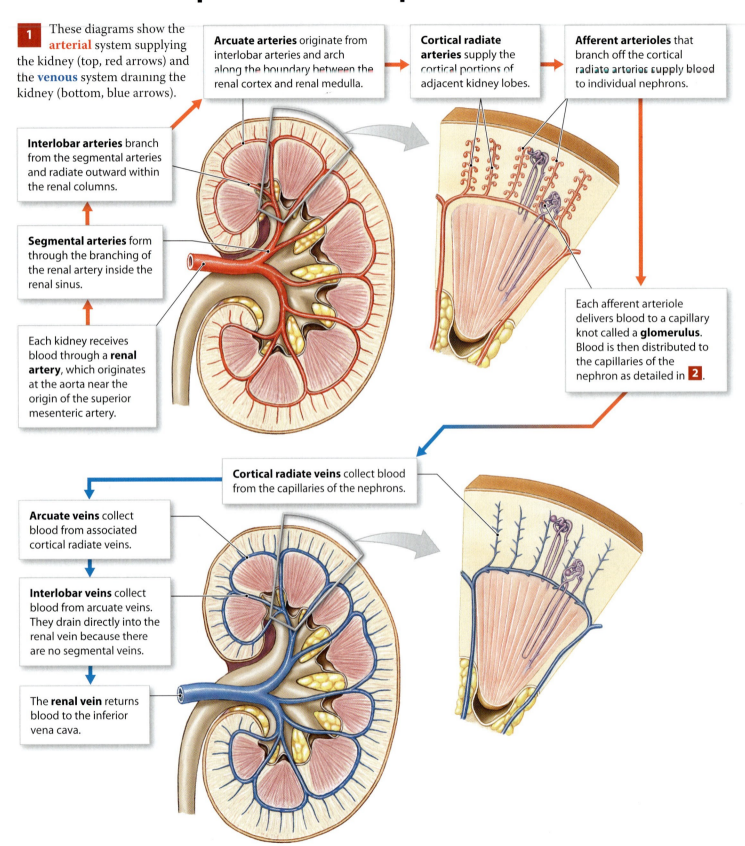

Cortical radiate veins collect blood from the capillaries of the nephrons.

Arcuate veins collect blood from associated cortical radiate veins.

Interlobar veins collect blood from arcuate veins. They drain directly into the renal vein because there are no segmental veins.

The **renal vein** returns blood to the inferior vena cava.

2 In a cortical nephron, the nephron loop is relatively short, and the **efferent arteriole** delivers blood to a network of **peritubular capillaries**, which surround the entire renal tubule. Both the nephrons and the peritubular capillaries are surrounded by interstitial fluid called peritubular fluid. These capillaries drain into small venules that carry blood to the cortical radiate veins.

3 In a juxtamedullary nephron, the peritubular capillaries are connected to the **vasa recta** (*vasa*, vessel + *recta*, straight)—long, straight capillaries that parallel the nephron loop.

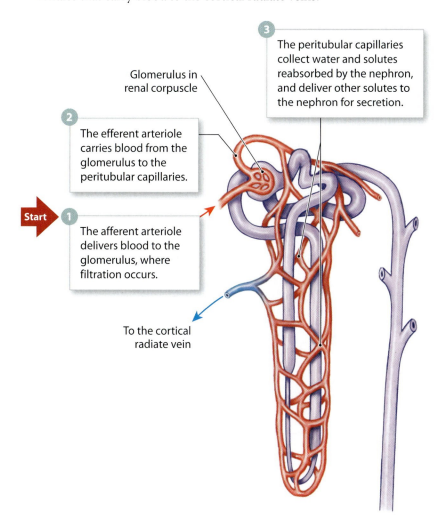

3 The peritubular capillaries collect water and solutes reabsorbed by the nephron, and deliver other solutes to the nephron for secretion.

Glomerulus in renal corpuscle

2 The efferent arteriole carries blood from the glomerulus to the peritubular capillaries.

Start **1** The afferent arteriole delivers blood to the glomerulus, where filtration occurs.

To the cortical radiate vein

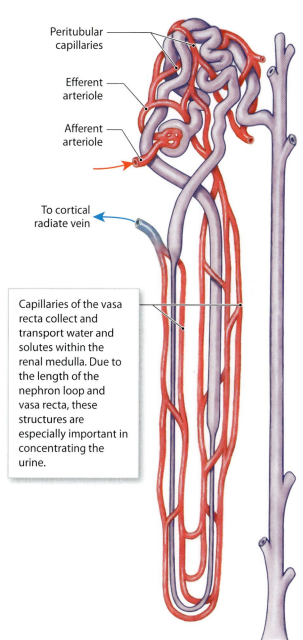

Peritubular capillaries

Efferent arteriole

Afferent arteriole

To cortical radiate vein

Capillaries of the vasa recta collect and transport water and solutes within the renal medulla. Due to the length of the nephron loop and vasa recta, these structures are especially important in concentrating the urine.

Each kidney has approximately 1.25 million nephrons, with a combined length of about 145 km (85 miles). Both the cortical and the juxtamedullary nephrons are innervated by renal nerves that enter at the hilum and follow the branches of the renal arteries. Most of the nerve fibers involved are sympathetic postganglionic fibers from the celiac plexus and the inferior splanchnic nerves. Sympathetic innervation adjusts blood flow and blood pressure at the glomeruli, and stimulates the release of the enzyme renin.

Module 17.3 Review

a. Trace the pathway of blood from the renal artery to the renal vein.

b. Describe how blood enters and leaves a glomerulus.

c. Describe the vasa recta.

1. Short answer

Label the kidney structures in the following diagram, and then provide a brief functional/anatomical description of each.

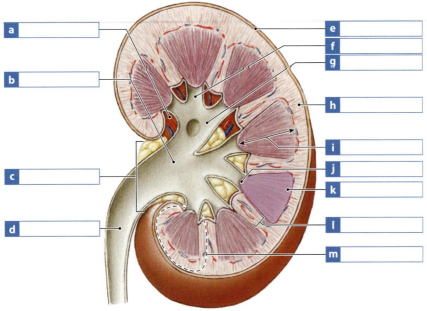

2. Concept map

Use each of the following terms once to fill in the blank boxes to correctly complete the urinary system concept map.

- ureter
- proximal convoluted tubule
- glomerulus
- urinary bladder
- renal tubules
- papillary duct
- major calyces
- renal medulla
- renal sinus
- nephrons

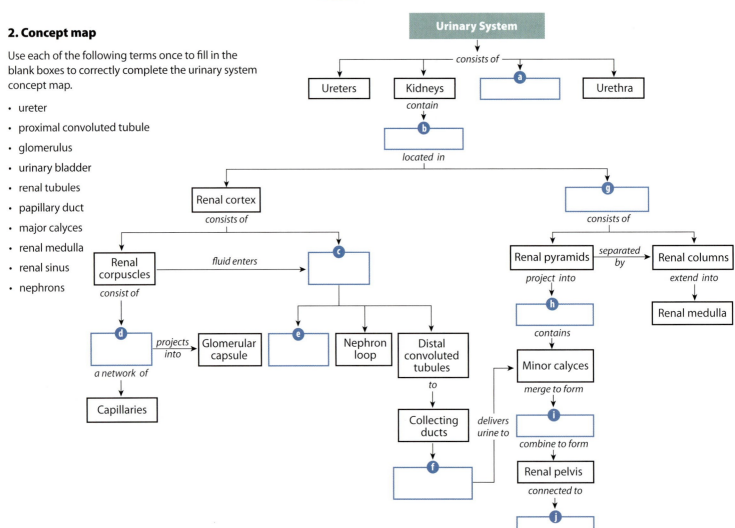

Overview of Renal Physiology

The goal of urine production is to maintain homeostasis by regulating the volume and composition of blood. This process involves excreting solutes—specifically, metabolic wastes. Three organic wastes in the blood plasma are noteworthy:

- **Urea** is the most abundant organic waste. Urea is a by-product of the breakdown of amino acids in the liver.
- **Creatinine** is generated in skeletal muscle tissue through the break-down of creatine phosphate, a high-energy compound that plays an important role in muscle contraction.
- **Uric acid** is a waste product formed during the recycling of the nitrogenous bases of RNA molecules.

The kidneys are usually capable of producing concentrated urine with an osmotic concentration of 1200–1400 mOsm/L, more than four times that of plasma.

1 The kidneys produce a fluid that is very different from other body fluids. This table demonstrates renal efficiency by comparing the concentrations of representative substances in plasma and urine.

Normal Laboratory Values for Solutes in Plasma and Urine		
Solute	**Plasma**	**Urine**
Ions (mEq/L)		
Sodium (Na⁺)	135–145	40–220
Potassium (K⁺)	3.5–5.0	25–100
Chloride (Cl⁻)	100–108	110–250
Bicarbonate (HCO₃⁻)	20–28	1.9
Metabolites and nutrients (mg/dL)		
Glucose	70–110	0.009
Lipids	450–1000	0.002
Amino acids	40	0.188
Proteins	6.0–8.0 g/dL	0.000
Nitrogenous wastes (mg/dL)		
Urea	8–25	1800
Creatinine	0.6–1.5	150
Ammonia	<0.1	60
Uric acid	2–6	40

Note: The values indicated are typical ranges. Specific numbers vary depending on the laboratory and methods used.

2 To perform their functions, the kidneys rely on three distinct physiological processes: filtration, reabsorption, and secretion.

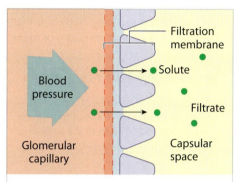

In **filtration**, blood pressure forces water and solutes across the membranes of the glomerular capillaries and into the capsular space. Solute molecules small enough to pass through the filtration membrane are carried by the surrounding water molecules.

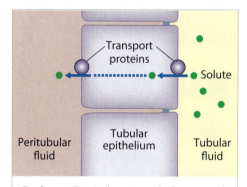

Reabsorption is the removal of water and solutes from the tubular fluid and their movement across the tubular epithelium and into the peritubular fluid.

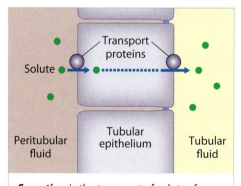

Secretion is the transport of solutes from the peritubular fluid, across the tubular epithelium, and into the tubular fluid.

Filtration, reabsorption, and secretion occur in specific regions of the nephron and collecting system

1 This diagram summarizes the functions of the various segments of the nephron and collecting system in the formation of urine. Most regions perform a combination of reabsorption and secretion, but the balance between the two processes varies from one region to another. Interaction between the collecting system and the nephron loops—especially the long loops of the juxtamedullary nephrons—regulates the final volume and solute concentration of urine.

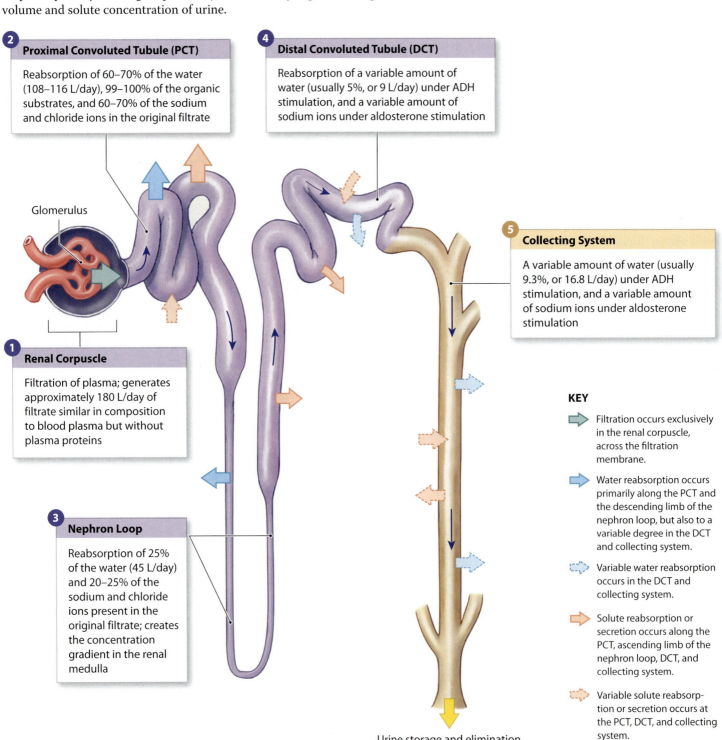

2 Proximal Convoluted Tubule (PCT)

Reabsorption of 60–70% of the water (108–116 L/day), 99–100% of the organic substrates, and 60–70% of the sodium and chloride ions in the original filtrate

4 Distal Convoluted Tubule (DCT)

Reabsorption of a variable amount of water (usually 5%, or 9 L/day) under ADH stimulation, and a variable amount of sodium ions under aldosterone stimulation

Glomerulus

1 Renal Corpuscle

Filtration of plasma; generates approximately 180 L/day of filtrate similar in composition to blood plasma but without plasma proteins

5 Collecting System

A variable amount of water (usually 9.3%, or 16.8 L/day) under ADH stimulation, and a variable amount of sodium ions under aldosterone stimulation

3 Nephron Loop

Reabsorption of 25% of the water (45 L/day) and 20–25% of the sodium and chloride ions present in the original filtrate; creates the concentration gradient in the renal medulla

KEY

Filtration occurs exclusively in the renal corpuscle, across the filtration membrane.

Water reabsorption occurs primarily along the PCT and the descending limb of the nephron loop, but also to a variable degree in the DCT and collecting system.

Variable water reabsorption occurs in the DCT and collecting system.

Solute reabsorption or secretion occurs along the PCT, ascending limb of the nephron loop, DCT, and collecting system.

Variable solute reabsorption or secretion occurs at the PCT, DCT, and collecting system.

Urine storage and elimination

2 The renal corpuscle, the start of the nephron, is responsible for filtering blood. This is the vital first step in forming urine. At the renal corpuscle, the capillary knot of the glomerulus projects into the capsular space like the heart projects into the pericardial cavity. Like the pericardium, the glomerular capsule has an outer parietal layer and an inner visceral layer.

The glomerular capsule forms the outer wall of the renal corpuscle and covers the glomerular capillaries.

The **capsular space** separates the inner and outer layers of the glomerular capsule.

Initial segment of renal tubule

The **efferent arteriole** delivers blood to peritubular capillaries. It has a smaller diameter than the afferent arteriole; this elevates the blood pressure within the glomerulus to support filtration.

DCT

The **juxtaglomerular complex** consists of specialized cells that secrete renin when glomerular blood pressure falls.

Outer layer

Inner layer

The **afferent arteriole** delivers blood from a cortical radiate artery.

3 The inner layer of the glomerular capsule consists of large cells with complex processes, or "feet," that wrap around the specialized dense layer of the glomerular capillaries. These unusual cells are called **podocytes** (PŌ-dō-sīts; *podos*, foot + *-cyte*, cell). Materials passing out of the blood at the glomerulus must be small enough to diffuse through the basement membrane and pass between the narrow gaps, or **filtration slits**, between adjacent processes.

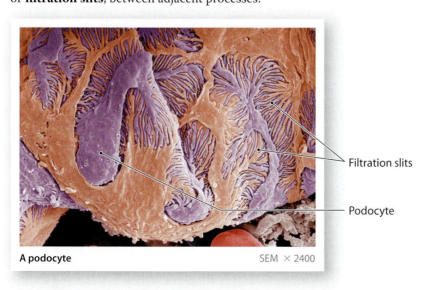

Filtration slits

Podocyte

A podocyte　　　　SEM × 2400

Module 17.4 Review

a. Identify the three distinct processes of urine formation in the kidney.

b. Where does filtration exclusively occur in the kidney?

c. Which hormone is responsible for regulating sodium ion reabsorption in the DCT and collecting system?

Urine volume and concentration are hormonally regulated

Urine volume and osmotic concentration are regulated through the control of water reabsorption. The water permeabilities of the PCT and descending limb of the nephron loop cannot be adjusted, and water is reabsorbed whenever the osmotic concentration of the peritubular fluid exceeds that of the tubular fluid. Because these water movements cannot be prevented, they represent **obligatory water reabsorption**. Obligatory reabsorption usually recovers 85 percent of the volume of filtrate produced. How much water is lost in urine? This depends on how much of the water in the remaining tubular fluid (15 percent of the filtrate volume, or roughly 27 liters per day) is reabsorbed along the DCT and collecting system. A process called **facultative water reabsorption** can precisely control the amount of water reabsorbed.

1 Without antidiuretic hormone (ADH), water is not reabsorbed in the distal collecting tubule or the collecting duct. All the fluid reaching the DCT is lost in urine. No facultative water reabsorption occurs, and the individual produces large amounts of very dilute urine.

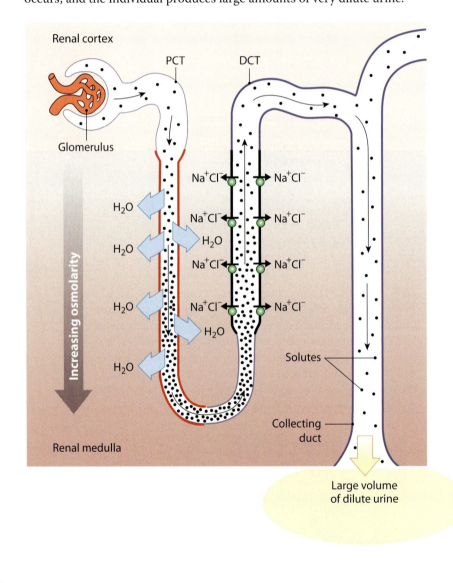

Obligatory Water Reabsorption

Glomerulus

Glomerular capsule

Proximal convoluted tubule

Nephron loop

Facultative Water Reabsorption

Distal convoluted tubule

Collecting duct

Urine storage and elimination

KEY

→ = Water reabsorption

⇢ = Variable water reabsorption

Renal cortex

PCT   DCT

Glomerulus

Na⁺Cl⁻   Na⁺Cl⁻

H_2O

H_2O Na⁺Cl⁻   Na⁺Cl⁻ H_2O

Na⁺Cl⁻   Na⁺Cl⁻

Increasing osmolarity

H_2O Na⁺Cl⁻   Na⁺Cl⁻

H_2O

H_2O

Renal medulla

Solutes

Collecting duct

Large volume of dilute urine

2 ADH causes special water channels, called **aquaporins**, to appear in the apical plasma membranes lining the DCT and collecting duct. Aquaporins dramatically enhance the rate of osmotic water movement. As ADH levels rise, the DCT and collecting system become more permeable to water, the amount of water reabsorbed increases, and the urine osmotic concentration climbs. Under maximum ADH stimulation, the DCT and collecting system become so permeable to water that the osmotic concentration of the urine is equal to that of the deepest portion of the renal medulla.

3 A healthy adult typically produces 1200 mL of urine per day (about 0.6 percent of the filtrate volume), with an osmotic concentration of roughly 1000 mOsm/L. However, normal values differ from individual to individual and from day to day, as the kidneys alter their function to maintain homeostatic conditions within body fluids.

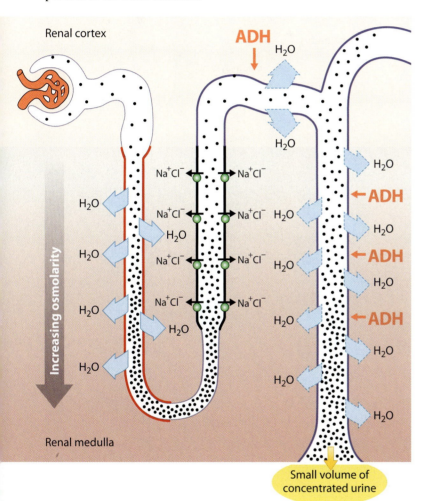

KEY

⬅ = Water reabsorption

⬅ = Variable water reabsorption

●→ = Na⁺/Cl⁻ transport

ADH = Antidiuretic hormone

General Characteristics of Normal Urine	
Characteristic	**Normal Range**
pH	4.5–8 (average: 6.0)
Specific gravity	1.003–1.030
Osmotic concentration (osmolarity)	855–1335 mOsm/L
Water content	93–97%
Volume	700–2000 mL/day
Color	Clear yellow
Odor	Varies with composition
Bacterial content	None (sterile)

Polyuria, the production of excessive amounts of urine, results from hormonal or metabolic problems, such as diabetes mellitus. (See Module 10.10 for a description of diabetes mellitus.)

Oliguria (a urine volume of 50–500 mL/day) and **anuria** (0–50 mL/day) are conditions that indicate serious kidney problems.

Module 17.5 Review

a. Can the permeability of the PCT to water ever change? Why or why not?

b. How would an increase in ADH levels affect the DCT?

c. When ADH levels in the DCT decrease, what happens to the urine osmotic concentration?

Renal function is an integrative process

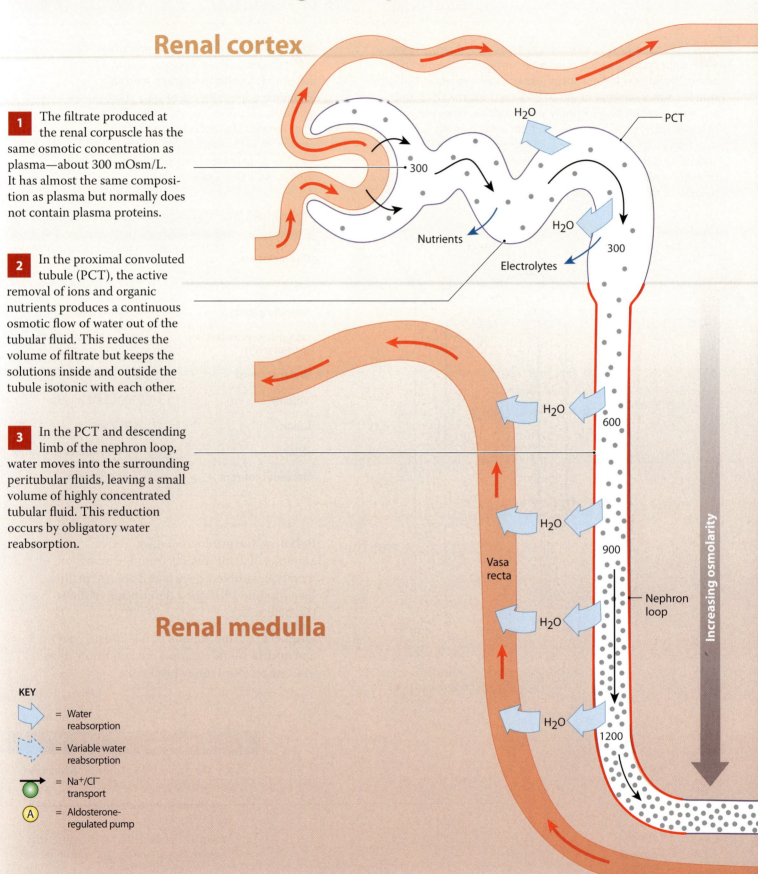

Renal cortex

1 The filtrate produced at the renal corpuscle has the same osmotic concentration as plasma—about 300 mOsm/L. It has almost the same composition as plasma but normally does not contain plasma proteins.

2 In the proximal convoluted tubule (PCT), the active removal of ions and organic nutrients produces a continuous osmotic flow of water out of the tubular fluid. This reduces the volume of filtrate but keeps the solutions inside and outside the tubule isotonic with each other.

3 In the PCT and descending limb of the nephron loop, water moves into the surrounding peritubular fluids, leaving a small volume of highly concentrated tubular fluid. This reduction occurs by obligatory water reabsorption.

Renal medulla

H_2O
PCT
300
H_2O
Nutrients
H_2O
300
Electrolytes

Increasing osmolarity

H_2O
600
H_2O
900
Vasa recta
Nephron loop
H_2O
H_2O
1200

KEY

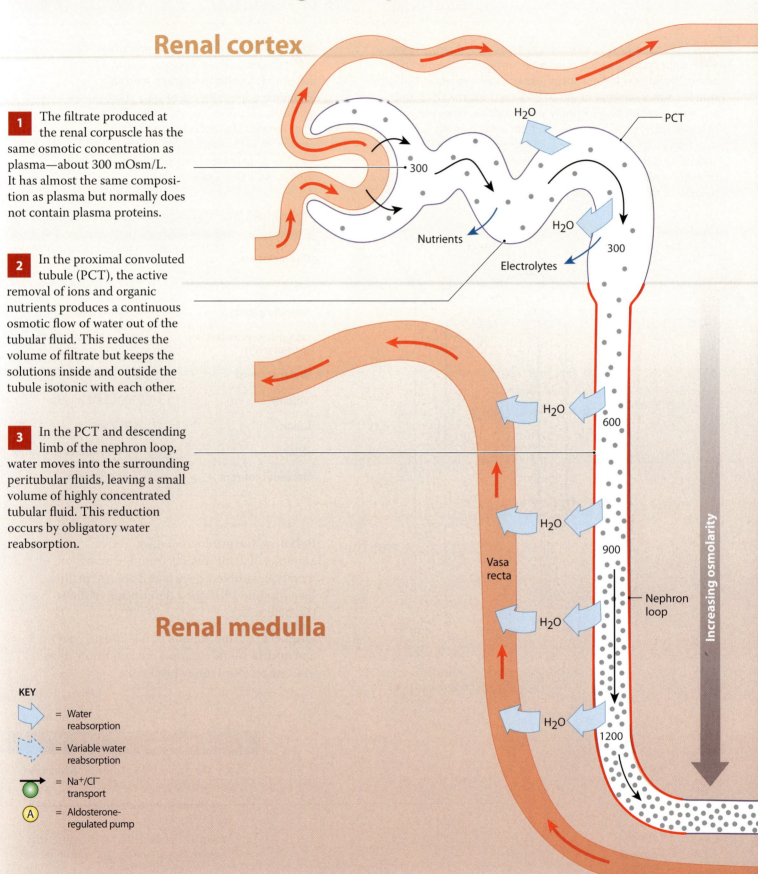 = Water reabsorption

= Variable water reabsorption

= Na^+/Cl^- transport

Ⓐ = Aldosterone-regulated pump

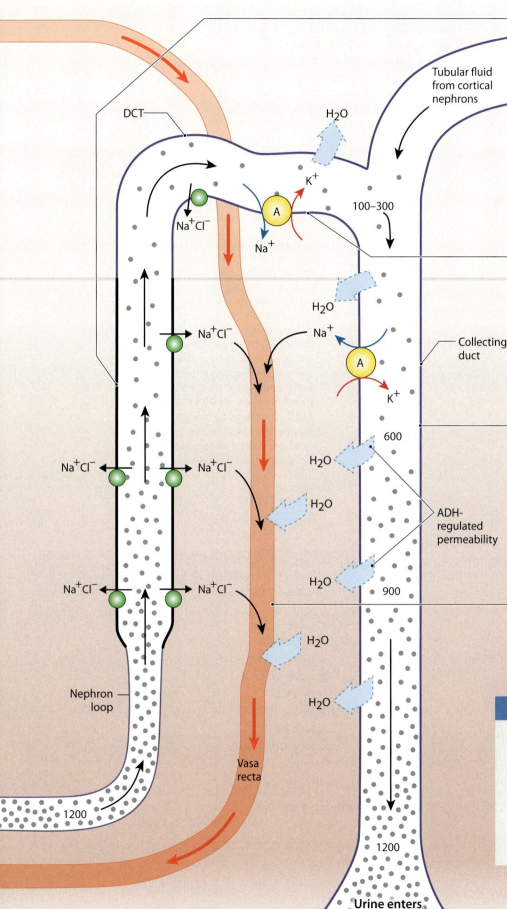

4 The thick ascending limb is impermeable to water and solutes. The tubule cells actively transport Na^+ and Cl^- out of the tubule, thereby lowering the osmotic concentration of the tubular fluid. Because just Na^+ and Cl^- are removed, urea accounts for a higher proportion of the total osmotic concentration at the end of the nephron loop.

5 Further alterations in the composition of the tubular fluid occur in the DCT and the collecting system. The osmotic concentration of the tubular fluid can be adjusted through active transport (reabsorption or secretion). For example, aldosterone promotes Na^+ reabsorption and K^+ secretion.

6 The final adjustments in the volume and osmotic concentration of the tubular fluid are made by controlling the water permeabilities of the distal portions of the DCT and the collecting system. The level of exposure to ADH determines the final urine concentration.

7 The vasa recta absorbs the solutes and water reabsorbed by the nephron loop and collecting ducts. By transporting these solutes and water into the venous system, the vasa recta maintains the concentration gradient of the renal medulla.

Module 17.6 Review

a. The filtrate produced at the renal corpuscle has the same osmotic pressure as _____.

b. In the PCT, ions and organic substrates are actively reabsorbed, thus causing what process to occur?

c. How is the concentration gradient of the renal medulla maintained?

Renal failure is a life-threatening condition

1 **Renal failure** occurs when the kidneys become unable to perform the excretory functions needed to maintain homeostasis. When kidney filtration slows for any reason, urine production declines. As the decline continues, symptoms of renal failure appear because water, ions, and metabolic wastes are retained. Renal failure impairs virtually all systems in the body. For example, it disturbs fluid balance, pH, muscular contraction, metabolism, and digestive function. The individual generally becomes hypertensive, anemia develops due to a decline in erythropoietin production, and central nervous system problems can lead to sleeplessness, seizures, delirium, and even coma.

Renal Failure

Acute Renal Failure

Acute renal failure occurs when exposure to toxic drugs, renal ischemia, urinary obstruction, or trauma causes filtration to slow suddenly or stop. Kidney function deteriorates rapidly, in just a few days, and may be impaired for weeks. Sensitized individuals can also develop acute renal failure after an allergic response to antibiotics or anesthetics. Individuals in acute renal failure may recover if they survive the incident. The kidneys may then regain partial or complete function. (With supportive treatment, the survival rate is approximately 50 percent.)

Chronic Renal Failure

In **chronic renal failure**, kidney function deteriorates gradually, and the associated problems accumulate over time. The management of chronic renal failure typically involves restricting water, salt, and protein intake. This combination reduces strain on the urinary system by minimizing (1) the volume of urine produced and (2) the amount of nitrogenous waste generated. Acidosis, a common problem in persons with renal failure, can be countered by ingesting bicarbonate ions. Chronic renal failure generally cannot be reversed, but its progression can be slowed.

2 **Hemodialysis** (hē-mō-dī-AL-i-sis) uses an artificial membrane to regulate the composition of blood by means of a dialysis machine. The basic principle involved in this process, called **dialysis**, is passive diffusion across a selectively permeable membrane. The patient's blood flows past an artificial dialysis membrane, which contains pores large enough to permit the diffusion of ions, nutrients, and organic wastes, but small enough to prevent the loss of plasma proteins. A special dialysis fluid flows on the other side of the membrane.

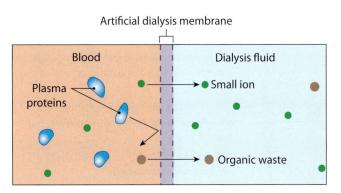

Artificial dialysis membrane

Blood · Dialysis fluid

Plasma proteins

Small ion

Organic waste

Comparative Composition of Plasma and Dialysis Fluid		
Component	**Plasma**	**Dialysis Fluid**
Electrolytes (mEq/L)		
Sodium (Na^+)	135–145	136–140
Potassium (K^+)	3.5–5.0	0–3.0
Calcium (Ca^{2+})	4.3–5.3	1.5
Magnesium (Mg^{2+})	1.4–2.0	0.5–1.0
Chloride (Cl^-)	100–108	99–110
Bicarbonate (HCO_3^-)	21–28	27–39
Phosphate (PO_4^{2-})	3	0
Sulfate (SO_4^{2-})	1	0
Nutrients (mg/dL)		
Glucose	70–110	100

As diffusion takes place across the dialysis membrane, the composition of the blood changes. Potassium ions, phosphate ions, sulfate ions, urea, creatinine, and uric acid diffuse across the membrane into the dialysis fluid. Bicarbonate ions and glucose diffuse from the dialysis fluid into the bloodstream. In effect, diffusion across the dialysis membrane takes the place of normal glomerular filtration, and the characteristics of the dialysis fluid ensure that important metabolites (substances necessary for a metabolic process) remain in the blood rather than diffusing across the membrane.

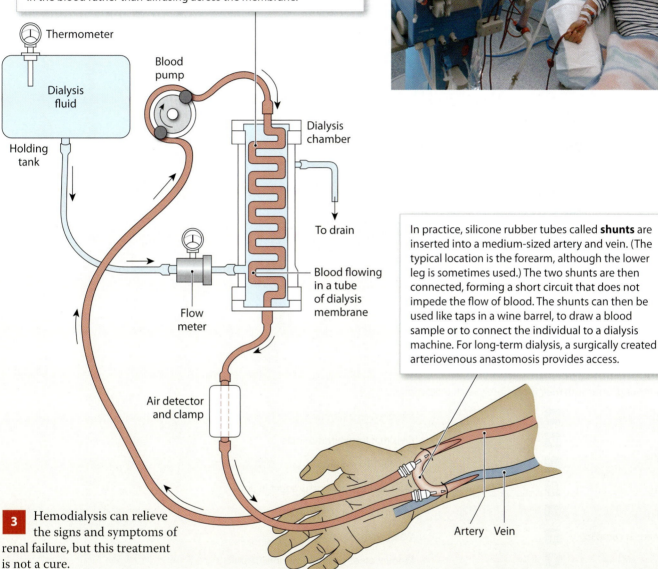

Thermometer

Dialysis fluid

Holding tank

Blood pump

Dialysis chamber

To drain

Blood flowing in a tube of dialysis membrane

Flow meter

Air detector and clamp

In practice, silicone rubber tubes called **shunts** are inserted into a medium-sized artery and vein. (The typical location is the forearm, although the lower leg is sometimes used.) The two shunts are then connected, forming a short circuit that does not impede the flow of blood. The shunts can then be used like taps in a wine barrel, to draw a blood sample or to connect the individual to a dialysis machine. For long-term dialysis, a surgically created arteriovenous anastomosis provides access.

Artery   Vein

3 Hemodialysis can relieve the signs and symptoms of renal failure, but this treatment is not a cure.

The only real cure for severe renal failure is a kidney transplant. This surgery involves implanting a new kidney obtained from a living donor or from a cadaver. In most cases, the damaged kidney is removed and its blood supply is connected to the transplant. The one-year success rate for transplantation is now 85–95 percent. The use of kidneys taken from close relatives significantly improves the success rate. Transplant recipients must take immunosuppressive drugs for the rest of their lives. This treatment reduces tissue rejection, but unfortunately it also lowers their resistance to infections.

Module 17.7 Review

a. Define hemodialysis.

b. Briefly explain the difference between chronic renal failure and acute renal failure.

c. Explain why patients on dialysis often receive Epogen or Procrit, a synthetic form of erythropoietin.

1. Short answer

Identify the collecting system and the various segments of the nephron in the following diagram, and describe the functions of each.

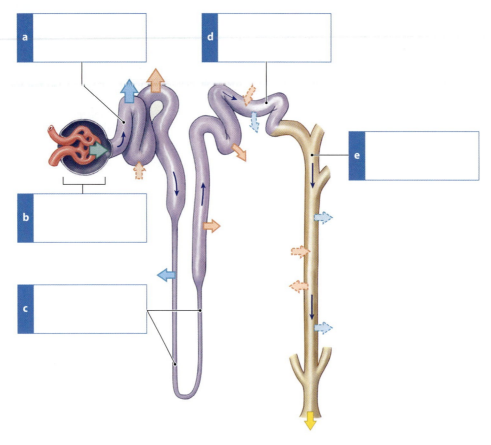

a _____

d _____

e _____

b _____

c _____

2. Matching

Match the following terms with their most closely related description.

- aquaporins
- ADH
- aldosterone
- PCT
- polyuria
- renal corpuscle
- juxtaglomerular complex
- filtrate
- urea
- podocytes

a _____ Site of plasma filtration

b _____ Cells that form filtration slits

c _____ Protein-free solution

d _____ Most abundant waste product in urine

e _____ Site of renin-secreting cells

f _____ Water channels

g _____ Production of excessive volume of urine

h _____ Ion pump—Na$^+$ reabsorbed and K$^+$ secreted

i _____ Primary site of nutrient reabsorption in the nephron

j _____ Regulates passive reabsorption of water from urine in the collecting system

3. Section integration

During hemodialysis, the composition of the blood changes.

a List six substances that diffuse out of the blood and two substances that diffuse into the blood during dialysis.

b What normal kidney function does a dialysis machine replace?

_____

_____

Urine Storage and Elimination

Filtrate modification and urine production end when the fluid enters the renal pelvis. The urinary tract (the ureters, urinary bladder, and urethra) is responsible for transporting, storing, and eliminating urine.

1 A **pyelogram** (PĪ-el-ō-gram) is an image of the urinary system, obtained by taking an x-ray of the kidneys after administering a contrast dye intravenously. A pyelogram shows the relative sizes and positions of the urinary system's main structures.

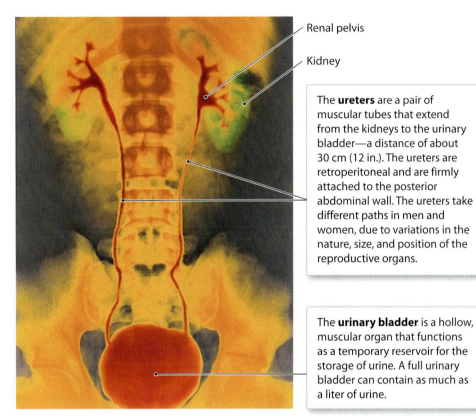

Renal pelvis

Kidney

The **ureters** are a pair of muscular tubes that extend from the kidneys to the urinary bladder—a distance of about 30 cm (12 in.). The ureters are retroperitoneal and are firmly attached to the posterior abdominal wall. The ureters take different paths in men and women, due to variations in the nature, size, and position of the reproductive organs.

The **urinary bladder** is a hollow, muscular organ that functions as a temporary reservoir for the storage of urine. A full urinary bladder can contain as much as a liter of urine.

2 This sectional view of a male pelvis shows the locations of the lower structures of the urinary tract.

3 This sectional view of a female pelvis shows the locations of the lower structures of the urinary tract.

Ureter

Urinary bladder

The **urethra** (plural, *urethrae*) extends from the neck of the urinary bladder and transports urine to the exterior of the body. The urethrae of males and females differ in length and in function. The male urethra is longer and transports semen as well as urine.

Male

Female

The ureters, urinary bladder, and urethra are specialized for conducting urine

1 The urinary bladder and urethra eliminate urine from the body. This process is called urination or **micturition** (mik-choo-RISH-un).

The wall of the urinary bladder contains mucosa, submucosa, and muscularis layers. The muscularis layer consists of inner and outer layers of longitudinal smooth muscle, with a circular layer between the two. Collectively, these layers form the powerful **detrusor** (dē-TROO-sor) muscle of the urinary bladder. Contraction of this muscle compresses the urinary bladder and expels its contents into the urethra.

The **ureters** penetrate the posterior wall of the urinary bladder without entering the peritoneal cavity. They pass through the bladder wall at an oblique angle.

The **ureteral openings** are slitlike rather than rounded. This shape helps prevent the backflow of urine toward the ureter and kidneys when the urinary bladder contracts.

The **trigone** (TRĪ-gōn) is the triangular area bounded by the ureteral openings and the entrance to the urethra.

The region surrounding the urethral opening, known as the **neck of the urinary bladder**, contains a muscular **internal urethral sphincter**. The smooth muscle fibers of this sphincter provide involuntary control over the discharge of urine from the bladder.

Rugae are folds of the urinary bladder lining that disappear as the bladder fills.

Prostate gland (males only)

Urethra

In both sexes, where the urethra passes through the urogenital diaphragm, a circular band of skeletal muscle forms the **external urethral sphincter**. This muscular band acts as a valve. The external urethral sphincter, which is under voluntary control via the perineal branch of the pudendal nerve, has a resting muscle tone and must be voluntarily relaxed to permit urination.

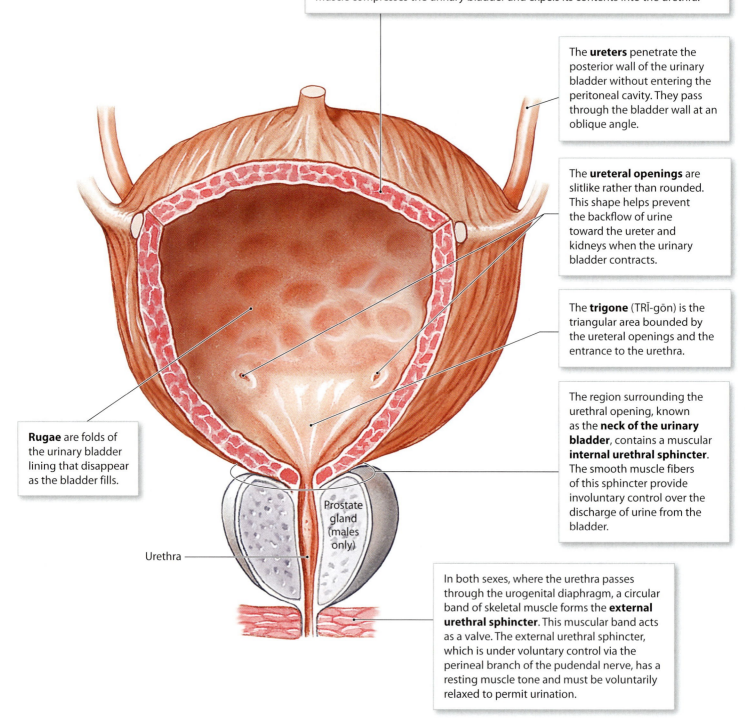

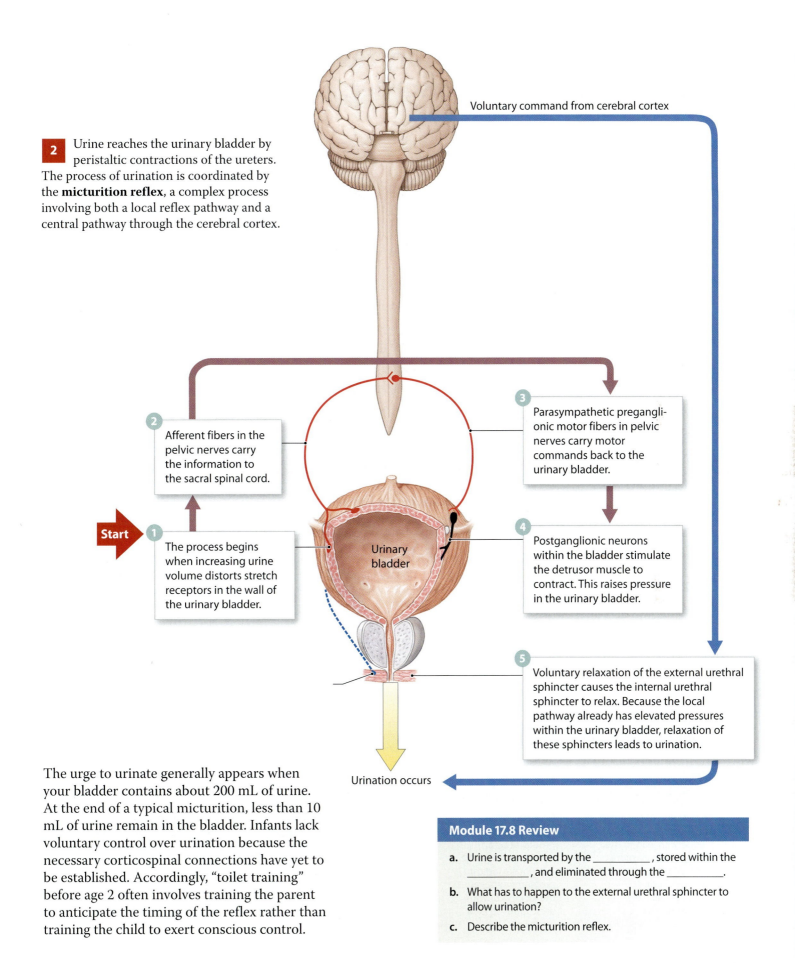

Voluntary command from cerebral cortex

2 Urine reaches the urinary bladder by peristaltic contractions of the ureters. The process of urination is coordinated by the **micturition reflex**, a complex process involving both a local reflex pathway and a central pathway through the cerebral cortex.

2 Afferent fibers in the pelvic nerves carry the information to the sacral spinal cord.

3 Parasympathetic preganglionic motor fibers in pelvic nerves carry motor commands back to the urinary bladder.

Start

1 The process begins when increasing urine volume distorts stretch receptors in the wall of the urinary bladder.

Urinary bladder

4 Postganglionic neurons within the bladder stimulate the detrusor muscle to contract. This raises pressure in the urinary bladder.

5 Voluntary relaxation of the external urethral sphincter causes the internal urethral sphincter to relax. Because the local pathway already has elevated pressures within the urinary bladder, relaxation of these sphincters leads to urination.

Urination occurs

The urge to urinate generally appears when your bladder contains about 200 mL of urine. At the end of a typical micturition, less than 10 mL of urine remain in the bladder. Infants lack voluntary control over urination because the necessary corticospinal connections have yet to be established. Accordingly, "toilet training" before age 2 often involves training the parent to anticipate the timing of the reflex rather than training the child to exert conscious control.

Module 17.8 Review

a. Urine is transported by the _____ , stored within the _____ , and eliminated through the _____ .

b. What has to happen to the external urethral sphincter to allow urination?

c. Describe the micturition reflex.

Physical exams and laboratory tests can often detect urinary disorders

1 The primary signs and symptoms of urinary system disorders include changes in volume and appearance of urine, frequency of urination, and pain. The nature and location of the pain provide clues to the source.

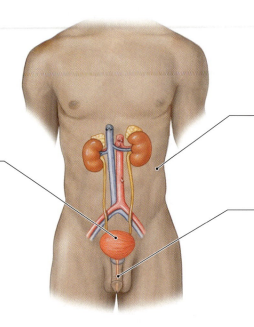

Pain in the superior pubic region may be associated with urinary bladder disorders.

Pain in the superior lumbar region or in the flank that radiates to the right upper quadrant or left upper quadrant can be caused by kidney infections such as **pyelonephritis**, or by kidney stones (**renal calculi**).

Dysuria (painful or difficult urination) can occur with cystitis or urethritis, or with urinary obstructions. In males, an enlarged prostate gland can compress the urethra and lead to dysuria.

2 Certain system-wide clinical signs are often associated with serious urinary system disorders.

Important Clinical Signs of Urinary System Disorders

- **Edema**. Renal disorders often lead to protein loss in the urine (**proteinuria**). If severe, they may cause generalized edema (swelling) in peripheral tissues. Facial swelling, especially around the eyes, is common.

- **Fever**. A fever commonly develops when pathogens infect the urinary system. Urinary bladder infections (**cystitis**) may result in a low-grade fever; kidney infections, such as pyelonephritis, can produce very high fevers.

3 Characteristic changes in urinary output or frequency give clues to underlying problems with the urinary system or other systems.

Abnormal Urine Output and Frequency

Increased Urgency or Increased Frequency	Changes in Urinary Output	Incontinence	Urinary Retention
An irritation of the lining of the ureters or urinary bladder can lead to the desire to urinate more often. However, the total amount of urine produced each day remains normal.	Changes in the volume of urine produced by a person with average fluid intake indicate problems either at the kidneys or with the control of renal function.	**Incontinence**, an inability to control urination voluntarily, may involve: (1) periodic involuntary leakage (stress incontinence); (2) inability to delay urination (urge incontinence); and (3) a continual, slow trickle of urine from a bladder that is always full (overflow incontinence).	In **urinary retention**, renal function is normal, at least initially, but urination does not occur. Urinary retention in males commonly results from an enlarged prostate gland that compresses the urethra.

4 Urinalysis is the clinical examination of a urine sample. Chemical analysis can be performed (see Module 17.5 for normal values), and there are also several screening tests that can be performed by recording changes in the color of test strips that are dipped in the sample. These tests can detect changes in urine pH and abnormal urinary concentrations of glucose, ketones, bilirubin, urobilinogen, plasma proteins, and hemoglobin. A test for one hormone in the urine, *human chorionic gonadotropin* (hCG), provides an early and reliable proof of pregnancy.

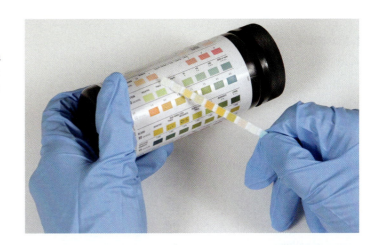

Potential Findings in Urinary Sediment

Crystals		Bacteria	Cells	Casts
Calcium carbonate	Calcium oxalate		RBCs	RBC casts
Calcium phosphate	Cystine crystals		WBCs	WBC casts
Triple phosphate	Uric acid crystals		Epithelial cells	Granular casts
Ammonium urate	Sodium urate	Assorted gram-negative and gram-positive bacteria		Hyaline casts

5 A urine sample may also be spun in a centrifuge, and any sediment examined under the microscope. Sediment may include mineral crystals, bacteria, or red or white blood cells. It may also contain deposits, known collectively as **casts**. Casts have a protein coat and form in DCTs and collecting ducts. Those containing RBCs or WBCs indicate glomerular damage or inflammation or infection (such as pyelonephritis), respectively. During a urinary tract infection, bacteria obtained from the urine can be cultured to determine their identities and sensitivity to specific antibiotics. The Gram stain is a method for differential staining of bacteria. Gram-positive bacteria stain purple, and gram-negative bacteria stain pink. Knowing the gram reaction enables clinicians to prescribe appropriate antibiotics.

Module 17.9 Review

a. What is the term for painful or difficult urination?

b. If a kidney stone obstructs a ureter, this would interfere with the flow of urine between which two points?

c. What types of casts might you find in urine sediment?

1. Matching

Match the following terms with their descriptions.

- urethra
- external urethral sphincter
- detrusor
- rugae
- internal urethral sphincter
- trigone
- renal calculi
- micturition
- parasympathetic postganglionic fibers
- prostate gland
- pyelogram

a	_____	The ring of smooth muscle in the neck of the urinary bladder
b	_____	Triangular area within the urinary bladder
c	_____	Voluntary relaxation of this muscle leads to urination
d	_____	Kidney stones
e	_____	Folds lining the surface of the empty urinary bladder
f	_____	Structure in males through which the urethra passes
g	_____	Contraction of this smooth muscle compresses the urinary bladder
h	_____	Image of the urinary system
i	_____	Tube that transports urine to the exterior
j	_____	Term for urination
k	_____	Stimulate muscle contractions that elevate pressure in the urinary bladder

2. Labeling

Use the following descriptions to fill in the boxes in the micturition reflex diagram below.

- individual relaxes external urethral sphincter, which causes internal sphincter muscle to relax
- afferent fibers carry information to sacral spinal cord
- stretch receptors stimulated
- detrusor muscle stimulated to contract
- parasympathetic preganglionic fibers carry motor commands

3. Short answer

List four primary signs and symptoms of urinary disorders.

4. Short answer

Briefly describe the similarities and differences in the following pairs of terms.

- cystitis/pyelonephritis
- stress incontinence/ overflow incontinence

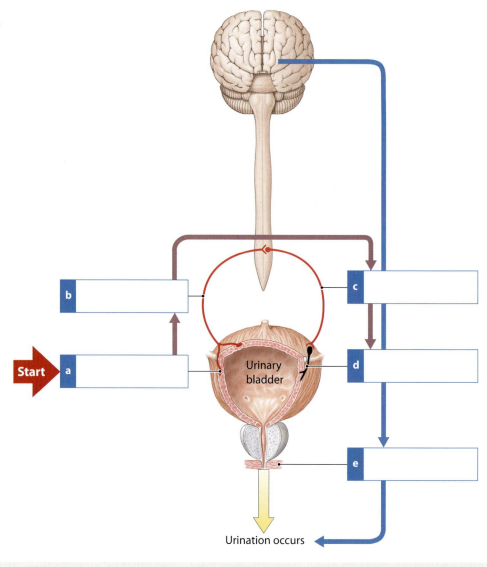

Start

a _____

b _____

c _____

d _____

e _____

Urinary bladder

Urination occurs

Fluid and Electrolyte Balance

Chapter 16 considered the metabolism of the organic components of the body. This chapter takes a broader look at the composition of the body as a whole. We will focus on the inorganic components: water and minerals. **Minerals** are the inorganic substances that dissociate in body fluids to form ions called electrolytes.

1 These pie charts compare the total body composition of adult males and females. The greatest variation is in the intracellular fluid (ICF), or cytosol, as a result of differences in the intracellular water content of fat versus muscle. Less striking differences occur in the extracellular fluid (ECF) values, due to variations in the interstitial fluid volume of various tissues and the larger blood volume in males versus females. The ECF and ICF are called **fluid compartments**, because they commonly behave as distinct entities. Because cells have a plasma membrane and active transport occurs at the membrane surface, cells are able to maintain internal environments quite distinct from that of the ECF.

2 Solid components account for only 40–50 percent of the mass of the body as a whole. This bar graph presents an overview of the solid components of a 70-kg (154-pound) individual with a minimum of body fat. This distribution averages values for males and females ages 18–40 years.

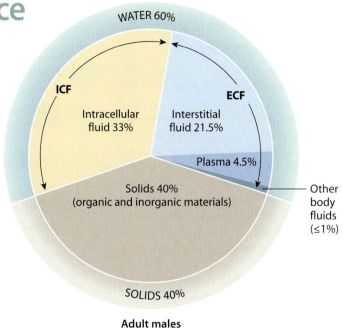

Adult males

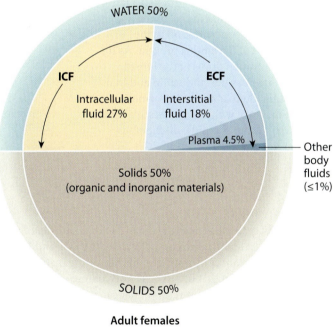

Adult females

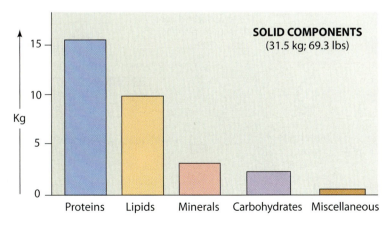

In this section we will consider the exchange of water and electrolytes between the ECF and ICF, and between the body and its external environment.

Fluid and mineral balance exists when gains equal losses

1 **Fluid balance** exists when the amount of water you gain each day is equal to the amount you lose. This diagram shows water balance in the body. Water enters the digestive tract through ingestion or secretion. Metabolic processes produce additional water. Only a small amount leaves the digestive tract in feces, while the rest is lost through evaporation and urination. The situation is complicated by the fact that the accessory digestive glands produce watery secretions that are mixed with arriving food. Most of that secreted water must be recovered along with water gained from food and drink.

All of this water movement involves passive water flow down osmotic gradients. Intestinal epithelial cells continuously absorb nutrients and ions, and these activities gradually lower the solute concentration in the lumen and elevate the solute concentration in the interstitial fluid of the lamina propria. As the solute concentration drops in the lumen, water moves across the epithelium and into the interstitial fluid, maintaining osmotic equilibrium. Once within the interstitial fluid, the absorbed water is rapidly distributed throughout the ECF.

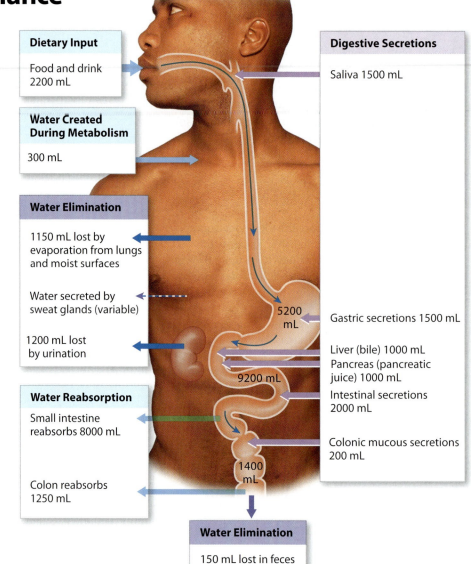

Dietary Input

Food and drink 2200 mL

Water Created During Metabolism

300 mL

Water Elimination

1150 mL lost by evaporation from lungs and moist surfaces

Water secreted by sweat glands (variable)

1200 mL lost by urination

Water Reabsorption

Small intestine reabsorbs 8000 mL

Colon reabsorbs 1250 mL

Digestive Secretions

Saliva 1500 mL

Gastric secretions 1500 mL

Liver (bile) 1000 mL
Pancreas (pancreatic juice) 1000 mL

Intestinal secretions 2000 mL

Colonic mucous secretions 200 mL

5200 mL

9200 mL

1400 mL

Water Elimination

150 mL lost in feces

2 This diagram illustrates the major factors that affect ECF volume. Although the composition of the ECF and ICF are very different, the two are at osmotic equilibrium. Note that the volume of the ICF is larger than that of the ECF. The volume of water held within cells represents a significant reserve that can prevent sudden changes in the solute and water concentrations in the ECF. A rapid water movement between the ECF and the ICF in response to an osmotic gradient is called a **fluid shift**. Fluid shifts occur rapidly if the osmotic concentration of the ECF changes, and reach equilibrium within minutes to hours. This can be an important factor when water intake is restricted but water losses are severe.

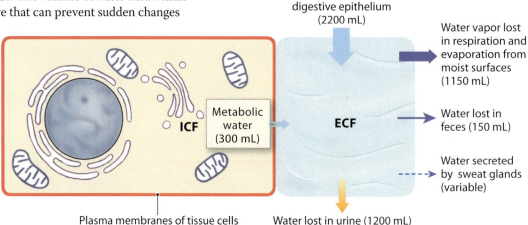

Water absorbed across digestive epithelium (2200 mL)

Water vapor lost in respiration and evaporation from moist surfaces (1150 mL)

Metabolic water (300 mL)

ICF

ECF

Water lost in feces (150 mL)

Water secreted by sweat glands (variable)

Plasma membranes of tissue cells

Water lost in urine (1200 mL)

3 **Mineral balance** is the balance between ion absorption, which occurs across the lining of the small intestine and colon, and ion excretion, which occurs primarily at the kidneys. Sweat glands are a potential source of both water and mineral loss, but the rate of secretion is extremely variable.

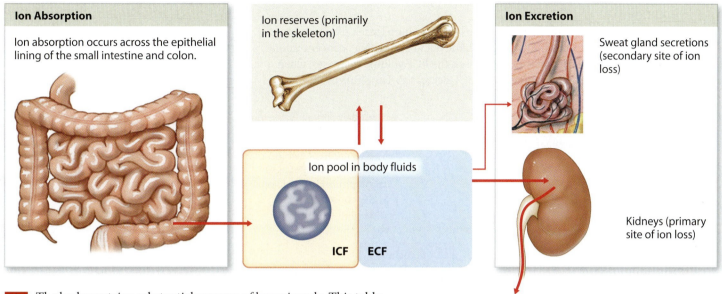

Ion Absorption

Ion absorption occurs across the epithelial lining of the small intestine and colon.

Ion reserves (primarily in the skeleton)

Ion pool in body fluids

ICF   ECF

Ion Excretion

Sweat gland secretions (secondary site of ion loss)

Kidneys (primary site of ion loss)

4 The body contains substantial reserves of key minerals. This table summarizes the functions of the various minerals and indicates the primary routes of ion excretion. The daily intake must on average equal the amount lost each day if the individual is to stay in mineral balance.

Minerals and Mineral Reserves

Mineral	Functions	Total Body Content	Primary Route of Excretion
Sodium	Essential for normal membrane function	110 g, primarily body fluids	Urine, sweat, feces
Potassium	Essential for normal membrane function	140 g, primarily cytoplasm	Urine
Chloride	Major anion of body fluids; forms HCl	89 g, primarily body fluids	Urine, sweat
Calcium	Essential for muscle and neuron function, and normal bone structure	1.36 kg, primarily skeleton	Urine, feces
Phosphorus	Required for bone mineralization, formation of high-energy compounds, and activation of enzymes	744 g, primarily skeleton	Urine, feces
Magnesium	Cofactor of enzymes	29 g	Urine
Iron	Component of hemoglobin and myoglobin	3.9 g	Urine (traces)
Zinc	Cofactor of enzyme systems	2 g	Urine, hair (traces)
Copper	Cofactor for hemoglobin synthesis	127 mg	Urine, feces (traces)
Cobalt	Cofactor for amino acid catabolism	1.1 g	Feces, urine

Module 17.10 Review

a. Identify routes of fluid loss from the body.

b. Describe a fluid shift.

c. Define fluid balance and mineral balance.

Water balance depends on sodium balance, and the two are regulated simultaneously

1 **Sodium balance** exists when sodium gains equal sodium losses. The regulatory mechanism involved changes the ECF volume but keeps the Na$^+$ concentration relatively stable. When sodium gains exceed losses, the ECF volume increases; when sodium losses exceed gains, the volume of the ECF decreases. The shifts in ECF volume occur without a significant change in the osmotic concentration of the ECF. These adjustments cause minor changes in ECF volume that do not cause adverse physiological effects.

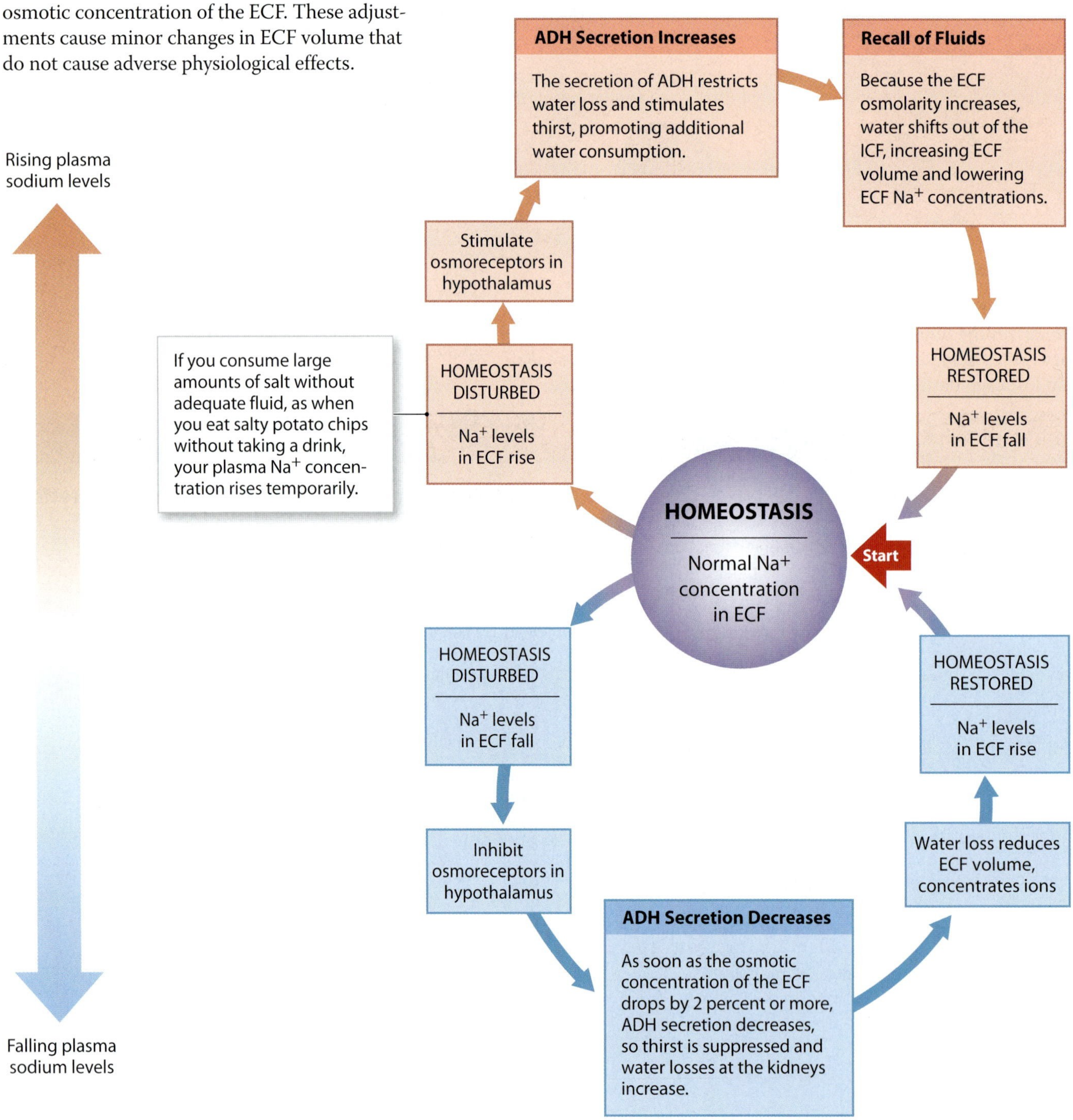

Rising plasma sodium levels

ADH Secretion Increases

The secretion of ADH restricts water loss and stimulates thirst, promoting additional water consumption.

Recall of Fluids

Because the ECF osmolarity increases, water shifts out of the ICF, increasing ECF volume and lowering ECF Na$^+$ concentrations.

Stimulate osmoreceptors in hypothalamus

If you consume large amounts of salt without adequate fluid, as when you eat salty potato chips without taking a drink, your plasma Na$^+$ concentration rises temporarily.

HOMEOSTASIS DISTURBED

Na$^+$ levels in ECF rise

HOMEOSTASIS RESTORED

Na$^+$ levels in ECF fall

HOMEOSTASIS

Normal Na$^+$ concentration in ECF

Start

HOMEOSTASIS DISTURBED

Na$^+$ levels in ECF fall

HOMEOSTASIS RESTORED

Na$^+$ levels in ECF rise

Inhibit osmoreceptors in hypothalamus

Water loss reduces ECF volume, concentrates ions

ADH Secretion Decreases

As soon as the osmotic concentration of the ECF drops by 2 percent or more, ADH secretion decreases, so thirst is suppressed and water losses at the kidneys increase.

Falling plasma sodium levels

2 If disturbances in sodium balance alter ECF volume significantly, the homeostatic mechanisms responsible for regulating blood volume and blood pressure will be activated. Why? Because when ECF volume changes, so does plasma volume and, in turn, so does blood volume. If ECF volume rises, blood volume goes up; if ECF volume drops, blood volume goes down.

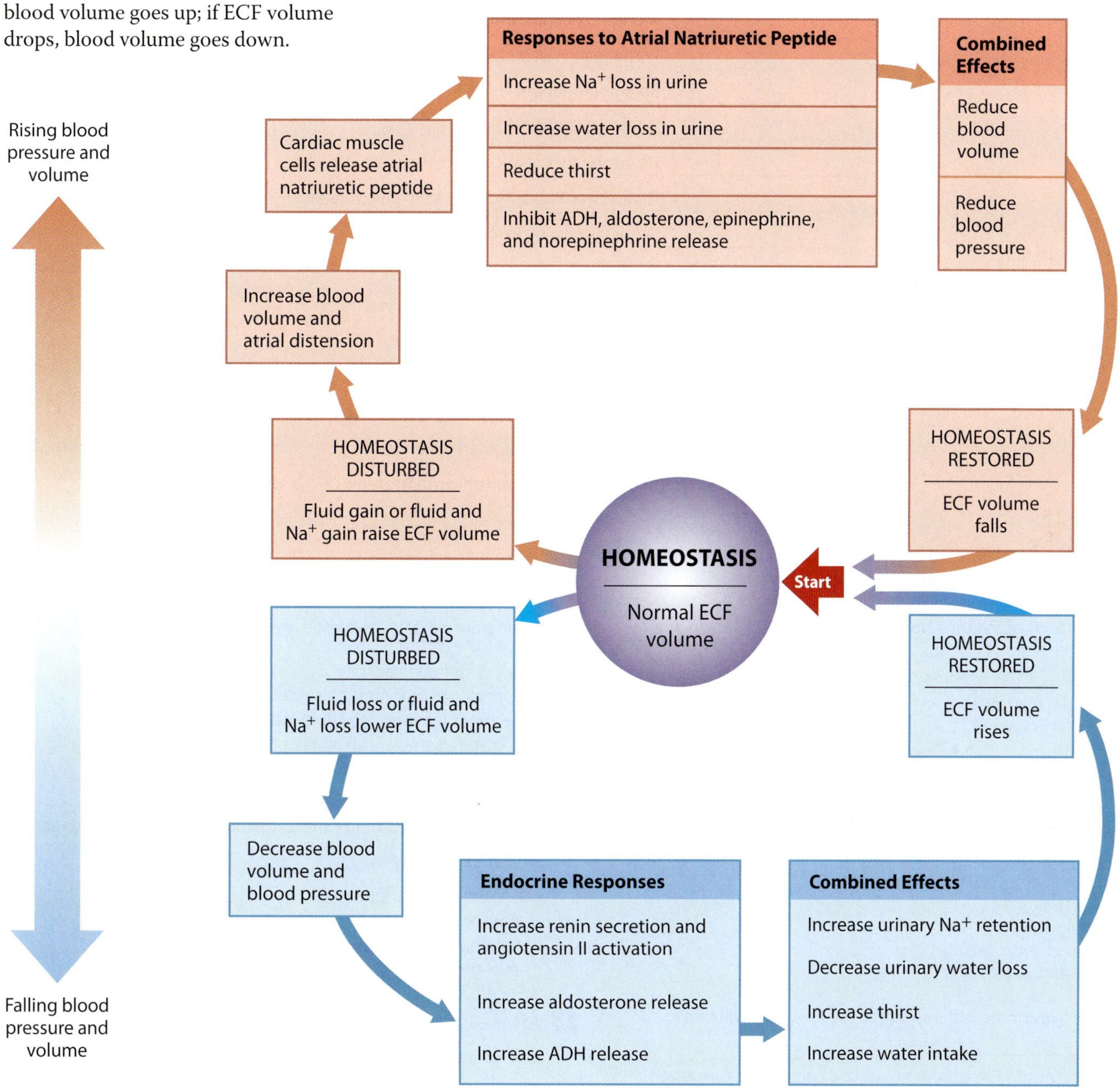

Rising blood pressure and volume

Responses to Atrial Natriuretic Peptide

Increase Na$^+$ loss in urine

Increase water loss in urine

Reduce thirst

Inhibit ADH, aldosterone, epinephrine, and norepinephrine release

Combined Effects

Reduce blood volume

Reduce blood pressure

Cardiac muscle cells release atrial natriuretic peptide

Increase blood volume and atrial distension

HOMEOSTASIS DISTURBED

Fluid gain or fluid and Na$^+$ gain raise ECF volume

HOMEOSTASIS RESTORED

ECF volume falls

HOMEOSTASIS

Normal ECF volume

Start

HOMEOSTASIS DISTURBED

Fluid loss or fluid and Na$^+$ loss lower ECF volume

HOMEOSTASIS RESTORED

ECF volume rises

Decrease blood volume and blood pressure

Falling blood pressure and volume

Endocrine Responses

Increase renin secretion and angiotensin II activation

Increase aldosterone release

Increase ADH release

Combined Effects

Increase urinary Na$^+$ retention

Decrease urinary water loss

Increase thirst

Increase water intake

Sustained abnormalities in the Na$^+$ concentration in the ECF occur only when there are severe problems with fluid balance.

Module 17.11 Review

a. What effect does inhibition of osmoreceptors have on ADH secretion and thirst?

b. What effect does aldosterone have on sodium ion concentration in the ECF?

c. Briefly summarize the relationship between sodium ion concentration and the ECF.

1. Matching

Match the following terms with the most closely related description.

- kidneys
- potassium
- fluid compartments
- fluid balance
- hypertonic plasma
- dehydration
- aldosterone
- plasma, interstitial fluid
- osmoreceptors
- fluid shift
- proteins
- ADH
- minerals
- sodium

a _____ Monitor plasma osmotic concentration

b _____ Water gain = water loss

c _____ Major components of ECF

d _____ Dominant cation in ECF

e _____ Hormone that restricts water loss and stimulates thirst

f _____ Dissociate in body fluids to form electrolytes

g _____ Dominant cation in ICF

h _____ ICF and ECF

i _____ Most important sites of sodium ion regulation

j _____ Water movement between ECF and ICF

k _____ Water moves from cells into ECF

l _____ Greatest contributor to the solid components of the body

m _____ Water losses greater than water gains

n _____ Regulates sodium ion absorption along DCT and collecting system

2. Multiple choice

Choose the bulleted item that best completes each statement.

a Nearly two-thirds of the total body water content is

_____ .

- extracellular fluid (ECF)
- intracellular fluid (ICF)
- tissue fluid
- interstitial fluid

b Electrolyte balance involves balancing the rates of absorption across the digestive tract with rates of loss at the _____ .

- heart and lungs
- stomach and liver
- kidneys and sweat glands
- pancreas and gallbladder

c If the ECF is hypertonic with respect to the ICF, water will move

_____ .

- from the ECF into the cells until osmotic equilibrium is restored
- from the cells into the ECF until osmotic equilibrium is restored
- in both directions until osmotic equilibrium is restored
- in response to the sodium–potassium exchange pump

d When pure water is consumed, the ECF

_____ .

- becomes hypotonic with respect to the ICF
- becomes hypertonic with respect to the ICF
- becomes isotonic with respect to the ICF
- electrolytes become more concentrated

e Physiological adjustments affecting fluid and electrolyte balance are regulated primarily by _____ .

- antidiuretic hormone
- aldosterone
- atrial natriuretic peptide
- all of these

f When water is lost but electrolytes are retained, the osmolarity of the ECF rises, and osmosis then moves water _____ .

- out of the ECF and into the ICF
- back and forth between the ICF and ECF
- out of the ICF and into the ECF
- none of the above

3. Short answer

What two primary steps are involved in the regulation of sodium ion concentrations?

_____

_____

_____

Acid-Base Balance

Your body is in **acid-base balance** when the production of hydrogen ions is precisely offset by their loss, and when the pH of body fluids remains within normal limits.

1 This diagram shows the major factors involved in maintaining acid-base balance. The primary challenge to homeostasis is that your body generates a variety of acids during normal metabolic operations, and a significant decline in body fluid pH must be prevented.

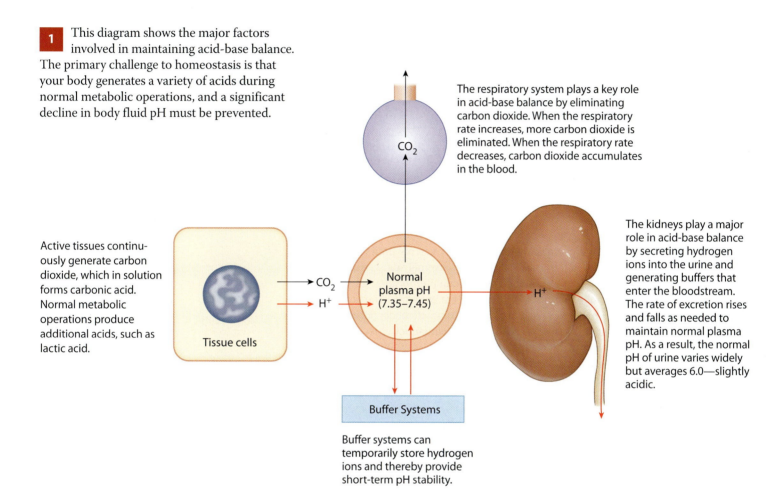

The respiratory system plays a key role in acid-base balance by eliminating carbon dioxide. When the respiratory rate increases, more carbon dioxide is eliminated. When the respiratory rate decreases, carbon dioxide accumulates in the blood.

Active tissues continuously generate carbon dioxide, which in solution forms carbonic acid. Normal metabolic operations produce additional acids, such as lactic acid.

Tissue cells

CO_2

H^+

Normal plasma pH (7.35–7.45)

H^+

The kidneys play a major role in acid-base balance by secreting hydrogen ions into the urine and generating buffers that enter the bloodstream. The rate of excretion rises and falls as needed to maintain normal plasma pH. As a result, the normal pH of urine varies widely but averages 6.0—slightly acidic.

Buffer Systems

Buffer systems can temporarily store hydrogen ions and thereby provide short-term pH stability.

To maintain homeostasis, the pH of body fluids must remain within specific ranges. Remember that pH is a measure of the hydrogen ion concentration. The normal balance between acid and base in the blood plasma exists as a relationship between materials produced and used by body metabolism and the relative amounts of acidic or basic materials excreted from the body. A normal state of acid-base balance in blood plasma is slightly alkaline, 7.35–7.45, indicating an excess of hydroxide ions. This section discusses the buffer systems and compensation mechanisms (respiratory and renal) that regulate hydrogen ions in body fluids to maintain acid-base balance.

Buffer systems oppose potentially dangerous disturbances in acid-base balance

1 The topic of pH and the chemical nature of acids, bases, and buffers was introduced in Module 2.7. This table reviews key terms important to the discussion that follows.

A Review of Important Terms Relating to Acid-Base Balance	
pH	The concentration of hydrogen ions [H$^+$] in a solution; the pH value is a number between 0 and 14
Neutral	A solution with a pH of 7; the solution contains equal numbers of hydrogen ions and hydroxide ions
Acidic	A solution with a pH below 7; in this solution, hydrogen ions [H$^+$] predominate
Basic, or alkaline	A solution with a pH above 7; in this solution, hydroxide ions [OH$^-$] predominate
Acid	A substance that dissociates to release hydrogen ions, decreasing pH
Base	A substance that dissociates to release hydroxide ions or to remove hydrogen ions, increasing pH
Salt	An ionic compound consisting of a cation other than a hydrogen ion and an anion other than a hydroxide ion
Buffer	A substance that tends to oppose changes in the pH of a solution by removing or replacing hydrogen ions; in body fluids, buffers maintain blood pH within normal limits (7.35–7.45)

2 The pH of the ECF normally remains within relatively narrow limits, usually 7.35–7.45. Any deviation from the normal range is extremely dangerous, because changes in H$^+$ concentrations disrupt the stability of plasma membranes, alter the structure of proteins, and change the activities of important enzymes. You could not survive for long with an ECF pH below 6.8 or above 7.7. In practice, decreases in pH are much more common than increases, because normal cellular activities generate several acids, including carbonic acid. Any shift in pH affects virtually all body systems, but the nervous and cardiovascular systems are particularly sensitive to pH fluctuations.

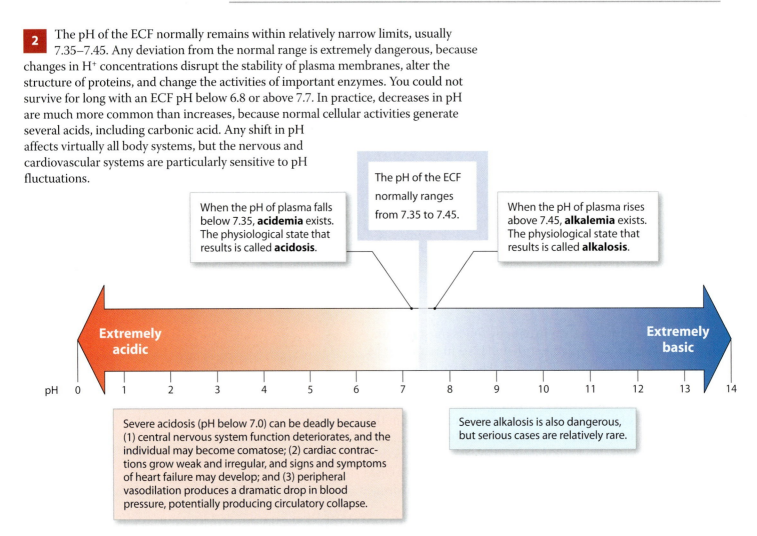

The pH of the ECF normally ranges from 7.35 to 7.45.

When the pH of plasma falls below 7.35, **acidemia** exists. The physiological state that results is called **acidosis**.

When the pH of plasma rises above 7.45, **alkalemia** exists. The physiological state that results is called **alkalosis**.

Extremely acidic

Extremely basic

pH 0 1 2 3 4 5 6 7 8 9 10 11 12 13 14

Severe acidosis (pH below 7.0) can be deadly because (1) central nervous system function deteriorates, and the individual may become comatose; (2) cardiac contractions grow weak and irregular, and signs and symptoms of heart failure may develop; and (3) peripheral vasodilation produces a dramatic drop in blood pressure, potentially producing circulatory collapse.

Severe alkalosis is also dangerous, but serious cases are relatively rare.

3 The carbon dioxide (CO_2) level is the most important factor affecting the pH of body tissues. This is because carbon dioxide combines with water to form **carbonic acid** (H_2CO_3). Since carbonic anhydrase converts most of the carbon dioxide in solution to carbonic acid, and most of the carbonic acid dissociates, there is an inverse relationship between the carbon dioxide level and pH.

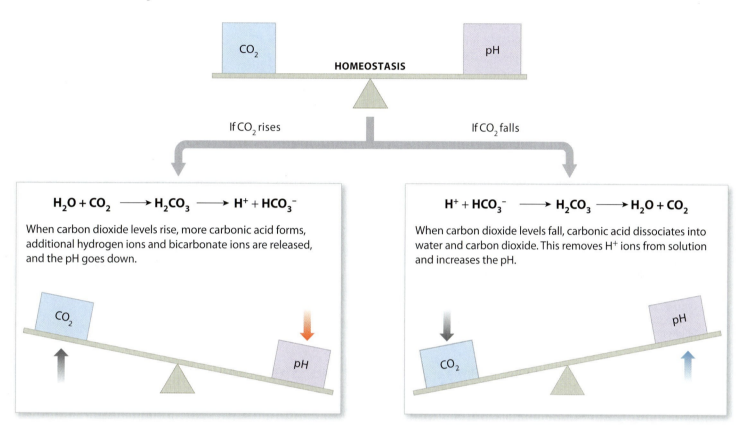

HOMEOSTASIS

If CO_2 rises

If CO_2 falls

$$H_2O + CO_2 \longrightarrow H_2CO_3 \longrightarrow H^+ + HCO_3^-$$

When carbon dioxide levels rise, more carbonic acid forms, additional hydrogen ions and bicarbonate ions are released, and the pH goes down.

$$H^+ + HCO_3^- \longrightarrow H_2CO_3 \longrightarrow H_2O + CO_2$$

When carbon dioxide levels fall, carbonic acid dissociates into water and carbon dioxide. This removes H^+ ions from solution and increases the pH.

4 A **buffer system** in body fluids generally consists of a combination of a weak acid (HY) and the anion (Y^-) released by its dissociation. The anion functions as a weak base. In solution, molecules of the weak acid exist in equilibrium with its dissociation products. In chemical notation, this relationship is represented as:

$$HY \rightleftharpoons H^+ + Y^-$$

Adding H^+ to the solution upsets the equilibrium and results in the formation of additional molecules of the weak acid.

$$H^+ + Y^- \xrightarrow{\ H^+\ } H^+ + HY$$

Removing H^+ from the solution also upsets the equilibrium and results in the dissociation of additional molecules of HY. This releases H^+.

$$H^+ + HY \xrightarrow[H^+]{} H^+ + Y^-$$

Buffer systems can temporarily compensate for shifts in pH, but ultimately the problem must be eliminated before all of the available buffers are tied up. A "fix" involves some combination of renal compensation and respiratory compensation. In **renal compensation** the kidneys secrete or generate either H^+ or HCO_3^-. In **respiratory compensation** the respiratory rate increases or decreases to control the rate at which carbon dioxide is eliminated.

Module 17.12 Review

a. Define acidemia and alkalemia.

b. What is the most important factor affecting the pH of the ECF?

c. Summarize the relationship between CO_2 levels and pH.

Buffer systems can delay but not prevent pH shifts in the ICF and ECF

1 The body has three major buffer systems, each with slightly different characteristics and distributions. Although buffer systems can tie up excess H$^+$, they provide only a temporary solution to acid-base imbalance. The hydrogen ions are not eliminated, but merely rendered harmless. In the process a buffer molecule is tied up, and the supply of buffers is limited.

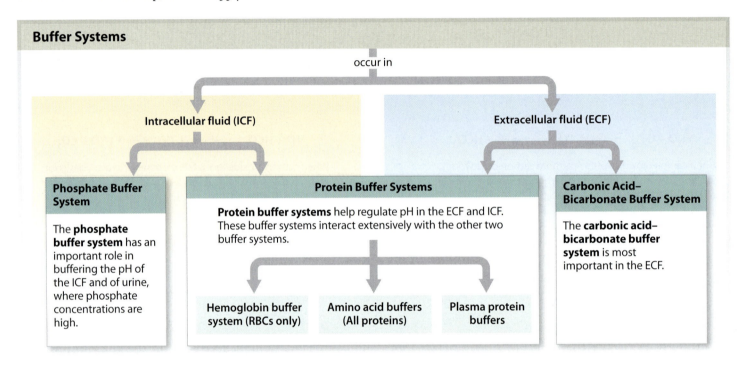

Buffer Systems

occur in

Intracellular fluid (ICF)   **Extracellular fluid (ECF)**

Phosphate Buffer System

The **phosphate buffer system** has an important role in buffering the pH of the ICF and of urine, where phosphate concentrations are high.

Protein Buffer Systems

Protein buffer systems help regulate pH in the ECF and ICF. These buffer systems interact extensively with the other two buffer systems.

Hemoglobin buffer system (RBCs only)   **Amino acid buffers (All proteins)**   **Plasma protein buffers**

Carbonic Acid–Bicarbonate Buffer System

The **carbonic acid–bicarbonate buffer system** is most important in the ECF.

2 The **hemoglobin buffer system** is the only intracellular buffer system that can have an immediate effect on the pH of body fluids. In the tissues, red blood cells absorb carbon dioxide from the plasma and convert it to carbonic acid. As the carbonic acid dissociates, hemoglobin proteins buffer the hydrogen ions. At the lungs, the entire reaction sequence proceeds in reverse and the CO_2 diffuses into the alveoli to be exhaled.

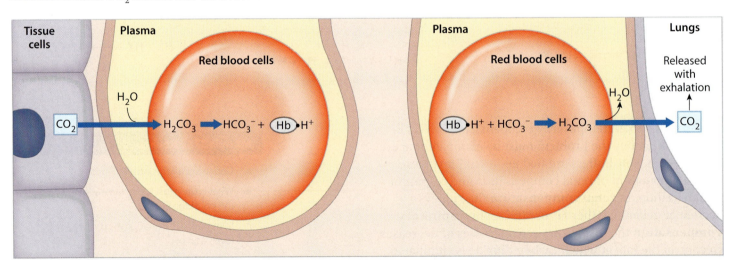

Tissue cells   Plasma   Red blood cells   Plasma   Red blood cells   Lungs

Released with exhalation

H_2O

$CO_2 \rightarrow H_2CO_3 \rightarrow HCO_3^- + \text{Hb} \cdot H^+$

$\text{Hb} \cdot H^+ + HCO_3^- \rightarrow H_2CO_3 \rightarrow CO_2$

H_2O

3 Protein buffer systems reduce the rate of pH change, usually by binding excess hydrogen ions. These buffer systems depend on the ability of amino acids to respond to pH changes by accepting or releasing H^+. The underlying mechanism is shown here. In a protein, most of the carboxyl and amino groups in the main chain are tied up in peptide bonds, leaving only the $-COO^-$ of the first amino acid and the $-NH_2$ of the last amino acid available as buffers. Thus, the R-groups of the component amino acids provide most of the buffering capacity of proteins.

Increasing acidity (decreasing pH)

Start

Normal pH
(7.35–7.45)

If pH drops, the carboxylate ion (COO^-) and the amino group ($-NH_2$) of a free amino acid can act as weak bases and accept additional hydrogen ions, forming a carboxyl group ($-COOH$) and an amino ion ($-NH_3^+$), respectively. Many of the R-groups can also accept hydrogen ions, forming RH^+.

At the normal pH of body fluids (7.35–7.45), the carboxyl groups of most amino acids have released their hydrogen ions.

4 The carbonic acid–bicarbonate buffer system involves freely reversible reactions. A change in the concentration of any participant affects the concentrations of all other participants and shifts the direction of the reactions under way.

BICARBONATE RESERVE

Body fluids contain a large reserve of HCO_3^-, primarily in the form of dissolved molecules of the weak base sodium bicarbonate ($NaHCO_3$). This readily available supply of HCO_3^- is known as the **bicarbonate reserve**.

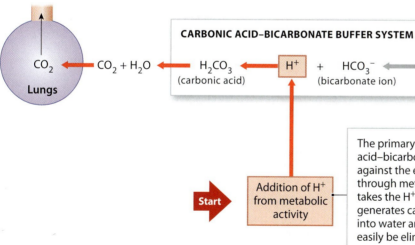

The primary function of the carbonic acid–bicarbonate buffer system is to protect against the effects of the acids generated through metabolic activity. In effect, it takes the H^+ released by these acids and generates carbonic acid that dissociates into water and carbon dioxide, which can easily be eliminated at the lungs.

Metabolic acid-base disorders result from the production or loss of excessive amounts of acids. The primary role of the carbonic acid–bicarbonate buffer system is to protect against such disorders. **Respiratory acid-base disorders** result from an imbalance between the rate at which CO_2 is generated and the rate at which CO_2 is eliminated at the lungs. The carbonic acid–bicarbonate buffer system cannot protect against respiratory disorders; the imbalances must be corrected by reflexive changes in the depth and rate of respiration.

Module 17.13 Review

a. Identify the body's three major buffer systems.

b. How do proteins and free amino acids act as buffers when pH drops below normal?

c. Describe the carbonic acid–bicarbonate buffer system.

Homeostatic responses to acidosis and alkalosis involve respiratory and renal mechanisms as well as buffer systems

Responses to Acidosis

1 **Metabolic acidosis** develops when acids release large numbers of hydrogen ions and the pH decreases. For homeostasis to be preserved, the excess H^+ must either be
(1) permanently tied up through the formation of water (linked to the formation of CO_2 that can be eliminated at the lungs), or
(2) removed from body fluids through secretion at the kidneys.

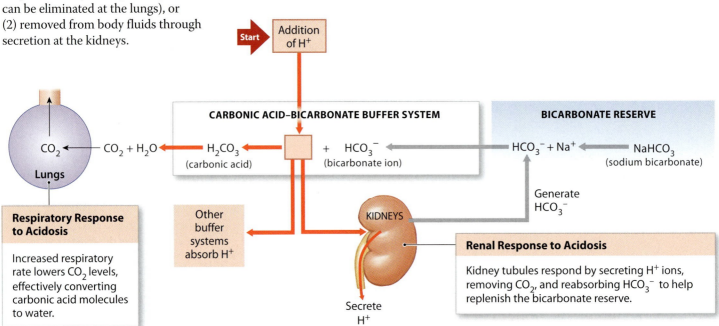

2 **Respiratory acidosis** develops when the rate of carbon dioxide removal by the lungs is less than the rate of carbon dioxide generation. This flowchart summarizes the integrated homeostatic responses to respiratory acidosis.

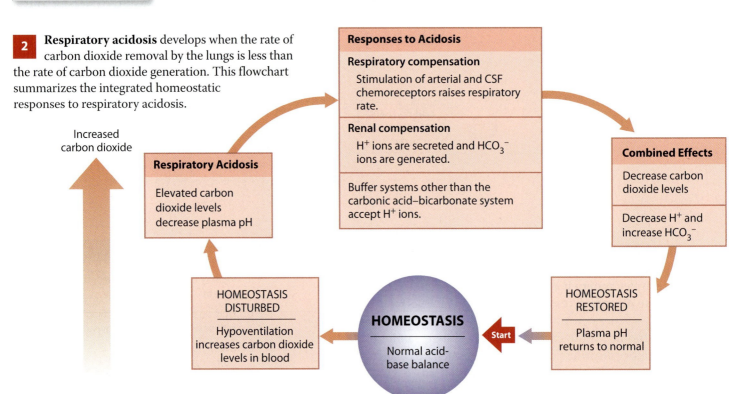

Responses to Alkalosis

3 **Metabolic alkalosis** develops when large numbers of hydrogen ions are removed from body fluids, raising pH. When this occurs, (1) the rate of H^+ secretion at the kidneys declines, (2) tubule cells do not reclaim the bicarbonates in tubular fluid, and (3) the collecting system transports HCO_3^- into tubular fluid while releasing a strong acid (HCl) into the ECF.

Start → Removal of H^+

Lungs

$CO_2 + H_2O$ →

CARBONIC ACID–BICARBONATE BUFFER SYSTEM

H_2CO_3 (carbonic acid) → ☐ + HCO_3^- (bicarbonate ion)

BICARBONATE RESERVE

$HCO_3^- + Na^+$ → $NaHCO_3$ (sodium bicarbonate)

Respiratory Response to Alkalosis

Decreased respiratory rate elevates CO_2, effectively converting CO_2 molecules to carbonic acid.

Other buffer systems release H^+

Generate H^+

KIDNEYS

Secrete HCO_3^-

Renal Response to Alkalosis

Kidney tubules respond by conserving H^+ ions and secreting HCO_3^-.

HOMEOSTASIS

Normal acid-base balance

HOMEOSTASIS DISTURBED

Hyperventilation decreases carbon dioxide levels in blood

Decreased carbon dioxide

Start ← **HOMEOSTASIS RESTORED**

Plasma pH returns to normal

Respiratory Alkalosis

Lower carbon dioxide levels raise plasma pH

Responses to Alkalosis

Respiratory compensation

Inhibition of arterial and CSF chemoreceptors decreases respiratory rate.

Renal compensation

H^+ ions are generated and HCO_3^- ions are secreted.

Buffer systems other than the carbonic acid–bicarbonate system release H^+ ions.

Combined Effects

Increase carbon dioxide levels

Increase H^+ and decrease HCO_3^-

4 **Respiratory alkalosis** develops if the rate of carbon dioxide elimination by the lungs exceeds the rate of carbon dioxide generation. This flowchart summarizes the homeostatic responses to respiratory alkalosis. Most cases are related to anxiety, and the resulting hyperventilation is self-limiting—the individual often faints and the respiratory rate then returns to normal levels.

Module 17.14 Review

a. Compare metabolic acidosis and metabolic alkalosis.

b. Compare respiratory acidosis and respiratory alkalosis.

c. If the kidneys are conserving HCO_3^- and eliminating H^+ in urine, to which condition are the kidneys responding?

1. Labeling

Use the following terms to label the boxes in the two flowcharts. Terms may be used more than once.

- plasma pH decrease
- plasma pH increase
- increased CO_2
- decreased CO_2
- increased
- decreased
- alkalosis
- acidosis
- generated
- secreted

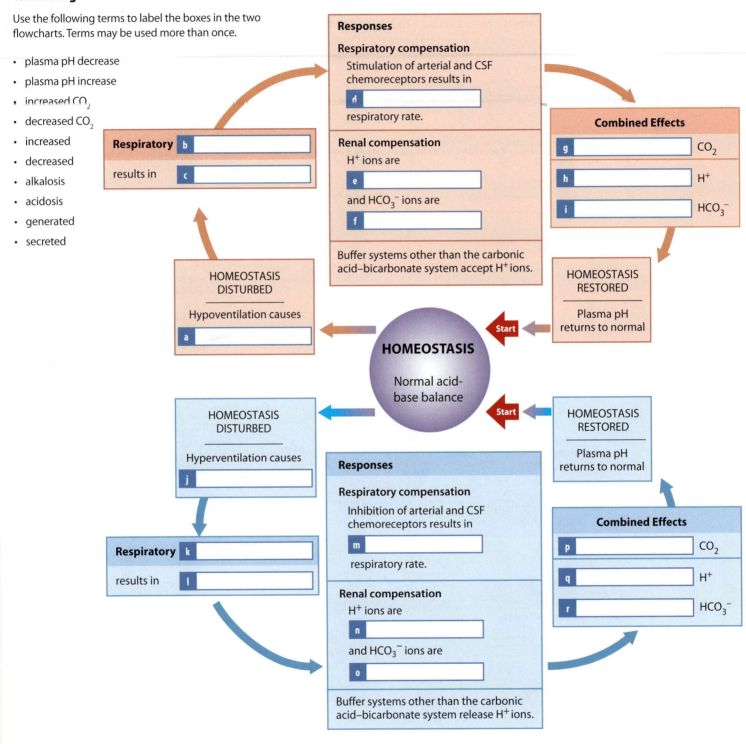

2. Section integration

After falling into an abandoned stone quarry filled with water and nearly drowning, a young boy is rescued. His rescuers assess his condition and find that his body fluids have high CO_2 and lactate levels, and low O_2 levels. Identify the underlying problem and recommend the necessary treatment to restore homeostatic conditions. _____

_____

_____

Visual Outline with Key Terms

Summarize the content of each module using the terms in the order provided.

SECTION 1

Anatomy of the Urinary System

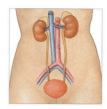

- urinary system
- urinary tract
- kidneys
- urine
- ureters
- urinary bladder
- urination
- urethra

17.1

The kidneys are paired retroperitoneal organs

- kidney
- hilum
- ureters
- retroperitoneal
- renal fascia
- fibrous capsule
- renal sinus
- renal cortex
- renal medulla
- renal pyramid
- renal papilla
- renal column
- kidney lobe
- minor calyx
- major calyx
- renal pelvis

17.2

A nephron can be divided into regions; each region has specific functions

- nephron
- renal corpuscle
- renal tubule
- capsular space
- filtrate
- glomerular capsule
- glomerulus
- proximal convoluted tubule (PCT)
- tubular fluid
- nephron loop
- distal convoluted tubule (DCT)
- collecting system
- collecting duct
- papillary duct
- cortical nephron
- juxtamedullary nephron

17.3

The kidneys are highly vascular, and the circulation patterns are complex

- renal artery
- segmental arteries
- interlobar arteries
- arcuate arteries
- cortical radiate arteries
- afferent arterioles
- glomerulus
- cortical radiate veins
- arcuate veins
- interlobar veins
- renal vein
- efferent arteriole
- peritubular capillaries
- vasa recta

SECTION 2

Overview of Renal Physiology

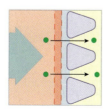

- urea
- creatinine
- uric acid
- filtration
- reabsorption
- secretion

17.4

Filtration, reabsorption, and secretion occur in specific regions of the nephron and collecting system

- efferent arteriole
- juxtaglomerular complex
- afferent arteriole
- capsular space
- podocytes
- filtration slits

17.5

Urine volume and concentration are hormonally regulated

- obligatory water reabsorption
- facultative water reabsorption
- distal convoluted tubule
- collecting duct
- antidiuretic hormone (ADH)
- aquaporins
- urine
- polyuria
- oliguria
- anuria

17.6

Renal function is an integrative process

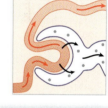

- osmotic concentration
- obligatory water reabsorption
- urea
- ADH
- vasa recta

17.7

Renal failure is a life-threatening condition

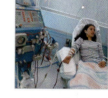

- renal failure
- acute renal failure
- chronic renal failure
- hemodialysis
- dialysis
- dialysis fluid
- shunts

SECTION 3

Urine Storage and Elimination

- pyelogram
- ureters
- urinary bladder
- urethra

17.8

The ureters, urinary bladder, and urethra are specialized for conducting urine

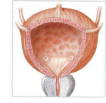

- micturition
- detrusor
- ureters
- ureteral openings
- trigone
- rugae
- neck of the urinary bladder
- internal urethral sphincter
- external urethral sphincter
- micturition reflex

• = *Term boldfaced in this module*

17.9

Physical exams and laboratory tests can often detect urinary disorders

- pyelonephritis
- renal calculi
- dysuria
- edema
- proteinuria
- fever
- cystitis
- incontinence
- urinary retention
- casts

SECTION 4

Fluid and Electrolyte Balance

- minerals
- electrolytes
- intracellular fluid (ICF)
- extracellular fluid (ECF)
- fluid compartments

17.10

Fluid and mineral balance exists when gains equal losses

- fluid balance
- fluid shift
- mineral balance
- ion absorption
- ion excretion

17.11

Water balance depends on sodium balance, and the two are regulated simultaneously

- sodium balance
- osmoreceptors
- ECF volume
- homeostasis

• = Term boldfaced in this module

SECTION 5

Acid-Base Balance

- acid-base balance
- hydrogen ions
- plasma pH

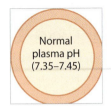

17.12

Buffer systems oppose potentially dangerous disturbances in acid-base balance

- acidemia
- acidosis
- alkalemia
- alkalosis
- carbonic acid
- buffer system
- renal compensation
- respiratory compensation

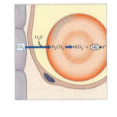

17.13

Buffer systems can delay but not prevent pH shifts in the ICF and ECF

- phosphate buffer system
- protein buffer systems
- carbonic acid–bicarbonate buffer system
- hemoglobin buffer system
- bicarbonate reserve
- metabolic acid-base disorders
- respiratory acid-base disorders

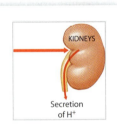

17.14

Homeostatic responses to acidosis and alkalosis involve respiratory and renal mechanisms as well as buffer systems

- metabolic acidosis
- carbonic acid–bicarbonate buffer system
- respiratory response to acidosis
- renal response to acidosis
- respiratory acidosis
- metabolic alkalosis
- respiratory response to alkalosis
- renal response to alkalosis
- respiratory alkalosis

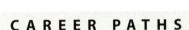

CAREER PATHS

Nurse Practitioner

"I'm able to interact with many different people in various states of health and effect a positive change on their lives."

— **Robin Andersen,**
Nurse Practitioner
UCSF Medical Center

Robin Andersen, a nurse practitioner (NP), always knew she would pursue a career in health care. "I wanted to choose a career that would make a difference in a tangible way. So I decided during high school to become an NP."

Robin works in acute care in a general surgery service in a major academic hospital.

Her day begins at 7 a.m., checking in with the physicians on her team for "sign-out" on each of the patients in their care.

As a nurse practitioner, Robin formulates a diagnosis and then initiates whatever treatment the patient may need. "I address cases of hemodynamic instability, fevers, nausea, pus, and pain—anything that may indicate a patient is having greater problems or may need a further workup." At the end of Robin's 9- to 11-hour day, she updates the team of physicians, so they can continue taking care of the patients.

A nurse practitioner must complete a bachelor's program in nursing and preferably work as a registered nurse for at least a year before applying to a master's level nurse practitioner program. Robin recommends that NP candidates make sure they complete all the prerequisite for the program, including biology, chemistry, and statistics. A good background in anatomy and physiology is also essential, both for obtaining the MSN (master of science in nursing) and succeeding as an NP.

According to Robin, "Every symptom we hear about or find in a patient has a basis in physiology. Physiology helps us understand why we are seeing certain lab results, hearing reports of pain, or finding changes in heart rate or blood pressure. Of course, anatomy goes hand in hand with physiology. If a patient complains of pain in a certain area, I have to know what is in that area to figure out what the source of the pain might be."

The job of nurse practitioner is uniquely rewarding. "One of the great things that being an NP offers over some of the other health provider roles is the two-piece combination of actual nursing experience with education in science and internal medicine. As a result, I'm able to interact with many different people in various states of health and effect a positive change in their lives. I help keep them safe, get them on an appropriate treatment plan, educate them about their illness, and help them take care of themselves."

For more information on nurse practitioners, including a list of accredited programs, contact the American Academy of Nurse Practitioners at http://www.aanp.org.

Think this is the career for you?

Key Stats:

- **Education and Training.** A master's degree is required, in addition to previous training as a registered nurse in a hospital.

- **Licensure.** All states require nurse practitioners to complete a national licensing examination.

- **Earnings.** Earnings vary but the median annual wage for registered nurses is $64,690. Earnings for nurse practitioners are higher depending on their specialty.

- **Expected Job Prospects.** Employment of registered nurses is expected to grow faster than the national average—22% from 2008 to 2018. There is even a higher demand for advanced practice specialties, including nurse practitioners.

Bureau of Labor Statistics, U.S. Department of Labor, *Occupational Outlook Handbook, 2010–11 Edition*, Registered Nurses, on the Internet at http://www.bls .gov/oco/ocos083.htm (visited September 14, 2011).

MasteringA&P®

Access more review material online in the Study Area at www.masteringaandp.com.

- Chapter guides
- Chapter quizzes
- Practice tests
- Art labeling activities
- Flashcards
- A glossary with pronunciations

- Practice Anatomy Lab™ (PAL™) 3.0 virtual anatomy practice tool
- Interactive Physiology® (IP) animated tutorials
- MP3 Tutor Sessions

PAL practice anatomy lab™ **For this chapter, follow these navigation paths in PAL:**

- Human Cadaver>Urinary System
- Anatomical Models>Urinary System
- Histology>Urinary System

iP For this chapter, go to these topics in the Urinary System and Fluids and Electrolytes in IP:

- Anatomy Review
- Glomerular Filtration
- Introduction to the Body Fluids
- Water Homeostasis
- Acid/Base Homeostasis

Chapter 17 Review Questions

1. Sylvia is suffering from severe edema (swelling) in her arms and legs. Her physician prescribes a diuretic (a substance that increases the volume of urine produced). Why might this help Sylvia's problems?

2. *Mannitol* is a sugar that is filtered, but not reabsorbed, by the kidneys. What effect would drinking a solution of mannitol have on the volume of urine produced?

3. How are proteins excluded from filtrate? Why is this important?

4. While visiting a foreign country, Milly inadvertently drinks some water, even though she had been advised not to. She contracts an intestinal disease that causes severe diarrhea. How would you expect her condition to affect her blood pH, urine pH, and pattern of breathing?

5. Differentiate among fluid balance, electrolyte balance, and acid-base balance, and explain why each is important to homeostasis.

For answers to all module, section, and chapter review questions, see the blue Answers tab at the back of the book.

18

The Reproductive System

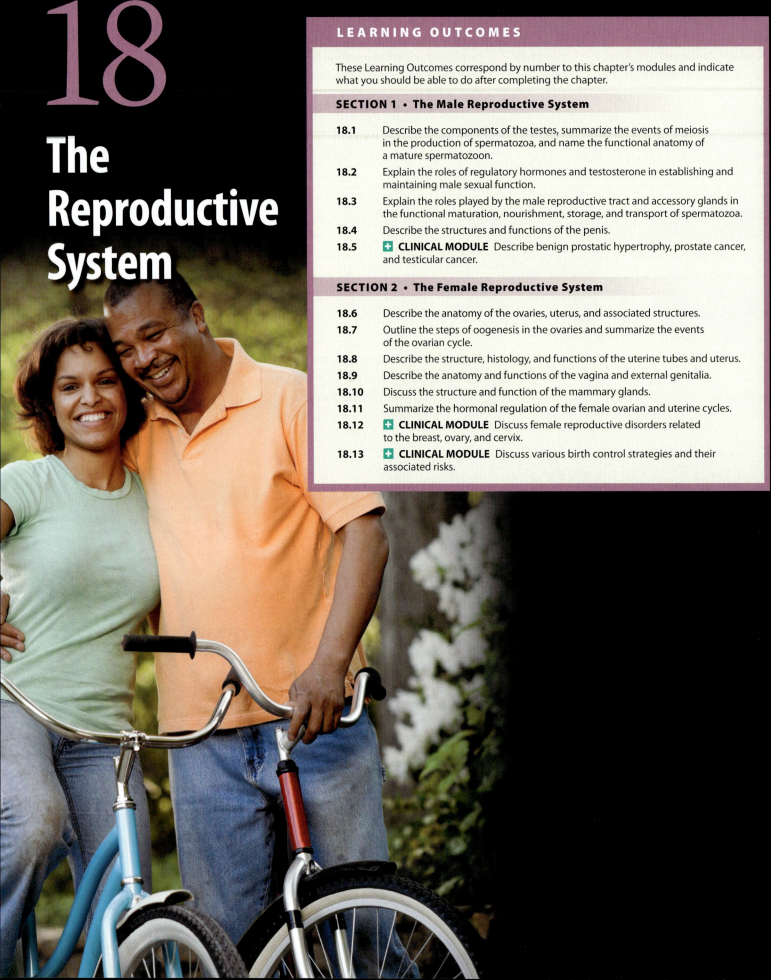

The Male Reproductive System

The reproductive system of both sexes includes the following basic components:

- **Gonads** (GŌ-nadz; *gone*, seed), or reproductive organs, which produce gametes and hormones.

- Accessory glands and organs that secrete fluids into the ducts of the reproductive system or into other excretory ducts.

- Perineal structures collectively known as the **external genitalia** (jen-i-TĀ-lē-uh).

1 This figure gives an overview of the components of the male reproductive system. The gonad is called a **testis** (plural, *testes*), and it produces male gametes called **spermatozoa** (sper-ma-tō-ZŌ-uh; singular, *spermatozoon*), or sperm. Mature spermatozoa travel along the **male reproductive tract**. As they proceed, they are mixed with the secretions of accessory glands to form a fluid known as **semen** (SĒ-men).

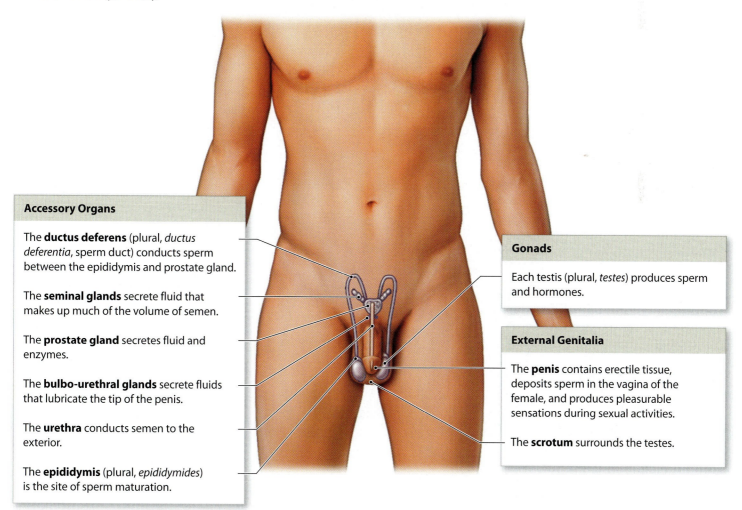

Accessory Organs

The **ductus deferens** (plural, *ductus deferentia*, sperm duct) conducts sperm between the epididymis and prostate gland.

The **seminal glands** secrete fluid that makes up much of the volume of semen.

The **prostate gland** secretes fluid and enzymes.

The **bulbo-urethral glands** secrete fluids that lubricate the tip of the penis.

The **urethra** conducts semen to the exterior.

The **epididymis** (plural, *epididymides*) is the site of sperm maturation.

Gonads

Each testis (plural, *testes*) produces sperm and hormones.

External Genitalia

The **penis** contains erectile tissue, deposits sperm in the vagina of the female, and produces pleasurable sensations during sexual activities.

The **scrotum** surrounds the testes.

The coiled seminiferous tubules of the testes are connected to the male reproductive tract

1 The main structures of the male reproductive system are shown in this sagittal section. Proceeding from a testis, the spermatozoa travel within the **epididymis** (ep-i-DID-i-mus), along the **ductus deferens** (DUK-tus DEF-e-renz), and then along the **ejaculatory duct** and the urethra before leaving the body. Accessory organs—the **seminal** (SEM-i-nal) **glands**, the **prostate** (PROS-tāt) **gland**, and the **bulbo-urethral** (bul-bō-ū-RĒ-thral) **glands**—secrete various fluids into the ejaculatory ducts and urethra. The external genitalia consist of the **scrotum** (SKRŌ-tum), which encloses the testes, and the **penis** (PĒ-nis), an erectile organ through which the distal portion of the urethra passes.

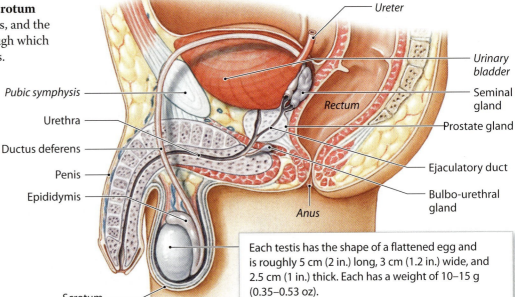

Ureter

Urinary bladder

Pubic symphysis

Seminal gland

Rectum

Prostate gland

Urethra

Ductus deferens

Ejaculatory duct

Penis

Bulbo-urethral gland

Epididymis

Anus

Each testis has the shape of a flattened egg and is roughly 5 cm (2 in.) long, 3 cm (1.2 in.) wide, and 2.5 cm (1 in.) thick. Each has a weight of 10–15 g (0.35–0.53 oz).

Scrotum

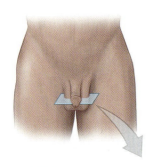

2 This horizontal section through the scrotum shows the internal organization of the testes.

Epididymis   Efferent ductule   Ductus deferens

A serous membrane lines the scrotal cavity and reduces friction between opposing surfaces.

Scrotum

Skin

A tough fibrous capsule covers the testis. This capsule is continuous with septa that subdivide the interior of the testis into separate lobules.

The **dartos muscle** is a layer of smooth muscle whose contractions wrinkle the skin of the scrotum.

Septa

The **cremaster** (krē-MAS-ter) **muscle** contracts to move the testes closer to the body.

Each lobule contains a coiled **seminiferous tubule**, which averages about 80 cm (32 in.) in length, and a typical testis contains nearly one-half mile of seminiferous tubules. Sperm production occurs within these tubules.

Scrotal cavity

Each seminiferous tubule is connected to a maze of passageways known as the **rete** (RĒ-tē; *rete*, a net) **testis**. Fifteen to 20 large **efferent ductules** connect the rete testis to the epididymis.

3 **Spermatogonia** (singular, *spermatogonium*) are the stem cells in the seminiferous tubules that divide to produce spermatozoa in a process called **spermatogenesis**. Each spermatogonium is diploid and contains 23 pairs of chromosomes. For clarity, the events shown here follow the fates of three representative chromosome pairs without identifying either maternal or paternal chromosomes.

Spermatogenesis

Mitosis of spermatogonium

Each mitotic division of a diploid spermatogonium produces two daughter cells. One is a spermatogonium that remains in contact with the basement membrane of the tubule, and the other becomes a **primary spermatocyte** that is displaced toward the lumen to begin the formation of haploid gametes.

Meiosis I

As meiosis I begins, each primary spermatocyte contains 46 individual chromosomes. At the end of meiosis I, the daughter cells are called **secondary spermatocytes**. Every secondary spermatocyte contains 23 chromosomes, each containing a pair of duplicate chromatids.

Meiosis II

The secondary spermatocytes soon enter meiosis II, which yields four haploid **spermatids**, each containing 23 chromosomes. For each primary spermatocyte that enters meiosis, four spermatids are produced.

Spermiogenesis (physical maturation)

In spermiogenesis, the last step of spermatogenesis, each spermatid matures into a single spermatozoon, or sperm. It takes about 9 weeks to complete both spermatogenesis (4 weeks) and spermiogenesis (5 weeks).

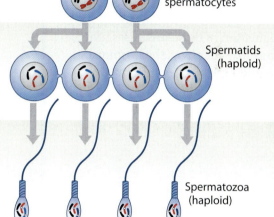

Primary spermatocyte (diploid)

DNA replication

Primary spermatocyte

Synapsis and tetrad formation

Secondary spermatocytes

Spermatids (haploid)

Spermatozoa (haploid)

Mature spermatozoa

4 Here you can see the distinctive, specialized features of a **spermatozoon**. Unlike other, less specialized cells, a mature spermatozoon lacks an endoplasmic reticulum, a Golgi apparatus, lysosomes, peroxisomes, inclusions, and many other intracellular structures. The loss of these organelles reduces the cell's size and mass; it is essentially a mobile carrier for the enclosed chromosomes, and extra weight would slow it down.

Structure of a Spermatozoon

The **acrosome** (ak-rō-SŌM) is a membranous compartment containing enzymes essential to fertilization.

The **head** is a flattened ellipse containing a nucleus with densely packed chromosomes.

The **neck** contains both centrioles of the original spermatid. The microtubules of the distal centriole are continuous with those of the middle piece and tail.

The **middle piece** contains mitochondria arranged in a spiral around the microtubules. Mitochondria provide the ATP required to move the tail.

The **tail** is a **flagellum**, a whiplike organelle that moves the sperm.

The tail of a spermatozoan is the only flagellum in the human body. Whereas cilia beat in a predictable, wavelike fashion, the flagellum of a spermatozoon has a complex, corkscrew motion. A normal adult male produces over 100-200 million sperm each day.

Module 18.1 Review

a. Name the male reproductive structures.

b. Define spermatogenesis, and list its major events.

c. Describe the functional anatomy of a typical spermatozoon.

Hormones play a key role in establishing and maintaining male sexual function

Testosterone is produced primarily by the **interstitial cells** of the testes. Interstitial cells lie outside the seminiferous tubules, the site where spermatogenesis occurs. (Small amounts of testosterone are also produced by the adrenal glands in both sexes.) Within the seminiferous tubules, spermatogenesis and spermiogenesis take place in a microenvironment between the plasma membranes of adjacent **nurse cells**. Nurse cells are also involved in the hormonal regulation of these processes. This flowchart diagrams the hormonal interactions that regulate male reproductive function.

HYPOTHALAMUS

When stimulated by GnRH from the hypothalamus, the anterior lobe of the pituitary gland releases **luteinizing hormone (LH)** and **follicle-stimulating hormone (FSH)**.

Anterior lobe of pituitary

Secretion of Luteinizing Hormone (LH)

LH targets the interstitial cells of the testes.

Secretion of Follicle-Stimulating Hormone (FSH)

FSH targets primarily the nurse cells of the seminiferous tubules.

TESTES

Interstitial Cell Stimulation

LH induces the secretion of testosterone and other androgens by the interstitial cells of the testes.

Nurse Cell Stimulation

Under FSH stimulation, and in the presence of testosterone from the interstitial cells, nurse cells (1) promote spermatogenesis and spermiogenesis and (2) secrete **inhibin**, which adjusts the rate of spermatogenesis by inhibiting FSH production.

Negative feedback

Testosterone

Inhibin

Negative feedback

Peripheral Effects of Testosterone

Maintenance of accessory glands and organs of the male reproductive system	Establishment and maintenance of male secondary sex characteristics, such as the distribution of facial hair, increased muscle mass and body size, and the quantity and location of characteristic adipose tissue deposits	Stimulation of bone and muscle growth	Effects on CNS, including maintaining libido (sexual drive) and related behaviors

By controlling the local environment around spermatocytes and spermatids, nurse cells facilitate both spermatogenesis and spermiogenesis.

KEY

→ Stimulation

⊣ Inhibition

Module 18.2 Review

a. Identify important regulatory hormones in the establishment and maintenance of male sexual function.

b. Identify the sources of hormones that control male reproductive functions.

c. What effect would low FSH levels have on sperm production?

The male reproductive tract receives secretions from the seminal, prostate, and bulbo-urethral glands

The testes produce physically mature spermatozoa; the other portions of the male reproductive system are responsible for the functional maturation, nourishment, storage, and transport of spermatozoa. The spermatozoa leaving the testes are physically mature, but immobile and incapable of fertilizing an oocyte. To become motile (actively swimming) and fully functional, spermatozoa must undergo a process called **capacitation**. Capacitation normally occurs in two steps: (1) Spermatozoa become motile when they are mixed with secretions of the seminal glands, and (2) they become capable of successful fertilization when exposed to conditions in the female reproductive tract.

1 This diagrammatic posterior view shows the urinary bladder, prostate gland, and other structures of the male reproductive system.

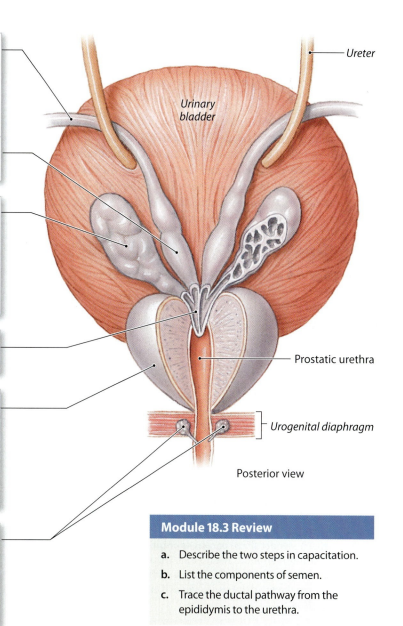

Each **ductus deferens** ascends through the inguinal canal, enters the abdominal cavity, and passes posteriorly to reach the prostate gland. In addition to transporting spermatozoa, the ductus deferens can store spermatozoa for several months. During this time, the spermatozoa remain in a temporary state known as suspended animation and have low metabolic rates.

The **ampulla** (am-PUL-uh) is the expanded distal portion of the ductus deferens that joins the duct of the seminal glands.

The **seminal glands**, also called the seminal vesicles, are on either side of the midline, between the posterior wall of the urinary bladder and the rectum. The seminal glands contribute about 60 percent of the volume of **semen**. When mixed with the secretions of the seminal glands, previously inactive but functional spermatozoa undergo the first step of capacitation and begin beating their flagella.

The **ejaculatory duct** carries fluid from the seminal gland and ampulla to the urethra.

The **prostate gland** is a small, muscular, rounded organ about 4 cm (1.6 in.) in diameter. The prostate gland encircles the proximal portion of the urethra as it leaves the urinary bladder. The prostate gland produces 20–30 percent of the volume of semen. Prostatic secretions may help prevent urinary tract infections in males. These secretions are ejected into the prostatic urethra by peristaltic contractions of the muscular prostate wall.

The paired **bulbo-urethral glands** are located at the base of the penis, covered by the fascia of the urogenital diaphragm. The duct of each gland empties into the urethra. These glands secrete a thick, alkaline mucus that helps neutralize any urinary acids that may remain in the urethra, and it also lubricates the tip of the penis.

Ureter

Urinary bladder

Prostatic urethra

Urogenital diaphragm

Posterior view

Module 18.3 Review

a. Describe the two steps in capacitation.

b. List the components of semen.

c. Trace the ductal pathway from the epididymis to the urethra.

The penis conducts semen and urine out of the body

1 The **penis** is a tubular organ through which the distal portion of the urethra passes. It conducts urine to the exterior and introduces semen into the female's vagina during sexual intercourse.

Pubic symphysis

The **root** of the penis is the fixed portion that attaches the penis to the body wall. This connection occurs within the urogenital triangle immediately inferior to the pubic symphysis.

The **body**, or shaft, of the penis is the tubular, movable portion of the organ.

The **neck** is the narrow portion of the penis between the shaft and the glans.

Ischial ramus

Corpus spongiosum

Corpora cavernosa

A fold of skin called the **prepuce** (PRĒ-poos), or foreskin, surrounds the tip of the penis. The prepuce attaches to the relatively narrow neck of the penis and continues over the glans.

Scrotum

The **glans** of the penis is the expanded distal end that surrounds the external urethral orifice.

External urethral orifice

2 This cross section shows that most of the body of the penis consists of three columns of vascular **erectile tissue.** Erectile tissue consists of a three-dimensional network with vascular spaces incompletely separated by partitions of elastic connective tissue and smooth muscle fibers. In the resting state, the arteries are constricted, and the muscular partitions are tense. This combination restricts blood flow into the erectile tissue. When smooth muscle fibers in arterial blood vessels dilate in response to sexual stimulation, blood flow increases, the vascular spaces engorge with blood, and erection of the penis occurs.

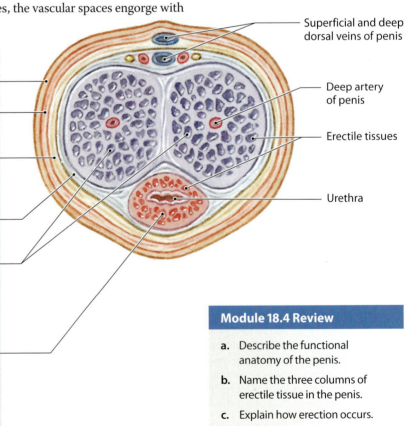

Tissue Layers of the Penis

The skin overlying the penis resembles that of the scrotum.

The dermis contains a layer of smooth muscle that is a continuation of the dartos muscle of the scrotum.

The underlying areolar tissue allows the thin skin to move without distorting deeper structures. The areolar tissue also contains superficial arteries, veins, and lymphatic vessels.

Deep to the areolar tissue, a dense network of elastic fibers encircles the internal structures of the penis.

The anterior surface of the flaccid penis covers two cylindrical masses of erectile tissue: the **corpora cavernosa** (KOR-por-a ka-ver-NŌ-suh; singular, corpus cavernosum). The corpora cavernosa extend along the length of the penis as far as its neck.

The relatively slender single **corpus spongiosum** (spon-jē-Ō-sum) surrounds the urethra. This erectile body extends from the superficial fascia of the urogenital diaphragm to the tip of the penis, where it expands to form the glans.

Superficial and deep dorsal veins of penis

Deep artery of penis

Erectile tissues

Urethra

Module 18.4 Review

a. Describe the functional anatomy of the penis.

b. Name the three columns of erectile tissue in the penis.

c. Explain how erection occurs.

Disorders of the prostate gland and testes are relatively common

1 Enlargement of the prostate gland, or **benign prostatic hypertrophy (BPH)**, typically occurs spontaneously in men over age 50. The increase in size occurs as testosterone production by the interstitial cells decreases. At the same time, the interstitial cells begin releasing small quantities of estrogen into the bloodstream. The combination of lower testosterone levels and the presence of estrogen probably stimulates prostatic growth. In severe cases, prostatic swelling constricts and blocks the urethra, producing urinary obstruction. **Prostate cancer**, a malignancy of the prostate gland, is the second most common cancer and the second most common cause of cancer deaths in males. The American Cancer Society estimates that approximately 217,730 new prostate cancer cases in 2010 will result in about 32,050 deaths. Blood tests are often used for screening purposes. The most sensitive is a blood test for **prostate-specific antigen (PSA)**. Elevated levels of this antigen, normally present in low concentrations, may indicate the presence of prostate cancer. Screening with periodic PSA tests is now being recommended for men over age 50. Treatment of localized prostate cancer often involves radiation therapy or surgical removal of the prostate gland—a **prostatectomy** (pros-ta-TEK-tō-mē).

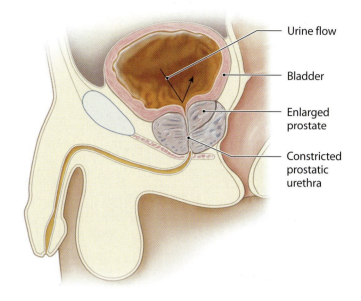

- Urine flow
- Bladder
- Enlarged prostate
- Constricted prostatic urethra

2 **Testicular cancer** occurs at a relatively low rate: about 3 cases per 100,000 males per year. Although only about 7900 new cases are reported each year in the United States, with less than 400 deaths, testicular cancer is the most common cancer among males aged 15–35. More than 95 percent of testicular cancers result from abnormal spermatogonia or spermatocytes, rather than abnormal nurse cells, interstitial cells, or other testicular cells. Treatment generally consists of a combination of orchiectomy (removal of the testes) and chemotherapy. The survival rate is now near 95 percent, primarily as a result of earlier diagnosis and improved treatment protocols. Cyclist Lance Armstrong won the grueling Tour de France six consecutive times after successful treatment for advanced testicular cancer.

Module 18.5 Review

a. Define benign prostatic hypertrophy.

b. What are two possible treatments for prostate cancer?

c. What cancer is the most common in males between the ages of 15 and 35?

1. Labeling

Label the structures of the male reproductive system in the accompanying diagram.

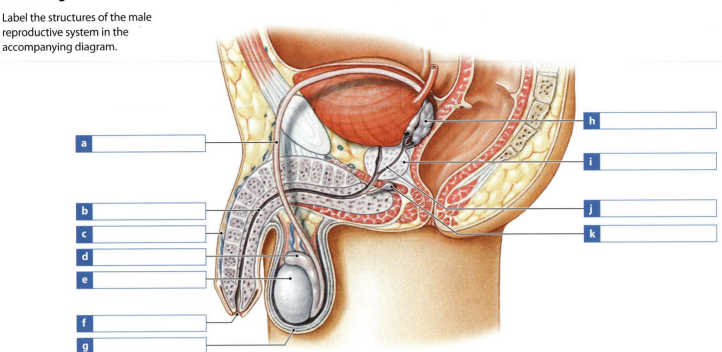

a _____

b _____

c _____

d _____

e _____

f _____

g _____

h _____

i _____

j _____

k _____

2. Matching

Match the following terms with the most closely related description.

- semen
- prepuce
- capacitation
- corpus spongiosum
- luteinizing hormone (LH)
- flagellum
- follicle-stimulating hormone (FSH)
- spermatogonia
- seminiferous tubules
- dartos muscle
- spermatogenesis
- interstitial cells
- spermiogenesis
- penis and scrotum

a _____ Scrotal smooth muscle

b _____ Sperm stem cells

c _____ Sites of sperm production

d _____ Produce testosterone

e _____ Physical maturation of spermatids

f _____ Sperm production

g _____ External genitalia

h _____ Spermatozoon tail

i _____ Process that enables spermatozoa to become fully functional

j _____ Spermatozoa, and secretions of seminal, prostate, and bulbo-urethral glands

k _____ Fold of skin surrounding tip of penis

l _____ Erectile tissue surrounding the urethra

m _____ Induces secretion of testosterone

n _____ Hormone that targets nurse cells

3. Section integration

In males, the endocrine disorder hypogonadism is primarily due to the underproduction of testosterone or the lack of tissue sensitivity to testosterone, and results in sterility. What are five primary functions of testosterone in males? _____

_____

_____

The Female Reproductive System

A woman's reproductive system produces sex hormones and functional gametes, and it must also be able to protect and support a developing embryo, maintain a growing fetus, and nourish a newborn infant. The main organs of the female reproductive system are the ovaries, uterine tubes, uterus, vagina, and components of the external genitalia. The female reproductive system also includes accessory organs—the mammary glands—and a variety of smaller accessory glands that secrete their products into the female reproductive tract.

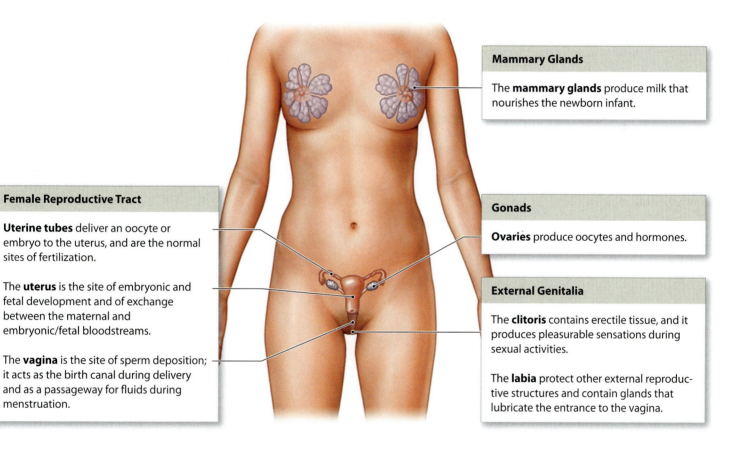

Mammary Glands

The **mammary glands** produce milk that nourishes the newborn infant.

Female Reproductive Tract

Uterine tubes deliver an oocyte or embryo to the uterus, and are the normal sites of fertilization.

The **uterus** is the site of embryonic and fetal development and of exchange between the maternal and embryonic/fetal bloodstreams.

The **vagina** is the site of sperm deposition; it acts as the birth canal during delivery and as a passageway for fluids during menstruation.

Gonads

Ovaries produce oocytes and hormones.

External Genitalia

The **clitoris** contains erectile tissue, and it produces pleasurable sensations during sexual activities.

The **labia** protect other external reproductive structures and contain glands that lubricate the entrance to the vagina.

This figure provides an overview of the components of the female reproductive system. The gonads in females, called ovaries (singular, *ovary*), produce immature female gametes called **oocytes** (Ō-ō-sīts), which later mature into **ova** (singular, *ovum*). Oocytes leave the ovary and then travel along the **female reproductive tract**. If fertilization occurs, it will occur in the uterine tubes and further embryonic development will occur within the uterus.

The ovaries and the female reproductive tract are in close proximity but are not directly connected

1 This sagittal section through the pelvic cavity shows the location of the female reproductive organs.

The paired **ovaries** are small, lumpy, almond-shaped organs near the lateral walls of the pelvic cavity. The ovaries have three main functions: (1) production of immature female gametes, or **oocytes**; (2) secretion of female sex hormones, including **estrogen** and **progesterone**; and (3) secretion of **inhibin**, involved in the feedback control of pituitary FSH production.

Each **uterine tube** begins with an expanded funnel, called an **infundibulum**, that is open into the pelvic cavity along the medial surface of the ovary. The other end of the uterine tube opens into the uterine cavity.

The **uterus** sits inferior to the ovary, usually angled forward above the urinary bladder.

The pocket formed between the posterior wall of the uterus and the anterior surface of the colon is the **rectouterine** (rek-tō-Ū-ter-in) **pouch**.

The **vagina** extends from the vaginal entrance, which opens to the exterior, to the base of the uterus.

Accessory glands lubricate the entrance to the vagina and the external genitalia.

External Genitalia

The **clitoris** contains erectile tissue; its stimulation produces pleasurable sensations during sexual activities.

The **labia** contain glands that lubricate the entrance to the vagina.

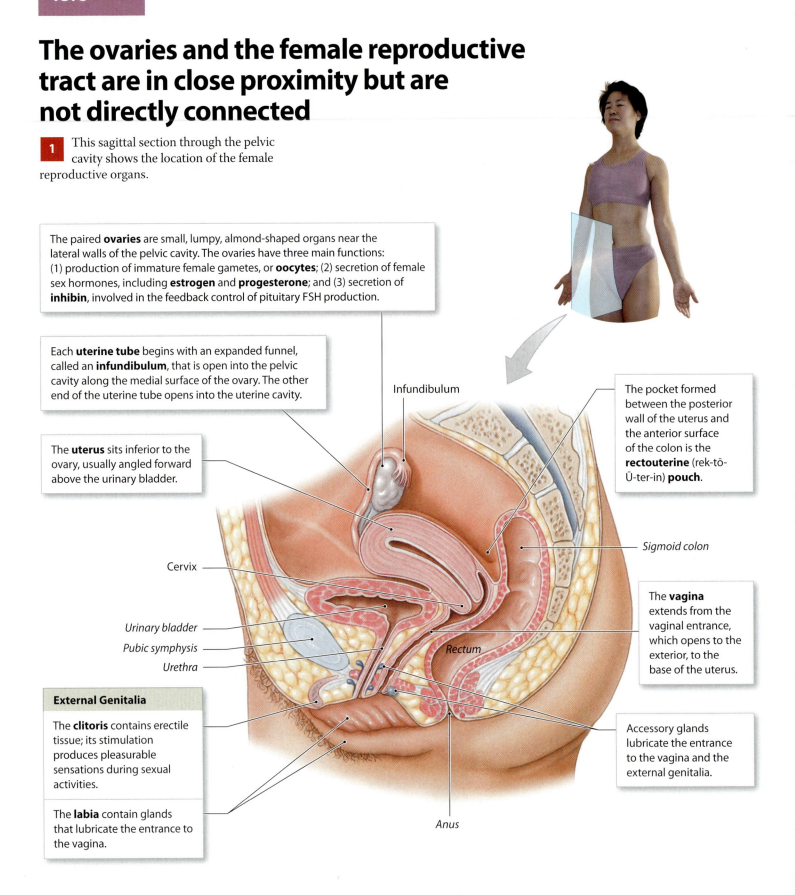

Infundibulum

Sigmoid colon

Cervix

Urinary bladder

Pubic symphysis

Urethra

Rectum

Anus

2 The position of each ovary is stabilized by several thickened peritoneal folds that are called *ligaments*. This is a view from above and behind, with the left uterine tube pulled away from the ovary to show the ligaments more clearly.

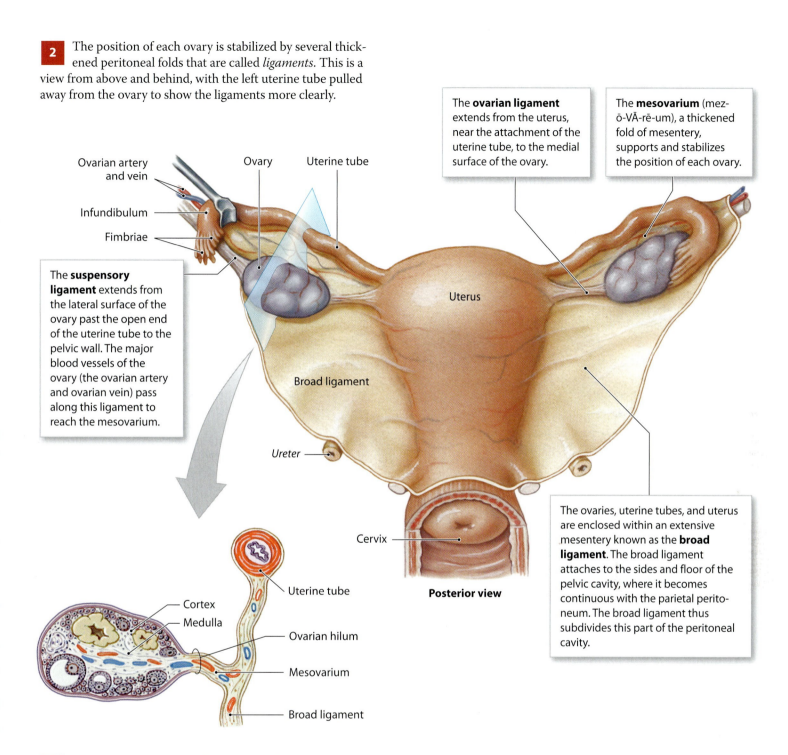

The **ovarian ligament** extends from the uterus, near the attachment of the uterine tube, to the medial surface of the ovary.

The **mesovarium** (mez-ō-VĀ-rē-um), a thickened fold of mesentery, supports and stabilizes the position of each ovary.

Ovarian artery and vein

Infundibulum

Fimbriae

Ovary

Uterine tube

The **suspensory ligament** extends from the lateral surface of the ovary past the open end of the uterine tube to the pelvic wall. The major blood vessels of the ovary (the ovarian artery and ovarian vein) pass along this ligament to reach the mesovarium.

Uterus

Broad ligament

Ureter

Cervix

Cortex

Medulla

Uterine tube

Ovarian hilum

Mesovarium

Broad ligament

Posterior view

The ovaries, uterine tubes, and uterus are enclosed within an extensive mesentery known as the **broad ligament**. The broad ligament attaches to the sides and floor of the pelvic cavity, where it becomes continuous with the parietal peritoneum. The broad ligament thus subdivides this part of the peritoneal cavity.

3 This is a cross section taken through the mesovarium and the broad ligament. A typical ovary is about 5 cm long, 2.5 cm wide, and about 8 mm thick (2 in. by 1 in. by 0.33 in.) and weighs 6–8 g (about 0.25 oz.). Blood vessels enter and leave the ovary at the ovarian hilum, where the ovary attaches to the mesovarium. The interior of the ovary can be divided into a superficial **cortex** and a deeper **medulla**. Gametes are produced in the cortex.

Module 18.6 Review

a. List the major organs of the female reproductive system.

b. Name the structures enclosed by the broad ligament, and cite the function of the mesovarium.

c. What roles do the ovaries perform?

Meiosis I in the ovaries produces a single haploid secondary oocyte that completes meiosis II only if fertilization occurs

Ovum production, or **oogenesis** (ō-ō-JEN-e-sis; oon, egg), begins before a woman's birth, accelerates at puberty, and ends at menopause. Between puberty and menopause, oogenesis occurs on a monthly basis as part of the ovarian cycle.

Oogenesis

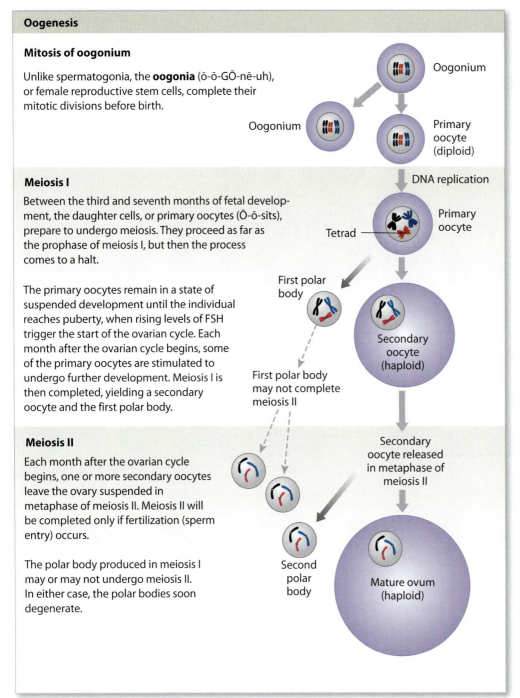

Mitosis of oogonium

Unlike spermatogonia, the **oogonia** (ō-ō-GŌ-nē-uh), or female reproductive stem cells, complete their mitotic divisions before birth.

Oogonium

Oogonium

Primary oocyte (diploid)

DNA replication

Meiosis I

Between the third and seventh months of fetal development, the daughter cells, or primary oocytes (Ō-ō-sīts), prepare to undergo meiosis. They proceed as far as the prophase of meiosis I, but then the process comes to a halt.

The primary oocytes remain in a state of suspended development until the individual reaches puberty, when rising levels of FSH trigger the start of the ovarian cycle. Each month after the ovarian cycle begins, some of the primary oocytes are stimulated to undergo further development. Meiosis I is then completed, yielding a secondary oocyte and the first polar body.

Tetrad

Primary oocyte

First polar body

First polar body may not complete meiosis II

Secondary oocyte (haploid)

Meiosis II

Each month after the ovarian cycle begins, one or more secondary oocytes leave the ovary suspended in metaphase of meiosis II. Meiosis II will be completed only if fertilization (sperm entry) occurs.

The polar body produced in meiosis I may or may not undergo meiosis II. In either case, the polar bodies soon degenerate.

Secondary oocyte released in metaphase of meiosis II

Second polar body

Mature ovum (haploid)

1 Although the nuclear events in the ovaries during meiosis are the same as those in the testes, the cytoplasm of the **primary oocyte** is unevenly distributed during the two meiotic divisions. Oogenesis produces one functional **ovum**, which contains most of the original cytoplasm, and two or three **polar bodies**, nonfunctional cells that later disintegrate. Another difference is that the ovary releases a **secondary oocyte** rather than a mature ovum. The secondary oocyte is suspended in metaphase of meiosis II; meiosis will not be completed unless fertilization occurs.

Not all primary oocytes produced during development survive until puberty. The ovaries have roughly 2 million **primordial follicles** at birth, each containing a primary oocyte. Primordial follicles exist at the earliest stages of development. By the time of puberty, the number has dropped to about 400,000. The rest of the primordial follicles degenerate in a process called **atresia** (a-TRĒ-zē-uh).

2 **Ovarian follicles** are specialized structures in the cortex of the ovaries where both oocyte growth and meiosis I occur. As the ovarian cycle proceeds, the structure of the follicle gradually changes. Although many primordial follicles develop into primary follicles, only a few of the primary follicles mature further in each ovarian cycle. Important events in the **ovarian cycle** are summarized here.

Formation of Primary Follicles

Follicle cells

Primary oocyte

Follicle cells enlarge, divide, and form several layers of cells around a primary oocyte. A clear region around the oocyte, called the **zona pellucida** (ZŌ-na pe-LOO-si-duh; *pellucidus*, translucent), develops. As the cells enlarge and multiply, they produce estrogen.

Formation of Secondary Follicles

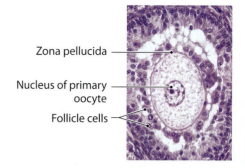

Zona pellucida

Nucleus of primary oocyte

Follicle cells

Secondary follicles develop as the wall of the follicle thickens and the deeper follicle cells begin secreting fluid that accumulates in small pockets. These pockets gradually expand and separate the inner and outer layers of the follicle.

Formation of a Tertiary Follicle

Antrum containing follicular fluid

Corona radiata

Secondary oocyte

By days 10–14 of the cycle, usually only one secondary follicle has become a **tertiary follicle**, or mature graafian (GRAF-ē-an) follicle, roughly 15 mm in diameter. The oocyte projects into the **antrum** (AN-trum), or expanded central chamber of the follicle. The granulosa cells associated with the secondary oocyte form a protective layer known as the **corona radiata** (kō-RŌ-nuh rā-dē-AH-tuh).

Primordial follicles

Corona radiata

Ruptured follicle

Degeneration of Corpus Luteum

If fertilization does not occur, after 12 days progesterone and estrogen levels fall markedly as the corpus luteum begins to degenerate. Fibroblasts invade the nonfunctional corpus luteum, producing a knot of pale scar tissue. The degeneration of the corpus luteum marks the end of the ovarian cycle. A new ovarian cycle then begins.

Formation of Corpus Luteum

The empty tertiary follicle initially collapses, and under LH stimulation the remaining granulosa cells proliferate to create the **corpus luteum** (LOO-tē-um; *lutea*, yellow), which secretes progesterone (prō-JES-ter-ōn) and estrogen. Progesterone prepares the uterus for pregnancy by stimulating the maturation of the uterine lining and the secretion by uterine glands.

Ovulation

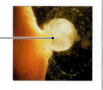

Secondary oocyte

At **ovulation**, the tertiary follicle releases the secondary oocyte and corona radiata into the pelvic cavity. Ovulation marks the end of the **follicular phase** of the ovarian cycle and the start of the **luteal phase**.

Module 18.7 Review

a. Define oocyte, and state the products resulting from its cell division.

b. What are the main differences in gamete production between males and females?

c. List the important events in the ovarian cycle.

The uterine tubes are connected to the uterus, a hollow organ with thick muscular walls

1 Each **uterine tube** is a hollow, muscular structure measuring roughly 13 cm (5.2 in.) in length. The distal portion of each uterine tube connects to the uterus. Fertilization—the penetration of the secondary oocyte by a sperm—occurs in the uterine tube. The **uterus** is a hollow muscular organ that is about 7.5 cm (3 in.) long with a maximum diameter of 5 cm (2 in.). It weighs 30–40 g (1–1.4 oz). The sectional illustration below shows the internal structure of the uterine tube and the connection between the lumen of the uterine tube and the large uterine cavity within the uterus.

The thickness of the smooth muscle layers in the wall of the **ampulla**, the middle segment of the uterine tube, gradually increases as the tube approaches the uterus.

The ampulla leads to the **isthmus** (IS-mus) of the uterine tube, a short segment connected to the uterine wall.

The **infundibulum** has numerous fingerlike projections that extend into the pelvic cavity. The projections are called **fimbriae** (FIM-brē-ē). Fimbriae drape over the surface of the ovary, but there is no physical connection between the two structures. The inner surfaces of the infundibulum are lined with cilia that beat toward the lumen of the uterine tube.

Infundibulum

Uterine cavity

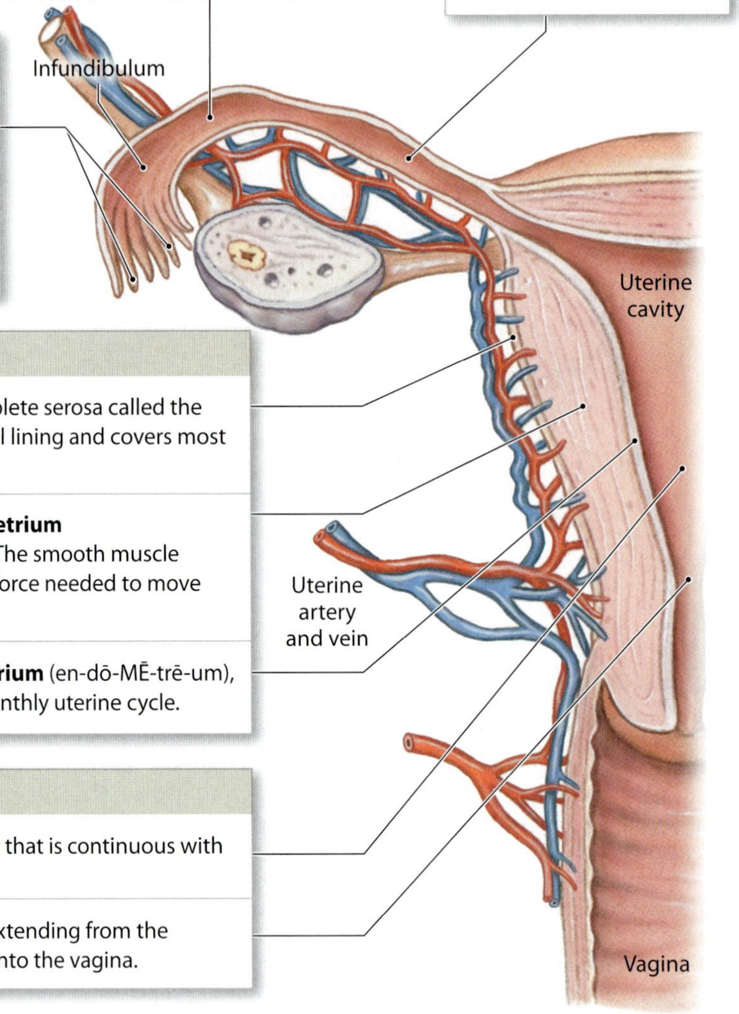

Layers of the Uterine Wall

The outer surface layer of the uterus is an incomplete serosa called the **perimetrium**. It is continuous with the peritoneal lining and covers most of the uterine surface.

The perimetrium covers a thick, muscular **myometrium** (mī-ō-MĒ-trē-um; *myo-*, muscle + *metra*, uterus). The smooth muscle tissue of the myometrium provides much of the force needed to move a fetus out of the uterus and into the vagina.

The inner lining consists of a glandular **endometrium** (en-dō-MĒ-trē-um), whose character changes in the course of the monthly uterine cycle.

Uterine artery and vein

The Uterine Lumen

The **uterine cavity** is the large, superior chamber that is continuous with the isthmus of the uterine tube on either side.

The **cervical canal** is a constricted passageway extending from the inferior end of the uterine cavity to the opening into the vagina.

Vagina

2 This colorized SEM shows the ciliated epithelium of the uterine tube. The cilia (colored yellow-green) beat toward the uterine cavity, causing fluid currents that help collect and transport the secondary oocyte after ovulation. Oocyte transport along the tube involves a combination of ciliary movement and peristaltic contraction stimulated by autonomic nerves.

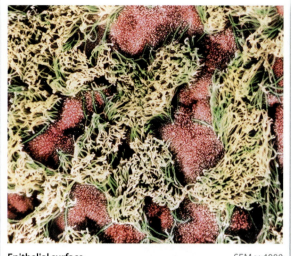

Epithelial surface   SEM × 4000

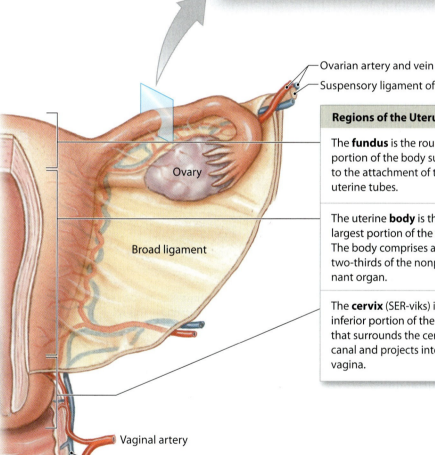

Ovarian artery and vein

Suspensory ligament of ovary

Ovary

Broad ligament

Vaginal artery

Vaginal vein

Vagina

Regions of the Uterus

The **fundus** is the rounded portion of the body superior to the attachment of the uterine tubes.

The uterine **body** is the largest portion of the uterus. The body comprises about two-thirds of the nonpregnant organ.

The **cervix** (SER-viks) is the inferior portion of the uterus that surrounds the cervical canal and projects into the vagina.

3 The uterus can be divided into three anatomical regions, as indicated here. The uterus, which is capable of great changes in size and shape, provides mechanical protection, nutritional support, and waste removal for the developing **embryo** (weeks 1–8) and **fetus** (week 9 through delivery). In addition, contractions in the muscular wall of the uterus are important in delivering the fetus at birth.

It normally takes 3-4 days for a secondary oocyte to travel from the infundibulum to the uterine cavity. If fertilization is to occur, the secondary oocyte must encounter spermatozoa during the first 12–24 hours of its passage along the uterine tube. After this period of time, the oocyte will begin to degenerate.

Module 18.8 Review

a. Name the regions of the uterus.

b. Describe the three layers of the uterine wall.

c. How do recently released secondary oocytes reach the uterine tube?

The entrance to the vagina is enclosed by external genitalia

1 The **vagina** is an elastic, muscular tube extending between the cervix and the **vestibule**, a space bounded by the female external genitalia. The vagina is typically 7.5–9 cm (3–3.6 in.) long, but its diameter varies because it is highly distensible. The internal passageway is called the **vaginal canal.** The vagina serves as a passageway for the elimination of menstrual fluids; receives the penis during sexual intercourse, and holds spermatozoa prior to their passage into the uterus; and forms the inferior portion of the birth canal, through which the fetus passes during delivery.

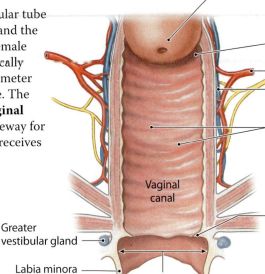

The cervix projects into the proximal end of the vagina. The shallow recess in the vagina surrounding the tip of the cervix is known as the **fornix** (FOR-niks).

Fornix

Vaginal artery

Vaginal vein

In the relaxed state, the vaginal lining forms folds called **rugae**. The vaginal canal is lined by a nonkeratinized stratified squamous epithelium.

Throughout childhood the vagina and vestibule are usually separated by the **hymen** (HĪ-men), an elastic epithelial fold of variable size that partially blocks the entrance to the vagina. An intact hymen is typically stretched or torn during sexual intercourse or tampon use.

Vaginal canal

Greater vestibular gland

Labia minora

Vestibule

The urethra opens into the vestibule just anterior to the vaginal entrance.

Vestibule

Labia minora

Hymen (torn)

Vaginal entrance

Anus

The bulge of the **mons pubis** is created by adipose tissue deep to the skin and superficial to the pubic symphysis.

Extensions of the labia minora encircle the body of the clitoris, forming its **prepuce**, or hood.

The **clitoris** (KLIT-ō-ris or kli-TŌR-is) projects into the vestibule. This small, rounded tissue projection contains erectile tissue comparable to the corpora cavernosa and corpus spongiosum of the penis.

The **labia majora** (singular, *labium majus*) are prominent folds of skin that encircle and partially conceal the labia minora and adjacent structures.

2 The area containing the female external genitalia is the **vulva** (VUL-vuh), or **pudendum** (pū-DEN-dum). The vagina opens into the vestibule, a central space bounded by small folds known as the **labia minora** (LĀ-be-uh mi-NOR-uh; singular, *labium minus*). A variable number of small **lesser vestibular glands** discharge their secretions onto the exposed surface of the vestibule, keeping it moist. During sexual arousal, a pair of ducts discharges the secretions of the **greater vestibular glands** into the vestibule. These mucous glands have the same embryological origins as the bulbo-urethral glands of males.

Module 18.9 Review

a. List the functions of the vagina.

b. Describe the anatomy of the vagina.

c. Cite the similarities that exist between certain structures in the reproductive systems of females and males.

The mammary glands nourish the infant after delivery

A newborn infant cannot fend for itself, and several of its key systems have yet to complete development. Over the initial period of adjustment to an independent existence, the infant obtains nourishment from the milk secreted by the maternal **mammary glands**. These organs are controlled mainly by hormones released by the reproductive system and the placenta, a temporary structure that nourishes the embryo and fetus. The interaction of these hormones results in milk production, or **lactation** (lak-TĀ-shun).

1 The mammary gland lies directly over the pectoralis major muscle. This figure shows the internal organization of the mammary tissue and its supporting structures.

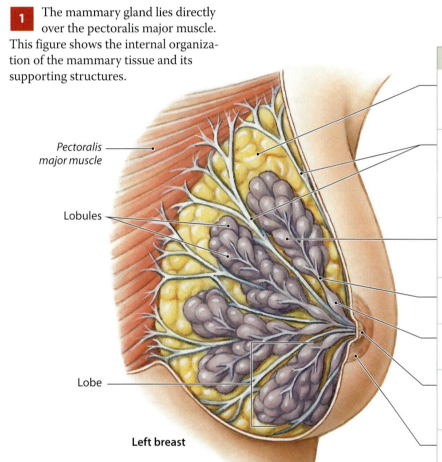

Pectoralis major muscle

Lobules

Lobe

Left breast

The Structure of a Mammary Gland

On each side, a mammary gland lies in the subcutaneous tissue of the **pectoral fat pad** deep to the skin of the chest.

Dense connective tissue surrounds the duct system and forms partitions that extend between the lobes and the lobules. These bands of connective tissue, the **suspensory ligaments of the breast**, originate in the dermis of the overlying skin.

The glandular tissue of the breast consists of separate **lobes**, each containing several secretory **lobules**. Each lobule is composed of many **secretory alveoli**.

Ducts leaving the lobules converge, giving rise to a single **lactiferous** (lak-TIF-er-us) **duct** in each lobe.

Near the nipple, each lactiferous duct enlarges, forming an expanded chamber called a **lactiferous sinus**.

Each breast has a **nipple**, a small conical projection where 15–20 lactiferous sinuses open onto the body surface.

The reddish-brown skin around each nipple is the **areola** (a-RĒ-ō-luh). Large sebaceous glands deep to the areolar surface give it a grainy texture.

Module 18.10 Review

a. Define lactation.

b. Explain whether the blockage of a single lactiferous sinus would or would not interfere with the delivery of milk to the nipple.

c. Trace the route of milk from its site of production to the outside of the woman's body.

The ovarian and uterine cycles are regulated by hormones of the hypothalamus, pituitary gland, and ovaries

Cyclical change in the gonadotropic hormone levels control the ovarian cycle. The ovarian and uterine cycles must operate in synchrony to ensure proper reproductive function. For example, a surge in LH levels triggers the rupture of the follicle wall and ovulation. This occurs after the repair and regeneration of the endometrial lining. If the two cycles are not properly coordinated, infertility results. A female who doesn't ovulate cannot conceive, even if her uterus is perfectly normal. A female who ovulates normally, but whose uterus is not ready to support an embryo, will also be infertile.

The uterine cycle, or **menstrual cycle**, begins at puberty. The first cycle, known as **menarche** (me-NAR-kē; *men*, month + *arche*, beginning), typically occurs at age 11–12. The uterine cycle averages 28 days in length, but it can range from 21 to 35 days in healthy women of reproductive age. The cycles continue until **menopause** (MEN-ō-pawz), the termination of the uterine cycle, at age 45–55. Over the interim, the regular appearance of uterine cycles is interrupted only by circumstances such as illness, stress, starvation, or pregnancy.

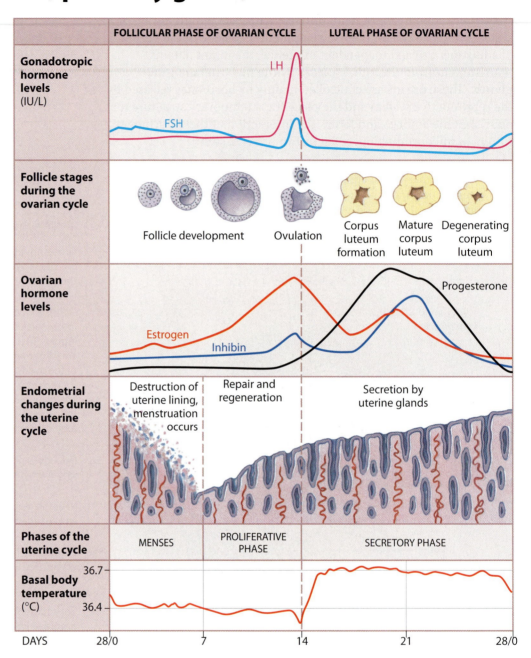

1 This illustration can aid your understanding of female reproductive physiology by integrating the key events in the ovarian and uterine cycles. The monthly hormonal fluctuations cause physiological changes that affect core body temperature. During the follicular phase—when estrogen is the dominant hormone—the **basal body temperature**, or the resting body temperature measured upon awakening in the morning, is about 0.3°C (0.5°F) lower than it is during the luteal phase, when progesterone dominates.

Module 18.11 Review

a. What event in the uterine cycle occurs when the levels of estrogen and progesterone have declined?

b. What changes would you expect to observe in the ovarian cycle if the LH surge did not occur?

c. Summarize the roles of the hormones in the ovarian and uterine cycles.

Female reproductive system disorders are relatively common and often deadly

1 The mammary glands are stimulated by the changing levels of circulating reproductive hormones that accompany the uterine cycle, and, late in the ovarian cycle, occasional discomfort or even inflammation of mammary gland tissues can occur. If inflamed lobules become walled off by scar tissue, **cysts** are created. Clusters of cysts can be felt in the breast as discrete masses, a condition known as **fibrocystic disease**. Biopsies may be needed to distinguish between this benign condition and **breast cancer**. Breast cancer, a malignant, metastasizing tumor of the mammary gland, is the leading cause of death in women between ages 35 and 45, but it is most common in women over age 50. An estimated 12 percent of U.S. women will develop breast cancer at some point in their lifetime. Notable risk factors include a family history of breast cancer, a first pregnancy after age 30, and early menarche or late menopause. Treatment of breast cancer begins with the removal of the tumor. Because in many cases cancer cells begin to spread before the condition is diagnosed, part or all of the affected mammary gland is surgically removed and usually the axillary lymph nodes on that side are biopsied to detect signs of metastasis. A combination of chemotherapy, radiation, and hormone treatments may be used to supplement the surgical procedures.

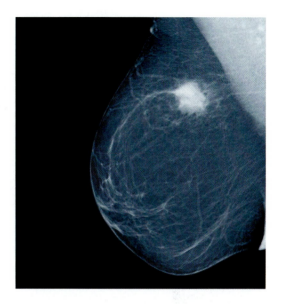

2 A woman in the United States has a 1-in-70 chance of developing **ovarian cancer** in her lifetime. In 2010, there were an estimated 21,880 new cases, and an estimated 13,850 deaths. Although ovarian cancer is the third most common reproductive cancer among women, it is the most dangerous because it is seldom diagnosed in its early stages. The prognosis is relatively good for cancers that originate in the general ovarian tissues or from abnormal oocytes. These cancers respond well to some combination of chemotherapy, radiation, and surgery. However, 85 percent of ovarian cancers are **carcinomas** (cancers derived from epithelial cells), and sustained remission occurs in only about one-third of the cases of this type.

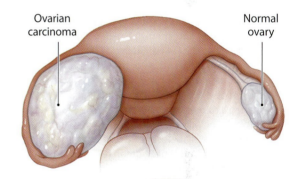

Ovarian carcinoma

Normal ovary

3 **Cervical cancer** is the most common cancer of the reproductive system in women ages 15–34. Each year roughly 13,000 U.S. women are diagnosed with invasive cervical cancer, and approximately one-third of them eventually die from the condition. Another 35,000 women are diagnosed with a less aggressive form of cervical cancer. Gardasil is a vaccine that helps protect against certain types of **human papillomavirus (HPV)**. HPV causes about 75 percent of cervical cancers. The HPV vaccine is a recent addition to the recommended immunization schedule for children. It is given in three doses, beginning at age 11 or 12.

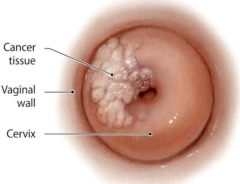

Cancer tissue

Vaginal wall

Cervix

Sexual activity carries the risk of infection with a variety of micro-organisms. Sexually transmitted diseases (STDs) are transferred from individual to individual, primarily or exclusively by sexual intercourse. At least two dozen bacterial, viral, and fungal infections are currently recognized as STDs. The bacterium *Chlamydia* can cause pelvic inflammatory disease (PID) and infertility; AIDS, caused by HIV, is a deadly viral disease. The incidence of STDs has been increasing in the United States since 1984. An estimated 15–24 million new cases occur each year, and almost 50 percent of those cases involve persons aged 19–24. Poverty, intravenous drug use, prostitution, and the appearance of drug-resistant pathogens all contribute to the problem.

Module 18.12 Review

a. Define sexually transmitted disease.

b. From which cell type does ovarian cancer usually arise?

c. Which pathogen is associated with most cases of cervical cancer?

Birth control strategies vary in effectiveness and in the nature of associated risks

Male Condom

Male condoms
(prophylactics or "rubbers") cover the glans and shaft of the penis during intercourse and keep spermatozoa from reaching the female reproductive tract. Of all the strategies described in this module, only latex condoms protect against **sexually transmitted diseases (STDs)**, such as syphilis, gonorrhea, human papillomavirus (HPV), and AIDS.

Diaphragm with Spermicide

A **diaphragm**, the most popular form of vaginal barrier in use today, consists of a dome of latex rubber with a small metal hoop supporting the rim. Because vaginas vary in size, women choosing this method must be individually fitted. Before intercourse, the diaphragm is inserted so that it covers the cervix's opening into the uterus. The diaphragm must be coated with a small amount of spermicidal (sperm-killing) jelly or cream to be an effective contraceptive. The failure rate of a properly fitted and used diaphragm is estimated at 5–6 percent.

Oral Contraceptives—Combined (Estrogen and Progesterone)

At least 20 brands of combination **oral contraceptives** are now available, and more than 200 million women are using them worldwide. In the United States, 33 percent of women under age 45 use a combination pill to prevent conception. The failure rate for combination oral contraceptives, when used as prescribed, is 0.24 percent over a 2-year period. (Failure for a birth control method is defined as a pregnancy.) Birth control pills are not risk free: Combination pills can worsen problems associated with severe hypertension, diabetes mellitus, epilepsy, gallbladder disease, heart trouble, and acne. Women taking oral contraceptives are also at increased risk of venous thrombosis, strokes, pulmonary embolism, and (for women over 35) heart disease. However, pregnancy itself has similar or higher risks.

Progesterone-Only Forms of Birth Control

Progesterone-only forms of birth control are now available: The progesterone-only pill and Depo-Provera are examples. The **progesterone-only pill** must be taken daily, and skipping even one pill may result in pregnancy. **Depo-Provera** is injected every 3 months. Uterine cycles are initially irregular and eventually cease in roughly 50 percent of women using this product. The most common problems with this contraceptive method are a tendency to gain weight and a slow return to fertility (up to 18 months) after injections are discontinued.

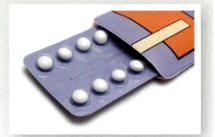

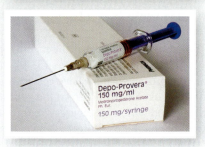

IUD (Intrauterine Device)

An **intrauterine device (IUD)** consists of a small plastic loop or a T that is inserted into the uterine cavity. The mechanism of action remains unclear, but IUDs are known to stimulate prostaglandin production in the uterus, and they are effective for years after insertion.

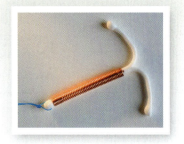

The Rhythm Method

"Natural Family Planning," also called the **rhythm method**, involves abstaining from sexual activity on the days ovulation might be occurring. The timing is estimated on the basis of previous patterns of menstruation; monitoring changes in indications of ovulation, including basal body temperature and cervical mucus texture; and, for some, urine tests for LH. Because of the irregularity of many women's uterine cycles, this method of contraception has a failure rate estimated to be 13–20 percent.

Post-Coital Contraceptives (Preven and Plan B)

Hormonal post-coital contraception, or the emergency "morning after" pill, involves taking either combination estrogen/progesterone birth control pills or progesterone-only pills in two large doses 12 hours apart within 72 hours of unprotected sexual intercourse. Particularly useful when barrier methods malfunction or coerced intercourse occurs, it reduces expected pregnancy rates by up to 89 percent. The progesterone-only version is considered safe for nonprescription use and is available for purchase over the counter for women over 18 years of age. (Purchase by women under age 18 is by prescription only.)

Surgical Sterilization—Male

In a **vasectomy** (va-SEK-tō-mē), each ductus deferens is cut and either a segment is removed (and the ends tied or cauterized) or silicone plugs are inserted (which makes it relatively easy to reverse the procedure). After a vasectomy, spermatozoa cannot pass from the epididymides to the distal portions of the reproductive tract. The surgery can be performed in a physician's office in a matter of minutes. The failure rate (due to incomplete closure/blockage of either ductus deferens) is 0.3 percent. After vasectomy, men experience normal sexual function, because the secretions of the epididymides and testes normally account for only about 5 percent of the volume of semen. Spermatozoa continue to develop, but they remain inactive and eventually degenerate.

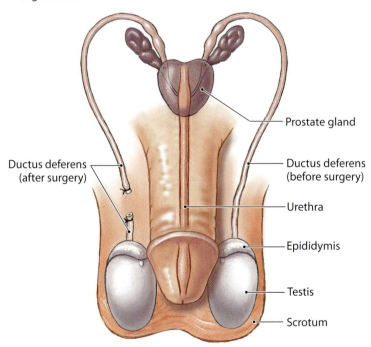

Ductus deferens (after surgery)

Prostate gland

Ductus deferens (before surgery)

Urethra

Epididymis

Testis

Scrotum

Surgical Sterilization—Female

The uterine tubes can be blocked by a surgical procedure known as a **tubal ligation**. The failure rate for this procedure is estimated at 0.45 percent. Because the surgery requires entry into the abdominopelvic cavity, most commonly by laparoscopy, general anesthetic is required and complications are more likely than with a vasectomy.

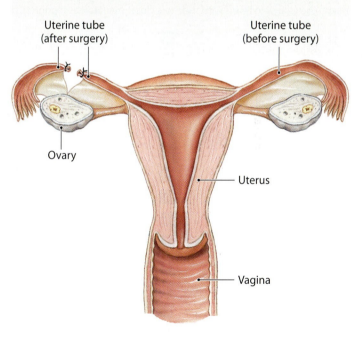

Uterine tube (after surgery)

Uterine tube (before surgery)

Ovary

Uterus

Vagina

Module 18.13 Review

a. Define vasectomy.

b. Which birth control method(s) provide some protection against sexually transmitted diseases?

c. The use of which birth control method often results in the cessation of the uterine cycle?

1. Labeling

Label the structures of the female reproductive system in the accompanying diagram.

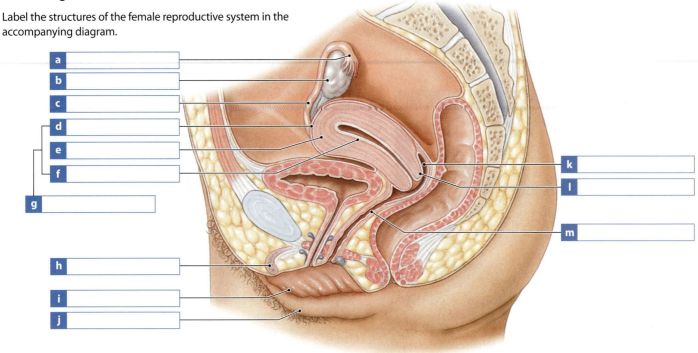

a _____

b _____

c _____

d _____

e _____

f _____

g _____

h _____

i _____

j _____

k _____

l _____

m _____

2. Matching

Match the following terms with the most closely related description.

- LH surge
- rectouterine pouch
- tubal ligation
- menarche
- ovaries
- corpus luteum
- vulva
- cervix
- broad ligament
- oocytes
- fertilization
- vasectomy
- lactation
- menstrual cycle

a _____ Immature female gametes

b _____ Begins puberty in female

c _____ Male surgical sterilization

d _____ Encloses the ovaries, uterine tubes, and uterus

e _____ Pocket posterior to the uterus

f _____ Endocrine structure

g _____ Averages 28 days

h _____ Milk production

i _____ Oocyte and hormone production

j _____ Triggers ovulation

k _____ Inferior portion of the uterus

l _____ Female surgical sterilization

m _____ Event that triggers meiosis II in a secondary oocyte

n _____ Contains female external genitalia

3. Section integration

In a condition known as endometriosis, endometrial cells are believed to migrate from the body of the uterus either into the uterine tubes or through the uterine tubes and into the peritoneal cavity, where they become established. Explain why periodic pain is a major symptom of endometriosis. _____

_____

_____

Visual Outline with Key Terms

Summarize the content of each module using the terms in the order provided.

SECTION 1

The Male Reproductive System

- gonads
- external genitalia
- testis
- spermatozoa
- male reproductive tract
- semen
  ○ accessory organs
- ductus deferens
- seminal glands
- prostate gland
- bulbo-urethral glands
- urethra
- epididymis
  ○ external genitalia
- penis
- scrotum

18.1

The coiled seminiferous tubules of the testes are connected to the male reproductive tract

- epididymis
- ductus deferens
- ejaculatory duct
- seminal glands
- prostate gland
- bulbo-urethral glands
- scrotum
- penis
- dartos muscle
- cremaster muscle
- seminiferous tubule
- rete testis
- efferent ductules
- spermatogonia
- spermatogenesis
- primary spermatocyte
- secondary spermatocytes
- spermatids
- spermiogenesis
- spermatozoon
- acrosome
- head
- neck
- middle piece
- tail
- flagellum

18.2

Hormones play a key role in establishing and maintaining male sexual function

- interstitial cells
- nurse cells
- luteinizing hormone (LH)
- follicle-stimulating hormone (FSH)
  ○ testosterone
- inhibin

Testosterone

18.3

The male reproductive tract receives secretions from the seminal, prostate, and bulbo-urethral glands

- capacitation
- ductus deferens
- ampulla
- seminal glands
- semen
- ejaculatory duct
- prostate gland
- bulbo-urethral glands

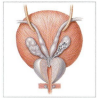

18.4

The penis conducts semen and urine out of the body

- penis
- root
- body
- neck
- glans
- prepuce
- erectile tissue
- corpora cavernosa
- corpus spongiosum

• = *Term boldfaced in this module*

18.5

Disorders of the prostate gland and testes are relatively common

- benign prostatic hypertrophy (BPH)
- prostate cancer
- prostate-specific antigen (PSA)
- prostatectomy
- testicular cancer

SECTION 2

The Female Reproductive System

- uterine tubes
- uterus
- vagina
- mammary glands
- ovaries
  ○ external genitalia
- clitoris
- labia
- oocytes
- ova
- female reproductive tract

18.6

The ovaries and the female reproductive tract are in close proximity but are not directly connected

- ovaries
- oocytes
- estrogen
- progesterone
- inhibin
- uterine tube
- infundibulum
- uterus
- clitoris
- labia
- rectouterine pouch
- vagina
- ovarian ligament
- mesovarium
- broad ligament
- suspensory ligament
- cortex (of ovary)
- medulla (of ovary)

18.7

Meiosis I in the ovaries produces a single haploid secondary oocyte that completes meiosis II only if fertilization occurs

- oogenesis
- primary oocyte
- ovum
- polar bodies
- secondary oocytes
- oogonia
- primordial follicles
- atresia
- ovarian follicles
- ovarian cycle
- zona pellucida
- secondary follicles
- tertiary follicle
- antrum
- corona radiata
- ovulation
- follicular phase
- luteal phase
- corpus luteum

18.8

The uterine tubes are connected to the uterus, a hollow organ with thick muscular walls

- uterine tube
- uterus
- infundibulum
- fimbriae
- ampulla (of the uterine tube)
- isthmus (of the uterine tube)
- perimetrium
- myometrium
- endometrium
- uterine cavity
- cervical canal
- embryo
- fetus
- fundus (of the uterus)
- body (of the uterus)
- cervix (of the uterus)

18.9

The entrance to the vagina is enclosed by external genitalia

- vagina
- vestibule
- vaginal canal
- fornix
- rugae
- hymen
- vulva
- pudendum
- labia minora
- lesser vestibular glands
- greater vestibular glands
- mons pubis
- prepuce
- clitoris
- labia majora

18.10

The mammary glands nourish the infant after delivery

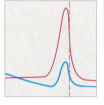

- mammary glands
- lactation
- pectoral fat pad
- suspensory ligaments of the breast
- lobes
- lobules
- secretory alveoli
- lactiferous duct
- lactiferous sinus
- nipple
- areola

18.11

The ovarian and uterine cycles are regulated by hormones of the hypothalamus, pituitary gland, and ovaries

- menstrual cycle
- menarche
- menopause
- basal body temperature

● = *Term boldfaced in this module*

18.12

Female reproductive system disorders are relatively common and often deadly

- cysts
- fibrocystic disease
- breast cancer
- ovarian cancer
- carcinomas
- cervical cancer
- human papillomaviruses (HPV)

18.13

Birth control strategies vary in effectiveness and in the nature of associated risks

- male condoms
- sexually transmitted diseases (STDs)
- diaphragm
- oral contraceptives
- progesterone-only pill
- Depo-Provera
- intrauterine device (IUD)
- rhythm method
- hormonal post-coital contraception
- vasectomy
- tubal ligation

CAREER PATHS

"I like talking to people, giving them information they might not have had before."

— **Tanya Bass**
Health Educator
CCFNC

Health Educator

Tanya Bass is the project coordinator for the statewide campaign Cervical Cancer-Free North Carolina (CCFNC). Tanya is a health educator, with the whole state as her classroom.

Health educators promote healthy lifestyles by providing communities and individuals with information about behaviors that can prevent illness or injury. For example, they work on anti-smoking campaigns, healthy eating initiatives, and, in Tanya's case, cancer prevention and screening.

"I like talking to people, giving them information they might not have had before," Tanya says. "If you like people, and if you're good at talking to people, this may be the career for you."

Tanya works on multiple fronts. She visits schools to promote a school health vaccination project, so that school nurses are able to administer the vaccine for the human papillomavirus (HPV)—the leading cause of cervical cancer. She also meets with community organizers and has helped produce a resource directory for healthcare providers and patients.

Because cervical cancer is treatable if caught early, CCFNC strives to ensure women have regular screenings. Tanya interviews women who have not been screened to find out why, and also urge them to do so.

If large numbers of women report similar obstacles to getting screenings, such as no affordable healthcare providers in an area, Tanya's findings can help advocates lobby for more clinics or lower-cost clinics in a given area. She also interviews providers who have had an increase in screenings to find out their best practices. CCFNC and other organizations have worked with and helped to influence lawmakers, ultimately leading to a state law in North Carolina that schools must provide info on the HPV vaccine and health classes must teach about HPV prevention.

Naturally, anatomy and physiology is important in Tanya's job. "You have to know your anatomy to even talk to people about

cervical cancer," she advises. Tanya's primary tool as a health educator, though, is personal contact. Communication is essential, but so too is cultural awareness and the ability to not pass judgment. Tanya emphasized that the important thing is that a woman is able to get tested and, if necessary, receive treatment.

For more information, visit the website of the National Commission for Health Educator Credentialing at http://nchec.org.

Think this is the career for you?
Key Stats:

- **Education and Training.** Entry-level positions require a bachelor's degree in health education. A master's degree is required for advancement or to work in public health.

- **Licensure.** Some states require certification to work in public health.

- **Earnings.** Earnings vary but the median annual salary is $45,830.

- **Expected Job Prospect.** Employment is expected to grow faster than the national average—by 18% through 2018.

Bureau of Labor Statistics, U.S. Department of Labor, *Occupational Outlook Handbook, 2010–11 Edition,* Health Educators, on the Internet at http://www.bls.gov/oco/ocos063.htm (visited September 14, 2011).

MasteringA&P®

Access more review material online in the Study Area at www.masteringaandp.com.

- Chapter guides
- Chapter quizzes
- Practice tests
- Art labeling activities
- Flashcards
- A glossary with pronunciations
- Practice Anatomy Lab™ (PAL™) 3.0 virtual anatomy practice tool
- Interactive Physiology® (IP) animated tutorials
- MP3 Tutor Sessions

practice anatomy lab™

For this chapter, follow these navigation paths in PAL:

- Human Cadaver>Reproductive System
- Anatomical Models>Reproductive System
- Histology>Reproductive System

For this chapter, go to this topic in the MP3 Tutor Sessions:

- Hormonal Control of the Menstrual Cycle

Chapter 18 Review Questions

1. What does amenorrhea (absence of normal menstrual cycles) in female athletes suggest about the relationship between body fat and menstruation?

2. How might exercise-induced amenorrhea (absence of normal menstrual cycles) be advantageous to human survival?

3. What are some health-related consequences of estrogen deficiency?

4. Diane has peritonitis (an inflammation of the peritoneum), which her physician says resulted from a sexually transmitted disease. Why might this condition occur more readily in females than in males?

5. Trace the pathway that a sperm follows from the male ejaculatory duct to the secondary oocyte.

For answers to all module, section, and chapter review questions, see the blue Answers tab at the back of the book.

19

Development and Inheritance

An Overview of Development

1 **Development** is the gradual modification of anatomical structures and physiological characteristics from fertilization to maturity. The changes that occur during development are truly remarkable. In a mere 9 months, all the tissues, organs, and organ systems we have studied so far take shape and begin to function. What begins as a single cell slightly larger than the period at the end of this sentence becomes an individual whose body contains trillions of cells organized into a complex array of highly specialized structures.

Prenatal Development

Embryological development comprises the events that occur during the first two months after **fertilization**, or conception, when the egg and sperm unite. The study of these events is called **embryology** (em-brē-OL-ō-jē).

Fetal development begins at the start of the ninth week and continues until birth. Embryological and fetal development are sometimes referred to collectively as **prenatal** (*natus*, birth) **development**, the primary focus of this chapter.

4 weeks

8 weeks

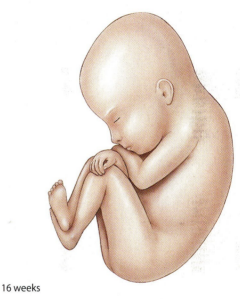

16 weeks

2 The time spent in prenatal development is known as **gestation** (jes-TĀ-shun). For convenience, we usually think of the gestation period as three integrated trimesters, each three months long.

Gestation

First Trimester
The **first trimester** is the period of embryological and early fetal development. During this period, the rudiments of all the major organ systems appear.

Second Trimester
The **second trimester** is dominated by the development of organs and organ systems, a process that nears completion by the end of the sixth month. During this period, body shape and proportions change. By the end of this trimester, the fetus looks distinctly human.

Third Trimester
The **third trimester** is characterized by the largest gain in fetal weight. Early in the third trimester, most of the fetus's major organ systems become fully functional. An infant born one month or even two months prematurely has a reasonable chance of survival if appropriate medical care is available.

Postnatal development begins at birth and continues to **maturity**, the state of full development or completed growth. A basic understanding of prenatal and postnatal development provides important insights into anatomical structures. In addition, many of the mechanisms of development and growth are similar to those responsible for the repair of injuries.

Cleavage continues until the blastocyst implants in the uterine wall

Fertilization involves the fusion of two haploid gametes, each containing 23 chromosomes, producing a **zygote** (ZĪ-gōt) that contains 46 chromosomes, the normal complement in a somatic cell. Fertilization most commonly takes place in the uterine tube (usually the ampulla). The zygote then undergoes a series of cell divisions, called cleavage. During cleavage, the cytoplasm of the zygote becomes subdivided among an ever-increasing number of progressively smaller blastomeres. A group of blastomeres created by cleavage divisions is called a **pre-embryo**.

1 Cleavage lasts roughly seven days. Within that period the pre-embryo travels the length of the uterine tube.

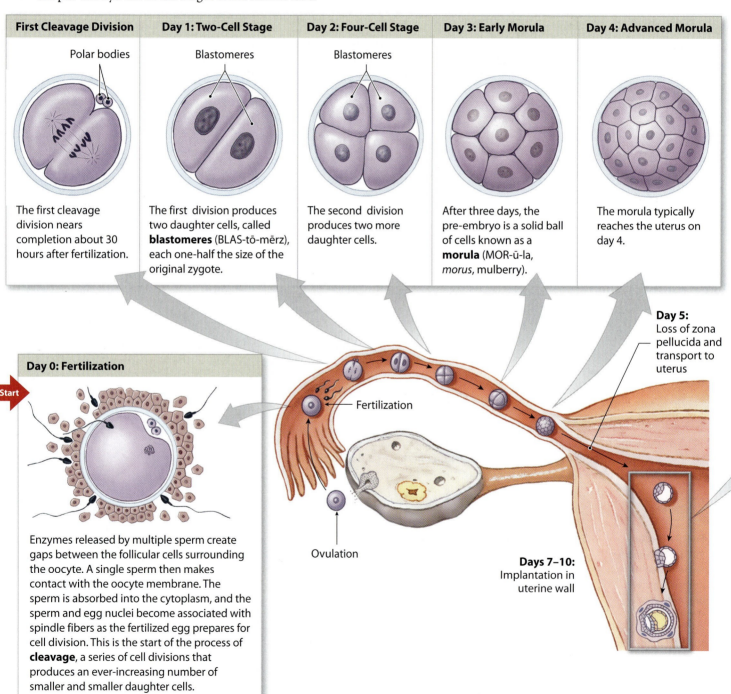

First Cleavage Division

Polar bodies

The first cleavage division nears completion about 30 hours after fertilization.

Day 1: Two-Cell Stage

Blastomeres

The first division produces two daughter cells, called **blastomeres** (BLAS-tō-mērz), each one-half the size of the original zygote.

Day 2: Four-Cell Stage

Blastomeres

The second division produces two more daughter cells.

Day 3: Early Morula

After three days, the pre-embryo is a solid ball of cells known as a **morula** (MOR-ū-la, *morus*, mulberry).

Day 4: Advanced Morula

The morula typically reaches the uterus on day 4.

Day 5:
Loss of zona pellucida and transport to uterus

Day 0: Fertilization

Start

Fertilization

Ovulation

Enzymes released by multiple sperm create gaps between the follicular cells surrounding the oocyte. A single sperm then makes contact with the oocyte membrane. The sperm is absorbed into the cytoplasm, and the sperm and egg nuclei become associated with spindle fibers as the fertilized egg prepares for cell division. This is the start of the process of **cleavage**, a series of cell divisions that produces an ever-increasing number of smaller and smaller daughter cells.

Days 7–10:
Implantation in uterine wall

Day 6: Blastocyst

Over the next two days, the blastomeres form a **blastocyst** (blas-tō-sist), a hollow ball with an inner cavity known as the **blastocoele** (BLAS-tō-sēl). At this stage the blastomeres are no longer identical in size and shape. The blastocyst is freely exposed to the fluid contents of the uterine cavity, which contains the glycogen-rich secretions of the uterine glands. The rate of growth and cell division now accelerates, and the blastocyst enlarges rapidly.

Day 7: Implantation

When fully formed, the blastocyst contacts the endometrium. **Implantation** begins with the attachment of the blastocyst to the endometrium of the uterus. Implantation proceeds as the blastocyst erodes the endometrial lining and becomes enclosed within the endometrium by day 10.

Day 8: Trophoblast Development

At the point of contact, the trophoblast cells divide rapidly, making the trophoblast several layers thick. The cells closest to the blastocoele remain intact, forming a layer of **cellular trophoblast**. Near the endometrial wall, the plasma membranes separating the trophoblast cells disappear, creating a layer of cytoplasm containing multiple nuclei. This layer is called the **syncytial** (sin-SISH-ul) **trophoblast**.

Day 9: Formation of Amniotic Cavity

As implantation proceeds, the trophoblast continues to enlarge and spread into the surrounding endometrium. The erosion of uterine glands releases nutrients that are absorbed by the trophoblast and distributed by diffusion to the inner cell mass. These nutrients provide the energy needed to support the early stages of embryo formation. Trophoblastic extensions (villi) grow around endometrial capillaries. As the capillary walls are destroyed, maternal blood begins to percolate through trophoblastic channels known as **lacunae**.

FUNCTIONAL ZONE OF ENDOMETRIUM

UTERINE CAVITY

Uterine glands

Blastocyst

The outer layer of cells, which separates the outside world from the blastocoele, is called the **trophoblast** (TRŌ-fō-blast, *trophos*, food + *blast*, precursor). The cells in this layer are responsible for providing nutrients to the developing embryo.

Blastocoele

The **inner cell mass** lies clustered at one end of the blastocyst. These cells are exposed to the blastocoele but are insulated from contact with the outside environment by the trophoblast. In time, the inner cell mass will form the embryo.

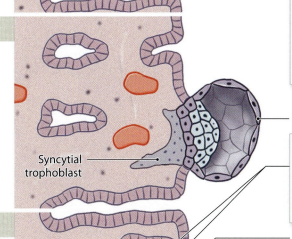

Cellular trophoblast

Syncytial trophoblast

Fingerlike **villi** extend away from the trophoblast into the surrounding endometrium, gradually increasing in size and complexity.

Lacuna

At the time of implantation, the inner cell mass has separated from the trophoblast. The separation gradually increases, creating a fluid-filled chamber called the **amniotic** (am-nē-OT-ik) **cavity.**

Endometrial capillary

Module 19.1 Review

a. Define cleavage.

b. What developmental stage begins once the zygote arrives in the uterine cavity?

c. Describe the blastocyst and its role in implantation.

Gastrulation produces three germ layers: ectoderm, endoderm, and mesoderm

Together, the three germ layers will form the body of the embryo.

Day 9: Formation of Amniotic Cavity (continued)

When the amniotic cavity first appears, the cells of the inner cell mass are organized into an oval sheet two layers thick: a superficial layer that faces the amniotic cavity, and a deeper layer that is exposed to the fluid contents of the blastocoele. Cells of the superficial layer migrate along the walls of the amniotic cavity and separate the amniotic cavity from the tropho-blast. This is the first step in the formation of the **amnion**, one of four **extra-embryonic membranes** we will consider further in Module 19.3. At this stage, nutrients released into the amnion and blastocoele by the advancing trophoblast are absorbed directly by the cells of the inner cell mass.

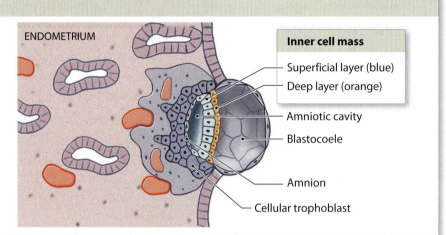

ENDOMETRIUM

Inner cell mass
- Superficial layer (blue)
- Deep layer (orange)

Amniotic cavity

Blastocoele

Amnion

Cellular trophoblast

Day 10: Yolk Sac Formation

While cells from the superficial layer of the inner cell mass migrate around the amniotic cavity, forming the amnion, cells from the deeper layer migrate around the outer edges of the blastocoele. This is the first step in the formation of the **yolk sac**, a second extra-embryonic membrane. For roughly the next two weeks, the yolk sac is the primary nutrient source for the inner cell mass. The yolk sac absorbs and distrib-utes nutrients released into the blastocoele by the trophoblast.

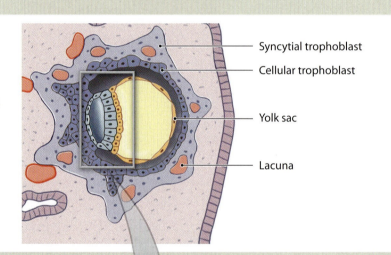

Syncytial trophoblast

Cellular trophoblast

Yolk sac

Lacuna

Day 12: Gastrulation

By day 12, superficial cells leave the surface and move between the two existing layers. This movement creates three distinct embryonic layers: (1) the **ectoderm**, consisting of superficial cells that did not migrate into the interior; (2) the **endoderm**, consist-ing of the cells that face the yolk sac; and (3) the **mesoderm**, consisting of the poorly organized layer of migrating cells between the ectoderm and the endoderm. Collectively, these three embryonic layers are called **germ layers**, and the cellular migration process that produces the germ layers is called **gastrulation** (gas-troo-LĀ-shun). Gastrulation produces an oval, three-layered sheet known as the **embryonic disc**. This disc will form the body of the embryo, whereas all other cells of the blastocyst will be part of the extra-embryonic membranes.

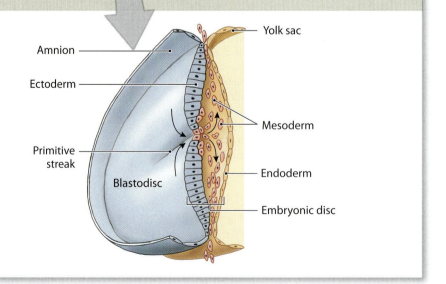

Amnion

Ectoderm

Primitive streak

Blastodisc

Yolk sac

Mesoderm

Endoderm

Embryonic disc

The different systems of the body originate from the cellular divisions of the three germ layers. The panels to the left illustrate the formation of the three germ layers, and the table below summarizes their importance to later development.

The Fates of the Germ Layers

Body System	Ectodermal Contributions
Integumentary system	Epidermis, hair follicles and hairs, nails, and glands communicating with the skin (sweat glands, mammary glands, and sebaceous glands)
Skeletal system	Pharyngeal cartilages and their derivatives in adults (portion of sphenoid, the auditory ossicles, the styloid processes of the temporal bones, the cornu and superior rim of the hyoid bone)
Nervous system	All neural tissue, including brain and spinal cord
Endocrine system	Pituitary gland and adrenal medullae
Respiratory system	Mucous epithelium of nasal passageways
Digestive system	Mucous epithelium of mouth and anus, salivary glands
Mesodermal Contributions	
Integumentary system	Dermis and hypodermis
Skeletal system	All components except some pharyngeal derivatives
Muscular system	All components
Endocrine system	Adrenal cortex, endocrine tissues of heart, kidneys, and gonads
Cardiovascular system	All components
Lymphatic system	All components
Urinary system	The kidneys, including the nephrons and the initial portions of the collecting system
Reproductive system	The gonads and the adjacent portions of the duct systems
Miscellaneous	The lining of the subdivisions of the ventral body cavity (pleural, pericardial, and peritoneal cavities) and the connective tissues that support all organ systems
Endodermal Contributions	
Endocrine system	Thymus, thyroid gland, and pancreas
Respiratory system	Respiratory epithelium (except nasal passageways) and associated mucous glands
Digestive system	Mucous epithelium (except mouth and anus), exocrine glands (except salivary glands), liver, and pancreas
Urinary system	Urinary bladder and distal portions of the duct system
Reproductive system	Distal portions of the duct system, stem cells that produce gametes

The trophoblast undergoes repeated nuclear divisions, shows extensive and rapid growth, has a very high demand for energy, invades and spreads through adjacent tissues, yet fails to activate the maternal immune system—in short, the trophoblast has many of the characteristics of cancer cells. In about 0.1 percent of pregnancies, something goes wrong with the regulatory mechanisms, and instead of developing normally, the trophoblast behaves like a tumor. This condition is called **gestational trophoblastic neoplasia**. Approximately 20 percent of gestational trophoblastic neoplasias metastasize to other tissues, with potentially fatal results. Consequently, prompt surgical removal of the mass is essential, and the surgery is sometimes followed by chemotherapy.

Module 19.2 Review

a. Define gestational trophoblastic neoplasia.

b. Define gastrulation and describe its importance.

c. What germ layer gives rise to nearly all body systems except the nervous and respiratory systems?

The extra-embryonic membranes form the placenta that supports fetal growth and development

Germ layers participate in the formation of four extra-embryonic membranes: the yolk sac (endoderm + mesoderm), the amnion (ectoderm + mesoderm), the allantois (endoderm + mesoderm), and the chorion (mesoderm + trophoblast). Although these membranes support embryological and fetal development, few traces of their existence remain in adult systems.

Formation of the Yolk Sac   Formation of the Amnion

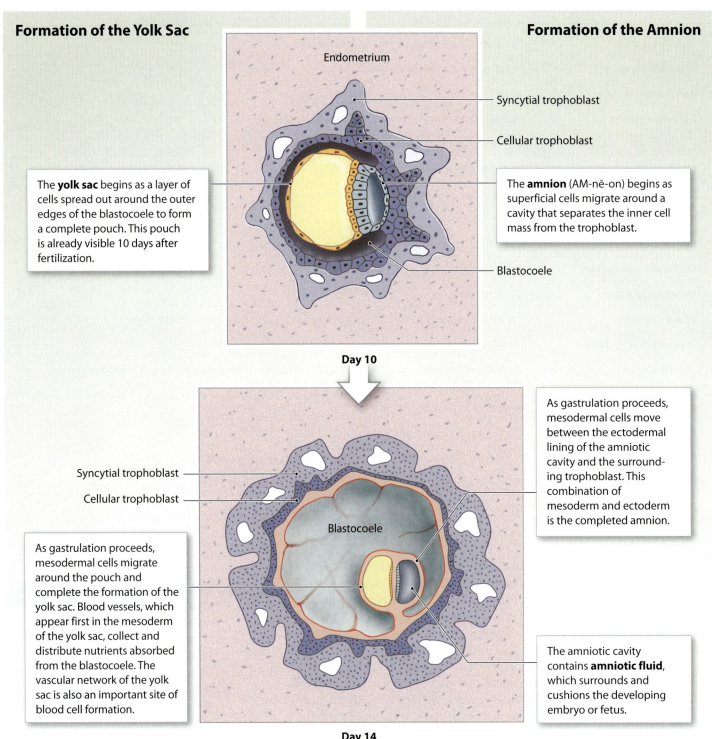

Endometrium

Syncytial trophoblast

Cellular trophoblast

The **yolk sac** begins as a layer of cells spread out around the outer edges of the blastocoele to form a complete pouch. This pouch is already visible 10 days after fertilization.

The **amnion** (AM-nē-on) begins as superficial cells migrate around a cavity that separates the inner cell mass from the trophoblast.

Blastocoele

Day 10

Syncytial trophoblast

Cellular trophoblast

Blastocoele

As gastrulation proceeds, mesodermal cells migrate around the pouch and complete the formation of the yolk sac. Blood vessels, which appear first in the mesoderm of the yolk sac, collect and distribute nutrients absorbed from the blastocoele. The vascular network of the yolk sac is also an important site of blood cell formation.

As gastrulation proceeds, mesodermal cells move between the ectodermal lining of the amniotic cavity and the surrounding trophoblast. This combination of mesoderm and ectoderm is the completed amnion.

The amniotic cavity contains **amniotic fluid**, which surrounds and cushions the developing embryo or fetus.

Day 14

Formation of the Allantois

Formation of the Chorion

The **allantois** (a-LAN-tō-is) begins as an outpocketing of the endoderm near the base of the yolk sac. The free endodermal tip then grows toward the wall of the blastocyst, surrounded by a mass of mesodermal cells.

The mesoderm associated with the allantois spreads around the blastocyst, separating the trophoblast from the blastocoele. This combination of mesoderm and trophoblast is the **chorion** (KŌ-rē-on), the beginning of the placenta.

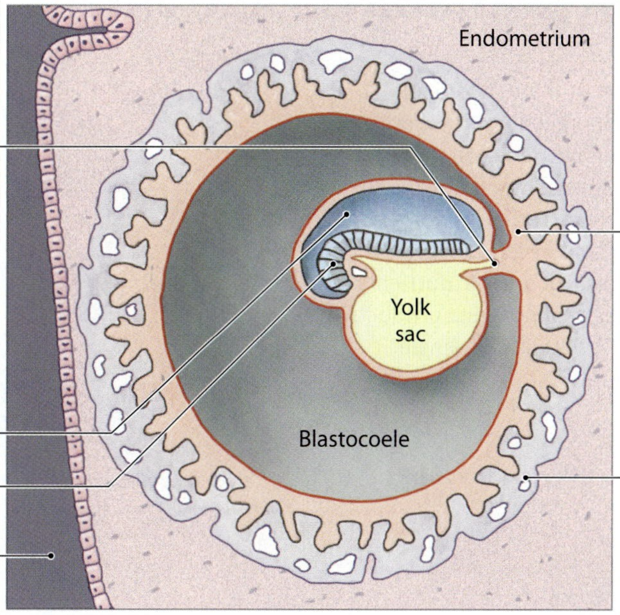

Endometrium

Yolk sac

Blastocoele

Amniotic cavity

Embryo

Uterine lumen

Syncytial trophoblast

Week 3

The allantois extends partway into the umbilical stalk. The base of the allantois will form the urinary bladder.

Umbilical stalk

The **placenta** forms the interface between fetal and maternal systems. It develops as villi of the chorion invade the endometrium and break down maternal blood vessels. The placenta becomes the primary support mechanism for the developing embryo; oxygen and nutrients are absorbed from the maternal bloodstream and exchanged for carbon dioxide and wastes.

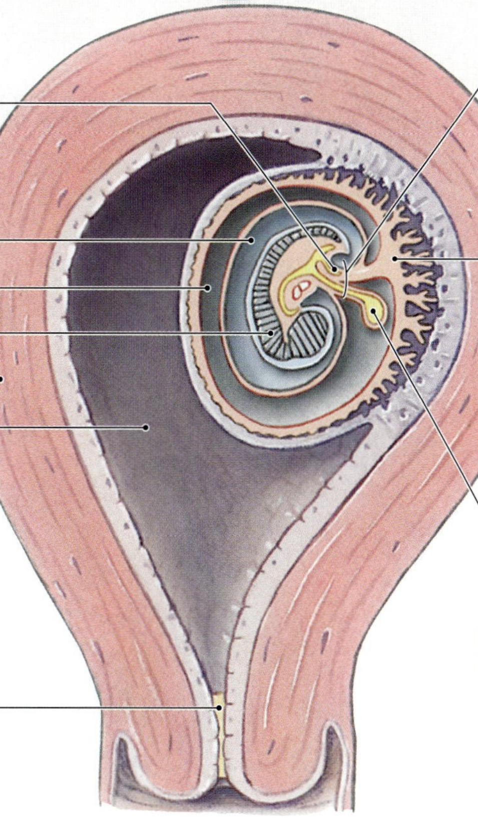

Amniotic cavity

Blastocoele

Embryo

Uterus

Uterine lumen

Yolk sac

Cervical (mucous) plug

Week 5

Module 19.3 Review

a. Name the four extra-embryonic membranes.

b. Which extra-embryonic membrane later gives rise to the urinary bladder?

c. From which germ layers do the extra-embryonic membranes form, and what are each membrane's functions?

The placenta performs many vital functions for the duration of prenatal development

1 This illustration gives a closer look at the structure of the placenta. The chorionic villi provide the surface area for active and passive exchanges of gases, nutrients, and wastes between the fetal and maternal bloodstreams. Blood flowing to the placenta through the paired **umbilical arteries** is deoxygenated and contains waste products generated by fetal tissues. At the placenta, oxygen supplies are replenished, organic nutrients are added, and carbon dioxide and other organic waste products are removed. Blood then returns to the fetus within a single **umbilical vein**.

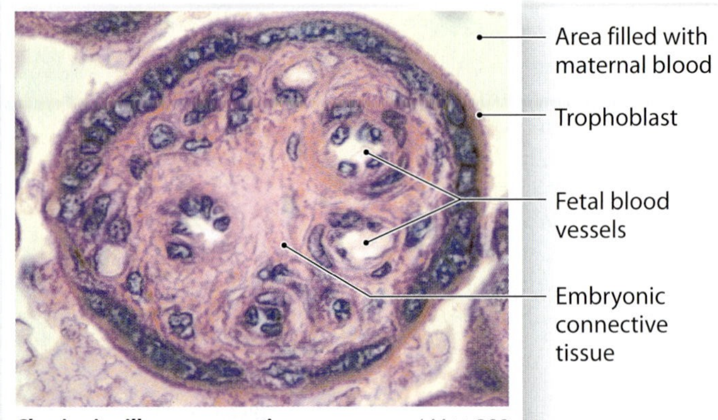

Area filled with maternal blood
Trophoblast
Fetal blood vessels
Embryonic connective tissue

Chorionic villus, cross section LM × 280

Umbilical cord (cut) Yolk sac Placenta

Amnion
Chorion

Endometrium
Myometrium
Uterine cavity
Cervical (mucous) plug in cervical canal
External os
Cervix
Vagina

Chorionic villi
Area filled with maternal blood
Maternal blood vessels

Umbilical vein Umbilical arteries Amnion Trophoblast (cellular and syncytial layers)

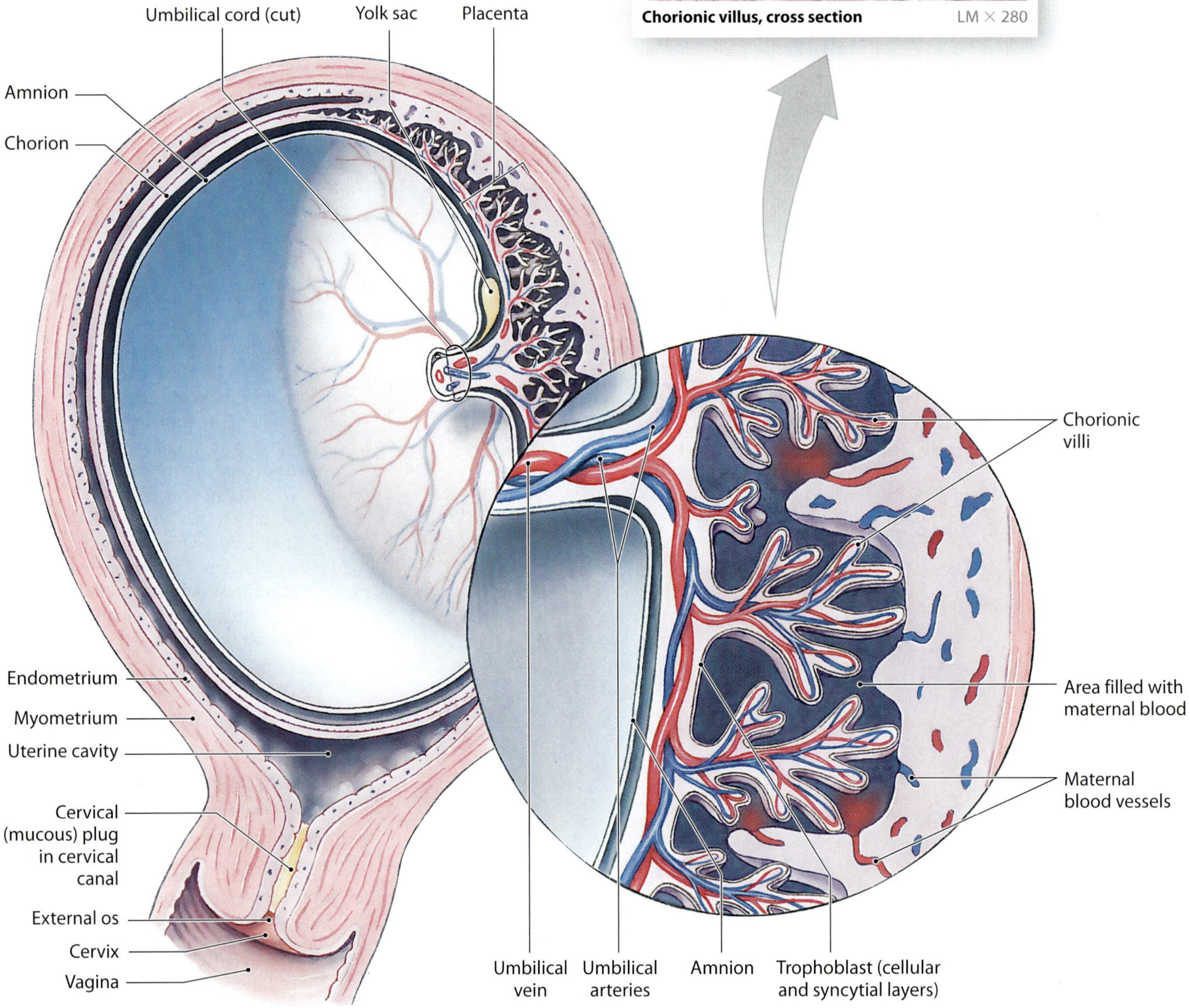

2 In addition to its role in nourishing the fetus, the placenta acts as an endocrine organ. Several hormones are synthesized by the syncytial trophoblast and released into the maternal bloodstream.

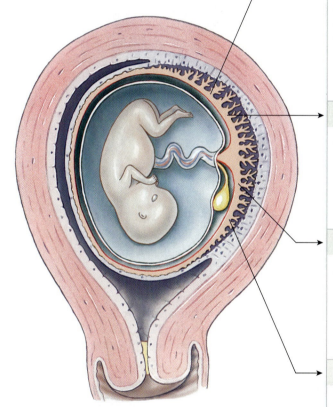

Placental Hormones

Human Chorionic Gonadotropin (hCG)

Human chorionic gonadotropin (hCG) appears in the maternal bloodstream soon after implantation has occurred. The presence of hCG in blood or urine samples provides a reliable indication of pregnancy. Kits sold for the early detection of pregnancy are sensitive to the presence of this hormone. In function, hCG resembles LH, because it maintains the integrity of the corpus luteum and promotes the continued secretion of progesterone. As a result, in pregnancy the endometrial lining remains perfectly functional, and menses does not occur. In the presence of hCG, the corpus luteum persists for three to four months before gradually decreasing in size and secretory function. The decline in luteal function does not trigger the return of uterine cycles, because by the end of the first trimester, the placenta is secreting both estrogen and progesterone.

Human Placental Lactogen (hPL)

Human placental lactogen (hPL) helps prepare the mammary glands for milk production. At the mammary glands, the conversion from inactive to active status requires the presence of placental hormones (hPL, estrogen, and progesterone) as well as several maternal hormones (GH, prolactin, and thyroid hormones).

Relaxin

Relaxin is a peptide hormone that is secreted by the placenta and the corpus luteum during pregnancy. Relaxin (1) increases the flexibility of the pubic symphysis, permitting the pelvis to expand during delivery; (2) causes dilation of the cervix, making it easier for the fetus to enter the vaginal canal; and (3) delays the onset of labor contractions until late in the pregnancy.

Progesterone and Estrogen

After the first trimester, the placenta produces sufficient amounts of progesterone to maintain the endometrial lining and continue the pregnancy. As the end of the third trimester approaches, estrogen production by the placenta accelerates. As we will see in a later module, the rising estrogen levels play a role in stimulating labor and delivery.

Module 19.4 Review

a. Name the hormones synthesized by the syncytial trophoblast.

b. The presence of which hormone in the urine provides a reliable indicator of pregnancy in home pregnancy tests?

c. When does the placenta become sufficiently functional to continue the pregnancy?

Organ systems are formed in the first trimester and become functional in the second and third trimesters

The first trimester is a critical period for development, because events in the first 12 weeks establish the basis for **organogenesis**, the process of organ formation. Over the next two trimesters the fetus grows larger and the organ systems increase in complexity to the stage at which they are capable of normal function.

1 This is a scanning electron micrograph of an embryo in the second week of development. The CNS is forming as a deep groove develops in a thick ectodermal band that lies along the posterior midline of the embryo.

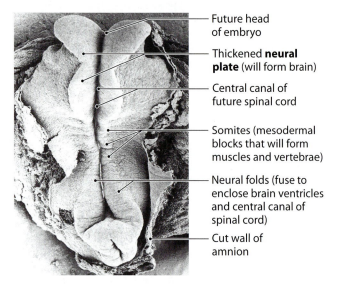

- Future head of embryo
- Thickened **neural plate** (will form brain)
- Central canal of future spinal cord
- Somites (mesodermal blocks that will form muscles and vertebrae)
- Neural folds (fuse to enclose brain ventricles and central canal of spinal cord)
- Cut wall of amnion

2 This photograph of a 4-week embryo shows many features you can probably recognize easily. The heart is beating, pushing blood to and from the placenta, providing the nutrients needed for growth and development.

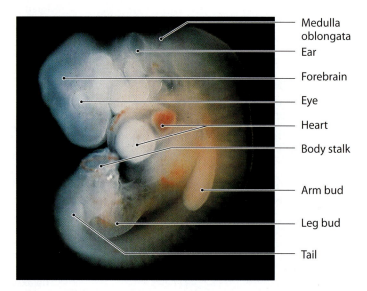

- Medulla oblongata
- Ear
- Forebrain
- Eye
- Heart
- Body stalk
- Arm bud
- Leg bud
- Tail

3 By week six, the placenta has formed and the embryo floats within the amniotic cavity. Body proportions are changing, the limbs are growing longer, and the skull bones are beginning to organize around the already-formed brain and eyes.

4 At the end of the first trimester, the fetus is considerably larger and its human features better defined. The axial and appendicular muscles are forming, and fetal movements will soon begin.

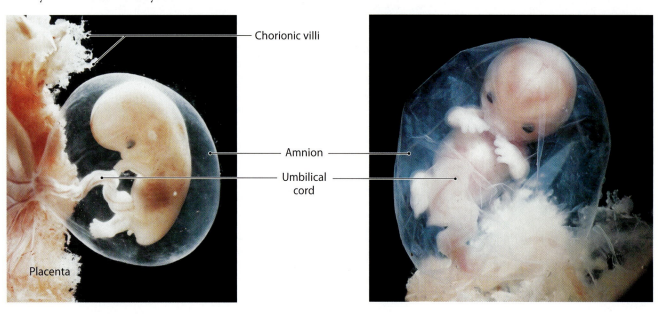

- Chorionic villi
- Amnion
- Umbilical cord
- Placenta

5 This is a fetus after four months of gestation. The face and palate have their proper form, and the cerebral hemispheres are rapidly enlarging. Hair follicles are present, and hair growth begins. Peripheral nerves have formed, sensory receptors are developing, and the fetus moves frequently. The first 8 weeks of fetal growth is the most rapid, with the fetus increasing in weight some twenty-five-fold. By the end of the second trimester, the fetus will have grown to a weight of about 0.64 kg (1.4 lb).

6 This is an ultrasound of a fetus after six months of gestation. During the third trimester, most of the organ systems become ready to perform their normal functions without maternal assistance. The rate of growth starts to slow, but in absolute terms the largest weight gain occurs in this trimester. In the final three months of gestation, the fetus gains about 2.6 kg (5.7 lb), reaching a full-term weight of approximately 3.2 kg (7 lb).

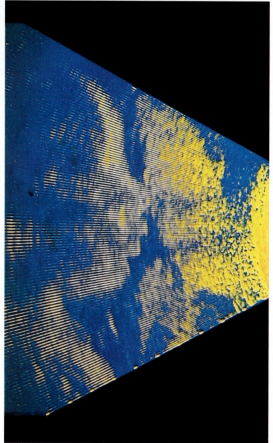

Module 19.5 Review

a. Define organogenesis.

b. Identify the main event in fetal development during the second trimester and third trimester.

c. During which trimester does the fetus undergo its largest absolute weight gain?

Pregnancy places anatomical and physiological stresses on maternal systems

Pregnancy places tremendous strains on the mother. The developing fetus is totally dependent on maternal organ systems for nourishment, respiration, and waste removal. Maternal systems perform these functions in addition to their normal operations. For example, the mother must absorb enough oxygen, nutrients, and vitamins for herself and for her fetus, and she must eliminate all the wastes that are generated. Although this is not a burden over the initial weeks of gestation, the demands placed on the mother become significant as the fetus grows.

1 The physical strains of pregnancy are considerable. It is not unusual for a woman to gain 6–7 kg (13–15 lb) in weight during pregnancy, and the weight is not aligned with the body axis. This means that moving and maintaining balance use additional energy. The sectional views below compare the organ positions in nonpregnant and pregnant women. When the pregnancy is at full term, the uterus and fetus are so large that they push many of the maternal abdominal organs out of their normal positions.

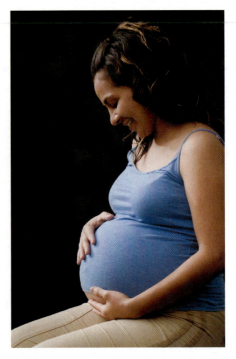

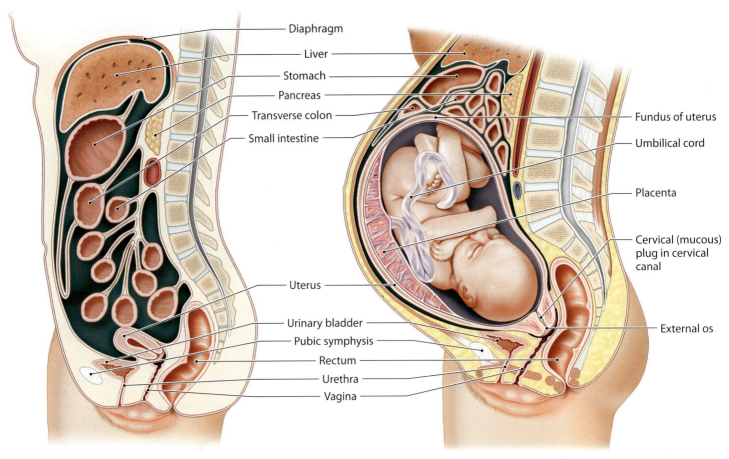

Diaphragm
Liver
Stomach
Pancreas
Transverse colon
Small intestine

Fundus of uterus
Umbilical cord
Placenta
Cervical (mucous) plug in cervical canal

Uterus
Urinary bladder
Pubic symphysis
Rectum
Urethra
Vagina

External os

Nonpregnant female

Pregnant female (full-term infant)

2 The physiological adjustments to pregnancy are even more extreme than the physical ones. This is a summary of the physiological changes that have occurred in maternal systems by the end of the third trimester.

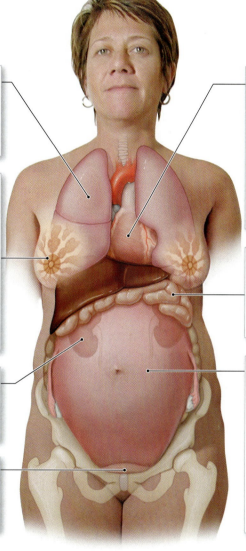

The mother's lungs must deliver the extra oxygen required, and remove the excess carbon dioxide generated, by the fetus. As a result, the maternal respiratory rate goes up and tidal volume increases.

Mammary gland development requires a combination of hormones, including human placental lactogen and placental prolactin (PRL), estrogen, progesterone, GH, and thyroxine from the maternal endocrine organs. By the end of the sixth month of pregnancy, the mammary glands are fully developed and begin to produce clear secretions that are stored in the duct system of those glands and may be expressed from the nipple.

The kidneys must eliminate the wastes produced by maternal and fetal systems. As a result, the maternal glomerular filtration rate (GFR) increases by roughly 50 percent.

Because the volume of urine produced increases and the weight of the uterus presses down on the urinary bladder, pregnant women need to urinate frequently.

The volume of blood flowing into the placenta reduces the volume in the systemic circuit of the mother. At the same time, fetal metabolic activity "steals" maternal oxygen and elevates maternal CO_2 levels. This combination stimulates the production of renin and erythropoietin, leading to an increase in maternal blood volume. By the end of gestation, maternal blood volume has increased by almost 50 percent.

Pregnant women must nourish both themselves and their fetus and so tend to have increased hunger sensations. Maternal requirements for nutrients can climb by up to 30 percent above normal.

At the end of gestation, a typical uterus has grown from 7.5 cm (3 in.) in length and 30–40 g (1–1.4 oz) in weight to 30 cm (12 in.) in length and 1100 g (2.4 lb) in weight. The uterus may then contain 2 liters of fluid, plus fetus and placenta, for a total weight of roughly 6–7 kg (13–15 lb). This remarkable expansion occurs through the enlargement (hypertrophy) of existing cells, especially smooth muscle fibers, rather than by an increase in the total number of cells.

Although pregnancy is a natural phenomenon, the physical and physiological demands on maternal systems make it potentially dangerous. At any age, the risks associated with pregnancy are significantly greater than those associated with the use of oral contraceptives. (The notable exception involves women who both take the pill and smoke.) For pregnant women over age 35, the chances of dying from pregnancy-related complications are almost twice as great as the chances of being killed in an automobile accident.

Module 19.6 Review

a. List the major changes that occur in maternal systems during pregnancy.

b. Why does a mother's blood volume increase during pregnancy?

c. Based on the illustrations showing the locations of the internal organs in nonpregnant and pregnant women, explain why some women experience difficulty breathing while pregnant.

Multiple factors initiate and accelerate the process of labor

The tremendous stretching of the uterus during pregnancy is associated with a gradual increase in the rate of spontaneous smooth muscle contractions in the myometrium. In the early stages of pregnancy, progesterone released by the placenta inhibits the uterine smooth muscle, preventing powerful contractions. Late in pregnancy, some women experience occasional spasms in the uterine musculature, but these contractions are neither regular nor persistent. Such contractions are called **false labor**. **True labor** begins when biochemical and mechanical factors reach a point of no return.

1 This sagittal section shows the position of the fetus at the onset of true labor. Most fetuses settle into this position.

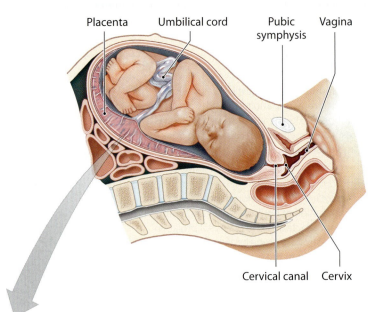

Placenta Umbilical cord Pubic symphysis Vagina

Cervical canal Cervix

2 After nine months of gestation, multiple factors interact to begin true labor. Once labor contractions have begun in the myometrium, positive feedback ensures that they will continue until delivery has been completed.

Placental Factors

Placental estrogen increases the sensitivity of the smooth muscle cells of the myometrium and makes contractions more likely. As delivery approaches, the production of estrogen accelerates. Estrogen also increases the sensitivity of smooth muscle fibers to oxytocin.

Relaxin produced by the placenta relaxes the pelvic articulations and dilates the cervix.

Maternal Oxytocin Release

Maternal oxytocin release is stimulated by high estrogen levels. The smooth muscle in a late-term uterus is 100 times more sensitive to oxytocin than the smooth muscle in a nonpregnant uterus.

Distortion of Myometrium

Distortion of the myometrium increases the sensitivity of the smooth muscle layers, promoting spontaneous contractions that get stronger and more frequent as the pregnancy advances.

Labor contractions move the fetus and further distort the myometrium. This distortion stimulates additional oxytocin and prostaglandin release. This **positive feedback** continues until delivery is completed.

LABOR CONTRACTIONS OCCUR

3 The goal of labor is **parturition** (par-toor-ISH-un), the forcible expulsion of the fetus and placenta. Labor has traditionally been divided into three stages: the dilation stage, the expulsion stage, and the placental stage.

Dilation Stage

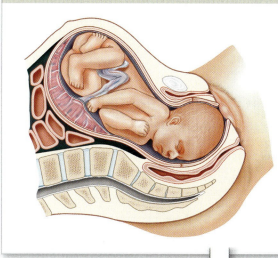

The **dilation stage** begins with the onset of true labor, as the cervix dilates and the fetus begins to shift toward the cervical canal, moved by gravity and uterine contractions. This stage typically lasts eight or more hours. At the start of the dilation stage, labor contractions last up to half a minute and occur once every 10–30 minutes; their frequency and duration increase steadily. Late in this stage, the amnion ruptures, an event sometimes referred to as "having one's water break."

Expulsion Stage

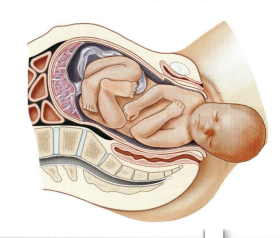

The **expulsion stage** begins as the cervix, pushed open by the approaching fetus, completes its dilation. In this stage, contractions reach maximum intensity, occurring at perhaps two- or three-minute intervals and lasting a full minute. Expulsion continues until the fetus has emerged from the vagina. In most cases, the expulsion stage lasts less than two hours. The arrival of the newborn infant into the outside world is called **delivery**, or birth.

Placental Stage

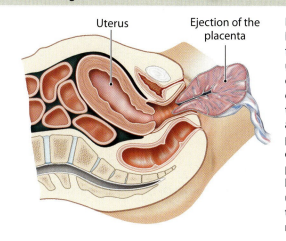

Uterus

Ejection of the placenta

During the **placental stage** of labor, muscle tension builds in the walls of the partially empty uterus, which gradually decreases in size. This uterine contraction tears the connections between the endometrium and the placenta, and the placenta, or **afterbirth**, is ejected. The disruption of the placenta is accompanied by a loss of blood, but associated uterine contractions compress the uterine vessels and usually restrict this flow.

Premature labor occurs when true labor begins before the fetus has completed normal development. The newborn's chances of surviving are directly related to its body weight at delivery. Even with massive supportive efforts, newborns weighing less than 400 g (14 oz) at birth will not survive, primarily because their respiratory, cardiovascular, and urinary systems are unable to support life without aid from maternal systems. Most fetuses born at 25–27 weeks of gestation (a birth weight under 600 g or 21.1 oz) die despite intensive neonatal care, and survivors have a high risk of developmental abnormalities. **Premature delivery** usually refers to birth at 28–36 weeks (a birth weight over 1 kg or 2.2 lb). With care, these newborns have a good chance of surviving and developing normally.

Module 19.7 Review

a. List and describe the factors involved in initiating labor contractions.

b. What chemicals are primarily responsible for initiating contractions of true labor?

c. Name the three stages of labor, and describe the events that characterize each stage.

After delivery, development initially requires nourishment by maternal systems

Developmental processes do not cease at delivery, because newborns have few of the anatomical, functional, or physiological characteristics of mature adults. The time immediately after birth and continuing through the first 28 days is known as the **neonatal period**. During the neonatal period, babies are called neonates or newborns. Infancy is the first year of life. Neonates and infants are dependent on the mother for nourishment, as well as for transportation and protection from environmental hazards such as extreme changes in temperature.

1 Milk is provided to infants through the **milk let-down reflex**, which is diagrammed here. By the end of the sixth month of pregnancy, the mammary glands are fully developed, and the gland cells begin to produce a secretion known as **colostrum** (kō-LOS-trum). Colostrum contains antibodies that help the infant ward off infections until the infant's own immune system becomes fully functional. After the first few days of nursing, the mammary gland produces breast milk, which has a much higher fat content than colostrum. Breast milk also contains antibodies and **lysozyme**, an enzyme with antibiotic properties.

3 **Stimulation of Hypothalamic Nuclei**

The stimulation of tactile receptors in the nipple leads to the stimulation of secretory neurons in the paraventricular nucleus of the maternal hypothalamus.

Posterior lobe of the pituitary gland

4 **Oxytocin Release**

The hypothalamic neurons release oxytocin at the posterior lobe of the pituitary gland. Oxytocin enters the circulation and is distributed throughout the body.

5 **Milk Ejected**

When circulating oxytocin reaches the mammary gland, this hormone causes the contraction of myoepithelial cells in the walls of the lactiferous ducts and sinuses. The result is milk ejection, or milk let-down.

Start

1 **Stimulation of Tactile Receptors**

Mammary gland secretion is triggered when the infant sucks on the nipple.

2 **Neural Impulse Transmission**

Impulses are propagated to the spinal cord of the mother and then to the brain.

2 Postnatal development includes five **life stages**: (1) the neonatal period, (2) infancy, (3) childhood, (4) adolescence, and (5) maturity. Once maturity has been reached, the individual is subject to the gradual changes that accompany senescence, or aging. Growth during infancy and childhood occurs under the direction of circulating hormones, notably growth hormone, adrenal steroids, and thyroid hormones. These hormones affect each tissue and organ in specific ways, depending on the sensitivities of the individual cells. Increasing amounts of hormones—mainly sex hormones—influence adolescent growth and development.

Postnatal Development

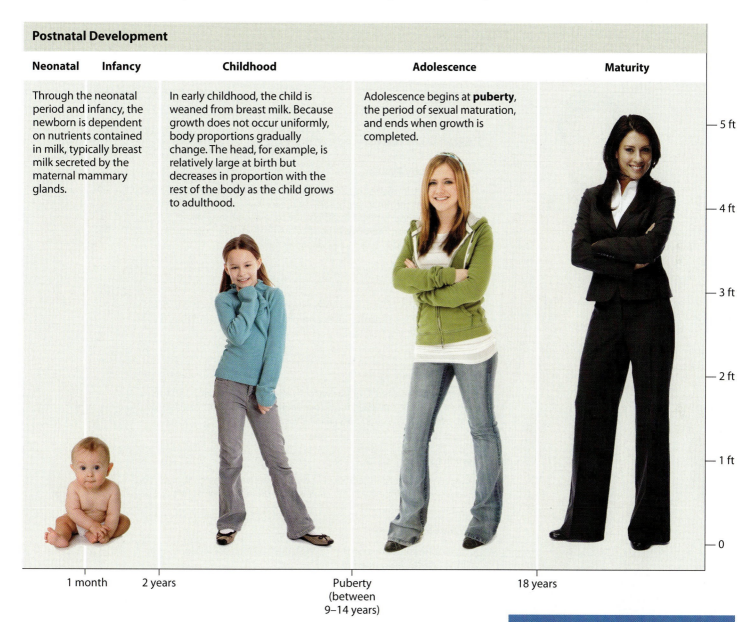

Neonatal	Infancy	Childhood	Adolescence	Maturity

Through the neonatal period and infancy, the newborn is dependent on nutrients contained in milk, typically breast milk secreted by the maternal mammary glands.

In early childhood, the child is weaned from breast milk. Because growth does not occur uniformly, body proportions gradually change. The head, for example, is relatively large at birth but decreases in proportion with the rest of the body as the child grows to adulthood.

Adolescence begins at **puberty**, the period of sexual maturation, and ends when growth is completed.

1 month   2 years   Puberty (between 9–14 years)   18 years

— 5 ft
— 4 ft
— 3 ft
— 2 ft
— 1 ft
— 0

Module 19.8 Review

a. What hormone causes the milk let-down reflex?

b. Explain the difference between colostrum and breast milk.

c. Name the stages of postnatal development, and describe the time frame involved for each of the stages.

At puberty, male and female sex hormones have differential effects on most body systems

Many body systems alter their activities in response to changes in circulating levels of sex hormones at puberty. The most important sex-related changes are summarized here. The effects of testosterone and estrogen are facilitated and enhanced by growth hormone, thyroid hormones, prolactin, and adrenocortical hormones.

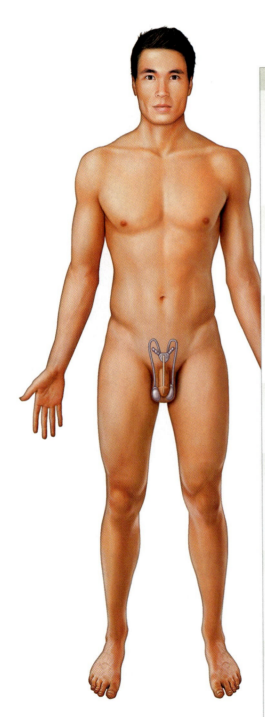

Responses to Testosterone in Males

Integumentary System

Testosterone stimulates the development of hairs on the face and chest, and stimulates terminal hair growth in the axillae and in the genital area. Adipose tissues respond differently to testosterone than to estrogen, and this difference produces the distinct distributions of subcutaneous body fat in males versus females.

Skeletal System

Testosterone accelerates bone deposition and skeletal growth. In the process, it promotes closure of the epiphyseal cartilages and thus places a limit on growth in height.

Muscular System

Testosterone stimulates the growth of skeletal muscle fibers, and the increased muscle mass accounts for significant sex differences in body mass, even for males and females of the same height.

Nervous System

A surge in testosterone secretion at puberty activates the central nervous system centers concerned with male sexual drive and sexual behaviors.

Cardiovascular System

Testosterone stimulates erythropoiesis, thereby increasing blood volume and the hematocrit.

Respiratory System

Testosterone stimulates disproportionate growth of the larynx and a thickening and lengthening of the vocal cords. These changes cause a gradual deepening of the voice in males.

Reproductive System

Testosterone stimulates the functional development of the accessory reproductive glands, such as the prostate gland and seminal glands, and helps promote spermatogenesis.

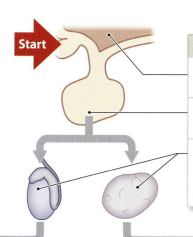

Start

At Puberty

The hypothalamus increases its production of gonadotropin-releasing hormone (GnRH).

Endocrine cells in the anterior lobe of the pituitary gland become more sensitive to the presence of GnRH, and circulating levels of FSH and LH rise rapidly.

Testicular or ovarian cells become more sensitive to FSH and LH, initiating (1) gamete production; (2) the secretion of sex hormones, which stimulate the appearance of secondary sex characteristics and behaviors; and (3) a sudden acceleration in the growth rate, culminating in closure of the epiphyseal cartilages.

Responses to Estrogen in Females

Integumentary System

Estrogen stimulates the hair follicles to continue to produce fine vellus hairs and stimulate terminal hair growth in the axillae and in the genital area. The combination of estrogen, prolactin, growth hormone, and thyroid hormones promotes the initial development of the mammary glands.

Skeletal System

Estrogen causes more rapid epiphyseal closure than does testosterone. In addition, the period of skeletal growth is briefer in females than in males, and so females generally do not grow as tall as males.

Muscular System

Estrogen stimulates the growth of skeletal muscle fibers, increasing strength and endurance, but not to the extent that testosterone does in males.

Nervous System

A surge in estrogen secretion at puberty activates central nervous system centers involved in female sexual drive and sexual behaviors.

Cardiovascular System

The iron loss associated with menses increases the risk of developing iron-deficiency anemia. Estrogen decreases plasma cholesterol levels and slows the formation of plaque within arteries. As a result, premeno-pausal women have a lower risk of atherosclerosis than do adult men.

Respiratory System

Estrogen does not cause excessive growth of the larynx and vocal cords, so females typically have higher-pitched voices than males.

Reproductive System

Estrogen targets the uterus, promoting a thickening of the myometrium and increasing blood flow to the endometrium. Estrogen also promotes the functional development of accessory reproductive structures in females.

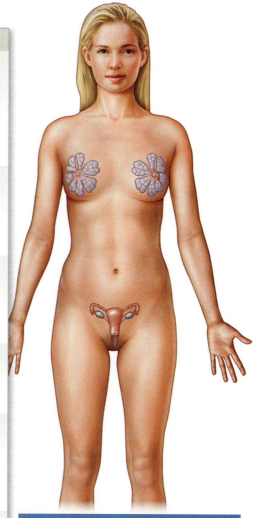

Module 19.9 Review

a. Name the three major interacting hormonal events associated with the onset of puberty.

b. Why does a male generally have a deeper voice and larger larynx than a female?

c. Why are premenopausal women at lesser risk of atherosclerosis than men?

1. Matching

Match the following terms with the most closely related description.

- hCG
- conception
- endometrium
- syncytial trophoblast
- colostrum
- amnion
- parturition
- morula
- embryonic disc
- neonatal period
- gestation
- inner cell mass
- relaxin
- blastocyst

a _____ Fertilization

b _____ First 28 days of life

c _____ Expulsion of the fetus and placenta

d _____ Period of prenatal development

e _____ Presence of this hormone indicates pregnancy

f _____ Softens pubic symphysis

g _____ Mammary gland secretion

h _____ Formed from mesoderm and ectoderm

i _____ Hollow ball of cells

j _____ Site of implantation

k _____ Forms the embryo

l _____ Cytoplasm with many nuclei

m _____ Solid ball of cells

n _____ Three-layered product of gastrulation

2. Multiple choice

Choose the bulleted item that best completes each statement.

a Fetal development begins at the start of the _____.

- implantation process
- second month after fertilization
- ninth week after fertilization
- sixth month after fertilization

b The umbilical _____ within the umbilical cord _____ from the placenta to the fetus.

- arteries; carry oxygenated blood
- arteries; carry deoxygenated blood
- vein; carries oxygenated blood
- vein; carries deoxygenated blood

c Organs and organ systems complete most of their development by the end of the _____.

- first trimester
- second trimester
- third trimester
- expulsion stage

d The four general processes that occur during the first trimester are _____.

- blastomere, blastocyst, morula, and trophoblast
- cleavage, implantation, placentation, and embryogenesis
- placentation, dilation, expulsion, and organogenesis
- fertilization, cleavage, placentation, dilation

e The most critical period in prenatal or neonatal life is the _____.

- first trimester
- second trimester
- third trimester
- expulsion stage

f The systems that were relatively nonfunctional during the fetal period that must become functional at birth are the _____ systems.

- cardiovascular, muscular, and skeletal
- integumentary, reproductive, and nervous
- respiratory, digestive, and urinary
- endocrine, nervous, and digestive

3. Section integration

Tina gives birth to a baby with a congenital deformity of the stomach. Tina believes that her baby's affliction is the result of a viral infection that she suffered during her third trimester. Is this a possibility? Explain. _____

_____

_____

Genetics and Inheritance

Although everyone goes through the same developmental stages, differences in both genetic structure and local environments produce distinctive individual characteristics. The term **inheritance** refers to the transfer of genetically determined characteristics from generation to generation. The study of the mechanisms responsible for inheritance is called **genetics**. We begin our study of genetics by examining two important concepts: *genotype* and *phenotype*.

1 One way to understand genotype and phenotype is to compare them to the architecture of a house. The house plan is the genotype and how the finished house looks is its phenotype.

Every nucleated somatic cell in your body carries copies of the original 46 chromosomes present when you were a zygote. Those chromosomes and their component genes constitute your **genotype** (JĒN-ō-tīp; *geno*, gene). In architectural terms, the genotype is a set of plans, like the blueprints for a house.

Through development and differentiation, the instructions contained in the genotype are expressed in many ways; specific genes may be activated or inactivated by interactions with the local environment. Collectively, the pattern of genetic expression within your genotype determines the anatomical and physiological characteristics that make you a unique individual. Those anatomical and physiological characteristics constitute your **phenotype** (FĒ-nō-tīp; *phaino*, to display). In architectual terms, the phenotype is the detailed structure of the completed building.

2 This photograph shows the **karyotype** (*karyon*, nucleus + *typos*, mark), or entire set of chromosomes, of a normal male. During fertilization, one member of each of the 23 chromosome pairs was contributed by the spermatozoon, and the other by the ovum. The two members of each pair are known as **homologous** (huh-MOL-ō-gus) **chromosomes**. Twenty-two of those pairs are called **autosomal** (aw-tō-SŌ-mul) **chromosomes**. Most of the genes of the autosomal chromosomes affect somatic characteristics, such as hair color and skin pigmentation. The chromosomes of the 23rd pair are called the **sex chromosomes**; one of their functions is to determine whether the individual is genetically male or female. The sex chromosomes of a male consist of an **X chromosome** and a shorter **Y chromosome**, whereas females have two X chromosomes.

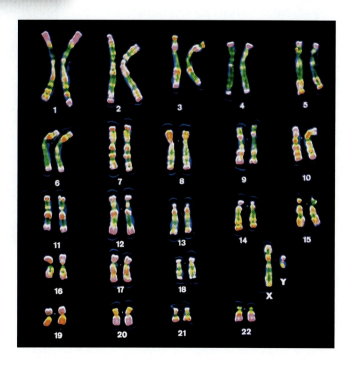

In this section we consider basic genetics as it applies to inherited characteristics, such as sex, hair color, and various clinical conditions.

Genes and chromosomes determine patterns of inheritance

1 The two chromosomes in a homologous autosomal pair have the same structure and carry genes that affect the same traits. The genes are also located at equivalent positions on their respective chromosomes. A gene's position on a chromosome is called a **locus** (LŌ-kus; plural, *loci*). The two chromosomes in a pair may carry the same form or different forms of each gene. The various forms of a given gene are called **alleles** (uh-LĒLZ).

2 The phenotype that results from a heterozygous genotype depends on the nature of the interaction between the corresponding alleles. The most common form of interaction among autosomal genes involves a pair of genes and is called **simple inheritance**. In one kind of simple inheritance—**strict dominance**—any dominant allele will be expressed in the phenotype, regardless of any conflicting instructions carried by the other allele. For instance, a person with only one allele for freckles will have freckles, because that allele is dominant over the "nonfreckle" allele. An allele that is **recessive** will be expressed in the phenotype only if that same allele is present on both chromosomes of a homologous pair.

If the two chromosomes of a homologous pair carry the same allele of a particular gene, you are **homozygous** (hō-mō-ZĪ-gus; *homos*, the same) for the trait affected by that gene. That allele will then be expressed in your phenotype. For example, if you receive a gene for curly hair from your father and a gene for curly hair from your mother, you will be homozygous for curly hair—and you will have curly hair.

When you have two different alleles for the same gene, you are **heterozygous** (het-er-ō-ZĪ-gus; *heteros*, other) for the trait determined by that gene.

The human genome contains roughly 23,000 genes. Thus an "average" autosomal pair contains about 1000 pairs of alleles.

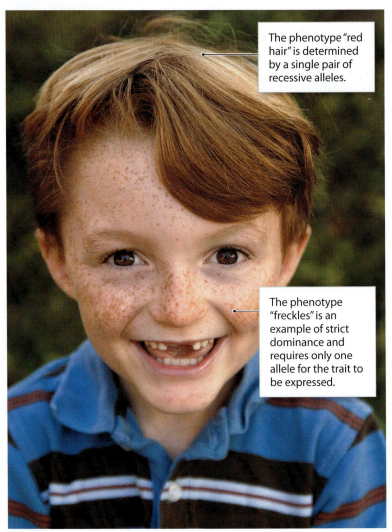

The phenotype "red hair" is determined by a single pair of recessive alleles.

The phenotype "freckles" is an example of strict dominance and requires only one allele for the trait to be expressed.

3 By convention, dominant alleles are indicated by capital letters, and recessive alleles by lowercase letters. So for a given trait *T*, the possible genotypes are *TT* (homozygous dominant), *Tt* (heterozygous), and *tt* (homozygous recessive). The **Punnett squares** shown in this figure indicate the possible offspring of an albino mother and a father with normal skin pigmentation. Because albinism is a recessive trait, the maternal alleles are designated *aa*. The father has normal pigmentation, a dominant trait, so his genotype could be either *AA* or *Aa*.

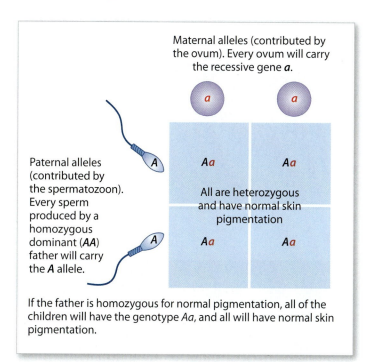

Maternal alleles (contributed by the ovum). Every ovum will carry the recessive gene *a*.

Paternal alleles (contributed by the spermatozoon). Every sperm produced by a homozygous dominant (*AA*) father will carry the *A* allele.

All are heterozygous and have normal skin pigmentation

If the father is homozygous for normal pigmentation, all of the children will have the genotype *Aa*, and all will have normal skin pigmentation.

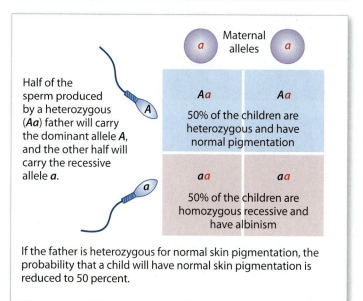

Maternal alleles

Half of the sperm produced by a heterozygous (*Aa*) father will carry the dominant allele *A*, and the other half will carry the recessive allele *a*.

50% of the children are heterozygous and have normal pigmentation

50% of the children are homozygous recessive and have albinism

If the father is heterozygous for normal skin pigmentation, the probability that a child will have normal skin pigmentation is reduced to 50 percent.

4 Many phenotypic characteristics are determined by interactions among several genes. Such interactions constitute **polygenic inheritance**. Because the resulting phenotype depends not only on the nature of the alleles but how those alleles interact, you cannot predict the presence or absence of phenotypic characters using a simple Punnett square. An example of polygenic inheritance is brown or black hair color.

Module 19.10 Review

a. Describe homozygous and heterozygous.

b. Differentiate between simple inheritance and polygenic inheritance.

c. The trait "curly hair" operates through strict dominance. What would be the phenotype of a person who is heterozygous for this trait?

Thousands of clinical disorders have been linked to abnormal chromosomes and/or genes

Chromosomal abnormalities can involve thousands of genes, and as a result they are usually lethal. Most of the time the embryo or fetus dies before delivery. However, there are a few autosomal chromosome abnormalities that do not result in prenatal mortality. In contrast, variations in the structure of individual genes are relatively common. Although more than 99 percent of human nucleotide bases are the same in all people, there are about 1.4 million single-base differences, or **single nucleotide polymorphisms (SNPs)**. Some of these SNPs are inconsequential, but others are associated with specific diseases.

1 **Trisomy 21**, or Down syndrome, is the most common viable chromosomal abnormality. Affected individuals exhibit mental retardation and physical malformations, including a characteristic facial appearance. The degree of mental retardation ranges from moderate to severe, and anatomical problems affecting the cardiovascular system often prove fatal during childhood or early adulthood. Although some individuals survive to moderate old age, many develop Alzheimer's disease while still relatively young (before age 40). For unknown reasons, there is a direct correlation between maternal age and the risk of having a child with trisomy 21. For a maternal age below 25, the incidence of Down's syndrome approaches 1 in 2000 births, or 0.05 percent. For maternal ages 30–34, the odds increase to 1 in 900, and during ages 35–44 they they go from 1 in 290 to 1 in 46, or more than 2 percent.

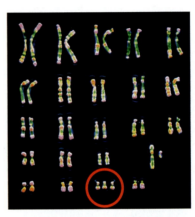
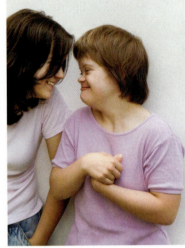

2 In **Klinefelter syndrome**, the individual carries the sex chromosome pattern XXY. The phenotype is male, but the extra X chromosome causes reduced androgen production. As a result, the testes fail to mature so the individuals are sterile, and the breasts are slightly enlarged. The incidence of this condition among newborn males averages 1 in 750 births.

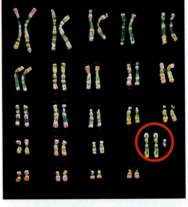
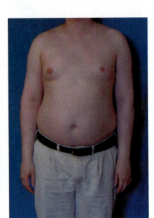

3 Individuals with **Turner syndrome** have only a single, female sex chromosome; their sex chromosome complement is designated XO. This kind of chromosomal deletion is known as **monosomy**. The incidence of this condition at delivery has been estimated as 1 out of every 2500 female live births worldwide. The condition may not be recognized at birth, because the phenotype is normal female. But maturational changes do not appear at puberty. The ovaries are nonfunctional, and estrogen production occurs at negligible levels. Some characteristic physical abnormalities that are often present include short stature, low-set ears, and a webbed neck.

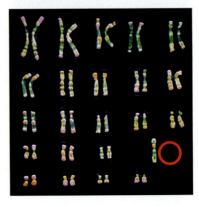
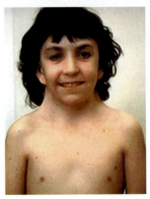

4 The X chromosome is much larger than the Y chromosome, and it carries genes that affect a variety of somatic structures. These characteristics are called **X-linked** (or sex linked), because in most cases there are no corresponding alleles on the Y chromosome. The best known single-allele characteristics are associated with identifiable diseases or functional deficits, such as color blindness.

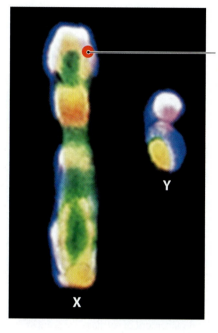

X-linked allele (allele not present on Y chromosome)

Y

X

5 The inheritance of color blindness exemplifies the differences between sex-linked inheritance and autosomal inheritance. In males, the presence of a dominant allele, *C*, on the X chromosome (X^C) results in normal color vision. A recessive allele, *c*, on the X chromosome (X^c) results in red–green color blindness. This Punnett square reveals that each son of a father with normal vision and a heterozygous (carrier) mother has a 50 percent chance of being red–green color blind, whereas any daughters will have normal color vision.

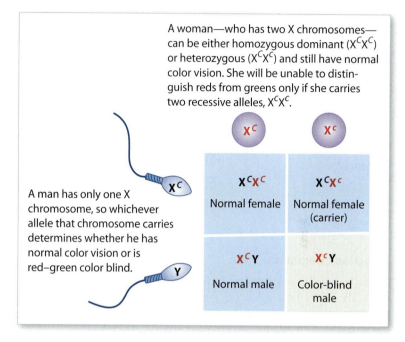

A woman—who has two X chromosomes—can be either homozygous dominant ($X^C X^C$) or heterozygous ($X^C X^c$) and still have normal color vision. She will be unable to distinguish reds from greens only if she carries two recessive alleles, $X^c X^c$.

A man has only one X chromosome, so whichever allele that chromosome carries determines whether he has normal color vision or is red–green color blind.

	X^c	X^c
X^C	$X^C X^c$ Normal female	$X^C X^c$ Normal female (carrier)
Y	X^cY Normal male	X^cY Color-blind male

Tests Performed during Pregnancy

Test	Remarks	Representative Uses
Amniocentesis	A needle, inserted through abdominal wall into uterine cavity, collects amniotic fluid for genetic diagnosis; performed in weeks 14–15 or closer to delivery if complications arise	Detects chromosomal abnormalities and birth defects such as spina bifida; late-pregnancy level of hemolysis in erythroblastosis fetalis; determines fetal lung maturity; evaluates fetal distress
Amniotic fluid analysis	Karyotyping to determine the presence of normal chromosomes in proper number	Detects chromosomal defects such as those in Down syndrome.
	Presence of bilirubin	Increased bilirubin values may indicate degree of hemolysis of fetal RBCs by mother's Rh antibodies.
Alpha-fetoprotein (AFP)	Week 16 of gestation: 5.7–31.5 ng/mL (lowers with increasing gestational age)	Increased values indicate possible neural tube defect, such as spina bifida.

6 A variety of clinical tests are used to monitor a developing fetus and detect signs of abnormal development.

Module 19.11 Review

a. Define single nucleotide polymorphism.

b. Name the disorder characterized by each of the following chromosome patterns: (1) XO and (2) XXY.

c. Why are X-linked traits expressed more frequently in males than females?

1. Matching

Match the following terms with the most closely related description.

- genotype
- heterozygous
- locus
- autosomes
- simple inheritance
- homozygous
- alleles
- polygenic inheritance
- homologous
- genetics
- X-linked
- phenotype

a	_____	Visible characteristics resulting from gene combinations
b	_____	Alternate forms of a gene
c	_____	Refers to the two members of a pair of chromosomes
d	_____	Inheritance of color blindness
e	_____	The particular gene combination for a specific trait
f	_____	Gene's position on a chromosome
g	_____	Two different alleles for the same gene
h	_____	Study of the mechanisms of inheritance
i	_____	Two identical alleles for the same gene
j	_____	Interactions between alleles on several genes
k	_____	Phenotype determined by a single pair of alleles
l	_____	Chromosomes affecting somatic characteristics

2. Section integration

Use the Punnett squares below to answer the questions concerning the following genetic conditions.

a Tongue rolling is inherited as a dominant trait (T). Explain how it is possible for two parents who are tongue rollers to have children who do not have the ability to roll the tongue.

b Achondroplasia dwarfism is an autosomal genetic disorder that results from problems with the replacement of cartilage by bone in the arms and legs. Assume that two dwarf parents have a normal-sized child. Determine the genotype of the parents and predict the probability that a second child would also be normal sized.

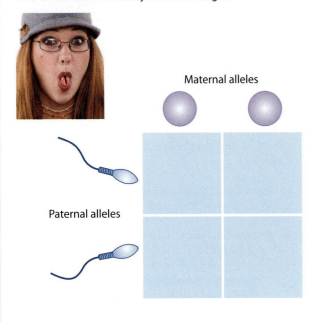

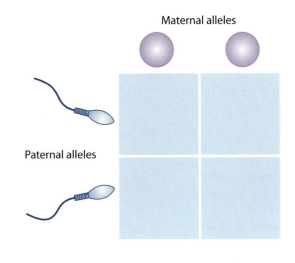

_____

_____

_____

Visual Outline with Key Terms

Summarize the content of each module using the terms in the order provided.

SECTION 1

An Overview of Development

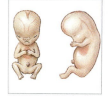

- development
- embryological development
- fertilization
- embryology
- fetal development
- prenatal development
- gestation
- first trimester
- second trimester
- third trimester
- postnatal development
- maturity

19.1

Cleavage continues until the blastocyst implants in the uterine wall

- fertilization
- zygote
- pre-embryo
- blastomeres
- morula
- cleavage
- blastocyst
- blastocoele
- implantation
- trophoblast
- inner cell mass
- cellular trophoblast
- syncytial trophoblast
- lacunae
- villi
- amniotic cavity

19.2

Gastrulation produces three germ layers: ectoderm, endoderm, and mesoderm

- amnion
- extra-embryonic membranes
- yolk sac
- ectoderm
- endoderm
- mesoderm
- germ layers
- gastrulation
- embryonic disc
- gestational trophoblastic neoplasia

19.3

The extra-embryonic membranes form the placenta that supports fetal growth and development

- yolk sac
- amnion
- amniotic fluid
- allantois
- chorion
- placenta

19.4

The placenta performs many vital functions for the duration of prenatal development

- umbilical arteries
- umbilical vein
- human chorionic gonadotropin (hCG)
- human placental lactogen (hPL)
- relaxin
- ○ progesterone
- ○ estrogen

19.5

Organ systems are formed in the first trimester and become functional in the second and third trimesters

- organogenesis
- neural plate

19.6

Pregnancy places anatomical and physiological stresses on maternal systems

- ○ pregnancy
- ○ anatomical changes
- ○ physiological stresses

19.7

Multiple factors initiate and accelerate the process of labor

- false labor
- true labor
- positive feedback
- parturition
- dilation stage
- expulsion stage
- delivery
- placental stage
- afterbirth
- premature labor
- premature delivery

19.8

After delivery, development initially requires nourishment by maternal systems

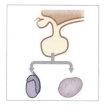

- neonatal period
- milk let-down reflex
- colostrum
- lysozyme
- ○ oxytocin
- ○ life stages
- ○ puberty

19.9

At puberty, male and female sex hormones have differential effects on most body systems

- ○ responses to testosterone in males
- ○ responses to estrogen in females

• = Term boldfaced in this module

SECTION 2

Genetics and Inheritance

- inheritance
- genetics
- genotype
- phenotype
- karyotype
- homologous chromosomes
- autosomal chromosomes
- sex chromosomes
- X chromosome
- Y chromosome

19.11

Thousands of clinical disorders have been linked to abnormal chromosomes and/or genes

- single nucleotide polymorphisms (SNPs)
- trisomy 21
  ○ Down syndrome
- Klinefelter syndrome
- Turner syndrome
- monosomy
- X-linked
  ○ carrier

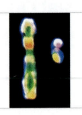

19.10

Genes and chromosomes determine patterns of inheritance

- locus
- alleles
- homozygous
- heterozygous
- simple inheritance
- strict dominance
- recessive
- Punnett squares
- polygenic inheritance

● = *Term boldfaced in this module*

CAREER PATHS

Pediatric Nurse

"You need to be able to explain why something is important. What can happen with a head injury? Why is it important to watch the child's neck? … A lot of what you do with nursing is teaching."

— **Walter Shaw**
Pediatric Nurse,
Children's Hospital of Alabama

For Walter Shaw, a pediatric nurse at a children's hospital in Alabama, working with children is a rewarding part of his job. He recalls a kidney transplant patient who, on a subsequent visit to the hospital, ran up and hugged him around the knees. "Kids don't know they're sick," he says. "That unconditional love you get from them is really special."

Pediatric nurses act as the link between the doctor and the patient in the hospital, following the patient throughout the process: from admitting, to assisting doctors with treatment, inserting central IV lines, all the way through to recovery. "Your job is to help the physicians do what they're supposed to do," Walter says. "A lot of what you do is monitoring the patient and letting the doctor know what's going on."

Walter has worked with adults and notes that, regarding nursing, the physiology skill

sets are similar, though there is a stronger focus on respiratory illness in pediatrics. Calculating appropriate dosages for medications can also be trickier, and "psychosocial skill" becomes more important. "It's especially needed when working with parents and patients of various age groups," Walter says. "You talk to a toddler in a different manner than an adolescent. Parents of newborns have different educational needs and wants than parents of older children."

Walter finds that the most challenging aspect of pediatric nursing is dealing with the patient's parents. Because parents will have many questions, knowing anatomy and physiology is crucial for nurses. "You need to be able to explain why something is important," Walter says. "What can happen with a head injury? Why is it important to watch the child's neck? Why is insulin important to a patient? A lot of what you do with nursing is teaching." Knowledge is important, but so is demeanor. "It takes a lot for a mother to walk away from her child, and to do that, she has to have confidence in you," Walter said. Pediatric nurses must be adept at communication, trust building, and general people skills. He also notes that the work can be physically demanding.

Pediatric nurses work wherever pediatricians work, which, in addition to hospitals, includes doctor's offices and clinics. Some work for schools or school districts. Walter points out that there is a push to get more men into nursing in general—he was one of 3 out of 60 students in his graduating class.

For more information, visit the website for the Pediatric Nursing Certification Board at http://www.pncb.org.

Think this is the career for you?
Key Stats:

- **Education and Training.** Most pediatric nurses are registered nurses (RNs), which require an associate's or bachelor's degree in nursing, though more and more are pursuing graduate degrees.

- **Licensure.** All states require nurses to graduate from an approved nursing program and pass a national licensing exam.

- **Earnings.** Earnings vary but the median annual salary is $64,690.

- **Expected Job Prospects.** Employment is expected to grow faster than the national average—by 22% through 2018.

Bureau of Labor Statistics, U.S. Department of Labor, *Occupational Outlook Handbook, 2010–11 Edition,* Registered Nurses, on the Internet at http://www.bls.gov/oco/ocos083.htm (visited September 14, 2011).

MasteringA&P®

Access more review material online in the Study Area at **www.masteringaandp.com.**

- Chapter guides
- Chapter quizzes
- Practice tests
- Art labeling activities
- Flashcards
- A glossary with pronunciations

- Practice Anatomy Lab™ (PAL™) 3.0 virtual anatomy practice tool
- Interactive Physiology® (IP) animated tutorials
- MP3 Tutor Sessions

Chapter 19 Review Questions

1. Explain the difference between cleavage and the cell division that occurs in cells after cleavage.

2. During true labor, what physiological mechanisms ensure that uterine contractions continue until delivery has been completed?

3. Which set of sex chromosomes results in a baby boy? What hormone must be present to stimulate male sexual characteristics?

4. Why would having a low sperm count interfere with the ability to conceive a child?

5. Hemophilia A, a condition in which blood does not clot properly, is a recessive trait located on the X chromosome (X^h). Suppose that a woman who is heterozygous for this trait (XX^h) mates with a normal male (XY). What is the probability that the couple will have hemophiliac daughters? What is the probability that the couple will have hemophiliac sons?

For answers to all module, section, and chapter review questions, see the blue Answers tab at the back of the book.

CHAPTER 1

Module 1.1 Review

a. Anatomy is the study of internal and external body structures; physiology is the study of how living organisms perform their vital functions.

b. Gross anatomy (also called macroscopic anatomy) is the study of body structures that can be seen with the unaided eye; microscopic anatomy is the study of body structures that cannot be seen without magnification.

c. The structures of body parts are closely related to their functions—that is, function follows form.

Module 1.2 Review

a. All specific functions are performed by specific structures, and the link between the two is always present, but not always understood.

b. The elbow joint moves in a single plane, like the opening and closing of a door on a hinge.

c. Twisting at the elbow joint is prevented by the ridges of the ulna that enclose the end of the humerus, along with ligaments and muscle that surround the joint.

Module 1.3 Review

a. An organ is a body structure composed of two or more tissues, and that has one or more specific functions.

b. The cellular level of organization.

c. The tissue level (a group of cells and cell products performing a specific function), the organ level (a combination of two or more tissues), and the organ system level (a group of different organs).

Module 1.4 Review

a. The body's 11 organ systems are the integumentary, skeletal, muscular, nervous, endocrine, cardiovascular, lymphatic, respiratory, digestive, urinary, and reproductive systems.

b. The digestive system provides nutrients and minerals for the skeletal system, which in turn protects soft tissues and organs of the digestive system.

c. A compound fracture could affect the (1) skeletal system (a broken bone), (2) integumentary system (disruption of skin integrity), (3) muscular system (broken bone tearing through a muscle), (4) cardiovascular system (blood loss at the site of injury), (5) lymphatic system (mobilization of specialized cells to defend against infection), and (6) nervous system (pain and nerve injury as a result of the trauma).

Section 1 Review

1. a. Approach the information in different ways; **b.** Set up a study schedule; **c.** Devote a block of study time each day; **d.** Practice memorization; **e.** Attend all lectures, labs, and study sessions; **f.** Read your lecture and lab assignments before coming to class; **g.** Do not procrastinate; **h.** Seek assistance as soon as you have a problem understanding the material.

2. Anatomy: right atrium, myocardium, left ventricle, endocardium, superior vena cava, ulna; Physiology: valve to aorta opens, valve between left atrium and left ventricle closes, pressure in left atrium, electrocardiogram, moving forearm toward shoulder

3. tissue: c; cell: b; organ: d; molecule: a; organism: f; organ system: e

4. a. skeletal system: support, protection of soft tissues, mineral storage, blood formation; **b.** digestive system: processing of food and absorption of nutrients, minerals, vitamins, and water; **c.** integumentary system: protection from environmental hazards, temperature control; **d.** urinary system: elimination of excess water, salts, and waste products, and control of pH; **e.** nervous system: directing immediate responses to stimuli, usually by coordinating the activities of other organ systems.

Module 1.5 Review

a. The components of homeostatic regulation include receptors, a control center, and effectors.

b. Negative feedback systems maintain homeostasis (and provide long-term control over the body's internal conditions and systems) by counteracting any stimulus that moves conditions outside their normal range.

c. Positive feedback is useful in processes such as blood clotting, which, once begun, must move quickly to completion. It is harmful in situations where stable conditions must be maintained, because it tends to exaggerate any departure from the desired condition. Thus positive feedback in the regulation of body temperature would cause a slight fever to spiral out of control, with fatal results.

Section 2 Review

1. a. positive feedback; **b.** homeostatic regulation; **c.** homeostasis; **d.** negative feedback; **e.** negative feedback

2. a. negative feedback; **b.** positive feedback; **c.** negative feedback; **d.** positive feedback

3. a. blood flow to skin increases, sweating increases, body surface cools, temperature declines; **b.** blood flow to skin decreases, shivering occurs, body heat is conserved, temperature rises

4. One reason your body temperature may have dropped is that your body may be losing heat faster than it is being produced. (This is more likely to occur on a cool day.) Perhaps hormones have caused a decrease in your metabolic rate, so your body is not producing as much heat as it normally would. Or you may have an infection that has temporarily reset the set point of the body's "thermostat." Infections normally cause a fever, but some can, oddly, cause a lower body temperature (hypothermia). For example, infection with Gram-negative bacteria (named for their reaction in a Gram stain test) can result in a lower body temperature. So, given the circumstances, it is likely that such a bacterial infection caused your body temperature to be 1.5 degrees below normal.

Module 1.6 Review

a. A person in the anatomical position is standing erect, facing the observer, arms at the sides with the palms facing forward, and the feet together.

b. Clinicians base their descriptions on four abdominopelvic quadrants (determined by the intersection of two imaginary perpendicular lines that cross at the navel, or umbilicus), whereas anatomists recognize nine abdominopelvic regions.

c. A person lying face down in the anatomical position is prone.

Module 1.7 Review

a. The purpose of directional and sectional terms is to provide a standardized frame of reference for describing the human body and its parts.

b. In the anatomical position, an anterior view shows the subject's face, whereas a posterior view shows the subject's back.

c. A midsagittal section would separate the two eyes.

Module 1.8 Review

a. Body cavities (1) protect internal organs and cushion them from shocks that occur during activity and (2) allow organs within them to change size and shape without disrupting the activities of nearby organs.

b. The ventral body cavity includes the thoracic cavity (which contains the pleural and pericardial cavities) and the abdominopelvic cavity (consisting of the peritoneal, abdominal, and pelvic cavities).

c. The body cavity inferior to the diaphragm is the abdominopelvic or peritoneal cavity.

Section 3 Review

1. a. superior; **b.** inferior; **c.** cranial; **d.** caudal; **e.** posterior or dorsal; **f.** anterior or ventral; **g.** lateral; **h.** medial; **i.** proximal; **j.** distal; **k.** proximal; **l.** distal

2. a. thoracic cavity; **b.** mediastinum; **c.** left lung; **d.** blood vessels; **e.** heart; **f.** diaphragm; **g.** abdominopelvic cavity; **h.** peritoneal cavity; **i.** digestive glands and organs; **j.** pelvic cavity; **k.** reproductive organs

Chapter Review

1. Homeostatic regulation refers to adjustments in physiological systems that are responsible for the preservation of homeostasis.

2. In negative feedback, a variation outside normal ranges triggers an automatic response that corrects the situation. In positive feedback, the initial stimulus produces a response that exaggerates the stimulus.

3. The stomach. (You would cut the pericardium to access the heart.)

4. When calcitonin is released in response to elevated calcium levels, it brings about a decrease in blood calcium levels, thereby decreasing the stimulus for its own release.

5. To see a complete view of the medial surface of each half of the brain, midsagittal sections are needed.

CHAPTER 2

Module 2.1 Review

a. An element is a pure substance made up only of atoms with the same atomic number.

b. Trace elements, found in very small amounts in the body, are chemical elements required for normal growth and maintenance. Some function as cofactors, but the functions of many have yet to be fully understood.

c. Hydrogen has three isotopes: hydrogen-1, with a mass of 1; deuterium, with a mass of 2; and tritium, with a mass of 3. The heavier sample must contain a higher proportion of one or both of the heavier isotopes.

Module 2.2 Review

a. The maximum number of electrons that can occupy an atom's first three electron shells is 2, 8, and 8, respectively.

b. Atoms of inert elements are unreactive because their outermost electron shell contains the maximum number of electrons possible.

c. A cation is formed when an atom loses one or more electrons from its outermost electron shell; it has an overall positive charge because it contains more protons than electrons. An anion is formed when an atom gains one or more electrons in its outermost electron

shell; it has an overall negative charge because it contains more electrons than protons.

Module 2.3 Review

a. The two most common types of chemical bonds are ionic bonds, which result from the attraction of oppositely charged atoms (ions), and covalent bonds, which result from the sharing of electrons.

b. The atoms in a water molecule are held together by polar covalent bonds, in which electrons are shared unequally.

c. The term *molecule* refers only to chemical structures held together by covalent bonds. Table salt is an ionic compound whose components—sodium ions and chloride ions—are held together by ionic bonds.

Section 1 Review

1. **a.** 2 **b.** 4 **c.** 1 **d.** 0 **e.** 6 **f.** 12 **g.** 7 **h.** 7 **i.** 20 **j.** 20

2. **a.** element; **b.** compound; **c.** element; **d.** compound

3. **a.** ions; **b.** neutrons; **c.** compound; **d.** atomic number; **e.** hydrogen bond; **f.** polar covalent bond; **g.** mass number; **h.** element; **i.** protons; **j.** isotopes; **k.** ionic bond; **l.** electrons

4. **a.** Each element consists of atoms containing a characteristic number of protons (atomic number). However, the outer energy level of inert elements is filled with electrons, whereas the outer energy level of reactive elements is not filled with electrons.

b. Both polar and nonpolar molecules are held together by covalent bonds. However, in a polar molecule the electrons are not shared equally, so it carries small positive and negative charges on its surface; in a nonpolar molecule the electrons are shared equally, so it is electrically neutral.

c. Both covalent bonds and ionic bonds bind atoms together, but covalent bonds involve the sharing of electrons between atoms, whereas ionic bonds involve the electrical attraction of oppositely charged atoms (ions).

Module 2.4 Review

a. The molecular formula for glucose, a compound composed of

6 carbon atoms, 12 hydrogen atoms, and 6 oxygen atoms, is $C_6H_{12}O_6$.

b. An enzyme is a protein that lowers the activation energy, which is the amount of energy required to start a chemical reaction.

c. Enzymes promote chemical reactions by lowering the activation energy requirements and making it possible for chemical reactions to proceed under conditions compatible with life.

Module 2.5 Review

a. Three types of chemical reactions important in human physiology are (1) decomposition reactions, in which a molecule is broken down into smaller fragments; (2) synthesis reactions, in which small molecules are assembled into larger ones; and (3) exchange reactions, in which parts of the reacting molecules are shuffled around to produce new products.

b. In hydrolysis reactions, water is a reactant, whereas in dehydration synthesis reactions, water is a product.

c. The source of the released energy was the potential energy stored in the covalent bonds of the glucose molecule. Energy was released when some of the covalent bonds were broken during this decomposition reaction.

Section 2 Review

1. **a.** H_2 **b.** 2 H **c.** 6 H_2O **d.** $C_{12}H_{22}O_{11}$

2. $C_6H_{12}O_6 + 6 O_2 \rightarrow 6 CO_2 + 6 H_2O$

3. **a.** hydrolysis reaction **b.** dehydration synthesis

4. **a.** enzyme; **b.** reactants; **c.** hydrolysis; **d.** endergonic; **e.** exchange reaction; **f.** products; **g.** exergonic; **h.** activation energy

5. A decreased amount of enzyme at the second step would limit the amount produced in the next two steps. This would cause a decrease in the amount of the final product.

Module 2.6 Review

a. Electrolytes are soluble inorganic molecules whose ions will conduct an electric current in solution.

b. Hydrophilic molecules are attracted to water molecules, whereas

hydrophobic molecules do not interact with water molecules.

c. Sodium chloride dissociates in water as the slightly positive poles of water molecules are attracted to the slightly negatively charged chloride ions, and the negative poles of water molecules are attracted to the positively charged sodium ions. The ions stay dissolved in solution because a layer of surrounding water molecules, or hydration sphere, separates them from each other.

Module 2.7 Review

a. pH is a measure of the hydrogen ion concentration in a solution. On the pH scale, 7 represents neutrality, values below 7 indicate acidic solutions, and values above 7 indicate alkaline (basic) solutions.

b. An acid is a compound whose dissociation in solution releases a hydrogen ion and an anion; a base is a compound whose dissociation releases a hydroxide ion (OH^-) into the solution or removes a hydrogen ion (H^+) from the solution; and a salt is an inorganic compound consisting of a cation other than H^+ and an anion other than OH^-.

c. The pH of various body fluids must remain relatively constant if the body is to maintain homeostasis and remain healthy.

Section 3 Review

1. Four important properties of water in the human body are effective lubrication (as between bony surfaces in a joint), reactivity (participates in chemical reactions), high heat capacity (readily absorbs and retains heat), and solubility (is a solvent for many substances).

2. **a.** alkalosis; **b.** solute; **c.** alkaline; **d.** hydrophilic; **e.** solvent; **f.** salt; **g.** buffers; **h.** acid; **i.** hydrophobic; **j.** water

3. **a.** acidic; **b.** neutral; **c.** alkaline; **d.** The pH 2 solution is 10,000 times more acidic than the pH 6 solution—it contains a ten thousand-fold (10^4) increase in the concentration of hydrogen ions (H^+). **e.** Three negative effects of abnormal fluctuations in pH are cell and tissue damage (due to broken

bonds), changes in the shapes of proteins, and altered cellular functions.

4. Table salt dissociates or dissolves in pure water but since it does not release either hydrogen (H^+) ions or (OH^-) ions, no change in pH occurs.

Module 2.8 Review

a. The three structural classes of carbohydrates are monosaccharides (glucose), disaccharides (sucrose), and polysaccharides (starch).

b. The C:H:O ratio of carbohydrates is 1:2:1. Carbohydrates are used in the body chiefly as energy sources.

c. Muscle cells make (synthesize) glycogen by linking numerous glucose molecules by a series of dehydration synthesis reactions.

Module 2.9 Review

a. Lipids are a diverse group of water-insoluble organic compounds that contain carbon, hydrogen, and oxygen in a ratio that does not approximate 1:2:1 (they contain less oxygen than do carbohydrates). Examples are fatty acids, fats (triglycerides), steroids, phospholipids, and glycolipids.

b. All fatty acids consist of a hydrocarbon chain and a carboxyl group. In saturated fatty acids, each carbon atom in the hydrocarbon chain has four single covalent bonds that bind the maximum number of hydrogen atoms possible. In unsaturated fatty acids, one or more of the carbon atoms in the hydrocarbon chain has double covalent bonds, so fewer hydrogen atoms are bonded.

c. In the hydrolysis of a triglyceride, the reactants are a triglyceride and three water molecules; and the products are a glycerol molecule and three fatty acids.

Module 2.10 Review

a. Cholesterol is a component of plasma membranes and is important for cell growth and division.

b. Steroids function as both components of cellular membranes (cholesterol) and sex hormones (testosterone and estrogens); phospholipids and glycolipids are important structural components of cellular membranes.

c. When phospholipids and glycolipids form a micelle, the hydrophobic tails of both molecules form a cluster inside the micelle, and their hydrophilic heads are outside the micelle.

Module 2.11 Review

a. Proteins are organic compounds formed from amino acids that contain a central carbon atom, a hydrogen atom, an amino group ($-NH_2$), a carboxyl group ($-COOH$), and a variable group, known as an R group or side chain.

b. During the dehydration synthesis of two amino acids, a peptide bond links the amino group of one amino acid with the carboxyl group of the other amino acid.

c. The heat of boiling breaks bonds that maintain the protein's tertiary structure, quaternary structure, or both. The resulting change in shape affects the ability of the protein molecule to perform its normal biological functions. These changes are known as denaturation.

Module 2.12 Review

a. The active site is the location on an enzyme where substrate binding occurs.

b. The reactants in an enzymatic reaction are called substrates.

c. An enzyme's specificity results from the unique shape of its active site, which permits only a substrate with a complementary shape to bind.

Module 2.13 Review

a. Cells obtain energy in the high-energy bonds of compounds such as ATP.

b. ATP (adenosine triphosphate) is a compound consisting of adenosine to which three phosphate groups are attached; high-energy bonds attach the second and third phosphates.

c. Adenosine monophosphate (AMP) is a nucleotide consisting of adenosine plus a phosphate group (PO_4^{3-}), and adenosine diphosphate (ADP) is a compound consisting of adenosine with two phosphate groups attached. Adding a phosphate group to AMP creates ADP.

Breaking one phosphate linkage of ATP, to form ADP, provides the primary source of energy for physiological processes.

Module 2.14 Review

a. Nucleic acids are large organic molecules, composed of carbon, hydrogen, oxygen, nitrogen, and phosphorus, that regulate the synthesis of proteins and make up the genetic material in cells.

b. The complementary strands of DNA are held together through complementary base pairing between adenine and thymine, and between guanine and cytosine.

c. The nucleic acid RNA (ribonucleic acid) contains the sugar, ribose. The nucleic acid DNA (deoxyribonucleic acid) contains the sugar, deoxyribose; both contain nitrogenous bases and phosphate groups.

Section 4 Review

1. a. glycogen; b. sucrose; c. polyunsaturated; d. active site; e. glycerol; f. cholesterol; g. ATP; h. RNA; i. peptide bond; j. nucleotide; k. monosaccharide

2. a. polysaccharide, polyunsaturated, polypeptide; b. triglyceride; c. disaccharide, diglyceride, dipeptide; d. glycogen, glycolipids

3. a. carbohydrates; b. polysaccharides; c. disaccharides; d. monosaccharides; e. lipids; f. fatty acids; g. glycerol; h. proteins; i. amino acids; j. nucleic acids; k. RNA; l. DNA; m. nucleotides; n. ATP; o. phosphate groups

Chapter Review

1. Ions may be larger or smaller than their atoms because the total electrical force between the protons and electrons are no longer equal. A loss of an electron would reduce the size of a sodium ion because the nucleus now has a greater overall positive charge (more positively-charged protons) pulling on fewer negatively-charged electrons. A gain of an electron would increase the size of a chloride ion because the protons

in the nucleus cannot hold the increased number of electrons as tightly as in a neutral atom.

2. If a person exhales large amounts of CO_2, the equilibrium will shift to the left and the level of H^+ in the blood will decrease. A decrease in the amount of H^+ will cause the pH to rise.

3. A nucleic acid. Carbohydrates and lipids do not contain the element nitrogen. Although both proteins and nucleic acids contain nitrogen, only nucleic acids contain phosphorus.

CHAPTER 3

Module 3.1 Review

a. Cytoplasm is the material between the plasma membrane and the nuclear membrane; cytosol is the fluid portion of the cytoplasm; the cytoskeleton provides strength and support and enables movement of cellular structures and materials.

b. Membranous organelles (and their functions) include endoplasmic reticulum (synthesis of secretory products, and intracellular storage and transport); nucleus (control of metabolism, storage and processing of genetic information, and control of protein synthesis); rough ER (modification and packaging of newly synthesized proteins); smooth ER (lipid and carbohydrate synthesis); Golgi apparatus (modification and packaging of proteins); lysosomes (removal of damaged organelles or intracellular pathogens); mitochondria (production of 95 percent of the ATP required by the cell); and peroxisomes (breakdown of organic compounds and neutralization of toxic compounds).

c. The nonmembranous organelles (and their functions) include the cytoskeleton (strengthen and support cell; move cellular structures and materials); microvilli (increase surface area to aid in absorption); centrioles (involved in the development of spindle fibers during cell division); cilia (beat rhythmically to aid fluid movement); ribosomes (synthesize proteins).

Module 3.2 Review

a. The general functions of the plasma membrane include physical isolation of the cell from its environment, regulation of exchange with the environment, sensitivity to the environment, and structural support.

b. The phospholipid bilayer of the plasma membrane is mostly responsible for isolating a cell from its external environment.

c. Cilia propel materials across cell surfaces.

Module 3.3 Review

a. Newly synthesized proteins from free ribosomes enter the cytosol; those from fixed ribosomes enter the ER.

b. The endoplasmic reticulum is a network of intracellular membranes connected to the nuclear envelope. The membranes of rough ER form sheets and contain ribosomes. Smooth endoplasmic reticulum (SER) lacks ribosomes, and its cisternae are tubular. A double membrane encloses a mitochondrion; the outer membrane surrounds the organelle, whereas the inner membrane contains folds called cristae and encloses a fluid, enzyme-filled matrix.

c. Mitochondria produce energy, in the form of ATP molecules, for the cell. A large number of mitochondria in a cell indicates a high demand for energy.

Module 3.4 Review

a. The Golgi apparatus (1) modifies and packages cellular secretions, such as hormones or enzymes; (2) renews or modifies the plasma membrane; and (3) packages special enzymes within vesicles for use within the cell.

b. Lysosomes contain digestive enzymes.

c. Lysosomes may (1) break down other intracellular organelles to enhance recycling and membrane flow; (2) fuse with vesicles containing fluids or solids from the external environment to obtain nutrients or destroy pathogens; and (3) break down and release their digestive enzymes, resulting in the destruction of the cell (autolysis).

Module 3.5 Review

a. The DNA in the nucleus stores the cell's instructions for synthesizing proteins.

b. The nucleus is a cellular organelle containing DNA, RNA, and proteins. Surrounding the nucleus is the double-membraned nuclear envelope; the gap within this double membrane is the perinuclear space. Nuclear pores allow for chemical communication between the nucleus and the cytosol.

c. The nuclei of typical human somatic cells contain 23 pairs of chromosomes.

Module 3.6 Review

a. The three types of RNA involved in protein synthesis are ribosomal RNA (rRNA), messenger RNA (mRNA), and transfer RNA (tRNA).

b. A gene is a portion of a DNA strand that functions as a hereditary unit and codes for a specific protein.

c. Translation is the synthesis of a protein using the information provided by the sequence of codons along an mRNA strand.

Module 3.7 Review

a. Interphase is the portion of a cell's life cycle during which the chromosomes are uncoiled and all normal cellular functions except mitosis are under way.

b. An important enzyme for DNA replication is DNA polymerase.

c. Apoptosis is the genetically controlled death of cells.

Module 3.8 Review

a. Mitosis is the division of a cell nucleus into two identical daughter cell nuclei, an essential step in cell division. Its four stages are prophase, metaphase, anaphase, and telophase.

b. A chromatid is a copy of a duplicated chromosome. Human cells normally contain 23 pairs of chromosomes, so during mitosis a human cell would contain 92 chromatids.

c. The appearance of a cell that completed mitosis but not cytokinesis would be one cell with two sets of chromosomes.

Module 3.9 Review

a. Cancer is an illness characterized by mutations that disrupt normal cell control mechanisms and produce malignant cells.

b. A tumor is a mass or swelling produced by the abnormal growth and division of cells. A benign tumor is a tumor in which the cells are not cancerous and typically remain in one place. A malignant tumor is a cancer that spreads from its original site to another in a process called metastasis. Metastasis leads to the establishment of secondary tumors.

c. Cancer is dangerous because the cells grow and multiply at the expense of healthy cells. Death may result when cancer cells kill or replace healthy cells, compress vital organs, or deprive normal tissues of essential nutrients.

Section 1 Review

1. a. microvilli: increased surface area to facilitate absorption of extracellular materials; **b.** Golgi apparatus: storage, alteration, and packaging of newly synthesized proteins; **c.** lysosome: intracellular removal of damaged organelles or pathogens; **d.** mitochondrion: production of 95 percent of the ATP required by the cell; **e.** peroxisome: neutralization of toxic compounds; **f.** nucleus: control of metabolism, storage and processing of genetic information, and control of protein synthesis; **g.** endoplasmic reticulum: synthesis of secretory products, and intracellular storage and transport; **h.** ribosomes: protein synthesis; **i.** cytoskeleton: provides strength and support, enables movement of cellular structures and materials

2. a. chromosomes; **b.** thymine; **c.** nuclear envelope; **d.** uracil; **e.** nucleoli; **f.** nuclear pore; **g.** transcription; **h.** gene; **i.** mRNA; **j.** tRNA; **k.** genetic information

3. The nucleus contains the information for synthesizing proteins in the nucleotide sequence of its DNA. Changes in the extracellular fluid can affect cells through the binding of molecules to plasma membrane receptors or by the diffusion of molecules through membrane channels. Such stimuli may result in alterations of genetic activity in the nucleus. These alterations may change biochemical processes and metabolic pathways through the synthesis of additional, fewer, or different enzymes. Altered genetic activity may also change the physical structure of the cell by synthesizing additional, fewer, or different structural proteins.

Module 3.10 Review

a. Diffusion is the passive movement of molecules from an area of higher concentration to an area of lower concentration until equilibrium is reached.

b. Factors that influence diffusion include distance, molecule size, lipid solubility, and concentration gradient.

c. Diffusion is driven by a concentration gradient. The larger the concentration gradient, the faster the rate of diffusion; the smaller the concentration gradient, the slower the rate of diffusion. If the concentration of oxygen in the lungs were to decrease, the concentration gradient between oxygen in the lungs and oxygen in the blood would decrease (as long as the oxygen level of the blood remained constant). So, oxygen would diffuse more slowly into the blood.

Module 3.11 Review

a. In both facilitated diffusion and active transport, a carrier protein transports materials across the plasma membrane.

b. Exocytosis is the ejection of cytoplasmic materials following the fusion of a membranous vesicle with the plasma membrane.

c. The engulfment of bacteria and their movement into the cell is called phagocytosis.

Section 2 Review

1. **a.** diffusion; **b.** facilitated diffusion; **c.** molecular size; **d.** diffusion of water; **e.** active transport; **f.** specific substances; **g.** vesicular transport; **h.** exocytosis; **i.** pinocytosis; **j.** "cell eating"
2. **a.** diffusion; **b.** diffusion; **c.** neither; **d.** diffusion; **e.** osmosis; **f.** osmosis

Module 3.12 Review

a. Histology is the study of tissues.

b. The body's four primary tissue types that form all body structures are epithelial, connective, muscle, and neural tissue.

c. Epithelial tissue covers external and internal surfaces and produces secretions; connective tissue fills internal spaces, provides support, and stores energy; muscle tissue is specialized to contract and produce movement; and neural tissue transmits information.

Module 3.13 Review

a. Epithelial tissue provides physical protection, controls permeability, provides sensations, and produces specialized secretions.

b. The three characteristic shapes of epithelial cells are squamous (flat), cuboidal (cube-shaped), and columnar (appear tall and rectangular). A single layer of epithelial cells is a simple epithelium, whereas multiple layers of epithelial cells make up a stratified epithelium.

c. The presence of many cilia on the free surface of epithelial cells indicates that substances are moved over the epithelial surface.

Module 3.14 Review

a. Epithelial intercellular connections include tight junctions, gap junctions, and desmosomes.

b. Epithelial tissues rely on blood vessels in underlying connective tissues to supply needed nutrients.

c. Gap junctions help coordinate the functioning of adjacent cells by allowing small molecules and ions to pass from cell to cell. In epithelial cells, gap junctions help coordinate functions such as the beating of cilia. In cardiac muscle tissue, they are essential for coordinated muscle cell contractions.

Module 3.15 Review

a. Keratinized epithelia are both tough (have mechanical strength) and water resistant.

b. All these sites are subject to mechanical stresses and abrasion—by food (pharynx and esophagus), by feces (anus), or during intercourse or childbirth (vagina).

c. No. A simple squamous epithelium provides too little protection against infection, abrasion, or dehydration to function effectively on the skin surface.

Module 3.16 Review

a. The epithelium that lines the urinary bladder and changes in appearance during stretching is a transitional epithelium. As the bladder stretches, the epithelium appears flattened.

b. In a sectional view, simple cuboidal epithelial cells are square and have central nuclei, and the distance between adjacent nuclei is roughly equal to the height of the epithelium.

c. Stratified cuboidal epithelia are associated with the ducts of sweat glands, mammary glands, and other exocrine glands.

Module 3.17 Review

a. In a sectional view, simple columnar epithelial cells appear as tall slender rectangles containing elongated nuclei close to the basement membrane. Cell height is several times the distance between adjacent nuclei.

b. A pseudostratified columnar epithelium is not truly stratified—even though its nuclei

make it appear so—because all of its epithelial cells contact the basement membrane.

c. The columnar epithelium lining the intestine typically has microvilli on its apical surface.

Module 3.18 Review

a. The two primary types of glands are endocrine glands and exocrine glands.

b. Apocrine secretion involves the loss of cytoplasm as well as the secretory products.

c. In holocrine secretion the secretory cells fill with secretions and then rupture, releasing their contents.

Module 3.19 Review

a. Cells found in areolar tissue are melanocytes, fixed macrophages, mast cells, fibroblasts, adipocytes (fat cells), plasma cells, free macrophages, stem cells, migrating phagocytic white blood cells, and lymphocytes.

b. A vitamin C deficiency would impair the production of collagen fibers, which add strength to connective tissue. The deficiency would result in tissue that is weak and prone to damage.

c. The three types of protein fibers in connective tissue are collagen fibers, elastic fibers, and reticular fibers. Collagen fibers are thick, strong fibers that resist stretching; elastic fibers are slender, stretchy fibers that can recoil to their original length after stretching; and reticular fibers are strong fibers that form a branching network.

Module 3.20 Review

a. The four types of membranes found in the body are synovial membranes, mucous membranes, serous membranes, and the cutaneous membrane.

b. The body cavities lined by serous membranes are the pleural, peritoneal, and pericardial cavities.

c. Synovial fluid lubricates joint cavities, and supplies oxygen and nutrients to cartilage cells.

Module 3.21 Review

a. Mature cartilage cells are called chondrocytes.

b. The fibers characteristic of cartilage supporting the ear are elastic fibers.

c. The type of cartilage damaged in a protruding intervertebral disc is fibrocartilage.

Module 3.22 Review

a. Mature bone cells in lacunae are called osteocytes.

b. The two types of bone are compact bone and spongy bone.

c. Exercise strengthens bone because it causes bones to grow thicker.

Module 3.23 Review

a. The three types of muscle tissue in the body are skeletal muscle tissue, cardiac muscle tissue, and smooth muscle tissue.

b. Muscle tissue containing small, tapering cells with single nuclei and no obvious striations is smooth muscle tissue.

c. These cells are most likely neurons.

Module 3.24 Review

a. The two processes in the response to tissue injury are inflammation and regeneration.

b. Inflammation produces swelling, redness, warmth, and pain.

c. Inflammation can occur in any organ in the body because all organs have connective tissues.

Section 3 Review

1. **a.** cardiac; **b.** smooth; **c.** nonstriated; **d.** multinucleate
2. **a.** perichondrium, periosteum, peritoneum, pericardium; **b.** osseous, osteocyte, periosteum, osteon; **c.** chondrocyte, perichondrium; **d.** lacunae; **e.** fibrocartilage; **f.** adipocytes; **g.** synovial membrane; **h.** cutaneous membrane; **i.** axon; **j.** intercalated disc; **k.** neuroglia; **l.** skeletal muscle tissue; **m.** smooth muscle tissue; **n.** regeneration; **o.** inflammation
3. Increased blood flow and blood vessel permeability

enhance the delivery of oxygen and nutrients and the migration of additional phagocytes into the area, and the removal of toxins and waste products from the area.

Chapter Review

1. Solution A must have initially had more solutes than solution B. As a result, water moved by osmosis across the selectively permeable membrane from side B to side A, increasing the fluid level on side A.

2. The isolation of the internal contents of membrane-bound organelles allows them to manufacture or store secretions, enzymes, or toxins that could adversely affect the cytoplasm in general. Another benefit is the increased efficiency of having specialized enzyme systems concentrated in one place. For example, the concentration of enzymes necessary for energy production in the mitochondrion increases the efficiency of cellular respiration.

3. Tight junctions block the passage of water or solutes between cells. In the digestive system, these junctions keep enzymes, acids, and waste products from damaging delicate underlying tissues.

4. Because apocrine secretions are released by pinching off a portion of the secreting cell, you could test for the presence of plasma membranes, specifically for the phospholipids in plasma membranes. Merocrine secretions do not contain a portion of the secreting cell, so they would lack membrane constituents.

5. Skeletal muscle tissue would be made up of densely packed fibers running in the same direction, but since muscle fibers are composed of cells, they would have many nuclei and mitochondria. Skeletal muscle also has an obvious banding pattern or striations due to the arrangement of the actin and myosin filaments within the cell. The student is probably looking at a slide of tendon (dense connective tissue). The small nuclei would be those of fibroblasts.

CHAPTER 4

Module 4.1 Review

a. The layers of the epidermis are the stratum basale, stratum spinosum, stratum granulosum, stratum lucidum, and stratum corneum.

b. Dandruff consists of cells from the stratum corneum.

c. A splinter that penetrates to the third epidermal layer of the palm is lodged in the stratum granulosum.

Module 4.2 Review

a. The two pigments in the epidermis are carotene (an orange-yellow pigment) and melanin (a brown, yellow-brown, or black pigment).

b. Exposure to sunlight or sunlamps darkens skin because the ultraviolet radiation they emit stimulates melanocytes in the epidermis and dermis to synthesize the pigment melanin.

c. From highest to lowest health risk, the rankings for skin cancer are (1) malignant melanoma, (2) squamous cell carcinoma, and (3) basal cell carcinoma.

Module 4.3 Review

a. The dermis (a connective tissue layer) lies between the epidermis and the hypodermis.

b. The capillaries and sensory neurons that supply the epidermis are located in the papillary layer of the dermis.

c. The presence of elastic fibers allows the dermis to undergo repeated cycles of stretching and recoiling (returning to its original shape).

Module 4.4 Review

a. One source is the epidermal cells in the stratum basale and stratum spinosum that synthesize vitamin D_3 when stimulated by ultraviolet (UV) radiation. Vitamin D_3 may also be obtained from the diet.

b. In the presence of ultraviolet radiation in sunlight, epidermal cells in the stratum spinosum and stratum basale convert a

cholesterol-related steroid into vitamin D_3.

c. Vitamin D_3 is needed to form strong bones and teeth. When the body surface is covered, ultraviolet radiation cannot reach the skin to stimulate vitamin D_3 production, so fragile bones can develop.

Section 1 Review

1. **a.** dermis: the connective tissue layer beneath the epidermis; **b.** epidermis: the protective epithelium covering the surface of the skin; **c.** papillary layer: vascularized areolar tissue containing capillaries, lymphatic vessels, and sensory neurons that supply the skin surface; **d.** reticular layer: interwoven meshwork of dense irregular connective tissue containing collagen fibers and elastic fibers; **e.** hypodermis (subcutaneous layer): layer of loose connective tissue, dominated by adipose tissue, below the dermis

2. **a.** accessory structures; **b.** epidermis; **c.** granulosum; **d.** papillary layer; **e.** nerves; **f.** reticular layer; **g.** collagen; **h.** hypodermis; **i.** loose connective; **j.** fat

3. **a.** Fingers (and toes) swell up because of the hypotonic osmotic flow of water into the dead, keratinized cells of the outer layer of the epidermis, the stratum corneum. Because the underlying strata and dermis do not expand, the larger surface area of the swollen stratum corneum must go somewhere, and it forms folds and creases, or wrinkles. Other areas of the body lack a thick stratum corneum, so little swelling and wrinkling result.

b. Epidermal cells in the stratum basale and stratum spinosum convert a cholesterol-related steroid to vitamin D_3 in the presence of UV radiation. The vitamin enters capillaries in the dermis and is carried to the liver. The liver produces an intermediate product that is released and carried to the kidneys. The kidneys convert that intermediate product into the hormone calcitriol. Calcitriol stimulates the intestinal absorption of calcium and phosphate, which are then available for bone growth and maintenance.

Module 4.5 Review

a. A typical hair is a keratinous strand produced by basal cells within a hair follicle.

b. When an arrector pili muscle contracts, it pulls the hair follicle erect, depressing the area at the base of the hair and making the surrounding skin appear higher. The overall effect is known as "goosebumps" or "chicken skin."

c. Pulling a hair is painful because its root is attached deep within the hair follicle, the base of which is surrounded by a root hair plexus consisting of sensory nerves. Cutting a hair is painless because a hair shaft contains no sensory nerves.

Module 4.6 Review

a. Two types of exocrine glands found in the skin are sebaceous (oil) glands and sweat glands.

b. Sebaceous secretions (sebum) lubricate and protect the keratin of the hair shaft, lubricate and condition the surrounding skin, and inhibit the growth of bacteria.

c. Deodorants are used to mask the odor of apocrine sweat gland secretions, which contain several kinds of organic compounds. Some of these compounds have an odor, or have the potential to produce an odor when metabolized by skin bacteria.

Module 4.7 Review

a. The nail bed is the area of epidermis covered by the nail, or nail body.

b. A fingernail is a keratinous structure that is produced by epithelial cells of the nail root and protects the underlying fingertip. Structures of the nail include a distal free edge, lunula, proximal nail fold, cuticle, nail root, and nail body.

c. Nail production occurs at the nail root, an epidermal fold that is not visible from the surface.

Module 4.8 Review

a. Common effects of the aging process on the skin include epidermal thinning due to a decline in basal cell activity; a decrease in the number of melanocytes; a reduction in sebaceous gland secretion; a decline in dendritic cell numbers; a reduction in vitamin D_3 production; a decline in glandular activity; reduced blood flow to the dermis; the stoppage of hair follicle functioning; and a slowing of the pace of skin repair.

b. With advancing age, melanocyte activity decreases, leading to gray or white hair.

c. As we age, the blood supply to the dermis decreases, and merocrine sweat glands become less active. Both changes make it more difficult for us to cool ourselves in hot weather as we get older.

Module 4.9 Review

a. The first step in the repair of tissue is inflammation.

b. Granulation tissue is the combination of blood clots, fibroblasts, and the extensive network of capillaries in healing tissue.

c. Skin can regenerate effectively even after undergoing considerable damage because stem cells persist in both the epithelial and connective tissue components of skin. In response to injury, cells of the stratum basale replace epithelial cells while connective tissue stem cells replace lost dermal cells.

Section 2 Review

1. **a.** merocrine sweat glands; **b.** nail root; **c.** apocrine sweat glands; **d.** scab; **e.** reticular layer of dermis; **f.** wrinkled skin; **g.** sebum; **h.** malignant melanoma; **i.** arrector pili; **j.** cuticle

2. **a.** free edge; **b.** nail body; **c.** lunula; **d.** proximal nail fold; **e.** cuticle; **f.** cuticle; **g.** proximal nail fold; **h.** nail root; **i.** lunula; **j.** nail body; **k.** phalanx; **l.** dermis; **m.** epidermis

3. **a.** hair shaft; **b.** sebaceous gland; **c.** arrector pili muscle; **d.** connective tissue sheath of hair follicle; **e.** root hair plexus

4. The puncture wound has a greater chance of becoming infected than the knife cut because the cut from the knife will bleed freely, washing away many of the bacteria from the wound site. In

a puncture wound, bacteria can be forced beneath the surface of the skin, where oxygen is limited, and past the skin's protective barriers, thus increasing the possibility of infection.

Chapter Review

1. The dermis consists of (1) the papillary layer, which consists of loose connective tissue and contains capillaries, lymphatic vessels, and sensory neurons, and (2) the reticular layer, which consists of dense irregular connective tissue and bundles of collagen and elastic fibers. Both layers contain networks of blood vessels, lymphatic vessels, nerve fibers, and accessory organs.

2. Regeneration of injured skin involves (1) inflammation and bleeding, (2) blood clot and scab formation, (3) clot dissolution and the appearance of collagen fibers and ground substance, and (4) scar tissue formation.

3. Fat-soluble substances easily pass through the permeability barrier, because it is composed primarily of lipids surrounding the epidermal cells. Water-soluble drugs are lipophobic ("fat-fearing") and thus do not readily penetrate the permeability barrier.

4. The child could eat more vitamin D-rich foods, such as fortified milk and salmon. Taking a vitamin D supplement could also help.

5. The chemicals in hair dyes break the protective covering of the cortex, allowing the dyes to stain the medulla of the shaft. Dying is not permanent because the cortex remains damaged, allowing shampoo and UV rays from the sun to enter the medulla and affect the color. Also, the living portion of the hair remains unaffected, so that when the shaft is replaced the color will be lost.

CHAPTER 5

Module 5.1 Review

a. A surface feature is a characteristic of a bone's surface that has a certain function, such as forming a joint, serving as a site of muscle attachment, or

allowing the passage of nerves and blood vessels.

b. The four broad categories for classifying bones according to shape are flat bones, long bones, irregular bones, and short bones.

c. A tubercle is a small, rounded projection on a bone, whereas a tuberosity is a small, rough projection that may occupy a broad area on the bone's surface.

Module 5.2 Review

a. The major parts of a long bone are the epiphysis, diaphysis, metaphysis, and marrow cavity.

b. The marrow cavity—the central space within a bone—contains the red bone marrow, the site of blood cell production, and the yellow bone marrow, adipose tissue that is an important site for energy reserves.

c. If the ratio of collagen to calcium phosphate in a bone increased, the bone's compressive strength would decrease, and it would also become more flexible.

Module 5.3 Review

a. Osteocytes are cells responsible for the maintenance and turnover of the mineral content of bone; osteoblasts are cells that produce the fibers and matrix of bone; and osteoclasts are cells that dissolve the fibers and matrix of bone.

b. An osteon is the basic functional unit of mature compact bone; it consists of osteocytes organized around a central canal and separated by concentric lamellae.

c. If the activity of osteoclasts (which demineralize bone) exceeded osteoblast activity (production of new bone), then the bone's mineral content (and thus its mass) would decline, making it weaker.

Module 5.4 Review

a. Appositional growth is enlargement of a bone by the addition of bone matrix at its surface.

b. The periosteum is the layer that surrounds a bone; it consists of an outer fibrous region and an inner cellular region. The endosteum is an incomplete

cellular lining on the inner (marrow) surfaces of bones.

c. The sample is likely from an epiphysis. The presence of lamellae that are not arranged in osteons is indicative of spongy bone, which occurs in epiphyses.

Module 5.5 Review

a. Endochondral ossification is the replacement of a cartilaginous model with bone.

b. During intramembranous ossification, fibrous connective tissue is replaced by bone.

c. X-rays of long bones, such as the femur, can reveal the presence or absence of the epiphyseal cartilage, which separates the epiphysis from the diaphysis so long as the bone is still lengthening. If the epiphyseal cartilage is still present, growth is still occurring; if it is not, the bone has reached its full length.

Module 5.6 Review

a. Marfan's syndrome is a hereditary disorder of connective tissue, resulting in abnormally long and thin limbs and digits. The condition usually causes life-threatening cardiovascular problems.

b. Gigantism results from overproduction of growth hormone *before* puberty, causing extreme height, whereas acromegaly results from overproduction of growth hormone *after* puberty, causing abnormally thick bones.

c. Pituitary growth failure is less common today in the United States, because children can be treated with synthetic growth hormone, which will provide adequate or near adequate amounts for normal growth and development.

Module 5.7 Review

a. An open fracture (also called a compound fracture) is a break in the bone in which bone pierces the skin; a closed fracture (also called a simple fracture) is a break in the bone in which no bone breaks the skin.

b. Immediately following a fracture, extensive bleeding occurs at the injury site. After several hours, a large blood clot called a fracture hematoma develops.

Next, an internal callus forms as a network of spongy bone unites the inner edges, and an external callus of cartilage and bone stabilizes the outer edges. The cartilaginous external callus is eventually replaced by bone, and the struts of spongy bone then unite the broken ends. With time, the swelling that initially marked the location of the fracture subsides and the fracture site is remodeled, leaving little evidence that a break occurred.

c. An external callus forms early in the healing process, when cells from the endosteum and periosteum migrate to the area of the fracture. These cells form an enlarged collar (external callus) that encircles the bone in the area of the fracture.

Section 1 Review

1. **a.** irregular bones; **b.** epiphyses; **c.** fossa; **d.** marrow cavity; **e.** trabeculae; **f.** osteoclasts; **g.** comminuted fracture; **h.** epiphyseal cartilage; **i.** osteon; **j.** appositional growth; **k.** endochondral ossification

2. **a.** intramembranous ossification; **b.** collagen; **c.** osteocytes; **d.** lacunae; **e.** hyaline cartilage; **f.** periosteum; **g.** compact bone

3. The fracture damaged the epiphyseal cartilage in Rebecca's right leg. Even though the bone healed properly, the damaged leg did not produce as much cartilage as did the undamaged leg. The result would be a shorter bone on the side of the injury.

Module 5.8 Review

a. The facial bones are the paired maxillae, palatine, nasal, inferior nasal conchae, zygomatic, and lacrimal and the unpaired vomer and mandible.

b. The bone that is fractured is the right parietal bone.

c. vomer: facial; ethmoid: cranial; sphenoid: cranial; temporal: cranial; inferior nasal conchae: facial

Module 5.9 Review

a. The temporal bone contains the carotid canal, through which

passes the internal carotid artery supplying the brain.

b. The foramen magnum is located in the occipital bone. This opening surrounds the connection between the brain and spinal cord.

c. The alveolar processes support the upper teeth in the maxillae, and the lower teeth in the mandible.

Module 5.10 Review

a. The ethmoid bone contains (1) the crista galli, which projects superior to the (2) cribiform plate that contains olfactory foramina, and (3) air cells, chambers that contribute to the paranasal sinuses. It also forms the (4) superior portion of the nasal septum, and its (5) nasal conchae project into the nasal cavity.

b. The paranasal sinuses are located in the sphenoid, ethmoid, and frontal bones, the paired palatine bones, and the maxillae.

c. The paranasal sinuses lighten the weight of the skull bones, and secrete mucus that flushes the surfaces of the nasal cavities.

Module 5.11 Review

a. There are seven associated bones of the skull.

b. Three auditory ossicles are located in each middle ear cavity, found within the petrous portion of the temporal bone. The ossicles play a key role in hearing by conducting vibrations produced by sound waves arriving at the tympanic membrane to the inner ear.

c. Your lab partner is correct: The hyoid bone does not directly attach to any other bone; instead, it supports the larynx and is the attachment site for muscles of the larynx, pharynx, and tongue.

Module 5.12 Review

a. A fontanelle is a relatively soft, flexible, fibrous connective tissue region between two flat bones in the developing skull.

b. The major fontanelles are the anterior fontanelle, occipital fontanelle, sphenoidal fontanelle, and mastoid fontanelle.

c. Because fontanelles are not ossified at birth, they permit flexibility of the skull during childbirth, and they allow for growth of the brain during infancy and early childhood.

Module 5.13 Review

a. The major components of a typical vertebra are the vertebral body, articular processes, and the vertebral arch, the last of which is composed of a spinous process, laminae, transverse processes, and pedicles.

b. The secondary curves of the spine allow us to balance our body weight to permit an upright posture with minimal muscular effort. Without the secondary curves, we would not be able to stand upright for extended periods.

c. The intervertebral discs attach to the body of the vertebra.

Module 5.14 Review

a. The dens is part of the axis, or second cervical vertebra, which is located in the cervical (neck) region of the vertebral column.

b. The presence of transverse foramina indicates that this vertebra is a cervical vertebra.

c. When you run your finger down a person's spine, you can feel the spinous processes of the vertebrae.

Module 5.15 Review

a. There are five vertebrae in the lumbar region, and five fused vertebrae in the sacrum.

b. The sacrum forms the posterior wall of the pelvic girdle.

c. The lumbar vertebrae must support a great deal more weight than do vertebrae that are more superior in the spinal column. The large vertebral bodies allow the weight to be distributed over a larger area.

Module 5.16 Review

a. True ribs (ribs 1-7) are attached directly to the sternum by their own costal cartilage. False ribs (ribs 8-12) either do not attach to the sternum (as in the floating ribs) or attach by means of a

common costal cartilage (as in ribs 8-10).

b. The three parts or regions of the sternum are, from superior to inferior, the manubrium, the body, and the xiphoid process.

c. In addition to the ribs and sternum, the thoracic vertebrae make up the thoracic cage.

Module 5.17 Review

a. Kyphosis may be caused by (1) osteoporosis with compression fractures affecting the anterior portions of vertebral bodies, (2) chronic contractions in muscles that insert on the vertebrae, or (3) abnormal vertebral growth.

b. The temporary lordosis that occurs during pregnancy is believed to stabilize the individual's center of gravity as the uterus enlarges.

c. Scoliosis is due to an abnormal lateral distortion or curvature of the spine. It is the most common condition related to distortions of vertebral curvature.

Module 5.18 Review

a. The bones of the pectoral girdle are two clavicles (collarbones) and two scapulae (shoulder blades).

b. In attaching the scapula to the sternum, the clavicle restricts the scapula's range of movement. A broken clavicle thus gives the scapula a greater range of movement but makes it less stable.

c. The humerus articulates with the scapula at the glenoid cavity.

Module 5.19 Review

a. The bone of the arm is the humerus, and the bones of the forearm are the radius and ulna.

b. The two rounded projections on either side of the elbow are the lateral and medial epicondyles of the humerus.

c. The radius is positioned laterally when the forearm is in the anatomical position.

Module 5.20 Review

a. Phalanges are bones of the fingers (or toes).

b. The carpal bones are the scaphoid, lunate, triquetrum, pisiform, trapezium, trapezoid, capitate, and hamate.

c. Bill has broken the tip of his thumb, also known as the pollex.

Module 5.21 Review

a. The acetabulum is the socket (fossa) on the lateral aspect of the pelvis that articulates with the head of the femur.

b. The three bones that fuse to make a hip bone (coxal bone) are the ilium, ischium, and pubis.

c. When you are seated, your body weight is borne by the ischial tuberosities.

Module 5.22 Review

a. The bones of the pelvis are the two hip (coxal) bones, the sacrum, and the coccyx.

b. The two pubic bones are joined anteriorly by the pubic symphysis.

c. The pelvis of females is adapted for supporting the weight of the developing fetus and enabling the newborn to pass through the pelvic outlet during delivery. Compared to males, the pelvis of females is smoother and lighter; has less-prominent markings; has an enlarged pelvic outlet; has a sacrum and coccyx with less curvature; has a pelvic inlet that is wider and more circular; is relatively broad and low; has ilia that project farther laterally; and has an inferior angle between the pubic bones that is greater than 100° (as opposed to 90° or less for males).

Module 5.23 Review

a. The bones of the lower limb are the femur (thigh), patella (kneecap), tibia and fibula (leg), tarsal bones, metatarsal bones, and phalanges.

b. The head of the femur articulates with the acetabulum.

c. The fibula both stabilizes the ankle joint and is an important point of attachment for muscles that move the foot and toes. When the fibula is fractured, those muscles cannot function properly, so walking becomes difficult—and painful.

Module 5.24 Review

a. The tarsal bones are the talus, calcaneus, cuboid, navicular, medial cuneiform, intermediate cuneiform, and lateral cuneiform.

b. The talus transmits the weight of the body from the tibia toward the toes.

c. Joey most likely fractured the calcaneus (heel bone).

Section 2 Review

1. a. axial; **b.** longitudinal; **c.** skull; **d.** mandible; **e.** lacrimal; **f.** occipital; **g.** temporal; **h.** hyoid; **i.** vertebral column; **j.** thoracic; **k.** lumbar; **l.** sacral; **m.** floating; **n.** sternum; **o.** xiphoid process

2. a. clavicle; **b.** scapula; **c.** humerus; **d.** radius; **e.** ulna; **f.** carpal bones; **g.** metacarpal bones; **h.** phalanges; **i.** hip bone (coxal bone); **j.** femur; **k.** patella; **l.** tibia; **m.** fibula; **n.** tarsal bones; **o.** metatarsal bones; **p.** phalanges

Module 5.25 Review

a. In a joint dislocation, the articulating surfaces of a joint are forced out of position.

b. Components of a synovial joint are a fibrous articular capsule, which surrounds the joint; articular cartilages, which resemble hyaline cartilages and cover the articulating bone surfaces; and synovial fluid, which is located within the joint cavity and provides lubrication, distributes nutrients, and absorbs shocks. Accessory structures include bursae, which are pockets filled with synovial fluid, that reduce friction and absorb shocks; fat pads, which protect the articular cartilages; menisci, which are fibrocartilage articular discs that allow for variation in the shapes of the articulating surfaces; ligaments, which are cords of fibrous tissue that support, strengthen, and reinforce the joint; and tendons, which pass across or around a joint, limit the range of motion, and provide mechanical support.

c. Articular cartilages lack a blood supply and thus rely on synovial fluid to supply nutrients and remove wastes. If the circulation of synovial fluid were impaired, the cartilages would no longer receive nutrients, and wastes would accumulate. The resulting conditions would cause the cartilages to degenerate, and cells in the tissue may die.

Module 5.26 Review

a. Based on the shapes of the articulating surfaces, synovial joints are classified as gliding, hinge, pivot, condylar, saddle, and ball-and-socket joints.

b. A ball-and-socket joint permits the widest range of motion.

c. shoulder: ball-and socket; elbow: hinge; thumb: saddle

Module 5.27 Review

a. The primary function of intervertebral discs is to separate and cushion the bodies of adjacent vertebrae.

b. A bulging disc is an intervertebral disc that is displaced or partly protruding as a result of a compressed gelatinous core distorting the outer fibrocartilage layer. In a herniated disc, the gelatinous core breaks through the outer fibrocartilage layer, causing it to protrude into the vertebral canal.

c. Osteoporosis is a condition in which there is a loss of bone mass to the point that normal function is compromised.

Module 5.28 Review

a. Rheumatism is a general term describing any painful condition of joints, muscles, or both that is not caused by infection or injury. Osteoarthritis is a form of rheumatism characterized by degeneration of the joint cartilage and the underlying bone. Osteoarthritis results from cumulative wear and tear or genetic factors affecting collagen formation.

b. An arthroscope is an instrument that uses thin, flexible optical fibers to view the interior structures of a joint. This instrument can also be modified to perform surgical procedures without the trauma of major surgery.

c. A person can slow the progression of arthritis by engaging in regular exercise, doing physical therapy, and taking anti-inflamatory drugs.

Section 3 Review

1. a. diarthrosis; **b.** articular cartilages; **c.** hip joint; **d.** fibrocartilage discs; **e.** synarthrosis; **f.** osteoporosis; **g.** fluid filled pouch; **h.** amphiarthrosis; **i.** arthritis; **j.** synovial fluid

2. a. marrow cavity; **b.** spongy bone; **c.** periosteum; **d.** synovial membrane; **e.** articular cartilage; **f.** joint cavity (containing synovial fluid); **g.** joint capsule; **h.** compact bone

3. a. condylar joint; **b.** pivot joint; **c.** gliding joint; **d.** hinge joint; **e.** ball-and-socket joint; **f.** saddle joint

4. Vertebral fractures are more likely in the elderly because of osteoporosis (a loss of bone mass that compromises normal function) and a reduction in cushioning by the intervertebral discs.

Chapter Review

1. Nutrients reach the osteocytes in spongy bone by diffusing along canaliculi that open onto the surface of the trabeculae.

2. The osteons are aligned parallel to the long axis of the shaft, which does not bend when forces are applied to either end. Stresses or impacts to the side of the shaft can lead to a fracture.

3. There are many long bones in the hand, each of which has two epiphyseal cartilages. Measuring the width of these epiphyseal cartilages will provide clues to the hormonal control of growth in the child.

4. The large bones of a child's cranium are not yet fused; they are connected by fontanelles, areas of fibrous tissue. By examining the bones, the anthropologist could readily see if sutures had formed. By knowing approximately how long it takes for the various fontanelles to close and by determining their sizes, she could estimate the age of the individuals at death.

5. In osteoporosis, a decrease in the calcium content of the bones leads to bones that are weak and brittle. Because the hip must

help support the body's weight, any weakening of the hip bones may result in their breaking under the weight of the body. The shoulder, by contrast, is not a load-bearing joint and is not subject to the same great stresses or strong muscle contractions as the hip joint. As a result, it is less likely to become broken.

CHAPTER 6

Module 6.1 Review

a. A tendon is a bundle of collagen fibers that connects a skeletal muscle to a bone. An aponeurosis is a broad collagenous sheet that takes the place of tendons and connects a skeletal muscle to a wider area of bone or more than one bone.

b. The epimysium is a dense layer of collagen fibers that surrounds the entire muscle; the perimysium divides the skeletal muscle into a series of compartments, each containing a bundle of muscle fibers called a fascicle. The endomysium surrounds individual skeletal muscle cells (fibers). The collagen fibers of the epimysium, perimysium, and endomysium come together to form either bundles known as tendons, or broad sheets called aponeuroses. Tendons and aponeuroses generally attach skeletal muscles to bones.

c. Muscle fibers are the same thing as muscle cells. They are called fibers, because their shape is elongated. Muscle fibers are enclosed in an endomysium, which contains capillaries, stem cells, and axons of neurons. Myofibrils are fine, longitudinal, "thread-like" structures consisting of myofilaments. Myofilaments are ultramicroscopic threads of protein that make up myofibrils in skeletal muscle.

Module 6.2 Review

a. A zone of overlap refers to the areas within a sarcomere where thin filaments are interspersed among thick filaments.

b. Thin filaments consist of actin, troponin, and tropomyosin. Thick filaments are composed of myosin molecules.

c. The sarcoplasmic reticulum is similar to the smooth endoplasmic reticulum found in other cells.

Module 6.3 Review

a. The neuromuscular junction is a specialized intercellular connection that enables a motor neuron to communicate with a skeletal muscle fiber.

b. Acetylcholine release is necessary for skeletal muscle contraction, because it serves as the first step in the process. A drug that blocks acetylcholine release would prevent ACh from binding with receptors on the motor end plate, so sodium ions would not rush into the muscle fiber's sarcoplasm, and no action potential would be generated in the sarcolemma. As a result, muscle contraction could not occur.

c. Without AChE (acetylcholinesterase), the motor end plate would be continuously stimulated by acetylcholine, locking the muscle in a state of contraction.

Module 6.4 Review

a. The sliding filament theory describes the process of sarcomere shortening caused by the sliding of thin and thick filaments past one another.

b. During a contraction, the sarcomere shortens as the Z lines become closer together. Both the H band and I bands also shorten, but the A band does not change length.

c. After a contraction, a muscle returns to its original length through a combination of gravity, opposing muscles contracting, and elasticity of tissues stretched by the contraction.

Module 6.5 Review

a. ATP is the molecule that supplies the energy for a muscle contraction.

b. Once the contraction process has begun, the steps that occur are (1) active-site exposure, (2) cross-bridge formation, (3) myosin head pivoting (power stroke), (4) cross-bridge detachment, and (5) myosin reactivation "recocking."

c. The breakdown of ATP into ADP + P enables myosin reactivation, because the energy released during this process is used to "recock" the myosin heads.

Section 1 Review

1. a. mitochondrion; b. sarcolemma; c. myofibril; d. thin filament; e. thick filament; f. sarcoplasmic reticulum; g. T tubules; h. sarcoplasm; i. myofibril

2. a. I band; b. A band; c. H band; d. Z line; e. zone of overlap; f. M line; g. thin filament; h. thick filament; i. sarcomere

3. a. myofibril, myofilament, myosin cell; b. sarcolemma, sarcoplasm, sarcomere, sarcoplasmic reticulum

Module 6.6 Review

a. A twitch is single stimulus-contraction-relaxation sequence in a muscle fiber.

b. During the relaxation phase, calcium levels are falling, which results in active sites being covered by tropomyosin and a decrease in the number of cross-bridges.

c. Complete tetanus occurs when action potentials arrive so quickly that the sarcoplasmic reticulum cannot reclaim calcium ions before the next stimulus is applied. The elevated Ca^{2+} concentration prolongs the contraction, making it continuous. Incomplete tetanus refers to a muscle producing near-peak tension during rapid cycles of contraction and relaxation.

Module 6.7 Review

a. A motor unit is a single motor neuron and all of the muscle fibers it innervates.

b. Recruitment is the increase in number of active motor units. Recruitment results in an increase in muscle tension.

c. The finer and more precise the movement produced by a particular muscle, the fewer the number of muscle fibers in the motor unit.

Module 6.8 Review

a. In an isotonic contraction, muscle tension rises and the skeletal muscle's length changes. In an isometric contraction, muscle tension rises but the muscle's length does not change, and the load does not move.

b. Yes, a skeletal muscle can contract without shortening, as occurs during an isometric contraction. Whether a contracting muscle shortens or elongates, or remains the same length (an isometric contraction) depends on the relationship between the resistance and the tension produced by actin–myosin interactions.

c. The peak muscle tension of the muscle undergoing the isometric contraction was twice as great as that produced by the muscle undergoing the isotonic contraction.

Module 6.9 Review

a. The two compounds in which energy is stored in resting muscle fibers are creatine phosphate (CP) and glycogen.

b. Muscle fibers produce lactate under conditions of anaerobic metabolism (when there is a lack of oxygen). These conditions occur at peak levels of muscle activity.

c. Oxygen debt (or excess postexercise oxygen consumption) is the amount of oxygen intake required after strenuous activity to produce the ATP needed to restore normal, pre-exertion conditions in the body.

Module 6.10 Review

a. Muscle hypertrophy is enlargement of fiber size without cell division (and thus of the entire muscle) stemming from increases in myofibrils, mitochondria, glycolytic enzymes, and glycogen reserves, often in response to activities such as bodybuilding. Muscle atrophy is the wasting away of muscle tissues from a lack of use, ischemia, or nutritional abnormalities.

b. While Fred's leg was immobilized, its muscles did not receive sufficient neural stimulation to maintain normal mass, tone, and strength—that is, his leg muscles atrophied—and the muscles were thus unable to support his weight.

c. A murder victim's time of death can be estimated according to the body's flexibility or rigidity because rigor mortis typically begins a few hours after death, reaches maximum rigidity some 2–7 hours after death, and subsides about 1–6 days later or when decomposition begins. Thus, for example, a victim whose body lacks any signs of rigor mortis likely died within the past few hours. At the molecular level, the membranes of the dead cells are no longer selectively permeable and the SR is no longer able to retain calcium ions. As calcium ions enter the sarcoplasm, a sustained contraction develops, making the body extremely stiff. Contraction persists because the dead muscle cells can no longer make the ATP required for cross-bridge detachment from the active sites. Rigor mortis lasts until the lysosomal enzymes released by autolysis break down the myofilaments.

Section 2 Review

1. a. isometric contraction; b. isotonic contraction; c. hypertrophy; d. atrophy; e. motor unit; f. rigor mortis; g. oxygen; h. lactic acid

2. a. fatty acids; b. O_2; c. glucose; d. glycogen; e. CP; f. creatine

3. a. Lactate results from the dissociation of lactic acid in muscles working at peak activity. The lactate diffuses into the bloodstream from where it is absorbed by the liver. The liver converts the lactate to pyruvate, much of which is then converted into glucose. The glucose is released into the bloodstream, where it is available for skeletal muscle fibers to absorb and rebuild glycogen reserves.

b. The muscle atrophy and paralysis characteristic of polio results from the destruction of motor neurons by viruses. Tetanus is characterized by sustained muscular contractions throughout the body. It is due to a bacterial toxin that suppresses the mechanism that inhibits motor neuron activity. Botulism and myasthenia gravis result from disruptions at the neuromuscular junctions (NMJs).

Botulism produces paralysis through a bacterial toxin that blocks ACh release at NMJs. Myasthenia gravis is a progressive muscular weakness due to a loss of ACh receptors at the motor end plates of the NMJs.

c. In rigor mortis, the membranes of the dead cells are no longer selectively permeable; the SR is no longer able to retain calcium ions. As calcium ions enter the sarcoplasm, a sustained contraction develops, making the body extremely stiff. Contraction persists because the dead muscle cells can no longer make the ATP required for cross-bridge detachment from the active sites. Rigor mortis begins a few hours after death and ends after 1–6 days, or when decomposition begins. Decomposition begins when the lysosomal enzymes released by autolysis break down the myofilaments.

Module 6.11 Review

a. A synergist is a muscle that helps a larger prime mover (or agonist—a muscle that is mainly responsible for a specific movement) perform its actions more efficiently.

b. The biceps brachii and the triceps brachii are antagonists, because they perform opposite actions.

c. The name flexor carpi radialis longus tells you that this muscle is a long muscle that lies next to the radius and flexes the wrist.

Module 6.12 Review

a. When doing jumping jacks, the lower limbs must perform abduction (when the limbs are spread apart) and adduction (when they are brought back together again).

b. Flexion and extension are the movements associated with hinge joints.

c. Dorsiflexion is upward movement of the foot through flexion at the ankle, whereas plantar flexion is ankle extension, as when pointing the toes.

Module 6.13 Review

a. Snapping your fingers involves opposition of the thumb and

flexion at the third metacarpophalangeal joint.

b. Pronation and supination of the hand are made possible by the rotation of the radius head.

c. Protraction, supination, and pronation occur while wriggling into tight-fitting gloves.

Module 6.14 Review

a. The axial muscles, which arise on the axial skeleton, position the head and spinal column and move the rib cage to make breathing possible.

b. biceps brachii: appendicular; external oblique: axial; temporalis: axial; vastus medialis: appendicular.

c. The following structures labeled in the figures in this module are not muscles: linea alba, flexor retinaculum, iliotibial tract, patella, tibia, clavicle, sternum, superior extensor retinaculum, inferior extensor retinaculum, lateral malleolus of fibula, medial malleolus of tibia, calcaneal tendon, and calcaneus.

Module 6.15 Review

a. The muscles of facial expression originate on the surface of the skull.

b. The masseter and temporalis muscles produce a biting action by elevating the mandible.

c. Contraction of the left and right zygomaticus muscles elevates the corners of the mouth into a smile.

Module 6.16 Review

a. The erector spinae from medial to lateral are the spinalis, longissimus, and iliocostalis muscles.

b. Both the iliocostalis and quadratus lumborum have origins on the iliac crest of the hip bone.

c. The splenius capitis and longissimus have insertions on the skull.

Module 6.17 Review

a. The abdominal muscles from superficial to deep are the external oblique, internal oblique, and transversus abdominis.

b. Damage to the external intercostal muscles would interfere with breathing.

c. A hit to the rectus abdominis muscle would cause that muscle to contract forcefully, resulting in flexion of the vertebral column. In other words, you would "double over."

Module 6.18 Review

a. Contractions of the levator scapulae and trapezius muscles elevate the scapulae and would produce a shoulder shrug.

b. The deltoid, pectoralis major, and latissimus dorsi are appendicular muscles that insert on the humerus. One of the actions of the deltoid is to raise (abduct) the arm.

c. deltoid: appendicular; external oblique: axial; gluteus maximus: appendicular; pectoralis major: appendicular

Module 6.19 Review

a. The triceps brachii inserts on the olecranon of the ulna and produces extension at the elbow. It also produces extension and adduction at the shoulder.

b. The muscles involved in flexion at the elbow are the biceps brachii, brachialis, and brachioradialis.

c. Injury to the flexor carpi ulnaris muscle would impair the ability to perform flexion and adduction of the wrist.

Module 6.20 Review

a. The quadriceps muscles are the rectus femoris, vastus intermedius, vastus lateralis, and vastus medialis muscles.

b. The muscles that flex the knee are the biceps femoris, semimembranosus, semitendinosus, sartorius, and popliteus muscles.

c. Sitting down on a chair requires flexion at the hip. Muscles that would produce that action include the adductor muscles, the gracilis, and the iliopsoas muscles.

Module 6.21 Review

a. The muscles involved in plantar flexion (extension at the ankle) include the gastrocnemius, soleus, and fibularis longus.

b. Dorsiflexion is the action performed and it occurs with contraction of the tibialis anterior.

c. A torn calcaneal tendon would make plantar flexion difficult, because this tendon attaches the soleus and gastrocnemius muscles to the calcaneus (heel bone).

1. **a.** frontalis; **b.** temporalis; **c.** orbicularis oculi; **d.** zygomaticus; **e.** buccinator; **f.** orbicularis oris; **g.** masseter; **h.** sternocleidomastoid

2. **a.** gluteus medius; **b.** tensor fasciae latae; **c.** gluteus maximus; **d.** an adductor muscle; **e.** gracilis; **f.** biceps femoris; **g.** semitendinosus; **h.** semimembranosus; **i.** sartorius

3. **a.** gastrocnemius; **b.** tibialis anterior; **c.** fibularis longus; **d.** soleus; **e.** extensor digitorum longus; **f.** superior extensor retinaculum; **g.** calcaneal tendon; **h.** inferior extensor retinaculum

1. Acetylcholine (ACh) released by the motor neuron at the neuromuscular junction changes the permeability of the plasma membrane at the motor end plate. The permeability change allows the influx of positive charges, which in turn trigger an electrical event called an action potential. The action potential spreads across the entire surface of the muscle fiber and into the interior along the T tubules. The cytoplasmic concentration of calcium ions (released from sarcoplasmic reticulum) increases, triggering the start of a contraction. The contraction ends when the ACh has been removed from the synaptic cleft and motor end plate by acetylcholinesterase (AChE).

2. ATP is generated from the anaerobic breakdown of glucose to pyruvate and from the aerobic breakdown of pyruvate in mitochondria.

3. The most obvious sign of organophosphate poisoning is uncontrolled tetanic contractions of skeletal muscles. Organophosphates block the action of the enzyme acetylcholinesterase, so acetylcholine released into the synaptic cleft of the neuromuscular junc-

tion would not be inactivated, causing a state of persistent contraction (spastic paralysis). If paralysis affected the muscles of respiration (which is likely), Terry would die of suffocation.

4. After death, the membranes of muscle cells are no longer selectively permeable, so calcium leaks in, triggering contractions that persist (rigor mortis) because the dead cells can no longer make the ATP required for cross-bridge detachment. The stiffness of rigor mortis begins 2-7 hours after death and disappears after 1-6 days or when decomposition begins. The timing depends on environmental factors.

5. Jared should do squatting exercises. Placing a barbell on his shoulders as he does squats will produce better results, because the quadriceps muscles will be working against a greater resistance.

a. Structural components of a typical neuron include a cell body (including the nucleus), an axon (including the axon hillock, collateral branches, and axon terminals), and dendrites.

b. According to structure, neurons are classified as bipolar, unipolar, or multipolar.

c. Most CNS neurons lack centrioles, so these cells cannot divide and replace themselves.

a. Central nervous system neuroglia are ependymal cells, microglia, astrocytes, and oligodendrocytes.

b. Astrocytes protect the CNS from circulating chemicals and hormones by maintaining the blood–brain barrier.

c. Microglia, cells that are embryologically related to monocytes and macrophages, would occur in increased numbers in a person with a CNS infection.

a. Schwann cells form a lipid-rich sheath around peripheral axons

and satellite cells surround neuron cell bodies in ganglia.

b. The neurilemma is the outer surface of a Schwann cell encircling an axon.

c. According to function, neurons are classified as sensory neurons, motor neurons, or interneurons.

a. The membrane potential is the unequal distribution of electrical charges across a cell's plasma membrane. In a resting cell, the inside of the plasma membrane is negatively charged, and the outside of the plasma membrane is positively charged.

b. The sodium–potassium exchange pump maintains the cell's resting potential by ejecting three sodium ions from the cell for every two potassium ions it recovers from the extracellular fluid.

c. Three body functions dependent upon changes in the membrane potential are muscle contraction, gland secretion, and information transfer in the nervous system.

a. Depolarization is a shift from the resting potential in which the membrane potential becomes less negative. Repolarization is the return of the membrane potential to the resting potential after the membrane has been depolarized.

b. Continuous propagation is the propagation of an action potential along an unmyelinated axon, wherein the action potential affects every portion of the membrane surface. Saltatory propagation is the relatively rapid propagation of an action potential between successive nodes of a myelinated axon.

c. The presence of myelin greatly increases the propagation speed of action potentials.

a. The parts of a synapse—the site where a neuron communicates with another neuron or with a cell of a different type—are a presynaptic cell and a postsynaptic cell, whose plasma membranes are separated by a

narrow gap called the synaptic cleft.

b. The events that occur when an action potential arrives at an axon terminal are (1) the depolarization of the presynaptic membrane; (2) the opening of calcium channels and an inflow of calcium ions that triggers the release of ACh into the synaptic cleft; (3) the binding of ACh to closed sodium channels of the postsynaptic membrane; (4) the opening of sodium channels and an inflow of sodium ions that leads to depolarization; and, if threshold is reached, (5) an action potential is initiated in the postsynaptic neuron.

c. Depolarization of the postsynaptic cell is temporary because the synaptic cleft contains an enzyme, called acetylcholinesterase (AChE), that breaks down the ACh released by the presynaptic cell.

1. **a.** dendrite; **b.** Nissl bodies; **c.** mitochondrion; **d.** nucleus; **e.** nucleolus; **f.** cell body; **g.** axon hillock; **h.** axon; **i.** collateral; **j.** axon terminal

2. **a.** axon of presynaptic cell; **b.** axon terminal; **c.** synaptic vesicles; **d.** synaptic cleft; **e.** postsynaptic cell; **f.** mitochondrion; **g.** presynaptic membrane; **h.** postsynaptic membrane

3. **a.** action potential; **b.** bipolar neurons; **c.** resting potential; **d.** oligodendrocytes; **e.** synapse; **f.** depolarization; **g.** interneurons; **h.** repolarization

4. In myelinated fibers, saltatory propagation transmits nerve impulses to the neuromuscular junctions rapidly enough to initiate muscle contractions and promote normal movements. In axons that have become demyelinated, nerve impulses cannot be propagated, and so the muscles are not stimulated to contract. Eventually, the muscles atrophy because of the lack of stimulation (a condition termed disuse atrophy).

a. The major regions of the brain are the cerebrum (composed of fissures, gyri, and sulci), the diencephalon (thalamus and hypothalamus), the cerebellum, and the brain stem (midbrain, pons, and medulla oblongata). Additionally, the brain contains four ventricles and some connecting passageways (the interventricular foramen and the cerebral aqueduct), plus the corpus callosum. Basal nuclei (gray matter) lie deep to the floor of the lateral ventricles within each cerebral hemisphere.

b. The medulla oblongata (the most caudal of the brain regions) relays sensory information to other parts of the brain stem and to the thalamus. It also contains centers that regulate autonomic function, such as heart rate and blood pressure.

c. A blocked cerebral aqueduct would prevent communication between the third and fourth ventricles.

Module 7.8 Review

a. The layers of the cranial meninges are the outer dura mater, the middle arachnoid mater, and the inner pia mater.

b. Cerebrospinal fluid (CSF) is produced at the choroid plexus in each of the ventricles. CSF reaches the subarachnoid space through the lateral and median apertures, and diffuses across the arachnoid granulations into the superior sagittal sinus.

c. If diffusion across the arachnoid granulations decreased, the volume of CSF in the ventricles would increase, because less CSF would reenter the bloodstream. The increased pressure within the brain due to accumulated CSF could damage the brain.

Module 7.9 Review

a. The lobes of the cerebrum—the frontal lobe, parietal lobe, occipital lobe, and temporal lobe—are named for the overlying bones of the skull.

b. The insula is an island of cortex located medial to the lateral sulcus.

c. Damage to the left postcentral gyrus would interfere with the awareness of sensory information from the right side of the body.

Module 7.10 Review

a. The primary motor cortex is located in the precentral gyrus of the frontal lobe of the cerebrum.

b. Damage to the temporal lobes of the cerebrum would interfere with the processing of olfactory (smell) and auditory (sound) sensations.

c. The stroke has damaged the speech center, located in the frontal lobe.

Module 7.11 Review

a. The main components of the diencephalon are the thalamus and hypothalamus.

b. The three regions of the brain stem are the midbrain, pons, and medulla oblongata.

c. The cerebellum adjusts postural muscles to maintain balance and programs and fine-tunes movements controlled at the conscious and subconscious levels.

Module 7.12 Review

a. The reticular formation extends from the medulla oblongata to the midbrain. This interconnected neural network regulates many vital involuntary functions of the body.

b. The reticular activating system (RAS) lies in the midbrain and is part of the reticular formation. The RAS directly affects the activity of the cerebral cortex and consciousness.

c. The reticular activating system (RAS) is responsible for rousing the cerebrum to a state of consciousness. If you are sleeping and your RAS were suddenly activated, you would wake up.

Module 7.13 Review

a. An electroencephalogram (EEG) is a graph of the electrical activity of the brain. The four wave types associated with an EEG are alpha waves (characteristic of normal resting adults), beta waves (characteristic of a person who is concentrating), theta waves (observed in children and frustrated adults), and delta waves (found in a person who is sleeping deeply or in individuals with certain pathological states).

b. While you are reading your textbook, EEG waves that might appear include alpha waves and beta waves.

c. A seizure is a temporary cerebral disorder accompanied by abnormal movements, unusual sensations, inappropriate behavior, or some combination of these signs and symptoms. Epilepsy is a clinical condition characterized by seizures.

Module 7.14 Review

a. A typical spinal cord has 31 pairs of nerves, and the spinal cord ends at the level of lumbar vertebra 1 or 2 (L_1 or L_2).

b. The gray matter in the spinal cord is composed of the cell bodies of neurons, neuroglia, and unmyelinated axons.

c. Gross anatomical features of the cross-sectioned spinal cord include the anterior median fissure (a deep groove along the anterior or ventral surface); the posterior median sulcus (a shallow longitudinal groove); white matter (composed of myelinated and unmyelinated axons); gray matter (composed of cell bodies of neurons, neuroglia, and unmyelinated axons); the central canal (a passageway containing cerebrospinal fluid); a dorsal root of each spinal nerve (axons of neurons whose cell bodies are in the dorsal root ganglion); a ventral root of each spinal nerve (the axons of motor neurons that extend into the periphery to control somatic and visceral effectors); dorsal root ganglia (contain cell bodies of sensory neurons); and spinal nerves (contain the axons of sensory and motor neurons).

Module 7.15 Review

a. The horns are projections of gray matter within the spinal cord. They include the anterior gray horn, lateral gray horn, and posterior gray horn.

b. Sensory nuclei receive and relay sensory information from peripheral receptors; motor nuclei issue motor commands to peripheral effectors.

c. The three columns of white matter in the spinal cord are the anterior white column, lateral white column, and posterior white column.

Module 7.16 Review

a. The three spinal meninges are the dura mater (the outermost component of the cranial and spinal meninges), arachnoid mater (the middle meninx that encloses cerebrospinal fluid), and pia mater (the innermost layer of the meninges bound to the underlying neural tissue).

b. The epidural space lies between the dura mater and the walls of the vertebral canal. It contains blood vessels and adipose tissue.

c. It is done below this level so that the spinal cord is not damaged. The spinal cord ends at the L_1 or L_2 level.

Section 2 Review

1. a. precentral gyrus; b. frontal lobe; c. lateral sulcus; d. temporal lobe; e. pons; f. central sulcus; g. postcentral gyrus; h. parietal lobe; i. occipital lobe; j. cerebellum; k. medulla oblongata

2. a. posterior median sulcus; b. central canal; c. sensory nuclei; d. motor nuclei; e. ventral root; f. anterior median fissure; g. posterior white column; h. dorsal root ganglion; i. lateral white column; j. posterior gray horn; k. lateral gray horn; l. anterior gray horn

3. a. columns; b. precentral gyrus; c. medulla oblongata; d. meninges; e. cauda equina; f. thalamus; g. dura mater; h. prefrontal cortex; i. cerebrospinal fluid

Chapter Review

1. Within the CNS, cerebrospinal fluid (CSF) fills the central canal, the ventricles, and the subarachnoid space. CSF acts as a shock absorber and a diffusion medium for dissolved gases, nutrients, chemical messengers, and waste products.

2. Transection of the spinal cord at C_7 would most likely result in paralysis from the neck down. Transection at T_{10} would produce paralysis and eliminate sensory input in the lower half of the body only.

3. The left hemisphere contains the general interpretive and speech centers and is responsible for performing analytical tasks, for logical decision making, and for language-based skills (reading, writing, and speaking). The right hemisphere analyzes sensory information and relates the body to the sensory environment. Interpretive centers in this hemisphere permit the identification of familiar objects by touch, smell, sight, taste, or feel. The right hemisphere is also important in understanding three-dimensional relationships and in analyzing the emotional context of a conversation.

4. Brain tumors do not result from uncontrolled division of neurons, but instead of neuroglial cells, which can divide. In addition, cells of the meningeal membranes can give rise to tumors.

5. Centers in the medulla oblongata are involved in respiratory and cardiac activity.

CHAPTER 8

Module 8.1 Review

a. The three layers are the outer epineurium, middle perineurium, and inner endoneurium. The major peripheral branches of a spinal nerve are the dorsal ramus and the ventral ramus.

b. A dermatome is a bilateral sensory region monitored by a single pair of spinal nerves.

c. Shingles is reactivation of the varicella-zoster virus (VZV), which is the same herpes virus that causes chickenpox. When active, the VZV causes painful inflammation of the nerve ganglia, with skin eruptions forming a girdle around the trunk or linear pattern down the limbs, because it follows a dermatome.

Module 8.2 Review

a. The cranial nerves by name and number are the following: olfactory (I), optic (II), oculomotor (III), trochlear (IV), trigeminal (V), abducens (VI), facial (VII), vestibulocochlear (VIII), glossopharyngeal (IX),

vagus (X), accessory (XI), and hypoglossal (XII).

b. The cranial nerves with motor functions only are the trochlear (IV), abducens (VI), accessory (XI), and hypoglossal (XII).

c. The cranial nerves that are mixed are the trigeminal nerve (V), the facial nerve (VI), the glossopharyngeal nerve (IX), and the vagus nerve (X).

Module 8.3 Review

a. A nerve plexus is a complex network of nerves. The major plexuses are the cervical, brachial, lumbar, and sacral.

b. Damage to the cervical plexus—or more specifically to the phrenic nerves, which originate in this plexus and innervate the diaphragm—would greatly interfere with the ability to breathe and might even be fatal.

c. Major nerves associated with the brachial plexus are the radial nerve, musculocutaneous nerve, median nerve, and ulnar nerve.

Module 8.4 Review

a. A sensory homunculus is a functional map of the primary sensory cortex showing the relative numbers of neurons associated with the sensory integration of each part of the body.

b. A corticospinal pathway carries motor commands from the cerebral cortex to the spinal cord, from which they continue to the skeletal muscles.

c. The right cerebral hemisphere controls the left side of the body.

Module 8.5 Review

a. All reflexes are rapid, unconscious responses to physical stimuli that restore or maintain homeostasis.

b. A reflex is a rapid, automatic response to a stimulus. A reflex arc includes the receptor, sensory neuron, motor neuron, and effector involved in a particular reflex; interneurons may be present, depending on the reflex considered.

c. In the patellar reflex, the response observed is extension of the leg at the knee and the effectors involved are the quadriceps femoris muscles.

Module 8.6 Review

a. Reinforcement is an enhancement of spinal reflexes. It occurs when the postsynaptic neuron enters a state of generalized facilitation caused by activity in other excitatory synapses.

b. Reflex testing provides information about the status of the nervous system.

c. A positive Babinski reflex is abnormal in adults; it indicates possible damage to descending tracts in the spinal cord.

Module 8.7 Review

a. Referred pain is a sensation felt in a part of the body other than its actual source.

b. After being bitten by a rabid animal, the rabies virus infects peripheral nerves. Retrograde flow then carries the viral particles into the CNS.

c. Amyotrophic lateral sclerosis (ALS), commonly called Lou Gehrig Disease, is a progressive degeneration of the motor neurons of the CNS, leading to muscle atrophy and eventual paralysis.

Section 1 Review

1. a. olfactory bulb (associated with N I, olfactory, S; **b.** oculomotor (III), M; **c.** trigeminal (V), B; **d.** facial (VII), B; **e.** glossopharyngeal (IX), B; **f.** vagus (X), B; **g.** optic (II), S; **h.** trochlear (IV), M; **i.** abducens (VI), M; **j.** vestibulocochlear (VIII), S; **k.** hypoglossal (XII), M; **l.** accessory (XI), M

2. a. receptor; **b.** sensory neuron; **c.** interneuron; **d.** spinal cord (CNS); **e.** motor neuron; **f.** effector

3. a. shingles; **b.** epineurium; **c.** plantar reflex; **d.** phrenic nerve; **e.** nerve plexus; **f.** corticospinal pathway; **g.** posterior column pathway; **h.** endoneurium; **i.** cerebral palsy; **j.** reinforcement

Module 8.8 Review

a. General responses to increased sympathetic activity include (1) heightened mental alertness, (2) increased metabolic rate, (3) reduced digestive and urinary functions, (4) activation of energy reserves, (5) increased respiratory rate and dilation of respiratory passageways, (6) increased heart rate and blood pressure, and (7) activation of sweat glands. General responses to increased parasympathetic activity include (1) decreased metabolic rate, (2) decreased heart rate and blood pressure, (3) decreased respiratory rate and constriction of the respiratory passageways, (4) increased secretion by salivary and digestive glands, (5) increased motility and blood flow in the digestive tract, and (6) stimulation of urination and defecation.

b. An intramural ganglion is a group of parasympathetic neurons embedded in the tissue of a target organ.

c. Spinal cord $\longrightarrow$ preganglionic neuron $(T_1-L_5) \longrightarrow$ collateral ganglia $\longrightarrow$ postganglionic fibers $\longrightarrow$ visceral effectors in abdominopelvic cavity.

Module 8.9 Review

a. Splanchnic nerves are preganglionic fibers to collateral ganglia in the viscera of internal organs, especially those of the abdomen.

b. The vagus nerve (X) is responsible for most of the outflow of the parasympathetic nervous system.

c. Sympathetic nerves are bundles of postganglionic fibers arising from the cervical sympathetic ganglia that innervate structures in the thoracic cavity.

Module 8.10 Review

a. Acetylcholine (ACh) is the neurotransmitter released by all parasympathetic neurons.

b. The parasympathetic division is sometimes referred to as the anabolic system because parasympathetic stimulation leads to a general increase in the nutrient content of the blood. Cells throughout the body respond to the increase by absorbing the nutrients and using them to support growth and other anabolic activities, such as the synthesis of glycogen.

c. In tense (anxious) individuals, an increase in sympathetic stimulation would probably cause an increase in some or all of the following: alertness, heart rate, blood pressure, and breathing.

Section 2 Review

1. **a.** cervical sympathetic ganglia; **b.** sympathetic chain ganglia; **c.** coccygeal ganglia; **d.** sympathetic nerves; **e.** cardiac and pulmonary plexuses; **f.** celiac ganglion; **g.** superior mesenteric ganglion; **h.** splanchnic nerves; **i.** inferior mesenteric ganglion
2. **a.** P; **b.** P; **c.** S; **d.** P; **e.** S; **f.** S; **g.** P; **h.** S; **i.** S; **j.** S; **k.** P; **l.** S; **m.** P
3. **a.** sympathetic division: lateral gray horns of spinal segments T_1–L_2, parasympathetic division: brain stem and spinal segments S_2–S_4; **b.** sympathetic division: near vertebral column, parasympathetic division: intramural, some terminal; **c.** sympathetic division: short; release ACh, parasympathetic division: long; release ACh; **d.** sympathetic division: long; usually release norepinephrine (NE), parasympathetic division: short; release ACh; **e.** sympathetic division: stimulates metabolism; increases alertness; prepares for emergency ("fight or flight"), parasympathetic division: promotes relaxation, nutrient uptake, energy storage ("rest and digest")

Chapter Review

1. N I: olfactory; N II: optic; N III: oculomotor; N IV: trochlear; N V: trigeminal; N VI: abducens; N VII: facial; N VIII: vestibulocochlear; N IX: glossopharyngeal; N X: vagus; N XI: accessory; N XII: hypoglossal
2. A sensory neuron that delivers sensations to the CNS is a first-order neuron. Within the CNS, the axon of the first-order neuron synapses on a second-order neuron, which is an interneuron located in the

spinal cord or brain stem. The second-order neuron synapses on a third-order neuron in the thalamus. The axons of third-order neurons synapse on neurons of the primary sensory cortex of the cerebral hemispheres.
3. A motor homunculus, a mapped-out area of the primary motor cortex, provides an indication of the degree of fine motor control available. A sensory homunculus indicates the degree of sensitivity of peripheral sensory receptors. Both of these maps show the relative numbers of neurons associated with each area of the body. The greater the number of neurons that innervate each area, the larger the part of the body appears on the homunculus.
4. Response time in the patellar reflex is faster than in a withdrawal reflex because with only one synapse, synaptic delay time is minimized. In a withdrawal reflex, total delay time is proportional to the number of synapses involved.
5. Sympathetic preganglionic fibers emerge from the thoracic and lumbar regions (T_1 through L_2) of the spinal cord. Parasympathetic fibers emerge from the brain stem (midbrain, pons, and medulla oblongata) and the sacral region of the spinal cord.

CHAPTER 9

Module 9.1 Review

a. The six types of tactile receptors are the following: free nerve endings (sensitive to touch and pressure), root hair plexuses (monitor distortions and movements across the body surface), tactile discs and specialized epithelial cells (detect fine touch and pressure), tactile corpuscles (detect fine touch and pressure), lamellated corpuscles (sensitive to pulsing or vibrating stimuli, such as deep pressure), and Ruffini corpuscles (sensitive to pressure and distortion of the skin).
b. Tactile receptors found only in the dermis are tactile cor-

puscles, lamellated corpuscles, and Ruffini corpuscles.
c. A Ruffini corpuscle is more sensitive to continuous deep pressure because, unlike a lamellated corpuscle, it undergoes little adaptation.

Module 9.2 Review

a. Baroreceptors are receptors that detect changes in pressure and chemoreceptors are receptors that detect changes in concentrations of dissolved chemical compounds or gases.
b. Chemoreceptors would be sensitive to changes in blood pH.
c. Baroreceptors are found along the digestive tract, within the walls of the urinary bladder, in the carotid sinus and aortic sinus, lungs, and colon.

Section 1 Review

1. **a.** free nerve ending; **b.** root hair plexus; **c.** tactile discs; **d.** tactile corpuscle; **e.** Ruffini corpuscle **f.** lamellated corpuscle
2. **a.** free nerve endings; **b.** chemoreceptors; **c.** tactile receptors; **d.** baroreceptors; **e.** thermoreceptors; **f.** chemoreceptors; **g.** thermoreceptors; **h.** chemoreceptors; **i.** baroreceptors; **j.** proprioceptors; **k.** nociceptors
3. **a.** The general senses refer to our sensitivity to temperature, pain, touch, pressure, vibration, and proprioception. **b.** Tactile receptors in the skin that may be stimulated are free nerve endings, tactile discs, tactile corpuscles, and root hair plexuses. Its bite would surely stimulate nociceptors (free nerve endings) that provide the sensation of pain.

Module 9.3 Review

a. Olfaction—the sense of smell—involves olfactory receptors in paired olfactory organs responding to chemical stimuli.
b. Gustation is the sense of taste, provided by taste receptors responding to chemical stimuli.
c. Axons from the olfactory epithelium collect into bundles that reach the olfactory bulb. Axons leaving the olfactory bulb then travel along the olfactory tract to the olfactory cortex,

hypothalamus, and portions of the limbic system.

Module 9.4 Review

a. The malleus, incus, and stapes are the three tiny bones in the middle ear.
b. The auditory tube is a passageway connecting the nasopharynx with the middle ear cavity that equalizes the pressure on the inside of the tympanic membrane; it is also called the Eustachian tube.
c. External ear infections are relatively uncommon because cerumen, secreted by the ceruminous glands, inhibits microbial growth, thereby lessening the chances of infection.

Module 9.5 Review

a. The bony labyrinth is composed of semicircular canals enclosing the semicircular ducts, the vestibule, and the cochlea.
b. The perilymph surrounds and separates the membranous labyrinth from the bony labyrinth.
c. Within the membranous labyrinth, receptors in the vestibule respond to gravity or acceleration, receptors in the semicircular ducts respond to rotation, and receptors in the cochlear duct respond to sound.

Module 9.6 Review

a. Otoliths ("ear stones") are calcium carbonate crystals that sit atop the gelatinous otolithic membrane of each macula.
b. Receptors in the saccule and utricle provide sensations of gravity (vertical movement) and linear (horizontal movement) acceleration.
c. Damage to the cupula of the lateral semicircular duct would interfere with the perception of horizontal rotation of the head.

Module 9.7 Review

a. The spiral organ is located in the cochlear duct of the cochlea of the internal ear.
b. Perilymph is located in the spaces of the scala vestibuli and scala tympani, and endolymph fills the cochlear duct.

c. Features visible on the sectional view of the cochlear spiral include the basilar membrane, vestibular membrane, scala vestibuli, spiral organ, cochlear duct, scala tympani, temporal bone (petrous part), and cochlear nerve.

Module 9.8 Review

a. A decibel is the unit used to measure the intensity of sound.

b. Sound waves enter the external acoustic meatus → travel to the tympanic membrane → auditory ossicles → oval window → scala vestibuli → basilar membrane to scala tympani → hair cells against tectorial membrane → cochlear branch of cranial nerve VIII.

c. If the round window could not move, the perilymph would not be moved by the vibration of the stapes at the oval window, reducing or eliminating the perception of sound.

Module 9.9 Review

a. Accessory structures associated with the eye include the eyelids (palpebrae), eyelashes, medial canthus, cornea, lateral canthus, lacrimal caruncle, conjunctiva, tarsal glands, and lacrimal apparatus.

b. Conjunctivitis, or pinkeye, is an inflammation of the conjunctiva, generally caused by a pathogenic infection, or by physical, allergic, or chemical irritation of the conjunctival surface. The most obvious sign, redness, results from the dilation of deep conjunctival blood vessels.

c. The conjunctiva would be the first layer of the eye affected by inadequate tear production.

Module 9.10 Review

a. The layers of the eye are the fibrous layer, the vascular layer, and the neural layer (retina).

b. The iris, the colored part of the eye, contains pigment cells that determine an individual's eye color.

c. The extrinsic eye muscles and their actions are the following: superior oblique (moves the eye downward and laterally), superior rectus (moves the eye upward), lateral rectus (moves the eye laterally), inferior oblique (moves the eye upward and laterally), inferior rectus (moves the eye downward), and medial rectus (moves the eye medially).

Module 9.11 Review

a. The cornea does not contain blood vessels.

b. Beginning with the cornea, the structures and fluids through which light passes are the aqueous humor, pupil of iris, lens, vitreous humor, and retina.

c. With decreased light intensity, the pupillary dilator muscles cause the pupils to enlarge.

Module 9.12 Review

a. The focal point is a specific point of light ray intersection on the retina.

b. When the ciliary muscles are relaxed you are viewing an object at a distance.

c. The elastic quality of the lens tends to decrease with age, thereby causing an increase in the near point of vision.

Module 9.13 Review

a. Rods are photoreceptors responsible for vision in dim lighting. Cones are photoreceptors requiring more light than rods. They provide us with color vision.

b. You will likely be unable to see at all. The light would be focused on the fovea that contains only cones, which need high-intensity light to be stimulated. The dimly lit room contains light that is too weak to stimulate the cones.

c. If you had been born without cones, you would still be able to see—so long as you had functioning rods—but you would see in black and white only, and with reduced visual sharpness (acuity).

Module 9.14 Review

a. Blue cones, green cones, and red cones are the three types of cones.

b. A camera's flash leaves a "ghost" image of the flash of light on your retina until the bleaching process is completed. Bleaching is an energy-consuming process that restores retinal to its original shape, permitting the photoreceptor to be again activated by light.

c. A vitamin A deficiency would reduce the quantity of retinal the body could produce, thereby interfering with night vision.

Module 9.15 Review

a. The optic radiation is the bundle of projection fibers linking the lateral geniculate nuclei of the thalamus with the visual cortex.

b. Visual images are perceived in the visual cortex of the occipital lobes.

c. The visual pathway is as follows: photoreceptors in the retina → bipolar cell → ganglion cell → axons from population of ganglion cells converge on optic disc → optic nerve (III) → optic chiasm → optic tract → lateral geniculate nucleus of the thalamus → occipital cortex of cerebral hemisphere.

Module 9.16 Review

a. Emmetropia is the term for normal vision.

b. Two refractive surgical procedures for correcting myopia and hyperopia are photorefractive keratectomy (PRK) and laser-assisted in-situ keratomileusis (LASIK). Both PRK and LASIK permanently change the cornea by using lasers to slice the corneal epithelium and reshape it.

c. A converging lens with a convex surface would correct hyperopia.

Module 9.17 Review

a. Cranial nerves VII (facial), IX (glossopharyngeal), and X (vagus) provide taste sensations from the tongue.

b. Conductive deafness, nerve deafness, and bacterial or viral infections are common—or lead to—hearing-related disorders.

c. Vertigo is caused by anything that alters the function of the internal ear receptor complex, the vestibular branch of the vestibulocochlear nerve, or sensory nuclei and pathways in the central nervous system. Other causes include anything that sets endolymph in motion, excessive alcohol consumption, exposure to certain drugs, and fever.

Section 2 Review

1. a. posterior cavity; b. choroid; c. fovea; d. optic nerve; e. optic disc; f. retina; g. sclera; h. conjunctiva; i. ciliary body; j. iris; k. lens; l. cornea; m. suspensory ligaments

2. a. external ear; b. middle ear; c. internal ear; d. auricle; e. external acoustic meatus; f. tympanic membrane; g. auditory ossicles; h. tympanic cavity; i. petrous part of temporal bone; j. vestibulocochlear nerve (VIII); k. cochlea; l. auditory tube

3. a. lingual papillae; b. taste bud; c. cones; d. cerebral cortex; e. optic disc; f. stapes; g. olfactory bulb; h. spiral organ; i. endolymph

4. When light falls on the eye, it passes through the cornea and strikes the photoreceptors of the retina, bleaching (breaking down) many molecules of the pigment rhodopsin into retinal and opsin. After an intense exposure to light, a photoreceptor cannot respond to further stimulation until its rhodopsin molecules have been regenerated by the conversion of retinal molecules to their original shape and recombination with opsin molecules. The "ghost" image remains until the rhodopsin molecules are regenerated.

Chapter Review

1. The general senses include both somatic and visceral sensations, such as temperature, pain, touch, pressure, vibration, and proprioception. The special senses include olfaction (smell), gustation (taste), hearing, equilibrium (balance), and vision. The receptors for the special senses are confined to the head.

2. Olfactory sensations are long lasting and linked to our memories and emotions because the olfactory system has extensive limbic system connections.

3. As you turn to look at your friend, your medial rectus muscles will contract, directing your gaze more medially.

In addition, your pupils will constrict and the lenses will become more spherical.

4. The loud noises from the fireworks have transferred so much energy to the endolymph in the cochlea that the fluid continues to move for a long while. As long as the endolymph is moving, it will vibrate the tectorial membrane and stimulate the hair cells. This stimulation produces the "ringing" sensation that Mona perceives. She finds it difficult to hear normal conversation because the vibrations associated with it are not strong enough to overcome the currents already moving through the endolymph, so the pattern of vibrations is difficult to discern against the background "noise."

5. The rapid descent in the elevator causes the maculae in the saccule of the vestibule to slide upward, producing the sensation of downward vertical motion. When the elevator abruptly stops, the maculae do not. It takes a few seconds for them to come to rest in the normal position. As long as the maculae are displaced, the perception of movement will remain.

CHAPTER 10

Module 10.1 Review

a. The endocrine system is composed of endocrine glands and tissues whose primary function is to produce hormones. These chemical messengers are transported in the bloodstream to target cells in other areas of the body.

b. Hormones are classified by chemical structure. Amino acid derivatives (thyroid hormones, catecholamines, and tryptophan derivatives) are small molecules structurally related to individual amino acids. Peptide hormones (short polypeptide chains and glycoproteins) are chains of amino acids; they are synthesized as prohormones. Lipid derivatives (eicosanoids and steroid hormones) are carbon rings and side chains that are built from fatty acids or cholesterol.

c. Primary organs of the endocrine system include the hypothala-

mus, pituitary gland, thyroid gland, parathyroid glands, adrenal glands, pancreas, and pineal gland. Organs and tissues with secondary endocrine functions include the heart, thymus, digestive tract, adipose tissue, kidneys, and gonads.

Module 10.2 Review

a. Hormones either bind to receptors on the plasma membrane and activate G proteins, or pass directly through the target cell's plasma membrane and bind to receptors in the nucleus.

b. Nonsteroid hormones bind to plasma membrane receptors because they are not lipid-soluble, and cannot penetrate the plasma membrane.

c. Steroid hormones bind to receptors in the cytoplasm.

Module 10.3 Review

a. Three mechanisms of hypothalamic control over endocrine function include (1) secretion of regulatory hormones to control activity of the anterior lobe of the pituitary gland, (2) production of antidiuretic hormone and oxytocin, and (3) control of sympathetic output to the adrenal medullae.

b. A regulatory hormone is a special hormone secreted by the hypothalamus to control endocrine cells in the pituitary gland.

c. Releasing hormones stimulate the synthesis and secretion of one or more hormones from the anterior lobe of the pituitary gland, and inhibiting hormones prevent the synthesis and secretion of hormones from the anterior lobe of the pituitary gland.

Section 1 Review

1. a. catecholamines; b. thyroid hormones; c. tryptophan derivatives; d. peptide hormones; e. short polypeptides; f. glycoproteins; g. small proteins; h. lipid derivatives; i. eicosanoids; j. steroid hormones; k. transport proteins

2. a. chemical messengers to relay information and instructions between cells

b. metabolic activities of many tissues and organs bearing the same receptors simultaneously

c. both systems rely on chemical messengers that bind to target cells with specific receptors

d. releasing and inhibiting

e. second messengers released when receptor binding occurs at the plasma membrane surface

f. binding to target receptors in the cytoplasm or nucleus

g. bound to specific transport proteins in the blood plasma

h. types, amounts, or activities of important enzymes and structural proteins

Module 10.4 Review

a. The two lobes of the pituitary gland are the anterior lobe and the posterior lobe. The anterior lobe contains endocrine cells that produce and secrete hormones. The hormones secreted by the posterior lobe are produced by neurons in the hypothalamus.

b. Anterior lobe hormones and their target tissues include the following: (1) Thyroid-stimulating hormone (TSH) targets the thyroid gland; (2) adrenocorticotropic hormone (ACTH) targets the adrenal cortex; (3) follicle-stimulating hormone (FSH) and (4) luteinizing hormone (LH) target the gonads (testes in males, ovaries in females); (5) growth hormone (GH) targets skeletal muscle cells, chondrocytes, epithelial cells, connective tissues (such as adipose tissue), and liver cells; and (6) prolactin (PRL) targets mammary glands in females. Posterior lobe hormones and their target tissues include the following: (1) Oxytocin (OXT) targets smooth muscles in the uterus and mammary glands, and (2) antidiuretic hormone (ADH) targets the kidneys. The pars intermedia of the anterior lobe secretes melanocyte-stimulating hormone (MSH), which targets melanocytes.

c. In dehydration, blood osmotic pressure is increased, which would stimulate the posterior lobe of the pituitary to release more ADH.

Module 10.5 Review

a. Thyroxine (T_4), triiodothyronine (T_3), and calcitonin are hormones associated with the thyroid gland.

b. Thyroid hormones elevate rates of oxygen consumption, increase heart rate and force of contraction, increase sensitivity to sympathetic stimulation, maintain normal sensitivity of respiratory centers to changes in oxygen and carbon dioxide concentrations, stimulate red blood cell formation, stimulate activity in other endocrine tissues, and accelerate mineral turnover in bone.

c. Most of the body's reserves of thyroid hormone, thyroxine, are bound to the blood-borne proteins called thyroid-binding globulins. Because these compounds represent such a large reservoir of thyroxine, it takes several days after removal of the thyroid gland for blood levels of thyroxine to decline.

Module 10.6 Review

a. The parathyroid glands are embedded in the posterior surfaces of the thyroid gland.

b. Parathyroid hormone (PTH) raises the blood calcium level by reducing calcium deposition in bones, increasing reabsorption of calcium from the blood by the kidneys, and increasing the kidneys' production of calcitriol which enhances the absorption of Ca^{2+} by the digestive tract.

c. Increased blood calcium levels would result in increased secretion of calcitonin.

Module 10.7 Review

a. The two regions of the adrenal gland are the cortex and medulla. The cortex secretes mineralocorticoids (primarily aldosterone), glucocorticoids (mainly cortisol, hydrocortisone, and corticosterone), and androgens; the medulla secretes epinephrine and norepinephrine.

b. The kidneys are the target tissues for aldosterone.

c. One function of cortisol is to decrease the cellular use of glucose while increasing both the available glucose (by promoting the breakdown of glycogen) and the conversion of amino acids to carbohydrates. Therefore, the net result of elevated cortisol levels would be an elevation of blood glucose.

Module 10.8 Review

a. The cells of the pancreatic islets (and their hormones) are alpha cells (glucagon) and beta cells (insulin).

b. Insulin is secreted to lower blood glucose concentrations.

c. Increased levels of glucagon stimulate the conversion of glycogen to glucose in the liver, which would in turn reduce the amount of glycogen stored in the liver.

Module 10.9 Review

a. The pineal gland lies in the posterior portion of the roof of the third ventricle.

b. Melatonin secretion is influenced by day–night patterns (light–dark cycles). Increased amounts of light would inhibit the production (and release) of melatonin from the pineal gland, which receives neural input from visual pathway collaterals.

c. Melatonin inhibits reproductive functions, protects against tissue damage by free radicals, and sets circadian rhythms.

Module 10.10 Review

a. Diabetes mellitus is a condition caused by inadequate production or diminished sensitivity to insulin, with a resulting elevation of blood glucose levels.

b. Two types of diabetes mellitus are Type 1 and Type 2. Type 1 is characterized by inadequate insulin production by the pancreatic beta cells. Type 2 is characterized by insulin resistance (failure of the body to use insulin properly).

c. Three clinical problems caused by diabetes are diabetic retinopathy (the proliferation of capillaries and hemorrhaging at the retina); diabetic nephropathy (degenerative changes in the kidneys, which can lead to kidney failure); and diabetic neuropathy (abnormal blood flow to neural tissues causing a variety of problems with peripheral nerves and autonomic function).

Module 10.11 Review

a. The alarm phase, the resistance phase, and the exhaustion phase are the three phases of the stress response.

b. The resistance phase is characterized by long-term metabolic adjustments, including mobilization of remaining energy reserves, conservation of glucose for neural tissues, elevation of blood glucose concentrations, and conservation of salts and water with loss of K^+ and H^+.

c. During the exhaustion phase of the stress response there is a collapse of vital systems.

Section 2 Review

1. **a.** pineal gland; **b.** pancreatic islets; **c.** adrenal glands; **d.** pituitary gland; **e.** thyroid gland; **f.** parathyroid glands

2. **a.** prolactin (PRL); **b.** growth hormone (GH); **c.** thyroid hormones (T_3, T_4); **d.** antidiuretic hormone (ADH); **e.** glucagon; **f.** oxytocin (OXT); **g.** calcitriol; **h.** cortisol; **i.** insulin; **j.** parathyroid hormone (PTH); **k.** calcitonin; **l.** aldosterone

3. **a.** osmoreceptors; **b.** pituitary gland; **c.** FSH; **d.** ACTH; **e.** hyperglycemia; **f.** homeostasis threat; **g.** diabetes mellitus; **h.** sympathetic activation; **i.** glucocorticoids; **j.** iodide ions

4. Calcitriol increases calcium ion absorption from the digestive system, and PTH inhibits osteoblast activity and enhances calcium ion reabsorption by the kidneys.

Chapter Review

1. Structurally, the hypothalamus is a part of the diencephalon. The hypothalamus (1) secretes ADH and oxytocin that target peripheral organs; (2) controls the secretions of the adrenal medullae; and (3) releases regulatory hormones that control the pituitary gland.

2. The adrenal medulla is controlled by the sympathetic nervous system, whereas the adrenal cortex is stimulated by the release of ACTH from the anterior lobe of the pituitary gland.

3. One benefit of a portal system is that it ensures that the controlling hormones will be delivered directly to the target cells. Second, because the hormones go directly to their target cells without first passing through the general circulation, they are not diluted. The hypothalamus can control the cells of the anterior lobe of the pituitary gland with much smaller amounts of releasing and inhibiting hormones than would be necessary if the hormones had to first go through the circulatory pathway before reaching the pituitary.

4. Julie will likely have elevated blood levels of parathyroid hormone. Because her poor diet does not supply enough calcium for her developing fetus, the fetus removes large amounts of calcium from the maternal blood. The resulting lowered maternal blood calcium levels lead to elevated blood parathyroid hormone levels and increased removal of stored calcium from the maternal skeleton.

5. **a.** *Patient A:* decreased insulin production indicating diabetes mellitus; *Patient B:* underproduction of thyroid hormones

   b. *Patient A:* pancreas disorder; *Patient B:* thyroid disorder

CHAPTER 11

Module 11.1 Review

a. The hematocrit, also called the packed cell volume (PCV), is the percentage of whole blood volume contributed by formed elements.

b. Blood consists of plasma (55%) and formed elements (45%). Plasma is made up of water, plasma proteins, and other solutes. The formed elements consist of RBCs, WBCs, and platelets.

c. During an infection, antibodies (also called immunoglobulins) and white blood cells would be elevated because they are part of the body's defense mechanisms.

Module 11.2 Review

a. It is important for RBCs to have a large surface area-to-volume ratio because this allows for a faster exchange of oxygen between the interior of the cell and the surrounding plasma.

b. Hemoglobin is a protein composed of four globular subunits, each bound to a heme molecule that contains an iron ion. Hemoglobin gives red blood cells the ability to transport oxygen in the blood.

c. Oxyhemoglobin (HbO_2) is hemoglobin in combination with oxygen, which is colored bright red; deoxyhemoglobin (Hb) is hemoglobin whose iron has not bound oxygen and is colored dark red.

Module 11.3 Review

a. Surface antigens on RBCs are proteins in the plasma membrane that determine blood type. The body recognizes these surface antigens as "normal" or "self."

b. Only Type O blood can be safely transfused into a person whose blood type is O.

c. A person with Type A blood has anti-B antibodies in his plasma, so if the person received a transfusion of Type B blood, the transfused red blood cells would clump, or agglutinate, potentially blocking blood flow to various organs and tissues.

Module 11.4 Review

a. Hemolytic disease of the newborn (HDN) is a condition in which maternal antibodies attack and destroy fetal red blood cells, resulting in fetal anemia. It occurs in a sensitized Rh^- mother who carries an Rh^+ fetus.

b. RhoGAM contains antibodies to the Rh factor. When RhoGAM is injected into a pregnant Rh- mother carrying a second Rh+ fetus. The injected anti-Rh antibodies bind to Rh antigens on any Rh+ fetal blood cells that find their way into the mother's blood. This prevents her immune system from making anti-Rh antibodies against the developing fetus' red blood cells.

c. An Rh^+ mother carrying an Rh^- fetus does not require a RhoGAM injection because the Rh surface antigen is already present and her blood does not contain anti-Rh antibodies.

The developing fetus is not at risk since there are no Rh surface antigens on its red blood cells.

Module 11.5 Review

a. The five types of white blood cells are neutrophils, eosinophils, basophils, monocytes, and lymphocytes.

b. An infected cut would contain a large number of neutrophils and monocytes, white blood cells that engulf pathogens or debris in infected tissues through phagocytosis.

c. During inflammation, basophils release histamine and other chemicals that promote inflammation.

Module 11.6 Review

a. A hemocytoblast is a multipotent stem cell whose divisions produce lymphoid and myeloid stem cells, which divide to form each of the various populations of formed elements.

b. Erythropoietin (EPO) is a hormone that is released when tissue oxygen concentrations are low. It stimulates erythropoiesis (red blood cell formation) in red bone marrow.

c. Both lymphoid stem cells and myeloid stem cells originate in the red bone marrow. Lymphoid stem cells only produce lymphocytes. Myeloid stem cells give rise to all types of blood formed elements, except lymphocytes.

Module 11.7 Review

a. Hemostasis is the stopping of blood loss that consists of three phases: the vascular phase, the platelet phase, and the coagulation phase.

b. The vascular phase is a period of local blood vessel constriction, or vascular spasm, at the injury site. The platelet phase follows as platelets are activated, aggregate at the site, and adhere to the damaged surfaces. The coagulation phase occurs as factors released by platelets and endothelial cells interact with clotting factors to form a blood clot. In this reaction sequence, fibrinogen is converted to insoluble fibers of fibrin.

c. An embolus is a drifting blood clot. A thrombus is a stationary blood clot that attaches to a vessel wall and obstructs blood flow.

Module 11.8 Review

a. Venipuncture is the piercing of a vein to obtain a blood sample.

b. The two types of leukemia are myeloid leukemia and lymphoid leukemia.

c. Iron deficiency anemia results from an iron deficiency. One may be deficient in iron if the dietary intake is inadequate or the iron reserves are depleted. Pernicious anemia results from a deficiency of vitamin B_{12}, which prevents normal stem cell division in the red bone marrow.

Section 1 Review

1. **a.** plasma; **b.** water; **c.** solutes; **d.** proteins **e.** electrolytes, glucose, urea; **f.** albumins; **g.** globulins; **h.** fibrinogen; **i.** formed elements; **j.** erythrocytes; **k.** leukocytes; **l.** platelets; **m.** neutrophils; **n.** eosinophils; **o.** basophils; **p.** lymphocytes; **q.** monocytes

2. **a.** drifting blood clot; **b.** mature RBCs; **c.** liquid matrix; **d.** monocyte; **e.** transport protein; **f.** cross-reaction; **g.** lymphocytes; **h.** RBCs rupture; **i.** venipuncture; **j.** pigment complex; **k.** erythropoietin; **l.** platelets

3. During differentiation, the red blood cells of humans (and other mammals) lose most of their organelles, including nuclei and ribosomes. As a result, mature circulating RBCs cannot divide or synthesize structural proteins and enzymes for cellular repairs.

Module 11.9 Review

a. The five general classes of blood vessels are arteries, arterioles, capillaries, venules, and veins.

b. A capillary is a small blood vessel, located between an arteriole and a venule, whose thin wall permits the diffusion of gases, nutrients, and wastes between plasma and interstitial fluids.

c. The blood vessels are veins. Arteries and arterioles have a large amount of smooth muscle in a thick, well-developed tunica media.

Module 11.10 Review

a. Arteriosclerosis is any thickening and toughening of arterial walls. Atherosclerosis is a type of arteriosclerosis characterized by changes in the endothelial lining and the formation of fatty deposits (plaque) in the tunica media.

b. Risk factors for the development of atherosclerosis include elevated cholesterol levels, high blood pressure, and cigarette smoking.

c. Balloon angioplasty is a procedure for treating small, soft plaques in which a catheter with an inflatable balloon is placed in a vessel wall. Once the catheter is in place, the balloon is inflated, pressing the plaque against the vessel wall.

Module 11.11 Review

a. Precapillary sphincters adjust the flow of blood into each capillary by alternately contracting and relaxing.

b. Vasomotion is the rhythmic cycling of blood vessel diameter. It is caused by the contraction and relaxation of smooth muscles at the entrance to each capillary. This results in a pulsed, rather than steady, flow through a capillary.

c. An arteriovenous anastomosis is a direct connection between an arteriole and a venule. This structure is regulated by the sympathetic nervous system and allows blood to bypass the capillary bed and flow directly into the venous circulation.

Module 11.12 Review

a. Varicose veins are swollen veins distorted by gravity and pooling of blood due to the failure of venous valves.

b. In the arterial system, pressures are high enough to keep the blood moving in one direction (away from the heart). In the venous system, blood pressure is too low to keep the blood moving on toward the heart. Valves in veins prevent blood from flowing backward whenever the venous pressure drops.

c. Blood flow is maintained in veins by the presence of valves, which prevent backflow of the blood, and the contraction of surrounding skeletal muscles squeezes venous blood toward the heart.

Module 11.13 Review

a. The two circulatory circuits of the cardiovascular system are the pulmonary circuit and the systemic circuit.

b. The three major patterns of blood vessel organization are the following: (1) The peripheral distributions of arteries and veins on the body's left and right sides are generally identical, except near the heart, where the largest vessels connect to the atria or ventricles; (2) a single vessel may have several names as it crosses specific anatomical boundaries, making accurate anatomical descriptions possible; and (3) tissues and organs are usually serviced by several arteries and veins.

c. right ventricle → right and left pulmonary arteries → pulmonary arterioles → alveolar capillaries → pulmonary venules → pulmonary veins → left atrium

Module 11.14 Review

a. The two large veins that collect blood from the systemic circuit are the superior vena cava and the inferior vena cava.

b. The aorta is the largest artery in the body.

c. A major difference between the arterial and venous systems is that in the neck and limbs there is dual venous drainage.

Module 11.15 Review

a. The two arteries formed by the division of the brachiocephalic trunk are the right common carotid and the right subclavian.

b. A blockage of the left subclavian artery would interfere with blood flow to the left arm.

c. The vein that is bulging is the external jugular vein.

Module 11.16 Review

a. The carotid sinus contains baroreceptors.

b. Branches of the external carotid artery are the superficial temporal artery, maxillary artery, occipital artery, facial artery, and lingual artery.

c. The external jugular vein and the internal jugular vein combine with the vertebral vein and subclavian vein to form the brachiocephalic vein.

Module 11.17 Review

a. The internal jugular veins drain the dural sinuses of the brain.

b. The internal carotid artery branches into the ophthalmic artery (supplies the eyes), anterior cerebral artery (supplies the frontal and parietal lobes of the brain), and middle cerebral artery (supplies the midbrain and lateral surfaces of the cerebral hemispheres).

c. The cerebral arterial circle, also known as the circle of Willis, is a ring-shaped anastomosis that encircles the infundibulum of the pituitary gland. Its anatomical arrangement creates alternate pathways in the cerebral circulation so that if blood flow is interrupted in one area, other blood vessels can continue to supply the brain with blood.

Module 11.18 Review

a. The inferior vena cava collects most of the venous blood inferior to the diaphragm.

b. Major tributaries of the inferior vena cava are the lumbar veins, gonadal veins, hepatic veins, renal veins, adrenal veins, and phrenic veins.

c. Rupture of the celiac trunk would most directly affect the stomach, spleen, pancreas, liver, gallbladder, and proximal portion of the small intestine.

Module 11.19 Review

a. The unpaired branches of the abdominal aorta that supply blood to the visceral organs are the celiac trunk, superior mesenteric artery, and inferior mesenteric artery.

b. The three veins that merge to form the hepatic portal vein are the superior mesenteric vein, inferior mesenteric vein, and splenic vein.

c. The celiac trunk divides into the common hepatic artery, the left gastric artery, and the splenic artery.

Module 11.20 Review

a. The common iliac artery branches to form the internal iliac artery and the external iliac artery.

b. The plantar venous arch delivers blood to the anterior tibial vein, posterior tibial vein, and the fibular veins.

c. A blockage of the popliteal vein would interfere with blood flow in the tibial and fibular veins (which form the popliteal vein) and the small saphenous vein (which joins the popliteal vein).

Section 2 Review Answers

1. **a.** common carotid; **b.** subclavian; **c.** brachiocephalic trunk; **d.** brachial; **e.** radial; **f.** popliteal; **g.** fibular; **h.** aortic arch of aorta; **i.** celiac trunk; **j.** renal; **k.** common iliac; **l.** external iliac; **m.** femoral; **n.** anterior tibial

2. **a.** vertebral; **b.** internal jugular; **c.** brachiocephalic; **d.** axillary; **e.** cephalic; **f.** median antebrachial; **g.** ulnar; **h.** great saphenous; **i.** fibular; **j.** superior vena cava; **k.** inferior vena cava; **l.** internal iliac; **m.** femoral; **n.** posterior tibial

Chapter Review

1. Artery walls are generally thicker and contain more smooth muscle and elastic fibers, enabling them to resist and adjust to the pressure generated by the heart. Venous walls are thinner; the pressure in veins is less than that in arteries. Arteries constrict more than veins do when not expanded by blood pressure, due to a greater degree of elastic tissue. Finally, the endothelial lining of an artery has a rippled appearance because it cannot contract and so forms folds. The lining of a vein looks like a typical endothelial layer and is smooth.

2. During running, the smooth muscle in arterioles relaxes, causing vasodilation and supplying more oxygen and nutrients to the active muscles.

3. Blood stabilizes and maintains body temperature by absorbing and redistributing the heat produced by active skeletal muscles.

4. A mature RBC does not contain ribosomes or a nucleus. As a result, RBCs cannot divide or synthesize structural proteins or enzymes necessary for cell maintenance.

5. Blood flow to the brain is relatively constant because a ring-shaped anastomosis within the cranium ensures that interruption in the flow in any of the four arteries (internal carotid and vertebral arteries) supplying the brain will not compromise its blood supply.

CHAPTER 12

Module 12.1 Review

a. From superficial to deep, the layers of the heart wall are the epicardium, myocardium, and endocardium.

b. The serous membrane lining the pericardial cavity is made up of the epicardium (also called the visceral pericardium), which attaches directly to the myocardium, and the parietal pericardium, which forms the outer wall of the pericardial cavity.

c. Cardiac tissue is highly metabolically active tissue that is totally dependent on aerobic metabolism, so mitochondria are necessary for aerobic respiration. This tissue also has a high demand for oxygen and nutrients, so the capillaries provide both.

Module 12.2 Review

a. The mediastinum is the region between the two pleural (lung) cavities that also contains the heart and great vessels (large arteries and veins linked to the heart), thymus, esophagus, and trachea.

b. The heart is located behind the sternum and within the mediastinum, surrounded by the pericardial sac, and positioned just superior to the diaphragm at the apex.

c. The anterior interventricular sulcus is found on the anterior surface and marks the boundary between the left and right ventricles. The coronary sulcus is a deep groove that marks the border between the atria and the ventricles. The shallower posterior interventricular sulcus marks the boundary between the left and right ventricles on the posterior surface.

Module 12.3 Review

a. The coronary arteries are the left coronary artery, anterior interventricular artery, right coronary artery, marginal arteries, circumflex artery, and posterior interventricular artery. The coronary veins are the great cardiac vein, anterior cardiac veins, coronary sinus, posterior cardiac vein, middle cardiac vein, and small cardiac vein.

b. During elastic rebound, some aortic blood is forced back toward the left ventricle, driving additional blood into the coronary arteries, while also forcing blood forward into the systemic circuit.

c. The great cardiac vein drains blood from the myocardial capillaries.

Module 12.4 Review

a. Damage to the semilunar valve on the right side of the heart would affect blood flow to the pulmonary trunk.

b. Contraction of the papillary muscles (just before the rest of the myocardium contracts) pulls on the chordae tendineae, which prevents the AV valves from opening back into the atria.

c. The left ventricle is more muscular than the right ventricle because the left ventricle must generate enough force to propel blood throughout the systemic circuit, except to the lungs, whereas the right ventricle must generate only enough force to propel blood a short distance to the lungs.

Module 12.5 Review

a. Coronary ischemia occurs when there is not an ample supply of blood through the coronary arteries, leading to tissue damage.

b. Stents are artificial mesh "tubes" that prop open the natural blood vessel, creating a conduit to restore blood flow. Without adequate blood flow to the cardiac muscle, the tissue would die.

c. The factors that cause the characteristic heart sounds that can be heard with a stethoscope are the closure of heart valves, the rushing of blood through the heart, and heart muscle contraction.

Section 1 Review

1. a. aortic arch; **b.** superior vena cava; **c.** right pulmonary arteries; **d.** ascending aorta; **e.** fossa ovalis; **f.** opening of coronary sinus; **g.** right atrium; **h.** pectinate muscles; **i.** tricuspid valve cusp; **j.** chordae tendineae; **k.** papillary muscle; **l.** right ventricle; **m.** inferior vena cava; **n.** pulmonary trunk; **o.** pulmonary valve; **p.** left pulmonary arteries; **q.** left pulmonary veins; **r.** interatrial septum; **s.** aortic valve; **t.** bicuspid valve cusp; **u.** left ventricle; **v.** interventricular septum; **w.** trabeculae carneae; **x.** moderator band

2. Deoxygenated blood flow: right atrium → right atrioventricular valve (tricuspid valve) → right ventricle → pulmonary semilunar valve. Oxygenated blood flow: left atrium → left atrioventricular valve (bicuspid valve) → left ventricle → aortic semilunar valve.

3. a. aorta; **b.** pericardial cavity; **c.** myocardium; **d.** coronary sinus; **e.** fossa ovalis; **f.** tricuspid valve; **g.** elastic recoil; **h.** intercalated discs; **i.** coronary arteries; **j.** endocardium; **k.** apex; **l.** aortic valve

Module 12.6 Review

a. The alternate term for contraction is systole, and the other term for relaxation is diastole.

b. The phases of the cardiac cycle are atrial systole (atria contract and the ventricles fill with blood), atrial diastole (the atria relax), ventricular systole (ventricles contract pushing the AV valves closed; as pressure in the ventricles rises, the semilunar

valves open and blood is forced out of the ventricles), and ventricular diastole (semilunar valves close and the ventricles fill about 70 percent full with blood).

c. No. When pressure in the left ventricle first rises, the heart is contracting but blood is not leaving the heart. During this initial phase of contraction, both the AV valves and the semilunar valves are closed. The increase in pressure is the result of increased tension as the cardiac muscle contracts. When the pressure in the ventricle exceeds the pressure in the aorta, the aortic semilunar valves are forced open, and blood is rapidly ejected from the ventricle.

Module 12.7 Review

a. The cardioacceleratory centers in the medulla oblongata activate sympathetic neurons to increase the heart rate. Also in the medulla oblongata, the cardioinhibitory centers control the parasympathetic neurons that slow the heart rate.

b. Bradycardia is a heart rate below 60 beats per minute. Tachycardia is a heart rate above 100 beats per minute.

c. Five important features of an ECG include the P wave (atrial depolarization), the QRS complex (ventricular depolarization), the T wave (ventricular repolarization), the P–R interval (time between the beginning of the P wave and the beginning of the next QRS complex), and the Q–T interval (time between the Q wave to the end of the T wave).

Module 12.8 Review

a. The amount of blood ejected by the left ventricle each minute is the cardiac output.

b. An increase in the venous return would fill the heart with blood, stretching the heart muscle. According to Starling's law of the heart, the more the heart muscle is stretched, the more forcefully it will contract. The more forceful the contraction, the more blood the heart will eject with each beat (stroke volume). So, increased venous

return would increase the stroke volume.

c. The heart pumps in proportion to the amount of blood that enters. A heart that beats too rapidly does not have sufficient time to fill completely between beats. So, when the heart beats too fast, very little blood leaves the ventricles and enters the circulation, so tissues suffer damage from inadequate blood supply.

Section 2 Review

1. a. systole, diastole; **b.** partially; **c.** left AV valve closes; **d.** less than; **e.** aortic valve is forced open; **f.** closed; **g.** closed

2. a. SA node (sinoatrial node). The SA node is the pacemaker of the heart. It generates the action potential for each heartbeat. **b.** Internodal pathways. The internodal pathways carry the action potential to atrial muscle cells and the AV node. **c.** AV node (atrioventricular node). The AV node receives the action potential carried by the internodal pathways. The AV node contains pacemaker cells that generate action potentials at a slower rate than the SA node. The AV node does not ordinarily affect the heart rate. **d.** AV bundle and bundle branches. The AV bundle receives the action potential next. It divides into right and left bundle branches that extend to the apex of the heart and then fan out into both ventricular walls. **e.** Purkinje fibers. The Purkinje fibers receive the action potential from the bundle branches. These fibers carry the action potential that will depolarize ventricular myocardial tissue and trigger ventricular systole.

3. a. contractility; **b.** automaticity; **c.** P wave; **d.** stroke volume; **e.** QRS complex; **f.** parasympathetic neurons; **g.** bradycardia; **h.** sympathetic neurons; **i.** Purkinje fibers; **j.** tachycardia; **k.** cardiac output; **l.** filling time

Module 12.9 Review

a. Blood flow is the volume of blood flowing per unit of time through a vessel or group of

vessels. Blood flow is directly proportional to blood pressure and inversely proportional to peripheral resistance.

b. In a healthy individual, blood pressure is higher at the aorta than at the inferior vena cava. Blood, like other fluids, moves along a pressure gradient from areas of high pressure to areas of low pressure. If the pressure were higher in the inferior vena cava than in the aorta, the blood would flow backward.

c. The water that is forced out of the capillary and not reabsorbed flows around the peripheral tissues and eventually returns to the venous system through vessels of the lymphatic system.

Module 12.10 Review

a. Tissue perfusion is blood flow through the tissues, which is meant to deliver adequate oxygen and nutrients to the body.

b. Autoregulation involves local factors changing the pattern of blood flow within capillary beds in response to chemical changes in interstitial fluids.

c. Baroreceptors involved in the homeostatic regulation of blood pressure are found in the carotid sinuses, aortic sinuses, and right atrium. They are stimulated by rising blood pressure and, in turn, stimulate the cardioinhibitory centers while inhibiting the cardioacceleratory centers and vasomotor center. The resulting decrease in cardiac output and vasodilation produces a drop in blood pressure. The baroreceptors are inhibited by falling blood pressure. Their effects are reversed from above, and an increased cardiac output and vasoconstriction increase blood pressure.

Module 12.11 Review

a. Epinephrine and norepinephrine from the adrenal medullae provide short-term regulation of falling blood pressure and blood volume.

b. Antidiuretic hormone (ADH), aldosterone, and erythropoietin (EPO) are hormones involved in the long-term regulation of blood pressure and blood volume.

c. Vasoconstriction of the renal artery would decrease both blood flow and blood pressure at the kidney. In response, the kidney would increase the amount of renin it releases, which in turn would lead to an increase in the level of angiotensin II. The angiotensin II would bring about increased blood pressure and increased blood volume partially through the increased secretion of ADH and aldosterone.

Module 12.12 Review

a. The respiratory pump is a mechanism by which a reduction of pressure in the thoracic cavity during inhalation assists venous return to the heart.

b. During exercise, cardiac output increases, and blood flow to skeletal muscles increases at the expense of blood flow to less essential organs.

c. Blood pressure increases during exercise because cardiac output increases. During exercise, blood flow to "nonessential" organs, such as visceral organs, decreases and is shunted to skeletal muscles to prevent systemic blood pressure from dropping too drastically.

Module 12.13 Review

a. Cardiac arrhythmias are abnormal patterns of cardiac electrical activity.

b. Ventricular fibrillation, responsible for cardiac arrest, is fatal because the heart merely quivers, and there is no effective heartbeat to pump blood systemically.

c. Paroxysmal atrial tachycardia (PAT) is characterized by premature atrial contraction that triggers a flurry of atrial activity.

Section 3 Review

1. The three primary factors influencing blood pressure and blood flow are cardiac output, blood volume, and peripheral resistance.

2. a. baroreceptors; **b.** vasodilation; **c.** venous return; **d.** autoregulation; **e.** local vasodilators; **f.** ACE; **g.** vasoconstriction; **h.** pulse pressure;

i. cardiac arrest; **j.** capillary hydrostatic pressure; **k.** atrial natriuretic peptide; **l.** medulla oblongata

3. a. arterioles; **b.** autonomic nervous system; **c.** increasing peripheral vasoconstriction; **d.** increased vasodilation, increased venous return, increased cardiac output; **e.** brain; **f.** nervous and endocrine

4. Arterial pressure is higher than venous pressure because it must overcome the peripheral resistance of the system of arteries and capillaries. Venous pressure is lower than arterial pressure because as blood moves to the heart, the venous vessels increase in diameter, which lowers the resistance to blood flow.

Chapter Review

1. The right atrium receives blood from the systemic circuit and passes it to the right ventricle, which pumps it into the pulmonary circuit. The left atrium collects blood returning from the lungs and passes it to the left ventricle, which ejects it into the systemic circuit.

2. Cardiac output cannot increase indefinitely because available filling time becomes shorter as the heart rate increases.

3. In this patient, only one of every two action potentials generated by the SA node is reaching the ventricles; thus there are two P waves but only one QRS complex and T wave. This condition frequently results either from damage to the internodal pathways of the atria or problems with the AV node.

4. Using CO = HR × SV, cardiac output for person A is 4500 mL, and for person B 8550 mL. According to the Frank-Starling principle, in a normal heart CO is directly proportional to venous return, so person B has the greater venous return. Ventricular filling increases as HR decreases, so person A has the longer ventricular filling time.

5. Given that the force of cardiac contraction is directly proportional to stroke volume, you would expect a reduced stroke volume with the administration of this drug.

CHAPTER 13

Module 13.1 Review

a. The function of lymphatic vessels, or lymphatics, is to transport lymph from peripheral tissues to the venous system.

b. Lymph is the fluid carried by lymphatic vessels. It is formed as interstitial fluid flows into lymphatic capillaries.

c. The regions of overlapping endothelial cells of a lymphatic capillary act as one-way valves. They permit the entry of fluids and solutes, but prevent their return to the intercellular spaces.

Module 13.2 Review

a. The lymphatic trunks empty into the thoracic duct and the right lymphatic duct.

b. The right lymphatic duct collects lymph from the right side of the body superior to the diaphragm. The thoracic duct collects lymph from the body inferior to the diaphragm and from the left side of the body superior to the diaphragm.

c. Lymphedema, an accumulation of interstitial fluid in a limb, results from blocked lymphatic drainage. If the condition does not resolve, connective tissues lose their elasticity and the swelling becomes permanent.

Module 13.3 Review

a. The three main classes of lymphocytes are T cells, B cells, and natural killer (NK) cells.

b. B cells are responsible for antibody-mediated immunity.

c. The red bone marrow, thymus, and peripheral lymphoid tissues are involved in lymphopoiesis.

Module 13.4 Review

a. Mucosa-associated lymphoid tissue (MALT) is a collection of lymphoid tissue that protects epithelial linings of the digestive, respiratory, urinary, and reproductive tracts.

b. Tonsils are large lymphoid nodules in the walls of the pharynx and include left and right palatine tonsils, a single pharyngeal tonsil (sometimes called the

adenoid), and a pair of lingual tonsils.

c. 1) The spleen removes abnormal red blood cells from the circulation (by phagocytosis), 2) stores iron recycled from red blood cells, and 3) initiates immune responses by B cells and T cells.

Section 1 Review

1. a. tonsil; **b.** cervical lymph nodes; **c.** right lymphatic duct; **d.** thymus; **e.** cisterna chyli; **f.** appendix; **g.** inguinal lymph nodes; **h.** axillary lymph nodes; **i.** thoracic duct; **j.** spleen; **k.** mucosa-associated lymphoid tissue (MALT)

2. a. lymphatic capillaries; **b.** lymphedema; **c.** right subclavian vein; **d.** thoracic duct; **e.** lymphoid organs; **f.** lymph nodes; **g.** efferent lymphatics; **h.** spleen; **i.** antigens; **j.** lymphopoiesis; **k.** cytotoxic T cells; **l.** tonsils; **m.** B cells; **n.** afferent lymphatics

Module 13.5 Review

a. The integumentary system is a physical barrier and serves as the first line of defense against pathogens and toxins entering body tissues. Skin secretions flush the surface, hair protects against mechanical abrasion, and the multiple layers of the skin create an interlocking barrier.

b. Neutrophils, eosinophils, monocytes, and macrophages are the different types of body phagocytes. There are two types of macrophages derived from monocytes: fixed and free. Fixed macrophages do not move and are found scattered among connective tissues. Free macrophages move and reach injury sites by migrating through adjacent tissues and traveling in the bloodstream.

c. Chemotaxis is movement of cells in response to chemicals, whereby the cells are attracted to or repelled by the chemicals.

Module 13.6 Review

a. Mast cell activation results in the release of histamine and heparin. Their presence stimulates the inflammatory response at the damaged area and also attract phagocytes.

b. The body's nonspecific defenses include physical barriers, phagocytes, immune surveillance, interferons, the complement system, inflammation, and fever.

c. A rise in the level of interferons suggests a viral infection.

Section 2 Review

1. **a.** physical barriers; **b.** phagocytes; **c.** immune surveillance; **d.** interferons; **e.** complement system; **f.** inflammation; **g.** fever

2. **a.** all of these; **b.** immune surveillance; **c.** the complement system; **d.** interferons; **e.** pyrogens; **f.** phagocytes

3. High body temperatures accelerate the body's metabolic processes, which may help to mobilize tissue defenses and speed the repair process. Additionally, the high body temperatures of a fever may inhibit some viruses and bacteria or speed their reproductive rates so that the disease runs its course more quickly.

Module 13.7 Review

a. Antigen presentation occurs when an antigen or antigenic fragment appears in the plasma membrane of an antigen-presenting cell (typically that of a phagocyte). T cells sensitive to this antigen are activated if they contact the membrane of the antigen-presenting cell.

b. Cytotoxic T cells and NK cells can be activated by direct contact with virus-infected cells.

c. Activated B cells called plasma cells produce antibodies.

Module 13.8 Review

a. The antigenic determinant site is the part of an antigen molecule to which an antibody attaches itself.

b. An antibody molecule consists of two parallel pairs of polypeptide chains: one pair of long heavy chains and one pair of short light chains. Each chain contains both constant segments and variable segments.

c. The five classes of immunoglobulins (Igs) are (1) IgG, responsible for resistance against many viruses, bacteria, and bacterial

toxins; (2) IgE, which releases chemicals that accelerate local inflammation; (3) IgD, located on the surfaces of B cells where it binds to antigens; (4) IgM, the first type of antibody secreted after an antigen arrives; and (5) IgA, which is found in glandular secretions and attacks pathogens before they enter tissues.

Module 13.9 Review

a. Mechanisms used by antigen-antibody complexes to destroy target antigens include neutralization, precipitation and agglutination, activation of complement, attraction of phagocytes, opsonization, stimulation of inflammation, and prevention of pathogen adhesion.

b. Opsonization is an effect of coating an object with antibodies leading to the attraction of phagocytes and enhanced phagocytosis.

c. Basophils and mast cells are involved in inflammation.

Module 13.10 Review

a. An allergy is an inappropriate or excessive immune response to an allergen, which is an antigen that triggers an allergic reaction.

b. Anaphylaxis is an allergic response in which a circulating allergen causes mast cells throughout the body to secrete histamine.

c. Histamines and other chemicals that cause pain and inflammation are released in response to massive stimulation of mast cells and basophils.

Module 13.11 Review

a. Autoimmune disorders are diseases that result from autoantibodies (self antibodies) produced against substances (self-antigens) naturally present in the body.

b. Immunosuppression is the partial or complete reduction of the immune response of an individual. It is also induced to help the survival of individuals after an organ transplant.

c. The increased incidence of cancer in the elderly may result from a decline in immune surveillance causing inadequate elimination of tumor cells.

Section 3 Review

1. **a.** antibody; **b.** graft rejection; **c.** acquired immunity; **d.** passive immunity; **e.** helper T cells; **f.** opsonization; **g.** agglutination; **h.** B cells; **i.** complete antigen; **j.** IgM; **k.** neutralization; **l.** IgG; **m.** anaphylaxis

2. **a.** viruses; **b.** macrophages; **c.** natural killer (NK) cells; **d.** helper T cells; **e.** B cells; **f.** antibodies; **g.** cytotoxic T cells; **h.** suppressor T cells; **i.** memory T cells and B cells

Chapter Review

1. Specificity: Each immune response is triggered by a specific antigen and defends against only that antigen. Versatility: The immune system can differentiate among tens of thousands of antigens it may encounter during an individual's normal lifetime. Immunologic memory: The immune response following a second exposure to a given antigen is stronger and lasts longer than the first exposure. Tolerance: Some antigens, such as those on an individual's own normal cells, do not elicit an immune response.

2. An antigen-antibody complex can eliminate the antigen by neutralization, agglutination, and precipitation; activation of complement; attraction of phagocytes; stimulation of inflammation; or prevention of bacterial or viral adhesion.

3. Examination of regional lymph nodes for the presence of cancer cells can help the physician determine if the cancer cells have remained localized in the lungs and were discovered at a relatively early stage, or have already spread (or metastasized) through the lymphatic system to establish tumors in other sites of the body. Both the stage of the disease and the sites involved help the physician decide about treatment options.

4. The presence of an elevated level of IgM antibodies, but very few IgG antibodies, suggests that Tony is in the early stages of a primary response to the measles virus, so he appears

to have contracted the disease. The subsequent levels of IgG antibodies will play a crucial role in the eventual control of the disease.

5. Yes, the crime lab could determine whether the sample is blood plasma, which contains IgM, IgG, IgD, and IgE, or semen, which contains only IgA.

CHAPTER 14

Module 14.1 Review

a. The cilia of the respiratory mucosa move mucus-trapped particles from the incoming air to the pharynx, where they will be swallowed and exposed to acids and enzymes in the stomach. This continuous process protects and cleans respiratory surfaces.

b. Cystic fibrosis is a lethal, inherited disease that results from the production of dense mucus that restricts respiratory passages and accumulates in the lungs. Massive chronic bacterial infection of the lungs can inhibit breathing, leading to death.

c. The blood flow in the nasal cavity delivers body heat to the area, so inhaled air is warmed before it leaves the nasal cavity. The heat also evaporates moisture from the epithelium to humidify the incoming air.

Module 14.2 Review

a. The esophagus is posterior to the trachea. The C-shaped tracheal cartilages, which are not complete posteriorly, allow room for the esophagus to expand when food or liquids are swallowed.

b. Objects are more likely to be lodged in the right primary bronchus because it is slightly wider and more vertical than the left primary bronchus.

c. The left lung has a cardiac notch to allow room for the heart.

Module 14.3 Review

a. The squamous epithelial cells of the alveolar walls are the gas exchange sites within the lungs, a second type of epithelial cell secretes surfactant, and alveolar macrophages protect the alveolar epithelium from particulate matter and infection.

b. The short distance across the respiratory membrane enhances the rate of diffusion of oxygen and carbon dioxide. It is composed of the alveolar epithelium, the capillary endothelium, and their fused basement membranes.

c. The alveoli would collapse.

Section 1 Review

1. a. nasal cavity; **b.** hard palate; **c.** pharynx; **d.** glottis; **e.** trachea; **f.** right lung; **g.** external nares; **h.** larynx; **i.** primary bronchus; **j.** bronchioles; **k.** alveolus

2. a. surfactant; **b.** trachea; **c.** simple squamous epithelial cells; **d.** bronchodilation; **e.** left bronchus; **f.** bronchoconstriction; **g.** right bronchus; **h.** respiratory membrane; **i.** soft palate; **j.** pharynx; **k.** larynx; **l.** cystic fibrosis; **m.** respiratory mucosa; **n.** internal nares

Module 14.4 Review

a. At a constant temperature, the pressure of a gas is inversely proportional to its volume—as volume increases, pressure drops, and as volume decreases, pressure rises.

b. The movements of the diaphragm and ribs affect the volume of the lungs.

c. The intrapulmonary pressure (the pressure inside the respiratory tract) and the atmospheric pressure (the pressure outside the respiratory tract) determine the direction of airflow. Air moves from the area with the higher pressure to the area with the lower pressure.

Module 14.5 Review

a. The measurable pulmonary volumes are the resting tidal volume, expiratory reserve volume (ERV), residual volume, and inspiratory reserve volume (IRV).

b. The primary inspiratory muscles are the diaphragm and the external intercostal muscles.

c. The accessory respiratory muscles become active whenever the primary respiratory muscles are unable to move enough air to meet the oxygen demands of tissues.

Module 14.6 Review

a. Driven by differences in concentration gradients, oxygen enters the blood at the lungs and leaves it in peripheral tissues; differences in concentration gradients also cause carbon dioxide to diffuse into the blood at the tissues and into the alveoli at the lungs.

b. Oxygen binds to hemoglobin in the red blood cells.

c. Carbon dioxide is transported in the bloodstream as bicarbonate ions, bound to hemoglobin, or dissolved in the plasma.

Module 14.7 Review

a. The respiratory centers that adjust the pace of respiration are located in the pons.

b. The medulla oblongata contains two groups of neurons that are involved in setting the pace of respiration. The dorsal respiratory group (DRG) is active in both quiet and forced breathing. The ventral respiratory (VRG) group is active only in forced breathing.

c. Hyperventilation causes hypocapnia, an abnormally low concentration of carbon dioxide in arterial blood. Hypoventilation causes hypercapnia, an abnormally high concentration of carbon dioxide in arterial blood.

Module 14.8 Review

a. Chronic obstructive pulmonary disease (COPD) is a general term for three progressive airway disorders—asthma, chronic bronchitis, and emphysema—that restrict airflow and ventilation.

b. Both cigarette smoking and exposure to secondhand smoke are important risk factors for developing lung cancer.

c. Aging results in deterioration of elastic tissue, decreased vital capacity, and some degree of emphysema.

Section 2 Review

1. a. inspiratory reserve volume (IRV): the amount of air that can be taken in above the resting tidal volume; **b.** resting tidal volume (V_T): the amount of air inhaled or exhaled during a single respiratory cycle while resting; **c.** expiratory reserve volume (ERV): the amount of air that can be expelled after a completely normal, quiet respiratory cycle; **d.** total lung capacity: the total volume of the lungs; **e.** vital capacity: the maximum amount of air that can be moved into or out of the lungs in a single respiratory cycle; **f.** residual volume: the amount of air remaining in the lungs after a maximal exhalation.

2. a. chloride shift; **b.** anoxia; **c.** DRG and VRG centers; **d.** internal intercostal muscles; **e.** bicarbonate ion; **f.** emphysema; **g.** external intercostal muscles; **h.** hemoglobin; **i.** hypocapnia; **j.** dead space of the lungs; **k.** centers in the pons; **l.** hypercapnia; **m.** atelectasis; **n.** respiratory cycle.

3. External respiration includes all the processes involved in the exchange of oxygen and carbon dioxide between the body's tissues and the external environment. Those processes include pulmonary ventilation, gas diffusion, and gas transport. Pulmonary ventilation, or breathing, is a process of external respiration that involves the physical movement of air into and out of the lungs. Internal respiration is the absorption of oxygen and the release of carbon dioxide by tissue cells.

Chapter Review

1. The air you were breathing while sleeping was so dry that it absorbed more than the normal amount of moisture as it passed through the nasal cavity. The nasal epithelia continued to secrete mucus, but the loss of moisture made the mucus quite viscous, so the cilia had difficulty moving it, causing nasal congestion. After taking a shower and drinking some juice, more moisture was transferred to the mucus, loosening it and making it easier to move—and easing the congestion.

2. Because the rib penetrated the chest wall, the thoracic cavity was damaged, as well as the inner, parietal pleura. Atmospheric air will then enter the pleural cavity (pneumothorax). This air breaks the fluid bond between the visceral and parietal pleural membranes. Breaking this fluid bond may lead to the collapse of the lung due to the natural elasticity of its walls. The resulting collapsed lung is called atelectasis.

3. As a result of emphysema, the larger air spaces and lack of elasticity will reduce the efficiency of both gas exchange at the pulmonary capillaries and pulmonary ventilation.

CHAPTER 15

Module 15.1 Review

a. The four layers of the digestive tract from superficial to deep are the mucosa (adjacent to the lumen), submucosa, muscularis externa, and serosa.

b. The mesenteries—sheets consisting of two layers of serous membrane separated by loose connective tissue—support and stabilize the organs in the abdominal cavity and provide a route for associated blood vessels, nerves, and lymphatic vessels.

c. The submucosal plexus is a nerve network that contains sensory neurons, parasympathetic ganglionic neurons, and sympathetic postganglionic fibers that innervate the mucosa and submucosa. The myenteric plexus is a network of parasympathetic neurons, interneurons, and sympathetic postganglionic fibers sandwiched between the circular and longitudinal muscle layers.

Module 15.2 Review

a. Smooth muscle fibers in the circular and longitudinal layers of the muscularis externa lie parallel to each other. In a longitudinal section of the digestive tract, the fibers of the superficial circular layer appear as little round balls, whereas the fibers of the deeper longitudinal layer are spindle-shaped.

b. A bolus is a moist, compact mass of food that enters the digestive tract.

c. Peristalsis is a wave of muscle contractions that moves a bolus along the digestive tract. Segmentation refers to a churning action that mixes intestinal contents with digestive fluids. Peristalsis is more efficient in propelling intestinal contents along the digestive tract.

Section 1 Review

1. **a.** mesenteric artery and vein; **b.** mesentery; **c.** circular folds; **d.** mucosa; **e.** submucosa; **f.** muscularis externa; **g.** serosa
2. **a.** esophagus; **b.** muscularis mucosae; **c.** lamina propria; **d.** circular folds; **e.** peristalsis; **f.** submucosa; **g.** myenteric plexus; **h.** muscularis externa; **i.** liver; **j.** segmentation; **k.** bolus; **l.** smooth; **m.** submucosal plexus; **n.** sphincter
3. A decrease in muscle tone in the digestive tract would cause a general decrease in motility and a weakening of peristaltic waves. This could result in constipation because undigested material would remain in the digestive tract longer, allowing more time for water reabsorption.

Module 15.3 Review

a. The hard palate forms the roof of the oral cavity. Posterior to the hard palate is the soft palate. The hard palate is supported by bone and the soft palate is composed of muscle with no bony support.

b. The oral cavity is lined by stratified squamous epithelium, which provides protection against friction and abrasion.

c. Lingual lipase begins digesting lipids in the oral cavity.

Module 15.4 Review

a. The four types of teeth are incisors, cuspids (canines), bicuspids (premolars), and molars. A typical tooth has a crown, neck, and root.

b. The third molars are sometimes called wisdom teeth.

c. The primary dentition is composed of deciduous teeth (primary teeth, milk teeth, or baby teeth), which are the temporary, first teeth to appear in children; there are usually 20 teeth comprising the primary dentition. The secondary dentition is composed of permanent teeth that appear subsequent to

the primary dentition; there are usually 32 teeth comprising the secondary dentition.

Module 15.5 Review

a. The pharynx is a membrane-lined cavity posterior to the nose and mouth that receives food and liquids and passes them to the esophagus as part of the swallowing process.

b. The esophagus is the structure connecting the pharynx to the stomach.

c. During the buccal phase, food is formed into a bolus. During the pharyngeal phase, the bolus contacts the palatal arches and the bolus moves into the esophagus. During the esophageal phase, swallowing begins as pharyngeal muscles contract, and the bolus is moved toward the stomach via peristaltic waves.

Module 15.6 Review

a. The four regions of the stomach are the fundus, cardia, body, and pylorus.

b. The inner lining of the stomach contains rugae, mucosal folds in the lining of the empty stomach that disappear as gastric distension occurs.

c. The three layers of the muscularis externa, whose muscle fibers are oriented in different directions—longitudinal, circular, and oblique—allow for the mixing and churning actions necessary for chyme formation.

Module 15.7 Review

a. Parietal cells secrete intrinsic factor and hydrochloric acid (HCl).

b. The alkaline mucous layer protects epithelial cells against the acid and enzymes in the gastric lumen.

c. Pepsinogen is an inactive proenzyme and pepsin is an active protein-digesting (proteolytic) enzyme. Hydrochloric acid in the stomach converts pepsinogen to pepsin.

Module 15.8 Review

a. The layers of the small intestine from superficial to deep are the mucosa, submucosa, muscularis externa, and serosa.

b. The intestinal mucosa bears transverse folds called circular folds that have small projections called intestinal villi. These folds and projections increase the surface area for absorption. Each villus contains a terminal lymphatic capillary called a lacteal. In between the bases of the villi are intestinal glands lined by mucous and stem cells.

c. Lacteals are special lymphatic capillaries of the small intestine that transport absorbed fatty acids that are too large to enter blood capillaries directly.

Module 15.9 Review

a. The three regions of the small intestine are the duodenum, jejunum, and ileum.

b. The proximal portion of the duodenum is found within the epigastric region.

c. The primary function of the duodenum is to receive chyme from the stomach and digestive secretions from the pancreas and liver.

Module 15.10 Review

a. Five major hormones involved with digestion are gastrin, secretin, gastric inhibitory peptide (GIP), cholecystokinin (CCK), and vasoactive intestinal peptide (VIP). Gastrin stimulates the stomach to produce acids and enzymes and increases movements of the stomach; Secretin stimulates the secretion of bile and buffers by the pancreas; GIP inhibits gastric activity and stimulates duodenum gland secretions, lipid synthesis by adipose tissue and glucose use by skeletal muscles; CCK speeds the production and secretion of all types of digestive enzymes; and, VIP stimulates secretions by intestinal glands, dilates regional capillaries, and inhibits acid production in the stomach.

b. A high-fat meal would raise the cholecystokinin (CCK) level in the blood.

c. The hormone secretin stimulates the pancreas to release a fluid high in buffers to neutralize the acids in chyme that enters the duodenum from the stomach. If the small intestine did not secrete

secretin, the pH of the intestinal contents would be lower than normal.

Module 15.11 Review

a. The three segments of the large intestine are the cecum, colon, and rectum. The four regions of the colon are the ascending colon, transverse colon, descending colon, and the sigmoid colon.

b. The main functions of the large intestine are to (1) reabsorb water and compact materials into feces, (2) absorb vitamins produced by bacteria, and (3) store fecal material prior to defecation.

c. Mass movements are powerful peristaltic contractions that push materials along the distal portion of the large intestine. Mass movements occur a few times daily in response to distension of the stomach and duodenum.

Module 15.12 Review

a. The enterogastric reflex and the defecation reflex are triggered by stretch receptors. The enterogastric reflex inhibits gastrin production and gastric contractions, thereby controlling the rate of chyme entering the duodenum. The defecation reflex initiates the ejection of feces from the rectum.

b. GIP stimulates the release of insulin from the pancreas.

c. Feces is composed of water (75%), bacteria (5%), and a mixture of indigestible materials, inorganic matter, and epithelial cell remains (20%).

Section 2 Review

1. **a.** crown; **b.** neck; **c.** root; **d.** enamel; **e.** dentin; **f.** pulp cavity; **g.** gingiva; **h.** cementum; **i.** periodontal ligament; **j.** root canal; **k.** bone of alveolus
2. **a.** material in jejunum; **b.** gastrin; **c.** GIP; **d.** secretin and CCK; **e.** VIP; **f.** inhibits; **g.** acid production; **h.** insulin; **i.** bile; **j.** intestinal capillaries; **k.** gallbladder; **l.** nutrient utilization by tissues
3. Both parietal cells and chief cells are secretory cells found in the gastric glands of the wall of the stomach. Parietal cells

secrete intrinsic factor and hydrochloric acid. Chief cells secrete pepsinogen, an inactive proenzyme.

a. The three pairs of salivary glands are the sublingual, submandibular, and parotid. The sublingual glands primarily secrete mucus, the submandibular glands secrete a mixture of buffers, mucus, salivary amylase and antibodies, and the parotid glands primarily secrete salivary amylase.

b. Saliva flushes the oral surfaces to help clean them; buffers in saliva keep the pH near 7.0 to prevent the buildup of acids produced by bacteria; saliva acts as a lubricant; antibodies in saliva protect against pathogens; and enzymes in saliva initiate complex carbohydrate digestion.

c. Damage to the parotid salivary glands, which secrete the carbohydrate-digesting enzyme salivary amylase, would interfere with the digestion of starches (complex carbohydrates).

a. The liver is divided into the left lobe, right lobe, caudate lobe, and quadrate lobe.

b. The falciform ligament marks the division between the left lobe and right lobe of the liver.

c. Degenerative changes in the liver restrict normal blood flow though the small veins and capillaries of the hepatic portal system. As the normally low blood pressures rise, these vessels become distended. If they rupture, extensive bleeding can take place.

a. The pathway of bile: hepatic ducts → common hepatic duct → common bile duct → duodenal ampulla and papilla → duodenal lumen

b. Pancreatic juice is an alkaline mixture of digestive enzymes, water, and ions secreted by the exocrine pancreas. Pancreatic juice enters the duodenum.

c. Major pancreatic enzymes include pancreatic alpha-amylase (breaks down starches), pancreatic lipase (breaks down lipids), nucleases (break down DNA and RNA), and proteolytic enzymes (break down proteins).

a. Periodontal disease is a condition characterized by the formation of dental plaque between the gums and teeth; this can result in tooth loss. Gingivitis, an inflammation of the gums, results from bacterial activity, leading to tooth decay and the eventual breakdown of the periodontal ligament and surrounding bone.

b. Cholecystitis is an inflammation of the gallbladder usually due to a blockage of the cystic duct or common bile duct by gallstones.

c. The bacterium responsible for most peptic ulcers is *Helicobacter pylori*.

1. a. liver; **b.** gallbladder; **c.** cystic duct; **d.** duodenum; **e.** duodenal papilla; **f.** hepatic ducts; **g.** common hepatic duct; **h.** common bile duct; **i.** hepatic portal vein; **j.** stomach; **k.** pancreas

2. a. starch; **b.** hepatopancreatic sphincter; **c.** common bile duct; **d.** saliva; **e.** mumps; **f.** peptic ulcer; **g.** portal hypertension; **h.** pancreatic lipase; **i.** liver; **j.** insulin; **k.** pancreatic juice; **l.** salivary reflex; **m.** gallbladder; **n.** gallstones

3. Salivary secretions continuously flush and clean oral surfaces; buffers in saliva prevent the buildup of acids produced by bacterial action; and saliva contains antibodies, which help control the growth of oral bacteria populations.

4. Such a blockage would interfere with the release of secretions into the duodenum by the pancreas, gallbladder, and liver. Each day the pancreas normally secretes about 1 liter of pancreatic juice, a mixture of a variety of digestive enzymes and buffer solution. The blockage of pancreatic juice would lead to pancreatitis, an inflammation of the pancreas. Extensive damage to exocrine cells by the blocked digestive enzymes would lead to the breakdown of cells that could destroy the pancreas and result in the individual's death. Blockage of bile secretion from the common bile duct could lead to damage of the wall of the gallbladder by the formation of gallstones and to jaundice (a yellow color to the skin and eyes) because bilirubin from the liver would not be excreted in the bile and, instead, would accumulate in body fluids.

1. The stomach performs four major digestive functions: bulk storage of ingested food, mechanical breakdown of ingested food, disruption of chemical bonds in the food through the actions of acids and enzymes, and production of intrinsic factor to aid in the absorption of Vitamin B_{12}.

2. If a gallstone is small enough, it can pass through the common bile duct and block the pancreatic duct. Enzymes from the pancreas then cannot reach the small intestine. As the enzymes accumulate, they irritate the duct and ultimately the exocrine pancreas, producing pancreatitis.

3. The small intestine, especially the jejunum and ileum, are likely involved. Barb's abdominal pain results from regional inflammation. Because inflamed intestines cannot absorb nutrients, she is deficient in iron and vitamin B_{12}, which are necessary for formation of hemoglobin and red blood cells; this accounts for her anemia. The underabsorption of other nutrients accounts for Barb's weight loss.

4. A root canal involves drilling through the enamel and the dentin. The infected pulp, nerve, and blood vessels can then be removed.

5. The primary effect of this surgery would be a reduction in the volume of food (and thus in the amount of calories) eaten because the person feels full after eating a small amount. This can result in significant weight loss.

a. The citric acid cycle is one of the reaction sequences that occur in the mitochondria. In the process, organic molecules are broken down, carbon dioxide molecules are released, and hydrogen atoms are transferred to coenzymes that deliver them to the electron transport system.

b. The ETS generates most of the ATP for the cell, consumes oxygen, and produces water as a by-product.

c. The ETS transfers electrons obtained from the H atoms of the citric acid cycle. As the electrons are transferred, at each step they release small amounts of energy. The energy is used to form ATP. Oxygen is the final acceptor in the transfer of the electrons. As long as oxygen is continually available, the electrons flow, and ATP is generated. A lack of oxygen stops the transfer of electrons along the ETS, the citric acid cycle, and the mitochondrial production of ATP.

a. The nutrient pool within a cell includes all the substrates in the cytosol that are available for anabolism or catabolism.

b. Cells carry out catabolism to release energy for use in cell growth, cell division, and tissue-specific activities.

c. Glycogenesis is the synthesis of glycogen from glucose. Glycogenolysis is the breakdown of glycogen to release glucose. Gluconeogenesis is the synthesis of glucose from smaller carbon chains. Glycolysis is the breakdown of glucose into two three-carbon chains.

1. a. fatty acids; **b.** glucose; **c.** proteins; **d.** two-carbon chains; **e.** citric acid cycle; **f.** coenzymes; **g.** ATP;

ANSWERS

h. electron transport system; i. O_2; j. CO_2; k. H_2O

2. During fasting or starvation, other tissues shift to fatty acid or amino acid catabolism, conserving glucose for neural tissue.

3. a. water; b. nutrient pool; c. anabolism; d. oxidative phosphorylation; e. glycogenolysis; f. coenzymes; g. oxygen; h. catabolism; i. nutrients abundant; j. triglycerides; k. citric acid cycle; l. glycogenesis; m. nutrients scarce; n. acetyl-CoA

Module 16.3 RevieAV w

a. Intestinal gas, or flatus, is a by-product of the breakdown of indigestible carbohydrates by bacteria in the colon.

b. Glycogen is synthesized from excess glucose molecules by liver and muscle cells, and serves as a glucose reserve.

c. Carbohydrates are a preferred energy source because proteins and fats are necessary for building the components of cells and tissues.

Module 16.4 Review

a. Micelles are lipid–bile salt complexes (containing fatty acids, glycerol, and monoglycerides) formed in the intestinal lumen. Chylomicrons are lipoproteins formed in intestinal epithelial cells and contain newly synthesized triglycerides, cholesterol, and other lipids surrounded by phospholipids and proteins.

b. The liver absorbs chylomicrons, removes the triglycerides, combines the cholesterol from the chylomicron with recycled cholesterol, and alters the surface proteins. The new complex is released into the bloodstream as a low-density lipoprotein (LDL).

c. Low-density lipoproteins (LDLs) deliver cholesterol to body tissues, and high-density lipoproteins (HDLs) absorb unused cholesterol from body tissues, returning it to the liver, where it may be packaged into new LDLs or excreted with bile salts in bile.

Module 16.5 Review

a. Enteropeptidase is an enzyme released by the duodenum that converts trypsinogen into trypsin.

b. The body either cannot synthesize essential amino acids, or the body cannot synthesize them in sufficient quantity; therefore, they must be supplied through the diet. The body does synthesize nonessential amino acids, so they do not need to be supplied through the diet.

c. Nutrients in the nutrient pool include glucose, fatty acids, and amino acids. Glucose is used for short-term energy use and excesses are stored as glycogen. Fatty acids are stored as energy and used as a long-term energy source; they are also used to build triglycerides. Amino acids are used for the synthesis of structural and functional proteins.

Module 16.6 Review

a. Nutrition is the ingestion and metabolism of nutrients from food.

b. Vitamins play an important role by serving as coenzymes in many metabolic pathways.

c. Minerals, such as sodium and chloride, determine the osmotic concentrations of body fluids. Minerals buffer body fluids and are essential in a variety of enzymatic reactions.

Module 16.7 Review

a. A balanced diet contains all the ingredients needed to maintain homeostasis and prevent malnutrition.

b. A complete protein meets the body's essential amino acid requirements. An incomplete protein is deficient in one or more essential amino acids.

c. The catabolism of lipids releases the greatest number of Calories per gram, approximately twice as much as a comparable weight of carbohydrates or proteins.

Module 16.8 Review

a. Ghrelin, a hormone secreted by the gastric mucosa when the stomach is not full, inhibits the satiety center and stimulates appetite.

b. Leptin is a peptide hormone produced by adipose tissue during the synthesis of triglycerides. It stimulates the satiety center and suppresses appetite.

c. Eating disorders are psychological problems that result in inadequate food consumption (anorexia nervosa) or excessive food consumption followed by purging (bulimia).

Module 16.9 Review

a. Phenylketonuria (PKU) is an inherited metabolic disorder resulting from an inability to convert phenylalanine to tyrosine. This results in a deficiency in some vital neurotransmitters, leading to severe brain damage.

b. Protein deficiency diseases are nutritional disorders resulting from a lack of one or more essential amino acids; kwashiorkor is an example of a protein deficiency disease.

c. Metabolic reserves are organic substrates in the body that can be catabolized to obtain ATP needed to sustain life.

Section 2 Review

1. a. leptin; b. skeletal muscle; c. B complex and C; d. ketone bodies; e. lipogenesis; f. hypovitaminosis; g. A,D,E,K; h. lipoproteins; i. pyruvate; j. lipolysis; k. uric acid; l. calorie; m. malnutrition; n. inhibits feeding center

2. a. jejunum; b. catabolized for energy or converted into glycogen for storage; c. venous circulation by the thoracic duct; d. the excess amount is readily excreted in the urine

3. a. Essential amino acids are necessary in the diet because the body cannot synthesize them, either in adequate amounts, or at all. The body can synthesize nonessential amino acids on demand. b. (1) Proteins are difficult to break apart because of their complex three-dimensional structure. (2) The energy yield of proteins (4.32 Cal/g) is less than that of lipids (9.46 Cal/g). (3) Proteins form the most

important structural and functional components of cells. Excessive protein catabolism would threaten homeostasis at the cellular to system levels of organization. c. Vitamins and minerals are types of cofactors. Sodium, potassium, calcium, and magnesium ions are important mineral cofactors. Several vitamins (B complex and C) are coenzymes in metabolic pathways. d. Liver cells can break down or synthesize most carbohydrates, lipids, and amino acids. The liver has an extensive blood supply, and so it can easily monitor blood composition of these nutrients and regulate accordingly. The liver also stores energy in the form of glycogen.

4. It appears that Darla is suffering from ketoacidosis as a consequence of her anorexia. Because she is literally starving herself, her body is metabolizing large amounts of fatty acids and amino acids to provide energy and in the process is producing large quantities of ketone bodies (normal metabolites from these catabolic processes). One of the ketones formed is acetone, which can be eliminated through the lungs. This accounts for the smell of aromatic hydrocarbons on Darla's breath. The ketones are also converted into keto acids. In large amounts this lowers the body's pH and begins to exhaust the alkaline reserves of the buffer system. This is probably the cause of her arrhythmias.

Module 16.10 Review

a. Basal metabolic rate (BMR) is the minimum resting energy expenditure of an awake, alert person.

b. Radiation accounts for about one-half of an individual's indoor heat loss.

c. Conduction is the direct transfer of heat through physical contact. Convection is the result of the conduction of heat to the air in contact with the skin. The air warmed by the skin rises and it is repeatedly replaced by cooler air until there is no difference in temperature.

Module 16.11 Review

a. The heat conservation mechanism that results in the conduction of heat from deep arteries to adjacent deep veins in the limbs is called countercurrent exchange.

b. Nonshivering thermogenesis involves the release of hormones that increase the metabolic activity of all tissues, resulting in an increase in body temperature.

c. The vasodilation of peripheral vessels would increase blood flow to the skin and thus the amount of heat the body can lose. As a result, body temperature would decrease.

Section 3 Review

1. a. energetics; b. radiation; c. thermoregulation; d. evaporation; e. nonshivering thermogenesis; f. 60 percent; g. peripheral vasoconstriction; h. countercurrent exchange; i. basal metabolic rate; j. conduction; k. 40 percent; l. peripheral vasodilation; m. hypothalamus; n. shivering thermogenesis

2. a. all of these; b. radiation, conduction, convection, and evaporation; c. physiological responses and behavioral modifications; d. peripheral vasoconstriction; e. triglycerides in adipose tissue; f. continuously

3. a. Energy use at rest is powered by mitochondrial energy production. Because mitochondrial energy production requires oxygen, the energy generated is proportional to oxygen consumption. b. The heat-gain center functions in preventing hypothermia, or below-normal body temperature, by conserving body heat and increasing the rate of heat production by the body. c. Nonshivering thermogenesis increases the metabolic rate of most tissues through the actions of two hormones, epinephrine and thyroid-stimulating hormone. In the short term, the heat-gain center stimulates the adrenal medullae to release epinephrine by the sympathetic division of the ANS. Epinephrine quickly increases the breakdown of glycogen (glycogenolysis) in liver and skeletal muscle, and the metabolic rate of most tissues. The long-term increase in metabolism occurs primarily in children as the heat-gain center adjusts the rate of thyrotropin-releasing hormone (TRH) release by the hypothalamus. When body temperature is low, additional TRH is released, which stimulates the release of thyroid-stimulating hormone (TSH) by the anterior lobe of the pituitary gland. The thyroid gland then increases its rate of thyroid hormone release, and these hormones increase rates of catabolism throughout the body.

Chapter Review

1. Vitamins and minerals are essential components of the diet because the body cannot synthesize most of the vitamins and minerals it requires.

2. Based just on the information given, Charlie would appear to be in good health, at least relative to his diet and probable exercise. Problems are associated with elevated levels of LDLs, which carry cholesterol to peripheral tissues and make it available for the formation of atherosclerotic plaques in blood vessels. High levels of HDLs indicate that a considerable amount of cholesterol is being removed from the peripheral tissues and carried to the liver for disposal. You would encourage Charlie not to change, and keep up the good work.

3. Vasoconstriction increases body temperature, and vasodilation decreases body temperature.

4. The brain region called the hypothalamus acts as the body's "thermostat" by regulating ANS control (through negative feedback) of such homeostatic mechanisms as sweating and shivering thermogenesis.

5. During starvation, the body must use its fat and protein reserves to supply the energy needed to sustain life. Among the proteins metabolized for energy are antibodies in the blood. The loss of antibodies coupled with a scarcity of amino acids needed to synthesize replacement antibodies, as well as protective molecules such as interferon and complement proteins, renders an individual more susceptible to contracting diseases, and less likely to recover from them.

CHAPTER 17

Module 17.1 Review

a. The kidneys are located in a retroperitoneal (behind the peritoneum) position and are deep to the abdominal organs.

b. The structural landmarks include the following: fibrous capsule (covering on the outer kidney surface), renal sinus (internal cavity), renal cortex (superficial region of the kidney), renal medulla (region between renal cortex and renal sinus), renal pyramid (conical structure), renal papilla (tip of renal pyramid), renal column (granular tissue between adjacent renal pyramids), kidney lobe (area consisting of a renal pyramid, overlying area of renal cortex, and adjacent tissues of the renal columns), minor calyx (collecting site for urine), major calyx (structure formed by the fusion of four or five minor calyces) and renal pelvis (funnel-shaped structure that collects urine from the major calyces).

c. The renal pyramid is a conical mass within the renal medulla that ends at the papilla.

Module 17.2 Review

a. The primary structures of the nephron are the renal corpuscle, proximal convoluted tubule, nephron loop, and distal convoluted tubule. The main structures of the collecting system are the collecting duct and papillary duct.

b. The renal corpuscle consists of the glomerular capsule and the glomerulus.

c. Glomerular fluid (also called filtrate) essentially lacks proteins, whereas blood plasma contains a variety of larger protein molecules.

Module 17.3 Review

a. Pathway of blood flow from the renal artery to the renal vein: renal artery → segmental arteries → interlobar arteries → arcuate arteries → cortical radiate arteries → afferent arterioles → glomerulus → efferent arterioles → peritubular capillaries → cortical radiate veins → arcuate veins → interlobar veins → renal vein

b. Blood enters the glomerulus by the afferent arteriole and leaves through the efferent arteriole.

c. The vasa recta are long, straight peritubular capillaries that parallel the nephron loop of a juxtamedullary nephron.

Section 1 Review

1. a. renal sinus: internal cavity within kidney that contains calyces and renal pelvis of the ureter, and blood vessels (segmental arteries, renal artery, and renal vein); b. renal pelvis: funnel-shaped expansion of the ureter that collects urine from the major calyces; c. hilum: depression on the medial border of kidney, entry point for renal artery and renal nerves, and exit point for renal vein and ureter; d. ureter: tube that conducts urine from the renal pelvis to the urinary bladder; e. fibrous capsule (outer layer): covering of the kidney's outer surface and lining of the renal sinus; f. minor calyx: subdivision of major calyces into which urine enters from the renal papillae; g. major calyx: primary subdivision of renal pelvis formed from the merging of four or five minor calyces; h. renal cortex: the outer portion of the kidney containing renal columns (extensions between the renal pyramids), and portions of nephrons (renal corpuscles, and the proximal and distal convoluted tubules); i. renal medulla: the inner portion of the kidney that contains the renal pyramids; j. renal papilla: tip of the renal pyramid that projects into a minor calyx; k. renal pyramid: conical mass of the kidney projecting into the medullary region containing part of the renal tubules and

collecting system ducts; **l.** renal columns: cortical tissue separating renal pyramids; **m.** kidney lobe: portion of kidney consisting of a renal pyramid and its associated tissue in the renal cortex

2. a. urinary bladder; **b.** nephrons; **c.** renal tubules; **d.** glomerulus; **e.** proximal convoluted tubule; **f.** papillary ducts; **g.** renal medulla; **h.** renal sinus; **i.** major calyces; **j.** ureter

Module 17.4 Review

a. The three distinct processes of urine formation in the kidney are filtration, reabsorption, and secretion.

b. Filtration exclusively occurs across the filtration membrane in the renal corpuscle.

c. Aldosterone is the hormone responsible for regulating sodium ion reabsorption in the DCT and collecting system.

Module 17.5 Review

a. No, the permeability of the PCT to water cannot change, because water reabsorption occurs whenever the osmotic concentration of the peritubular fluid exceeds that of the tubular fluid.

b. Increased ADH levels cause the appearance of more water channels, or aquaporins, in the DCT. As a result, more water is reabsorbed into the peritubular fluid, which reduces the volume of water in the urine, making the urine more concentrated.

c. Decreased ADH levels in the DCT reduce the urine osmotic concentration due to the presence of more water in the urine. The result is a larger volume of more dilute urine.

Module 17.6 Review

a. The filtrate produced at the renal corpuscle has the same osmotic pressure as plasma.

b. The active reabsorption of ions and organic substrates causes osmosis to occur.

c. The concentration gradient of the renal medulla is maintained

by the removal of solutes and water from the area by the vasa recta, which transport them to the venous system.

Module 17.7 Review

a. Hemodialysis is the process of using an artificial semipermeable membrane to remove waste products from the blood of a person whose kidneys are not functioning properly.

b. Chronic renal failure involves a gradual loss of renal function, whereas acute renal failure involves a sudden loss of renal function.

c. Patients on dialysis are often given Epogen or Procrit (a synthetic form of erythropoietin) to treat anemia, which occurs because their malfunctioning kidneys produce too little erythropoietin (EPO). EPO is the hormone that stimulates the development of red blood cells in the red bone marrow.

Section 2 Review

1. a. proximal convoluted tubule: reabsorbs water, ions, and all organic substrates (nutrients) from the filtrate; **b.** renal corpuscle: expanded chamber that encloses the glomerulus and where the original filtrate is formed; **c.** nephron loop: portion of the nephron that reabsorbs water and sodium and chloride ions, and produces the concentration gradient in the renal medulla; **d.** distal convoluted tubule: reabsorbs variable amount of water (under ADH control) and variable amount of Na$^+$ (under aldosterone control), and secretes variable amount of solutes; **e.** collecting system: performs variable reabsorption of water (under ADH control) and reabsorption of sodium ions (under aldosterone control), and receives and delivers urine for storage and elimination

2. a. renal corpuscle; **b.** podocytes; **c.** filtrate; **d.** urea; **e.** juxtaglomerular complex; **f.** aquaporins; **g.** polyuria; **h.** aldosterone; **i.** PCT; **j.** ADH

3. a. Six substances that diffuse from the blood are potassium ions, phosphate ions, sulfate

ions, urea, creatinine, and uric acid. Two substances that diffuse into the blood are bicarbonate ions and glucose. **b.** A dialysis machine replaces normal glomerular filtration.

Module 17.8 Review

a. Urine is transported by the ureters, stored within the urinary bladder, and eliminated through the urethra.

b. The external urethral sphincter (a voluntary muscle) must be relaxed for urination to occur.

c. The urge to urinate usually appears when the urinary bladder contains about 200 mL of urine. The micturition reflex begins to function when the stretch receptors have provided adequate stimulation to the parasympathetic motor neurons. The activity in the motor neurons generates action potentials that reach the smooth muscle in the wall of the urinary bladder. These efferent impulses travel over the pelvic nerves, producing a sustained contraction of the urinary bladder.

Module 17.9 Review

a. Dysuria is the term for painful or difficult urination.

b. Obstruction of a ureter by a kidney stone would interfere with the flow of urine between the kidney and the bladder.

c. You might find casts of red blood cells, white blood cells, granular casts, and hyaline casts in urine sediment.

Section 3 Review

1. a. internal urethral sphincter; **b.** trigone; **c.** external urethral sphincter; **d.** renal calculi; **e.** rugae; **f.** prostate gland; **g.** detrusor; **h.** pyelogram; **i.** urethra; **j.** micturition; **k.** parasympathetic postganglionic fibers

2. a. stretch receptors stimulated; **b.** afferent fibers carry information to the sacral spinal cord; **c.** parasympathetic preganglionic fibers carry motor commands; **d.** detrusor muscle contraction stimulated;

e. individual relaxes external urethral sphincter, which causes internal urethral sphincter to relax

3. Four primary signs and symptoms of urinary disorders are (1) changes in the volume of urine, (2) changes in the appearance of urine, (3) changes in the frequency of urination, and (4) pain.

4. cystitis/pyelonephritis: both conditions involve inflammation and infections of the urinary system, but cystitis refers to the urinary bladder, whereas pyelonephritis refers to the kidney; stress incontinence/overflow incontinence: both conditions involve an inability to control urination, but stress incontinence involves periodic involuntary leakage, whereas overflow incontinence involves a continual, slow trickle of urine

Module 17.10 Review

a. The major routes of fluid loss are urination (kidneys), evaporation at the skin, evaporation at the lungs, and water loss in feces.

b. A fluid shift is a rapid movement of water between the ECF and ICF in response to an osmotic gradient.

c. Fluid balance is when the amount of water you gain each day is equal to the amount you lose. Mineral balance is the state of the body in which ion gains (ion absorption) and losses (ion excretion) are equal.

Module 17.11 Review

a. When osmoreceptors are inhibited, ADH release is decreased, and thirst is suppressed.

b. Aldosterone causes increased urinary sodium retention and thus increases the sodium ion concentration in the ECF.

c. Shifts in sodium balance result in expansion or contraction of the ECF. Large variations in ECF volume are corrected by homeostatic mechanisms triggered by changes in blood volume. If the blood volume becomes too low, ADH and aldosterone are secreted, increasing the sodium ion concentration in the ECF. If the volume becomes too high, natriuretic peptides are secreted.

Section 4 Review

1. **a.** osmoreceptors; **b.** fluid balance; **c.** plasma, interstitial fluid; **d.** sodium; **e.** ADH; **f.** minerals; **g.** potassium; **h.** fluid compartments; **i.** kidneys; **j.** fluid shift; **k.** hypertonic plasma; **l.** proteins; **m.** dehydration; **n.** aldosterone

2. **a.** intracellular fluid (ICF); **b.** kidneys and sweat glands; **c.** from the cells into the ECF until osmotic equilibrium is restored; **d.** becomes hypotonic with respect to the ICF; **e.** all of these; **f.** out of the ICF and into the ECF

3. Two primary steps involved in the regulation of sodium ion concentrations are (1) changes in the circulating levels of ADH, and (2) a fluid shift into or out of the ICF.

Module 17.12 Review

a. Acidemia is the condition in which plasma pH falls below 7.35. Alkalemia exists when the plasma pH is above 7.45.

b. The dissociation of carbonic acid is the most important factor affecting the pH of the ECF.

c. An inverse relationship exists between pH and CO_2 levels.

Module 17.13 Review

a. The body's three major buffer systems are the phosphate buffer system, the protein buffer system, and the carbonic acid–bicarbonate buffer system.

b. Proteins and free amino acids can accept hydrogen ions on their exposed amino groups and many R groups. Hydrogen ions can also be accepted by their exposed carboxylate ions, which re-form as carboxyl groups.

c. The carbonic acid–bicarbonate buffer system prevents pH changes caused by acids generated by metabolic activity. It uses the H^+ released by these acids to generate carbonic acid,

which dissociates into H_2O and CO_2, the latter of which is exhaled from the lungs.

Module 17.14 Review

a. Metabolic acidosis results from the depletion of the bicarbonate reserve, caused by an inability to excrete hydrogen ions at the kidneys, the production of large numbers of acids, or bicarbonate loss. Metabolic alkalosis results when bicarbonate ion concentrations become elevated.

b. Respiratory acidosis results from excessive levels of carbon dioxide in body fluids. Respiratory alkalosis develops when respiratory activity lowers plasma carbon dioxide levels to below-normal levels.

c. The kidneys conserve HCO_3^- and eliminate H^+ in the urine during metabolic acidosis.

Section 5 Review

1. **a.** increased CO_2; **b.** acidosis; **c.** plasma pH decrease; **d.** increased; **e.** secreted; **f.** generated; **g.** decreased; **h.** decreased; **i.** increased; **j.** decreased CO_2; **k.** alkalosis; **l.** plasma pH increase; **m.** decreased; **n.** generated; **o.** secreted; **p.** increased; **q.** increased; **r.** decreased

2. The young boy has metabolic and respiratory acidosis. The metabolic acidosis resulted primarily from the large amounts of lactic acid generated by the boy's muscles as he struggled in the water. (The dissociation of lactic acid releases hydrogen ions and lactate ions.) Sustained hypoventilation during drowning contributed to both tissue hypoxia (low O_2) and respiratory acidosis. Respiratory acidosis developed as the CO_2 increased in the ECF, increasing the production of carbonic acid and its dissociation into H^+ and HCO_3^-. Prompt emergency treatment is essential; the usual procedure involves some form of artificial or mechanical respiratory assistance (to increase the respiratory rate and decrease CO_2 in the ECF) coupled with the intravenous infusion of a buffered isotonic

solution that would absorb the hydrogen ions in the ECF and increase body fluid pH.

Chapter Review

1. Increasing the volume of urine produced decreases the total blood volume of the body, which in turn leads to a decreased blood pressure. Edema frequently results when the pressure of the blood exceeds the opposing reabsorption forces at the capillaries in the affected area. Depending on the actual cause of the edema, decreasing the blood pressure would decrease edema formation and possibly cause some of the fluid to move from the interstitial spaces back to the blood.

2. Because mannitol is filtered but not reabsorbed, drinking a mannitol solution would lead to an increase in the osmolarity of the filtrate. Less water would be reabsorbed, and an increased volume of urine would be produced.

3. Proteins are excluded from filtrate because they are too large to fit through the slit pores. Keeping proteins in the plasma ensures that the osmolarity of the blood will oppose filtration and return water to the plasma.

4. Digestive secretions contain high levels of bicarbonate, so individuals with diarrhea can lose significant amounts of this important ion, leading to acidosis. We would expect Milly's blood pH to be lower than 7.35, and that of her urine to be low (due to increased renal excretion of hydrogen ions). We would also expect an increase in the rate and depth of breathing as the respiratory system tries to compensate by eliminating carbon dioxide.

5. Fluid balance is a state in which the amount of water gained each day is equal to the amount lost to the environment. It is vital that the water content of the body remain stable, because water is an essential ingredient of cytoplasm and accounts for about 99 percent of ECF volume. Electrolyte balance exists when there is neither a net gain nor a net loss of any ion in body fluids. It is impor-

tant that the ionic concentrations in body water remain within normal limits; if levels of calcium or potassium become too high, for instance, cardiac arrhythmias can develop. Acid-base balance exists when the production of hydrogen ions precisely offsets their loss. The pH of body fluids must remain within a relatively narrow range; variations outside this range can be life threatening.

CHAPTER 18

Module 18.1 Review

a. Male reproductive structures are the scrotum, two testes, two epididymides, a pair of ductus deferens, two ejaculatory ducts, a urethra, two seminal glands, a prostate gland, two bulbo-urethral glands, and a penis.

b. Spermatogenesis is the production of spermatozoa and involves mitosis, meiosis I, meiosis II, and spermiogenesis.

c. A typical spermatozoon has an acrosome that contains enzymes essential to fertilization; a head with a nucleus that is packed with chromosomes; a neck that contains centrioles; a middle piece containing mitochondria to provide ATP for propulsion; and a flagellum, the whiplike structure that moves the cell.

Module 18.2 Review

a. Important regulatory hormones in the establishment and maintenance of male sexual function are GnRH (gonadotropin-releasing hormone) from the hypothalamus, LH (luteinizing hormone), and FSH (follicle-stimulating hormone) from the anterior lobe of the pituitary gland. Testosterone is an important hormone that exerts its effects on peripheral tissues and aids in spermatogenesis and spermiogenesis by stimulating nurse cell activities.

b. The testes, hypothalamus, and the anterior lobe of the pituitary gland secrete the hormones that control male reproductive functions.

c. Low FSH levels would lead to reduced sperm production.

Module 18.3 Review

a. The two steps of capacitation involve motility and fertilization capability. In the first step, spermatozoa become motile when they are mixed with seminal gland secretions. In the second step, the spermatozoa become capable of successful fertilization when they are exposed to conditions in the female reproductive tract.

b. Semen is male reproductive fluid. It contains sperm plus the secretions of the seminal, prostate, and bulbo-urethral glands.

c. The tail of each epididymis connects with the ductus deferens, which passes through the inguinal canal as part of the spermatic cord. Near the prostate gland, each ductus deferens enlarges to form an ampulla. The junction of the base of the seminal gland and the ampulla creates the ejaculatory duct, which empties into the urethra. The urethra then provides an exit route.

Module 18.4 Review

a. The parts of the penis include the root, body, neck, glans, and prepuce. The root attaches the penis to the body wall; the body (shaft) is the tubular, movable portion; the neck is the narrow portion between the shaft and glans; the glans is the expanded distal end that surrounds the external urethral orifice; and the prepuce (foreskin) surrounds the tip of the penis.

b. The three columns of erectile tissue are the corpus spongiosum and the paired corpora cavernosa.

c. When smooth muscles in the arterial walls of the erectile tissue relax, vessels dilate, and blood flow increases. This results in engorgement of the vascular spaces with blood, causing erection of the penis.

Module 18.5 Review

a. Benign prostatic hypertrophy is the medical term for an enlarged prostate gland that occurs spontaneously.

b. Two possible treatments for prostate cancer are radiation therapy and prostatectomy (removal of the prostate gland).

c. Testicular cancer is the most common cancer among males aged 15–35.

Section 1 Review

1. a. ductus deferens; **b.** urethra; **c.** penis; **d.** epididymis; **e.** testis; **f.** external urethral orifice; **g.** scrotum; **h.** seminal gland; **i.** prostate gland; **j.** ejaculatory duct; **k.** bulbo-urethral gland

2. a. dartos muscle; **b.** spermatogonia; **c.** seminiferous tubules; **d.** interstitial cells; **e.** spermiogenesis; **f.** spermatogenesis; **g.** penis and scrotum; **h.** flagellum; **i.** capacitation; **j.** semen; **k.** prepuce; **l.** corpus spongiosum; **m.** luteinizing hormone (LH); **n.** follicle-stimulating hormone (FSH)

3. Normal levels of testosterone (1) promote the functional maturation of spermatozoa, (2) maintain the accessory glands and organs of the male reproductive system, (3) establish and maintain male secondary sex characteristics, (4) stimulate bone and muscle growth, and (5) stimulate sexual behaviors and sexual drive (libido).

Module 18.6 Review

a. The major organs of the female reproductive system are the ovaries, uterine tubes, uterus, vagina, and external genitalia.

b. The ovaries, uterine tubes, and uterus are enclosed within the broad ligament. The mesovarium supports and stabilizes each ovary.

c. The ovaries produce immature female gametes called oocytes; secrete female sex hormones, including estrogen and progesterone; and secrete inhibin, which is involved in the feedback control of FSH production.

Module 18.7 Review

a. An oocyte is an immature female gamete whose meiotic divisions will produce a single secondary oocyte and three polar bodies.

b. Males produce gametes from puberty until death; females produce gametes only from puberty to menopause. Males produce many gametes at a time; females typically produce one or two per each ovarian cycle. Males release mature gametes that have completed meiosis; females release secondary oocytes suspended in metaphase of meiosis II.

c. Important events in the ovarian cycle are (1) the formation of primary follicles, (2) the formation of secondary follicles, (3) the formation of a tertiary follicle, (4) ovulation, (5) the formation of a corpus luteum, and (6) the degeneration of the corpus luteum.

Module 18.8 Review

a. The regions of the uterus are the fundus, body, and cervix.

b. The outer layer is the perimetrium, composed of an incomplete serosa; the middle layer is the muscular myometrium; and, the endometrium is the inner, glandular layer.

c. Recently released secondary oocytes reach the uterine tube with the aid of the beating action of cilia on the inner surfaces of the fimbriae of the infundibulum.

Module 18.9 Review

a. The vagina (1) serves as a passageway for the elimination of menstrual fluids; (2) receives the penis during sexual intercourse, and holds spermatozoa prior to their passage into the uterus; and (3) forms the inferior portion of the birth canal, through which the fetus passes during delivery.

b. The vagina is a muscular tube extending between the uterus and external genitalia; its lining forms folds called rugae. The proximal portion of the vagina is marked by the cervix, which dips into the vaginal canal, and the shallow recess known as the fornix. The hymen, a thin epithelial fold, partially blocks the entrance to the vagina until physical distortion ruptures it.

c. The greater vestibular glands in females are similar to the bulbo-urethral glands in males, and both the penis and the clitoris have characteristic erectile tissue.

Module 18.10 Review

a. Lactation is the secretion of milk by the mammary glands.

b. Blockage of a single lactiferous sinus would not interfere with the delivery of milk to the nipple, because each breast generally has 15–20 lactiferous sinuses.

c. Route of milk flow: secretory alveoli of the secretory lobules of the mammary gland → ducts within a lobe → lactiferous duct of the lobe → lactiferous sinus → surface of the nipple

Module 18.11 Review

a. Completion of the decline in the levels of estrogen and progesterone signals the beginning of menses and the start of a new uterine cycle.

b. If the LH surge did not occur during an ovarian cycle, ovulation and corpus luteum formation could not occur.

c. The pituitary secretes FSH and LH. FSH stimulates follicle development, and activated follicles and ovarian cells produce estrogen. High estrogen levels stimulate LH secretion and a massive release of LH. Progesterone is produced by the corpus luteum and is the principal hormone of the luteal phase of the uterine cycle. The changes in estrogen and progesterone levels are responsible for the maintenance of the uterine cycle.

Module 18.12 Review

a. A sexually transmitted disease (STD) is a disease that is transferred from one individual to another primarily or exclusively through sexual contact.

b. Ovarian cancer usually arises from epithelial cells.

c. The human papillomaviruses (HPV) cause most cases of cervical cancer.

Module 18.13 Review

a. A vasectomy is the surgical removal of a segment of each ductus deferens and the tying, cauterizing, or inserting of silicone plugs into the cut ends to prevent spermatozoa from reaching the distal portions of the male reproductive tract.

b. Condoms provide some protection against sexually transmitted diseases.

c. Depo-Provera injections (a progesterone-only form of birth control) result in the cessation of the uterine cycle in 50 percent of women using this product. (Uterine cycles eventually resume after use of the product is discontinued.)

Section 2 Review

1. a. infundibulum; **b.** ovary; **c.** uterine tube; **d.** perimetrium; **e.** myometrium; **f.** endometrium; **g.** uterus; **h.** clitoris; **i.** labium minus; **j.** labium majus; **k.** fornix; **l.** cervix; **m.** vagina

2. a. oocytes; **b.** menarche; **c.** vasectomy; **d.** broad ligament; **e.** rectouterine pouch; **f.** corpus luteum; **g.** menstrual cycle; **h.** lactation; **i.** ovaries; **j.** LH surge; **k.** cervix; **l.** tubal ligation; **m.** fertilization; **n.** vulva

3. The endometrial cells have receptors for estrogens and progesterone and respond to these hormones as if the cells were still in the body of the uterus. Under the influence of estrogens, the endometrial cells proliferate at the beginning of the uterine (menstrual) cycle and begin to develop glands and blood vessels, which then further develop under the control of progesterone. This dramatic increase in tissue size exerts pressure on neighboring tissues or in some other way interferes with their function. It is the recurring expansion of tissue in an abnormal location that causes periodic pain.

Chapter Review

1. The presence of exercise-induced amenorrhea suggests that a certain amount of body fat is necessary for menstrual cycles to occur. If body fat levels fall below some set point, menstruation ceases.

2. Because a woman lacking adequate body fat might not have the energy reserves needed to have a successful pregnancy, compensatory mechanisms in her body prevent pregnancy by shutting down the ovarian cycle, and thus the menstrual cycle. When her body subsequently accumulates sufficient energy reserves in body fat, the cycles begin again.

3. Estrogen deficiency can lead to infertility, vaginal and breast atrophy, and osteoporosis. It may also increase the risk of heart attacks later in life.

4. Women more frequently experience peritonitis stemming from a urinary tract infection because infectious organisms exiting the urethral orifice can readily enter the nearby vagina. From there, they can then proceed to the uterus, into the uterine tubes, and finally into the peritoneal cavity. No such direct path of entry into the abdominopelvic cavity exists in men.

5. From the male ejaculatory duct, the sperm travel into the urethra as a component of semen. (Semen is a viscid fluid containing sperm and secretions of the testes, seminal glands, prostate gland, and bulbo-urethral glands.) The semen is then ejaculated into the female vagina. Sperm then continue through the vaginal canal, through the uterus, to the uterine tubes. Roughly half of the ejaculate enters the left uterine tube and the other half enters the right uterine tube. The sperm continue swimming toward the ovary until they encounter the waiting secondary oocyte.

CHAPTER 19

Module 19.1 Review

a. Cleavage is the series of cell divisions occurring in the oocyte immediately after fertilization. Cleavage produces an ever-increasing number of smaller and smaller daughter cells.

b. The blastocyst stage begins once the zygote arrives in the uterine cavity.

c. The blastocyst consists of an outer trophoblast and an inner cell mass. Implantation occurs when the blastocyst adheres to and then becomes enclosed within the uterine lining about 7–10 days after fertilization.

Module 19.2 Review

a. Gestational trophoblastic neoplasia is a tumor formed when something goes wrong with the trophoblastic cell division regulatory mechanisms. If untreated, the neoplasm may become malignant.

b. Gastrulation is the cellular migration process that forms the three primary germ layers—the endoderm, ectoderm, and mesoderm—from the embryonic disc. It is from these germ layers that the body systems differentiate.

c. The mesoderm gives rise to nearly all body systems except the nervous and respiratory systems.

Module 19.3 Review

a. The four extra-embryonic membranes are the yolk sac, amnion, allantois, and chorion.

b. The base of the allantois gives rise to the urinary bladder.

c. The yolk sac forms from endoderm and mesoderm; it is an important site of blood cell formation. The amnion forms from the ectoderm and mesoderm; it encloses the fluid that surrounds and cushions the developing embryo and fetus. The allantois forms from endoderm and mesoderm; its base gives rise to the urinary bladder. The chorion forms from mesoderm and trophoblast; it surrounds the blastocoele.

Module 19.4 Review

a. The hormones synthesized by the syncytial trophoblast are human chorionic gonadotropin (hCG), human placental lactogen (hPL), relaxin, progesterone, and estrogen.

b. The presence of human chorionic gonadotropin (hCG) in the urine provides a reliable indicator of pregnancy.

c. After the first trimester, the placenta is sufficiently functional to maintain the pregnancy.

Module 19.5 Review

a. Organogenesis is the process of organ formation.

b. In the second trimester, the organ systems increase in complexity. During the third trimester, many of the organ systems become fully functional.

c. During the third trimester, the fetus undergoes its largest weight gain.

Module 19.6 Review

a. The major changes that occur in maternal systems during pregnancy are increases in respiratory rate and tidal volume, blood volume, nutrient requirements, glomerular filtration rate (GFR), and the size of the uterus and mammary glands.

b. A mother's blood volume increases during pregnancy to compensate for the reduction in maternal blood volume resulting from blood flow through the placenta.

c. Difficulty breathing in pregnant women results from the enlarged uterus pressing against the diaphragm and crowding the lungs. This makes it more difficult for them to expand their lungs.

Module 19.7 Review

a. Relaxin, produced by the placenta, relaxes the pelvic articulations and dilates the cervix, and the weight of the fetus distorts the uterus. Rising estrogen levels promote the release of oxytocin; increased estrogen enhances the sensitivity of the myometrium to the effects of oxytocin; and the already stretched smooth muscles of the myometrium contract more.

b. Estrogen and oxytocin stimulate the production of prostaglandins, which are then primarily responsible for the initiation of true labor.

c. The dilation stage begins with the onset of true labor, as the cervix dilates and the fetus begins to move toward the cervical canal; late in this stage, the amnion ruptures. The expulsion stage begins as the cervix dilates completely and continues until the fetus has completely emerged from the vagina (delivery). In the placental stage, the uterus gradually contracts, tearing the connections between the endometrium and the placenta and ejecting the placenta.

Module 19.8 Review

a. Oxytocin causes the milk let-down reflex.

b. Colostrum is produced by the mammary glands from the end of the sixth month of pregnancy until a few days after birth and contains antibodies. After that, the glands begin producing breast milk, which contains antibodies and lysozyme but has a higher fat content than colostrum.

c. The postnatal stages of development are the neonatal period, from birth to one month; infancy, from one month to age 2; childhood, from two until sexual maturation begins; adolescence, which begins with the onset of sexual maturation (puberty) between ages 9–14 and ends when growth in body size ends (around 18 years); and maturity, which includes the rest of the individual's life. A final stage called senescence, or aging, overlaps with maturity.

Module 19.9 Review

a. The three interacting hormonal events associated with the onset of puberty are (1) increased GnRH production by the hypothalamus; (2) increased sensitivity to GnRH by the anterior lobe of the pituitary gland, and a rapid increase in circulating levels of FSH and LH; and (3) increased sensitivity to FSH and LH by ovarian and testicular cells.

b. Males have deeper voices and larger larynges than do females because testosterone stimulates laryngeal development in males to a greater extent than estrogen stimulates laryngeal development in females.

c. The higher estrogen levels in premenopausal women (compared to adult men) cause reductions in plasma cholesterol levels, thereby slowing plaque formation and lowering the risk of atherosclerosis.

Section 1 Review

1. a. conception; **b.** neonatal period; **c.** parturition; **d.** gestation; **e.** hCG; **f.** relaxin; **g.** colostrum; **h.** amnion; **i.** blastocyst; **j.** endometrium; **k.** inner cell mass; **l.** syncytial trophoblast; **m.** morula; **n.** embryonic disc

2. a. ninth week after fertilization; **b.** vein; oxygenated blood; **c.** second trimester; **d.** cleavage, implantation, placentation, embryogenesis; **e.** first trimester; **f.** respiratory, digestive, and urinary

3. It is very unlikely that the baby's condition is the result of a viral infection contracted during the third trimester. The development of organ systems occurs during the first trimester, and by the end of the second trimester, most organ systems are fully formed. During the third trimester, the fetus undergoes tremendous growth, but very little new organ formation occurs.

Module 19.10 Review

a. Homozygous means that homologous chromosomes carry the same allele of a given gene. Heterozygous means that

homologous chromosomes carry different alleles of a given gene.

b. In simple inheritance, phenotypic characteristics are determined by interactions between a single pair of alleles. Polygenic inheritance involves interactions among alleles of several genes.

c. The phenotype of a person who is heterozygous for curly hair—that is, a person with one dominant allele and one recessive allele for that trait—would be "curly hair."

Module 19.11 Review

a. A single nucleotide polymorphism is a variation in a single base pair in a DNA sequence.

b. (1) XO: Turner's syndrome; (2) XXY: Klinefelter's syndrome

c. X-linked traits are expressed more frequently in males than females because males have a Y chromosome that has no corresponding allele.

Section 2 Review

1. a. phenotype; **b.** alleles; **c.** homologous; **d.** X-linked; **e.** genotype; **f.** locus; **g.** heterozygous; **h.** genetics; **i.** homozygous; **j.** polygenic inheritance; **k.** simple inheritance; **l.** autosomes

2. a. Children who cannot roll the tongue must be homozygous recessive for the condition, and the only way a child can receive two recessive alleles from tongue-rolling parents is to have parents who are both heterozygous. **b.** The allele for achondroplasia dwarfism must be dominant. In order for two dwarf parents to produce a normal child, the child must be

homozygous recessive, and thus both parents must be heterozygous. The probability that their next child (and all subsequent children) will be normal sized is 1 in 4, or 25 percent.

Chapter Review

1. Cleavage is a series of cell divisions within the zygote in which the cells get smaller with each division. Cell division involves cells growing after each division.

2. Positive feedback mechanisms between increasing levels of oxytocin and increased uterine distortion ensure that labor contractions continue until delivery has been completed.

3. The sex chromosomes XY result in a baby boy. Testosterone must be present to stimulate male sexual characteristics.

4. Although technically it takes only one sperm to fertilize a secondary oocyte, having a low sperm count means that most sperm entering the female reproductive tract are killed or disabled before they reach the uterus. So, too few sperm reach the secondary oocyte in a uterine tube to produce sufficient acrosomal enzymes to enable one sperm to penetrate the waiting oocyte.

5. The probability that this couple's daughters will have hemophilia is zero, because each daughter will receive a dominant normal allele from her father. There is a 50 percent chance that a son will have hemophilia, because each son has a 50 percent chance of receiving the mother's normal allele, and a 50 percent chance of receiving the mother's recessive allele.

A

abdomen: The region of the trunk between the inferior margin of the rib cage and the superior margin of the pelvis.

abdominopelvic cavity: The term used to refer to the general region bounded by the abdominal wall and the pelvis; it contains the peritoneal cavity and visceral organs.

abducens: Cranial nerve VI, which innervates the lateral rectus muscle of the eye.

abduction: Movement away from the midline of the body, as viewed in the anatomical position.

abscess: A localized collection of pus within a damaged tissue.

absorption: The active or passive uptake of gases, fluids, or solutes.

accommodation: An alteration in the curvature of the lens of the eye to focus an image on the retina.

acetabulum: The fossa on the lateral aspect of the pelvis that accommodates the head of the femur.

acetylcholine (ACh): A chemical neurotransmitter in the brain and peripheral nervous system; the dominant neurotransmitter in the peripheral nervous system, released at neuromuscular junctions and synapses of the parasympathetic division.

acetylcholinesterase (AChE): An enzyme found in the synaptic cleft, bound to the postsynaptic membrane, and in tissue fluids; breaks down and inactivates acetylcholine molecules.

acetyl-CoA: An acetyl group bound to coenzyme A, a participant in the anabolic and catabolic pathways for carbohydrates, lipids, and many amino acids.

acetyl group: $-CH_3CO$.

acid: A compound whose dissociation in solution releases a hydrogen ion and an anion; an acidic solution has a pH below 7.0 and contains an excess of hydrogen ions.

acidosis: An abnormal physiological state characterized by a plasma pH below 7.35.

acinus/acini: A histological term referring to a blind pocket, pouch, or sac.

acoustic: Pertaining to sound or the sense of hearing.

acromion: A continuation of the scapular spine that projects superior to the capsule of the shoulder joint.

acrosome: A membranous sac at the tip of a spermatozoon that contains hyaluronidase.

actin: The protein component of microfilaments that forms thin filaments in skeletal muscles and produces contractions of all muscles through interaction with thick (myosin) filaments.

action potential: A propagated change in the membrane potential of excitable cells, initiated by a change in the membrane permeability to sodium ions.

active transport: The ATP-dependent absorption or secretion of solutes across a plasma membrane.

acute: Sudden in onset, severe in intensity, and brief in duration.

adaptation: A change in pupillary size in response to changes in light intensity; a decrease in receptor sensitivity or perception after chronic stimulation.

adduction: Movement toward the axis or midline of the body, as viewed in the anatomical position.

adenine: One of the nitrogenous bases in the nucleic acids RNA and DNA.

adenosine: A compound consisting of adenine and ribose.

adenosine diphosphate (ADP): A compound consisting of adenosine with two phosphate groups attached.

adenosine monophosphate (AMP): A nucleotide consisting of adenosine plus a phosphate group (PO_4^{3-}); also called *adenosine phosphate*.

adenosine triphosphate (ATP): A high-energy compound consisting of adenosine with three phosphate groups attached; the third is attached by a high-energy bond.

adenylate cyclase: An enzyme bound to the inner surfaces of plasma membranes that can convert ATP to cyclic-AMP.

adipocyte: A fat cell.

adipose tissue: Loose connective tissue dominated by adipocytes.

adrenal cortex: The superficial portion of the adrenal gland that produces steroid hormones; also called *suprarenal cortex*.

adrenal gland: A small endocrine gland that secretes steroids and catecholamines and is located superior to each kidney; also called *suprarenal gland*.

adrenal medulla: The core of the adrenal gland; a modified sympathetic ganglion that secretes catecholamines (E and NE) into the blood during sympathetic activation; also called *suprarenal medulla*.

adrenocortical hormone: Any steroid produced by the adrenal cortex.

adrenocorticotropic hormone (ACTH): The hormone that stimulates the production and secretion of glucocorticoids by the adrenal cortex; released by the anterior lobe of the pituitary gland in response to corticotropin-releasing hormone.

adventitia: The superficial layer of connective tissue surrounding an internal organ; fibers are continuous with those of surrounding tissues, providing support and stabilization.

aerobic: Requiring the presence of oxygen.

aerobic metabolism: The complete breakdown of organic substrates into carbon dioxide and water, via pyruvate; a process that yields large amounts of ATP but requires mitochondria and oxygen.

afferent: Toward a center.

afferent arteriole: An arteriole that carries blood to a glomerulus of the kidney.

afferent fiber: An axon that carries sensory information to the central nervous system.

agglutination: The aggregation of red blood cells due to interactions between surface antigens and plasma antibodies.

aggregated lymphoid nodules: Lymphoid nodules beneath the epithelium of the small intestine; also called *Peyer patches*.

agonist: A muscle responsible for a specific movement; also called a *prime mover*.

agranular: Without granules; *agranular leukocytes* are monocytes and lymphocytes.

alba: White.

albicans: White.

albuginea: White.

aldosterone: A mineralocorticoid produced by the adrenal cortex; stimulates sodium and water conservation at the kidneys; secreted in response to the presence of angiotensin II.

alkalosis: The condition characterized by a plasma pH greater than 7.45; associated with a relative deficiency of hydrogen ions or an excess of bicarbonate ions.

alveolar sac: An air-filled chamber that supplies air to several alveoli.

alveolus/alveoli: Blind pockets at the end of the respiratory tree, lined by a simple squamous epithelium and surrounded by a capillary network; sites of gas exchange with the blood; a bony socket that holds the root of a tooth.

Alzheimer's disease: A disorder resulting from degenerative changes in populations of neurons in the cerebrum, causing dementia characterized by problems with attention, short-term memory, and emotions.

amino acids: Organic compounds whose chemical structure can be summarized as R—$CHNH_2$—COOH.

amino group: $-NH_2$.

amnion: One of the four extra-embryonic membranes; surrounds the developing embryo or fetus.

amniotic fluid: Fluid that fills the amniotic cavity; cushions and supports the embryo or fetus.

amphiarthrosis: An articulation that permits a small degree of independent movement; *see* **interosseous membrane** and **pubic symphysis.**

ampulla/ampullae: A localized dilation in the lumen of a canal or passageway.

amygdaloid body: A basal nucleus that is a component of the limbic system and acts as an interface between that system, the cerebrum, and sensory systems.

amylase: An enzyme that breaks down polysaccharides; produced by the salivary glands and pancreas.

anabolism: The synthesis of complex organic compounds from simpler precursors.

anaerobic: Without oxygen.

anal triangle: The posterior subdivision of the perineum.

anaphase: The mitotic stage in which the paired chromatids separate and move toward opposite ends of the spindle apparatus.

anaphylaxis: A hypersensitivity reaction due to the binding of antigens to immunoglobulins (IgE) on the surfaces of mast cells; the release of histamine, serotonin, and prostaglandins by mast cells then causes widespread inflammation; a sudden decline in blood pressure may occur, producing anaphylactic shock.

anastomosis: The joining of two tubes, usually referring to a connection between two peripheral vessels without an intervening capillary bed.

anatomical position: An anatomical reference position; the body viewed from the anterior surface with the palms facing forward.

anatomy: The study of the structure of the body.

androgen: A steroid sex hormone primarily produced by the interstitial cells of the testis and manufactured in small quantities by the adrenal cortex in both sexes.

anemia: The condition marked by a reduction in the hematocrit, the hemoglobin content of the blood, or both.

angiotensin I: The hormone produced by the activation of angiotensinogen by renin; angiotensin-converting enzyme converts angiotensin I into angiotensin II in lung capillaries.

angiotensin II: The hormone produced by the renin-angiotensin system that causes an elevation in systemic blood pressure, stimulates the secretion of aldosterone, promotes thirst, and causes the release of antidiuretic hormone; angiotensin-converting enzyme in lung capillaries converts angiotensin I into angiotensin II.

angiotensinogen: The blood protein produced by the liver that is converted to angiotensin I by the enzyme renin.

anion: An ion bearing a negative charge.

anoxia: Tissue oxygen deprivation.

antagonist: A muscle that opposes the movement of an agonist.

antebrachium: The forearm.

anterior: On or near the front, or ventral surface, of the body.

antibiotic: A chemical agent that selectively kills pathogens, primarily bacteria.

antibody: A globular protein produced by plasma cells that will bind to specific antigens and promote their destruction or removal from the body.

antibody-mediated immunity: The form of immunity resulting from the presence of circulating antibodies produced by plasma cells; also called *humoral immunity*.

anticholinesterase: A chemical compound that blocks the action of acetylcholine and causes prolonged and intensive stimulation of postsynaptic membranes.

anticodon: Three nitrogenous bases on a tRNA molecule that interact with a complementary codon on a strand of mRNA.

antidiuretic hormone (ADH): A hormone synthesized in the hypothalamus and secreted at the neurohypophysis (posterior lobe of the pituitary gland); causes water retention by the kidneys and an elevation of blood pressure.

antigen: A substance capable of inducing the production of antibodies.

antigen-antibody complex: The combination of an antigen and a specific antibody.

antigenic determinant site: A portion of an antigen that can interact with an antibody molecule.

antigen-presenting cell (APC): A cell that processes antigens and displays them, bound to MHC proteins; essential to the initiation of a normal immune response.

antihistamines: A chemical agent that blocks the action of histamine on peripheral tissues.

antrum: A chamber or pocket.

anulus: A cartilage or bone shaped like a ring; also spelled *annulus*.

anus: The external opening of the anal canal.

aorta: The large, elastic artery that carries blood away from the left ventricle and into the systemic circuit.

apocrine secretion: A mode of secretion in which the glandular cell sheds portions of its cytoplasm.

aponeurosis/aponeuroses: A broad tendinous sheet that may serve as the origin or insertion of a skeletal muscle.

appendicular: Pertaining to the upper or lower limbs.

appendix: A blind tube connected to the cecum of the large intestine.

appositional growth: The enlargement of a cartilage or bone by the addition of cartilage or bony matrix at its surface.

aqueous humor: A fluid similar to perilymph or cerebrospinal fluid that fills the anterior chamber of the eye.

arachidonic acid: One of the essential fatty acids.

arachnoid granulations: Processes of the arachnoid mater that project into the superior sagittal sinus; sites where cerebrospinal fluid enters the venous circulation.

arachnoid mater: The middle meninx that encloses cerebrospinal fluid and protects the central nervous system.

arbor vitae: The central, branching mass of white matter inside the cerebellum.

arcuate: Curving.

areolar tissue: Loose connective tissue with an open framework.

arrector pili: Smooth muscles whose contractions force hairs to stand erect.

arrhythmias: Abnormal patterns of cardiac contractions.

arteriole: A small arterial branch that delivers blood to a capillary network.

artery: A blood vessel that carries blood away from the heart and toward a peripheral capillary.

articular: Pertaining to a joint.

articular capsule: The dense collagen fiber sleeve that surrounds a joint and provides protection and stabilization.

articular cartilage: The cartilage pad that covers the surface of a bone inside a joint cavity.

articulation: A joint; the formation of words.

ascending tract: A tract carrying information from the spinal cord to the brain.

association areas: Cortical areas of the cerebrum that are responsible for the integration of sensory inputs and/or motor commands.

astrocyte: One of the four types of neuroglia in the central nervous system; responsible for maintaining the blood–brain barrier by the stimulation of endothelial cells.

atherosclerosis: The formation of fatty plaques in the walls of arteries, restricting blood flow to deep tissues.

atom: The smallest stable unit of matter.

atomic number: The number of protons in the nucleus of an atom.

atomic weight: Roughly, the average total number of protons and neutrons in the atoms of a particular element.

atria: Thin-walled chambers of the heart that receive venous blood from the pulmonary or systemic circuit.

atrial natriuretic peptide (ANP): Hormone released by specialized atrial cardiocytes when they are stretched by an abnormally large venous return; promote fluid loss and reductions in blood pressure and in venous return.

atrioventricular (AV) node: Specialized cardiocytes that relay the contractile stimulus to the bundle of His, the bundle branches, the Purkinje fibers, and the ventricular myocardium; located at the boundary between the atria and ventricles.

atrioventricular (AV) valve: One of the valves that prevents backflow into the atria during ventricular systole.

atrophy: The wasting away of tissues from a lack of use, ischemia, or nutritional abnormalities.

auditory: Pertaining to the sense of hearing.

auditory ossicles: The bones of the middle ear: malleus, incus, and stapes.

auditory tube: A passageway that connects the nasopharynx with the middle ear cavity.

auricle: A broad, flattened process that resembles the external ear; in the ear, the expanded, projecting portion that surrounds the external auditory meatus, also called *pinna*; in the heart, the externally visible flap formed by the collapse of the outer wall of a relaxed atrium.

autoantibodies: Antibodies that react with antigens on the surfaces of a person's own cells and tissues.

autoimmunity: The immune system's sensitivity to normal cells and tissues, resulting in the production of autoantibodies.

autolysis: The destruction of a cell due to the rupture of lysosomal membranes in its cytoplasm.

automaticity: The spontaneous depolarization to threshold, characteristic of cardiac pacemaker cells.

autonomic ganglion: A collection of visceral motor neurons outside the central nervous system.

autonomic nerve: A peripheral nerve consisting of preganglionic or postganglionic autonomic fibers.

autonomic nervous system (ANS): Centers, nuclei, tracts, ganglia, and nerves involved in the unconscious regulation of visceral functions; includes components of the central nervous system and the peripheral nervous system.

autopsy: The detailed examination of a body after death.

autoregulation: Changes in activity that maintain homeostasis in direct response to changes in the local environment; does not require neural or endocrine control.

autosomal: Chromosomes other than the X or Y sex chromosome.

avascular: Without blood vessels.

axilla: The armpit.

axon: The elongate extension of a neuron that conducts an action potential.

axon hillock: In a multipolar neuron, the portion of the cell body adjacent to the initial segment.

axoplasm: The cytoplasm within an axon.

B

bacteria: Single-celled microorganisms, some pathogenic, that are common in the environment and in and on the body.

baroreception: The ability to detect changes in pressure.

baroreceptor reflex: A reflexive change in cardiac activity in response to changes in blood pressure.

baroreceptors: The receptors responsible for baroreception.

basal nuclei: Nuclei of the cerebrum that are important in the subconscious control of skeletal muscle activity.

base: A compound whose dissociation releases a hydroxide ion (OH^-) or removes a hydrogen ion (H^+) from the solution.

basement membrane: A layer of filaments and fibers that attach an epithelium to the underlying connective tissue.

basophils: Circulating granulocytes (white blood cells) similar in size and function to tissue mast cells.

B cells: Lymphocytes capable of differentiating into plasmocytes (plasma cells), which produce antibodies.

benign: Not malignant.

beta cells: Cells of the pancreatic islets that secrete insulin in response to elevated blood sugar concentrations.

beta oxidation: Fatty acid catabolism that produces molecules of acetyl-CoA.

bicarbonate ions: HCO_3^-; anion components of the carbonic acid–bicarbonate buffer system.

bicuspid: Having two cusps or points; refers to a premolar tooth, which has two roots, or to the left AV valve, which has two cusps.

bicuspid valve: The left atrioventricular (AV) valve, also called *mitral valve*.

bifurcate: To branch into two parts.

bile: The exocrine secretion of the liver; stored in the gallbladder and ejected into the duodenum.

bile salts: Steroid derivatives in bile; responsible for the emulsification of ingested lipids.

bilirubin: A pigment that is the by-product of hemoglobin catabolism.

biopsy: The removal of a small sample of tissue for pathological analysis.

bladder: A muscular sac that distends as fluid is stored and whose contraction ejects the fluid at an appropriate time; used alone, the term usually refers to the urinary bladder.

blastocyst: An early stage in the developing embryo, consisting of an outer trophoblast and an inner cell mass.

blood–brain barrier: The isolation of the central nervous system from the general circulation; primarily the result of astrocyte regulation of capillary permeabilities.

blood pressure: A force exerted against vessel walls by the blood in the vessels, due to the push exerted by cardiac contraction and the elasticity of the vessel walls; usually measured along one of the muscular arteries, with systolic pressure measured during ventricular systole and diastolic pressure during ventricular diastole.

bolus: A compact mass; usually refers to compacted ingested material on its way to the stomach.

bone: *See* **osseous tissue.**

bowel: The intestinal tract.

brachial: Pertaining to the arm.

brachial plexus: A network formed by branches of spinal nerves C_5–T_1 en route to innervating the upper limb.

brachium: The arm.

bradycardia: An abnormally slow heart rate, usually below 50 bpm.

brain stem: The brain minus the cerebrum, diencephalon, and cerebellum.

brevis: Short.

bronchial tree: The trachea, bronchi, and bronchioles.

bronchodilation: The dilation of the bronchial passages; can be caused by sympathetic stimulation.

bronchus/bronchi: A branch of the bronchial tree between the trachea and bronchioles.

buccal: Pertaining to the cheeks.

buffer: A compound that stabilizes the pH of a solution by removing or releasing hydrogen ions.

buffer system: Interacting compounds that prevent increases or decreases in the pH of body fluids; includes the carbonic acid–bicarbonate buffer system, the phosphate buffer system, and the protein buffer system.

bulbo-urethral glands: Mucous glands at the base of the penis that secrete into the penile urethra; the equivalent of the greater vestibular glands of females; also called *Cowper glands.*

bundle branches: Specialized conducting cells in the ventricles that carry the contractile stimulus from the bundle of His to the Purkinje fibers.

bundle of His: Specialized conducting cells in the interventricular septum that carry the contracting stimulus from the AV node to bundle branches and then to Purkinje fibers.

bursa: A small sac filled with synovial fluid that cushions adjacent structures and reduces friction.

C

calcaneal tendon: The large tendon that inserts on the calcaneus; tension on this tendon produces extension (plantar flexion) of the foot; also called *Achilles tendon.*

calcaneus: The heel bone, the largest of the tarsal bones.

calcification: The deposition of calcium salts within a tissue.

calcitonin: The hormone secreted by C cells of the thyroid when calcium ion concentrations are abnormally high; restores homeostasis by increasing the rate of bone deposition and the rate of calcium loss by the kidneys.

calculus/calculi: A solid mass of insoluble materials that form within body fluids, especially the gallbladder, kidneys, or urinary bladder.

callus: A localized thickening of the epidermis due to chronic mechanical stresses; a thickened area that forms at the site of a bone break as part of the repair process.

canaliculi: Microscopic passageways between cells; bile canaliculi carry bile to bile ducts in the liver; in bone, canaliculi permit the diffusion of nutrients and wastes to and from osteocytes.

cancer: An illness caused by mutations leading to the uncontrolled growth and replication of the affected cells.

capacitation: The activation process that must occur before a spermatozoon can successfully fertilize an oocyte; occurs in the vagina after ejaculation.

capillary: A small blood vessel, located between an arteriole and a venule, whose thin wall permits the diffusion of gases, nutrients, and wastes between plasma and interstitial fluids.

capitulum: A general term for a small, elevated articular process; refers to the rounded distal surface of the humerus that articulates with the head of the radius.

caput: The head.

carbaminohemoglobin: Hemoglobin bound to carbon dioxide molecules.

carbohydrase: An enzyme that breaks down carbohydrate molecules.

carbohydrate: An organic compound containing carbon, hydrogen, and oxygen in a ratio that approximates 1:2:1.

carbon dioxide: CO_2; a compound produced by the citric acid cycle reactions of aerobic metabolism.

carbonic anhydrase: An enzyme that catalyzes the reaction $H_2O + CO_2 \rightarrow H_2CO_3$; important in carbon dioxide transport, gastric acid secretion, and renal pH regulation.

carcinogenic: Stimulating cancer formation in affected tissues.

cardia: The area of the stomach surrounding its connection with the esophagus.

cardiac: Pertaining to the heart.

cardiac cycle: One complete heartbeat, including atrial and ventricular systole and diastole.

cardiac output: The amount of blood ejected by the left ventricle each minute.

cardiac reserve: The potential percentage increase in cardiac output above resting levels.

cardiac tamponade: A compression of the heart due to fluid accumulation in the pericardial cavity.

cardiocyte: A cardiac muscle cell.

cardiovascular: Pertaining to the heart, blood, and blood vessels.

cardiovascular centers: Poorly localized centers in the reticular formation of the medulla oblongata of the brain; includes cardioacceleratory, cardioinhibitory, and vasomotor centers.

cardium: The heart.

carotene: A yellow-orange pigment, found in carrots and in green and orange leafy vegetables, that the body can convert to vitamin A.

carotid artery: The principal artery of the neck, servicing cervical and cranial structures; one branch, the internal carotid, provides a major blood supply to the brain.

carotid body: A group of receptors, adjacent to the carotid sinus, that are sensitive to changes in the carbon dioxide levels, pH, and oxygen concentrations of arterial blood.

carotid sinus: A dilated segment at the base of the internal carotid artery whose walls contain baroreceptors sensitive to changes in blood pressure.

carotid sinus reflex: Reflexive changes in blood pressure that maintain homeostatic pressures at the carotid sinus, stabilizing blood flow to the brain.

carpus/carpal: The wrist.

cartilage: A connective tissue with a gelatinous matrix that contains an abundance of fibers.

catabolism: The breakdown of complex organic molecules into simpler components, accompanied by the release of energy.

catalyst: A substance that accelerates a specific chemical reaction but that is not altered by the reaction.

catecholamine: Epinephrine, norepinephrine, dopamine, and related compounds.

catheter: A tube surgically inserted into a body cavity or along a blood vessel or excretory passageway for the collection of body fluids, monitoring of blood pressure, or introduction of medications or radiographic dyes.

cation: An ion that bears a positive charge.

cauda equina: Spinal nerve roots distal to the tip of the adult spinal cord; they extend caudally inside the vertebral canal en route to lumbar and sacral segments.

caudal/caudally: Closest to or toward the tail (coccyx).

caudate nucleus: One of the basal nuclei involved with the subconscious control of skeletal muscular activity.

cavernous tissue: Erectile tissue that can be engorged with blood; located in the penis (males) and clitoris (females).

cell: The smallest living unit in the human body.

cell-mediated immunity: Resistance to disease through the activities of sensitized T cells that destroy antigen-bearing cells by direct contact or through the release of lymphotoxins; also called *cellular immunity.*

center of ossification: The site in a connective tissue where bone formation begins.

central canal: Longitudinal canal in the center of an osteon that contains blood vessels and nerves, a passageway along the longitudinal axis of the spinal cord that contains cerebrospinal fluid.

central nervous system (CNS): The brain and spinal cord.

centriole: A cylindrical intracellular organelle composed of nine groups of microtubules, three in each group; functions in mitosis or meiosis by organizing the microtubules of the spindle apparatus.

centromere: The localized region where two chromatids remain connected after the chromosomes have replicated; site of spindle fiber attachment.

cephalic: Pertaining to the head.

cerebellum: The posterior portion of the metencephalon, containing the cerebellar hemispheres; includes the arbor vitae, cerebellar nuclei, and cerebellar cortex.

cerebral cortex: An extensive area of neural cortex covering the surfaces of the cerebral hemispheres.

cerebral hemispheres: A pair of expanded portions of the cerebrum covered in neural cortex.

cerebrospinal fluid (CSF): Fluid bathing the internal and external surfaces of the central nervous system; secreted by the choroid plexus.

cerebrovascular accident (CVA): The occlusion of a blood vessel that supplies a portion of the brain, resulting in damage to the dependent neurons; also called *stroke.*

cerebrum: The largest portion of the brain, composed of the cerebral hemispheres; includes the cerebral cortex, the basal nuclei, and the internal capsule.

cerumen: The waxy secretion of the ceruminous glands along the external acoustic meatus.

ceruminous glands: Integumentary glands that secrete cerumen.

cervix: The inferior portion of the uterus.

chemoreception: The detection of changes in the concentrations of dissolved compounds or gases.

chemotaxis: The attraction of phagocytic cells to the source of abnormal chemicals in tissue fluids.

chloride shift: The movement of plasma chloride ions into red blood cells in exchange for bicarbonate ions generated by the intracellular dissociation of carbonic acid.

cholecystokinin (CCK): A duodenal hormone that stimulates the contraction of the gallbladder and the secretion of enzymes by the exocrine pancreas.

cholesterol: A steroid component of plasma membranes and a substrate for the synthesis of steroid hormones and bile salts.

choline: A breakdown product or precursor of acetylcholine.

cholinesterase: The enzyme that breaks down and inactivates acetylcholine.

chondrocyte: A cartilage cell.

chondroitin sulfate: The predominant proteoglycan in cartilage, responsible for the gelatinous consistency of the matrix.

chordae tendineae: Fibrous cords that stabilize the position of the AV valves in the heart, preventing backflow during ventricular systole.

chorion/chorionic: An extra-embryonic membrane, consisting of the trophoblast and underlying mesoderm, that forms the placenta.

choroid: The middle, vascular layer in the wall of the eye.

choroid plexus: The vascular complex in the roof of the third and fourth ventricles of the brain, responsible for the production of cerebrospinal fluid.

chromatid: One complete copy of a DNA strand and its associated nucleoproteins.

chromatin: A histological term referring to the grainy material visible in cell nuclei during interphase; the appearance of the DNA content of the nucleus when the chromosomes are uncoiled.

chromosomes: Dense structures, composed of tightly coiled DNA strands and associated histones, that become visible in the nucleus when a cell prepares to undergo mitosis or meiosis; normal human somatic cells each contain 46 chromosomes.

chronic: Habitual or long term.

chylomicrons: Relatively large droplets that may contain triglycerides, phospholipids, and cholesterol in association with proteins; synthesized and released by intestinal cells and transported to the venous blood by the lymphatic system.

ciliary body: A thickened region of the choroid that encircles the lens of the eye; includes the ciliary muscle and the ciliary processes that support the suspensory ligaments of the lens.

cilium/cilia: A slender organelle that extends above the free surface of an epithelial cell and generally undergoes cycles of movement; composed of a basal body and microtubules in a 9 + 2 array.

circulatory system: The network of blood vessels and lymphatic vessels that facilitate the distribution and circulation of extracellular fluid.

circumduction: A movement at a synovial joint in which the distal end of the bone moves in a circular direction, but the shaft does not rotate.

circumvallate papilla: One of the large, dome-shaped papillae on the superior surface of the tongue that forms a V, separating the body of the tongue from the root.

cisterna: An expanded or flattened chamber; associated with the Golgi apparatus.

citric acid cycle: The reaction sequence that occurs in the matrix of mitochondria; in the process, organic molecules are broken down, carbon dioxide molecules are released, and hydrogen atoms are transferred to coenzymes that deliver them to the electron transport system.

clot: A network of fibrin fibers and trapped blood cells; also called a *thrombus* if it occurs within the cardiovascular system.

clotting factors: Plasma proteins, synthesized by the liver, that are essential to the clotting response.

clotting response: The series of events that results in the formation of a clot.

coccyx: The terminal portion of the spinal column, consisting of relatively tiny, fused vertebrae.

cochlea: The spiral portion of the bony labyrinth of the internal ear that surrounds the organ of hearing.

cochlear duct: The central membranous tube within the cochlea that is filled with endolymph and contains the spiral organ (*organ of Corti*); also called *scala media*.

codon: A sequence of three nitrogenous bases along an mRNA strand that will specify the location of a single amino acid in a peptide chain.

coelom: The ventral body cavity, lined by a serous membrane and subdivided during fetal development into the pleural, pericardial, and abdomino-pelvic (peritoneal) cavities.

coenzymes: Complex organic cofactors; most are structurally related to vitamins.

cofactor: Ions or molecules that must be attached to the active site before an enzyme can function; examples include mineral ions and several vitamins.

collagen: A strong, insoluble protein fiber common in connective tissues.

collateral ganglion: A sympathetic ganglion situated anterior to the spinal column and separate from the sympathetic chain.

colliculus/colliculi: A little mound; in the brain, refers to one of the thickenings in the roof of the mesencephalon; the superior colliculi are associated with the visual system, and the inferior colliculi with the auditory system.

colloid/colloidal suspension: A solution containing large organic molecules in suspension.

colon: The large intestine.

comminuted: Broken or crushed into small pieces.

commissure: A crossing over from one side to another.

common bile duct: The duct formed by the union of the cystic duct from the gallbladder and the bile ducts from the liver; terminates at the duodenal ampulla, where it meets the pancreatic duct.

compact bone: Dense bone that contains parallel osteons.

complement: A system of 11 plasma proteins that interact in a chain reaction after exposure to activated antibodies or the surfaces of certain pathogens; complement proteins promote cell lysis, phagocytosis, and other defense mechanisms.

compliance: Distensibility; the ability of certain organs to tolerate changes in volume; indicates the presence of elastic fibers and smooth muscles.

compound: A molecule containing two or more elements in combination.

concentration: The amount (in grams) or number of atoms, ions, or molecules (in moles) per unit volume.

concentration gradient: Regional differences in the concentration of a particular substance.

conception: Fertilization.

concha/conchae: Three pairs of thin, scroll-like bones that project into the nasal cavities; the superior and middle conchae are part of the ethmoid, and the inferior conchae articulate with the ethmoid, lacrimal, maxilla, and palatine bones.

condyle: A rounded articular projection on the surface of a bone.

congenital: Present at birth.

congestive heart failure (CHF): The failure to maintain adequate cardiac output due to cardiovascular problems or myocardial damage.

conjunctiva: A layer of stratified squamous epithelium that covers the inner surfaces of the eyelids and the anterior surface of the eye to the edges of the cornea.

connective tissue: One of the four primary tissue types; provides a structural framework that stabilizes the relative positions of the other tissue types; includes connective tissue proper, cartilage, bone, and blood; contains cell products, cells, and ground substance.

continuous propagation: The propagation of an action potential along an unmyelinated axon or a muscle plasma membrane, wherein the action potential affects every portion of the membrane surface.

contractility: The ability to contract; possessed by skeletal, smooth, and cardiac muscle cells.

conus medullaris: The conical tip of the spinal cord that gives rise to the filum terminale.

convergence: In the nervous system, the innervation of a single neuron by axons from several neurons; most common along motor pathways.

coracoid process: A hook-shaped process of the scapula that projects above the anterior surface of the capsule of the shoulder joint.

Cori cycle: The metabolic exchange of lactate from skeletal muscle for glucose from the liver; performed during the recovery period after muscular exertion.

cornea: The transparent portion of the fibrous layer of the anterior surface of the eye.

corniculate cartilages: A pair of small laryngeal cartilages.

cornu: Horn-shaped.

coronoid: Hooked or curved.

corpora quadrigemina: The superior and inferior colliculi of the mesencephalic tectum (roof) in the brain.

corpus/corpora: Body.

corpus callosum: A large bundle of axons that links centers in the left and right cerebral hemispheres.

corpus luteum: The progestin-secreting mass of follicle cells that develops in the ovary after ovulation.

cortex: The outer layer or portion of an organ.

corticospinal tracts: Descending tracts that carry motor commands from the cerebral cortex to the anterior gray horns of the spinal cord.

corticosteroid: A steroid hormone produced by the adrenal cortex.

corticosterone: A corticosteroid secreted by the zona fasciculata of the adrenal cortex; a glucocorticoid.

corticotropin-releasing hormone (CRH): The releasing hormone, secreted by the hypothalamus, that stimulates secretion of adrenocorticotropic hormone by the anterior lobe of the pituitary gland.

cortisol: A corticosteroid secreted by the zona fasciculata of the adrenal cortex; a glucocorticoid.

costa/costae: A rib/ribs.

cotransport: The membrane transport of a nutrient, such as glucose, in company with the movement of an ion, normally sodium; transport requires a carrier protein but does not involve direct ATP expenditure and can occur regardless of the concentration gradient for the nutrient.

countercurrent exchange: The transfer of heat, water, or solutes between two fluids that travel in opposite directions.

countercurrent multiplication: Active transport between two limbs of a loop that contains a fluid moving in one direction; responsible for the concentration of urine in the kidney tubules.

covalent bond: A chemical bond between atoms that involves the sharing of electrons.

coxal bone: Hip bone.

cranial: Pertaining to the head.

cranial nerves: Peripheral nerves originating at the brain.

cranium: The braincase; the skull bones that surround and protect the brain.

creatine: A nitrogenous compound, synthesized in the body, that can form a high-energy bond by connecting to a phosphate group and that serves as an energy reserve.

creatine phosphate: A high-energy compound in muscle cells; during muscle activity, the phosphate group is donated to ADP, regenerating ATP; also called *phosphorylcreatine*.

creatinine: A breakdown product of creatine metabolism.

crenation: Cellular shrinkage due to an osmotic movement of water out of the cytoplasm.

cribriform plate: A portion of the ethmoid bone that contains the foramina used by the axons of olfactory receptors en route to the olfactory bulbs of the cerebrum.

cricoid cartilage: A ring-shaped cartilage that forms the inferior margin of the larynx.

crista ampullaris: The region in the wall of the ampulla of a semicircular duct containing hair cells; a receptor complex sensitive to movement along the plane of the semicircular canal.

cross-bridge: A myosin head that projects from the surface of a thick filament and that can bind to an active site of a thin filament in the presence of calcium ions.

cuneiform cartilages: A pair of small cartilages in the larynx.

cupula: A gelatinous mass that is located in the ampulla of a semicircular duct in the internal ear and whose movement stimulates the hair cells of the crista ampullaris.

cutaneous membrane: The epidermis and papillary layer of the dermis.

cuticle: The layer of dead, keratinized cells that surrounds the shaft of a hair; a narrow zone of stratum corneum that extends across the surface of a nail at its exposed base.

cyanosis: A bluish coloration of the skin due to the presence of deoxygenated blood in vessels near the body surface.

cystic duct: A duct that carries bile between the gallbladder and the common bile duct.

cytochrome: A pigment component of the electron transport system.

cytokinesis: The cytoplasmic movement that separates two daughter cells at the completion of mitosis.

cytology: The study of cells.

cytoplasm: The material between the plasma membrane and the nuclear membrane; cell contents.

cytosine: One of the nitrogenous bases in the nucleic acids RNA and DNA.

cytoskeleton: A network of microtubules and microfilaments in the cytoplasm.

cytosol: The fluid portion of the cytoplasm.

cytotoxic: Poisonous to cells.

cytotoxic T cells: Lymphocytes involved in cell-mediated immunity that kill target cells by direct contact or by the secretion of lymphotoxins; also called *killer T cells* and *T_C cells*.

D

daughter cells: Genetically identical cells produced by somatic cell division.

deamination: The removal of an amino group from an amino acid.

decomposition reaction: A chemical reaction that breaks a molecule into smaller fragments.

defecation: The elimination of fecal wastes.

degradation: Breakdown, catabolism.

dehydration: A reduction in the water content of the body that threatens homeostasis.

dehydration synthesis: The joining of two molecules associated with the removal of a water molecule.

demyelination: The loss of the myelin sheath of an axon, normally due to chemical or physical damage to Schwann cells or oligodendrocytes.

denaturation: A temporary or permanent change in the three-dimensional structure of a protein.

dendrite: A sensory process of a neuron.

deoxyribonucleic acid (DNA): A nucleic acid consisting of a double chain of nucleotides that contains the sugar deoxyribose and the nitrogenous bases adenine, guanine, cytosine, and thymine.

deoxyribose: A five-carbon sugar resembling ribose but lacking an oxygen atom.

depolarization: A change in the membrane potential from a negative value toward 0 mV.

depression: Inferior (downward) movement of a body part.

dermatome: A sensory region monitored by the dorsal rami of a single spinal segment.

dermis: The connective tissue layer beneath the epidermis of the skin.

detrusor muscle: Collectively, the three layers of smooth muscle in the wall of the urinary bladder.

development: Growth and the acquisition of increasing structural and functional complexity; includes the period from conception to maturity.

diabetes mellitus: Polyuria and glycosuria, most commonly due to the inadequate production or diminished sensitivity to insulin with a resulting elevation of blood glucose levels.

diapedesis: The movement of white blood cells through the walls of blood vessels by migration between adjacent endothelial cells.

diaphragm: Any muscular partition; the respiratory muscle that separates the thoracic cavity from the abdominopelvic cavity.

diaphysis: The shaft of a long bone.

diarthrosis: A synovial joint.

diastolic pressure: Pressure measured in the walls of a muscular artery when the left ventricle is in diastole (relaxation).

diencephalon: A division of the brain that includes the epithalamus, thalamus, and hypothalamus.

differential count: The determination of the relative abundance of each type of white blood cell on the basis of a random sampling of 100 white blood cells.

differentiation: The gradual appearance of characteristic cellular specializations during development as the result of gene activation or repression.

diffusion: Passive molecular movement from an area of higher concentration to an area of lower concentration.

digestion: The chemical breakdown of ingested materials into simple molecules that can be absorbed by the cells of the digestive tract.

digestive system: The digestive tract and associated glands.

digestive tract: An internal passageway that begins at the mouth, ends at the anus, and is lined by a mucous membrane; also called *gastrointestinal tract*.

dilate: To increase in diameter; to enlarge or expand.

disaccharide: A compound formed by the joining of two simple sugars by dehydration synthesis.

dissociation: *See* ionization.

distal: A direction away from the point of attachment or origin; for a limb, away from its attachment to the trunk.

distal convoluted tubule (DCT): The portion of the nephron closest to the connecting tubules and collecting duct; an important site of active secretion.

diuresis: Fluid loss at the kidneys; the production of unusually large volumes of urine.

divergence: In neural tissue, the spread of information from one neuron to many neurons; an organizational pattern common along sensory pathways of the central nervous system.

diverticulum: A sac or pouch in the wall of the colon or other organ.

DNA molecule: Two DNA strands wound in a double helix and held together by hydrogen bonds between complementary nitrogenous base pairs.

dorsal: Toward the back, posterior.

dorsal root ganglion: A peripheral nervous system ganglion containing the cell bodies of sensory neurons.

dorsiflexion: Upward movement of the foot through flexion at the ankle.

Down's syndrome: A genetic abnormality resulting from the presence of three copies of chromosome 21; individuals with this condition have characteristic physical and intellectual deficits.

duct: A passageway that delivers exocrine secretions to an epithelial surface.

ductus arteriosus: A vascular connection between the pulmonary trunk and the aorta that functions throughout fetal life; normally closes at birth or shortly thereafter and persists as the ligamentum arteriosum.

ductus deferens: A passageway that carries spermatozoa from the epididymis to the ejaculatory duct; also called the vas deferens.

duodenal ampulla: A chamber that receives bile from the common bile duct and pancreatic secretions from the pancreatic duct.

duodenal papilla: A conical projection from the inner surface of the duodenum that contains the opening of the duodenal ampulla.

duodenum: The proximal 25 cm (9.8 in.) of the small intestine that contains short villi and submucosal glands.

dura mater: The outermost component of the cranial and spinal meninges.

E

eccrine glands: Sweat glands of the skin that produce a watery secretion.

ectoderm: One of the three primary germ layers; covers the surface of the embryo and gives rise to the nervous system, the epidermis and associated glands, and a variety of other structures.

ectopic: Outside the normal location.

effector: A peripheral gland or muscle cell innervated by a motor neuron.

efferent: Away from an organ or structure.

efferent arteriole: An arteriole carrying blood away from a glomerulus of the kidney.

efferent fiber: An axon that carries impulses away from the central nervous system.

ejaculation: The ejection of semen from the penis as the result of muscular contractions of the bulbospongiosus and ischiocavernosus muscles.

ejaculatory ducts: Short ducts that pass within the walls of the prostate gland and connect the ductus deferens with the prostatic urethra.

elastase: A pancreatic enzyme that breaks down elastin fibers.

elastin: Connective tissue fibers that stretch and recoil, providing elasticity to connective tissues.

electrocardiogram (ECG, EKG): A graphic record of the electrical activities of the heart, as monitored at specific locations on the body surface.

electroencephalogram (EEG): A graphic record of the electrical activities of the brain.

electrolytes: Soluble inorganic compounds whose ions will conduct an electrical current in solution.

electron: One of the three fundamental subatomic particles; bears a negative charge and normally orbits the protons of the nucleus.

electron transport system (ETS): The cytochrome system responsible for aerobic energy production in cells; a complex bound to the inner mitochondrial membrane.

element: All the atoms with the same atomic number.

GLOSSARY

elevation: Movement in a superior, or upward, direction.

embryo: The developmental stage beginning at fertilization and ending at the start of the third developmental month.

embryology: The study of embryonic development, focusing on the first two months after fertilization.

endocardium: The simple squamous epithelium that lines the heart and is continuous with the endothelium of the great vessels.

endochondral ossification: The replacement of a cartilaginous model by bone; the characteristic mode of formation for skeletal elements other than the bones of the cranium, the clavicles, and sesamoid bones.

endocrine gland: A gland that secretes hormones into the blood.

endocrine system: The endocrine (ductless) glands/organs of the body.

endocytosis: The movement of relatively large volumes of extracellular material into the cytoplasm via the formation of a membranous vesicle at the cell surface; includes pinocytosis and phagocytosis.

endoderm: One of the three primary germ layers; the layer on the undersurface of the embryonic disc; gives rise to the epithelia and glands of the digestive system, the respiratory system, and portions of the urinary system.

endogenous: Produced within the body.

endolymph: The fluid contents of the membranous labyrinth (the saccule, utricle, semicircular ducts, and cochlear duct) of the internal ear.

endometrium: The mucous membrane lining the uterus.

endomysium: A delicate network of connective tissue fibers that surrounds individual muscle cells.

endoneurium: A delicate network of connective tissue fibers that surrounds individual nerve fibers.

endoplasmic reticulum: A network of membranous channels in the cytoplasm of a cell that function in intracellular transport, synthesis, storage, packaging, and secretion.

endosteum: An incomplete cellular lining on the inner (medullary) surfaces of bones.

endothelium: The simple squamous epithelial cells that line blood and lymphatic vessels.

enteroendocrine cells: Endocrine cells scattered among the epithelial cells that line the digestive tract.

enterogastric reflex: The reflexive inhibition of gastric secretion; initiated by the arrival of chyme in the small intestine.

enterokinase: An enzyme in the lumen of the small intestine that activates the proenzymes secreted by the pancreas.

enzyme: A protein that catalyzes a specific biochemical reaction.

eosinophil: A microphage (white blood cell) with a lobed nucleus and red-staining granules; participates in the immune response and is especially important during allergic reactions.

ependyma: The layer of cells lining the ventricles and central canal of the central nervous system.

epicardium: A serous membrane covering the outer surface of the heart; also called *visceral pericardium*.

epidermis: The epithelium covering the surface of the skin.

epididymis: A coiled duct that connects the rete testis to the ductus deferens; site of functional maturation of spermatozoa.

epidural space: The space between the spinal dura mater and the walls of the vertebral foramen; contains blood vessels and adipose tissue; a common site of injection for regional anesthesia.

epiglottis: A blade-shaped flap of tissue, reinforced by cartilage, that is attached to the dorsal and superior surface of the thyroid cartilage; folds over the entrance to the larynx during swallowing.

epimysium: A dense layer of collagen fibers that surrounds a skeletal muscle and is continuous with the tendons/aponeuroses of the muscle and with the perimysium.

epineurium: A dense layer of collagen fibers that surrounds a peripheral nerve.

epiphyseal cartilage: The cartilaginous region between the epiphysis and diaphysis of a growing bone.

epiphysis: The head of a long bone.

epithelium: One of the four primary tissue types; a layer of cells that forms a superficial covering or an internal lining of a body cavity or vessel.

equilibrium: A dynamic state in which two opposing forces or processes are in balance.

erection: The stiffening of the penis due to the engorgement of the erectile tissues of the corpora cavernosa and corpus spongiosum.

erythema: Redness and inflammation at the surface of the skin.

erythrocyte: A red blood cell; has no nucleus and contains large quantities of hemoglobin.

erythropoietin: A hormone released by most tissues, and especially by the kidneys, when exposed to low oxygen concentrations; stimulates erythropoiesis (red blood cell formation) in bone marrow.

Escherichia coli: A normal bacterial resident of the large intestine.

esophagus: A muscular tube that connects the pharynx to the stomach.

essential amino acids: Amino acids that cannot be synthesized in the body in adequate amounts and must be obtained from the diet.

essential fatty acids: Fatty acids that cannot be synthesized in the body and must be obtained from the diet.

estrogens: A class of steroid sex hormones that includes estradiol.

evaporation: A movement of molecules from the liquid state to the gaseous state.

eversion: A turning outward.

excitable membranes: Membranes that propagate action potentials, a characteristic of muscle cells and nerve cells.

excretion: The removal of waste products from the blood, tissues, or organs.

exocrine gland: A gland that secretes onto the body surface or into a passageway connected to the exterior.

exocytosis: The ejection of cytoplasmic materials by the fusion of a membranous vesicle with the plasma membrane.

expiration: Exhalation; breathing out.

extension: An increase in the angle between two articulating bones; the opposite of flexion.

external acoustic meatus: A passageway in the temporal bone that leads to the tympanic membrane of the middle ear.

external ear: The auricle, external acoustic meatus, and tympanic membrane.

external nares: The nostrils; the external openings into the nasal cavity.

external receptors: General sensory receptors in the skin, mucous membranes, and special sense organs that provide information about the external environment and about our position within it.

external respiration: The diffusion of gases between the alveolar air and the alveolar capillaries and between the systemic capillaries and peripheral tissues.

extracellular fluid: All body fluids other than that contained within cells; includes plasma and interstitial fluid.

extra-embryonic membranes: The yolk sac, amnion, chorion, and allantois.

extrinsic pathway: A clotting pathway that begins with damage to blood vessels or surrounding tissues and ends with the formation of tissue thromboplastin.

F

facilitated diffusion: The passive movement of a substance across a plasma membrane by means of a protein carrier.

falciform ligament: A sheet of mesentery that contains the ligamentum teres, the fibrous remains of the umbilical vein of the fetus.

falx: Sickle-shaped.

fasciculus: A small bundle; usually refers to a collection of nerve axons or muscle fibers.

fatty acids: Hydrocarbon chains that end in a carboxylic acid group.

fauces: The passage from the mouth to the pharynx, bounded by the palatal arches, the soft palate, and the uvula.

feces: Waste products eliminated by the digestive tract at the anus; contains indigestible residue, bacteria, mucus, and epithelial cells.

fenestra: An opening.

fertilization: The fusion of a secondary oocyte and a spermatozoon to form a zygote.

fetus: The developmental stage lasting from the start of the third developmental month to delivery.

fibrin: Insoluble protein fibers that form the basic framework of a blood clot.

fibrinogen: A plasma protein that is the soluble precursor of the insoluble protein fibrin.

fibroblasts: Cells of connective tissue proper that are responsible for the production of extracellular fibers and the secretion of the organic compounds of the extracellular matrix.

fibrocartilage: Cartilage containing an abundance of collagen fibers; located around the edges of joints, in the intervertebral discs, the menisci of the knee, and so on.

fibrous layer: The outermost layer of the eye, composed of the sclera and cornea.

fibula: The lateral, slender bone of the leg.

filiform papillae: Slender conical projections from the dorsal surface of the anterior two-thirds of the tongue.

filtrate: The fluid produced by filtration at a glomerulus in the kidney.

filtration: The movement of a fluid across a membrane whose pores restrict the passage of solutes on the basis of size.

fimbriae: Fringes; the fingerlike processes that surround the entrance to the uterine tube.

fissure: An elongate groove or opening.

flagellum/flagella: An organelle that is structurally similar to a cilium but is used to propel a cell through a fluid; found on spermatozoa.

flatus: Intestinal gas.

flexion: A movement that reduces the angle between two articulating bones; the opposite of extension.

flexor: A muscle that produces flexion.

flexor reflex: A reflex contraction of the flexor muscles of a limb in response to a painful stimulus.

flexure: A bending.

follicle: A small secretory sac or gland.

follicle-stimulating hormone (FSH): A hormone secreted by the anterior lobe of the pituitary gland; stimulates oogenesis (female) and spermatogenesis (male).

fontanelle: A relatively soft, flexible, fibrous region between two flat bones in the developing skull; also spelled *fontanel.*

foramen/foramina: An opening or passage through a bone.

forearm: The distal portion of the upper limb between the elbow and wrist.

forebrain: The cerebrum.

fossa: A shallow depression or furrow in the surface of a bone.

fourth ventricle: An elongate ventricle of the pons, cerebellum, and medulla oblongata; the roof contains a region of choroid plexus.

fovea: The portion of the retina that provides the sharpest vision because it has the highest concentration of cones; within the *macula.*

fracture: A break or crack in a bone.

frenulum: A bridle; usually referring to a band of tissue that restricts movement (e.g., *lingual frenulum*).

frontal plane: A sectional plane that divides the body into an anterior portion and a posterior portion; also called *coronal plane.*

fructose: A hexose (six-carbon simple sugar) in foods and in semen.

fundus: The base of an organ such as the stomach, uterus, or gallbladder.

G

gallbladder: The pear-shaped reservoir for bile after it is secreted by the liver.

gametes: Reproductive cells (spermatozoa or oocytes) that contain half the normal chromosome complement.

gametogenesis: The formation of gametes.

ganglion/ganglia: A collection of neuron cell bodies outside the central nervous system.

gap junctions: Connections between cells that permit electrical coupling.

gaster: The stomach; the body, or belly, of a skeletal muscle.

gastric: Pertaining to the stomach.

gastric glands: The tubular glands of the stomach whose cells produce acid, enzymes, intrinsic factor, and hormones.

gastrointestinal (GI) tract: *See* **digestive tract.**

gene: A portion of a DNA strand that functions as a hereditary unit, is located at a particular site on a specific chromosome, and codes for a specific protein or polypeptide.

genetics: The study of mechanisms of heredity.

genitalia: The reproductive organs.

germinal centers: Pale regions in the interior of lymphoid tissues or lymphoid nodules, where cell divisions occur that produce additional lymphocytes.

gestation: The period of intrauterine development; pregnancy.

gland: Cells that produce exocrine or endocrine secretions.

glenoid cavity: A rounded depression that forms the articular surface of the scapula at the shoulder joint.

glial cells: *See* **neuroglia.**

globular proteins: Proteins whose tertiary structure makes them rounded and compact.

glomerular capsule: The expanded initial portion of the nephron that surrounds the glomerulus.

glomerular filtration rate: The rate of filtrate formation at the glomerulus.

glomerulus: A ball or knot; in the kidneys, a knot of capillaries that projects into the enlarged, proximal

end of a nephron; the site of filtration, the first step in the production of urine.

glossopharyngeal nerve: Cranial nerve IX.

glucagon: A hormone secreted by the alpha cells of the pancreatic islets; elevates blood glucose concentrations.

glucocorticoids: Hormones secreted by the adrenal cortex to modify glucose metabolism; cortisol and corticosterone are important examples.

gluconeogenesis: The synthesis of glucose from protein or lipid precursors.

glucose: A six-carbon sugar, $C_6H_{12}O_6$; the preferred energy source for most cells and normally the only energy source for neurons.

glycerides: Lipids composed of glycerol bound to fatty acids.

glycogen: A polysaccharide that is an important energy reserve; a polymer consisting of a long chain of glucose molecules.

glycogenesis: The synthesis of glycogen from glucose molecules.

glycogenolysis: Glycogen breakdown and the liberation of glucose molecules.

glycolipids: Compounds created by the combination of carbohydrate and lipid components.

glycolysis: The anaerobic cytoplasmic breakdown of glucose into two 3-carbon molecules of pyruvate, with a net gain of two ATP molecules.

glycoprotein: A compound containing a relatively small carbohydrate group attached to a large protein.

glycosuria: The presence of glucose in urine.

Golgi apparatus: A cellular organelle consisting of a series of membranous plates that give rise to lysosomes and secretory vesicles.

gomphosis: A fibrous synarthrosis that binds a tooth to the bone of the jaw.

gonadotropin-releasing hormone (GnRH): A hypothalamic releasing hormone that causes the secretion of both follicle-stimulating hormone and luteinizing hormone by the adenohypophysis (anterior pituitary gland).

gonadotropins: Follicle-stimulating hormone and luteinizing hormone, hormones that stimulate gamete development and sex hormone secretion.

gonads: Reproductive organs that produce gametes and hormones.

granulocytes: White blood cells containing granules that are visible with the light microscope; includes eosinophils, basophils, and neutrophils; also called *granular leukocytes.*

gray matter: Areas in the central nervous system that are dominated by neuron cell bodies, neuroglia, and unmyelinated axons.

greater omentum: A large fold of the dorsal mesentery of the stomach; hangs anterior to the intestines.

groin: The inguinal region.

gross anatomy: The study of the structural features of the body without the aid of a microscope.

growth hormone (GH): An adenohypophysis (anterior pituitary gland) hormone that stimulates tissue growth and anabolism when nutrients are abundant and restricts tissue glucose dependence when nutrients are in short supply.

growth hormone–inhibiting hormone (GH–IH): A hypothalamic regulatory hormone that inhibits growth hormone secretion by the anterior lobe of the pituitary gland.

guanine: One of the nitrogenous bases in the nucleic acids RNA and DNA.

gustation: Taste.

gyrus: A prominent fold or ridge of neural cortex on the surfaces of the cerebral hemispheres.

H

hair: A keratinous strand produced by epithelial cells of the hair follicle.

hair cells: Sensory cells of the internal ear.

hair follicle: An accessory structure of the integument; a tube lined by a stratified squamous epithelium that begins at the surface of the skin and ends at the hair papilla.

hallux: The big toe.

haploid: Possessing half the normal number of chromosomes; a characteristic of gametes.

hard palate: The bony roof of the oral cavity, formed by the maxillae and palatine bones.

helper T cells: Lymphocytes whose secretions and other activities coordinate cell-mediated and antibody-mediated immunities; also called T_H *cells.*

hematocrit: The percentage of the volume of whole blood contributed by cells; also called *volume of packed red cells (VPRC)* or *packed cell volume (PCV).*

hematoma: An abnormal collection of clotted or partially clotted blood outside a blood vessel.

hematuria: The abnormal presence of red blood cells in urine.

heme: A porphyrin ring containing a central iron atom that can reversibly bind oxygen molecules; a component of the hemoglobin molecule.

hemocytoblasts: Stem cells whose divisions produce each of the various populations of blood cells.

hemoglobin: A protein composed of four globular subunits, each bound to a heme molecule; gives red blood cells the ability to transport oxygen in the blood.

hemolysis: The breakdown of red blood cells.

hemopoiesis: Blood cell formation and differentiation.

hemorrhage: Blood loss; to bleed.

hemostasis: The cessation of bleeding.

heparin: An anticoagulant released by activated basophils and mast cells.

hepatic duct: The duct that carries bile away from the liver lobes and toward the union with the cystic duct.

hepatic portal vein: The vessel that carries blood between the intestinal capillaries and the sinusoids of the liver.

hepatocyte: A liver cell.

heterozygous: Possessing two different alleles at corresponding sites on a chromosome pair; the individual's phenotype is determined by one or both of the alleles.

hiatus: A gap, cleft, or opening.

high-density lipoprotein (HDL): A lipoprotein with a relatively small lipid content; responsible for the movement of cholesterol from peripheral tissues to the liver.

hilum: A localized region where blood vessels, lymphatic vessels, nerves, and/or other anatomical structures are attached to an organ.

hippocampus: A region, beneath the floor of a lateral ventricle, involved with emotional states and the conversion of short-term to long-term memories.

histamine: The chemical released by stimulated mast cells or basophils to initiate or enhance an inflammatory response.

histology: The study of tissues.

histones: Proteins associated with the DNA of the nucleus; the DNA strands are wound around them.

holocrine: A form of exocrine secretion in which the secretory cell becomes swollen with vesicles and then ruptures.

homeostasis: The maintenance of a relatively constant internal environment.

hormone: A compound that is secreted by one cell and travels through the bloodstream to affect the activities of cells in another portion of the body.

human chorionic gonadotropin (hCG): The placental hormone that maintains the corpus luteum for the first three months of pregnancy.

human immunodeficiency virus (HIV): The infectious agent that causes acquired immune deficiency syndrome (AIDS).

human placental lactogen (hPL): The placental hormone that stimulates the functional development of the mammary glands.

humoral immunity: *See* **antibody-mediated immunity**.

hyaluronan: A carbohydrate component of proteoglycans in the matrix of many connective tissues.

hyaluronidase: An enzyme that breaks down the bonds between adjacent follicle cells; produced by some bacteria and found in the acrosomal cap of a spermatozoon.

hydrogen bond: A weak interaction between the hydrogen atom on one molecule and a negatively charged portion of another molecule.

hydrolysis: The breakage of a chemical bond through the addition of a water molecule; the reverse of dehydration synthesis.

hydrophilic: Freely associating with water; readily entering into solution; water-loving.

hydrophobic: Incapable of freely associating with water molecules; insoluble; water-fearing.

hydrostatic pressure: Fluid pressure.

hydroxide ion: OH^-.

hypercapnia: High plasma carbon dioxide concentrations, commonly as a result of hypoventilation or inadequate tissue perfusion.

hyperpolarization: The movement of the membrane potential away from the normal resting potential and farther from 0 mV.

hypersecretion: The overactivity of glands that produce exocrine or endocrine secretions.

hypertension: Abnormally high blood pressure.

hypertonic: In comparing two solutions, the solution with the higher osmolarity.

hypertrophy: An increase in tissue size without cell division.

hyperventilation: A rate of respiration sufficient to reduce plasma carbon dioxide concentrations to levels below normal.

hypocapnia: An abnormally low plasma carbon dioxide concentration; commonly a result of hyperventilation.

hypodermic needle: A needle inserted through the skin to introduce drugs into the subcutaneous layer.

hypodermis: The layer of loose connective tissue below the dermis; also called *subcutaneous layer* or *superficial fascia*.

hypophyseal portal system: The network of vessels that carries blood from capillaries in the hypothalamus to capillaries in the anterior lobe of the pituitary gland.

hyposecretion: Abnormally low rates of exocrine or endocrine secretion.

hypothalamus: The floor of the diencephalon; the region of the brain containing centers involved with the subconscious regulation of visceral functions, emotions, drives, and the coordination of neural and endocrine functions.

hypotonic: In comparing two solutions, the solution with the lower osmolarity.

hypoventilation: A respiratory rate that is insufficient to keep plasma carbon dioxide concentrations within normal levels.

hypoxia: A low tissue oxygen concentration.

I

ileum: The distal 2.5 m of the small intestine.

ilium: The largest of the three bones whose fusion creates a coxal bone (hip bone).

immunity: Resistance to injuries and diseases caused by foreign compounds, toxins, or pathogens.

immunization: The production of immunity by the deliberate exposure to antigens under conditions that prevent the development of illness but stimulate the production of memory B cells.

immunoglobulin: A circulating antibody.

implantation: The attachment of a blastocyst into the endometrium of the uterine wall.

inclusions: Aggregations of insoluble pigments, nutrients, or other materials in cytoplasm.

incus: The central auditory ossicle, situated between the malleus and the stapes in the middle ear cavity.

infarct: An area of dead cells that results from an interruption of blood flow.

infection: The invasion and colonization of body tissues by pathogens.

inferior: Below, in reference to a particular structure, with the body in the anatomical position.

inferior vena cava: The vein that carries blood from the parts of the body inferior to the heart to the right atrium.

infertility: The inability to conceive; also called *sterility*.

inflammation: A nonspecific defense mechanism that operates at the tissue level; characterized by swelling, redness, warmth, pain, and some loss of function.

infundibulum: A tapering, funnel-shaped structure; in the brain, the connection between the pituitary gland and the hypothalamus; in the uterine tube, the entrance bounded by fimbriae that receives the oocytes at ovulation.

ingestion: The introduction of materials into the digestive tract by way of the mouth; eating.

inguinal canal: A passage through the abdominal wall that marks the path of testicular descent and that contains the testicular arteries, veins, and ductus deferens.

inguinal region: The area of the abdominal wall near the junction of the trunk and the thighs that contains the external genitalia; the groin.

inhibin: A hormone, produced by nurse (sustentacular) cells of the testes and follicular cells of the ovaries, that inhibits the secretion of follicle-stimulating hormone by the adenohypophysis (anterior lobe of the pituitary gland).

injection: The forcing of fluid into a body part or organ.

inner cell mass: Cells of the blastocyst that will form the body of the embryo.

innervation: The distribution of sensory and motor nerves to a specific region or organ.

insertion: A point of attachment of a muscle; the end that is easily movable.

insoluble: Incapable of dissolving in solution.

inspiration: Inhalation; the movement of air into the respiratory system.

insulin: A hormone secreted by beta cells of the pancreatic islets; causes a reduction in plasma glucose concentrations.

integument: The skin.

intercalated discs: Regions where adjacent cardiocytes interlock and where gap junctions permit electrical coupling between the cells.

intercellular fluid: *See* **interstitial fluid**.

interferons: Peptides released by virus-infected cells, especially lymphocytes, that slow viral replication and make other cells more resistant to viral infection.

interleukins: Peptides, released by activated monocytes and lymphocytes, that assist in the coordination of cell-mediated and antibody-mediated immunities.

internal ear: The membranous labyrinth that contains the organs of hearing and equilibrium.

internal nares: The entrance to the nasopharynx from the nasal cavity.

internal receptors: Sensory receptors monitoring the functions and status of internal organs and systems.

internal respiration: The diffusion of gases between interstitial fluid and cytoplasm.

interneuron: An association neuron; central nervous system neurons that are between sensory and motor neurons.

interosseous membrane: The fibrous connective tissue membrane between the shafts of the tibia and fibula and between the radius and ulna; an example of a fibrous amphiarthrosis.

interphase: The stage in the life cycle of a cell during which the chromosomes are uncoiled and all normal cellular functions except mitosis are under way.

interstitial fluid: The fluid in the tissues that fills the spaces between cells.

interstitial growth: A form of cartilage growth through the growth, mitosis, and secretion of chondrocytes in the matrix.

interventricular foramen: The opening that permits fluid movement between the lateral and third ventricles of the brain.

intervertebral disc: A fibrous cartilage pad between the bodies of successive vertebrae that absorbs shocks.

intestine: The tubular organ of the digestive tract.

intracellular fluid: The cytosol.

intramembranous ossification: The formation of bone within a connective tissue without the prior development of a cartilaginous model.

intrinsic factor: A glycoprotein, secreted by the parietal cells of the stomach, that aids the intestinal absorption of vitamin B_{12}.

intrinsic pathway: A pathway of the clotting system that begins with the activation of platelets and ends with the formation of platelet thromboplastin.

inversion: A turning inward.

in vitro: Outside the body, in an artificial environment.

in vivo: In the living body.

involuntary: Not under conscious control.

ion: An atom or molecule bearing a positive or negative charge due to the donation or acceptance, respectively, of one or more electrons.

ionic bond: A molecular bond created by the attraction between ions with opposite charges.

ionization: Dissociation; the breakdown of a molecule in solution to form ions.

iris: A contractile structure, made up of smooth muscle, that forms the colored portion of the eye.

ischemia: An inadequate blood supply to a region of the body.

ischium: One of the three bones whose fusion creates a coxal bone.

islets of Langerhans: *See* **pancreatic islets**.

isotonic: A solution with an osmolarity that does not result in water movement across plasma membranes.

isotopes: Forms of an element whose atoms contain the same number of protons but different numbers of neutrons (and thus differ in atomic weight).

isthmus: A narrow band of tissue connecting two larger masses.

J

jejunum: The middle part of the small intestine.

joint: An area where adjacent bones interact; also called *articulation*.

juxtaglomerular complex: Specialized cells in between the walls of the DCT and afferent and efferent arterioles adjacent to the glomerulus; a complex responsible for the release of renin and erythropoietin.

keratin: The tough, fibrous protein component of nails, hair, calluses, and the general integumentary surface.

keto acid: The carbon chain that remains after the catabolism of lipids and some amino acids.

ketoacidosis: A reduction in the pH of body fluids due to the presence of large numbers of ketone bodies.

ketone bodies: Keto acids produced during the catabolism of lipids and some amino acids; specifically, acetone, acetoacetate, and beta-hydroxybutyrate.

kidney: A component of the urinary system; an organ functioning in the regulation of plasma composition, including the excretion of wastes and the maintenance of normal fluid and electrolyte balances.

killer T cells: See **cytotoxic T cells.**

Krebs cycle: See **citric acid cycle.**

Kupffer cells: Phagocytic cells of the liver.

labium/labia: Lip; the labia majora and labia minora are components of the female external genitalia.

labrum: A lip or rim.

labyrinth: A maze of passageways; the structures of the internal ear.

lacrimal gland: A tear gland on the dorsolateral surface of the eye.

lactase: An enzyme that breaks down the milk sugar lactose.

lactate: An anion released by the dissociation of lactic acid, produced from pyruvate under anaerobic conditions.

lactation: The production of milk by the mammary glands.

lacteal: A terminal lymphatic within an intestinal villus.

lacuna: A small pit or cavity.

lambdoid suture: The synarthrosis between the parietal and occipital bones of the cranium.

lamellae: Concentric layers; the concentric layers of bone within an osteon.

lamellated corpuscle: A receptor sensitive to vibration.

lamina: A thin sheet or layer.

lamina propria: The reticular tissue that underlies a mucous epithelium and forms part of a mucous membrane.

large intestine: The terminal portions of the intestinal tract, consisting of the colon, the rectum, and the anal canal.

laryngopharynx: The division of the pharynx that is inferior to the epiglottis and superior to the esophagus.

larynx: A complex cartilaginous structure that surrounds and protects the glottis and vocal cords; the superior margin is bound to the hyoid bone, and the inferior margin is bound to the trachea.

latent period: The time between the stimulation of a muscle and the start of the contraction phase.

lateral: Pertaining to the side.

lateral apertures: Openings in the roof of the fourth ventricle that permit the circulation of cerebrospinal fluid into the subarachnoid space.

lateral ventricle: A fluid-filled chamber within a cerebral hemisphere.

lens: The transparent refractive structure that is between the iris and the vitreous humor.

lesser omentum: A small pocket in the mesentery that connects the lesser curvature of the stomach to the liver.

leukocyte: A white blood cell.

ligament: A dense band of connective tissue fibers that attaches one bone to another.

ligamentum arteriosum: The fibrous strand in adults that is the remnant of the ductus arteriosus of the fetal stage.

ligamentum teres: The fibrous strand in the falciform ligament of adults that is the remnant of the umbilical vein of the fetal stage.

ligate: To tie off.

limbic system: The group of nuclei and centers in the cerebrum and diencephalon that are involved with emotional states, memories, and behavioral drives.

lingual: Pertaining to the tongue.

lipid: An organic compound containing carbons, hydrogens, and oxygens in a ratio that does not approximate 1:2:1; includes fats, oils, and waxes.

lipogenesis: The synthesis of lipids from nonlipid precursors.

lipolysis: The catabolism of lipids as a source of energy.

lipoprotein: A compound containing a relatively small lipid bound to a protein.

liver: An organ of the digestive system that has varied and vital functions, including the production of plasma proteins, the excretion of bile, the storage of energy reserves, the detoxification of poisons, and the interconversion of nutrients.

lobule: Histologically, the basic organizational unit of the liver.

local hormone: See **prostaglandin.**

loop of Henle: See **nephron loop.**

loose connective tissue: A loosely organized, easily distorted connective tissue that contains several fiber types, a varied population of cells, and a viscous ground substance.

lumbar: Pertaining to the lower back.

lumen: The central space within a duct or other internal passageway.

lungs: The paired organs of respiration, situated in the pleural cavities.

luteinizing hormone (LH): A hormone produced by the adenohypophysis (anterior lobe of the pituitary gland). In females, it assists FSH in follicle stimulation, triggers ovulation, and promotes the maintenance and secretion of endometrial glands. In males, it stimulates testosterone secretion by the interstitial cells of the testes.

lymph: The fluid contents of lymphatic vessels, similar in composition to interstitial fluid.

lymphatic vessels: The vessels of the lymphatic system; also called *lymphatics.*

lymph nodes: Lymphoid organs that monitor the composition of lymph.

lymphocyte: A cell of the lymphatic system that participates in the immune response.

lymphokines: Chemicals secreted by activated lymphocytes.

lymphopoiesis: The production of lymphocytes from lymphoid stem cells.

lysis: The destruction of a cell through the rupture of its plasma membrane.

lysosome: An intracellular vesicle containing digestive enzymes.

lysozyme: An enzyme, present in some exocrine secretions, that has antibiotic properties.

macrophage: A phagocytic cell of the monocyte–macrophage system.

macula: The region of the eye containing a high concentration of cones and no rods. A receptor complex, located in the saccule or utricle of the internal ear, that responds to linear acceleration and gravity.

major histocompatibility complex: See **MHC protein.**

malignant tumor: A form of cancer characterized by rapid cell growth and the spread of cancer cells throughout the body.

malleus: The first auditory ossicle, bound to the tympanic membrane and the incus.

malnutrition: An unhealthy state produced by inadequate dietary intake or absorption of nutrients, calories, and/or vitamins.

mamillary bodies: Nuclei in the hypothalamus that affect eating reflexes and behaviors; a component of the limbic system.

mammary glands: Milk-producing glands of the female breast.

manus: The hand.

marrow: A tissue that fills the internal cavities in bone; dominated by hemopoietic cells (red bone marrow) or by adipose tissue (yellow bone marrow).

marrow cavity: The space within a bone that contains the marrow.

mast cell: A connective tissue cell that, when stimulated, releases histamine, serotonin, and heparin, initiating the inflammatory response.

mastication: Chewing.

mastoid sinus: Air-filled spaces in the mastoid process of the temporal bone.

matrix: The extracellular fibers and ground substance of a connective tissue.

maxillary sinus: One of the paranasal sinuses; an air-filled chamber lined by a respiratory epithelium that is located in a maxilla and opens into the nasal cavity.

meatus: An opening or entrance into a passageway.

mechanoreception: The detection of mechanical stimuli, such as touch, pressure, or vibration.

medial: Toward the midline of the body.

mediastinum: The central tissue mass that divides the thoracic cavity into two pleural cavities.

medulla: The inner layer or core of an organ.

medulla oblongata: The most caudal of the brain regions, also called the *myelencephalon.*

medullary rhythmicity center: The center in the medulla oblongata that sets the background pace of respiration; includes inspiratory and expiratory centers.

megakaryocytes: Bone marrow cells responsible for the formation of platelets.

meiosis: Cell division that produces gametes with half the normal somatic chromosome complement.

melanin: The yellow-brown pigment produced by the melanocytes of the skin.

melanocyte: A specialized cell in the deeper layers of the stratified squamous epithelium of the skin; responsible for the production of melanin.

melanocyte-stimulating hormone (MSH): A hormone, produced by the pars intermedia of the adenohypophysis (anterior lobe of the pituitary gland), that stimulates melanin production.

melatonin: A hormone secreted by the pineal gland.

membrane: Any sheet or partition; a layer consisting of an epithelium and the underlying connective tissue.

membrane potential: The potential difference, measured across a plasma membrane and expressed in millivolts, that results from the uneven distribution of positive and negative ions across the plasma membrane.

membranous labyrinth: Endolymph-filled tubes that enclose the receptors of the internal ear.

meninges: Three membranes that surround the surfaces of the central nervous system; the dura mater, the arachnoid, and the pia mater.

meniscus: A fibrous cartilage pad between opposing surfaces in a joint.

menses: The first portion of the uterine cycle in which the endometrial lining sloughs away; menstrual period.

merocrine: A method of secretion in which the cell ejects materials from secretory vesicles through exocytosis.

mesentery: A double layer of serous membrane that supports and stabilizes the position of an organ in the abdominopelvic cavity and provides a route for the associated blood vessels, nerves, and lymphatic vessels.

mesoderm: The middle germ layer, between the ectoderm and endoderm of the embryo.

messenger RNA (mRNA): RNA formed at transcription to direct protein synthesis in the cytoplasm.

metabolic turnover: The continuous breakdown and replacement of organic materials within cells.

metabolism: The sum of all biochemical processes under way within the human body at any moment; includes anabolism and catabolism.

metabolites: Compounds produced in the body as a result of metabolic reactions.

metacarpal bones: The five bones of the palm of the hand.

metaphase: The stage of mitosis in which the chromosomes line up along the equatorial plane of the cell.

metaphysis: The region of a long bone between the epiphysis and diaphysis, corresponding to the location of the epiphyseal cartilage of the developing bone.

metarteriole: A vessel that connects an arteriole to a venule and that provides blood to a capillary plexus.

metastasis: The spread of cancer cells from one organ to another, leading to the establishment of secondary tumors.

metatarsal bone: One of the five bones of the foot that articulate with the tarsal bones (proximally) and the phalanges (distally).

MHC protein: A surface antigen that is important to the recognition of foreign antigens and that plays a role in the coordination and activation of the immune response; also called *human leukocyte antigen (HLA)*.

micelle: A droplet with hydrophilic portions on the outside; a spherical aggregation of bile salts, monoglycerides, and fatty acids in the lumen of the intestinal tract.

microfilaments: Fine protein filaments visible with the electron microscope; components of the cytoskeleton.

microglia: Phagocytic neuroglia in the central nervous system.

microtubules: Microscopic tubules that are part of the cytoskeleton and are a component in cilia, flagella, the centrioles, and spindle fibers.

microvilli: Small, fingerlike extensions of the exposed plasma membrane of an epithelial cell.

micturition: Urination.

midbrain: Part of the brain stem.

middle ear: The space between the external and internal ears that contains auditory ossicles.

midsagittal plane: A plane passing through the midline of the body that divides it into left and right halves.

mineralocorticoid: Corticosteroids produced by the adrenal cortex; steroids such as aldosterone that affect mineral metabolism.

mitochondrion: An intracellular organelle responsible for generating most of the ATP required for cellular operations.

mitosis: The division of a single cell nucleus that produces two identical daughter cell nuclei; an essential step in cell division.

mitral valve: *See* **bicuspid valve**.

mixed gland: A gland that contains exocrine and endocrine cells, or an exocrine gland that produces serous and mucous secretions.

mixed nerve: A peripheral nerve that contains sensory and motor fibers.

molecular weight: The sum of the atomic weights of all the atoms in a molecule.

molecule: A chemical structure containing two or more atoms that are held together by covalent chemical bonds.

monocytes: Phagocytic agranulocytes (white blood cells) in the circulating blood.

monoglyceride: A lipid consisting of a single fatty acid bound to a molecule of glycerol.

monosaccharide: A simple sugar, such as glucose or ribose.

motor unit: All of the muscle cells controlled by a single motor neuron.

mucins: Proteoglycans responsible for the lubricating properties of mucus.

mucosa: A mucous membrane; the epithelium plus the lamina propria.

mucosa-associated lymphoid tissue (MALT): The extensive collection of lymphoid tissues linked with the epithelia of the digestive, respiratory, urinary, and reproductive tracts.

mucous (adjective): Indicating the presence or production of mucus.

mucous cell: A goblet-shaped, mucus-producing, unicellular gland in certain epithelia of the digestive and respiratory tracts; also called goblet cells.

mucus (noun): A lubricating fluid that is composed of water and mucins and is produced by unicellular and multicellular glands along the digestive, respiratory, urinary, and reproductive tracts.

multipolar neuron: A neuron with many dendrites and a single axon; the typical form of a motor neuron.

muscle: A contractile organ composed of muscle tissue, blood vessels, nerves, connective tissues, and lymphatic vessels.

muscle tissue: A tissue characterized by the presence of cells capable of contraction; includes skeletal, cardiac, and smooth muscle tissues.

muscularis externa: Concentric layers of smooth muscle responsible for peristalsis.

muscularis mucosae: The layer of smooth muscle beneath the lamina propria; responsible for moving the mucosal surface.

mutagens: Chemical agents that induce mutations and may be carcinogenic.

mutation: A change in the nucleotide sequence of the DNA in a cell.

myelin: An insulating, lipid-rich sheath around an axon; consists of multiple layers of neuroglial plasma membrane; significantly increases the impulse propagation rate along the axon.

myelination: The formation of myelin.

myenteric plexus: Parasympathetic motor neurons and sympathetic postganglionic fibers located between the circular and longitudinal layers of the muscularis externa.

myocardial infarction: A heart attack; damage to the heart muscle due to an interruption of regional coronary circulation.

myocardium: The cardiac muscle tissue of the heart.

myofibril: Organized collections of myofilaments in skeletal and cardiac muscle cells.

myofilaments: Fine protein filaments composed primarily of the proteins actin (thin filaments) and myosin (thick filaments).

myoglobin: An oxygen-binding pigment that is especially common in slow skeletal muscle fibers and cardiac muscle cells.

myometrium: The thick layer of smooth muscle in the wall of the uterus.

myosin: The protein component of thick filaments.

N

nail: A keratinous structure produced by epithelial cells of the nail root.

nares, external: The entrance from the exterior to the nasal cavity.

nares, internal: The entrance from the nasal cavity to the nasopharynx.

nasal cavity: A chamber in the skull that is bounded by the internal and external nares.

nasolacrimal duct: The passageway that transports tears from the nasolacrimal sac to the nasal cavity.

nasolacrimal sac: A chamber that receives tears from the lacrimal ducts.

nasopharynx: A region that is posterior to the internal nares and superior to the soft palate and ends at the oropharynx.

necrosis: The death of cells or tissues from disease or injury.

negative feedback: A corrective mechanism that opposes or negates a variation from normal limits.

neonate: A newborn infant, or baby.

neoplasm: A tumor, or mass of abnormal tissue.

nephron: The basic functional unit of the kidney.

nephron loop: The portion of the nephron that creates the concentration gradient in the renal medulla; also called *loop of Henle*.

nerve impulse: An action potential in a neuron plasma membrane.

neural cortex: An area of gray matter at the surface of the central nervous system.

neurilemma: The outer surface of a neuroglia that encircles an axon.

neuroglandular junction: A cell junction at which a neuron controls or regulates the activity of a secretory (gland) cell.

neuroglia: Cells of the central nervous system and peripheral nervous system that support and protect neurons; also called *glial cells*.

neuromuscular junction: A synapse between a neuron and a muscle cell.

neuron: A cell in neural tissue that is specialized for intercellular communication through (1) changes in membrane potential and (2) synaptic connections.

neurotransmitter: A chemical compound released by one neuron to affect the membrane potential of another.

neutron: A fundamental particle that does not carry a positive or a negative charge.

neutrophil: A microphage that is very numerous and normally the first of the mobile phagocytic cells to arrive at an area of injury or infection.

nipple: An elevated epithelial projection on the surface of the breast; contains the openings of the lactiferous sinuses.

Nissl bodies: The ribosomes, Golgi apparatus, rough endoplasmic reticulum, and mitochondria of the perikaryon of a typical neuron.

nitrogenous wastes: Organic waste products of metabolism that contain nitrogen, such as urea, uric acid, and creatinine.

nociception: Pain perception.

node of Ranvier: The area between adjacent neuroglia where the myelin covering of an axon is incomplete.

noradrenaline: *See* **norepinephrine**.

norepinephrine (NE): A catecholamine neurotransmitter in the peripheral nervous system and central nervous system, released at most sympathetic neuromuscular and neuroglandular junctions, and a hormone secreted by the adrenal medulla; also called *noradrenaline*.

nucleic acid: A polymer of nucleotides that contains a pentose sugar, a phosphate group, and one of four nitrogenous bases that regulate the synthesis of proteins and make up the genetic material in cells.

nucleolus: The dense region in the nucleus that is the site of ribosomal RNA synthesis.

nucleoplasm: The fluid content of the nucleus.

nucleoproteins: Proteins of the nucleus that are generally associated with DNA.

nucleotide: A compound consisting of a nitrogenous base, a simple sugar, and a phosphate group.

nucleus: A cellular organelle that contains DNA, RNA, and proteins; in the central nervous system, a mass of gray matter.

nurse cells: Supporting cells of the seminiferous tubules of the testis; responsible for the differentiation of spermatids, and the secretion of inhibin, and androgen-binding protein; also called *sustentacular cells.*

nutrient: A part of food (vitamin, mineral, carbohydrate, lipid, protein, or water) that is necessary for normal physiologic function.

O

occlusal surface: The opposing surfaces of the teeth that come into contact when chewing food.

ocular: Pertaining to the eye.

oculomotor nerve: Cranial nerve III, which controls the extra-ocular muscles other than the superior oblique and the lateral rectus muscles.

olecranon: The proximal end of the ulna that forms the prominent point of the elbow.

olfaction: The sense of smell.

olfactory bulb: The expanded ends of the olfactory tracts; the sites where the axons of the first cranial nerves (I) synapse on central nervous system interneurons that lie inferior to the frontal lobes of the cerebrum.

oligodendrocytes: Central nervous system neuroglia that maintain cellular organization within gray matter and provide a myelin sheath in areas of white matter.

oocyte: A cell whose meiotic divisions will produce a single ovum and three polar bodies.

oogenesis: Ovum production.

opsonization: An effect of coating an object with antibodies; the attraction and enhancement of phagocytosis.

optic chiasm: The crossing point of the optic nerves.

optic nerve: The second cranial nerve (II), which carries signals from the retina of the eye to the optic chiasm.

optic tract: The tract over which nerve impulses from the retina are transmitted between the optic chiasm and the thalamus.

orbit: The bony recess of the skull that contains the eyeball.

organelle: An intracellular structure that performs a specific function or group of functions.

organic compound: A compound containing carbon, hydrogen, and in most cases oxygen.

organogenesis: The formation of organs during embryological and fetal development.

organs: Combinations of tissues that perform complex functions.

origin: In a skeletal muscle, the point of attachment that does not change position when the muscle contracts; usually defined in terms of movements from the anatomical position.

oropharynx: The middle portion of the pharynx, bounded superiorly by the nasopharynx, anteriorly by the oral cavity, and inferiorly by the laryngopharynx.

osmolarity: The total concentration of dissolved materials in a solution, regardless of their specific identities, expressed in moles; also called *osmotic concentration.*

osmoreceptor: A receptor sensitive to changes in the osmolarity of plasma.

osmosis: The movement of water across a selectively permeable membrane from one solution to another solution that contains a higher solute concentration.

osmotic pressure: The force of osmotic water movement; the pressure that must be applied to prevent osmosis across a membrane.

osseous tissue: A strong connective tissue containing specialized cells and a mineralized matrix of crystalline calcium phosphate and calcium carbonate; also called bone.

ossicles: Small bones.

ossification: The formation of bone.

osteoblast: A cell that produces the fibers and matrix of bone.

osteoclast: A cell that dissolves the fibers and matrix of bone.

osteocyte: A bone cell responsible for the maintenance and turnover of the mineral content of the surrounding bone.

osteolysis: The breakdown of the mineral matrix of bone.

osteon: The basic histological unit of compact bone, consisting of osteocytes organized around a central canal and separated by concentric lamellae.

otic: Pertaining to the ear.

otolith: The aggregation of calcium carbonate crystals on top of a gelatinous otolithic membrane; located above one of the maculae of the vestibule.

oval window: An opening in the bony labyrinth where the stapes attaches to the membranous wall of the scala vestibuli.

ovarian cycle: The monthly chain of events that leads to ovulation.

ovary: The female reproductive organ that produces gametes.

ovulation: The release of a secondary oocyte, surrounded by cells of the corona radiata, after the rupture of the wall of a tertiary follicle; in females, the periodic release of an oocyte from an ovary.

ovum/ova: The functional product of meiosis II, produced after the fertilization of a secondary oocyte.

oxytocin (OXT): A hormone produced by hypothalamic cells and secreted into capillaries at the posterior lobe of the pituitary gland; stimulates smooth muscle contractions of the uterus or mammary glands in females and the prostate gland in males.

P

pacemaker cells: Cells of the sinoatrial node that set the pace of cardiac contraction.

palate: The horizontal partition separating the oral cavity from the nasal cavity and nasopharynx; divided into an anterior bony (hard) palate and a posterior fleshy (soft) palate.

palatine: Pertaining to the palate.

palpate: To examine by touch.

palpebrae: Eyelids.

pancreas: A digestive organ containing exocrine and endocrine tissues; the exocrine portion secretes pancreatic juice, and the endocrine portion secretes hormones, including insulin and glucagon.

pancreatic duct: A tubular duct that carries pancreatic juice from the pancreas to the duodenum.

pancreatic islets: Aggregations of endocrine cells in the pancreas; also called *islets of Langerhans.*

pancreatic juice: A mixture of buffers and digestive enzymes that is discharged into the duodenum under the stimulation of the enzymes secretin and cholecystokinin.

papilla: A small, conical projection.

paralysis: The loss of voluntary motor control over a portion of the body.

paranasal sinuses: Bony chambers, lined by respiratory epithelium, that open into the nasal cavity; the frontal, ethmoidal, sphenoidal, and maxillary sinuses.

parasagittal: A section or plane that parallels the midsagittal plane but that does not pass along the midline.

parasympathetic division: One of the two divisions of the autonomic nervous system; generally responsible for activities that conserve energy and lower the metabolic rate.

parathyroid glands: Four small glands embedded in the posterior surface of the thyroid gland; secrete parathyroid hormone.

parathyroid hormone (PTH): A hormone secreted by the parathyroid glands when plasma calcium levels fall below the normal range; causes increased osteoclast activity, increased intestinal calcium uptake, and decreased calcium ion loss at the kidneys.

parietal: Relating to the parietal bone; referring to the wall of a cavity.

parietal cells: Cells of the gastric glands that secrete hydrochloric acid and intrinsic factor.

parotid salivary glands: Large salivary glands that secrete a saliva containing high concentrations of salivary (alpha) amylase.

patella: The bone of the kneecap.

pathogen: A disease-causing organism.

pathogenic: Disease-causing.

pathologist: A physician specializing in the identification of diseases on the basis of characteristic structural and functional changes in tissues and organs.

pelvic cavity: The inferior subdivision of the abdominopelvic cavity; encloses the urinary bladder, the sigmoid colon and rectum, and male or female reproductive organs.

pelvis: A bony complex created by the articulations among the coxal bones, the sacrum, and the coccyx.

penis: A component of the male external genitalia; a copulatory organ that surrounds the urethra and serves to introduce semen into the female vagina; the developmental equivalent of the female clitoris.

peptide: A chain of amino acids linked by peptide bonds.

peptide bond: A covalent bond between the amino group of one amino acid and the carboxyl group of another.

pericardial cavity: The space between the parietal pericardium and the epicardium (visceral pericardium) that covers the outer surface of the heart.

pericardium: The fibrous sac that surrounds the heart; its inner, serous lining is continuous with the epicardium.

perichondrium: The layer that surrounds a cartilage, consisting of an outer fibrous region and an inner cellular region.

perikaryon: The cytoplasm that surrounds the nucleus in the cell body of a neuron.

perilymph: A fluid similar in composition to cerebrospinal fluid; located in the spaces between the bony labyrinth and the membranous labyrinth of the internal ear.

perimysium: A connective tissue partition that separates adjacent fasciculi in a skeletal muscle.

perineum: The pelvic floor and its associated structures.

perineurium: A connective tissue partition that separates adjacent bundles of nerve fibers in a peripheral nerve.

periodontal ligament: Collagen fibers that bind the cementum of a tooth to the periosteum of the surrounding alveolus.

periosteum: The layer that surrounds a bone, consisting of an outer fibrous region and inner cellular region.

peripheral nervous system (PNS): All neural tissue outside the central nervous system.

peripheral resistance: The resistance to blood flow; primarily caused by friction with the vascular walls.

peristalsis: A wave of smooth muscle contractions that propels materials along the axis of a tube such as the digestive tract, the ureters, or the ductus deferens.

peritoneum: The serous membrane that lines the peritoneal cavity.

peritubular capillaries: A network of capillaries that surrounds the proximal and distal convoluted tubules of the kidneys.

permeability: The ease with which dissolved materials can cross a membrane; if the membrane is freely permeable, any molecule can cross it; if impermeable, nothing can cross; most biological membranes are selectively permeable.

peroxisome: A membranous vesicle containing enzymes that break down hydrogen peroxide (H_2O_2).

pes: The foot.

petrous: Stony; usually refers to the thickened portion of the temporal bone that encloses the internal ear.

pH: The negative exponent of the hydrogen ion concentration, expressed in moles per liter; solutions may be acidic, neutral, or basic.

phagocyte: A cell that performs phagocytosis.

phagocytosis: The engulfing of extracellular materials or pathogens; the movement of extracellular materials into the cytoplasm by enclosure in a membranous vesicle.

phalanx/phalanges: Bone(s) of the finger(s) or toe(s).

pharmacology: The study of drugs, their physiological effects, and their clinical uses.

pharynx: The throat; a muscular passageway shared by the digestive and respiratory tracts.

phenotype: Physical characteristics that are genetically determined.

phosphate group: PO_4^{3-}; a functional group that can be attached to an organic molecule; required for the formation of high-energy bonds.

phospholipid: An important membrane lipid whose structure includes both hydrophilic and hydrophobic regions.

phosphorylation: The addition of a high-energy phosphate group to a molecule.

photoreception: Sensitivity to light.

physiology: The study of how living things function.

pia mater: The innermost layer of the meninges bound to the underlying neural tissue.

pineal gland: Neural tissue in the posterior portion of the roof of the diencephalon; secretes melatonin.

pinna: See **auricle.**

pinocytosis: The introduction of fluids into the cytoplasm by enclosing them in membranous vesicles at the cell surface.

pituitary gland: An endocrine organ that is situated in the sella turcica of the sphenoid and is connected to the hypothalamus by the infundibulum; includes the posterior lobe and the anterior lobe; also called the *hypophysis.*

placenta: A temporary structure in the uterine wall that permits diffusion between the fetal and maternal circulatory systems.

plantar: Referring to the sole of the foot.

plantar flexion: Ankle extension; toe pointing.

plasma: The fluid ground substance of whole blood; what remains after the cells have been removed from a sample of whole blood.

plasma cell: An activated B cell that secretes antibodies; plasmocyte.

plasma membrane: A cell membrane; plasmalemma.

platelets: Small packets of cytoplasm that contain enzymes important in the clotting response; manufactured in bone marrow by megakaryocytes.

pleura: The serous membrane that lines the pleural cavities.

pleural cavities: Subdivisions of the ventral body cavity that enclose the lungs.

plexus: A network or braid.

polar body: A nonfunctional packet of cytoplasm that contains chromosomes eliminated from an oocyte during meiosis.

polar bond: A covalent bond in which electrons are shared unequally.

pollex: The thumb.

polypeptide: A chain of amino acids strung together by peptide bonds; those containing more than 100 peptides are called *proteins.*

polyribosome: Several ribosomes linked by their translation of a single mRNA strand.

polysaccharide: A complex sugar, such as glycogen or a starch.

polyunsaturated fats: Fatty acids containing carbon atoms that are linked by double bonds.

pons: The portion of the metencephalon that is anterior to the cerebellum.

popliteal: Pertaining to the back of the knee.

positive feedback: A mechanism that increases a deviation from normal limits after an initial stimulus.

postcentral gyrus: The primary sensory cortex, where touch, vibration, pain, temperature, and taste sensations arrive and are consciously perceived.

posterior: Toward the back; dorsal.

postganglionic neuron: An autonomic neuron in a peripheral ganglion, whose activities control peripheral effectors.

postsynaptic membrane: The portion of the plasma membrane of a postsynaptic cell that is part of a synapse.

potential difference: The separation of opposite charges; requires a barrier that prevents ion migration.

precentral gyrus: The primary motor cortex of a cerebral hemisphere, located anterior to the central sulcus.

prefrontal cortex: The anterior portion of each cerebral hemisphere; thought to be involved with higher intellectual functions, predictions, calculations, and so forth.

preganglionic neuron: A visceral motor neuron in the central nervous system whose output controls one or more ganglionic motor neurons in the peripheral nervous system.

premotor cortex: The motor association area between the precentral gyrus and the prefrontal area.

preoptic nucleus: The hypothalamic nucleus that coordinates thermoregulatory activities.

presynaptic membrane: The synaptic surface where neurotransmitter release occurs.

prevertebral ganglion: See **collateral ganglion.**

prime mover: A muscle that performs a specific action.

proenzyme: An inactive enzyme secreted by an epithelial cell.

progesterone: The most important progestin secreted by the corpus luteum after ovulation.

progestins: Steroid hormones structurally related to cholesterol; progesterone is an example.

prognosis: A prediction about the possible course or outcome from a specific disease.

prolactin: The hormone that stimulates functional development of the mammary glands in females; a secretion of the adenohypophysis (anterior lobe of the pituitary gland).

pronation: The rotation of the forearm that makes the palm face posteriorly.

prone: Lying face down with the palms facing the floor.

pronucleus: An enlarged ovum or spermatozoon nucleus that forms after fertilization but before amphimixis.

prophase: The initial phase of mitosis; characterized by the appearance of chromosomes, the breakdown of the nuclear membrane, and the formation of the spindle apparatus.

proprioception: The awareness of the positions of bones, joints, and muscles.

prostaglandin: A fatty acid secreted by one cell that alters the metabolic activities or sensitivities of adjacent cells; also called a *local hormone.*

prostate gland: An accessory gland of the male reproductive tract, contributes roughly one-third of the volume of semen.

prosthesis: An artificial substitute for a body part.

protease: See **proteinase.**

protein: A large polypeptide with a complex structure.

proteinase: An enzyme that breaks down proteins into peptides and amino acids.

proteoglycan: A compound containing a large polysaccharide complex attached to a relatively small protein; examples include hyaluronan and chondroitin sulfate.

proton: A fundamental particle bearing a positive charge.

protraction: Movement anteriorly in the horizontal plane.

proximal: Toward the attached base of an organ or structure.

proximal convoluted tubule (PCT): The portion of the nephron that is situated between the glomerular capsule and the nephron loop; the major site of active reabsorption from filtrate.

pseudopodia: Temporary cytoplasmic extensions typical of mobile or phagocytic cells.

pseudostratified epithelium: An epithelium that contains several layers of nuclei but whose cells are all in contact with the underlying basement membrane.

puberty: A period of rapid growth, sexual maturation, and the appearance of secondary sexual characteristics; normally occurs at ages 10–15 years.

pubic symphysis: The fibrocartilaginous amphiarthrosis between the pubic bones of the coxal bones.

pubis: The anterior, inferior component of the hip bone.

pudendum: The external genitalia.

pulmonary circuit: Blood vessels between the pulmonary semilunar valve of the right ventricle and the entrance to the left atrium; the blood flow through the lungs.

pulmonary ventilation: The movement of air into and out of the lungs.

pupil: The opening in the center of the iris through which light enters the eye.

Purkinje cell: A large, branching neuron of the cerebellar cortex.

Purkinje fibers: Specialized conducting cardiocytes in the ventricles of the heart.

pyloric sphincter: A sphincter of smooth muscle that regulates the passage of chyme from the stomach to the duodenum.

pylorus: The gastric region between the body of the stomach and the duodenum; includes the pyloric sphincter.

pyruvate: The ion formed by the dissociation of pyruvic acid, a three-carbon compound produced during glycolysis.

Q

quaternary structure: The three-dimensional protein structure produced by interactions between protein subunits.

R

ramus/rami: A branch/branches.

raphe: A seam.

receptive field: The area monitored by a single sensory receptor.

rectum: The inferior 15 cm (6 in.) of the digestive tract.

rectus: Straight.

red blood cell (RBC): *See* **erythrocyte**.

reductional division: The first meiotic division, which reduces the chromosome number from 46 to 23.

reflex: A rapid, automatic response to a stimulus.

reflex arc: The receptor, sensory neuron, motor neuron, and effector involved in a particular reflex; interneurons may be present, depending on the reflex considered.

relaxation phase: The period after a contraction when the tension in the muscle fiber returns to resting levels.

relaxin: A hormone that loosens the pubic symphysis; secreted by the placenta.

renal: Pertaining to the kidneys.

renal corpuscle: The initial portion of the nephron, consisting of an expanded chamber that encloses the glomerulus.

renin: The enzyme released by cells of the juxtaglomerular complex when renal blood flow declines; converts angiotensinogen to angiotensin I.

repolarization: The movement of the membrane potential away from a positive value and toward the resting potential.

respiration: The exchange of gases between cells and the environment; includes pulmonary ventilation, external respiration, internal respiration, and cellular respiration.

respiratory pump: A mechanism by which changes in the intrapleural pressures during the respiratory cycle assist the venous return to the heart.

resting potential: The membrane potential of a normal cell under homeostatic conditions.

rete: An interwoven network of blood vessels or passageways.

reticular activating system (RAS): The mesencephalic portion of the reticular formation; responsible for arousal and the maintenance of consciousness.

reticular formation: A diffuse network of gray matter that extends the entire length of the brain stem.

retina: The innermost layer of the eye, lining the vitreous chamber; also called *neural layer*.

retinal: A visual pigment derived from vitamin A.

retraction: Movement posteriorly in the horizontal plane.

retroperitoneal: Behind or outside the peritoneal cavity.

reverberation: A positive feedback along a chain of neurons such that they remain active once stimulated.

rheumatism: A condition characterized by pain in muscles, tendons, bones, or joints.

Rh factor: A surface antigen that may be present (Rh-positive) or absent (Rh-negative) from the surfaces of red blood cells.

rhodopsin: The visual pigment in the membrane disks of the distal segments of rods.

rhythmicity center: A center in the medulla oblongata responsible for the pace of respiration; includes inspiratory and expiratory centers.

ribonucleic acid: A nucleic acid consisting of a chain of nucleotides that contain the sugar ribose and the nitrogenous bases adenine, guanine, cytosine, and uracil.

ribose: A five-carbon sugar that is a structural component of RNA.

ribosome: An organelle that contains rRNA and proteins and is essential to mRNA translation and protein synthesis.

rod: A photoreceptor responsible for vision in dim lighting.

rough endoplasmic reticulum (RER): A membranous organelle that is a site of protein synthesis and storage.

round window: An opening in the bony labyrinth of the internal ear that exposes the membranous wall of the tympanic duct to the air of the middle ear cavity.

rugae: Mucosal folds in the lining of the empty stomach that disappear as gastric distension occurs; folds in the urinary bladder.

S

saccule: A portion of the vestibular apparatus of the internal ear; contains a macula sensitive to changes in vertical movements and gravity.

sagittal plane: A sectional plane that divides the body into left and right portions.

salt: An inorganic compound consisting of a cation other than H^+ and an anion other than OH^-.

saltatory propagation: The relatively rapid propagation of an action potential between successive nodes of a myelinated axon.

sarcomere: The smallest contractile unit of a striated muscle cell.

sarcoplasm: The cytoplasm of a muscle cell.

scala media: *See* **cochlear duct**.

scala tympani: The perilymph-filled chamber of the internal ear, adjacent to the basilar membrane; pressure changes there distort the round window.

scala vestibuli: A coiled tube filled with perilymph that lies within the bony labyrinth; it is continuous with the scala tympani at the tip of the cochlear spiral.

scar tissue: The thick, collagenous tissue that forms at an injury site.

Schwann cells: Neuroglia responsible for the neurilemma that surrounds axons in the peripheral nervous system.

sciatic nerve: A nerve innervating the posteromedial portions of the thigh and leg.

sclera: The fibrous, outer layer of the eye that forms the white area of the anterior surface; a portion of the fibrous layer of the eye.

sclerosis: A hardening and thickening that commonly occurs secondary to tissue inflammation.

scrotum: The loose-fitting, fleshy pouch that encloses the testes of the male.

sebaceous glands: Glands that secrete sebum; normally associated with hair follicles.

sebum: A waxy secretion that coats the surfaces of hairs.

secondary sex characteristics: Physical characteristics that appear at puberty in response to sex hormones but are not involved in the production of gametes.

secretin: A hormone, secreted by the duodenum, that stimulates the production of buffers by the pancreas and inhibits gastric activity.

semen: The fluid ejaculate that contains spermatozoa and the secretions of accessory glands of the male reproductive tract.

semicircular ducts: The tubular components of the membranous labyrinth of the internal ear; contain receptors sensitive to rotations of the head.

semilunar valve: A three-cusped valve guarding the exit from one of the cardiac ventricles; the pulmonary and aortic valves.

seminal glands: Glands of the male reproductive tract that produce roughly 60 percent of the volume of semen; also called *seminal vesicles*.

seminiferous tubules: Coiled tubules where spermatozoon production occurs in the testes.

senescence: Aging.

septa: Partitions that subdivide an organ.

serous cell: A cell that produces a serous secretion.

serous membrane: A squamous epithelium and the underlying loose connective tissue; the lining of the pericardial, pleural, and peritoneal cavities.

serous secretion: A watery secretion that contains high concentrations of enzymes.

serum: The ground substance of blood plasma from which clotting agents have been removed.

sesamoid bone: A bone that forms within a tendon.

sigmoid colon: The S-shaped 18-cm (7.1 in.)-long portion of the colon between the descending colon and the rectum.

sign: The visible, objective evidence of the presence of a disease.

simple epithelium: An epithelium containing a single layer of cells above the basement membrane.

sinoatrial (SA) node: The natural pacemaker of the heart; situated in the wall of the right atrium.

sinus: A chamber or hollow in a tissue; a large, dilated vein.

skeletal muscle: A contractile organ of the muscular system.

skeletal muscle tissue: A contractile tissue dominated by skeletal muscle fibers; characterized as striated, voluntary muscle.

sliding filament theory: The concept that a sarcomere shortens as the thick and thin filaments slide past one another.

small intestine: The duodenum, jejunum, and ileum; the digestive tract between the stomach and the large intestine.

smooth endoplasmic reticulum (SER): A membranous organelle in which lipid and carbohydrate synthesis and storage occur.

smooth muscle tissue: Muscle tissue in the walls of many visceral organs; characterized as nonstriated, involuntary muscle.

soft palate: The fleshy posterior extension of the hard palate, separating the nasopharynx from the oral cavity.

solute: Any materials dissolved in a solution.

solution: A fluid containing dissolved materials.

somatic: Pertaining to the body.

somatic nervous system (SNS): The efferent division of the nervous system that innervates skeletal muscles.

somatomedins: Compounds stimulating tissue growth; released by the liver after the secretion of growth hormone; also called *insulin-like growth factors*.

sperm: *See* **spermatozoon**.

spermatic cord: Collectively, the spermatic vessels, nerves, lymphatic vessels, and the ductus deferens, extending between the testes and the proximal end of the inguinal canal.

spermatocyte: A cell of the seminiferous tubules that is engaged in meiosis.

spermatogenesis: Spermatozoon production.

spermatozoon/spermatozoa: A male gamete; also called *sperm*.

sphincter: A muscular ring that contracts to close the entrance or exit of an internal passageway.

spinal nerve: One of 31 pairs of nerves that originate on the spinal cord from anterior and posterior roots.

spindle apparatus: Microtubule-based structure that distributes duplicated chromosomes to opposite ends of a dividing cell during mitosis.

spinous process: The prominent posterior projection of a vertebra; formed by the fusion of two laminae.

spiral organ: A receptor complex in the scala media of the cochlea that includes the inner and outer hair cells, supporting cells and structures, and the tectorial membrane, also called the organ of Corti; provides the sensation of hearing.

spleen: A lymphoid organ important for the phagocytosis of red blood cells, the immune response, and lymphocyte production.

spongy bone: Composed of a network of bony struts; within epiphyses and lines marrow cavity.

squamous: Flattened.

squamous epithelium: An epithelium whose superficial cells are flattened and platelike.

stapes: The auditory ossicle attached to the tympanic membrane.

stenosis: A constriction or narrowing of a passageway.

stereocilia: Elongate microvilli characteristic of the epithelium of the epididymis, portions of the ductus deferens, and the internal ear.

steroid: A ring-shaped lipid structurally related to cholesterol.

stimulus: An environmental change that produces a change in cellular activities; often used to refer to events that alter the membrane potentials of excitable cells.

stratified: Containing several layers.

stratum: A layer.

stretch receptors: Sensory receptors that respond to stretching of the surrounding tissues.

subarachnoid space: A meningeal space containing cerebrospinal fluid; the area between the arachnoid membrane and the pia mater.

subclavian: Pertaining to the region immediately posterior and inferior to the clavicle.

subcutaneous layer: *See* **hypodermis.**

submucosa: The region between the muscularis mucosae and the muscularis externa.

substrate: A participant (product or reactant) in an enzyme-catalyzed reaction.

sulcus: A groove or furrow.

summation: The temporal or spatial addition of contractile force or neural stimuli.

superior: Above, in reference to a portion of the body in the anatomical position.

superior vena cava (SVC): The vein that carries blood to the right atrium from parts of the body that are superior to the heart.

supination: The rotation of the forearm such that the palm faces anteriorly.

supine: Lying face up, with palms facing anteriorly.

suppressor T cells: Lymphocytes that inhibit B cell activation and the secretion of antibodies by plasma cells.

surfactant: A lipid secretion that coats the alveolar surfaces of the lungs and prevents their collapse.

sustentacular cells: See nurse cells.

sutural bones: Irregular bones that form in fibrous tissue between the flat bones of the developing cranium; also called *Wormian bones.*

suture: A fibrous joint between flat bones of the skull.

sympathetic division: The division of the autonomic nervous system that is responsible for "fight or flight" reactions; primarily concerned with the elevation of metabolic rate and increased alertness.

symphysis: A fibrous amphiarthrosis, such as that between adjacent vertebrae or between the pubic bones of the coxal bones.

symptom: An abnormality of function as a result of disease; subjective experience of patient.

synapse: The site of communication between a nerve cell and some other cell; if the other cell is not a neuron, the term *neuromuscular* or *neuroglandular junction* is often used.

synaptic delay: The period between the arrival of an impulse at the presynaptic membrane and the initiation of an action potential in the postsynaptic membrane.

syncytium: A multinucleate mass of cytoplasm, produced by the fusion of cells or repeated mitoses without cytokinesis.

syndrome: A discrete set of signs and symptoms that occur together.

synergist: A muscle that assists a prime mover in performing its primary action.

synovial cavity: A fluid-filled chamber in a synovial joint.

synovial fluid: The substance secreted by synovial membranes that lubricates joints.

synovial joint: A freely movable joint where the opposing bone surfaces are separated by synovial fluid; a diarthrosis.

synovial membrane: An incomplete layer of fibroblasts facing the synovial cavity, plus the underlying loose connective tissue.

synthesis: Manufacture; anabolism.

system: An interacting group of organs that performs one or more specific functions.

systemic circuit: The vessels between the aortic valve and the entrance to the right atrium; the system other than the vessels of the pulmonary circuit.

systole: A period of contraction in a chamber of the heart, as part of the cardiac cycle.

systolic pressure: The peak arterial pressure measured during ventricular systole.

T

tactile: Pertaining to the sense of touch.

tarsal bones: The bones of the ankle (the talus, calcaneus, navicular, and cuneiform bones).

tarsus: The ankle.

T cells: Lymphocytes responsible for cell-mediated immunity and for the coordination and regulation of the immune response; includes regulatory T cells (helpers and suppressors) and cytotoxic (killer) T cells.

tectospinal tracts: Descending tracts of the medial pathway that carry involuntary motor commands issued by the colliculi.

telodendria: Terminal axonal branches that end in synaptic terminals.

telophase: The final stage of mitosis, characterized by the disappearance of the spindle apparatus, the reappearance of the nuclear membrane, the disappearance of the chromosomes, and the completion of cytokinesis.

temporal: Pertaining to time (temporal summation) or to the temples (temporal bones).

tendon: A collagenous band that connects a skeletal muscle to an element of the skeleton.

teres: Long and round.

terminal: Toward the end.

tertiary structure: The protein structure that results from interactions among distant portions of the same molecule; complex coiling and folding.

testes: The male gonads, sites of gamete production and hormone secretion.

testosterone: The principal androgen produced by the interstitial cells of the testes.

tetraiodothyronine: T_4, or thyroxine, a thyroid hormone.

thalamus: The walls of the diencephalon.

therapy: The treatment of disease.

thermoreception: Sensitivity to temperature changes.

thermoregulation: Homeostatic maintenance of body temperature.

thick filament: A cytoskeletal filament in a skeletal or cardiac muscle cell; composed of myosin, with a core of titin.

thin filament: A cytoskeletal filament in a skeletal or cardiac muscle cell; consists of actin, troponin, and tropomyosin.

thorax: The chest.

threshold: The membrane potential at which an action potential begins.

thrombin: The enzyme that converts fibrinogen to fibrin.

thymine: One of the nitrogenous bases in the nucleic acid DNA.

thymosins: Thymic hormones essential to the development and differentiation of T cells.

thymus: A lymphoid organ, the site of T cell formation.

thyroglobulin: A circulating transport globulin that binds thyroid hormones.

thyroid gland: An endocrine gland whose lobes are lateral to the thyroid cartilage of the larynx.

thyroid hormones: Thyroxine (T_4) and triiodothyronine (T_3), hormones of the thyroid gland; stimulate tissue metabolism, energy utilization, and growth.

thyroid-stimulating hormone (TSH): The hormone, produced by the anterior lobe of the pituitary gland, that triggers the secretion of thyroid hormones by the thyroid gland.

thyroxine: A thyroid hormone; also called T_4 or *tetraiodothyronine.*

tidal volume: The volume of air moved into and out of the lungs during a normal quiet respiratory cycle.

tissue: A collection of specialized cells and cell products that performs a specific function.

tonsil: A lymphoid nodule in the wall of the pharynx; the palatine, pharyngeal, and lingual tonsils.

topical: Applied to the body surface.

toxic: Poisonous.

trabecula: A connective tissue partition that subdivides an organ.

trachea: The windpipe; an airway extending from the larynx to the primary bronchi.

tract: A bundle of axons in the central nervous system.

transcription: The encoding of genetic instructions on a strand of mRNA.

transection: The severing or cutting of an object in the transverse plane.

translation: The process of peptide formation from the instructions carried by an mRNA strand.

transudate: A fluid that diffuses across a serous membrane and lubricates opposing surfaces.

transverse tubules: The transverse, tubular extensions of the sarcolemma that extend deep into the sarcoplasm, contacting the membranes of the sarcoplasmic reticulum; also called *T tubules.*

tricuspid valve: The right atrioventricular valve, which prevents the backflow of blood into the right atrium during ventricular systole.

trigeminal nerve: Cranial nerve V, which provides sensory information from the lower portions of the face (including the upper and lower jaws) and delivers motor commands to the muscles of mastication.

triglyceride: A lipid that is composed of a molecule of glycerol attached to three fatty acids.

trochanter: Large process near the head of the femur.

trochlea: A pulley; the spool-shaped medial portion of the condyle of the humerus.

trochlear nerve: Cranial nerve IV, controlling the superior oblique muscle of the eye.

trunk: The thoracic and abdominopelvic regions; a major arterial branch.

T tubules: *See* **transverse tubules**.

tuberculum: A small, localized elevation on a bony surface.

tuberosity: A large, roughened elevation on a bony surface.

tumor: A tissue mass formed by the abnormal growth and replication of cells.

tunica: A layer or covering.

twitch: A single stimulus–contraction–relaxation cycle in a skeletal muscle.

tympanic duct: *See* **scala tympani**.

tympanic membrane: The membrane that separates the external acoustic meatus from the middle ear; the membrane whose vibrations are transferred to the auditory ossicles and ultimately to the oval window; also called *eardrum* or *tympanum*.

U

umbilical cord: The connecting stalk between the fetus and the placenta; contains the allantois, the umbilical arteries, and the umbilical vein.

umbilicus: The navel.

unipolar neuron: A sensory neuron whose cell body is in a dorsal root ganglion or a sensory ganglion of a cranial nerve.

unmyelinated axon: An axon whose neurilemma does not contain myelin and along which continuous propagation occurs.

uracil: One of the nitrogenous bases in the nucleic acid RNA.

ureters: Muscular tubes, lined by transitional epithelium, that carry urine from the renal pelvis to the urinary bladder.

urethra: A muscular tube that carries urine from the urinary bladder to the exterior.

urinary bladder: The muscular, distensible sac that stores urine prior to micturition.

urination: The voiding of urine; micturition.

uterus: The muscular organ of the female reproductive tract in which implantation, placenta formation, and fetal development occur.

utricle: The largest chamber of the vestibular apparatus of the internal ear; contains a macula sensitive to changes in horizontal movements and gravity.

V

vagina: A muscular tube extending between the uterus and the vestibule.

vascular: Pertaining to blood vessels.

vasoconstriction: A reduction in the diameter of arterioles due to the contraction of smooth muscles in the tunica media; elevates peripheral resistance; may occur in response to local factors, through the action of hormones, or from the stimulation of the vasomotor center.

vasodilation: An increase in the diameter of arterioles due to the relaxation of smooth muscles in the tunica media; reduces peripheral resistance; may occur in response to local factors, through the action of hormones, or after decreased stimulation of the vasomotor center.

vasomotion: Rhythmic changes in vessel diameter by precapillary sphincters; results in changes in the pattern of blood flow through a capillary bed.

vasomotor center: The center in the medulla oblongata whose stimulation produces vasoconstriction and an elevation of peripheral resistance.

vein: A blood vessel carrying blood from a capillary bed toward the heart.

vena cava: One of the major veins delivering systemic blood to the right atrium; superior and inferior venae cavae.

ventilation: Air movement into and out of the lungs.

ventral: Pertaining to the anterior surface.

ventricle: A fluid-filled chamber; in the heart, one of the large chambers discharging blood into the pulmonary or systemic circuits; in the brain, one of four fluid-filled interior chambers.

venule: Thin-walled veins that receive blood from capillaries.

vertebral canal: The passageway that encloses the spinal cord; a tunnel bounded by the neural arches of adjacent vertebrae.

vertebral column: The cervical, thoracic, and lumbar vertebrae, the sacrum, and the coccyx.

vesicle: A membranous sac in the cytoplasm of a cell.

vestibular duct: *See* **scala vestibuli**.

villus/villi: A slender, finger-shaped projection of a mucous membrane.

virus: A noncellular pathogen.

viscera: Organs in one of the subdivisions of the ventral body cavity.

visceral: Pertaining to viscera or their outer coverings.

viscosity: The resistance to flow that a fluid exhibits as a result of molecular interactions within the fluid.

viscous: Thick, syrupy.

vitamin: An essential organic nutrient that functions as a coenzyme in vital enzymatic reactions.

vitreous humor: The gelatinous mass in the vitreous chamber of the eye.

voluntary: Controlled by conscious thought processes.

W

white blood cells (WBCs): The granulocytes and agranulocytes of whole blood.

white matter: Regions in the central nervous system that are dominated by myelinated axons.

X

xiphoid process: The slender, inferior extension of the sternum.

Y

Y chromosome: The sex chromosome whose presence indicates that the individual is a genetic male.

Z

zygote: The fertilized ovum, prior to the start of cleavage.

Illustration Credits

All illustrations by Medical & Scientific Illustration, except for the following by Imagineering: p. 18, p. 19, p. 22 top, p. 24 second from top left, p. 96–97, p. 105 left, p. 214 right, p. 215, p. 249 second from top right, p. 385, p. 392 top, p. 393 top, p. 427 top left, p. 467, p. 476, p. 493 top left, p. 512 top, p. 633, p. 641, p. 655 top left, p. 655 second from top right, p. 670, p. 672–673, p. 676 far left, p. 677 far right, p. 685 second and third from top right

Photo Credits

Frontmatter p. vi top: Johannes de Ketham. Fasiculo de Medicina. (Venice, 1495); p. vi bottom: zoomstudio/iStockphoto; p. vii: Andresr/ Shutterstock; p. xix: Leigh Schindler/iStockphoto

Chapter 1 Chapter Opener: AYAKOVLEV.COM/ Shutterstock; p. 3 top: Andreas Reh/iStockphoto; middle: Larry Mulvehill/Photo Researchers, Inc.; p. 6: Ralph T. Hutchings; p. 15: Andreas Vesalius. De humani corporis fabrica (Liber VII) from *Images from the History of Medicine*; p. 17 all; Custom Medical Stock Photo, Inc.; p. 23 top left: Andreas Reh/iStockphoto; bottom right: Andreas Vesalius. De humani corporis fabrica (Liber VII) from *Images from the History of Medicine*; p. 24 top left: Custom Medical Stock Photo, Inc.; middle: Bork/ Shutterstock; bottom: Reflekta/Shutterstock; p. 25: Dule964/Dreamstime

Chapter 2 Chapter Opener: Lomachevsk/ Dreamstime; p. 35: Valua Vitali/iStockphoto; p. 41 top: Brandon Parry/Shutterstock; bottom: Ryerson Clark/iStockphoto; p. 47 left: Denis Pepin/ iStockphoto; right: Camilla Wisbauer/iStockphoto; p. 62: Denis Pepin/iStockphoto; p. 63 top: Donald R. Swartz/Shutterstock; bottom: Blend Images/Fotolia

Chapter 3 Chapter Opener: Henrik Jonsson/ iStockphoto; p. 72: Dr. Birgit Satir; p. 74: Don W. Fawcett; p. 80 left and middle: Ed Reschke/ PhotoLibrary; right: James Solliday/Biological Photo Service; p. 81 all: Ed Reschke/PhotoLibrary; p. 83: Dr. Riaz Agha/From: The International Journal of Surgery Image Library (theijs.com); p. 98: Robert B. Tallistch; p. 99: Frederic H. Martini; p. 100 top: Robert B. Tallistch; bottom: Gregory N. Fuller; p. 101 both: Robert B. Tallistch; p. 102: Frederic H. Martini; p. 103 top: Robert B. Tallistch; bottom: Gregory N. Fuller; p. 105: Dr. Holger Jastrow; p. 107 top and middle: Robert B. Tallistch; bottom: Frederic H. Martini; p. 109 top: Robert Brons/Biological Photo Service; middle: Science Source/Photo Researchers, Inc.; bottom: Robert B. Tallistch; p. 111: Robert B. Tallistch; p. 112 all: Robert B.Tallistch; p. 114 top: iStockphoto; bottom: Leah-Anne Thompson/iStockphoto; p. 120 top: Newstockimages/iStockphoto; bottom left: lightpoet/Shutterstock; bottom right: DNY59/ iStockphoto; p. 121: Pali Rao/iStockphoto

Chapter 4 Chapter Opener: Don Bayley/ iStockphoto; p. 124 all: Robert B. Tallistch; p. 125 top: Robert B. Tallistch; p. 125 bottom: Cloud Hill Imaging/www.lastrefuge.co.uk; p. 126: Robert B. Tallistch; p. 127 all: Elizabeth A. Abel; p. 128 left: Steve Gschmeissner/Photo Researchers, Inc.; right: Prof. P. Motta, Dept. of Anatomy, University "La Sapienza," Rome/Science Photo Library/Photo Researchers, Inc.; p. 129: Biophoto Associates/ Photo Researchers, Inc.; p. 135 both: Frederic H. Martini; p. 136: Jim Thomas M.D./Dermnet.com, Interactive Medical Media; p. 138: William C. Ober; p. 139: Kristine Hannon; p. 141: Biophoto Associates/Photo Researchers, Inc.; p. 142 top: Stephen Coburn/Shutterstock; bottom: Nancy Louie/iStockphoto; p. 143: stefanolunardi/ Shutterstock

Chapter 5 Chapter Opener: Jana Blašková/ iStockphoto; p. 149: seelevel.com; p. 151: Robert B. Tallistch; p. 155: Ralph T. Hutchings; p. 156 top left: Dennis MacDonald/Alamy; bottom left: Gary Parker Photography; right: Joeff Davis; p. 157 left: MUSTAFA OZER/AFP/Getty Images/Newscom; top right: Frederick Kaplan M.D./From: *Overexpression of an Osteogenic Morphogen in Fibrodysplasia Ossificans Progressiva*. Adam B. Shafritz et al. New England Journal of Medicine. 1996 Aug 22;335(8):555–561. Courtesy of the Mütter Museum, College of Physicians of Philadelphia. Copyright ©1996 Massachusetts Medical Society. All rights reserved; bottom right: Laura Gould/ From: *Mosby's Dental Dictionary*, 2nd edition, 2008. © Elsevier; p. 158: iStockphoto; p. 159 top left: Dr. Kathleen Welch; top middle left: SIU Biomed com/Custom Medical Stock Photo, Inc.; top middle: Custom Medical Stock Photo, Inc.; top middle right: Zephyr/Photo Researchers, Inc.; top right: Frederic H. Martini; bottom left: Mark Aiken (chezlark.com); bottom middle left: Image reprinted with permission from eMedicine.com, 2009. Available at: http://emedicine.medscape.com/article/ 1260633-overview.; bottom middle right: LIVING ART ENTERPRISES, LLC; bottom right: Scott Camazine/Photo Researchers, Inc.; p. 168: right Based on Life-size Auditory Ossicles, model LT-E13 by 3B Scientific; bottom: Ralph T. Hutchings; p. 172 both: Ralph T. Hutchings; p. 173 both: Ralph T. Hutchings; p. 174 all: Ralph T. Hutchings; p. 175 both: Ralph T. Hutchings; p. 177 top left: Caters News Agency; right: Dreamstime LLC; bottom: Princess Margaret Rose Orthopaedic Hospital/ Science Photo Library/Photo Researchers, Inc.; p. 178 both: Ralph T. Hutchings; p. 179 all: Ralph T. Hutchings; p. 180 both: Ralph T. Hutchings; p. 181 both: Ralph T. Hutchings; p. 182: Ralph T. Hutchings; p. 184: Ralph T. Hutchings; p. 186 both: Ralph T. Hutchings; p. 187 both: Ralph T. Hutchings; p. 188: Ralph T. Hutchings; p. 189 both: Ralph T. Hutchings; p. 195 left and middle: Prof. P. Motta/SPL/Photo Researchers, Inc.; right: Michael J. Timmons; p. 196 top: Steve Rabin/ iStockphoto; middle left and right: Abramson Lab and Rheumatology Research Laboratory, NYU School of Medicine; bottom left: Sanjiv Jari (www .thekneedoc.co.uk) (www.sportsmedclinic.com);

bottom right: Stanley S. Tao, Scott Orthopedic Center, Huntington, West Virginia (www .scottorthopedic.com); p. 197 top: Mark D. Miller, S. Ward Casscells Professor of Orthopaedic Surgery, University of Virginia; bottom left, middle and right: Smith & Nephew, Inc.; p. 199: Gary Parker Photography; p. 200: Ralph T. Hutchings; p. 201 top left and right: Ralph T. Hutchings; middle left: Dreamstime LLC; bottom right: Ralph T. Hutchings; p. 202 top left: Ralph T. Hutchings; middle right: Mark D. Miller, S. Ward Casscells Professor of Orthopaedic Surgery, University of Virginia; bottom: Yuri Arcurs/Shutterstock; p. 203 top: sunil menon/ iStockphoto; middle: aceshot1/Shutterstock

Chapter 6 Chapter Opener: Juanmonino/ iStockphoto; p. 224 right: Terry Wilson/ iStockphoto; left: Chris Schmidt/iStockphoto; p. 250 bottom left: Terry Wilson/iStockphoto; p. 252 top left: Stylephotographs/Dreamstime; right: Dean Mitchell/Shutterstock; p. 253: Catherine Yeulet/iStockphoto

Chapter 7 Chapter Opener: Sebastian Kaulitzki/ iStockphoto; p. 274: Ralph T. Hutchings; p. 275: Ralph T. Hutchings; p. 281: Jodi Jacobson/ iStockphoto; p. 284: Michael J. Timmons; p. 285: Ralph T. Hutchings; p. 289 left: Steve Cole/ iStockphoto; right: Ryan McVay/Lifesize/ Thinkstock

Chapter 8 Chapter Opener: Jana Blašková/ iStockphoto; p. 293: William C. Ober; p. 304 top: Lisa F. Young/iStockphoto; middle: University of Rochester Department of Pathology; bottom: Len Tillim/iStockphoto; p. 305 top: Cecilia Magill/ Photo Researchers, Inc.; middle: Bettmann/Corbis Images; bottom: Gunilla Elam/Photo Researchers, Inc.; p. 315: University of Rochester Department of Pathology; p. 316 top: Konstantin Sutyagin/ Shutterstock; bottom: Susy56/Dreamstime; p. 317: Design Pics Inc./Alamy

Chapter 9 Chapter Opener: Lana K/Shutterstock; p. 319 left: iStockphoto; right: Ankya/Shutterstock; p. 334: Michael J. Timmons; p. 338: Ralph T. Hutchings; p. 339: iStockphoto; p. 343 both: Diane Hirsch/Fundamental Photographs NYC; p. 346: Custom Medical Stock Photo, Inc.; p. 351: David Kevitch/iStockphoto; p. 352 top left: iStockphoto; top right: Sharon Dominick/iStockphoto; bottom left: Olga Ekatarincheva/iStockphoto; bottom right: Micheline Dubé/iStockphoto; p. 353 top: iStock- photo; bottom: Konrad Lange/iStockphoto; p. 356 top right: Custom Medical Stock Photo, Inc.; bottom right: Micheline Dubé/iStockphoto; p. 357 top: StockLite/Shutterstock; bottom: Oliver Hoffmann/Shutterstock

Chapter 10 Chapter Opener: Michael Svobada/iStockphoto; p. 360 top left: Miha Krivic/iStockphoto; bottom left: Kelly Cline/ iStockphoto; right: James Steidl/iStockphoto; p. 368: Robert B. Tallistch; p. 370: Frederic H. Martini; p. 372: Ward's Natural Science Establishment; p. 376: Lutz Slomianka;

p. 377: Mikhail Kokhanchikov/iStockphoto; p. 378: Stephanie Horrocks/iStockphoto; p. 379: Ruhani Kaur/UNICEF India Country Office; p. 382 top: Mikhail Kokhanchikov/iStockphoto; middle: Galina Barskaya/Shutterstock.com; bottom: Aspen Photo/Shutterstock; p. 383: Joshua Hodge Photography/iStockphoto

Chapter 11 Chapter Opener: Bogdan Pop/iStockphoto; p. 388 top left: Frederic H. Martini; top right: Cheryl Power/Photo Researchers, Inc.; bottom: Ed Reschke/Photolibrary; p. 391: Karen E. Petersen, Dept of Biology, University of Washington; p. 394: Robert B. Tallistch; p. 400 top: Shevelev Vladimir/Shutterstock; bottom left: iStockphoto; bottom right: Eye of Science/Photo Researchers, Inc.; p. 401 top left: National Medical Slide Bank/Custom Medical Stock Photo, Inc.; top right: Wikipedia; bottom: Christopher Badzioch/iStockphoto; p. 403: Kyu Oh/iStockphoto; p. 404: Biophoto Associates/Photo Researchers, Inc.; p. 406 top: B&B Photos/Custom Medical Stock Photo, Inc.; bottom: Peter Arnold, Inc./Photolibrary; p. 409: Timothy Kosachev/iStockphoto; p. 427: Robert B. Tallistch; p. 428 top: Shevelev Vladimir/Shutterstock; bottom: B&B Photos/Custom Medical Stock Photo, Inc.; p. 430 top: Alexraths (Alexander Raths)/Dreamstime; bottom: dlewis33/iStockphoto; p. 431: Malota/Dreamstime

Chapter 12 Chapter Opener: Mads Abildgaard/iStockphoto; p. 435: Robert B. Tallistch; p. 442 top left and right: Howard Sochurek/Corbis; p. 442 bottom: ICVI-CCN/Photo Researchers, Inc.; p. 443 top left: Science Photo Library/Photo Researchers, Inc.; top right: Biophoto Associates/Science Source/Photo Researchers, Inc.; middle top: Margaret Grimes; middle bottom: Patty Quinehan; bottom: iStockphoto; p. 449: Larry Mulvehill/Photo Researchers, Inc.; p. 450: William C. Ober; p. 459: iStockphoto; p. 463 top: Margaret Grimes; bottom: William C. Ober; p. 464: Du Cane Medical Imaging Ltd./Photo Researchers, Inc.; p. 465 top: Mark Kostich/iStockphoto; bottom: Adivin/iStockphoto

Chapter 13 Chapter Opener: Sebastian Kaulitzki/iStockphoto; p. 469: Frederic H. Martini; p. 471: Barkley Fahnestock/iStockphoto; p. 474: Ralph T. Hutchings; p. 477: iStockphoto; p. 478: Josh Hodges/iStockphoto; p. 479: CC-BY-SA Photo:Etxrge;

p. 480: Ashok Rodrigues/iStockphoto; p. 489: Monika Wisniewska/iStockphoto; p. 490 top left: Stan Rohrer/iStockphoto; top right: Courtesy of Sentara Heathcare; bottom: Elwynn/Dreamstime.com; p. 491 left: C. Goldsmith, P. Feorino, E. L. Palmer, & W. R. McManus/CDC; right: Steve Code/iStockphoto; p. 494 top: Stan Rohrer/iStockphoto; middle: zilli/iStockphoto; bottom: Andrew Gentry/Shutterstock

Chapter 14 Chapter Opener: Niderlander/Dreamstime; p. 498: iStockphoto; p. 514 top: Dana Spiropoulou/iStockphoto; middle: Jaren Wickland/iStockphoto; bottom: iStockphoto; p. 518 top: Dana Spiropoulou/iStockphoto; middle left: Danijelm/Dreamstime; middle right: WavebreakMedia-Micro/Fotolia; p. 519: muratseyit/iStockphoto

Chapter 15 Chapter Opener: Sebastian Kaulitzki/Shutterstock; p. 524: Frederic H. Martini; p. 531: Ralph T. Hutchings; p. 532 middle: Alfred Pasieka/Peter Arnold, Inc./Photolibrary; bottom: Astrid and Hanns-Frieder Michler/Science Photo Library/Photo Researchers, Inc.; p. 535: Ralph T. Hutchings; p. 539: M.I. Walker; p. 552 top: Dr. Richard Nejat; middle: Barbara Rice/CDC; bottom left: Gastrolab/Photo Researchers, Inc.; bottom right: PD Image Stell98/Wikimedia Commons; p. 553 top: David M. Martin, M.D./Photo Researchers, Inc.; middle: Ronald Bleday M.D./BWH; bottom: Joel Mancuso, University of California, Berkeley; p. 557: Tracy Whiteside/Shutterstock

Chapter 16 Chapter Opener: Blend_Images/iStockphoto.com; p. 571: Celso Pupo Rodrigues/iStockphoto; p. 572: Ilya Genkin/iStockphoto; p. 573: Olga Lyubkina/iStockphoto; p. 574: choosemyplate.gov/USDA; p. 575 top: Kelly Cline/iStockphoto; bottom: JoopHoek/iStockphoto; p. 577 top left: B-d-s/Dreamstime; bottom left: katsgraph-icslv/iStockphoto; right: David Gaylor/Shutterstock; p. 578 top: William C. Ober; bottom left: Lyle Conrad/CDC; bottom right: Thalia Oster, COE; p. 582: Phanie Agency/Photo Researchers, Inc.; p. 583 left: Madlen/Shutterstock; right: Fotandy/Dreamstime; p. 587 top: Ilya Genkin/iStockphoto; bottom: choosemyplate.gov; p. 588 top left: Lyle Conard/CDC; top right: Phanie Agency/Photo Researchers, Inc.; middle: KLH49/iStockphoto; bottom: Ironrodart/Dreamstime; p. 589: esolla/iStockphoto

Chapter 17 Chapter Opener: Luis Pedrosa/iStockphoto; p. 593: Ralph T. Hutchings; p. 601: Steve Gschmeissner/Photo Researchers, Inc.; p. 607: Beranger/Photo Researchers, Inc.; p. 609: Photo Researchers, Inc.; p. 613: David Gold/iStockphoto; p. 629 top: Berenger/Photo Researchers, Inc.; bottom: Photo Researchers, Inc.; p. 630: Sean Locke/iStockphoto

Chapter 18 Chapter Opener: Steve Cole/iStockphoto; p. 635: Eye of Science/Photo Researchers, Inc.; p. 639: Elizabeth Kreutz/Lance Armstrong Foundation; p. 645 top and middle: Frederic H. Martini; bottom: C. Edelmann/La Villete/Photo Researchers, Inc.; p. 647: Custom Medical Stock Photo, Inc.; p. 651: iStockphoto; p. 652 top left: Kim Gunkel/iStockphoto; top right: Getty Images Inc. RF; middle left: Brent Melton, iStockphoto; middle right: Shutterstock; bottom left: Alamy; bottom right: Alamy; p. 653 left: Jennifer Trechard/iStockphoto; right: Maureen Spuhler; p. 656 top: Brent Melton/iStockphoto; bottom; Leigh Schindler/iStockphoto; p. 657: Mark Bowden/iStockphoto

Chapter 19 Chapter Opener: Alberto Pomares/iStockphoto; p. 666: Frederic H Martini; p. 668 top left: Dr. Arnold Tamarin; top right, bottom left and right: Lennart Nilsson/Scanpix Sweden AB; p. 669 left: Lennart Nilsson/Scanpix Sweden AB; right: Photo Researchers, Inc.; p. 670: Aldo Murillo/iStockphoto; p. 675 left & middle left: Jaroslav Wojcik/iStockphoto; middle right: Justin Horrocks/iStockphoto; right: Jacob Wackerhausen/iStockphoto; p. 679 top right: Valentin Casarsa/iStockphoto; top left: Bob Ainsworth/iStockphoto; bottom: Science Photo Library/Photo Researchers, Inc.; p. 680: Lisa McCorkle/iStockphoto; p. 681: Noriko Cooper/iStockphoto; p. 682 top, middle and bottom left: Science Photo Library/Photo Researchers, Inc.; top right: Fotolia; middle right: Malkolm Gin/Wikimedia commons; bottom right: Kumar Vikram/From: Robbins Pathologic Basis of Disease, 6th edition, by Kumar and Hagler, from the Interactive Case Study Companion, ©1999, W. B. Saunders Company, Elsevier; p. 683: Science Photo Library/Photo Researchers, Inc.; p. 684: iStockphoto; p. 685: Lennart Nilsson/Scanpix Sweden AB; p. 686 top left and right: Science Photo Library/Photo Researchers, Inc.; middle: Ruslan Dashinsky/iStockphoto; bottom: PhotoAlto/SuperStock; p. 687 Flirt/SuperStock

double helix of DNA, 59, 75
Down syndrome (trisomy 21), 682, 683
DRG (dorsal respiratory group), 512
DSA (digital subtraction angiography), 442
ducts of exocrine glands, 104
ductus deferens, 633, 634, 637
duodenal ampulla, 550, 551
duodenal glands, 313
duodenal papilla, 550, 551
duodenal ulcer, 553
duodenum, 540
  arteries supplying, 420, 422
  digestion in, 544–545, 565, 570
  hormones produced by, 541
  stomach in relation to, 534
  veins draining, 423
dura mater
  cranial, 272, 273
  spinal, 284, 285
dural folds, 272
dural sinuses, 272, 417, 419
dysentery, 553
dysuria, 612

E

E (epinephrine). See epinephrine
ear (otic region), 16
  anatomical regions of, 328
  auditory ossicles and, 161, 168, 328, 329, 337, 353
  external, 328
  hearing and, 325, 328, 336–337. See also hearing sense
  infections of, 328
  internal, 328. See also internal ear
  middle, 163, 168, 328, 329
eardrum, 328. See also tympanic membrane
eating, facial muscles used in, 236–237
eating disorders, 577
ECF, 66. See also extracellular fluid
ECG/EKG (electrocardiogram), 5, 449
ectoderm, 662
  body systems derived from, 663
ectopic bones (heterotopic bones), 157
ectopic pacemaker, 461
edema, 612
EEG (electroencephalogram), 281
effector, 11–13
  in homeostatic regulation, 11–13
  negative feedback and, 12
  in nervous system (overview), 255
  positive feedback and, 13
  in reflex arcs, 300–301
  somatic, 261
  visceral, 261, 307
efferent arteriole, 597, 601
efferent ductules of seminiferous tubule, 634
efferent fibers, 261
efferent lymphatics, 474
eicosanoids, 360, 362
ejaculatory duct, 634, 637
EKG/ECG (electrocardiogram), 5
elastase, 551, 570
elastic arteries, 405
elastic cartilage, 109
  of external ear, 109, 328
elastic fibers, 106
  of alveoli, 502–503
  in arteries, 404, 405
  in cartilage, 109, 110
  in dermis, 128, 137
elastic membrane of arteries, 404
elastic rebound (coronary circulation), 439
elastin, wrinkled skin and, 128
elbow joint, 6, 228
  antecubital region of, 16
  bones articulating at, 180, 272

  dislocation and, 272
  extension at, 228, 242
  flexion at, 228, 243
  as hinge synovial joint, 194, 272
  muscles that move, 234–235, 242–243
  olecranal region of, 17, 181, 272
  radial fossa and, 180
  range of motion and, 272
  reflex testing and, 303
  reinforcing ligaments of, 272
  structure/function, 6
electrical charge of atoms, 28, 30–31
  chemical notation to describe, 36
electrical currents/forces, 28
  electrolytes and, 42–43
  ion diffusion rate and, 88
electrocardiogram (ECG/EKG), 5, 449
  cardiac arrhythmias and, 461
electroencephalogram (EEG), 281
electrolytes, 42–43, 387
  major, in plasma, 387
  in sebum, 134
  in sweat gland secretions, 135
electron cloud, 27, 28
  energy levels and, 30–31
electron shells, 28, 30–31
electron transport system (ETS), 560–561
  cellular metabolism and, 562–563
electrons, 27
  electron cloud and, 27, 28
  energy levels and, 30–31
elements, 28, 29
  chemical notation rules to describe, 36
  chemical reactions and, 35–40
  of human body (chart), 29
  inert, 30–31
  reactive, 30–31
  trace, 29
elevation (movement), 233
  facial expression muscles and, 236–237
  trunk/back muscles and, 240, 241
embolism, 398
embolus, 398
embryo
  defined, 647
  pre-embryos and, 660
embryology, 659
embryonic development, 659
  amniotic cavity formation, 661, 662
  bone formation, 154–155
  of brain, 269
  cleavage and implantation, 660–661
  extra-embryonic membranes and, 662, 664–665
  germ layer formation and, 662–663
embryonic disc, 662
emmetropia, 351
emotional responses, to smells, 326
emotions
  hypothalamus and, 270
  limbic system and, 278
emphysema, 514
EMT/paramedic, 142–143
emulsification, 568
enamel of teeth, 530
endergonic reactions, 37
endocardium, 4, 434
endochondral ossification, 154–155
endocrine functions, of skin, 129
endocrine glands, 94, 104
  intercellular communication and, 359, 362
endocrine pancreas, 374–375
endocrine system, 8, 9, 358–383
  body weight percentage, 227
  cardiovascular regulation by, 458
  disorders of, 156–157
  embryonic germ layers derived from, 663
  functions of (summary), 8, 9, 361
  hormones of, 359–363. See also hormones
  hypothalamic coordination of, 278, 363
  intercellular communication and, 359, 362

  nervous system similarities with, 359
  organs of, 361, 365–376.
endocytosis, 89
  ATP and, 559
endoderm, 662
  body systems derived from, 663
endolymph, 330, 331
  ampulla of semicircular duct and, 332
  in rotational head movements, 330, 331
endolymphatic duct, macula and, 333
endometrium, 646, 665
endomysium, 206
endoneurium of spinal nerves, 292
endoplasmic reticulum (ER), 66–67, 70
  of epithelial cells, 125
endosteum, 153
endothelium
  of arteries, 404, 405
  of veins, 404, 405
energetics, 582
energy, 35
  activation, 37
  anabolism and, 39, 559, 562–563
  ATP and, 57, 71, 222. See also ATP
  basal metabolic rate and, 582–583
  body temperature and, 582–583
  carbohydrates and, 48, 566–567
  catabolism and, 38, 559–563
  cellular, 57, 71, 562–563, 571
  dietary sources of, 48–49, 574–575
  endergonic reactions, 37
  exergonic reactions, 37
  kinetic, 35, 38
  light, 348–349
  metabolism and, 581–586
  potential, 35, 38
  stored (as potential energy), 35
  study of (energetics), 582
  work and, 35
energy levels of electrons, 30–31
energy reserves
  fight-or-flight response and, 308
  hypodermis as storage site for, 128
  lipid deposits as, 51
  skeletal muscle as, 205
  yellow bone marrow and, 148
energy sources
  carbohydrates, 48–49, 566–567
  fatty acids/fats, 51
  glucose as primary, 48
  lipids, 50–53
energy storage
  in adipose tissue of hypodermis, 128
  glycerides and, 51
  glycogen and, 49
enlargements of spinal cord, 282
enteritis, 553
enteroendocrine cells, 537
enterogastric reflex, 544
enteropeptidase, 570
enzyme-substrate complex, 56
enzymes, 37, 56–57
  acetylcholinesterase, 210–211
  activation energy and, 37
  active site and, 56
  DNA polymerase, 79
  equilibrium and, 51
  Golgi apparatus secretory vesicles and, 72
  high body temperature and, 55
  lingual lipase, 529
  in lysosomes, 72–73
  magnesium as cofactor for many, 29
  metabolic, in mitochondrial matrix, 71
  pancreatic, 551
  protein-digesting, 189
  proteolytic, 189
  specificity of, 56
eosinophils, 387, 394, 395, 397, 479, 481
  formation of, 396–397
  in nonspecific immunity, 479, 481
ependyma, 258

ependymal cells, 258
  brain ventricles lined by, 271
  cerebrospinal fluid and, 273
epicardium (visceral pericardium), 4, 434
epicondyles (lateral/medial)
  of femur, 186
  of humerus, 180, 272
epidermal ridges
  fingerprints and, 125
  of papillary layer of dermis, 128
epidermis, 123, 124–126, 131
  age-related changes in, 137
  avascularity of, 125
  dermis in relation to, 123, 131
  exocrine glands lubrication of, 131
  fingerprints and, 125
  hypodermis in relation to, 123, 131
  as part of cutaneous membrane, 123, 131
  pigmentation and skin color, 126–127
  stratified squamous epithelium of, 99, 123
  sunlight, vitamin D_3 synthesis and, 123, 129, 131
  thick/thin, 124
epididymis, 633, 634
epidural space, 285
epigastric region, 17, 540
epiglottis, 532
  taste receptors and, 327
epilepsies, 281
epimysium, 206
epinephrine (E), 266
  as a neurotransmitter, 266, 312
  adrenal medulla secretion of, 365, 373
  blood pressure changes and, 457, 458
  derived from amino acid tyrosine, 360
  effects of, 266, 373, 458
  in fight-or-flight response, 308
  heart rate affected by, 451
  receptors for, 362
  second messenger system and, 362
  in stress response alarm phase, 378–379
  sympathetic activation and, 308, 373, 458
epineurium of spinal nerve, 292
epiphyseal artery, of humerus, 149
epiphyseal cartilage, 155
  achondroplasia and, 156
  Marfan's syndrome and, 156
epiphyseal fractures, 159
epiphyseal line, 155
epiphyseal vein, of humerus, 149
epiphysis, 148
  in bone formation, 154–155
epithalamus, pineal gland of, 275
epithelial tissue (epithelia), 65, 91, 92, 94–105
  avascular nature of, 97
  cell differentiation and, 65
  columnar, 95, 102–103
  cuboidal, 95, 100
  gland cells in, 94, 104–105
  growth hormone and, 367
  intercellular attachments of, 96–97
  layers of cells in, 95
  regenerative abilities of, 115
  shapes of cells in, 95, 98–103
  simple, 95
  squamous, 95, 98–99
  stratified, 95, 99, 100, 103, 123
EPO (erythropoietin), 397, 458
eponyms, anatomical terms and, 15
equations, chemical, 36
equilibrium (balance), 325, 330–333
  axial skeleton abnormalities affecting, 176
  disorders of, 353
  ear (internal) structures of, 328, 330–333
  gravity sensations and, 330, 331, 333
  innervation of, 295, 328, 333
  linear acceleration and, 330, 331, 333

third ventricle of brain, 271, 273
   cerebrospinal fluid and, 273
   diencephalon and, 278
thirst
   antidiuretic hormone and, 618–619
   hypothalamus and, 278
   low blood pressure/volume and, 458, 459
thoracic aorta, 420
thoracic artery (internal), 414, 420
thoracic cage, 161, 176
thoracic cavity, 20, 21
   arteries of, 412, 414
   components of, 21, 176
   innervation of, 295, 297
   muscular walls of, 239
   parasympathetic innervation and,
      309, 311
   pressure changes during breathing,
      506–507
   sympathetic innervation and, 308, 310
   thoracic cage and protection of, 176
   veins of, 413, 415
thoracic curve, 170
   kyphosis and, 177
thoracic duct, 467, 470–471
thoracic lymph nodes, 471
thoracic nerves, 296
   reflexes and, 303
thoracic region (chest/thorax), 16
   cavity of. *See* **thoracic cavity**
   muscle attachment sites and, 176
   muscles of, 239
   skeleton of. *See* **thoracic cage**
   vertebrae located in, 170, 173
thoracic spinal nerves, 282, 291
   sympathetic division of PNS and, 292
thoracic vein (internal), 415
thoracic vertebrae, 170, 173
   spinal nerves of, 282
**thoracodorsal fascia, vertebral column
   muscles and, 238**
thoracolumbar division. *See* **sympathetic
   division of ANS**
thoracolumbar fascia, 238, 241
thorax/chest (thoracic region), 16
thrombin, 399
thrombus, 398
thrombus (blood clot), 442
thumb (pollex), 16, 182
   muscles that move, 242
   opposition movement of, 233
   saddle joint of, 194
thymine (T), 58–59
   genetic code, protein synthesis and,
      76–77
thymosins, 475
thymus, 396, 467, 475
   hormones secreted by, 361, 475
   lymphocyte production in, 473
thyroglobulin, 368
   thyroiditis and, 490
thyroid-binding globulins (TBGs), 369
thyroid follicles, 368
thyroid gland, 361, 365, 368–369
   hormones of, 360, 368–369, 667
   iodine and, 29
   pituitary gland regulation of, 363,
      365, 366
   in resistance phase of stress
      response, 379
   simple cuboidal epithelium of, 100
**thyroid-stimulating hormone (TSH/
   thyrotropin), 360, 366**
thyroiditis, 490
thyroxine (T₄), 369
   derived from tyrosine, 360
   during pregnancy, 667, 671
   heart rate affected by, 451
tibia, 161, 186, 187
   articulation with fibula, 187, 191
   articulation with talus, 188
   foot-moving muscles and, 246–247
   gracilis muscle and, 244
   at knee joint, 193

leg-moving muscles and, 244, 245
   medial malleolus and, 187, 234
   muscles originating on, 234, 246
**tibial arteries (anterior/posterior),
   412, 424**
tibial nerve, reflexes and, 303
tibial tuberosity, 187, 245
tibial veins (anterior/posterior), 413, 425
tibialis anterior muscle, 234, 247
tibiofibular joints, 187, 191
tidal volume (V_T), 507, 508, 509
tight junctions, 96–97
tin, 29
tissue, 7
tissue factor (clotting factor), 399
tissue injuries, inflammatory response, 115
tissue level of organization, 7, 91–119
tissue perfusion, 456
tissue plasminogen activator (t-PA), 399
tissue repair, 114–115
   hemostasis and, 398–399
   in inflammatory response, 480
   skin injuries and, 137, 138–139
tissues, 7, 91–119
   carbon dioxide and, 511
   cells in relation to, 7, 65, 91, 92
   connective, 65, 91, 92, 106–107
   epithelial, 65, 91, 92, 94–105
   histology as study of, 92
   hypoxia/anoxia and, 505
   injury response of, 114–115
   muscle, 65, 91, 112
   neural, 65, 93, 113, 262–263
   organs in relation to, 7, 8, 91
   oxygen absorption and, 502–503,
      505, 511
   regeneration abilities and, 115
   scar, 115
   types of, 65, 91, 92
toenails, functions of, 131, 136
toes, 16
   arteries supplying, 424
   bones (phalanges) of, 188–189
   great (hallux), 16, 188
   long bones of, 146
   movements of, 247
   muscles that move, 246, 247
   plantar reflex and, 303
   veins draining, 425
tongue, 521, 527, 528, 529
   functions of, 529
   hyoid bone and, 168
   stratified squamous epithelium of, 99
   taste buds and, 327
   taste receptors of, 327
tongue muscles
   hyoid bone as attachment site for, 168
   innervation of, 294, 295
tonsillitis, 474
tonsils, 467, 474
tooth decay, 531. *See also* **teeth**
total lung capacity, 509
touch receptors, 255, 319
   fine touch/pressure, 320, 321
   in skin, 123, 125, 128, 131, 320–321
   tactile corpuscles and, 321
   tactile discs and, 320
touch sense
   cerebral cortex area of, 277
   epithelial tissue and, 125
   exteroceptors and, 261
   hair/hair follicles on skin and, 131
   Merkel cells and, 125
   posterior column pathway to CNS, 298
   skin and, 123, 125, 131
   somatic sensory receptors and, 255
toxic chemicals
   lysosomes and, 72–73
   skin and protection against, 123, 131
toxins, tissue injury response to, 114–115
trabeculae, 152
trabeculae carneae, 440
trabecular bone. *See* **spongy bone**
trace elements, 29

trachea, 499, 500
   cartilages of, 500
   cilia of, 103
   mucous cells of, 104
   thyroid gland in relation to, 368
   trachealis muscle of, 500
tracheal cartilages, 500
trachealis muscle, 500
tracts (ascending/descending), 284
transcription (in protein synthesis), 77
transfer RNA (tRNA), 59, 77
transfusion reactions, 391
transitional epithelium, 101
translation (in protein synthesis), 77
transplant rejection, 490
transport globulins, 387
transport vesicles, 72
   vesicular transport and, 85, 89
transudate, 108
transverse arch of foot, 189
transverse colon, 540, 543
**transverse foramen of cervical vertebrae,
   170, 172**
transverse fractures, 159
transverse (horizontal) plane, 19
transverse (horizontal) section, 19
transverse processes of vertebral arch, 171
   of cervical vertebrae, 170, 172
   of lumbar vertebrae, 170, 174
   muscles of vertebral column and, 238
   regional differences in, compared, 170
   rib articulation and, 171
   of thoracic vertebrae, 170, 173
transverse sinuses, 419
transverse tubules (T tubules), 209
   in muscle contraction, 210–211
transversus abdominis muscle, 239, 240
transversus thoracic muscle, 508
trapezium bone, 182
   saddle joint of, 194
trapezius muscle, 234, 235, 240, 241
   innervation of, 294
trapezoid bone, 182
triceps brachii muscle, 234, 242, 243
   as agonist muscle, 228
   lateral/long heads of, 235, 242
   tendon connective tissue and, 107
triceps reflex, 302, 303
**tricuspid valve (right AV valve),
   440, 441**
trigeminal nerve (V), 295
triglycerides, 51
   cellular metabolism and, 562, 563
   digestion of, 568–569
   growth hormone and, 367
   metabolism of, 568–569
   in sebum, 134
   synthesized by smooth ER, 70
trigone, 610
triiodothyronine (T₃), 369
   heart rate affected by, 451
**trimesters of pregnancy (first/second/
   third), 659**
   organ system development during,
      668–669
tripeptides, 54
triplet (DNA), 76
triquetrum bone, 182
trisomy 21 (Down syndrome), 682
tritium, 28
tRNA (transfer RNA), 59, 77
**trochanters (surface feature
   of bone), 147**
   of femur (greater/lesser), 186
trochlea of eye, 341
trochlea (surface feature of bone), 147
   of ankle, 188
   of humerus, 147, 180, 181, 272
trochlear nerve (IV), 294, 341
trochlear notch of ulna, 181, 272
trophoblast, 661, 666
   gestational trophoblastic neoplasia
      and, 663
tropic hormones, 366

tropomyosin, 208
   in muscle contraction cycle, 214–215
troponin, 208
   in muscle contraction cycle, 214–215
true labor, 672
true pelvis, 185
true ribs, 176
trunk region, 16
   appendicular muscles originating on,
      240–241
   arteries of, 412
   limb muscles and, 242–243
   muscles of, 239, 240, 241
   nerve supply of, 292
   pectoral girdle and, 178–179
   rotation (left/right), 232
   sympathetic innervation of, 308
   veins of, 413
   vertebral column support of, 170
trunks (large arteries), 4, 411
tryosine, of thyroglobulin molecules, 368
trypsin, 551, 570
tryptophan, 360
**TSH (thyroid-stimulating hormone/
   thyrotropin), 366**
   derived from glycoproteins, 360
tubal ligation, 653
tubercles (surface feature of bone), 147
   of humerus (greater/lesser), 147, 180
tuberosity (surface feature of bone), 147
tubular fluid, 594, 595, 604–605
tumors, 82–83
**tunica externa (tunica adventitia) of
   blood vessels, 404–405**
**tunica intima (tunica interna) of blood
   vessels, 404–405**
tunica media of blood vessels, 404–405
Turner syndrome, 682
twisting mechanoreceptors, 319
twitch, muscle, 218–219
tympanic cavity (middle ear), 328, 329
tympanic membrane (eardrum), 328, 329
   auditory ossicles and, 168, 328, 329, 337
   external acoustic meatus and, 164, 328
   pitch/volume of sound sensations and,
      336–337
   sound waves and, 329, 336, 337
tympanum. *See* **tympanic membrane**
**Type 1 (insulin dependent) diabetes,
   377, 490**
**Type 2 (non-insulin dependent)
   diabetes, 377**
Type A blood, 390, 391
Type A fibers of pain receptors, 319
Type AB blood, 390, 391
Type B blood, 390, 391
Type C fibers of pain receptors, 319
Type O blood, 390, 391
tyrosine (amino acid), 126, 130

U

ulcers, stomach, 553
ulna, 161, 181
   articulations of, 180, 181
   condylar joint of, 194
   at elbow joint, 6, 180, 181, 272
   hinge joint of, 194
   humerus articulation and, 180
   radio-ulnar joint and, 181
   rotational movements and, 232
ulnar artery, 412, 414
ulnar collateral arteries, 414
ulnar collateral ligament, 272
ulnar head, 181
ulnar nerve, 296, 297
ulnar notch of radius, 181
ulnar vein, 413, 415
ultrasound diagnostics, careers in, 24–25
ultraviolet (UV) radiation
   cancer and overexposure to, 127
   elastin in dermis reduced by, 128
   melanin production by skin and, 123

skin color changes and, 126
sunburn and, 137
vitamin D₃ synthesis and, 129

INDEX